Name Reactions for Homologations
Part I

Name Reactions for Homologations

Part I

Edited by
Jie Jack Li
Bristol-Myers Squibb Company

Foreword by
E. J. Corey
Harvard University

WILEY

A JOHN WILEY & SONS, INC., PUBLICATION

Published by John Wiley & Sons, Inc., Hoboken, New Jersey.
Published simultaneously in Canada.

Library of Congress Cataloging-in-Publication Data:

Name reactions for homologations / edited by Jie Jack Li ; foreword by E. J. Corey.
v. cm.
Includes index.
ISBN 978-0-470-08507-3 (pt. 1 : cloth) — ISBN 978-0-470-46498-4 (pt. 2 : cloth)
1. Organic compounds—Synthesis. 2. Chemical reactions. 3. Chemical tests and reagents.
I. Li, Jie Jack.

QD262.N363 2009
547'.2—dc22 2008055902

Printed in the United States of America

10 9 8 7 6 5 4 3 2 1

Dedicated to

Chris Limberakis, John Montgomery, and Derek A. Pflum

for the good ol'days in Ann Arbor

Foreword

Part of the charm of synthetic organic chemistry derives from the vastness of the intellectual landscape along several dimensions. First, there is the almost infinite variety and number of possible target structures that lurk in the darkness waiting to be made. Then, there is the vast body of organic reactions that serve to transform one substance into another, now so large in number as to be beyond credibility to a non-chemist. There is the staggering range of reagents, reaction conditions, catalysts, elements, and techniques that must be mobilized in order to tame these reactions for synthetic purposes. Finally, it seems that new information is being added to that landscape at a rate that exceeds the ability of a normal person to keep up with it. In such a troubled setting any author, or group of authors, must be regarded as heroic if through their efforts, the task of the synthetic chemist is eased.

These two volumes on methods for the extension of carbon chains by the use of coupling reactions brings to the attention of practicing synthetic chemists and students of chemistry a wide array of tools for the synthesis of new and useful molecules. It is a valuable addition to the literature by any measure and surely will prove its merit in years to come. The new knowledge that arises with its help will be impressive and of great benefit to humankind.

E. J. Corey
October 1, 2008

Preface

This book is the third volume of the series *Comprehensive Name Reactions*, an ambitious project conceived by Prof. E. J. Corey of Harvard University in the summer of 2002. Volume 1, *Name Reactions in Heterocyclic Chemistry*, was published in 2005 and was warmly received by the organic chemistry community. Volume 2, *Name Reactions for Functional Group Transformations* was published in 2007. After publication of the current Volume 3 and 4 on homologations in 2009, we plan to roll out Volume 5, *Name Reactions on Ring Formation* in 2010; and Volume 6, *Name Reactions in Heterocyclic Chemistry-2*, in 2011, respectively.

Continuing the traditions of the first two volumes, each name reaction in Volume 3 is also reviewed in seven sections:

1. *Description;*
2. *Historical Perspective;*
3. *Mechanism;*
4. *Variations and Improvements;*
5. *Synthetic Utility;*
6. *Experimental; and*
7. *References.*

I also introduced a symbol [R] to highlight review articles, book chapters, and books dedicated to the respective name reactions.

I have incurred many debts of gratitude to Prof. E. J. Corey. What he once told me — "*The desire to learn is the greatest gift from God*" — has been a true inspiration. Furthermore, it has been my great privilege and a pleasure to work with a collection of stellar contributing authors from both academia and industry. Some of them are world-renowned scholars in the field; some of them have worked intimately with the name reactions that they have reviewed; some of them even discovered the name reactions that they authored in this series. As a consequence, this book truly represents the state-of-the-art for *Name Reactions for Homologations*.

I welcome your critique.

Jack Li
October 1, 2008

Contributing authors:

Dr. Nadia M. Ahmad
Takeda Cambridge
418 Cambridge Science Park
Cambridge
CB4 0PA
United Kingdom

Dr. Marco A. Biamonte
Medicinal Chemistry
BIOGEN IDEC
5200 Research Place
San Diego, CA 92122

Dr. Victor Cee
Amgen, Inc.
Mailstop 29-1-B
One Amgen Center Drive
Thousand Oaks, CA 91320

Dr. Timothy T. Curran
Chemical Development
Vertex Pharmaceuticals
130 Waverly Street
Cambridge, MA 02139

Dr. Matthew J. Fuchter
Department of Chemistry
Imperial College London
London SW7 2AZ, UK

Dr. Paul Galatsis
Department of Chemistry
Pfizer Global Research & Development
Eastern Point Rd
Groton, CT 06340

Prof. Brian Goess
Department of Chemistry
Plyler Hall
Furman University
Greenville, SC 29613

Dr. David L. Gray
Pfizer Global Research & Development
Eastern Point Rd
Groton, CT 06340

Prof. Gordon W. Gribble
Department of Chemistry
6128 Burke Laboratory
Dartmouth College
Hanover, NH 03755

Dr. Timothy J. Hagen
Medicinal Chemistry
deCODE
2501 Davey
Woodridge, IL 60517

Dr. Xiaojun Han
Discovery Chemistry
Bristol-Myers Squibb Company
5 Research Parkway
Wallingford, CT 06492

Dr. Chris Limberakis
Pfizer Global Research and Development
Eastern Point Road
Groton, CT 06340

Dr. Ivar M. McDonald
Discovery Chemistry
Bristol-Myers Squibb Company
5 Research Parkway
Wallingford, CT 06492

Dr. John Mancuso
Department of Medicinal Chemistry
MethylGene Inc.
7220 Frederick-Banting Rd.
Montreal, PQ H4S 2A1, Canada

Dr. Vincent Mascitti
Pfizer Global Research and Development
Eastern Point Road
Groton, CT 06340

Prof. Richard J. Mullins
Department of Chemistry
Xavier University
3800 Victory Parkway
Cincinnati, OH 45207-4221

Dr. Jeremy M. Richter
Discovery Chemistry
Bristol–Myers Squibb Company
311 Pennington–Rocky Hill Road,
Pennington, NJ 08534

Dr. Frank Rong
ChemPartner
No.2 Building, 998 Halei Road
Zhangjiang Hi-Tech Park, Pudong
Shanghai, China 201203

Prof. Kevin M. Shea
Department of Chemistry
Clark Science Center
Smith College
Northampton, MA 01063

Xiangdong Alan Wang
Discovery Chemistry
Bristol-Myers Squibb Company
5 Research Parkway
Wallingford, CT 06492

Prof. John P. Wolfe
Department of Chemistry
University of Michigan
930 N. University Ave.
Ann Arbor, MI 48109

Dr. Larry Yet
Albany Molecular Research, Inc.
21 Corporate Circle
P.O. Box 15098
Albany, NY 12212-5098

Dr. Ji Zhang
Process Research & Development
Pharmaceutical Research Institute
Bristol-Myers Squibb
Princeton, NJ 08543-4000

Table of Contents

Foreword		vi
Preface		vii
Contributing Authors		viii
Chapter 1.	**Organometallics**	**1**
Section 1.1	***Palladium Chemistry***	***2***
1.1.1	Heck reaction	2
1.1.2	Hiyama cross-coupling reaction	33
1.1.3	Kumada cross-coupling reaction	47
1.1.4	Negishi cross-coupling reaction	70
1.1.5	Sonogashira reaction	100
1.1.6	Stille coupling	133
1.1.7	Suzuki coupling	163
1.1.8	Tsuji–Trost reaction	185
Section 1.2	***Organocopper Reagents***	***212***
1.2.1	Castro–Stephens coupling	212
1.2.2	Glaser coupling	236
1.2.3	Ullmann reaction	258
Section 1.3	***Other Organometallic Reagents***	***268***
1.3.1	McMurry coupling	268
1.3.2	Nicholas reaction	284
1.3.3	Nozaki–Hiyama–Kishi reaction	299
1.3.4	Tebbe olefination	319
Chapter 2.	**Carbon-Chain Homologations**	**335**
2.1	Arndt–Eistert homologation	336
2.2	Morita–Baylis–Hillman reaction	350
2.3	Benzoin condensation	381
2.4	Corey–Fuchs reaction	393
2.5	Henry reaction	404
2.6	Horner–Wadsworth–Emmons reaction	420
2.7	Julia–Lythgoe olefination	447
2.8	Knoevenagel condensation	474
2.9	Mukaiyama aldol reaction	502
2.10	Peterson olefination	521
2.11	Sakurai allylation reaction	539

2.12 Stetter reaction 576
2.13 Wittig reaction 588

Chapter 3. Radical Chemistry 613

3.1 Barton–McCombie deoxygenation 614
3.2 Barton nitrite photolysis 633
3.3 Sandmeyer reaction 648
3.4 Wohl–Ziegler reaction 661

Appendixes

Appendix 1, Table of Contents for Volume 1: *Name Reactions in Heterocyclic Chemistry* 675
Appendix 2, Table of Contents for Volume 2: *Name Reactions for Functional Group Transformations* 678
Appendix 3, Table of Contents for Volume 4: *Name Reactions for Homologations-2* 680
Appendix 4, Table of Contents for Volume 5: *Name Reactions for Ring Formations* 682
Appendix 5, Table of Contents for Volume 6: *Name Reactions in Heterocyclic Chemistry-II* 684

Subject index 687

Chapter 1. Organometallics 1

Section 1.1	***Palladium Chemistry***	***2***
1.1.1	Heck reaction	2
1.1.2	Hiyama cross-coupling reaction	33
1.1.3	Kumada cross-coupling reaction	47
1.1.4	Negishi cross-coupling reaction	70
1.1.5	Sonogashira reaction	100
1.1.6	Stille coupling	133
1.1.7	Suzuki coupling	163
1.1.8	Tsuji–Trost reaction	185
Section 1.2	***Organocopper Reagents***	***212***
1.2.1	Castro–Stephens coupling	212
1.2.2	Glaser coupling	236
1.2.3	Ullmann reaction	258
Section 1.3	***Other Organometallic Reagents***	***268***
1.3.1	McMurry coupling	268
1.3.2	Nicholas reaction	284
1.3.3	Nozaki–Hiyama–Kishi reaction	299
1.3.4	Tebbe olefination	319

1.1.1 Heck Reaction

Mathew J. Fuchter

1.1.1.1 Description

The Heck reaction is the palladium-catalyzed alkenylation or arylation of olefins.[1–24] It has become one of the most widely used C–C bond forming tools in organic synthesis.

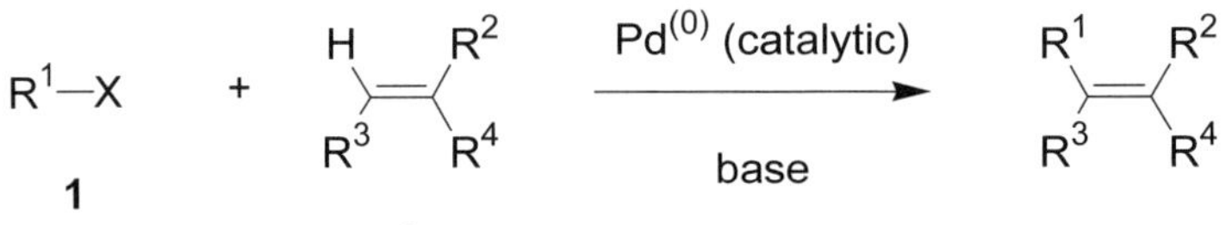

1 2 3

R^1 = aryl, alkenyl, alkyl (with no β-hydrogen)

X = Cl, Br, I, OTf, OTs, N_2^+

An extensive range of functional groups and substitution patterns on the olefin **2** are tolerated and aryl, alkenyl and some alkyl (lacking β-hydrogen atoms) electrophiles **1** are suitable reaction partners. The active catalyst is generated *in situ* from a variety of available palladium(0) or palladium(II) precatalysts ($Pd(OAc)_2$, $Pd_2(dba)_3$, *etc.*). A large number of ligands have been employed including phosphines, palladacycles and carbenes and even "ligand-free" conditions are commonly exploited.[16] By using enantiomerically pure chiral ligands, the reaction can be rendered stereoselective (at centres adjacent to the newly formed olefin). A stoichiometric amount of base is needed, but in practice 3-5 molar equivalents are often used. Tertiary amine bases (for example Et_3N or PMP) or inorganic bases such as K_2CO_3 can be employed. Halide-scavenging additives (such as Ag_3PO_4) can be useful for aryl/alkenyl halide substrates, especially in the case of asymmetric Heck reactions. The reaction tolerates a range of solvents, however polar aprotic solvents such as DMF or NMP are most frequently utilized. The reaction most commonly takes place at elevated temperatures.

1.1.1.2 Historical Perspective

In the early 1970s, T. Mizoroki and R. F. Heck independently discovered that aryl, benzyl and styryl halides react with olefinic compounds and elevated temperature in the presence of a hindered amine base and a catalytic amount of palladium.[25,26] This was based on the previous work of Heck when he was at the Hercules Powder Company, Delaware in 1968. He discovered that when palladium(II) chloride (interestingly the use of palladium was inspired by a colleague studying the Wacker reaction) was dissolved in acetonitrile with phenylmercuric chloride (**4**) and ethylene gas (**5**), the presumed transient phenylpalladium chloride rapidly absorbed one equivalent of ethylene to produce styrene (**6**) in high yield.[27]

4 **5** **6**

As reports began to appear on the formation of halo(aryl)palladium-phosphine complexes, Heck hypothesized that these intermediates could replace the arylmercurial-palladium combination. Crucially, he also reasoned that the use of a base to quench the hydrogen halide generated would render the reaction catalytic. In the 1970s the use of palladium was considered exotic and the reaction a mere curiosity, which meant its importance was underestimated for decades. In 1982, Heck published a review that contained all known examples in a mere 45 pages.

Today however, the paramount importance of organopalladium chemistry has propelled the Heck reaction into one of the most widely used catalytic C–C bond forming reactions. This was driven by its operational simplicity, unprecedented functional group compatibility and wide applicability. Indeed, from materials science to enantioselective organic synthesis, nearly every sub-discipline of modern organic chemistry has embraced the Heck reaction. It is hard to overstate its importance and in fact, this reaction may also be considered as a forerunner to all other widely used palladium-catalyzed couplings (Stille, Suzuki, Negeshi, Hiyama, *etc.*). Perhaps the greatest social impact of the Heck reaction has been its use in the coupling of alkynes to aryl halides; a reaction which was used to couple fluorescent dyes to DNA bases, allowing the automation of DNA sequencing and the elucidation of the human genome.

1.1.1.3 Mechanism

The general mechanism for the Heck reaction has been accepted for many years however numerous recent studies have shown the active catalytic species to vary dramatically depending on the ligands, reaction conditions and substrates.

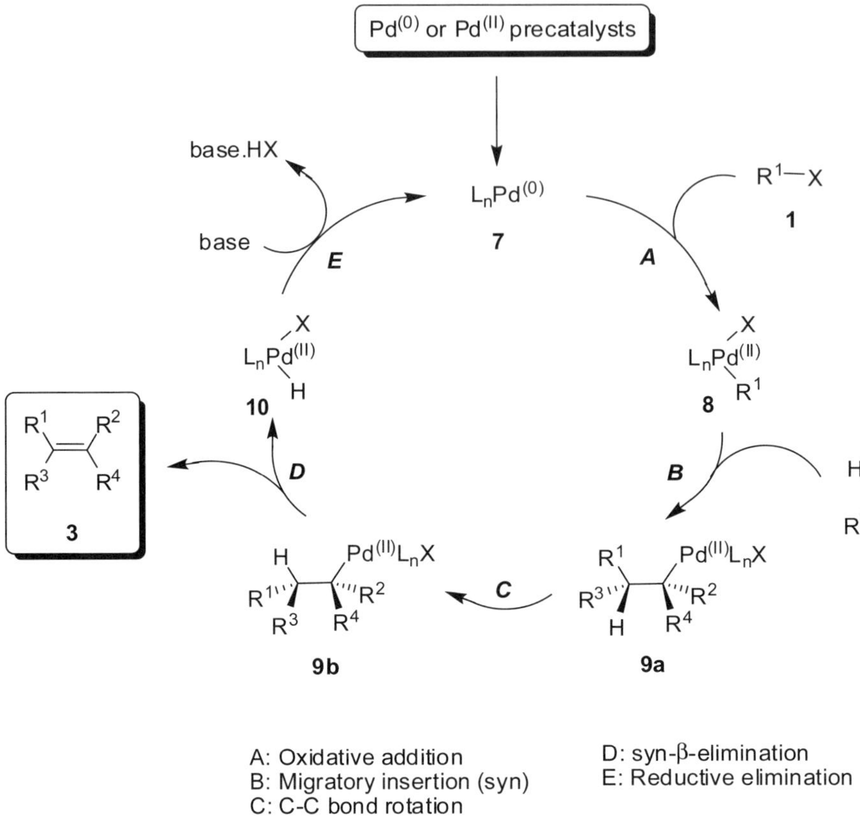

There are several extensive reviews regarding mechanistic studies on the Heck reaction which should be consulted for further detail.[16,28] In the basic mechanism, aryl (or alkenyl) halides or perfluorosulfonates **1** undergo oxidative addition to a palladium(0) catalyst **7** to afford a σ-arylpalladium(II) complex **8**. All palladium precatalysts are converted to the active palladium(0) catalyst **7** *in situ,* most commonly by phosphine in phosphine assisted catalytic cycles.[16] The order of reactivity for the oxidative addition is X = I > OTf > Br >> Cl.[29] Alkene coordination, followed by *syn* addition provides a σ-alkylpalladium(II) complex **9a**. It is widely accepted that this carbopalladation of an alkene is the rate-determining step for Heck reactions.[30] Rapid β-hydride elimination releases alkene product **3**. In order to undergo *syn*-β-hydride elimination, the palladium and hydrogen atoms must be co-planar (as in conformer **9b**). Finally, a base is required for the conversion of the hydridopalladium(II) complex **10** to the active palladium(0) catalyst **7**, completing the catalytic cycle. While useful in explaining the discreet mechanistic steps of the Heck reaction, this generalized scheme ignores the precise coordination number, geometry and formal charge on the palladium and therefore several more detailed mechanistic scenarios have been reported.

Cationic versus "Neutral" Pathways

Historically, the Heck reaction was the functionalization of olefins by aryl iodides, bromides, aroyl chlorides, or the corresponding vinyl halides, carried out without ligands in the case of aryl iodides or in the presence of monodentate phosphines (e.g., Ph_3P).[2] Under these reaction conditions a square planar palladium(II) oxidative addition complex with a weak Pd–PR_3 bond (or Pd–solvent in the case of iodides) and a strong Pd–X bond is generated.[6] Dissociation of one of these neutral ligands gives a free coordination site to which the alkene can bind. In this context, Heck reported that chelating phosphines "in general do not form useful catalysts".[31] Indeed, in the case of reactions employing aryl halide substrates and bidentate ligands, suppression of the reaction is observed due to competitive coordination of the chelating ligand, shifting the equilibrium of **11**/**12** to the left.[6]

Neutral mechanism

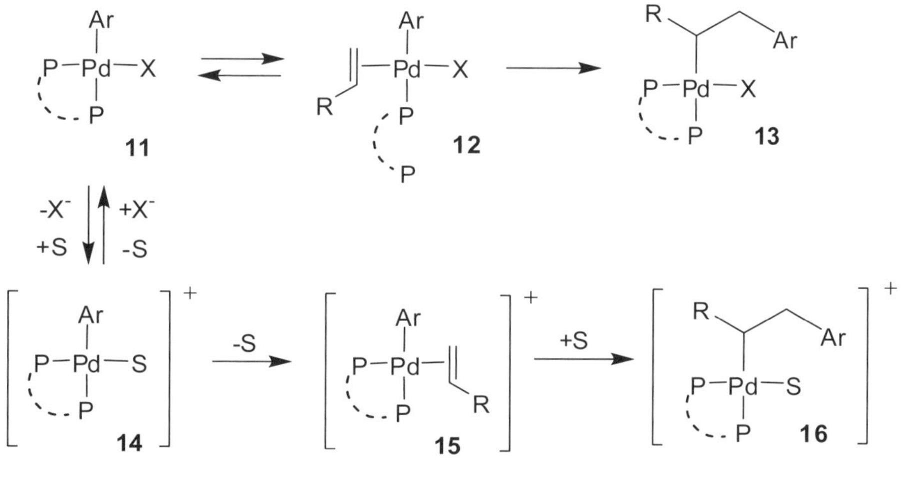

Cationic mechanism

The cationic reaction manifold was first reported independently by Capri[32] and Hayashi[33] to describe the Heck reaction of aryl triflates in the presence of palladium-diphosphine catalytic systems. This scenario arises from the lability of the Pd–OTf bond present in complex **11** (X = OTf).[6] Dissociation of the anionic counterion ($^{-}$OTf) affords cationic complex **14** with a vacant coordination site (transiently occupied by a solvent molecule,

S), thus allowing binding of the olefin without decomplexation of either phosphorus atom of the bidentate ligand. The ability to use chelating ligands was crucial to the development of asymmetric Heck reactions employing chiral diphosphines, first pioneered independently by Shibasaki and Overman in 1989:[17] Partial dissociation of the chiral bidentate ligand under "neutral" conditions would diminish the rigidity of the ligand and could lead to erosion of the enantioselectivity. As well as the use of aryl (alkenyl) triflates, aryl (alkenyl) halides can be used in the presence of Ag(I) or Tl(I) additives. These additives mediate halide extraction from complex **11**,[17] facilitating the cationic pathway. Furthermore, the reactivity of complexes **11** and **14** depends on the charge density of the unsaturated system. Competition studies have shown electron-poor olefins (good π-acceptors and poor σ-donors) react faster with neutral complex **11**, whereas electron-rich olefins (poor π-acceptors and good σ-donors) react faster with cationic complex **14**.[6]

Cabri has used the cationic pathway to explain the regioselectivity of the Heck reaction. Moreover, the use of triflates or halide scavenging additives (cationic conditions) in the asymmetric Heck reaction has become widespread. However, recent studies have shown the employment of either "neutral" conditions or cationic conditions may not lead to the expected reaction pathway. In 1992, Overman reported a Pd/BINAP catalyzed Heck cyclization of aryl halides in high enantioselectivity *without halide scavengers* (i.e. "neutral" conditions).[34] Since monodentate analogues used to mimic a partially dissociated BINAP gave products of low enantiopurity, it was rationalized that both phosphorus atoms remain coordinated to the palladium in the enantio-discriminating step, despite the "neutral" reaction conditions. Whilst Overman has suggested the reaction proceeds via an associative process involving a pentacoordinated palladium species, theoretical and experimental data has largely dismissed this due to high activation energies of the subsequent migratory insertion.[17]

Recent work by Amatore and Jutand on the Heck reaction of aryl palladium complexes ligated by 1,3-bis(diphenylphosphino)propane) (dppp) offers some fascinating new insight into the "neutral" *versus* cationic pathways.[30] Based on kinetic studies, they suggest the reactions of electron-rich alkenes (isobutyl vinyl ether) *always* proceed via a cationic mechanism, despite the use of so-called "neutral" conditions. The ability of the of the base to deliver anions (acetate, carbonate) is often overlooked, and regardless of the medium cationic complex **22** is always the most intrinsically reactive. As such, the regioselectivity of the Heck reaction of electron rich alkenes, as well as asymmetric Heck reactions under "neutral" conditions can be rationalized by considering the rates and equilibria constants of all palladium species under a given set of reaction conditions (concentration, ionic strength of the solvent, additives, counterions, substrates, *etc.*).[30]

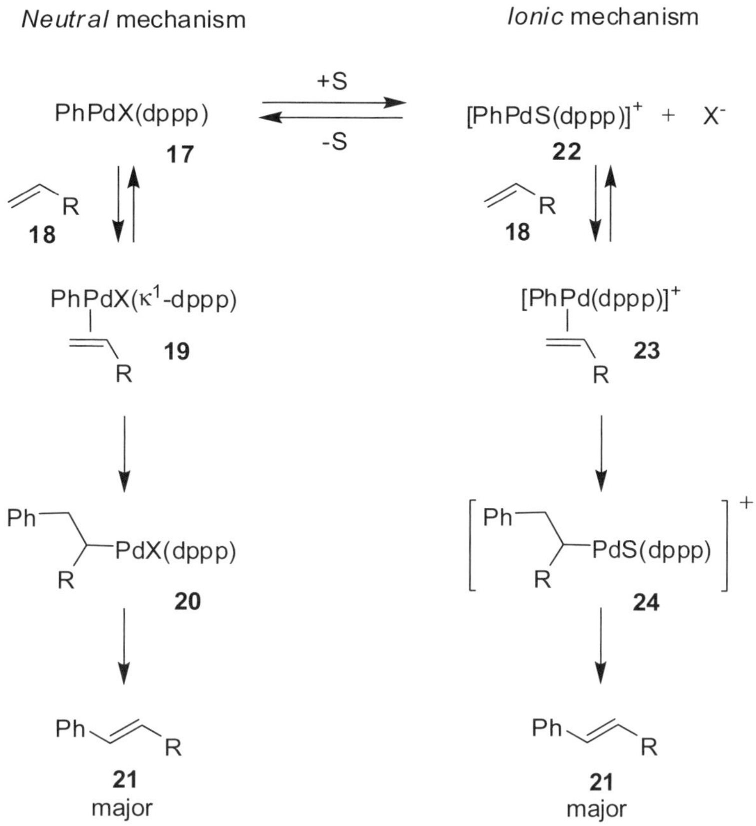

Anionic Pathways

In many palladium-mediated reactions, the exact role of the precatalyst is ignored and simply seen as a means to provide the active palladium(0) catalytic species. However, studies by Amatore and Jutand have shown the counterions of the precatalyst can be non-innocent and dramatically influence the reaction mechanism.[35] $Pd(OAc)_2$ is the most common precatalyst used in the Heck reaction and previous studies have considered the acetate anion as an innocent bystander. Experimental evidence now suggests that the $Pd(OAc)_2$/phosphine systems initiate a catalytic cycle involving anionic palladium(0) and palladium(II) complexes.[35] The active catalyst generated is anionic species **26**, which undergoes oxidative addition to afford a pentacoordinated palladium species **28**, where both the acetate and iodide anions remain ligated to the palladium(II) centre. This short-lived species rapidly loses the halide ion to yield a new palladium(II) complex, *trans*-$[ArPd(OAc)(PPh_3)_2]$ (**29**). The increased reactivity of complex **29** compared to $[ArPdI(PPh_3)_2]$ has been attributed to the bidentate nature of the acetate ligand, which may assist in phosphine release to open a coordination site for the alkene substrate. Migratory insertion, followed by β-hydride elimination provides olefin **31** and hydridopalladium complex **32**. Base-mediated conversion to palladium(0) species **26** completes the catalytic cycle. This proposed mechanism not only details the crucial role of acetate ions in many Heck reactions, but also provides and explanation of the beneficial effects of additives such as KOAc in certain cases.[17]

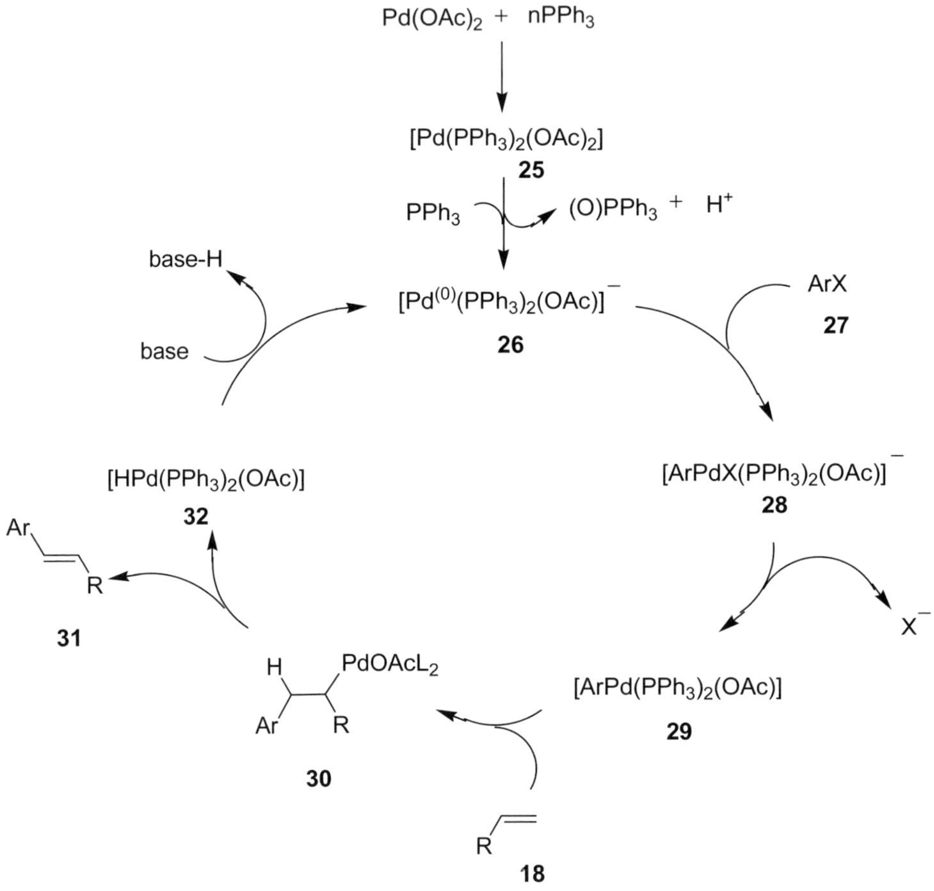

Related studies on the formation of active catalytic species' derived from $Pd(OAc)_2$ and bidentate phosphine ligands has also been reported.[36] A stable palladium(0) complex is formed in the presence of $Pd(OAc)_2$, dppp, water and triethylamine. In this case oxidative addition to PhI gives the cationic complex $[PhPd(dppp)(dppp(O))]^+$, in which the oxidized dppp(O) ligand is monodentate. The complex [PhPd(OAc)(dppp)] is only formed on addition of excess acetate anions. These results suggest that the anionic pathway could be relevant to systems employing chelating phosphines, in the presence of acetate additives.

Regioselectivity and Stereoselectivity

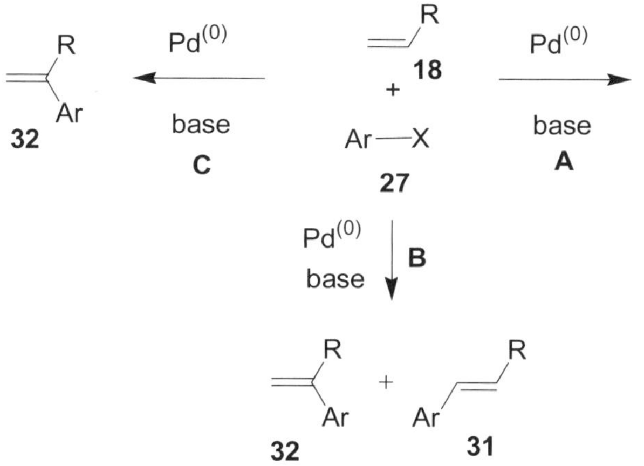

A: Where R = Aryl, alkyl, CO_2R, CN, etc.
B: Where R = OR, NR_2, etc.
C: Where R = OR, NR_2, etc. and X = OTf

The direction of addition of the organopalladium species to the olefin is almost exclusively sterically controlled, i.e. addition will take place at the least substituted carbon to provide the linear product **31**.[2] In the case of alkenes containing electron withdrawing double bonds, once again, addition predominantly gives the linear product **31**. For alkenes appended with electron donating groups however, mixtures are often obtained with the sterically favoured isomer predominating. Work by Cabri using aryl triflates or aryl halides with halide scavengers (cationic pathway) has demonstrated, under these conditions, the branched product **32** is obtained in high selectivity for electron-rich olefins.[6] A recent mechanistic rationale for this effect has been reported.[30] While this is a useful guide, it is possible to override this intrinsic regioselectivity bias using other factors such as chelation control. Electron-rich olefins bearing pendant heteroatom functionalized substituents can form linear products exclusively by exploiting neighbouring-group effects.[23]

Other selectivity issues can arise during β-hydride elimination. If there is more than one sp^3-bonded hydrogen atom beta to the palladium group in the olefin adduct, then a mixture of geometric isomers may result.[2] Also, if the hydridopalladium(II) species is not scavenged fast enough by the base, re-addition to the double bond may occur and once again, a mixture of geometric isomers may result.[16] The hydridopalladium(II) species can also be potentially scavenged by the starting olefin; a process which results in isomerization of the starting alkene and therefore leads to the formation of isomeric Heck products.[16] In certain cases it has been demonstrated that the use of low temperatures[37] or additives such as silver salts can minimize this type of alkene isomerization.[38]

The stereoselectivity of the Heck reaction is governed by *syn*-β-hydride elimination. In the majority of cases, the elimination obeys the Curtin–Hammett kinetic control principle[39] and the ratio of *E*- and *Z*-isomers reflects the relative energy of the respective transition states. Unless R (see **18**) is very small (for example CN), the *E*-isomer is predominant and the reaction is highly stereoselective.[16]

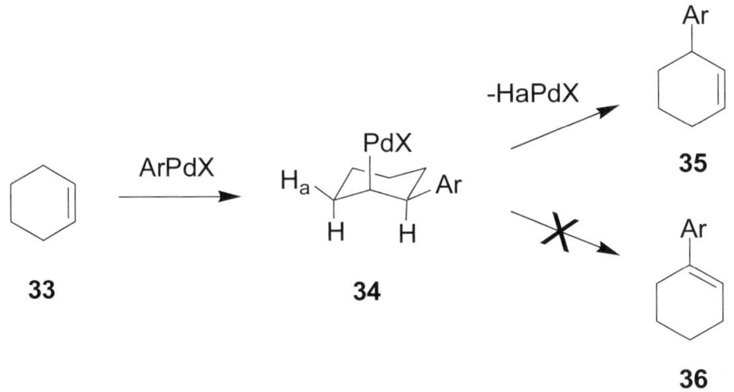

Selectivity issues noted above are largely irrelevant for intramolecular reactions. In these cases, regiocontrol in the migratory insertion is largely governed by the size of the ring being formed with 5-*exo* and 6-*exo* cyclizations being particularly favoured.[17] The use of cyclic olefin substrates also aids the regioselectivity of the reaction. Stereospecific *syn* addition of an arylpalladium species to a cyclic alkene, such as cyclohexene (**33**) produces σ-alkylpalladium(II) intermediate **34**, bearing a single *syn*-β-hydrogen (H_a). *Syn* elimination of this hydrogen provides product **35** exclusively (providing no isomerization of the product occurs under the reaction conditions, *vide supra*).[17] As a notable alternative, Tietze has used allyl silanes to control β-elimination in acyclic systems.[40] The additional elements of control in the intramolecular Heck reaction of cyclic substrates is the reason for its huge success in asymmetric, complex molecule synthesis.[17]

"Ligand-free" Catalysts

In some of his original work, Heck demonstrated that the reaction of aryl iodides can be carried out using $Pd(OAc)_2$ in the absence of additional ligands.[2] It was subsequently shown by Jeffery that this works particularly well in the presence of tetraalkyl ammonium salts.[7] A detailed mechanism for the "ligand-free" Heck reaction has been reported by de Vries.[24]

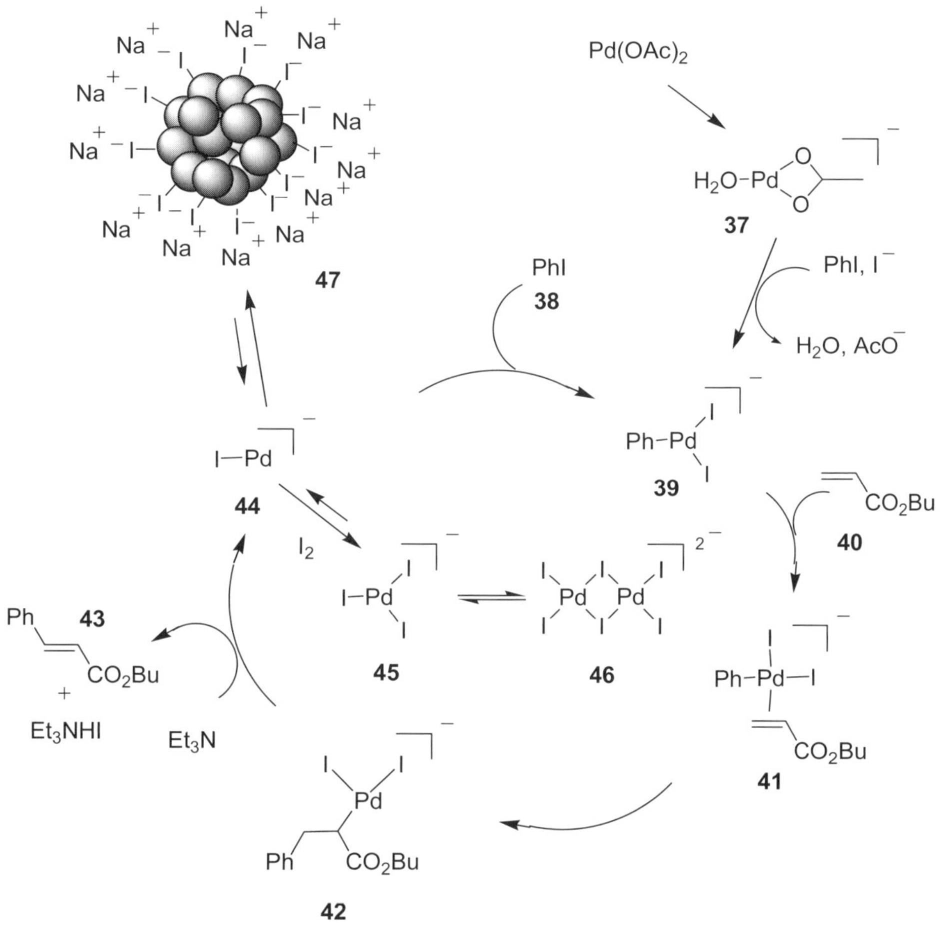

The reaction of iodobenzene (**38**) and olefin **40** was studied under "ligand-free" conditions in the presence of sodium halide additives. As in the generalized mechanism, $Pd(OAc)_2$ is reduced to transient palladium(0) complex **37**. This undergoes oxidative addition and ligand exchange to form anionic species **39** (potentially containing additional solvent ligands). Olefin complexation, migratory insertion and β-hydride elimination furnish the product **43** along with highly underligated species **44**. At this stage complex **44** can do one of three things: (a) react with traces of iodine (arsing from aerobic oxidation), to give **45** or dimer **46**; (b) form soluble palladium nanoparticles **47**; or (c) react with iodobenzene (**38**). Since oxidative addition is fast for reactive aryl iodides, (c) is the primary pathway observed. Once the substrate is consumed, rapid formation of palladium nanoparticles **47** occurs, which in turn conglomerate to form palladium black.[24]

For the less-reactive aryl bromides, the situations change. Since oxidative addition of these substrates is slower, formation of palladium nanoparticles is prevalent.[24] If these particles grow beyond a certain size, they precipitate as palladium black and the reaction stops. This is the reason aryl iodides were initially reported to be the only substrates to undergo the Heck reaction under "ligand-free" conditions. Therefore, one explanation of the success of Jeffery's conditions[7] is that the tetraalkyl ammonium additives stabilize the palladium nanoparticles/colloids, preventing formation of palladium black. Indeed, the pioneering work of Reetz[70] and Hermann[71] has shown that pre-formed stabilized palladium colloids can be used as active catalysts in the reaction. This unifying mechanism can be extended to other high-temperature Heck reactions (solid-supported palladium, palladacycles) in the absence of strongly coordinating ligands.[24] Of perhaps the most experimental significance is that, while most groups have sort stabilizing agents to prevent aggregation of colloidal palladium in these reactions, de Vries has demonstrated that simply maintaining a low substrate/catalyst ratio allows the Heck reaction to compete with colloid formation. Using these conditions he has successful used aryl bromides under "ligand-free" Heck reactions, in the absence of any additional stabilising agents.[24]

1.1.1.4 Synthetic Utility

In his initial review on the scope of the reaction, Heck reported the use of a variety of relatively simple aryl, heteroaryl and vinyl halides.[2] For example, exposure of bromopyridine **48** to alcohol **49** gave ketone **51** in good yield, following tautomerization of the initial Heck adduct **50**.

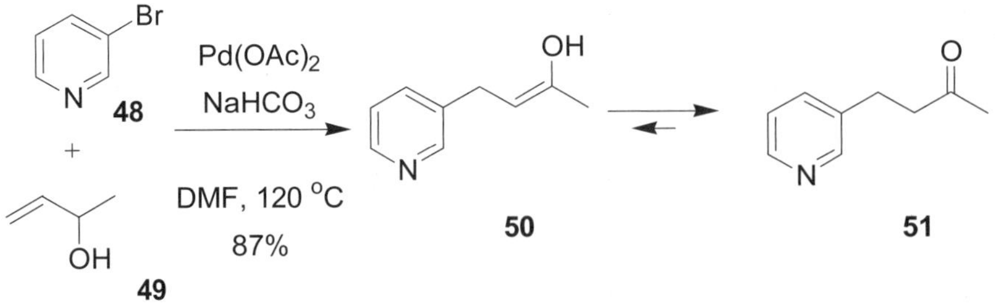

Nowadays however, the Heck reaction is one of the most widely used catalytic C–C bond forming reactions and there are numerous examples in nearly every sub-discipline of modern organic chemistry. The proceeding section will highlight some of the most accomplished uses of this flexible synthetic method.

Asymmetric Intramolecular Heck reaction

Perhaps one of the most challenging aspects of complex molecule synthesis is control of the absolute sense of stereochemistry for the preparation of optically-active compounds. In 1989, Shibasaki and Overman independently reported the first examples of asymmetric Heck reactions.[17] These efforts focused on intramolecular cyclization reactions, which display extra elements of regiocontrol. To date, the asymmetric intramolecular Heck has been exploited in the synthesis of terpenoids, akaloids and polyketides, forging key tertiary and quaternary stereocentres.[17]

Some of the most spectacular examples have come from the laboratories of Overman at the University of California, Irvine. In his synthesis of (–)-quadrigemine C (**57**), Overman noted that the C3 and C3‴ quaternary stereocentres have the same absolute stereochemistry. Therefore, following a Stille reaction to prepare key substrate **54**, a double asymmetric Heck reaction was performed, yielding decacyclic system **56** in good yield and in 90% *ee*.[41] This example truly displays the synthetic power of the Heck reaction in forging quaternary, crowded stereocentres.

Interestingly, the Heck cyclization of anilides, such as **54**, has constituted a frequent strategy in the asymmetric synthesis of alkaloids.[17] While it is often assumed that migratory insertion of the arylpalladium(II) species into the carbon-carbon double bond is the stereocontrolling step, recent studies by Curran have offered an alternative explanation.[42]

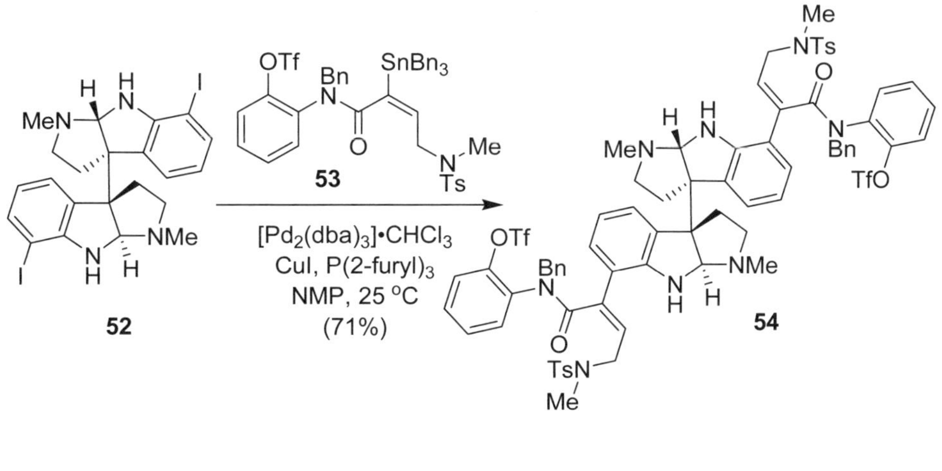

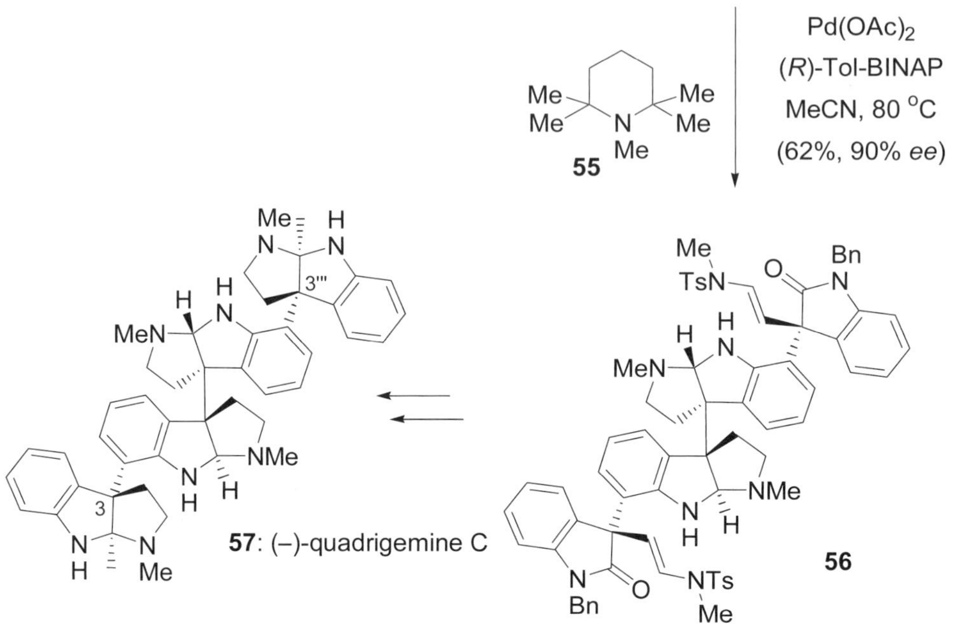

For iodoanilides, such as **58**, hindered rotation around the N–Ar bond renders these molecules axially chiral. Curran showed that low-temperature Heck reactions of chiral anilines **58** with an achiral palladium catalyst occur with efficient transfer of chirality from the chiral axis of the precursor to the stereocentre of the product. Since at high temperature the two axially chiral enantiomers will be rapidly equilibrating, this suggests that the stereocontrolling step in the asymmetric Heck reaction of similar substrates is, in fact, a dynamic kinetic resolution (oxidative addition to the aryl–X bond).

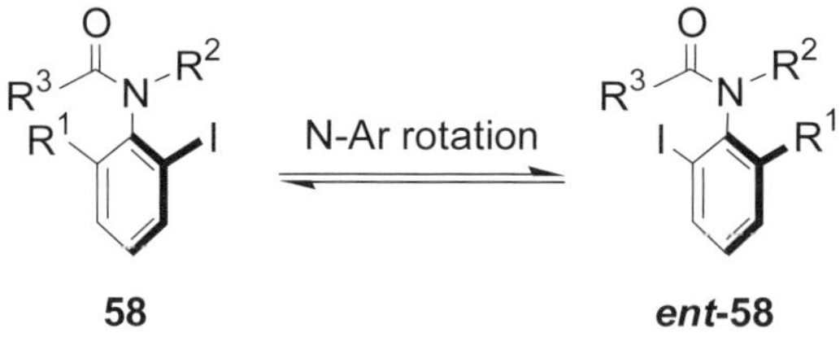

Exposure of prochiral cyclopentadiene **59** to catalytic $Pd(OAc)_2$, (*S*)-BINAP and *n*-BuNOAc furnished diquinane **64** in 89% yield and 80% *ee*. The mechanism presumably involves oxidative addition, followed by coordination of either enantiotopic double bond to yield diastereomeric intermediates **60** and **62**. The energetically favoured complex **62** undergoes insertion followed by rapid σ–π isomerization to generate the π-allyl palladium species **63**. Trapping of the intermediate with an acetate anion proceeds with good control of the regioselectivity (attack at the least hindered terminus of the π-allyl complex **63**) and stereoselectivity (attack on the opposite face to palladium) to yield **64**.[43]

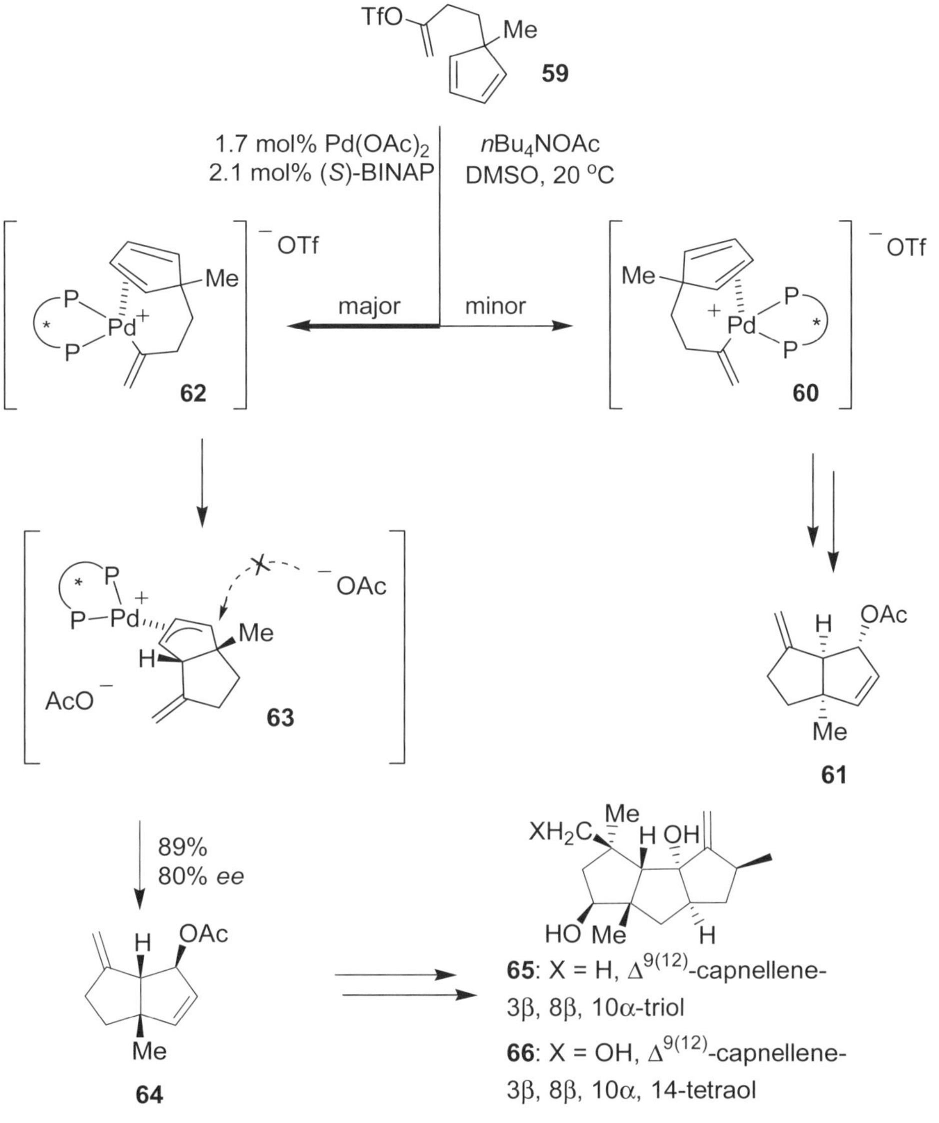

Shibasaki has also reported impressive applications of the asymmetric intramolecular Heck reaction. For example, the Shibasaki group have applied their chemistry to the synthesis of compound **64**, a key intermediate in the total synthesis of two complex triquinane sesquiterpenes **65** and **66**, by making use of a Heck reaction/anion capture cascade sequence.[43]

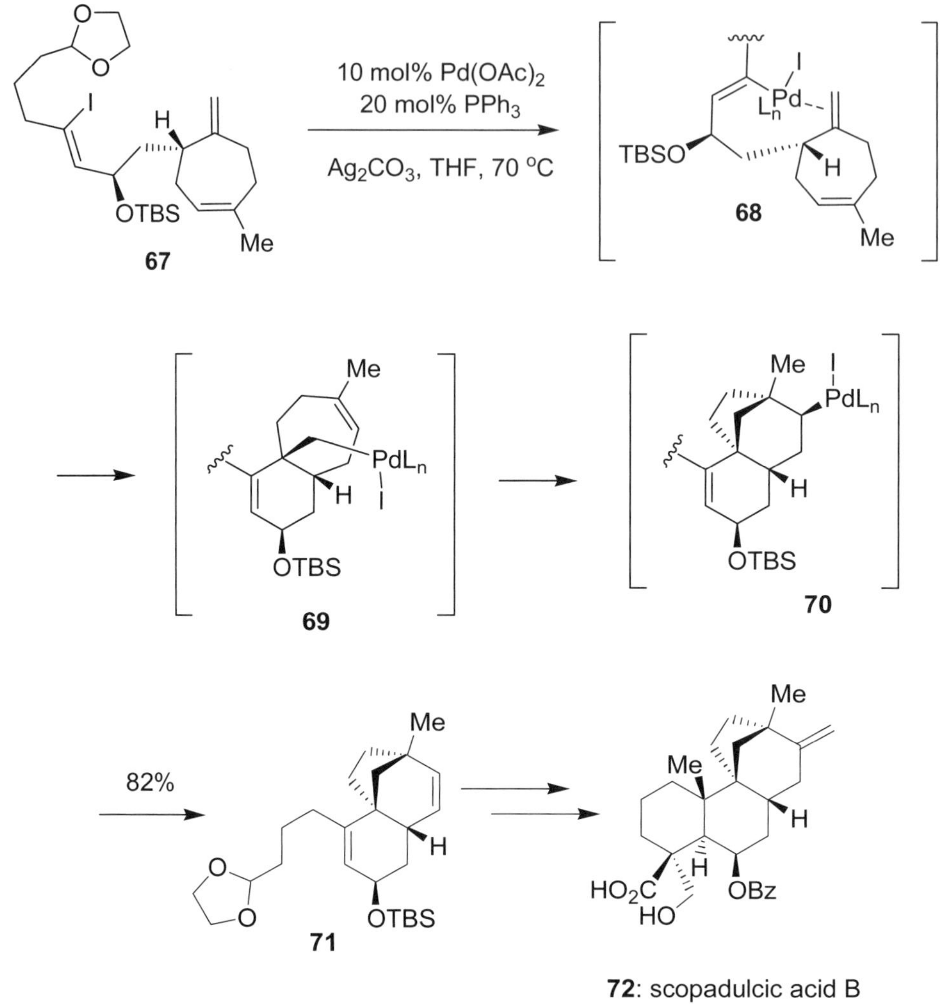

One of the most enabling features of the Heck reaction is the ability to facilitate polyene cyclizations in the synthesis of complex multiple ring systems. In the context of natural product synthesis, Overman pioneered this approach towards the synthesis of the scopadulcic acid family of diterpenes.[44] This inventive strategy formed three out of the four ring systems, including a sterically congested bridged bicycle and two of the three quaternary stereocentres from a simple monocyclic precursor, employing an Heck cyclization cascade. In this case the stereochemistry of the product was under substrate control (i.e. there was no need for chiral ligands). Thus, compound **67** was converted into tricyclic intermediate **71** in one step, using catalytic $Pd(OAc)_2$, PPh_3 and a silver additive. Oxidative addition of the palladium(0) species into the C–I bond of compound **67** followed by the first cyclization, gave intermediate **69**, which was unable to undergo β-hydride elimination due to the lack of suitably disposed hydrogen atoms. A second migratory insertion furnished intermediate **70**, which rapidly underwent β-hydride elimination to yield key intermediate **71**. Compound **71** was subsequently elaborated to scopadulcic acid B (**72**).

Asymmetric intermolecular Heck reaction

Perhaps the ultimate goal in the application of the asymmetric Heck reaction is to control both regio- and stereoselectivity in *inter*molecular reactions, which lack the extra elements of regiocontrol, compared to their intramolecular counterparts. Hayashi et al. reported the first example of an asymmetric intermolecular Heck reaction in 1991.[45] In their study, the use of aryl triflates was essential to achieve high levels of enantioinduction. Numerous examples have subsequently been reported in the literature with limited success.[19] In one impressive example, Hayashi reported the Heck reaction of dihydropyrrole **73** and alkenyl triflate **74** using (*R*)-BINAP as the chiral ligand. The product **75** was isolated in high yield and excellent enantioselectivity as the sole regioisomer.[46]

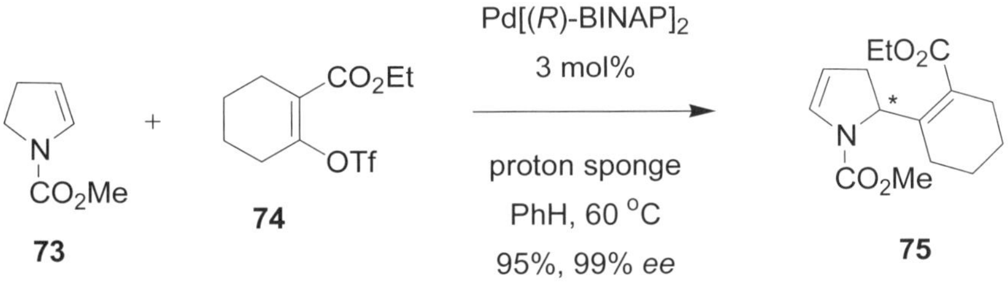

73 **74** **75**

The majority of asymmetric Heck reactions reported employ BINAP as the as the chiral ligand, however initial reports suggest other ligating molecules may offer some benefits.[19] A dramatic example of this was the introduction of oxazoline-based *P,N*-ligands by Pfaltz. Using these ligands, several previously reported cases of asymmetric Heck reactions were improved, giving excellent enantioselectivities (> 99% *ee*). Other studies have reported the use of alternative *P,N*-ligands, *N,N*-ligands and bidentate phosphines.[19]

Zipper reactions

In 1990, Negishi reported an impressive example of a domino Heck cascade reaction, where an acyclic polyunsaturated precursor was transformed into the tetracyclic steroidal ring framework in a single step.[48] Following this work, the Trost group reported several alternative examples of impressive cascade cyclizations.[49] For example, heptacyclic compound **78** was generated in a single synthetic step from precursor **76**. The first step of this cyclization cascade differs slightly from the traditional Heck mechanism, involving hydropalladation of an alkyne to give intermediate **77**. Subsequently, seven consecutive intramolecular Heck reactions furnished the polycyclic product **78** in good yield (76%) These cascade cyclizations were termed "zipper reactions" and nicely illustrate the power of this methodology.

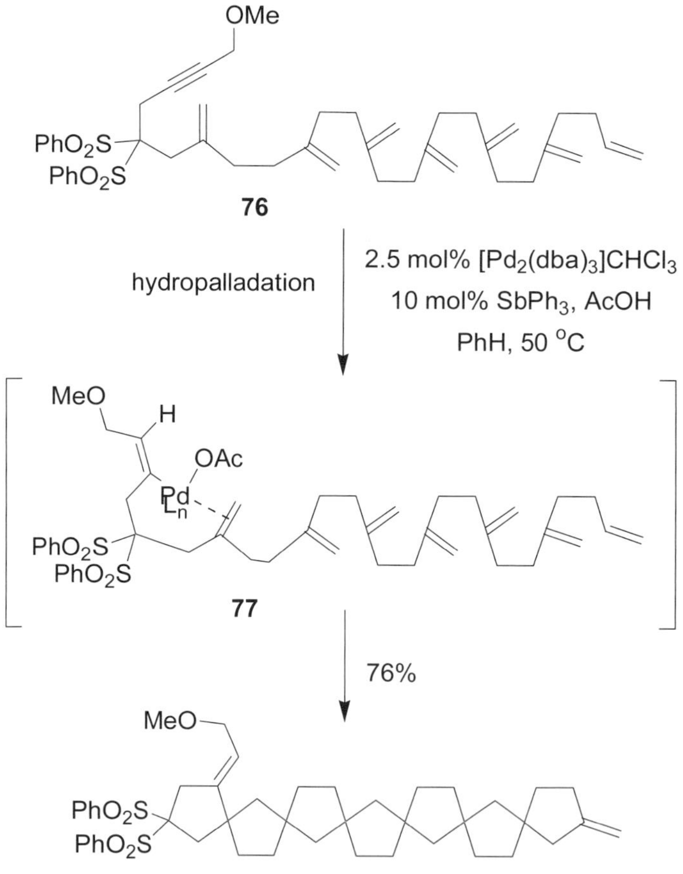

78

Highly conjugated systems

The Heck reaction has also proved extremely useful in the synthesis of highly conjugated organic materials.[4,16] For example, using Jeffery's conditions, extended derivatives of paracyclophane **79** were prepared.[50]

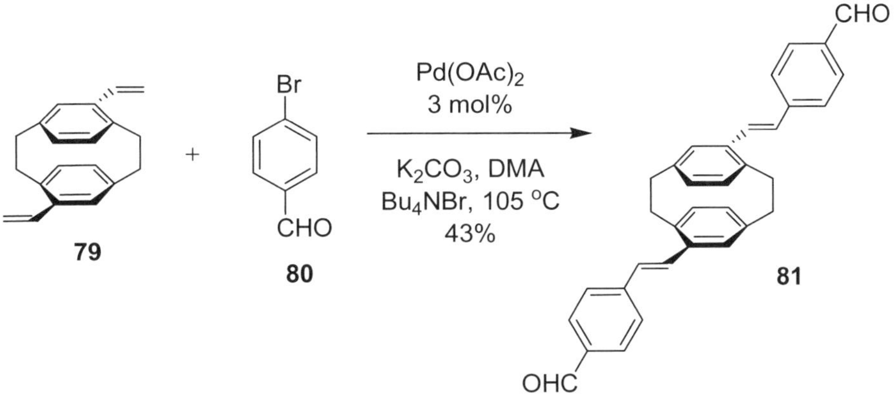

Although the preparation of hexaalkynylbenzenes had been reported in the 1980s, the synthesis of the corresponding hexaalkenyl derivatives was largely unknown. Studies by de Meijere demonstrated that the Heck reaction could be used to prepare conjugated molecules, such as **84** in good yields.[4] Interestingly once again the application of Heck's original procedure failed to initiate the reaction, whereas Jeffery's modified conditions successfully furnished the desired product **84**.

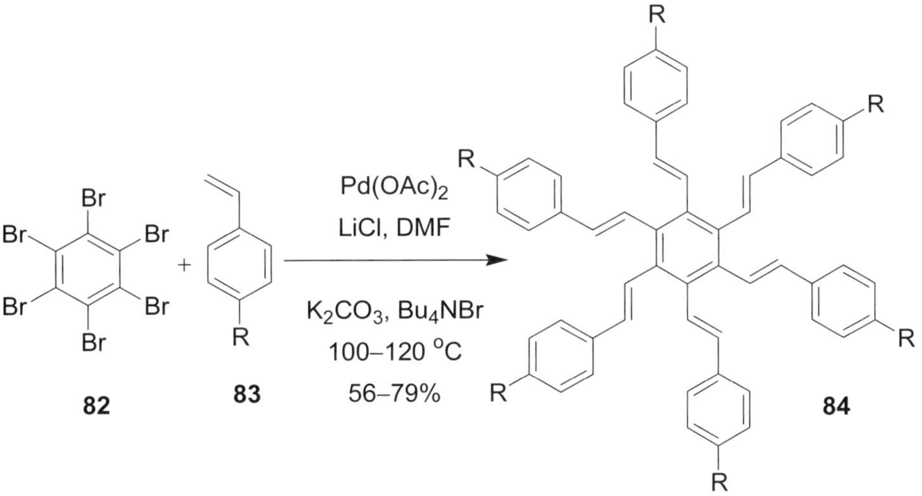

Polymers

Due to its mildness and efficiency the Heck reaction has been explored for polymerization chemistry.[4] For example, oligomeric stilbenes, such as **87** were successfully prepared and studied in the context of photochemically induced reactions.[51]

Likewise, poly(*p*-phenylenevinylene) polymers are of particular interest in the field of organoelectronics and the Heck reaction has proved a useful tool in their construction.[52]

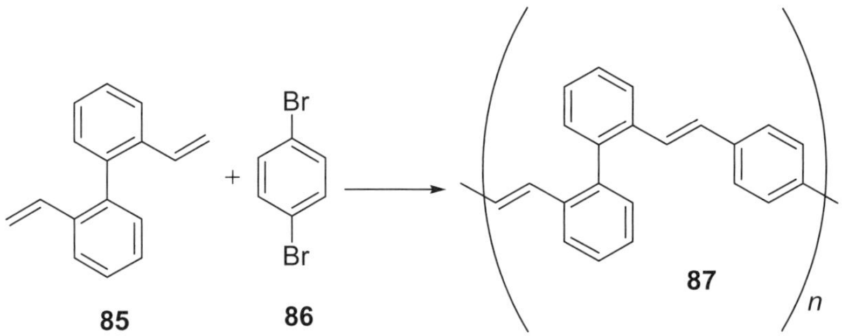

85 **86** **87**

The use of resin-bound substrates in combinatorial chemistry is often essential to creating diverse chemical libraries. Such polymer-supported reagents are suitable for Heck-type chemistry; for example resin-bound iodobenzyl alcohol **88** was found to undergo reaction with a large range of olefins.[53]

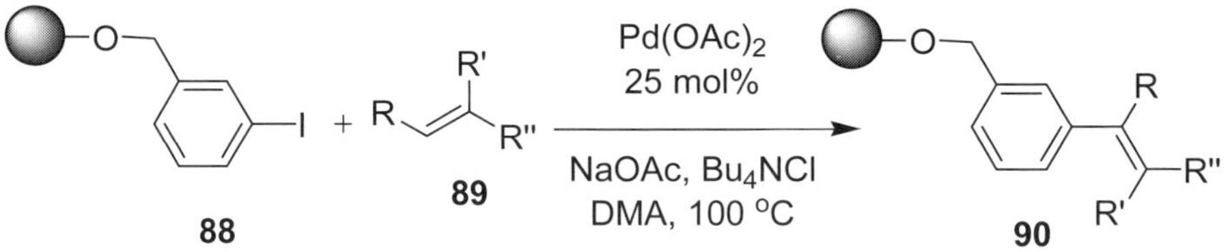

88 **89** **90**

1.1.1.5 Variations and Improvements

Ligands with the highest activity

Numerous ligands have been applied successfully to the Heck reaction, including phosphines, phosphites, palladacycles and carbenes.[16] Of the reported ligands however, several stand out for their high activity and substrate scope. Perhaps the greatest advance of this chemistry was the discovery of ligands which enable the Heck reaction of aryl chlorides.[16] Aryl chlorides are more readily available and less expensive than their aryl bromide and aryl iodide counterparts and therefore the ability to use them in palladium-mediated transformations vastly improves the industrial relevance of these processes. Several of the most active catalysts are highlighted below.

Phosphines: Spencer demonstrated vast improvements in the phosphine-assisted Heck reaction in the early 1980s.[54] He showed that the use of $P(o\text{-Tol})_3$ in polar aprotic solvents can give high turn-over number (TON) processes. Heck first reported the use of $P(o\text{-Tol})_3$[2] and interestingly this ligand is a key precursor to a highly active palladacycle (*vide supra*). Following these early efforts, in 1992 Milstein reported the use of bulky, electron-rich chelating phosphines (1,3-bis(diisopropylphosphino)propane) in the Heck reaction of aryl chlorides at high temperatures.[55] In this case a highly reactive monochelated palladium(0) species is formed which can undergo oxidative addition to aryl chlorides. Following subsequent de-ligation of the chloride ligand, the complex undergoes migratory insertion via the cationic route. Perhaps the greatest discovery however, was the use of the electron-rich, bulky monophosphine, tri-*tert*-butylphosphine ($P(t\text{-Bu})_3$) by G. Fu at MIT in 1999.[56] In this study, he demonstrated that aryl chlorides could be effectively used in the Heck reaction at 100 °C. He subsequently reported a slightly modified procedure in 2001 where aryl bromides and aryl chlorides could undergo the Heck reaction at room temperature.[57] Amongst the impressive results in this publication, Fu demonstrated that bulky, *ortho*-substituted aryl bromides (for example **91**) and activated (electron-poor) aryl chlorides, such as **96** can undergo the Heck reaction at room temperature in high yield. He also demonstrated unactivated (electron-rich) aryl chlorides can undergo Heck coupling at relatively low, elevated temperatures (70 °C), and at higher temperatures (120 °C), the loading of palladium can be reduced to 0.1 mol%.[57] Hartwig has also reported active bulky phosphine catalysts for the Heck reaction under mild conditions.[58]

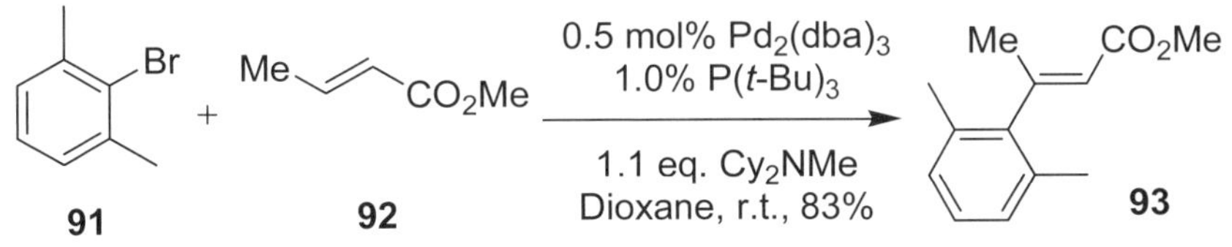

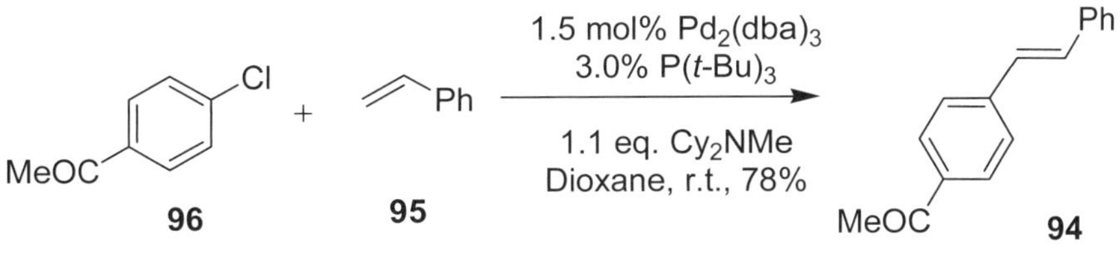

Traditionally, aryl chlorides are usually regarded as unreactive due to their reluctance to oxidatively add to palladium(0) complexes.[59] While intuitively it would seem bulky, electron-rich phosphines such as ($P(t\text{-}Bu)_3$) facilitate this reaction by increasing the electron density on palladium, studies have shown the most likely explanation due to the formation of a highly reactive monoligated palladium species (PdL).[60] These heavily underligated complexes show a vastly increased reactivity towards oxidative addition. Additional studies by Fu have highlighted a dichotomy however, since other bulky phosphines, such as tricyclohexylphosphine (PCy_3), do not furnish active catalysts in the Heck reaction (whereas they do in the case of other palladium-mediated cross coupling reactions).[61] He has attributed this to the base-mediated palladium(0) regeneration step ($L_2Pd^{(II)}HX \rightarrow Pd^{(0)}L_2$) of the catalytic cycle being kinetically slow and thermodynamically unfavourable in certain cases. This suggests the difference in reactivity between $P(t\text{-}Bu)_3$ and PCy_3 is due to highly sensitive steric effects effecting the ability of the corresponding L_2PdHX complexes to reductively eliminate.

Palladacycles: As mentioned previously, the use of $P(o\text{-}Tol)_3$ as a ligand was pioneered by Spencer, however it was Herrmann who demonstrated its true potential.[62] He discovered that treatment of $Pd(OAc)_2$ with $P(o\text{-}Tol)_3$ in toluene actually affords a cyclometallated palladacycle **97**, now known as Herrmann's catalyst. When this complex was employed in the Heck reaction, efficient coupling of aryl bromides and activated aryl chlorides was observed at high temperatures.

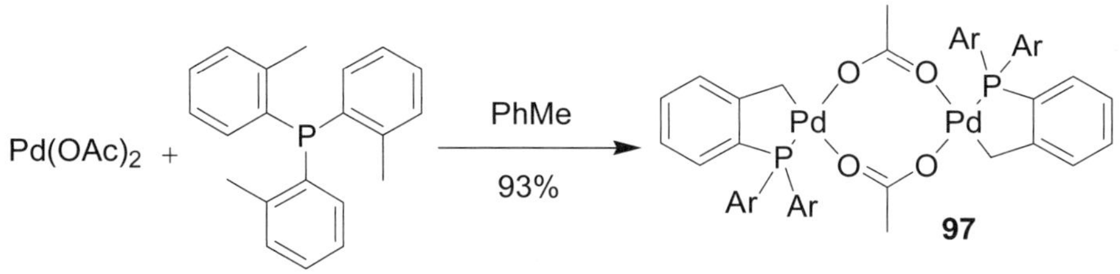

97

This highly active catalyst was the cause of much controversy however, in regards to a mechanistic understanding of its chemistry.[20] In Herrmann's original paper he hypothesized that since the catalyst is recovered unchanged from the reaction, the active species must be palladium(II), therefore invoking a Pd(II)/Pd(IV) catalytic cycle. This was further compounded by Shaw's speculations on exactly how this mechanistic cycle could operate.[63,64] However, despite these suggestions, elegant studies by Hartwig set the record straight.[65] He demonstrated that upon treatment with an amine base, palladacycle **97** fragmented to form a palladium(0) species, $Pd[(o\text{-}Tol)_3P]_2$. It therefore became apparent that the original speculations on the reactivity of $P(o\text{-}Tol)_3$ were most likely correct; the active catalyst is an underligated palladium species caused by the large cone angle of the ligand.[20] Hartwig's experiments also suggest that species **97** may lead, upon activation to colloidal palladium(0), and the mechanism therefore follows an analogous path to "ligand-free" reactions.[24] Further support for this mechanism can be gleamed from the fact that the Heck reaction of aryl chlorides with Herrmann's catalyst **97** required a tetraalkyl ammonium salt additive ($[Bu_4N]Br$),[62] and such additives are frequently employed to facilitate "ligand-free" reactions. Herrmann's pioneering studies sparked the synthesis of numerous other palladacycles and pincer-type complexes.[16]

Carbenes: The use of stable carbene ligands in palladium mediated processes was also pioneered by Herrmann,[66] however mechanistically, their role in the catalysis is unclear. Heterocyclic carbenes are strong σ-donor ligands and have stronger binding to palladium(II) than phosphines, therefore dissociative ligand processes are unlikely.[20] For this reason, olefin insertion has been postulated to occur via either a pentacoordinated complex or a cationic pathway (following halide decomplexation), both of which are unfavoured.[16,20] The enhancement of TONs by tetraalkyl ammonium salts however, suggests the presence of palladium colloids.[16] Therefore it would seem that, as in the case of Herrmann's palladacycle (**97**), the carbene complexes simply provide a source of palladium(0) colloids, which undergo the Heck reaction under "ligand-free" conditions.

"Ligand-free" and Supported Catalysts

As mentioned previously, Heck reported the ability to carry out the reaction of aryl iodides using $Pd(OAc)_2$ in the absence of additional ligands.[2] Indeed, it has been known for several decades that the Heck reaction using "naked" palladium can exhibit very high turnover frequencies (TOF), but short lifetimes, due to the lack of stabilizing ligands to prevent the formation of palladium black.[21] It was the seminal work of Jeffery which established the use of tetraalkyl ammonium salts to expand the applicability of "ligand-free" Heck reactions.[7] In fact the use of quaternary ammonium salts in the Heck reaction is often now referred to as Jeffery's conditions. The protocol gained popularity when, in 1987, Larock utilized these conditions in the intramolecular cyclization to form nitrogen heterocycles, demonstrating its potential in complex molecule synthesis.[67]

There have been numerous explanations for the beneficial effect of the quaternary ammonium salts including their ability to act as phase-transfer agents solubilizing inorganic bases, their role as a source of anions which can promote the reaction and their capacity to act as ion exchangers.[16] While these processes may be of relevance, their primary role appears to be as a stabilizing agent of the palladium nanoparticles formed in "ligand-free" reactions. Indeed, these salts are known to be active colloid stabilizers.[21] Further work by Reetz demonstrated that there is an induction period of approximately one hour in the Heck reaction of iodobenzene and ethyl acrylate.[68] After this period both catalysis and the formation of palladium colloids was observed.

Regardless of the active species, importantly, Jeffery's conditions allowed the reaction to be carried out at < 100 °C, at least in the case of aryl iodides.[16] This protocol has therefore been embraced in the synthesis of a diverse range of molecular targets. For example, the preparation of pyrroloquinoxaline **99**, described by the editor of this book, in high yield using Jeffery's conditions was achieved in just two hours.[69] As a comparison, the phosphine-assisted conditions only gave marginal yields due to the strong binding of the reagents and product to the catalyst.

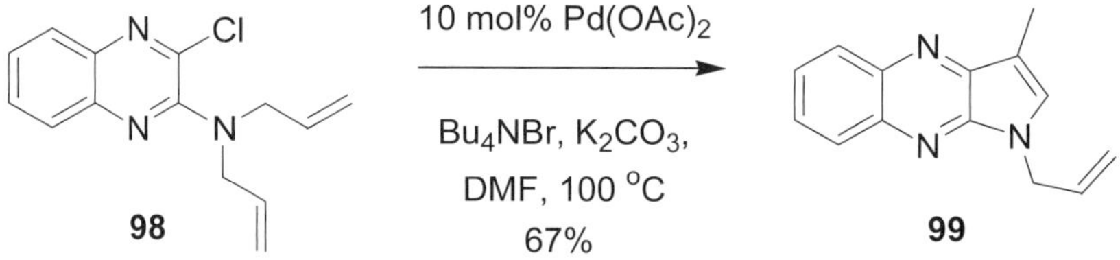

Since it is hypothesized that Jeffery's conditions are a means of utilizing palladium nanoparticles/colloids in the Heck reaction, it seems reasonable to suggest that pre-formed palladium colloids should also be catalytically active. Indeed, this idea has mostly been developed thanks to the seminal works of Reetz.[70] In 1996 he reported the preparation of uniform propylene carbonate stabilized palladium clusters (diameter 8–10 nm) which were stable up to 155 °C and catalytically active in the Heck reaction.[70] While Reetz's early colloids displayed some catalytic activity, it was the colloids reported by Beller and Herrmann, stabilized by tetraoctylammonium bromide, that started to show good activity.[71] Further dramatic improvements were made by Reetz, who reported the use of the stabilizing ligand *N,N*-dimethylglycine (DMG).[72] Impressive TON of 106,700 were achieved for the Heck reaction of bromobenzene with styrene using only 0.0009 mol% of palladium precatalyst and a 20-fold amount of DMG. In his continuing studies, Reetz later noted the use of tetraarylphosphonium salts as an additional additive, enabling the coupling of chlorobenzene with styrene in 96% conversion.[73,16] Also using these colloidal systems, the challenging olefination of **100** with ethylene was achieved in good yield.

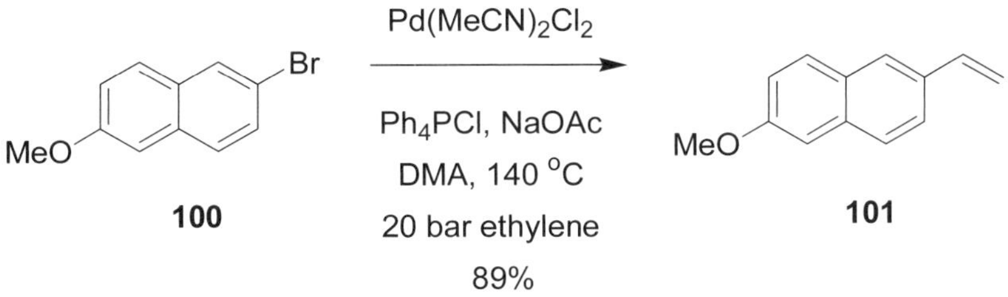

100 **101**

In his preliminary work, Heck reported that the heterogeneous palladium on carbon (Pd/C) could function as an effective catalyst at high temperatures.[2] Since this report, there have been numerous examples of palladium immobilized on a plethora of supports including polystyrene, silica, carbon, porous glass or encapsulated in zeolites and dedrimers.[16,20,24] Unfortunately in many cases, relatively low TON are observed. The main mechanistic issue in all these cases is whether catalysis is truly heterogeneous (i.e. the whole catalytic cycle takes place on the supported catalyst) or whether it operates in solution, through leaching of palladium(0). Indeed, there is a growing body of evidence suggesting in almost all cases, it is homogenous catalysis via active palladium(0) anionic species resulting from palladium nanoparticles/colloids.[24] The main goal for using such supported systems is to prevent leaching of palladium into commercially valuable products. However, on the contrary, it would appear a small degree of leaching is necessary for effective catalysis. This therefore fundamentally limits this approach.

Alternative Solvents

Aqueous media: Following the discovery that the Heck reaction could be accomplished under phase transfer conditions, it was demonstrated that it could also be efficiently carried out in the aqueous phase using palladium salts in the presence of inorganic bases.[74] In these reactions, miscible co-solvents such as DMF or HMPA are often employed as a means to solubilize lipophilic organic reactants. Whereas the reaction of bromobenzene with acrylic acid catalyzed by $Pd(OAc)_2$, in the presence of $P(o\text{-Tol})_3$ and K_2CO_3 gave only 12% yield of cinnamic acid in DMF, the addition of 10 % (v/v) or more of water increased the yield to quantitative.[16] Indeed, for water-soluble iodoarene substrates, full conversion was observed with 0.0005 mol% palladium, which corresponds to 200,000 catalytic cycles. It is thought the addition of water promotes the formation of palladium nanoparticles, which mediate the reaction under "ligand-free" conditions.[16]

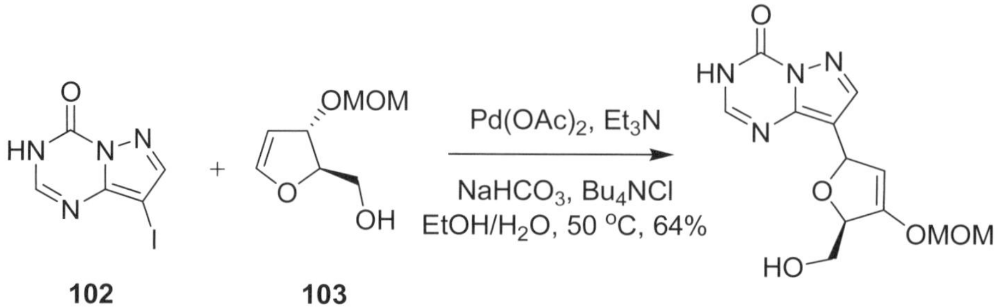

102 **103** **104**

Another beneficial effect of water was observed for the synthesis of C-nucleosides, for example **104**. Whereas the Heck reaction could not be achieved under standard conditions, the use of "ligand-free" conditions in aqueous ethanol (1:1, v/v) furnished the product in good yield.[75] Interestingly suppression of double bond migration in the product was also observed (a common issue in these substrates).

Phosphine-assisted catalysis under aqueous conditions requires the use of water-soluble phosphines, for example sulfonated derivatives.[21] Of particular note is the fact that for several cyclization reactions, a reversal of regioselectivity was observed; whereas standard Heck conditions furnish the *exo* cyclization adduct preferentially, the phosphine-assisted aqueous conditions gave the *endo* product predominantly.[76,16]

Ionic liquids: The use of ionic liquids (or molten salts) in the Heck reaction has mainly been to allow easy recycling of the catalytic system. There is some evidence however that this medium can activate the catalyst in certain cases.[16] Ionic liquids are highly polar and therefore facilitate the cationic mechanism, and can also contribute to the stabilization of underligated palladium(0) species through the formation of anionic complexes with halide ions.[16] Indeed, in some examples which generally require halide scavenging agents for the Heck reaction to proceed efficiently, the use of ionic liquids has been shown to negate the need for these additives. This has been attributed to the ease of dissociation of the halide, to give a cationic palladium complex in this highly polar medium.[30] Numerous ionic liquids have been employed including *n*-Bu_4NBr, Ph_3MePCl, Ph_3MePBr, *n*-Bu_3-*n*-$C_{16}H_{33}NBr$ and 1-methylimidazolium bromide. It must be noted however, that in general, imidazolium salts cannot be used in the absence of phosphine ligands below 100 °C due to their ability to form carbene complexes (which release catalytically active palladium only at higher temperatures, *vide infra*).[16,21] In the presence of Ph_3P, a system of 1-butyl-3-methylimidazolium•PF_6 and Et_3N gave quantitative yield in the coupling of iodobenzene with ethyl acrylate at 100 °C.[77] Importantly, the reaction could be rejuvenated without any loss of activity by extracting triethylammonium iodide with water, the product with cyclohexane and recharging the system with fresh reagents and base.

Supercritical fluids: Carbon dioxide forms a supercritical phase under relatively low temperatures and pressures and it therefore has been explored as a solvent for the Heck reaction.[16,21] Following the reaction, the gas can be collected and recycled, allowing the design of true waste-free technology. The dielectric constant of supercritical CO_2 is similar to pentane and therefore modified catalytic systems are often required to aid solubility.[16,21] Unusually, one of the best catalytic precursors in supercritical CO_2 is $Pd(OCOF_3)_2$, a strong electrophile and oxidant. Heck reactions in supercritical water have also been investigated.[16,21]

Fluorous Systems: Fluorous systems employ fluorinated compounds in perfluorinated solvents, which are immiscible with organic solvents. This allows the design of biphasic systems, where product is extracted from a reusable catalytic fluorous phase with organic solvents. Such systems have recently been applied to the Heck reaction with limited success.[16,21]

Less Usual Leaving Groups

Besides halides and triflates, other electrophiles can be applied to Heck reactions. The first classical alternative was diazonium salts.[20] Reactions proceed in the absence of phosphine (partly due to the fact that phosphines result in uncontrolled decomposition of the diazonium salt). The Heck reaction using these species can be useful in cases when mild conditions are required. Alternatively, iodonium salts behave in a similar manner to diazonium salts and show better tolerance to bases.[20] The reactions take place at ambient temperature and so are once again most useful in situations when mild conditions are required. Some main group metallic compounds such as lead(IV) and thallium(III) have also been shown to undergo Heck-type chemistry and can be useful in specific cases.[20] Of particular interest is the fact that acid chlorides and anhydrides can be employed in Heck chemistry, the use which was pioneered by Blaser and Spencer in 1982.[78,20] The process involves oxidative addition of palladium into the C–X bond followed by decarbonylation to yield the intermediate ArPdX species. de Vries has exploited this reaction, demonstrating the use of benzoic anhydride (**105**) as an effective arylating agent.[79,24]

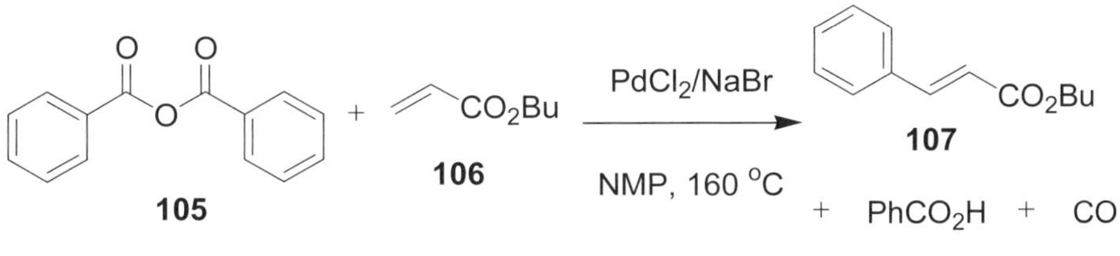

Other Metal Catalysts

Other metals can catalyze Heck-type reactions, although none thus far match the versatility of palladium. Copper salts have been shown to mediate the arylation of olefins, however this reaction most probably differs from the Heck mechanistically.[20] Likewise, complexes of platinum(II), cobalt(I), rhodium(I) and iridium(I) have all been employed in analogous arylation chemistry, although often with disappointing results.[20] Perhaps the most useful alternative is the application of nickel catalysis. Unfortunately, due to the persistence of the nickel(II) hydride complex in the catalytic cycle, the employment of a stoichiometric reductant, such as zinc dust is necessary, however the nickel-catalyzed Heck reaction does offer one distinct advantage. Unlike its palladium counterpart, it is possible to use aliphatic halides.[20] For example, cyclohexyl bromide (**108**) was coupled to styrene to yield product **110**.

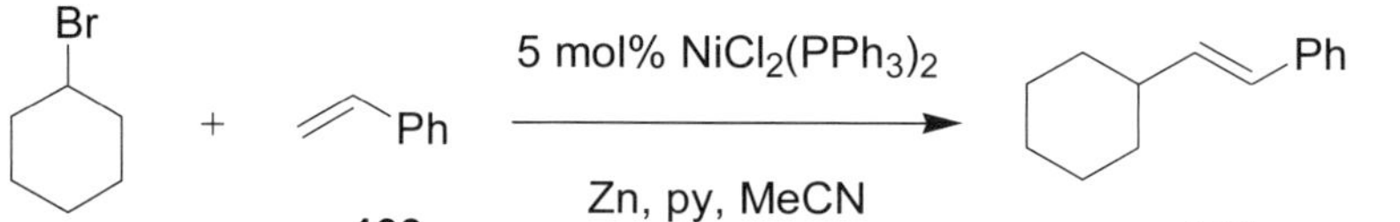

108 **109** **110**

Reductive Heck Reaction

In 1987, Larock published a paper on the palladium-mediated synthesis of nitrogen heterocycles.[67] For one example, he noted the cyclization of iodoaniline **111** in the presence of one equivalent of sodium formate gave indoline **116** in good yield. The use of sodium formate was based on previous precedent for the reduction of aryl halides.[80] Mechanistically he proposed the reaction proceeded via oxidative addition and migratory insertion to give intermediate **113**, which lacks a β-hydrogen atom, and therefore cannot undergo β-hydride elimination. Instead, ligand exchange with the formate anion gives species **114**, which can decarboxylate, providing hydridopalladium complex **115**. Reductive elimination provides indoline **116,** regenerating the palladium(0) catalyst. Similar reports on the use of formate as a hydride source in Heck-type reactions was published soon after by Grigg.[81]

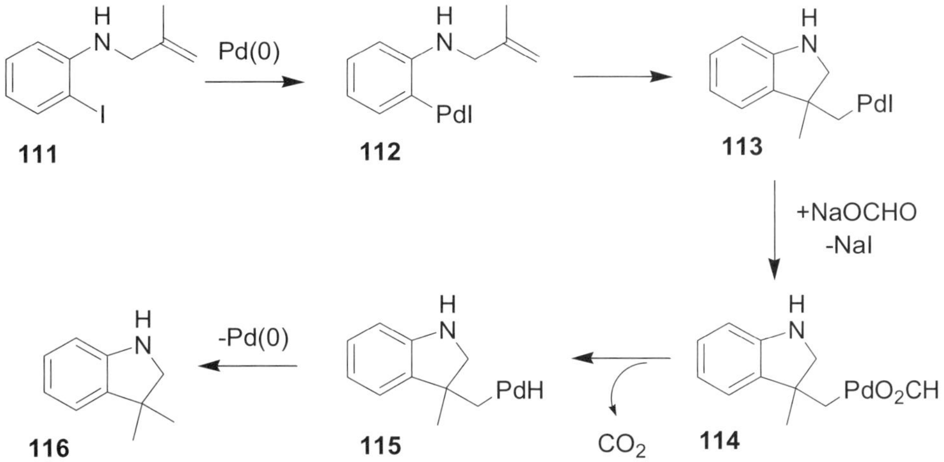

This reductive Heck reaction has become a useful tool in cyclization reactions for complex molecule synthesis. In his "protecting group free" synthesis of ambiguine H (**119**), Baran successfully applied this strategy.[82] Slow addition of Herrmann's catalyst (**97**)[62] to substrate **117** provided intermediate **118** in a reliable 65% yield. This chemistry proved both robust and scalable, providing gram quantities of **118**.

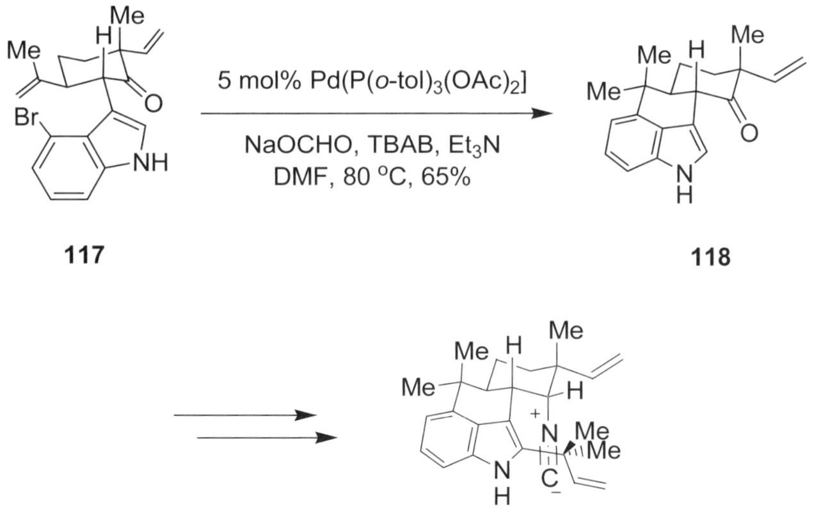

119: (+)-ambiguine H

1.1.1.6 Experimental

The proceeding examples highlight active modifications of Heck's original conditions. For the preliminary conditions, readers are directed to Heck's initial review article.[2]

Heck Reaction Using Fu's Conditions

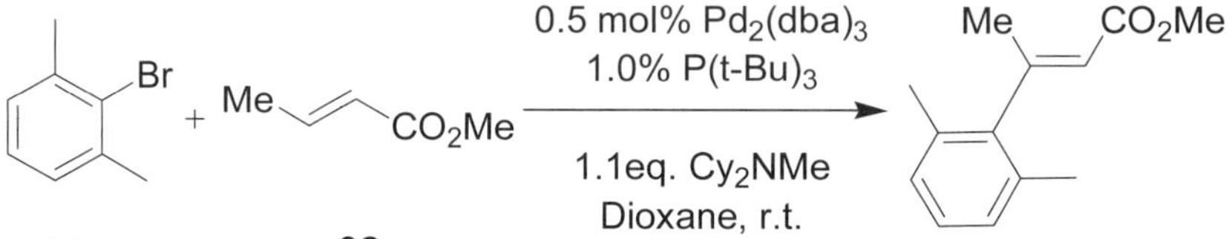

91 **92** **93**

(*E*)-3-(2,6-Dimethylphenyl)-3-methyl Acrylic Acid Methyl Ester (93).[57]
To a solution of $Pd_2(dba)_3$ (11.9 mg, 0.013 mmol), 2-bromo-*m*-xylene (**91**, 0.115 mL, 0.863 mmol), methyl crotonate (**92**, 0.100 mL, 0.943 mmol) and Cy_2NMe (0.200 mL, 0.934 mmol) in anhydrous dioxane (0.60 mL) was added P(*t*-Bu)$_3$ (0.10 M solution in dioxane; 0.26 mL, 0.026 mmol) under argon. The resulting mixture was stirred at ambient temperature for 49 h. The mixture was diluted with EtOAc, filtered through a pad of silica gel with copious washings and concentrated *in vacuo*. Flash column chromatography (SiO_2; Et_2O/hexanes, 5 : 95) furnished compound **93** (150 mg, 85%) as a clear, colorless liquid.

Heck Reaction Using Jeffery's Conditions

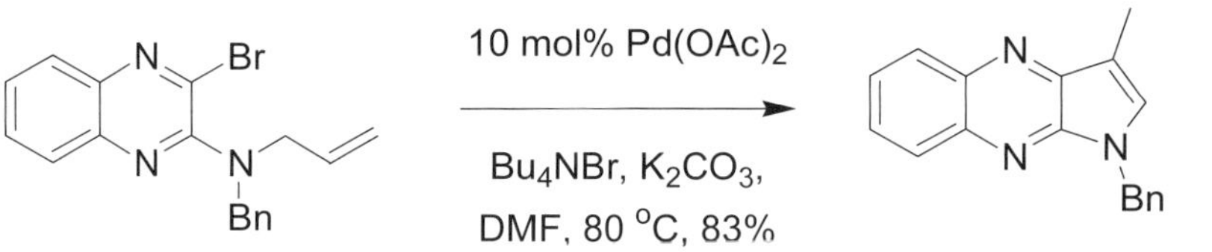

120 **121**

1-Benzyl-3-methyl-1H-pyrrolo[2,3-b]quinoxaline (121).[69]
To a solution of allylbenzyl-(3-bromoquinoxalin-2-yl)amine (**120**, 500 mg, 1.42 mmol) in DMF (15 mL) was added Pd(OAc)$_2$ (32 mg, 0.14 mmol), K_2CO_3 (580 mg, 4.25 mmol), and Bu_4NBr (456 mg, 1.42 mmol). The resulting mixture was stirred at 80 °C for 30 min and cooled to room temperature. The mixture was diluted with CH_2Cl_2 (100 mL), washed with water (3 × 20 mL) and brine (20 mL), dried (Na_2SO_4), and filtered, and the filtrate was concentrated *in vacuo*. The residue was purified by flash

chromatography to furnish 1-benzyl-3-methylpyrrolo[2,3-*b*]quinoxaline (**121**) as a yellow solid (321 mg, 83% yield).

1.1.1.7 References

1. [R] Heck, R. F. In *Comprehensive Organic Synthesis*; Trost, B. M., Fleming, I., Eds.; Pergamon Press: Oxford, 1991; Vol. 4.
2. [R] Heck, R. F. *Org. React. (N. Y.)* **1982**, *27*, 345.
3. [R] Daves, G. D.; Hallberg, A. *Chem Rev.* **1989**, *89*, 1433.
4. [R] de Meijere, A.; Meyer, F. E. *Angew. Chem., Int. Ed. Engl.* **1994**, *33*, 2379.
5. [R] de Meijere, A, Braese, S. In *Transition Metal Catalyzed Reactions*; Davies S. G., Murahashi, S.-I., Eds.; Blackwell Science: Oxford, 1999.
6. [R] Cabri, W.; Candiani, I. *Acc. Chem. Res.* **1995**, *28*, 2.
7. [R] Jeffery, T. In *Advances in Metal-Organic Chemistry*; Liebeskind, L. S., Ed.; Jai Press Inc: Greenwich, CT, 1996; Vol. 5.
8. [R] Crisp, G. T. *Chem. Soc. Rev.* **1998**, *27*, 427.
9. [R] Brase, S.; de Meijere, A. In *Metal-Catalyzed Cross-Coupling Reactions*; Diedrich, F., Stang, P. J., Eds.; Wiley–VCH: New York, 1998.
10. [R] Shibasaki, M.; Boden, C. D. J.; Kojima, A. *Tetrahedron*, **1997**, *53*, 7371.
11. [R] Overman, L. E. *Pure Appl. Chem.* **1994**, *66*, 1423.
12. [R] Link, J. T.; Overman, L. E. In *Metal-Catalyzed Cross-Coupling Reactions*; Diedrich, F., Stang, P. J., Eds.; Wiley-VCH: New York, 1998.
13. [R] Link, J. T.; Overman, L. E. *Chemtech* **1998**, *28*, 19.
14. [R] Gibson, S. E.; Middleton, R. J. *Contemp. Org. Synth.* **1996**, *3*, 447.
15. [R] Reetz, M. T. In *Metal-Catalyzed Reactions*; Davies, S. G., Murahashi, S.-I., Eds.; Blackwell Sci.: Oxford, 1999.
16. [R] Beletskaya, I. P.; Cheprakov, A. V. *Chem. Rev.* **2000**, *100*, 3009.
17. [R] Dounay, A. B.; Overman, L. E. *Chem. Rev.* **2003**, *103*, 2945.
18. [R] Biffs, A.; Zecca, M.; Basato, M. *J. Mol. Catal. A: Chem.* **2001**, *173*, 249.
19. [R] Shibasaki, M.; Vogl, E. M.; Ohshima, T. *Adv. Synth. Catal.* **2004**, *346*, 1533.
20. [R] Farina, V. *Adv. Synth. Catal.* **2004**, *346*, 1553.
21. [R] Alonso, F.; Beletskaya, I. P.; Yus, M. *Tetrahedron*, **2005**, *61*, 11771.
22. [R] Trzeciak, A. M.; Ziółkowski, J. J. *Coord. Chem. Rev.* **2005**, *249*, 2308.
23. [R] Oestreich, M. *Eur. J. Org. Chem.* **2005**, 783.
24. [R] de Vries, J. G. *Dalton Trans.* **2006**, 421.
25. Mizoroki, T.; Mori, K.; Ozaki, A. *Bull. Chem. Soc. Jpn.* **1971**, *44*, 581.
26. Heck, R. F.; Nolley, J. P. Jr. *J. Org. Chem.* **1972**, *37*, 2320.
27. Heck, R. F. *J. Am. Chem. Soc.* **1968**, *90*, 5518.
28. [R] Amatore, C.; Jutand, A. *J. Organomet. Chem.* **1999**, 254.
29. [R] Hegedus, L. S. In *Transition Metals in the Synthesis of Complex Organic Molecules*, 1st ed.; University Science Books: Mill Valley, CA, 1994; p 87.
30. Amatore, C.; Godin, B.; Jutand, A.; Lemaître, F. *Organometallics*, **2007**, *26*, 1757.
31. [R] Heck, R. F. *Acc. Chem. Res.* **1979**, *12*, 146.
32. Cabri, W.; Candiani, I.; DeBernardis, S.; Francalance, F.; Penco, S. *J. Org. Chem.* **1991**, *56*, 5796.
33. Ozawa, F.; Kubo, A.; Hayashi, T. *J. Am. Chem. Soc.* **1991**, *113*, 1417.
34. Ashimori, A.; Overman, L. E. *J. Org. Chem.* **1992**, *57*, 4571.
35. Amatore, C.; Jutand, A. *Acc. Chem. Res.* **2000**, *33*, 314.
36. Amatore, C.; Jutand, A.; Thuilliez, A. *Organometallics*, **2001**, *20*, 3241.
37. Overman, L. E.; Rucker, P. V. *Tetrahedron Lett.* **1998**, *39*, 4643.
38. Abelman, M. M.; Oh, T.; Overman, L. E. *J. Org. Chem.* **1987**, *52*, 4130.
39. [R] Hammett, L. P. *Physical Organic Chemistry: Reaction Rates, Equilibria and Mechanisms*, 2nd ed.; McGraw-Hill: New York, 1970.
40. Tietze, L. F.; Schimpf, R. *Angew. Chem., Int. Ed. Engl.* **1994**, *33*, 1089.

41. Lebsack, A. D.; Link, J. T.; Overman, L. E.; Stearns, B. A. *J. Am. Chem. Soc.* **2002**, *124*, 9008.
42. Lapierre, A. J. B.; Geib, S. J.; Curran, D. P. *J. Am. Chem. Soc.* **2007**, *129*, 494.
43. a) Kagechika, K.; Shibasaki, M. *J. Org. Chem.* **1991**, *56*, 4093; b) Kagechika, K.; Ohshima, T.; Shibasaki, M. *Tetrahedron* **1993**, *49*, 1773.
44. a) Overman, L. E.; Ricca, D. J.; Tran, V. D. *J. Am. Chem. Soc.* **1993**, *115*, 2042; b) Kucera, D. J.; O'Connnor, S. J.; Overman, L. E. *J. Org. Chem.* **1993**, *58*, 5304.
45. Ozawa, F.; Kubo, A.; Hayashi, T. *J. Am. Chem. Soc.* **1991**, *113*, 1417.
46. Ozawa, F.; Kobatake, Y.; Hayashi, T. *Tetrahedron Lett.* **1993**, *34*, 2505.
47. Loiseleur, O.; Meier, P.; Pfaltz, A. *Angew. Chem., Ind. Ed. Engl.* **1996**, *35*, 200.
48. Zhang, Y.; Wu, G.; Angel, G.; Negishi, E. *J. Am. Chem. Soc.* **1990**, *112*, 8590.
49. a) Trost, B. M.; Shi, Y. *J. Am. Chem. Soc.* **1991**, *113*, 701; b) Trost, B. M. *Angew. Chem., Int. Ed. Engl.* **1995**, *34*, 259.
50. a) Koenig, B.; Knieriem, B.; de Meijere, A. *Chem. Ber.* **1993**, *126*, 1643; b) Bazan, G. C.; Oldham, W. J.; Lachicotte, R. J.; Tretiak, S.; Chernyak, V.; Mukamel, S. *J. Am. Chem. Soc.* **1998**, *120*, 9188.
51. Scherf, U.; Müllen, K. *Synthesis*, **1992**, 23.
52. Lee, Y.; Liang, Y.; Yu, L. *Synlett* **2006**, 2879.
53. Berteina, S.; Wendeborn, S.; Brill, W. K. D.; Mesmaeker, A. D. *Synlett* **1998**, *6*, 676.
54. Spencer, A. *J. Organomet. Chem.* **1983**, *258*, 101.
55. Portnoy, M.; Ben-David, Y.; Milstein, D. *Organometallics* **1993**, *12*, 4734.
56. Littke, A. F.; Fu, G. C. *J. Org. Chem.* **1999**, *64*, 10.
57. Littke, A. F.; Fu, G. C. *J. Am. Chem. Soc.* **2001**, *123*, 6989.
58. Stambuli, J. P.; Stauffer, S. R.; Shaughnessy, K. H.; Hartwig, J. F. *J. Am. Chem. Soc.* **2001**, *123*, 2677.
59. [R] Grushin, V. V.; Alper, H. *Chem. Rev.* **1994**, *94*, 1047.
60. Christmann, U.; Vilar, R. *Angew. Chem., Int. Ed.* **2005**, *44*, 366.
61. Hills, I. D.; Fu, G. C. *J. Am. Chem. Soc.* **2004**, *126*, 13178.
62. Herrmann, W. A.; Brossmer, C.; Öfele, K.; Reisinger, C.-P.; Priermeier, T.; Beller, M.; Fischer, H. *Angew. Chem., Int. Ed. Engl.* **1995**, *34*, 1844.
63. Shaw, B. L. *New. J. Chem.* **1998**, *22*, 77.
64. Shaw, B. L.; Petera, S. D.; Staley, E. A. *Chem. Commun.* **1998**, 1361.
65. Louie, J.; Hartwig, J. F. *Angew. Chem., Int. Ed. Engl.* **1996**, *35*, 2359.
66. Herrmann, W. A.; Elison, M.; Fischer, J.; Koecher, G. R.; Artus, J. *Angew. Chem., Int. Ed. Engl.* **1995**, *34*, 2371.
67. Larock, R. C.; Babu, S. *Tetrahedron Lett.* **1987**, *28*, 5291.
68. Reetz, M. T.; Westermann, *Angew. Chem., Int. Ed. Engl.* **2000**, *39*, 165.
69. Li, J. J. *J. Org. Chem.* **1999**, *64*, 8425.
70. Reetz, M. T.; Lohmer, G. *Chem. Commun.* **1996**, 1921.
71. Beller, M.; Fischer, H.; Kuehlein, K.; Reisinger, C.-P.; Herrmann, W. A. *J. Organomet. Chem.* **1996**, *520*, 257.
72. Reetz, M. T.; Westermann, E.; Lohmer, R.; Lohmer, G. *Tetrahedron Lett.* **1998**, *39*, 8449.
73. Reetz, M. T.; Lohmer, G.; Schwickardi, R. *Angew. Chem., Int. Ed. Engl.* **1998**, *37*, 481.
74. Zhao, F.; Shirai, M.; Arai, M.; *J. Mol. Cat. (A)* **2000**, *154*, 39.
75. Zhang, H. C.; Daves, G. D.; *Organometallics*, **1993**, *12*, 1499.
76. Lemaire–Audoire, S.; Savignac, M.; Dupuis, C.; Genet, J. P. *Tetrathedron Lett.* **1996**, *37*, 2003.
77. Carmichael, A. J.; Earle, M. J.; Holbrey, J. D.; McCormac, P. B.; Seddon, K. R. *Org. Lett.* **1999**, *1*, 997.
78. Blaser, H.-U.; Spencer, A. *J. Organomet. Chem.* **1982**, *233*, 267.
79. Stephan, M. S.; Teunissen, A. J. J. M.; Verzijl, G. K. M.; de Vries, J. G. *Angew. Chem., Int. Ed.* **1998**, *37*, 662.
80. Pri-Bar, I.; Buchman, O. *J. Org. Chem.* **1986**, *51*, 734.
81. Burns, B.; Grigg, R.; Ratananukul, P.; Visuvanathar, S.; Stevenson, P.; Tanachat, W. *Tetrahedron Lett.* **1988**, *29*, 4329.
82. Baran, P. S.; Maimone, T. J.; Richter, J. M. *Nature*, **2007**, *446*, 404.

1.1.2 Hiyama Cross-Coupling Reaction

Larry Yet

1.1.2.1 Description

The palladium-catalyzed reaction of alkenyl-, aryl-, alkynyl- and alkylsiloxanes with aryl, alkyl, and alkenyl halides and triflates in the presence of activators is known as the Hiyama cross-coupling reaction and several reviews have been published.[1–8] This chapter will present major developments and examples of recent carbon–carbon bond formation methodology and improvements as well as their use in natural products synthesis in the last few years.

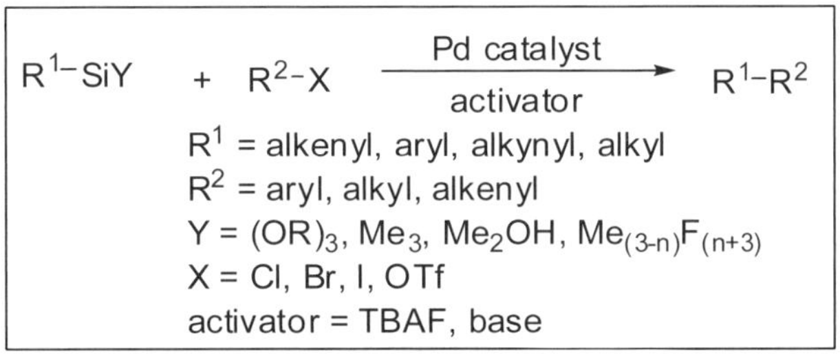

Organotin, organoboron, and organozinc reagents have become useful reagents for palladium-catalyzed cross-coupling reactions, which have found widespread applications in modern synthetic organic chemistry applications. The low molecular weight of organosilanes as well as the lack of toxicity, ease of activation, and high stability in most chemical reactions makes them ideal for use as nucleophilic partners in the cross-coupling reaction with organic halides and pseudohalides.

1.1.2.2 Historical Perspective

In the early 1980's, Hiyama started his academic research at the Sagami Chemical Research Center focusing on the activation of C–Si and Si–Si bonds using fluoride ion to generate the corresponding naked anion species that were otherwise destabilized by metallic counter ions.[9] Typical examples included synthetic reactions of $Me_3Si–SiMe_3$ with 1,3-butadienes[10], reduction of ketones with $HSiMe_2Ph$,[11] and carbonyl addition of carbenoid-type carbanions $^{-}CX_nR_{3-n}$,[12] all mediated at room temperature by tetrabutylammonium fluoride (TBAF) or by $[(Et_2N)_3S^+(Me_3SiF_2^-)]$ (TASF).

Hiyama then wondered what would happen if a palladium complex was present in the reaction mixture. The questions he asked, "Does a fluoride ion simply attack the Pd(II) to deactivate the catalyst?" Hiyama then found through experimentation that a fluoride ion was found to preferentially attack the silicon to generate the anionic species that were later proved to be the pentacoordinated silicates, and the involvement of this species was shown to be essential for smooth transmetallation of the organosilicon reagents to complete the catalytic cycle of the cross-coupling reaction.[13–15] These initial observations then created a slew of palladium-catalyzed silicon-based reactions known today as the Hiyama cross-coupling reaction.

1.1.2.3 Mechanism

The most commonly accepted mechanism for this coupling was initially proposed by Hiyama and Hatanaka, which involves three steps.[16] The first step is the oxidative addition of the aryl halide to the palladium(0) catalyst to give arylpalladium complex **1**. The second step involves the transmetallation of the arylpalladium complex **1** with the anionic arylsilicate **2** to give bis(aryl)palladium complex **3**. Finally, the cross-coupled product **4** is produced and the palladium(0) catalyst is regenerated through reductive elimination of the bis(aryl)palladium(II) complex **3**. The key intermediate to this process is the requirement for the pentacoordinate arylsilicate anion **2**, typically formed by treatment of the tetracoordinate silane with the activating anion, such as tetrabutylammonium fluoride (TBAF).

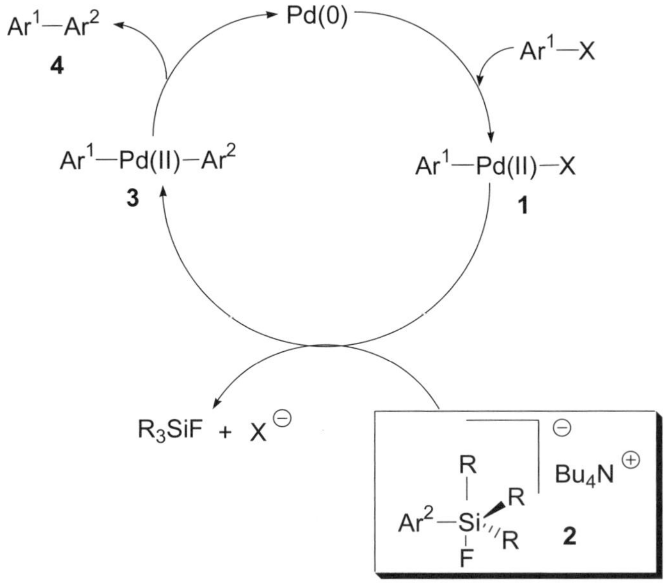

To overcome the limitations of using TBAF as an activating reagent in sensitive compounds containing silicon protecting groups, Denmark reported that simple deprotonation of the silanol might open a new pathway for activation.[17] In this mechanism, the conjugate base of the silanol served two roles. First, the silanolate **5** displaces the halide ion on the organopalladium–X species to generate a palladium silanolate complex **6**. Then another silanolate molecule activates the palladium silanolate complex through the formation of a pentacoordinate siliconate complex **7**, which undergoes transmetallation to give **8**. Finally, reductive elimination of **8** gives the product **4**. Bases that can be used include potassium trimethylsilanoate, cesium carbonate, sodium hydride, and sodium *tert*-butoxide.

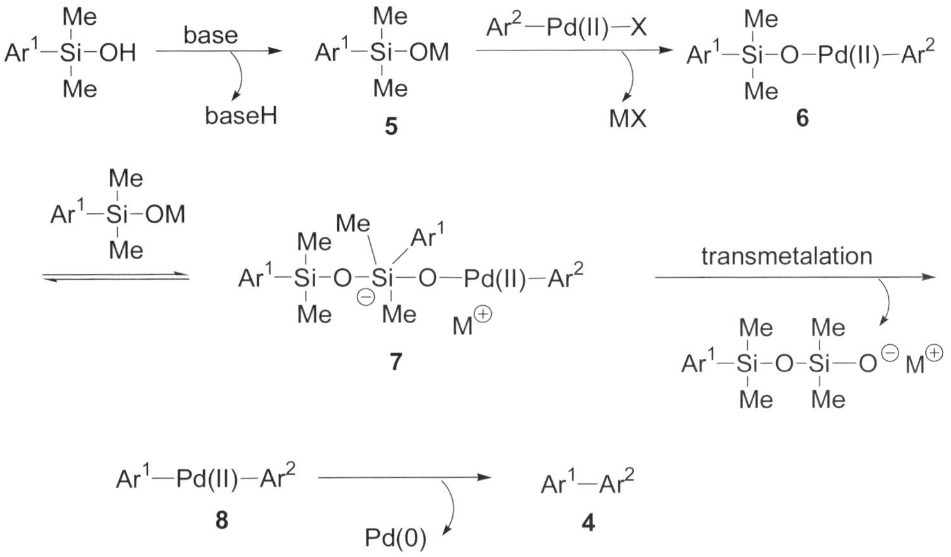

1.1.2.4 Variations and Improvements

Several reviews have shown variations and improvements in the scope and limitations of the Hiyama cross-coupling reaction.[1–8] This section reports the recent advancements in this area over the last few years.

Preparation of the Organosiloxanes

Classical synthetic routes to arylsilanes consist of the reaction of aryl Grignard or aryllithium compounds with silicon electrophiles and many of these routes are presented in the various reviews. Recently, metal-catalyzed reactions of aryl halides with silanes have proven to be useful routes to functionalized arylsilanes. Alami reported the first platinum-catalyzed selective silylation of aryl iodides and bromides **9** having electron-withdrawing group to give arylsilanes **10** with triethylsilane and sodium acetate in NMP.[18] Heteroaromatic halides were also readily silylated. Previous palladium-catalyzed methods were limited due to the difficult synthesis where only electron-rich, para-substituted aryl iodides afforded good yields of the arylsiloxanes. Denmark demonstrated a mild and general palladium-catalyzed insertion of 1,2-diethoxy-1,1,2,2-tetramethyldisilane to a variety of para-substituted aryl bromides **11** to afford the aryldimethylsilanols **12**.[19]

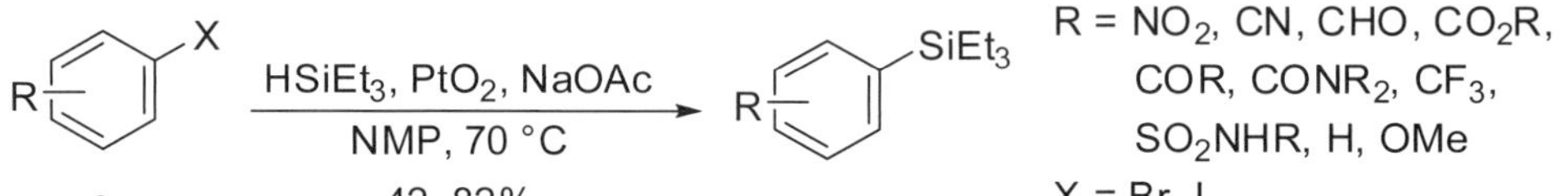

9 **10**

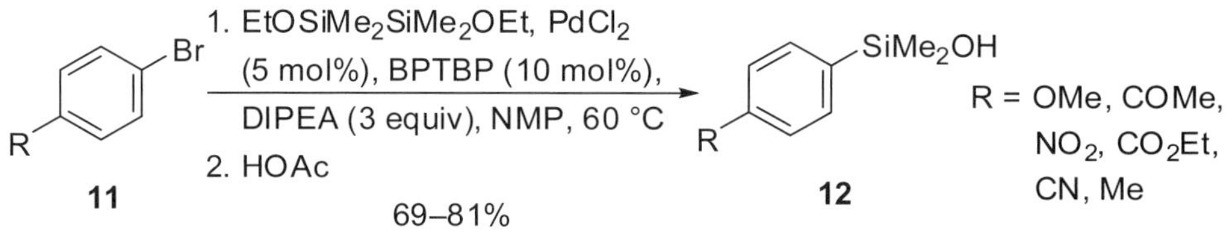

11 **12**

A series of aryltriethoxysilanes **14** were synthesized from the reaction of aryl bromides **13** with magnesium powder, 1,2-dibromoethane and tetraethylorthosilicate in tetrahydrofuran via sonochemical Barbier-type conditions.[20] DeShong previously reported the preparation of aryltrialkoxysilanes via treatment of aryl Grignard or lithium reagents with tetralkylorthosilicates.[21] DeShong also synthesized a selection of *ortho*-substituted aryltriethyoxysilanes by directed orthometallation protocols.[22]

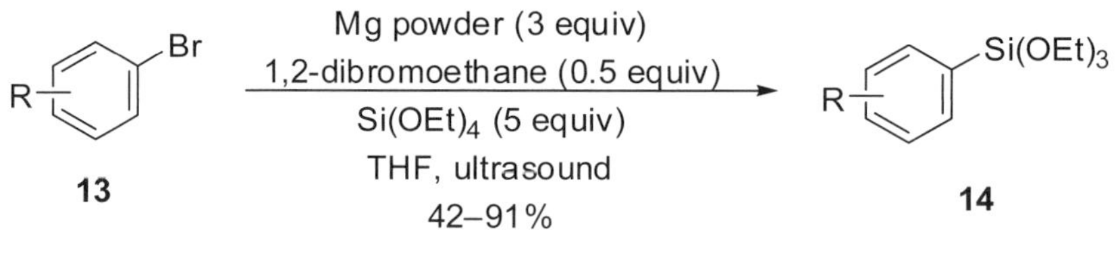

13 **14**

Hiyama Cross-Coupling Improvements

Zhang developed mild conditions for the palladium-catalyzed fluoride-free cross-coupling between aryl bromides **15** and aryltrimethoxysilanes **16** in good to high yields to afford biaryls **17** in the presence of poly(ethylene glycol) and sodium hydroxide.[23] Significant increase in substrate reactivity and reduction in reaction times were noted. Hiyama reported the use of triallyl(aryl)silanes as stable and easily accessible arylsilanes for the cross-coupling reactions of aryl bromides in the presence of palladium catalyst ($PdCl_2/PCy_3$) and TBAF.[24]

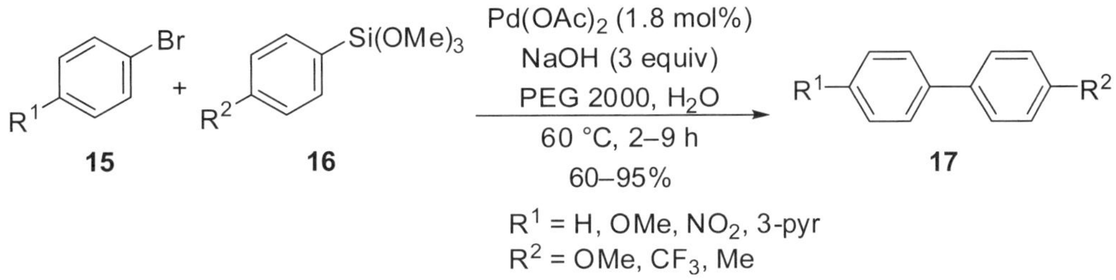

To date, nearly all studies of the Hiyama reaction have focused on couplings of C_{sp^2}–X electrophiles. Fu developed the first method for achieving the room-temperature Hiyama couplings of unactivated alkyl bromides and iodides.[25] Palladium-catalyzed reactions of alkyl bromides and iodides **18** with aryltrimethoxysilanes **19** in the presence of phosphorus ligand and TBAF afforded coupled products **20** in moderate to good yields.

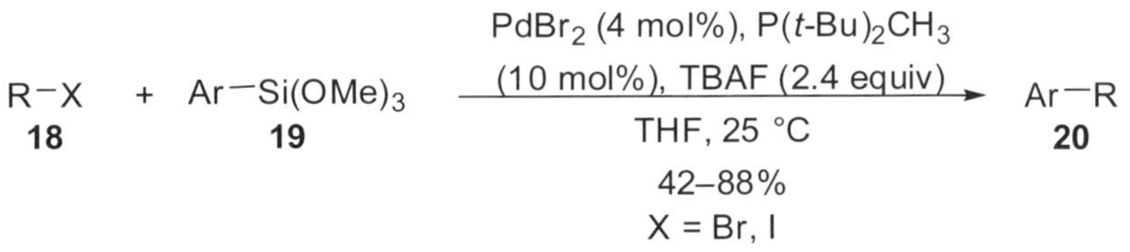

The use of ligands has found beneficial use in the promotion of Hiyama cross-coupling reactions. Li has showed that palladium(II) acetate/DABCO was an inexpensive and efficient catalytic system for the Hiyama cross-coupling reactions of aryl halides **21** with aryltrimethoxy-silanes **22** to give biaryls **23**.[26] He also reported the improved palladium-catalyzed Hiyama cross-coupling reaction of aryl halides with aryltrimethoxysilanes under solvent-free conditions with $PdCl_2(CH_3CN)_2$ and $P(o\text{-tol})_3$ and tetrabutylammonium fluoride.[27] Phosphite ligand **24** was utilized in the palladium-catalyzed Hiyama cross-coupling reactions of trimethoxysilylbenzene with aryl bromides and chlorides in the presence of $Pd(acac)_2$ in *p*-xylene at 80 °C with tetrabutylammonium fluoride.[28] Hydrazone **25** was a good ligand in the $PdCl_2$-catalyzed Hiyama reaction of aryl bromides with aryltriethoxysilanes with tetrabutylammonium fluoride in toluene at 80 °C.[29]

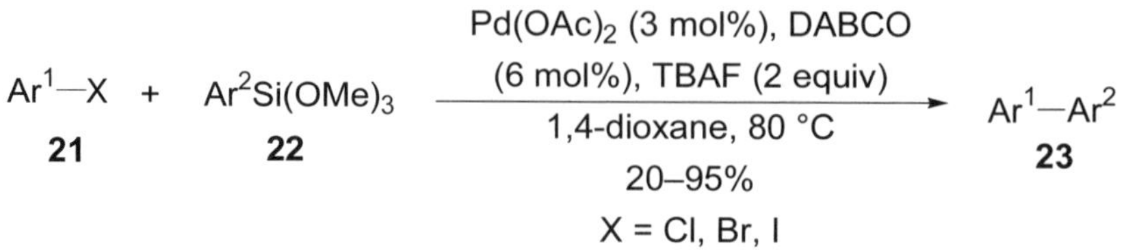

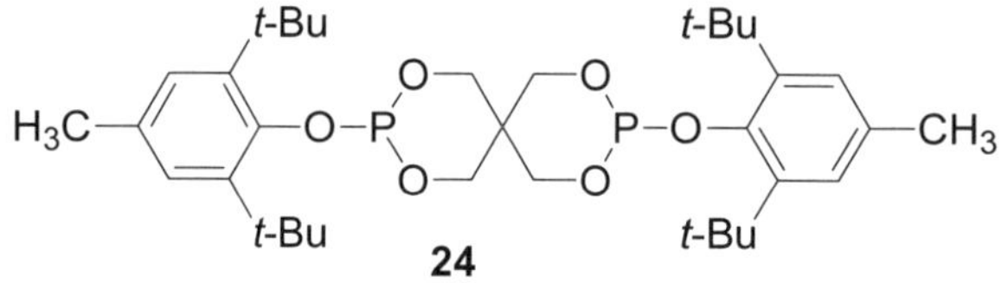

24

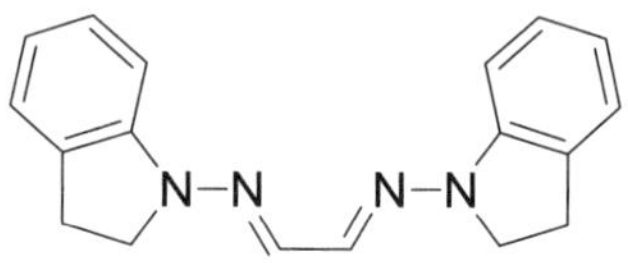

25

A colloidal palladium nanoparticle prepared from a Fischer carbene complex of tungsten with K_2PdCl_4 as the reductant and PEG as the capping agent, efficiently catalyzed the Hiyama cross-coupling reactions in air.[30]

1.1.2.5 Synthetic Utility

Many reviews have been published in the last few years on the applications of the Hiyama cross-coupling reaction in organic synthesis.[1–8] In general, preparation of biaryls or heterobiaryls is the most common application of this reaction since new methodology development is based on examples of these. The most prevalent applications of the aryl and heteroaryl halides with organosilane precursors involve reactions such as vinylations, alkenylations, and alkynylations. The reader is encouraged to consult these key reviews. This section will describe some of the more recent applications published in the last few years.

Range of Organosilyl Precursors

The organosilyl precursors for the Hiyama cross-coupling reaction is not limited to the aryltrialkoxysilanes but can include the following structures shown below, where their applications are numerous.[1–8]

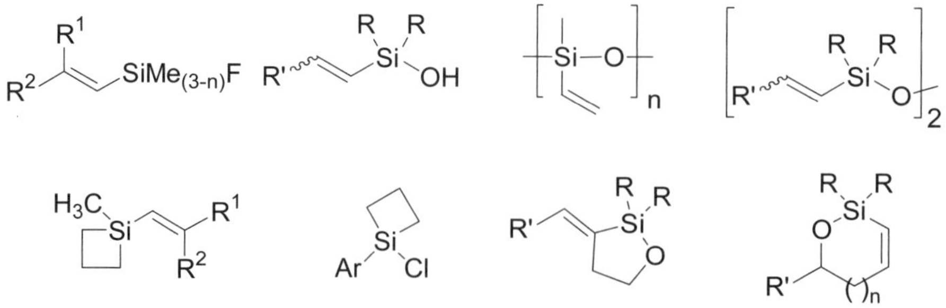

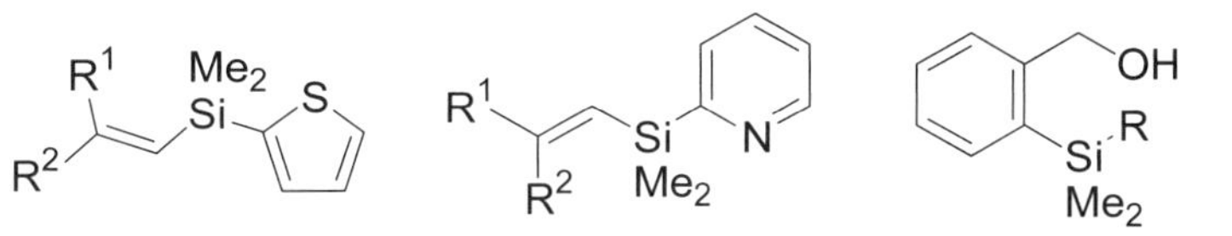

Heterocycle Examples

The applications of aryltrialkoxysilanes in Hiyama cross-couplings with heteroaryl halides have been documented in a review by DeShong.[7] The Hiyama cross-coupling has now been widened to include heterocyclic silanolates.

Denmark reported a mild and general palladium-catalyzed cross-coupling reaction of 2-indolylsilanols **26** with aryl iodides in the presence of stoichiometric copper(I) iodide and sodium *tert*-butoxide to afford coupled products **27** in good yields.[31] In a different protocol, Denmark showed that sodium silanoates derived from a number of heterocyclic (pyrrole, thiophene, furan, indole) silanols **28** underwent cross-coupling reactions with a variety of aryl iodides and bromides to give products **30** with palladium catalyst **29**.[32] Denmark reported that isoxazolylsilanols **33**, prepared from a [3 + 2] cycloaddition reaction between alkynyldimethylsilyl ether **32** and aryl and alkyl nitrile oxides **31**, cross-coupled with aryl iodides to give 3,4,5-trisubstituted isoxazoles **34**.[33] Yoshida demonstrated that (2-pyridyl)allyldimethylsilanes **35** were found to be novel pyridyl transfer reagents in the palladium-catalyzed reactions of aryl iodides to give 2-arylpyridines **36** in the presence of silver(I) oxide as an activator.[34]

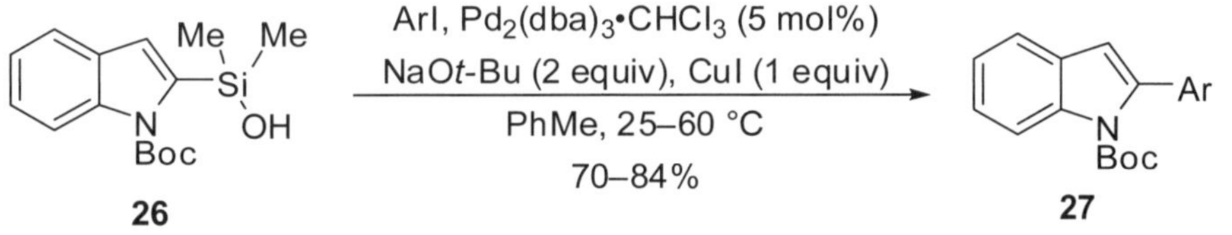

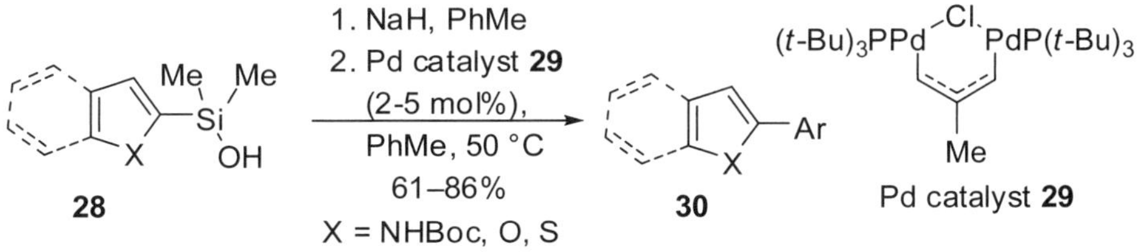

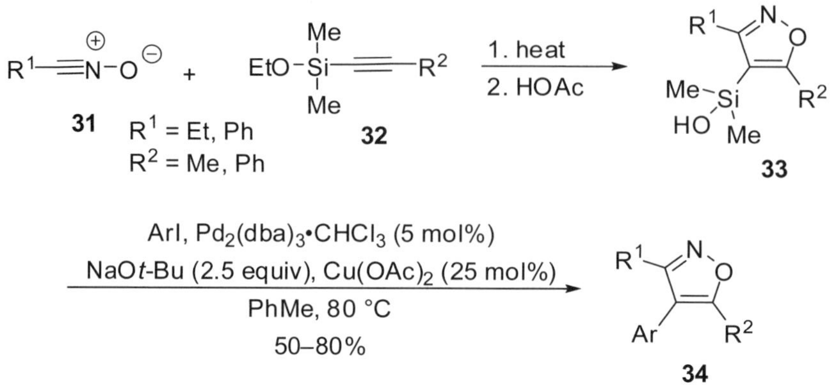

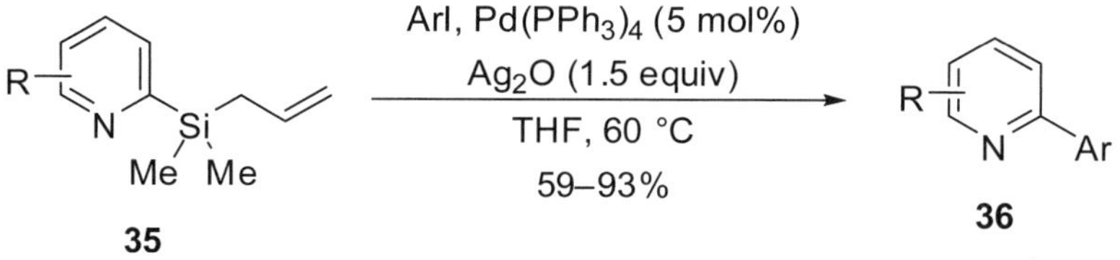

Conjugate Additions

Aryltrialkoxysilanes can participate not only in palladium-catalyzed cross-coupling reactions but also in palladium-catalyzed conjugate addition reactions to α,β-unsaturated compounds. Denmark showed that addition of aryltrialkoxysilanes to α,β-unsaturated compounds (ketones, aldehydes) and nitroalkenes **37** in the presence of $SbCl_5$, TBAF, acetic acid, and palladium(II) acetate in acetonitrile gave the conjugate addition product **38** in variable yields depending on the substrates.[35] Wolf also reported the conjugate addition of aryltrimethoxysilanes to α,β-unsaturated compounds (ketones, aldehydes, esters, nitroalkanes, and nitriles) **39** in the presence of POPd1 gave conjugate product **40** in water under microwave irradiation.[36] This method eliminated the need for stoichiometric additives such as TBAF and an excess of arylsiloxane, and does not require an inert atmosphere.

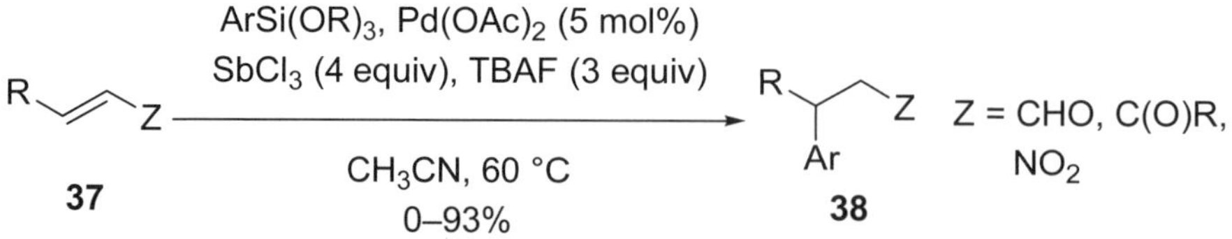

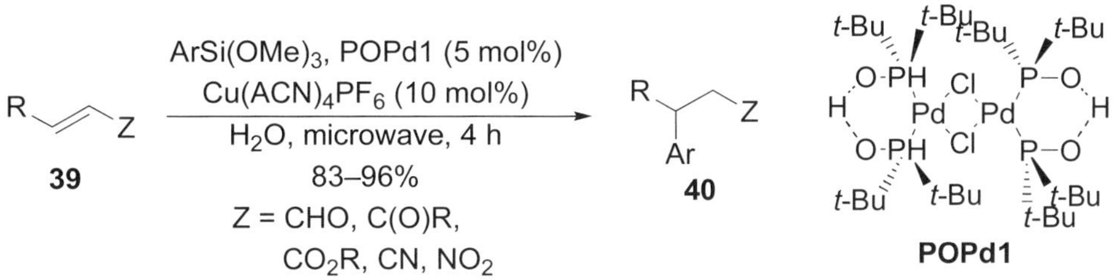

Tandem Reactions

Multistep or tandem reactions have been employed in the use of the organosilicon moiety to create complex structures rapidly. These tandem reactions terminate with the palladium-catalyzed cross-coupling reactions. Tandem reactions have been involved in processes such as intermolecular and intramolecurlar hydrosilylation/cross-coupling, silylformylation/cross-coupling, Mizoroki–Heck reaction/cross-coupling, ring-closing metathesis/cross-coupling, and Alder–ene/cross-coupling reactions. Examples of these tandem reactions are represented in Denmark's review.[5]

Denmark published a sequential rhodium-catalyzed silylcarbocyclization of enynes parlayed with a palladium-catalyzed silicon-based cross-coupling reaction for the synthesis of highly substituted cyclopentanes.[37] 1,6-Enynes **41** reacted with benzyldimethylsilane in the presence of rhodium catalysts to afford five-membered rings **42** bearing a (*Z*)-alkylidenylbenzylsilyl group. A variety of substitution patterns and heteroatom substituents were compatible. Cyclopentenes **42** then underwent Hiyama cross-coupling reaction with aryl iodides in the presence of TBAF to give coupled products **43** in good yields.

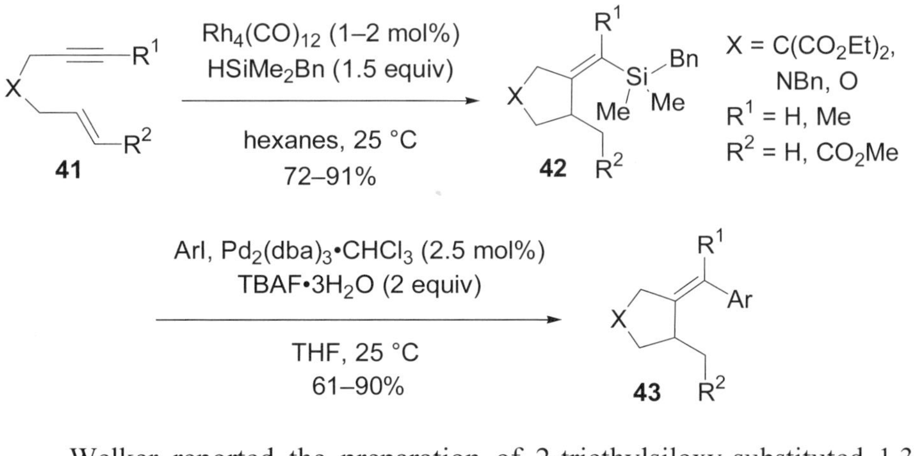

Welker reported the preparation of 2-triethylsiloxy-substituted 1,3-butadienes **45** from 2-chlorobutadiene (**44**) via a Grignard reaction, addition to triethoxysilyl chloride and alcoholysis.[38] These dienes **45** then participated in a Diels–Alder/cross-coupling reaction, respectively with dienophiles **46** to give cycloadducts **47** and then Hiyama cross-couplings with aryl iodides to give final product **48**.

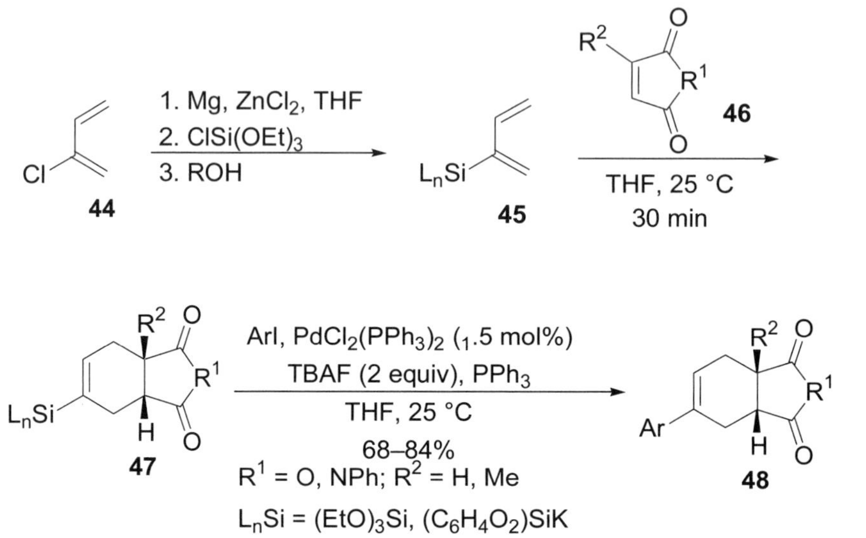

Marciniec demonstrated that 9-vinycarbazole (**49**) can undergo cross-metathesis with vinyltriethoxysilane with the ruthenium catalyst to give vinylsiloxane carbazole **50** which then participated in a Hiyama palladium-catalyzed reaction with aryl iodides to furnish (*E*)-*N*-styrylcarbazoles **51**.[39]

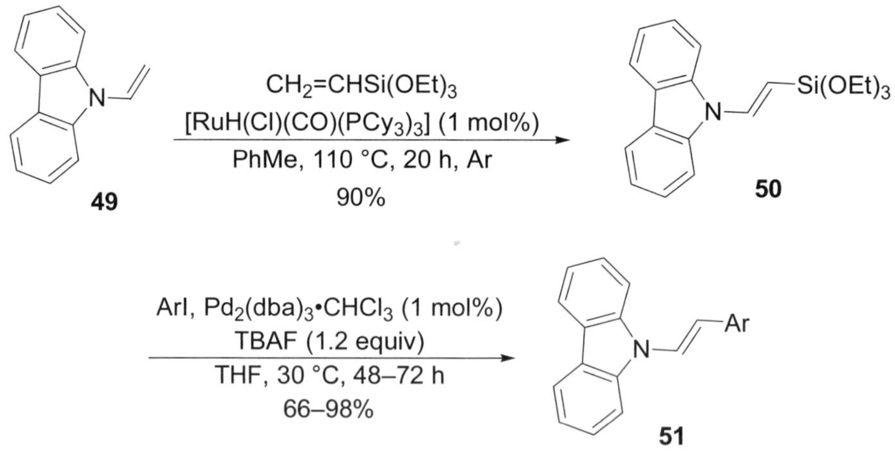

Applications in Natural Products

The Hiyama cross-coupling reaction has not been highly represented in the preparation of natural products compared to the Negishi, Stille, and Suzuki reactions. DeShong utilized this key step in synthetic studies towards several natural products. For example, the Hiyama cross-coupling reaction was employed in the early synthesis of sterically-hindered biaryl **54** from 2-bromo-3,4,5-trimethoxybenzaldehyde (**52**) and arylsiloxane **53**.[40] Elaboration of **54** to phenanthrol **55** was accomplished in six steps. Finally phenanthrol ring expansion provided racemic *N*-acetyl colchinol-*O*-methyl ether (**56**). Similarly, DeShong approached the studies towards colchicine via palladium-catalyzed siloxane cross-coupling of 5-bromotropolone (**57**) to give aryltropolone **58**.[41] Siloxane-based cross-coupling of highly functionalized 4-bromopyridines **59** with aryltrialkylsilanes furnished sterically demanding biaryls **60** towards studies for the synthesis of streptonigrin and lavendamycin.[42]

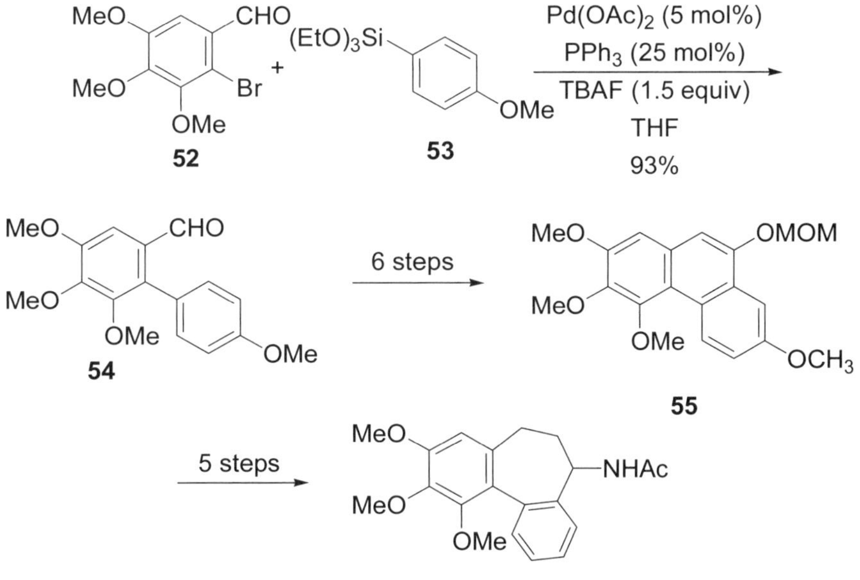

N-acetyl colchinol-*O*-methyl ether (**56**)

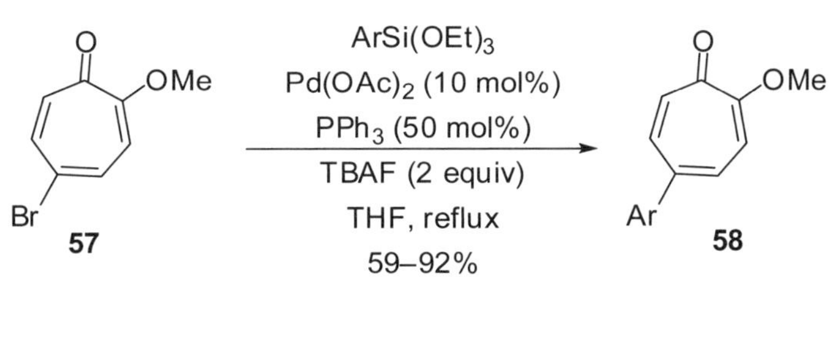

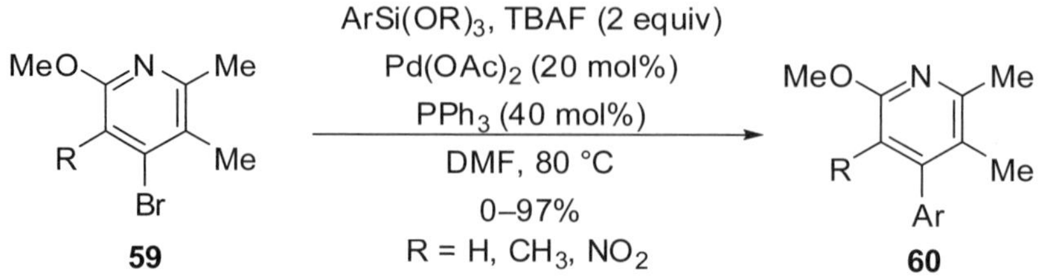

Denmark showcased the intramolecular silicon-assisted cross-coupling reaction of vinyl iodide **61** to give oxecene **62** in the first total synthesis of (+)-brasilenyne (**63**).[43]

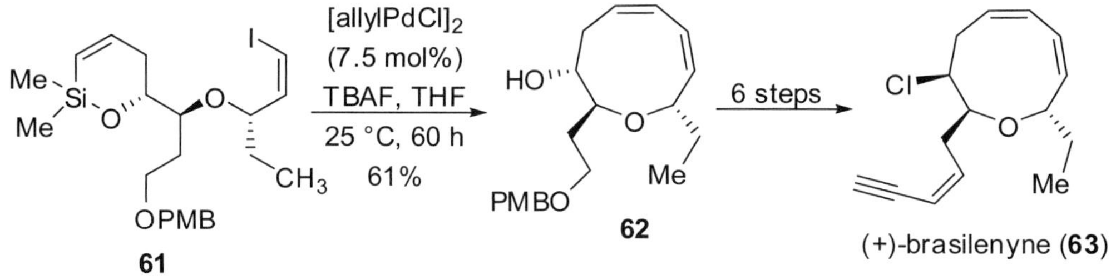

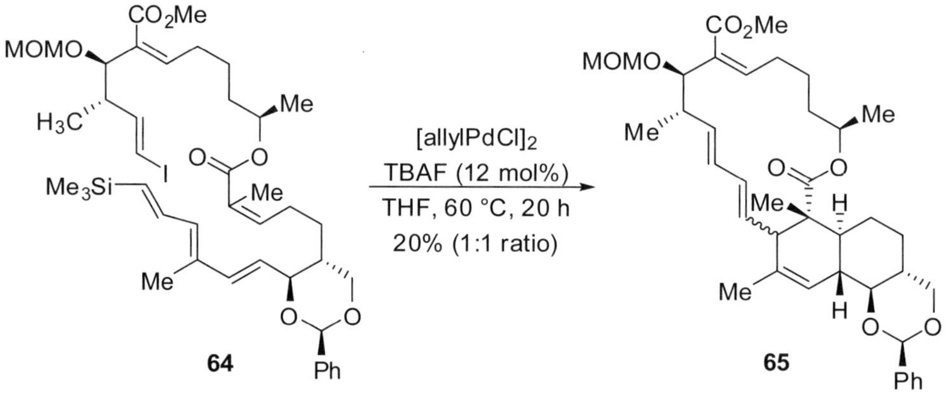

Tadano utilized the Hiyama reaction in a formal synthesis of the antimicrobial tricyclic macrolides tubelactomicins.[44] The intramolecular Hiyama cross-coupling reaction of **64** aided in the synthesis of a 24-membered macrolactone equipped with all the requisite functionalities, which then triggered the transannular Diels–Alder reaction to give **65**. The 24-membered lactone formation was also achieved by the intramolecular ring-closing metathesis reaction.

1.1.2.6 Experimental

2,6-Dimethylbiphenyl (67)[45]

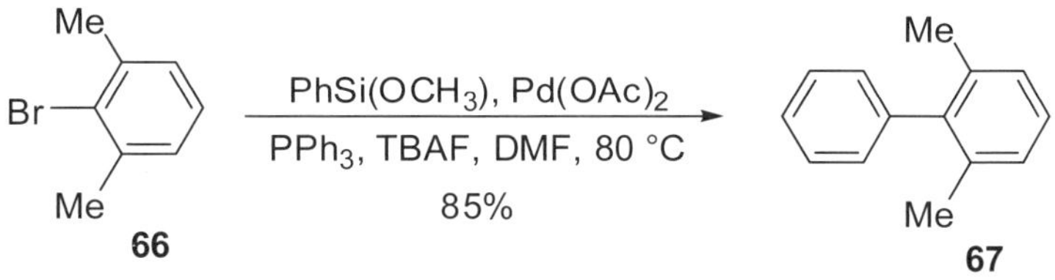

Phenyltrimethoxysilane (2.18 g, 10.9 mmol) was added to 2-bromo-*m*-xylene (**66**, 1.00 g, 5.43 mmol), $Pd(OAc)_2$ (119 mg, 0.530 mmol) and triphenylphosphine (283 mg, 1.08 mmol) in DMF (40 mL). TBAF (10.8 mL, 10.8 mmol, 1.0 M in THF) was added dropwise via a syringe. The reaction mixture was degassed with argon and was heated at 80 °C for 24 h. The reaction was quenched with water (50 mL) and extracted with diethyl ether (4 × 50 mL). The organic layers were dried over $MgSO_4$ and concentrated in vacuo. Purification by flash chromatography eluting with pentane afforded **67** (825 mg, 85%) as a colorless oil.

(*E*)-(Hept-1-enyl)naphthalene (69).[46]

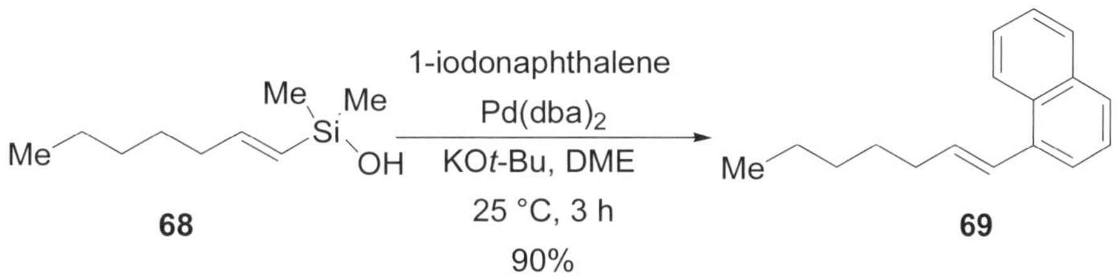

A mixture of potassium *tert*-butoxide (449 mg, 4.0 mmol), (*E*)-**68** (344 mg, 2.0 mmol), 1-iodonaphthalene (292 μL, 2.0 mmol) and $Pd(dba)_2$ (58 mg, 0.1 mmol) was stirred in DME (4 mL) at room temperature for 3 h, and then was filtered through a pad of silica gel. Purification by column chromatography (RP C18, $MeOH/H_2O$, 9/1) afforded 403 mg (90%) of (*E*)-**69** (403 mg, 90%) as a colorless oil.

1.1.2.7 References

1. [R] Hiyama, T. In *Metal Catalyzed Cross-coupling Reactions*, Chapter 10, Ed. By Diederich, F.; Stang, P. J., Wiley-VCH, Weinhein, 1998, pg 421.
2. [R] Hiyama, T.; Shirakwa, E., *Topics in Current Chemistry* **2002**, *219*, 61.
3. [R] Denmark, S. E.; Sweis, R. F. *Acc. Chem. Res.* **2002**, *35*, 835.
4. [R] Denmark, S. E.; Sweis, R. F. *Chem. Pharm. Bull.* **2002**, *50*, 1531.

5. [R] Denmark, S. E.; Ober, M. H. *Aldrichimica Acta* **2003**, *36*, 75.
6. [R] Spivey, A. C.; Gripton, C. J. G.; Hannah, J. P. *Curr. Org. Synth.* **2004**, *1*, 211.
7. [R] Handy, C. J.; Manoso, A. S.; McElroy, W. T.; Seganish, W. M.; DeShong, P. *Tetrahedron* **2005**, *61*, 12201.
8. [R] Denmark, S. E.; Baird, J. D. *Chem. Eur. J.* **2006**, *12*, 4954.
9. Hiyama, T. *J. Organometal. Chem.* **2002**, *653*, 58.
10. Hiyama, T.; Obayashi, M.; Mori, H.; Nozaki, H. *J. Org. Chem.* **1983**, *48*, 912.
11. Fujita, M.; Hiyama, T. *J. Org. Chem.* **1988**, *53*, 5415.
12. Fujita, M. Obayashi, M.; Hiyama, T. *Tetrahedron* **1988**, *44*, 4135.
13. Hatanaka, Y.; Hiyama, T. *J. Org. Chem.* **1988** *53,* 918.
14. Hatanaka, Y.; Hiyama, T. *Synlett* **1994**, 845.
15. Hiyama, T.; Hatanka, Y. *Pure Appl. Chem.* **1994**, *66*, 1471.
16 Hatanaka, Y.; Goda, K.; Hiyama, T. *J. Organomet. Chem.* **1994**, *465*, 97.
17. Denmark, S. E.; Sweis, R. F. *J. Am. Chem. Soc.* **2001**, *123*, 6439.
18. Hamze, A.; Provot, O.; Alami, M.; Brion, J.-D. *Org. Lett.* **2006**, *8*, 931.
19. Denmark, S. E.; Kallemeyn, J. M. *Org. Lett.* **2003**, *5*, 3483.
20. Lee, A. S.-Y.; Chang, Y.-T.; Chu, S.-F.; Tsao, K.-W. *Tetrahedron Lett.* **2007**, *47*, 7085.
21. Manoso, A. S.; Ahn, C.; Soheili, A.; Handy, C. J.; Correia, R.; Seganish, W. M.; DeShong, P. *J. Org. Chem.* **2004**, *69*, 8305.
22. Seganish, W. M.; DeShong, P. *J. Org. Chem.* **2004**, *69*, 6790.
23. Shi, S.; Zhang, Y. *J. Org. Chem.* **2007**, *72*, 5927.
24. Nakao, Y.; Oda, T.; Sahoo, A. K.; Hiyama, T. *J. Organomet. Chem.* **2003**, *687*, 570.
25. Lee, J.-Y.; Fu, G. C. *J. Am. Chem. Soc.* **2003**, *125*, 5616.
26. Li, J.-H.; Deng, C.-L.; Liu, W.-J.; Xie, Y.-X. *Synthesis* **2005**, 3039.
27. Li, J.-H.; Deng, C.-L.; Xie, Y.-X. *Synthesis* **2006**, 969.
28. Ju, J.; Nam, H.; Jung, H. M.; Lee, S. *Tetrahedron Lett.* **2006**, *47*, 8673.
29. Mino, T.; Shirae, Y.; Saito, T.; Sakamoto, M.; Fujita, T. *J. Org. Chem.* **2006**, *71*, 9499.
30. Srimani, D.; Sawoo, S.; Sarkar, A. *Org. Lett.* **2007**, *9*, 3639.
31. Denmark, S. E.; Baird, J. D. *Org. Lett.* **2004**, *6*, 3649.
32. Denmark, S. E.; Baird, J. D. *Org. Lett.* **2006**, *8*, 793.
33. Denmark, S. E.; Kallemeyn, J. M. *J. Org. Chem.* **2005**, *70*, 2839.
34. Nokami, T.; Tomida, Y.; Kamei, T.; Itami, K.; Yoshida, J.-i. *Org. Lett.* **2006**, *8*, 729.
35. Denmark, S. E.; Amishiro, N. *J. Org. Chem.* **2003**, *68*, 6997.
36. Lerebours, R.; Wolf, C. *Org. Lett.* **2007**, *9*, 2737.
37. Denmark, S. E.; Liu, J. H.-C. *J. Am. Chem. Soc.* **2007**, *129*, 3737.
38. Pidaparthi, R. R.; Welker, M. E.; Day, C. S.; Wright, M. W. *Org. Lett.* **2007**, *9*, 1623.
39. (a) Marciniec, B.; Majchrzak, M.; Prukala, W.; Kubicki, M.; Chadyniak, D. *J. Org. Chem.* **2005**, *70*, 8550. (b) Prukala, W.; Marciniec, B.; Majchrzak, M.; Kubicki, M. *Tetrahedron* **2007**, *63*, 1107.
40. Seganish, W. M.; DeShong, P. *Org. Lett.* **2006**, *8*, 3951.
41. Seganish, W. M.; Handy, C. J.; DeShong, P. *J. Org. Chem.* **2005**, *70*, 8948.
42. McElroy, W. T.; DeShong, P. *Org. Lett.* **2003**, *5*, 4779.
43. (a) Denmark, S. E.; Yang, S.-M. *J. Am. Chem. Soc.* **2002**, *124*, 15196. b) Denmark, S. E.; Yang, S.-M. *J. Am. Chem. Soc.* **2004**, *126*, 12432.
44. Anzo, T.; Suzuki, A.; Sawamura, K.; Motozaki, T.; Hatta, M.; Takao, K.-i.; Tadano, K.-i. *Tetrahedron Lett.* **2007**, *48*, 8442.
45. Mowery, M. E.; DeShong, P. *Org. Lett.* **1999**, *1*, 2137.
46. Denmark, S. E.; Sweis, R. F. *J. Am. Chem. Soc.* **2001**, *123*, 6439.

1.1.3 Kumada Cross-Coupling Reaction

Mathew J. Fuchter

1.1.3.1 Description

The Kumada cross-coupling reaction was originally reported as the nickel-catalyzed cross-coupling of Grignard reagents with aryl- or alkenyl halides. It has subsequently been developed to encompass the coupling of organolithium or organomagnesium compounds with aryl-, alkenyl or alkyl halides, catalyzed by nickel or palladium.[1–8]

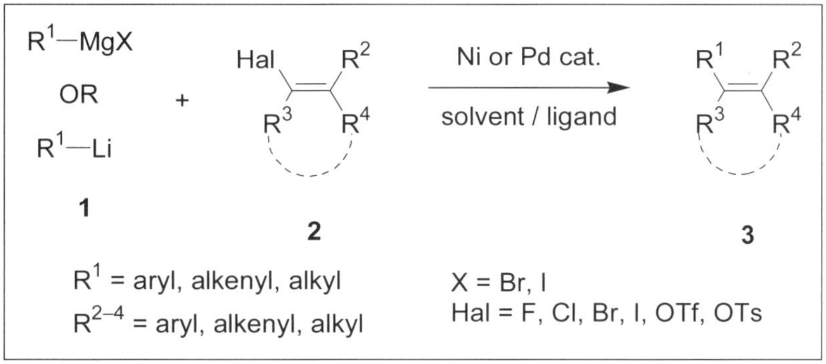

In the nickel-catalyzed process, catalytic activity depends on the phosphine ligand employed and the following trend is observed: $Ni(dppp)Cl_2$ > $Ni(dppe)Cl_2$ > $Ni(PPR_3)_2Cl_2$ ~ $Ni(dppb)Cl_2$. Even alkyl Grignard reagents (with β-hydrogen atoms) can undergo nickel-catalyzed cross coupling without β-hydride elimination. Chlorinated aromatic compounds (**2**, Hal = Cl) react with ease and even fluorobenzene (**2**, Hal = F) can be utilized. The coupling is stereoselective and the stereochemistry of the alkenyl halide reagent **2** is preserved. While organolithium reagents cannot be used in the nickel-catalyzed reaction, they are suitable reaction partners for the palladium-catalyzed process, which is more stereo- and chemoselective. More reactive aryl halide substrates **2** are required (Hal = Br, I, and under modified conditions Cl) however, for the palladium-catalyzed process. By using enantiomerically pure chiral ligands, the reaction can be rendered stereoselective (for saturated coupling partners). The reaction tolerates a range of non-protic solvents including Et_2O, THF, DME and toluene. The reaction most commonly takes place at ambient or slightly elevated temperatures. Side-reactions including homocoupling and reduction can be avoided by: 1) slow addition of organolithiums (to avoid rearrangement of transient α-bromo alkenyllithiums into lithium acetylides); 2) the use of high purity catalyst; and 3) avoiding the use of excess reagents.[5]

1.1.3.2 Historical Perspective

The discovery of the stereoselective cross-coupling reaction between aryl- or alkenyl halides and Grignard reagents under nickel catalysis is attributed to two independent publications in 1972. The first report, from the laboratories of R. J. P. Corriu in Montpellier, France, detailed the coupling of β-bromostyrene with phenylmagnesium bromide in the presence of several nickel salts, for example nickel(II) acetylacetonate.[9] The second, from the laboratories of M. Kumada in Kyoto, Japan, described the coupling of Grignard reagents, such as phenylmagnesium bromide (**4**) with aryl- or vinyl chloride (**5**), catalyzed by $NiCl_2$(dppe).[10] Interestingly, this research was carried out by a graduate student, K. Sumitani, alongside M. Kumada's research associate K. Tamao, who has made significant contributions in the field of organosilicon chemistry (for example, the Tamao–Kumada–Flemming oxidation). In the following years, Kumada and co-workers fully explored the scope of the reaction and thus the transformation is now referred to as the Kumada cross-coupling.[11,4]

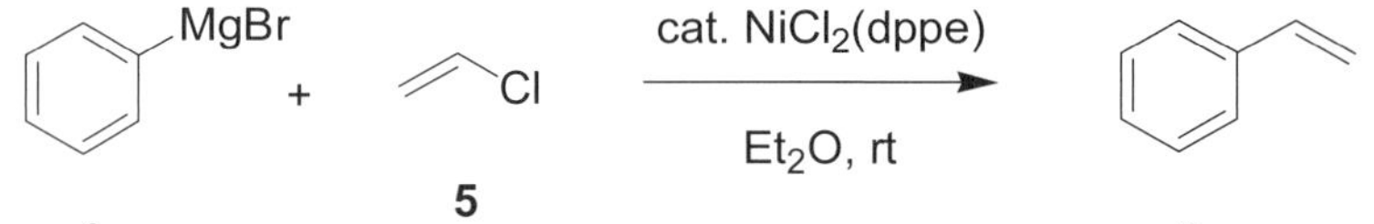

4 **5** **6**

It would be incorrect however, to state that these initial reports were the first examples of this type of chemistry being performed. Indeed, as early as 1923, a French chemist, largely unrecognized today, A. Job and co-workers reported that "a solution of C_6H_5MgBr in diethyl ether gives in the presence of $NiCl_2$ a derivative able to absorb CO, NO, C_2H_4, C_2H_2 and H_2".[12] He subsequently went on to detail the catalytic effect of the nickel salt and its use in several chemical reactions.[13] Several years later in 1939, H. Gilman and co-workers from the chemical laboratory of Iowa State College reported the homocoupling of phenylmagnesium iodide in the presence of catalytic nickel(II) bromide, to yield biphenyl in quantitative yield.[14] Most importantly however, a publication by M. S. Kharasch and co-workers from the George Herbert Jones Laboratory of the University of Chicago in 1941, reported the coupling of phenylmagnesium bromide with bromobenzene using 4 mol% of nickel(II) chloride, amongst a variety of other salts.[15] Perhaps in part due to his work on organocobalt chemistry,[16,17] the mechanism of this reaction was attributed to homocoupling of phenyl radicals derived from the Grignard reagent, however, it remains uncertain whether a radical or polar pathway is occurring in the case of the nickel salt. Even the independent reports by Corriu and Kumada, published in 1972, followed a flurry of activities on the transition metal-catalyzed C–C bond forming reactions of unreactive alkenyl and aryl halides.[18–23]

Key to the publication of Kumada and co-workers, was the use of a nickel(II)–phosphine complex. Indeed, they subsequently demonstrated that the catalytic activity of the complex strongly depends on the nature of the phosphine ligand.[4] In 1973, it was reported that the use of an optically active phosphine ligand could induce asymmetric induction in the related reaction of secondary alkyl Grignard reagents and vinyl chloride, under nickel catalysis.[24,6] Perhaps the most notable development however, was a publication by S.-I. Murahasi and co-workers, which detailed the ability of the Kumada cross-coupling to be carried out under palladium, as apposed to nickel catalysis.[25,5] One distinct advantage of this procedure, was that it allowed versatile organolithium reagents to be used as an alternative to Grignard reagents.

Following this early activity, application of the Kumada cross-coupling reaction in synthesis dwindled somewhat in latter years. This was a result of issues with functional group compatibility: The high basicity of Grignard and organolithium reagents prohibits the use of base-sensitive functional groups. Alternative protocols using less nucleophilic (and basic) coupling reagents such as organozinc (Negishi), organoboron (Suzuki–Miyaura), organotin (Stille) and organosilicon (Hiyama) became more widespread.[26] However, since many of the alternative coupling reagents are synthetically prepared from Grignard or organolithium reagents, these alternative procedures are less direct and experimentally straightforward than the Kumada coupling. Thus this procedure has seen a slight renaissance lately.[27,28]

1.1.3.3 Mechanism

The general mechanism of the nickel-catalyzed process can be described as follows: The initial nickel(II) complex **7** undergoes transmetallation with the Grignard reagent **1** to form a diorganonickel species **8**. Reaction of complex **8** with an organohalide **9** forms the homocoupled product **10** and the nickel(II) species **11**, which enters the catalytic cycle. This preliminary sequence can be viewed as initiation of the nickel complex, and is negligible in the overall reaction due to the catalytic quantity of nickel. The first step of the catalytic cycle is transmetallation of the active nickel(II) complex **11** with Grignard reagent **1**. Coordination of the organohalide **9** to the diorganonickel species **12,** gives complex **13** which subsequently undergoes oxidative addition of organohalide **9**, releasing the coupled product **14**.[29]

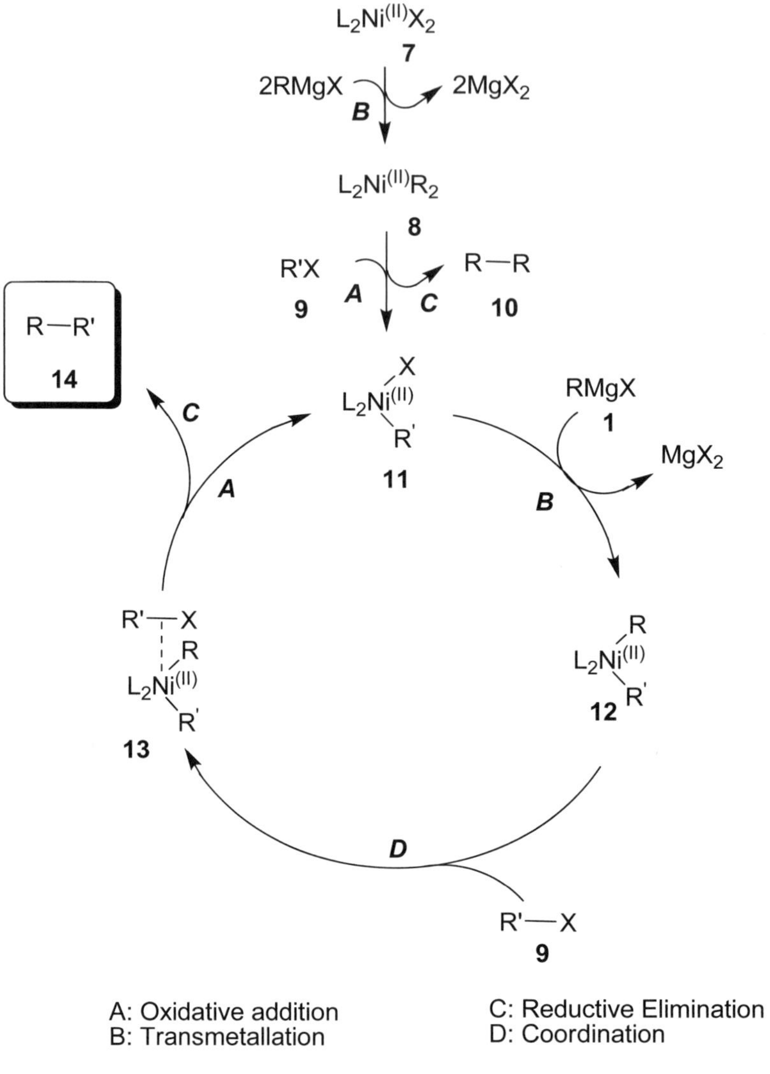

In the basic mechanism for the palladium-catalyzed process, organohalide **9** undergoes oxidative addition to a palladium(0) catalyst **15** to afford a σ-organopalladium(II) complex **16**. All palladium precatalysts are converted to the active palladium(0) catalyst **15** *in situ,* most commonly by phosphine in phosphine assisted catalytic cycles. Transmetallation of the palladium(II) complex **16** with organometallic reagent **17** gives diorganopalladium complex **19**. Reductive elimination of the product **14** regenerates the active palladium(0) catalyst.[29]

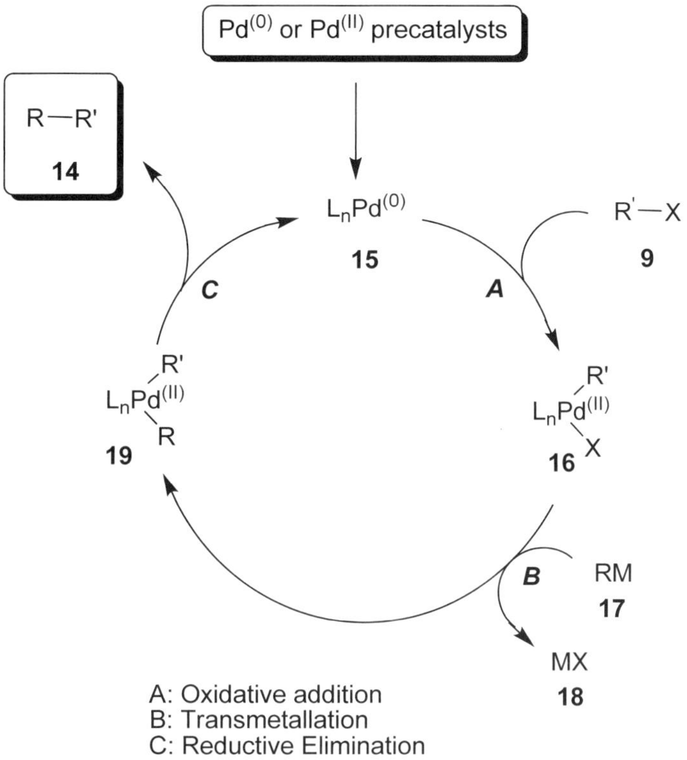

While useful in explaining the discrete mechanistic steps of the Kumada cross-coupling reaction this generalized scheme ignores the precise coordination number, geometry and formal charge of the nickel or palladium centres. Indeed, readers are referred to more detailed mechanistic studies for further information.[30,31] For example, in the palladium catalyzed reaction, Murahashi has pointed out the benefit of using $Pd(PPh_3)_2LiCl$, formed upon addition of methyl lithium to $PdCl_2(PPh_3)_2$.[5] Following their publication of this synthetic method in 1984,[32] Negishi confirmed the presence of $Pd(PPh_3)_2LiCl$ in the reaction.[33] It has since been established in elegant studies by Amatore and Jutand, that anions can play a vital role in palladium mediated cross-couplings, and that in many cases, the active catalytic species may involve a formally anionic palladium centre.[31] This situation is most likely further complicated by the presence of strongly nucleophilic Grignard or organolithium reagents. Indeed, Knochel has postulated that an organopalladate of the type $[MgX]^+[RPdL_2]^-$ may be involved in certain coupling processes.[34]

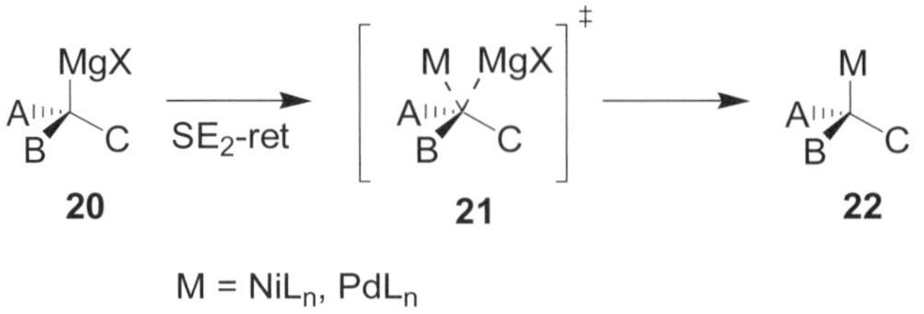

For given systems it is possible to render the Kumada cross-coupling stereoselective. Elegant studies by Hoffmann and co-workers have provided details on the stereochemical course of the transmetallation step when using saturated reaction partners.[35] Utilizing a chiral Grignard regent as a probe, they determined that transmetallation of the Grignard reagent **20** by nickel or palladium, proceeds with retention of configuration to give **22**; a concerted S_E2-ret process.[36] Since virtually all examples of asymmetric Kumada cross-coupling reactions use racemic secondary Grignard reagents (the Grignard reagents usually undergo racemization on a rate comparable to cross-coupling), they can be viewed as a dynamic kinetic resolution, with transmetallation as the enantiodiscriminating step.[6]

1.1.3.4 Synthetic Utility

Despite its historical significance, the application of the Kumada cross-coupling reaction in synthesis dwindled somewhat in recent years due to due issues with functional group compatibility, and thus alternative cross-coupling protocols have flourished.[26] In certain cases however, the Kumada cross-coupling offers distinct advantages. For example, in many palladium-mediated cross-coupling reactions, the use of aryl or alkenyl bromides, iodides or tosylates are required as a cross-coupling partner. The use of readily available and cheap aryl or alkenyl chloride reagents is problematic, due to their low reactivity towards oxidative addition to Pd(0).[26] On the other hand, for the nickel-catalyzed Kumada cross-coupling, aryl or alkenyl chlorides are the reagent of choice in view of their high reactivity and high yields.[4] Indeed, the Hokko Chemical Industry Company Ltd., Japan, industrialized a process involving a Kumada coupling of aryl Grignard reagents (for example, **23**) and vinyl chloride (**5**) to give styrene compounds in good yields. In 2002, it was reported that styrene **24** was being prepared at approximately 5000 kg per year using this process.[37]

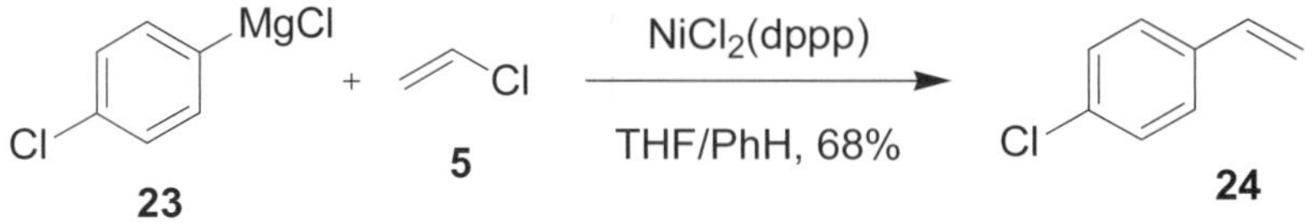

Kumada, Tamao and co-workers demonstrated another important aspect of the Kumada cross-coupling reaction prior to 1982.[38] They demonstrated that even alkyl Grignard reagents containing β-hydrogen atoms can selectivity undergo cross-coupling. For example, *o*-dibutylbenzene (**27**) can be prepared in good yield from butyl Grignard reagent **26** and dichlorobenzene **25**.

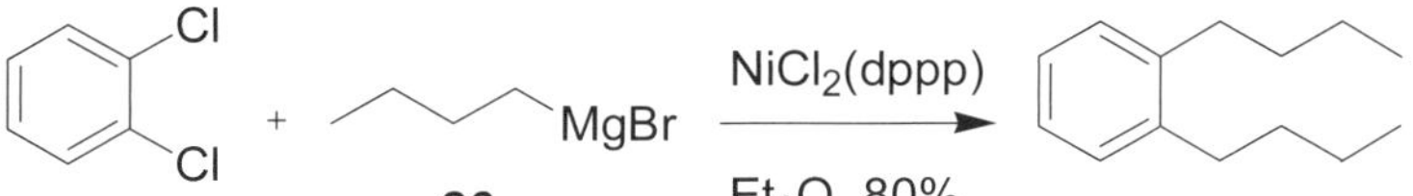

25 **26** **27**

The use of secondary alkyl Grignard reagents however, can result in product mixtures. This is due to alkyl group isomerization from secondary to primary, and has been shown to be highly dependent on the basicity of the phosphine ligands, as well as the electronic nature of the aromatic halides.[39,40] For example, cross-coupling of *sec*-propyl Grignard regent **29** with chlorobenzene (**28**), results in the desired product **30** in a 96 : 4 ratio when 1,3-bis(diphenylphosphino)propane (dppp) is used as a ligand, whereas isomerization is the major mechanistic pathway when 1,3-bis(diphenyl-phosphino)ethane (dppe) is employed.

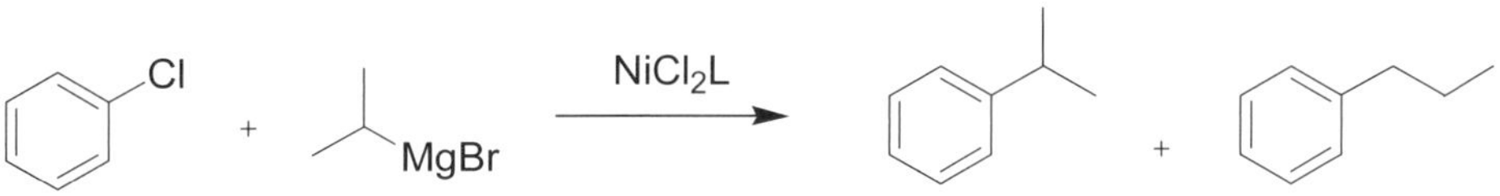

28 **29** **30** **31**

dppp	96	:	4
dppe	9	:	84

The importance of the seminal paper by Kumada, Tamao and co-workers lies in the use of nickel-phosphine complexes, rather than nickel salts in the Kumada cross-coupling reaction.[10] As the example above already demonstrates, the ability to tune the reactivity of the system by modifying the group which ligates the nickel is of prime importance, and results in a versatile catalytic procedure. As already discussed, the use of aryl or alkenyl chloride reagents is problematic in palladium-mediated cross-coupling reactions,[26] and aryl fluorides are highly inert due to the strength of the C–F bond. While Kumada and Tamao first demonstrated that aryl fluorides are suitable reagents for the nickel-catalyzed Kumada cross-coupling (albeit in poor yields),[4] Herrmann and co-workers demonstrated, that in the presence of suitably stabilizing ligands (*N*-heterocyclic carbenes in this case), aryl fluorides and aryl Grignard reagents can be coupled at ambient temperature.[38] Similarly, Ackermann and co-workers have disclosed a phosphine oxide ligand which can mediate an analogous reactivity profile.[39] In an alternative, but related approach, a fascinating report by Nakamura and co-workers detailed the use of bimetallic cooperation to realize this transformation. They reported the efficient coupling of aryl fluorides **32** with phenylmagnesium bromide, catalyzed by nickel ligated to **34**.[40] Computational studies indicated that a bimetallic synergy facilitates C–F bond activation in the oxidative addition step, and the oxidative addition transition state can be represented by structure **35**.

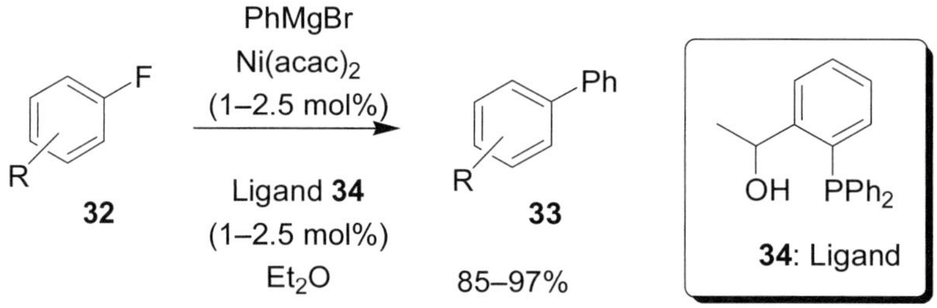

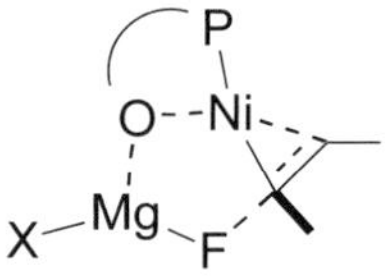

35

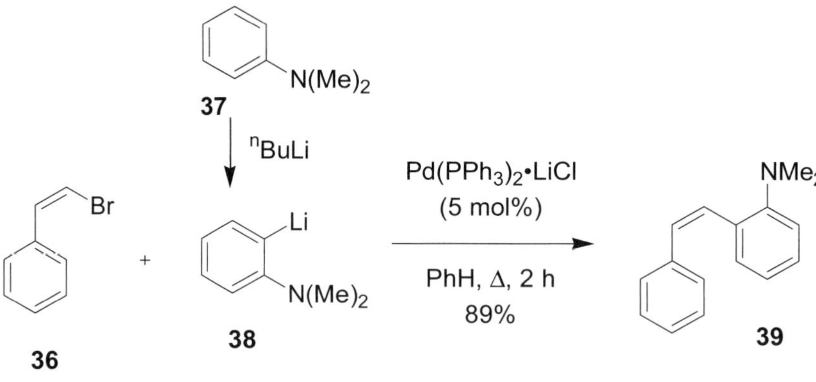

While the nickel-catalyzed reaction facilitates oxidative addition of substrates problematic to palladium-mediated processes, there is still a great deal of interest in the palladium-catalyzed Kumada cross-coupling reaction due to its enhanced chemoselectivity.[25,5] Also, as mentioned previously, one distinct advantage of the palladium-catalyzed procedure, is that it allows versatile organolithium reagents to be used as an alternative to Grignard reagents. The strongest merit of organolithium reagents is their preparation by direct lithiation of hydrocarbons, particularly when directed by neighbouring heteroatoms. Thus, lithiated aniline derivative **38**, prepared *in situ* from *ortho*-lithiation of **37** can be coupled directly to vinyl bromide **36** in good yield.[5] Clearly, this procedure is extremely experimentally straightforward and therefore highly useful for given cases.

As discussed above however, a switch from nickel to palladium renders the use of aryl or alkenyl chloride reagents problematic, due to their low reactivity towards oxidative addition to Pd(0).[26] To overcome this issue, in 1999 Nolan and co-workers reported the palladium-catalyzed Kumada cross-coupling of aryl chlorides using *N*-heterocyclic carbenes as ligands for the palladium centre.[41] Using *N*-heterocyclic carbene **42** (prepared *in situ* from the imidazolium salt, IPr•HCl), the biaryl products **41** were isolated in moderate to good yields. A subsequent publication from Li in 2002 detailed the use of a phosphine oxide ligand to mediate an analogous process.[42] Finally, other *N*-heterocyclic carbene-based precatalysts have been reported to mediate this transformation.[43]

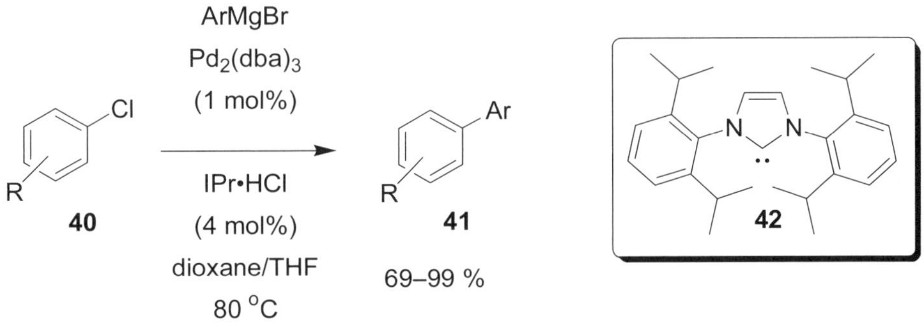

Asymmetric Kumada Cross-Coupling

Asymmetric synthesis using the Kumada cross-coupling reaction has most frequently been studied using racemic secondary Grignard reagents.[3] Since such reagents usually undergo racemization on a rate comparable to cross-coupling, the reaction process can be viewed as a dynamic kinetic resolution to produce enantioenriched products. The first example of an asymmetric, nickel-catalyzed Kumada cross-coupling reaction utilized (–)-DIOP as a ligand, although the products were only isolated in 13–17% *ee*.[24,3] Subsequent studies have surveyed a large number of optically active phosphines in the quest for improved enantioselectivities. Two of the best ligand systems are the ferrocenylphosphines containing (dialkylamino)alkyl sidechains and β-(dialkylamino)alkylphosphines.[3] For example, in the latter class of ligands, Valphos (**46**) was shown to mediate the asymmetric Kumada cross-coupling of racemic Grignard reagent **43** with vinyl bromide to give **44** in 80% *ee*. Oxidative cleavage of the double bond gave a short synthesis of the anti-inflammatory agent Ibuprofen (**45**).[44]

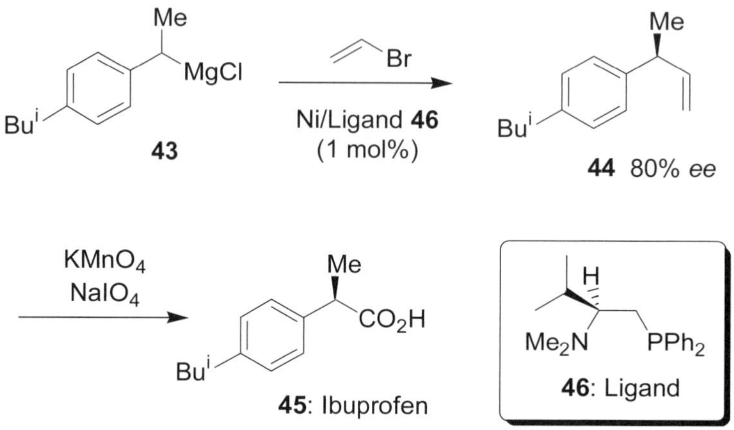

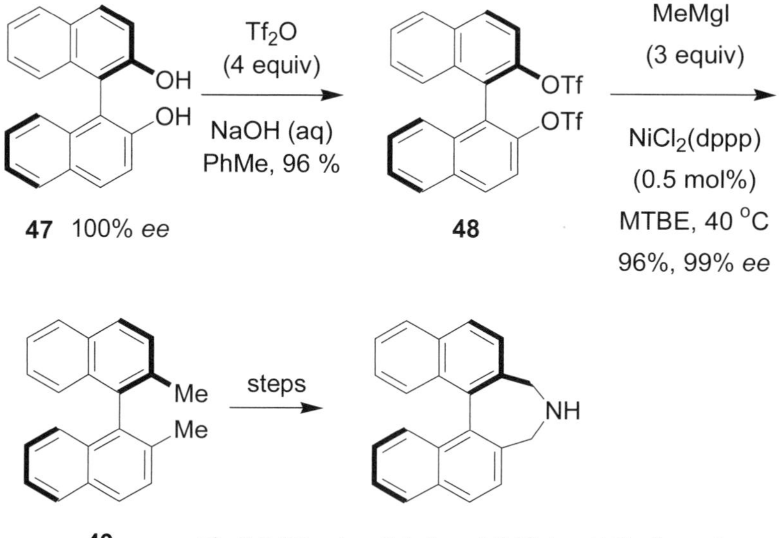

50: 3,5-Dihydro-4H-dinaphth[2,1-c:1',2'-e]azepine

Unfortunately, despite significant work,[3] there a very few reports detailing the asymmetric Kumada cross-coupling of racemic secondary Grignard reagents, to give truly exception levels of enantioselection (> 95% *ee*). One area that has shown promise however, is in the synthesis of axially chiral biaryls.[3] Indeed, control of atropisomeric selectivity in asymmetric cross coupling reactions is an exciting and challenging field.[45] In order to prepare azepine **50**, a precursor to C_2-symmetric chiral quaternary ammonium salts, that can serve as asymmetric phase-transfer catalysts,

Ikunaka *et al.* employed a Kumada cross-coupling reaction.[46] Conversion of commercially available (*R*)-BINOL (**47**) to the bistriflate **48**, followed by Kumada cross-coupling with methylmagnesium iodide, gave access to methyl derivative **49** in good yield. The synthesis proved reliable and scalable as apposed to other cited reports.

For the above example however, the Kumada cross-coupling is not stereoselective and the source of chiral information lies in the commercially available, but relatively expensive (*R*)-BINOL (**47**). An impressive alternative was reported 15 years previous to this by Hayashi and co-workers.[47] Asymmetric cross-coupling of aryl Grignard **51** with napthyl bromide **52**, gave the desired product **49** in an impressive 95% *ee*. While other ligands had proved less successful in mediating an analogous transformation,[3] the use of ferrocenylphosphine (*S*)-(*R*)-**53** dramatically improved the selectivity.[47]

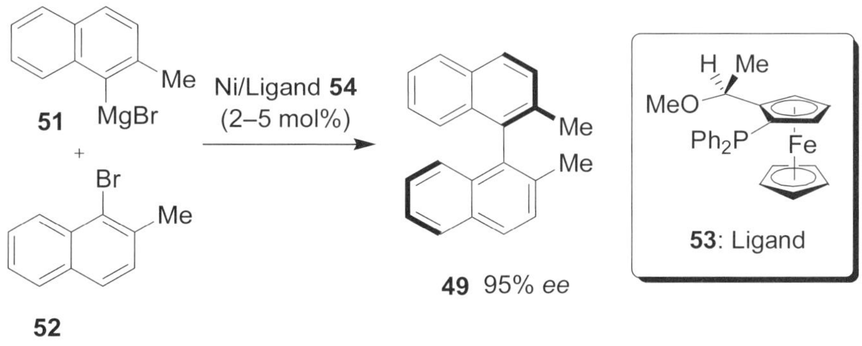

In another example of the enantioselective synthesis of axially chiral molecules, Hayashi and co-workers reported the successful enantioposition-selective asymmetric Kumada cross-coupling, for the preparation of **55**. Indeed, using (*S*)-phenphos (**56**) as a ligand, the Kumada cross-coupling of bistriflate **54** gave the desired product **55** in good yield and a high level of asymmetric induction.[48]

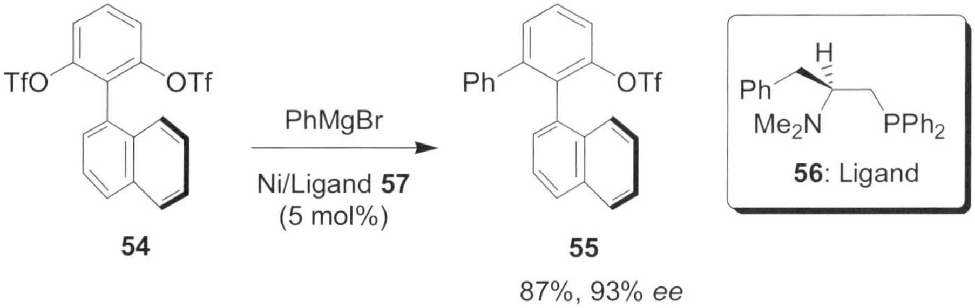

Natural Product Synthesis

There are several examples of the synthesis of naturally isolated, complex molecular architectures utilizing the Kumada cross-coupling as a key step. For example, the enantioselective synthesis of (+)-ambruticin (**59**) was reported from the laboratories of Jacobsen.[49] Conversion of an (*E*)-vinyl iodide **57** to diene **58** was achieved in good yield using a Kumada cross-coupling. The stereochemistry of the vinyl iodide **57** was conserved in the transformation.

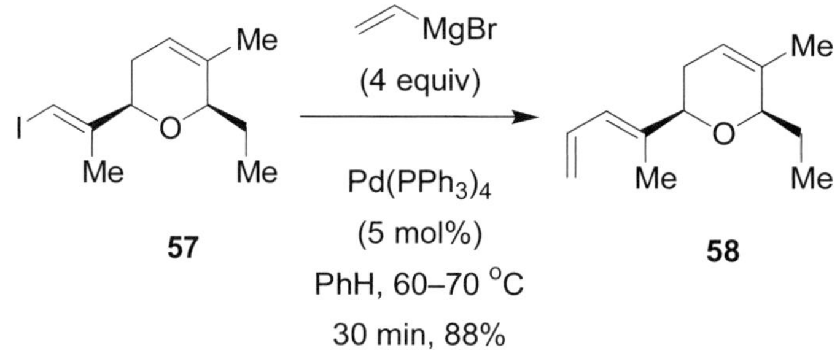

57 **58**

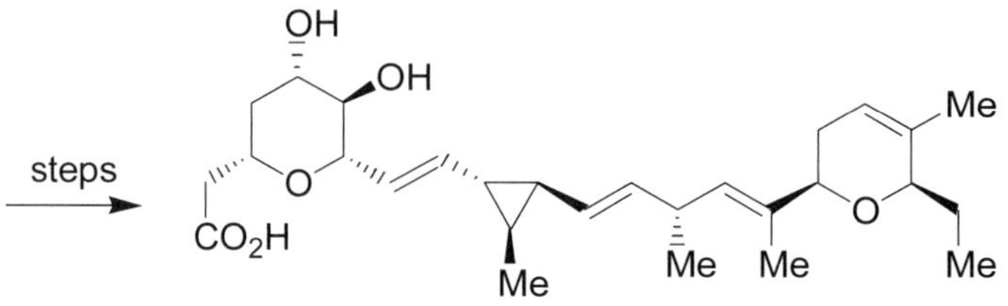

59: (+)-ambruticin

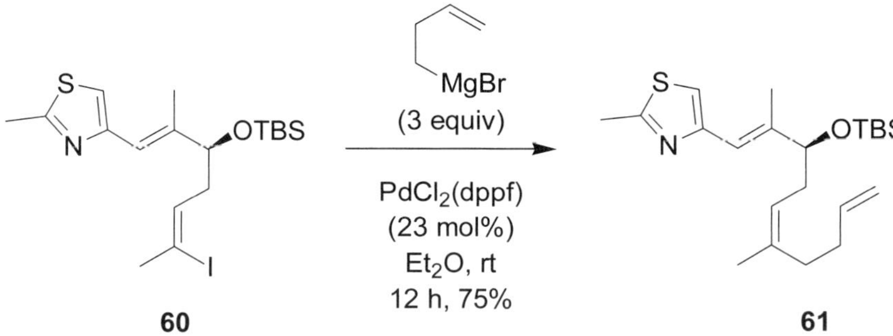

60 **61**

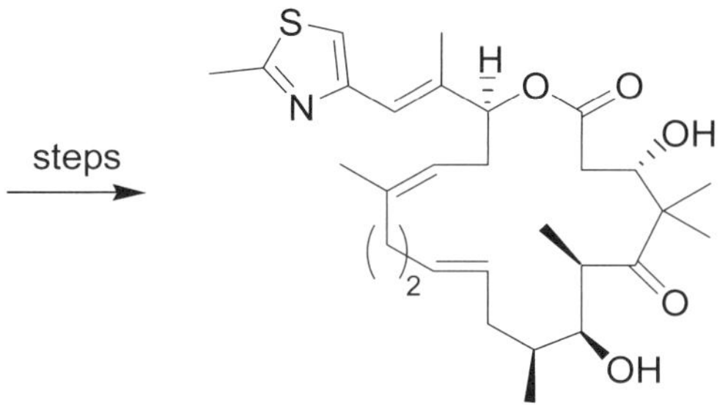

62: [18]-dehydrodesoxyepothilone B

In 2002, Danishefsky and co-workers reported a highly concise synthesis of [18]-dehydrodesoxyepothilone B (**62**).[50] The synthetic approach was based on a key ring-closing metathesis reaction, and the precursor to this was prepared by a Kumada cross-coupling reaction. Coupling of vinyl iodide **60** with allylmagnesium bromide, under palladium-catalyzed conditions gave key intermediate **61** in good yield.

The synthesis of tetrahydrocannabinol **65** by Kobayashi and co-workers employed a Kumada cross-coupling in its final stages, utilizing an unusual enol phosphate as the substrate.[51] The synthetic sequence employed a three-step 1,4-addition strategy to functionalize an α-iodinated cyclohexanone via the conjugate addition of a cuprate. The resulting enolate was trapped as the corresponding phosphate **63** and engaged in a Kumada cross-coupling reaction with methylmagnesium chloride.

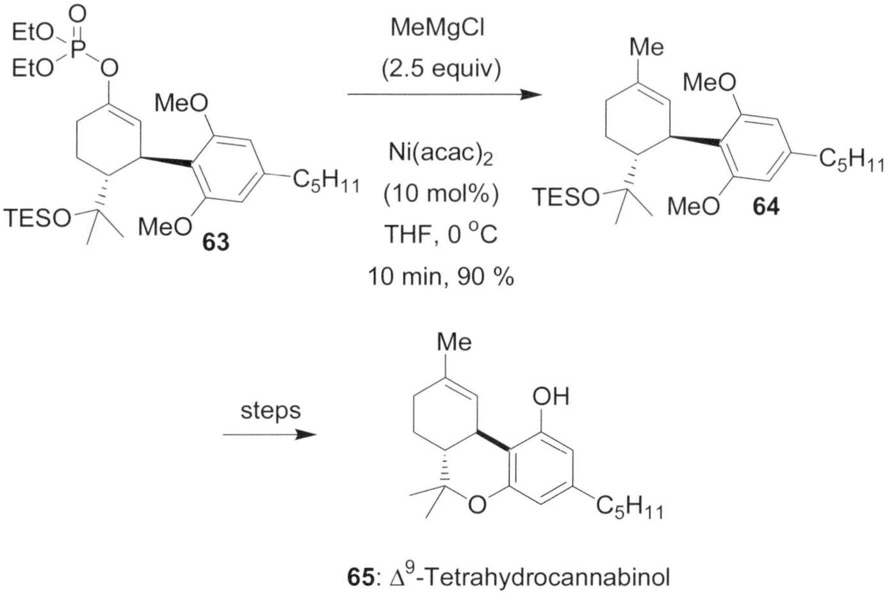

65: Δ^9-Tetrahydrocannabinol

1.1.3.5 Variations and Improvements

Kumada Cross-Coupling of Aryl Tosylates

As described above, many electrophiles can be utilized in the Kumada cross-coupling including aryl and alkenyl iodides, bromides, chlorides and fluorides. To enable oxidative addition of a carbon–oxygen bond, it is usual to convert it to an aryl or vinyl triflate. Aryl or vinyl tosylates on the other hand are used much less frequently employed, despite the fact that they are readily prepared and cheaper than the corresponding triflates. They are also more stable to water and often crystalline. This greater stability however, renders them less-reactive towards oxidative addition, and therefore catalytic procedures with an enhanced activity must be applied. Studies by Hartwig and co-workers have demonstrated that aryl and alkenyl tosylates are suitable reaction partners in the palladium-catalyzed Kumada cross-coupling.[52,53] Using sterically hindered ligands from the Josiphos family **69**, efficient coupling of aryl or alkenyl tosylates **67** with Grignard reagents **66** was observed in good to excellent yields at ambient temperature. Additional studies by Ackmermann and co-workers have demonstrated that phosphine oxides are also suitable ligands for the Kumada cross-coupling of aryl tosylates.[54]

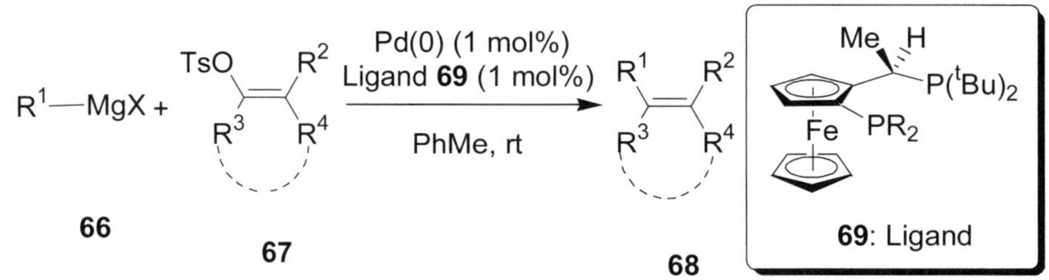

66 **67** **68**

Cross-Coupling of Knochel-type Grignard Reagents and Triarylmagnesiates

Traditionally, the palladium-catalyzed Kumada cross-coupling was developed to allow the use of organolithium reagents. The attractiveness of these reagents in synthesis stems from their ease of preparation, either by direct lithiation of hydrocarbons or low temperature lithium halogen exhange.[55] The downside of their use however, is that the high polarity of the carbon-lithium bond precludes the presence of sensitive functional groups. Classically, Grignard reagents are prepared by direct reaction of magnesium metal with organic halides at elevated temperatures. Unfortunately, this method is also not compatible with sensitive functionality. In recent years, pioneering work by Knochel has demonstrated the power of the magnesium-halogen exchange reaction. Using *i*-propylmagnesium chloride at temperatures below 0 °C, Knochel has published extensively on the synthesis of functionalized Grignard reagents **71** from the magnesium–halogen exchange of aryl iodides **70**.[56,57] Since only reactive electrophiles such as aldehydes and ketones react with Grignard reagents rapidly at temperatures below 0 °C, a whole host of Grignard reagents can be prepared bearing sensitive functional groups.[56,57]

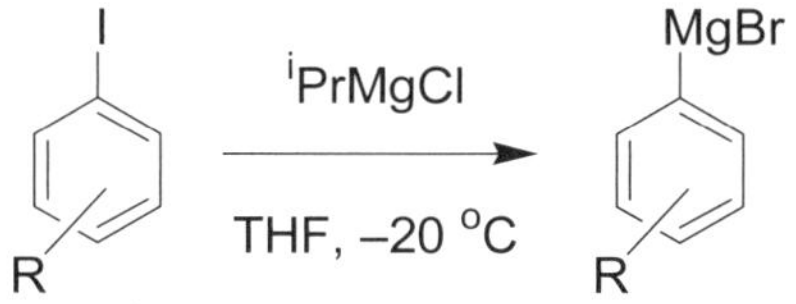

70 **71**

These elegant studies by Knochel and co-workers have reinvigorated the use of Grignard reagents in synthesis, due to the ready availability of reagents bearing previously unobtainable functionality. Also, due to the more covalent nature of the carbon-magnesium bond, these reagents are, in general, more functional group compatible than the corresponding organolithium reagents.[57] A large body of Knochel's work focuses on the functionalization of heteroaromatics.[56,57] For example, during their studies Knochel and co-workers noticed and interesting selectivity in the synthesis of pyridine derivatives using a Kumada cross-coupling. Exposure of bromopyridine derivative **72** to phenylmagnesium chloride results in the direct substitution of the phenyl sulfonyl group in 77% yield, whereas under Kumada conditions, smooth cross-coupling of the Grignard reagent with the aryl bromide moiety results in the production of biaryl **74**.[58]

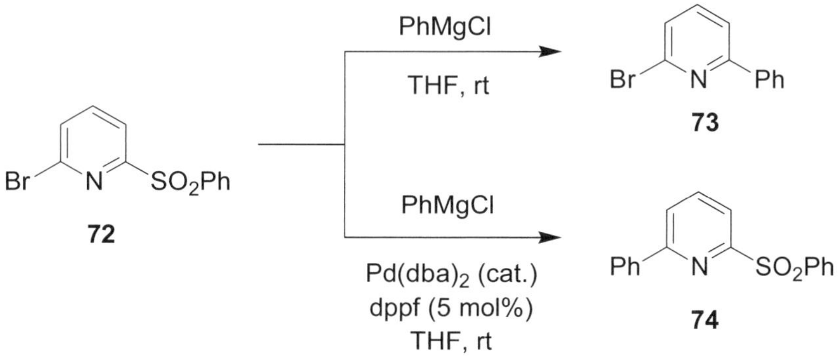

72 **73** **74**

Despite the synthetic flexibility of the magnesium–halogen exchange using *iso*-propylmagnesium chloride, the reaction is still considerably slow, especially with substrates other than organic iodides. Knochel and co-workers have recently reported an improved system however, which employs one equivalent of lithium chloride.[59] Thus, magnesium–halogen exchange is mediated by *i*-PrMgCl•LiCl, which has a greatly enhanced reactivity profile, allowing the exchange of aryl bromides at low temperature. It is postulated that enhanced reactivity upon addition of LiCl stems from breaking down of the aggregated *i*-PrMgCl, in addition to the reactive nature of the magnesiate [*i*-$PrMgCl_2^-Li^+$].[59] This method has allowed the preparation of Grignard reagents bearing highly sensitive functionality. Despite this success, in the context of the Kumada cross-coupling, temperatures above 0 °C are required, at which unfortunately, these sensitive magnesiate reagents are often unstable. Recent work by Buchwald and co-workers however, has recently opened up the possibility of utilizing Knochel's magnesiates in the Kumada cross-coupling reaction. Utilizing the most active ligand systems developed in his laboratories (such as **78**), Buchwald and co-workers demonstrated the effective Kumada cross coupling of aryl iodides and Knochel-type magnesiates at temperatures ranging from –20 to –62 °C.[60] For example, Kumada cross-coupling of functionalized thiophene **76** with aryl iodide **75** gave access to **77** in good yield.

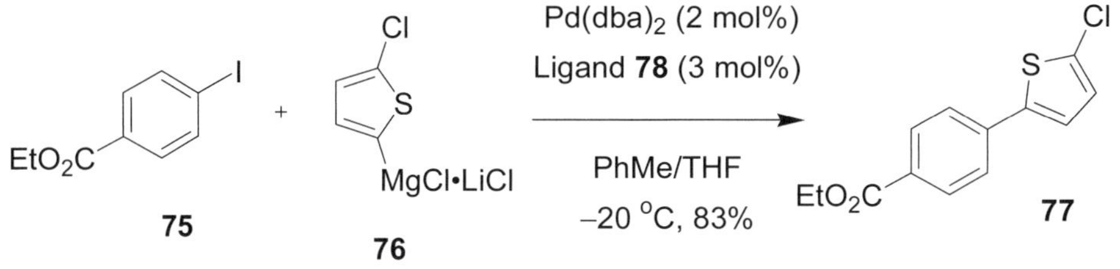

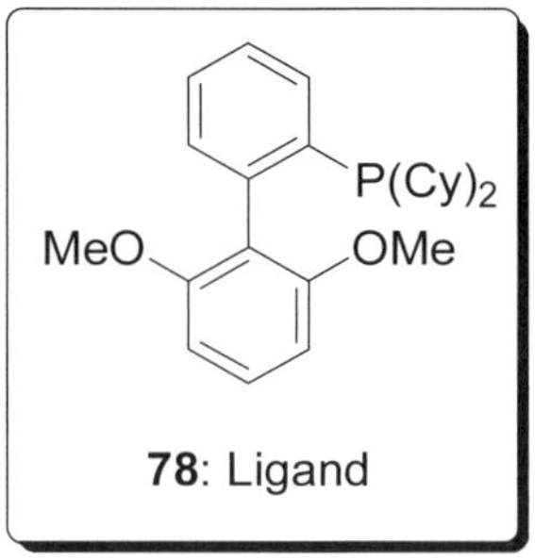

One of the remaining major drawbacks in the synthesis of functionalized aryl Grignard reagents is that magnesium insertion or halogen–magnesium exchange on electron-rich aryl halides is often problematic.

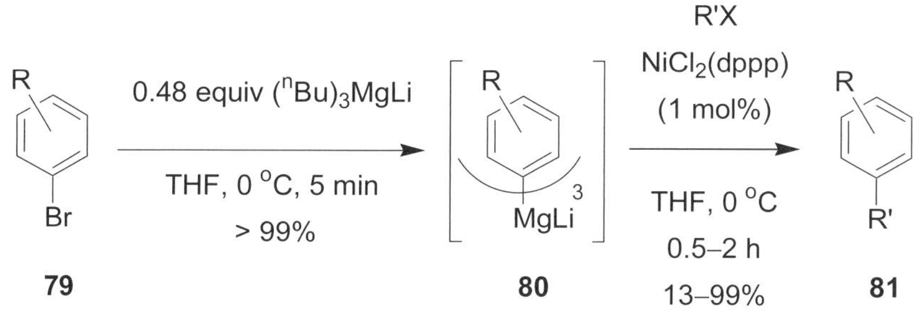

Recent work by Lau and co-workers has addressed this issue in the synthesis of electron-rich biaryls. While magnesium–halogen exchange of **79** using Knochel's conditions failed to give the desired Grignard reagents, Lau and co-workers found that conversion to the triaryl magnesiate **80** proceeded in high yield in five minutes.[61] They subsequently found that such magnesiates undergo efficient nickel-catalyzed Kumada cross-coupling reactions in mostly high yield.

Kumada Cross-Coupling of Alkyl Halides

While aryl and vinyl electrophiles have been thoroughly investigated as cross-coupling partners over the last 30 years or so, the use of alkyl halides as electrophiles has remained largely unknown until recently.[62] Firstly, since the $C(sp^3)$–X bond in alkyl halides is more electron rich than the $C(sp^2)$–X bond in aryl and vinyl halides, these substrates are much less reactive to oxidative addition to a low-valent transition metal complex. Furthermore, the resultant alkyl metal complex is highly reactive owing to the absence of stabilizing electronic interactions with the metal d orbitals.[62] As such, the fast and thermodynamically favoured β-hydride elimination leads predominantly to olefinic by-products in the majority of catalytic systems. Finally, the relatively slow reductive elimination of the cross-coupling product increases the likelihood of further side-reactions (elimination, hydrodehalogenation).[62]

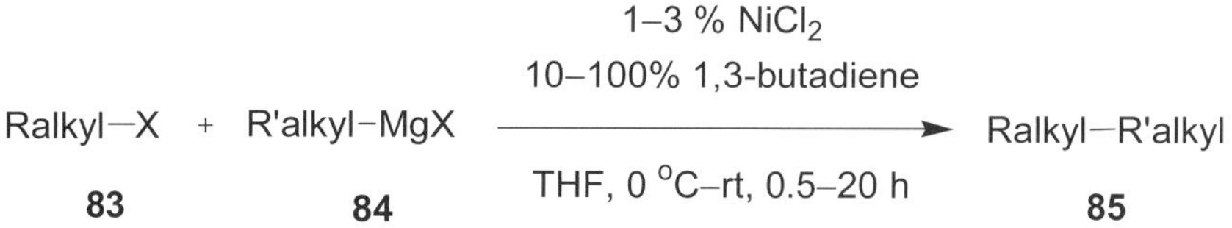

Although the first report of this type of coupling appeared in the 1970s from the laboratories of J. K. Kochi,[21,22] only a few subsequent publications emerged until the early 1990s. In 1992, Suzuki demonstrated the ability of alkyl halides to be used in the cross-coupling of organoboron reagents,[63] whereas in 1995 Knochel reported the use of organozinc reagents.[64] In terms of the Kumada cross-coupling however, Kambe and co-workers first reported the nickel-catalyzed cross-coupling of alkyl bromides, chlorides and tosylates **83** with Grignard reagents **84** in 2002.[65] It is important to note that prior to this study, several other publications on a similar theme had emerged using other metallic catalysts. Fascinatingly, Kambe and co-workers reported the beneficial addition of 1,3-butadiene as opposed to phosphine ligands. Cross-coupling of alkyl bromides and tosylates was observed quantitatively at 0 °C in the presence of the diene, whereas reduction and/or elimination of the electrophile were mainly observed in its absence. Mechanistically, they postulated that nickel (0), generated from reduction of nickel(II) by the Grignard reagent, undergoes reaction with 2 equivalents of butadiene to generate **87**. Transmetallation with the Grignard reagent gives the formally anionic species **88**, which undergoes alkylation and reductive elimination to give the product and regenerate the catalyst.[65] It is important to note however, that this postulation, inferring nickel(IV) intermediate, contradicts mechanistic interpretations of similar nickel-catalyzed cross coupling reactions.[62]

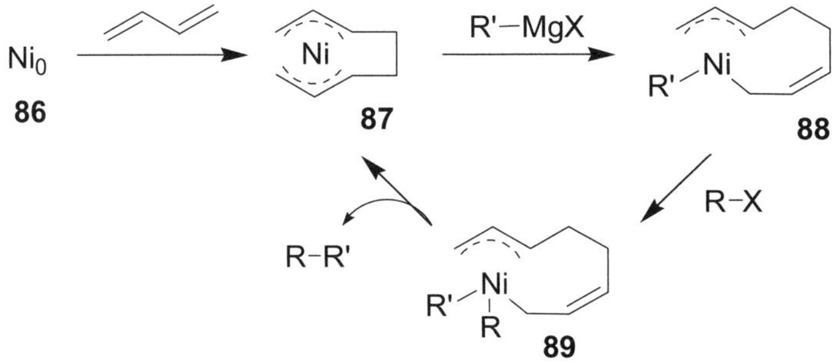

The following year, Kambe and co-workers reported an analogous strategy using a palladium catalyst, [Pd(acac)$_2$].[66] Once again, the need for 1,3-butadiene was highlighted and the palladium-catalyzed system exhibited higher chemoselectivity. Another report emerged from the same group highlighting the use of nickel or copper catalysts in the alkyl–alkyl cross-coupling reaction of alkyl fluorides.[67] As well as these pioneering efforts, reports have emerged from the laboratories of M. Beller which highlight the Kumada cross-coupling of alkyl chlorides under palladium catalysis.[68,69]

Other Metallic Catalysts

Numerous other metallic salts have been shown to exhibit similar reactivity profiles in the Kumada coupling, although not necessarily via the same mechanistic pathway. Even in the early studies of Kharasch, Kochi, Gilman and a variety of others, cobalt and iron salts displayed similar reactivity to nickel and palladium. Perhaps as a result of this, the use of cobalt and iron salts has been resurrected in cross-coupling reactions. For example, Knochel has demonstrated that catalytic cobalt(II) chloride is able to mediate a Kumada cross-coupling of chloroheteroaromatics with aryl Grignard reagents in good yield.[70] Likewise, Oshima and co-workers have reported the use of cobalt(III) acetylacetonate in the catalyzed Kumada cross-coupling of alkyl halides with 1-(trimethylsilyl)ethenylmagnesium, although large catalytic loadings of the cobalt complex were required.[71]

With the growing importance of sustainable chemistry however, iron-catalyzed cross-coupling reactions are becoming more-widespread, owing to the fact iron is inexpensive and more environmentally friendly than palladium, cobalt, or nickel. While important breakthroughs have been made in the laboratories of Cahiez and Nakamura, the use of iron in the Kumada cross-coupling has been pioneered by A. Fürstner and co-workers.[72] Using FeX_n (n = 2, 3; X = Cl, acac) as a precatalyst, Fürstner and co-workers observed the effective cross-coupling of a variety of substrates with Grignard reagents. It was postulated that the active catalysts were highly reduced iron-magnesium clusters of formal composition $[Fe(MgX)_2]_n$, in which the iron is in its –2 oxidation state.[72] To probe this hypothesis, they utilized a structurally defined iron(–II) complex, $[Li(tmeda)]_2[Fe(C_2H_4)_4]$ in the Kumada cross-coupling of alkyl halides with Grignard reagents.[72] Pleasingly, this catalyst was extremely effective, and since the iron-catalyzed reaction turned out to be significantly faster than the uncatalyzed version, the process tolerated other polar groups present in the substrates. For example, α-bromoester **90** was converted to its phenylated derivative in high yield within minutes.[72]

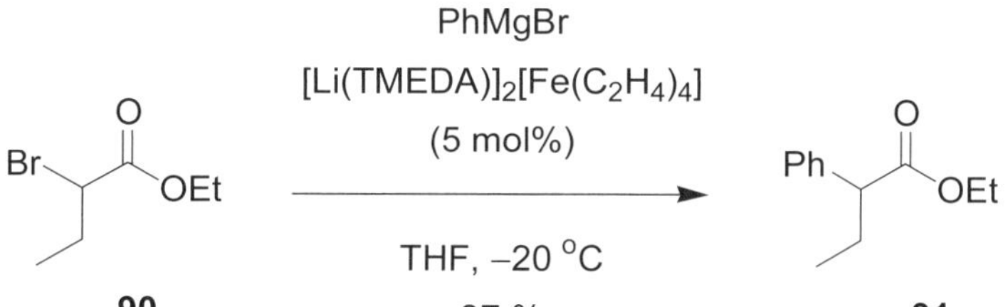

90 **91**

Although the mechanism of this reaction is not entirely understood, studies by Hoffmann and co-workers have provided some interesting insight.[35] Utilizing a chiral Grignard regent as a probe, they observed a significant loss of optical purity of the cross-coupling product, when mediated by low-valent iron or cobalt. They attribute this to the transmetallation step, which may involve a radical pathway, i.e. pathway B. However, since the actual oxidation state of the metal is unknown (and hence it's ability to oxidize a Grignard reagent), they also mention another potential mechanistic scenario, whereby reversible carbon–metal bond homolysis could lead to racemisation.[35]

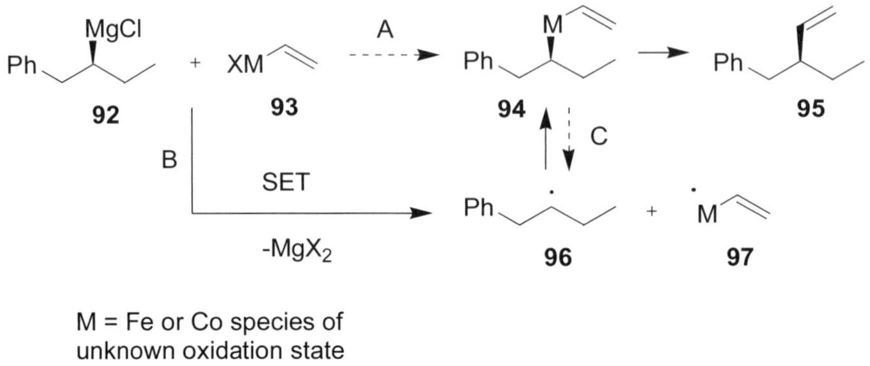

Fürstner has also reported the loss of optical purity in the iron-catalyzed cross-coupling of optically-pure secondary alkyl halides, as well as the cyclization of alkyl iodides bearing unsaturation, both of which suggest the presence of radical intermediates.[72] Numerous substrates however, do not undergo analogous 5-*exo*-trig cyclizations and tertiary halides remain unchanged and thus caution should be taken in generalizing the exact mechanistic pathway for all substrates.[72] Mechanistic studies are ongoing in the laboratories of A. Fürstner. For example, recently he has reported the isolation and structural characterization of a homoleptic "super-ate" complex of iron, which has implications in the iron-catalyzed Kumada cross-coupling reaction.[73]

1.1.3.6 Experimental

Standard Nickel-Catalyzed Procedure

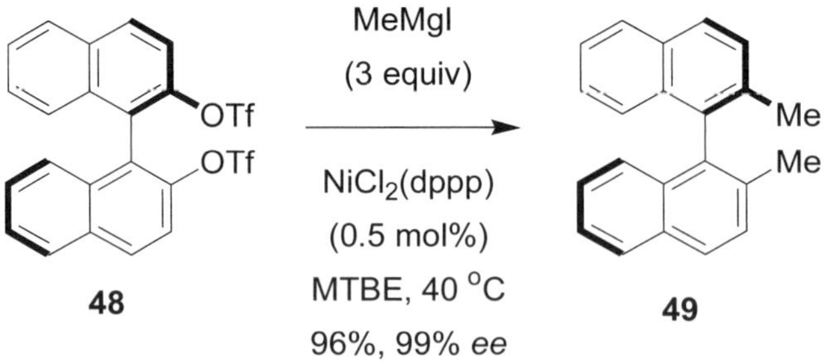

(*R*)-2,2′-Dimethyl-1,1′-binaphthyl (49).[46]

Under an atmosphere of nitrogen, a solution of MeI (3.90 g, 27.5 mmol) in *tert*-butyl methyl ether (MTBE, 4.0 mL) was added dropwise to a stirred suspension of Mg turnings (660 mg, 27.1 mmol) in MTBE (7.0 mL) such that gentle reflux was maintained throughout the addition. The mixture was allowed to cool to 30 °C where MTBE (5.0 mL) and $NiCl_2$(dppp) (250 mg, 0.46 mmol) were added in sequence. A solution of crude (*R*)-**48** (5.00 g, 9.08 mmol) in MTBE (20 mL) was added dropwise, and the mixture was stirred and heated under reflux (at 55 °C) for 30 min. Consumption of (*R*)-**48** was confirmed by TLC [AcOEt/*n*-hexane (1:4); *Rf* 0.46 for (*R*)-**48**, 0.79 for (*R*)-**49**]. The mixture was allowed to cool to room temperature (20–25 °C). PhMe (30 mL) was added, and the mixture was poured into ice-chilled water (50 mL). To the mixture was added 35% aqueous HCl (50 mL). The layers were separated, and the organic layer was washed with H_2O (30 mL × 2) and saturated aqueous NaCl solution (30 mL × 1). The organic solution was dried ($MgSO_4$) and concentrated in vacuo [40–50 °C (bath temperature), 50–60 mmHg]. The solid residue (2.60 g) was mounted on a short pad of silica gel (Merck Kieselgel 60, 7.8 g). Elution with AcOEt/*n*-hexane (1 : 4; 200 mL) gave (*R*)-**49** (2.46 g, 96.1%) as white crystals: 99.6% *ee*.

Highly-Active Palladium-Catalyzed Procedure.

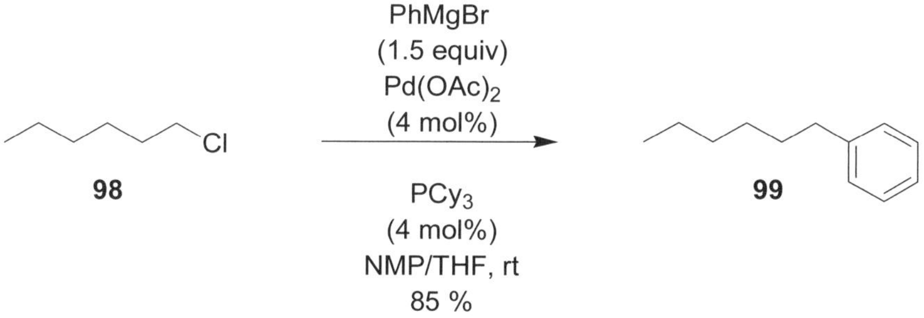

Hexylbenzene (99).[68]

A 25-mL Schlenk flask was charged with $Pd(OAc)_2$ (0.0180 g, 0.080 mmol) and PCy_3 (0.0224 g, 0.080 mmol), sealed with a septum, and purged with argon for 15 min. NMP (5 mL) and **98** (0.27 mL, 2 mmol) were added by syringe. Then, phenylmagnesium bromide (3 mL, 3 mmol, 1 M in THF) was added dropwise over 1 min to the stirred mixture. After 20 h at room temperature, the reaction was quenched with MeOH (1 mL) and water (1 mL). The solution was concentrated to about 6 mL by rotary evaporation, and subjected to silica gel column chromatography (heptane) to give a colorless liquid (0.276 g, 1.7 mmol, 85% yield).

1.1.3.7 References

1. [R] Negishi, E.-I. Liu, F. *Palladium- or Nickel-Catalyzed Cross-Coupling With Organometals Containing Zinc, Magnesium, Aluminium and Zirconium.* In *Metal-Catalyzed Cross-Coupling Reactions*: Diederich, F., Stang, P. J., Eds.; Wiley-VCH: Weinheim, Germany, 1998, 1.
2. [R] Anastasia, L., Negishi, E.-I. *Palladium-catalyzed Aryl-Aryl Coupling.* In *Handbook of Organopalladium Chemistry for Organic Synthesis*: 2002, Vol. 1, 311.
3. [R] Hayashi, T. *Palladium-catalyzed Asymmetric Cross-Coupling.* In *Handbook of Organopalladium Chemistry for Organic Synthesis*: 2002, Vol. 1, 791.
4. [R] Tamao, K. *J. Organomet. Chem.* **2002**, *653*, 23.
5. [R] Murahasi, S.-I. *J. Organomet. Chem.* **2002**, *653*, 27.
6. [R] Hayashi, T. *J. Organomet. Chem.* **2002**, *653*, 41.
7. [R] Hiller, A. C.; Grasa, G. A.; Viciu, M. S.; Lee, H. M.; Yang, C.; Nolan, S. P. *J. Organomet. Chem.* **2002**, *653*, 69.
8. [R] Huo, S., Negishi, E.-I. *Palladium-catalyzed Alkenyl-Aryl, Aryl-Alkenyl, And Alkenyl-Alkenyl Coupling Reactions.* In *Handbook of Organopalladium Chemistry for Organic Synthesis*: 2002, Vol. 1, 335.
9. Corriu, R. J. P.; Masse, J. P. *J. Chem. Soc., Chem. Commun.* **1972**, 144.
10. Tamao, K.; Sumitani, K.; Kumada, M. *J. Am. Chem. Soc.* **1972**, *94*, 4374.
11. Tamao, K.; Sumitani, K.; Kiso, Y.; Zemayashi, M.; Fujioka, A.; Kodama, S.; Nakajima, I.; Minato, A.; Kumada, M. *Bull. Chem. Soc. Jpn.* **1976**, *49*, 1958.
12. Job, A.; Reich, R. *Compt. Rend.* **1923**, *177*, 1439.
13. Job, A.; Reich, R. *Compt. Rend.* **1924**, *179*, 330.
14. Gilman, H.; Lichtenwalter, M. *J. Am. Chem. Soc.* **1939**, *61*, 957.
15. Kharasch, M. S.; Fields, E. K. *J. Am. Chem. Soc.* **1941**, *63*, 2316.
16. Kharasch, M. S.; Fuchs, C. F. *J. Am. Chem. Soc.* **1943**, *65*, 504.
17. Kharasch, M. S.; Lewis, D. W.; Reynolds, W. B. *J. Am. Chem. Soc.* **1943**, *65*, 493.
18. Chuit, C. Felkin, H.; Frajerman, C.; Roussi, G.; Swierczewski, G.; *Chem. Commun.* **1968**, 1604.
19. [R] Collman, J. P. *Acc. Chem. Res.* **1968**, *1*, 126.
20. Uchino, M.; Yamamoto, A.; Ikeda, S. *J. Organomet. Chem.* **1970**, *24*, C63.
21. Kochi, J. K.; Tamura, M. *J. Am. Chem. Soc.* **1971**, *93*, 1483.
22. Kochi, J. K.; Tamura, M. *J. Am. Chem. Soc.* **1971**, *93*, 1485.
23. Tamura, M.; Kochi, J. K. *J. Organomet. Chem.* **1972**, *42*, 205.
24. Consiglio, G.; Botteghi, C. *Helv. Chim. Acta.* **1973**, *56*, 460.
25. Yamamura, M.; Moritani, I.; Murahashi, S.-I. *J. Organomet. Chem.* **1975**, *91*, C39.
26. [R] *Metal-Catalyzed Cross-Coupling Reactions*: Diederich, F., Stang, P. J., Eds.; Wiley-VCH: Weinheim, Germany, 2004.
27. Limmert, M. E.; Roy, A. H.; Hartwig, J. F. *J. Org. Chem.* **2005**, *70*, 9364.
28. Martin, R.; Buchwald, S. L. *J. Am. Chem. Soc.* **2007**, *129*, 3822.
29. [R] *Strategic Applications of Named Reactions in Organic Synthesis*: Kurti, L., Czako, B.; Academic Press, 2005.
30. Fauvarque, J.-F.; Pflüger, F.; Troupel, M. *J. Organomet. Chem.* **1981**, *208*, 419.
31. Amatore, C.; Jutand, A. *J. Organomet. Chem.* **1999**, *576*, 254.
32. Murahashi, S.-I.; Naota, T.; Tanigawa, Y. *Org. Synth.* **1984**, *62*, 39.
33. Negishi, E.-I.; Takahashi, T.; Akiyoshi, K. *J. Chem. Soc., Chem. Commun.* **1986**, 1338.
34. Knochel, P.; Dohle, W.; Gommermann, N.; Kneisel, F. F.; Kopp, F.; Korn, T.; Sapountiz, I.; Vu, V. A. *Angew. Chem., Int. Ed.* **2003**, *42*, 4302.
35. Hölzer, B.; Hoffman, R. W. *Chem. Commun.* **2003**, 732.
36. Gawler, R. E. *Tetrahedron Lett.* **1999**, *40*, 4297.
37. [R] Banno, T.; Hayakawa, Y.; Umeno, M. *J. Organomet. Chem.* **2002**, *653*, 288.
38. Böhm, V. P.; Gstöttmayr, C. W. K.; Weskamp, T.; Herrmann, W. A. *Angew. Chem., Int. Ed.* **2001**, *40*, 3387.
39. Ackermann, L.; Born, R.; Spatz, J. H.; Meyer, D. *Angew. Chem., Int. Ed.* **2005**, *44*, 7216.
40. Yoshikai, N.; Mashima, H.; Nakamura, E. *J. Am. Chem. Soc.* **2005**, *127*, 17978.

41. Huang, J.; Nolan, S. P. *J. Am. Chem. Soc.* **1999**, *121*, 9889.
42. Li, G. Y. *J. Organomet. Chem.* **2002**, *653*, 63.
43. Organ, M. G.; Abdel-Hadi, M.; Avola, S.; Hadei, N.; Nasielski, J.; O'Brien, C. J.; Valente, C. *Chem. Eur. J.* **2007**, *13*, 150.
44. Hayashi, T.; Konishi, M.; Fukushima, M.; Kanehira, T.; Hioki, M.; Kumada, M. *J. Org. Chem.* **1983**, *48*, 2195.
45. Bringmann, G.; Price Mortimer, A. J.; Keller, P. A.; Gresser, M. J.; Garner, J.; Breuning, M. *Angew. Chem., Int. Ed.* **2005**, *44*, 5384.
46. Ikunaka, M.; Maruoka, K.; Okuda, Y.; Ooi, T. *Org. Process Res. Dev.* **2003**, *7*, 644.
47. Hayashi, T.; Hayashizaki, K.; Kiyoi, T.; Ito, Y. *J. Am. Chem. Soc.* **1988**, *110*, 8153.
48. Hayashi, T.; Niizuma, S.; Kamikawa, T.; Suzuki, N. Uozumi, Y. *J. Am. Chem. Soc.* **1995**, *117*, 9101.
49. Liu, P.; Jacobsen, E. N. *J. Am. Chem. Soc.* **2001**, *123*, 10772.
50. Rivkin, A.; Njardarson, J. T.; Biswas, K.; Chou, T.-C.; Danishefsky, S. J. *J. Org. Chem.* **2002**, *67*, 7737.
51. William, A. D.; Kobayashi, Y. *J. Org. Chem.* **2002**, *67*, 8771.
52. Roy, A. H.; Hartwig, J. F. *J. Am. Chem. Soc.* **2003**, *125*, 8704.
53. Limmert, M. E.; Roy, A. H.; Hartwig, J. F. *J. Org. Chem.* **2005**, *70*, 9364.
54. Ackermann, L.; Althammer, A. *Org. Lett.* **2006**, *8*, 3457.
55. Wittig, G.; Pockels, U.; Dröge, H. *Chem. Ber.* **1938**, *71*, 1903.
56. Knochel, P.; Dohle, W.; Gommermann, N.; Kneisel, F. F.; Kopp, F.; Korn, T.; Sapountzis, I.; Vu, V. A. *Angew. Chem., Int. Ed.* **2003**, *42*, 4302.
57. Ila, H.; Baron, O.; Wagner, A. J.; Knochel, P. *Chem. Commun.* **2006**, 583.
58. Bonnet, V.; Mongin, F.; Trécourt, F.; Quéguiner, G.; Knochel, P. *Tetrahedron,* **2002**, *58*, 4429.
59. Krasovskiy, A.; Knochel, P. *Angew. Chem., Int. Ed.* **2004**, *43*, 3333.
60. Martin, R.; Buchwald, S. L. *J. Am. Chem. Soc.* **2007**, *129*, 3844.
61. Lau, S. Y. W.; Hughes, G.; O'Shea, P. D.; Davies, I. W. *Org. Lett.* **2007**, *9*, 2239.
62. Frisch, A. C.; Beller, M. *Angew. Chem., Int. Ed.* **2005**, *44*, 674.
63. Ishiyama, T.; Abe, S.; Miyaura, N. Suzuki, A. *Chem. Lett.* **1992**, 691.
64. Devasagayaraj, A.; Stüdemann, T.; Knochel, P. *Angew. Chem., Int. Ed. Engl.* **1995**, *34*, 2723.
65. Terao, J.; Watanabe, H.; Ikumi, A.; Kuniyasu, H.; Kambe, N. *J. Am. Chem. Soc.* **2002**, *124*, 4222.
66. Terao, J.; Naitoh, Y.; Kuniyasu, H.; Kambe, N. *Chem. Lett.* **2003**, *32*, 890.
67. Terao, J.; Ikumi, A.; Kuniyasu, H.; Kambe, N. *J. Am. Chem. Soc.* **2003**, *125*, 5646.
68. Frisch, A. C.; Shaikh, N.; Zapf, A.; Beller, M. *Angew. Chem., Int. Ed.* **2002**, *114*, 4056.
69. Frisch, A. C.; Rataboul, F.; Zapf, A.; Beller, M. *J. Organomet. Chem.* **2003**, *687*, 403.
70. Korn, T. J.; Cahiez, G.; Knochel, P. *Synlett*, **2003**, 1893.
71. Ohmiya, H.; Yorimitsu, H.; Oshima, K. *Org. Lett.* **2006**, *8*, 3093.
72. Martin, R.; Fürstner, A. *Angew. Chem., Int. Ed.* **2004**, *43*, 3955. And references therein.
73. Fürstner, A.; Krause, H.; Lehmann, C. W. *Angew. Chem., Int. Ed.* **2006**, *45*, 440.

1.1.4 Negishi Cross-Coupling Reaction

Larry Yet

1.1.4.1 Description

The Negishi cross-coupling reaction is the versatile nickel- or palladium-catalyzed coupling of organozinc compounds with various halides or triflates (aryl, alkenyl, alkynyl, acyl).[1]

$$R^1\text{—}X \quad + \quad R^2Zn\text{—}Y \xrightarrow[\text{solvent}]{NiL_n \text{ or } PdL_n} R^1\text{—}R^2$$

R^1 = aryl, alkenyl, alkynyl, acyl
R^2 = aryl, heteroaryl, alkenyl, allyl, Bn, homoallyl, homopropargyl
X = Cl, Br, I, OTf
Y = Cl, Br, I
L_n = PPh_3, dba, dppe

Both nickel- and palladium-phosphine complexes work well as catalysts. However, Pd catalysts tend to give somewhat higher yields and better stereoselectivity, and their functional group tolerance is better. The active catalysts are relatively unstable Ni(0)- and Pd(0)-complexes but these can be generated in situ from more stable Ni(II)- and Pd(II)-complexes with a reducing agent such as 2 equivalents of DIBAL-H or *n*-BuLi. The most widely used ligand is PPh_3, but other achiral and chiral phosphine ligands have been successfully used. The various organozinc reagents can be prepared either by direct reaction of the organic halide with zinc metal or activated zinc metal or by transmetallation of the corresponding organolithium or Grignard reaction with a zinc halide. The use of organozinc reagents allow for a much greater functional group tolerance in both coupling partners than in the Kumada cross-coupling where organolithiums and Grignard reagents are utilized as coupling partners. Other advantages of the use of organozincs include: high regio- and stereoselectivity, wide scope and applicability, few side reactions and almost no toxicity. The reaction is mostly used for the coupling of two $C(sp^2)$ carbons but $C(sp^2)$–$C(sp)$, $C(sp^2)$–$C(sp^3)$, and $C(sp^3)$–$C(sp^3)$ couplings are well-known. Of all the various organometals (Al, Zr, B, Sn, Cu, Zn), organozincs are usually the most reactive in palladium-catalyzed cross-coupling reactions and do not require the use of additivies such as bases as in Suzuki reactions to boost the reactivity.

1.1.4.2 Historical Perspective

In 1972, after the discovery of nickel-catalyzed coupling of alkenyl and aryl iodides with Grignard reagents (Kumada cross-coupling), it became apparent in order to improve the functional group tolerance of the process, the organometallic coupling partners should contain less electropositive metals than lithium and magnesium. In 1976, E. Negishi reported the first stereospecific nickel-catalyzed alkenyl–alkenyl and alkenyl–aryl cross-coupling of alkenylalanes (organoaluminums) with alkenyl- or aryl halides.[2] Extensive research by Neghishi showed that the best results (reaction rate, yield, and stereoselectiviity) were obtained when organozincs are coupled in the presence of Pd(0) catalysts.[3–5]

Negishi's group published seminal papers on the palladium- or nickel-catalyzed cross-coupling reactions between 1976–1978.[6–13] The palladium-catalyzed reaction of alkynylzinc chlorides with alkenyl halides[8] along with the related alkynyl–aryl,[9] aryl–aryl,[10] and benzyl–aryl[10] coupling reactions provided some of the earliest examples of the palladium-catalyzed cross-coupling of organozincs, and showed superior reactivity of organozincs under the palladium-catalyzed cross-coupling conditions relative to the ten or so other types of organometals. The first examples of the palladium-catalyzed carboalumination–cross-coupling tandem reaction were also reported in 1978.[13] The use of Zn salts, such as $ZnCl_2$ or $ZnBr_2$, as additives or cocatalysts in the coupling step of this tandem reaction was shown to be highly desirable or even essential to observing satisfactory results. This study demonstrated, for the first time, the concept of double metal catalysis and the favorable effects of additives on the palladium- or nickel-catalyzed cross-coupling.[13] These findings reported established that the palladium- or nickel-catalyzed cross-coupling can be achieved with organometals containing various metal counteractions other than Mg, which had previously been used almost exclusively.

1.1.4.3 Mechanism

The palladium-catalyzed Negishi cross-coupling reaction follows a general pathway common to other palladium-catalyzed type reactions.[14] The active catalyst in this reaction is the Pd(0) species which proceeds through an oxidative addition step of the organic halide to give the Pd(II) species **1**, which then undergoes transmetalation with the zinc halides to afford *trans*-adduct Pd(II) species **2**. Isomerization of **2** to the *cis*-adduct Pd(II) species **3** followed by reductive elimination affords the product R^1–R^2 and the regenerated Pd(0) catalyst, which then continues with the catalytic cycle.

Compared to other palladium-catalyzed cross-coupling reactions such as Stille and Suzuki, mechanistic proposals on the transmetalation step in

Negishi cross-couplings are purely speculative, and the essential stereochemical and kinetic aspects remain obscure. Casares and Espinet have studied the *cis-trans* isomerization of this transmetallation reaction with $ZnMe_2$ and ZnMeCl using $[PdRfCl(PPh_3)_2]$ (Rf = 3,5-dichloro-2,4,6-trifluorophenyl) with ^{19}F NMR spectroscopy as a tool for the mechanistic probes.[15] They found that each methylating reagent afforded stereoselectively a different isomer (*cis* or *trans*) of the $[PdRfCl(PPh_3)_2]$ coupling intermediate. The key point they found was that the choice of the organozinc reagent could strongly affect the outcome of the Negishi cycle. The mechanism of the nickel-catalyzed Negishi cross-coupling reactions mirrors that of the palladium-catalyzed processes.[16]

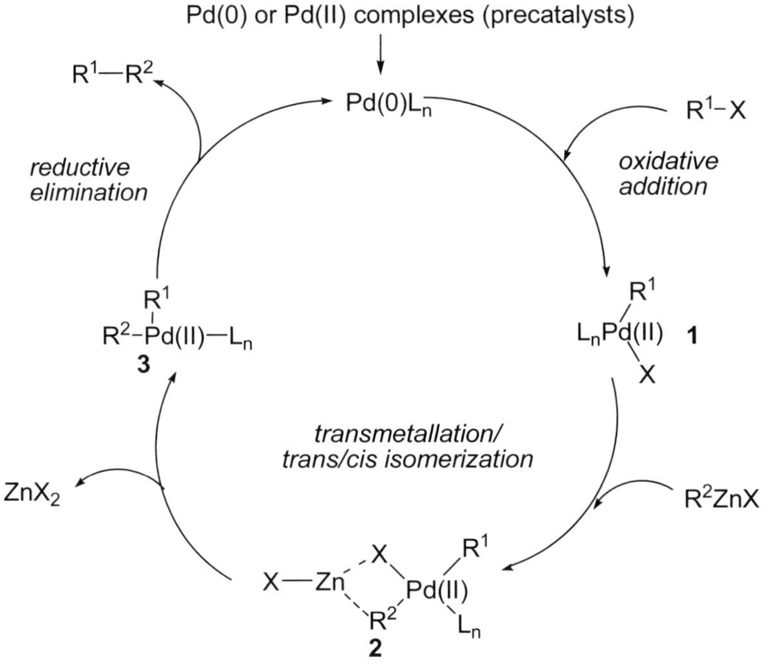

1.1.4.4 *Variations and Improvements*

Phosphine-type ligands have been utilized in the acceleration of Negishi cross-coupling reactions. Santelli published a procedure where the system combining the tetraphosphine *cis,cis,cis*-1,2,3,4-tetrakis(diphenylphosphino-methyl)cyclopentane (**4**, Tedicyp) and $[Pd(C_3H_5)Cl]_2$ was found to be a very active catalyst for the cross-coupling of aryl bromides with alkyl- or arylzinc derivatives.[17] Buchwald reported a new catalyst system for the palladium-catalyzed cross-coupling of organozinc reagents with aryl halides with $Pd_2(dba)_3$ and biphenyl ligand **5**.[18] This system permitted efficient preparation of hindered biaryls (tri- and tetra-*ortho*-substituted) and functioned effectively at low levels of catalyst, and tolerated a wide range of functional groups and heterocyclic substrates. Fu showed that commercially available $Pd(P(t\text{-}Bu)_3)_2$ effected the first general method for the Negishi cross-coupling of a wide range of aryl and vinyl chlorides with aryl- and alkylzinc reagents.[19] The process tolerated nitro groups, and it efficiently generated sterically hindered biaryls. In addition, a high turnover number (> 3000) could be achieved.

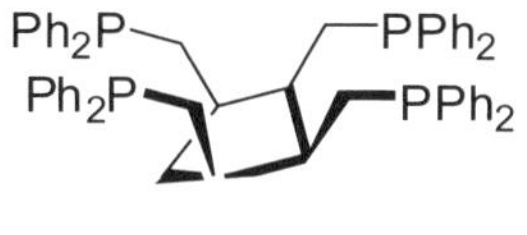

4

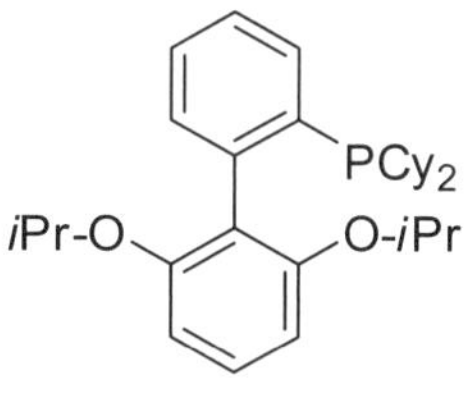

5

In the last few years, the use of microwave irradiation has appeared in several reports for the Negishi cross-coupling reactions. Suna reported that arylzinc reagents **7** could be readily prepared from aryl iodides **6** using a zinc–copper couple in a microwave environment followed by a Negishi cross-coupling with 4-bromobenzaldehyde under palladium- or nickel-catalyzed conditions to give biphenyl aldehydes **8**.[20] Furthermore, Suna published a report where arylmagnesium species could be efficiently generated from magnesium turnings and aryl chlorides or aryl bromides under microwave irradiation, followed by transmetallation with $ZnCl_2$–TMEDA to give the corresponding arylzinc reagents.[21] Finally these arylzinc reagents underwent efficient Negishi cross-coupling reactions with aryl bromides. Kappe reported a general and efficient protocol for high-speed microwave-assisted Negishi cross-coupling reactions of aryl chlorides.[22] A range of electron-rich and electron-deficient aryl chlorides **9** participated in this accelerated reaction to give aryls **10**. The use of $Ni(acac)_2$ catalyst was also successful in these reactions. An example of a solid-phase Negishi cross-coupling reaction was also shown.

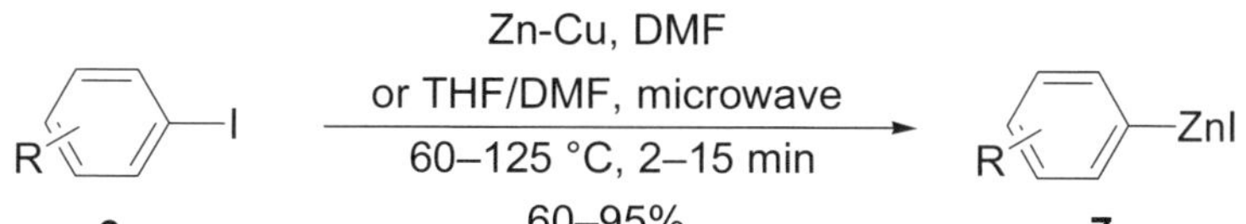

6 **7**

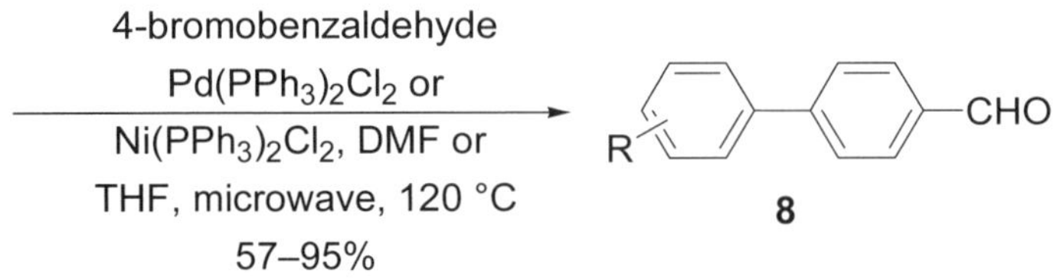

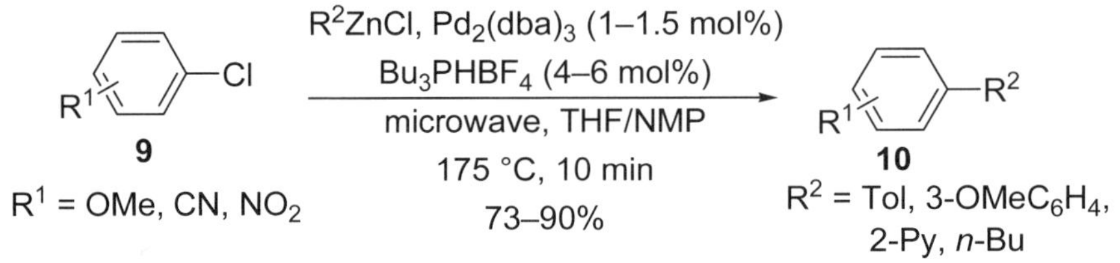

Knochel reported that a combination of diethyl phosphite–DMAP and Ni(II) salts formed a very effective catalytic system for the cross-coupling reactions of arylzinc halides with aryl, heteroaryl, alkenyl bromides, chlorides, triflates, and nonaflates.[23] The choice of solvent was quite important and the mixture of THF–*N*-ethylpyrrolidinone (NEP) (8:1) was found to be optimal in the synthesis of biaryls **13** with arylzinc bromides **11** with aryl halides **12**.

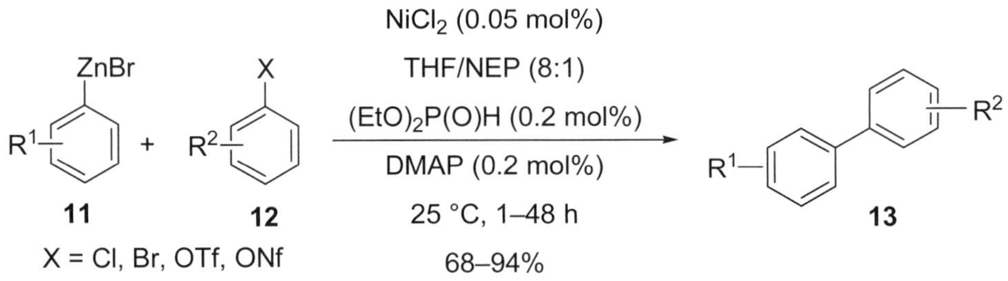

Organ reported the high-yielding Negishi cross-coupling reaction of unactivated alkyl bromides possessing β-hydrogens with alkylzinc bromides utilizing a Pd/*N*-heterocyclic carbene (NHC) catalyst **20** at room temperature.[24] Under optimal conditions, a number of alkyl bromides **14** and alkylzinc bromides **15** possessing common functional groups such as amide, nitrile, ester, acetal, and alkyne were effectively cross-coupled to give **16**. It was noteworthy that β-substituted alkyl bromides and alkylzinc bromides successfully underwent cross-coupling. Organ has also developed the first user-friendly Negishi protocol capable of routinely cross-coupling all combinations of alkyl and aryl centers.[25] The use of an easily synthesized, air stable, highly active, well-defined precatalyst PEPPSI–IPr (**21**; PEPPSI = pyridine-enhanced precatalyst preparation, stabilization and initiation; IPr =

diisopropylphenylimidazolium derivative) substantially increased the scope, reliability, and ease-of-use of the Negishi reaction. All organohalides and routinely used pseudohalides **17** were excellent coupling partners such as the use of chlorides, bromides, iodides, triflates, tosylates, and mesylates with aryl chlorides **18** resulting in high yield of the coupled product **19**. Furthermore, all reactions were performed by using general laboratory techniques with no glovebox necessary as the precatalyst was weighed and stored in air. Utilization of this methodology allowed for the easy synthesis of an assortment of sterically encumbered biaryls and druglike heteroaromatics, demonstrating the value of the PEPPSI–IPr system. Furthermore, this was also the first time Pd–NHC methodology surpassed the related phosphine-ligated Negishi processes both in activity and use.

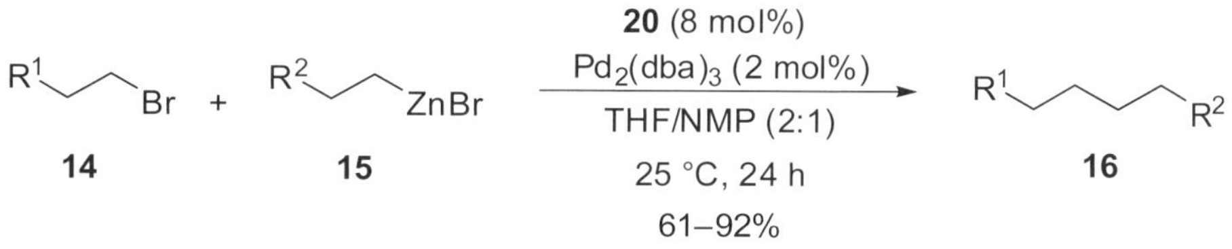

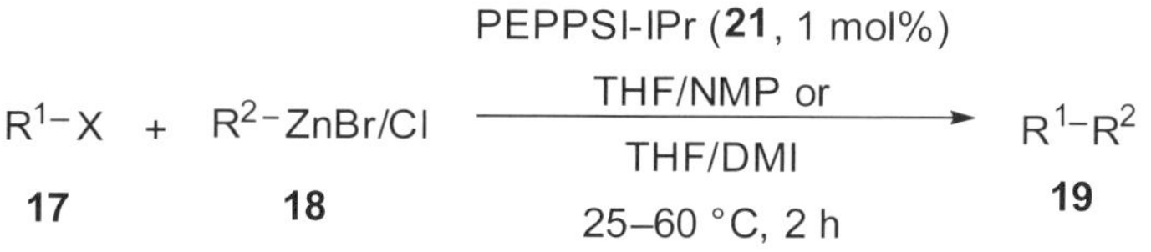

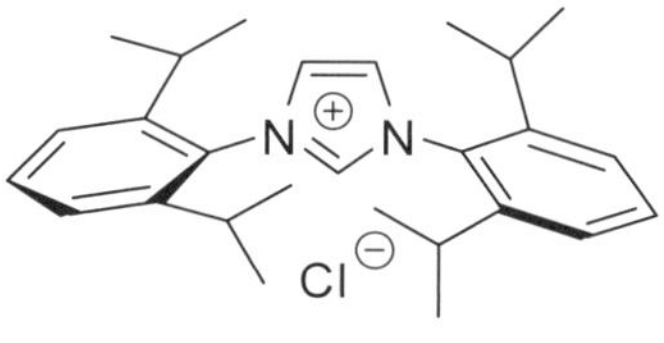

20

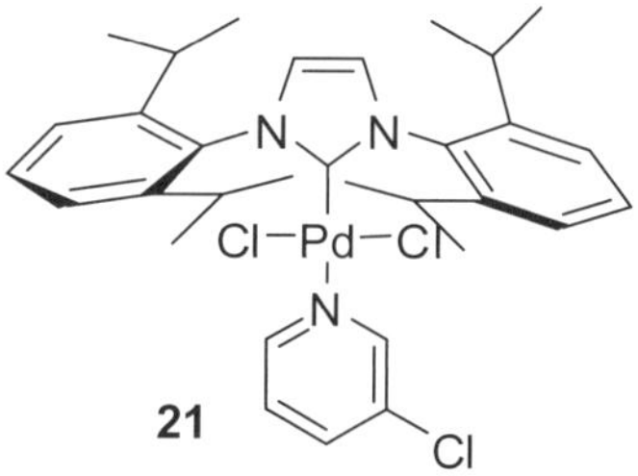

21

Fu reported the first catalytic enantioselective cross-couplings of secondary alkyl electrophiles such as the Negishi cross-coupling reaction of a range of α-bromo amides **22** with an array of organozinc reagents with a pyridinyl oxazoline ligand **26** to afford α-substituted amides **23**.[26] This asymmetric carbon-carbon bond formation proceeded smoothly in the presence of groups such as an olefin, a benzyl ether, an acetal, an imide and a nitrile. Fu described also the first highly enantioselective Negishi cross-coupling reactions of racemic secondary benzylic bromides and chlorides **24** with organozinc reagents with the opposite pyridinyl oxazoline ligand **27** to give substituted indanes **25** in high enantiomeric excesses. Functionalized organozinc reagents, including those that bear a cyano or a chloride group, coupled with 1-bromoindanes in very good enantiomeric excesses.[27] Fu also established that Ni(cod)$_2$/*s*-Bu-Pybox catalyzed Negishi cross-coupling reactions of an array of functionalized alkyl bromides and iodides at room temperature.[28] This represented the first nickel- or palladium-catalyzed method for cross-coupling of unactivated, *β*-hydrogen-containing secondary alkyl halides. A single method (2% Pd$_2$(dba)$_3$/8% PCy$_3$/NMI in THF/NMP at 80 °C) achieved the cross-coupling of a range of *α*-hydrogen-containing primary alkyl iodides, bromides, chlorides, and tosylates with an array of alkyl-, alkenyl-, and arylzinc halides.[29] The process was compatible with a variety of functional groups, which included esters, amides, imides, nitriles, and heterocycles.

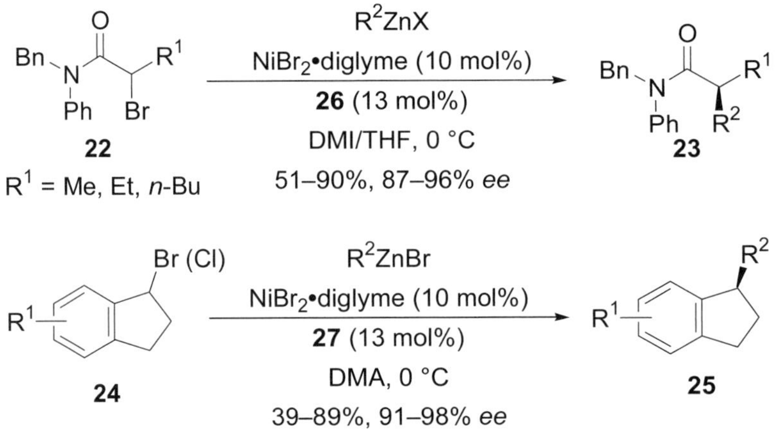

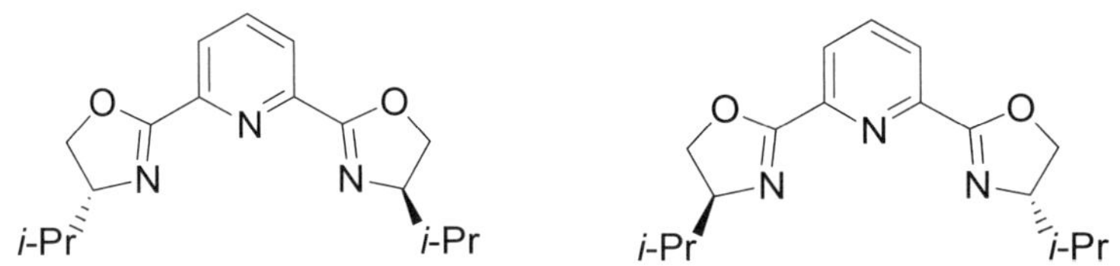

26, (*R*)-(*i*-Pr)-Pybox **27**, (*S*)-(*i*-Pr)-Pybox

Lipshutz showed that nickel on charcoal ("Ni/C") was found to be an efficient heterogeneous catalyst for mediating carbon-carbon bond constructions involving chloroarenes **28** and functionalized organozinc reagents to give **29**.[30] Importantly retention of nickel on the solid support offers control over such critical parameters as waste disposal and toxicity.

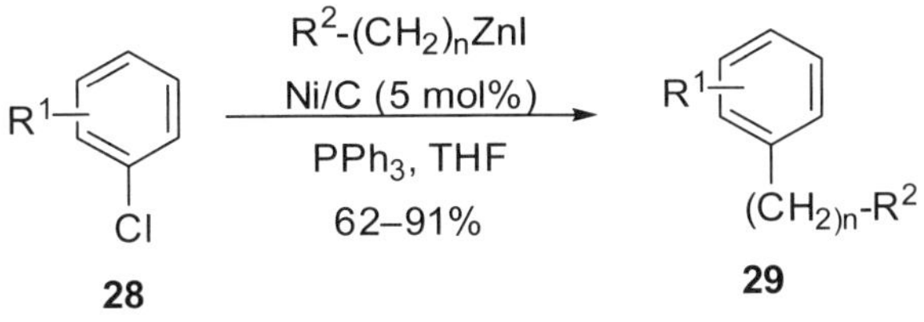

1.1.4.5 Synthetic Utility

Five-Membered Heterocycles

1-(Benzyloxy)-4-iodopyrazole (**30**) could be metalated by using *iso*-propylmagnesium chloride, transmetallated with zinc chloride and Negishi cross-coupling of intermediate **31** with aryl and heteroaryl iodides to give the corresponding products **32** in good to excellent yields.[31] The reaction worked well with both electron-donating and electron-withdrawing substituents. Experiments with 1-(benzyloxy)-4-(tributylstannyl)pyrazole (prepared via pyrazolylmagnesium bromide) in a Stille reaction or the Grignard intermediate directly in a Kumada–Corriu reaction failed. Cross-coupling reactions of zinc organyls **34** in the 5-position with acid chlorides to **35** and aryl halides to **36** were accomplished in generally good yields.[32]

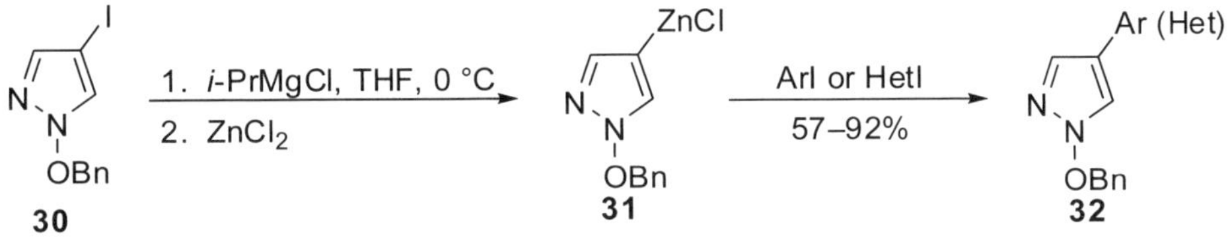

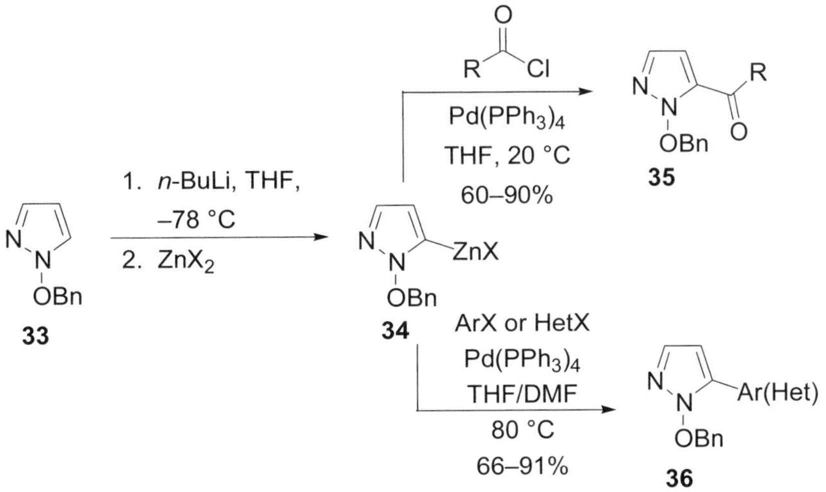

5-Iodoimidazole **37** readily underwent halogen–magnesium exchange followed by transmetallation with zinc bromide and subsequent Negishi cross-coupling with iodobenzene to give product **38**.[33] A Negishi reaction in the 2-position of imidazole has been reported on the solid phase.[34]

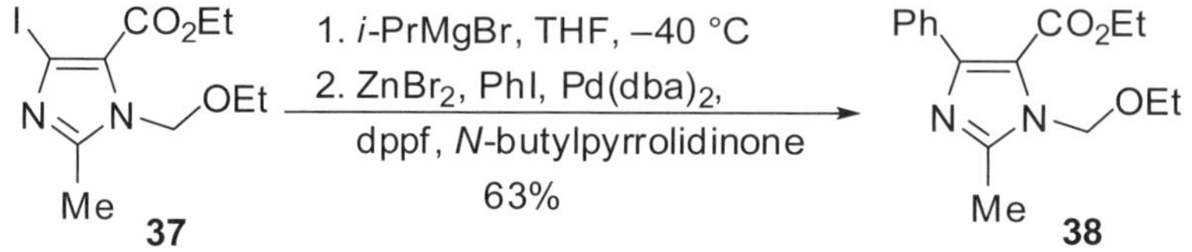

37 38

The Merck process chemists found that 2-oxazolylzinc species **40,** successfully prepared from oxazole **39** subsequently cross-coupled with aryl or heteroaryl bromides to oxazoles **41**.[35] The use of solid zinc chloride helped the transmetallation; however, long reaction times were required for good yields. Anderson also reported 2-oxazolylzinc species **40** was successfully cross-coupled with aryl iodides and triflates to compounds **41**.[36] This methodology was employed in the synthesis of oxazole-containing partial ergot alkaloids.[37]

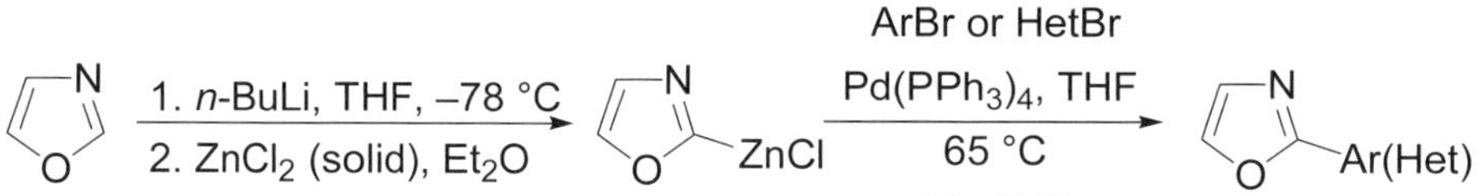

39 40 41

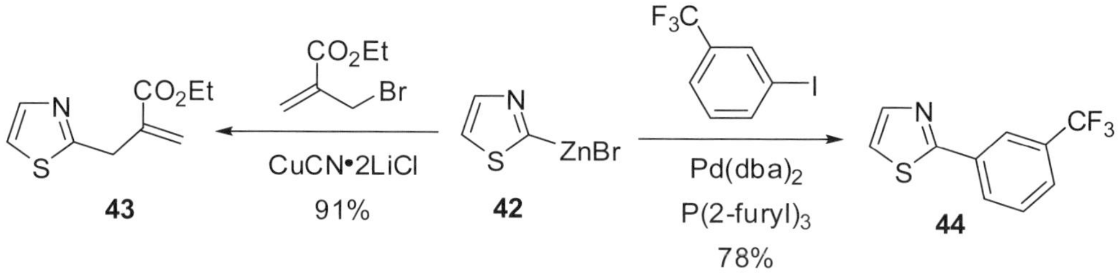

43 42 44

Knochel showed that 2-thiazolylzinc species **42** cross-coupled with aromatic or aliphatic electrophiles to give the corresponding products **43** and **44**, respectively, in the presence of copper and palladium catalysts.[38] 2,4-Dibromothiazole has been used as halide in Negishi reactions; the cross-coupling took place selectively in the 2-position in 50–62% yield.[39] The formation of zinc species in the 5-position of thiazole can also be prepared if the 2-position was protected with a trimethylsilyl group as in **45** before the organozinc derivative **46** was formed in the 5-position. The subsequent

Negishi cross-coupling reaction and deprotection gave the desired products **47** with substituted aryl and heteroaryl halides.[40]

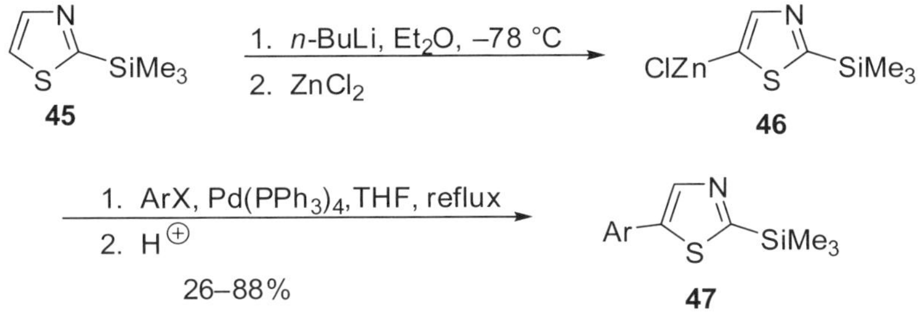

Several derivatives of 4-aryl-1,5-disubstituted-1,2,3-triazole **51** were synthesized in good yields via 1,3-dipolar cycloaddition of aryl azides **48** and alkynylzinc species **49** followed by Negishi reaction of **50** under new conditions.[41] Of all the combinations screened, Pd/xantphos was found to be superior. Kumada coupling was not as effective as the Negishi coupling in this reaction.

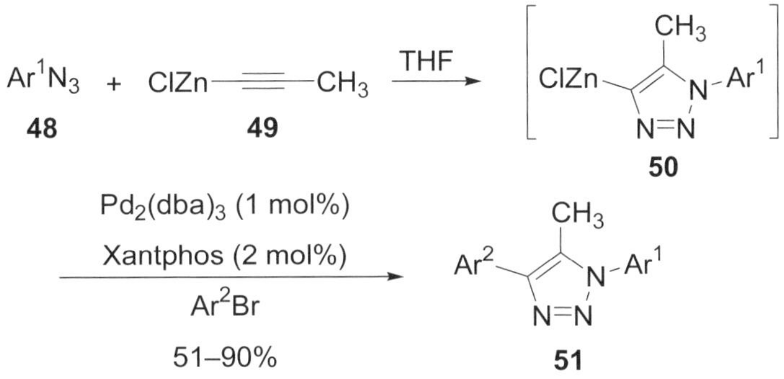

Six-Membered Rings

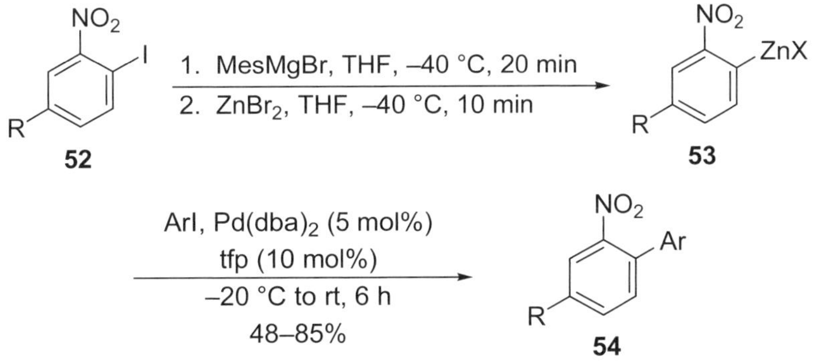

Knochel reported that various ortho nitro-substituted arylmagnesium reagents prepared by an iodine–magnesium exchange reaction starting from the corresponding aryl iodides **52** and mesitylmagnesium bromide are readily transmetallated to the corresponding organozinc compounds **53**.[42] These nitro-containing organometallics underwent a smooth Negishi cross-coupling reaction with various aryl iodides to give highly functionalized nitrosubstituted biphenyls **54**.

Pyridines have been employed as substrates for Negishi cross-coupling reactions. An efficient procedure for cross-coupling of 3-bromopyridine (**55**) with various mono- and dichloroheteroaryl compounds giving the corresponding biaryls **56** has been described.[43] The Negishi reaction also gave a mild and efficient method for the conversion of bromopyridines **58** into functionalized 2,2′-bipyridines **59** using commercially available 2-pyridylzinc bromide (**57**).[44] This method was also extended to the conversion of dibromopyridines to 5- and 6-bromobipyridines, which are powerful synthons for incorporation into larger supramolecular systems. A similar protocol for the efficient, modified Negishi cross-coupling strategy for substituted 2,2′-bipyridines from 2-bromo- and 2-chloropyridines has also been reported.[45] A convenient scalable synthesis of 6,6′-dimethyl-2,2′-bipyridine-4-ester based on modified Negishi cross-coupling conditions from substituted 2-chloro and 2-bromopyridines has been disclosed.[46]

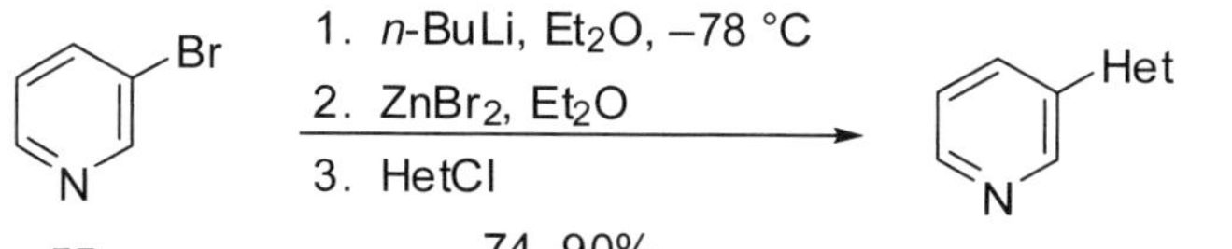

55 **56**

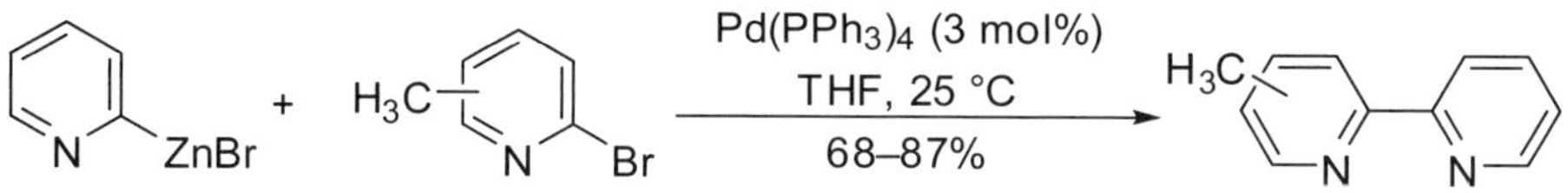

57 **58** **59**

Queguiner prepared the first Negishi cross-coupling reaction of a pyridazine. 3,6-Dimethoxypyridazine (**60**) was lithiated and transmetallated with zinc chloride at the 4-position to give zinc species **61**, which underwent smooth cross-couplings with a few halides to give pyridazinyl compounds **62**.[47] Negishi cross-coupling reactions of pyridazines have also been utilized in the synthesis of analogs of nicotinic acetylcholine receptor agonists[48] and anxiolytic drugs with improved side-effect profiles.[49]

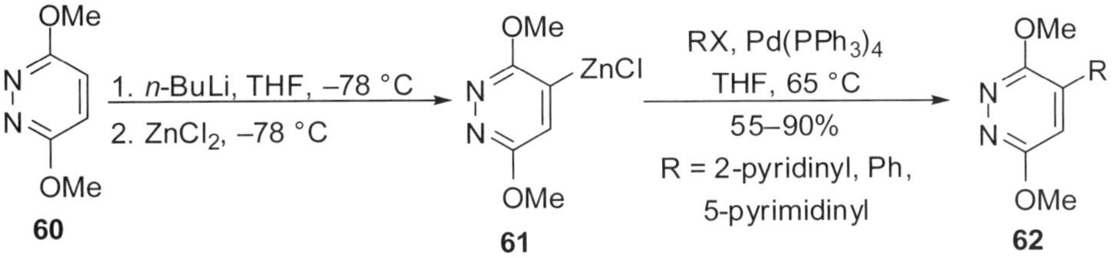

A general procedure for the synthesis of 4-arylpiperidines **64** via the coupling of 4-(*N*-BOC-piperidyl)zinc iodide, generated from zinc insertion of **63** with improved conditions, with aryl halides and triflates in the presence of palladium/copper cocatalysis has been reported.[50]

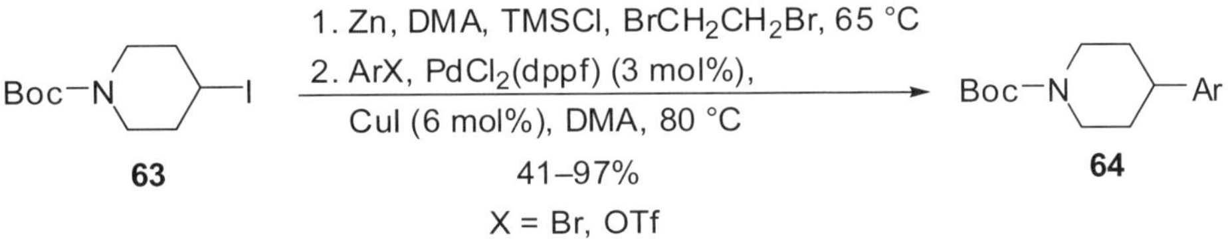

Chiral binaphthalenes are amongst the most useful chiral ligands and auxiliaries employed in asymmetric synthesis. Although the Stille and Suzuki–Miyaura reactions are very efficient for aryl–aryl carbon bond formation, sterically congested coupling partners such as naphthyls have the problems of poor yields and extensive deboronation in the Suzuki couplings. The naphthyl–naphthyl Negishi cross-coupling reaction was a possible alternative to overcome this problem. A new synthetic approach affording for the first time chiral binaphthalene derivatives **67** via an asymmetric Negishi reaction with 1-naphthyl bromides **65** and 2-naphthyl zinc species **66** with ligand **68** in good yields and good enantioselectivities was reported.[51]

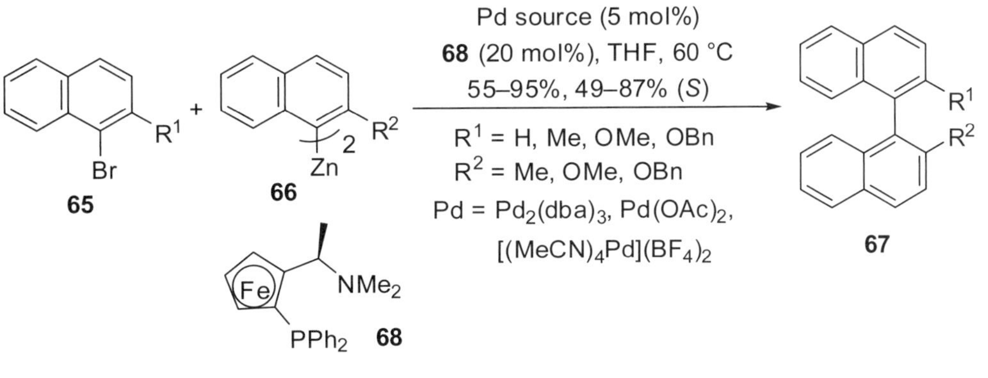

Sequential cross-coupling reactions of 2,4-dichloropurine **69** have been conducted.[52a] 2,6-Disubstituted purine **69** was cross-coupled with MeZnBr to give monosubstituted purine **70**, which underwent Suzuki cross-coupling with phenylboronic acid to furnish disubstituted purine **71**. Recently, the synthesis of 6-(2-hydroxyethyl)purines was developed based on the Negishi palladium-catalyzed cross-coupling reactions of 6-chloropurines with the Reformatsky reagent followed by reduction by sodium borohydride and treatment with manganese(IV) oxide.[52b]

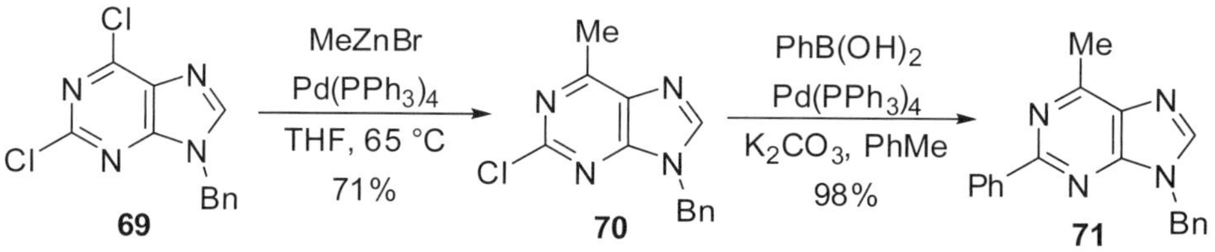

69 **70** **71**

Strategies for controlling the regioselective reactions between 2,4-dichloroquinoline (**72**) and organozinc reagents have been reported.[53] 2,4-Dichloroquinoline (**72**) has been found to react with benzylic zinc reagent **73** in the presence of catalytic amounts of palladium complexes to exclusively give α-substituted products such as **75**. Several metal salts were examined as an additive for χ-selective coupling reactions. The most effective additive for selective coupling reaction at the χ-position has been found to be LiCl to give products such as **74**.

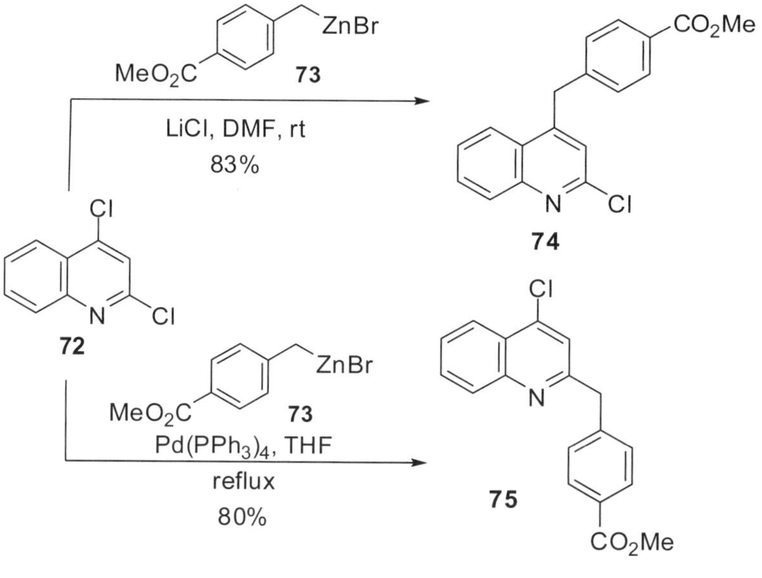

Alkenes, Alkynes, Polyenes, Enynes

Polysubstituted alkenes, alkynes, polyenes, and enynes are present in many naturally occurring biologically active compounds such as terpenoids, pheromones, *etc.* They are also key intermediates in a number of transformations leading to natural products and have remained an active area of research for organic chemists.

The Negishi cross-coupling reaction of arylzinc chlorides and bromides with functionalized vinylic tellurides **76** in the presence of a catalytic amount of $PdCl_2$ in THF at room temperature was reported.[54] This cross-coupling reaction was general and permitted the synthesis of functionalized substituted alkenes **77** in good yields and high stereoselectivity. In this way, there were some advantages to use vinylic tellurides instead of the other methods, such as the easy access by stereoselective reactions to either (*Z*)- or (*E*)-vinylic tellurides, no isomerization of the double bond and the enhanced stability of these compounds. The use of vinylic tellurides in cross-coupling reactions tolerated many sensitive functional groups and provided mild reaction conditions. The Negishi cross-coupling reaction of vinylic- and aryltellurides with heteroarylzinc chlorides catalyzed by $PdCl_2$/CuI was reported. This cross-coupling reaction was general and permitted the formation of a new sp^2–sp^2 carbon bond in good yields and high stereoselectivity.

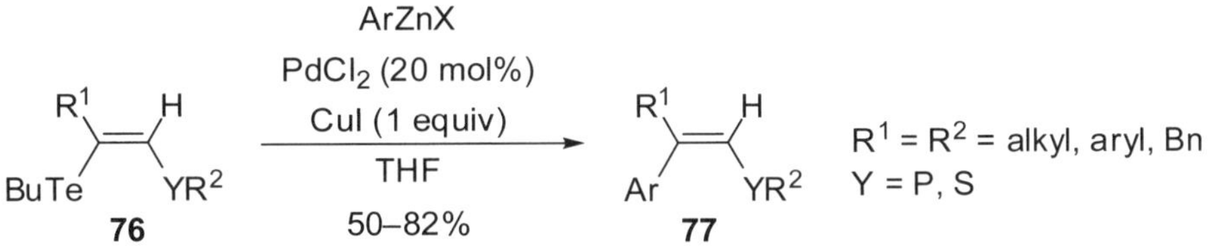

The application of this methodology in the synthesis of 1-(*Z*)-atractylodinol (**81**), a natural product isolated from the dried rhizomes of *Atractylodes lancea De Candolle* widely used in China and Japan against rheumatic diseases, digestive disorders, night blindness, and influenza.[55] The total synthesis of 1-(*Z*)-atractylodinol, a natural polyacetylenic alcohol with several biological activities, has been achieved using a newly developed telluride synthon and a novel use for the Negishi type coupling reaction employing vinyl tellurides. The protection of the alkyne terminus as its trimethylsilyl derivative **79** followed by a cross-coupling reaction with an excess 2-furylzinc chloride from furan (**78**) gave **80**, precursor to **81**.

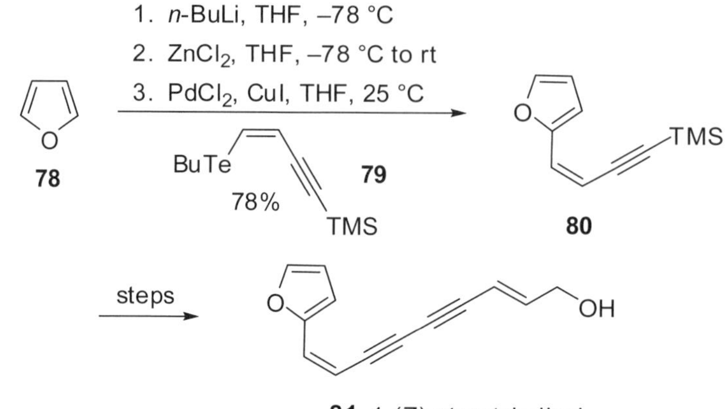

78 **79** **80**

81, 1-(*Z*)-atractylodinol

Tetrasubstituted olefins containing a 1,4-diene structural unit as in **85** can be regio- and stereoselectively constructed in one pot by the allylzincation of acetylenic sulfones **82** with **83**, followed by Negishi cross-coupling of intermediate **84** with halohydrocarbons in the presence of catalytic nickel.[56]

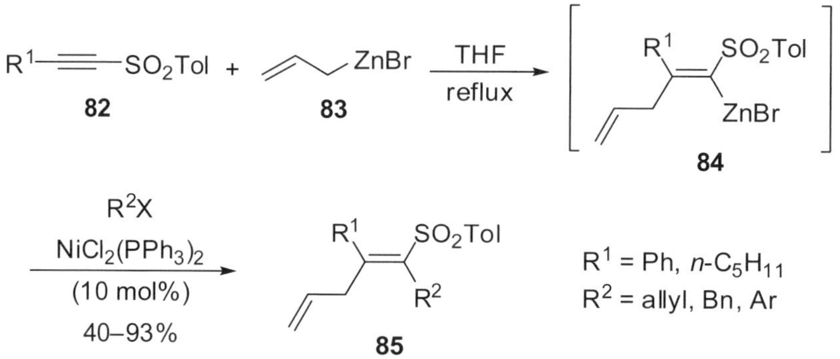

82 **83** **84** **85**

The 1-fluoro-1-haloalkenes **86** undergo palladium-catalyzed Negishi cross-couplings with primary alkylzinc bromides to give multisubstituted fluoroalkenes.[57] The alkylation was transselective giving pure *Z*-fluoroalkenes **87** in most cases. The highest yields were obtained with $Pd_2(dba)_3$ and $PdCl_2(dppb)$ catalysts but the best stereochemical outcome was obtained with less reactive $Pd(PPh_3)_4$. The tertiary alkylzincs also produced the desired fluoroalkenes in high yields.

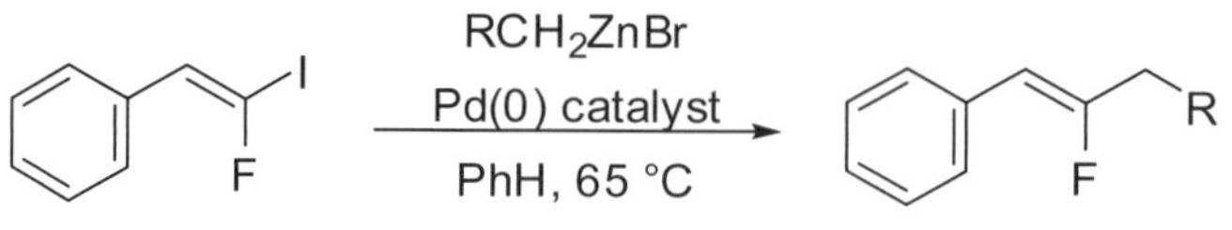

86 **87**

A highly stereospecific synthesis of (*E*)- or (*Z*)-α-fluoro-α,*β*-unsaturated ketones **90**, via a kinetically controlled Negishi palladium-catalyzed coupling reaction was developed providing an easy and general access to valuable fluorinated intermediates.[58] The synthesis involved a reaction between *E*/*Z* *gem*-bromofluoroolefins **88** and alkoxyvinylzinc species **89** under controlled reaction temperature. At 10 °C, **(*Z*)-90** was obtained along with unreacted **(*Z*)-88**. At THF reflux, the recovered olefin was transformed into **(*E*)-90**.

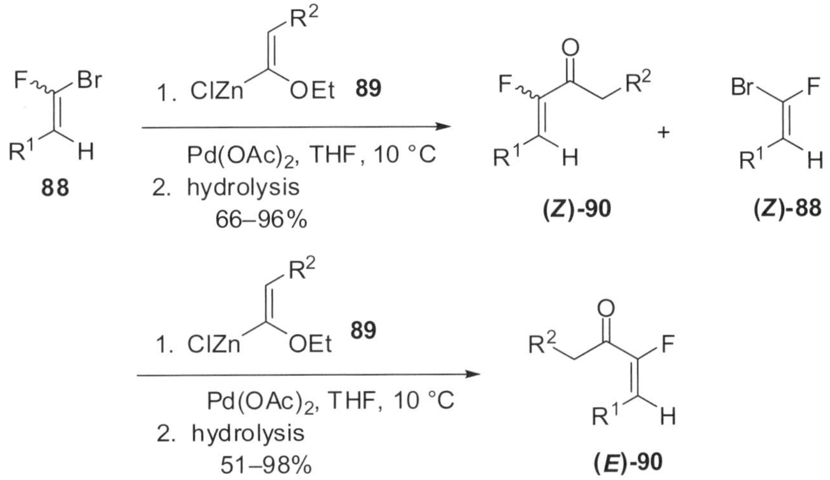

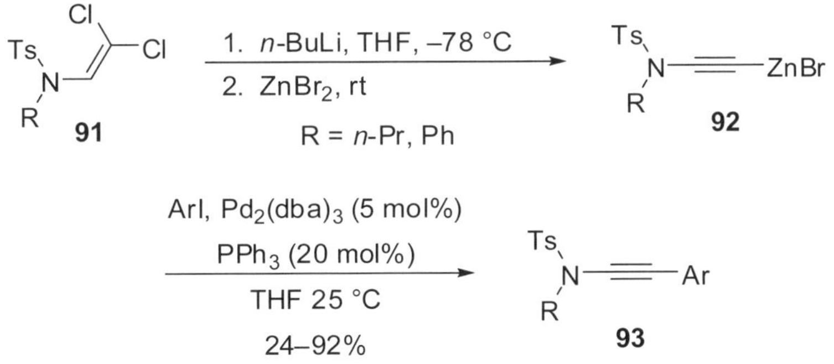

Negishi coupling of *N*-ethynylzinc tosylamides derivatives **92**, prepared from 1,1-dichlorovinylamide **91** with aryl iodides in the presence of $Pd_2(dba)_3$ and triphenylphosphine afforded *N*-aryl and *N*-alkyl arylynamides **93**.[59]

New conjunctive reagents **94** can be used, after transmetallation, in Negishi cross-couplings with vinyl and aryl iodides to give **95**.[60] The subsequently unmasked terminal alkynes could be further manipulated to obtain retinoid-like products.

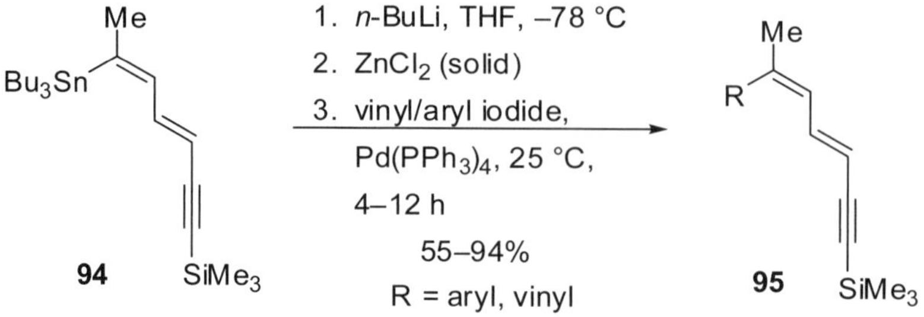

A one-pot synthesis and derivatization of diynes and triynes is reported.[61] The polyyne framework was formed from a dibromoolefin precursor **96** based on a carbenoid rearrangement, and the resulting Li-acetylide is then transmetallated with zinc chloride which then allowed for the divergent preparation of aryl polyynes **97** via Negishi palladium-catalyzed cross-coupling reactions.

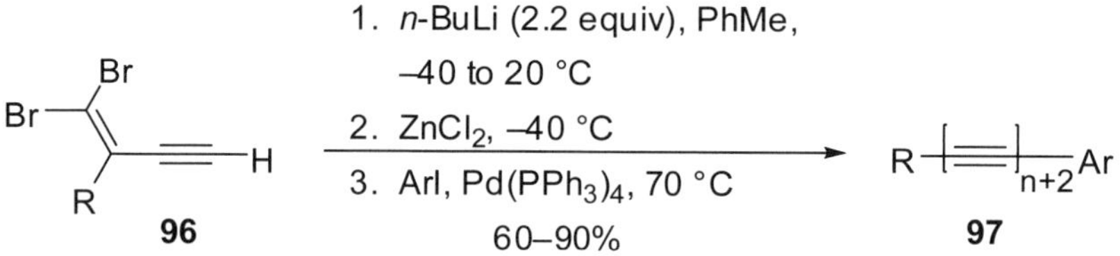

Three isomers of tetraethynylethene (**98**, $C_{10}H_4$) have been prepared by palladium-catalyzed Negishi coupling of a trimethylsilylbutadiynyl zinc reagent with a bromoalkene, followed by mild deprotection with potassium carbonate in methanol.[62] The unsubstituted enynes, 3-ethynyloct-3-ene-1,5,7-triyne (**99**), *trans*-dec-5-ene-1,3,7,9-tetrayne (**100**), and *cis*-dec-5-ene-1,3,7,9-tetrayne (**101**), exhibit modest stability at –20 °C but decomposed rapidly at room temperature.

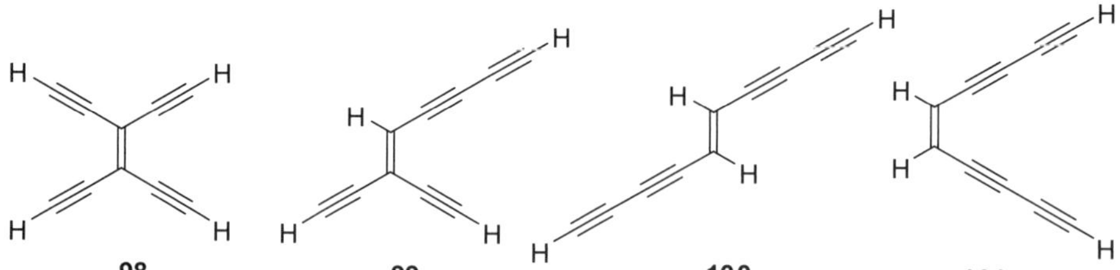

98 **99** **100** **101**

Carbohydrates

The Negishi cross-coupling reactions could also be applied to functionalized carbohydrate precursors. For example, Negishi and Stille coupling reactions of 3,4,6-tri-*O*-benzyl-2-(tri-*n*-butylstannyl)-*D*-glucal (**102**) with (*Z*)-vinyl iodide **103** provided access to functionalized spiroketal **104**.[63] With the Stille reaction, erosion of alkene chemistry (3 : 1) was noticed plus it was sluggish requiring higher temperatures but no erosion of alkene geometry occurred when the Negishi reaction was performed.

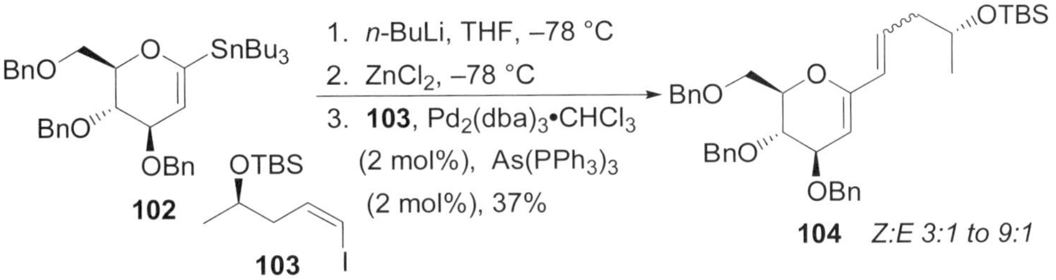

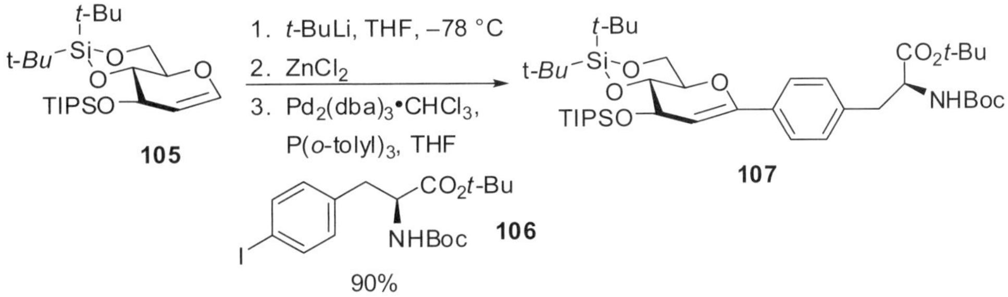

A *p*-(*C*-glucopyranosyl)-*L*-phenylalanine derivative, protected to be directly incorporated into a peptidic chain, was prepared from glucose on a gram scale, with a Negishi cross-coupling reaction as the key step.[64] The zincated glucal **105** and *p*-iodo-*L*-phenylalanine **106** were involved in this organometallic coupling, which gave rise to a link between the sugar and amino acid moieties; the β-*gluco* configuration of the *C*-glycopyranosyl amino acid **107** was ascertained by a stereoselective hydroboration of the double bond of the glucal.

Amino Acids

Protected amino acids have also participated in Negishi cross-coupling reactions. Protected *L*-serine iodides were converted under Rieke conditions to iodozinc species of **108**, which were efficiently coupled to give phenylalanine-derived chiral amino acids **109** in the presence of palladium and copper catalysts.[65]

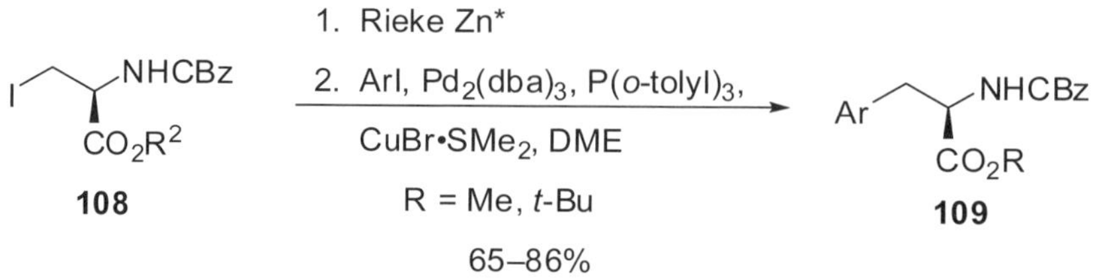

108 **109**

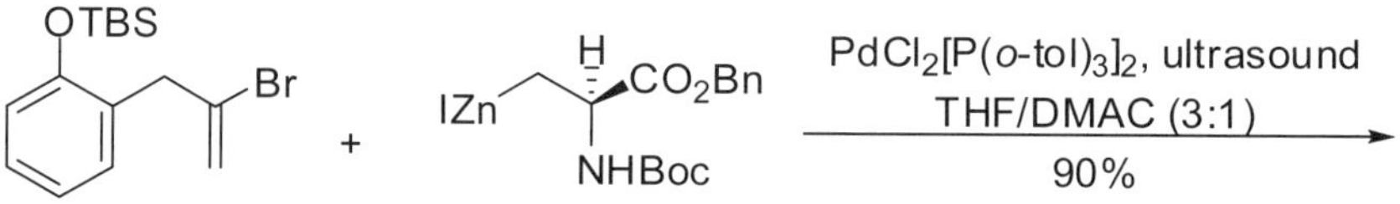

110 **111**

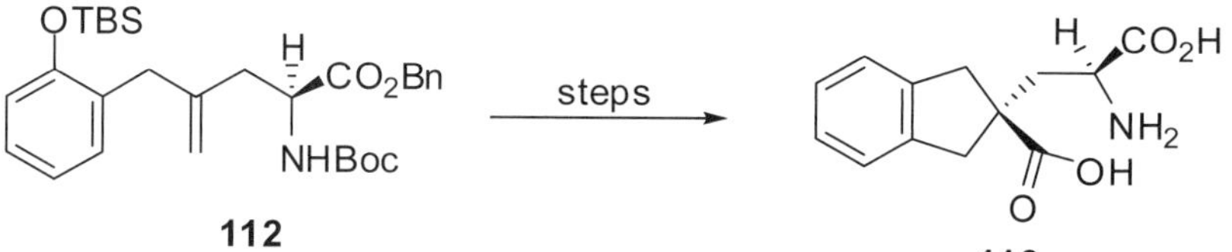

112 **113**

A general route to excitatory amino acid analogues has been developed. The key reactions involved were a Negishi coupling of Jackson's organozinc reagent **111** with vinyl bromide **110** and subsequent ring closure of **112** using the Mitsunobu reaction as the key step to **113**.[66] The efficient and direct synthesis of protected biaryl amino acids, including dityrosine by Negishi cross-coupling of the Jackson's reagent **111** with iodo- and di-iodobiaryls, is also reported.[67]

Quaternary, α-vinyl amino acids are potential mechanism-based inactivators of pyridoxal phosphate (PLP) dependent enzymes, particularly amino acid decarboxylases (AADC's). Protected α-formyl amino acids, themselves available from the corresponding α-vinyl amino acids, are stereoselectively transformed into the (*Z*)-configured α-(2′-fluoro)vinyl amino acids via a three-step sequence.[68] Palladium-mediated Negishi-type coupling with Me_2Zn or the analogous Ni-mediated procedure of Knochel could be applied to the synthesis of **115** from **114**, an unusual analogue of α-methylleucine in which a fluorine atom takes the place of a methyl group.

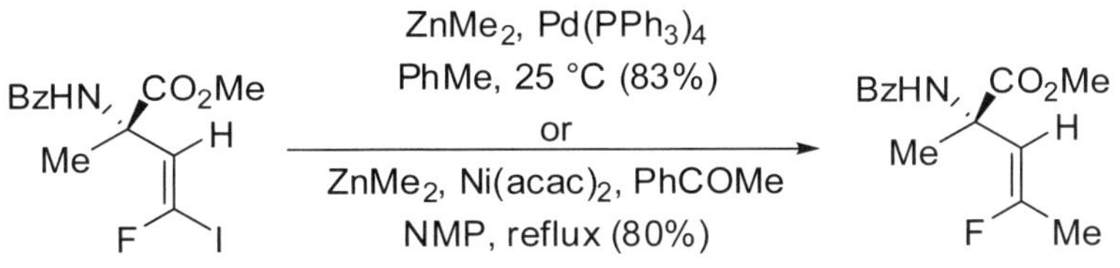

114 115

Miscellaneous

Simple amides and esters **116** were conveniently deprotonated by $Zn(tmp)_2$ (tmp = 2,2,6,6-tetramethylpiperidinyl anion) to generate zinc enolates **117**.[69] The zinc enolates **117** were readily coupled with aryl bromides using typical palladium-catalyzed Negishi cross-coupling reactions to give arylketones **118**. Enolates formed by this method were suitable for use in aldol reactions that tolerate base-sensitive functional groups.

116 117 118

Arene-, phenylmethane- and alkenesulfonyl chlorides are suitable electrophilic reagents in desulfinylative carbon–carbon bond formation cross-coupling reactions with organozinc reagents.[70] Organozinc reagents **120** underwent desulfinylative Negishi C–C cross-coupling reactions with sulfonyl chlorides **119**. However, in the presence of 1–5 mol% of Pd(0) catalyst such as $Pd[P(t\text{-}Bu)_3]_2$, smooth reactions occurred with elimination of SO_2 and formation of products **121** of C–C cross-coupling. All reactions were accompanied with the concurrent formation of homocoupling product R^1–R^1 of organozinc chloride as minor by-products.

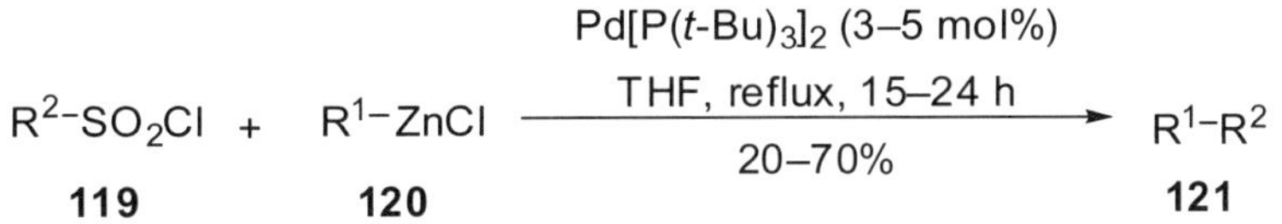

119 120 121

Natural Products

The Negishi cross-coupling reactions have been showcased as key steps in the synthesis of many natural products and in structure-activity relationship drug discovery programmes.

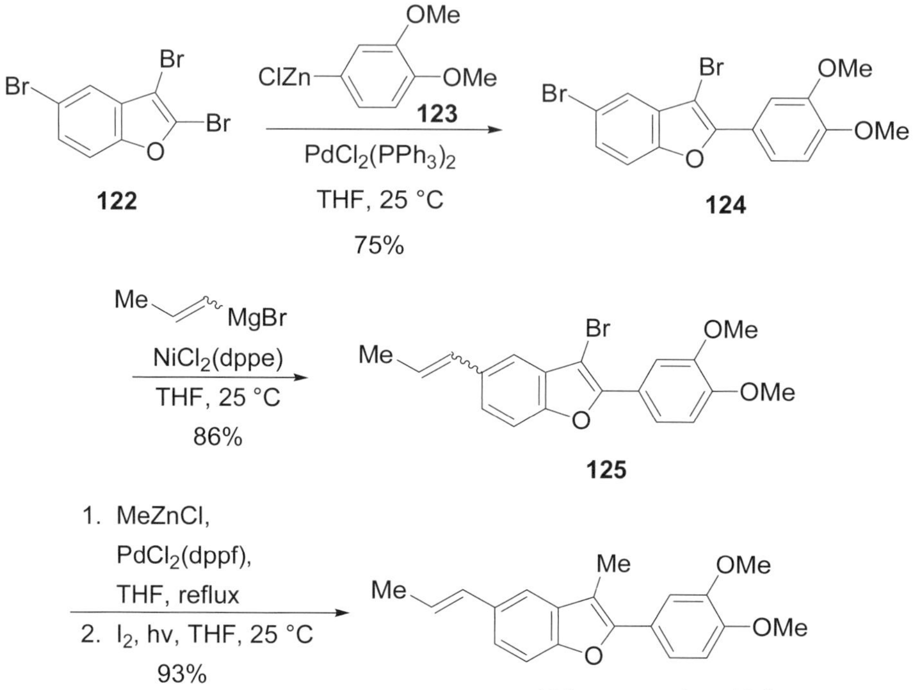

126, eupomatenoid 4

Bach has developed a synthetic approach to the trisubstituted benzofurans of the neolignan eupomatenoid family of natural products based on the regioselective transition metal-catalyzed cross-coupling reactions on the easily available 2,3,5-tribromobenzofuran (**122**).[71] The sequence commenced with 2,3,5-tribromobenzofuran (**122**) with zinc species **123**, which underwent regioselective Negishi cross-coupling to yield dibromide **124**. The Kumada cross-coupling was the method of choice to differentiate between the sterically different, but electronically similar, positions C-3 and C-5. A 1-prop-1-enyl group was established by this means to yield the monobromide **125**. Finally, the least reactive position at carbon atom C-3 was addressed in a Negishi cross-coupling with an excess of methylzinc chloride. Quantitative isomerization of the double bond in compounds **125** led exclusively to the more stable naturally occurring (*E*)-configurated eupomatenoid (**126**).

Scyphostatin (**130**) has been isolated as a potent inhibitor of neutral sphingomyerinase (N-Smase) from the mycleial extract of *Dasyscyphus mollissima.* The hydrophobic side chain of scyphostatin was synthesized by the construction of the C12′-C13′ trisubstituted *E*-olefin moiety by Negishi coupling.[72] Treatment of iodide **127** with *tert*-butyllithium (3 equiv) in the presence of zinc chloride afforded the organozinc species *in situ* and Negishi palladium-catalyzed cross-coupling with vinyl iodide **128** afforded the hydrophoboic side chain of scyphostatin **129**. Another similar report utilizing the Negishi cross-coupling reaction for the side-chain has been disclosed.[73]

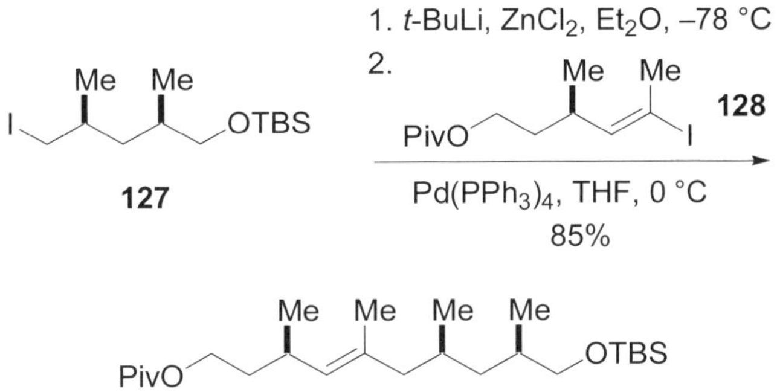

129

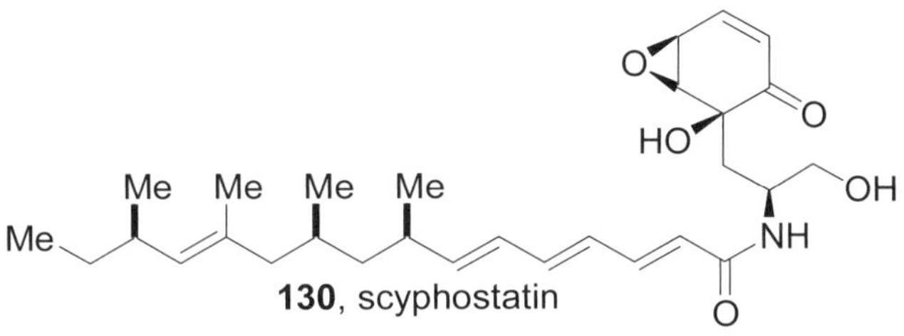

130, scyphostatin

2′,4-Disubstituted-2,4′-bithiazoles are prevelant in several natural products which exhibit antiinfective and cytotoxic properties. Regioselective Negishi cross-coupling of 2,4-dibromothiazole (**131**) delivered 4-bromothiazole **132**, which was converted into a nucleophile by bromine–lithium exchange and transmetallation to zinc.[74] The 4-thiazolylzinc chloride underwent another cross-coupling with another equivalent of 2,4-dibromothiazole (**131**) to yield the bithiazole **133**, which bore a residual bromine atom at the 4-position. Subsequent cross-coupling with boronic acid **134** led to **135**, an immediate precursor to the natural product, cystothiazole E (**136**).

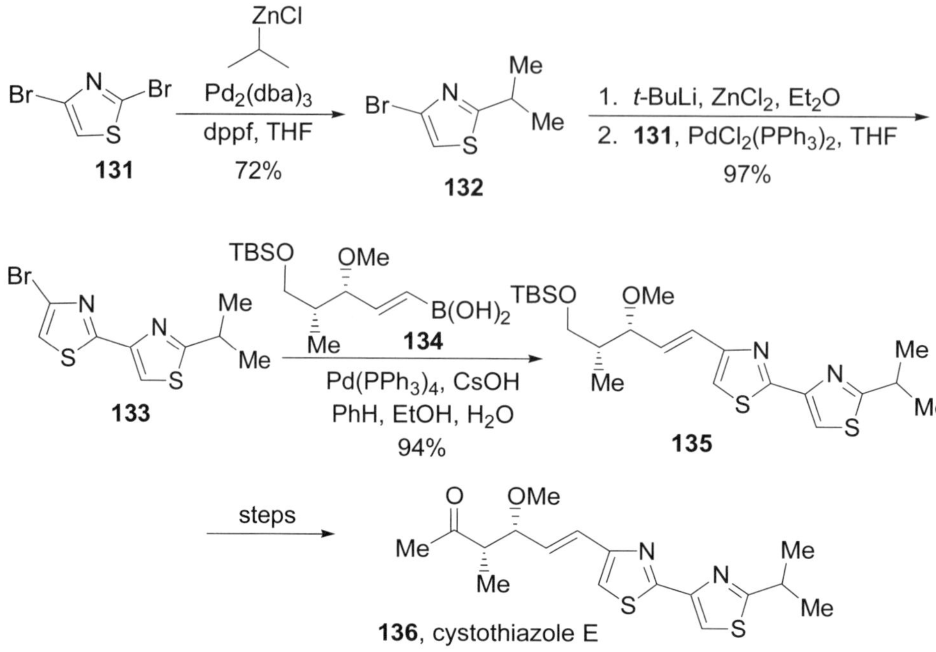

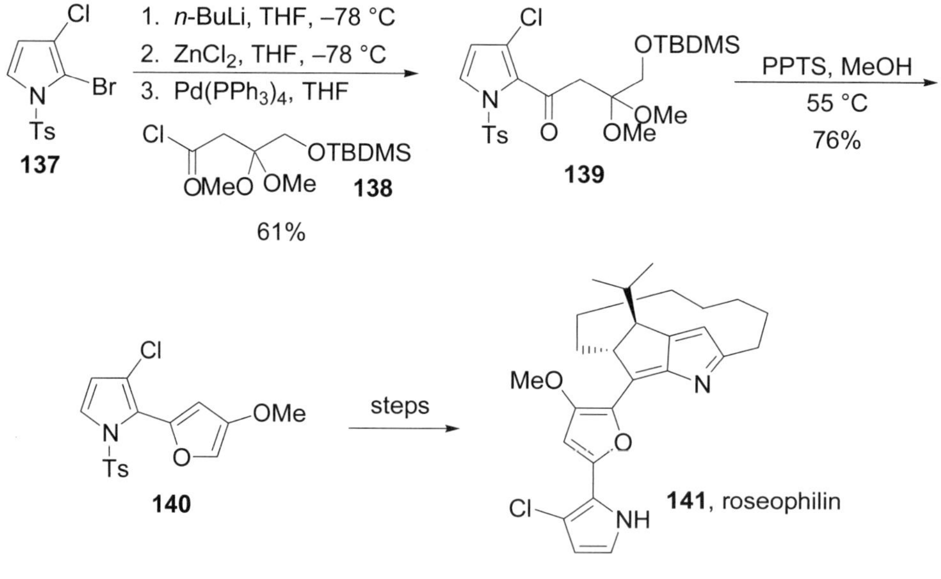

The first total synthesis of roseophilin (**141**), a novel antibiotic isolated from *Streptomyces griseoviridis*, employed the Negishi cross-coupling reaction in the early steps.[75] Thus, Fürstner demonstrated that the selective conversion of the C2-bromo–C3-chloro system **137**, via successive formation of the zinc intermediate, then cross-coupling with acid chloride **138**, into the corresponding C2-acylated product **139**. The silyl group of **139**

was cleaved upon exposure to pyridinium *p*-toluenesulfonate (PPTS) in methanol followed by spontaneous cyclization to the pyrrolylfuran derivative **140**, precursor to roseophilin (**141**).

Lactonamycin (**146**) showed significant levels of antimicrobial activity toward Gram-positive bacteria, being especially effective against methicillin-resistant *Staphylococcus aureus* (MRSA) and vancomycin-resistant *Enterococcus* (VRE). In addition, lactonamycin (**146**) showed significant levels of cytotoxicity against various tumor cell lines. The synthesis of **146** was carried out using a high-yielding Negishi coupling of organozinc species of benzyl bromide **142** with triflate **143** to obtain coupled product **144**, which was further elaborated to the tetracyclic CDEF ring system **145** of lactonamycin (**146**).[76]

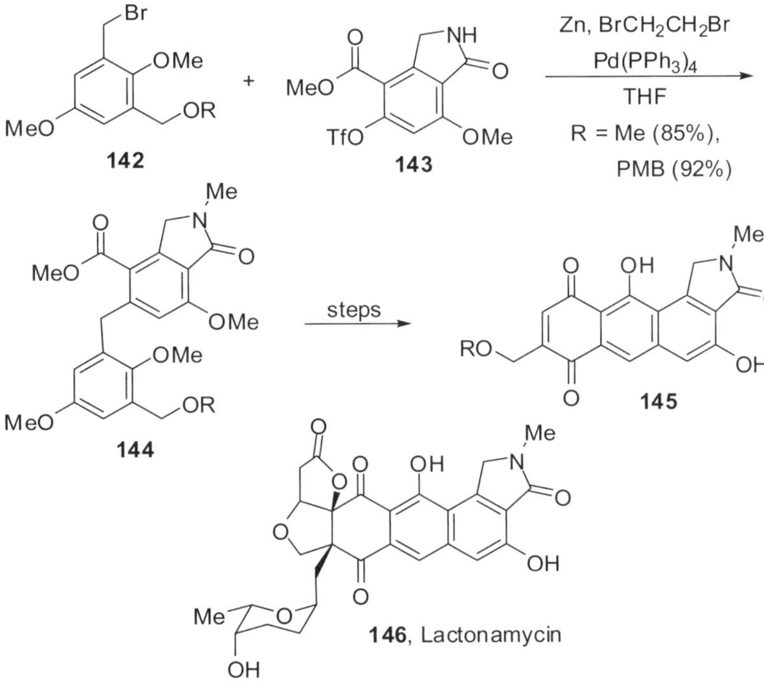

Xerulinic acid (**151**) inhibited the biosynthesis of cholesterol in HeLa S3 cells by blocking HMG-CoA synthase. Bisstannane **147** was transmetallated to the zincate intermediate and was Negishi cross-coupled with butenolide **148** to give the all *trans*-polyene **149**.[77] The Stille reaction of **149** with enediyne **150** followed by deprotection of the silyl group with tetrabutylammonium fluoride afforded xerulinic acid (**151**).

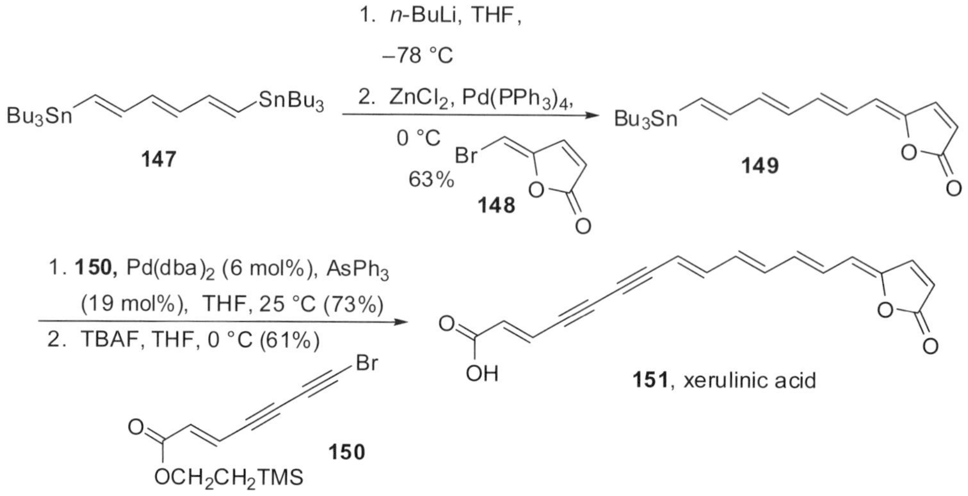

The tetracyclic A/B/C/D ring core of (20*S*)-camptothecin, one of the most potent antitumor natural products isolated from *Camptotheca acuminate*, was prepared by a combined directed ortho metallation/cross-coupling strategy.[78] Snieckus showed that 2-bromo-6-methoxypyridine (**152**) could be sequentially treated with *tert*-butyllithium (2 equiv) at –78 °C and anhydrous zinc bromide. The resulting organozinc species was then subjected to palladium-catalyzed cross-coupling reaction with triflate **153** to afford the biaryl **154**, a precursor to the tetracyclic A/B/C/D ring core **155**.

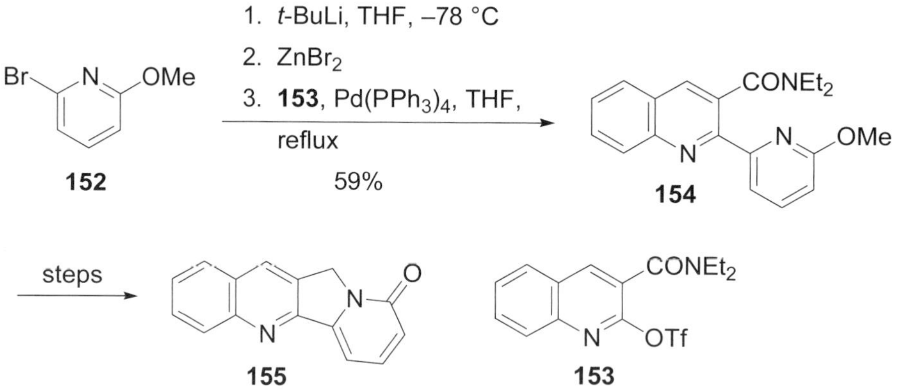

The steroid hormone 1*R*,25-dihydroxyvitamin D_3 (1*R*,25-$(OH)_2$-D_3, calcitriol, **158**) is the bioactive metabolite of vitamin D_3. This B-ring-seco-steroid plays an important role in the regulation of mineral metabolism and finds application in the treatment of osteodystrophy due to renal failure, rickets, osteoporosis, and psoriasis. The bromoolefin **156** was subjected to sequential metallation and transmetallation to give the corresponding organozinc derivative, which was treated with vinyl triflate **157** to give, by the palladium-catalyzed cascade and deprotection, the desired hormone calcitriol (**158**).[79]

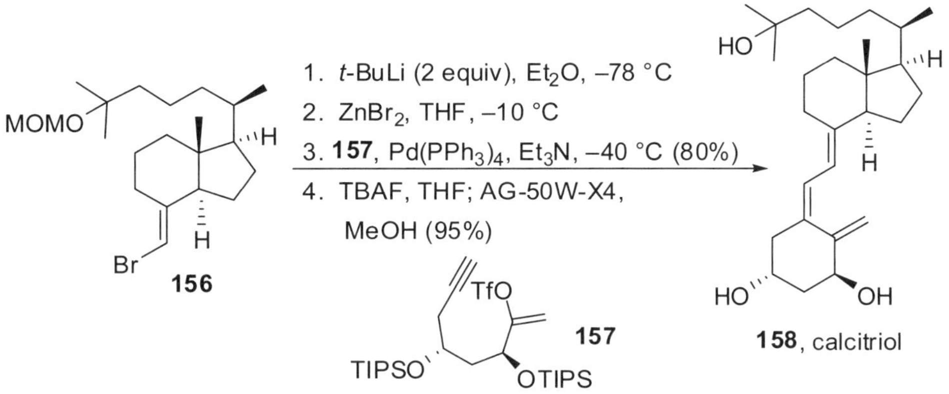

One of the current approaches to the treatment of cancer involves the disruption of kinase activity and signal transduction pathways. For example, protein kinase C (PKC) plays a crucial role in signal transductions, cellular proliferation, and differentiation. PKC is the term for a whole family of cytosolic serine/threonine kinases. Phenylamino-pyrimidines like 3-{4-[2-(3-chlorophenylamino)-pyrimidin-4-yl]pyridin-2-yl-amino}propanol (**159**, CGP 60474) represent a promising class of inhibitors of PKC with a high degree of selectivity versus other serine/threonine and tyrosine kinases and show competitive kinetics relative to ATP.[80] Analogs of CGP 60474 were synthesized as useful models for the evaluation of structure–activity relationships of phenylamino-pyrimidine-type protein kinase C inhibitors.

The signal transduction pathway in which Raf kinase operates has long been implicated in oncogenesis and is critical for proliferation, survival, and angiogenesis in various cancer models. [4,7′]Bis-isoisoquinolinyl-1-yl-(2-*tert*-butyl-pyrimidine-5-yl)amine (**160**) was identified as effective inhibitors of B-Raf kinase and was ultimately promoted for development as a drug candidate for the treatment of melanoma.[81] The key step in the synthesis was the palladium-catalyzed Negishi coupling of 4-bromo-1-chloroisoquinoline with trifluoromethanesulfonic acid isoquinoline-7-yl ester to yield 1-chloro-[4,7′]bis-isoquinolinyl.

Cyclin-dependent kinases (CDKs) play a key role in regulating cell cycle machinery. This family of kinases requires association with a cyclin regulatory subunit for activity. Different CDK/cyclin pairs are active during each phase of the cell cycle. Negishi cross-coupling reactions were the key steps in the synthesis of [1,3,5]triazine-pyridine biheteroaryls **161** as a novel series of potent cyclin-dependent kinase inhibitors.[82]

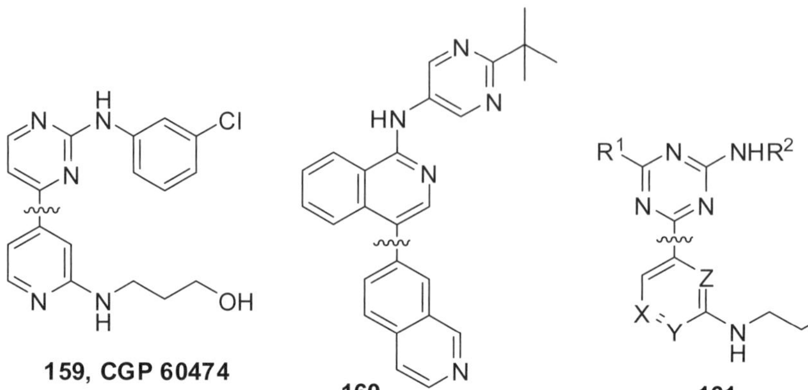

159, CGP 60474

160

161

1.1.4.7 Experimental

Negishi Cross-Coupling in the Preparation of a Furanoenyne

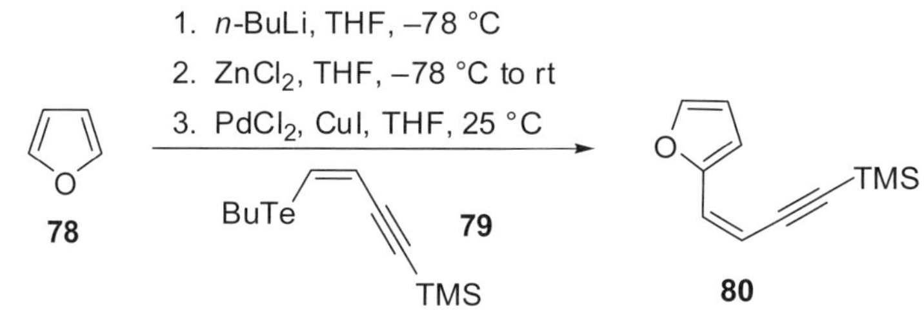

(*Z*)-[4-(Furan-2-yl)buten-3-en-1-ynyl]trimethylsilane (80).[55]
n-Butyllithium (12 mmol, 1.43 M in hexane, 8.40 mL) was added to a solution of freshly distilled furan (**78**, 0.87 mL, 12 mmol) in THF (12 mL) at –78 °C and stirred for 45 min. After this time, a suspension of anhydrous $ZnCl_2$ (1.22 g, 9.0 mmol) in THF (9 mL) was added and the mixture was warmed up to room temperature. A pale yellow solution was observed. Another two-necked, round-bottomed flask equipped with a magnetic stirring bar, rubber septum, under argon was charged sequentially with $PdCl_2$ (0.105 g; 0.60 mmol), CuI (0.57 g; 3.0 mmol), THF (3 mL), and compound **79** (0.924 g; 3.0 mmol). The mixture was stirred at room temperature for 10 min; then 2-furylzinc chloride was transferred dropwise from other flask via cannula. The dark solution was stirred at room temperature for 32 h. After this time, the mixture was filtered through a pad of silica gel/Celite and treated with aqueous ammonium chloride (30 mL) and CH_2Cl_2 (3 × 20 mL). The organic phase was separated, dried over $MgSO_4$ and concentrated under vacuum. The residue was purified by flash chromatography to give **80** (0.40 g, 78%) as a yellow oil.

Negishi Cross-Coupling in the Preparation of a Bipyridine

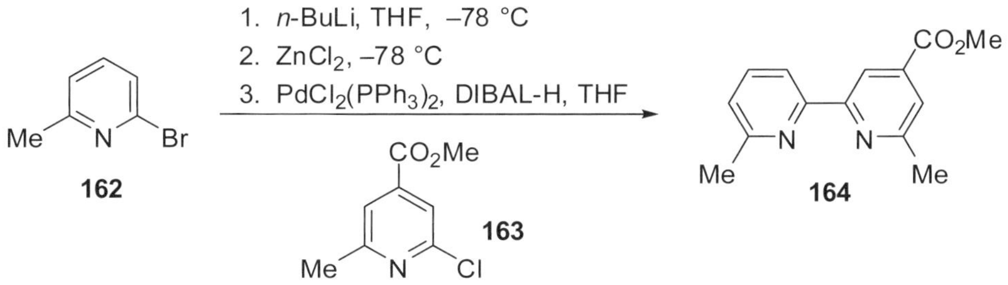

Methyl 6,6′-Dimethyl-2,2′-bipyridine-4-carboxylate (164).[83]
A solution of *n*-BuLi (1.5 M in hexane, 1.1 equiv) was slowly added to a stirred solution of 2-bromo-6-picoline (1.02 g, 5.93 mmol) in anhydrous THF (10 mL) at –78 °C, and the resulting mixture was stirred for 15 min at this temperature. Then, a 0.44 M solution of anhydrous $ZnCl_2$ (1.1 equiv) in THF was added and the stirring was continued for 30 min at room temperature. In a separate flask, a solution of methyl 3-chloro-5-methylbenzoate (**163**, 715 mg, 3.85 mmol) in anhydrous THF (5 mL) was added to a solution containing 5 mol % of a catalyst prepared by reaction of a 0.014 M solution of $PdCl_2(PPh_3)_2$ with diisobutylaluminium hydride (1.0 M in hexane, 2 equiv) and the mixture was stirred at room temperature for 10 min. The pyridylzinc chloride solution **162** prepared above was then added dropwise, and the resulting mixture was heated at reflux for 1.5 h, cooled, and poured into saturated aqueous $NaHCO_3$. The aqueous phase was extracted with Et_2O and the organic extracts were concentrated to give a solid residue which was purified by flash chromatography over alumina (petroleum ether/ether 90 : 10), to yield **164** (827 mg, 84%) as a white solid.

1.1.4.8 References

1. (a) [R] Negishi, E. *Acc. Chem. Res.* **1982**, *15*, 340. (b) [R] Negishi, E. In *Metal-Catalyzed Cross Coupling*, Eds. Diedrerich, F; Stang, P. J., Wiley, New York, **1998**. (c) Negishi, E. *J. Organomet. Chem.* **2002**, *653*, 34. (d) [R] Negishi, E.; Zeng, X.; Tan, Z.; Mingxing, Q.; Hu, Q.; Huang, Z. In *Metal-Catalyzed Cross-Coupling Reactions*, 2nd ed., Vol. 2; de Meijere, A.; Diederich, F., Eds.; Wiley-VCH: Weinheim, **2004**, 815. (e) [R] Negishi, E; Hu, Z.; Huang, M. Q.; Wang, G. *Aldrichimica Acta* **2005**, *38*, 71.
2. Baba, S.; Negishi, E. *J. Am. Chem. Soc.* **1976**, *98*, 6729.
3. King, A. O.; Okukado, N.; Negishi, E. *J. Chem. Soc., Chem. Commun.* **1977**, 683.
4. Negishi, E.; King, A. O.; Okukado, N. *J. Org. Chem* **1977**, *42*, 1821.
5. Negishi, E. In *Aspects Mech. Organometal. Chem., [Proc. Symp.]*, Ed. Brewster, J. H., Plenum, New York, **1978**, 285.
6. Negishi, E.; Baba, S. *J. Chem. Soc., Chem. Commun.* **1976**, 597.
7. Baba, S.; Negishi, E. *J. Am. Chem. Soc.* **1976**, *98*, 6729.
8. King, A. O.; Okukado, N.; Negishi, E. *J. Chem. Soc., Chem.Commun.* **1977**, 683.
9. King, A. O.; Negishi, E.; Villani, F. J., Jr.; Silveira, A., Jr. *J. Org. Chem.* **1978**, *43*, 358.
10. Negishi, E.; King, A. O.; Okukado, N. *J. Org. Chem.* **1977**, *42*, 1821.

11. Negishi, E.; van Horn, D. E. *J. Am. Chem. Soc.* **1977**, *99*, 3168.
12. Okukado, N.; van Horn, D. E.; Klima, W. L.; Negishi, E. *Tetrahedron Lett.* **1978**, *19*, 1027.
13. Negishi, E.; Okukado, N.; King, A. O.; van Horn, D. E.; Spiegel, B. I. *J. Am. Chem. Soc.* **1978**, *100*, 2254.
14. (a) Espinet, P.; Echavarren, A. *Angew. Chem., Int. Ed.* **2004**, *43*, 4704. (b) Casado, A. L.; Espinet, P. *Organometallics* **1998**, *31*, 852.
15. Casares, J. A.; Espinet, P.; Fuentes, B.; Salas, G. *J. Am. Chem. Soc.* **2007**, *129*, 3508.
16. Takahashi, T.; Kanno, K.-i. In *Modern Organonickel Chemistry*; Tamaru, Y., Ed.; Wiley-VCH: Weinheim, **2005**, 41.
17. Kondolff, I.; Doucet, H.; Santelli, M. *Organometallics* **2006**, *25*, 5219.
18. Milne, J. E.; Buchwald, S. L. *J. Am. Chem. Soc.* **2004**, *126*, 13028.
19. Dai, C.; Fu, G. C. *J. Am. Chem. Soc.* **2001**, *123*, 2719.
20. Mutule, I.; Suna, E. *Tetrahedron Lett.* **2004**, *45*, 3909.
21. Mutule, I.; Suna, E. *Tetrahedron* **2005**, *61*, 11168.
22. Walla, P.; Kappe, C. O. *J. Chem. Soc., Chem. Commun.* **2004**, 564.
23. (a) Gavryushin, A.; Kofink, C.; Manolikakes, G.; Knochel, P. *Org. Lett.* **2005**, *7*, 4871. (b) Gavryushin, A.; Kofink, C.; Manolikakes, G.; Knochel, P. *Tetrahedron* **2006**, *62*, 7521.
24. (a) Hadei, N.; Kantchev, E. A. B; O'Brien, C. J.; Organ, M. G. *J. Org. Chem.* **2005**, *70*, 8503. (b) Hadei, N.; Kantchev, E. A. B; O'Brien, C. J.; Organ, M. G. *Org. Lett.* **2005**, *7*, 3805. (c) O'Brien, C. J.; Kantchev, E. A. B.; Chass, G. A.; Hadei, N.; Hopkinson, A. C.; Organ, M. G.; Setiadi, D. H.; Tang, T.-H.; Fang, D.-C. *Tetrahedron* **2005**, *61*, 9723.
25. Organ, M. G.; Avola, S.; Dubovyk, I.; Hadei, N.; Kantchev, E. A. B.; O'Brien, C. J.; Valente, C. *Chem.-Eur. J.* **2006**, *12*, 4749.
26 Fischer, C.; Fu, G. C. *J. Am. Chem. Soc.* **2005**, *127*, 4594.
27. Arp, F. O.; Fu, G. C. *J. Am. Chem. Soc.* **2005**, *127*, 10482.
28. Zhou, J.; Fu, G. C. *J. Am. Chem. Soc.* **2003**, *125*, 14726.
29. Zhou, J.; Fu, G. C. *J. Am. Chem. Soc.* **2003**, *125*, 12527.
30. Lipshutz, B. H.; Blomgren, P. A. *J. Am. Chem. Soc.* **1999**, *121*, 5819.
31. (a) Felding, J.; Kristensen, J.; Bjerregaard, T. B.; Sander, L.; Vedsø, P.; Begtrup, M. *J. Org. Chem.* **1999**, *64*, 4196. (b) Pawlas, J.; Vedsø, P.; Jakobsen, P.; Huusfeldt, P. O.; Begtrup, M. *J. Org. Chem.* **2000**, *65*, 9001.
32. Kristensen, J.; Begtrup, M.; Vedsø, P. *Synthesis* **1998**, 1604.
33. Abarbri, M.; Thibonnet, J; Berillon, L.; Dehmel, F; Rottlander, M.; Knochel, P. *J. Org. Chem.* **2000**, *65*, 4618.
34. Havez, S.; Begtrup, M.; Vedsø, P.; Andersen, K; Ruhland, T. *Synthesis* **2001**, 909.
35. Reeder, M. R.; Gleaves, H. E.; Hoover, S. A.; Imbordino R. J; Pangborn, J. J. *Org. Process Res. Dev.* **2003**, *7*, 696.
36. Anderson, B. A.; Harn, N. K. *Synthesis* **1996**, 583.
37. Anderson, B. B.; Becke, L. M.; Booher, R. N.; Flaugh, M. E.; Harn, N. K.; Kress, T. J.; Varie, D. L.; Wepsiec, J. P. *J. Org. Chem.* **1997**, *62*, 8634.
38. Prasad, A. S. B.; Stevenson, T. M.; Citineni, J. R.; Nyzam, V.; Knochel, P. *Tetrahedron* **1997**, *53*, 7237.
39. Bach, T.; Heuser, S. *Tetrahedron Lett.* **2000**, *41*, 1707.
40. Jensen, J.; Skjaerbaek, N.; Vedsø, P. *Synthesis* **2001**, 128.
41. Akao, A.; Tsuritani, T.; Satoshi K. S; Sato, K.; Nonoyama, N.; Mase, T.; Yasuda, N. *Synlett* **2007**, 31.
42. Sapountzis, I.; Dube, H.; Knochel, P. *Adv. Synth. Catal.* **2004**, *346*, 709.
43. Simkovsky, N. M.; Ermann, M.; Roberts, S. M.; Parry, D. M.; Baxter, A. D. *J. Chem. Soc., Perkin Trans. 1*, **2002**, 1847.
44. Fang, Y.-Q.; Hanan, G. S. *Synlett* **2003**, 852.
45. Kiehne, U.; Bunzen, J.; Staats, H.; Lutzen, A. *Synthesis* **2007**, 1061.
46. Havas, F.; Danel, M.; Galaup, C.; Tisnes, P.; Picard, C. *Tetrahedron Lett.* **2007**, *48*, 999.
47. Turck, A.; Ple, N.; Lepretre-Gaquere, A.; Queguiner, G. *Heterocycles* **1998**, *49*, 205.
48. Sharples, C. G. V.; Karig, G.; Simpson, G. L.; Spencer, J. A.; Wright, E.; Millar, N. S.; Wonnacott, S.; Gallager, T. *J. Med. Chem.* **2002**, *45*, 3235.
49. Collins, I.; Castro, J. L.; Street, L. J. *Tetrahedron Lett.* **2000**, *41*, 781.

50. Gorley, E. G.; Conrad, K.; Murry, J. A.; Savarin, C.; Holko, J.; Boice, G. *J. Org. Chem.* **2004**, *69*, 5120.
51. (a) Genov, M.; Fuentes, B.; Espinet, P.; Pelaz, B. *Tetrahedron: Asymmetry* **2006**, *17*, 2593. (b) Genov, M.; Almorin, A.; Espinet, P. *Tetrahedron: Asymmetry* **2007**, *18*, 625.
52. (a) Hocek, M.; Dvorakova, H. *J. Org. Chem.* **2003**, *68*, 5773. (b) Hasnik, Z.; Silhar, P.; Hocek, M. *Tetrahedron Lett.* **2007**, *48*, 5589.
53. Shiota, T.; Yamamori, T. *J. Org. Chem.* **1999**, *64*, 453.
54. (a) Alves, D.; Schumacher, R. F.; Brandao, R.; Nogueira, C. W.; Zeni, G. *Synlett* **2006**, 1035. (b) Zeni, G.; Alves, D.; Braga, A. L.; Stefani, H. A.; Nogueira, C. W. *Tetrahedron Lett.* **2004**, *45*, 4823.
55. Oliveira, J. M.; Zeni, G.; Malvestiti, I.; Menezes, P. H. *Tetrahedron Lett.* **2006**, *47,* 8183.
56. Xie, M.; Wang, J.; Gu, X.; Sun, Y.; Wang, S. *Org. Lett.* **2006**, *8*, 431.
57. Andrei, D.; Wnuk, S. F. *J. Org. Chem.* **2006**, *71*, 405.
58. Dutheuil, G.; Paturel, C.; Lei, X.; Couve-Bonnaire, S.; Pannecoucke, X. *J. Org. Chem.* **2006**, *71*, 4316.
59. (a) Rodriguez, D.; Castedo, L.; Saa, C. *Synlett* **2004**, 783. (b) Martinez-Esperon, M. F.; Rodriguez, D.; Castedo, L.; Saa, C. *Tetrahedron* **2006**, *62*, 3843.
60. Lipshutz, B. H.; Clososki, G. C.; Chrisman, W.; Chung, D. W.; Ball, D. B.; Howell, J. *Org. Lett.* **2005**, *7*, 4561.
61. Morisaki, Y.; Luu, T.; Tykwinski, R. R. *Org. Lett.* **2006**, *8*, 689.
62. Bowling, N. P.; McMahon, R. J. *J. Org. Chem.* **2006**, *71*, 5841.
63. Conway, J. C.; Urch, C. J.; Quayle, P.; Xu, J. *Synlett* **2006**, 776.
64. Ousmer, M.; Boucard, V.; Lubin-Germain, N.; Uziel, J.; Auge, J. *Eur. J. Org. Chem.* **2006**, 1216.
65. Kruppa, M.; Imperato, G.; Konig, B. *Tetrahedron* **2006**, *62*, 1360.
66. Goundry, W. R. F.; Lee, V.; Baldwin, J. E. *Synlett* **2006**, 2407.
67. Moreno, E.; Nolasco, L. A.; Caggiano, L.; Jackson, R. F. W. *Org. Biomol. Chem.* **2006**, *4*, 3639.
68. Berkowitz, D. B.; de la Salud-Bea, R.; Jahng, W.-J. *Org. Lett.* **2004**, *6*, 1821.
69. Hlavinka, M. L.; Hagadorn, J. R. *Tetrahedron Lett.* **2006**, *47*, 5049.
70. Dubbaka, S. R.; Vogel, P. *Tetrahedron Lett.* **2006**, *47*, 3345.
71. Bach, T.; Bartels, M. *Tetrahedron Lett.* **2002**, *43*, 9125.
72. Takagi, R.; Tsuyumine, S.; Nishitani, H.; Miyanaga, W.; Ohkata, K. *Aust. J. Chem.* **2004**, *57*, 439.
73. Inoue, M.; Yokota, W.; Katoh, T. *Synthesis* **2007**, 622.
74. (a) Bach, T.; Heuser, S. *Agnew. Chem. Int. Ed.* **2001**, *40*, 3184. (b) Bach, T.; Heuser, S. *Chem.-Eur. J.* **2002**, *8*, 5585.
75. Fürstner, A.; Weintritt, H. *J. Am. Chem. Soc.* **1998**, *120*, 2817.
76. Wehlan, H.; Jezek, E.; Lebrasseur, N.; Pave, G.; Roulland, E.; White, A. J. P.; Burrows, J. N.; Barrett, A. G. M. *J. Org. Chem.* **2006**, *71*, 8151.
77. Sorg, A.; Bruckner, R. *Angew. Chem., Int. Ed.* **2004**, *43*, 4523.
78. Nguyen, T.; Wicki, M. A.; Snieckus, V. *J. Org. Chem.* **2004**, *69*, 7816.
79. Gomez-Reino, C.; Vitale, C.; Maestro, M.; Mourino, A. *Org. Lett.* **2005**, *7*, 5885.
80. (a) Stanetty, P.; Rohrling, J.; Stanetty, P.; Hattinger, G.; Schnurch, M.; Mihovilovic, M. D. *J. Org. Chem.* **2005**, *70*, 5215. (b) Stanetty, P.; Rohrling, J.; Hatinger, G.; Schnurch, M.; Mihovilovic, M. D. *Tetrahedron* **2006**, *62*, 2380.
81. Denni-Dischert, D.; Marterer, W.; Banziger, M.; Yusuff, N.; Batt, D.; Ramsey, T.; Geng, P.; Michael, W.; Wang, R.-M.; Taplin, F., Jr.; Versace, R.; Cesarz, D.; Perez, L. B. *Org. Proc. Res. & Dev.* **2006**, *10*, 70.
82. Kuo, G.-H.; DeAngelis, A.; Emanuel, S.; Wang, A.; Zhang, Y.; Connolly, P. J.; Chen, X.; Gruninger, R. H.; Rugg, C.; Fuentes-Pesquera, A.; Middleton, S. A.; Jolliffee, L.; Murray, W. V. *J. Med. Chem.* **2005**, *48*, 4535.
83. Havas, F.; Danel, M.; Galaup, C.; Tisnes, P.; Ficard, C. *Tetrahedron Lett.* **2007**, *48*, 999.

1.1.5 Sonogashira Reaction

David L. Gray

1.1.5.1 Description

The palladium catalyzed C–C bond formation processes that couples the terminal sp hybridized carbon from an alkyne (**1**) with an sp^2 carbon of an aryl or vinyl halide **2** to afford a disubstituted alkyne (**3**) is commonly referred to as a Sonogashira coupling.[1–6] As with many palladium-mediated coupling processes, there are numerous variants of this name reaction, including the analogous entirely copper-mediated process (see Stevens–Castro reaction, section 1.2.1), but the most common version uses catalytic copper as a co-promoter. The Sonogashira coupling has found broad utility in complex molecule synthesis, and truly excels in certain transformations for which it is particularly suited. The generic reaction equation does not visually capture the impressive diversity of compounds which have successfully undergone Sonogashira coupling. Among the over 1000 articles which report on Sonogashira couplings are a number of recent and comprehensive reviews of this reaction, its application, and ongoing advancements in the field.[7–21]

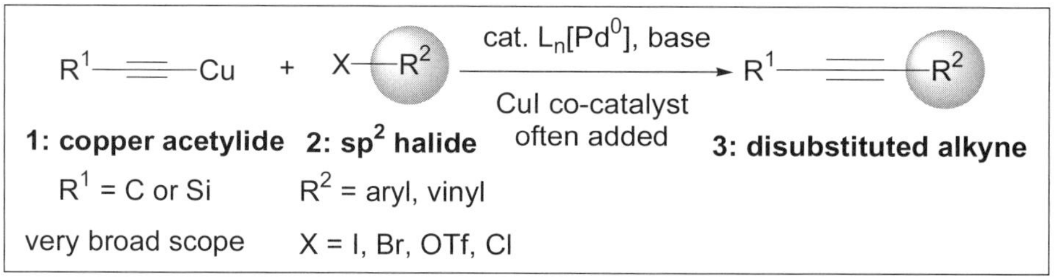

A difference between the Sonogashira coupling and many other palladium-mediated C–C bond forming reactions covered in this chapter is the fact that one of the coupling partners (the terminal alkyne) does not need to be activated with a halide or other metal, the alkyne itself being a sufficiently good ligand for the palladium by virtue of its *d*-orbital structure and inherent lack of steric bulk. In early embodiments, the coupling was primarily observed with activated (electron deficient) aryl/vinyl iodides or bromides, and functional group compatibility was shown to be excellent. An examination of successful substrates reveals examples of Sonogashira coupling in the presence of nearly any common functional group.[7,22–25] In fact, this coupling process is among the most tolerant in terms of other functionalities, making it particularly effective for the synthesis of complex molecules and large supramolecular constructs.[26,27] Key requirements for reliable reactivity include an alkyne that is not highly electron deficient (conjugation with a carbonyl dramatically slows the reaction rates, though this can be overcome) and the familiar activation constraints on the halide coupling partner, with electron rich aryl bromides, and particularly aryl chlorides, necessitating more forcing conditions and specially tuned catalyst systems.[15] Steric bulk around the reacting alkyl halide and significant bulk on the alkyne are normally tolerated. The Sonogashira coupling reaction is usually complete within 8 hours and often within 30 minutes. Palladium loading varies widely, however in a large percentage of synthetic application, between 0.5 and 5 mol% of the metal is used.

In its early form, the reactions were generally run in amine bases as solvent with co-catalytic CuI.[1] The functional palladium catalyst can be added to the reaction mixtures either as a Pd(0) species such as $[Pd(PPh_3)_4]$, or as a Pd(II) precatalyst like $[Pd(PPh_3)_2Cl_2]$. This latter pre-catalyst was the palladium source for the cross-coupling of 1-bromopropene (**4**) and propargyl alcohol (**5**) to form ene–yne **6**, which was demonstrated in Sonogashira's original disclosure.[1] More recent advances include powerful "copper-free" conditions capable of coupling notoriously unreactive aryl chlorides, and a host of developments in the areas of heterogeneous catalyst systems, novel solvents, ligands, and palladium sources, though the early CuI, Pd/PPh_3-based systems remain as the methods chosen for the majority of synthetic application.[7,28–36] One caveat to the use of the Sonogashira coupling is the ready formation of copper-promoted alkyne homo-dimers. A number of protocols have been developed to minimize the formation of these products and some of these techniques will be discussed in section 1.1.5.3.[37–39]

The union of an sp^2 and an sp center does not create any new stereochemistry, making for an attractive retrosynthetic disconnection in certain situations. In cases where *E*- or *Z*-geometry exists in the aryl halide component (e.g., **4**), the stereochemical integrity of the olefin is retained during a Sonogashira coupling.[9]

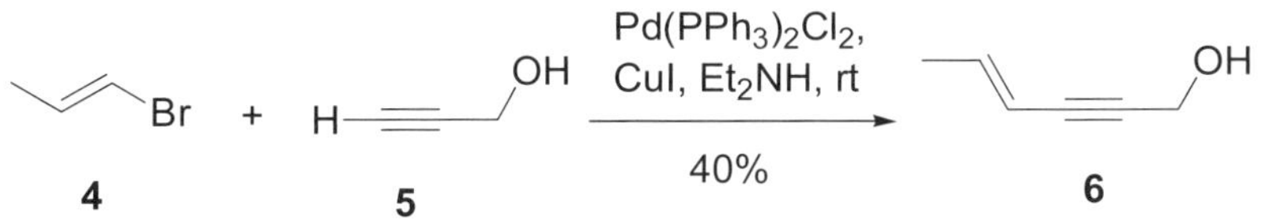

A large number of ligands for palladium have been employed, though the largest and most important class is the monodentate phosphines. Multidentate phosphanes, palladacycles, and *N*-heterocyclic carbenes are emerging areas of ligand development and application.[8] A stoichiometric amount of base is necessary to turn the catalytic system over, but in practice two or more molar equivalents are normally used. Organic amine bases (for example Et_3N or *i*-Pr_2NH) were used as solvent in much of the early investigation of this reaction, and later, inorganic bases such as K_2CO_3 or Cs_2CO_3 were shown to be effective in many cases. While reaction solvent is a parameter that has been varied in many studies with effect, when taken collectively, any of the solvents used in Pd-based cross coupling chemistry have functioned well in the appropriate Sonogashira system. Apart from the aforementioned components, (Pd^0, Cu^I, ligand for Pd, solvent, and base), additional additives are uncommon save for phase transfer agents used in aqueous systems, and tetra-alkyl amines which accelerate ligand-free protocols.[42–44] Many successful Sonogashira transformations occur at room temperature, but reaction have been reported as low as –20 °C and in excess of 150 °C.[7–45]

1.1.5.2 Historical Perspective

The Sonogashira coupling has its origins in the extension of copper-catalyzed processes that were being described in the late 1950's.[6] Conceptually, it is related to the Castro–Stephens coupling (see 1.2.1) and shares mechanistic connection to other productive Pd-mediated cross couplings in common use today (See sections 1.1.1–1.1.4), with many of the same pioneers in organometallic chemistry making important contributions to the advancement of this reaction in parallel with their studies of other cross-coupling methodologies. In consecutive *Journal of Organometallic Chemistry* articles in 1975, Heck and Cassar independently described Pd-catalyzed conditions for the union of alkynes and aryl iodides under basic conditions at high temperature, Heck's procedure using amine bases as solvent, and Cassar's protocol featuring sodium methoxide in DMF.[2,3] As part of a program directed toward characterizing alkyne reactivity with metals, Kenkichi Sonogashira and co-workers had previously observed that copper acetylides effectively transmetalated to platinum and that the resulting Pt-alkyne intermediates went on to react at the unactivated alkyne terminus.[7,8] Sonogashira's insight was to combine the copper-mediated transmetalation of alkynes that his group was studying with a metal that offered more in terms of catalyst tenability, namely palladium. Several months after the Heck and Cassar disclosures, Sonogashira reported on 15 examples of Pd-catalyzed cross coupling between terminal alkynes and aryl/alkenyl halides that proceeded *at room temperature* in amine solvent when co-catalytic CuI was added. It was this extension of Heck and Cassar's results that yielded a productive and robust C–C bond formation with especially mild reaction conditions.[1]

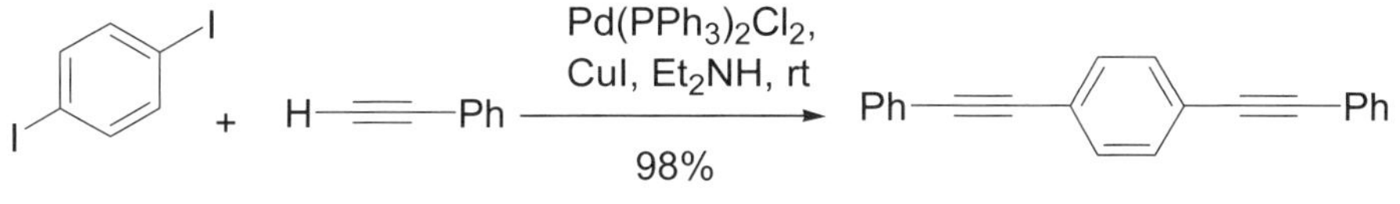

Typical of the reactions exemplified in Sonogashira, Todha, and Hagihara's 1975 report is the double coupling of 1,4-diiodobenzene (**7**) with phenylacetylene (**8**), which proceeded to give a remarkably high yield of linear triphenyl compound **9** using 0.5 mol% of a Pd(II) precatalyst and 1 mol% cuprous iodide in diethylamine. The same initial publication also described the formation of symmetrically disubstituted alkynes from aryl iodides and bubbled-in acetylene gas under the same conditions (coupling at both acetylene termini).[1] The striking mildness of the copper-promoted coupling conditions (when compared to the similar Heck and Cassar palladium-only conditions for the same transformation) spurred significant research into the application of this reaction and later, into the broadening of its scope. It was quickly established that the original Sonogashira protocol was very tolerant of functional groups and steric bulk within the reacting partners, and could therefore be adapted to complex molecule synthesis with little modification. It is likely these features which contribute the current popularity of the Sonogashira reaction.[49] Of course, while the addition of copper yielded tangible benefits in terms of reaction mildness, there were a number of challenges with the methodology which provided fertile ground for organometallic research.[8,10,13,15] These initial shortcomings included relatively high catalyst loadings with aryl bromide coupling partners, significant alkyne homocoupling (Glaser reaction – see 1.2.2), inertness of unactivated aryl bromide and chlorides, complexity of two-metal systems, and the need for a large excess of amine base. These challenges have all been addressed to a certain extent, and in some cases, a solution has been to eliminate the copper in so called "copper-free" Sonogashira conditions (which could just as easily have been termed Heck–Cassar couplings). The term "Sonogashira coupling" however, is now a blanket description commonly applied to the Pd(0) mediated union of a terminal alkyne and an sp^2 (or even sp^3) halide or triflate, regardless of whether copper(I) salts are present. The ene-yne moiety of Sonogashira products has found applicability in diverse and highly useful ways spanning everything from self-assembly and guest-host constructs, to dyes, sensors, biomolecule conjugates, polymers, and heterocycle synthesis.[27,49–63]

1.1.5.3 Mechanism

Mechanism of Sonogashira Reaction with Copper Co-Catalyst

It has been generally accepted that the copper co-catalyzed Sonogashira coupling proceeds via a dual catalytic cycle- a palladium cycle which is similar to that which is postulated for the Heck and Suzuki couplings, and an ancillary copper cycle which facilitates the transfer of an un-activated acetylynic group to the palladium metal center.[6] There are similarities between key steps of the postulated Pd catalytic cycle for the Sonogashira coupling, and those elements in Heck and Suzuki couplings which unify many of the considerations that have driven progress in the palladium-mediated C–C bond formation area. The difficulty of unequivocally "proving" the details of a complex catalytic cycle leave room for additional insight into the precise nature of the various intermediates, however, as a basis for informing progress in the field, the basic 3-step catalytic cycle consisting of A) oxidative addition, B) transmetalation, and C) reductive elimination has been fruitful.[9] There is more uncertainty surrounding the Cu cycle, as the putative intermediates (e.g., **10** and **11**) have not been directly characterized and there is less analogy with other well-studied systems. There are excellent resources which provide insight into mechanistic subtleties as presently understood, however this scheme outlines a suitable framework for practical mechanistic discussion.[8–10,16]

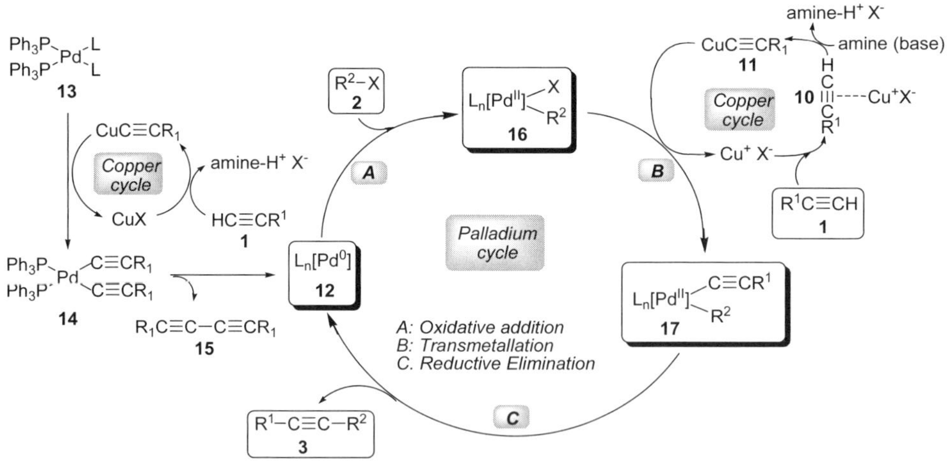

The requisite entry into the main catalytic cycle is a Pd(0) species (**12**) which can either be added as $[Pd^0(PPh_3)_4]$ or similar, or generated *in situ* from a Pd(II) pre-catalyst and excess phosphine or alkyne. In Sonogashira's original condition, it is likely that the active Pd(0) catalyst **12** was generated from [$Pd(PPh_3)_2Cl_2$] (**13**, L = Cl) via reductive coupling of sacrificial alkyne as outlined.[6] Thus, the "copper cycle" and a base facilitate the sequential transfer of two alkyne ligands (**1**) onto Pd(II) to afford intermediate **14**. Reductive C–C bond formation then expels diacetylene species **15** and generates low valent palladium **12**. The classical Pd cycle then begins with the oxidative addition (step A) of this Pd(0) species into the C–X bond of the aryl or vinyl halide component **2** to afford a Pd(II) complex (**16**). The L_n designation within **16** is purposefully ambiguous in that a number of factors are thought to determine the coordination chemistry and charge state of this Pd(II) intermediate complex, with counterions, main ligand properties, and solvent effects typically proposed as the major players. Regardless of its exact structure, progress towards eventual product occurs in step B when the electron deficient metal center accepts a terminal alkyne donor ligand in a transmetalation event, leading to intermediate **17**. A co-catalytic copper cycle is postulated to form a transient copper acetylide **11**, which transfers alkyne to palladium. Activation of the alkyne starting material **1** likely precedes formation of **11** via the intermediacy of the π-alkyne copper complex **10**. In step C, complex **17** reductively eliminates disubstituted alkyne product **3**, forming a new carbon-carbon bond and re-generating the Pd(0) **12**. A minimum of one equivalent of base (frequently an amine) is required to remove the net acid (HX) generated in the reductive elimination step, and this action is often thought to be focused within the copper cycle when copper salts are present.

Similar to other Pd-promoted C–C bond formations, the ligand plays a key role in funnelling the course of the reaction into a productive manifold. In favourably biased reactions, most any Pd(0)-based catalyst/ligand system can be coerced to afford useful yields of product. However, the importance of the ligand is particularly apparent with more recalcitrant systems, where slow steps need to be accelerated by ligand effects in order to avoid side reactions and the plating out of palladium black, which limits turnover.[13] Certainly, the optimization of phosphine ligands is the single biggest development contributing to the increasing scope of the Sonogashira coupling. Not surprisingly, the order of reactivity for halide substrates **2** parallels their propensity to undergo oxidative addition, namely X = I > OTf > Br >> Cl.[9] For iodides and electron deficient bromides or triflates, the oxidative addition is generally fast and is not rate limiting.[16] Transmetalation from the copper acetylide is typically the slower event unless electron rich bromides, and particularly aryl chlorides are employed, in which case the situation changes considerably and the oxidative addition may be rate limiting.[15]

Under Sonogashira reaction conditions, the propensity of terminal alkynes to form dimerized products (**15**) is often attributed to oxidation of copper(I), which is well known to reductively dimerize terminal alkynes in its higher oxidation state (see section 1.2.2, Glaser coupling). The mechanistic scheme also points to a second, minor pathway for the formation of **15** mediated by palladium. The homocoupled side product is a unique issue to the Sonogashira coupling among the reactions covered in this chapter.

As is the case with many other Pd-mediated coupling processes, there can be varying coordination states at the metal center at each stage of the catalytic cycle, depending upon the conditions. Familiar Pd(0)L_2 (**12**) seems likely with ligands like PPh_3, but evidence suggests that very bulky phosphanes shift **12** to a singly ligated Pd(0)L catalyst species, a condition which promotes oxidative addition to less reactive aryl bromides and chlorides.[15] On the other end of that spectrum, heterogeneous catalyst conditions lead to a weakly or transiently coordinated Pd(0) catalyst **12**.[13,64] Jutand advocates that **12** can also be an anionic Pd(0)L_2X^- complex, particularly with certain precatalysts like $Pd(OAc)_2$ or with aryl chloride substrates, where the acetate or Cl anion can serve as a ligand to drive formation of pentacoordinate palladium species.[16]

Mechanism of "Copper-Free" Sonogashira Reaction

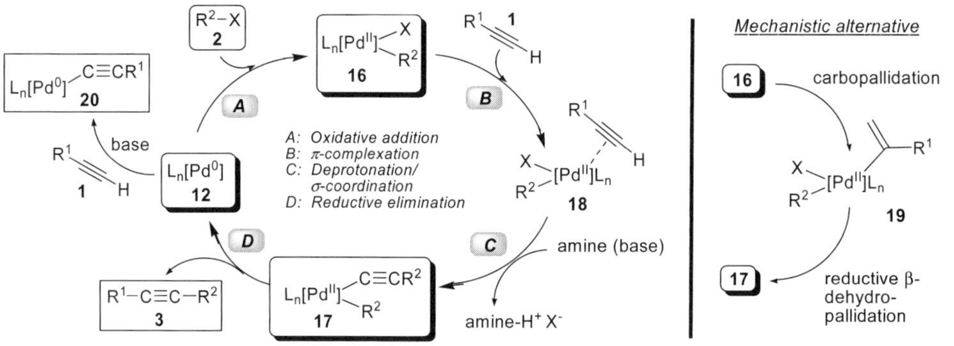

Recent publications concerning "Ligand-free" and "copper-free" conditions for the Sonogashira coupling raise additional points about the reactivity pathway.[65] In copper free variants, several mechanistic possibilities have been proposed, one of which is presented here. Laying aside specifics about the Pd(0) entry into the catalytic cycle, palladium complex **12** again oxidatively inserts into sp^2 halide **2** (step A) to give rise to complex **16**. In step B, π-complexation of **1** with **16** would acidify the acetylinic proton in species **18** and facilitate its removal by an amine base (step C) with coordination of the now (formally) anionic acetylene ligand to the metal center. Palladium(II) species **17** is now set up for step D – reductive elimination of product **3** and reformation of active catalyst **12**.

One might envision that the base plays a direct role in deprotonating the alkyne, with the resulting acetylide anion coordinating to the palladium, however, activation of the acetylynic proton in an η2 complex **18** seems to be required, because the employed amine and inorganic bases are not sufficiently strong to directly deprotonate the alkyne, though the pKa of the acetylinic proton has been shown to influence reaction outcome and rate.[66] When amine bases are used under "copper-free" conditions, the influence of specific amines is more pronounced than with standard copper co-promoted reactions.[68] Perhaps surprisingly, while copper(I) salts have robust acceleration effects most Sonogashira couplings with iodides and electron–poor bromides regardless of ligand/catalysts system, the protocols that have demonstrated the capability to successfully couple deactivated aryl bromides and chlorides are mainly "copper-free" and the addition of CuI to these reactions has a deleterious effect on both rate and yield.[8,28,67] Clearly, the copper-free Sonogashira coupling is less well characterized, though surely it proceeds via oxidative addition and reductive elimination. A detailed understanding of the process for coordination of the alkyne remains elusive. In fact, at least one leading researcher recently suggested that a carbopallidation/reductive β-dehydropallidation sequence advancing through transient vinyl palladium **19** cannot be ruled out owing to the lack of palladium-derived structural information translating to Sonogashira products. This latter pathway may be more relevant when there is no amine base in solution.[9,69]

Additional Mechanistic Considerations

One interesting observation is that the oxidative addition of palladium(0)-tetrakistriphenylphosphine to iodobenzene is slowed when performed in the presence of phenylacetlyene, with the rate of the overall reaction having an inversely proportional relationship to alkyne concentration.[16] This strongly suggests that a portion of the active Pd(0) catalyst **12** is coordinated to the alkyne in the form of deactivated $[(\eta 2\text{-}R^1C{\equiv}CH)Pd^0L_2]$ **20**, and helps to explain inconsistent effects on rate due to alkyne concentration. With an alkyne ligand on Pd(0), the oxidative addition will be slowed. Oxidative addition is typically not considered to be the rate-limiting step in this postulated catalytic cycle, but this slowing can have divergent consequences. Slowing of step A can actually lead to favourable effects on turnover numbers (TONs) when there is a significant rate mismatch between a fast oxidative addition and a slower transmetalation. The coordination of the alkyne to the Pd can move the relative rates closer to unity, the ideal situation for maximum catalytic efficiency. Much effort has been devoted to the coupling of propiolic acid derivatives because the products would have broad utility.[70,71] Success in this area has been limited, likely due to the affinity of Pd(0) for propiolates, which affinity would increase the contribution of a significantly deactivated species **20** [(η2-RC≡CH)Pd^0L_2], R = CO_2H.[16] To further obscure the details, many of the amine bases used in the Sonogashira reaction can themselves be transient ligands for the Pd metal, displacing phosphine to form [Pd^0L_n(amine)] complexes in a reversible process that should be a more significant contributor to reaction outcome when the amine base is used as solvent.[13,16,69]

Regioselectivity and Stereoselectivity

The Sonogashira coupling is regio- and stereochemically straightforward. The aryl or vinyl halide component always couples with retention of any *E,Z* stereochemical information contained within the starting materials, and no additional stereocenters are generated, nor are migrations observed. The alkyne component couples at the unsubstituted terminus and affords and expected product. Information derived from stereochemical outcomes of other Pd coupling reactions has been invaluable in establishing the mechanistic underpinnings of those processes. Conversely, the lack of such clues in the product has hampered efforts to more fully define the mechanistic course of the Sonogashira coupling.

"Ligand-free" Catalysis

Background and general mechanistic considerations for "ligand-free" catalysis are covered in previous sections, and many of the same concepts apply to the Sonogashira coupling. The term "ligand-free" is meant to convey the absence of traditional "strong" ligands (like PPh_3) rather than a truly naked Pd(0) species. Basic amines or solvents are assumed to serve as weak ligands for the palladium center in these protocols. Owing to the transient nature of the interactions, it is difficult to glean much structural information on weakly coordinated palladium species or to make mechanistic generalizations across brand of catalysis. Nanoparticulate or soluble molecular palladium has been shown to be the operant catalyst in some "ligand-free" systems. Aryl iodides (e.g., **7**) couple so readily with terminal alkynes that $Pd(OAc)_2$ has been shown to catalyze this transformation without the addition of any traditional ligand.[42–44,72,73] There is an ongoing debate as to the nature of the active species when classical ligands are absent, but it is clear that poorly ligated palladium can demonstrate interesting catalytic activity.[74]

Heterogeneous Catalysis

Heterogeneous catalysis encompasses a broad and active area of cross coupling research. Similar to ligand-free catalysis, the chemistries and mechanistic proposals surrounding productively harnessed heterogeneous palladium-based cross coupling are complex and evolving.[75] For palladium nanoparticles, a well accepted relationship exists between particle size, accessible surface area, and activity. Preparations that maximize accessible metal surface area lead to enhanced reactivity. At present, mechanistic work is primarily focused on the characterization of reactive surface metal.[76] The strategies used to increase turnover with these reactive, but inherently unstable catalyst species are somewhat generally applicable to Heck, Stille, or Sonogashira protocols and include addition of ionic liquids or quaternary ammonium salts, reservoir strategies where soluble palladium is thought to slowly release into solution, capture of palladium in peroskovites or other mesoporous matrices, and the use of palladium coated particles which maximize the amount of surface-accessible reactive catalyst.[74] Each of these approaches aims to maximize operant Pd(0) while mitigating against the expected aggregations of the low valent metal. One must be careful to differentiate between truly insoluble metal, and catalysis which occurs via low ppm leaching of palladium into solution. A great challenge in this area is proving unequivocally that catalysis is occurring exclusively in the heterogeneous phase. Even so, the expected progression for Sonogashira couplings in the heterogeneous phase is via intermediates like **14** and **15** with undefined ligands L_n.

Alkyne Homocoupling and Reactivity of Acetylene

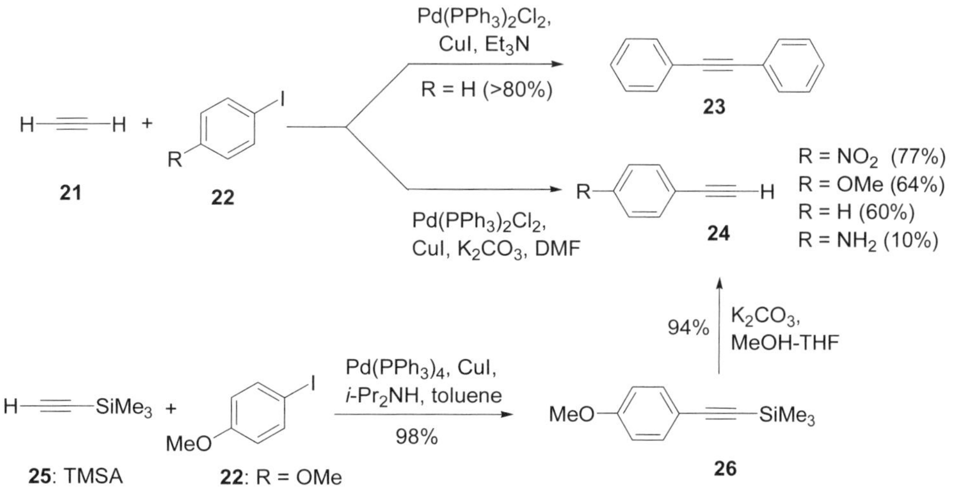

Acetylene is ideally a very useful reagent for Sonogashira coupling, however, this gas poses selectivity problems in this reaction owing to the propensity for the product to couple with additional halide to form symmetrical constructs. In fact, under the typical conditions, the bis-coupled acetylene **23** has been reported to be the major isolated product from reaction of acetylene (**21**) and iodobenzene (**22**: R = H).[39,77]

The common work-around for this reactivity problem is to use TMS acetylene (**25**). This reagent is easily handled and the silyl group is readily removed upon treatment with mild base (**26** → **24**: R = OMe).[78] Productive monocoupling of **22** and acetylene (**21**) to afford the terminal alkynes (**24**) has been achieved by judicious choice of DMF as a reaction solvent. The proposal is that acetylene has high solubility in DMF, and that the high molarity of acetylene compared to starting halide statistically drives the reaction equilibrium toward formation of the desired monocoupled alkyne **24**. Use of TMS acetylene is a general solution to the synthesis of monosubstituted alkynes, while the reactions with acetylene itself afforded products (**24**) in good yield only when aryl iodides were coupled.[77]

Separate from the issue with acetylene, the homocoupling of alkynes is a reaction shunt that is usually undesired and arises chiefly via copper redox chemistry. This end to end union of two of the same alkynes is also known as a Glaser coupling (see 1.2.2). In order to minimize the formation of these dimers, the alkyne is often added slowly to keep its relative concentration low.[37] Oxygen has been identified as a promoter for the unwanted homocoupling of alkynes, and most procedures now actively remove this gas from reaction solvents and are run under inert atmospheres to minimize Glaser products. Running Sonogashira coupling under an atmosphere of H_2 suppresses the oxidative pathway, but most experimental procedures simply call for degassed solvents and reaction under nitrogen.[38] A more recent development is the "sila-Sonogashira" wherein the Glaser pathway is completely avoided by effecting direct coupling of trimethylsilylacetylenes (reaction occurring on the silyl-bearing carbon).[79,80]

1.1.5.4 Synthetic Utility

The Sonogashira finds frequent application in nearly every area of synthetic chemistry, and consequently, a dizzying array of diverse compound classes are represented among the successful products of this reaction. Though the product structures may look quite distinct, the reason that the Sonogashira coupling is typically chosen is often the same – either the mild reaction conditions and corresponding functional group tolerance or incorporation of the useful alkyne moiety. Despite the significant work on copper-free conditions, heterogeneous catalyst systems, advanced phosphanes, aqueous reaction, and other "improved" protocols, the vast majority of the literature on actual application of the Sonogashira coupling still uses the CuI promoted conditions with simple Pd/PPh_3-based catalyst systems and reaction conditions that are not too dissimilar from early versions of the reaction, the main difference being that amines are rarely used neat as solvent.[7] Despite the preponderance of the "old" conditions in synthetic application, there is recently a trend of uptake which indicates that newer procedures are indeed more powerful, and general. Strategically, some larger classes of applications emerge:

- Cross-coupling of advanced and precious fragments in total synthesis
- Construction of long saturated or polyunsaturated alkyl chains in target molecules
- Synthesis of rigid polyalkynl–polyaryl products
- Synthesis of organomaterials where the electrooptical properties of arylalkynes themselves are of interest
- Heterocycle synthesis

Complex Fragment Coupling

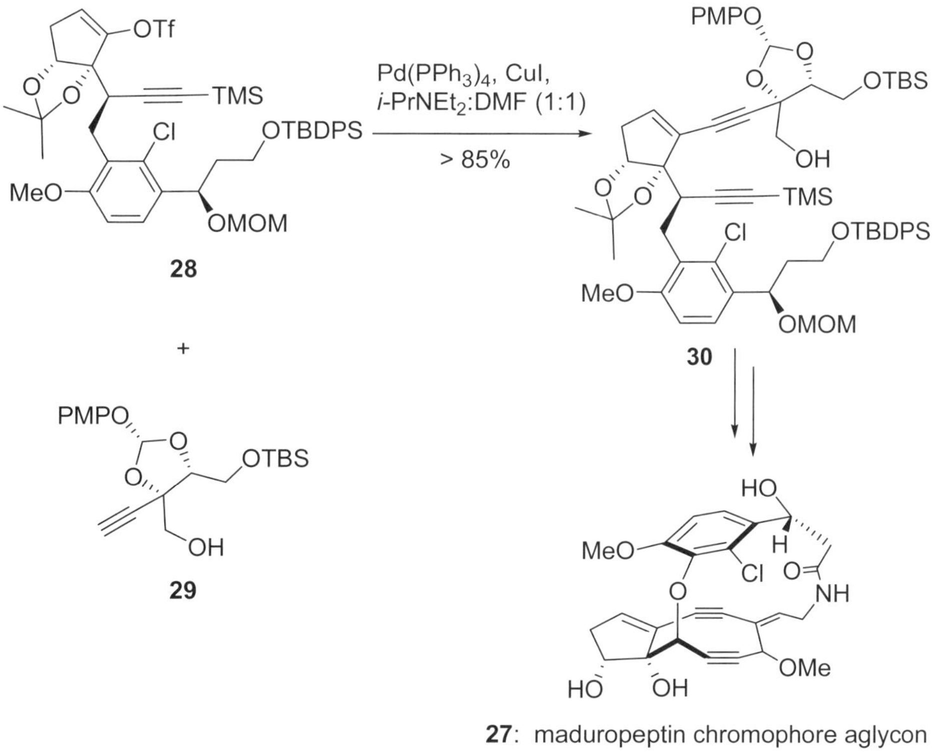

27: maduropeptin chromophore aglycon

Each year, dozens of published syntheses of natural products feature a Sonogashira coupling. A recent example of coupling large and precious (and potentially sensitive) fragments can be found in Hirama and Inoue's synthesis of the maduropeptin chromophore.[81] Some of the more spectacular applications of this reaction have come in the area of ene–dyine natural products of which maduropeptin chromophore (**27**) is an example.[82] Sonogashira coupling gives direct access to the namesake ene–yne functionality and maintains the mildness necessary for handling such energetically loaded systems. The coupling between vinyl triflate **28** and hindered alkyne **29** was effected using 5 mol% Pd(0) tetrakistriphenylphosphine and 10 mol% CuI in DMF and Hunigs base (1 : 1). This union proceeds to give an 85% yield of **30**, despite the presence of a free hydroxyl and multiple protecting groups. The high yield is particularly gratifying when the individual fragments represent dozens of linear steps.

Another example which demonstrates the high tolerance of the Sonogashira coupling for heteroatoms is the joining of substituted iodoindole **31** and imidazole alkyne **32** in Baran's elegant biomimetic synthesis of chartelline C (**33**). In this case, the cross coupling to synthetic intermediate **34** is accomplished in a 1:1 mixture of DME and Et_3N and is complete after 7 hours at 50 °C.[83]

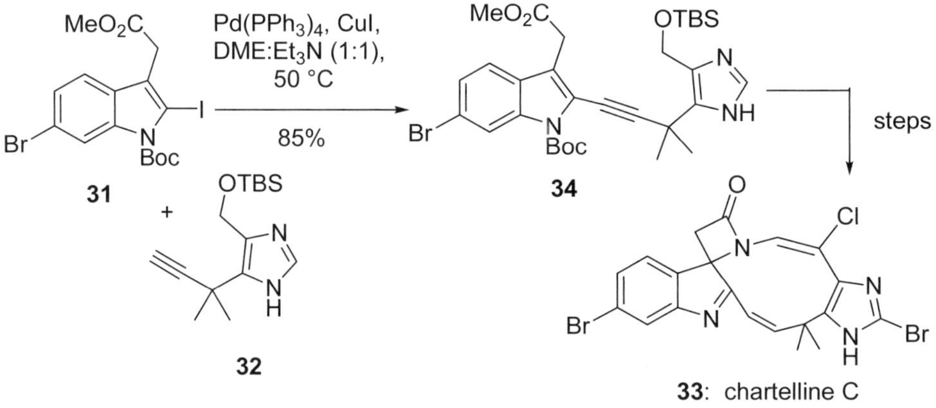

Fukuyama *et al.* accomplished penultimate step analoging of vinblastine via Sonogashira coupling. In this example, the union is carried out on the completed molecule arraying nearly a dozen different functional groups.[84] Three aryl iodides were coupled to **35** in good to excellent yields and the derived vinblastine analogs **36** screened for anti-cancer activity after removal of the TMS protecting group. When *p*-nitroiodobenzene was the coupling partner, standard copper-promoted Sonogashira conditions were adequate to afford high yield of product **36** (Ar = *p*-$PhNO_2$). In contrast, attempted coupling with *p*-methoxyiodobenzene and 2-iodopyridine afforded significant quantities of homodimerized **35**. An effective solution which conserved precious alkyne **35** was to switch to a more active catalytic system. The ability to execute this reaction at room temperature without the need for protection of these moieties showcases the mildness and generality of the Sonogashira coupling process and the effectiveness of modern conditions using tri-2-furyl-phosphine and $Pd_2(dba)_3$ with catalytic CuI in a mixture acetonitrile and triethylamine.

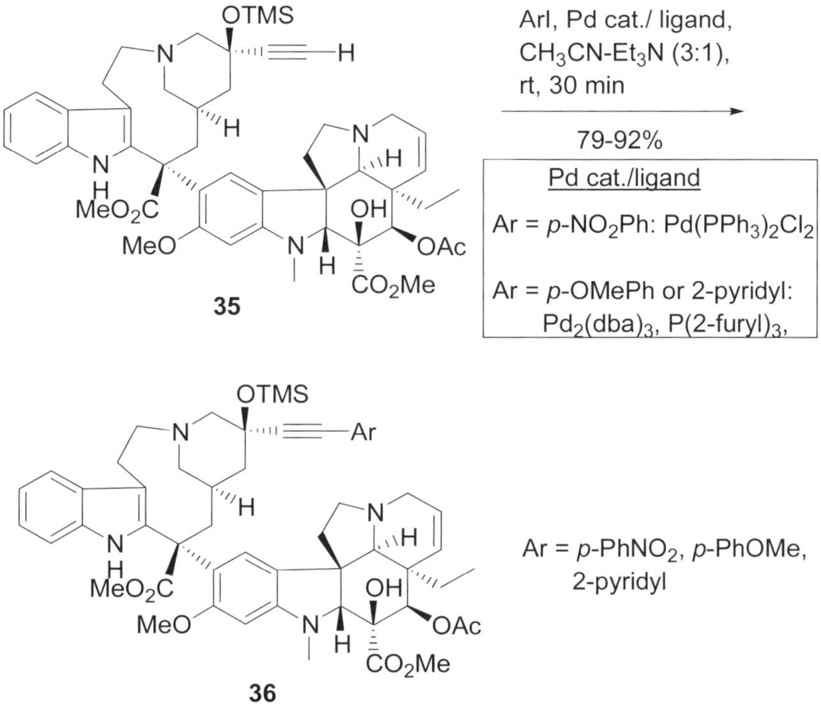

The Sonogashira coupling also finds regular use in combinatorial library synthesis protocols. Takahashi's library of clavulone analogs (**37**) was synthesized beginning with the union terminal alkyne **38** to a collection of 12 halide monomers (**39**) while the alkyne portion was bound to polystyrene resin via a THP linker. Standard Sonogashira coupling conditions in DMF at 40 °C were successful in this case, and after 24 hours, the functionalized alkynes **40** were isolated by filtration and rinse of the resin. Additional diversity was added to the immobilized cross coupling products **40** in a subsequent step, and the derived analogs cleaved from resin and purified to yield a set of 74 compounds with general structure **37**.[85]

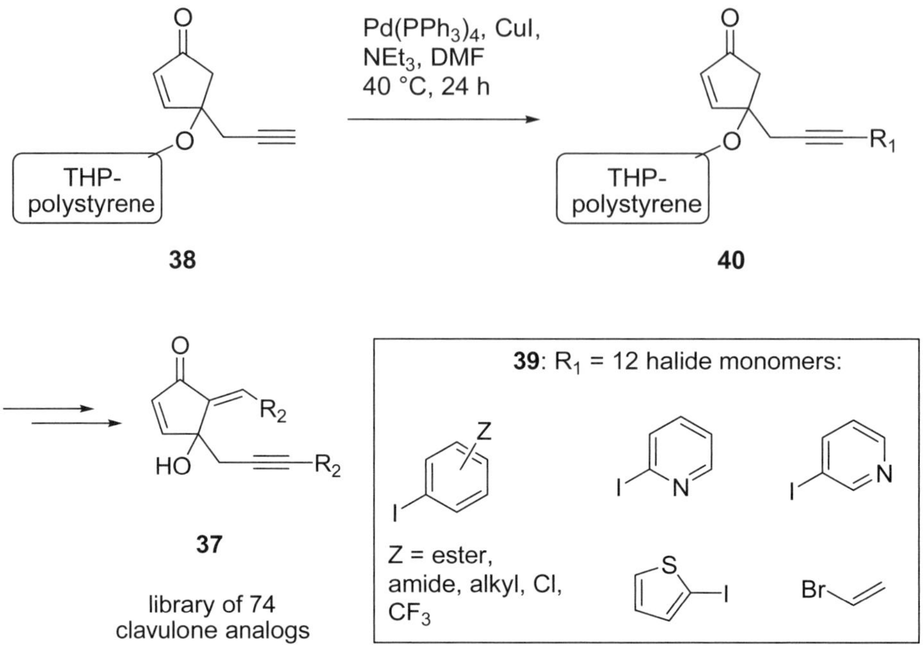

Fatty Acids and Polyunsaturated Systems

From the early 1980s to the present day, the Sonogashira coupling has constituted a frequent strategy for the synthesis of long saturated and polyunsaturated alkyl chains by coupling vinyl halides and alkynes. The product ene–yne (e.g., **41**) can be reduced to afford dienes of defined geometry, controlled by choice appropriate reduction reagents. Hydrogenation of alkynes to *Z*-olefins is often done with poisoned palladium catalysts such as Lindlars catalyst.[86] The *E*-olefins can be accessed with various complementary chemical reductions, with one of the more recent methodologies being hydrosilation/protodesilation reactions employing ruthenium catalyst and a trialkoxysilane.[87,88] Nicolaou's group completed several synthesis of biologically important members of the eicosatetraenoic acid family. A representative portion of this work, published within a decade Sonogashira's initial report, features two key ene–yne couplings using modified original conditions ([$Pd(PPh_3)_4$], CuI, amine base) in benzene at room temperature. Initially, bis-alkyne **42** was united with *Z*-vinyl bromide **43** to deliver **44**. Removal of the terminal TMS set the stage for the second high-yielding Sonogashira, this time with ester-bearing vinyl bromide **45**. From bis-alkyne **41**, a double Lindlar reduction generated the targeted *cis*-diene system and further manipulation led to the arachadonic acid metabolite **46**.[89]

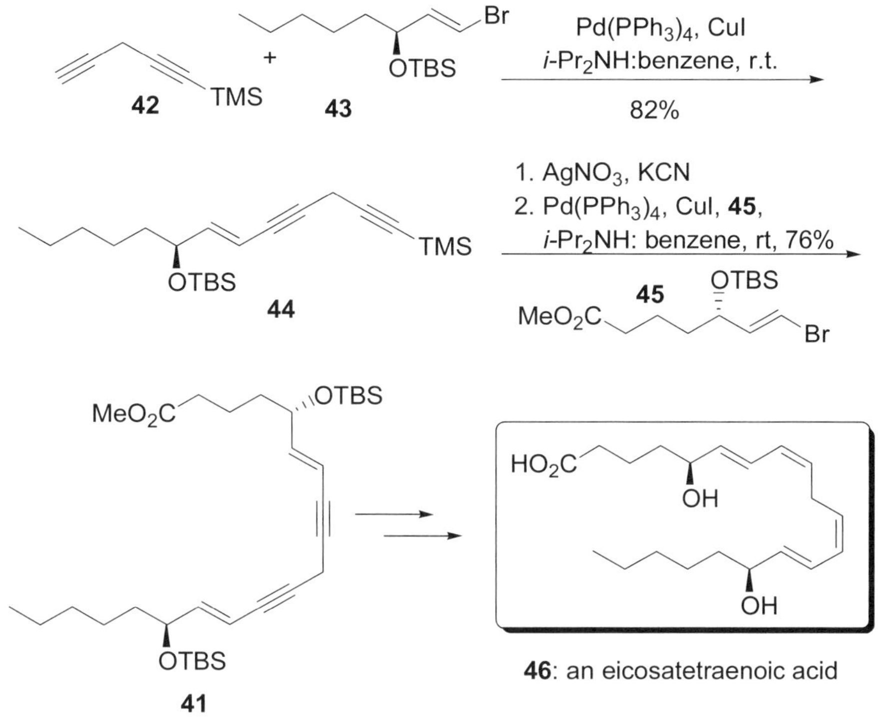

Self-assembly Subunits, Dendrimers, and Biopolymers

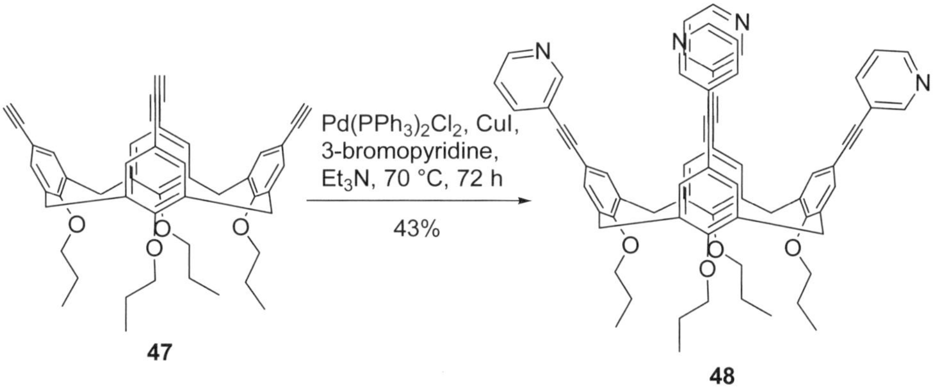

Dyker *et al.* studied the properties of cone calixarenes with aryl and pyridyl head groups. The fourfold Sonogashira coupling of **47** with several different aryl bromides was more sluggish than most, and went to completion over 72 hours under standard conditions in 24–95% yield depending on the aryl halide employed. In one example **47** was tetra-coupled with 3-bromopyridine, which reaction yielded 43% yield of host **48**, and this molecule was studied for its ability to sequester *N*-methylpyridinium and tetramethyl ammonium ions.[90]

Nanostructures formed from spontaneous assembly of extended arylene-ethylene macrocycles (e.g., **49**) have special properties by virtue of the non-collapsible and highly delocalized nature of this bond connection.[91] The individual macrocycle units can get quite complex, containing a dozen or more alkynes, so not surprisingly, many synthetic routes to molecules in this classification feature the Sonogashira coupling. In a relatively simple example, She, Pan, and co-workers found that the strongly deactivated aryl iodide **50** would only afford unwanted alkyne–alkyne coupling material in THF and DMF. Switching solvents to the ionic liquid 1-butyl-3-methylimidazolium tetrafluoroborate **51** had a pronounced accelerating effect on the cross coupling and allowed the copper loading to be reduced to 1 mol%, which, in turn, reduced the formation of alkyne dimers and allowed for isolation of desired cyclic product **49**, albeit in low yield. With less electron-rich aryl iodides, coupling yields were greatly improved, and in such cases, THF as solvent delivered the product in modest yields in the absence of the molten salt.[92]

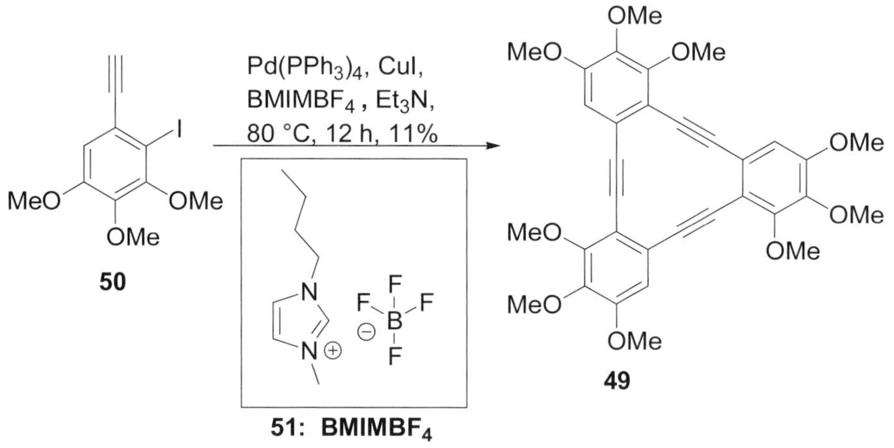

The compatibility of the Sonogashira coupling with multiple functionalities has provided some unique applications for conjugating organic compounds with biopolymers. One of the motivations for an ongoing push to extend Sonogashira coupling to systems which function well in water, is the desire to apply this reaction to the derivatization of biological molecules.[19] The large ring in the 21 amino acid cyclic peptide **52** was closed via an intramolecular Sonogashira coupling of **53**. The peptide (**53**) was synthesized on RINK-resin using a combination of solution phase and resin-bound peptide synthesis techniques. Immobilization of **53** served an additional purpose of preventing intermolecular reactivity in the coupling of the large ring. In fact, attempted off-resin macrocyclization of the same peptide was unsuccessful. Exposure of **53** to typical Sonogashira conditions in refluxing triethylamine followed by global peptide deprotection (with concomitant release off the bead) and HPLC purification, gave the desired product **52**. This compound (**52**) was designed to be a loop mimic of human immunoglobulin E with potential therapeutic application.[93]

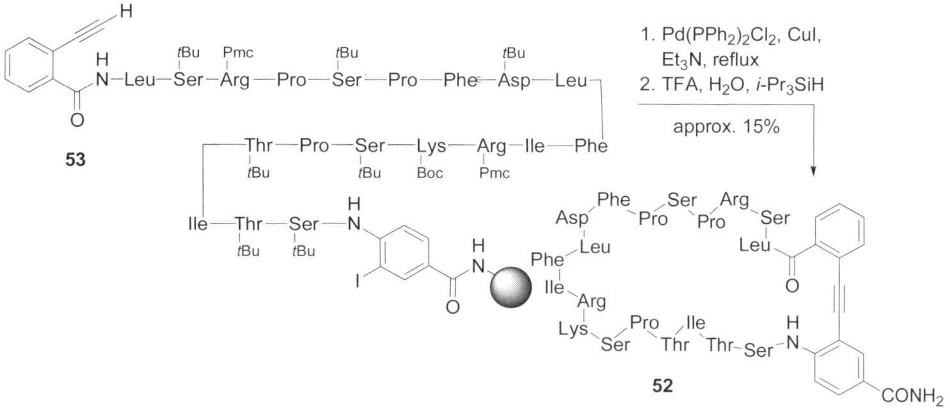

Compounds with Electronic and Optical Properties

A highly enabling aspect of the Songashira reaction is its ready applicability to classes of rigid, extended phenylene alkynes. These highly conjugated compounds have intriguing electronic and optical characteristics deriving from the extended delocalization of the rigid π-system. Frequently, multiple arene groups are arrayed in series to obtain specific properties. A representative example is the synthesis of polythiophene **54**, which relies upon two Sonogashira couplings for key bond formations. In the first coupling, dibromophenol **55** is a challenging substrate, but it was cleanly bis-coupled to TIPS-acetylene **56** at room temperature using Fu's phosphane salt conditions—tri-*t*-butyl phosphine tetrafluoroborate salt and [$Pd(PPh_3CN)_2Cl_2$)] with CuI and *i*-Pr_2NH in dioxane, providing phenol intermediate **57**. The advantage of the phosphine salt is that it is air stable and more easily handled than the oxygen-sensitive tri-*t*-butyl phosphine.[94] Following base-promoted removal of the silyl groups, a second double Sonogashira with polythiophene iodide **58** was accomplished using more traditional conditions. Compound **54** and several similar constructs were then characterized for their absorption, fluorescence, quantum yield, and lifetime.[95] Facile modular construction is advantageous when looking to fine tune these semi-empirical properties.

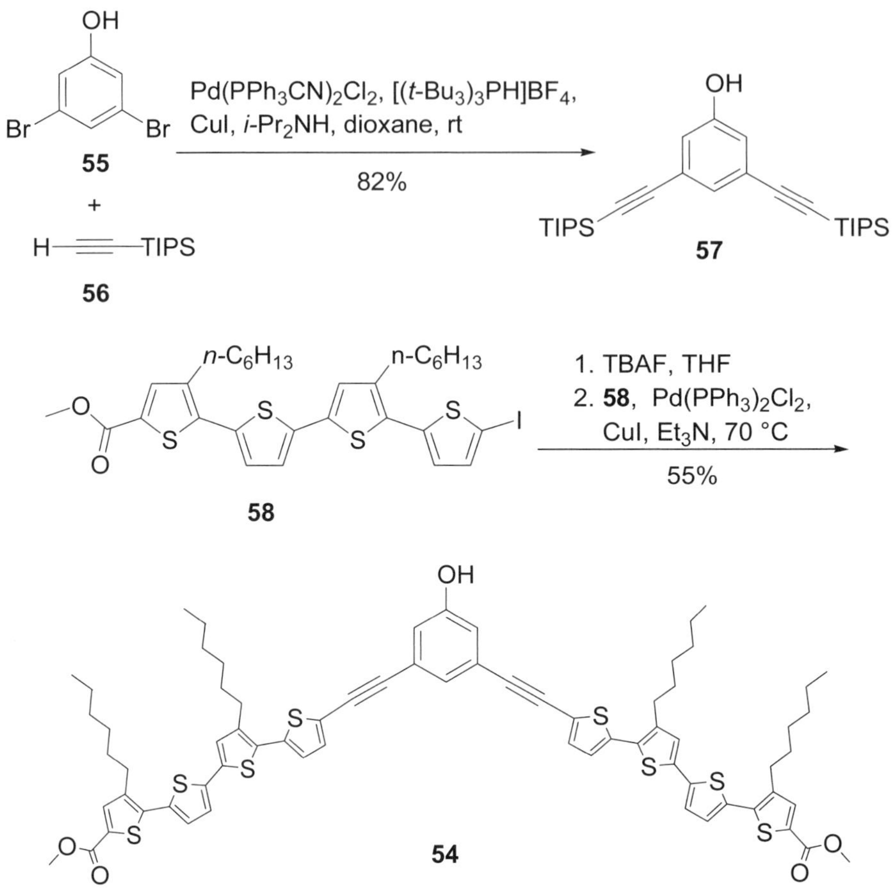

Heterocycle Synthesis

The Sonogashira coupling is firmly entrenched as an integral reaction for heterocycle formation.[52–54] The coupling of an alkyne to an appropriately functionalized molecule can set it up for intermolecular heterocyclization (**61** → **65**) where the employed metals can serve to activate the alkyne toward the cyclization event and yield operationally simple one-pot procedures. The most general application of this approach forms a new 5-membered aromatic system, often fused to another ring system. After standard reaction of **59** and **60**, the transformation from **61** to thiazolo-1,2,4-triazole **65** occurs in one pot following an initial standard condition Sonogashira coupling with **62**, and in this instance, the heterocyclization did not occur in the absence of copper salt. A plausible mechanism might involve engagement of a metal-activated alkyne (**63**), followed by favourable base-promoted isomerisation of **64**.[96] Owing to broad toleration for unprotected heteroatoms, the Sonogashira is well suited for the task of installing the requisite alkyne with a minimal protecting groups.

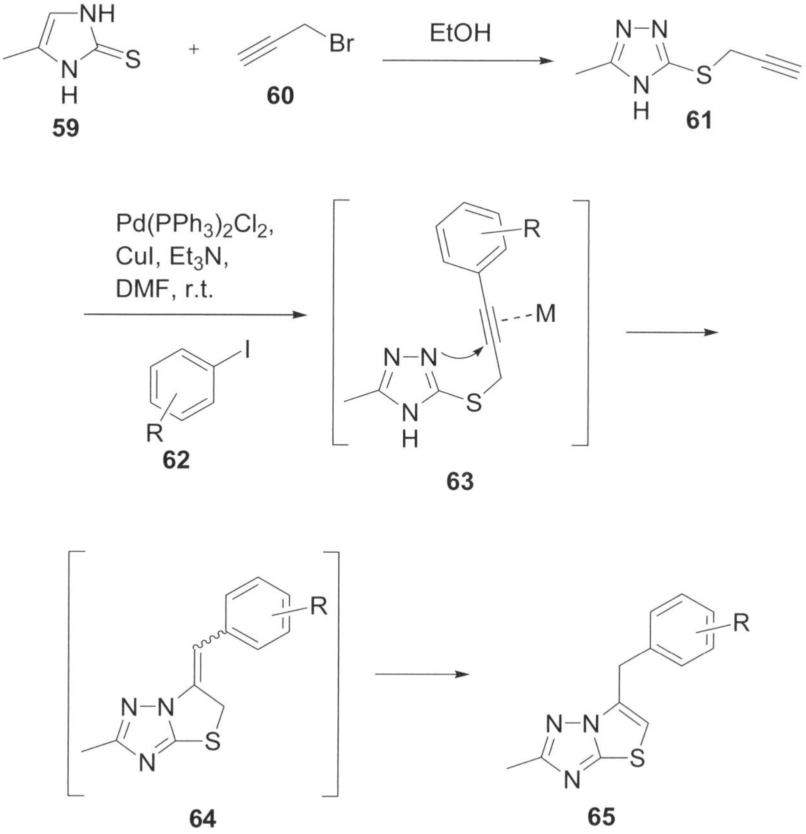

For some of the fused heterocycles, the coupling/cyclization protocol affords one of the easiest entries into that particular substituted heterocycle.

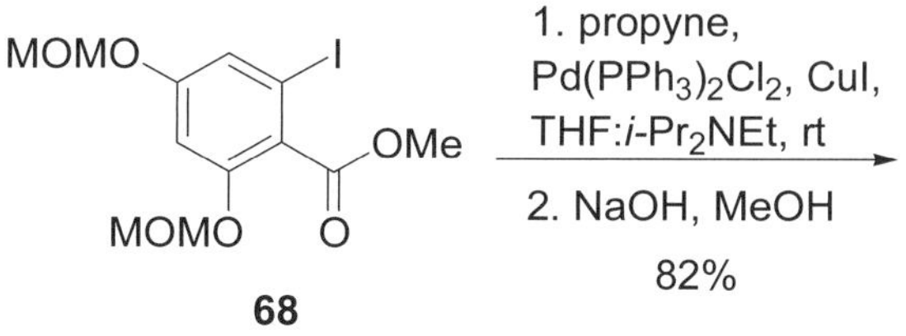

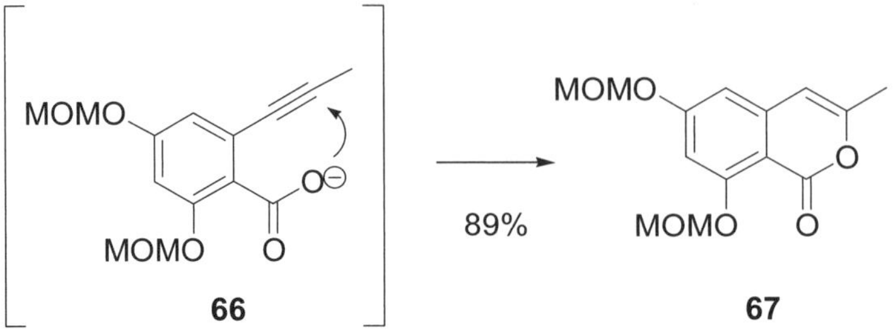

Recent work on the synthesis of Cassarin demonstrates the formation of an isocoumarin (**67**) in a base-promoted 6-*endo*-dig cyclization of acid **68** onto the just-coupled alkyne. The stage is set for this reaction when aryl iodide **68** reacts with *in situ* generated propyne in a Sonogashira coupling catalyzed by [$Pd(PPh_3)_2Cl_2$] and cuprous iodide in THF–*i*-Pr_2NH at room temperature.[97]

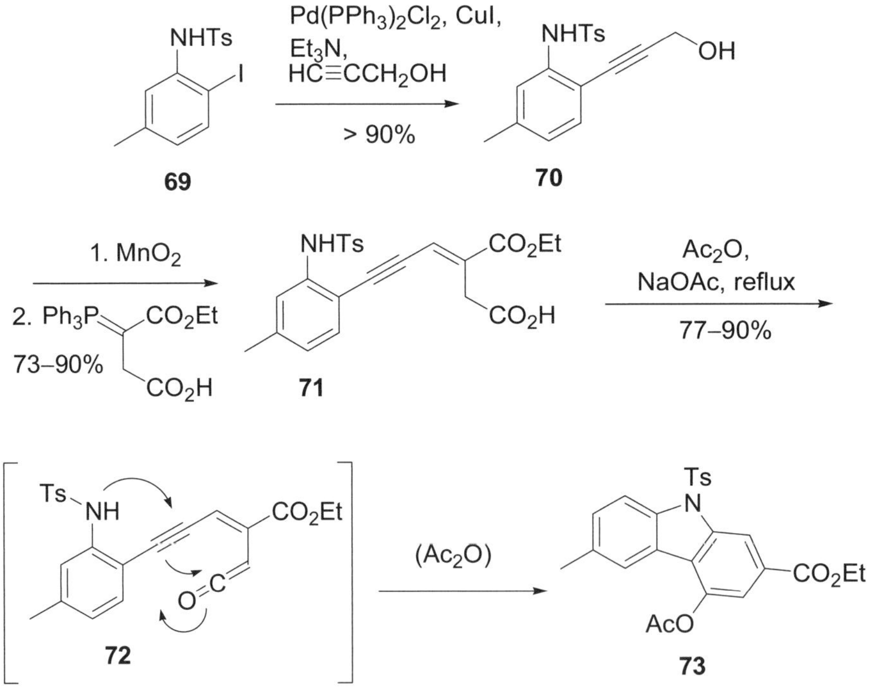

A nice substituted carbazole synthesis begins with installation of propargyl alcohol via its Sonogashira coupling with aryl iodide **69**. Subsequent oxidation of propargylic alcohol **70** to the aldehyde, followed by stabilized Wittig reaction, affords a conjugated intermediate ene–yne **71**, which undergoes a cascade cyclization upon treatment with acetic anhydride, proceeding via ketene ene–yne **72** to carbazole **73** in 76% overall yield.[98]

Emerging Applications – Bioconjugates

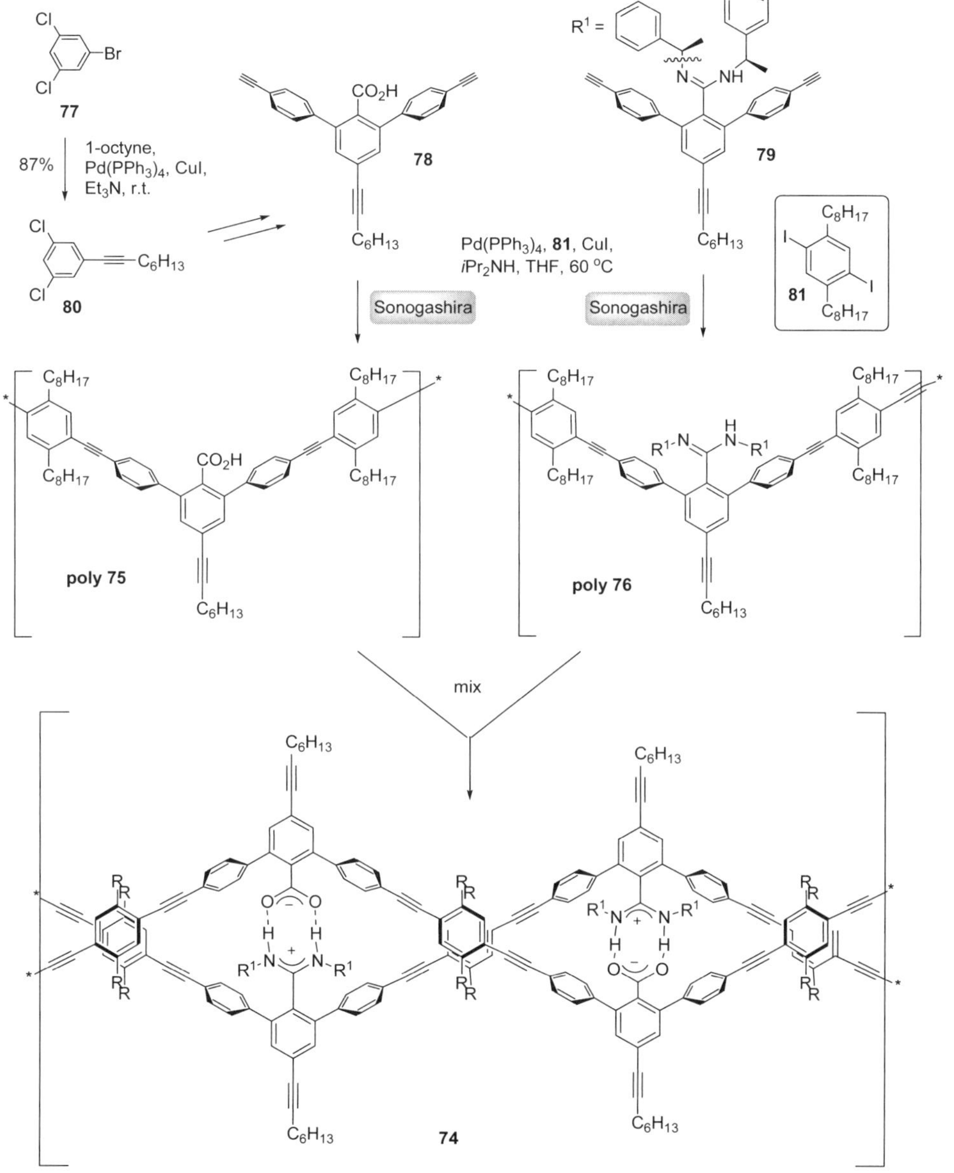

The structurally impressive DNA mimic **74** was designed around a pair of oligomeric constructs (poly-**75** and poly-**76**) that hold complementary sets of lipophilic and hydrogen bonding interactions. The Sonogashira reaction features prominently in the construction of these large molecules, serving first to attach octynyl chains to trihalo precursor **77**. The dichloro product **78** is further functionalized to monomers **79** and chiral monomer **80**. The oligomerization is accomplished by cross-linking these monomers in a series of double Sonogashira couplings between diiodophenyl linker **81** and bisacetylene **79** (and, in a separate pot, **80**). The oligomers **75** and **76** were found to have average molecular weights of 24,000 and 38,000 respectively and formed a chiral helical structure (**74**) upon mixing under controlled conditions. The Sonogashira couplings were performed under very typical reaction conditions employing Pd-tetrakistriphenylphosphine and CuI with mixtures of triethylamine and either toluene or THF as solvent.[99]

There are numerous examples of harnessing ene–yne coupling chemistry to assist with the characterization a biological system such as the Sonogashira coupling of ferrocene acetylene chromophore **82** to a uracil-derived bromide **83**. This reaction proceeded smoothly in DMF at room temperature under otherwise standard conditions. After isolation, the tagged uracil–ferrocene conjugate **84** could conceivably serve as a component of a bioelectronic gene-sensing system.[100]

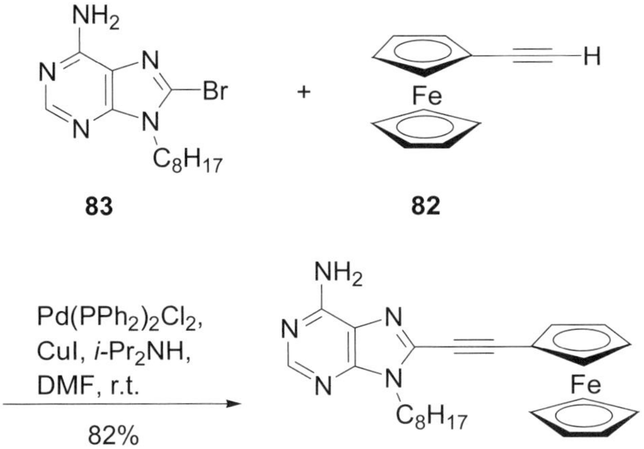

83 **82** **84**

Polymers

Designer polyene/yne oligomers of the types exemplified by compounds **54** and poly-**75** and poly-**76** are often synthesized in sequences containing Sonogashira couplings. Applications for rigid, polymeric structures based on polyaryleneethynylenes are emerging, however, true industrial scale polymer synthesis has not been achieved, due either to lack of demand or unavailability of the extremely high TON catalyst systems and ultra clean reactions that are typically necessary for cost-effective large-scale polymerization. Polyphenylene–ethynylene with average molecular weight of 4300 was synthesized by polymerization of diiodobenzene and acetylene gas in a 3:1 mixture of acetonitrile and water under $Pd(OAc)_2/PPh_3/CuI/Et_3N$ conditions, but overall, these conditions are not ready for serious polymer synthesis.[101]

1.1.5.5 Variations and Improvements

For the majority of applications, the chemist is looking to the Sonogashira coupling to afford mild, reliable bond construction. In this context, the basic CuI-promoted conditions worked out early in the evolution of this reaction remain a viable option. Key areas for additional development are increasing catalyst turnover, maintaining lower reaction temperatures with less activated systems, and developing generally effective recyclable catalysts.[8]

Ligands with the Highest Activity

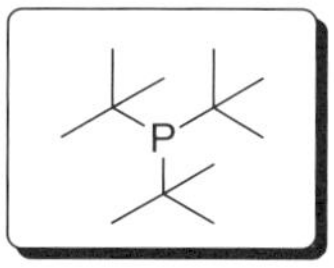

85

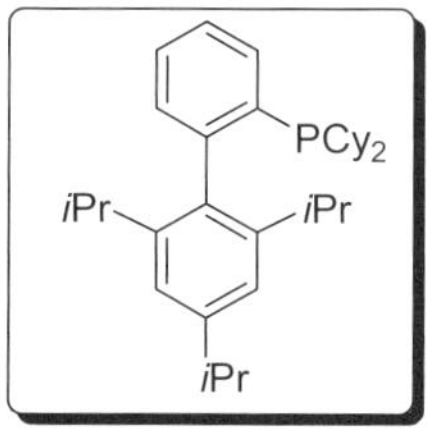

86

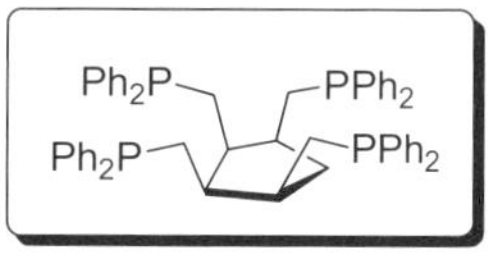

87

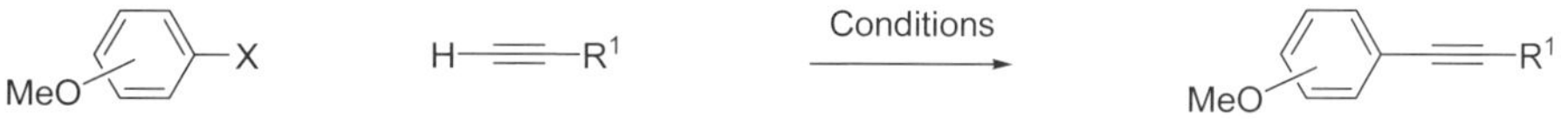

88 89 90

Entry	Ligand	X	88	89	Pd	CuI?	Base	Solvent	Temp °C	%Yield	TON
1	**85**	Br	*p*-OMe	*n*Bu	$Na_2[PdCl_4]$	Yes	$i Pr_2NH$	$i Pr_2NH$	80	85	850
2	**85**	Br	*o*-OMe	Ph	$Na_2[PdCl_4]$	Yes	$i Pr_2NH$	$i Pr_2NH$	80	87	17400
3	**85**	Cl	*p*-OMe	Ph	$Na_2[PdCl_4]$	Yes	Na_2CO_3	xylene	120	75	41
4	**86**	Cl	*p*-OMe	*t*Bu	$PdCl_2(MeCN)_2$	No	Cs_2CO_3	MeCN	95	89	870
5	**86**	Cl	*o*-OMe	Ph	$PdCl_2$(MeCN)2	No	Cs_2CO_3	MeCN	95	93	950
6	**87**	Br	*p*-OMe	$CH(OEt)_2$	$(Pd(C_3H_5)Cl)_2$	Yes	K_2CO_3	DMF	130	>50	8200
7	**87**	Cl	*p*-OMe	Ph	$(Pd(C_3H_5)Cl)$2	No	K_2CO3	DMF	140	31	155

Much of the ligand evolution that has occurred in the field of Pd catalyzed cross couplings was done without, perhaps, the Sonogashira coupling as a main focus. The mechanistic insights which led to the introduction of phosphanes with large cone angles and high σ-donating ability for promoting Heck and Suzuki couplings of aryl chlorides at room temperature were also successfully applied to the similarly recalcitrant Sonogashira with aryl chlorides.[28,102–104] Among the more successful of these monodentate phosphanes for the Sonogashira coupling is tri-*t*-butylphosphine (**85**) which is able to handle traditionally difficult electron-rich aryl bromides like 4-bromoanisole (entry 1) as well as aryl chlorides at elevated reaction temperatures (entries 2 and 3). Buchwald's hindered biphenyl dicyclohexyl phosphine (X-Phos, **86**) is perhaps even more capable, giving good turnover even with electron rich aryl chlorides (entries 4 and 5) and the first example of Sonogashira reaction with aryl tosylates.[28] With this latter catalyst, the addition of CuI was found to hamper the reaction, a finding which has been duplicated in other catalytic systems for the Sonogashira couplings of aryl chlorides (see "CuI?" column in the table).

The effectiveness of ligands **85** and **86** derives from acceleration of slow oxidative addition to aryl halides. As an additional benefit, these powerful ligands form product rapidly and can essentially shut down the homodimerization pathway even in the presence of CuI.

Apart from the ability to couple previously inaccessible substrates, a valuable measure of ligand effectiveness is derived by comparing turn-over numbers (TONs) for a benchmark reaction such as the Sonogashira coupling of 4-bromoanisole and phenylacetylene. It is by this measure that these two catalysts really distinguish themselves. An examination of the table highlights the precipitous drop in TON that occurs with nearly all catalytic systems in going from more reactive halides to less reactive substrates. It important to recognize that very few catalytic systems are capable of accessing **90** via Sonogashira coupling of substrates **88** (X = Br, Cl) and **89**. For favourable reactions such as the coupling of diiodobenzene and phenylacetylene (**7** and **8**), a number of catalytic systems including mono- and multidentate phosphanes, palladacycles, carbene ligands, and ligandless procedures are able to reach TON's of 100,000 or more. With more difficult substrates, however, the validated bulky monophosphines such as **85** and **86**, together with a few similar ligands stand in a class by themselves.[8]

The tetradentate phosphine **87** is of note, recording high TONs for the coupling of 4-bromo- and 4-chloroanisole (see table entries 6 and 7). Beyond the various phosphanes, *N*-heterocyclic carbenes ligands (NHC's) have also received much attention and have been successfully demonstrated in both typical and challenging Sonogashira couplings, though in general, these ligands have not demonstrated advantages over **85** or **86**.[105,106] One notable example of the budding potential of the strongly donating carbene ligands is the reaction of *alkyl* bromide **91**, and 1-hexyne (**92**) in a DMF–Et_2O mixture to afford alkynol **93**. To achieve this impressive result, 7.5 mol% of Pd and CuI were required. Interestingly, Fu showed that imidazolium carbene ligand **94** was uniquely able to minimize competing (undesired) β-hydride elimination from the Pd(II) complex where other high-performance phosphanes ligands failed.[29] Glorius similarly showed that the NHC ligand **95** also promotes Sonogashira coupling of alkyl halides under mild conditions.[107]

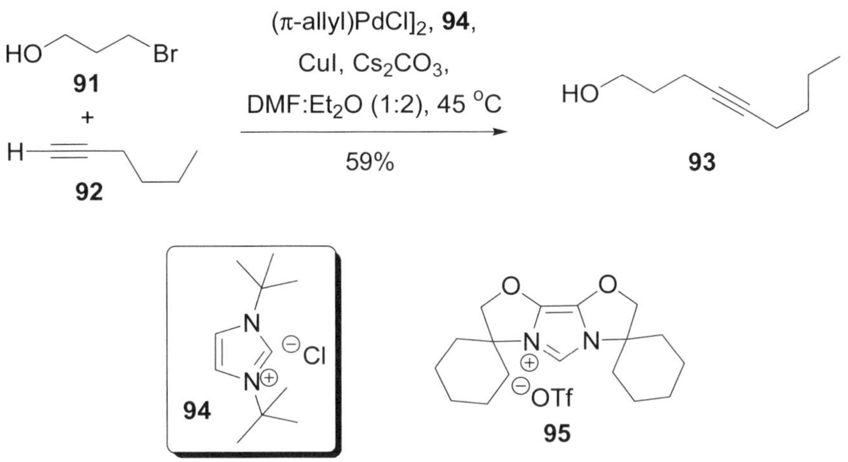

"Ligand-free" Catalysts

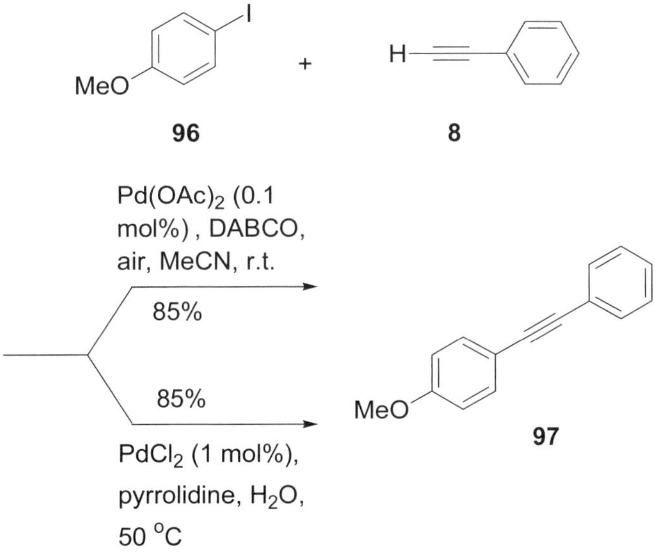

While there are open questions about the exact nature and processing of the molecular or colloidal Pd which may operate under the so called "ligand-free" conditions, there is not doubt that stabilized Pd nanoparticles are highly capable of catalyzing Sonogashira couplings. Li reported that DABCO is a superior amine additive when compared to tetrabutylammonium salts in widespread use for stabilizing Pd(0) in "ligand-free" cross couplings.[33,34] For example, *p*-iodoanisole (**96**) reacts with phenylacetylene (**8**) in the presence of 0.1 mol% $Pd(OAc)_2$ and DABCO in acetonitrile to afford the coupled product **97** in 85% yield. Similarly, 1 mol% $PdCl_2$ catalyzes the same reaction in water at 50 °C when 5 equivalents of pyrrolidine are added.[32] This result if of interest, however, the reaction fails with more challenging aryl bromides. The potential advantages of lower cost and simpler work-up procedures are mitigated by the limited scope, higher palladium loadings, and the stabilizing additives (PPh_3, tetraalkylammonium salts) that have been necessary for with many ligand-free protocols reported to date.

Heterogeneous "ligand-free" reagent systems have been employed with some success, though they too have yet to demonstrate the power and generality of the bulky phosphanes in coupling alkynes to unreactive halides. Common palladium sources including Pd/C and $Pd(OH)_2$ provide an environmentally friendlier way to accomplish the Sonogashira transformation because the majority of the catalyst can be recovered and often recycled after filtration from the reaction mixture. For the coupling of 3-bromopyridine (**98**) and butyne-ol **99**, a mixture of $Pd(OH)_2$, CuI, K_2CO_3, and PPh_3 in DME-water at 80°C for 21 hours resulted in an 84% yield of the pyridyl alkyne **100**. The elevated temperature and longer reaction time is evidence of the lower level of activity often seen with the heterogeneous catalysts. In this system, the role of the copper salt and phosphane have not been fully elucidated.[34]

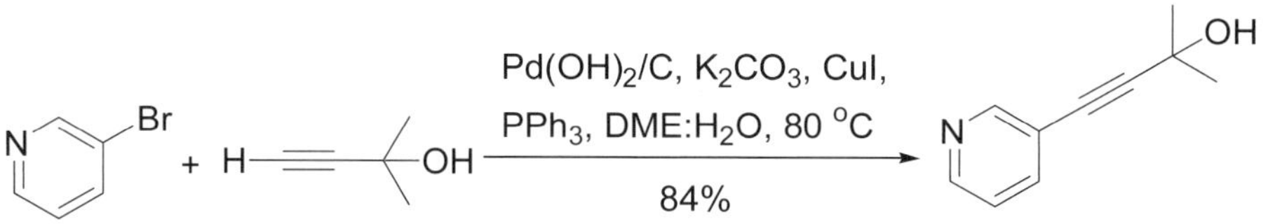

98 **99** **100**

One approach towards more efficient use of precious metals is to engineer surfaces coated with palladium. In one embodiment of this concept, chemically-controlled deposition of Pd and Ni results in defined particles with Pd-rich shells built up on inexpensive nickel cores (**101**).[108]

102 **8**

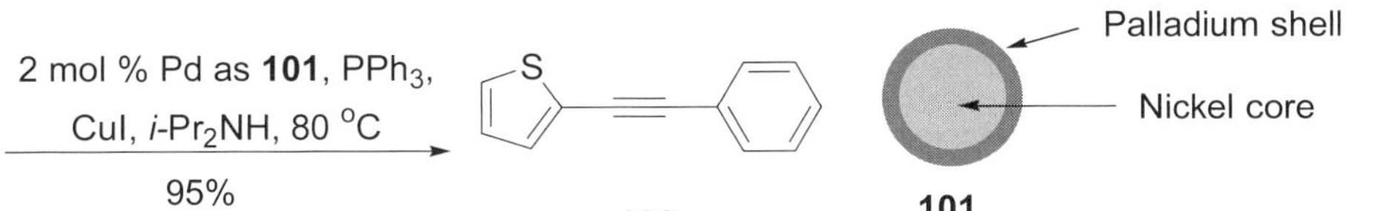

103 **101**
palladium/nickel bimetallic nanoparticle

This technology has been applied to aryl-alkyne cross coupling, where thiophene bromide **102** couples with phenylacetylene (**8**) in a reaction mix consisting of Pd/Ni bimetallic nanoparticulate catalyst **101** (approximtely 2 mol% palladium), PPh_3, and CuI in *i*-Pr_2NH. Coupling occurs in two hours at 80 °C in high yield, and the nanoparticules can be filtered and recycled. It is not always clear which heterogeneous catalysts merely constitute a reservoir for slow release of colloidal/soluble Pd into solution and which actually have catalytically active sites immobilized.[74] An important application of truly immobilized catalyst would be in the synthesis of active drug ingredients, with the potential to eliminate contamination of product by low ppm quantities of toxic metal. If, however, if the true active catalysts are some type of soluble palladium, there will be limited pharmaceutical application for those ligand-free protocols.

Alternative Solvents – Aqueous Reactions

The Sonogashira reaction proceeds in an admirably wide range of solvents given the correct catalytic system, and water is no exception. Numerous reports have surfaced showcasing this coupling in water using phase transfer agents, or microwave heating. To date, the best reactivity is seen in mixed solvent systems and substrate scope if limited, however, the potential advantage of facile catalyst recovery continues to drive progress in this area. An interesting extension of substrate scope came with the introduction of biarylphosphine ligand **104** bearing a solubilising sulphate salt. The non-sulfonlylated version of this ligand was able to promote Sonogashira couplings with aryl chlorides, and the addition of the sulphate allows for the coupling of water soluble alkynes in aqueous mixtures. Thus, acid **105** and 3-chloroanisole (**106**) unite to generate acid **107** in high yield when heated to 100 °C in a 1:1 MeCN–H_2O mixture with ligand **104**, [$PdCl_2(MeCN)_2$], and Cs_2CO_3.[109]

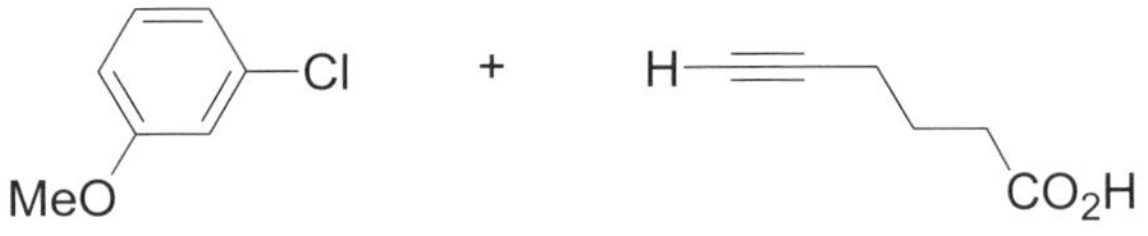

106 **105**

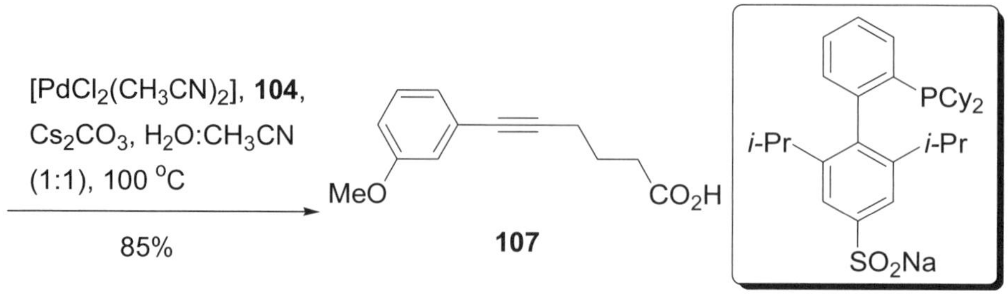

107 **104**

Non-traditional Coupling Partners

Under the broad umbrella of the Sonogashira coupling, less common substrates for both the alkyne and halide partners have been described. There are examples of Sonogashira-type coupling of terminal acetylenes to benzylic and primary bromides and iodides, and even secondary bromides. There are also limited examples of aryl tosylates and sulfonates (so called pseudo-halides) entering into Sonogashira reaction.[107,110] The union of phenylacetylene and acid chlorides (e.g., benzoyl chloride) catalyzed by Pd/C in refluxing toluene–Et_3N has been reported to proceed in excellent yield.[111,112] For the alkyne component, Mori demonstrated that terminal silanes can couple with a range of aryl halides using 5 mol% [$Pd(PPh_3)_4$] and CuCl in DMF at moderately elevated temperature.[113–115] One can envision an effective strategy for sequential coupling of different halides using this sila-Sonogashira. Molander's alkyne tetrafluoroborate salts afford an alternative related entry into mild reaction systems.[116] Beyond the examples discussed here, there are additional isolated examples of unusual reaction partners.[8]

Other Metal Catalysts

Aluminium, Zinc, Nickel, Ruthenium, and other metal acetylides will transmetalate to Pd to afford a species which can productively couple to aryl and vinyl halides.[8] There is a surge in methods which do not require expensive palladium, in particular, copper- and nickel-based methodologies.[117] The so called "palladium-free" Sonogashira begins to look like a Castro–Stephens coupling when promoted by copper. Indeed a host of other methods are available to effect the net transformation of generating an ene-yne from an acetylide and aryl or vinyl halide. These processes begin to stray from the already large categorical umbrella of the Sonogashira coupling and will not be discussed in this chapter.

1.1.5.6 Experimental

Sonogashira Reaction using Modification of Original Conditions

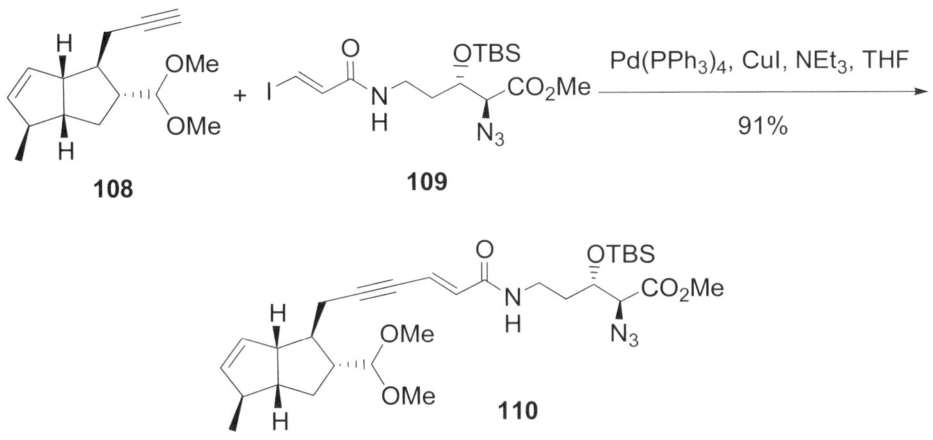

Cylindramide intermediate (110).[118]

Under argon gas, a solution of compound **108** (630 mg, 1.3 mmol) and compound **109** (276 mg, 1.18 mmol) in anhydrous THF (1.6 mL) was added to a stirred suspension of [Pd(PPh_3)$_4$] (19 mg, 0.016 mmol, 1.3 mol%) and copper(I) iodide (9 mg, 0.047 mmol, 4 mol %) in Et_3N (2.8 mL) and the reaction mixture stirred at room temperature. After 1 hour, the solvent was removed under vacuum and the residue purified by silica gel chromatography (petroleum ether/EtOAc, 3:1) to give compound **110** (630 mg, 91%, > 95% NMR purity) as a yellow resin.

Sonogashira Reaction using Bulky Phosphine Ligand and Copper-Free Conditions

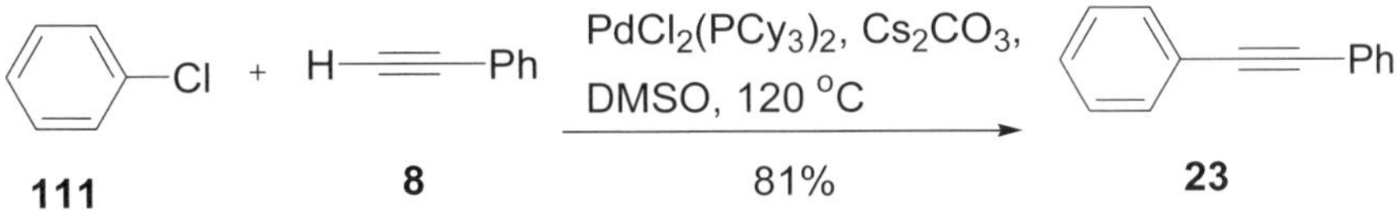

Diphenylacetylene (23).[119]

A mixture of chlorobenzene (**111**, 75.0 mg, 0.66 mmol), phenyl acetylene (**8**, 61.5 mg, 0.6 mmol), Cs_2CO_3 (230.0 mg, 0.66 mmol), [$PdCl_2(PCy_3)_2$] (15.4 mg, 0.02 mmol), and DMSO (0.8 mL) under nitrogen in a sealed tube was heated with stirring at 120 °C for 12 h. After cooling to room temperature, the reaction mixture was diluted with CH_2Cl_2 to 1.5 mL total volume and octadecane (45 mg, 0.18 mmol) was added as an internal standard for GC analysis. After GC–MS analysis, the solvents and volatiles were removed under vacuum and the residue was subjected to preparative TLC isolation (silica, eluted with cyclohexane). Compound **23** was obtained (85.5 mg, 0.48 mmol, 81%) as a white solid.

1.1.5.7 References

1. Sonogashira, K.; Tohda, Y.; Hagihara, N. *Tetrahedron Lett.* **1975**, *16*, 4467.
2. Cassar, L. *J. Organomet. Chem.* **1975**, *93*, 253.
3. Dieck, H.; Heck, F. *J. Organomet. Chem.* **1975**, *93*, 259.
4. [R] Sonogashira, K. In *Metal-Catalyzed Cross-Coupling Reactions*; Diederich, F.; de Meijere, A., Eds.; Wiley-VCH: Weinheim, **2004**; Vol. 1, 319.
5. [R] Sonogashira, K. In *Comprehensive Organic Synthesis*; Trost, B. M., Fleming, I., Eds.; Pergamon: Oxford, 1991; Vol. 3, p 521.
6. [R] Sonogashira, K. *J. Organomet. Chem.* **2002**, *653*, 46.
7. [R] Chinchilla, R.; Najera, C. *Chem. Rev.* **2007**, *107*, 874.
8. [R] Doucet, H.; Hierso, J-C. *Angew. Chem., Int. Ed.* **2007**, *46*, 834.
9. [R] Negishi, E.; Anastasia, L. *Chem. Rev.* **2003**, *103*, 1979.
10. [R] Tykwinski, R. R. *Angew. Chem., Int. Ed.* **2003**, *42*, 1566.
11. [R] Brandsma, L. *Synthesis of Acetylenes, Allenes and Cumulenes: Methods and Techniques*; Elsevier: Oxford, 2004; p293.
12. [R] Rossi, R.; Carpita, A.; Bellina, F. *Org. Prep. Proced. Int.* **1995**, *27*, 127.
13. [R] Farina, V. *Adv. Synth. Catal.* **2004**, *346*, 1553.
14. [R] Herrmann, W.; Öfele, K.; von Preysing, D.; S.; Schneider, S. *J. Organomet. Chem.* **2003**, *687*, 229.
15. [R] Fu, G.; Littke, A. *Angew. Chem., Int. Ed.* **2002**, *41*, 4176.
16. [R] Jutand, A. *Pure Appl. Chem.* **2004**, *76*, 565.
17. [R] Shi Shun, A.; Tykwinski, R. *Angew. Chem., Int. Ed.* **2006**, *45*, 1034.
18. [R] Hermann, W. *Angew. Chem., Int. Ed.* **2002**, *41*, 1290.
19. [R] Gauthier, M.; Klok, H.-A. *Chem. Commun.* **2008**, *23*, 2591.
20. [R] Singh, B.; Kaval, N.; Tomar, S.; Van der Eycken, E.; Parmar, V. *Org. Proc. R&D.* **2008**, *12*, 468.
21. [R] Peris, E.; Crabtree, R-H. *Coord. Chem. Rev.* **2004**, *248*, 2239.
22. Zheng, S.-L.; Reid, S.; Lin, N.; Wang, B. *Tetrahedron Lett.* **2006**, *47*, 2331.
23. Organ, M. G.; Ghasemi, H. *J. Org. Chem.* **2004**, *69*, 695.
24. Raju, S.; Batchu, V. R.; Swamy, N. K.; Dev, R. V.; Babu, J. M.; Kumar, P. R.; Mukkanti, K.; Pal, M. *Tetrahedron Lett.* **2006**, *47*, 83.
25. Huang, Q.; Hunter, J.; Larock, R. *J. Org. Chem.* **2002**, *67*, 3437.
26. [R] Nicolaou, K. C.; Bulger, P.; Sarlah, D. *Angew. Chem., Int. Ed.* **2005**, *44,* 4442.
27. [R] Bunz, U. *Chem. Rev.* **2000**, *100*, 1605.
28. Gelman, D.; Buchwald, S. *Angew. Chem., Int. Ed.* **2003**, *42*, 5993.
29. Eckhardt, M.; Fu, G. *J. Am. Chem. Soc.* **2003**, *125*, 13642.
30. Shi, S.; Zhang, Y. *Synlett* **2007**, *12*, 1843
31. Liang Bo; D.; Mingji; C.-J.; Yang, Z. *J. Org. Chem.* **2005**, *70*, 391.
32. Urgaonkar, S.; Verkade, J. *J. Org. Chem.* **2004**, *69*, 5752.
33. Li, J.; Zhang, X.; Xie, Y. *Synthesis* **2005**, 804.
34. Li, J.; Hu, X.; Liang, Y.; Xie, Y. *Tetrahedron* **2006**, *62*, 31.
35. Alami, M.; Crousse, B.; Ferri, F. *J. Organomet. Chem.* **2001**, *624*, 114.
36. Mori, Y.; Seki, M. *J. Org. Chem.* **2003**, *68*, 1571.
37. Thorand, S.; Krause, N. *J. Org. Chem.* **1998**, *63*, 8551.
38. Elangovan, A.; Wang, Y.-H.; Ho, T.-I. *Org. Lett.* **2003**, *5*, 1841.
39. Traschsel, D. *Helv. Chim. Acta* **2003**, *86*, 2754.
40. Nguefack, J.-F.; Bolitt, V.; Sinou, D. *Tetrahedron Lett.* **1996**, *37*, 5527
41. Rossi, R.; Carpita, A.; Lezzi, A. *Tetrahedron* **1984**, *40*, 2773
42. Bhattacharya, S.; Sengupta, S. *Tetrahedron Lett.* **2004**, *45*, 8733

43. Rossi, R.; Carpita, A.; Lezzi, A. *Tetrahedron* **1984**, *40*, 2773.
44. Bhattacharya, S.; Sengupta, S. *Tetrahedron Lett.* **2004**, *45*, 8733.
45. Nakamura, K.; Okubo, H.; Yamaguchi, M. *Synlett*, **1999**, 549.
46. Masai, H.; Sonogashira, K.; Hagihara, N. *J. Organomet. Chem*, **1971**, *26*, 271.
47. Sonogashira, K.; Takahashi, S.; Hagihara, N. *Macromolecules* **1977**, *10*, 879
48. Hong, P.; Sonogashira, K.; Hagihara, N. *Tetrahedron Lett.* **1970**, 1633
49. Böhm, V.; Herrmann, W. *Eur. J. Org. Chem.* **2000**, 3679.
50. Genêt, J.-P.; Savignac, M. *J. Organomet. Chem.* **1999**, *576*, 305.
51. [R] (a) *Electronic Materials: The Oligomer Approach*; Müllen, K., Wegner, G., Eds.; Wiley-VCH: Weinheim, 1998. [R] (b) Nalwa, H. S.; Miyata, S. *Nonlinear Optics of Organic Molecules and Polymers*; CRC Press: Boca Raton, FL, 1997.
52. Sorensen, U. S.; Pombo-Villar, E. *Tetrahedron* **2005**, *61*, 2697
53. Bach, T.; Bartels, M. *Synthesis* **2003**, 925.
54. Zoppellaro, G.; Baumgarten, M. *Eur. J. Org. Chem.* **2005**, 2888.
55. Martin, R. E.; Diederich, F. *Angew. Chem., Int. Ed.* **1999**, *38*, 1350
56. Breitenkamp, R. B.; Arnt, L.; Tew, G. N. *Polym. Adv. Technol.* **2005**, *16*, 189.
57. Zhou, N.; Merschrod, E. F.; Zhao, Y. *J. Am. Chem. Soc.* **2005**, *127*, 14154.
58. Sugiura, H.; Nigorikawa, Y.; Saiki, Y.; Nakamura, K.; Yamaguchi, M. *J. Am. Chem. Soc.* **2004**, *126*, 14858.
59. Meier, H.; Mühling, B.; Oehlhof, A.; Theisinger, S.; Kirsten, E. *Eur. J. Org. Chem.* **2006**, 405.
60. Nagy, A.; Novak, Z.; Kotschy, A. *J. Organomet. Chem.* **2005**, *690*, 4453.
61. Corona, C.; Bryant, B. K.; Arterburn, J. B. *Org. Lett.* **2006**, *8*, 1883
62. Toyota, S.; Iida, T.; Kunizane, C.; Tanifuji, N.; Yoshida, Y. *Org. Biomol. Chem.* **2003**, *1*, 2298
63. [R] Testero, S.; Mata, E. *J. Comb. Chem.,* **2008**, *10*, 487
64. Urgaonkar, S.; Verkade, J. G. *J. Org. Chem.* **2004**, *69*, 5752.
65. Ljungdahl, T; Bennur, T; Dallas, A; Emtenaes, H; Maartensson, J. *Organometallics* **2008**, *27*, 2490.
66. Jutand, A.; Negri, S.; Principaul, A. *Eur. J. Org. Chem.* **2005**, 631.
67. Feuerstein, M.; Doucet, H.; Santelli, M. *Tetrahedron Lett.* **2004**, *45*, 8443
68. Amatore, C.; Bensalem, S.; Ghalem, S.; Jutand, A. *J. Organomet. Chem.* **2004**, *689*, 4642
69. Amatore, C.; Bensalem, S.; Ghalem, S.; Jutand, A.; Medjour, Y. *Eur. J. Org. Chem.* **2004**, 366.
70. Moon, J.; Jeong, M.; Nam, H.; Ju, J.; Moon, J.-H.; Jung, H-.M,; Lee, S. *Org. Lett.,* **2008**, *10*, 945.
71. Dobler, M. R. *Tetrahedron Lett.* **2003**, *44*, 7115.
72. Li, J.-H.; Liang, Y.; Xie, Y.-X. *J. Org. Chem.* **2005**, *70*, 4393.
73. Jana, S.; Dutta, B.; Bera, R.; Koner, S. *Inorg. Chem.***2008**, *12*, 5512.
74. Reetz, R. T.; Westermann, E. *Angew. Chem., Int. Ed.* **2000**, *39*, 165.
75. Reetz, M. T.; Masse, M. *Adv. Mater.* **1999**, *11*, 773.
76. Choudary, B. M. Madhi, S.; Chowdari, N. S.; Kantam, M. L.; Sreedhar, B. *J. Am. Chem. Soc.* 2002, **124**, 14127.
77. Vasilevsky, S.; Klyatskaya, S.; Elguero, J. *Tetrahedron*, **2004**, *60*, 6685
78. Walker, W. H., IV; Rokita, S. E. *J. Org. Chem.* **2003**, *68*, 1563
79. Alonso, D.; Botella, L.; Najera, C.; Pacheco, M. *Synthesis*, **2004**, *10*, 1713
80. Nishihara, Y.; Ikegashira, K.; Mori, A.; Hiyama, T. *Chem. Lett.,* **1997**, *12*, 1233.
81. Komano, K.; Shimamura, S.; Inoue, M.; Hirama, M. *J. Am. Chem. Soc.* **2007**, *129*, 14184.
82. Nicolaou, K. C. *Angew. Chem., Int. Ed. Engl.* **1993**, *32*, 1377.
83. Baran, P.; Shenvi, R. *J. Am. Chem. Soc.* **2006**, *128*, 14028.
84. Miyazaki, T.; Yokoshima, S.; Simizu, S.; Osada, H.; Tokuyama, H.; Fukuyama, T. *Org. Lett.* **2007**, *23*, 4737.
85. Kitade, M.; Tanaka, H.; Oe, S.; Iwashima, M.; Iguchi, K.; Takahashi, T. *Chem. Eur J.* **2006**, *12*, 1368.
86. Lindlar, H.; Dubuis, R. *Org. Syn.* **1966**, *46*, 89.
87. Fürstner, A.; Radkowski, K. *Chem. Commun.* 2002, 2182.
88. Trost, B. M.; Ball, Z. T.; Jöge, T. *J. Am. Chem. Soc.* 2002, **124**, 7922.

89. Nicolaou, K. C.; Webber, S. *J. Am. Chem. Soc.* **1984**, *106*, 5734.
90. Dyker, G.; Mastalerz, M.; Müller, I. M. *Eur. J. Org. Chem.* **2005**, 3801.
91. Xia, Y.; Yang, P.; Sun, Y.; Wu, Y.; Mayers, B.; Gates, B.; Yin, Y.; Kim, F.; Yan, H. *Adv. Mat.*, **2003**, *15*, 353.
92. Li, Y.; Zhang, J.; Wang, W.; Miao, Q.; She, X.; Pan, X. *J. Org. Chem.* **2005**, *70*, 3285.
93. Spivey, A.; McKendrick, J.; Srikaran. *J. Org. Chem.* **2003**, *68*, 1843.
94. Netherton, M.; Fu, G. *Org. Lett.* **2001**, *3*, 4295.
95. Zhang, Y.; Zhao, C.; Yang, J.; Kapiamba, M.; Haze, O.; Rothberg, L.; Ng, M.-K. *J. Org. Chem.* **2006**, *71*, 9473.
96. Heravi, M.; Kivanloo, A.; Rahimzadeh, M.; Bakavoli, M.; Ghassemzadeh, M.; Neumuller, B. *Tetrahedron Lett.* **2005**, *46*, 1607.
97. Rudyanto, M.; Tomizawa, Y.; Morita, H.; Honda, T. *Org. Lett.*, **2008**, *10*, 1921.
98. Serra, S.; Fuganti, C. *Synlett*, **2005**, 809.
99. Maeda, T.; Furusho, Y.; Sakurai, S.; Kumaki, J.; Okoshi, K.; Yashima, E. *J. Am. Chem. Soc.* **2008**, *130*, 7938.
100. Coutouli-Argyropoulou, E.; Tsitabani, M.; Petrantonakis, G.; Terzis, A.; Raptopoulou, C. *Org. Biomol. Chem.* **2003**, *1*, 1382.
101. Li, C.-J.; Slaven, W.; John, V.; Banerjee, S. *J. Chem. Soc., Chem. Commun.* **1997**, 1569.
102. Hundertmark, T.; Littke, A., Buchwald, S.; Fu, G. *Org. Lett.* **2000**, *2*, 1729.
103. Hierso, J-C.; Fihri, A.; Amardeil, R.; Meunier, P.; Doucet, H.; Santelli, M.; Ivanov, V. *Org. Lett.* **2004**, *6*, 3473.
104. (a) Köllhofer, A.; Plenio, H. *Adv. Synth. Catal.* **2005**, *347*, 1295. (b) A. Köllhofer, H. Plenio, *Chem. Eur. J.* **2003**, *9*, 1416.
105. Fukuyama, T.; Shinmen, M.; Nishitani, S.; Sato, M.; Ryu, I. *Org. Lett.* **2002**, *4*, 1691.
106. Hierso, J.-C.; Boudon, J.; Picquet, M.; Meunier, P. *Eur. J. Org. Chem.* **2007**, *4*, 583.
107. Altenhoff, G.; Wurtz, S.; Glorius, F. *Tetrahedron Lett.* **2006**, *47*, 2925.
108. Son, S.; Jang, Y.; Park, J.; Na, H.-B.; Park, H-M.; Yun, H.-J.; Lee, J.; Hyeon, T. *J. Am. Chem. Soc.* **2004**, *126*, 5026.
109. Anderson, K.; Buchwald, S. *Angew. Chem., Int. Ed.* **2005**, *44*, 6173.
110. Dubbaka, S. R.; Vogel, P. *Adv. Synth. Catal.* **2004**, *346*, 1793.
111. (a) Likhar, P.; Subhas, M.; Roy, M.; Roy, S.; Kantam, M. *Helvetica. Chim. Acta*, **2008**, *91*, 259; (b) Collett, M.; Jones, D.; Renyard, S. *J. Chem. Soc. Perkin Trans. I*, **1986**, 1471.
112. Karpov, A. S.; Müller, T. J. J. *Org. Lett.* **2003**, *5*, 3451.
113. Nishihara, Y.; Ikegashira, K.; Mori, A.; Hiyama, T. *Chem. Lett.* **1997**, 1233.
114. Yang, C.; Nolan, S. P. *Organometallics* **2002**, *21*, 1020.
115. Nishihara, Y.; Ikegashira, K.; Hirabayashi, K.; Ando, J.; Mori, A.; Hiyama, T. *J. Org. Chem.* **2000**, *65*, 1780.
116. Molander, G.; Katona, B. W.; Machroubi, F. *J. Org. Chem.* **2002**, *67*, 8416.
117. Wang, L.; Pinhua, L.; Zhang, Y.; *Chem Commun.* **2004**, 514.
118. Cramer, N.; Buchweitz, M.; Laschat, S.; Frey, W.; Baro, A.; Mathieu, D.; Richter, C.; Schwalbe, H. *Chem. Eur. J.* **2006**, *12*, 2488.
119. Hua, R.; Yi, C. *J. Org. Chem.* **2006**, *71*, 2536.

1.1.6 Stille Coupling

Vincent Mascitti

1.1.6.1 Description

The reaction between an organic electrophile **1** and an organostannane **2** mediated by a transition metal catalyst (originally palladium) to form a new sigma carbon carbon bond is referred to as the Stille cross-coupling reaction (equation 1).

$$\underset{\mathbf{1}}{R_1{-}X} + \underset{\mathbf{2}}{R_2{-}Sn(alkyl)_3} \xrightarrow{\text{Pd(0) catalytic}} R_1{-}R_2 + X{-}Sn(alkyl)_3 \quad (1)$$

Commonly used organic electrophiles involve $C(sp^2)$ hybridized carbon as coupling partners like in acid chlorides[1], (hetero)aryl halides (Cl, Br, I) and triflates, alkenyl halides and triflates; activated $C(sp^3)$ hybridized carbon like allyl halides and acetates, benzyl halides are also used. Recently the use of unactivated alkyl halides has also been reported[2]. Aryl sulphonyl chlorides[3] and arenediazonium salts[4] have also been used as organic eletrophiles.

Organotin reagents involving $C(sp^2)$ or C(sp) hybridized carbons, like in alkenyl, aryl, heteroaryl, alkynyl organostannanes, are the most widely used. Examples of use of allyl and alkyl organotin compounds are also reported. The relative order of ligand transfer from the organostannane is: alkynyl > alkenyl > aryl > allyl ~ benzyl >> alkyl.

The catalyst used is often palladium (0) (like $Pd(PPh_3)_4$, $Pd_2(dba)_3$), or a source of palladium (II) (like $Pd(OAc)_2$, $BnPdCl(PPh_3)_2$ to name a few), that gets reduced to the active species palladium(0) in situ. Methods using other metals like manganese, copper, and nickel have been reported; the latter has been applied for instance in the successful Stille coupling of unreactive aryl chlorides as well as in the coupling of unactivated primary and secondary alkyl halides.[2]

The Stille coupling is usually carried out in a dipolar solvent (like DMF, DMSO or NMP) or in an ethereal solvent (like THF or dioxane).

When the above reaction is performed under an atmosphere of carbon monoxide, CO insertion occurs with the concomitant formation of two sigma carbon carbon bonds to give a ketone as product. This reaction is referred to as the Stille carbonylative coupling reaction (see equation 2 below).

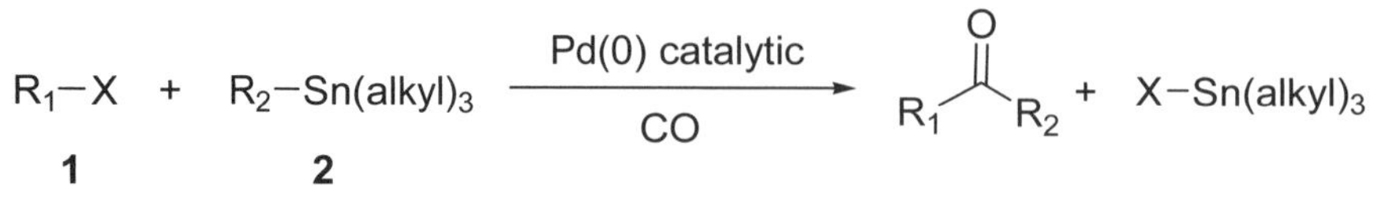

(2)

The Stille coupling is one of the most powerful tools available to synthetic chemists to date for the formation of sigma carbon carbon bonds. This is clearly demonstrated by the impressive amount of syntheses involving this transformation.[5] One advantage of this reaction is that it can be performed under very mild and neutral conditions (on the contrary to other cross coupling reactions, like the Suzuki coupling, which are done under basic conditions) compatible with sensitive motifs and functional groups found for instance in natural products. Furthermore, the organostannane precursors are usually easily accessible, easily purified and stored under normal conditions (on the contrary to other air- or moisture-sensitive reagents required for other types of palladium mediated cross couplings) and tolerate many functional groups[6] (properties mainly due to the low polarity of the carbon tin bond compared to other organometallic reagents like Grignards and organozincs respectively used in Kumada and Negishi cross couplings). Last but not least, the reaction has usually a high rate of success and the numerous precedents found in literature as well as the numerous options available to tune a particular coupling (solvent, catalyst, ligand, additives, ability to change the pair electrophile/organostannane to the other possible combination, *etc.*) allow for a quick screening of the most optimal set of conditions for a particular motif.[7] The extensive use of the Stille coupling in total syntheses, not only to build small fragments but also to combine together advanced intermediates harboring very elaborated carbon frameworks is no stranger to that and should be regarded as a testimony of the power of this method. One drawback however is the generation of toxic tin-containing by-products (often in stoechiometric amounts), which renders the reaction potentially problematic on scale-up. Elegant solutions to this problem have emerged in the literature.[8]

1.1.6.2 Historical Perspective

The reaction between an organic electrophile and a tin reagent (hexaalkyldistannane) was first reported by Colin Eaborn and co-workers[9] in 1976. For instance reaction of a stoechiometric amount of bromobenzene and hexabutyldistannane **3** in toluene in a sealed tube heated at 120 °C for 40 hours in presence of a catalytic amount of $Pd(PPh_3)_4$ led to the formation of tributyl(phenyl)stannane **4** and biphenyl **5** in respectively 28 and 31% yield.

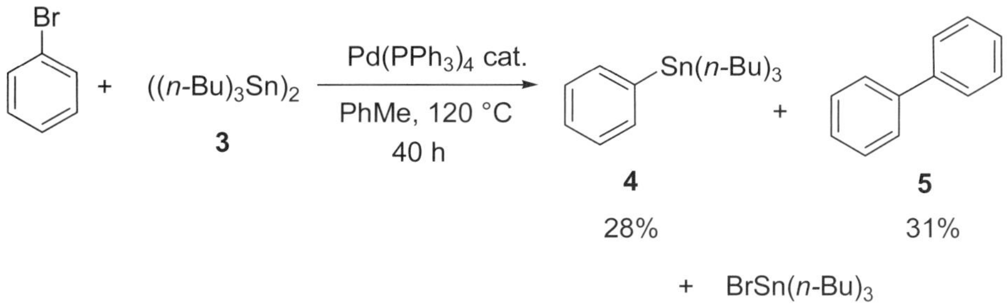

A year later, Kosugi, Shimizu and Migita reported the first palladium-catalysed sigma carbon carbon bond formation from an organic electrophile (acid chloride[10] or aryl halide[11]) and an organostannane. Thus, heating a mixture of allyltributyltin **6** and bromobenzene in a sealed tube at 100 °C for 20 hours in presence of a catalytic amount of $Pd(PPh_3)_4$ led to the clean formation of allylbenzene **7** in high yield.[11]

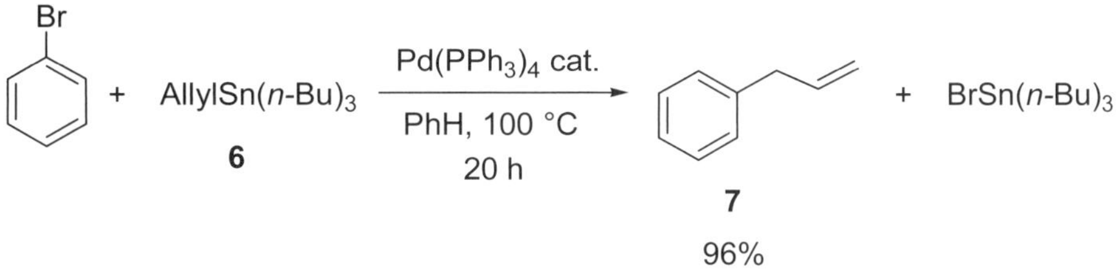

The first report by Stille was published in 1978 and dealt with the formation of ketones, mediated by a palladium(II) catalyst, starting from acid chlorides and organostannanes.[12] For instance acetophenone was produced in 89% isolated yield upon treatement of a solution of benzoyl chloride and tetramethylstannane in HMPA at 65 °C for 10–15 minutes in presence of benzylchlorobis(triphenylphosphine)palladium(II) (**8**). The experimental conditions reported in this article proved to be milder and higher yielding than the ones reported by Kosugi, Shimizu and Migita (see below).

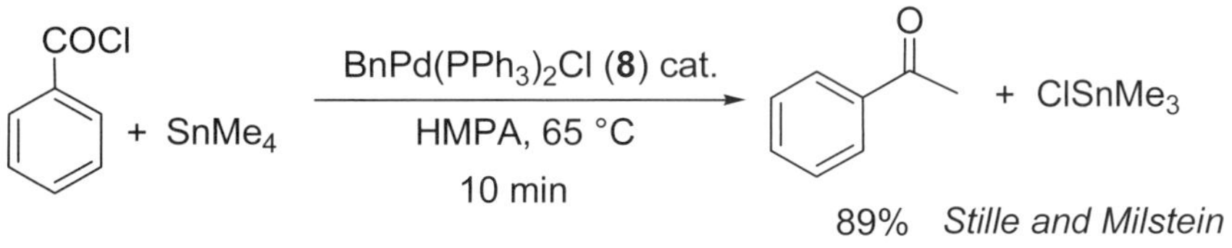

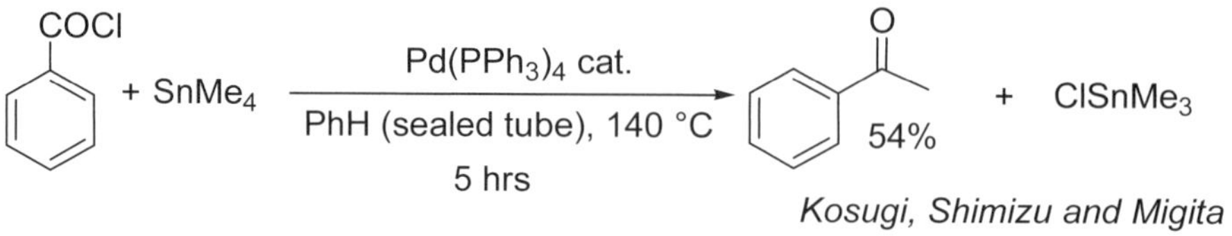

Stille then studied extensively this reaction[13] and in recognition of Stille's contribution, this reaction is now referred to as the Stille coupling.[14]

1.1.6.3 Mechanism

The mechanism of the Stille coupling presented by Stille as a working model in his 1986 review,[13] is represented below.

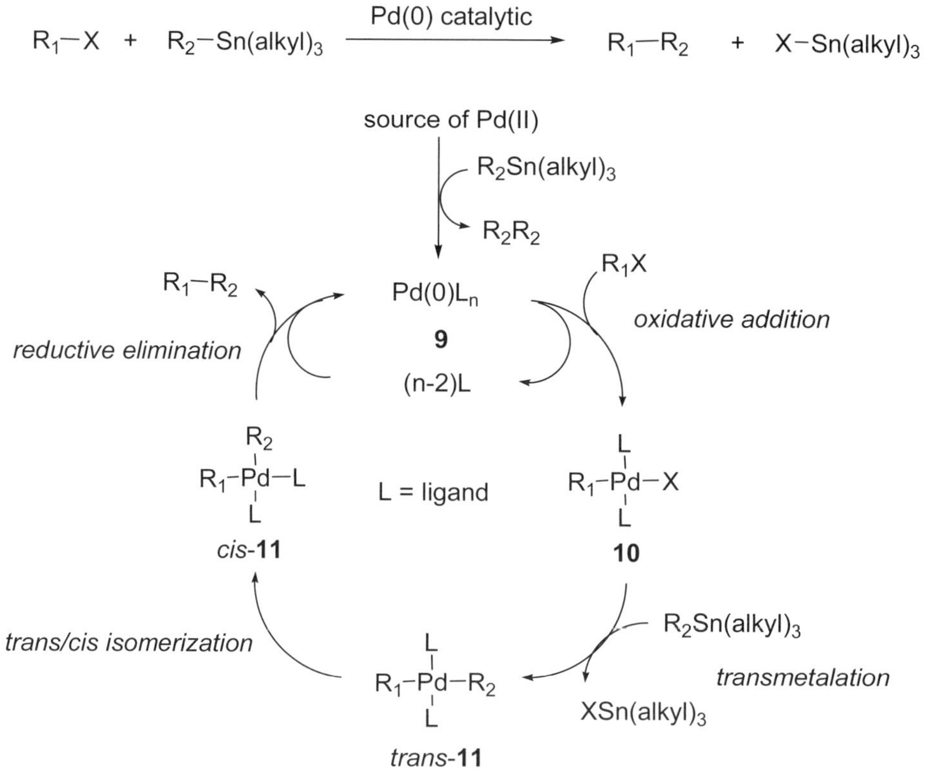

If using a palladium(II) catalyst, the catalytic active species palladium(0) **9** is generated in situ from reduction of the palladium(II) precursor with the organostannane present in the medium. Oxidative addition of the organic electrophile then generates a 16-electron palladium(II) complex intermediate **10**, which then undergoes a transmetalation step to produce *trans*-**11**. A rapid *trans*/*cis*-isomerization to produce *cis*-**11** followed by a reductive elimination to generate the carbon carbon bond in the product and to regenerate the active catalytic palladium(0) species **9** complete the catalytic cycle.

The carbonylative Stille coupling obeys to the same catalytic loop except that there is a carbon monoxide insertion into intermediate **10** to produce **12**, which undergoes the transmetalation step (carbonyl insertion into **12** is faster than transmetalation).

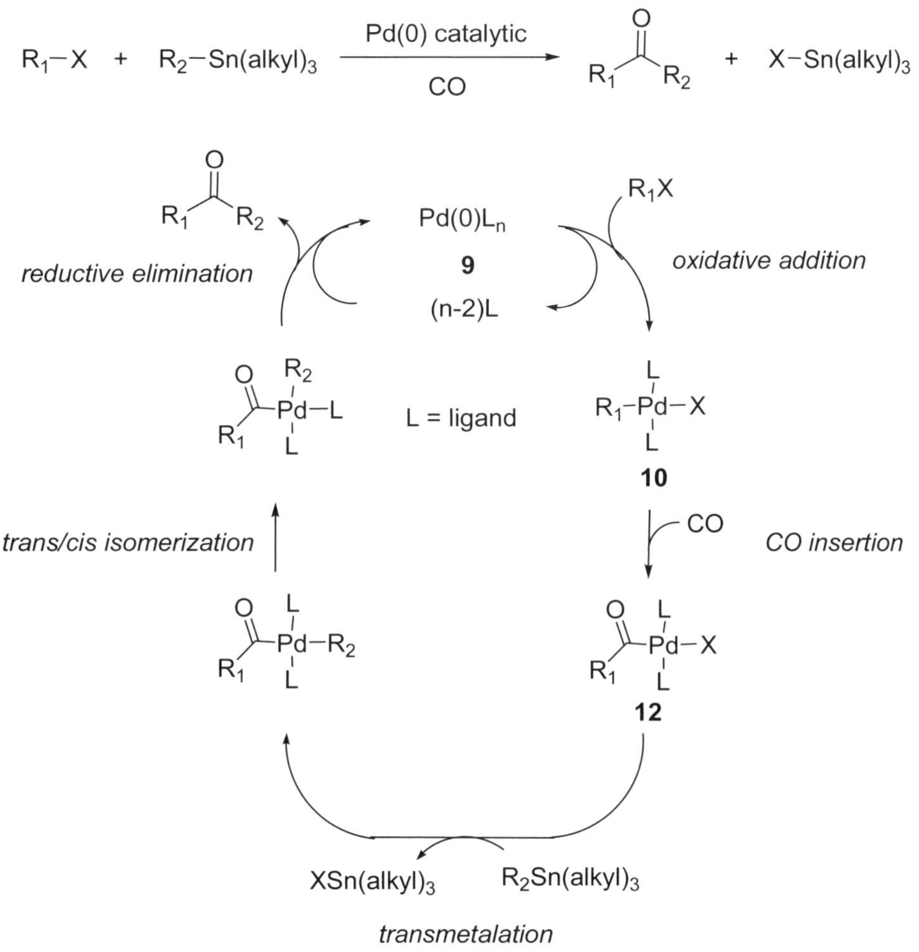

The mechanism deserves some additional comments. It is admitted that the transmetalation step is the rate-limiting step in most cases.[15] However, at times either the oxidative addition[16] or the reductive elimination[17] could be the rate-limiting step. As it will be seen in the next paragraph, knowing that the transmetalation step is the rate-determining step in most cases is of key importance when trying to fine-tune the reaction.

The mechanism of the reaction has been the subject of considerable attention over the past twenty years and very detailed physicochemical studies aiming at decoding this mechanistic black box have been published[18]. A very detailed review on the subject has been published in 2004 by Espinet and Echavarren[15] in which a more detailed view of the pathways that may be involved in the catalytic loop of the Stille coupling was proposed. It is presented below.

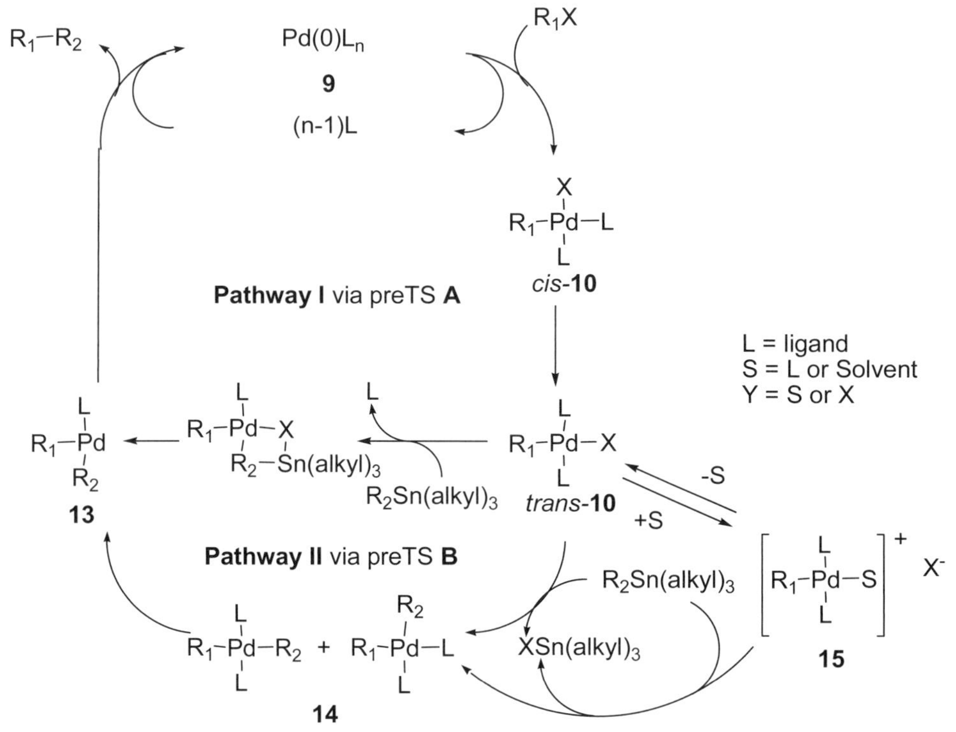

pre Transition State assembly:

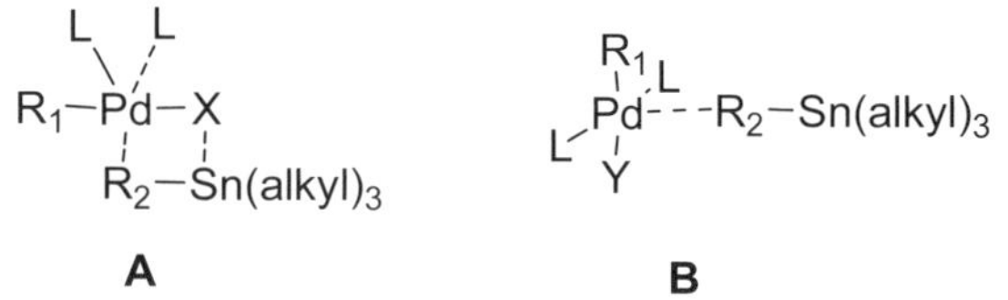

Oxidative addition of the organic electrophile leads to *cis*-**10** (relative to R_1 and X), which is the kinetic addition product. This product rapidly isomerises to the thermodynamically more stable *trans*-**10** (by virtue of the transphobia effect[19] between a phosphane ligand and a *trans* C-donor ligand);

two transmetalation pathways are then suggested depending of the reaction conditions.

In **pathway I**, *trans*-**10** leads to **13** via a cyclic transition state (see **A**) in an *associative L-for-R_2* substitution. Reductive elimination then completes the catalytic loop.

In **pathway II**, *trans*-**10** leads to **14** (either directly or by the intermediary of **15**) via an acyclic transition state (see **B**) in an *associative Y-for-R_2* substitution. Reductive elimination then completes the catalytic loop. Espinet and Echavarren have proposed that **pathway I** is favored for conditions involving organic halides as electrophiles and a solvent with moderate coordinative ability towards palladium. **Pathway II** would be favored for good leaving groups like triflates and in presence of solvents with good coordinative ability for palladium.

Postulating a universal mechanism more detailed than the working models presented above seems to be a very ambitious endeavour since clearly reaction mechanisms are very dependent of each parameter (catalyst, nature of the ligand, nature of the counter ion in the electrophile, nature of the organostannane, presence of additives like lithium chloride[20] or copper salts[21]). The next paragraph will give us an additional flavour of this.

1.1.6.4 Variations and Improvements

The variations and improvements of the Stille coupling are numerous in literature and each of the parameters that can potentially influence the mechanism of the reaction has been studied. In this paragraph, we do not have the pretension to give an exhaustive list of the numerous modifications that have been published but instead we will focus on presenting a representative set of examples. This will allow emphasis on the logical path taken to bring the reaction to where it stands now, 30 years after its discovery.

Influence of the ligand

Based on the working models presented in the mechanism section it is easily conceivable that each parameter prone to modify the sphere of coordination of the metal will have an influence on the course of the reaction.

The ligand is no exception to this. Indeed, it was found by Farina and Krishnan that changing the ligand triphenylphosphine PPh_3 for ligands like tri-2-furylphosphine (TFP) or triphenylarsine $AsPh_3$ led to large rate enhancements in the Stille coupling[22] (see below).

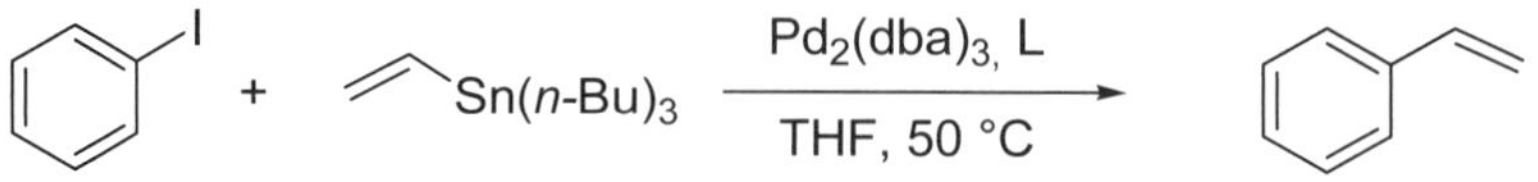

Ligand[a] L	rel rate	yield[b] (%)
PPh_3	1	15
TFP	105	> 95
$AsPh_3$	1100	> 95

a. Pd:L ratio = 1:4. b. HPLC yield after 72 hrs.

It was proposed that in this case, ligands like TFP and $AsPh_3$ were increasing the rate of the reaction because of their ability to readily dissociate from Pd(II) intermediate and therefore increase the rate of the rate limiting transmetalation step.

Influence of additives

It is reported in literature that additives like LiCl[20] or Cu(I)[23] salts can have a dramatic influence on the coupling. The "copper effect" in Stille coupling reactions was investigated by Farina and Liebeskind and coworkers[21]. For instance in the reaction of iodobenzene and vinyltributyltin in dioxane at 50 °C catalyzed by $Pd_2(dba)_3$ in presence of a strong ligand like PPh_3, it was found that the addition of 2 molar equivalents of CuI per mol of catalyst led to a > 100 fold increase in reaction rate.

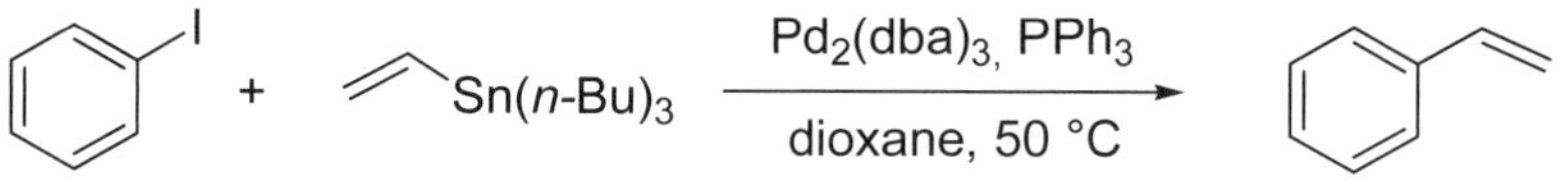

Pd:L:CuI	rel rate	yield[a] (%)
1:4:0	1	85
1:4:1	5	91
1:4:2	**114**	**> 95**

a. HPLC yield determined after catalyst had decomposed.

Based on NMR studies it was proposed that in this case the rate increase was due to the ability of CuI to scavenge the strong ligand PPh_3 free in solution. Indeed, strong ligands in solution are known to inhibit the rate-limiting transmetalation step.

Following on this discovery, in 1996 Liebeskind reported that copper(I) thiophene-2-carboxylate (CuTC) was able to promote the Stille coupling between organic iodides (mainly vinyl and activated aryl) and organostannanes in the absence of palladium, at or below room temperature in NMP as solvent.[24]

CuTC proved to be crucial in the successful late stage Stille coupling involved in the recent total synthesis of (–)-archazolid B reported by the Trauner group[25] (**16** + **17** → **18**). Palladium and copper were required to promote this coupling.

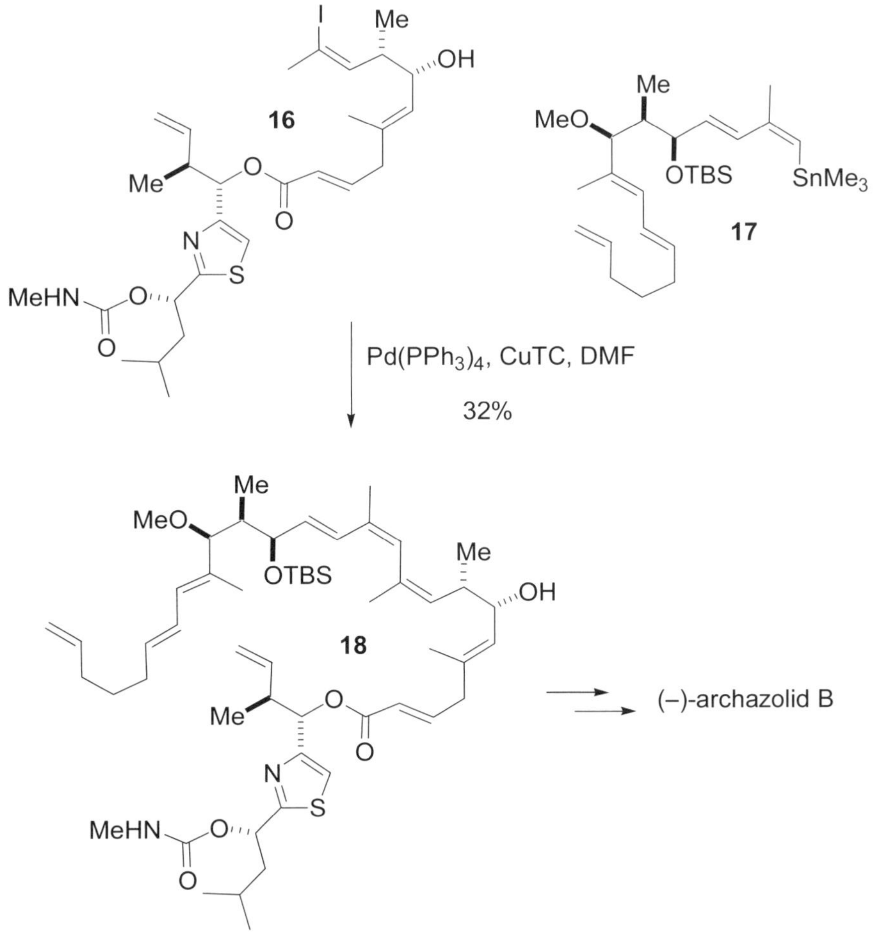

Cuprous chloride proves to be an ally of choice in fine tuning tricky Stille coupling reactions. This was demonstrated by the Corey group while working towards the total synthesis of natural product nicandrenone.[26] One step in the synthetic scheme required the Stille coupling between a 1-substituted vinylstannane and an aryl triflate (or nonaflate). However, 1-substituted vinylstannes are notoriously known for being poor coupling partners in Stille reactions mainly for steric reasons, leading to poor yields of desired coupling product with competing formation of products resulting from *cine* substitution. This prompted the study of new conditions for this particular type of systems.

It was found that the coupling between vinylstannane **19** and aryltriflate **20** mediated by $Pd(PPh_3)_4$ proceeded in high yield when performed in polar solvent DMSO at 60 °C for 48 hours in the presence of 6 equivalents of LiCl and 5 equivalents of CuCl to produce **21** in 95% isolated yield.

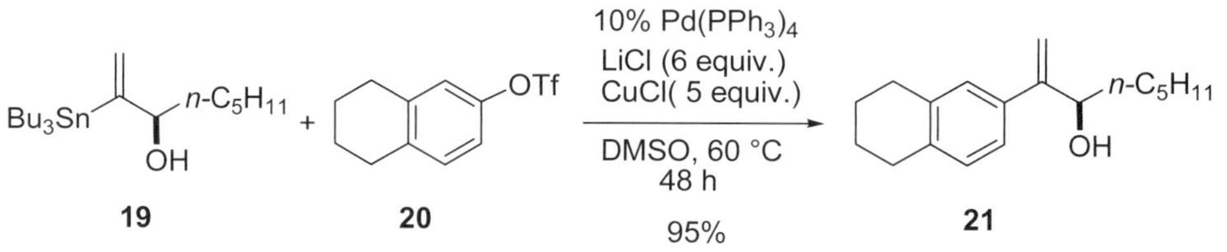

It was proposed that the influence of CuCl was to promote the formation of a more reactive metalloid (RCu or RCuLiCl) than the starting organostannane. This reactive metalloid then participates in the transmetalation step with palladium (II) intermediate to yield to the desired product after reductive elimination. The conditions developed by Corey were successfully applied later on by the Tanner group in their studies towards the total synthesis of zoanthamine.[27]

Other transition metals, like manganese[28] and nickel,[29] have been used in the Stille coupling (as additive or as replacement of palladium). As we will see later, nickel proved to be very powerful in replacing palladium as catalyst and is very promising in terms of expanding the scope of the reaction.

Activation of the organostannane

As seen with the previous example, activation of the organostannane to promote the transmetalation step has a dramatic and positive influence on the outcome of the reaction. In 1992, Vedejs *et al.* showed that intramolecular coordination of tin by a nucleophilic center like a nitrogen can considerably increase the rate of the transmetalation step in the Stille coupling[30]. Derivatives like **22** proved to be more reactive than their analogue trialkyl tin derivatives and the by product halide **23** can be easily removed and recycled from the crude since it crystallizes in the course of the reaction.

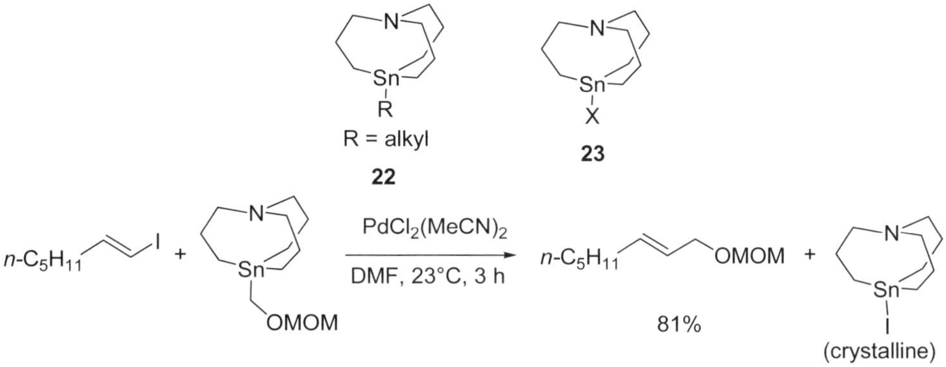

Other methods of activating organostannanes have been reported in literature. These methods capitalize on the use of fluoride ions and the formation in situ or not of activated hypervalent stannate intermediates. For instance, Garcia Martinez has used the hypervalent tin reagent tetrabutylammonium difluorotribenzylstannate **24** (prepared in quantitative yield from Bn_3SnF) in the synthesis of unsymmetrical diarylmethanes starting from aryl triflates (e.g., **24** + **25** → **26**).[31]

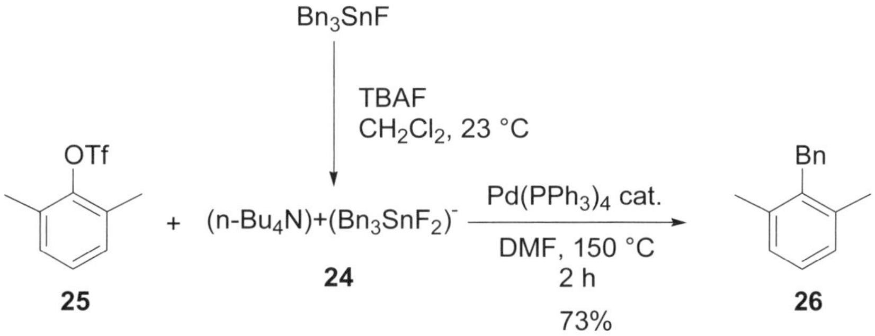

Nolan has also taken advantage of hypervalent organostannate (generated in situ) in Stille coupling catalyzed by a Palladium/imidazolium salt system (e.g., **27** + **28** → **29**).[32]

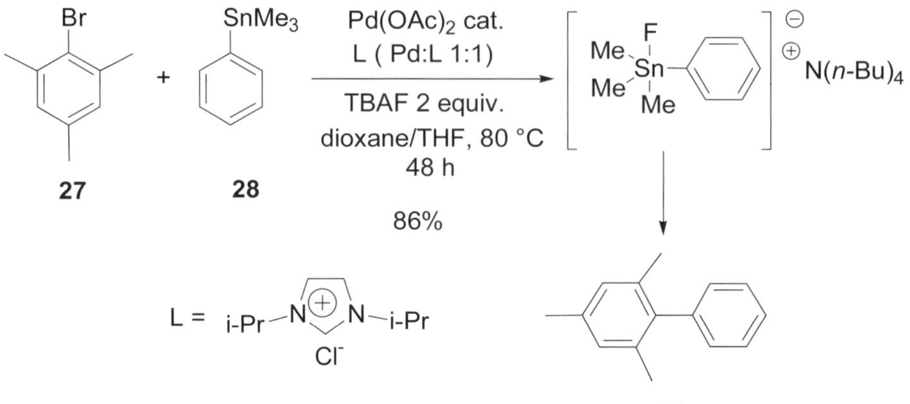

27 **28** **29**

Four examples were also reported where aryl chlorides were used as substrate, best yields and shorter reaction times being obtained with activated aryl chlorides (like 1-(4-chlorophenyl)ethanone).

Fouquet developed an interesting methodology using activated alkyl tin reagents which allows for the formation of C(sp^3) C(sp^2) sigma carbon carbon bonds starting from primary alkyl bromides or iodides.[33] The reaction is limited to the use of primary alkyl halides (β and γ substitution is tolerated).

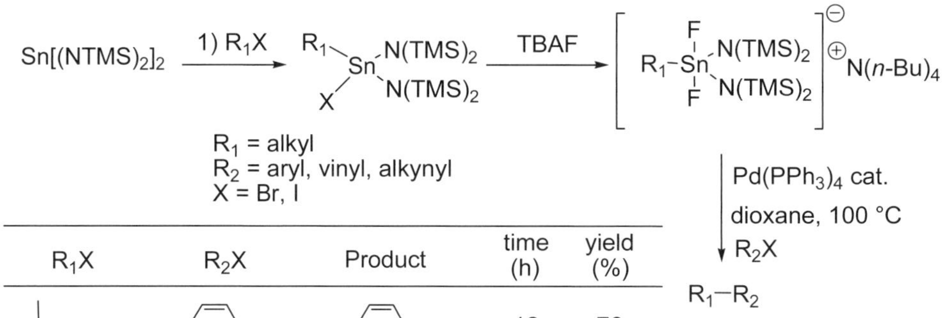

R_1X	R_2X	Product	time (h)	yield (%)
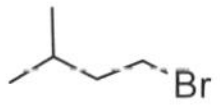	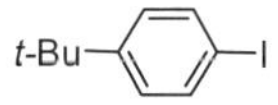	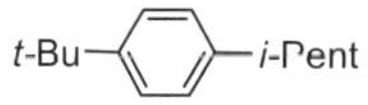	12	76
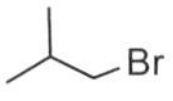	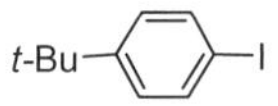		36	63
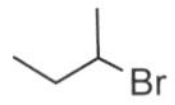	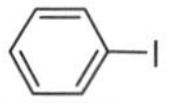	iodobenzene unreacted	72	-

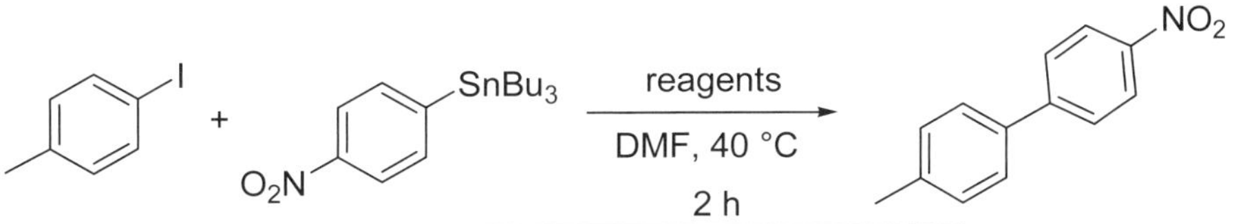

reagents[a]	yield[b] (%)
$Pd(PPh_3)_4$	2
$Pd(PPh_3)_4$, CsF	8
$Pd(PPh_3)_4$, CuI	46
$Pd(PPh_3)_4$, CsF, CuI	**98**
CsF, CuI	0

a. $Pd(PPh_3)_4$ (10 mol%), CuI (20 mol%), CsF (2 equiv.)
b. isolated yield.

Taking advantage of the "copper" and "fluoride" effects, Baldwin and co-workers have developed conditions ($Pd(PPh_3)_4$/CuI/CsF) that allow the Stille coupling of electronically unfavourable and/or sterically hindered substrates.[34]

Development of powerful catalytic systems: the aryl chloride case.

In 1999, the Fu group reported the first general method for the Stille cross coupling of aryl chlorides[35] (**30**→**31**). On the contrary to previously reported methods involving the coupling of aryl chlorides, the Fu conditions allowed for the coupling in good yield of not only electron poor but also electron neutral and electro rich aryl chlorides. The conditions involved the use of 1.5 mol% of $Pd_2(dba)_3$, 6 mol% of the electron rich and sterically demanding phosphine $P(t\text{-}Bu)_3$ as ligand and 2.2 equivalents of CsF.

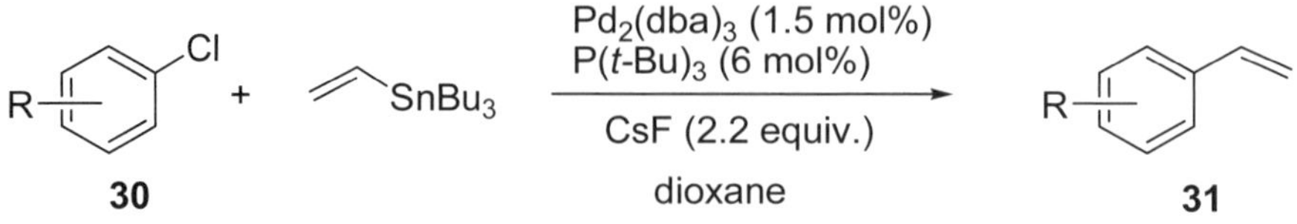

aryl chloride	temp. (°C)	time (h)	yield[a] (%)
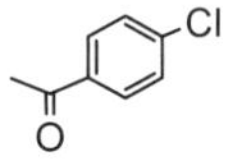	80	12	87
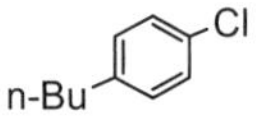	100	23	80
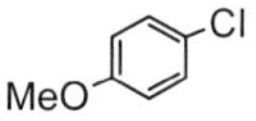	100	48	82
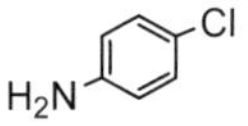	100	48	61
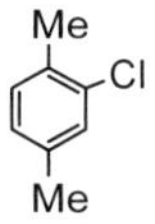	100	36	71

a. isolated yield.

Additional advantages of this methodology are:
- Allow Stille coupling with aryl bromides at room temperature;
- Very hindered biaryl can be synthesized (like tetra ortho substituted biaryls;
- Aryl chlorides can be coupled in presence of aryl triflates.[36]

This method should be regarded as a method of choice when looking to perform a Stille coupling on a sensitive substrate potentially prone to decomposition under the more classical Stille conditions which require higher temperature.

In 2004 Verkade and co-workers reported a catalyst system of general applicability for the Stille coupling of aryl chlorides at 110 °C in dioxane and the coupling of aryl bromides at room temperature in THF. The catalyst system uses $Pd_2(dba)_3$, the bulky proazaphosphatrane ligand **32** in presence of CsF (or Me_4NF).

The same year Baldwin also reported conditions capable of coupling electron rich aryl chlorides.

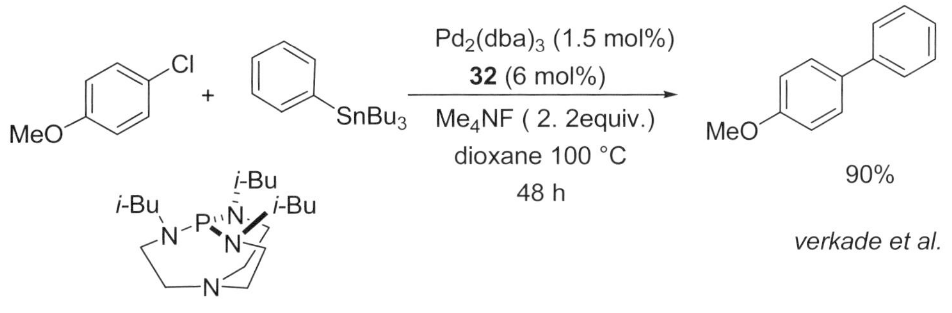

32

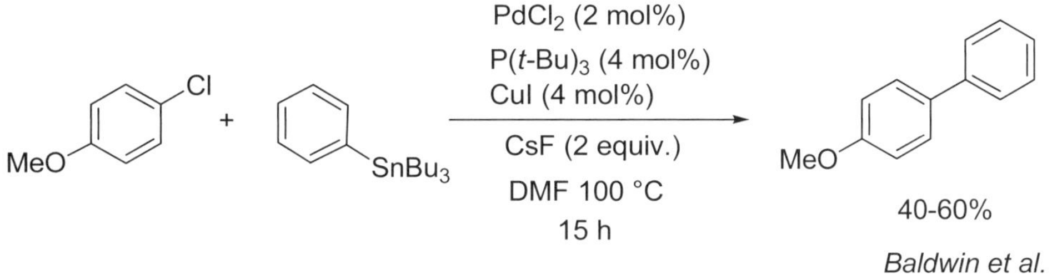

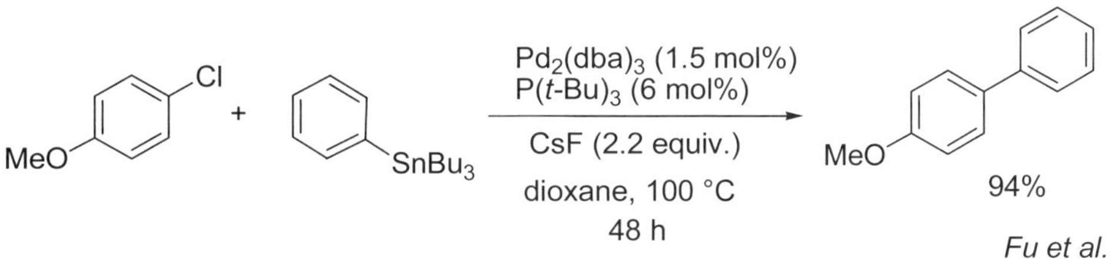

Development of powerful catalytic systems: the alkyl halide case.

After having proposed a general solution to the Stille coupling of unactivated aryl chlorides, the Fu laboratories reported in 2003 a general method for the coupling of unactivated primary alkyl bromides with vinylstannanes[37] (**33** + **34** → **35**). It is a real tour de force since these substrates harbor β-hydrogens and therefore the spectre of undesired β-hydride elimination loomed as a real threat.[38] At that time only 5 successes had been reported in literature but only on specific substrates.[39]

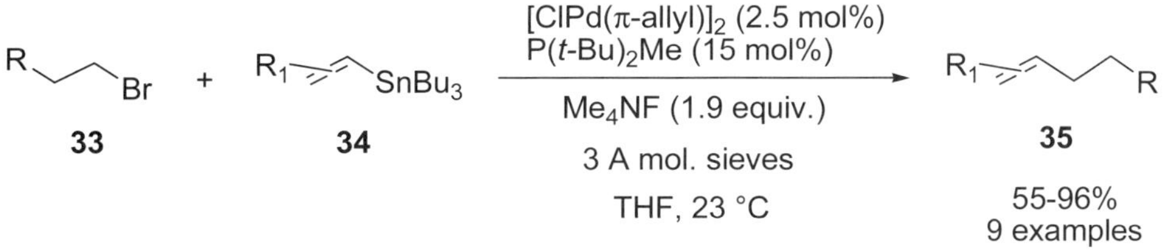

It is worth noting from a practical point of view that the air sensitive phosphine P(*t*-Bu)$_2$Me can be replaced by the commercially available, air- and moisture- stable salt [HP(*t*-Bu)$_2$Me]BF$_4$ without decrease in yield. The drawback of this methodology was that it is limited to the use of vinylstannanes and that arylstannanes were ineffective partners. The solution came a few months later with the discovery that electron-rich alkyldiaminophosphane PCy(pyrrolidinyl)$_2$ was a superior ligand compared to P(*t*-Bu)$_2$Me (**36** + **37** → **38**).[40]

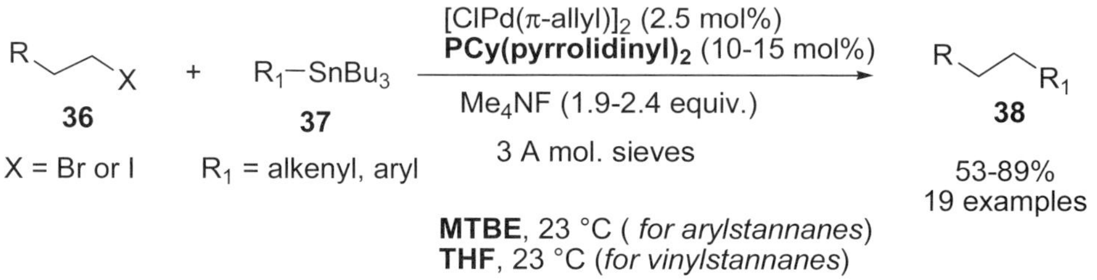

Besides the role of the ligand, the solvent plays a key role, methyl *tert*-butyl ether (MTBE) being the solvent of choice for the coupling with arylstannanes. Other solvents like acetonitrile, *tert*-amyl alcohol, or dichloromethane proved to be not suitable. From a synthetic point of view it is worth mentioning that these conditions are compatible with a wide range of functionalities like, ester, amide, nitrile, benzyl, terminal olefin, tetrahydropyran, 1,3-dioxolane. For an alternative method, see also the work of Fouquet mentioned above.[33,41]

The Stille coupling involving secondary alkyl electrophiles still remained an unmet challenge…but not for long since in 2005 the Fu group published a landmark communication reporting the first catalytic system able to perform Stille cross couplings of unactivated secondary alkyl halides (bromide and iodide) with electronically and sterically diverse monoorganotin reagents[42] (**39** + **40** → **41**).

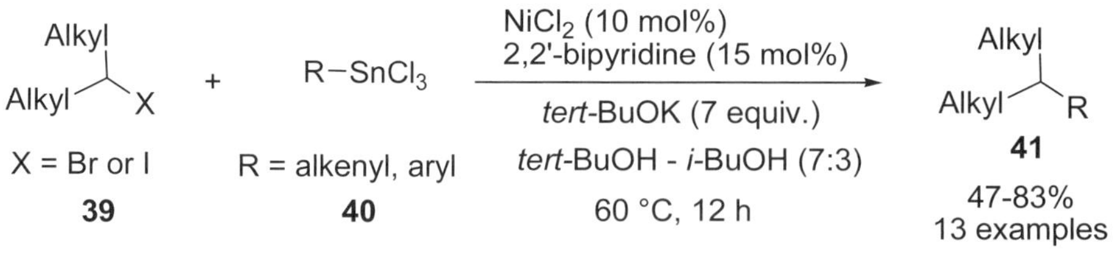

A key was the realization that Ni, the neighbour of Pd in the groupVIII, was superior in such coupling. Nickel catalysts have already been used in the Stille coupling for the generation of C(sp^2)–C(sp^2) sigma carbon–carbon bonds. For instance Shirakawa and Hiyama had shown that Ni(0) catalyst (formed by the in situ reduction of $Ni(acac)_2$ by DIBAL-H) combined with PPh_3 could catalyze the Stille coupling of aryl halides with vinyl-, aryl- or allyl-stannanes[43].

Fu's methodology is impressive for several reasons:
- It uses unactivated secondary alkyl halides (bromide and iodide, cyclic or acyclic).
- It works equally well using unactivated primary alkyl halides (bromide and iodide).
- It uses inexpensive and air stable $NiCl_2$ as catalyst.
- Palladium ($Pd_2(dba)_3$ or $Pd(OAc)_2$) was ineffective as a catalyst.
- Reaction conditions are mild.
- Preliminary results suggest that the coupling may be done with retention of configuration.
- Mechanism may proceed through the initial formation of an alkyl radical.
- It works on electronically and sterically diverse aryl- and alkenyl-monoorganostannanes (easily accessible starting from alkenyl- or aryl- trialkylstannanes by redistribution with $SnCl_4$).
- This process does not generate toxic triorganotin compound ($XSn(alkyl)_3$) in stoechiometric amount but instead inorganic tin-based by-products which are less toxic and easier to eliminate in the work-up[44, 45].

Reduction of the potential toxicity associated with tin compounds.

Tin reagents particularly trialkyl organostannanes are toxic chemicals[44,45] and extreme care should be exercised when manipulating them (avoiding inhalation and contact as well as working in a well ventilated hood is very important). In most cases the Stille reaction will produce toxic halotrialkylstannane $XSn(alkyl)_3$ (in stoichiometric amount) which are also difficult to remove (for instance $ClSnBu_3$ is high boiling, non polar and tends to streak when purifying a crude by flash chromatography over silica gel). Fortunately methods have appeared in literature that generate less toxic and easier to eliminate tin-containing by-products. The Fu methodology using monoorganostannanes of general formula $RSnCl_3$ is one of them.

Other methods have appeared in literature that capitalize on the use of alkyltrichlorostannanes to perform Stille couplings in aqueous solutions (via an organostanoate intermediate **42**)[46]. The by products are inorganic tin by products, less toxic and easily eliminated in the work-up.[44,45]

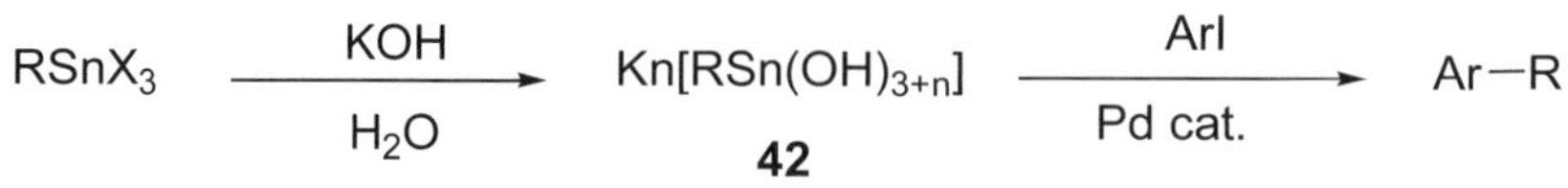

Solid Phase chemistry has also been used to help solving this problem by immobilizing one of the partners (organic electrophile or organostannane) on the solid support to help purification[47]. An example of this approach is found in Nicolaou's total synthesis of (S)-zearalenone[48] (**43** → **44**).

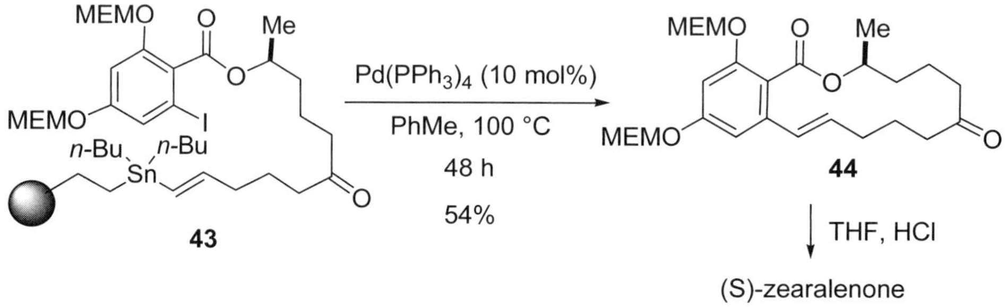

The application of modern solid-phase chemistry to metal catalyzed cross coupling reaction (including the Stille coupling) has been the subject of a very recent review by Testero and Matta.[47a] Ionic liquid supported tin reagents have also been developed.[49]

Curran has used fluorous tin reactants (**45**) to facilitate the separation of the desired product from tin by-products via a three-phase (aqueous/organic/fluorous) extraction protocol as presented below.[50] This also allows for easy isolation and recycling of the tin fluorous compound.

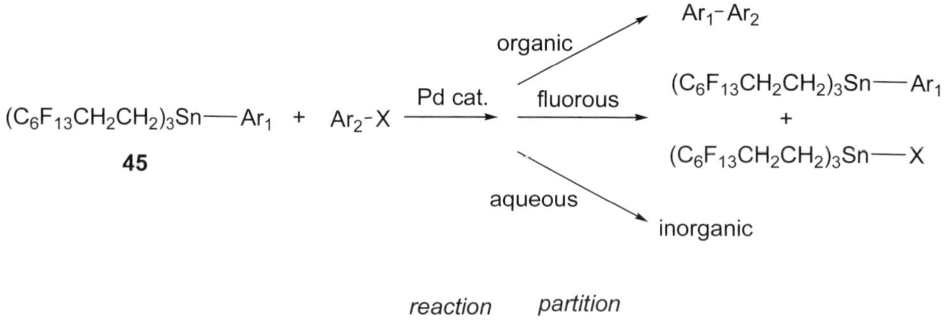

Curran and Hallberg also applied this method in combination with the rapid microwave assisted Stille coupling of simple organic halides or triflates.[51] Stille couplings under microwave irradiation conditions usually leads to shorter reaction time compared to classical conditions and have been the subject of a very recent review.[52]

Maleczka proposed an ingenious approach (the 'Sn–F' route), catalytic in tin, for the Stille coupling involving alkenylstannanes.[53,54] The alkenylstannane is formed in situ from the corresponding terminal alkyne and a catalytic amount of trialkyltinhydride as depicted on the catalytic loop below.

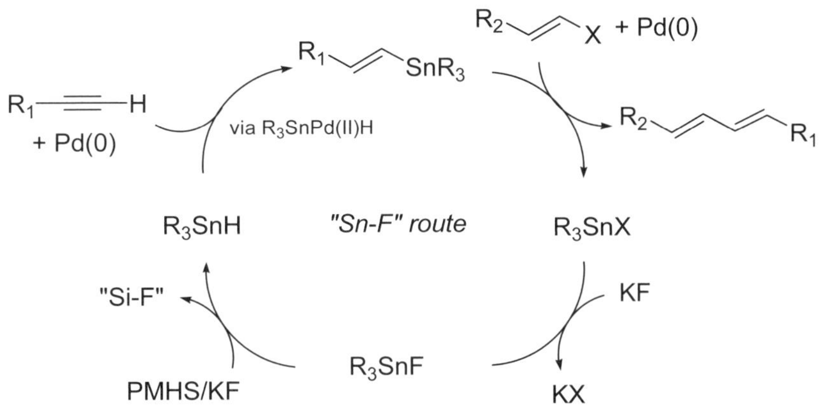

Polymethylhydrosiloxane (PMHS) made hypercoordinate by KF is responsible for the regeneration of the tin hydride. These conditions were successfully applied to the $C(sp^2)$–$C(sp^2)$ carbon–carbon bond formation involving aryl iodides, alkenyl halides (bromide or iodide) or benzyl bromide. Aryl nonaflate and allyl bromide proved to be ineffective partners under these conditions.

The Stille–Kelly reaction.

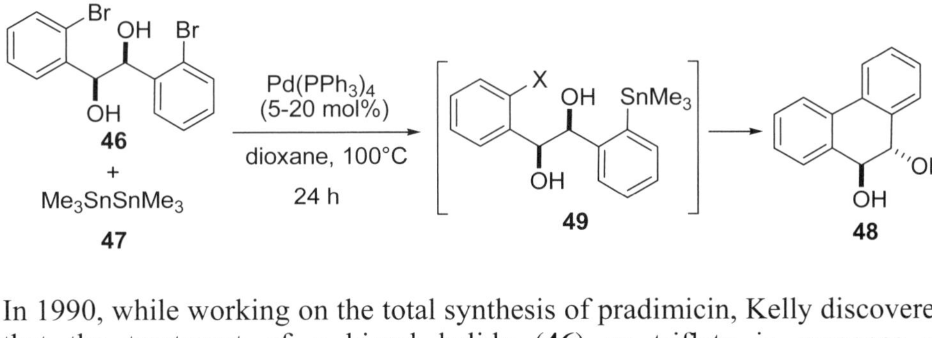

In 1990, while working on the total synthesis of pradimicin, Kelly discovered that the treatment of a biaryl halide (**46**) or triflate in presence of hexaalkyldistannane (**47**) and a catalytic amount of $Pd(PPh_3)_4$ led to the clean intramolecular biaryl coupling to produce **48** in 80% isolated yield.[55] This reaction can be regarded as the transmetalation with the distannane mediated by palladium to afford the intermediate organostannane **49** which then intramolecularly reacts in a classical Stille fashion to produce the desired biaryl product.

1.1.6.5 Synthetic Utility

As mentioned already in the description, the Stille coupling is one of the most powerful **and reliable** tool available to synthetic chemists to form a sigma carbon carbon bond (C(Sp^2) C(sp^2) or C(sp^2) C(Sp^3)). There are countless examples reported in literature over the past 30 years. Not only this reaction found an abundant application in the synthesis of small molecules but it has also been widely used in total syntheses of complex natural products; sometimes as a key step! The "stitching" approaches used by the Nicolaou and Danishefsky groups to respectively close at a late stage of the synthesis the 29-membered ring macrocycle found in rapamycin,[56] and to form the enediyne motif found in dynemicin[57] constitute perfect examples of the power of the Stille coupling.

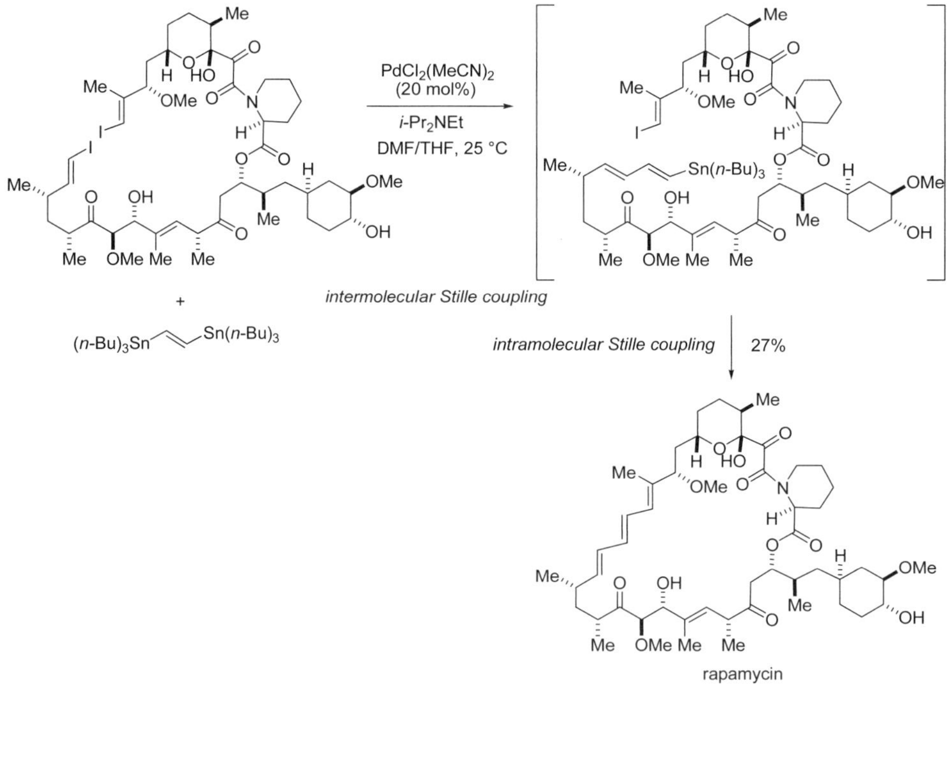

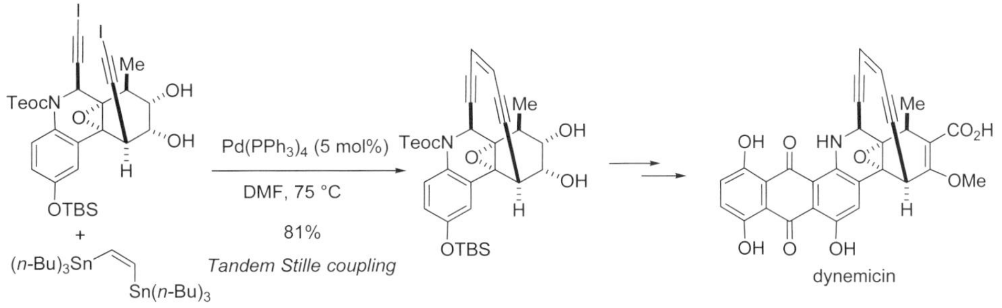

The use of the Stille reaction (cross coupling and carbonylative variation) has been the subject of numerous excellent reviews[58] and books[7] over the past few years and we will mention in this part only a few examples to illustrate the power and broad applicability of the method.

The Stille–Hegedus synthesis of the diterpene jatrophone took advantage of an intramolecular carbonylative Stille coupling between a vinyl triflate and a vinylstannane motif to forge the macrocycle found in the natural product (see below).[59]

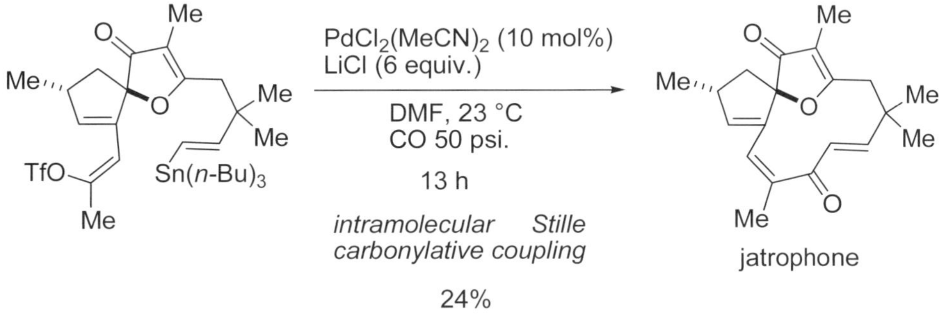

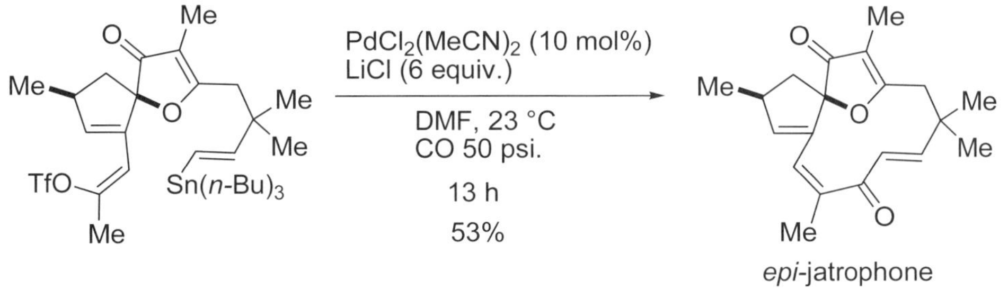

As often in organic synthesis, a remotely placed stereocenter had a profound influence on the course of the reaction (probably due to a conformational change).

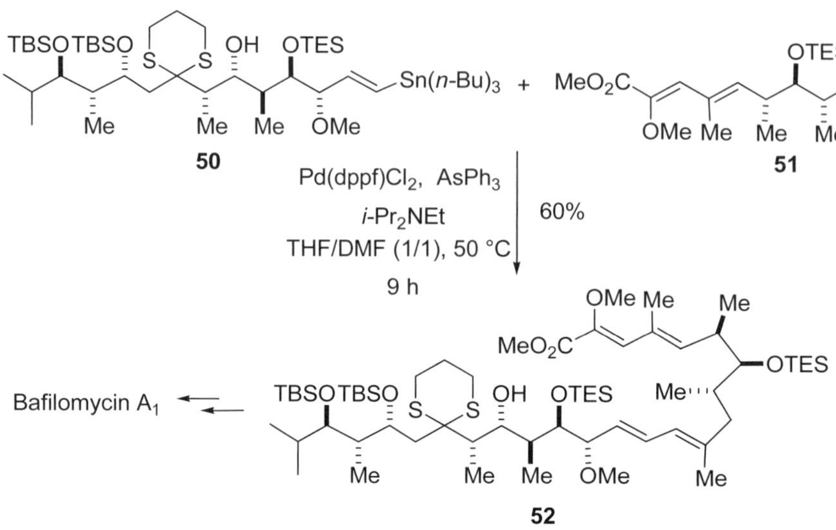

In their total synthesis of bafilomycin A_1 Hanessian and co-workers used a Stille coupling to forge the carbon carbon bond between two advanced intermediates **50** and **51**.[60] The coupling was catalyzed by Pd(dppf)Cl_2 in presence of triphenylarsine as ligand and proceeded cleanly at 50 °C in a mixture of THF/DMF as solvent to afford the desired advanced intermediate **52** possessing all the carbon backbone of the natural product.

In their total synthesis of palominol, Corey and Snyder capitalized on a π-allyl Stille reaction early in the synthesis between 1-substituted vinylstannane **53** and allyl acetate **54** to produce intermediate **55** in excellent yield.[61]

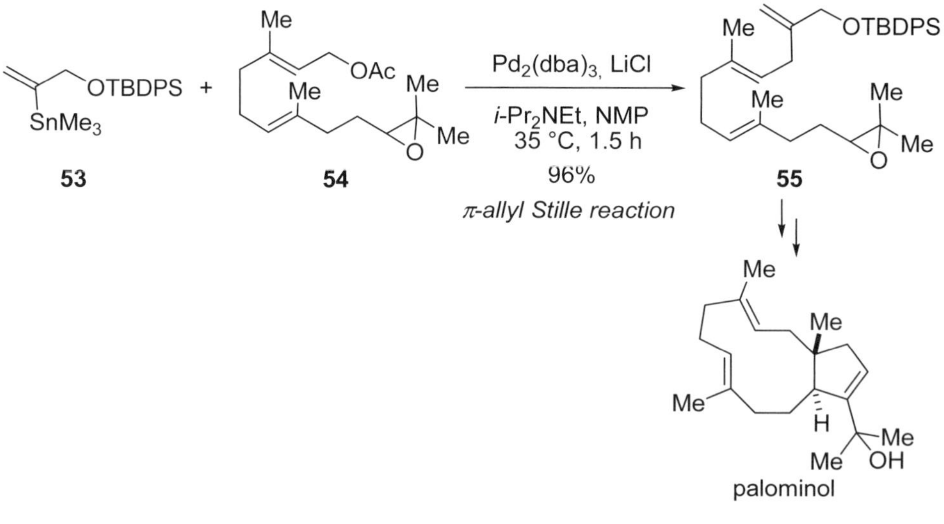

It is worth noting that under these conditions, the 1-substituted vinylstannane behaved well (1-substituted vinylstannanes tend to behave poorly in Stille reactions due to steric hindrance) and that no isomeric mixture of products was obtained (palladium mediated couplings on farnesyl substrates had led in the past to isomeric mixtures of products). The newly formed carbon $C(sp^3)–C(sp^2)$ bond resulted from the attack at the less hindered terminus of the π-allyl intermediate (which is a usual trend in this type of couplings).

Recently, in the course of studies directed towards the total synthesis of saudin, the Stoltz group developed a tandem Stille-oxa-electrocyclization reaction to access substituted pyran systems (**56** + **57** → **58**)[62].

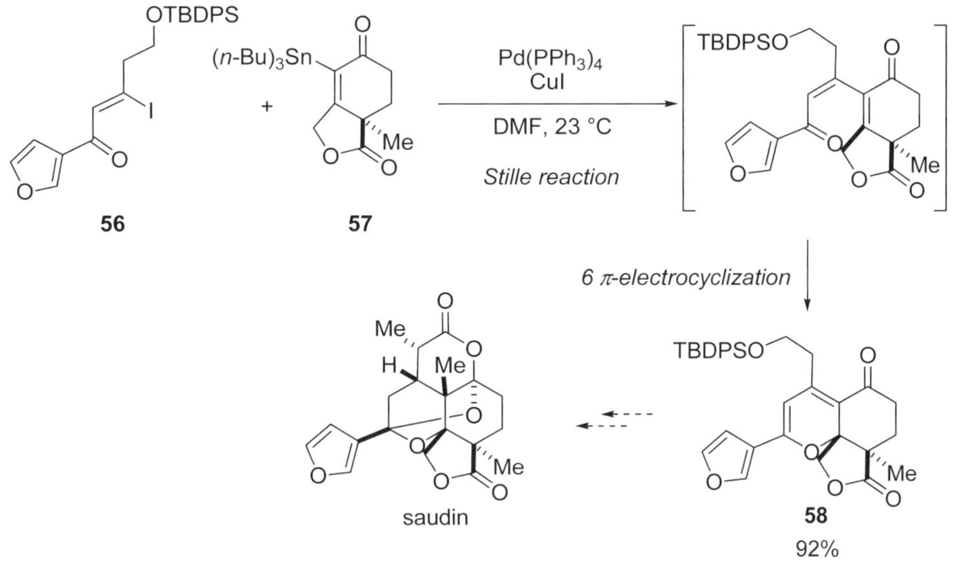

It is interesting to note that if the coupling partners were reversed (**59** + **60** → **58**), the coupling was more tricky; the use of a stoechiometric amount of CuI under anaerobic conditions proved to be crucial to get high yields of the desired product.

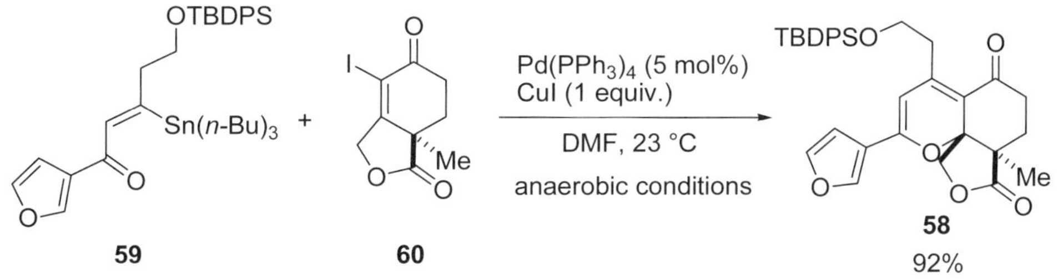

It was postulated that in presence of air copper (I) is oxidized to copper (II) and promotes the homocoupling[63] of the vinylstannane **59** to the detriment of the formation of the desired product.

One pot cascade sequences involving a Stille coupling (Heck/Stille, Heck/ carbonylative Stille, Stille/Diels Alder, Stille/electrocyclization, *etc.*) have been reported in literature.[58,64]

Process chemists at Pfizer took advantage of the Stille coupling reaction between an iodothienopyridine **61** and 5-(tributylstannyl)-1-imidazole **62** to produce **63**, an intermediate in the synthesis of a cGMP bulk lot of VEGFR kinase inhibitor **64**.[65]

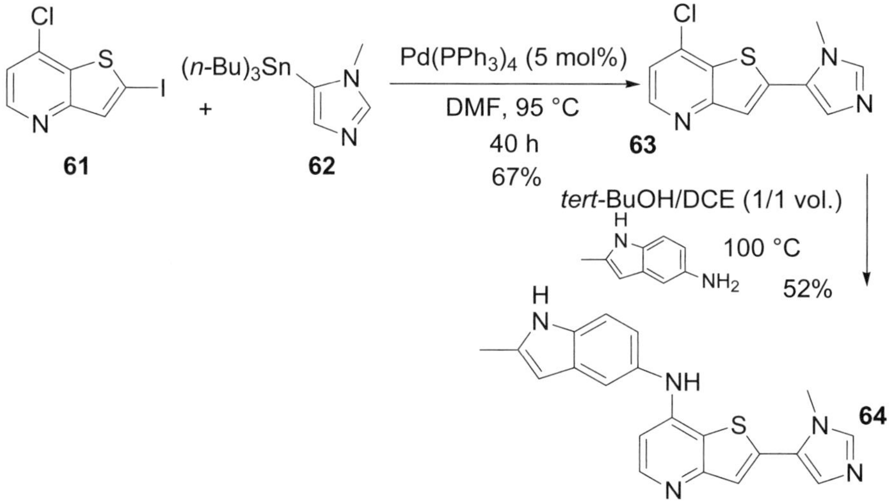

From all the palladium cross coupling conditions tried (Suzuki–Miyaura, Heck, Negishi, Hiyama, Kumada–Tamao, Kobayashi, DeShong), only the Stille coupling proved to be reliable on scale over 50 g. It is also interesting to note that adequate work-up allowed for getting below the acceptable upper limit of 20 ppm of stannane content (analytically determined by ICP).

1.1.6.6 Experimental

Caution: Tin reagents particularly trialkyl organostannanes are toxic chemicals[44,45] and extreme care should be exercised when manipulating them (avoiding inhalation and contact as well as working in a well ventilated hood is very important).

A useful trick when doing the work up of a Stille coupling (or of any reaction involving formation of non polar tin by-products) is to do a biphasic

MeCN/hexanes extraction if solubility of the desired organic product allows. (n-Bu_3)SnCl is soluble in hexanes but not in MeCN.

(n-Bu)$_3$SnX can also be removed by washing with aqueous KF solution to form polymeric (n-Bu)$_3$SnF which can be removed by filtration.

Washing with a dilute aqueous solution of ammonium hydroxide also helps making trialkyltin halides more water soluble.

General procedure for the Stille coupling of unactivated secondary alkyl halides reported by Fu:[42]

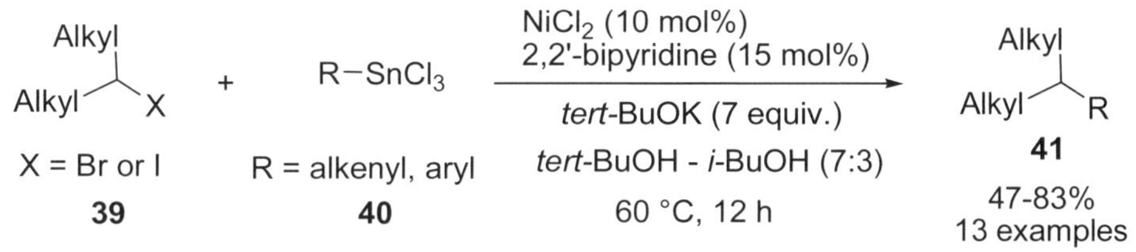

Into a 20 mL sample vial was added $NiCl_2$ (13 mg, 0.10 mmol) and 2,2′-bipyridine (23 mg, 0.15 mmol); these compounds were weighed in air, and no special handling precautions were employed. The vial was sealed with a septum screw-cap and purged under a steady stream of argon (nitrogen may also be used) for 20 min. A 1.0 M stock solution of *tert*-BuOK in *tert*-BuOH:*i*-BuOH (7.0 mL, 7.0 mmol of *tert*-BuOK) was added, followed by the aryltrichlorotin reagent (**40**, 1.2 mmol) and the alkyl halide (**39**, 1.0 mmol). The reaction mixture was heated in an oil bath to 60 °C for 12 h, with vigorous stirring under an argon atmosphere (the reaction mixture typically turns deep purple within 1 h of heating). Then, the cooled reaction mixture was poured into a separatory funnel that contained an aqueous 1 M HCl solution (50 mL), and the mixture was extracted with Et_2O (3 × 20 mL). The combined organic layers were washed with brine (40 mL), dried over $MgSO_4$, filtered and concentrated under reduced pressure. The resulting crude reaction mixture was purified by column chromatography.

Note: the reaction is air- and moisture-sensitive. The *tert*-BuOK solution and the tin reagent should be handled/stored under nitrogen or argon.

Stille carbonylative coupling:[59]

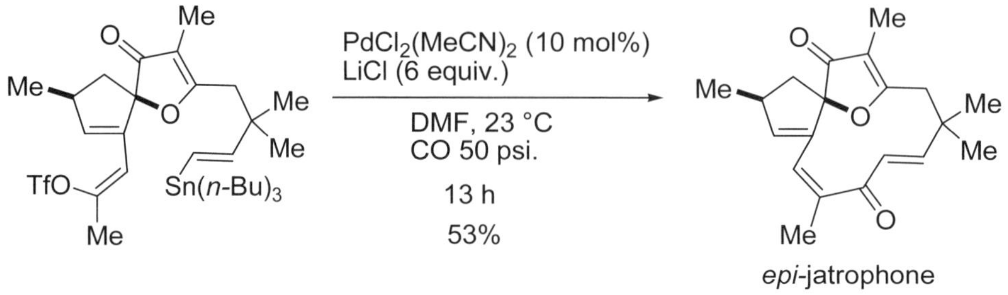

To a Fisher–Porter tube were added the vinylic triflate (30 mg, 0.042 mmol) in 10 mL of DMF and LiCl (11 mg, 0.26 mmol). The solution was bubbled with carbon monoxide for 30 min. This was followed by the addition of $PdCl_2(MeCN)_2$ (1 mg, 0.004 mmol) in 3 mL of DMF. The tube was then pressurized to 50 psi and the mixture stirred at room temperature for 13 h, after which time palladium black had precipitated out of solution. The tube was vented and the solution taken up in ether and washed with water. The organic layer was dried over Na_2SO_4. Filtration and removal of solvent under reduced pressure afforded an oily residue that was purified by column chromatography on silica gel with 25% EtOAc/hexanes to give 7 mg (53%) of product as a white solid.

π-Allyl Stille coupling:[61]

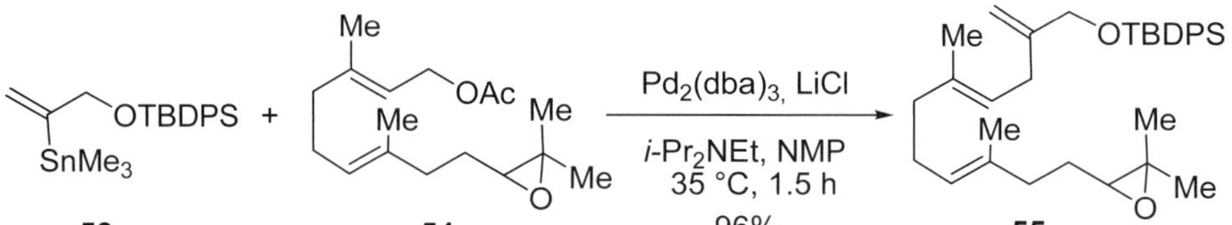

53 **54** **55**

A solution of epoxy acetate **54** (2.07 g, 7.39 mmol, 1.2 equiv.) in 2-methyl-*N*-pyrrolidinone (17 mL) was added via cannula to a reaction vessel containing $Pd_2(dba)_3$ (677 mg, 0.739 mmol, 0.12 equiv) and flame dried LiCl (1.25 g, 29.6 mmol, 4.7 equiv.) at 25 °C. A 3 mL wash of 2-methyl-*N*-pyrrolidinone was used to quantitate the transfer. A solution of stannane **53** (2.90 g, 6.31 mmol, 1.0 equiv) and *i*-Pr_2NEt (2.58 mL, 14.8 mmol, 2.3 equiv.) in 2-methyl-*N*-pyrrolidinone (17 mL) was then added via cannula, again using a 3 mL wash of 2-methyl-*N*-pyrrolidinone to quantitate the transfer. The resultant dark red solution was stirred at 25 °C for 10 min, and then was warmed to 35 °C and stirred for an additional 1.5 h. Upon completion, the reaction contents were cooled to 25 °C, poured into saturated aqueous NH_4Cl (100 mL), and extracted with EtOAc (3 × 50 mL). The combined organic layers were then washed with water (50 mL) and brine (50 mL), dried over $MgSO_4$ and concentrated. The resultant red-brown residue was purified by flash chromatography over silica gel (hexanes/EtOAc, 1:0 to 4:1) to give Stille coupling product **55** (3.12g, 96% yield).

Large scale Stille coupling:[65]

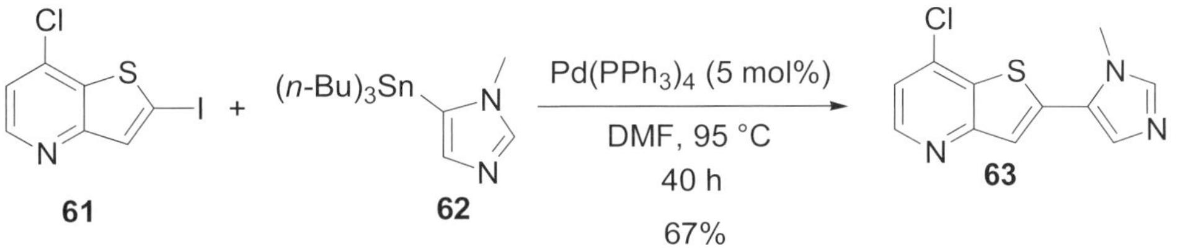

A 22-L, three-neck, round-bottom flask equipped with a mechanical stirrer was charged with stannane **62** (672 g, 1.81 mol), **61** (535 g, 1.81 mol), $Pd(PPh_3)_4$ (105 g, 0.091 mol, 5 mol%) and DMF (2.7 L), and heated to 95 °C under nitrogen. After 40 h, HPLC analysis indicated complete conversion. The reaction mixture was cooled to 10 °C and quenched by the addition of 1 N HCl (5.3 L). EtOAc (4.2 L) was added, and the mixture was filtered. The layers were seprarated, and the aqueous phase was extracted with two 4-L portions of EtOAc. The combined organic layers were washed with water (3 L). The aqueous extracts were combined, the pH was adjusted to 10–10.5 as needed. HPLC analysis indicated essentially complete extraction of product from the aqueous phase at this point. The organic extracts were combined, washed with water (three portions of 4 L each) and brine (2 L), and concentrated under vacuum to provide a tacky solid. MTBE (4 L) was added, and the mixture was concentrated under vacuum. An additional 3 L of MTBE was then added, and the resulting slurry was stirred for 2 h. The solids were collected by filtration, rinsing with MTBE. After drying at 40 °C under vacuum, the product (**63**) was obtained as an off-white solid (302 g, 1.21 mol, 67% yield).

1.1.6.7 References

1. The original report by Stille was between an acid chloride and an organostannane; see section 1.1.6.2.
2. See section 1.1.6.4.
3. Dubbaka, S. R.; Vogel, P. *J. Am. Chem. Soc.* **2003**, *125*, 15292.
4. (a) Kikiukawa, K.; Kono, K.; Wada, F.; Matsuda, T. *J. Org. Chem.* **1983**, *48*, 1333. (b) Kikiukawa, K.; Idemoto, T.; Katayama, A.; Kono, K.; Wada, F.; Matsuda, T. *J. Chem. Soc. Perkin Trans. 1* **1987**, 1511. (c) Dughera, S. *Synthesis* **2006**, 1117. For a review on the use of diazonium salts in palladium catalyzed cross-coupling reactions see [R] Roglans, A.; Pla-Quintana, A.; Moreno-Manas, M. *Chem. Rev.* **2006**, *106*, 4622.

5. See section 1.1.6.5.
6. [R] (a) Farina, V.; Krishnamurthy, V.; Scott, W. K. *Organic Reactions, vol. 50*, Wiley, New York **1997**. [R] (b) Lee, A. S.-Y.; Dai, W.-C. *Tetrahedron* **1997**, *53*, 859.
7. [R] A very thorough investigation organized by type of substrate has been reported in the following book: Farina, V.; Krishnamurthy, V.; Scott, W. J. *The Stille Reaction*; John Wiley & Sons: New York **1998**.
8. See section 1.1.6.4.
9. Azarian, D; Dua, S. S.; Eaborn, C.; Walton, D. R. M. *J. Organomet. Chem.* **1976**, *117*, C55–C57.
10. (a) Kosugi, M.; Shimizu, Y.; Migita, T. *Chem. Lett.* **1977**, 1423. (b) Kosugi, M.; Shimizu, Y.; Migita, T. *J. Organomet. Chem.* **1977**, *129*, C36. For a historical note, see also Kosugi, M.; Fugami, K. *J. Organomet. Chem.* **2002**, *653*, 50.
11. Kosugi, M.; Sasazawa, K.; Shimizu, Y.; Migita, T. *Chem. Lett.* **1977**, 301.
12. Milstein, D.; Stille, J. K. *J. Am. Chem. Soc.* **1978**, *100*, 3636.
13. For a review, see Stille, J. K. *Angew. Chem., Int. Ed. Engl.* **1986**, *25*, 508.
14. Also referred sometimes as the Migita-Kosugi-Stille coupling.
15. (a) Collman, J. P.; Hegedus, L. S.; Norton, J. R.; Finke, R. G. *Principles and Applications of Organotransition Metal Chemistry*; University Science Books; Mill Valley, CA **1987**. (b) Negishi, E.; Takahashi, T.; Baba, S.; Van Horn, D. E.; Okukado, N. *J. Am. Chem. Soc.* **1987**, *109*, 2393. For an excellent review on the mechanisms of the Stille reaction see: Espinet, P.; Echavarren, A. M. *Angew. Chem., Int. Ed. Engl.* **2004**, *43*, 4704. See also: Casares, J. A.; Espinet, P.; Salas, G. *Chem. Eur. J.* **2002**, *8*, 4843.
16. Casado, A. L.; Espinet, P.; Gallego, A. M. *J. Am. Chem. Soc.* **2000**, *122*, 11771.
17. Casado, A. L.; Espinet, P.; Gallego, A. M.; Martinez-Ilarduya, J. M. *Chem. Commun.* **2001**, 339.
18. (a) Casado, A. L.; Espinet, P. *J. Am. Chem. Soc.* **1998**, *120*, 8978. (b) Casado, A. L.; Espinet, P.; Gallego, A. M. *J. Am. Chem. Soc.* **2000**, *122*, 11771. (c) Casado, A. L.; Espinet, P.; Gallego, A. M.; Martinez-Ilarduya, J. M. *Chem. Commun.* **2001**, 339. (d) Amatore, C.; Bahsoun, A. A.; Jutand, A.; Meyer, G.; Ntepe, A. N.; Ricard, L. *J. Am. Chem. Soc.* **2003**, *125*, 4212. (e) Santos, L. S.; Rosso, G. B.; Pilli, R. A.; Eberlin, M. N. *J. Org. Chem.* **2007**, *72*, 5809. (f) Perez-Temprano, M H.; Nova, A.; Casares, J. A.; Espinet, P. *J. Am. Chem. Soc.* **2008**, *130*, 10518.
19. Vicente, J.; Arcas, A.; Bautista, D.; Jones, P. G. *Organometallics* **1997**, *16*, 2127.
20. For a review on the influence of chloride ions in palladium catalysed reaction see: Jutand, A. *Appl. Organometal. Chem.* **2004**, *18*, 574.
21. Farina, V. Kapadia, S.; Krishnan, B.; Wang, C.; Liebeskind, L. S. *J. Org. Chem.* **1994**, *59*, 5905.
22. Farina, V.; Krishnan, B. *J. Am. Chem. Soc.* **1991**, *113*, 9585.
23. (a) Falck, J. R.; Bhatt, R. K.; Ye, J. *J. Am. Chem. Soc.* **1995**, *117*, 5973. (b) Takeda, T.; Matsunaga, K.-I.; Kabasawa, Y.; Fujiwara, T. *Chem Lett.* **1995**, 771. (c) Piers, E. Romero, M. A. *J. Am. Chem. Soc.* **1996**, *118*, 1215. (d) Kang, S. K.; Kim, J. S.; Choi, S. C. *J. Org. Chem.* **1997**, *62*, 4208. (e) Kang, S. K.; Kim, J. S.; Yoon, S. K.; Lim, K. H.; Yoon, S. S. *Tetrahedron Lett.* **1998**, *39*, 3011. (f) Kang, S. K.; Kim, W. Y.; Jiao, X. G. *Synthesis* **1998**, 1252. (g) Naso, F.; Babudri, F.; Farinola, G. M. *Pure Appl. Chem.* **1999**, *71*, 1485.
24. Allred, G. D.; Liebeskind, L. S. *J. Am. Chem. Soc.* **1996**, *118*, 2748.
25. Roethle, P. A.; Chen, I. T.; Trauner, D. *J. Am. Chem. Soc.* **2007**, *129*, 8960.
26. Han, X.; Stoltz, B. M.; Corey, E. J. *J. Am. Chem. Soc.* **1999**, *121*, 7600.
27. Nielsen, T. E.; Le Quement, S.; Juhl, M.; Tanner, D. *Tetrahedron* **2005**, *61*, 8013.
28. Kang, S. K.; Kim, J. S.; Choi, S. C. *J. Org. Chem.* **1997**, *62*, 4208.
29. (a) Percec, V.; Bae, J. Y.; Hill, D. H. *J. Org. Chem.* **1995**, *60*, 6895. (b) Cui, D.-M.; Hashimoto, N.; Ikeda, S.-I.; Sato, Y. *J. Org. Chem.* **1995**, *60*, 5752. (c) Shirakawa, E.; Yamasaki, K.; Hiyama, T. *J. Chem. Soc., Perkin Trans. 1* **1997**, 2449. (d) Shirakawa, E.; Yamasaki, K.; Hiyama, T. *Synthesis* **1998**, 1544.
30. Vedejs, E.; Haight, A. R.; Moss, W. O. *J. Am. Chem. Soc.* **1992**, *114*, 6556. See also Brown, J. M.; Pearson, M.; Jastrzebski, J. T. B. H.; van Koten, G. *J. Chem. Soc., Chem. Commun.* **1992**, 1440.

31. Garcia Martinez, A.; Osio Barcina, J.; del Rosario Colorado Heras, M.; de Fresno Cerezo, A. *Org. Lett.* **2000**, *2*, 1377.
32. Grasa, G. A.; Nolan, S. P. *Org. Lett.* **2001**, *3*, 119.
33. (a) Herve, A.; Rodriguez, A. L.; Fouquet, E. *J. Org. Chem.* **2005**, *70*, 1953. See also Bourdier, T.; Huiban, M.; Huet, A.; Sobrio, F.; Fouquet, E.; Perrio, C.; Barré, L. *Synthesis* **2008**, 978.
34. (a) Mee, S. P. H.; Lee, V.; Baldwin, J. E. *Angew. Chem., Int. Ed. Engl.* **2004**, *43*, 1132. (b) Mee, S. P. H.; Lee, V.; Baldwin, J. E. *Chem. Eur. J.* **2005**, *11*, 3294.
35. Littke, A. F.; Fu, G. C. *Angew. Chem., Int. Ed. Engl.* **1999**, *38*, 2411. For a review on the palladium catalyzed coupling reactions of aryl chlorides see [R] Littke, A. F.; Fu, G. C. *Angew. Chem., Int. Ed. Engl.* **2002**, *41*, 4176.
36. Littke, A. F.; Schwarz, L.; Fu, G. C. *J. Am. Chem. Soc.* **2002**, *124*, 6343.
37. Menzel, K.; Fu, G. C. *J. Am. Chem. Soc.* **2003**, *125*, 3718.
38. [R] (a) Luh, T.-Y.; Leung, M.-K.; Wong, K.-T. *Chem. Rev.* **2000**, *100*, 3187. (b) Cardenas, D. J. *Angew. Chem., Int. Ed. Engl.* **2003**, *42*, 384.
39. (a) for α-bromoketones see: Kosugi, M.; Takano, I.; Sakurai, M.; Sano, H.; Migita, T. *Chem. Lett.* **1984**, 1221. (b) for α-halolactones see: Simpson, J. H.; Stille, J. K. *J. Org. Chem* **1985**, *50*, 1759. (c) for 1-bromo-1-phenylethane see: Sustmann, R.; Lau, J.; Zipp, M. *Tetrahedron Lett.* **1986**, *27*, 5207. (d) α-chloroether see: Bhatt, R. K.; Shin, D.-S.; Falck, J. R.; Mioskowski, C. *Tetrahedron Lett.* **1992**, *33*, 4885. (e) for fluorinated alkyl iodides see: Shimizu, R.; Fuchikami, T. *Tetrahedron Lett.* **1996**, *37*, 8405 and Shimizu, R.; Fuchikami, T. *Tetrahedron Lett.* **2001**, *42*, 6891.
40. Tang, H.; Menzel, K.; Fu, G. C. *Angew. Chem. Int. Ed. Engl.* **2003**, *42*, 5079.
41. For a review on the catalysts developed for the cross coupling of non activated alkyl halides see also: [R] Frisch, A. C.; Beller, M. *Angew. Chem., Int. Ed. Engl.* **2005**, *44*, 674.
42. Powell, D. A.; Maki, T.; Fu, G. C. *J. Am. Chem. Soc.* **2005**, *127*, 510.
43. (a) Shirakawa, E.; Yamasaki, K.; Hiyama, T. *J. Chem. Soc., Perkin Trans. 1* **1997**, 2449. (b) Shirakawa, E.; Yamasaki, K.; Hiyama, T. *Synthesis* **1998**, 1544. See also, Cui, D.-M.; Hashimoto, N.; Ikeda, S.-I.; Sato, Y. *J. Org. Chem.* **1995**, *60*, 5752.
44. [R] Bulten, E. J.; Meinema, H. A. *Metals and their Compounds in the Environment*; Merian E.; Ed.; VCH: New York, **1991**; chapter II, 30, p1243.
45. [R] Davie, A. G.; Smith, P. G. *Comprehensive Organometallic Chemistry*; Wilkinson, G.; Stone, F. G. A., Ed.; Pergamon, New York, **1982**, vol. 2, chapter 11.
46. See for example: (a) Roshchin, A.; Bumagin, N. A.; Beletskaya, I. P. *Tetrahedron Lett.* **1995**, *36*, 125. (b) Rai, R.; Aubrecht, K. B.; Collum, D. B. *Tetrahedron Lett.* **1995**, *36*, 3111. For a review on the developments of palladium catalyzed reactions in aqueous media see also: Genet, J.-P.; Savignac, M. *J. Organomet. Chem.* **1999**, *576*, 305.
47. [R] (a) For a recent review see: Testero, S. A.; Mata, E. G. *J. Comb. Chem.* **2008**, *10*, 487. (b) Gerlach, M.; Jordens, F.; Kuhn, H.; Neumann, W. P.; Peterseim, M. *J. Org. Chem.* **1991**, *56*, 5971. (c) Kuhn, H.; Neumann, W. P. *Synlett* **1994**, 123. (d) Deshpande, M. S. *Tetrahedron Lett.* **1994**, *35*, 5613. (e) Hernan, A. G.; Guillot, V.; Kuvshinov, A.; Kilburn, J. D. *Tetrahedron Lett.* **2003**, 44, 8601. (f) Chrétien, J.-M.; Mallinger, A.; Zammattio, F.; Le Grognec, E.; Paris, M.; Montavon, G.; Quintard, J.-P. *Tetrahedron Lett.* **2007**, 48, 1781. (g) Lau, K. C. Y.; Chiu, P. *Tetrahedron Lett.* **2007**, 48, 1813. (h) Zaho, H.; wang, Y.; Sha, J.; Sheng, S.; Mingzhong, C. *Tetrahedron* **2008**, 64, 7517. (i) Charette, A. B.; Roy, M.-N. *Synthesis and Applications of New and Improved Tetraarylphosphonium-supported Tin Reagents*; abstract of the 236th ACS national meeting, Philadelphia, PA, August 17-21 **2008**.
48. Nicolaou, K. C., Winssinger, N.; Pastor, J.; Murphy, F. *Angew. Chem., Int. Ed. Engl.* **1998**, *37*, 2534.
49. Vitz, J.; Mac, D. H.; Legoupy, S. *Green Chem.* **2007**, *9*, 431. For a review on the use of ionic liquid in organometallic catalysis see: [R] Dupont, J.; de Souza, R. F.; Suarez, P. A. Z. *Chem. Rev.* **2002**, *102*, 3667.
50. Hoshino, M.; Degenkolb, P.; Curran, D. P. *J. Org. Chem.* **1997**, *62*, 8341.
51. Larhed, M.; Hoshino, M.; Hadida, S.; Curran, D. P.; Hallberg, A. *J. Org. Chem.* **1997**, *62*, 5583.
52. Appukkuttan, P.; Van der Eycken, E. *Eur. J. Org. Chem.* **2008**, 1133.

53. (a) Gallagher, W. P.; Maleczka, Jr. R. E. *J. Org. Chem.* **2005**, *70*, 841. (b) Maleczka, Jr. R. E.; Gallagher, W. P. *Org. Lett.* **2001**, *3*, 4173.
54. For the Sn–O route see: (a) Maleczka, Jr. R. E.; Gallagher, W. P.; Terstiege, I. *J. Am. Chem. Soc.* **2000**, *122*, 384. (b) Gallagher, W. P.; Terstiege, I.; Maleczka, Jr. R. E.; *J. Am. Chem. Soc.* **2001**, *123*, 3194.
55. Kelly, T. R.; Li, Q.; Bhushan, V. *Tetrahedron Lett.* **1990**, *31*, 161. For additional examples see: Mori, M.; Kaneta, N.; Shibasaki, M. *J. Org. Chem.* **1991**, *56*, 3486. Fukuyama, Y.; Yaso, H.; Mori, T.; Takahashi, H.; Minami, H.; Kodama, M. *Heterocycles* **2001**, *54*, 259. Yue, W. S.; Li, J. J. *Org. Lett.* **2002**, *4*, 2201.
56. Nicolaou, K. C.; Chakraborty, T. K.; Piscopio, A. D.; Minowa, N.; Bertinato, P. *J. Am. Chem. Soc.* **1993**, *115*, 4419.
57. Shair, M. D.; Yoon, T.-Y.; Mosny, K. K.; Chou, S. J.; Danishefsky, S. J. *J. Am. Chem. Soc.* **1996**, *118*, 9509.
58. [R] (a) For a review on the Stille reaction in the synthesis of natural products see: De Souza, M. V. N. *Current Organic Synthesis* **2006**, *3*, 313. [R] (b) For a review on the palladium catalyzed cross coupling reactions in total synthesis see: Nicolaou, K. C.; Bulger, P. G.; Sarlah, D. *Angew. Chem., Int. Ed. Engl.* **2005**, *44*, 4442. [R] (c) For reviews on the intramolecular Stille reaction in synthesis see: Duncton, M. A. J.; Pattenden, G. *J. Chem. Soc., Perkin Trans. 1* **1999**, 1235. Pattenden, G.; Sinclair, D. J. *J. Organomet. Chem.* **2002**, *653*, 261.
59. Gyorkos, A. C.; Stille, J. K.; Hegedus, L. S. *J. Am. Chem. Soc.* **1990**, *112*, 8465.
60. Hanessian, S.; Ma, J.; Wang, W. *J. Am. Chem. Soc.* **2001**, *123*, 10200.
61. Snyder, S. A.; Corey, E. J. *J. Am. Chem. Soc.* **2006**, *128*, 740.
62. Tambar, U. K.; Kano, T.; Zepernick, J. F.; Stoltz, B. M. *Tetrahedron Lett.* **2007**, *48*, 345.
63. Han, X.; Corey, E. J. *Org. Lett.* **1999**, *1*, 1871.
64. For examples see: (a) Kosugi, M.; Tamura, H.; Sano, H.; Migita, T. *Chem. Lett.* **1987**, 193. (b) Larock, R. C.; Lee, N. H. *J. Org. Chem.* **1991**, *56*, 6253. (c) Grigg, R.; Redpath, J.; Sridharan, V.; Wilson, D. *Tetrahedron Lett.* **1994**, *35*, 4429.
65. Ragan, J. A.; Raggon, J. W.; Hill, P. D.; Jones, B. P.; McDermott, R. E.; Munchhof, M. J.; Marx, M. A.; Casavant, J. M.; Cooper, B. A.; Doty, J. L.; Lu, Y. *Org. Proc. Res. & Dev.* **2003**, *7*, 676.

1.1.7 The Suzuki Reaction

John P. Wolfe and Josephine S. Nakhla

1.1.7.1 Description

The Suzuki reaction (also referred to as the Suzuki–Miyaura reaction) is the palladium-catalyzed cross-coupling of an alkenyl-, aryl-, or heteroaryl halide or pseudohalide (R–X) with an alkyl, alkenyl, aryl, or heteroaryl boron reagent (R′–BY_2). This transformation leads to formation of a carbon-carbon bond with stereospecific and regiospecific replacement of X with R′.[1–9] In a representative example, boronic acid **1** was coupled with bromobenzene in the presence of aqueous Na_2CO_3 and 3 mol% $Pd(PPh_3)_4$ in refluxing DME to provide biaryl derivative **2** in 85% yield.[10]

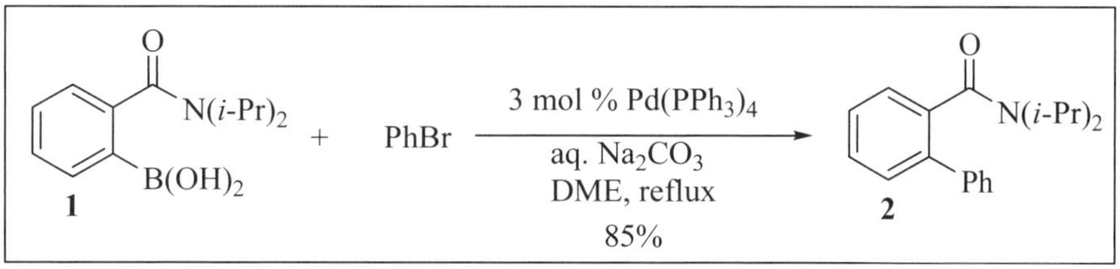

1.1.7.2 General Trends

Through extensive experimentation, several reactivity trends have been established that hold true for most Suzuki coupling reactions.[1–9] In general, the order of reactivity of the electrophilic component is alkenyl–X > aryl–X > alkyl–X. The nature of the halogen also has an impact on reactivity, with the general trend of I > Br ~ OTf >> Cl. In addition, sterically hindered electrophiles are typically less reactive than unhindered derivatives. These trends can be exploited for selective coupling of substrates bearing two or more different halogens, and in some instances selectivity can be obtained in reactions of substrates that contain two or more of the same halogen.[11] In terms of the nucleophilic component, steric hindrance also leads to decreased reactivity, and the order of reactivity with respect to carbon hybridization is spC–B > sp^2C–B > sp^3C–B.

1.1.7.3 Comparison with Other Cross-Coupling Reactions

The Suzuki reaction is a member of a class of transformations that are typically referred to as cross-coupling reactions,[12–15] which involve the coupling of an aryl halide or related electrophile with a main-group organometallic reagent. Many of these processes are also named reactions, and the most common variations employ Grignard reagents (Kumada coupling-chapter 1.1.3),[16,17] organotin reagents (Stille Coupling-chapter 1.1.6),[18] organozinc reagents (Negishi coupling-chapter 1.1.4),[19,20] or organosilicon reagents (Hiyama coupling-chapter 1.1.2)[21,22] as the main group organometallic species.

The Suzuki reaction has many advantages over these related cross-coupling reactions. A large number of organoboron coupling partners are commercially available, including a vast array of heteroaromatic derivatives. The preparation of organoboron compounds can also be accomplished through several different methods, the most common of which involves treatment of a Grignard or organolithium reagent with a trialkoxyboron derivative.[6,23] The synthesis of organoboron reagents can be achieved under mild conditions through Pd-catalyzed coupling reactions of aryl/heteroaryl halides with bis(pinacolato)diboron,[24] and alkylboron reagents can be generated *in situ* from readily available alkene precursors via hydroboration.[3] In addition, the synthesis of organoboron reagents has also been accomplished through C–H bond functionalization.[25,26] The organoboron reagents are relatively non-toxic, particularly in comparison to organotin reagents. Moreover, the boron-containing byproducts of the reactions can easily be separated from the desired product by extraction with aqueous base. In contrast, removal of tin-containing byproducts is frequently a problem in Stille coupling reactions.[18] The organoboron reagents do not react with most common functional groups, and Suzuki reactions can be conducted under relatively mild conditions. Thus, the functional group tolerance of Suzuki reactions is usually higher than related Kumada coupling or Negishi coupling processes. These features make the Suzuki reaction one of the most valuable and extensively used methods for C–C bond formation in organic synthesis.[1–21]

Although the Suzuki reaction has many advantages over other cross-coupling reactions, there are three principal limitations of this method. The Suzuki reaction requires the use of either an inorganic base or fluoride, which can be problematic with some functionalized substrates. However, use of anhydrous conditions can help to mitigate some of these complications. The organoboron reagents can be difficult to purify, although in many cases use of high-purity reagents is not essential. Finally, in contrast to organostannanes, which are relatively nonpolar, and can be carried several steps through a synthetic sequence, organoboron reagents are fairly polar, and react with many commonly used reagents (e.g., oxidants, nucleophiles, bases). Recent studies described below have led to the development of new organoboron reagents that are more robust, and more amenable to use in multistep synthetic sequences.

1.1.7.4 Historical Perspective

Two significant examples of Pd-catalyzed or mediated cross-coupling type reactions of organoboron compounds were reported prior to Suzuki's 1979 disclosure of what has become known as the Suzuki reaction. The first example was described by Heck in 1975 in the context of studies on the mechanism of Pd-catalyzed coupling reactions between alkenyl halides and activated alkenes. As shown below, treatment of alkenyl boronic acid **3** with methyl acrylate in the presence of stoichiometric $Pd(OAc)_2$ generated conjugated diene **4** in 70% yield.[27] This reaction likely proceeds via transmetalation of the boronic acid with $Pd(OAc)_2$, insertion of the alkene into the Pd–C bond of the intermediate alkenyl palladium complex, and β-hydride elimination to provide the diene product. Importantly, the stereochemistry of the alkenylboronic acid was transferred to the diene products. A second example, which involved a Pd-catalyzed coupling of an alkynyl(tributyl)borate with iodobenzene was described at an American Chemical Society meeting by Negishi in 1977 and subsequently published in a book chapter on transition-metal catalyzed C–C bond formation.[28]

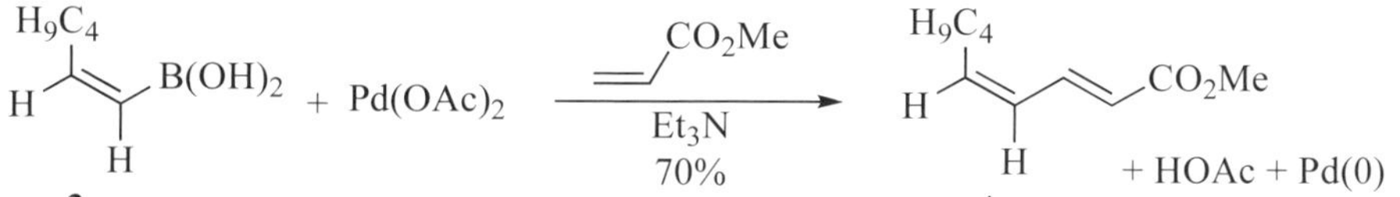

3 **4**

In 1979, Miyaura, Yamada, and Suzuki demonstrated that 1-alkenylboranes react with 1-alkenyl or 1-alkynyl halides in the presence of a base using $Pd(PPh_3)_4$ as a catalyst to afford diene or enyne products.[29] For example, treatment of (*E*)-β-bromostyrene with 1.1. equiv of alkenylboron reagent **5** in the presence of 2 equiv NaOEt and 1 mol% $Pd(PPh_3)_4$ afforded diene **6** in 80% yield. The use of base was essential for the success of these transformations, as the weakly nucleophilic organoboronic acid or ester derivatives do not undergo transmetalation to $L_nPd(Ar)(X)$ complexes at sufficient rates to facilitate catalysis.[29,30] In contrast, the reaction of the organoboron reagent with base leads to generation of a highly nucleophilic boron "ate" complex, which undergoes facile transmetalation with Pd(II).

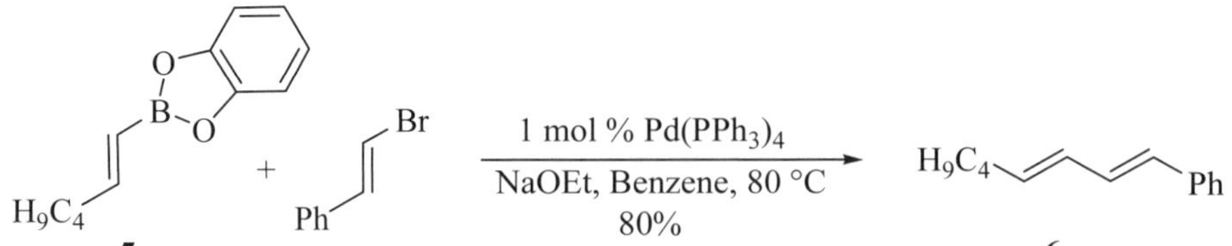

5 **6**

Since the initial communication by Suzuki and coworkers, the Suzuki reaction has evolved into one of the most commonly used methods for the construction of C–C bonds. The Suzuki reaction has found many applications in the synthesis of biologically active molecules[3,31] and useful materials.[32,33] The reactions are amenable to scale-up, and are broadly employed in the pharmaceutical industry.[34]

1.1.7.5 Mechanism

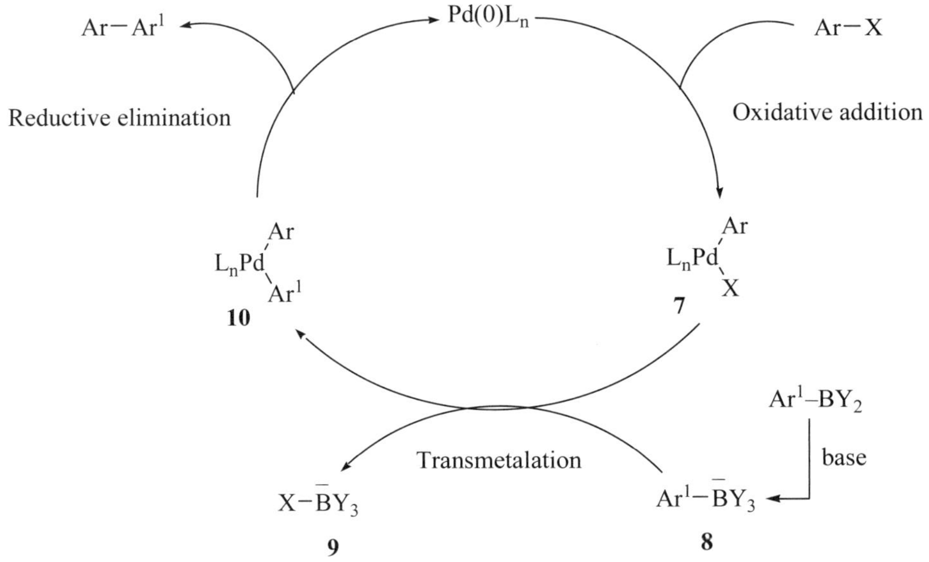

The mechanism of the Suzuki reaction is believed to be similar to that of other well-established Pd-catalyzed cross-coupling reactions.[12–14,35] The active $L_nPd(0)$ catalyst is often generated *in situ* by coordination of added phosphine ligands to a stable source of Pd(0), such as $Pd_2(dba)_3$. In cases where Pd(II) complexes such as $Pd(OAc)_2$ are used as precatalysts, these species are converted to the active $L_nPd(0)$ species under the reaction conditions. The catalytic cycle then commences with oxidative addition of the aryl halide (or related electrophile) to Pd(0), which generates Pd(II) intermediate **7**.[36] Reaction of the organoboron reagent with base leads to formation of borate **8**,[37] which undergoes transmetalation with **7** to afford Pd(Ar)(Ar[1]) complex **10** and the boron-containing byproduct **9**. Reductive elimination from **10** forms the C–C bond with concomitant regeneration of the Pd(0) catalyst. Key intermediates **7** and **10** in this catalytic cycle have been observed in the coupling reaction of 3-bromopyridine with phenylboronic acid.[38] Detailed kinetic studies on the Pd-catalyzed Suzuki coupling reaction have also been conducted, and indicate that the turnover-limiting step of the catalytic cycle is substrate-dependent.[39]

As noted above, the principal difference between the Suzuki reaction and other Pd-catalyzed cross coupling processes (e.g., Stille coupling of organostannanes with aryl/alkenyl halides or Negishi coupling of organozinc reagents with similar electrophiles) is the use of an organoboron reagent as the main-group coupling partner. Although the mechanisms of the various Pd-catalyzed cross-coupling reactions share many common features, differences between these processes are apparent in the transmetalation steps of the catalytic cycles.

Several groups have conducted studies to elucidate specific details concerning the mechanism of transmetalation from boron to palladium. Evidence suggests that two different pathways may be operable in this process. The first involves exchange of hydroxide with halogen to generate palladium hydroxide **11**, which then undergoes transmetalation with the neutral organoboron reagent through a four-centered transition state (**12**).[40] Alternatively, transmetalation may occur through reaction of the borate complex **13** with a cationic Pd(II) intermediate that derives from dissociation of the halide from the oxidative addition complex **7** shown above.[40,41] The preference for reactivity via one path over another may depend on the Lewis acidity of the organoboron reagent, along with other parameters of individual reactions. When alkylboron reagents are employed transmetalation occurs with retention of stereochemistry.[40,42]

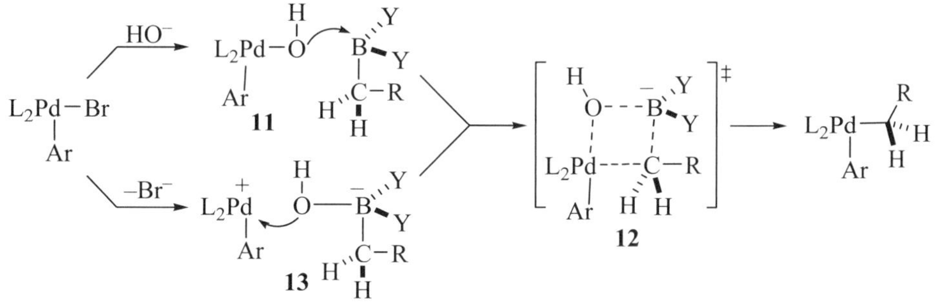

Interestingly, the relative reactivity of aryl bromides vs. aryl triflates in Suzuki coupling reactions differs from other Pd-catalyzed cross-coupling processes.[43] Brown has demonstrated that 2- and 3-bromophenyl triflate derivatives undergo selective substitution of the triflate group in Stille, Negishi, and Kumada coupling reactions, as well as Pd-catalyzed Heck reactions and *N*-arylation reactions. In contrast, the bromide group is selectively replaced in the Suzuki coupling. The exact origin of this effect remains unclear, but could arise from differences in oxidative addition, or differences in transmetalation if oxidative addition is reversible under the reaction conditions.[44]

1.1.7.6 Variations and Improvements

New, Highly Active Catalysts for Suzuki Coupling Reactions

Since the initial report by Suzuki in 1979, the optimization of reaction parameters, such as base and solvent, has been explored by many different groups.[1–6,31] Replacement of the base with KF or CsF has been demonstrated to improve functional group tolerance in some systems.[45] Use of microwave heating,[46–48] water[49] or ionic liquids[50,51] as solvents, solid supports,[52,53] or other non-traditional conditions[54] can also provide advantages. However, the research that has arguably led to the broadest expansion of scope has been focused on catalyst development.

Although many Suzuki coupling reactions can be effected with triphenylphosphine-based catalysts, which were traditionally used in these transformations, these catalysts suffer from several key limitations. In general, the reactivity of $Pd(PPh_3)_4$ and related triphenylphosphine palladium complexes is relatively low, and it is generally not feasible to employ these catalysts for transformations of unreactive substrates such as aryl chlorides, or very hindered aryl halides or boronic acids. In addition, it is difficult to carry out Suzuki reactions at low catalyst loadings with triphenylphosphine-derived catalysts, and reactions of many interesting heterocyclic compounds are also challenging. In order to address these limitations, many studies over the past ten years have focused on the development of new ligands and/or catalysts for palladium-catalyzed Suzuki coupling reactions. The description of every single ligand or catalyst that has been surveyed falls beyond the scope of this chapter. However, representative examples of several new catalysts and their utility in traditionally difficult Suzuki coupling reactions are outlined below.

In addition to recent work on the development of new homogenous catalysts, several groups have also explored the utility of heterogeneous catalyst systems for Suzuki coupling reactions.[55,56] In some instances these catalysts can be recovered and reused after catalytic reactions. However, in many cases the use of added phosphine ligands is required, and it is not clear these catalysts are effective with a similarly broad range of substrates as the homogenous catalysts described below.

Suzuki Coupling Reactions of Aryl Chlorides

The low reactivity of aryl chloride substrates, which are cheaper and often more readily available than the analogous aryl iodides or bromides, has been a longstanding problem in cross-coupling chemistry.[57,58] This problem arises from the fact that aryl chlorides undergo oxidative addition to Pd(0) much more slowly than aryl iodides or bromides. Although it has been known for some time that the rate of oxidative addition to Pd(0) can be accelerated by using electron-rich ligands, many of these ligands also slow the rate of the key C–C bond forming reductive elimination step of the catalytic cycle. A solution to this problem involves the use of electron-rich ligands that are also sterically bulky, as the size of the ligand can increase the rate of the reductive elimination step.[59] This concept has led to the development of several different new catalysts for Suzuki coupling reactions.

The first general catalyst for Suzuki coupling reactions of aryl chlorides was reported by Buchwald in 1998, and featured a new type of biaryl(dialkyl)phosphine ligand (**15**).[59] In subsequent years, Buchwald has expanded on this general structure to develop a family of ligands with broad utility in Suzuki coupling reactions (**14**–**19**).[60–63]

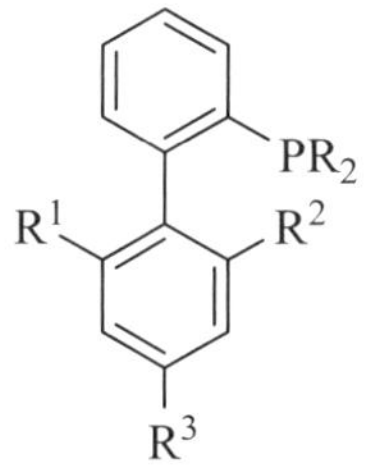

14: R = Cy; R^1, R^2, R^3 = H
15: R = Cy; R^1, R^3 = H; R^2 = NMe_2
16: R = Cy; R^1, R^2 = OMe; R^3 = H
17: R = Cy; R^1, R^2, R^3 = *i*Pr
18: R = *t*Bu; R^1, R^2, R^3 = H

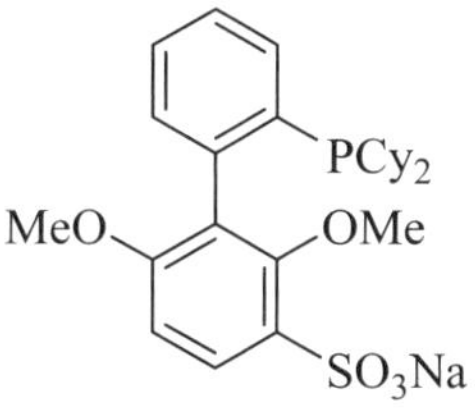

19

Ligand **19** is water soluble, and has been employed in Suzuki coupling reactions conducted in aqueous media.[64] Ligand **17** provides optimal results in Suzuki coupling reactions of aryl tosylates and aryl benzenesulfonates.[65] Ligand **16** is thought to have the greatest substrate scope of the Buchwald biaryl-phosphine derivatives, and is widely used in both academia and industry.[62] In a representative example, the reaction of 4-chloroanisole (**20**) with 2-methoxyphenyl boronic acid (**21**) using 0.5 mol% of a palladium catalyst ligated by **16** provided **22** in 98% yield in only 3 h at rt.[62]

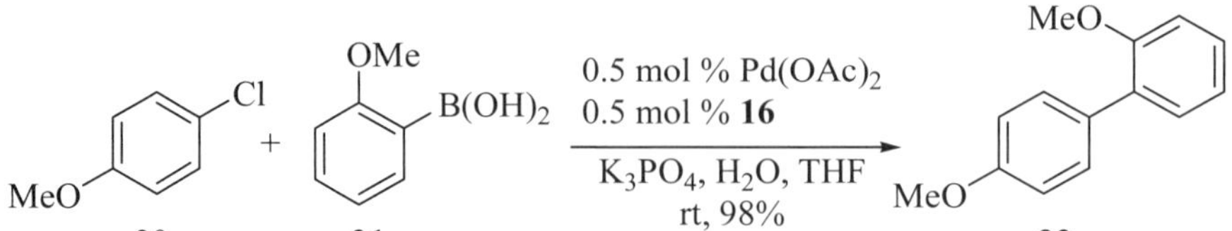

20 **21** **22**

Several other research groups have also explored the use of bulky, electron-rich phosphine ligands for Suzuki coupling reactions. Ligands **23** (Hartwig),[66] **24** (Beller),[67] **25** (Guram),[68] and P*t*-Bu$_3$ (Fu)[69] also exhibit good reactivity in transformations involving aryl chloride substrates. Hartwig has also demonstrated that dimeric (LPdBr)$_2$ complexes (L = P*t*-Bu$_3$ or PAd$_3$) are effective catalysts for Suzuki coupling reactions of aryl chlorides.[70] In addition, *N*-heterocyclic carbene ligands (e.g., **26**) have been successfully employed in Pd-catalyzed Suzuki coupling reactions of aryl chlorides,[71–73] as have palladacyclic catalysts (e.g., **27**).[74–76]

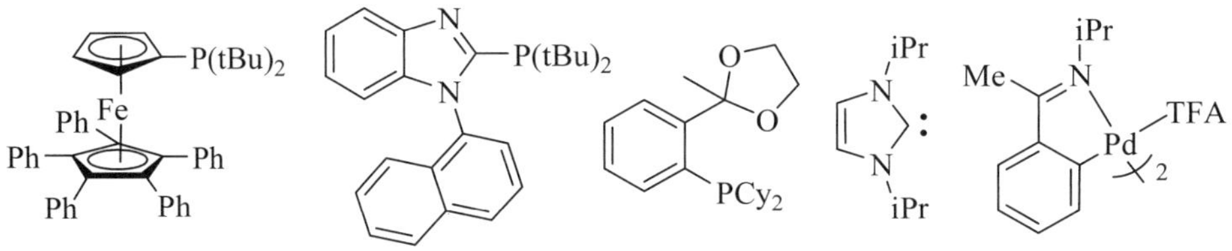

23 **24** **25** **26** **27**

The new catalysts and ligands described above are also generally effective in Suzuki coupling reactions of aryl bromides and iodides. In addition, several of the catalysts derived from these ligands are sufficiently active and long-lived to allow for use of low levels of palladium (< 0.1 mol%), which is of great significance for large scale applications.[60,73,74]

Suzuki Coupling Reactions of Heteroaromatic Compounds

Suzuki coupling reactions of heteroaromatic halides and/or heteroaryl boronic acids are often challenging transformations. Although some transformations can be effected with triphenylphosphine-derived catalysts,[77] many heteroarenes can coordinate strongly to palladium, which often leads to catalyst poisoning or low reactivity. Fortunately, several of the catalyst systems/ligands described above are also highly effective for Suzuki coupling reactions of heteroaromatic systems, and the Buchwald ligands appear to have particularly good generality.[78,79] For example, the coupling of boronic acid **29** with **28** provided **30** in 96% yield when a catalyst composed of Pd(OAc)$_2$/**16** was employed. In addition to the ligands noted above,

PCy_3,[80] *p*-substituted phenyl(PR_2) ligands (R = Cy or *t*-Bu),[81] sulfonated fluorenyl dialkylphosphines,[82] and diaryl- or dialkylphosphine oxides[83] have also shown good reactivity in Suzuki reactions of heteroaryl halides/boronic acids.

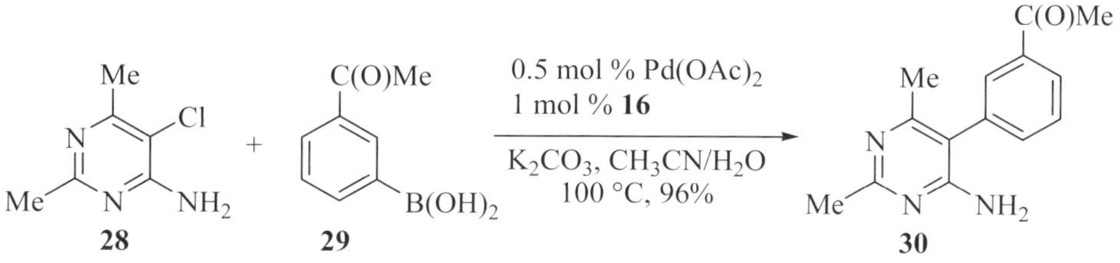

Organotrifluoroborate Reagents for Suzuki Coupling Reactions

Classical Suzuki coupling reactions typically employ either boronic acids, or boronate esters or trialkylboranes (derived from hydroboration reactions) as the nucleophilic coupling partner. However, there are several problems associated with the use, purification, and/or handling of these reagents. For example, boronic acids can be quite difficult to purify, and can undergo relatively facile dehydration to generate less reactive cyclic trimers (e.g., **31**). Boronic acids can also undergo protiodeboronation or homocoupling under typical Suzuki reaction conditions, and vinylboronic acid readily polymerizes. Some of these limitations may be overcome by use of pinacol boronate esters, which are generated by reaction of a boronic acid with pinacol. However, these reagents tend to be more expensive and less reactive than boronic acids, and their use is less atom-economical. Finally, it is challenging to carry trivalent boron compounds through multi-step syntheses due to their sensitivity to a variety of common reagents, such as oxidizing agents and Lewis bases.

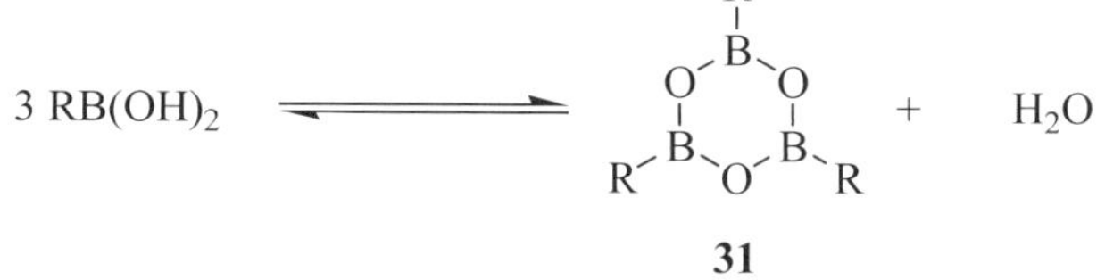

In recent years, the use of organotrifluoroborate reagents (RBF_3K) in Suzuki coupling reactions been shown to solve many of the problems described above. These reagents were originally developed by Vedejs in 1995 as air-stable precursors to aryl(difluoro)boron Lewis acids,[84] and are typically prepared in high yield by treatment of boronic acids or esters with

KHF_2.[84–87] These reagents are free-flowing solids that can easily be purified by crystallization. In addition, they have greatly increased stability to oxidants, nucleophiles, and bases, and are resistant to trimerization.

Organotrifluoroborate reagents were first employed in Suzuki coupling reactions of aryl diazonium salts by Genet[86,88] and diaryl iodonium salts by Xia and Chen.[89] Although these electrophiles can be efficiently transformed,[90] their use is relatively uncommon due to cost and safety issues. More recent studies by Molander have led to a significant expansion of the scope of this chemistry to include aryl/alkenyl halide and sulfonate electrophiles.[87] In representative examples, the coupling of **32** with **33** and **35** with **36** afforded **34** and **37** in good yields.[86,91] Interestingly, the coupling reactions of aryl halides require addition of base (e.g., Et_3N or K_2CO_3) to the reaction mixture, whereas transformations of the apparently more reactive aryl diazonium salts or diaryl iodonium salts are effective in the absence of base.

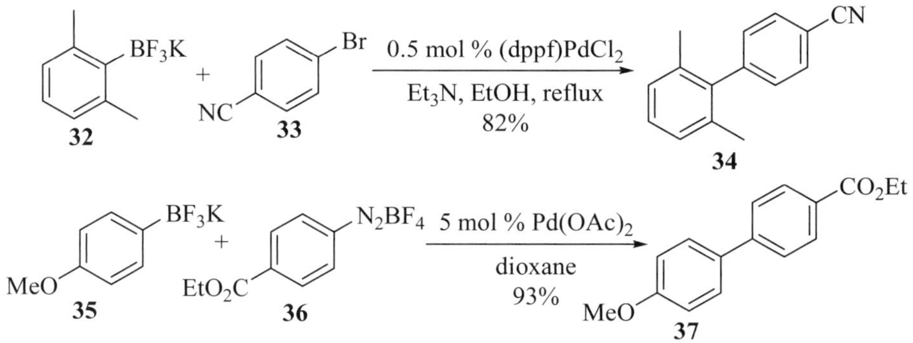

The scope of the Suzuki coupling reactions of organotrifluoroborate salts is quite broad, and examples of aryl–aryl, aryl–alkenyl, aryl–benzyl, aryl–alkyl, alkenyl–aryl, and alkenyl–alkenyl couplings have been described. These reactions can also be conducted with heteroaryl halides and/or heteroaryl trifluoroborates, and are highly tolerant of functional groups, such as ethers, nitriles, carbonyls, and halogens.[87] It is important to note that silyl ethers are also stable to the reaction conditions, even though fluoride ions may be present. The use of potassium vinyltrifluoroborate, which does not undergo polymerization that prohibits use of the analogous boronic acid, has also been realized. In addition, the transformations can be conducted using air stable, isolable alkyl trifluoroborate salts.[92] This allows for the installation of methyl groups via Suzuki coupling, which is not easily accomplished with other readily available organoboron reagents.

In addition to exhibiting excellent functional group tolerance, the organotrifluoroborate reagents are also stable towards many commonly used reagents and transformations, including nucleophilic substitution reactions,[93,94] azide dipolar cycloaddition reactions,[95] lithiation/alkylation reactions,[96] oxidations,[97] epoxidations,[98] dihydroxylations,[99] and Wittig or Horner–Wadsworth–Emmons olefinations.[100] For example, treatment of bromomethylpotassium trifluoroborate with *N*-methylpiperazine afforded **38** in 72% yield. Subsequent cross-coupling of **38** with 4-bromoanisole provided **39** in 68% yield.[94] Similarly, treatment of potassium 3-butenyltrifluroborate with OsO_4/NMO gave diol **40**, which was coupled with alkenyl bromide **41** to generate **42** in 60% yield.[99] These transformations have great potential utility in multistep complex molecule synthesis.

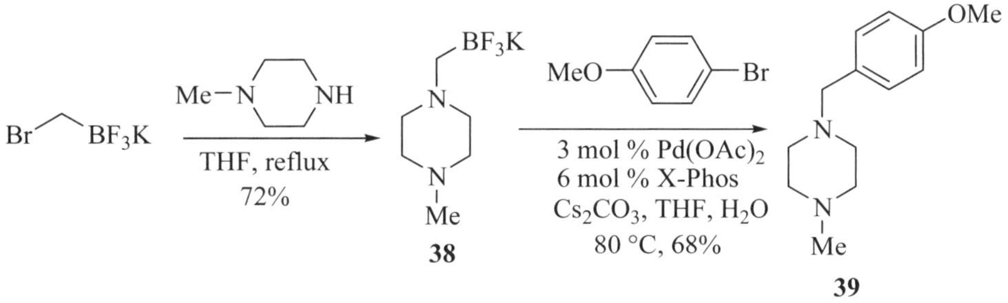

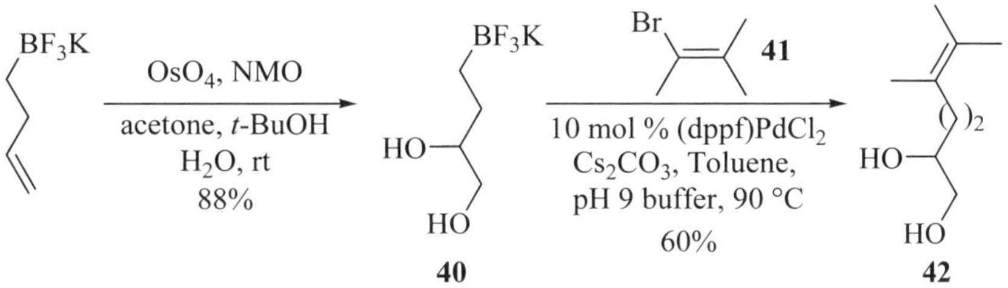

Suzuki Coupling Reactions of Unactivated Alkyl Halides and Sulfonates

Although cross-coupling reactions of aryl and alkenyl halides have been broadly studied and applied to numerous synthetic challenges over the past thirty years, the use of alkyl halides as electrophiles in these reactions was seldom reported. Cross-coupling reactions of alkyl halides are quite challenging to effect due to the fact that they are less reactive than aryl/alkenyl halides towards oxidative addition to Pd(0), and the resulting alkylpalladium(II)halide species undergo relatively facile competing β-hydride elimination that leads to alkene formation rather than cross coupling. Prior to 2001, most examples of cross-coupling reactions of alkyl halides involved substrates that did not contain β-hydrogens and/or were activated towards oxidative addition, such as benzylic halides or iodocyclopropanes.[101,102] In 1992 the first examples of Suzuki coupling reactions of alkyl halides were reported by Suzuki and Miyaura,[103] although the scope of these transformations was limited to primary alkyl iodide substrates.

In 2001 Fu described the first examples of Pd-catalyzed Suzuki coupling reactions of primary alkyl bromides with alkyl-9-BBN reagents.[104] The following year the scope of these transformations was extended to allow for use of aryl and alkenylboronic acids as nucleophiles,[105] and alkyl tosylates as electrophiles.[106] The key for the success of these reactions was use of either PCy_3 (alkyl-9-BBN reagents) or $P(t\text{-}Bu)_2Me$ (boronic acids) as ligands for palladium; other ligands provided unsatisfactory yields. In a representative example, *n*-octyl bromide (**43**) was coupled with phenylboronic acid to afford **44** in 85% yield using a catalyst composed of $Pd(OAc)_2$ and $P(t\text{-}Bu)_2Me$. Subsequent studies have indicated that NHC-ligated palladium catalysts also provide good results for some of these transformations.[107,108]

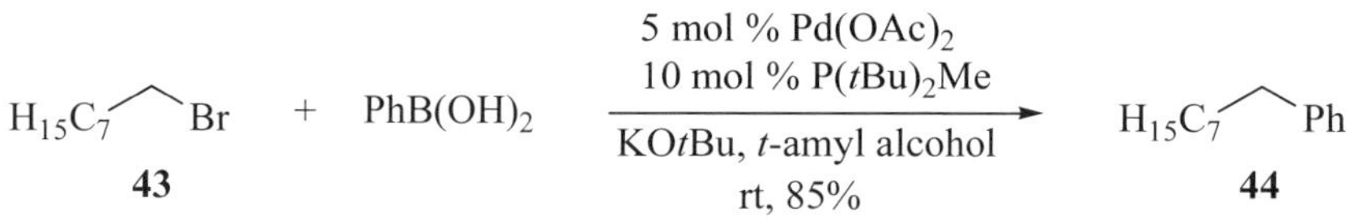

Suzuki Coupling reactions of secondary alkyl halides were not effective using the palladium catalysts described above. However, these latter transformations can be achieved using nickel catalyst systems.[109–111] For example the Ni/prolinol-catalyzed coupling of **45** with phenylboronic acid gave **46** in 84% yield.

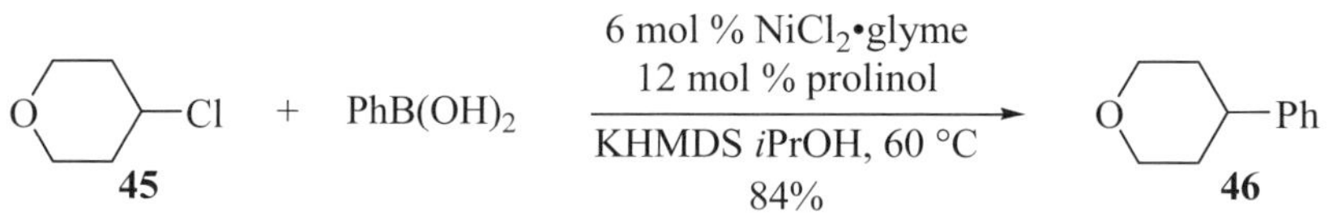

Enantioselective Suzuki Coupling Reactions

The asymmetric construction of axially chiral biaryl derivatives has been accomplished through enantioselective Suzuki coupling reactions, although these transformations remain relatively rare.[112] Two different strategies have been employed, the first of which involves diastereoselective reactions of chiral aryl halides. For example, Uemura has reported stereoselective couplings of $Cr(CO)_3$-complexed aryl halides such as **47**, which yield diastereomerically pure biaryls (e.g., **48**).[113] Subsequent photolytic removal of the Cr-moiety affords enantiomerically pure compounds. Asymmetric coupling reactions of arenes bearing sp^3-stereocenters have also been reported.[114–116]

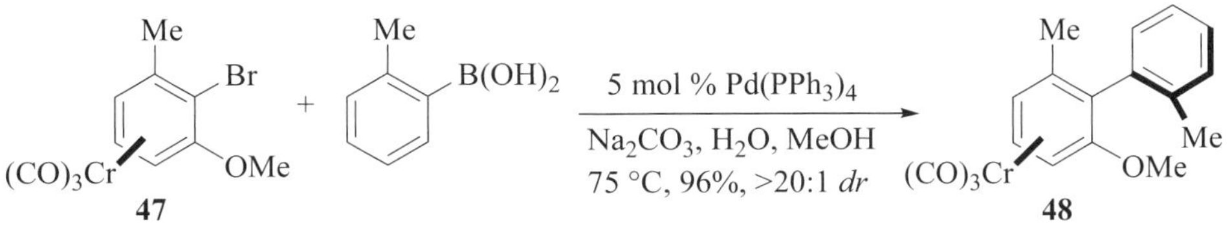

Enantioselective Suzuki coupling reactions that employ chiral catalysts for the synthesis of nonracemic biaryl derivatives have also been explored by several groups.[112] The first examples of these transformations were reported independently by Buchwald[117] and Cammidge[118] in 2000, although related asymmetric desymmetrizations[119,120] and the use of chiral catalysts for diastereoselective reactions of chiral substrates had been previously described.[121] In a representative example,[117] boronic acid **50** was coupled with aryl bromide **49** to provide **51** in 93% yield and 88% *ee*; selectivities of up to 92% *ee* were obtained in some systems.

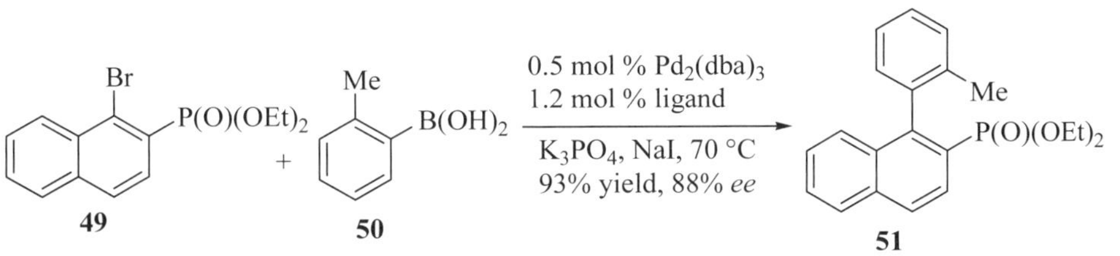

A very recent paper described the first examples of asymmetric Suzuki coupling reactions of unactivated alkyl halides.[122] These transformations take advantage of the fact that oxidative addition of secondary alkyl halides to Ni(0) proceeds through radical intermediates. This leads to scrambling of stereochemistry when achiral catalysts are employed, but can be exploited to achieve dynamic kinetic resolution with chiral catalysts. For example, use of a catalyst composed of Ni(COD)$_2$ and chiral 1,2-diamine ligand **55** for the coupling of **52** with **53** gave **54** in 78% yield and 90% *ee*.

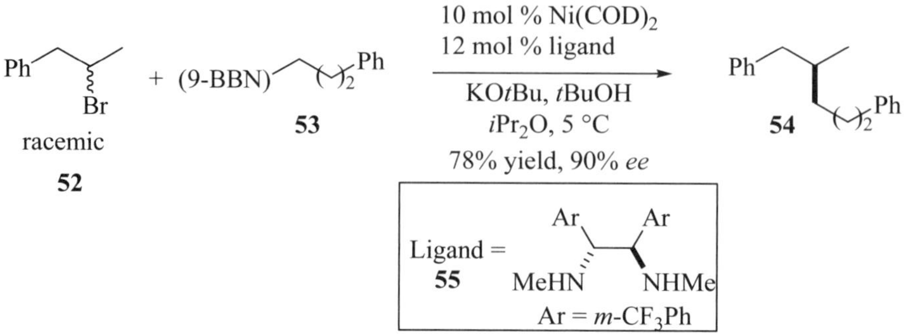

1.1.7.7 *Synthetic Utility*

The Suzuki reaction has been frequently employed in the synthesis of natural products and other complex molecules.[3,31] A few recent, representative examples that illustrate the chemoselectivity and functional group tolerance of these transformations are described below.

Coleman and coworkers have reported a lynchpin synthesis of (6*E*)-2′-*O*-methylmyxalamide D (**60**) that employs a Suzuki coupling of pinacolboronate **58** with alkenyl iodide **59**.[123] The key intermediate alkenylboronate (**58**) was generated through chemoselective Stille coupling of vinylstananne **56** with **57**. This example further highlights the importance of base in the Suzuki coupling, and illustrates the utility of substrates bearing two different main-group elements that can undergo selective transmetalation under appropriate conditions.

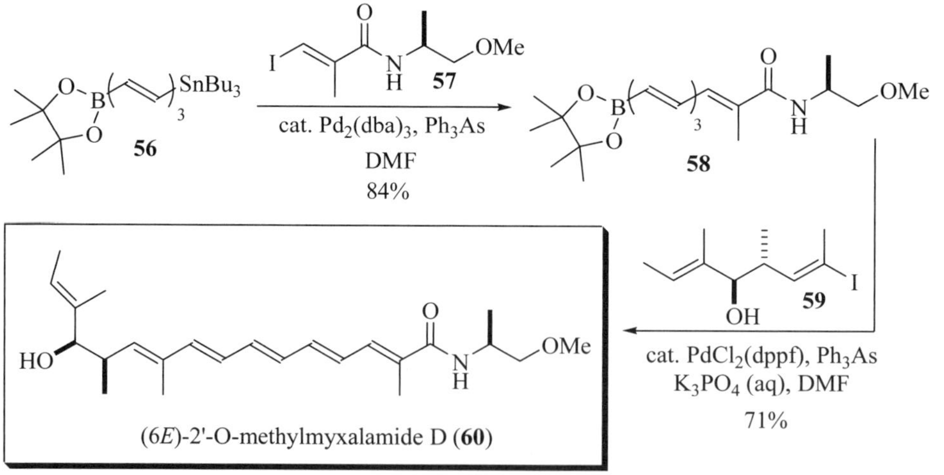

A different Suzuki-Coupling strategy for the lynchpin synthesis of polyene-containing natural products was recently described by Burke.[124] This strategy is illustrated in the construction of the polyene fragment of amphotericin B, and employs a MIDA-protected β-bromovinylboronate ester (**62**, MIDA = *N*-methyliminodiacetic acid) as a key building block. The MIDA-protected boronate is unreactive under anhydrous conditions, which allows for chemoselective coupling of the alkenyl bromide with alkenylboronic acid **61**. The MIDA group was cleaved from the resulting product **63** with aqueous NaOH, and the resulting boronic acid was then treated with alkenyl chloride **64** to afford **65**. A second sequence involving *in situ* deprotection of **65** followed by cross coupling with dienyl chloride **66** provided **67**, which is a fragment of amphotericin B, in moderate overall yield. This strategy was also employed for the synthesis of retinal and β-parinaric acid.

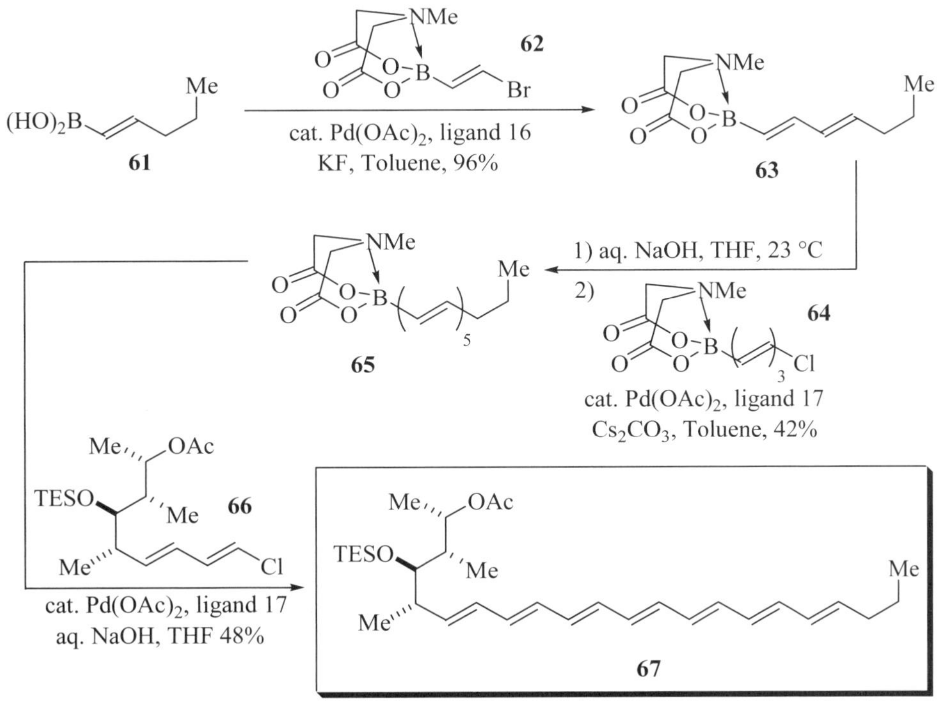

The utility of potassium organotrifluroborate salts in natural product synthesis was demonstrated by Molander through a formal total synthesis of the macrolide oximidine II. Alkyne **68** was selectively hydroborated with di(isopropylprenyl)borane and then converted to the potassium trifluoroborate salt **69**. Formation of the macrocyclic ring was achieved through intramolecular Suzuki coupling of **69**, which generated **70** in a 42% yield.[125] Intermediate **70** was transformed to **71** in two steps to complete the formal synthesis.

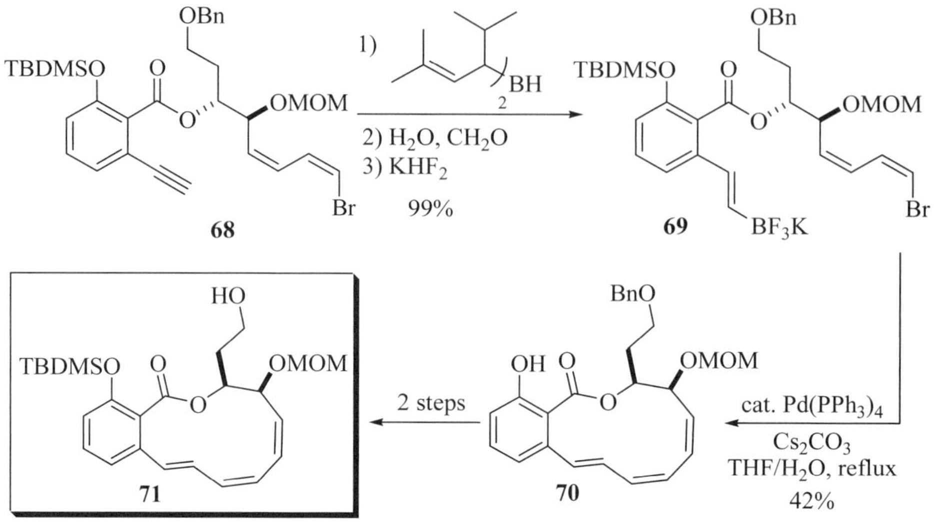

Nicolaou and coworkers recently employed an intramolecular Suzuki macrocyclization reaction in their total synthesis of the antibiotic marinomycin A (**73**).[126] The requisite organoboron reagent was generated *in situ* through treatment of enyne ester **72** with catecholborane in the presence of catalytic dicyclohexylborane, and the Pd-catalyzed Suzuki coupling effected closure of the 44-membered ring. The natural product was obtained in 23% yield over three steps from **72** after TBAF-mediated global deprotection.

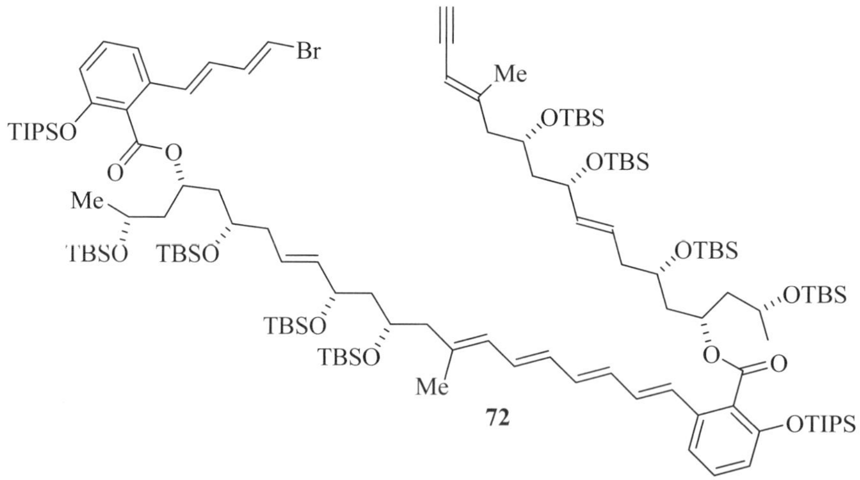

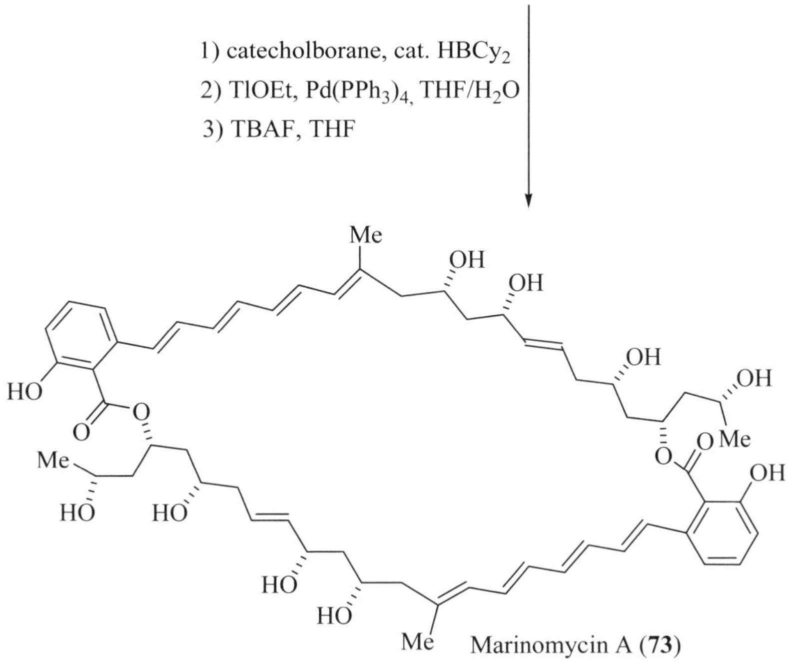

Marinomycin A (**73**)

The Suzuki coupling of an *in situ*-generated alkyl-9-BBN reagent with a vinyl phosphonate was employed by Sasai for the synthesis of the skipped ladder polyether brevenal.[127] As shown below, treatment of **74** with 9-BBN followed by addition of **75**, Cs_2CO_3, and a palladium catalyst provided **76**. Hydroboration of the crude alkene product afforded alcohol **77** in 84% total yield over two steps. This intermediate was transformed to the natural product after several additional steps.

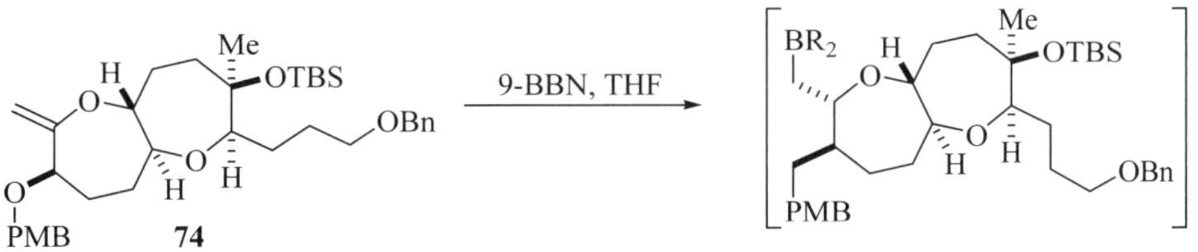

74

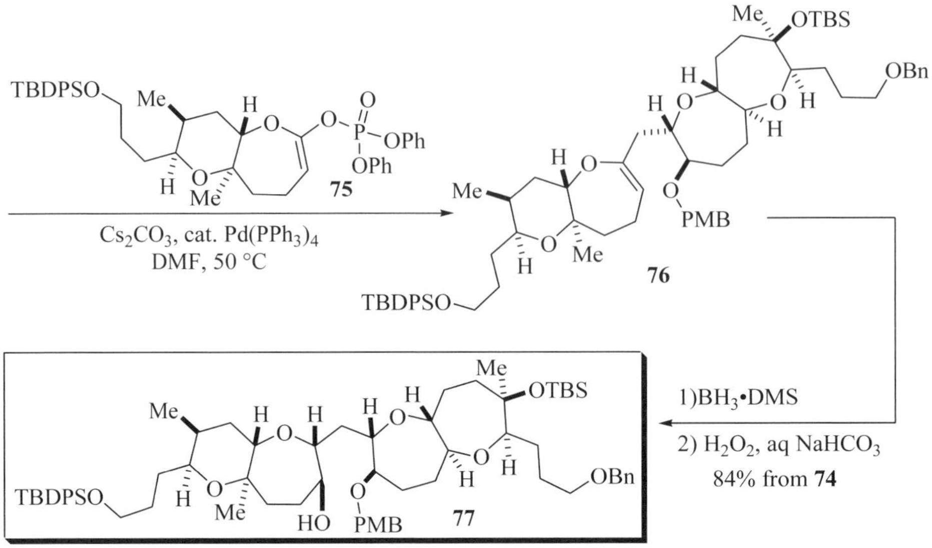

1.1.7.8 Experimental

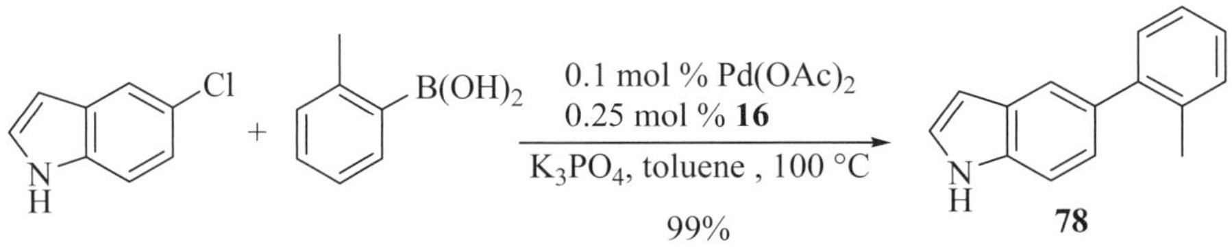

5-(2′-Methylphenyl)-1*H*-indole (78).[62]

An oven-dried resealable Schlenk tube was purged with argon and charged with 5-chloroindole (152 mg, 1.0 mmol), 2-methylphenylboronic acid (204 mg, 1.5 mmol), K_3PO_4 (424 mg, 2.0 mmol). The tube was capped with a rubber septum, evacuated and backfilled with argon three times, and then toluene (3 mL), and 200 μL of a catalyst solution composed of $Pd(OAc)_2$ (2.2 mg, 0.1 mmol), **16** (10.3 mg, 0.025 mmol), and THF (2 mL) were added via syringe. The septum was replaced with a Teflon screwcap, the tube was sealed, and the mixture was heated to 100 °C with stirring for 15 h. The mixture was then cooled to rt, diluted with ether (10 mL), filtered through a plug of silica gel, and concentrated. The crude product was purified by flash chromatography on silica gel using 9:1 hexanes:ether as the eluent to afford 207 mg (99%) of the title compound as a colorless oil.

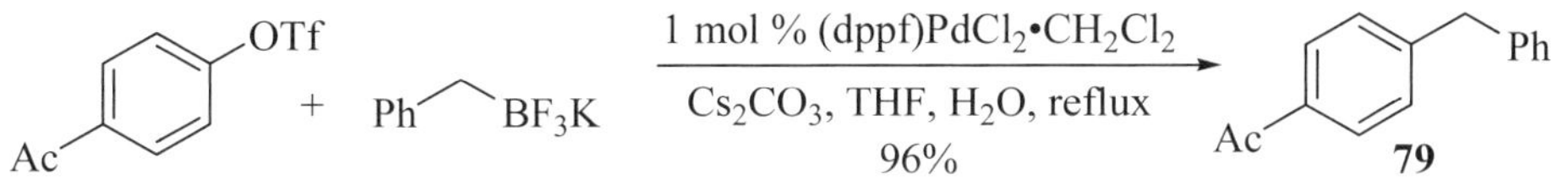

79

1-(4-Benzylphenyl)ethanone (79).[128]

A round-bottom flask equipped with a stirbar and a reflux condenser was purged with argon and charged with potassium benzyltrifluoroborate (106 mg, 0.5 mmol), Cs_2CO_3 (489 mg, 1.5 mmol), $PdCl_2(dppf)\cdot CH_2Cl_2$ (36 mg, 0.045 mmol), 4-acetylphenyltriflate (134 mg, 0.5 mmol), and THF (5 mL). Water (0.5 mL) was added, and the resulting mixture was heated to reflux for 18 h. The mixture was cooled to rt and diluted with water (10 mL). The organic layer was extracted with ether (50 mL × 3) and the ethereal extracts were washed with 1 M HCl (10 mL), and brine (20 mL). The organic layer was dried over anhydrous $MgSO_4$, filtered, and concentrated *in vacuo*. The crude product was purified by flash chromatography on silica gel using 20:1 hexanes:ether as the eluent to afford 108 mg (96%) of the title compound as a colorless oil.

1.1.7.8 References

1. [R] Suzuki, A. in *Modern Arene Chemistry*, Astruc, D., Ed., Wiley, VCH: Weinheim, 2002, pp 53–106.
2. [R] Miyaura, N.; Suzuki, A. *Chem. Rev.* **1995**, *95*, 2457–2483.
3. [R] Chemler, S. R.; Trauner, D.; Danishefsky, S. J. *Angew. Chem., Int. Ed.* **2001**, *40*, 4544–4568.
4. [R] Suzuki, A. in *Metal-Catalyzed Cross-Coupling Reactions*, Diederich, F.; Stang, P. J., Eds.; Wiley-VCH: Weinheim, 1998, Ch 2.
5. [R] Bellina, F.; Carpita, A.; Rossi, R. *Synthesis* **2004**, 2419–2440.
6. [R] Suzuki, A. Brown, H. C. in *Organic Syntheses via Boranes*, Aldrich Chemical Co: Milwaukee, WI, 2003, Vol 3.
7. [R] Kotha, S.; Lahiri, K.; Kashinath, D. *Tetrahedron* **2002**, *58*, 9633–9695.
8. [R] Suzuki, A. *J. Organomet. Chem.* **1999**, *576*, 147–168.
9. [R] Sasaki, M.; Fuwa, H. *Synlett* **2004**, 1851–1874.
10. Alo, B. I.; Kandil, A.; Patil, P. A.; Sharp, M. J.; Siddiqui, M. A.; Snieckus, V.; Josephy, P. D. *J. Org. Chem.* **1991**, *56*, 3763–3768.
11. [R] Schroter, S.; Stock, C.; Bach, T. *Tetrahedron* **2005**, *61*, 2245–2267.
12. [R] Negishi, E. -i. *B. Chem. Soc. Jpn.* **2007**, *80*, 233–257.
13. [R] Corbet, J.-P.; Mignani, G. *Chem. Rev.* **2006**, *106*, 2651–2710.
14. [R] Hassan, J.; Sevignon, M.; Gozzi, C.; Schulz, E.; Lemaire, M. *Chem. Rev.* **2002**, *102*, 1359–1469.
15. [R] Stanforth, S. P. *Tetrahedron* **1998**, *54*, 263–303.
16. [R] Murahashi, S.-i. *J. Organomet. Chem.* **2002**, *653*, 27–33.
17. [R] Tamao, K. *J. Organomet. Chem.* **2002**, *653*, 23–26.
18. [R] Farina, V.; Krishnamurthy, V.; Scott, W. J. *Org. React.* **1997**, *50*, 1–652.
19. [R] Negishi, E.-i. *Acc. Chem. Res.* **1982**, *15*, 340–348.
20. [R] Knochel, P.; Singer, R. D. *Chem. Rev.* **1993**, *93*, 2117–2188.
21. [R] Hiyama, T.; Shirakwa, E. *Top. Curr. Chem.* **2002**, *219*, 61–85.
22. [R] Denmark, S. E.; Sweis, R. F. *Acc. Chem. Res.* **2002**, *35*, 835–846.
23. [R] Chauder, B.; Green, L.; Snieckus, V. *Pure. Appl. Chem.* **1999**, *71*, 1521–1529.
24. Murata, M.; Watanabe, S.; Masuda, Y. *J. Org. Chem.* **1997**, *62*, 6458–6459.

25. Cho, J.-Y.; Tse, M. K.; Holmes, D.; Maleczka, R. E., Jr.; Smith, M. R., III. *Science* **2002**, *295*, 305–308.
26. Chen, H.; Schlecht, S.; Semple, T. C.; Hartwig, J. F. *Science* **2000**, *287*, 1995–1997.
27. Dieck, H. A.; Heck, R. F. *J. Org. Chem.* **1975**, *40*, 1083–1090.
28. Negishi, E. i. in *Aspects of Mechanism and Organometallic Chemistry*, Brewster, J. H., Ed., Plenum Press: New York, 1978, pp 285–317.
29. Miyaura, N.; Yamada, K.; Suzuki, A. *Tetrahedron Lett.* **1979**, *20*, 3437–3440.
30. Baba, S.; Negishi, E.-i. *J. Am. Chem. Soc.* **1976**, *98*, 6729–6731.
31. [R] Nicolaou, K. C.; Bulger, P. G.; Sarlah, D. *Angew. Chem., Int. Ed.* **2005**, *44*, 4442–4489.
32. [R] Schluter, A. D. *J. Polym. Sci. A.* **2001**, *39*, 1533–1556.
33. [R] Schiedel, M.-S.; Briehn, C. A.; Bauerle, P. *J. Organomet. Chem.* **2002**, *653*, 200–208.
34. [R] Doucet, H.; Hierso, J.-C. *Curr. Opin. Drug Disc. Dev.* **2007**, *10*, 672–690.
35. [R] Phan, N. T. S.; Van Der Sluys, M.; Jones, C. W. *Adv. Synth. Catal.* **2006**, *348*, 609–679.
36. [R] Stille, J. K.; Lau, K. S. Y. *Acc. Chem. Res.* **1977**, *10*, 434–442.
37. [R] Leadbeater, N. E. *Chem. Commun.* **2005**, 2881–2902.
38. Aliprantis, A. O.; Canary, J. W. *J. Am. Chem. Soc.* **1994**, *116*, 6985–6986.
39. Smith, G. B.; Dezeny, G. C.; Hughes, D. L.; King. A. O.; Verhoeven, T. R. *J. Org. Chem.* **1994**, *59*, 8151–8156.
40. Matos, K.; Soderquist, J. A. *J. Org. Chem.* **1998**, *63*, 461–470.
41. Braga, A. A. C.; Morgon, N. H.; Ujaque, G.; Maseras, F. *J. Am. Chem. Soc.* **2005**, *127*, 9298–9307.
42. Ridgway, B. H.; Woerpel, K. A. *J. Org. Chem.* **1998**, *63*, 458–460.
43. Espino, G.; Kurbangalieva, A.; Brown, J. M. *Chem. Commun.* **2007**, 1742–1744.
44. Roy, A. H.; Hartwig, J. F. *J. Am. Chem. Soc.* **2003**, *125*, 13944–13945.
45. Wright, S. W.; Hageman, D. L.; McClure, L. D. *J. Org. Chem.* **1994**, *59*, 6095–6097.
46. [R] Appukkuttan, P.; Van der Eycken, E. *Eur. J. Org. Chem.* **2008**, 1133–1155.
47. [R] Singh, B. K.; Kaval, N.; Tomar, S.; Van der Eycken, E.; Parmar, V. S. *Org. Proc. Res. Dev.* **2008**, *12*, 468–474.
48. Leadbeater, N. E.; Williams, V. A.; Barnard, T. M.; Collins, M. J., Jr. *Org. Proc. Res. Dev.* **2006**, *10*, 833–837.
49. [R] Genet, J. P.; Savignac, M. *J. Organomet. Chem.* **1999**, *576*, 305–317.
50. [R] DuPont, J.; de Souza, R. F.; Suarez, P. A. Z. *Chem. Rev.* **2002**, *102*, 3667–3691.
51. Mathews, C. J.; Smith, P. J.; Welton, T. *Chem. Commun.* **2000**, 1249–1250.
52. [R] Franzen, R. *Can. J. Chem.* **2000**, *78*, 957–962.
53. [R] Lorsbach, B. A.; Kurth, M. J. *Chem. Rev.* **1999**, *99*, 1549–1581.
54. [R] Alonso, F.; Beletskaya, I. P.; Yus, M. *Tetrahedron* **2008**, *64*, 3047–3101.
55. [R] Yin, L. X.; Liebscher, J. *Chem. Rev.* **2007**, *107*, 133–173.
56. [R] Felpin, F.-X.; Ayad, T.; Mitra, S. *Eur. J. Org. Chem.* **2006**, 2679–2690.
57. [R] Bedford, R. B.; Cazin, C. S. J.; Holder, D. *Coord. Chem. Rev.* **2004**, *248*, 2283–2321.
58. [R] Littke, A. F.; Fu, G. C. *Angew. Chem., Int. Ed.* **2002**, *41*, 4176–4211.
59. Old, D. W.; Wolfe, J. P.; Buchwald, S. L. *J. Am. Chem. Soc.* **1998**, *120*, 9722–9723.
60. [R] Martin, R.; Buchwald, S. L. *Acc. Chem. Res.* **2008**, Articles ASAP; DOI: 10.1021/ar800036s.
61. Wolfe, J. P.; Singer, R. A.; Yang, B. H.; Buchwald, S. L. *J. Am. Chem. Soc.* **1999**, *121*, 9550–9561.
62. Walker, S. D.; Barder, T. E.; Martinelli, J. R.; Buchwald, S. L. *Angew. Chem., Int. Ed.* **2004**, *43*, 1871–1876.
63. Barder, T. E.; Walker, S. D.; Martinelli, J. R.; Buchwald, S. L. *J. Am. Chem. Soc.* **2005**, *127*, 4685–4696.
64. Anderson, K. W.; Buchwald, S. L. *Angew. Chem., Int. Ed.* **2005**, *44*, 6173–6177.
65. Nguyen, H. N.; Huang, X.; Buchwald, S. L. *J. Am. Chem. Soc.* **2003**, *125*, 11818–11819.
66. Kataoka, N.; Shelby, Q.; Stambuli, J. P.; Hartwig, J. F. *J. Org. Chem.* **2002**, *67*, 5553–5556.
67. Harkal, S.; Rataboul, F.; Zapf, A.; Fuhrmann, C.; Riermeier, T.; Monsees, A.; Beller, M. *Adv. Synth. Catal.* **2004**, *346*, 1742–1748.
68. Bei, X.; Turner, H. W.; Weinberg, W.; H.; Guram, A. S.; Petersen, J. L. *J. Org. Chem.* **1999**, *64*, 6797–6803.
69. Littke, A. F.; Fu, G. C. *Angew. Chem., Int. Ed* **1998**, *37*, 3387–3388.

70. Stambuli, J. P.; Kuwano, R.; Hartwig, J. F. *Angew. Chem., Int. Ed.* **2002**, *41*, 4746–4748.
71. [R] Hiller, A. C.; Grasa, G. A.; Viciu, M. S.; Lee, H. M.; Yang, C.; Nolan, S. P. *J. Organomet. Chem.* **2002**, *653*, 69–82.
72. Fürstner, A.; Leitner, A. *Synlett* **2001**, 290–292.
73. Marion, N.; Navarro, O.; Mei, J.; Stevens, E. D.; Scott, N. M.; Nolan, S. P. *J. Am. Chem. Soc.* **2006**, *128*, 4101–4111.
74. [R] Bedford, R. B. *Chem. Commun.* **2003**, 1787–1796.
75. Schnyder, A.; Indolese, A. F.; Studer, M.; Blaser, H.-U. *Angew. Chem., Int. Ed.* **2002**, *41*, 3668–3671.
76. Roca, F. X.; Richards, C. J. *Chem. Commun.* **2003**, 3002–3003.
77. Thompson, A. E.; Hughes, G.; Batsanov, A. S.; Bryce, M. R.; Parry, P. R.; Tarbit, B. *J. Org. Chem.* **2005**, *70*, 388–390.
78. Billingsley, K. L.; Anderson, K. W.; Buchwald, S. L. *Angew. Chem., Int. Ed.* **2006**, *45*, 3484–3488.
79. Billingsley, K.; Buchwald, S. L. *J. Am. Chem. Soc.* **2007**, *129*, 3358–3366.
80. Kudo, N.; Perseghini, M.; Fu, G. C. *Angew. Chem., Int. Ed.* **2006**, *45*, 1282–1284.
81. Guram, A. S.; Wang, X.; Bunel, E. E.; Faul, M. M.; Larsen, R. D.; Martinelli, M. J. *J. Org. Chem.* **2007**, *72*, 5104–5112.
82. Fleckenstein, C. A.; Plenio, H. *Chem. Eur. J.* **2008**, *14*, 4267–4279.
83. Billingsley, K.; Buchwald, S. L. *Angew. Chem., Int. Ed.* **2008**, *47*, 4695–4698.
84. Vedejs, E.; Chapman, R. W.; Fields, S. C.; Lin, S.; Schrimpf, M. R. *J. Org. Chem.* **1995**, *60*, 3020–3027.
85. Murphy, J. M.; Tzschucke, C. C.; Hartwig, J. F. *Org. Lett.* **2007**, *9*, 757–760.
86. Darses, S.; Michaud, G.; Genet, J.-P. *Eur. J. Org. Chem.* **1999**, 1875–1883.
87. [R] Molander, G. A.; Ellis, N. *Acc. Chem. Res.* **2007**, *40*, 275–286.
88. Darses, S.; Genet, J.-P.; Brayer, J.-L.; Demoute, J.-P. *Tetrahedron Lett.* **1997**, *38*, 4393–4396.
89. Xia, M.; Chen, Z.-C. *Synth. Commun.* **1999**, *29*, 2457–2465.
90. [R] Roglans, A.; Pla-Quintana, A.; Moreno-Manas, M. *Chem. Rev.* **2006**, *106*, 4622–4643.
91. Molander, G. A.; Biolatto, B. *J. Org. Chem.* **2003**, *68*, 4302–4314.
92. Molander, G. A.; Yun, C. –S.; Ribagorda, M.; Biolatto, B. *J. Org. Chem.* **2003**, *68*, 5534–5539.
93. Molander, G. A.; Canturk, B. *Org. Lett.* **2008**, *10*, 2135–2138.
94. Molander, G. A.; Gormisky, P. E.; Sandrock, D. L. *J. Org. Chem.* **2008**, *73*, 2052–2057.
95. Molander, G. A.; Ham, J. *Org. Lett.* **2006**, *8*, 2767–2770.
96. Molander, G. A.; Ellis, N. M. *J. Org. Chem.* **2006**, *71*, 7491–7493.
97. Molander, G. A.; Petrillo, D. E. *J. Am. Chem. Soc.* **2006**, *128*, 9634–9635.
98. Molander, G. A.; Ribagorda, M. *J. Am. Chem. Soc.* **2003**, *125*, 11148–11149.
99. Molander, G. A.; Figueroa, R. *Org. Lett.* **2006**, *8*, 75–78.
100. Molander, G. A.; Figueroa, R. *J. Org. Chem.* **2006**, *71*, 6135–6140.
101. [R] Doucet, H. *Eur. J. Org. Chem.* **2008**, 2013–2030.
102. [R] Netherton, M. R.; Fu, G. C. *Top. Organomet. Chem.* **2005**, *14*, 85–108.
103. Ishiyama, T.; Abe, S.; Miyaura, N.; Suzuki, A. *Chem. Lett.* **1992**, *21*, 691–694.
104. Netherton, M. R.; Dai, C.; Neuschutz, K.; Fu, G. C. *J. Am. Chem. Soc.* **2001**, *123*, 10099–10100.
105. Kirchhoff, J. H.; Netherton, M. R.; Hills, I. D.; Fu, G. C. *J. Am. Chem. Soc.* **2002**, *124*, 13662–13663.
106. Netherton, M. R.; Fu, G. C. *Angew. Chem., Int. Ed.* **2002**, *41*, 3910–3912.
107. Arentsen, K.; Caddick, S.; Cloke. F. G. N.; Herring, A. P.; Hitchcock, P. B. *Tetrahedron Lett.* **2004**, *45*, 3511–3515.
108. Valente, C.; Baglione, S.; Candito, D.; O'Brien, C. J.; Organ, M. G. *Chem. Commun.* **2008**, 735–737.
109. Gonzalez-Bobes, F.; Fu, G. C. *J. Am. Chem. Soc.* **2006**, *128*, 5360–5361.
110. Zhou, J.; Fu, G. C. *J. Am. Chem. Soc.* **2004**, *126*, 1340–1341.
111. Saito, B.; Fu, G. C. *J. Am. Chem. Soc.* **2007**, *129*, 9602–9603.
112. [R] Baudoin, O. *Eur. J. Org. Chem.* **2005**, 4223–4229.
113. [R] Kamikawa, K.; Uemura, M. *Synlett* **2000**, 938–949.

114. Lipshutz, B. H.; Keith, J. M. *Angew. Chem., Int. Ed.* **1999**, *38*, 3530– 3533.
115. Baudoin, O.; Décor, A.; Cesario, M.; Gueritte, F. *Synlett* **2003**, 2009–2012.
116. Broutin, P.-E.; Colobert, F. *Org. Lett.* **2003**, *5*, 3281–3284.
117. Yin, J.; Buchwald, S. L. *J. Am. Chem. Soc.* **2000**, *122*, 12051–12052.
118. Cammidge, A. N.; Crepy, K. V. L. *Chem. Commun.* **2000**, 1723–1724.
119. Uemura, M.; Nishimura, H.; Hayashi, T. *Tetrahedron Lett.* **1993**, *34*, 107–110.
120. Suk, Y. C.; Shibasaki, M. *Tetrahedron: Asymmetry* **1998**, *9*, 3751–3754.
121. Nicolaou, K. C.; Li, H.; Boddy, C. N. C.; Ramanjulu, J. M.; Yue, T.-Y.; Natarajan, S.; Chu, X.-J.; Brase, S.; Rubsam, F. *Chem. Eur. J.* **1999**, *5*, 2584–2601.
122. Saito, B.; Fu, G. C. *J. Am. Chem. Soc*. **2008**, *130*, 6694–6695.
123. Coleman, R. S.; Lu, X.; Modolo, I. *J. Am. Chem. Soc.* **2007**, *129*, 3826–3827.
124. Lee, S. J.; Gray, K. C.; Paek, J. S.; Burke, M. D. *J. Am. Chem. Soc.* **2008**, *130*, 466–468 and references therein.
125. Molander, G. A.; Dehmel, F. *J. Am. Chem. Soc.* **2004**, *126*, 10313–10318.
126. Nicolaou, K. C.; Nold, A. L.; Milburn, R. R.; Schindler, C. S.; Cole, K. P.; Yamaguchi, J. *J. Am. Chem. Soc.* **2007**, *129*, 1760–1768.
127. Fuwa, H.; Ebine, M.; Sasaki, M. *J. Am. Chem. Soc.* **2006**, *128*, 9648–9650.
128. Molander, G. A.; Ito, T. *Org. Lett.* **2001**, *3*, 393–396.

1.1.8 Tsuji–Trost Reaction

Mathew J. Fuchter

1.1.8.1 Description

The Tsuji–Trost reaction is the palladium-catalyzed substitution of allylic leaving groups by carbon nucleophiles. These reactions proceed via π-allylpalladium intermediates **2**.[1–18]

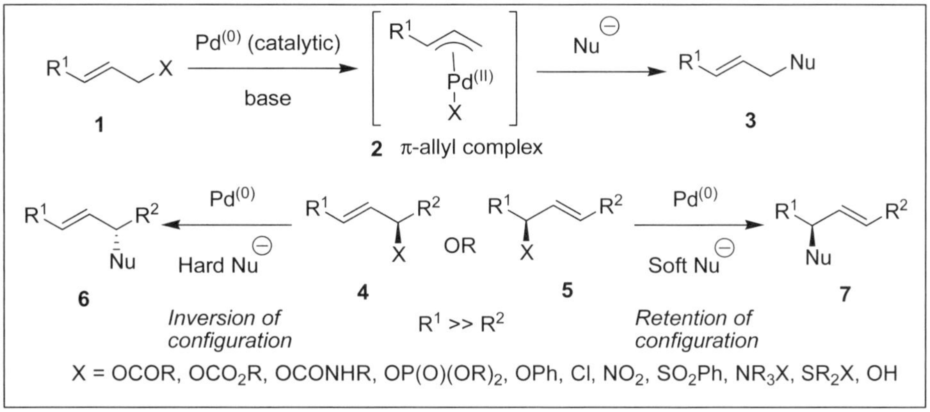

A wide range of leaving groups (X) on the allylic reagent **1** can be utilized including halides, acetates, ethers, sulfones, carbonates, carbamates, epoxides and phosphates. The order of reactivity of the various leaving groups follows the trend: Cl > OCO_2R > OAc >> OH. For most substrates **1**, a stoichiometric amount of base is required (to deprotonate the "soft" nucleophile). However, allylic carbonates undergo decarboxylation, generating a sufficiently basic alkoxide and therefore can be used under neutral conditions. Traditionally, only "soft" carbon nucleophiles were employed in the Tsuji–Trost reaction, the most common being active methylene compounds with two electron-withdrawing groups (RR′CH, where R,R′ = CN, CO_2R, NO_2, *etc.*), enamines and enolates. It is equally possible however, to employ nitrogen-, oxygen-, and sulfur-based "soft" nucleophiles, as well as "hard" organometallic nucleophiles. In general, addition of the nucleophile occurs to the least hindered terminus of the π-allyl intermediate **2**, regardless of the position of the leaving group. For optically-active reagents **4** or **5**, substitution with "hard" nucleophiles occurs with overall inversion of configuration (to yield **6**). In the case of "soft" nucleophiles, substitution takes place with overall retention of configuration (to yield **7**).

1.1.8.2 Historical Perspective

In 1965, Jiro Tsuji and co-workers demonstrated that π-allylpalladium chloride (**8**) could be substituted with several nucleophiles including enamines and the anions derived from diethyl malonate.[19] This was an extension of their studies on the reaction of olefins activated by palladium(II) with "soft" nucleophiles.[20]

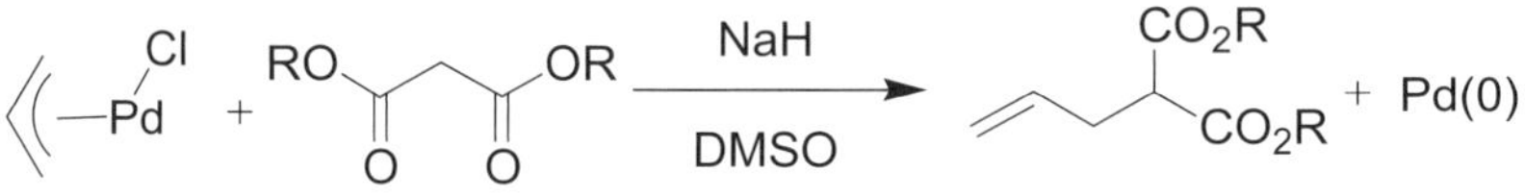

8 **9** **10**

The reaction was found to require DMSO as the solvent and resulted in the formation of allylated product **10** and palladium black. Importantly, these studies constituted a conceptual shift in organometallic chemistry. Many organometallic compounds known at the time (for example Grignard reagents) were established as nucleophiles, however these studies demonstrated certain organometallic regents could function as electrophiles.[21] The formation of palladium(0) in the reaction highlighted the potential for it to be rendered catalytic. Indeed, previously in 1964, Tsuji had developed the palladium-catalyzed carbonylation of allylic compounds.[22] In 1967, several other groups reported the palladium-catalyzed telomerization of butadiene with nucleophiles, which constituted the first examples of catalytic allylation reactions.[23,24] It was not until 1970 however the first examples of palladium-catalyzed allylation of nucleophiles using allylic compounds appeared.[25,26] In 1973, B. M. Trost reported that alkyl-substituted π-alkylpalladium complexes (often derived from the corresponding olefins) could by alkylated by "soft" carbon nucleophiles with complete regio- and stereoselectivity.[27] He subsequently published extensively on the use of this chemistry in complex molecule synthesis.[28] Nowadays this reaction is a powerful method of forging C–C bonds, particularly in the synthesis of intricate molecular architectures.

1.1.8.3 Mechanism

The basic mechanism of the Tsuji–Trost reaction is as follows: All palladium precatalysts are converted to the active palladium(0) catalyst **11** *in situ,* most commonly by phosphine in phosphine assisted catalytic cycles. Following coordination of the allylic reagent **1** to the palladium(0) catalyst **11**, oxidative addition occurs to give π-allylpalladium(II) complexes **13/14** (this step is also known as ionization). Complexes **13/14** can interconvert via ligand exchange reactions. As mentioned previously, for most substrates **1**, a stoichiometric amount of base is required to deprotonate the "soft" nucleophile. However, allylic carbonates undergo decarboxylation, generating a sufficiently basic alkoxide and therefore can be used under neutral conditions. Nucleophilic attack of the anionic nucleophile on complex **13**/**14**, followed by reductive elimination gives the complexed product **15**. Ligand exchange regenerates **12** and releases the product **3**.

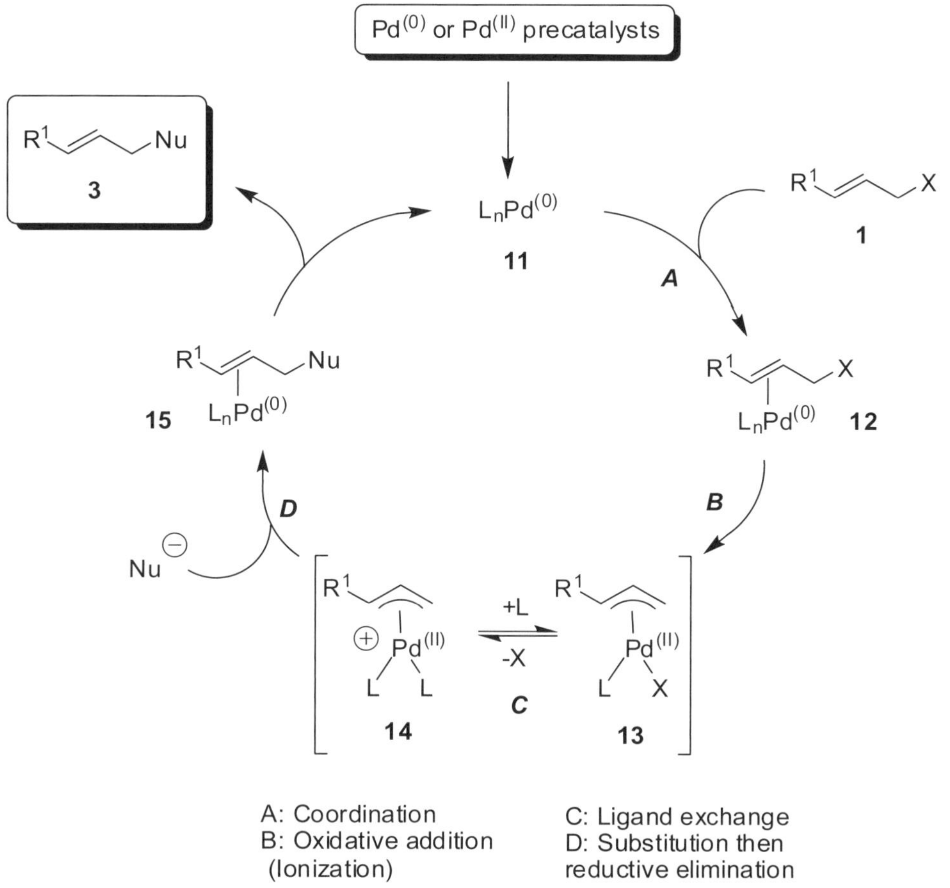

The regioselectivity of the process depends on several factors: (1) The charge distribution in the intermediate π-allyl palladium species, which favours attack at the more substituted allyl terminus; (2) steric hindrance to the approach of the nucleophiles, which favours attack at the less substituted allyl terminus; (3) electronic effects of substituents of the allyl unit; (4) the stability of the initial olefin–palladium(0) complex.[7,29] As a general rule of thumb however, in the case of palladium catalyzed reactions, steric control dominates and attack predominately occurs at the least hindered terminus.[29]

Another important consideration is the stereoselectivity of the reaction. Oxidative addition occurs under stereoelectronic control and one can think of this step as an S_N2-like displacement of the leaving group by the incoming palladium "nucleophile".[30] In the case of a substrate such as **4**, oxidative addition occurs with inversion of stereochemistry, to give complex **16**.

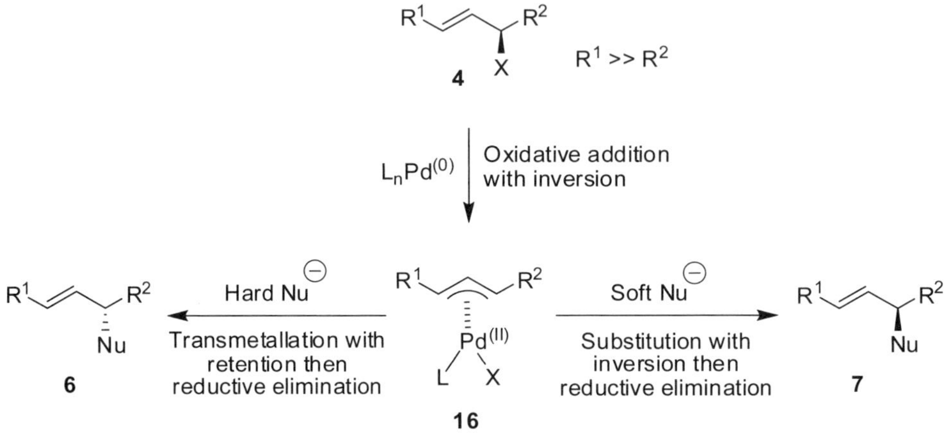

Two different pathways subsequently occur for "soft" or "hard" nucleophiles. "soft" nucleophiles, such as those derived from conjugate acids with a $pK_a < 25$ and most heteroatoms, directly the attack the π-allyl unit (i.e. from outside the coordination sphere of the metal) from the opposite face of the palladium, resulting in a second inversion to give product **7**.[12] On the other hand, "hard" nucleophiles such as conjugate acids with a $pK_a > 25$, attack the metal centre directly (transmetallation), followed by reductive elimination. This gives products with inversion of configuration, **6**.[12]

The Asymmetric Allylic Alkylation

In 1977, Trost published the first example of an asymmetric variant of the Tsuji–Trost reaction, termed the asymmetric allylic alkylation reaction (AAA).[31] Much of the subsequent development of the AAA reaction can be attributed to the dedicated work of Trost and co-workers.[17,18] There was a substantial time lag however, in the development of processes where high enantioselectivities were realized in a predictable fashion. This was due, in part, to the fact that chiral, asymmetrically pure ligands must create a chiral environment on the opposite face of the allyl fragment to the metal centre (a stereoelectronic requirement, *vide infra*).[12] This obviously represents a significant design challenge in the production of effective ligand systems.

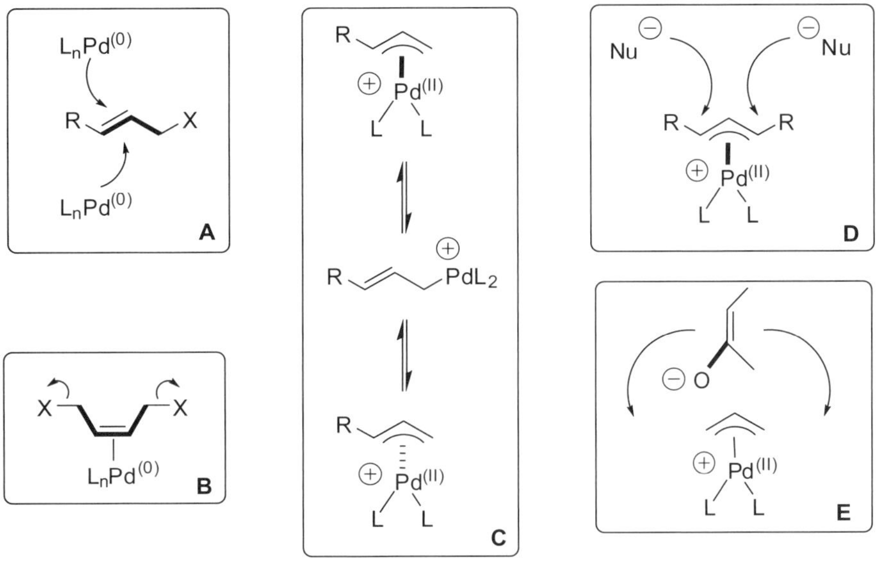

A: Enantiotopic faces of olefin
B: Enantiotopic leaving groups
C: Enantiotopic faces of the allyl complex
D: Enantiotopic termini of the allyl complex
E: Enantiotopic faces of the nucleophile

Further intricacies of the AAA present both a sizeable challenge, but also a unique opportunity.[12,17,18] The general catalytic cycle offers at least five opportunities for enantiodiscrimination: In the olefin complexation step, if one complex leads to oxidative addition at a rate significantly faster than the other, and nucleophilic capture of that diastereomer is fast relative to π–σ–π equilibration, then enantiotopic olefin face coordination becomes the enantiodetermining step (mechanism A). In a case where there are two potential leaving groups on a *meso* or on an achiral *gem*-disubstituted system, enantiotopic ionization of the leaving groups is the enantiodetermining step (mechanism B). Where initial olefin coordination is rapid and reversible, two diastereomeric palladium complexes can form, which can interchange through a π–σ–π equilibration step. This form of equilibration involves a change in hapticity of the allyl ligand (from η^3 to η^1), carbon–carbon bond rotation, and a second change in hapticity (from η^1 to η^3).[12] Either the more abundant or the more reactive diastereomeric complex leads to the product (mechanism C). If the starting material is a chiral racemic moiety, but ionization leads to a *meso* π-allyl intermediate, differentiation of enantiotopic allyl termini is the enantioselection event (mechanism D). Finally, in the case of a prochiral nucleophile, enantioface discrimination of an achiral allyl complex can be enantiodetermining (mechanism E).

Trost has published several extensive reviews, which elegantly demonstrate examples of all these modes of enantioselection in operation.[12,17,18] Additives can be employed to distinguish between different selection modes. For example, in the AAA of substrate **17**, chloride ion additives rapidly increase the rate of π–σ–π equilibration and this, in turn results in a decrease in the enantioselectivity.[17] Since this is not a base-catalyzed effect and lowering the concentration, increases the %*ee*, this suggests enantiotopic olefin face coordination is the enantiodetermining event in this AAA reaction. Further evidence is gained from the fact that changing to the equivalent chiral racemic branched substrate gives a branched product with low enantioselectivity.[17]

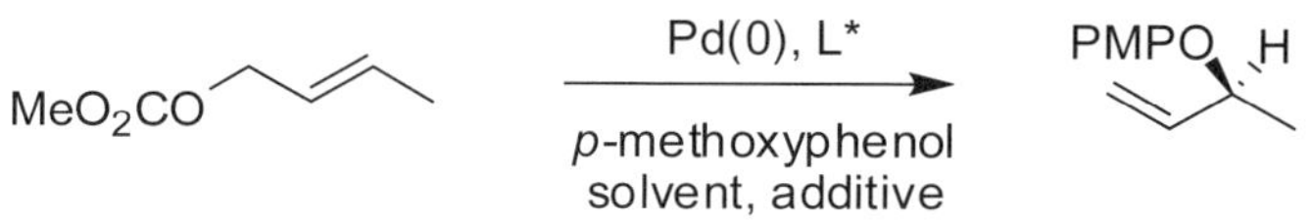

17 **18**

Perhaps the most important mechanistic implication of all is the very fact that the allylpalladium complexes can interconvert via π–σ–π equilibration. This enables chiral racemic material to be transformed into products of enantiopurity through a dynamic kinetic asymmetric transformation (DYKAT).[17,18] This powerful strategy has facilitated the construction of numerous complex, asymmetric molecules from simple racemic starting materials. Dynamic kinetic asymmetric transformations are extremely rare in other asymmetric reactions, highlighting the importance of the AAA reaction.

1.1.8.4 Synthetic Utility

"Soft" carbon nucleophiles

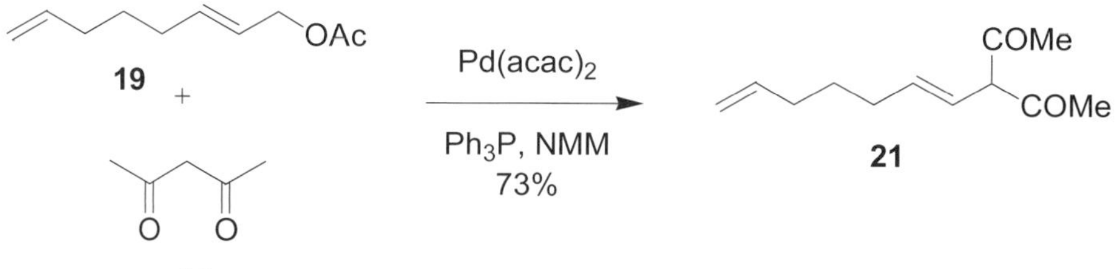

π-Allyl palladium cations can be regarded as "soft" electrophiles, and react smoothly with "soft" nucleophiles. For carbon-based nucleophiles, "soft" methylene compounds (conjugate acids with a pK_a < 25) with two electron-withdrawing groups (RR′CH, where R,R′ = CN, CO_2R, NO_2, *etc.*) are most commonly employed.

In the first catalytic Tsuji–Trost reaction, allylic acetate **19** was readily converted into product **21** in good yield.[26] Following this precedent, numerous examples of this allylation reaction have been reported using activating groups such as carbonyl, sulfonyl, cyano, nitro, aryl, olefinic, imino, *etc.* Readers are referred to the many comprehensive reviews on the topic for extensive examples of the Tsuji–Trost reaction in synthesis.[1–18]

One interesting example of the Tsuji–Trost reaction is the ring-opening of a vinyl cyclopropane bearing two electron-withdrawing groups. Under palladium catalysis, substrate **22** undergoes ring-opening oxidative addition (cleaving a C–C bond in the process) to give π-allyl intermediate **23**. Subsequent cyclization furnishes the bicyclic compound **24** in good yield.[32]

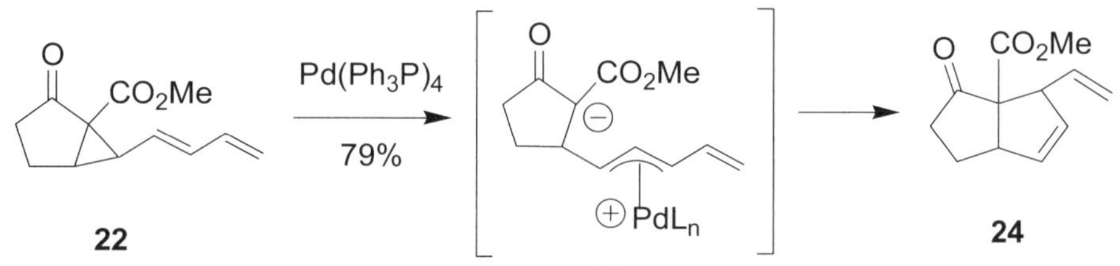

22 **23** **24**

The regiochemistry of the Tsuji–Trost reaction warrants further consideration (see **1.1.8.3**). While steric control often dominates, with nucleophilic attack occurring at the least hindered terminus of the allyl moiety, examples of branched adducts have been reported. In certain cases the electronic bias of a given substrate can give rise to attack at the more hindered terminus.[12] Other cases are less straightforward. It has been proposed that at low temperature and with short reaction times the reaction is under kinetic control, but at elevated temperature and prolonged reaction times the reaction is under thermodynamic control. Indeed, allyl malonates have been shown to rearrange to the more thermodynamically-stable (branched) regioisomer in the presence of palladium(0).[33] Regioselectivity can also be influenced by leaving groups, nucleophiles and ligands. For example, in studies of the Tsuji–Trost reaction of allylic acetates **25** and **26** in refluxing THF, the branched isomer **27** predominates.[34]

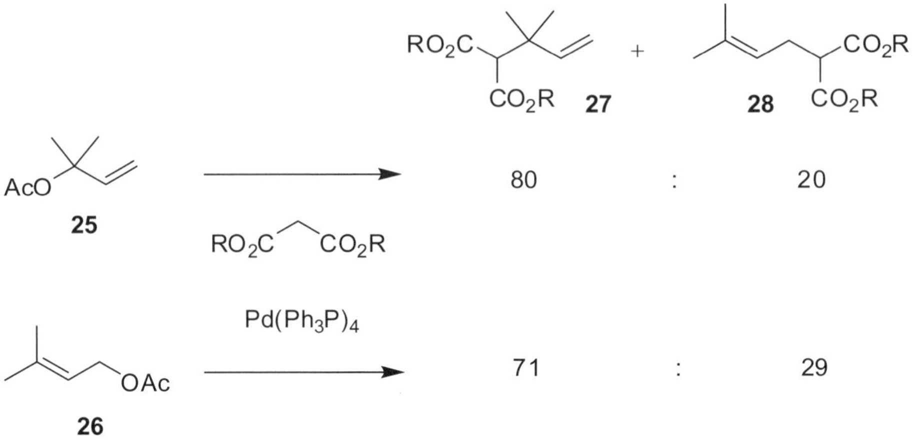

Intramolecular allylation has proved extremely useful in the synthesis of macrocyclic compounds. In 1977, Nozaski and co-workers reported the first example in the synthesis of humulene **31**.[35] Intramolecular Tsuji–Trost allylation provided key intermediate **30** in moderate yield.

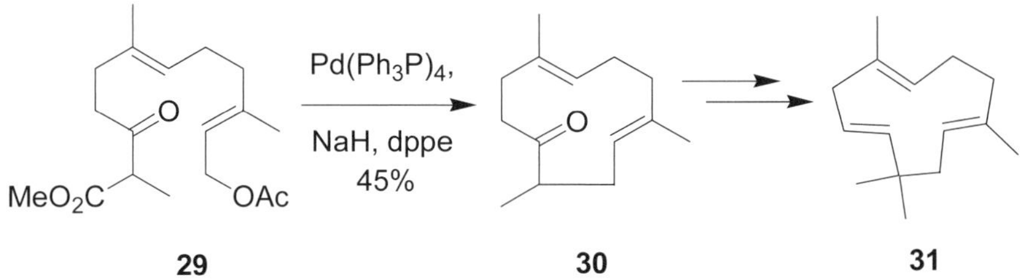

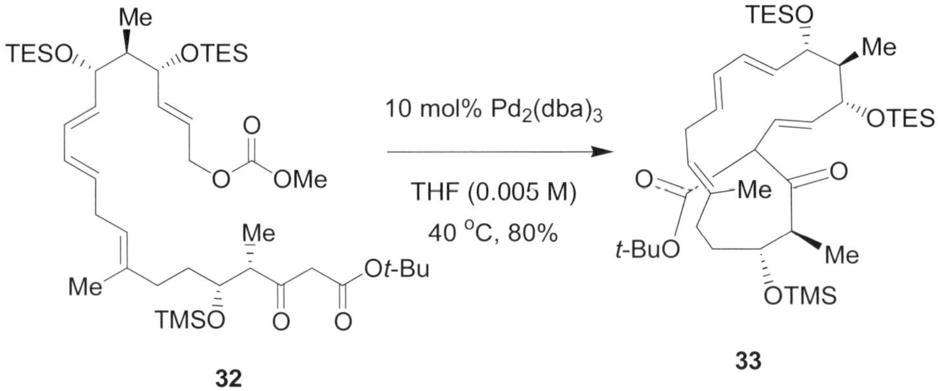

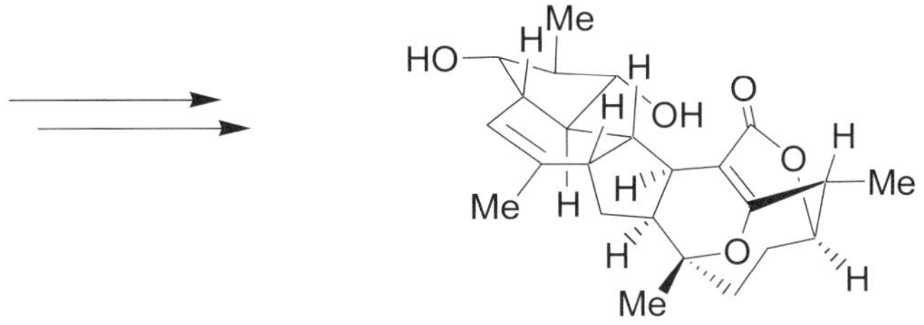

34: (+)-FR182877

More recently, in 2003, Sorensen demonstrated an impressive, scalable synthesis of cytotoxic natural product (+)-FR182877 **34**, which employed an intramolecular Tsuji–Trost allylation reaction to prepare the 19-membered macrocycle **33**.[36] Exposure of allylic carbonate **32** to 10 mol% palladium catalyst under high dilution formed the key bond in good yield and complete diastereoselectivity. This key intermediate was subsequently converted to (+)-FR182877 **34** via an intramolecular Diels–Alder reaction.

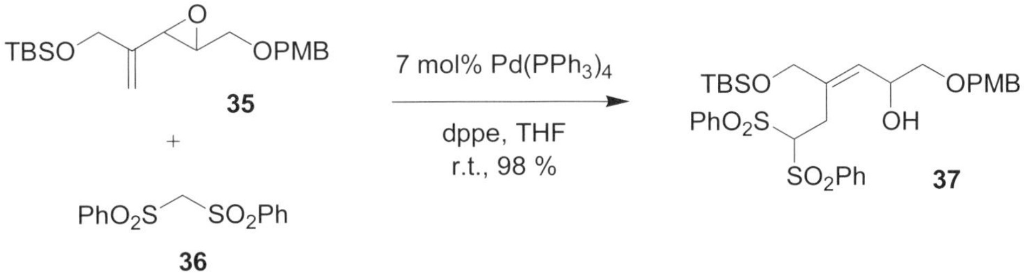

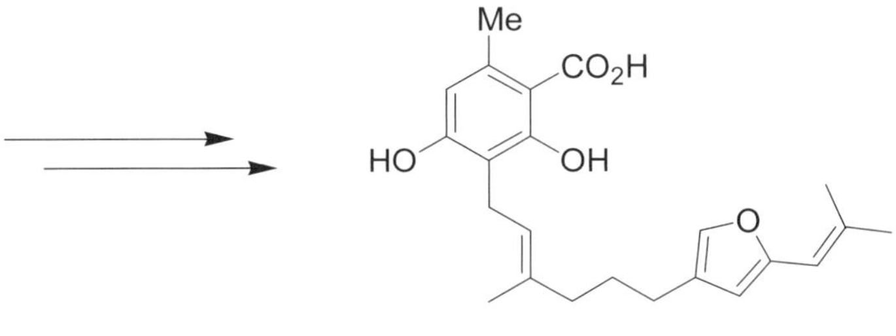

38: cristatic acid

Another potent cytotoxic compound cristatic acid (**38**) was prepared for the first time by A. Fürstner.[37] Allylation of vinyl epoxide **35** with bis(phenylsulfonyl)methane (**36**) gave 1,4-diol **37** in almost quantitative yield. This substrate was subsequently elaborated to cristatic acid (**38**). This example nicely highlights the use of the vinyl epoxide activating group in Tsuji–Trost reactions.

Nitrogen nucleophiles

Amines are suitable nucleophiles in the Tsuji–Trost reaction, with the use of simple amines, imides, azides, sulfonamides and heterocyclic amines having been reported.[11] In fact, one of the first examples of the catalytic version of the Tsuji–Trost reaction involved the use of aliphatic amine nucleophiles.[25] Subsequently these important nucleophiles have been used extensively in synthesis.

The water soluble vitamin (+)-biotin (**41**) was prepared by Seki and co-workers from *L*-cysteine in only 11 steps.[38] The key ring-forming reaction was an intramolecular allylic amination of a *cis*-allylic carbonate **39**. As expected, the allylation took place with net retention and furnished key intermediate **40** in good yield.

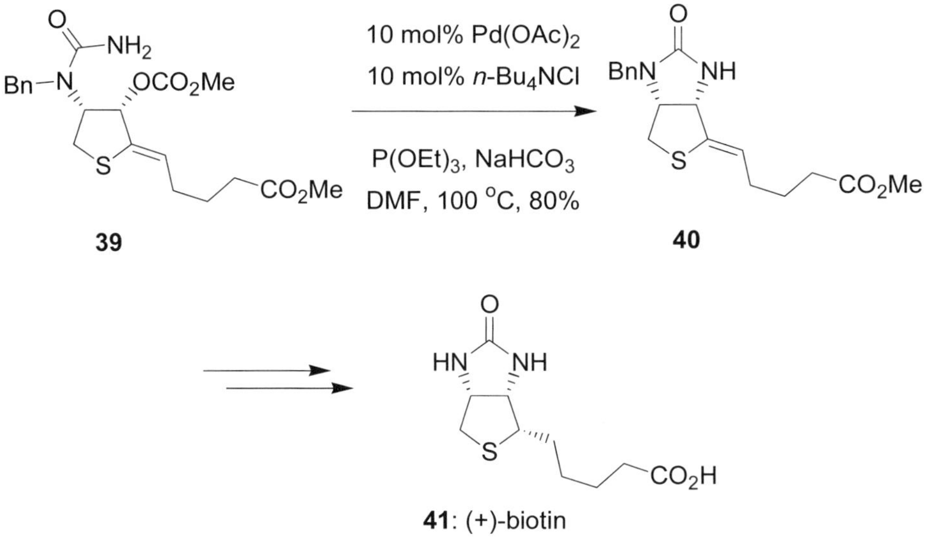

Oxygen nucleophiles

Aliphatic alcohols are poor *O*-nucleophiles, and the allylation of alcohols to form alkyl allyl ethers is somewhat sluggish.[11] One method to overcome this issue is to use aliphatic carbonates as the leaving group. These substrates are decarboxylated upon oxidative addition, releasing an alkoxide moiety, which gives an allyl ether in the absence of any other nucleophiles.[11] Phenolic alcohols and carboxylates are far better substrates for the Tsuji–Trost reaction however, and have been used extensively. For example, the monoepoxide of cyclopentadiene **42** is readily attacked by acetic acid to give *cis*-dibsubtituted cyclopentene **43** in good yield.[39] Note how the reaction proceeds both regio- and stereoselectively (net retention) via a π-allyl intermediate.

While the reaction of aliphatic alcohols can be problematic, there are several examples reported in the synthesis of complex molecular targets. For example, the use of Tsuji–Trost reaction in the synthesis of *cis*-2,5-disubstituted tetrahydrofurans was reported by Williams and co-workers. They used a "soft" oxygen nucleophile in an intramolecular reaction to prepare the C7–C22 core of amphidinolide K **45**.[40] It was found the addition of Me_3SnCl was necessary to both suppress acyl migration and ensure the oxygen was strongly nucleophilic.

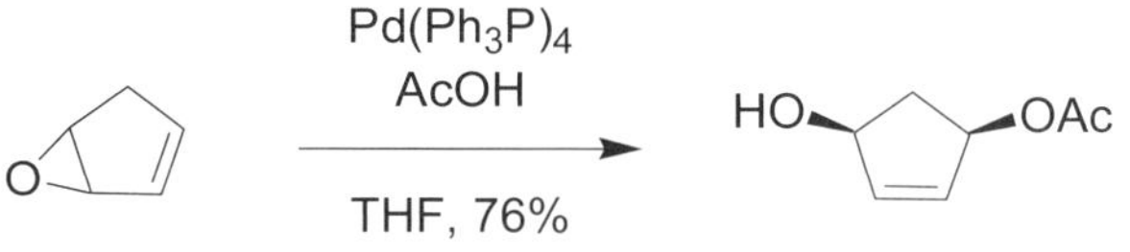

42 **43**

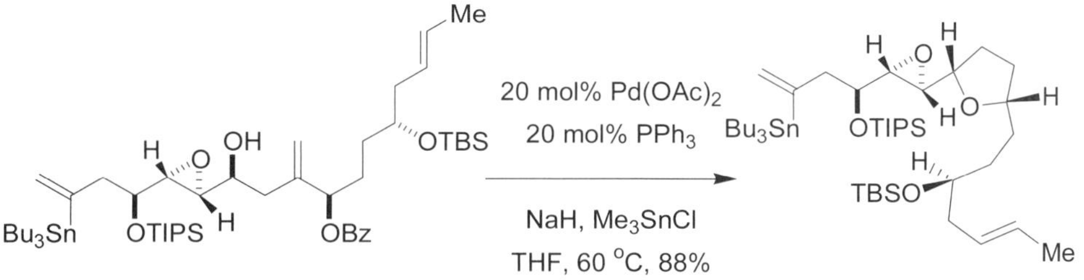

44 **45**

"Hard" carbon nucleophiles

Various allylic compounds react with "hard" carbon nucleophiles ($pK_a > 25$) including organometallic compounds of Zn, B, Al, Sn and Si via a transmetallation pathway (see **1.1.8.3**).[11] For example, the reaction of allylic lactone **46** with phenylzinc chloride gave the product **48** in excellent yield.[41] Note the inversion of the stereochemistry at the allylic carbon (see **1.1.8.3**).

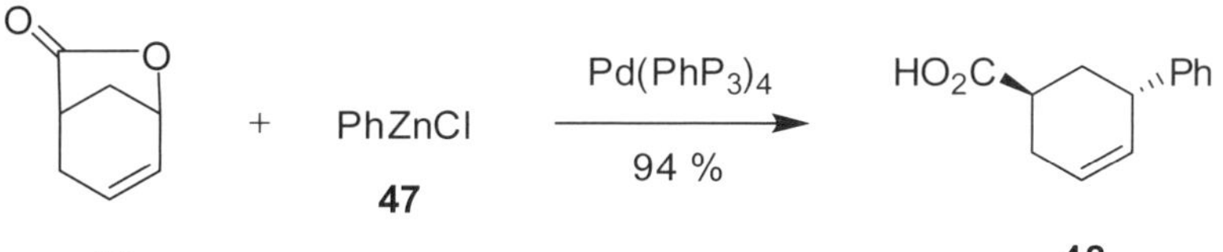

46 **47** **48**

Simple ketones cannot be allylated under standard Tsuji–Trost conditions however transmetallation via tin enolates has proven to be a useful modification. For example, Trost reported the use of tin enolate **49** in the preparation of intermediate **51**, which was subsequently annulated to give bicycle **52**.[42] Additionally by using enol acetates in the presence of Bu_3SnOMe, the quantity of tin can be substoichiometric, allowing the allylation of simple ketones under bimetallic tin and palladium catalysis.[11]

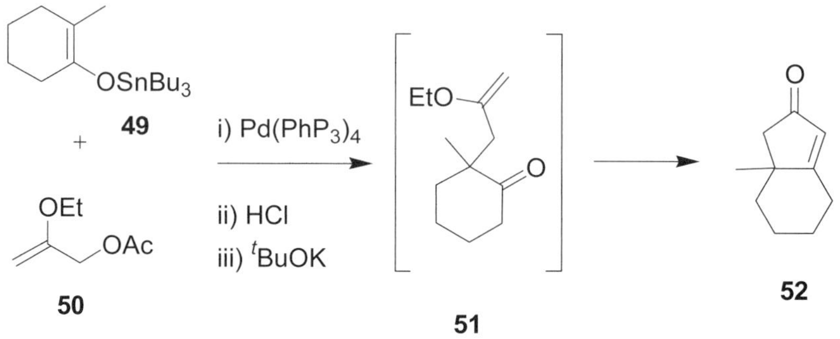

AAA Reaction: Carbon Nucleophiles

Asymmetric carbon-carbon bond formation is one of the ongoing challenges in synthetic organic chemistry. Good enantioselectivity at the nucleophile and electrophile has been achieved with "soft" carbon nucleophiles on both cyclic and acyclic electrophiles.[12,17,18] The majority of the reported examples in the AAA using carbon nucleophiles utilize cyclic electrophiles. One elegant example from the Trost laboratory was the synthesis of L-showdomycin (**58**).[43] The synthesis employed a desymmetrization of *meso*-substrate **53** as the first step (see mechanism B). Thus *meso*-dihydrofuran **53** was transformed into adduct **55** using imidosulfone **54** and ligand **59** in 67% yield and 92% *ee*. The resulting adduct was further alkylated using an achiral palladium complex with 1,3-bisdiphenyl-phosphinopropane (dppp) to yield key intermediate **57**, which was transformed to L-showdomycin (**58**) in eight further steps.

It is worth mentioning the huge success diphenylphosphino benzoic acid-based ligands (such as **59**), first pioneered by Trost and co-workers, have had in AAA reactions.[12,17,18] While many C_2-symmetrical ligands such as BINAP, DIOP and CHIRAPHOS have proved extraordinarily successful in other asymmetric transformations, their performance in AAA reactions is somewhat lacklustre.[12] Trost has published a working model to act as a predictive tool on the outcome of a given AAA reaction using ligands such as **59**.[44] Due to the numerous potential coordination modes of ligand **59** however, the exact catalytic species for a given reaction may differ.[45] Other successfully employed ligands include C_2-symmetric diamine ligands,[46] bisoxazoline ligands,[47] ferrocenyl ligands,[48,49] electronically differentiated *P,S*-[50] and *P,N*-based[51] ligands.

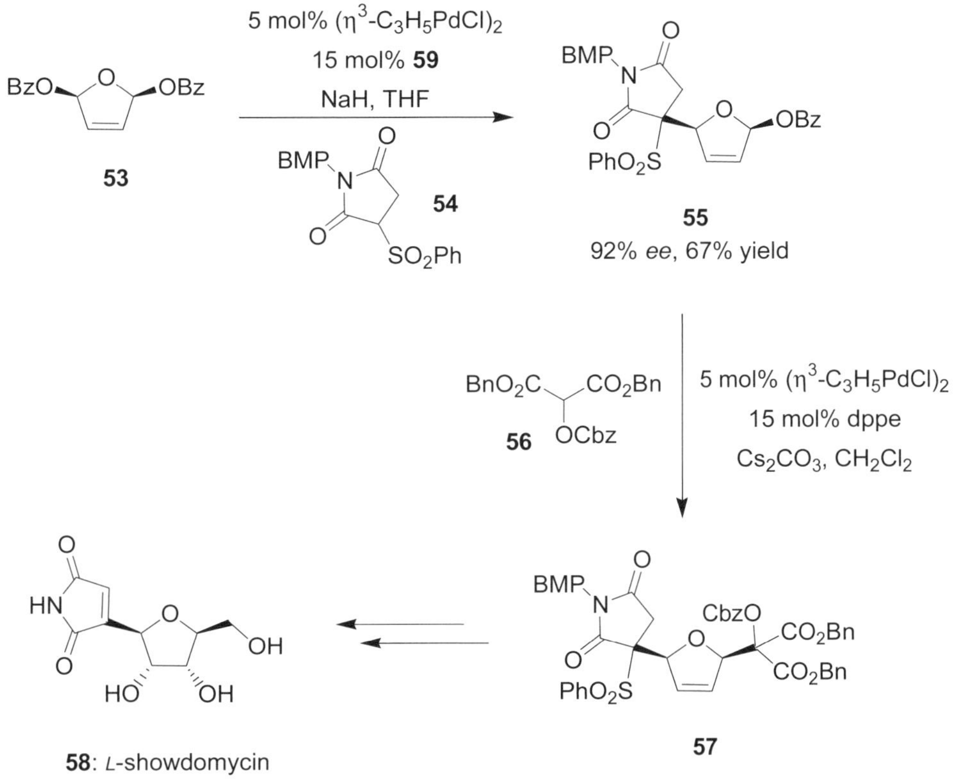

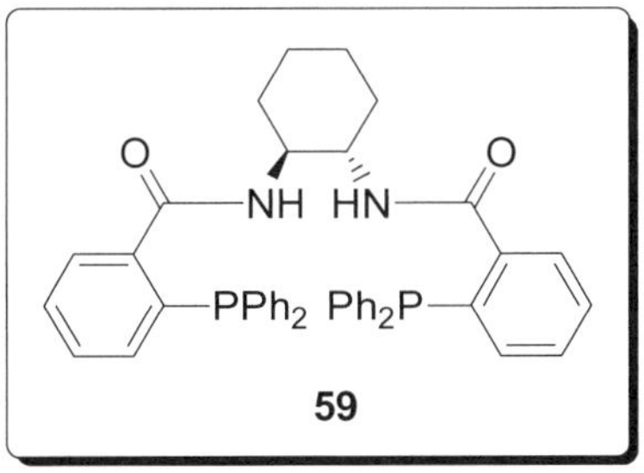

AAA reaction: Oxygen Nucleophiles

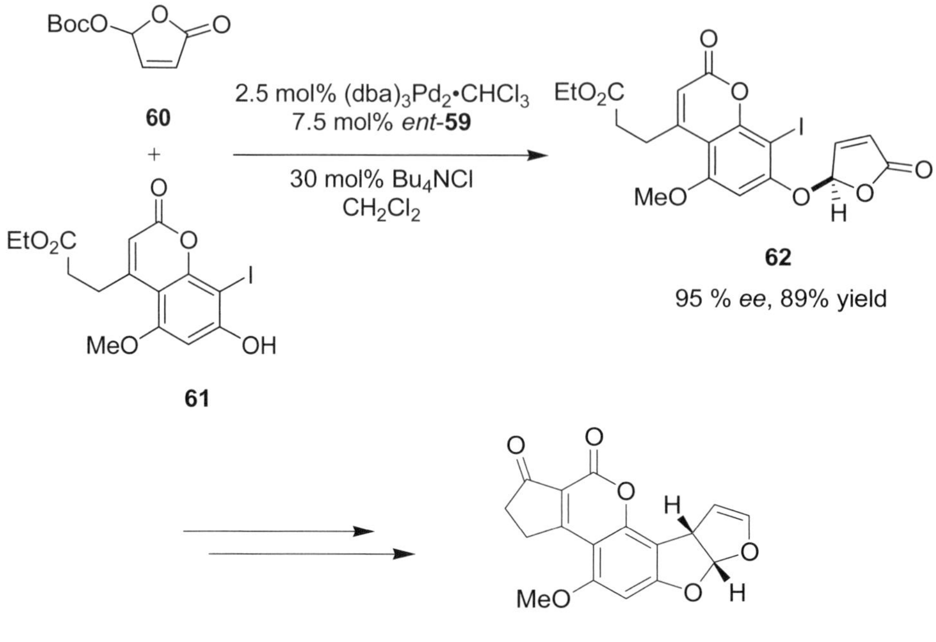

63: aflatoxin B_1

With over a dozen biologically-active, complex molecules prepared using oxygen nucleophiles in the AAA reaction, asymmetric formation of carbon-oxygen bonds is well precedented.[12,17,18] Phenolic oxygens are the most thoroughly explored nucleophiles and have been employed in a number of syntheses. For example, AAA of racemic butenolide **60** with coumarin **61** gave the product in high yield and enantiopurity.[52] Product **62** was a key intermediate in the synthesis of mycotoxin (–)-aflatoxin B (**63**). Facile π–σ–π equilibration (see mechanism C) of the allylpalladium species formed upon ionization of butenolide **60** enables a dynamic kinetic asymmetric transformation (DYKAT), furnishing product **62** in high enantiopurity from a racemic reagent.

Carboxylates have also proven to be suitable nucleophiles in the AAA reaction, especially with cyclic electrophiles.[17,18] For example, DYKAT of conduritol B tetracarboxylate facilitated a synthesis of D-*myo*-inositol 1,4,5-triphosphate (**66**).[53] Racemic substrate **64** was transformed to the enantiopure disubstituted product **65** in 80% yield. This was a key intermediate in the preparation of D-*myo*-inositol 1,4,5-triphosphate (**66**), a key component of intracellular signalling.

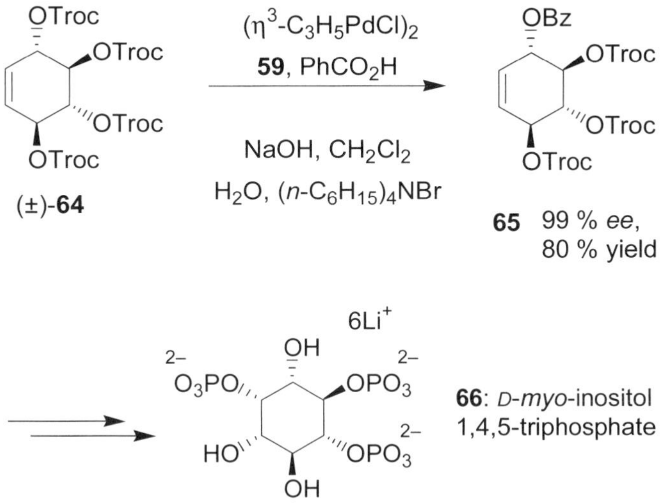

Troc = 2,2,2,-trichloroethoxycarbonyl

Aliphatic, primary alcohols can also be utilized in the AAA reaction and several examples have been reported.[17,18]

AAA Reaction: Amine Nucleophiles

Amines as nucleophiles in AAA have proven a challenge due to mono- vs. bisalkylation, regioselectivity issues and rate of nucleophilic addition vs allyl equilibration.[17] While no AAA-based total synthesis has been reported utilizing an alkylamine nucleophile, novel ligand systems are being developed which provide unprecedented selectivity. For example, novel *P,N*-ferrocene ligand **69** mediated a regio- and enantioselective amination of allylic acetates.[54] The product **68** was isolated predominantly as the branched isomer in 90% yield, 98% *ee*.

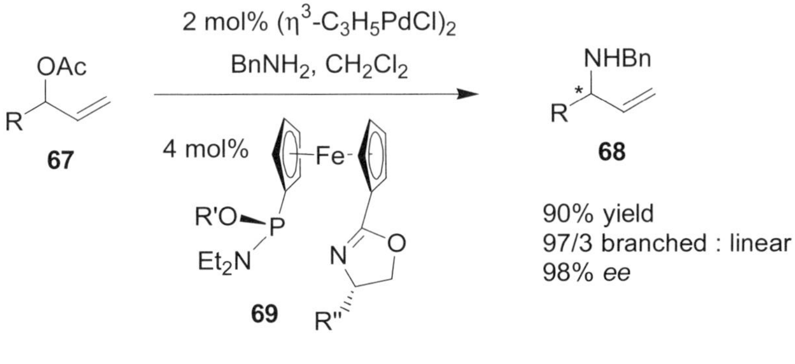

Numerous other amine nucleophiles have been used with great success in the AAA reaction including azides, imides, sulfonamides, and heterocycles.

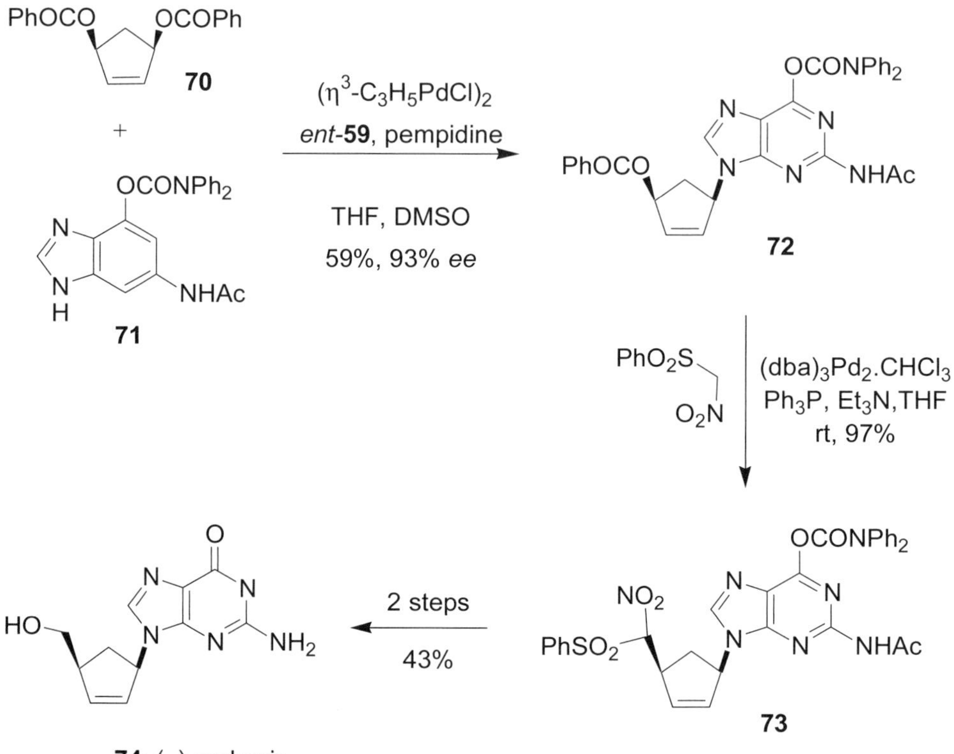

74: (–)-carbovir

Trost reported an elegant and expedient preparation of the antiviral agent (–)-carbovir (**74**).[55] Desymmetrization of dibenzoate **70** with a purine base **71** gave product **72** in moderate yield and high enantioselectivity. A second Tsuji–Trost allylation with a carbon nucleophile proceeded with net retention to give adduct **73**. This was converted to (–)-carbovir (**74**) in a two further steps.

In another example of the power of amine nucleophiles in the AAA reaction, Trost reported a highly enantioselective route to mannostatin A (**77**), a specific nanomolar inhibitor of α-*D*-mannosidase. Once again a desymmetrization strategy was employed, cyclizing meso substrate **75** in an intramolecular AAA reaction to give adduct **76**.[56] While other ligands provided inefficient in this task, the diphenylphosphino benzoic acid-based ligand **59** delivered key intermediate **76** in 97% *ee*, which was subsequently converted to mannostatin A (**77**).

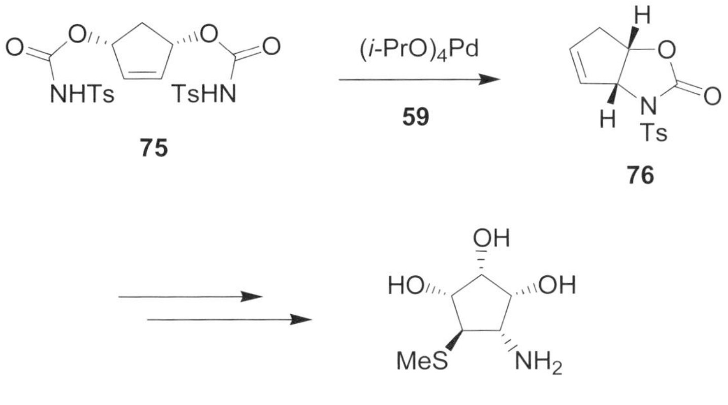

77: mannostatin A

AAA Reaction: Sulfur Nucleophiles

Allylic sulfones are important and versatile intermediates in organic synthesis due to the ability of the sulfone to impart both nucleophilic and electrophilic properties to the α-carbon. Pleasingly, the use of sodium benzenesulfinate in an AAA reaction of *meso* substrate **78** furnished the chiral adduct **79** in 85% yield and essentially asymmetrically pure.[57]

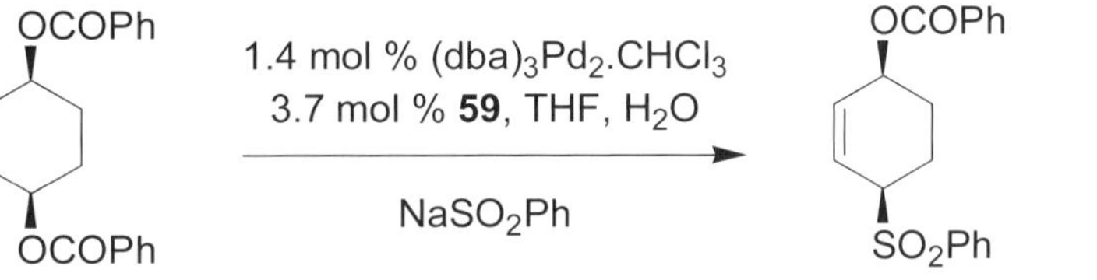

78 **79** 85% yield
> 99% *ee*

1.1.8.5 *Variations and Improvements*

There are many variations on the Tsuji–Trost reaction whereby transient π-allylpalladium compounds are harnessed in a variety of pathways. Some examples are highlighted below.

2-(Trimethylsilylmethyl)allyl Acetate.

2-(Trimethylsilylmethyl)allyl acetate (**80**), which has a silyl group in the allylic position undergoes a formal [3 + 2] cycloaddition reaction with electron deficient alkenes to give methylenecyclopentane derivatives **84**.[11] Following oxidative addition, elimination of the TMS group, facilitated by the proximal positive charge, generates dipolar intermediate **82**. The cyclization of reactive intermediate **82** proceeds by a Michael addition to the activated double bond to give **83**, followed by intramolecular allylation.[11] In general, tri-isopropyl phosphite is a particularly good ligand for this process a numerous cyclic compounds have been prepared.[11]

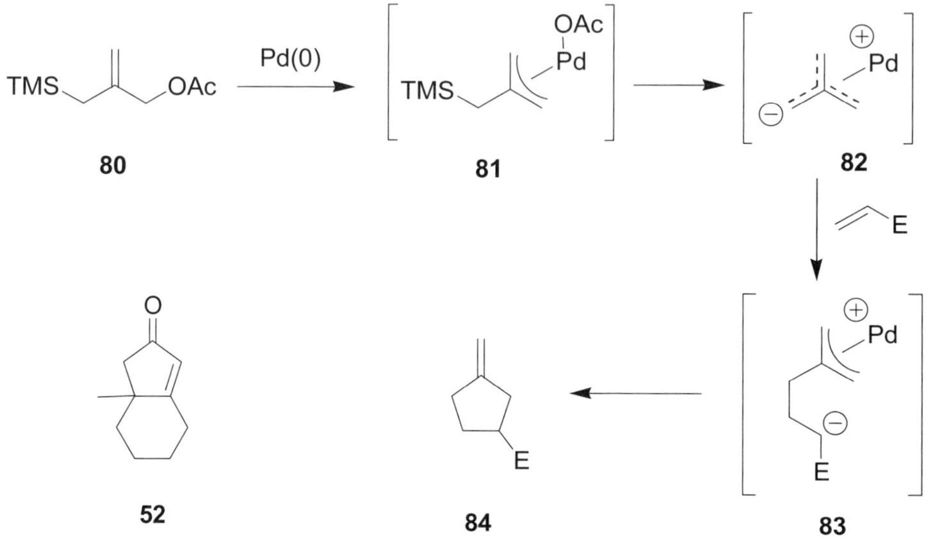

In their synthesis of the naturally occurring product loganin, Trost and co-workers prepared bicycle **87** as a key intermediate. Exposure of substrate **85** to cyclopentenone **86** under palladium catalysis, furnished the desired [3 + 2] cycloaddition adduct **87** in moderate yield.[58]

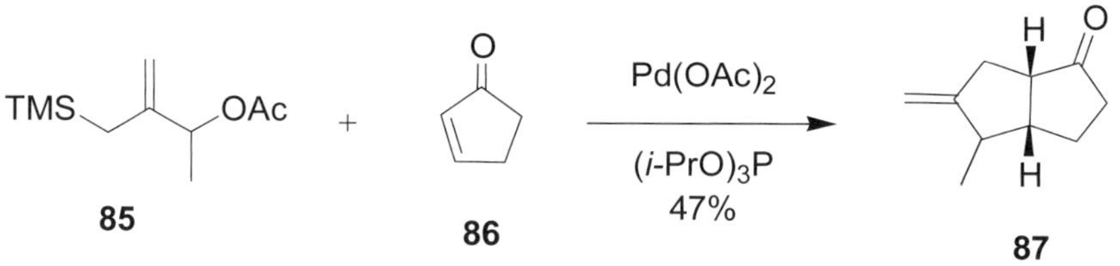

Carbonylation

Carbonylation of allylic compounds in alcoholic solvent gives β,γ-unsaturated esters, however, these substrates are far less reactive to palladium-catalyzed carbonylation than aryl or alkenyl halides.[11] In general, a large positive pressure of carbon monoxide is necessary to drive these reactions forward.

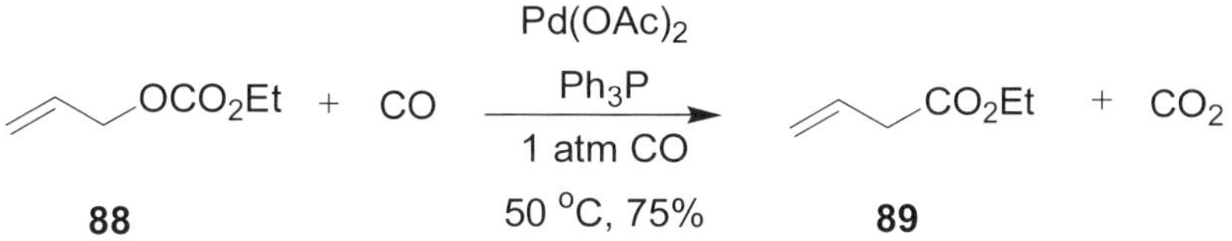

88 **89**

While carbonylation of allylic halides occurs with reasonable rates, allylic carbonates are the most reactive substrates. For example, carbonylation of carbonate **88** proceeds under relatively mild conditions to furnish allylic ester **89** in good yield.[59]

Since it is known that π-allylpalladium acetate is converted to allyl acetate by reductive elimination when treated by CO, it is understandable that carbonylation of allylic acetates is problematic.[11] Forcing conditions and additives such as NaBr can aid this reaction however. For example, under a high pressure of carbon monoxide and with NaBr as an additive, allylic acetate **90** was converted to allylic ester **91** in good yield.[60] It has been suggested this reaction could proceed via the allylic bromide.[11,60]

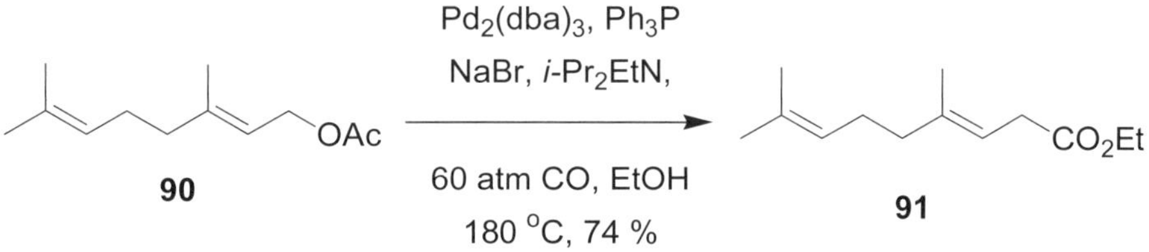

90 **91**

Elimination to Form Conjugated Dienes

When allylic compounds are exposed to palladium(0) in the absence of any suitable nucleophiles, 1,4-elimination occurs to yield conjugated dienes.[11] Following oxidative addition, hydride elimination from π-allylpalladium compound **93** gives conjugated diene **94**, often as a mixture of *E*- and *Z*-isomers.[11]

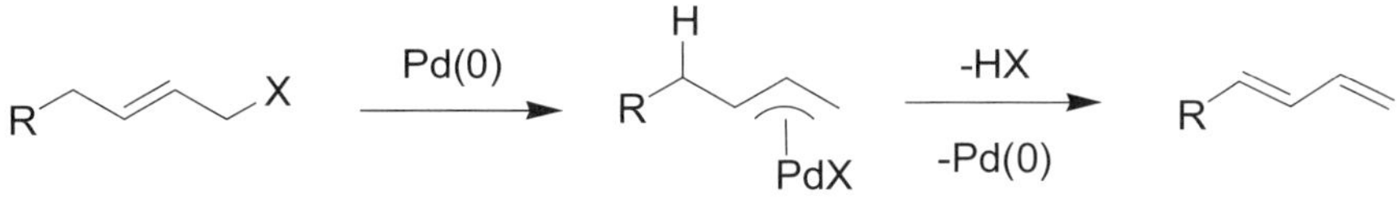

92 **93** **94**

This strategy has been exploited in the synthesis of variety of naturally occurring compounds such as pheromones and steroids.[11] For example, selective elimination of the acetate moiety from cyanohydrin derivative **95** was used to prepare key intermediate **96**. Diene **96** was subsequently elaborated to pellitorine (**97**).[61]

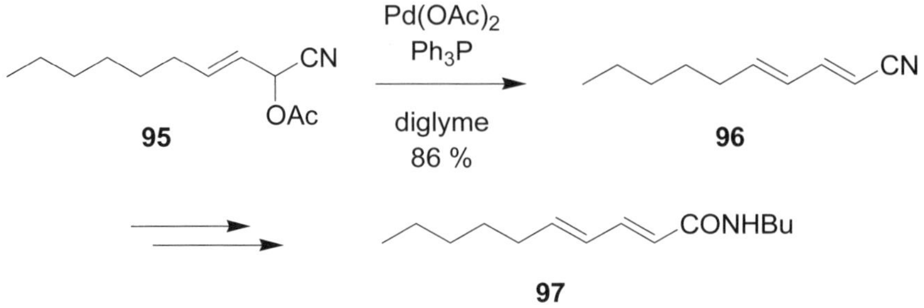

Oxidation of Alcohols

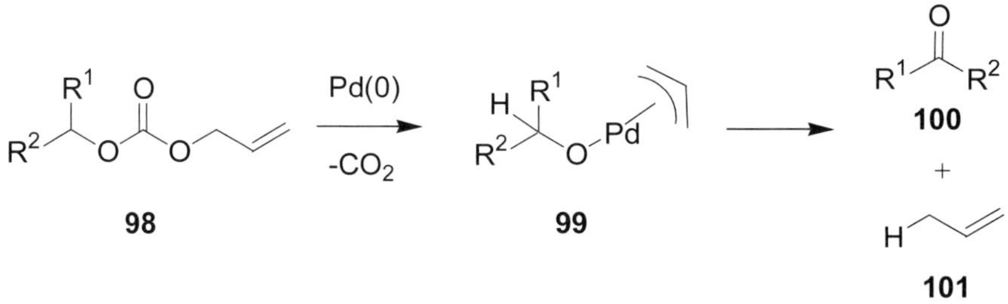

Smooth oxidation of alcohols under mild, neutral conditions can be achieved via π-allylpalladium intermediates. The reaction of aliphatic allyl carbonate **98** with a palladium(0) catalyst generates palladium alkoxide species **99**, which subsequently undergoes β-hydride elimination to give carbonyl compound **100**. This reaction is very clean since the by-products are CO_2 and propylene (**101**).[62] The reaction can readily be applied to secondary alcohols, allylic primary and secondary alcohols and benzylic alcohols. The oxidation of primary alcohols is prohibitively slow.

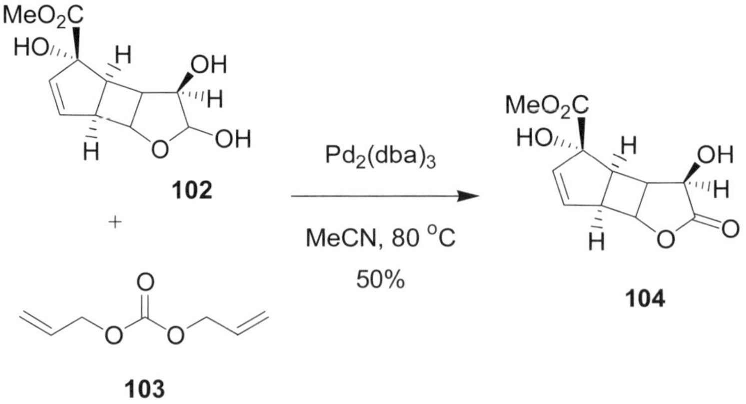

Indeed the slow oxidation of primary alcohols enables the use of external allylating agents as oxidants. For example, unprotected lactol **102** was selectively oxidized with diallyl carbonate **103** to yield lactone **104**.[63] Lactone **104** was subsequently used in the synthesis of echinosporin.

Hydrogenolysis of Allylic Compounds.

Palladium-catalyzed hydrogenolysis of allylic compounds gives access to alkene products. While initial studies focused on the use of ammonium formate as a hydride source, subsequent studies showed a plethora of metal hydride agents could be employed, including $LiAlH_4$, $NaBH_4$, hydrosilanes, tin hydrides, *etc.*[11] These studies highlighted an interesting regiochemical issue. For all reducing agents apart from formate, the reduction of terminal allylic compounds gave the more substituted 2-alkenes, whereas for formate, the terminal 1-alkene is produced.[11]

Mechanistically, this can be rationalized as follows.[11] Exposure of π-allylpalladium complex **105** to a metal hydride species results in transmetallation to give palladium species **106**. The hydride is then transferred to the less hindered side of the allylic system by reductive elimination, furnishing the internal alkene **107**. In the presence of formate however, ligand exchange gives palladium complex **108**, which can undergo concerted decarboxylation and hydride transfer to give terminal alkene **110**. Formation of π-allylpalladium formate **108** in the hydrogenolysis was confirmed by NMR studies.[64]

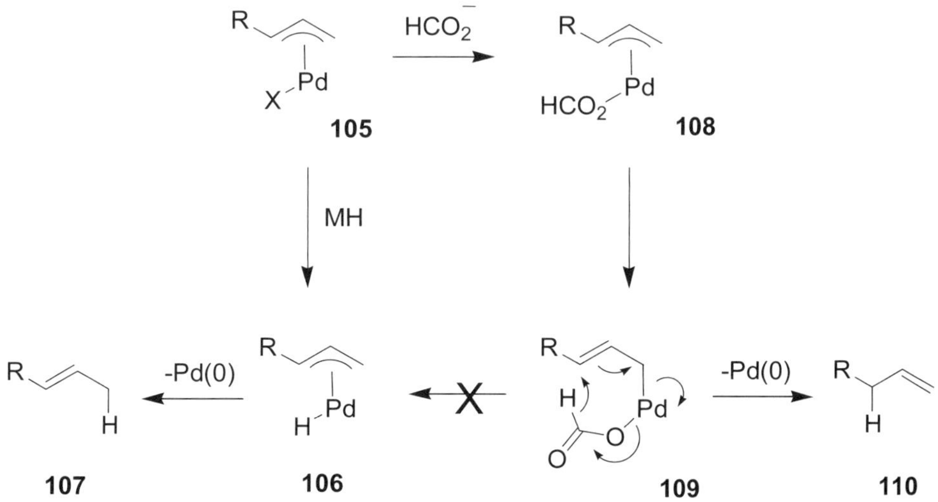

There are many examples of this reaction in the preparation of both cyclic and acyclic compounds.[11] For example, regioselective hydrogenolysis of compound **111** furnished the desired product **112** bearing an *exo*-methylene group in almost quantitative yield. This compound was subsequently used in the preparation of gomisin A and schizandrin.[65] Asymmetric reductions have also been reported using this reaction pathway.[66]

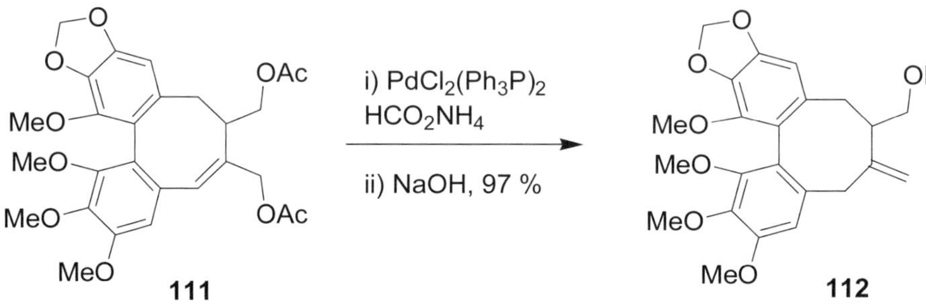

Other Metal Catalysts

Several other metal catalysts have been shown to mediate the Tsuji–Trost reaction, with molybdenum being the most developed. Trost first reported the use of molybdenum for allylic alkylation in 1982.[67] The most important aspect of the use of this metal is its regiocomplimentary with the palladium-catalyzed process. While palladium preferentially gives linear adducts (in the absence of electronic bias), molybdenum gives preferentially branched adducts.[67,17]

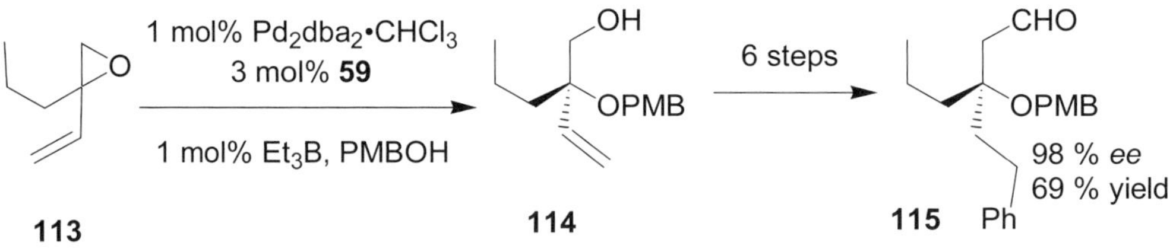

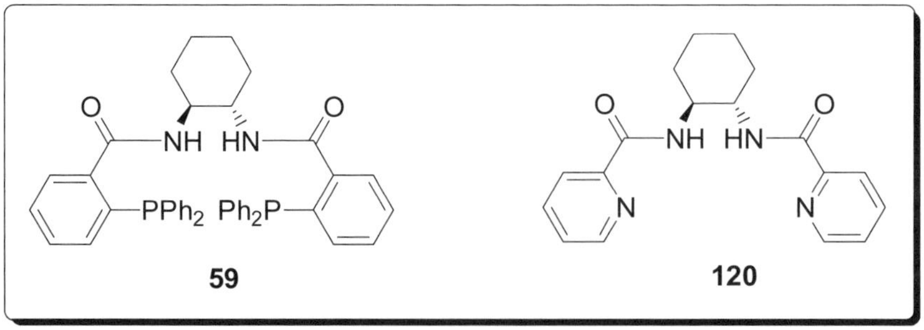

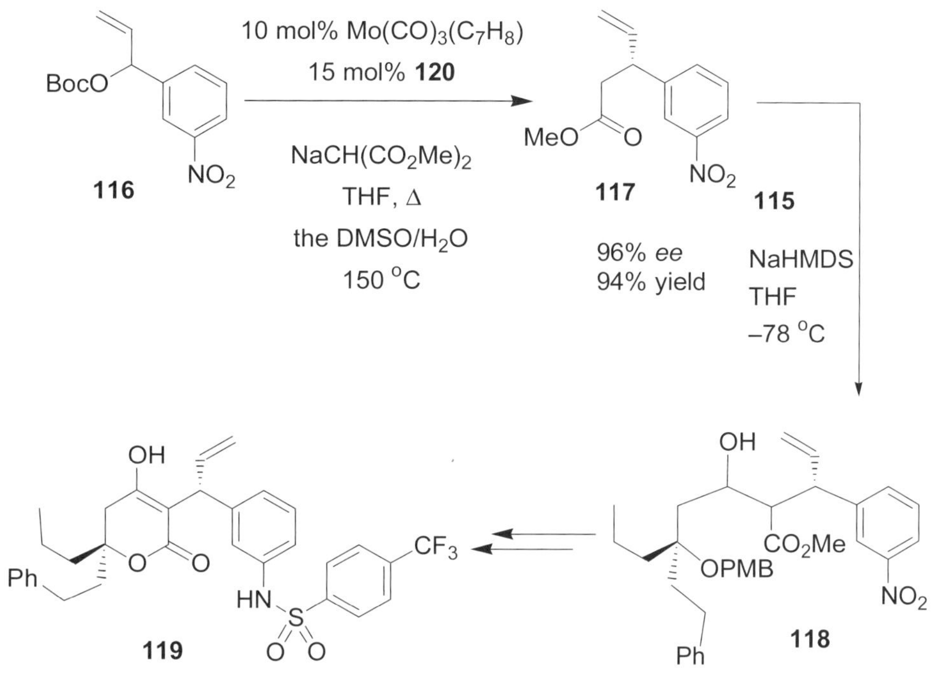

Just like palladium, the molybdenum-catalyzed reaction has been extended to an asymmetric variant. Following development of the reaction by Trost and co-workers,[68] they reported a concise and efficient synthesis of the HIV therapeutic tipranavir (**119**).[69] While the key tetrasubstituted centre in compound **115** was constructed by a palladium–boron co-catalyzed DYKAT reaction, the stereogenic centre in nitroaromatic **117** was installed in a molybdenum-catalyzed DYKAT reaction. Both compounds were subsequently elaborated to tipranavir (**119**). While it has been determined that the molybdenum-catalyzed reaction proceeds with overall retention, the stereochemistry of each step has not been extensively studied. Recent studies have suggested a retention–retention pathway may be in operation.[70]

Progress has also been made in the use of other metals. For example, iridium-catalyzed processes have been developed for AAA reactions.[17] The regioselectivity of this reaction is analogous to molybdenum. Allylation of "soft" carbon nucleophiles has been reported using chiral iridium complexes, with the products isolated in high enantiomeric purity (> 91% *ee*).[71,72] Hartwig has reported the use of chiral iridium complexes derived from ligand **123** in the formation of chiral allylic amines.[73] The branched amines **122** were isolated in moderate to good yield and excellent enantiomeric purity (> 94% *ee*).

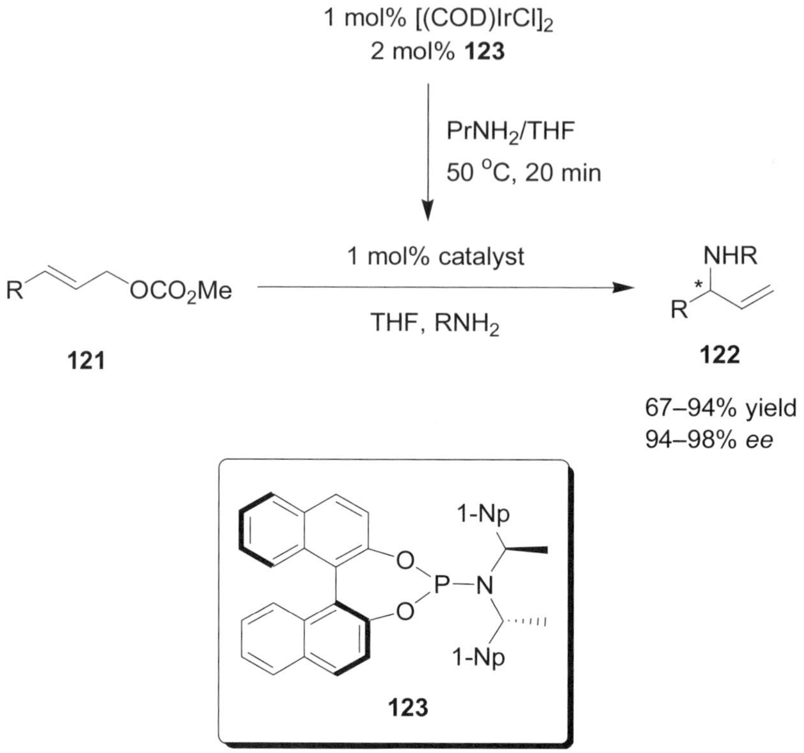

Nickel and platinum mediated allylation reactions have been reported and in terms of AAA reactions, perhaps the most useful processes have involved "hard" nucleophiles.[17] For example, high enantioselectivities have been realized in nickel-catalyzed processes employing Grignard reagents.[74] Tungsten has also been used in Tsuji–Trost type reactions, however it has thus far not been applied to complex molecule synthesis.[17]

1.1.8.6 Experimental

Tsuji–Trost Reaction

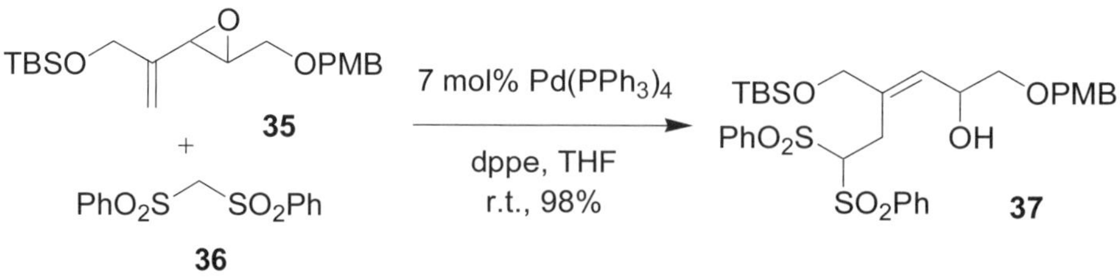

6,6-Bis-phenylsulfonyl-4-(tert-butyldimethylsilanyloxymethyl)-1-(4-methoxy-benzyloxy)-hex-3-en-2-ol (37).[37]

To a solution of $Pd(PPh_3)_4$ (280 mg, 0.24 mmol), dppe (182 mg, 4.6 mmol) and bis(phenylsulfonyl)methane (**36**) (1.007 g, 3.40 mmol) in THF (100 mL) was added epoxide **35** (1.25 g, 3.43 mmol) and the resulting mixture was stirred for 14 h at ambient temperature. An aqueous extractive work-up followed by flash chromatography of the crude material (hexane/EtOAc, 10/1→4/1) afforded product **37** as a colorless syrup (2.20 g, 98%, mixture of diastereoisomers).

AAA reaction

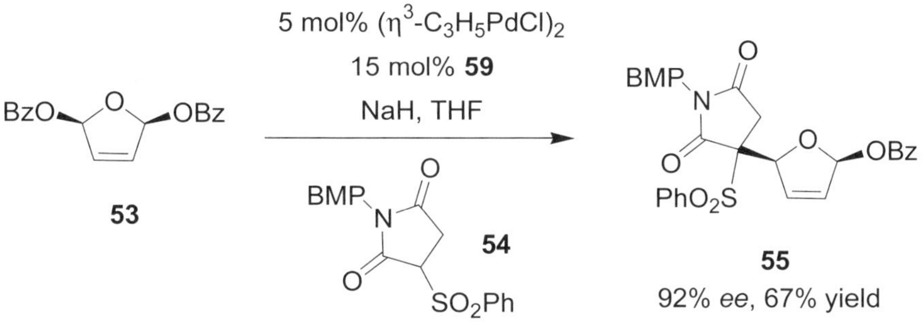

53 **54** **55**

92% *ee*, 67% yield

Preparation of (*2S*,*5S*)-2-[3′-Phenylsulfonyl-*N*′-(4-methoxybenzyl) succinimid-3′-yl]-5-benzoyloxy-2,5-dihydrofuran (55).[43]

To sodium hydride (68% oil, 3.9 mg, 0.097 mmol) in a cooled sonicator bath (0 °C) was added **54** (46 mg, 0.13 mmol) in THF (300 μL). The mixture was degassed with argon while bubbling ensued for 3–5 min. A prestirred solution of **53** (20 mg, 0.064 mmol), bis(η^3-allyl)di-μ-chlorodipalladium (0.5 mg, 5 mol% Pd), and ligand **59** (3.0 mg, 15%) in THF (300 μL) under argon was cannulated into the solution of nucleophile in the sonicator. Additional THF was added to help wash all the material into the reactive flask (200 μL, 0.08M overall). The reaction mixture was degassed by bubbling with argon with sonication for 10 min. The mixture was sealed and stirred for 4 h at 0 °C. Direct application to flash chromatography (50 : 50 petroleum ether/ethyl acetate) gave a crude oil with a ratio of starting material to product of 8 : 92. The product was purified by flash chromatography (75 : 25 petroleum ether/ethyl acetate) to give 24 mg (67%) of **55** as a colorless oil. The diastereomeric ratio was 7 : 3 by proton NMR. The diastereomers were separated by flash chromatography (80 : 20 petroleum ether/ethyl acetate) for enantiomeric excess determination by chiral HPLC. The enantiomeric excess for both the major and minor diastereomers was 92%.

1.1.8.7 References

1. [R] Trost, B. M. In *Stereochemistry of Organic and Bioorganic Transformations*; Bartmann, W., Sharpless, K. B., Eds.; VCH: New York, 1986.
2. [R] Trost, B. M. *Pure Appl. Chem.* **1981**, *53*, 2357.
3. [R] Trost, B. M.; Verhoeven, T. R. In *Comprehensive Organometallic Chemistry*; Wilkinson, G., Ed.; Pergamon Press: Oxford, 1982; Vol. 8, Chapter 57.
4. [R] Tsuji, J. *Pure Appl. Chem.* **1982**, *54*, 197.
5. [R] Tsuji, J. *Tetrahedron*, **1986**, *42*, 4361.
6. [R] Trost, B. M. *Chemtracts: Org. Chem.* **1988**, *1*, 415.
7. [R] Trost, B. M. *Angew. Chem., Int. Ed. Engl.* **1989**, *28*, 1173.
8. [R] Trost, B. M. *ACS Adv. Chem.* **1992**, *230*, 463.
9. [R] Godleski, S. A. In *Comprehensive Organic Synthesis*; Trost B. M., Fleming, I., Semmerlhack, M. F., Eds.; Pergamon Press: Oxford, 1991; Vol. 4, Chapter 3.3.
10. [R] Frost, C. G.; Howart, J.; Williams, J. M. J.; *Tetrahedron: Asymmetry*, **1992**, *3*, 1089.
11. [R] Tsuji, J. In *Palladium Reagents and Catalysts*: John Wiley & Sons: New York, 1995; Chapter 4.2.
12. [R] Trost, B. M.; Van Vranken, D. L. *Chem. Rev.* **1996**, *96*, 395.
13. [R] Moberg, C.; Bremberg, U.; Hallman, K.; Svensson, M.; Norrby, P.-O.; Hallberg, A.; Larhed, M.; Csoregh, I. Pure Appl. Chem. **1999**, *71*, 1477.
14. [R] Acemoglu, L.; Williams, J. M. J. In *Handbook of Organopalladium Chemistry for Organic Synthesis*; Negishi, E.-I., Eds.; John Wiley & Sons: New York, 2002; Vol. 2, 1689–1705.
15. [R] Mandai, T. In *Handbook of Organopalladium Chemistry for Organic Synthesis*; Negishi, E.-I., Eds.; John Wiley & Sons: New York, 2002; Vol. 2, 1845–1858.
16. [R] Tsuji, J. In *Handbook of Organopalladium Chemistry for Organic Synthesis*; Negishi, E.-I., Eds.; John Wiley & Sons: New York, 2002; Vol. 2, 1669-1687.
17. [R] Trost, B. M.; Crawley, M. L. *Chem. Rev.* **2003**, *103*, 2921.
18. [R] Trost, B. M. *J. Org. Chem.* **2004**, *69*, 5813.
19. Tsuji, J.; Takahashi, H.; Morikawa, M. *Tetrahedron Lett.* **1965**, 4387.
20. Tsuji, J.; Takahashi, H. *J. Am. Chem. Soc.* **1965**, *87*, 3275.
21. Tsuji, J. *New. J. Chem.* **2000**, *24*, 127.
22. Tsuji, J.; Kiji, J.; Morikawa, M. *J. Am. Chem. Soc.* **1964**, *86*, 4350.
23. Takahashi, S.; Shibano, T.; Hagihara, N. *Tetrahedron Lett.* **1967**, 2451.
24. Smutny, E. J. *J. Am. Chem. Soc.* **1967**, *89*, 6793.
25. Hata, G.; Takahashi, K.; Miyaka, *J. Chem. Soc., Chem. Commun.* **1970**, 1392.
26. Atkins, K. E.; Walker, W. E.; Manyik, R. M. *Tetrahedron Lett.* **1970**, 3821.
27. Trost, B. M.; Fullerton, T. J. *J. Am. Chem. Soc.* **1973**, *95*, 292.
28. [R] Trost, B. M. *Acc. Chem. Res.* **1980**, *13*, 385.
29. Trost, B. M.; Hung, M.-H. *J. Am. Chem. Soc.* **1984**, *106*, 6837.
30. [R] Trost, B. M.; Machacek, M. R.; Aponick, A. *Acc. Chem. Res.* **2006**, *39*, 747.
31. Trost, B. M.; Strege, P. E. *J. Am. Chem. Soc.* **1977**, *99*, 1650.
32. Morizawa, Y.; Oshima, K.; Nozaki, H. *Tetrahedron Lett.* **1982**, *23*, 2871.
33. Nilsson, Y. I. M.; Andersson, P. G.; Bäckvall, J. E. *J. Am. Chem. Soc.* **1993**, *115*, 6609.
34. Cuvigny, T.; Julia, M.; Rolando, C. *J. Organomet. Chem.* **1985**, *285*, 395.
35. Kitagawa, Y.; Itoh, A.; Hashimoto, S.; Yamamoto, H.; Nozaki, H. *J. Am. Chem. Soc.* **1977**, *99*, 3865.
36. Vanderwal, C. D.; Vosburg, D. A.; Weiler, S.; Sorenson, E. J. *J. Am. Chem. Soc.* **2003**, *125*, 5393.
37. Fürstner, A.; Gastner, T. *Org. Lett.* **2000**, *2*, 2467.
38. Seki, M.; Mori, Y.; Hatsuda, M.; Yamada, S. *J. Org. Chem.* **2002**, *67*, 5527.
39. Deardorff, D. R.; Myles, D. C.; MacFerrin, K. D. *Tetrahedron Lett.* **1985**, *26*, 5615.
40. Williams, D. R.; Meyer, K. G. *Org. Lett.* **1999**, *1*, 1303.
41. Matsushita, H.; Negishi, E. *J. Chem. Soc., Chem. Commun.* **1982**, 160.
42. Trost, B. M.; Keinan, E. *Tetrahedron Lett.* **1980**, *21*, 2591.
43. Trost, B. M.; Kallander, L. S. *J. Org. Chem.* **1999**, *64*, 5427.

44. [R] Trost, B. M.; Machacek, M. R.; Aponick, A. *Acc. Chem. Res.* **2006**, *39*, 747.
45. Amatore, C.; Jutand, A.; Mensah, L.; Ricard, L. *J. Organomet. Chem.* **2007**, *692*, 1457.
46. Andersson, P. G.; Harden, A.; Tanner, D.; Norrby, P. D. *Chem. Eur. J.* **1995**, *1*, 12.
47. Pfaltz, A. *Acc. Chem. Res.* **1993**, *26*, 339.
48. Hayashi, T.; Yamamoto, A.; Ito, Y.; Hagihara, T. *Tetrahedron Lett.* **1986**, *27*, 191.
49. Togni, A.; Breutel, C.; Schnyder, A.; Spindler, F.; Landert, H.; Tijani, A. *J. Am. Chem. Soc.* **1994**, *116*, 4062.
50. Evans, D. A.; Campos, K. R. Tedrow, J. S.; Michael, F. E.; Gagne, M. R. *J. Am. Chem. Soc.* **2000**, *122*, 7905.
51. Sprinz, J.; Kiefer, M.; Helmchen, G.; Reggelin, M.; Huttner, G.; Walter, O.; Zsolnai, L. *Tetrahedron Lett.* **1994**, *35*, 1523.
52. (a) Trost, B. M.; Toste, F. D. *J. Am. Chem. Soc.* **2003**, *125*, 3090; (b) Trost, B. M.; Toste, F. D. *J. Am. Chem. Soc.* **1999**, *121*, 3543.
53. Trost, B. M.; Patterson, D. E.; Hembre, E. *J. Am. Chem. Soc.* **1999**, *121*, 10834.
54. You, S. L.; Zhu, X.-Z.; Luo, Y.-M.; Hou, X.-L.; Dai, L.-X. *J. Am. Chem. Soc.* **2001**, *123*, 7471.
55. (a) Trost, B. M.; Madsen, R.; Guile, S. D.; Brown, B. *J. Am. Chem. Soc.* **2000**, *122*, 5947; (b) Trost, B. M.; Madsen, R.; Guile, S. D.; Elia, A. E. H. *Angew. Chem., Int. Ed. Engl.* **1996**, *35*, 1569.
56. Trost, B. M.; Van Vranken, D. L. *J. Am. Chem. Soc.* **1993**, *115*, 444.
57. Trost, B. M.; Organ, M. G.; O'Doherty, G. A.; *J. Am. Chem. Soc.* **1995**, *117*, 9662.
58. Trost, B. M.; Nanninga, T. N.; *J. Am. Chem. Soc.* **1985**, *107*, 1293.
59. Tsuji, J.; Sato, K.; Okumoto, H. *Tetrahedron Lett.* **1982**, *23*, 5189.
60. Murahashi, S.; Imada, Y.; Taniguchi, Y.; Higashiura, *Tetrahedron Lett.* **1988**, *29*, 4945.
61. Mandai, T.; Gotoh, J.; Otera, J.; Kawada, M. *Chem. Lett.* **1980**, 313.
62. Tsuji, J.; Minami, I.; Shimizu, I. *Tetrahedron Lett.* **1984**, *25*, 279.
63. Smith, A. S.; Sulikpwski, G. A.; Fujimoto, K. *J. Am. Chem. Soc.* **1989**, *111*, 8039.
64. Oshima, M.; Shimizu, I.; Yamamoto, A.; Ozawa, F. *Organometallics* **1991**, *10*, 1221.
65. Mukaiyama, C.; Mitsuhashi, H.; Wakamatsu, T. *Tetrahedron Lett.* **1992**, *33*, 4165.
66. Hayashi, T.; Iwamura, H.; Naito, M.; Matsumoto, Y.; Uozumi, Y.; Miki, M.; Yanagi, K. *J. Am. Chem. Soc.* **1994**, *116*, 775.
67. Trost, B. M.; Lautens, M. *J. Am. Chem. Soc.* **1982**, *104*, 5543.
68. Krsa, S. W.; Hughes, D. L.; Reamer, R. A.; Mathre, D. J.; Sun, Y.; Trost, B. M. *J. Am. Chem. Soc.* **2002**, *124*, 12656. And references therein.
69. Trost, B. M.; Anderson, N. G. *J. Am. Chem. Soc.* **2002**, *124*, 14320.
70. Hughes, D. L.; Lloyd-Jones, G.C.; Krsa, S. W.; Gourioua, L.; Bonnet, V. D.; Sun, Y.; Mathre, D. J.; Reamer, R. A. *Proc. Natl. Acad. Sci. USA.* **2004**, *101*, 5379.
71. Takeuchi, R.; Kashio, M. *J. Am. Chem. Soc.* **1998**, *120*, 8647.
72. Jannsen, J. P.; Helmchen, G. *Tetrahedron Lett.* **1997**, *38*, 8025.
73. Leitner, A.; Shu, C.; Hartwig, J. F. *Org. Lett.* **2005**, *7*, 1093.
74. Gomez-Bengoa, E.; Heron, N. M.; Didiuk, M. T.; Luchaco, C. A.; Hoveyda, A. H. *J. Am. Chem. Soc.* **1998**, *120*, 7649.

1.2.1 Castro–Stephens Reaction

David L. Gray

1.2.1.1 Description

The Castro–Stephens reaction is the cross coupling of a copper acetylide (**1**) and an aryl or vinyl halide (**2**) to give a disubstituted alkyne (**3**).[1] The reaction, which shares some common elements with the Sonogashira, Cadiot–Chodkiewicz, Rosenmund–von Braun, Hay, and Glaser coupling reactions, was discovered by Stephens and Castro in the early 1960s and has found some applications in synthesis.[2–5]

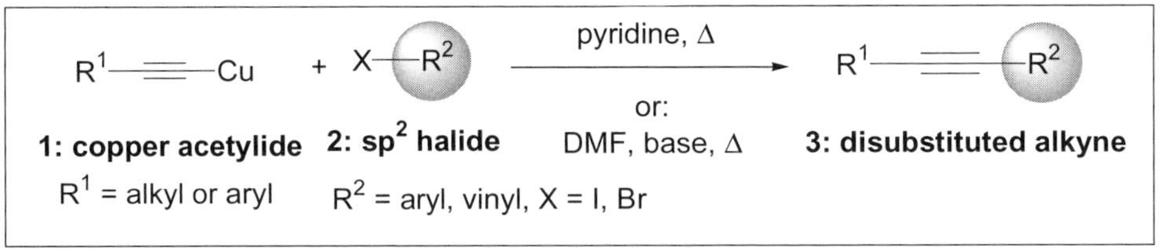

In many ways, the utility of the Castro–Stephens coupling has been supplanted by the more convenient Sonogashira coupling because the latter palladium-based reaction accesses the same target space with broader scope and functional group toleration, milder conditions, and greater operational simplicity (see 1.1.5).[6] There are a very few instances where the Castro–Stephens reaction is a superior option, but the need to pre-generate a air-sensitive and potentially explosive copper acetylide in the traditional protocol had made it a less popular choice for disubstituted alkyne synthesis. Perhaps 70% of published traditional Castro–Stephens couplings involve the union of copper (I) phenylacetylide (**4**) and an aryl iodide (**5**) to give linearly linked biphenyl compounds like **6**. Advancements in Castro–Stephens methodology have given the chemist simpler procedures for this reaction and have marginally broadened the scope of this process.

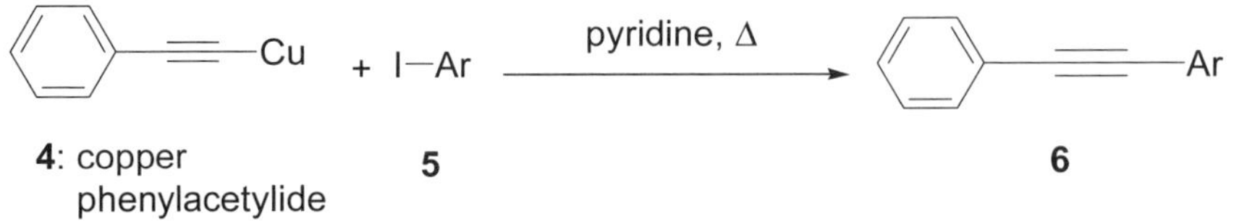

A somewhat limited range of coupling partners participates in Castro–Stephens coupling chemistry. Substrates tend to be simple, robust molecules that can stand up to basic copper salt solutions during prolonged heating to 110 °C or more and many common functional groups are not tolerated. For the halide partner, examples of coupling to aryl bromides are extremely rare – limited to cases where the aryl ring is further activated by an electron withdrawing group.[7] The two published classical Castro–Stephens couplings of **4** to aryl or vinyl chlorides were exemplified in Castro and Owsley's follow-up work on the reaction's scope, and are the special cases where the aryl chloride is either *ortho* to a carboxylic acid (e.g., 2-chlorobenzoic acid, **7**, X = Cl) or extremely activated (picryl chloride, **8**). Reaction of **8** with **4** in refluxing pyridine gave the expected biarylalkyne **9**, while the same protocol with **7** led to phthalide **11**, by way of alkynylated intermediate **11**.[8]

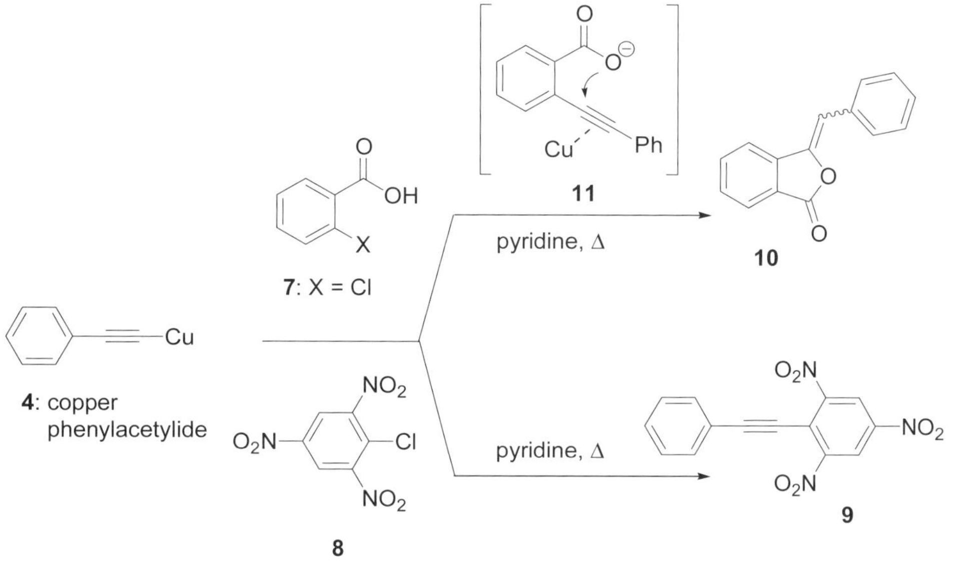

By definition, the reaction of a pre-generated copper acetylide requires stoichiometric copper, and the process generates one equivalent of a by-product copper(I) salt (usually cuprous iodide). The majority of successful reactions are carried out either in refluxing pyridine, or in DMF at elevated temperature, and reaction times often extend to 24 hours. One of the reasons for the narrow solvent range is that copper acetylides are insoluble polymeric species which are unavailable for any coupling chemistry until heated in particular polar solvents which have affinity for copper(I). Oxygen must be rigorously excluded to avoid the copper-mediated Glaser homocoupling of alkynes (see 1.2.2). An important advancement in the methodology introduced *in situ* formation of the copper acetylide, and later, reaction conditions that are catalytic in copper.

There is a great deal of nomenclature confusion in the Sonogashira/Castro–Stephens literature, with some authors calling palladium-based couplings "Castro–Stephens " reactions, and others calling copper-only procedures "palladium-free Sonogashira couplings".[9,10] Presently, the term "Sonogashira reaction" has morphed to cover any coupling of a terminal alkyne and a halide, irrespective of metal catalyst or pre-activation requirements. A useful rule for distinction of the two name reactions is that the classical Castro–Stephens coupling is a copper-only process that utilizes a discrete, pre-formed copper acetylide, while the Sonogashira, as first conceived, features palladium (with co-catalytic cuprous halide) and starts from an unactivated alkyne.[2] Advances in both methodologies, particularly the *in situ* formation of copper acetylides for Castro–Stephens reactions, have blurred the lines between these two methodologies, and under some conditions, they share key mechanistic features. The substrate scope, reaction conditions, and postulated intermediates for the pre- and the *in situ*-generated copper acetylide coupling reactions are similar, therefore, the latter procedures will be considered Castro–Stephens-like couplings in this section. Owing to heightened sensitivity within the chemical community toward the depletion of non-renewable resources (like precious noble metals) and also because of cost considerations, there has been renewed interest in copper-based cross coupling for C–N, C–O, and C–C bond formation including a number of advancements in so-called "palladium-free" Sonogashira couplings.

1.2.1.2 Historical Perspective

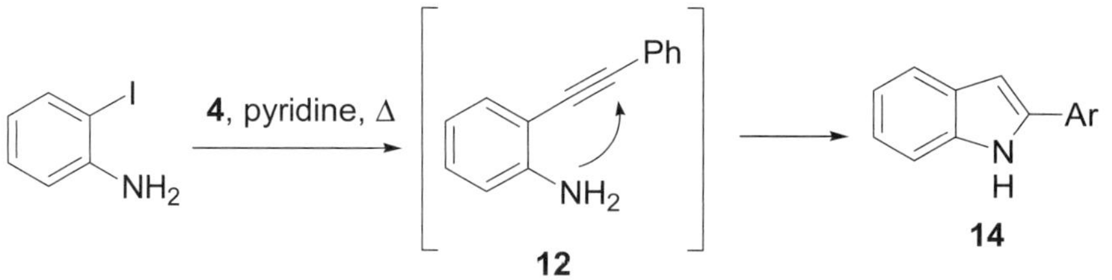

12

14

C. E. Castro and R. D. Stephens were working at the University of California, Riverside in the Chemistry and Nematology Departments where they were studying metal-promoted reductions of alkynes as well as heterocycle formation from *ortho*-heteroatom substituted biaryl acetylenes. They found that some of the required precursors for this work could be prepared by "exposing aryl iodides to cuprous acetylides in refluxing pyridine". In many cases, the coupled biaryl acetylenes (**11**–**13**) went on to cyclize with the *ortho*-oxygen or *ortho*-nitrogen atoms under those same conditions, thus the initial publications on this chemistry highlighted both the uncovered reactivity, and its utility in the synthesis of 2-arylindoles (**14**), 2-arylbenzofurans (**15**) and 3-benzylidenephthalides (**10**).

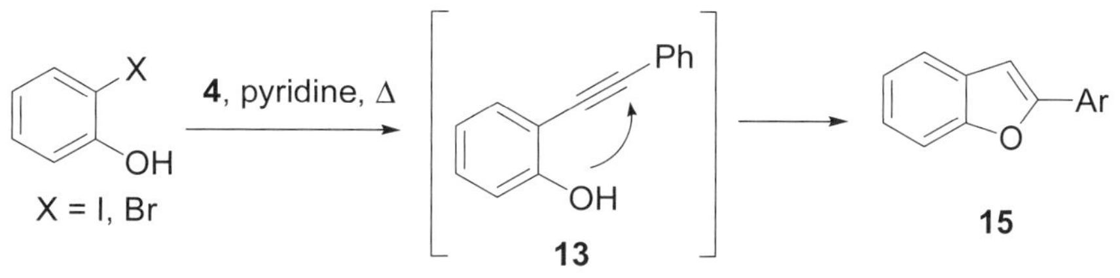

In their initial 1963 *Journal of Organic Chemistry* publication, Stephens and Castro proposed some potential reaction pathways and also noted the similarity of the reactivity they had uncovered to the much older Rosenmund von Braun displacement of aryl halides by copper cyanide.[1] In the next few years, Castro's laboratory did a relatively complete study of reaction scope for both the alkyne and halide components, exemplifying over 40 successful substrate pairs.[8] For the halide component, they included successful reactions with activated and deactivated aryl iodides, a pair of bromophenols, *o*-bromobenzoic acid, *o*-chlorobenzoic acid, picryl chloride, and an iodothiophene. On the alkyne side, they demonstrated that primarily phenyl, but also substituted phenyl, alkyl, pyridyl, propargyl alcohol, and propiolate acetylides would undergo productive reaction in pyridine or occasionally DMF. For some of the more reactive halides, room temperature reaction was demonstrated. Additionally, Castro showed that *in situ* formation of copper acetylides was possible for a half-dozen examples when *N*-ethylpiperidine was added to the reactions. This obviated the hassle of pre-forming the oxygen sensitive copper reagent for those substrates. Interestingly, in the intervening years, the scope they outlined in the mid 1960's has not expanded very much. Perhaps some of Castro's success can be attributed to a proficiency in generating, purifying, and handling cuprous acetylide reagents.

Copper acetylide (Cu≡Cu) itself is highly explosive and other metal acetylides have explosive character.[11] Copper acetylides are generally synthesized by exposure of a terminal acetylene to CuCl in aqueous ammonia solution, and this treatment will cause the newly formed reagent to precipitate from the reaction mixture as a colourful, polymeric solid within minutes of adding the copper salt (see 1.2.1.6 for a preparation of copper(I) phenylacetylide **4**). It has been shown that successful Castro–Stephens coupling requires that this insoluble metal acetylide be carefully purified in an often lengthy process of rinsing and filtering. In 1993, Miura reported on a version of this reaction that was catalytic in copper salt and generated the required copper-bound alkyne *in situ* with a mixture of CuI, K_2CO_3, and PPh_3

in DMF at 120 °C.[12,13] The scope of this variant at least equalled what was available from pre-formed copper acetylide coupling, with aryl and vinyl iodides giving reliable coupling to copper acetylides generated *in situ*. For example, Miura reported that phenylacetylene (**16**) and *E*-vinyl iodide **17** unite to give **18** in 72% isolated yield at only 80 °C. Running the same reaction under carbon monoxide and in dimethylimidazolidinone with a slightly different phosphine/copper ratio yielded an appreciable amount (33%) of the carbonylated product **19** along with similar yields of expected ene-yne **18**.

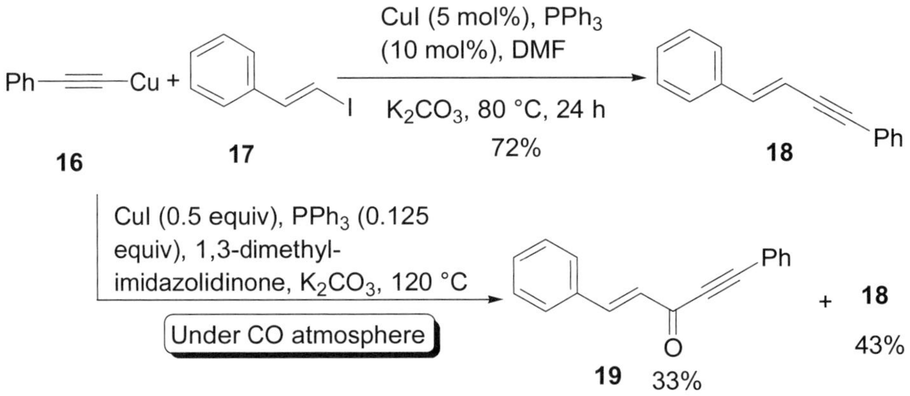

Modern extensions of Miura's work have yielded similarly simple *in situ* procedures which are also catalytic in copper and do not require pre-formation of a copper acetylide.[14,15] As a rule, regardless of Cu-acetylide origin, the scope of cross coupling with sp^2 halides remains limited to activated (electron deficient) aryl bromides and vinyl, aryl, or activated heteroaryl iodides, and reaction temperatures below 100 °C or reaction times less than 3 hours are unusual. There are situations where the coupling of acetylene itself would be of interest. The formal product of acetylene coupling is obtained in other cross coupling methodology (e.g., Sonogashira reaction) by use of TMS–acetylene and later deprotection. That sequence is ineffective in the Castro–Stephens reaction because the copper acetylide of TMS–acetylene is inherently unstable and decomposes before it can react productively. It is uncertain if major advances in the traditional Castro–Stephens coupling methodology are on the horizon, however, there is significant contemporary effort going into developing copper-catalyzed reactions to synthesize disubstituted acetylenes (**3**) from **1** and **2**.

1.2.1.3 Mechanism

Mechanism of Classical Castro–Stephens Reaction

Copper is able to participate in cross coupling chemistry in ways that are similar to palladium. Indeed there are copper-promoted analogs to many of the common Pd-based coupling procedures. Distinct from palladium, however, is the more complex and dynamic redox chemistry of the copper metal center.[16,17] Copper exists in four stable oxidation states from 0 to +3, and is able to readily shuttle among oxidation states and access several coordination geometries in each oxidation state. For instance, Cu(I), is regularly observed in linear (2 ligand), trigonal (3 or 4 ligand), or T-shaped (3 ligand) geometries. To further complicate matters at the metal center, solvent effects on copper complexes are significant because organocopper compounds are found to be highly aggregated and frequently engage in three-center (bridging) copper-copper bonding. Where the discrete geometry and binary oxidation state of palladium has led to isolable intermediates and testable hypothesis for some palladium-based cross-coupling reaction mechanisms, the aforementioned fluidity and complexity of copper has hampered rigorous mechanistic studies of reactions like the Castro–Stephens.[11]

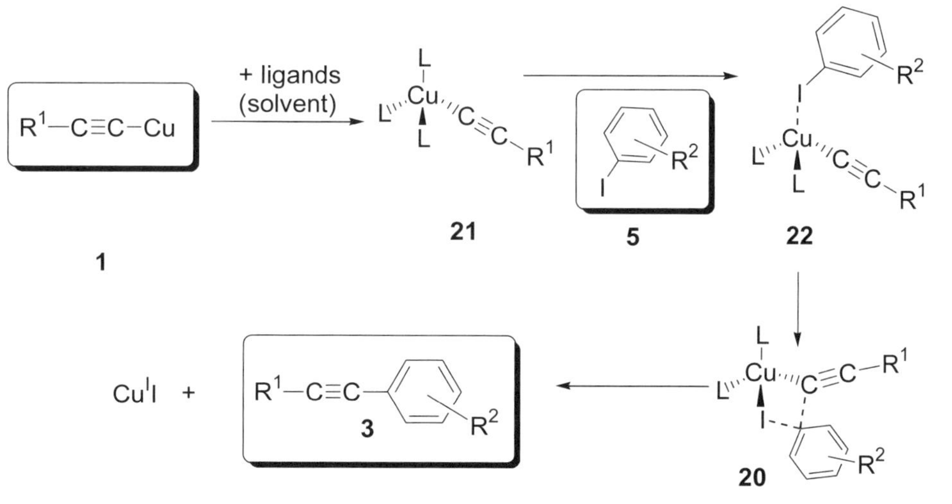

Copper acetylides have been characterized as polymeric species. They have notoriously poor solubility in organic solvents and are not completely soluble even in DMF at 100 °C, though this may be advantageous for cross-coupling because there is evidence that the reaction proceeded best under partially heterogenous conditions. The pKa of the acetylinc proton does not seem to exert influence on reaction outcome, and this observation may make the Castro–Stephens coupling more attractive (vs. Sonogashira coupling) in certain systems. Stephens and Castro put forth a mechanistic hypothesis in 1963 that still has merit, and they provided some rationale for their proposed key transition state intermediate **20**. Building upon that framework, when a copper acetylide disassociates from its polymer state and goes into solution (**1**), it takes on additional solvent ligands to afford a tetrahedral complex **21**.

It is proposed that the coordination of this species with the halide component **5** occurs via a fleeting initial interaction of the copper and the halogen atom (complex **22**), which progressed to a bridging three-center bond with the halogen-bearing sp^2 carbon in the transition state **20**. At this point, coupling of the two carbon ligands liberates the product **3** along with the favourable formation of CuI.

Hammett studies measuring the effect of ring substituents for copper promoted carbon substitution of aryl halides have demonstrated a weak positive correlation (ρ-values +0.1 to +1.1) which confirm the nucleophilic character of the copper reagent. The smaller ρ-values could indicate a weaker, through-atom-communication of the aryl ring's electronic information in the transition state. The three-center bonding represented in **20** is entirely consistent with the Hammett data as well as with the relative rate data for different halides, where the relative rates follow with the polarizability of the halides (e.g., I > Br > Cl), and aryl triflates do not react.[11,16,17] The very strong *ortho* effect of chelating heteroatom groups that Stephens and Castro noted (e.g., reaction of **7** → **10**) is also in line with intermediates like **20** and argues against mechanistically relevant π-complexation of copper to the arene.

Catalytic Mechanism and in situ Acetylide Formation

When Miura published the first example of a catalytic Castro–Stephens coupling, he made a number of careful mechanistic observations and proposed a catalytic cycle which incorporates many of the same intermediates (e.g., **20** and **21**) postulated for the non-catalytic reaction.[13] The entry into this catalytic cycle, copper iodide (**23**) is converted to the soluble, solvent ligated copper(I) species **24**. In this scheme, with the ligand PPh_3 present and no pyridine solvent to ligate copper, the ligands (L) could be phosphines. The metal activates a terminal alkyne (**25**), acidifying the proton for removal by base via a transient complex **26.** The mechanistic requirement for at least one equivalent of base is a difference between the stoichiometric copper reactions and the *in situ* catalytic methods. Following deprotonation, the copper acetylide **21** is proposed to lose a ligand to generate a coordinatively unsaturated complex **27** which is predisposed to coordinate with the aryl halide component **5** in a halogen-centred 3-atom bond between copper and the aryl ring. Within this complex (**20**), bonding occurs between the alkyne and the aryl ring, ultimately leading to C–C bond formation, creation of coupled product **5**, and regeneration of the copper entry catalyst **24**. This proposal was based upon several important experimental results. First, when phenylacetylene (**16**) was treated with 0.5 equivalent of CuI in the presence of excess K_2CO_3 in DMF at room temperature, the characteristic polymeric copper phenylacetylene **4** was quantitatively formed. This precipitate was a competent reagent and reacted with iodobenzene only at elevated temperature (where the polymer begins to dissolve into its soluble monomeric form) to deliver the desired coupling product. This result and the surrounding observations provided clear evidence that *in situ* formation of copper(I) acetylides was occurring under the reaction conditions, and that these acetylides were highly similar to pre-formed copper acetylide.

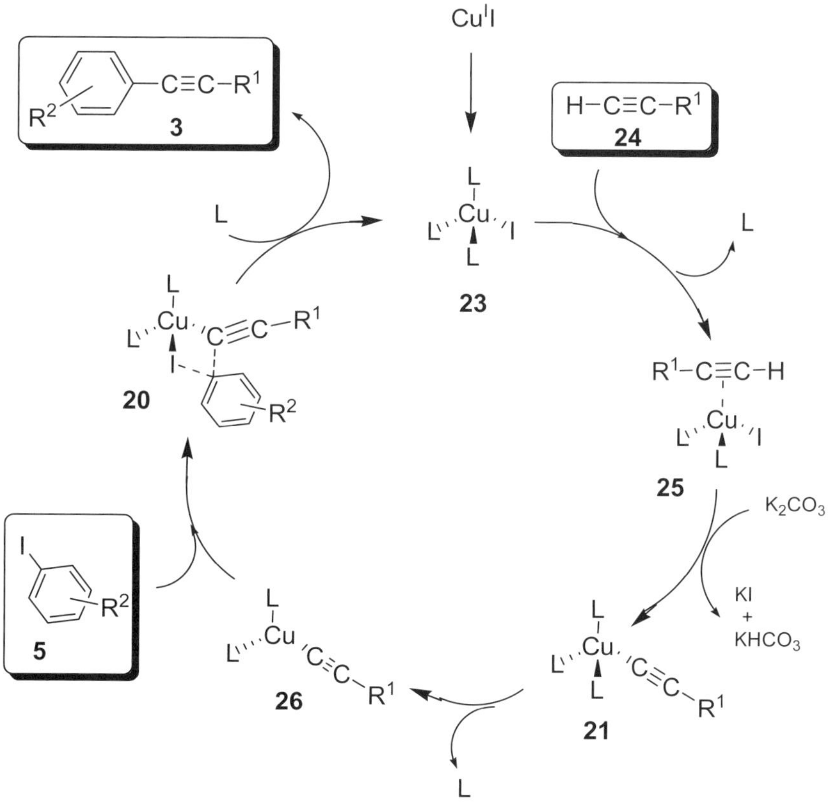

Second, Miura did a set of experiments to conclusively show that addition of PPh_3 greatly accelerated the coupling reaction. This result was attributed to the ability of PPh_3 to solubilise the copper acetylide (as species **21**). Using this observation, Miura reasoned that PPh_3 must be coordinating to a copper(I) acetylide. He was able to isolate and partially characterize a labile complex which had spectral (FT–IT and X-ray fluorescence) properties consistent with **21** (L = PPh_3). This complex was able to react with iodobenzene to afford coupling product **3** when heated with iodobenzene in DMF. In the course of studying this process, Miura examined the effects of varied ligands and found that stronger σ-donor ligands (stronger than PPh_3) shut the reaction down. He hypothesized that accessing the coordinatively unsaturated species **27** might be difficult with very strongly donating ligands, thus providing some support for the putative complex **27**. Following Miura's work, publications using bidentate bipyridine and phenanthroline-based ligand systems have emerged.[14] The protocols that are catalytic in copper often use inorganic bases, and it is interesting to note that amine bases retard cross coupling in many of these catalytic reaction systems. Relative to the original Castro–Stephens conditions, all of these catalytic, *in situ* methods are more convenient, however, the reaction appears to have reached a reactivity wall with electron rich aryl bromides as these substrates are not coupled effectively by any of the newer procedures.

Regioselectivity and Stereoselectivity

Similar to the Sonogashira and related sp–sp^2 couplings, there are no regio- or stereoselectivity complications in the Castro–Stephens coupling. Reaction conditions are quite vigorous, however, and where vinyl halides with defined geometry are used, the potential for olefin scrambling must be considered because there are rare cases where olefin isomerisation has been observed.[18]

Ligand Effects

Traditional Castro–Stephens couplings with pre-generated copper acetylides have not been aided by the addition of ligands for copper, and pyridine or other reaction solvent serves to fill coordination sites on the metal center. The general necessity to solubilise pre-formed (polymeric) copper acetylide with pyridine leads to a situation where a strong ligand for the copper (pyridine) is abundant and probably overrides any added ligand. For catalytic protocols, where copper acetylide can be accessed in a more soluble form (e.g., **21**, L = PPh_3) , monodentate phosphine, bidentate pyridine-based ligands, and low molecular weight PEG (polyethyleneglycol) have all been reported to be effective ligands in DMF or DMSO.[14,19] The most interesting development along these lines is the methodology developed by J.-H. Li, which uses DABCO as a ligand under otherwise fairly typical Miyura conditions.[20]

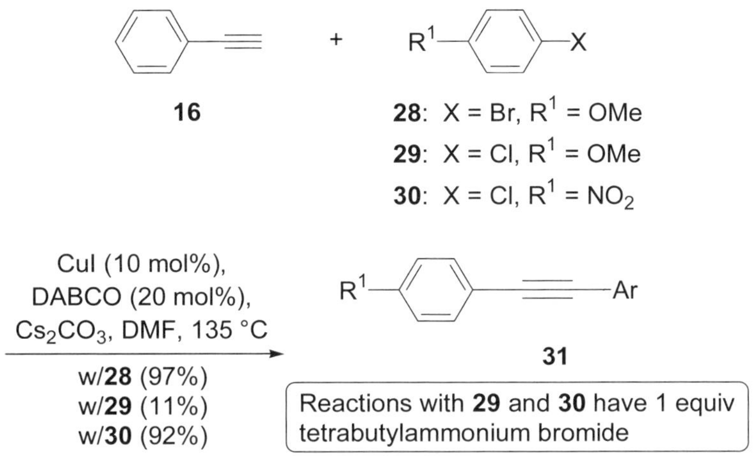

When 20 mol% of this base is added to 10 mol% CuI in DMF with 2 equivalents of Cs_2CO_3, the resulting catalyst is highly active for Castro–Stephens coupling of terminal alkynes and aryl bromides. For the reaction of **16** with *p*-bromoanisole **28**, the product **31** (R^1 = OMe) is obtained in excellent yield after 24 hours at 140 °C. Most impressively, with the addition of one equivalent of tetrabutylammonium bromide, even *p*-methoxychlorobenzene **29** couples to **16** under these conditions, affording the desired product in 11% yield, while *p*-nitrochlorobenzene **30** couples with phenylacetylene in 92% yield (55% in the absence of the tetraalkylammonium salt). Alkyl alkynes also couple effectively to both aryl and vinyl iodides and bromides in this methodology. The DABCO ligand is highly effective for Castro–Stephens coupling, displaying a substrate scope not seen with any other copper-promoted Csp–Csp^2 coupling methodology. Interestingly, the same catalytic system (CuI, DABCO) is also effective for Suzuki couplings of aryl halides and boronic acids.[20]

Cadiot–Chodkiewicz Reaction

One of the reactions that has some mechanistic relationship to the Castro–Stephens coupling is the Cadiot–Chodkiewicz synthesis of unsymmetrical bis-alkynes (**32**).[3] This coupling reaction finds occasional use in modern synthesis. Like the Castro–Stephens, this cross coupling is promoted by cuprous halides and proceeds through a putative copper acetylide. The Cadiot–Chodkiewicz reaction requires a terminal alkyne (**33**), an alkynyl halide (**34**), and a base, and typically goes to product under conditions which are considerably milder than Castro–Stephens couplings. Similar to most other copper-based homologation reactions, it is important to exclude oxygen to prevent the formation of copper(II).

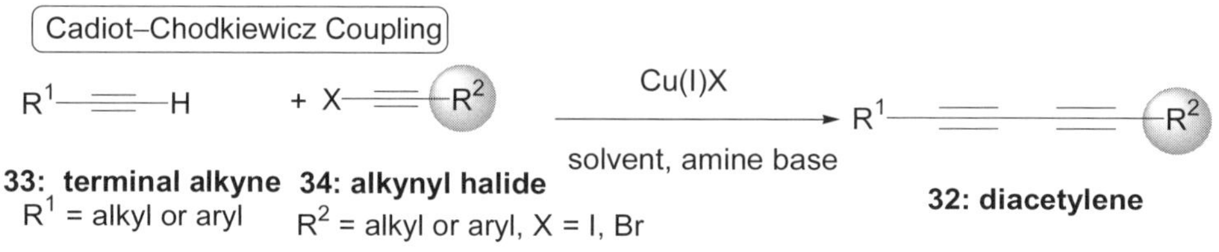

This reaction process takes advantage of the ease with which a copper acetylide will oxidatively insert into an alkynyl halide bond. The postulated mechanism begins with and *in situ* base- and Cu(I)-induced formation of a copper acetylide (**1**) from a terminal alkyne (**33**). This intermediate undergoes oxidative addition into the activated C–X bond of an alkynyl halide (**34**) to afford the copper(III) species **35**. Reductive elimination of the bis-alkyne **32** from complex **35** delivers the reaction product and regenerates the copper(I) halide **36** which may re-enter the catalytic cycle.

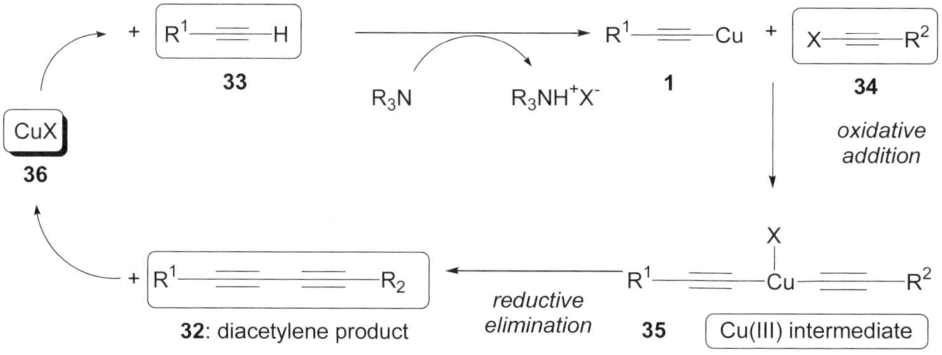

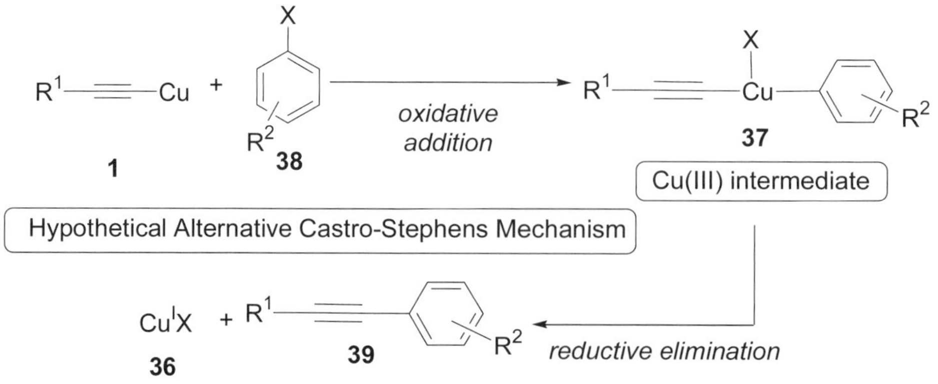

It is conceivable that a copper(III) intermediate like **35** could be involved in Castro–Stephens coupling, with copper acetylide **1** advancing through complex **37** after oxidative addition of the aryl halide (**5**). By analogy to the Cadiot–Chodkiewicz reaction, a reductive elimination event would liberate the disubstituted alkyne **39**. The evidences for the Castro–Stephens mechanism outlined in 1.2.1.3 are much more compelling at present, and experimental support for this alternate mechanism is lacking, but chemists working on copper-catalyzed cross coupling methodology would certainly benefit from additional mechanistic studies in this area.

Rosenmund–von Braun Reaction

The Rosenmund–von Braun synthesis of aryl nitriles (**40**) from aryl halides (**38**) also shares some commonality with the Castro–Stephens reaction.[4] This transformation was discovered nearly a century ago, and was mentioned in the initial Castro–Stephens publication where the authors noted potential mechanistic similarity.

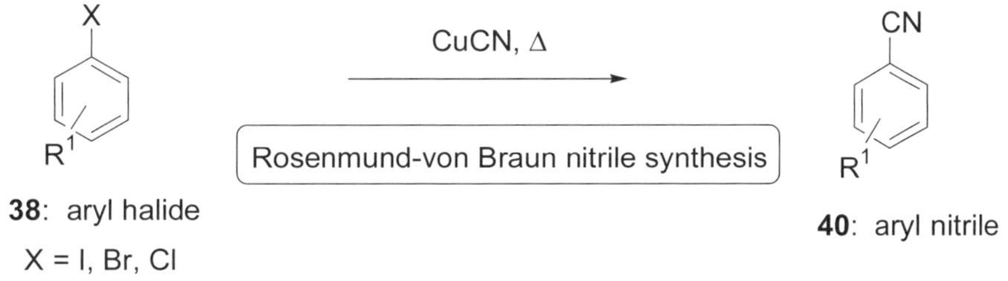

The typical Rosenmund–von Braun conditions are very harsh (200 °C neat), and product purification is generally troublesome, but the reaction is of historical interest and is occasionally used in the synthesis of simple aryl nitriles.

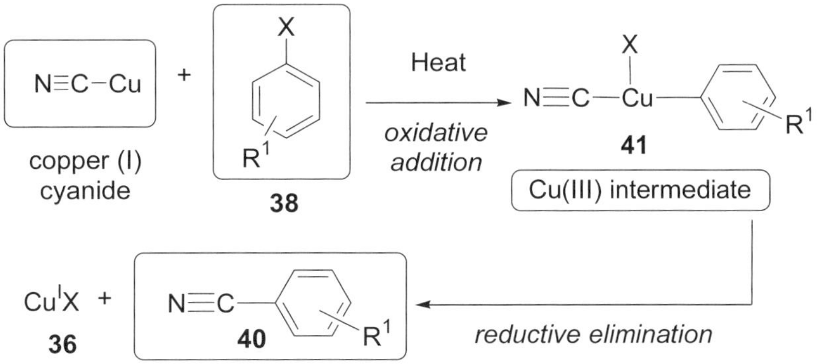

A possible mechanism for the conversion of aryl halides **38** to aryl nitriles **40** invokes a copper(III) intermediate **41**, which is reminiscent of the postulated key complex **35** in the Cadiot–Chodkiewicz mechanism outlined above. Despite the gross similarities among these three transformations, there is no clear evidence that the oxidative addition/elimination pathway and copper(III) intermediates which define the Cadiot–Chodkiewicz and Rosenmund–von Braun reaction mechanisms are operant in the typical Castro–Stephens coupling.

1.2.1.4 Synthetic Utility

Alkynlation of Aryl Halides

One example where Castro–Stephens reaction proved to be superior to other methods is the synthesis of the binaphthyl compound **43**. In this case, Sonogashira reaction on **42** afforded only an undesired helicene product, where the Castro–Stephens coupling with **4** gave the desired bis-alkyne **43** in 40% yield. Later a Negishi coupling approach was even more successful, delivering the same product in 90% yield.[21]

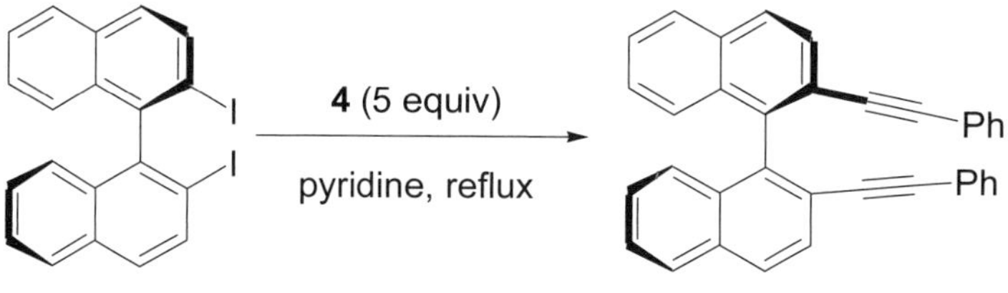

42 43

Copper acetylides of protected propargyl alcohols have been used as partners in Castro–Stephens couplings on multiple occasions. Coupling of iodopyrimidine **44** with the pre-formed copper reagent **45** was carried out in refluxing pyridine was complete in 2.5 hours, and gave the desired pyrimidine **46** in high yield. This intermediate was used to synthesize uracils with novel carbon substituents.[22]

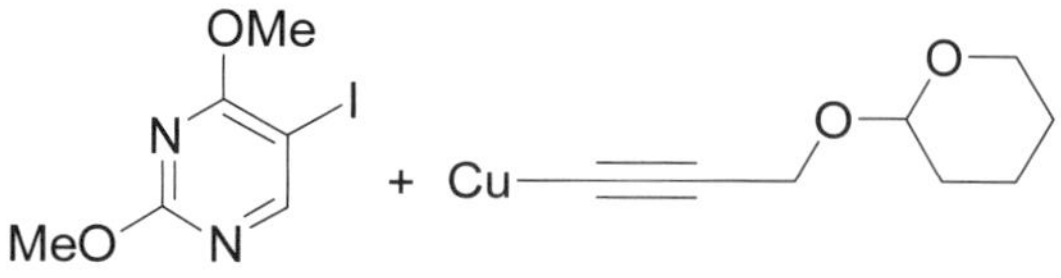

44 45

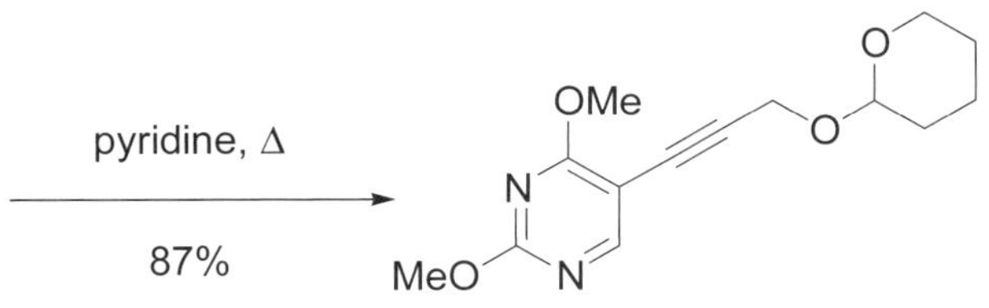

46

Macrocycle Synthesis

Compounds like tribenzocyclotriyne **47** are of interest because they have the ability to form conducting complexes with low-valent nickel by virtue of their planar, anti-aromatic character and cavity size. Youngs *et al.* published a synthesis of **47** where cyclotrimerization of precursor **48** is accomplished by synthesizing and then purifying the copper acetylide of the latter iodo alkyne and then refluxing said material in pyridine for 24 hours. In this way, the desired cyclotrimer **47** is obtained in an impressive 80% yield. The palladium-based Sonogashira reaction was attempted for this same substrate (**48**) and was much less effective (5% yield of **47**). There are several examples where the Castro–Stephens approach was superior to other methods for creating cyclic arene/yne macrocycles.[23,24]

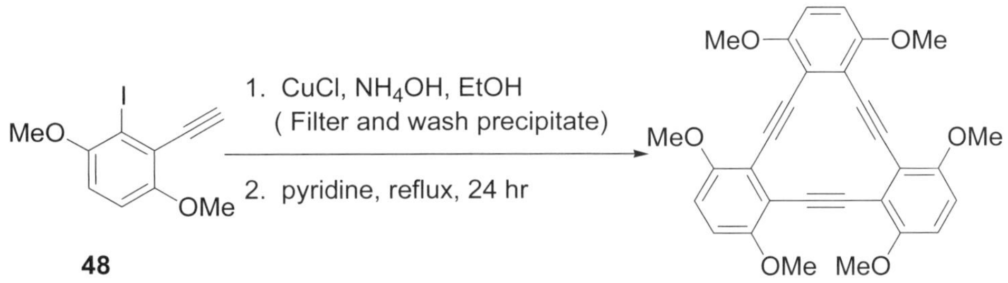

48

47

One of the few applications of the Castro–Stephens reaction to natural product synthesis is another example of the apparent suitability of the copper sp–sp^2 coupling for macrocyclic ring closure. As part of studies directed towards the synthesis of oximidine I, the copper-initiated coupling was mildly successful for the formation of a large ring via intramolecular closure of an *in situ*-formed copper acetylide and an *E*-vinyl iodide within **49**.[18,25] Thus following Miura's conditions, adding CuI, PPh_3, K_2CO_3, and compound **49** together in DMF and heating this mixture to 110 °C for 26 hours, provided a 31% yield of **50**. The obtained *cis/cis* diene geometry was postulated to arise via an isomerisation event caused by the high reaction temperature. Computational modelling of the **50** provided evidence that the all *cis* arrangement was lower in energy than the initially formed *cis*/*trans* product.

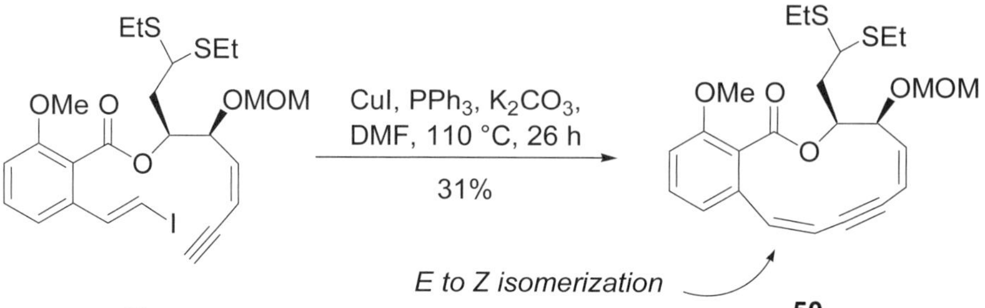

49 50

Heterocycle Synthesis

In their first full disclosure of their reaction, Stephens and Castro reported that coupling acetylides to *ortho*-halobenzoic acids (**7**) led, via intermediate **11**, to isocoumarins **51**.[1b] Several years later, they corrected the structure of the product of this reaction, realizing that the post-coupling cyclization had occurred in a 5-*exo*-dig arrangement to yield the benzylidene phthalide **52**. This reaction typifies the utility of the Castro–Stephens reaction in heterocycle synthesis, where examples are essentially limited to closure of *ortho*-heteroatoms onto a newly coupled alkyne, perhaps facilitated by the presence of copper. Castro did find one example where the small alkyl acetylide of propyne (**1**, R = Me) gave a quantity of the isocoumarin **51** in a mixture with **52**. The profound activation mediated by the *ortho* heteroatoms might come from a coordination of the heteroatom to the metal center, which would facilitate (by proximity) the formation of transition-state complex **20**. With iodobenzoic acid (**7**, X = I), unique examples of room temperature Castro–Stephens couplings were observed. Ethyl propiolate (**1**, R = CO_2Et), 7, and *N*-ethyl piperidine catalyst **53** in pyridine gave a 37% yield of phthalide **52** in a room temperature reaction.[8]

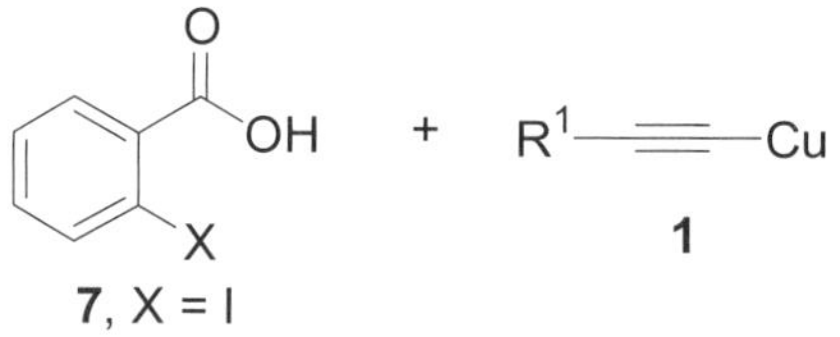

7, X = I **1**

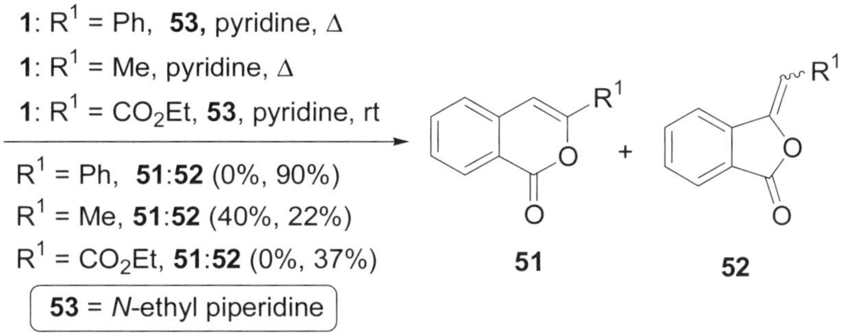

While attempting to couple iodopyrazine **54** to a terminal acetylene under Sonogashira conditions, Vasilevsky noted that the Castro–Stephens reaction was successful in some instances where Sonogashira reaction only gave starting material or side products, particularly with less acidic alkynes.[26] With limited examples, he concluded that for pyrazine iodides, the Sonogashira coupling doesn't work well above a pKa threshold for the alkyne. Others have noted pKa effects for the Sonogashira reaction.[27] A series of substituted copper phenylacetylides were reacted with **54** in refluxing pyridine. Good yields of products **55** were obtained with all groups. This synthesis could also be adapted to afford the heterocyclic derivative **56** from reaction with propiolic copper acetylide **57**. This cyclization is most likely occurring via homologated intermediate **58**, with ring closure aided by the presence of the ketone.

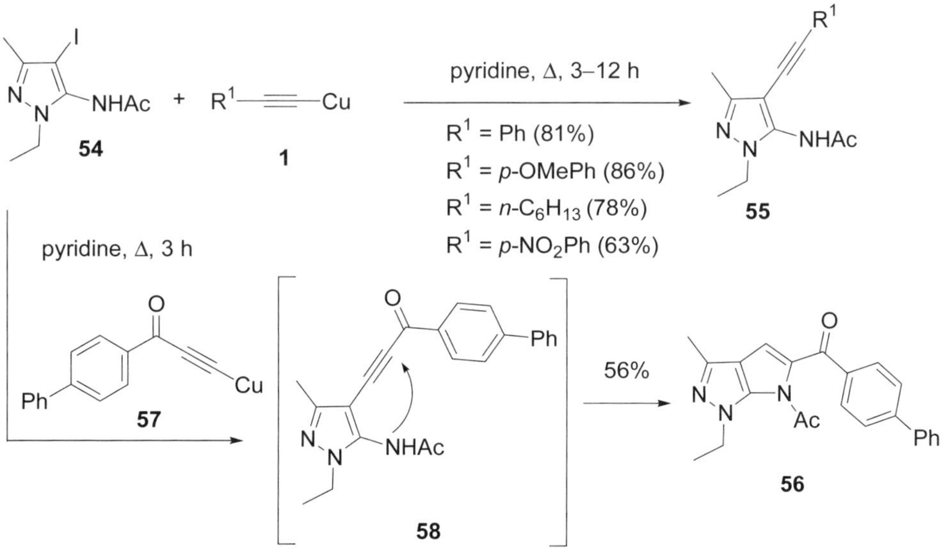

1.2.1.5 Variations and Improvements

Copper Nanoclusters – Heterogeneous Catalysis

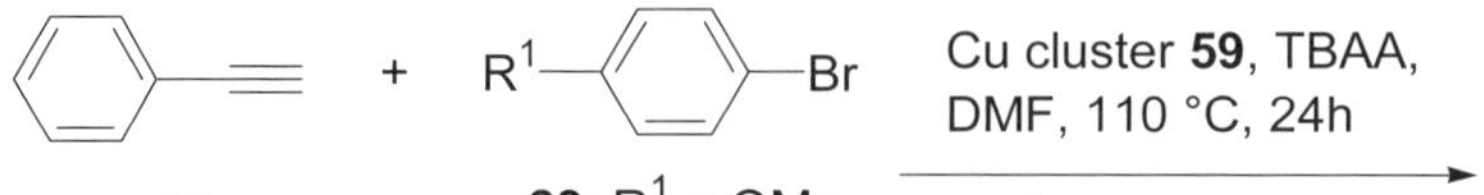

16 **28**: R^1 = OMe **61**: R^1 = CN

TBAA = tetrabutylammonium acetate

Cu cluster **59**, TBAA, DMF, 110 °C, 24h

R^1 = CN, (91%)
R^1 = OMe, (16%)

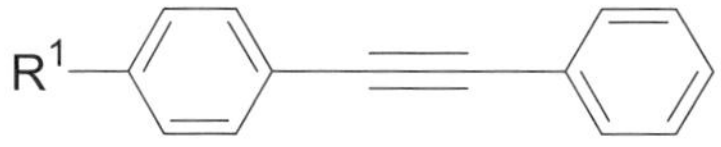

60: R^1 = OMe
62: R^1 = CN

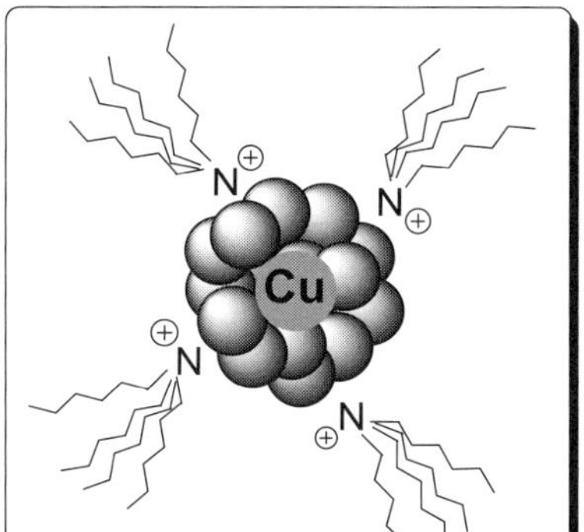

59

One of the driving forces behind the substitution of palladium for copper is a growing awareness of environmental and sustainable chemistry issues. Accordingly, the substitution of rare metals for more common elements and the use of protocols which allow for facile re-use of metal catalysts are important goals. The use of copper nanoclusters of 2–10 nM diameter (**59**) to catalyze Castro–Stephens reactions is a foothold into "green" arene–alkyne cross coupling.[10] The catalytic system consists of microscopic nanoclusters which are stabilized against self-aggregation by tetraalkylammonium salt additives. It is possible to conceptualize the entire nanoparticle as a substitute for a single copper atom, with developed charges shared across multiple copper centers and substrate ligands perhaps residing on different atoms. The nature of reacting species is poorly understood for nanoparticules, but the reactivity seen with these catalysts is promising. When alkyne **16** and *p*-bromoanisole (**28**) are heated to 110 °C for 24 hours with 5 mol% of these tiny copper clusters and tetrabutylammonium acetate as a base, a 16% yield of **60** was obtained. This reaction only proceeded with an alkylammonium aggregation inhibitor/stabilizer (tetraoctlyammonium formate). The yield for the traditionally challenging reaction of **28** → **60** is unimpressive, however, with a handful of more favourable substrate pairs, such as the conversion of **61** to **62**, yields were excellent. One advantage for such heterogeneous catalysts is that they can often be filtered and re-used. In this case, the catalysts began losing some activity after 3 recycles.

Copper-Catalyzed Sonogashira

As efforts have accelerated to find copper-based methods for Csp^2–O and Csp^2–N bond formation, the developed ligands have found some application in Castro–Stephens type couplings of aryl iodides and terminal acetylenes. For example, phenylacetylene (**16**) was united to **63** under the influence of $Cu(acac)_2$ and the diketone ligand **64** in DMF. The product (**65**) was isolated in 65% yield. The more active iodoarene **66** yielded better yields of the expected disubstituted acetylene **67** when it was reacted under the some conditions.[9] Among the handful of bidentate ligands used to date in so called "palladium-free" Sonogashira coupling approaches, none have demonstrated robust coupling of aryl bromides to alkynes.

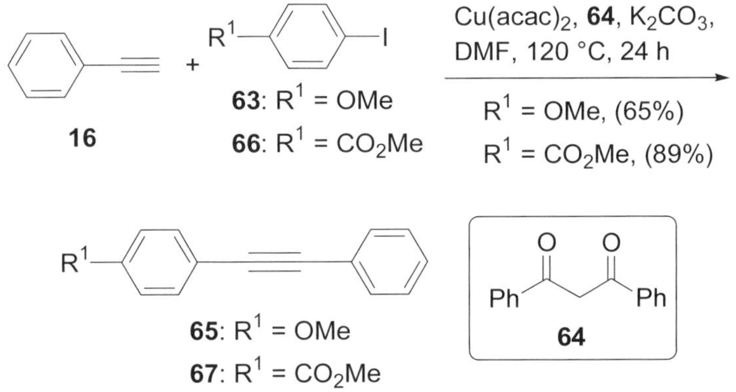

Castro–Stephens-like Coupling with Allylic Halides

One of the other complex molecule examples which showcase a coupling under Castro–Stephens-like conditions appeared in the synthesis of dehydroepothilone D (**68**), an anti-mitotic agent of interest for its anti-proliferation activity.

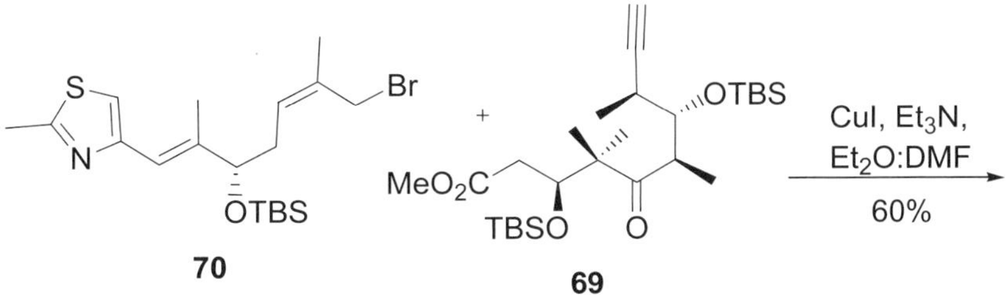

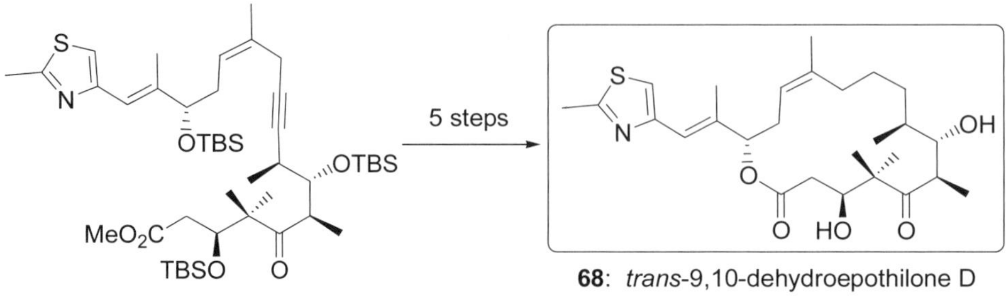

71

68: *trans*-9,10-dehydroepothilone D

This synthesis included a Castro–Stephens-type coupling of two highly functionalized fragments, where **69**, and allylic bromide **70**, together with Et_3N and CuI, were heated to 50 °C in an Et_2O/DMF solvent system. With this treatment, a 60% yield of the coupled product **71** was obtained.[28] There are several other examples of allylic substitution with copper acetylides, and these reactions clearly occur under milder conditions than typical Castro–Stephens couplings.[29] This reaction may be more of an S_N2 displacement, however, the net transformation and reaction conditions bear some similarity to the Castro–Stephens reaction.

Silylated Alkynes

Until recently, there has not been much of a role for silyl alkynes in Castro–Stephens reactions though they are ubiquitous in Pd-based alkyne couplings due to the extreme instability of TMS copper acetylide under typical reaction conditions. The stability of the acetylide increases as the alkyl group on silicon become larger. Advances in C–Si bond cleavage methodology has led to a few methods where silyl alkynes like **72** can be used as a reagent to generate reaction-competent copper(I) acetylides. In one such example, cuprous chloride in DMF is reported to cleave the C–Si bond in TMS–alkyne **72** to afford a discrete copper species **73**. This material was characterized and is not reactive as an acetylide, however, following additional processing, the desired reagent **4** is formed in good overall yield. The reagent thus obtained is viable in Castro–Stephens couplings.[15,30]

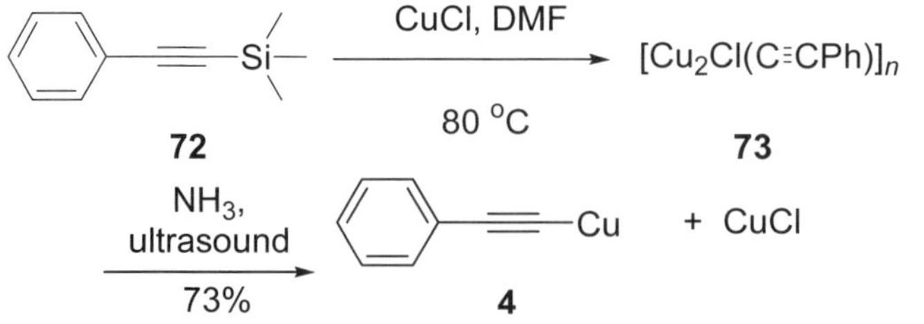

72 **73** **4**

Related Processes—Cadiot–Chodkiewicz Reaction

The Cadiot–Chodkiewicz coupling typically proceeds under conditions which are considerably milder than Castro–Stephens reactions. Triethylsilylacetylene **74** rapidly undergoes Cadiot–Chodkiewicz coupling with alkynyl bromide **75** to generate the unsymmetrical bisalkyne **76** in nearly quantitative yield when those two reactants are treated with catalytic cuprous chloride and catalytic ammonium hydroxide in *n*-butylamine solution. This coupling process affords one of the best entries into compounds such as **76** and is permissive of TES and larger silylated copper acetylide species because of the lower reaction temperature.[31]

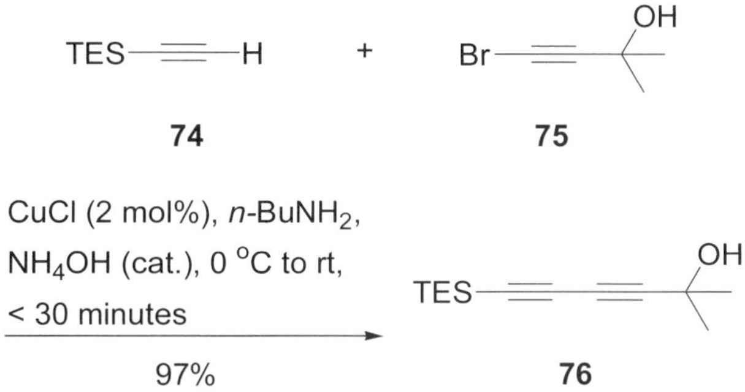

Similarly, exposure of propargyl alcohol **77** or alkyne **78** to hydroxylamine in aqueous MeOH with two equivalents of copper(I) chloride and excess diethylamine with under nitrogen atmosphere forms *in situ* copper acetylides that react with **79** to efficiently yield bisacetylenes **80** and **81**. These intermediate were used in a total synthesis of falcarindiol.[32]

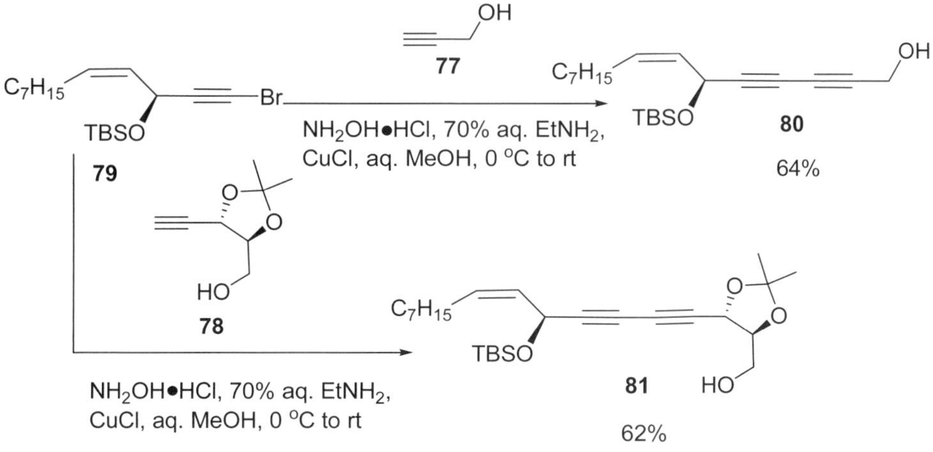

Related Processes: Rosenmund–von Braun Reaction

The application of ionic liquid solvent has led to Rosenmund–von Braun conditions which are considerably milder than previous variants.[33,34]

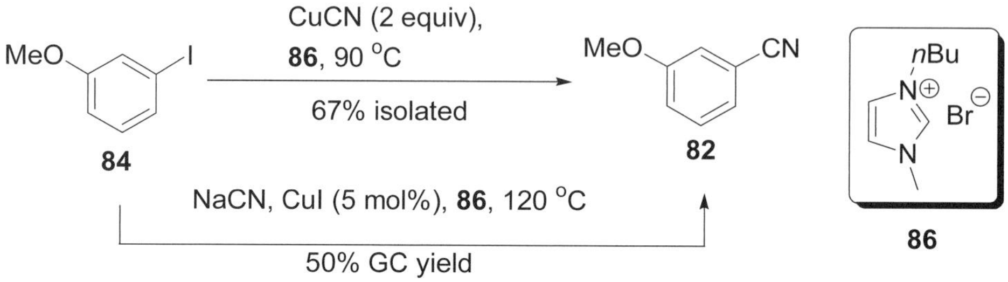

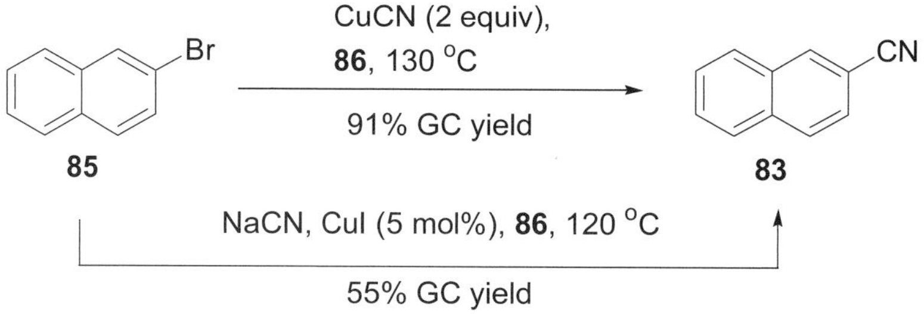

Aryl nitriles **82** and **83** were synthesized from the corresponding iodide **84** and bromide **85** in imidazolium bromide ionic solvent (**86**). The described protocol using either excess copper cyanide or catalytic copper(I) iodide and sodium cyanide, allows for these reaction to be run at between 90 and 130 °C. Isolation of the products is still somewhat problematic, however, the catalyst can be recycled from this medium and effectively re-used.[35]

1.2.1.6 Experimental

Preparation of Copper(I) Acetylide

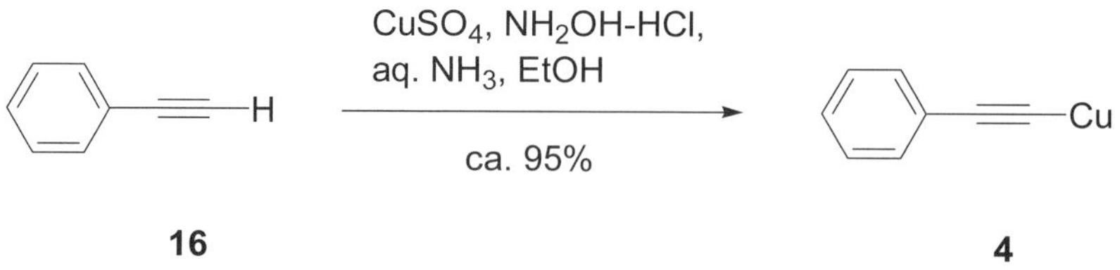

Copper(I) phenylacetylide (4).[36]

A solution of 25.0 g. (0.100 mole) of copper(II) sulfate pentahydrate in 100 mL of concentrated aqueous ammonia was placed in a large Erlenmeyer flask fitted with an effective magnetic stirbar and cooled to 0 °C. Stirring was maintained through the course of the reaction and the reaction is maintained under a constant stream of N_2. Water (400 mL) is added followed by addition of solid hydroxylamine hydrochloride (13.9 g, 0.200 mole) over 10 minutes. A solution of 10.25 g (0.1005 mole) of phenylacetylene (**16**) in 500 mL of 95% ethanol is then added rapidly to the pale blue solution. The reaction flask is swirled by hand, copper(I) phenylacetylide separates as a copious yellow precipitate, and an additional 500 mL of water is added. After the mixture has been allowed to stand for 5 minutes, the precipitate is collected on a sintered glass filter and washed successively with five 100 mL portions of water, five 100 mL portions of absolute ethanol, and five 100 mL portions of anhydrous diethyl ether. The copper(I) acetylide **4** is dried in a 250 mL, round-bottom flask heated to 65 °C for 4 hours under reduced pressure on a rotary evaporator, yielding 14.8–16.4 g (90–99%) of a bright yellow solid. The dry acetylide may be stored under nitrogen in a brown bottle.

Castro–Stephens Reaction and Heterocyclization

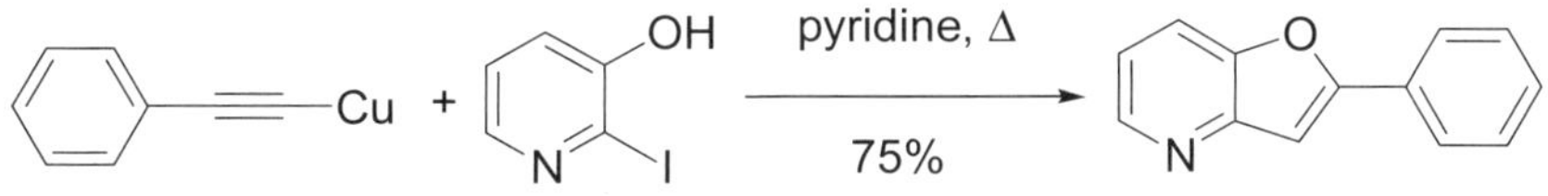

4 **87** **88**

2-Phenylfuro[3,2-b]pyridine (88).[36]

A 300-mL, three-necked flask fitted with a nitrogen inlet stopcock, a magnetic stirring bar, and a condenser attached to a nitrogen outlet bubbler is charged with 2.47 g (0.0150 mole) of copper(I) phenylacetylide (**4**). The system is purged with nitrogen for 20 minutes before 80 mL of pyridine is added. The resulting mixture is stirred for 20 minutes under a nitrogen atmosphere, and 3.30 g (0.0149 mole) of 3-hydroxy-2-iodopyridine (**87**) is added. The mixture, which changes in color from yellow to dark green as the acetylide dissolves, is warmed in an oil bath at 110–120 °C for 9 hours with continuous stirring under a nitrogen atmosphere. The reaction solution is transferred to a 500-mL, round-bottom flask and concentrated to a volume of 20 mL at 60–70 °C (20–80 mmHg) with a rotary evaporator. The pyridine solution is treated with 100 mL of concentrated aqueous ammonia, and the resulting deep-blue mixture is stirred for 10 minutes and extracted with five 100 mL portions of ether. The combined ethereal extracts are washed with three 250 mL portions of water, dried over anhydrous magnesium sulfate, and concentrated with a rotary evaporator. The crude product, 2.6–2.76 g of orange semisolid, is dissolved in 100 mL of boiling cyclohexane. The solution is filtered, concentrated to a volume of about 30 mL, and cooled in an ice bath. The partially purified product **88** crystallizes as 2.3–2.7 g of orange solid, m.p. 83–89 °C. Further purification is effected by sublimation at 110–120° (0.01–0.2 mmHg), yielding 2.2–2.4 g (75–82%) of a yellow solid, m.p. 90–91°C.

Castro–Stephens Reaction Using In situ Modification

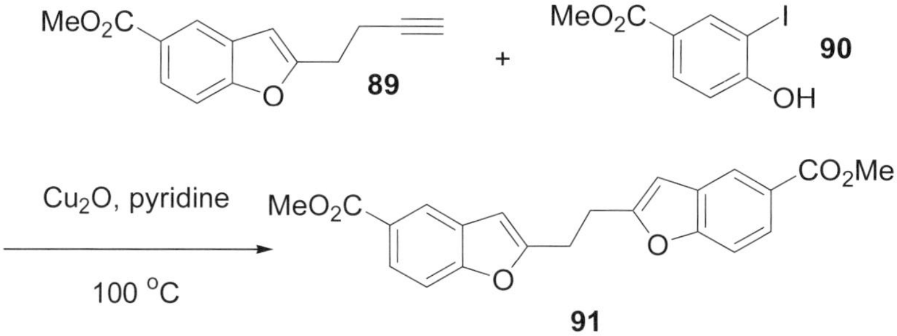

Methyl 2-(2-(5-(methoxycarbonyl)benzofuran-2-yl)ethyl)benzofuran-5-carboxylate (91).[37]

A mixture of **89** (7.6 g, 33.5 mmol), **90** (9.2 g, 33 mmol), and copper(I) oxide (3.2 g, 22.5 mmol) in dry pyridine (40 mL) was stirred under N_2 at 110 °C overnight. The mixture was allowed to cool to ambient temperature, diluted with EtOAc (50 mL), filtered though a 5 cm Celite pad, and concentrated. The residue was dissolved in EtOAc (75 mL), washed with 2 M HCl (12.5 mL) and brine (25 mL), dried over Na_2SO_4, and concentrated. The residue was purified by column chromatography, eluting with (1 : 1) hexanes/diethyl ether to produce **91** as a white solid (5.24 g, 43%): mp 165–167 °C (hexanes/EtOAc).

1.2.1.7 References

1. (a) Stephens, R. D.; Castro, C. E. *J. Org. Chem.* **1963**; *28*; 2163; (b) Stephens, R. D.; Castro, C. E. *J. Org. Chem.* **1963**; *28;* 3313–3315.
2. Sonogashira, K.; Tohda, Y.; Hagihara, N. *Tetrahedron Lett.* **1975**, *16*, 4467.
3. (a) Chodkiewicz, W. *Ann. Chim. Paris* **1957**, *2*, 819–869; [R] (b) Cadiot, P.; Chodkiewicz, W. In *Chemistry of Acetylenes*; Viehe, H. G., Ed.; Marcel Dekker: New York, **1969**.
4. Koelsch, C.; Whitney, A. *J. Org. Chem.* **1941**, *6*, 597–647.
5. Hay, A. *J. Org. Chem.* **1962**, *27*, 3320–3321.
6. Siemsen, P.; Livingston, R.; Diederich, F. *Angew. Chem.* **2000**, *39*, 2632–2657.
7. Gonzalez-Rojas, C.; Oprunenko, Y.; Morales, R. *Russ. Chem. Bull. Int. Ed.* **2004**, *53*, 474–475.
8. Castro, C. E.; Gaughan, J.; Owsley, D. *J. Org. Chem.* **1966**, *31*, 4071–4078.
9. Monnier, F.; Turtaut, F.; Duroure, L.; Taillefer, M. *Org. Lett.* **2008**, *10*, 3203–3206.

10. Thathagar, M.; Beckers, J.; Rothenberg, G. *Green Chem.* **2004**, 215–218.
11. [R] (a) Jastrzebski, J.; van Koten, G. in *Modern Organocopper Chemistry* Ed. N. Krause. **2002** Wiley-VCH. Pp. 1–44; [R] (b) Mori, S.; Nakamura, E. in Modern Organocopper Chemistry Ed. N. Krause. 2002 Wiley-VCH pp 315–346.
12. Okura, K.; Furuune, M.; Enna, M.; Miura, M.; Nomura, M. *J. Org. Chem.* **1993**, *58*, 4716–4721.
13. Okura, K.; Furuune, M.; Miura, M.; Nomura, M. *Tetrahedron Lett.* **1992**, *37*, 5363–5364.
14. Gujadhur, R.; Bates, C.; Venkataraman, R. *Org. Lett.* **2001**, *3*, 4315–4317.
15. Marshall, J.; Chobanian, H.; Yanik, M. *Org. Lett.* **2001**, *3*, 4107–4110.
16. [R] Beletskaya, I.; Cheprakov, A. *Coord. Chem. Rev.* **2004**, *248*, 2337–2364.
17. Lindey, J. *Tetrahedron* **1984**, *40*, 1433–1456.
18. Haack, T.; Kurtkaya, S.; Snyder, J.; Georg, G. *Org. Lett.* **2003**, *5*, 5019–5022.
19. Colacino, E.; Daich, L.; Martinez, J.; Lamaty, F. *Synlett* **2007**, 1279–1283.
20. Li, J.-H.; Li, J.-L.; Wang, D.-P.; Pi, S.-F.; Xie, Y.-X.; Zhang, M.-B.; Hu, X.-C. *J. Org. Chem.* **2007**, *72*, 2053–2057.
21. Juricek, M.; Kasak, P.; Stach, M.; Putala, M. *Tetrahedron Lett.* **2007**, *48*, 8869–8873.
22. Woon, E.; Dhami, A.; Mahon, M.; Threadgill, M. *Tetrahedron* **2006**, *62*, 4829–4837.
23. Kinder, J.; Tessier, C.; Youngs, W. *Synlett* **1993**, 149–150.
24. Staab, H.; Neunhoeffer, K. *Synthesis* **1974**, 424.
25. Coleman, R.; Garg, R. *Org. Lett.* **2001**, *3*, 3487–3490.
26. Tretyakov, E.; Knight, D.; Vasilevsky, S. *J. Chem Soc., Perkin Trans. I* **1999**, 3713–3720.
27. (a) Jutand, A.; Negri, S.; Principaul, A. *Eur. J. Org. Chem.* **2005**, 631; (b) Suo, Q-L.; Ma, Y-Q.; Wang, Y-B.; Han, L.-M.; Bai, Y.-G.; Weng, L.-H. *J. Coord. Chem.,* **2008**, 1234–1243.
28. White, J.; Carter, R.; Sundermann, K.; Wartmann, M. *J. Am. Chem. Soc.* **2001**, *123*, 5407–5413.
29. Mignani, G.; Chevalier, C.; Grass, R.; Allmang, G.; Morel, D. *Tetrahedron Lett.* **1990**, *36*, 5161–5164.
30. Nishihara, Y.; Takemura, M.; Mori, A.; Osakada, K. *J. Organomet. Chem.* **2001**, *620*, 282–286.
31. Marino, J. P.; Nguyen, H. N. *J. Org. Chem.* **2002**, *67*, 6841–6844.
32. Bandyopadhyay, A.; Varghese, B.; Sankararaman, S. *J. Org. Chem.* **2006**, *71*, 4544–4548.
33. Steffen, W.; Laskoski, M.; Collins, G.; Bunz, U. H. F. *J. Organomet. Chem.* **2001**, *630*, 132–138.
34. Zanon, J.; Klapars, A.; Buchwald, S. *J. Am. Chem. Soc.* **2003**, *125*, 2890–2891.
35. Wu, J.; Beck, B.; Ren, R. *Tetrahedron Lett.* **2002**, *43*, 387–389.
36. Owsley, D.; Castro, C. *Org. Synth.* **1988**, *52*, 128–131.
37. Bakunova, A.; Bakunov, S.; Wenzler, T.; Barszcz, T.; Werbovetz, K.; Brun, R.; Hall, J.; Tidwell, R. *J. Med. Chem.* **2007**, *50*, 5807–5823.

1.2.2 Glaser Coupling

Gordon W. Gribble

1.2.2.1 Description

The Glaser Coupling reaction describes the oxidative coupling of terminal acetylenes, **1**→**2**, under the influence of copper(I) and base.[1–6] Several closely related acetylenic homocoupling variations are also discussed in this chapter.

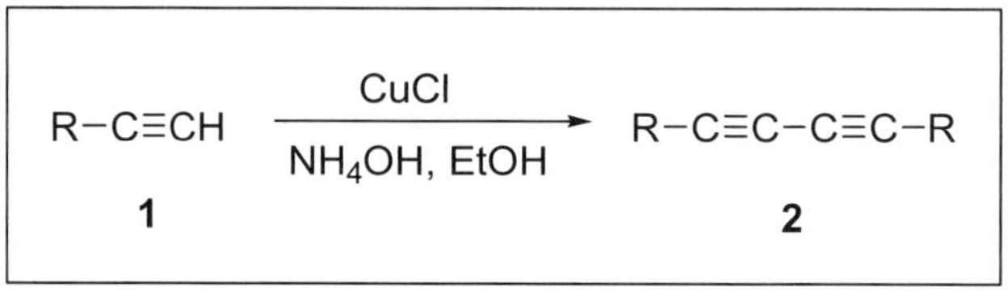

1.2.2.2 Historical Perspective

In 1869, Carl Glaser observed that an ethanolic ammonia solution of copper(I) phenylacetylide in the presence of air undergoes oxidative coupling to afford diphenyldiacetylene.[7] In 1956, Eglinton and Galbraith described an acetylenic oxidative dimerization using copper(II) acetate in methanolic pyridine.[8] In 1957, Cameron and Bennett demonstrated that some amines (*t*-butylamine, ethylenediamine, and pyridine) can substitute for ammonia in the original Glaser protocol.[9] In 1960, Hay reported a fourth variation involving copper(I) chloride, oxygen, and the bidentate amine *N,N,N',N'*-tetramethylethylene diamine (TMEDA) or pyridine.[10] Examples of all four methods — Glaser, Eglinton, Cameron, and Hay — are presented in the following sections, as will be other lesser well-known and newer modifications and variations. Whereas the Glaser, Cameron, and Hay coupling reactions are catalytic with copper(I), the Eglinton modification is stoichiometric (or with excess) in copper(II). The more recent palladium-catalyzed terminal alkyne homocouplings (e.g., Cadiot–Chodkiewicz and Sonogashira) are covered elsewhere.

1.2.2.3 Mechanism

Glaser originally invoked a copper phenylacetylide dimer (Ph-C≡C-Cu-Cu-C≡C-Ph) as the species that reacts with oxygen to form diphenyldiacetylene and Cu_2O.[7a] Subsequent studies by Salkind,[11] Vaitiekunas,[12] Bohlmann,[13] and others[2] provide further details on the mechanism of this alkyne coupling reaction. Diederich *et al.*[2] conclude that the mechanism proposed by Bohlmann, which does not involve acetylenic radicals, is most consistent with the experimental data. Thus, copper–acetylene π-complex **3** (or a copper acetylide π-complex) and copper(II) complex **4** are in equilibrium with dimeric copper acetylide **5**. Subsequent dimerization to **6** is followed by collapse to diacetylene **7**.[13]

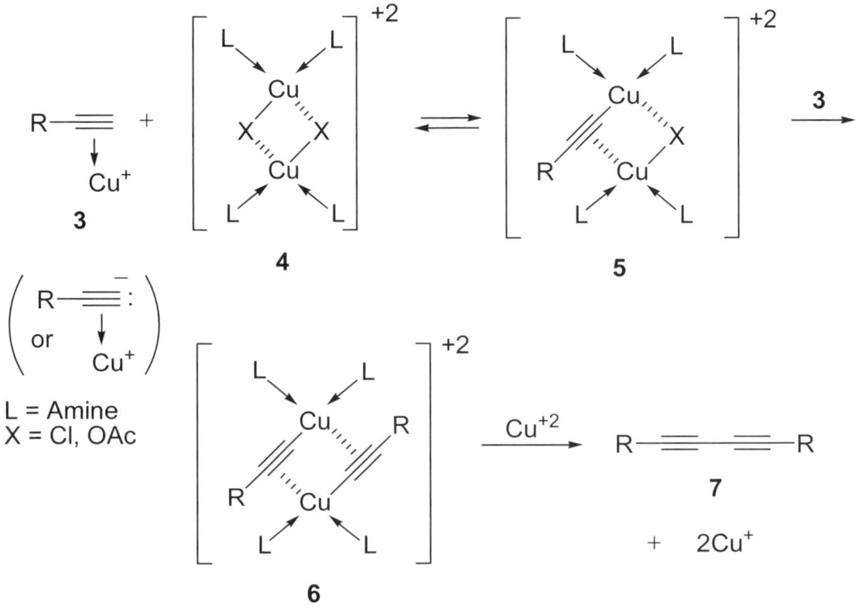

It should be noted that most presentations of the Glaser and related acetylene homocouplings show a simpler mechanism involving base-catalyzed formation of a copper(I) acetylide **8**, oxidation to copper(II) acetylide **9**, and homocoupling of the resultant acetylenic radical **10** to afford **7**.[2,14]

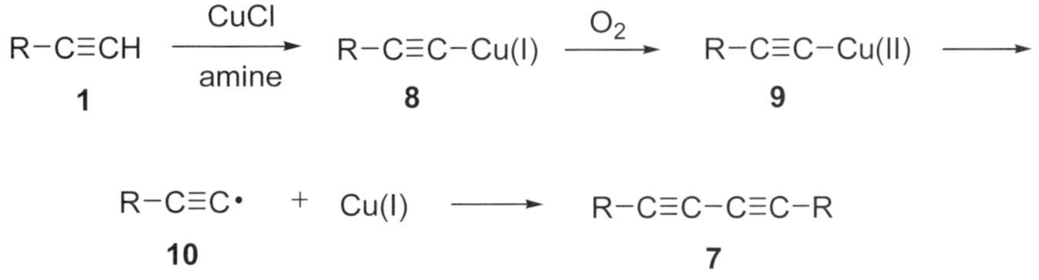

1.2.2.4 Variations and Improvements

Several minor variations and practical improvements on the Glaser and related acetylenic coupling reactions have been developed. For example, when CuI is substituted for CuCl in the standard Hay conditions (TMEDA, O_2, acetone) a greatly improved yield of **12** is obtained, which was ascribed to the more soluble CuI·2TMEDA in acetone.[15] The homocoupling of 2-ethynyl-1-methylpyrrole (**13**) only proceeds in the presence of 1,8-diazabicyclo[5.4.0]undec-7-ene (DBU), which may be due to the decreased acidity of the acetylenic hydrogen.[16]

$$\underset{\mathbf{11}}{\text{RO–C≡C–H}} \xrightarrow[\text{acetone, } O_2\text{, rt, 65–95\%}]{\text{CuI•2TMEDA}} \underset{\mathbf{12}}{\text{RO–C≡C–C≡C–OR}} \quad \text{(ref. 15)}$$

R = *t*-Bu, *n*-$C_{10}H_{21}$, 1-adamantyl, *c*-C_6H_{13}, L-menthyl, 2,6-diMeOPh

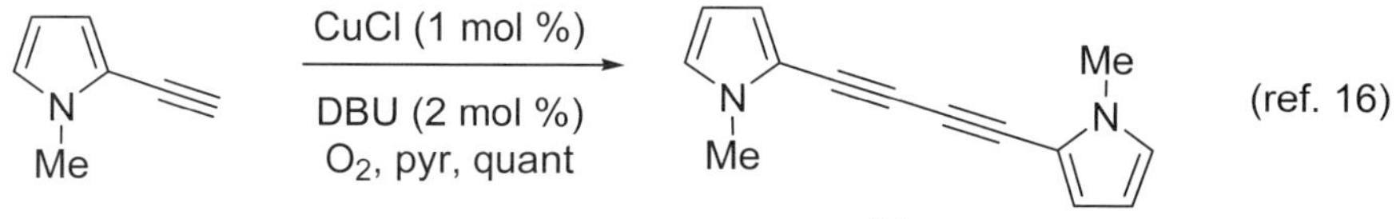

13 **14**

Glaser coupling can be carried out both in supercritical CO_2 using NaOAc as base, e.g., **15** to **16**,[17] and in ionic liquids, e.g., **17** to **18**.[18] Likewise, water near its critical point serves as a solvent for Glaser coupling, **19**→**20**.[19] Terminal alkynes **21** are effectively coupled to **22** with the recyclable system $Cu(OAc)_2$–polyethyleneglycol (PEG)–NaOAc.[20] PEG 6000 afforded the highest efficiency and this catalyst system could be recycled more than five times. Recycling of the resulting Cu_2O was accomplished by heating in acetic acid in air. A microwave solvent-free Glaser coupling in the presence of KF–Al_2O_3 forms diacetylenes **24**.[21]

$$\underset{\mathbf{15}}{R^1\text{–C≡C–H}} \xrightarrow[\text{sc } CO_2\text{, MeOH, 40 °C, 71–100\%}]{CuCl_2\text{, NaOAc}} \underset{\mathbf{16}}{R^1\text{–C≡C–C≡C–}R^1} \quad \text{(ref. 17)}$$

R = Ph, *n*-C_5H_{11}, *n*-C_6H_{13}, CH_2OH, CH_2OAc

$$\underset{\mathbf{17}}{R^2\text{–C≡C–H}} \xrightarrow[\text{[bmin]PF}_6\text{, rt; 85–95\%}]{\text{CuCl-TMEDA (cat), O}_2} \underset{\mathbf{18}}{R^2\text{–C≡C–C≡C–}R^2} \quad \text{(ref. 18)}$$

R^2 = Ph, CH_2OH, *n*-Bu, CH_2CH_2CN, CH_2CH_2OTHP, CH_2OMe, CH_2OTs, $(CH_2)_4OH$, others

$$\underset{\mathbf{19}}{R^3\text{–C≡C–H}} \xrightarrow[\text{H}_2\text{O, 200 °C; 59–82\%}]{\text{CuCl}_2\text{ (30 mol \%)}} \underset{\mathbf{20}}{R^3\text{–C≡C–C≡C–}R^3} \quad \text{(ref. 19)}$$

R^3 = Ph, 4-MePh, 2-ClPh, 4-FPh, *n*-C_8H_{17}, *n*-C_6H_{13}

$$\underset{\mathbf{21}}{R^4\text{–C≡C–H}} \xrightarrow[\text{NaOAc, 120 °C; 86–98\%}]{\text{Cu(OAc)}_2\text{-PEG}} \underset{\mathbf{22}}{R^4\text{–C≡C–C≡C–}R^4} \quad \text{(ref. 20)}$$

R^4 = Ph, 4-FPh, 4-EtPh, cyclohex-1-enyl, *n*-Bu, *n*-Hex, CH_2OH

$$\underset{\mathbf{23}}{R^5\text{–C≡C–H}} \xrightarrow[\text{microwave; 53–75\%}]{\text{CuCl}_2\text{, KF-Al}_2\text{O}_3} \underset{\mathbf{24}}{R^5\text{–C≡C–C≡C–}R^5} \quad \text{(ref. 21)}$$

R^5 = *n*-C_8H_{17}, *n*-C_6H_{13}, Ph, 4-MePh, 2-ClPh, 2-FPh, 4-FPh

The novel catalytic system AgOTs–$CuCl_2$–TMEDA is reported to effect homocoupling in high yield of a series of benzo[*b*]furan alkynes on the solid phase.[22] For example, compound **25** is obtained after cleavage from the bead in high purity. The conventional Glaser, Hay, and Eglinton methods fared badly. This seems to be the first report of Ag(I) promoting a copper-mediated alkyne homocoupling.

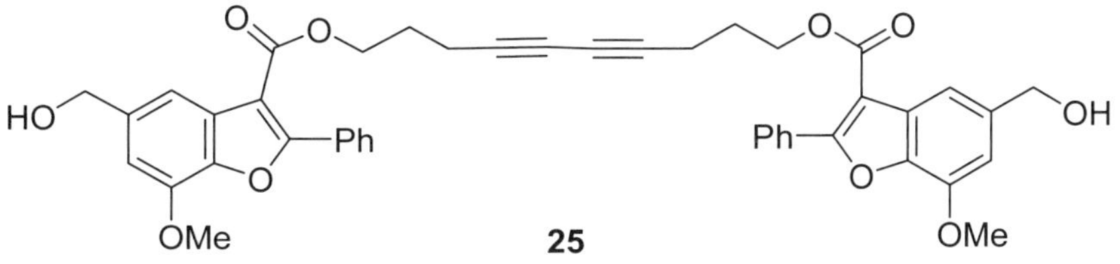

25

Several methods have been reported involving Glaser-type homocoupling of alkynyltrialkylsilanes. Thus, exposure of various alkynyltrimethylsilanes **26** to CuCl in a polar solvent such as DMF affords good to excellent yields of dialkynes **27**.[23] Interestingly, whereas other silanes give good to excellent yields of coupled acetylenes **27** when R = *n*-C_6H_{13} (e.g., $SiPhMe_2$, 83%; $Si(OMe)_3$, > 99%; $SiMe_2OSiMe_3$, 92%; $SiMe_2(OH)$, 89%), both $SiEt_3$ and Si*i*-Pr_3 fail completely.[23b] In this study it was also possible to isolate the intermediate alkynylcopper complexes **28**, which upon further heating in DMF in air afford diacetylenes.[23c]

$$\underset{\mathbf{26}}{R{-}C{\equiv}C{-}SiMe_3} \xrightarrow[\text{DMF, 60 °C; 70–99\%}]{\text{CuCl}} \underset{\mathbf{27}}{R{-}C{\equiv}C{-}C{\equiv}C{-}R} \quad \text{(ref. 23)}$$

R = Ph, 4-MePh, 4-AcPh, 2-thienyl, *n*-C_6H_{13}

$$[Cu_2Cl(C{\equiv}C{-}R)]_n$$

28 (40–67%)

R = 4-AcPh, 4-MeOPh, 2-thienyl, 4-ClPh, 4-*n*-C_8H_{17}Ph, *n*-C_6H_{13}

The use of $Cu(OAc)_2$ under the Eglinton conditions also effects homocoupling of both alkynyltrimethylsilanes, **26**→**27**,[24] and dialkynyltrialkylsilanes, **29**→**30**.[25]

$$\underset{\mathbf{26}}{R{-}C{\equiv}C{-}SiMe_3} \xrightarrow[\text{K}_2\text{CO}_3\text{, pyr; MeOH, 50 °C; 82–98\%}]{\text{Cu(OAc)}_2\bullet\text{H}_2\text{O}} \underset{\mathbf{27}}{R{-}C{\equiv}C{-}C{\equiv}C{-}R} \quad \text{(ref. 24)}$$

R = Ph, 2-BrPh, 2,6-diBrPh, PhC≡C

$$\underset{\mathbf{29}}{R{-}(C{\equiv}C)_2{-}Si\textit{i}{-}Pr_3} \xrightarrow[\text{TBAF, pyr; Et}_2\text{O; 82–100\%}]{\text{Cu(OAc)}_2} \underset{\mathbf{30}}{R{-}(C{\equiv}C)_2{-}(C{\equiv}C)_2{-}R} \quad \text{(ref. 25)}$$

R = Ph, 2-BrPh, 3-Me_2NPh, 1-naphthyl, *n*-C_5H_{11}, 2-thienyl

The former study was successfully applied to the synthesis of benzoannulenes **33** relying on the selective reactivity of the trialkylsilyl groups.[24]

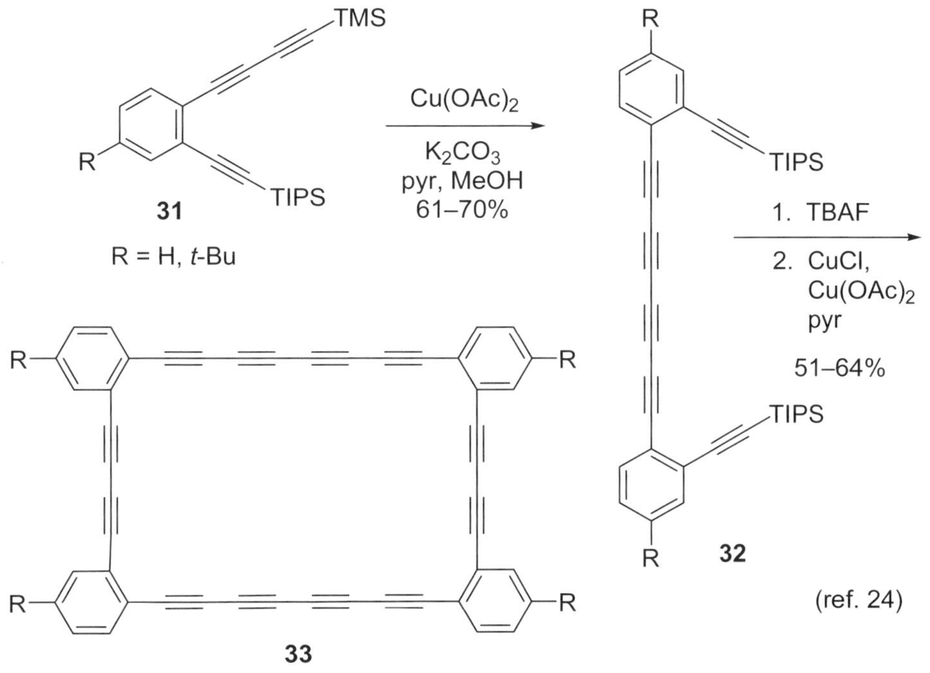

Alkynylboronates may also be used in Glaser-type homocoupling as summarized in **34**→**35**.[26] Homocoupling of terminal alkynes that do not employ copper include $Co_2(CO)_8$[27] and $NiCl_2$,[28] but these are beyond the scope of the present chapter.

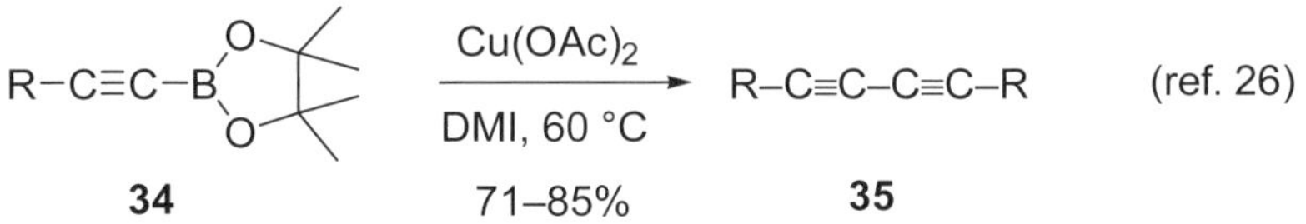

R = Ph, 4-MePh, 2-MePh, 4-MeOPh, 3-CF_3Ph, 2-thienyl, 2-propenyl, CH_2OMe, $(EtO)_2CH$, *n*-C_6H_{13}, *t*-$BuMe_2SiO(CH_2)_4$

1.2.2.5 Synthetic Utility

Virtually from the onset of its discovery by Glaser in 1869, the copper-catalyzed coupling of terminal acetylenes has seen enormous applications, far too many to be fully documented herein. Therefore, emphasis is on recent examples. In addition to the synthesis of straight-chain (literally!) dialkynes, the Glaser coupling reaction and its variants have been extensively employed in the synthesis of cyclic polyalkynes (annulenes, catenanes, knots, and others), and in the synthesis of polymeric alkynes.

A selection of diacetylenes that have been synthesized using Glaser coupling methodology is presented here. As can be seen, a myriad of diacetylenes are available, running the gamut of heterocyclic structures. Compound **46** was prepared from the corresponding trimethylsilyl acetylene using $CuBr_2$ under Eglinton conditions.[38] Interestingly, the attempted homocoupling of the ethylene precursor corresponding to that which afforded **48** failed completely, which may be due to copper–alkene complexation.[40b] In contrast, diacetylene **49** is obtained in modest yield.[41] The 1-alkynyl tosylamides **51** are best synthesized using CuI under classic Hay conditions.[43] Alkynes **52** are prepared from the corresponding trimethylsilyl acetylenes.[44]

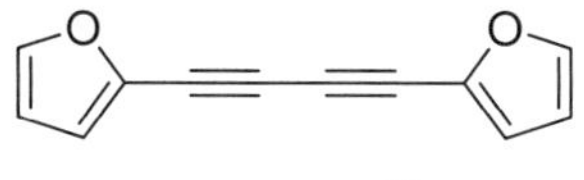

36 (88%)[29]

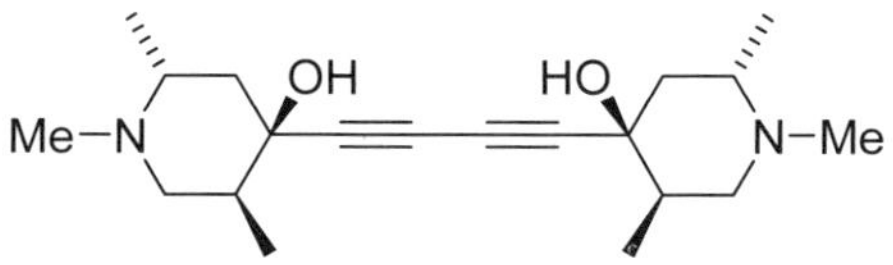

37 (96%)[30]

38 (83%)[31]

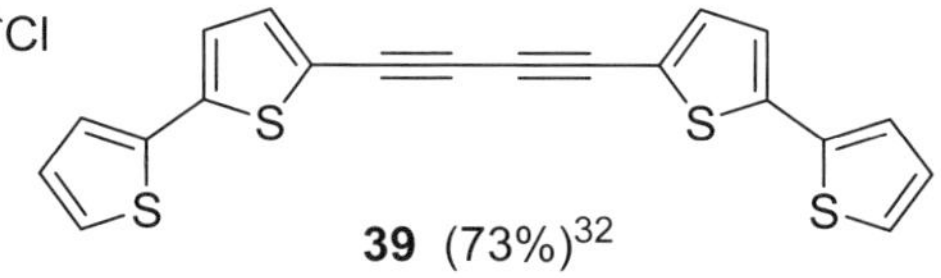

39 (73%)[32]

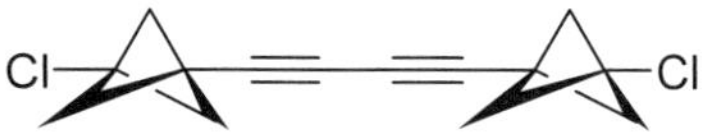

40 (69%)[33]

41 (93%)[34]

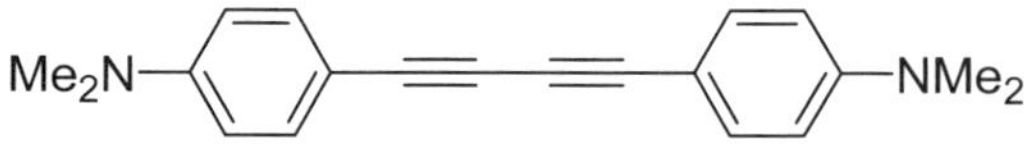

42 (60%)[35]

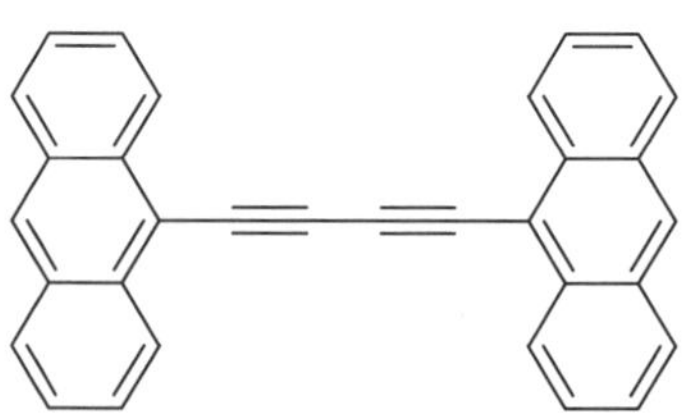

43 (91%)[36]

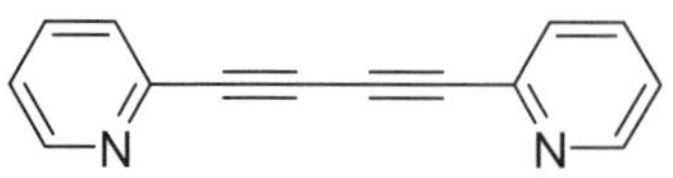

44 (79%)[37]

45 (90%)[37]

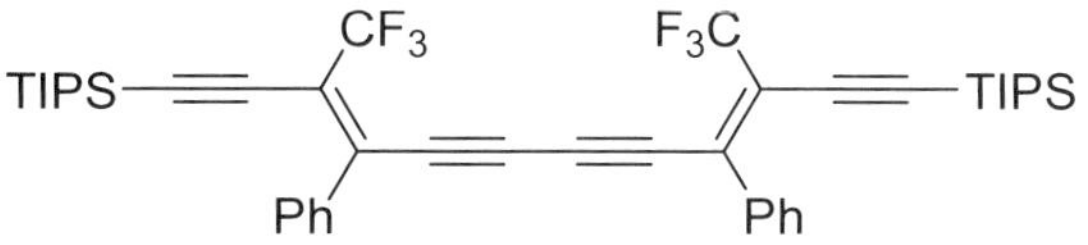

46 (77%)[38]

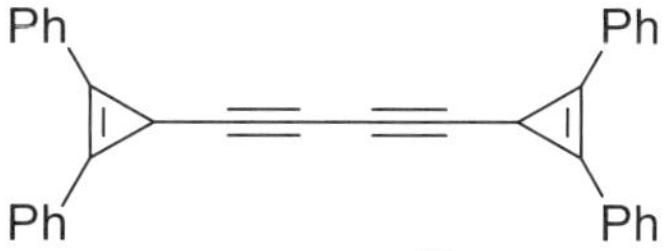

47 (72%)[39]

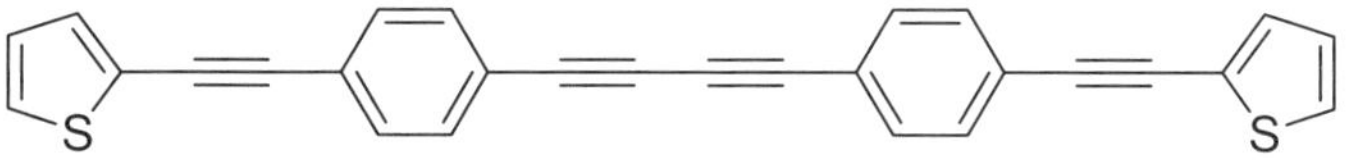

48 (71%)[40]

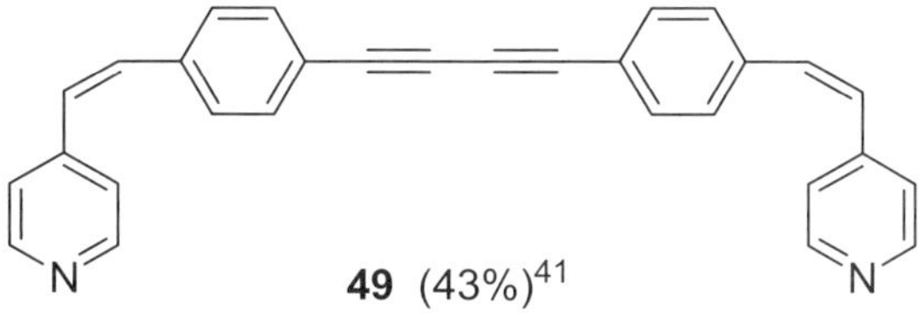

49 (43%)[41]

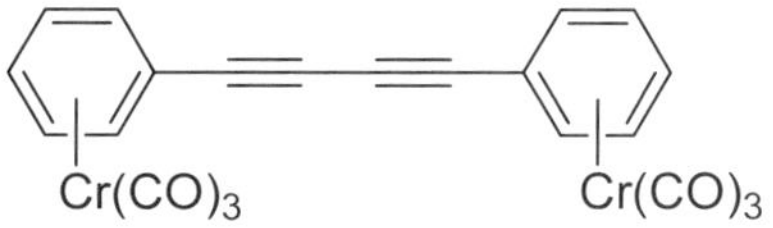

50 (95%)[42a]

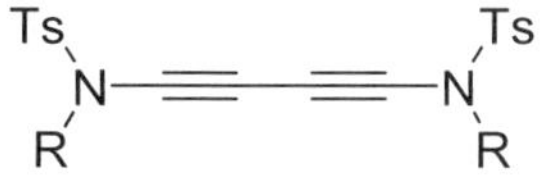

R = Ph, 4-MePh, *n*-Pr, allyl, Bn

51 (84–100%)[43]

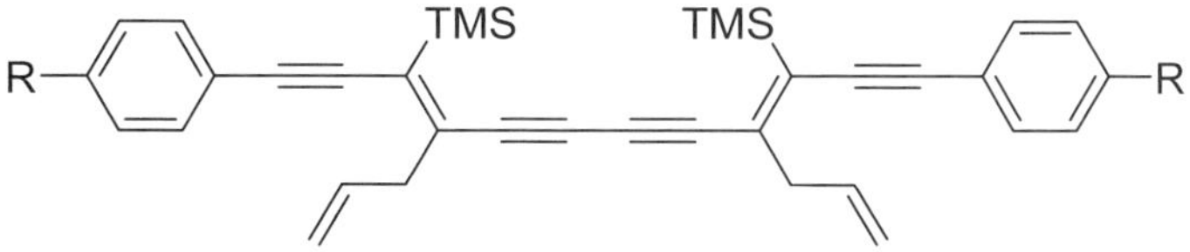

52, R = H, Cl, OMe (49–55%)[44]

Other substituted diacetylenes that have been made via Glaser coupling methods include simple carbohydrates,[45] cyclooctaamylose analogues,[46] nucleosides,[47] ferrocene-labeled amino acids,[48] phthalocyaninato complexes,[49] β-cyclodextrins,[50] novel lithium ion binding ionophores,[51] ruthenium "dumbbells",[52] dendritic 1,1'-binaphthalene carbohydrate receptors,[53] calix[4]arenes,[54] a highly fluorescent pyrene,[55] trityl-based chemical sensors,[56] and novel phosphocholines.[57] A few examples of the Glaser homocoupling of diacetylenes to tetraacetylenes are known. For example, **53–55** can be fashioned in this manner.

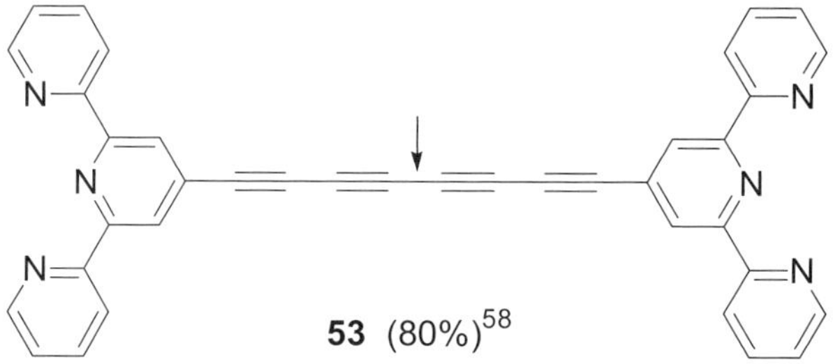

53 (80%)[58]

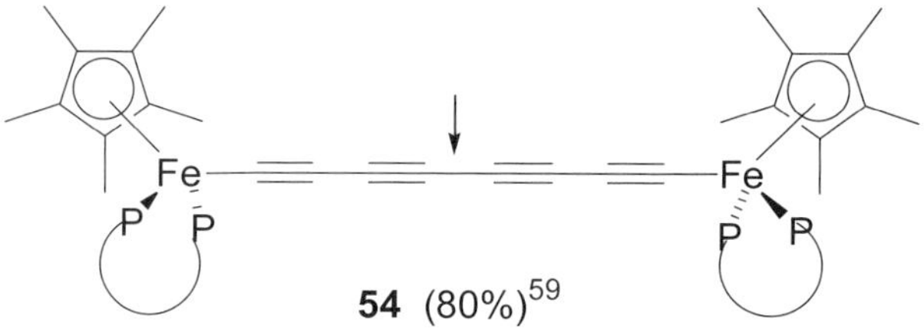

54 (80%)[59]

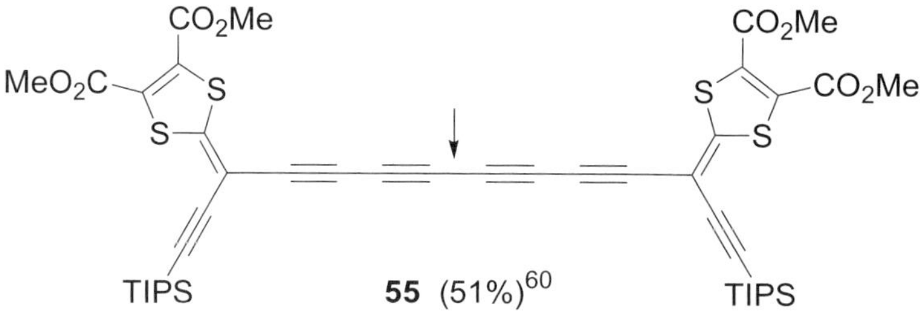

55 (51%)[60]

Rare are "mixed" Glaser heterocouplings, but an example is **56** plus **57** to give **58** in low yield.[61] An interesting, but not unexpected, side reaction in the attempted homocoupling of alkyne **59** is the copper-catalyzed formation of pyrrole **60**, which is the major product under all Glaser conditions.[62] An elegant application of copper-catalyzed homocoupling is the synthesis of diacetylene **61** (83% yield) and its conversion to bipyridine **62** via a double intramolecular Diels–Alder reaction.[63] Other examples were also described.

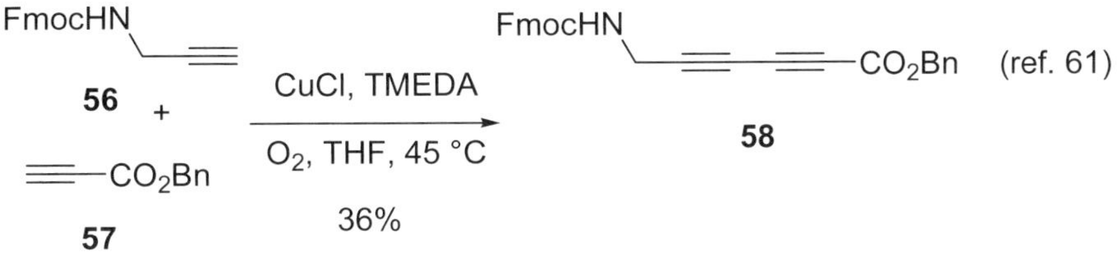

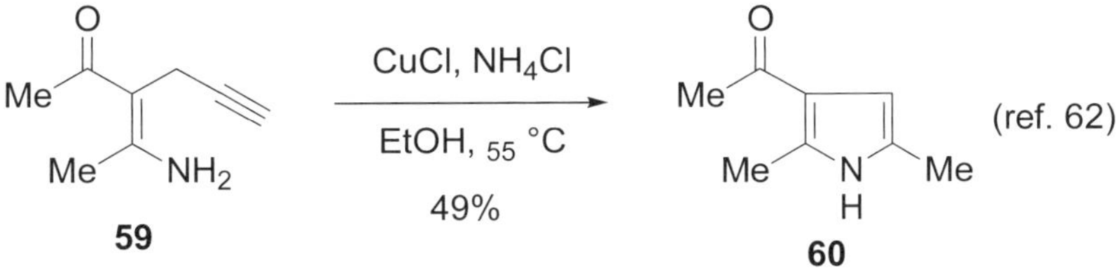

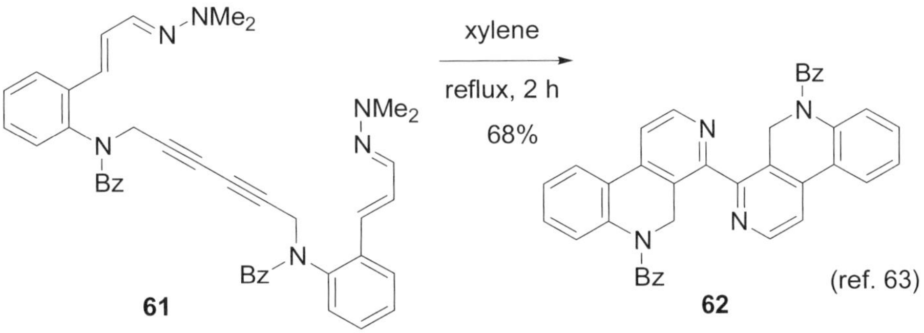

Another major application of Glaser couplings is the construction of cyclic polyalkynes, most notably annulenes[64] and catenanes.[65] The pioneering work of Sondheimer,[66] Eglinton,[67] Nakagawa,[68] and others,[69,70] e.g., **63–68**, suggested that the Glaser coupling would be a general route to cyclic polyacetylenes.

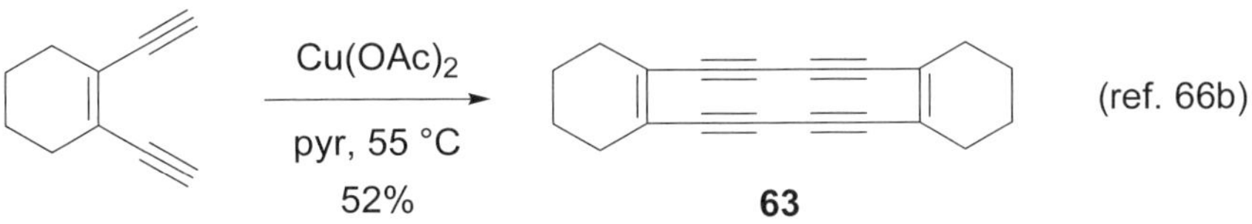

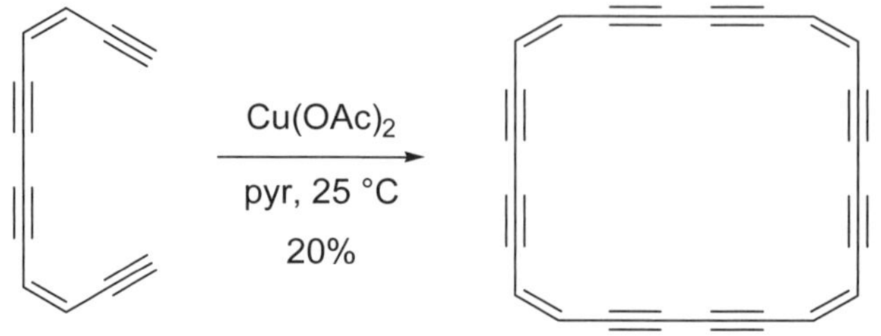

64

(ref. 66c)

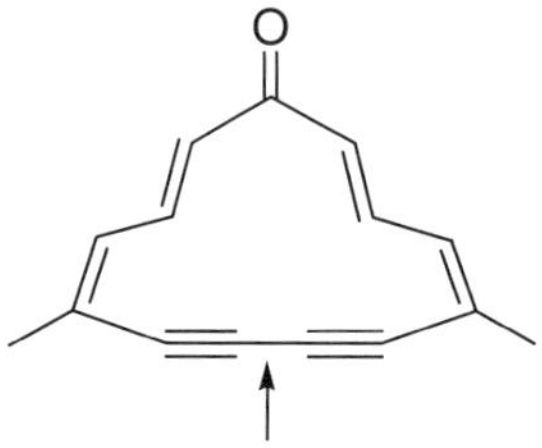

65 (80%)[66d]

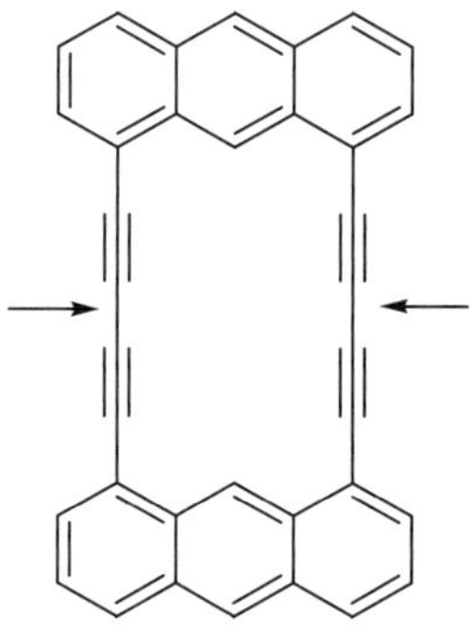

66 (95%)[68a]

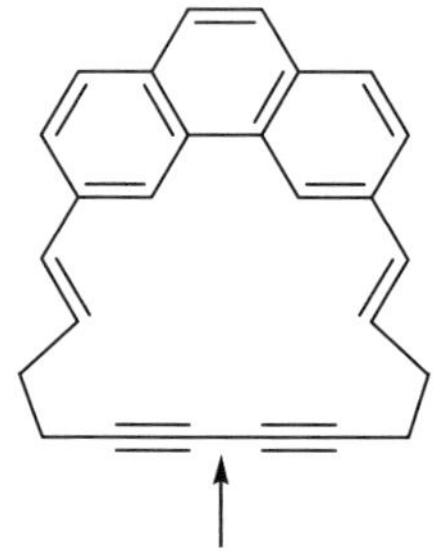

67 (50%)[69]

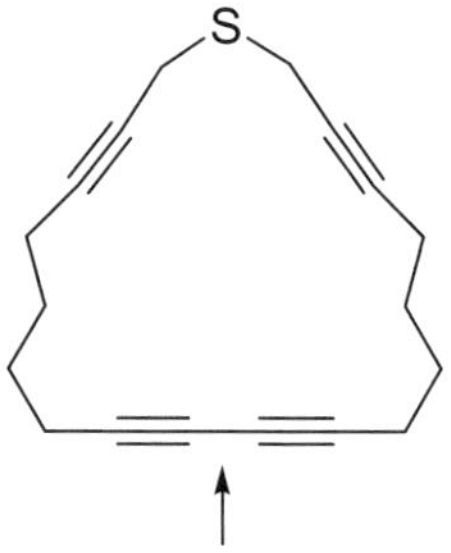

68 (24%)[70]

The recent work of Diederich,[71] Haley,[72] Höger,[73] de Meijere and Scott,[74] Fallis,[75] Bunz,[76] and several others[77–85] has established Glaser-type couplings as the preeminent synthetic route to a multitude of complex cyclic polyacetylenes. A limited selection of these molecules is illustrated here, **69–76**. Höger has pointed out that copper salts may act as a template in the intermolecular dimerization of functionalized bisacetylenes.[73c] The trimer corresponding to **74** is isolated in 9%.[80] Annulene **76** is unusually stable, being sterically protected by the bulky trimethylsilyl groups.[84]

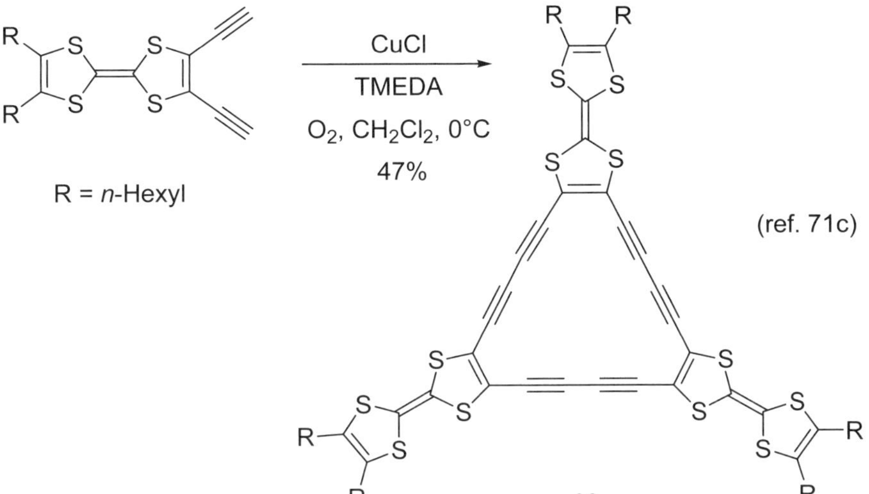
R = n-Hexyl
CuCl
TMEDA
O2, CH2Cl2, 0°C
47%
(ref. 71c)

69

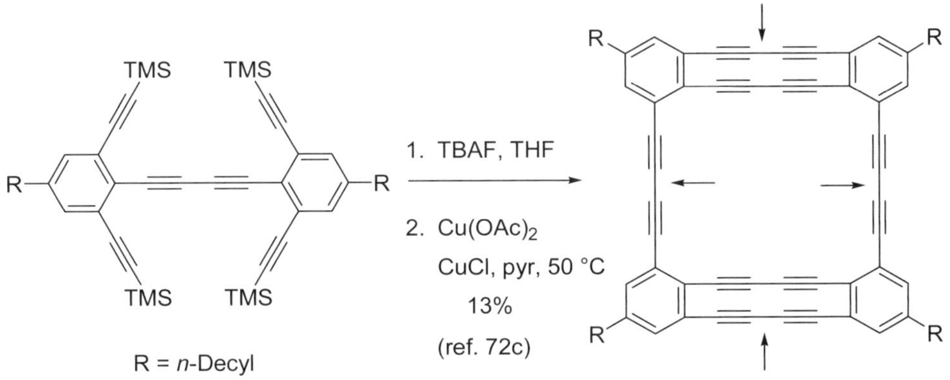
TMS
R = n-Decyl
1. TBAF, THF
2. Cu(OAc)2
CuCl, pyr, 50 °C
13%
(ref. 72c)

70

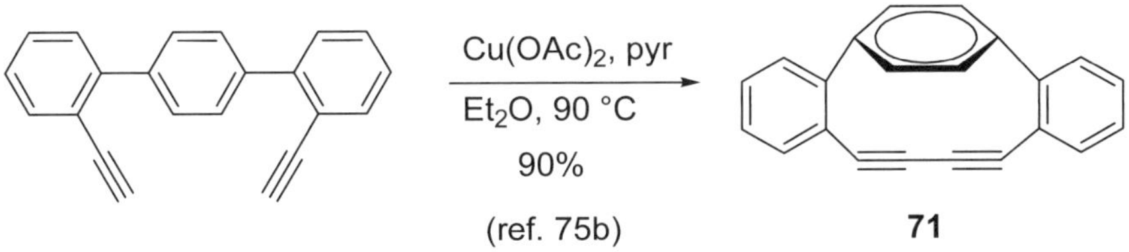
Cu(OAc)2, pyr
Et2O, 90 °C
90%
(ref. 75b)

71

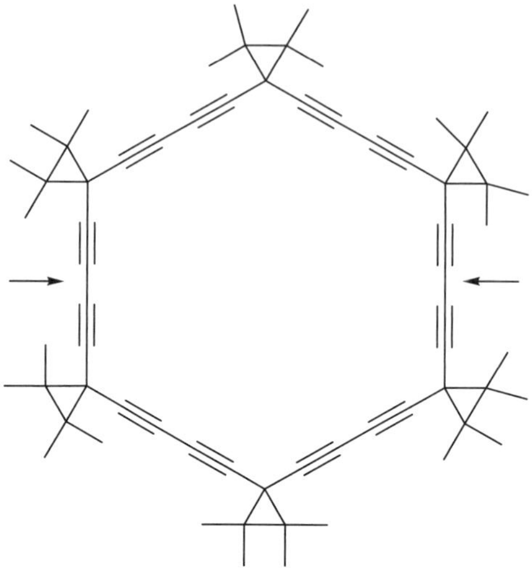

72 (49%)[74c]

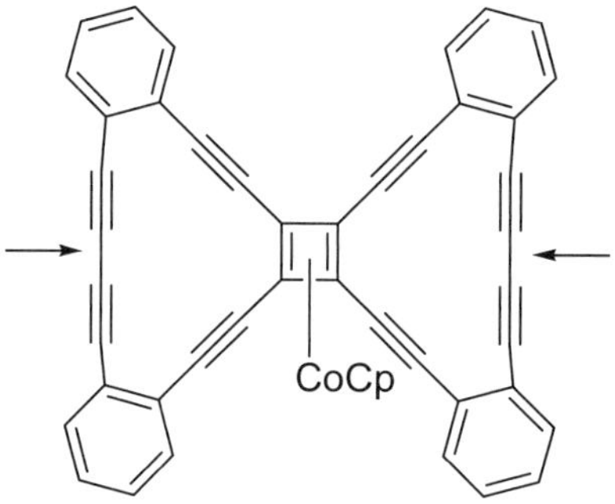

73 (94%)[76b]

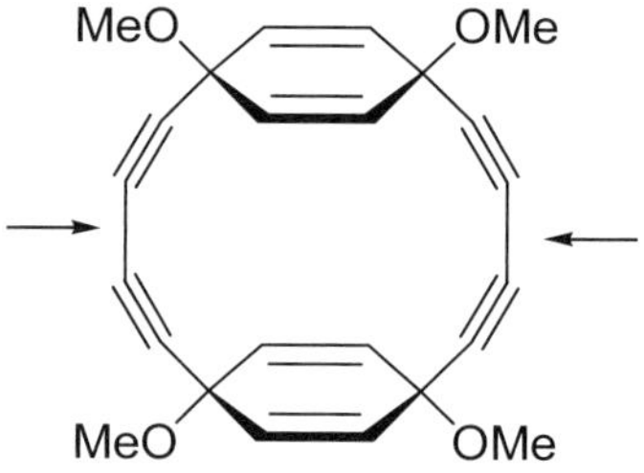

74 (2.5%)[80]

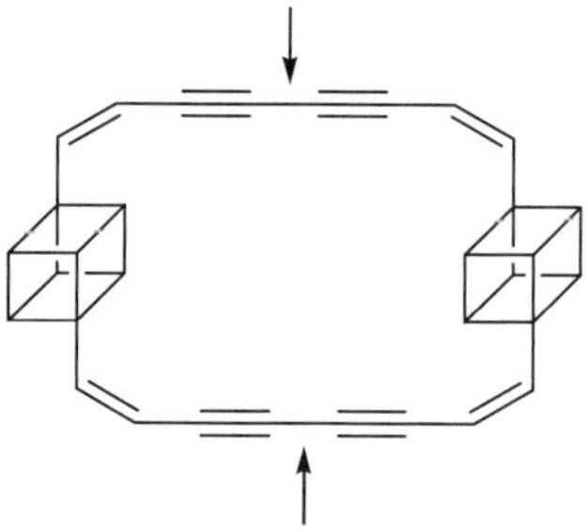

75 (50%)[81]

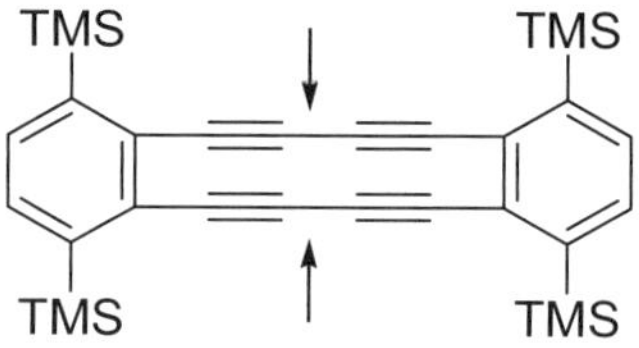

76 (61%)[84]

Glaser coupling methodologies have also been instrumental in the synthesis of cyclic polyacetylenes linked to porphyrins,[86] lipids,[87–90] carbohydrates,[91] cyclophane ethers (e.g., **77**, **78**),[92–94] dicationic aromatic ether dyestuffs,[95] aromatic ether liquid crystals,[96] and macrocyclic glycerol lactones (e.g., **79**).[97] Several aromatic heterocycle-based macrocyclic polyacetylenes are available via Glaser coupling techniques. These include the 1,3,4-thiadiazole polyether **80**[98] and diethynylcarbazoles.[99]

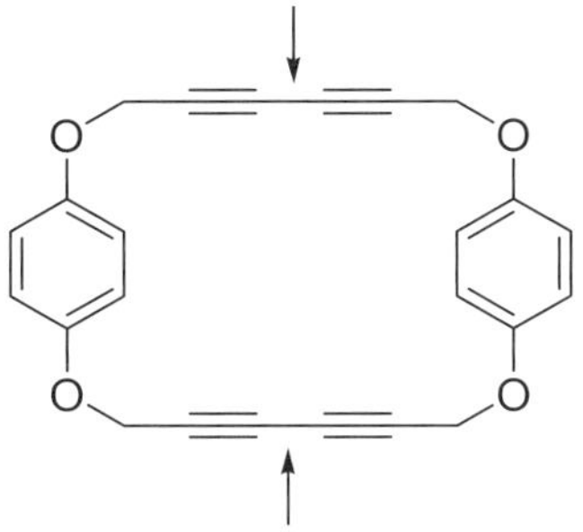

77 (11%)[93a]

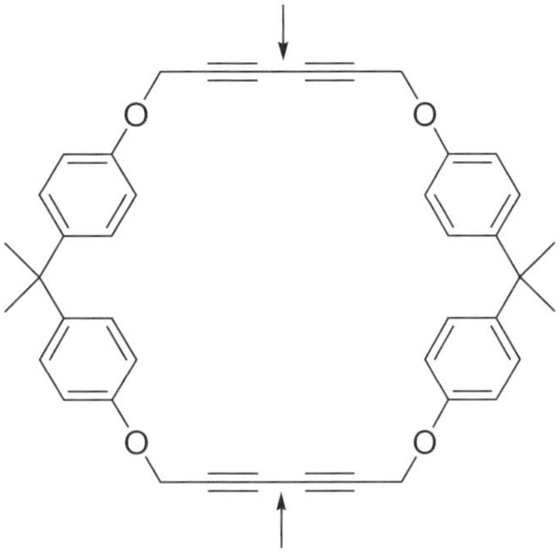

78 (7%)[93b]

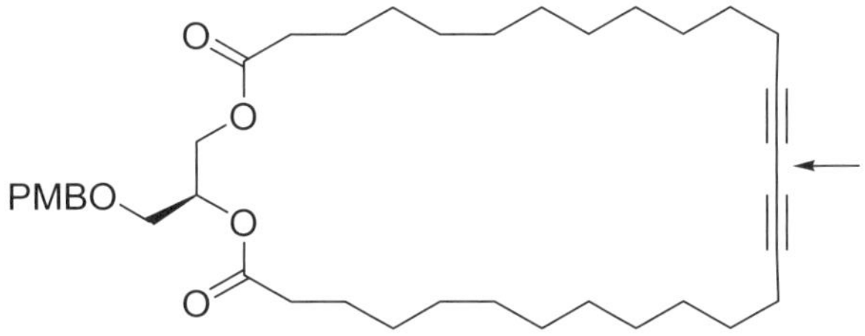

79 (60%)[97a]

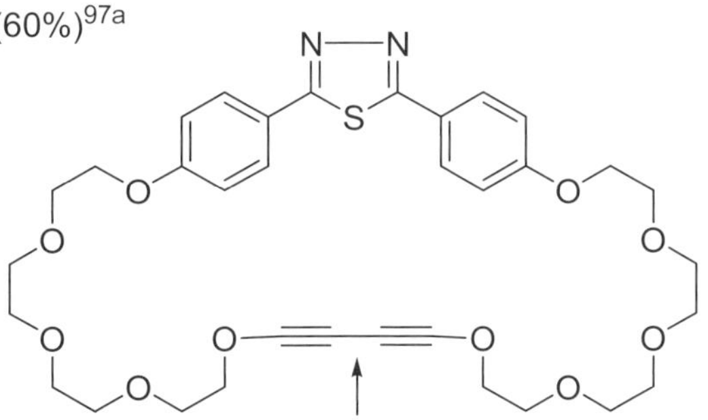

80 (43%)[98]

The Glaser homocoupling of terminal alkynes has proven a powerful method for constructing catenanes, molecules consisting of two ([2]catenanes) or more interlocking rings (e.g., **81**, **82**), rotaxanes (e.g., **83**), and knots (e.g., a trefoil knot **84** (not shown)).[100] Catenanes containing up to 147-membered rings are known using Glaser coupling to close the second ring.[101] Asymmetric [2]catenanes,[102] porphyrin [2]catenanes,[103] donor–

acceptor [2]catenanes,[104] and molecular composite knots[105] have also been synthesized via Glaser coupling.

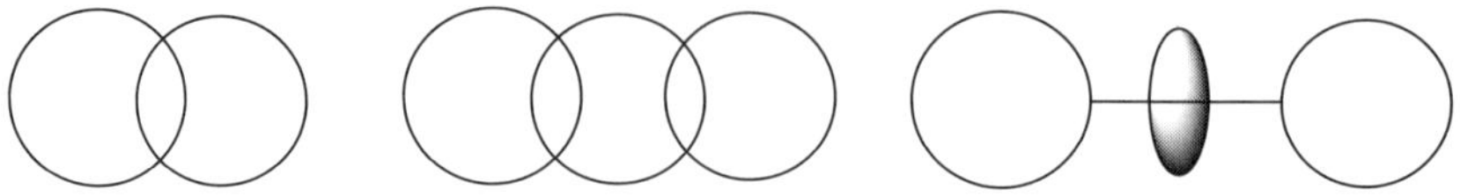

81 ([2]catenane) **82** ([3]catenane) **83** (rotaxane)

The final major application of Glaser-Type coupling of terminal dialkynes is the synthesis of polymeric acetylenes.[106] Following the seminal work of Hay[107] and others,[106] polymeric coupling of terminal bis-acetylenes has led to a myriad of polymeric acetylenes. Some pioneering examples are **85** (polymerization of *m*-diethynylbenzene),[107] **86** (polymerization of 1,8-nonadiyne),[108] α-carbyne (polyyne) (**87**) (polymerization of dicopper acetylide),[109] **88** (polymerization of 2,5-diethynylthiophene),[110] and porphyrins (e.g., **89**).[111]

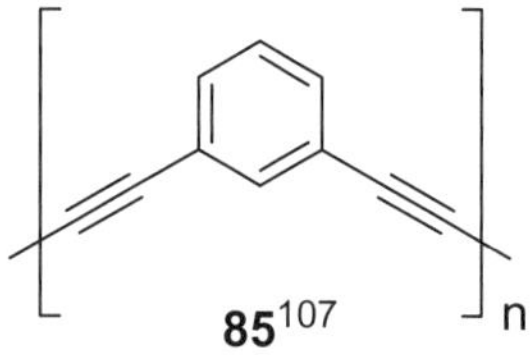

85[107]

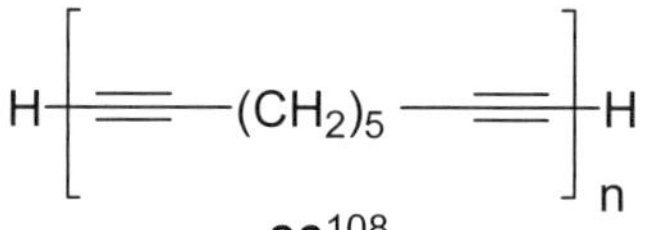

86[108]

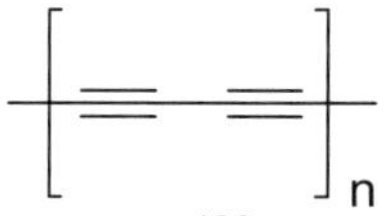

87[109]

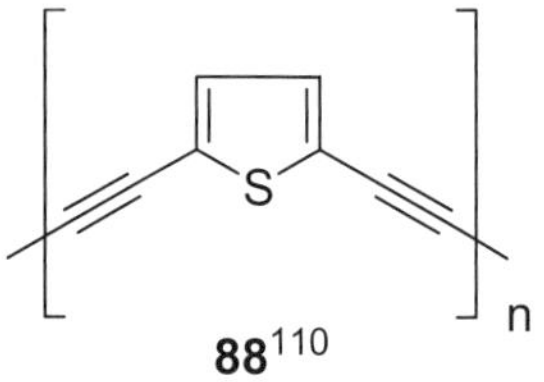

88[110]

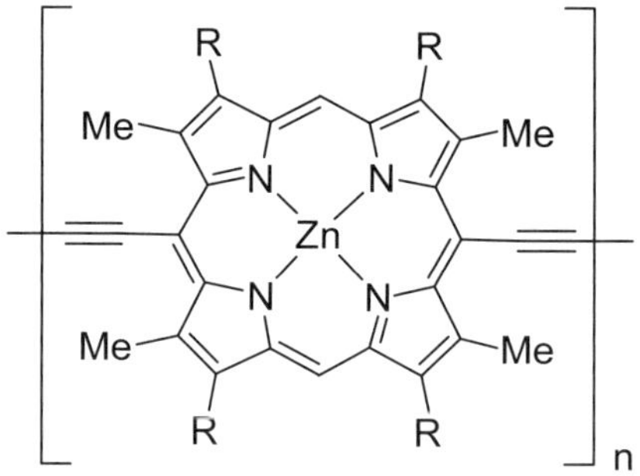

89[111a]

Mixed acetylenic polymerizations are also achieveable via Glaser couplings. For example, coupling between 4,4'-diethynylbenzophenone and phenylacetylene affords polymer **90**.[112] Likewise, several soluble polymers of type **91** are prepared in quantitative yield using standard Hay conditions (CuCl, O_2, TMEDA, pyridine, *o*-dichlorobenzene, 75 °C).[112] End-capping with phenylacetylene or 3,5-di(*tert*-butyl)phenylacetylene was also used to prepare "molecular rods" (e.g., **92**, **93**).[113]

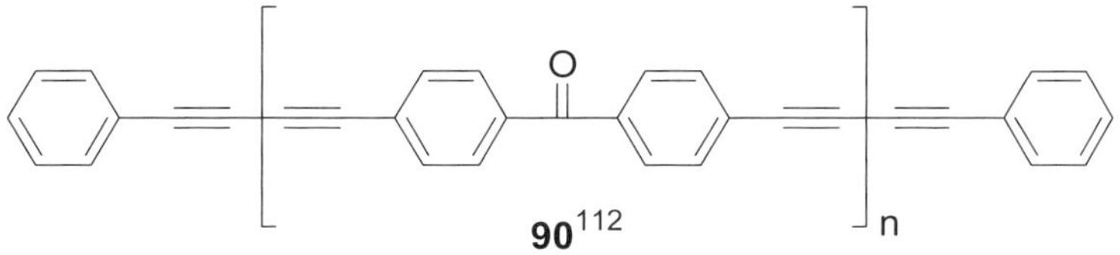

90[112]

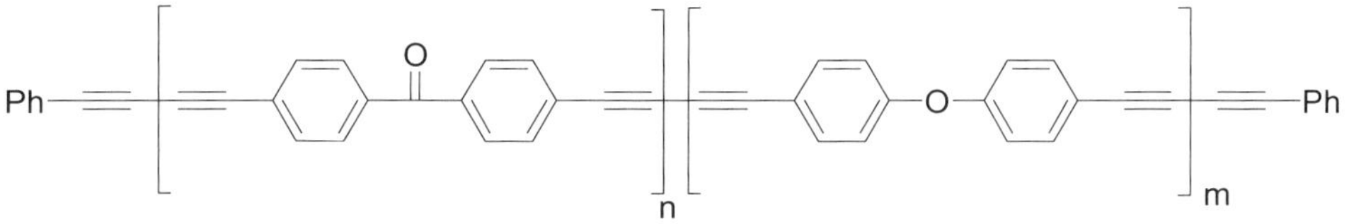

91[112]

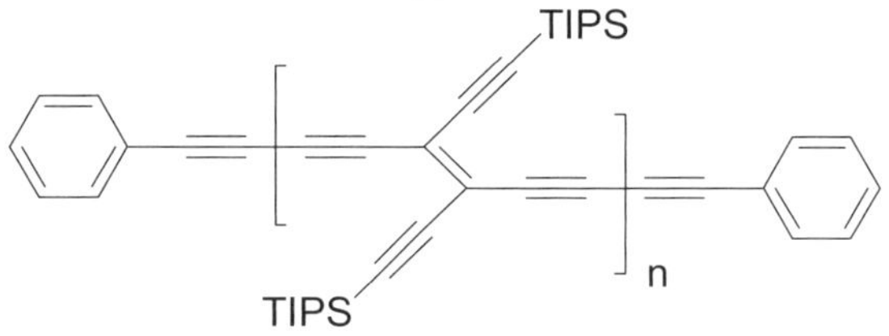

92[113a]

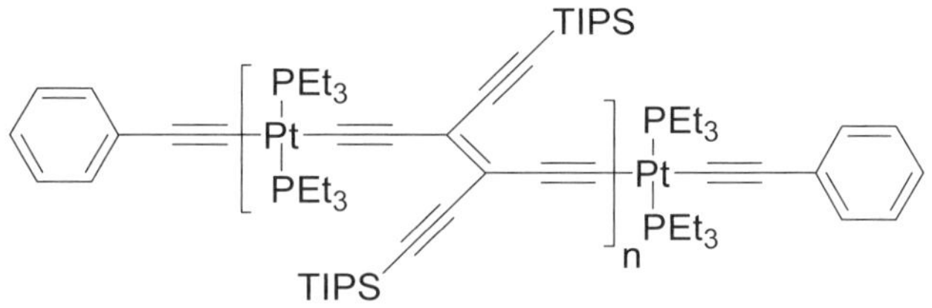

93[113c]

Other polymeric acetylenic materials constructed using Glaser methodology include water-soluble rotaxanes as insulated molecular wires,[114] exceptionally long polymeric thiophenes as novel nanomaterials,[115] macrocyclic oligothiophenes with cavities in the nanometer region,[116] ruthenium-capped polyacetylenes,[117] crown-ether-capped polyacetylenes,[118] asymmetric polyacetylenic binaphthols,[119] hyperbranched polyacetylenes,[120] end-capped carbynes,[121] polyacetylenic oligoazulenes,[112] acetylenic polycarbosilanes (e.g., **94**),[123] and polymeric naphthylacetylenes (e.g., **95**).[124]

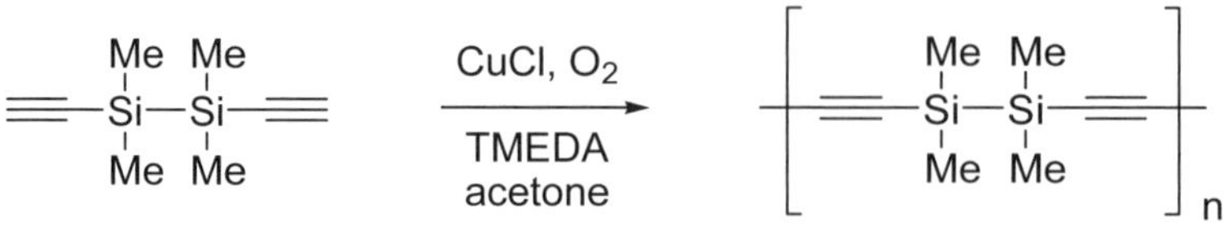

18% **94** (ref. 123)

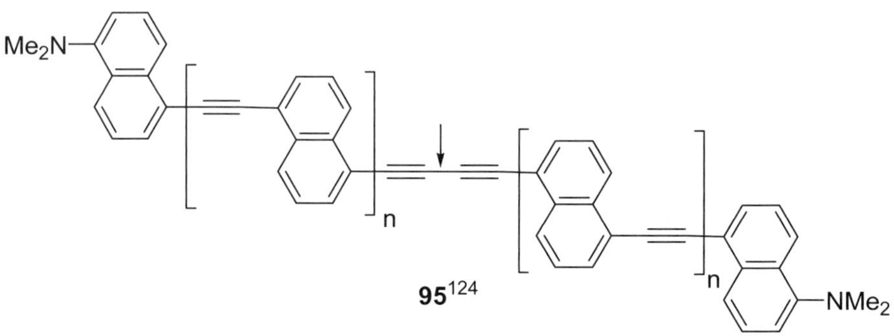

95[124]

1.2.2.6 Experimental

The reader is referred to the syntheses of diphenyldiacetylene,[125] cyclooctadeca-1,3,7,9,13,15-hexayne,[126] and 1,4-bis(trimethylsilyl)buta-1,3-diyne[127] published in *Organic Syntheses*, 2,7-dimethyl-3,5-octadiyn-2,7-diol published in an undergraduate laboratory text,[128] and 3,5-octadiyne and 2,4-hexadiyn-1,6-diol published in *Preparative Acetylenic Chemistry*.[129]

2,4-Hexadiyne-1,6-diol (Glaser Method):[9]
Propargyl alcohol (11.2 g, 0.20 mol) was added with stirring to a mixture of copper(I) chloride (25 g, 0.13 mol), ammonium chloride (40 g, 0.75 mol), concentrated ammonium hydroxide (12.5 mL, 0.18 mol NH_3), and water (200 mL). The mixture was stirred under a slight (30 Torr) positive pressure of oxygen for 20 h. The blue-green reaction mixture was acidified with dilute hydrochloric acid, diluted to 750 mL with water and extracted with ether in a continuous extractor for 24 h. Evaporation of ether from the extract left a solid which, when recrystallized from hot water, gave 9.1 g (83%) of the title compound, mp 111.5–112 °C.

1,4-Di-(2′-quinolyl)-1,3-butadiyne (Cameron Modification):[130]
A solution of cuprous chloride (10 mg, 0.1 mmol) and 2-ethynylquinoline (85 mg, 0.55 mmol) in freshly distilled pyridine (30 mL), under oxygen at 40 °C was stirred for 150 min. Then, the solvent was removed at reduced atmosphere giving a brown solid, which was washed with an aqueous ammonium chloride solution and extracted with dichloromethane. The

combined extracts were dried with anhydrous sodium sulfate, filtered, and the solvent was evaporated affording a brown solid that was crystallized from acetonitrile-hexane (1:1). The title compound was isolated as a white solid (76 mg, 89%).

_N,N′_-Phenylbuta-1,3-diyne-1,4-tosylamide (51, R = Ph) (Hay Modification with CuI):[43]

TMEDA (5 μL, 0.033 mmol) was added to a suspension of copper(I) iodide 93 mg, 0.017 mmol) in dry acetone (4 mL) under oxygen at rt. After 15 min, a solution of *N*-phenyl-*N*-tosyl ynamide (45 mg, 0.177 mmol) in acetone (4 mL) was added and the mixture was vigorously stirred until TLC showed complete consumption of the starting material (3 h). After removal of the solvent, the crude residue was purified by column chromatography on silica gel using hexanes-ethyl acetate (1 : 3) as eluent, yielding 41 mg (91%) of **51** (R = Ph) as white prisms.

9,9′-Dianthryldiacetylene (43) (Eglinton Modification):[36]

A mixture of 9-ethynylanthracene (0.02 g, 1.1 mmol), cupric acetate monohydrate (5.0 g, 25 mmol), pyridine (10 mL) and methanol (1 mL) was stirred for 3 h at 50 °C. The insoluble material was collected by filtration and washed with a small amount of methanol, water, and a small amount of ethanol, successively. The orange tiny cubes (0.20 g, 91%, mp 287–291 °C), thus obtained, were dissolved in toluene, and passed through a short column of alumina to give **43** as orange cubes, mp 290–292 °C, which was identical with an authentic sample.

1.2.2.7 References

1. [R] Cadiot, P.; Chodkiewicz, W. "Chemistry of Acetylenes," Ed. Viehe, H. G., Dekker, New York, 1979, pp 597–647.
2. [R] Siemsen, P.; Livingston, R. C.; Diederich, F. *Angew. Chem., Int. Ed.* **2000,** *39*, 2632–2657.
3. [R] Hay, A. S. *J. Polymer Sci. Part A: Polym. Chem.* **1998**, *36*, 505–517.
4. [R] Eglinton, G.; McCrae, W. *Adv. Org. Chem.* **1963**, *4*, 225–328.
5. [R] Nakagawa, M. "The Chemistry of the Carbon-Carbon Triple Bond, Part 2," Ed. Patai, S., John Wiley, New York, 1978, pp 635–712.
6. [R] Simándi, L. I. "The Chemistry of Triple-Bonding Functional Groups, Supplement C, Part 1," Ed. Patai, S.; Rappoport, Z., John Wiley, New York, 1983, pp 529–534.
7. a) Glaser, C. *Ber. Dtsch. Chem. Ges.* **1869**, *2*, 422–424. b) Glaser, C. *Ann. Chem. Pharm.* **1870**, *154*, 137–171.
8. a) Eglinton, G.; Galbraith, A. R. *Chem. Ind. (London)* **1956**, 737–738. b) Eglinton, G.; Galbraith, A. R. *J. Chem. Soc.* **1959**, 889–896.
9. Cameron, M. D.; Bennett, G. E. *J. Org. Chem.* **1957**, *22*, 557–558.
10. a) Hay, A. S. *J. Org. Chem.* **1960**, *25*, 1275–1276. b) Hay, A. S. *J. Org. Chem.* **1962**, *27*, 3320–3321.
11. a) Zal'kind, Yu. S.; Aizikovich, M. A. *J. Gen. Chem. USSR*, **1937**, *7*, 227–233. b) Zal'kind, Yu. S.; Gverdtsiteli, I. M. *J. Gen. Chem. USSR* **1939**, *9*, 971–974. c) Zal'kind, Yu. S.;

Fundyler, B. M. *J. Gen. Chem. USSR* **1939**, *9*, 1725–1728. d) Salkind, J. S. Fundyler, Fr. B. *Ber.* **1936**, *69*, 128–130.
12. Vaitiekunas, A.; Nord, F. F. *J. Am. Chem. Soc.* **1954**, *76*, 2733–2736.
13. Bohlmann, F.; Schonowsky, H.; Inhoffen, E.; Grau, G. *Chem. Ber.* **1964**, *97*, 794–800.
14. a) Clifford, A. A.; Waters, W. A. *J. Chem. Soc.* **1963**, 3056–3062. b) For a recent theoretical study of the coupling mechanism using DFT calculations, see Fomina, L.; Vazquez, B.; Tkatchouk, E.; Fomine, S. *Tetrahedron* **2002**, *58*, 6741–6747.
15. Valenti, E.; Pericãs, M. A.; Serratosa, F. *J. Am. Chem. Soc.* **1990**, *112*, 7405–7406.
16. Vasilevsky, S. F.; Verkruijsse, H. D.; Brandsma, L. *Recl. Trav. Chim. Pays-Bas.* **1992**, *111*, 529–530.
17. Li, J.; Jiang, H. *Chem. Commun.* **1999**, 2369–2370.
18. Yadav, J. S.; Reddy, B. V. S.; Reddy, K. B.; Gayathri, K. U.; Prasad, A. R. *Tetrahedron Lett.* **2003**, *44*, 6493–6496.
19. Li, P.-H.; Yan, J.-C.; Wang, M.; Wang, L. *Chin. J. Chem.* **2004**, *22*, 219–221.
20. Lu, X.; Zhang, Y.; Luo, C.; Wang, Y. *Synth. Commun.* **2006**, *36*, 2503–2511.
21. a) Kabalka, G. W.; Wang, L.; Pagni, R. M. *Synlett* **2001**, 108–110. For related conditions, see b) Sharifi, A.; Mirzaei, M.; Naimi-Jamal, M. R. *J. Chem. Res. (S)* **2002**, 628–630. c) Sharifi, A.; Mirzaei, M.; Naimi-Jamal, M. R. *Monatsh.* **2006**, *137*, 213–217.
22. Liao, Y.; Fathi, R.; Yang, Z. *Org. Lett.* **2003**, *5*, 909–912.
23. a) Ikegashira, K.; Nishihara, Y.; Hirabayashi, K.; Mori, A.; Hiyama, T. *Chem. Commun.* **1997**, 1039–1040. b) Nishihara, Y.; Ikegshira, K.; Hirabayashi, K.; Ando, J.; Mori, A.; Hiyama T. *J. Org. Chem.* **2000**, *65*, 1780–1787. c) Nishihara Y.; Takemura, M.; Mori, A.; Osakada, K. *J. Organomet. Chem.* **2001**, *620*, 282–286.
24. Haley, M. M.; Bell, M. L.; Brand, S. C.; Kimball, D. B.; Pak, J. J.; Wan, W. B. *Tetrahedron Lett.* **1997**, *38*, 7483–7486.
25. Heuft, M. A.; Collins, S. K.; Yap, G. P. A.; Fallis, A. G. *Org. Lett.* **2001**, *3*, 2883–2886.
26. Nishihara, Y.; Okamoto, M.; Inoue, Y.; Miyazaki, M.; Miyasaka, M.; Takagi, K. *Tetrahedron Lett.* **2005**, *46*, 8661–8664.
27. Krafft, M. E.; Hirosawa, C.; Dalal, N.; Ramsey, C.; Stiegman, A. *Tetrahedron Lett.* **2001**, *42*, 7733–7736.
28. Li, P. H.; Wang, L.; Wang, M.; Yan, J. C. *Chin. Chem. Lett.* **2004**, *15*, 1295–1298.
29. Carpita, A.; Rossi, R.; Veracini, C. A. *Tetrahedron* **1985**, *41*, 1919–1929.
30. Abdulganeeva, S. A.; Grishina, G. V.; Potapov, V. M.; Erzhanov, K. B.; Shapovalov, A. A. *Chem. Het. Cpds.* **1983**, 408–411.
31. Sen, B. K.; Majumdar, K. C. *J. Indian Chem. Soc.* **LX**, 409–410.
32. Kagan, J.; Arora, S. K. *J. Org. Chem.* **1983**, *48*, 4317–4320.
33. Bunz, U.; Szeimies, G. *Tetrahedron Lett.* **1989**, *30*, 2087–2088.
34. Müller, T. J. J. *Tetrahedron Lett.* **1999**, *40*, 6563–6566.
35. Rodríguez, J. G.; Ramos, S.; Martín-Villamil, R.; Fonseca, I.; Albert, A. *J. Chem. Soc., Perkin Trans. 1* **1996**, 541–543.
36. Akiyama, S.; Ogura, F.; Nakagawa, M. *Bull. Chem. Soc. Japan* **1971**, *44*, 3443–3445.
37. Fritzsche, U.; Hunig, S. *Tetrahedron Lett.* **1972**, *13*, 4831–4834.
38. Jeon, H. H.; Son, J. B.; Choi, J. H.; Jeong, I. H. *Tetrahedron Lett.* **2007**, *48*, 627–631.
39. Gilbertson, R. D.; Wu, H-P.; Gorman-Lewis, D.; Weakley, T. J. R.; Weiss, H-C.; Boese, R.; Haley, M. M. *Tetrahedron* **2004**, *60*, 1215–1223.
40. a) Rodríguez, J. G.; Lafuente, A.; Rubio, L. *Tetrahedron Lett.* **2004**, *45*, 5685–5688. b) Rodríguez, J. G.; Esquivias, J.; Lafuente, A.; Rubio, L. *Tetrahedron* **2006**, *62*, 3112–3122.
41. Rodríguez, J. G.; Martín-Villamil, R.; Lafuente, A. *Tetrahedron* **2003**, *59*, 1021–1032.
42. a) Müller, T. J. J.; Blümel, J. *J. Organomet. Chem..* **2003**, *683*, 354–367. b) Yagi, S.; Kitayama, H.; Takagishi, T. *J. Chem. Soc., Perkin Trans. 1* **2000**, 925–932.
43. Rodríguez, D.; Castedo, L.; Saá, C. *Synlett* **2004**, 377–379.
44. Liu, Y.; Gao, H. *Org. Lett.* **2006**, *8*, 309–311.
45. a) Murty, K. V. S. N., Vasella, A. *Helv. Chim. Acta* **2001**, *84*, 939–963. b) Gan, Z.; Roy, R. *Tetrahedron Lett.* **2000**, *41*, 1155–1158.
46. Hoffmann, B.; Zanini, D.; Ripoche, I.; Bürli, R.; Vasella, A. *Helv. Chim. Acta* **2001**, *84*, 1862–1888.
47. Jung, F.; Burger, A.; Biellmann, J-F. *Org. Lett.* **2003**, *5*, 383–385.

48. Brosch, O.; Weyhermüller, T.; Metzler-Nolte, N. *Eur. J. Inorg. Chem.* **2000**, 323–330.
49. Maya, E. M.; Vásquez, P.; Torres, T.; Gobbi, L.; Diederich, F.; Pyro, S.; Echegoyen, L. *J. Org. Chem.* **2000**, *65*, 823–830.
50. Faiz, J. A.; Spencer, N.; Pikramenou, Z. *Org. Biomol. Chem.* **2005**, *3*, 4239–4245.
51. Paquette, L. A.; Tae, J. *J. Am. Chem. Soc.* **2001**, *123*, 4974–4984.
52. Chen, W-Z.; Ren, T. *Inorg. Chem.* **2006**, *45*, 9175–9177.
53. Bähr, A.; Felber, B.; Schneider, K.; Diederich, F. *Helv. Chim. Acta* **2000**, *83*, 1346–1376.
54. Al-Saraierh, H.; Miller, D. O.; Georghiou, P. E. *J. Org. Chem.* **2005**, *70*, 8273–8280.
55. Benniston, A. C.; Harriman, A.; Lawrie, D. J.; Rostron, S. A. *Eur. J. Org. Chem.* **2004**, 2272–2276.
56. Raker, J.; Glass, T. E. *J. Org. Chem.* **2001**, *66*, 6505–6512.
57. a) Patwardhan, A. P.; Thompson, D. H. *Org. Lett.* **1999**, *1*, 241–243. b) Patwardhan, A. P.; Thompson, D. H. *Langmuir* **2000**, *16*, 10340–10350.
58. Ziessel, R.; Suffert, J.; Youinou, M-T. *J. Org. Chem.* **1996**, *61*, 6535–6546.
59. Coat, F.; Lapinte, C. *Organometallics* **1996**, *15*, 477–479.
60. Nielsen, M. B.; Utesch, N. F.; Moonen, N. N. P.; Bouden, C.; Gisselbrecht, J-P.; Concilio, S.; Piotto, S. P.; Seiler, P.; Günter, P.; Gross, M.; Diederich, F. *Chem. Eur. J.* **2002**, *8*, 3610–3613.
61. van Swieten, P. F.; Maehr, R.; van den Nieuwendijk, A. M. C. H.; Kessler, B. M.; Reich, M.; Wong, C-S.; Kalbacher, H.; Leeuwenburgh, M. A.; Driessen, C.; van der Marel, G. A.; Ploegh, H. L.; Overkleeft, H. S. *Bioorg. Med. Chem. Lett.* **2004**, *14*, 3131–3134.
62. Vizer, S. A..; Yerzhanov, K.B.; Dembitsky, V. M. *Heteroatom. Chem.* **2006**, *17*, 66–73.
63. Bushby, N.; Moody, C. J.; Riddick, D. A.; Waldron, I. R. *J. Chem. Soc., Perkin Trans. 1* **2001**, 2183–2193.
64. [R] a) Spitler, E. L.; Johnson, II, C. A.; Haley, M. M. *Chem. Rev.* **2006**, *106*, 5344-5386. b) Haley, M. M.; Pak, J. J.; Brand, S. C. *Top. Curr. Chem.* **1999**, *201*, 81–130. c) Diederich, F.; Gobbi, L. *Top. Curr. Chem.* **1999**, *201*, 43–79.
65. [R] Godt, A. *Eur. J. Org. Chem.* **2004**, 1639–1654.
66. a) Sondheimer, F.; Amiel, Y. *J. Am. Chem. Soc.* **1957**, *79*, 5817–5820. b) Pilling, G. M.; Sondheimer, F. *J. Am. Chem. Soc.* **1971**, *93*, 1970–1977. c) McQuilkin, R. M.; Garratt, P. J.; Sondheimer, F. *J. Am. Chem. Soc.* **1970**, *92*, 6682–6683. d) Cresp, T. M.; Ojima, J.; Sondheimer, F. *J. Org. Chem.* **1977**, *42*, 2130–2134 e) [R] Huang, N. Z.; Sondheimer, F. *Acc. Chem. Res.* **1982**, *15*, 96–102.
67. Behr, O. M.; Eglinton, G.; Galbraith, A. R.; Raphael, R. A. *J. Chem. Soc.* **1960**, 3614–3625.
68. a) Akiyama, S.; Misumi, S.; Nakagawa, M. *Bull. Chem. Soc. Japan* **1960**, *33*, 1293–1298. b) Fukui, K.; Okamoto, T.; Nakagawa, M. *Tetrahedron Lett.* **1971**, *12*, 3121–3124.
69. Meissner, U.; Meissner, B.; Staab, H. A. *Angew. Chem. Internat. Ed.* **1973**, *12*, 916–918.
70. Carruthers, W.; Pellatt, M. G. *J. Chem. Soc. (C)* **1971**, 1485–1488.
71. a) Boldi, A. M.; Diederich, F. *Angew. Chem., Int. Ed. Engl.* **1994**, *33*, 468–471. b) Kivala, M.; Mitzel, F.; Boudon, C.; Gisselbrecht, J.-P.; Seiler, P.; Gross, M.; Diederich, F. *Chem. Asian J.* **2006**, *1*, 479–489. c) Andersson, A. S.; Kilsa, K.; Hassenkam, T.; Gisselbrecht, J.-P.; Boudon, C.; Gross, M.; Nielsen, M. B.; Diederich, F. *Chem. Eur. J.* **2006**, *12*, 8451–8459.
72. a) Haley, M. M.; Langsdorf, B. L. *Chem. Commun.* **1997**, 1121–1122. b) Pak, J. J.; Weakley, T. J. R.; Haley, M. M. *J. Am. Chem. Soc.* **1999**, *121*, 8182–8192. c) Marsden, J. A.; O'Connor, M. J.; Haley, M. M. *Org. Lett.* **2004**, *6*, 2385–2388.
73. a) Höger, S.; Meckenstock, A.-D.; Pellen, H. *J. Org. Chem.* **1997**, *62*, 4556–4557. b) Höger, S.; Meckenstock, A.-D. *Chem. Eur. J.* **1999**, *5*, 1686–1691. c) Höger, S.; Bonrad, K.; Karcher, L.; Meckenstock, A.-D. *J. Org. Chem.* **2000**, *65*, 1588–1589. d) Ziegler, A.; Mamdouh, W.; Ver Heyen, A.; Surin, M.; Uji-i, H.; Abdel-Mottaleb, M. M. S.; De Schryver, F. C.; De Feyter, S.; Lazzaroni, R.; Höger, S. *Chem. Mater.* **2005**, *17*, 5670–5683.
74. a) de Meijere, A.; Kozhushkov, S.; Puls, C.; Haumann, T.; Boese, R.; Cooney, M. J.; Scott, L. T. *Angew. Chem., Int. Ed. Engl.* **1994**, *33*, 869–871. b) de Meijere, A.; Kozhushkov, S.; Haumann, T.; Boese, R.; Puls, C.; Cooney, M. J.; Scott, L. T. *Chem. Eur. J.* **1995**, *1*, 124–131. d) de Meijere, A.; Kozhushkov, S. I. *Chem. Eur. J.* **2002**, 8, 3195–3202.
75. a) Romero, M. A.; Fallis, A. G. *Tetrahedron Lett.* **1994**, *35*, 4711–4714. b) Collins, S. K.; Yap, G. P. A.; Fallis, A. G. *Angew. Chem., Int. Ed.* **2000**, *39*, 385–388.

76. a) Altmann, M.; Friedrich, J.; Beer, F.; Reuter, R.; Enkelmann, V.; Bunz, U. H. F. *J. Am. Chem. Soc.* **1997**, *119*, 1472–1473. b) Laskoski, M.; Roidl, G.; Ricks, H. L.; Morton, J. G. M.; Smith, M. D.; Bunz, U. H. F. *J. Organomet. Chem.* **2003**, *673*, 13–24.
77. Ohkita, M.; Ando, K.; Suzuki, T.; Tsuji, T. *J. Org. Chem.* **2000**, *65*, 4385–4390.
78. García-Frutos, E. M.; Fernández-Lázaro, F.; Maya, E. M.; Vázquez, P.; Torres, T. *J. Org. Chem.* **2000**, *65*, 6841–6846.
79. Tobe, Y.; Utsumi, N.; Kawabata, K.; Nagano, A.; Adachi, K.; Araki, S.; Sonoda, M.; Hirose, K.; Naemura, K. *J. Am. Chem. Soc.* **2002**, *124*, 5350–5364.
80. Srinivasan, M.; Sankararaman, S.; Hopf, H.; Varghese, B. *Eur. J. Org. Chem.* **2003**, 660–665.
81. Moriarty, R. M.; Pavlović, D. *J. Org. Chem.* **2004**, *69*, 5501–5504.
82. a) Werz, D. B.; Gleiter, R.; Rominger, F. *J. Org. Chem.* **2004**, *69*, 2945–2952. b) Baier, M.; Gleiter, R.; Rominger, F. *Eur. J. Org. Chem.* **2006**, 5264–5278.
83. Iyoda, M.; Kuwatani, Y.; Yamagata, S.; Nakamura, N.; Todaka, M.; Yamamoto, G. *Org. Lett.* **2004**, *6*, 4667–4670.
84. Setaka, W.; Kanai, S.; Kabuto, C.; Kira, M. *Chem. Lett.* **2006**, *35*, 1364–1365.
85. a) Goichi, M.; Toyota, S. *Chem. Lett.* **2006**, *35*, 684–685. b) Goichi, M.; Miyahara, H.; Toyota, S. *Chem. Lett.* **2006**, *35*, 920–921.
86. a) Anderson, H. L. *Inorg. Chem.* **1994**, *33*, 972–981. b) Marvaud, V.; Vidal-Ferran, A.; Webb, S. J.; Sanders, J. K. M. *J. Chem. Soc., Dalton Trans.* **1997**, 985–990. c) Nakash, M.; Clyde-Watson, Z.; Feeder, N.; Davies, J. E.; Teat, S. J.; Sanders, J. K. M. *J. Am. Chem. Soc.* **2000**, *122*, 5286–5293.
87. a) Menger, F. M.; Chen, X. Y.; Brocchini, S.; Hopkins, H. P.; Hamilton, D. *J. Am. Chem. Soc.* **1993**, *115*, 6600–6608. b) Menger, F. M.; Chen, X. Y. *Tetrahedron Lett.* **1996**, *37*, 323–326.
88. Ladika, M.; Fisk, T. E.; Wu, W. W.; Jons, S. D. *J. Am. Chem. Soc.* **1994**, *116*, 12093–12094.
89. Taguchi, K.; Arakawa, K.; Eguchi, T.; Kakinuma, K.; Nakatani, Y.; Ourisson, G. *New J. Chem.* **1998**, 63–69.
90. a) Miyawaki, K.; Takagi, T.; Shibakami, M. *Synlett* **2002**, 1326–1328. b) Miyawaki, K.; Goto, R.; Takagi, T.; Shibakami, M. *Synlett* **2002**, 1467–1470.
91. a) Bukownik, R. R.; Wilcox, C. S. *J. Org. Chem.* **1988**, *53*, 463–471. b) Belghiti, T.; Joly, J.-P.; Didierjean, C.; Dahaoui, S.; Chapleur, Y. *Tetrahedron Lett.* **2002**, *43*, 1441–1443.
92. a) Whitlock, B. J.; Jarvi, E. T.; Whitlock, H. W. *J. Org. Chem.* **1981**, *46*, 1832–1835. b) Jarvi, E. T.; Whitlock, H. W. *J. Am. Chem. Soc.* **1982**, *104*, 7196–7204. c) Cloninger, M. J.; Whitlock, H. W. *J. Org. Chem.* **1998**, *63*, 6153–6159.
93. a) Srinivasan, M.; Sankararaman, S.; Hopf, H.; Dix, I.; Jones, P. G. *J. Org. Chem.* **2001**, *66*, 4299–4303. b) Sankararaman, S.; Srinivasan, M.; Narayanan, V.; Varghese, B. *Ind. J. Chem.* **2004**, *43B*, 1499–1503.
94. Lehn, J.-M.; Schmidt, F.; Vigneron, J.-P. *Tetrahedron Lett.* **1988**, *29*, 5255–5258.
95. Berscheid, R.; Vögtle, F. *Synthesis* **1992**, 58–62.
96. Godt, A.; Duda, S.; Ünsal, O.; Thiel, J.; Härter, A.; Roos, M.; Tschierske, C.; Diele, S. *Chem. Eur. J.* **2002**, *8*, 5094–5106.
97. a) Hébert, N.; Beck, A.; Lennox, R. B.; Just, G. *J. Org. Chem.* **1992**, *57*, 1777–1783. b) Bhattacharya, S.; Ghosh, S.; Easwaran, K. R. K. *J. Org. Chem.* **1998**, *63*, 9232–9242.
98. Hegmann, T.; Neumann, B.; Wolf, B.; Tschierske, C. *J. Mater. Chem.* **2005**, *15*, 1025–1034.
99. Zhao, T.; Liu, Z.; Song, Y.; Xu, W.; Zhang, D.; Zhu, D. *J. Org. Chem.* **2006**, *71*, 7422–7432.
100. [R] a) Schill, G. *Catenanes, Rotaxanes, and Knots*, Academic, New York, 1971. b) Dietrich-Buchecker, C. O.; Sauvage, J.-P. *Chem. Rev.* **1987**, *87*, 795–810. c) Amabilino, D. B.; Stoddart, J. F. *Chem. Rev.* **1995**, *95*, 2725–2828.
101. a) Duda, S.; Godt, A.; *Eur. J. Org. Chem.* **2003**, 3412–3420. b) Ünsal, Ö.; Godt, A. *Chem. Eur. J.* **1999**, *5*, 1728–1733.
102. Hamilton, D. G.; Prodi, L.; Feeder, N.; Sanders, J. K. M. *J. Chem. Soc., Perkin Trans. 1* **1999**, 1057–1065.
103. Gunter, M. J.; Farquhar, S. M. *Org. Biomol. Chem.* **2003**, *1*, 3450–3457.
104. Miljanić, O. S.; Dichtel, W. R.; Mortezaei, S.; Stoddart, J. F. *Org. Lett.* **2006**, *8*, 4835–4838.
105. Carina, R. F.; Dietrich-Buchecker, C.; Sauvage, J.-P. *J. Am. Chem. Soc.* **1996**, *118*, 9110–9116.

106. [R] a) Ogawa, T. *Prog. Polym. Sci.* **1995**, *20*, 943–985. b) Szafert, S.; Gladysz, J. A. *Chem. Rev.* **2006**, *106*, 1–33.
107. Hay, A. S. *J. Polym. Sci. A.* **1969**, *7*, 1625–1634.
108. Knol, K. E.; van Horssen, L. W.; Challa, G.; Havinga, E. E. *Polym. Commun.* **1985**, *26*, 71–73.
109. Cataldo, F. *Polym. Int.* **1997**, *44*, 191–200.
110. Rutherford, D. R.; Stille, J. K.; Elliott, C. M.; Reichert, V. R. *Macromolecules* **1992**, *25*, 2294–2306.
111. a) Anderson, H. L.; Martin, S. J.; Bradley, D. D. C. *Angew. Chem., Int. Ed. Engl.* **1994**, *33*, 655–657. b) Screen, T. E. O.; Lawton, K. B.; Wilson, G. S.; Donley, N.; Ispasoiu, R.; Goodson III, T.; Martin, S. J.; Bradley, D. D. C.; Anderson, H. L. *J. Mater. Chem.* **2001**, *11*, 312–320.
112. Kwock, E. W.; Baird, Jr., T.; Miller, T. M. *Macromolecules* **1993**, *26*. 2935–2940.
113. a) Boldi, A. M.; Anthony, J.; Gramlich, V.; Knobler, C. B.; Boudon, C.; Gisselbrecht, J.-P.; Gross, M.; Diederich, F. *Helv. Chim. Acta* **1995**, *78*, 779–796. b) Schenning, A. P. H. J.; Arndt, J.-D.; Ito, M.; Stoddart, A.; Schreiber, M.; Siemsen, P.; Martin, R. E.; Boudon, C.; Gisselbrecht, J.-P.; Gross, M.; Gramlich, V.; Diederich, F. *Helv. Chim. Acta* **2001**, *84*, 296–334. c) Siemsen, P.; Gubler, U.; Bosshard, C.; Günter, P.; Diederich, F. *Chem. Eur. J.* **2001**, *7*, 1333–1341.
114. a) Anderson, S.; Aplin, R. T.; Claridge, T. D. W.; Goodson III, T.; Maciel, A. C.; Rumbles, G.; Ryan, J. F.; Anderson, H. L. *J. Chem. Soc., Perkin Trans. 1* **1998**, 2383–2397. b) Taylor, P. N.; Hagan, A. J.; Anderson, H. L. *Org. Biomol. Chem.* **2003**, *1*, 3851–3856.
115. a) Sumi, N.; Nakanishi, H.; Ueno, S.; Takimiya, K.; Aso, Y.; Otsubo, T. *Bull. Chem. Soc. Jpn.* **2001**, *74*, 979–988. b) Inouchi, K.; Kobashi, S.; Takimiya, K.; Aso, Y.; Otsubo, T. *Org. Lett.* **2002**, *4*, 2533–2536.
116. a) Krömer, J.; Rios-Carreras, I.; Fuhrmann, G.; Musch, C.; Wunderlin, M.; Debaerdemaeker, T.; Mena-Osteritz, E.; Bäuerle, P. *Angew. Chem., Int. Ed.* **2000**, *39*, 3481–3486. b) Fuhrmann, G.; Krömer, J.; Bäuerle, P. *Synth. Met.* **2001**, *119*, 125–126.
117. Xu, G.-L.; Wang, C.-Y.; Ni, Y.-H.; Goodson III, T. G.; Ren, T. *Organometallics* **2005**, *24*, 3247–3254.
118. Lee, L.-H.; Lynch, V.; Lagow, R. J. *J. Chem. Soc., Perkin Trans. 1* **2000**, 2805–2809.
119. Xie, X.; Phuan, P.-W.; Kozlowski, M. C. *Angew. Chem., Int. Ed.* **2003**, *42*, 2168–2170.
120. Häussler, M.; Zheng, R.; Lam, J. W. Y.; Tong, H.; Dong, H.; Tang, B. Z. *J. Phys. Chem. B* **2004**, *108*, 10645–10650.
121. Gibtner, T.; Hampel, F.; Gisselbrecht, J.-P.; Hirsch, A. *Chem. Eur. J.* **2002**, *8*, 408–432.
122. Elwahy, A. H. M.; Hafner, K. *Eur. J. Org. Chem.* **2006**, 3910–3916.
123. Kim, J. H.; Park, Y. T. *Bull. Korean Chem. Soc.* **2006**, *27*, 869–874.
124. Rodríguez, J. G.; Tejedor, J. L. *Tetrahedron* **2005**, *61*, 3033–3043.
125. Campbell, I. D.; Eglinton, G. *Org. Syn. Coll. Vol. V* **1973**, 517–520.
126. Stöckel, K.; Sondheimer, F. *Org. Syn. Coll. Vol. VI* **1991**, 68–75.
127. Jones, G. E. Kendrick, D. A.; Holmes, A. B. *Org. Syn. Coll. Vol. VIII* **1993**, 63–67.
128. [R] Williamson, K. L.; Minard, R.; Masters, K. M. *Macroscale and Microscale Organic Experiments, 5th Ed.* Houghton Mifflin, Boston, 2007, pp 399–403.
129. [R] Brandsma, L. *Preparative Acetylenic Chemistry, 2nd Edition*, Elsevier, Amsterdam, 1988, pp 220–223.
130. Rodríguez, J. G.; de los Rios, C.; Lafuente, A. *Tetrahedron* **2005**, *61*, 9042–9051.

1.2.3 Ullmann Coupling

Nadia M. Ahmad

1.2.3.1 Description

The Ullmann reaction can be taken to refer to two different transformations. The first is the copper mediated coupling of two aryl groups to give a biaryl compound; this is the "classic" Ullmann reaction. The second, the Ullmann-type reaction, is the nucleophilic aromatic substitution between aryl nucleophiles and aryl halides, the most common of which is the Ullmann ether synthesis.[1,2] The classic Ullmann reaction will be reviewed in this chapter; the reader is refered to several excellent reviews for details on the Ullmann-type reaction.[3,4]

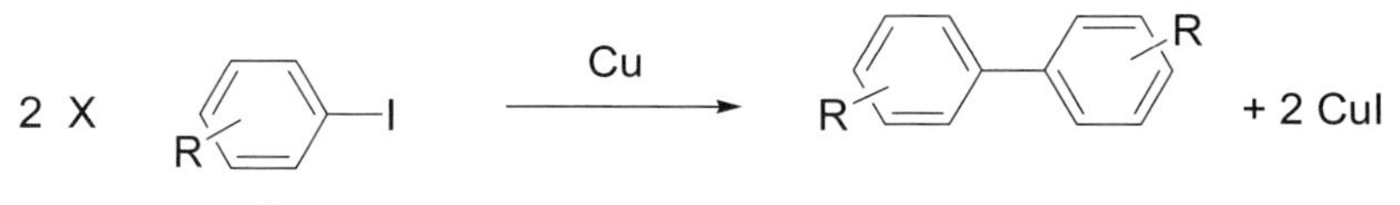

1.2.3.2 Historical Perspective

The reaction of aryl halides with stoichiometric copper to produce aryl–aryl bonds was reported by Fritz Ullmann in 1901 in series of papers in Chemische Bericht.[5–8] Ullman, a German chemist, studied chemistry in Nurenberg and received his Ph.D. from the University of Geneva under the tutelage of Carl Gräbe in 1895. Ullmann has the honour of several reactions being named after him, including the Graebe–Ullmann reaction,[9a,b] and the Jourdan–Ullmann–Goldberg synthesis,[8,10,11] a reaction discovered with his wife, Irma Goldberg.

The reaction employing copper was utilized almost exclusively in the following decades for biaryl formation until the rising popularity of nickel, in an Ullmann-type coupling. This was followed by the use of zinc, tin, boron, and eventually, palladium, arguably the most commonly used transition metal in organic synthesis today for such transformations.

1.2.3.3 Mechanism

The mechanism proceeds by oxidative addition of the copper into the aryl halide **3**. The copper(I)–aryl species **4** then undergoes another oxidative coupling with another equivalent of the aryl halide resulting in a biaryl copper compound 7. Reductive elimination follows resulting in the formation of the carbon–carbon bond and a biaryl compound **8** is produced.

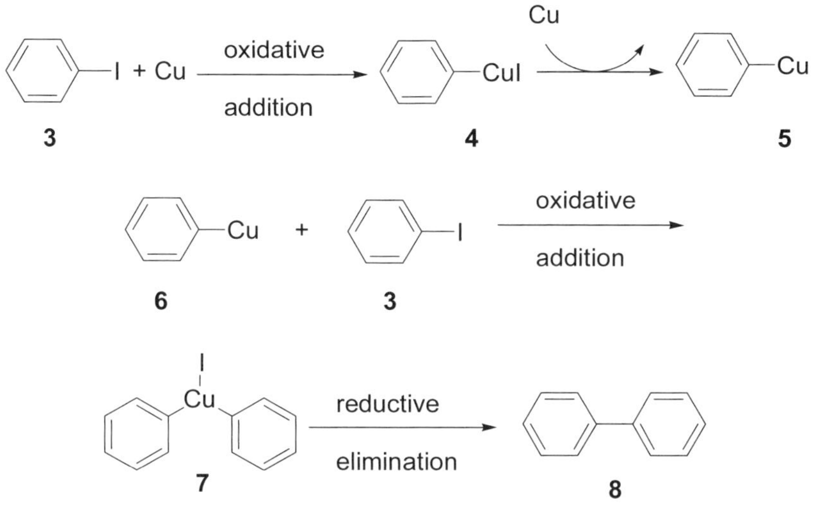

Thus it is clear that electron-withdrawing groups on the aromatic ring, particularly in the ortho position to the halogen, result in an activating effect on the reaction. Electron-donating groups on the other hand, hinder the reaction if not inhibiting it altogether. For obvious reasons, bulky substituents in the ortho positions can also decrease the effectiveness of the reaction, although exceptions exist.

1.2.3.4 Variations, Improvements or Modifications

Variations in the Ullmann reaction centre mainly on the catalysts used to carry out the transformation and modifications to the conditions in order to improve yields. Additionally, although the Ullmann reaction is traditionally the reaction of aryl halides with copper, other metals have also been utilized, often to inhibit the formation of by-products.[12–15]

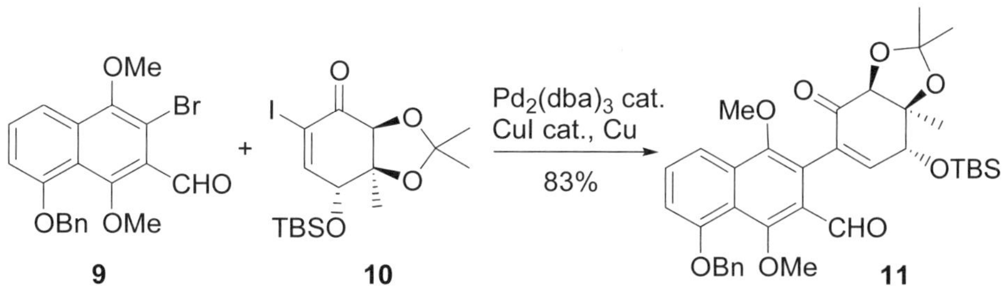

The replacement of copper by other metals in the Ullmann reaction usually results in milder and more efficient pathways. The common use of Pd in conjunction with copper in the Ullmann coupling can be seen in many examples. Nicolaou *et al.* utilized such modifications in their total synthesis of kinamycins C, F, and J.[16] Bromide **9** underwent coupling with iodide **10** to give aldehyde **11** in a satisfactory 83% yield.

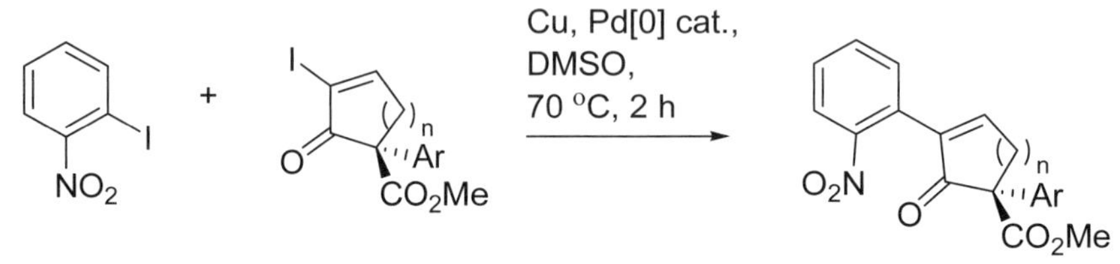

12 **13a-f** **14**

Substrate	n	Ar	Yield %
13a	1	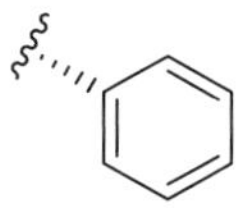	60
13b	1	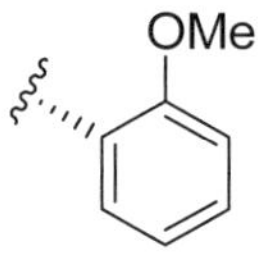	58
13c	1	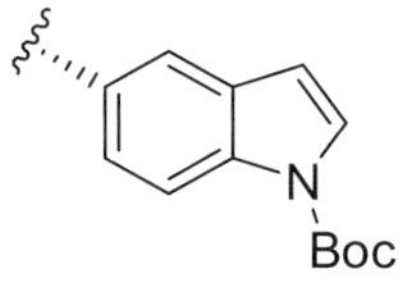	86
13d	2	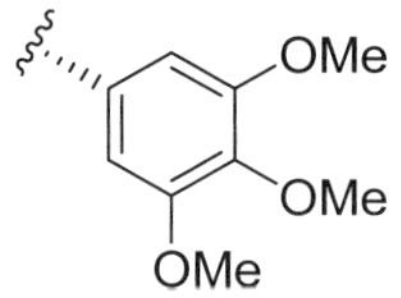	56
13e	22	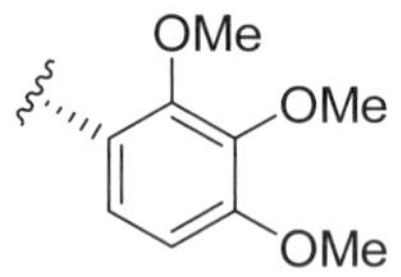	56
13f	2	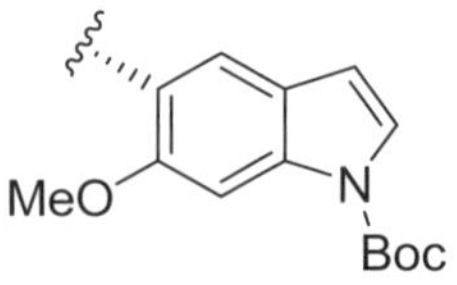	91

The synthesis of vincristine and vinblastine has received much attention over the years due to their use in cancer therapy, especially breast and testicular cancer, and acute leukemia. However due to their neurotoxicological side effects the identification and effective synthesis of analogues displaying improved therapeutic properties has been a pursuit of several research groups.[17–19] Accordingly, Banwell and coworkers have reported the use of a Pd[0]-catalyzed Ullmann cross-coupling between 2-iodonitrobenzene **12** and iodides **13** in their synthesis of the indole–indoline analogues.[20] Good to excellent yields were obtained. The Ullmann products then underwent reductive cyclisations to give the desired indole-indoline structures.

The use of copper bronze was used in the Ullmann coupling of protected iodoresorcinol **15** to give the symmetrical biphenyl **16**.[21] Despite the obvious steric hinderance afforded by the *ortho*-methoxy groups, the reaction took place in a good yield of 72%.

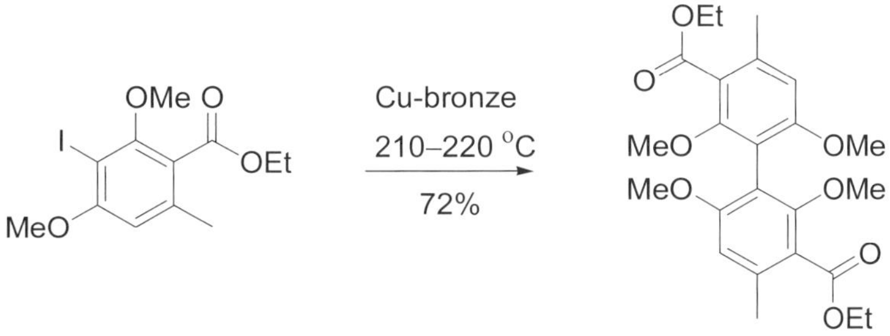

15 **16**

Nickel is a popular alternative to copper in the Ullmann reaction as the reaction conditions are mild and the nickel reagent is inexpensive. You and coworkers have reported the use of a nickel(0)-mediated Ullmann coupling in their reactions of *ortho*-carbonyl-substituted aryl halides **17**, with Zn powder used *in situ* as a reductant, to form *trans*-9,10-dihydroxy-9,10-dihydrophenanthrenes **18**.[22] These phenanthrenes could then be used as chiral ligands in asymmetric synthesis.

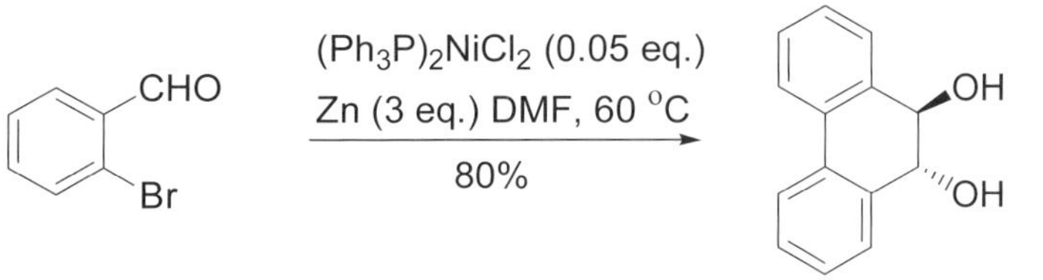

17 **18**

1.2.3.5 Synthetic Utility

Occasionally, the Ullmann reaction is the only successful method of coupling two aryl groups when other, more prevalent coupling reactions fail. Broady and coworkers for example, attempted to synthesize the core of colchicine-based natural products (**19**) initially utilizing a Suzuki reaction to carry out the key intermolecular biaryl coupling (**20** and **21** → **22**).[23] However, despite significant efforts to optimise the reaction between **20** and **21**, the yields remained below 10%.

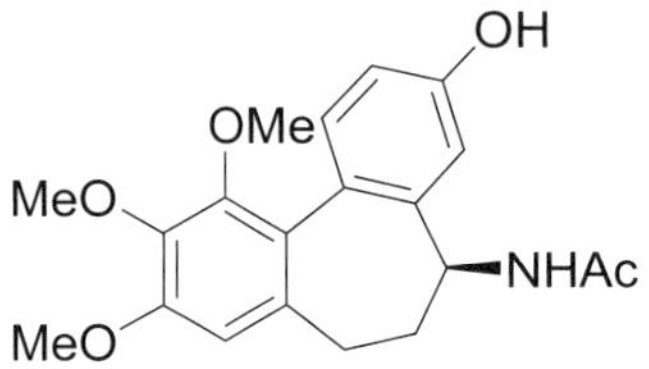

19

(*S*)-(−)-*N*-acetylcolchinol

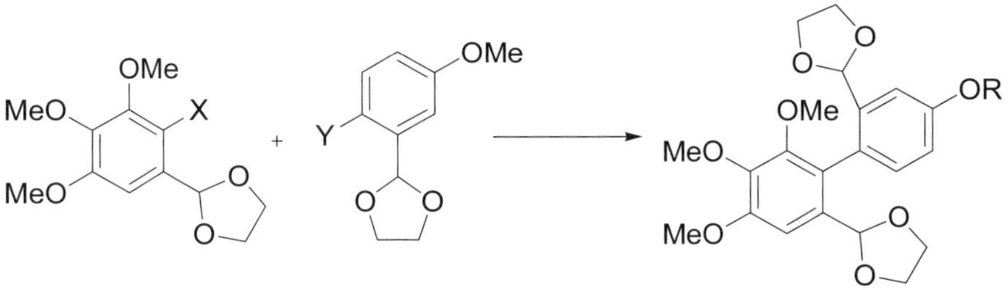

20 **21** **22**, < 10% yield

X = Br and Y = $B(OH)_2$ or $B(OAlk)_2$
or Y = Br and X = $B(OH)_2$ or $B(OAlk)_2$

The Ziegler–Ullmann reaction was then attempted. This is the reaction devised by Ziegler in which an Ullmann reaction between aryl groups bearing certain substituents is particularly favoured.[24–26] Electron-rich aromatic systems bearing *ortho*-substituents capable of coordinating to copper work best, and thus a sulfur atom is typically preferred. Thus, aryl bromide **24** was converted to the diaryl system **25** in a satisfactory 70% yield, with concommitant conversion of the imine to the aldehyde during work-up. The thioacetal group was then further manipulated to form the 7-membered central colchicine ring. The method illustrates the synthetic utility of the Ullmann reaction which can be employed when other, perhaps more known procedures, fail.

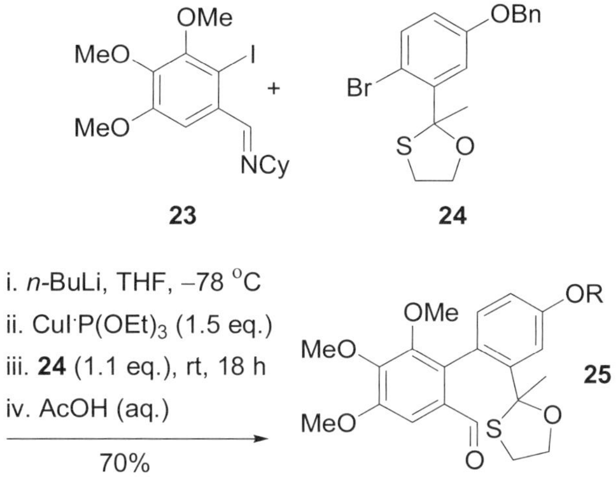

The coordinating effect of an *ortho*-substituent is further exemplified by Kelly and coworkers' synthesis of santiagonamine.[27] Although the yield of the reaction of imine **26** with iodide **27** was still poor, it was still significantly better than the reaction with the corresponding aldehyde (7%). This was partly attributed to the coordinating N lone pair and partly due to the precedent of success Ullmann reactions carried out with imines.[28–29]

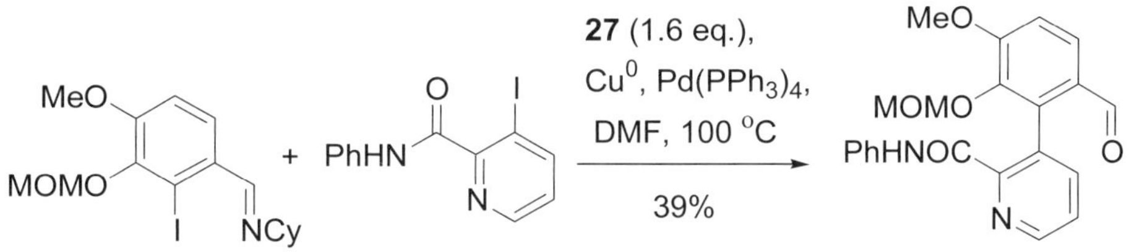

26 **27** **28**

The Ullmann reaction can be used to selectively couple at one aromatic halogen than another depending on the substituents around it. As previously mentioned *ortho*-substituents can be used to direct the coupling and indeed, facilitate the reaction over a less hindered halogen substitute. Crudden and coworkers investigated the Ullmann reaction for the synthesis of biaryl **30**.[30] The Ullmann homocoupling of the iodonitro derivative **29** proceeded with a high yield of 82%; however, the nitro group then had to be reduced and converted to the halogen *via* a Sandmeyer reaction. Conversely the Ullmann reaction of the dibromo derivative **31** gave the desired product **32** – due to the coordinating *ortho*-ester only one product was formed, albeit in lower yield (55%). However, the shortness of the second route compensated for the lower yield. The authors found the Ullman to be a particularly useful reaction in this case as cross-coupling reactions forming tetrasubstituted biaryl systems are rarely effective. The biaryl systems were synthesized for use as chiral materials.

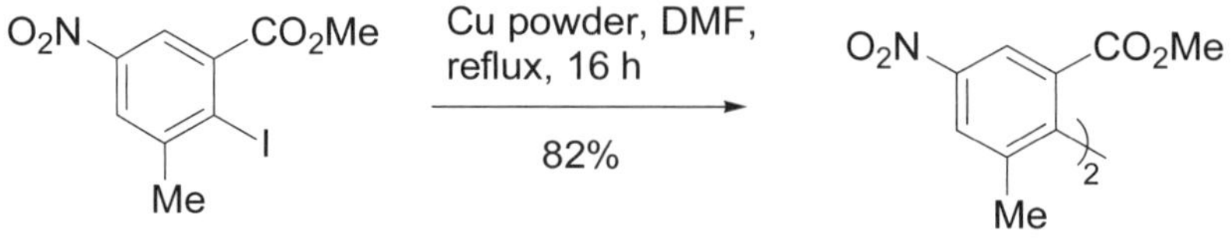

29 30

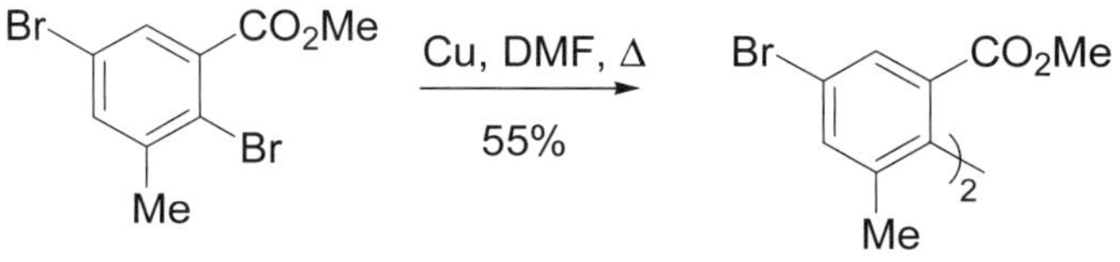

31 32

Symmetrical biaryl systems have been synthesised using high-intensity ultrasound (US). The zinc-mediated Ullmann coupling reaction of iodide **33** was carried out in aqueous solvent under bubbling CO_2 with Pd/C to give biaryl **34**. The procedure has the advantage of fast reaction times and green credentials.[31]

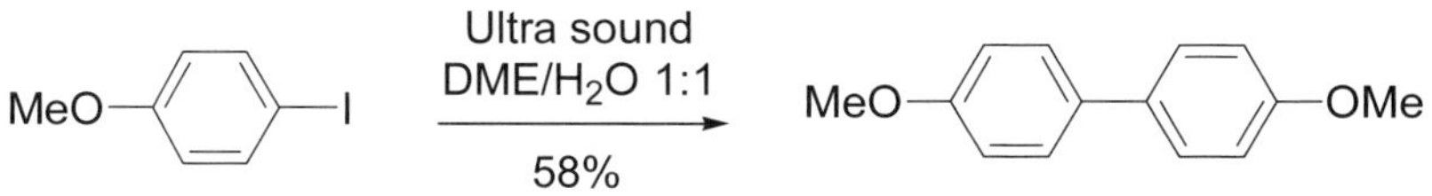

33 34

Similarly, homocoupling reactions can also be carried out using microwave irradiation. Kappe and coworkers demonstrate such procedures, synthesizing bisquinolones **36** from quinolones **35** *via* both one-pot borylation/Suzuki reactions and Ni(0)-mediated Ullmann homocouplings.[32] In these examples, the yields and product distributions are comparable although the Ullmann benefits as the cheaper alternative to using a diboron reagent.

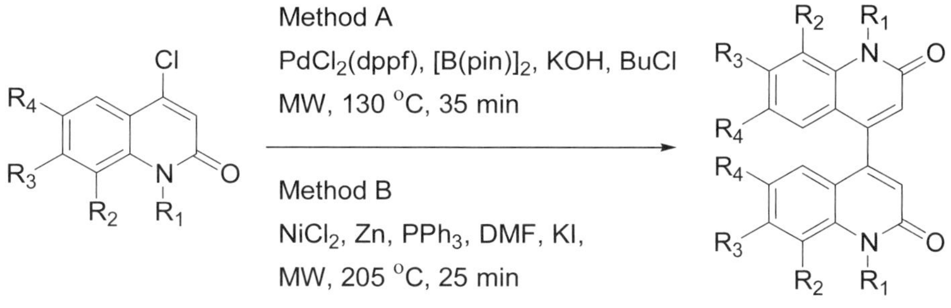

35 **36**

Substrate		Yield %	
		Method A	Method B
xa	R_1 = Me, R_2 = R_3 = R_4 = H	85	90
xb	R_1-R_2 = $(CH_2)_3$, R_3 = R_4 = H	68	68
xc	R_1 = Ph, R_2 = R_3 = R_4 = H	83	39
xd	R_1 = Me, R_2 = R_3 = H, R_4 = OMe	83	70
xe	R_1 = Me, R_2 = R_4 = H, R_3 = OMe	70	74

1.2.3.6 Experimental

Asymmetric Ullmann Coupling Using Copper

Synthesis of (*S*)-[6,6′-(2*R*,4*R*-Pentadioxyl)]-(2,2′)-bis(diphenylphosphoryl)-(1,1′)-biphenyl (38)[33]

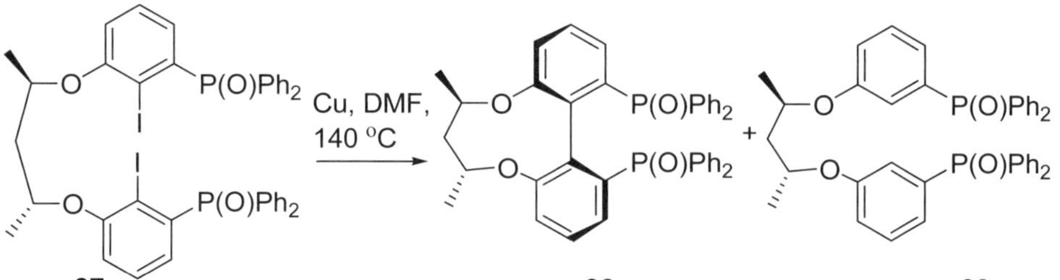

37 **38** 70% **39** 18%

DMF (5 mL) was added into a flask containing Cu powder (0.215 g, 3.36 mmol) and 37 (0.382 g, 0.42 mmol). The resulting mixture was stirred at 140 °C for 12 h under a nitrogen atmosphere. After removal of the DMF solvent under reduced pressure, the residue was boiled for 5 min. with hot CH_2Cl_2 (10 mL × 3). The insoluble solid was removed by filtration and washed with hot CH_2Cl_2 (5 mL × 3). The combined filtrate was washed successively with saturated aqueous ammonium chloride and brine and was dried over anhydrous Na_2SO_4. After removal of the solvent, the residue was purified by silica gel chromatography to give (*SRR*)-**38** as a white solid (194 mg, 0.296 mmol, 70.5%) and a recovered compound **39** (50 mg, 0.076 mmol, 18.1%).

Intramolecular Ullmann Coupling Using Palladium

5,7-Dihydro-dibenzo[a,c]cycloheptane-6,6-dicarboxylic acid diethyl ester 41[34]

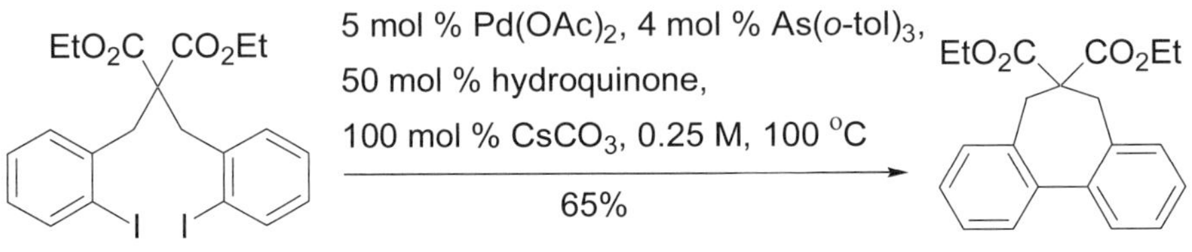

40 **41**

To a mixture of **40** (592 mg, 1 mmol), hydroquinone (55 mg, 0.5 mmol) and Cs_2CO_3 (0.326 g, 1 mmol) was added a homogenous pre-stirred DMA solution (2.50 mL) of $Pd(OAc)_2$ (11.2 mg, 0.05 mmol) and tri-*o*-tolylarsine (17.4 mg, 0.05 mmol). The reaction mixture darkened immediately upon addition of the catalyst solution to the solid reagents. The mixture was degassed, using N_2 and house vacuum, and heated under N_2 at 100 °C for 24 h. The reaction mixture was cooled to room temperature, quenched with HCl (20 mL, 2 M), diluted with water (20 mL), and extracted with EtOAc (3 × 25 mL). The combined organics were washed with 10% NaOH (4 × 20 mL), brine, dried over $MgSO_4$, concentrated under reduced pressure, and purified using flash column chromatography on silica gel (elution with hexanes:toluene:ether 30:10:1) to afford biaryl **41** as a colourless oil (221 mg, 65%) and recovered **40** (81 mg, 14%).

1.2.3.7 References

1. [R] Frlan, R.; Kikelj, D. *Synthesis* **2006**, *14*, 2271–2285.
2. [R] Sawyer, J. S. *Tetrahedron* **2000**, *56*, 5045–5065.
3. [R] Liang, Y.; Li, J-H. *Youji Huaxue* **2005**, *25*, 147–151.
4. [R] Beletskaya, I. P.; Cheprakov, A. V. *Coordination Chem. Rev.* **2004**, *248*, 2337–2364.
5. Ullmann, F.; Bielecki, J. *Chem. Ber.* **1901**, *34*, 2174–2178.
6. Ullmann, F. *Chem. Ber.* **1906**, *39*, 1693–1696.
7. Ullmann, F. *Chem. Ber.* **1905**, *38*, 2111–2112.
8. Ullmann, F. *Chem. Ber.* **1903**, *36*, 2382–2384.
9. (a) Graebe. C.; Ullmann, F. *Ann.* **1896**, *16*, 291; (b) Ullmann, F. *Ann.* **1904**, *82*, 332.
10. Jourdan, F. *Chem. Ber.* **1885**, *18*, 1444.
11. Goldberg, I. *Chem. Ber.* **1906**, *39*, 1691.
12. [R] Nelson, T. D.; Crouch, R. D. *Org. React.* **2004**, *63*, 265–555.

13. [R] Hassan, J.; Sévignon, M.; Gozzi, C.; Schulz, E.; Lemaire, M. *Chem. Rev.* **2002**, *102*, 1359–1469.
14. [R] Fanta, P. E. *Chem. Rev.* **1964**, 613–632.
15. [R] Fanta, P. E. *Synthesis* **1974**, *1*, 9–21.
16. Nicolaou, K. C.; Li, H.; Nold, A. L.; Pappo, D.; Lenzen, A. *J. Am. Chem. Soc.* **2007**, *129*, 10356–10357.
17. Miyazaki, T.; Yokoshima, S.; Simizu, S.; Osada, H.; Tokuyama, H.; Fukuyama, T. *Org. Lett.* **2007**, *9*, 4737–4740.
18. Shao, Y.; Ding, H.; Tang, W.; Lou, L.; Hu, L. *Bioorg Med. Chem.* **2007**, *15*, 5061–5075.
19. Li, W.; Shao, Y.; Hu, L.; Zhang, X.; Chen, Y.; Tong, L.; Li, C.; Shen, X.; Ding, J. *Cancer Biol. Ther.* **2007**, *6*, 787–794.
20. Harvey, M. J.; Banwell, M. G.; Lupton, D. W. *Tetrahedron Lett.* **2008**, *49*, 4780–4783.
21. Hauser, F. M.; Gauuan, P. J. F. *Org. Lett.* **1999**, *1*, 671–672.
22. You, T-p.; Chen, Q-a.; Lin, S-z. *Synlett.* **2007**, *13*, 2101–2105.
23. Broady, S. D.; Golden, M. D.; Leonard, J.; Muir, J. C.; Maudet, M. *Tetrahedron Lett.* **2007**, *48*, 4627–4630.
24. Ziegler F. E.; Chiliwner, I.; Fowler, K. W.; Kanfer, S. J.; Kuo, S. J.; Sinha, N. D. *J. Am. Chem. Soc.* **1980**, *102*, 790–798.
25. Ziegler, F. E.; Fowler, K. W.; Rodgers, W. B.; Wester, R. T. *Org. Synth.* **1987**, *65*, 108–118.
26. Stark, L. M.; Lin, X.-F.; Flippin, L. A. *J. Org. Chem.* **2000**, *65*, 3227–3230.
27. Markey, M. D.; Fu, Y.; Kelly, T. R. *Org. Lett.* **2007**, *9*, 3255–3257.
28. Sainsbury, M. *Tetrahedron* **1980**, *36*, 3327–3359.
29. Stark, L. M.; Lin, X.-F.; Flippin, L. A. *J. Org. Chem.* **2000**, *65*, 3227–3230.
30. Montoya-Pelaez, P. J.; Uh, Y-S.; Lata, C.; Thompson, M. P.; Lemieux, R. P.; Crudden, C. M. *J. Org. Chem.* **2006**, *71*, 5921–5929.
31. Cravotto, G.; Beggiato, M.; Penoni, A.; Palmisano, G.; Tollari, S.; Leveque, J-M.; Bonrath, W. *Tetrahedron Lett.* **2005**, *46*, 2267–2271.
32. Hashim, J.; Glasnov, T. N.; Kremsner, J. M.; Kappe, C. O. *J. Org. Chem.* **2006**, *71*, 1707–1710.
33. Qui, L.; Kwong, F. Y.; Wu, J.; Wai, H. L.; Chan, S.; Yu, W.-Y.; Li, Y.-M.; Guo, R.; Zhou, Z.; Chan, A. S. C. *J Am. Chem. Soc.* **2006**, *128*, 5955–5965.
34. Hennings, D. D.; Iwama, T.; Rawal, V. H. *Org. Lett.* **1999**, *1*, 1205–1208.

1.3.1 McMurry Coupling

Brian Goess

1.3.1.1 Description

Alkenes may be generated via the intra- or intermolecular reductive coupling of carbonyl compounds in a titanium-mediated process known as the McMurry coupling.[1]

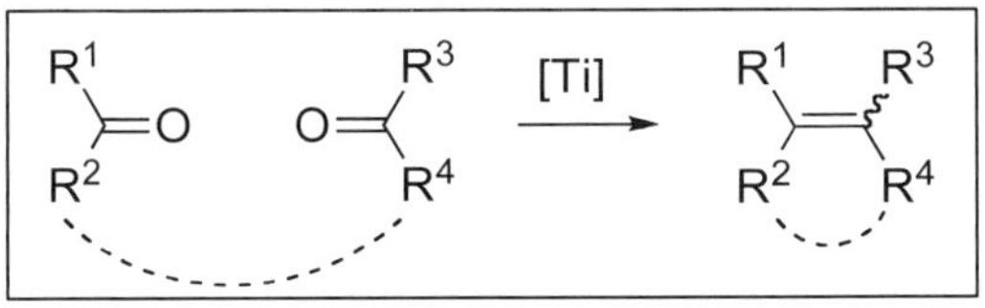

The reaction is a versatile strategy for carbon–carbon bond construction as evidenced by the large number of natural and non-natural compounds that have been synthesized using the McMurry reaction as a key step. Furthermore, the reaction has stimulated a significant number of theoretical studies, including examinations of the unique and often highly-strained molecules that can be prepared using the McMurry reaction and of the mechanism of the reaction itself.

1.3.1.2 Historical Perspective

A titanium-induced carbonyl coupling reaction was discovered serendipitously by the McMurry group in 1974 during their search for a new, high-yielding reduction of α,β-unsaturated carbonyl compounds such as **1** to the corresponding alkene (**2**).[2] Their plan was to conduct a hydride reduction of the carbonyl in the presence of an oxophilic transition metal, which they had hoped might coordinate strongly to the alkoxide intermediate and facilitate a second hydride reduction to yield the deoxygenated product (**2**). However, their experimental conditions, which employed $LiAlH_4$ as the strong hydride donor and $TiCl_3$ as the oxophile, instead generated the product of reductive dimerization (**3**).

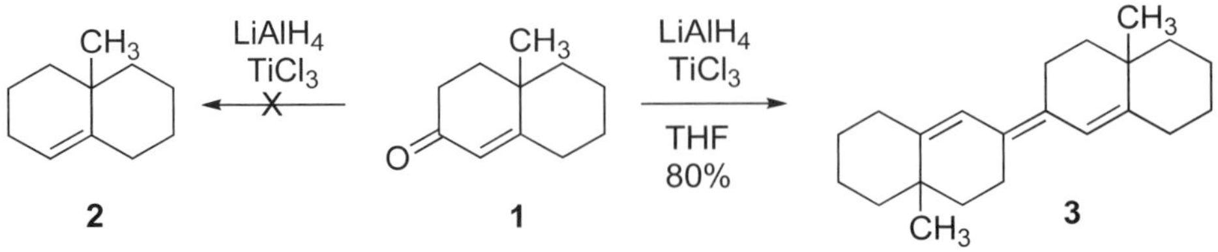

Two other groups made a similar discovery almost simultaneously,[3,4] and the potential value of these transformations was immediately apparent to the synthetic community. Early investigations into substrate scope led to the discovery that the reductive coupling was successful for a wide range of aldehydes and ketones, including saturated and unsaturated ketones and aldehydes, aryl ketones and aldehydes, and diaryl ketones.[5] In these early stages of reaction development, there were few techniques for overcoming the problems inherent with achieving selective reductive heterocoupling of two different carbonyl-containing compounds. Therefore, acyclic products were limited to symmetrical alkenes, although this limitation was overcome in specific cases when an excess of one reaction partner was used.[5]

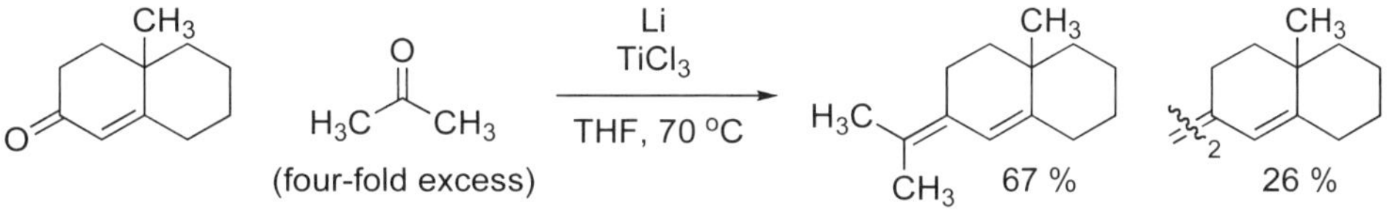

If both carbonyl substrates have sufficiently different redox potentials, a reasonable yield of crossed product is obtained without resorting to using one substrate in excess. This result was unexpected given the relative ease that a bisaromatic ketone should undergo ketyl formation followed by self pinacol coupling and was one of the first results that indicated more than one mechanism may be operative in McMurry couplings.[5] In this case, the diaryl ketone may be reduced selectively to a dianion, which then undergoes nucleophilic addition to the saturated ketone.

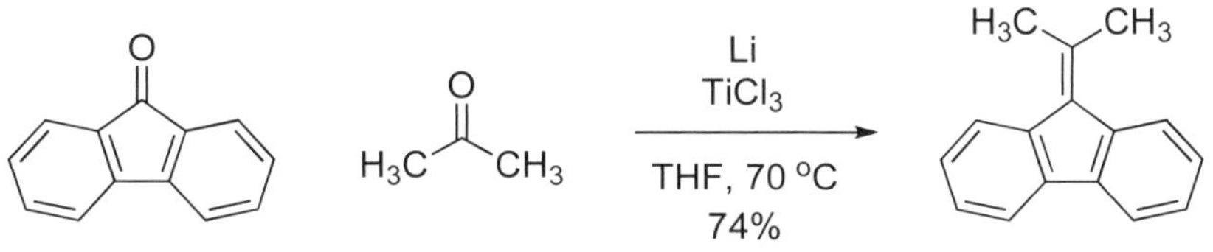

A more recent study has provided a strategy for heterocoupling structurally similar aryl or diaryl ketones.[6] Heteroatom directing groups are appended to one of the ketone reaction partners, thus giving it an enhanced affinity for the titanium surface on which the reaction proceeds. Binding of these substrates deactivates the titanium surface and retards the corresponding homocoupling reaction. The result is increased selectivity for the heterocoupled product (**4**). Removal of the directing group can then lead to more symmetrical heterocoupled products (**5**).

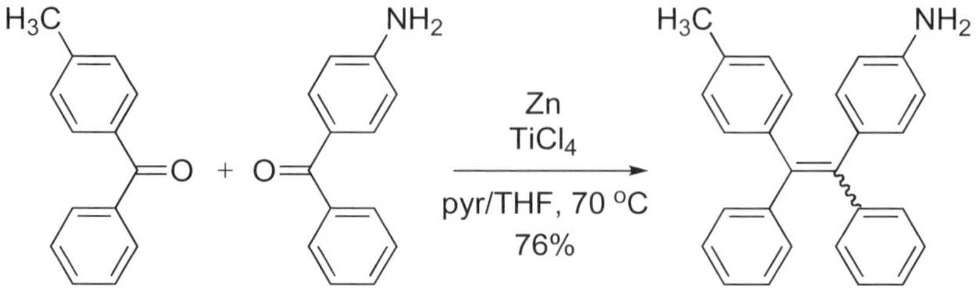

4

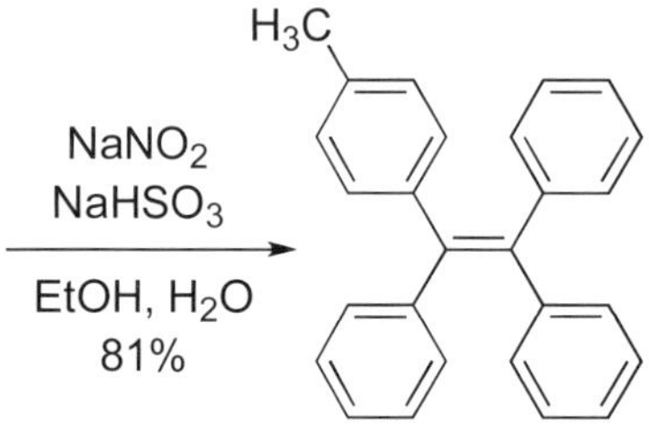

5

In cases where the formation of alkene geometrical isomers was possible the more stable isomer usually predominated. Notably, it was demonstrated that the stereoselectivity was kinetically-controlled; no alkene isomerization was observed when single geometrical isomers were resubmitted to the reaction conditions.[2]

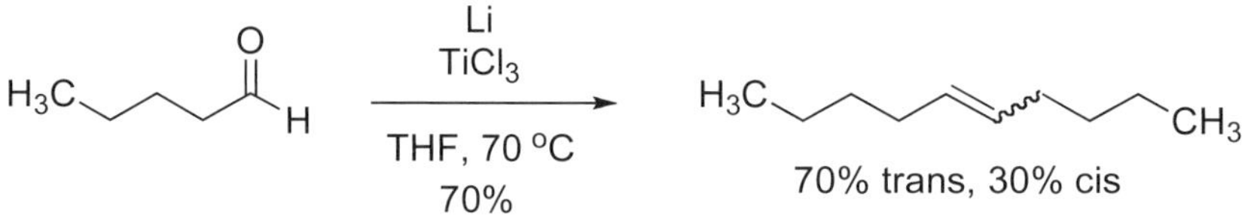

The robust nature of the coupling reaction was repeatedly demonstrated through the synthesis of unusual ring structures as exemplified by the key bond-forming step of Hopf's synthesis of triply-bridged cyclophanes.[7]

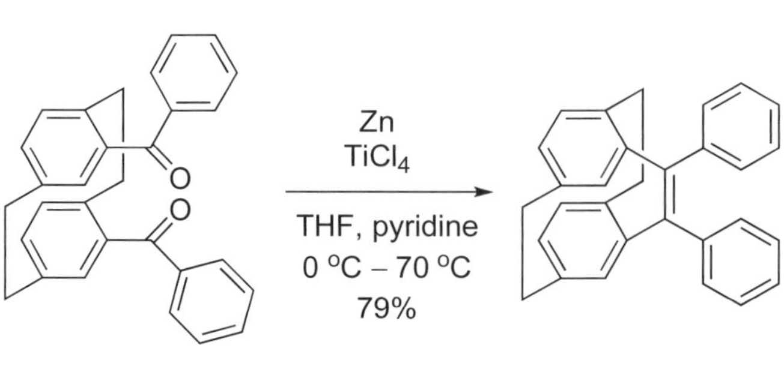

These examples also demonstrate another hallmark of the development of the McMurry coupling, namely the ongoing search for the best source of low-valent metal for reductive couplings. Early efforts by McMurry and Geise[8] to substitute other transition metals for titanium failed to uncover a suitable replacement. Various sources of low-valent titanium were also intensively investigated. McMurry's initial combination of $LiAlH_4/TiCl_3$ proved difficult to reproduce consistently. He later reported improved variations on this reagent combination including $K/TiCl_3$,[9] $Li/TiCl_3$,[10] and $Zn–Cu/TiCl_3(DME)_{1.5}$.[11] In each of these cases, the reaction was performed in two steps wherein the titanium chloride and reducing agent were combined prior to introduction of the carbonyl-containing substrate. A breakthrough in the convenience of low-valent titanium preparation came in 1994 when Fürstner described a single-step "instant" method for the purpose of indole synthesis.[12] Here, the active titanium species is generated upon coordination of the substrate carbonyl to $TiCl_3$ followed by reduction of this complex in situ with zinc dust. This one-step procedure is effective only when the reducing agent is not strong enough to reduce the carbonyl. Using this methodology, indole (**7**) was prepared from oxo amide (**6**) in high yield. The method is also applicable to furans[13] and a number of other heterocycles.

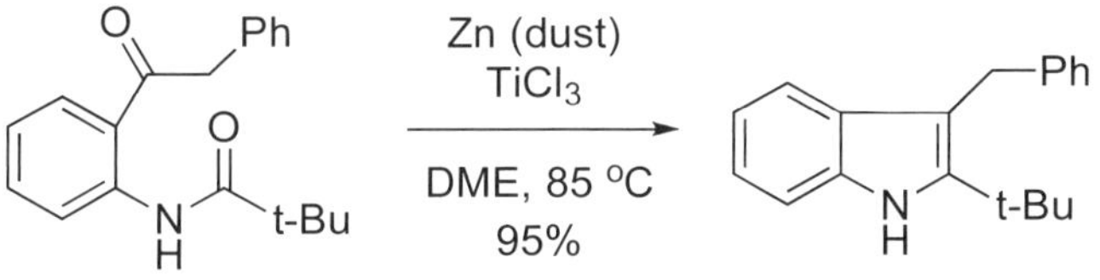

6 **7**

Importantly, this method was shown to be effective for many conventional McMurry reactions as well, and a catalytic variant, using commercially-available titanium powder, followed one year later.[14] The catalytic reaction relied on an admixed chlorosilane, which both activated the commercial titanium powder by destroying the tightly bound oxide layer and regenerated, via ligand exchange, the active titanium chloride from the inactive titanium oxychloride product.

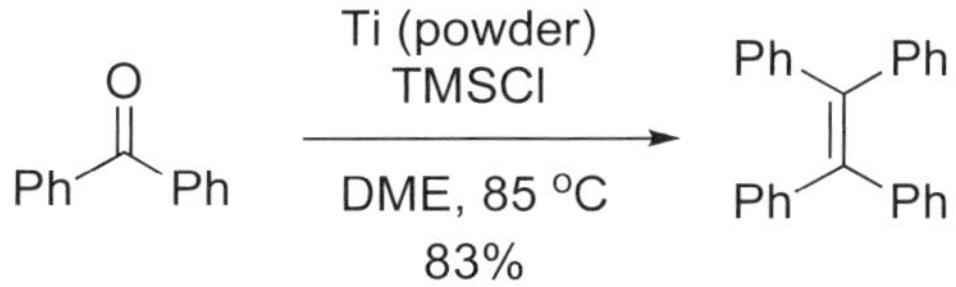

The high temperatures required for most McMurry reactions have prompted recent studies into activating the low-valent titanium reagents for use in room temperature couplings with correspondingly increased functional group tolerance. For instance, addition of alkaline or alkaline earth metal salts was found to enhance the activity of the activated titanium reagent.[15] The authors suggest that exchange of lithium for more electropositive metal cations such as potassium increases the electron density on titanium, which may lead to higher activity for the titanium intermetallic complex.

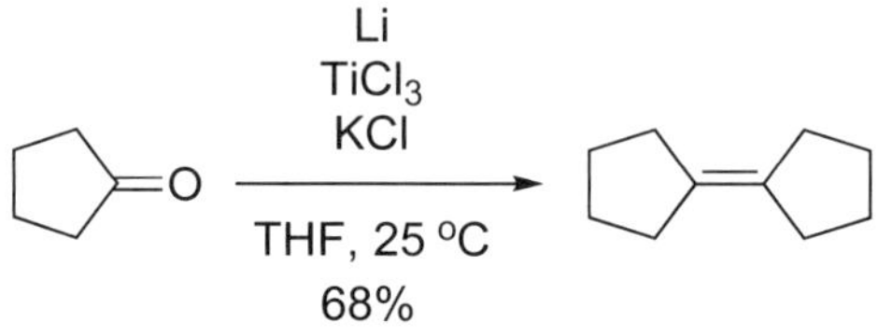

Microwave heating was found to accelerate reaction times for conventional McMurry reactions, which ordinarily require refluxing for more than two hours.[16]

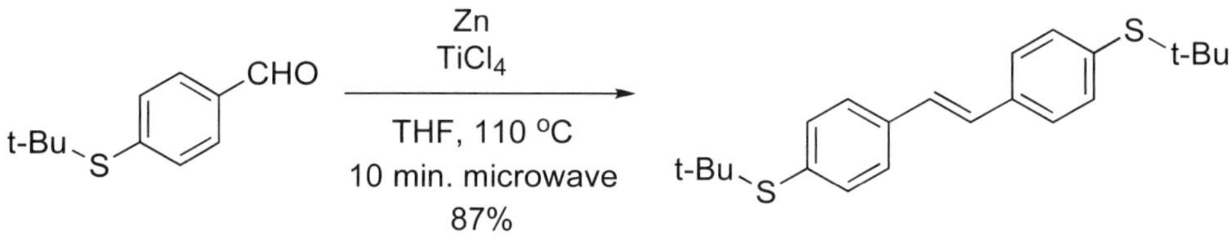

Sonication allows coupling aromatic aldehydes and ketones at lower temperatures.[17]

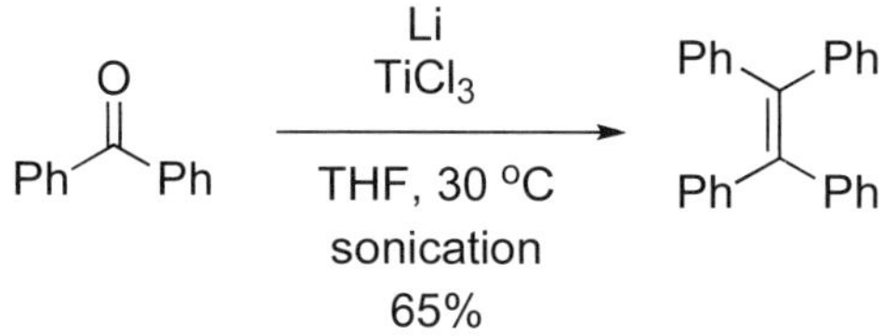

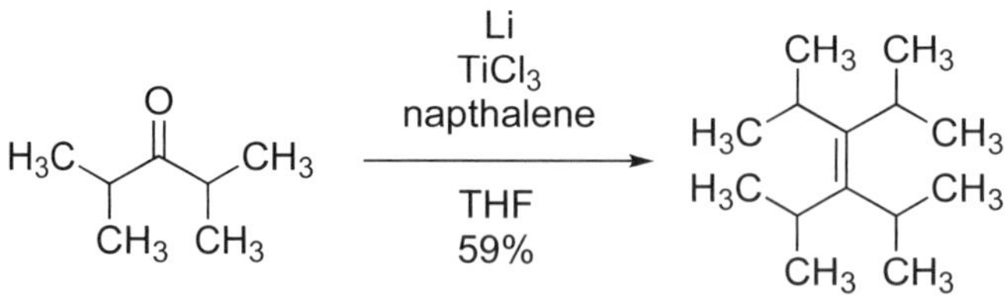

And the use of metal-arenes as single-electron reducing agents for $TiCl_3$ allows the McMurry reaction between aromatic and aliphatic aldehydes and ketones to proceed at room temperature.[18]

1.3.1.3 Mechanism

The precise mechanism of the McMurry coupling is not known, though the details have been the subject of intense speculation since the discovery of the reaction. The reagent combinations used to perform McMurry couplings result in heterogeneous solutions, thus complicating the determination of both the structure of the reaction intermediates and of the oxidation states of titanium, which vary as the reaction progresses. Nonetheless, in recent years significant progress has been made in determining the identity of early reaction intermediates, and the results of these studies point to a context-dependent mechanism accommodating elements from a variety of mechanisms proposed since the early 1970s.

McMurry's original mechanistic proposal involved an initial pinacol coupling followed by a deoxygenation step.[2a] The pinacol step, involving a one-electron reduction from a titanium complex to generate a ketyl (**8**), followed by a carbon–carbon bond-forming dimerization to yield an intermediate pinacolate dianion (**9**), had significant circumstantial support. Reducing metals were known to react with aldehydes and ketones to generate radical anions that subsequently dimerized to form pinacols. Furthermore, pinacols could be isolated in high yields as products if the reaction mixtures were hydrolyzed before completion at temperatures well below reflux.[2]

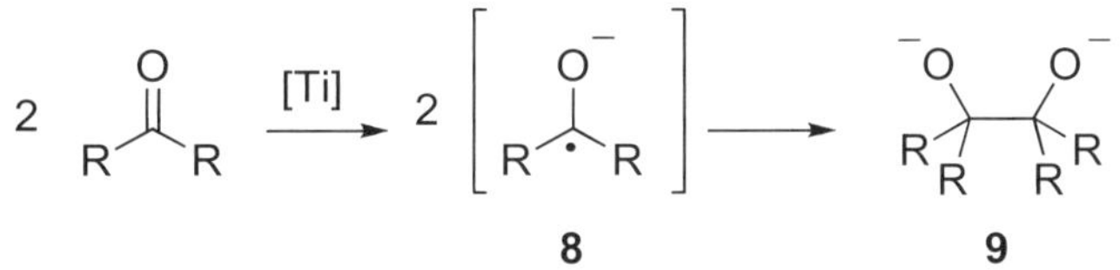

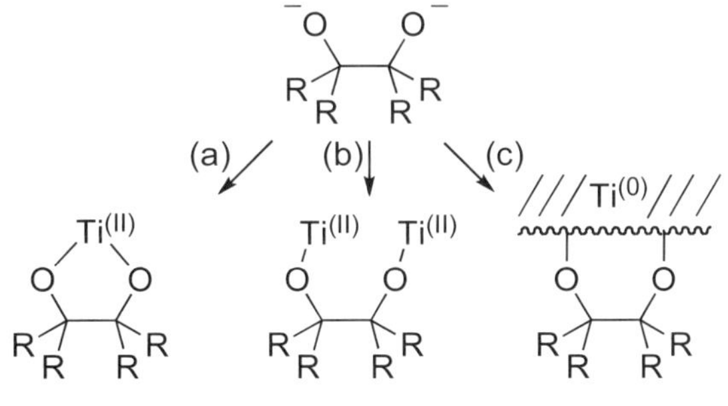

The deoxygenation step was both unprecedented at the time and unique to titanium, thus its mechanism was more challenging to probe. McMurry postulated three general mechanisms: (a) titanacycle formation, (b) an acyclic variant with two coordinated titaniums, and (c) a hybrid mechanism where both oxyanions coordinate to the surface of a finely divided titanium particle presumably formed from the reduction of $TiCl_3$.[5] In each case, cleavage of the two carbon-oxygen bonds then occurs via a stepwise radical process to yield the product alkene and a titanium oxide. That this cleavage must be non-concerted was demonstrated when configurationally pure diols were found to yield mixtures of *E* and *Z* alkenes when treated under the reaction conditions.[5]

Two mechanistic probes were designed to distinguish the three pinacol pathways. When it was discovered that both *cis*-camphanediol (**10**), which can easily form a 5-membered metallacycle, and *trans*-camphanediol (**11**), which cannot, reacted at a similar rate to yield camphene, pathway (a) was ruled out.

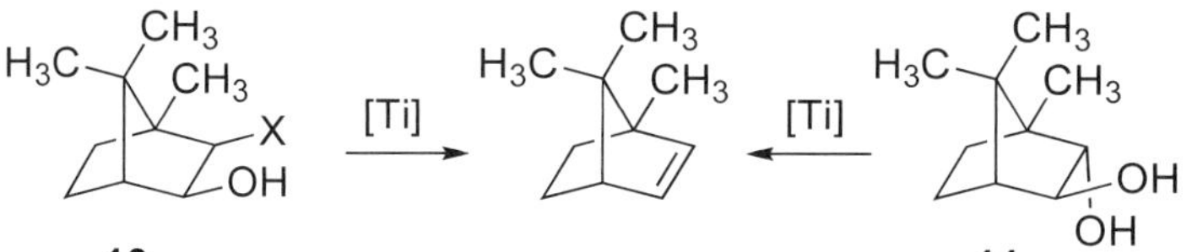

10 **11**

When it was discovered that *trans*-9,10-decalindiol (**13**) was inert to titanium under all conditions whereas *cis*-9,10-decalindiol (**12**) reacted as expected, pathway (b) was ruled out in favor of a pathway that required a common titanium surface for reaction.

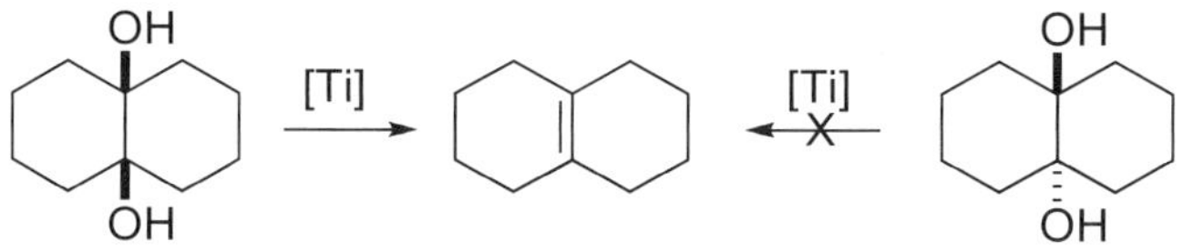

12 **13**

This left pathway (c) as the only reasonable alternative, and this mechanistic proposal received early independent experimental support.[19] The templating effect inherent in pathway (c) also nicely explained the unusual ease with which large rings could be formed using the McMurry protocol.[20] It was later suggested that coordination to the titanium surface might instead occur at the ketyl stage, in which case the carbon-carbon bond formation would also be facilitated by templating of the ketyl radicals.[19]

More recent studies have demonstrated that Ti(0) is not required for the McMurry reaction to proceed. Indeed, the assumption that zero-valent titanium is prerequisite for the McMurry reaction persisted even in spite of early evidence that pinacol couplings could be performed with well-characterized Ti(II) species and without intervention of Ti(0) intermediates.[21] Furthermore, the reductive coupling of gas-phase benzaldehyde was effected on TiO_2 surfaces on which X-ray photoelectron spectroscopy detected no Ti(0).[22] Finally, when the McMurry reagent was prepared by one of the traditional methods, Ti(II) chlorohydride complex **14** (one possible structure shown) was produced according to the following equation:[23]

$$[TiCl_3(THF)_3] + 0.5\ LiAlH_4 \longrightarrow 0.5\ H_2 + 0.5\ LiCl + 0.5\ AlCl_3 + \mathbf{14}$$

14 = 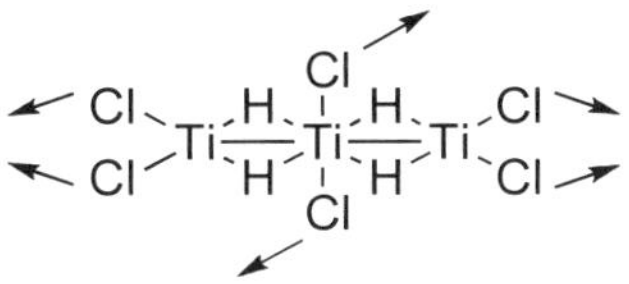

Complex **14** proved to be active in the reductive coupling of benzophenone according to the following equation, yielding a Ti(III) oxychloride by-product:

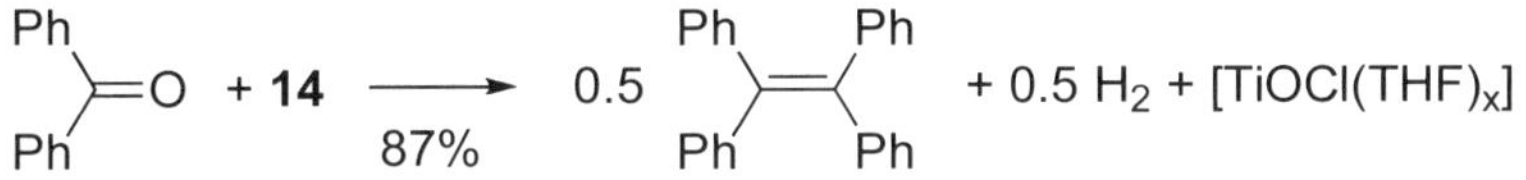

These results imply that the McMurry reaction, when performed with one of the most commonly applied reagent sets, $TiCl_3$–$LiAlH_4$–THF, involves changes in the oxidation state of titanium only between Ti^{1+}, Ti^{2+} and Ti^{3+}. The necessity of finely divided Ti(0) seems to be precluded by these results.[24]

The established mechanism for pinacolate formation has also been challenged by an analysis of the reaction between acetophenone and $TiCl_3$(DME)–Zn(Cu).[25] Aliquots quenched at various stages of the reaction revealed products from which a nucleophilic addition mechanism could be inferred. DFT calculations indicated that this reaction pathway is more energetically favorable than the corresponding ketyl pathway.

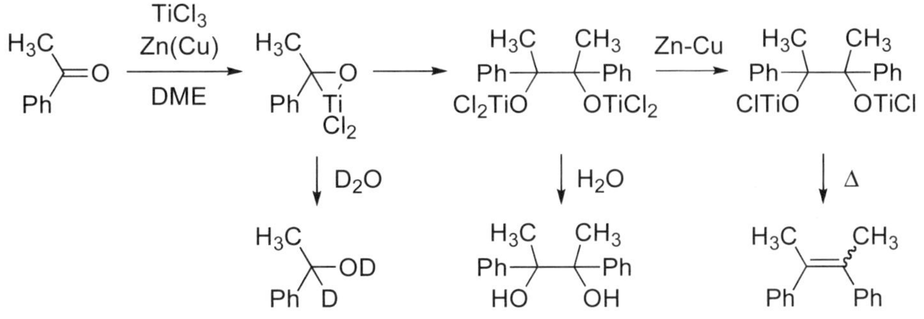

When sterically hindered systems are investigated, evidence has been found for a mechanistic pathway that proceeds not through pinacolate intermediates but through carbenes.[26] In the reaction between diisopropyl ketone and $TiCl_4$–Li(Hg) two products were observed, the expected coupling product (**15**) and a product (**16**) that could only be explained through the existence of a carbenoid intermediate. Data supporting this mechanistic hypothesis include the lack of pinacol product and the observation that the titanium pinacolate intermediate is unstable and readily decomposes to carbonyl and titanium chloride. Similar evidence using di-*tert*-buytl ketone has also been obtained.[27]

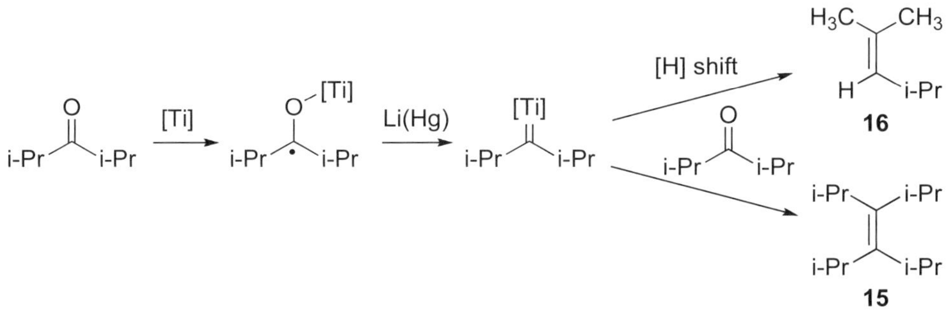

1.3.1.4 Variations, Improvements and Modifications

As reflux temperatures are usually required to effect the deoxygenation step of the McMurry reaction, pinacol intermediates can be isolated if the McMurry reaction is run below room temperature. Though a mixture of diol diastereomers is formed in intermolecular reactions, intramolecular reactions can be highly diastereoselective.[28]

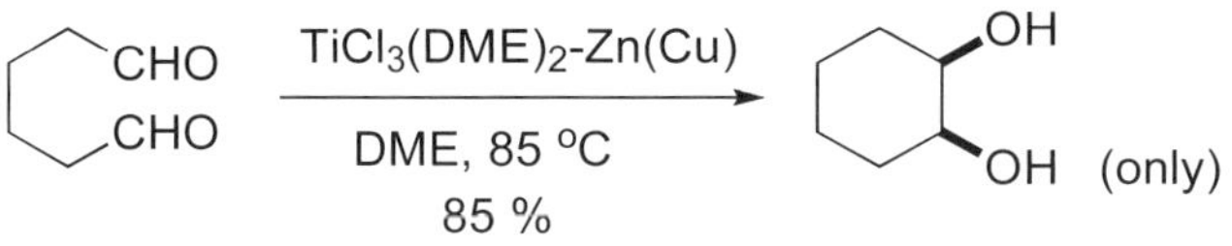

A catalytic version of this reaction has been developed, but diastereoselectivities are low to moderate.[29] Diastereoselectivities for catalytic pinacol coupling of aromatic aldehydes are usually much higher.[30] Recent work on catalytic, enantioselective aromatic aldehyde pinacol couplings have produced enantioselectivities above 90%.[31]

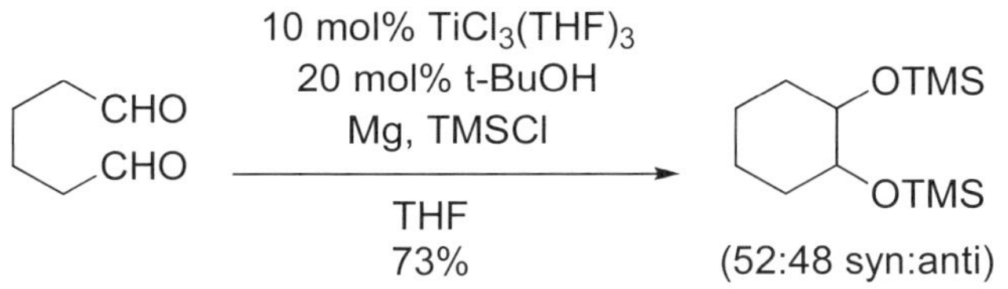

This reaction has been extended to the synthesis of vicinal arylamines,[32] and a one-pot reductive coupling of aryl aldehydes and amines has also been reported.[33]

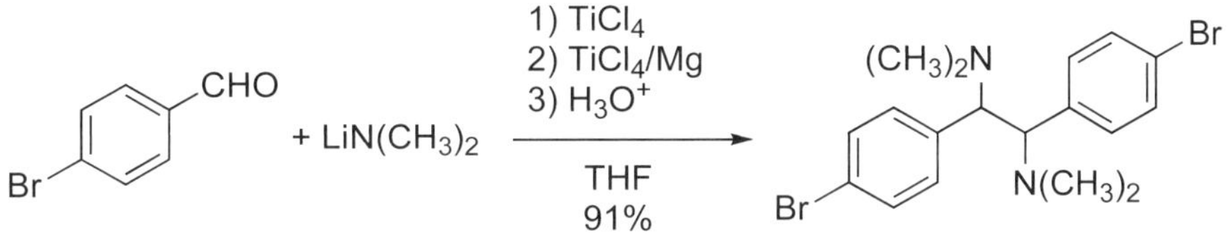

One of the earliest variations on the traditional McMurry coupling is the McMurry keto ester coupling for the synthesis of enol ethers. Large rings can be prepared by this method, albeit in lower yields, and the cyclic enol ether can be hydrolyzed to yield a cyclic alkanone.[34]

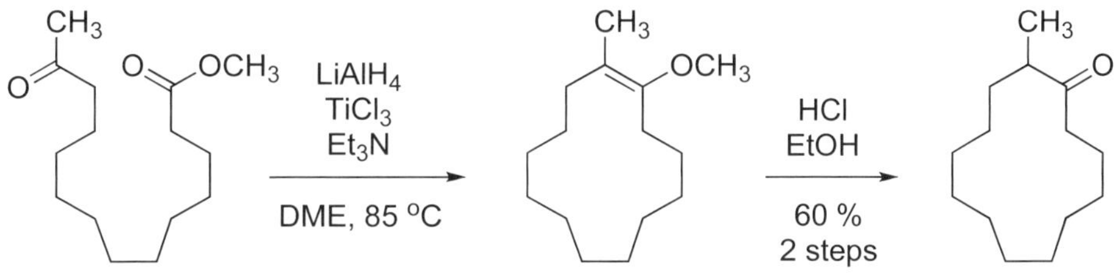

The titanium-mediated deoxygenation step has also been made more general. For example, allylic and benzylic 1,4-diols and 1,4-dimethyl ethers are transformed into more highly unsaturated products under McMurry conditions.[35]

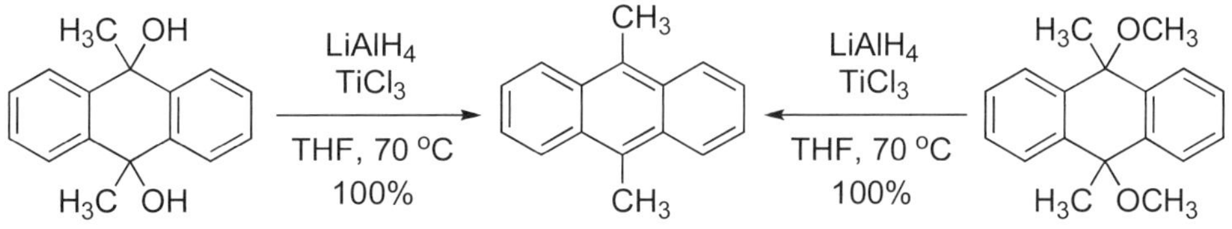

Furthermore, epoxides[36] and halohydrins[37] are deoxygenated to alkenes, nitro compounds[38] and oximes[39] are reduced to imines, and sulfoxides are reduced to sulfides.[40]

McMurry conditions have been used to self-couple allylic and benzylic alcohols, though this process has not been as successful in the intramolecular synthesis of large rings.[41]

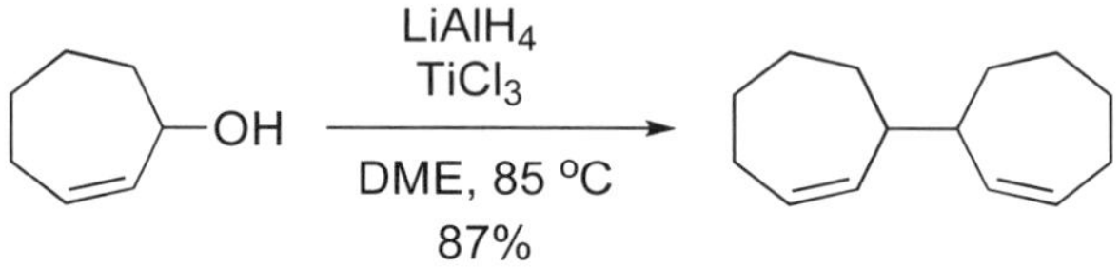

1.3.1.5 Synthetic Utility

In addition to the numerous examples of the McMurry couplings described above, the following examples illustrate the power of the transformation in the synthesis of a broad range of natural products and non-natural molecules. Of particular note in the following examples is the impressive functional group tolerance of the McMurry conditions. In general, alcohols, tosyl alcohols, alkyl ethers, silyl ethers, alkyl silanes, vinyl silanes, amines, sulfides, and alkenes are inert to McMurry conditions. Acteals, halides, alkynes, nitriles, and carboxylic acids are semi-compatible.

In his recent synthesis of 19-hydroxytaxoid **18**, an intermediate in the synthesis of the anti-cancer agent 19-hydroxytaxol, Mukaiyama performed an intermolecular McMurry pinacol coupling on diketone **17**.[42] Note that the reaction temperature was sufficiently low to allow isolation of the diol intermediate.

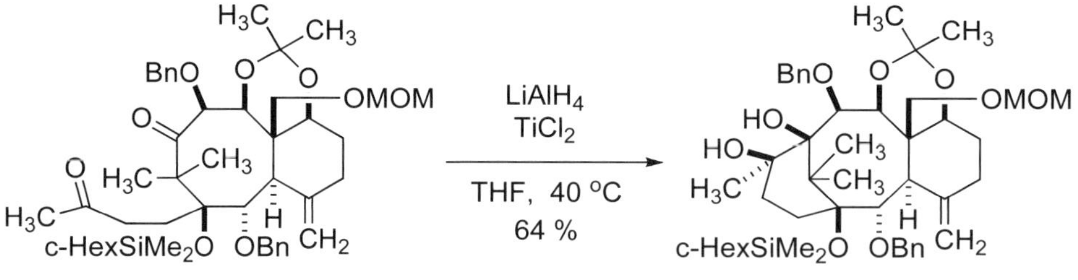

17 **18**

McMurry pinacol couplings can proceed with stereochemical induction from nearby stereogenic carbons. For example, diastereomer **21** was the major heterocoupled stereoisomer observed when aldehyde **19** and ketone **20** were treated under McMurry conditions.[43] A diastereomer of **21** was also produced in 22% yield along with the product of homocoupling of

19. This route provides access to 2-*C*-trifluoromethyl-substituted-*D*-ribose (**22**).

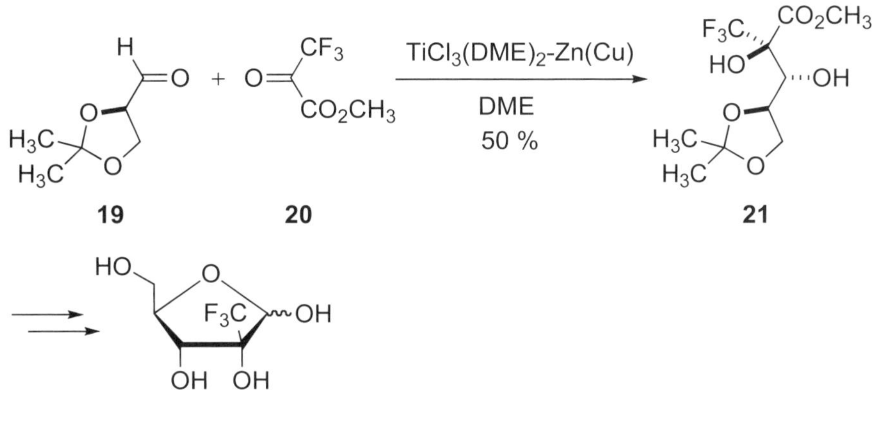

22

Calogeropoulou took advantage of the stereoselectivity of the McMurry coupling to prepare **(*E*)-23**, a synthetic intermediate en route to (*Z*)-4-hydroxytamoxifen (**24**), the active metabolite of the anti-cancer drug Tamoxifen.[44] A three-fold excess of propiophenone was used to account for unproductive reductive homocoupling.

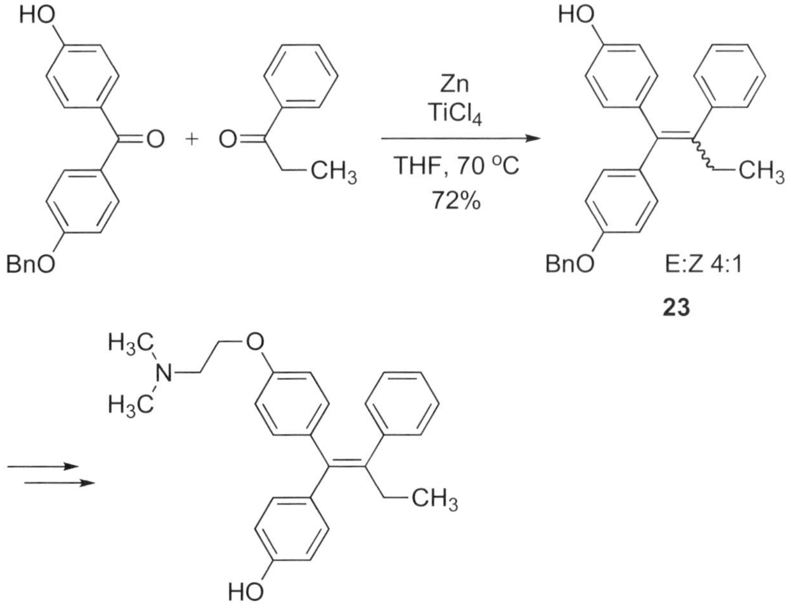

24

The use of the McMurry reaction to form large rings was aptly demonstrated by Eguchi in his synthesis of this 36-membered ring-containing precursor to an archaebacterial diether lipid (**25**).[45]

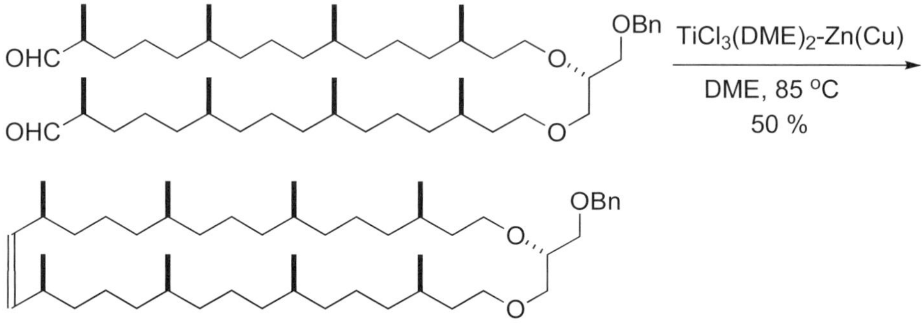

25

The McMurry reaction is often employed in the synthesis of highly-strained alkenes of theoretical interest. For instance, the rotational barrier of the cyclohexyl groups in **26** is 18.7 kcal/mol as determined by dynamic NMR.[46]

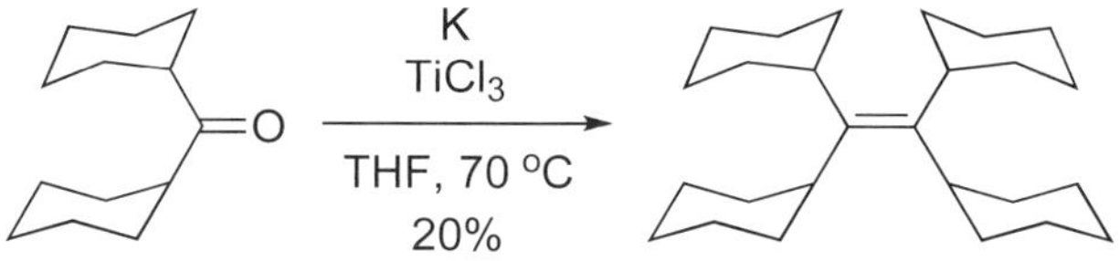

26

Spherand **27** has an inner surface of high electron density and was found to bind silver(I) ions in a static, square-planar, d^{10} organometallic complex.[47] Receptor **28** was found to bind a single silver(I) ion which hops intramolecularly between the two adjoined cavities.[48]

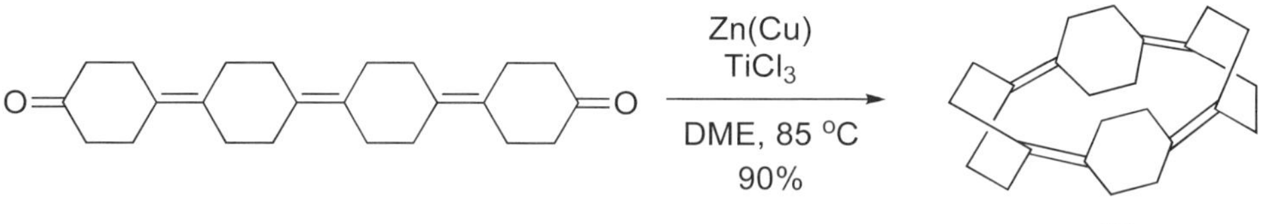

27

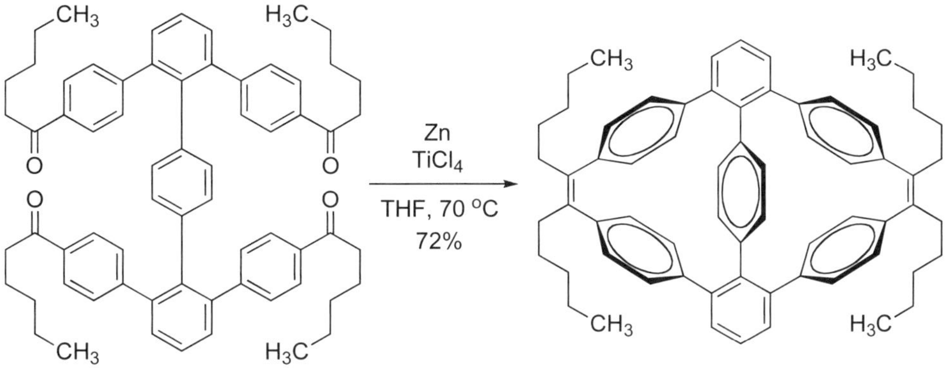

28 **29**

1.3.1.6 Experimental

The examples presented illustrate two of the common ways McMurry reactions are run. The first uses the instant method and is intramolecular. The second uses a different order of addition and is intermolecular.

Indole (7).[12]

To a 100 mL two-necked flask, equipped with a Teflon-coated magnetic stirring bar and a reflux condenser connected to an argon line, was added keto-amide **6** (0.395 g, 1.34 mmol), $TiCl_3$ (0.772 g, 5.00 mmol), and zinc dust (0.654 g, 10.00 mmol). The mixture was suspended in DME (50 mL), refluxed for 92 h, cooled to room temperature, and filtered through a short plug of silica. The inorganic residues were washed with ethyl acetate (50 mL), the filtrate was evaporated, and the residue was purified by flash column chromatography using 10% ethyl acetate in *n*-hexanes to yield indole **7** as a viscous oil (0.334 g, 95 %).

Receptor (29).[48]

To a Schlenck flask containing chilled (~ 0 °C) anhydrous THF (100 mL) was added $TiCl_4$ (3 mL, 27 mmol) via a dropping funnel under an argon atmosphere. To this mixture was added Zn dust (2.2 g, 34 mmol) and dry pyridine (0.1 g, 1.3 mmol), and the resulting black suspension was warmed to room temperature and refluxed for 2 h. A solution of **28** (2.8 g, 3 mmol) in THF (200 mL) was added dropwise to the black reaction mixture over 4 h while refluxing, and the resulting mixture was refluxed for an additional 12 h. The reaction mixture was cooled to room temperature and quenched with 10 % aqueous K_2CO_3 (50 mL). The organic layer was separated and the aqueous suspension was extracted with dichloromethane (5 × 50 mL). The combined organic extracts were dried over anhydrous $MgSO_4$, filtered and evaporated to afford a syrupy liquid which was purified by flash chromatography on silica gel using 1: 9 mixture of ethyl acetate and hexanes to afford pure **29** as a crystalline solid (1.86 g, 72%).

1.3.1.7 References

1. a) [R] McMurry, J. E. *Chem. Rev.* **1989**, *89*, 1513–1524. b) [R] Lenoir, R. *Synthesis* **1989**, 883–897. c) [R] Fürstner, A.; Bogdanović, B. *Angew. Chem., Int. Ed.* **1996**, *35*, 2443–2469. d) [R] Fürstner, A. *Pure App. Chem.* **1998**, *70*, 1071–1076. e) [R] Ephritikhine, M. *Chem. Commun.* **1998**, *6*, 2549–2554. f) [R] Hirao, T. *Synlett.* **1999**, 175–181. g) [R] Ladipo, F. T. *Curr. Org. Chem.* **2006**, *10*, 965–980.
2. [R] a) McMurry, J. E.; Fleming, M. P. *J. Am. Chem. Soc.* **1974**, *96*, 4708–4709. [R] b) McMurry, J. E. *Acc. Chem. Res.* **1983**, *16*, 405–411.
3. Mukaiyma, T.; Sato, T.; Hanna, J. *Chem. Lett.* **1973**, 1041–1044.
4. Tyrlik, S.; Wolochowicz, I. *Bull. Soc. Chim. Fr.* **1973**, 2147–2148.
5. McMurry, J. E.; Fleming, M. P.; Kees, K. L.; Krepski, L. R. *J. Org. Chem.* **1978**, *43*, 3255–3266.
6. Duan, X-F.; Zeng, J.; Lü. J-W.; Zhang, Z-B. *J. Org. Chem.* **2006**, *71*, 9873–9876.
7. Hopf, H.; Mlynek, C. *J. Org. Chem.* **1990**, *55*, 1361–1363.
8. Dams, R.; Malinowski, M.; Geise, H. J. *Bull. Soc. Chim. Belg.* **1982**, *91*, 149–152.
9. McMurry, J. E.; Fleming, M. P. *J. Org. Chem.* **1976**, *41*, 896–897.
10. McMurry, J. E.; Krepski, L. R. *J. Org. Chem.* **1976**, *41*, 3929–3930.
11. McMurry, J. E.; Lectka, T.; Rico, J. G. *J. Org. Chem.* **1989**, *54*, 3748–3749.
12. Fürstner, A.; Hupperts, A.; Ptock, A.; Janssen, E. *J. Org. Chem.* **1994**, *59*, 5215–5229.
13. Fürstner, A.; Jumbam, D. N. *Tetrahedron* **1992**, *48*, 5991–6010.
14. Fürstner, A.; Hupperts, A. *J. Am. Chem. Soc.* **1995**, *117*, 4468–4475.
15. Rele, S.; Chattopadhyay, S.; Nayak, S. K. *Tetrahedron Lett.* **2001**, *42*, 9093–9095.
16. Stuhr-Hansen, N. *Tetrahedron Lett.* **2005**, *46*, 5491–5494.
17. Nayak, S. K.; Banerji, A. *J. Org. Chem.* **1991**, *56*, 1940–1942.
18. Rele, S.; Talukdar, S.; Banerji, A.; Chattopadhyay, S. *J. Org. Chem.* **2001**, *66*, 2990–2994.
19. Dams, R.; Malinowski, M.; Westdorp, I.; Geise, H. Y. *J. Org. Chem.* **1982,** *47*, 248–259.
20. McMurry, J. E.; Kees, K. L. *J. Org. Chem.* **1977**, *42*, 2655–2656.
21. Corey, E. J.; Danheiser, R. L.; Chandrasekaran, S. *J. Org. Chem.* **1976**, *41*, 260–265.
22. Idriss, H.; Pierce, K. G.; Barteau, M. A. *J. Am. Chem. Soc.* **1994**, *116*, 3063–3074.
23. Aleandri, L. E.; Becke, S.; Bogdanović, B.; Jones, D. J. *J. Organomet. Chem.* **1994**, *472*, 97–112.
24. Bogdanović, B.; Bolte, A. *J. Organomet. Chem.* **1995**, *502*, 109–121.
25. Stahl, M.; Pidun, U.; Frenking, G. *Angew. Chem., Int. Ed.* **1997**, *36*, 2234–2237.
26. Villiers, C.; Ephritikhine, M. *Angew. Chem., Int. Ed.* **1997**, *36*, 2380–2382.
27. Villiers, C.; Vandais, A.; Ephritikhine, M. *J. Organomet. Chem.* **2001**, *617–618*, 744–747.
28. McMurry, J. E.; Rico, J. G. *Tetrahedron Lett.* **1989**, *30*, 1169–1172.
29. Lipski, T. A.; Hilfiker, M. A.; Nelson, S. G. *J. Org. Chem.* **1997**, *62*, 4566–4567.
30. Bandini, M.; Cozzi, P. G.; Morganti, S.; Umani-Ronchi, A. *Tetrahedron Lett.* **1999**, *40*, 1997–2000.
31. a) Chatterjee, A.; Bennur, T.H.; Joshi, N. N. *J. Org. Chem.* **2003**, *68*, 5668–5671. b) Li, Y.-G.; Tian, Q.-S.; Zhao, J.; Feng, Y.; Li, M.-J.; You, T. *Tetrahedron: Asymmetry* **2004**, *15*, 1707–1710.
32. Mangeney, P.; Grojean, F.; Alexakis, A.; Normant, J. F. *Tetrahedron Lett.* **1988**, *29*, 2675–2676.
33. Betschart, C.; Schmidt, B.; Seebach, D. *Helv. Chim. Acta* **1988**, *71*, 1999–2021.
34. McMurry, J. E.; Miller, D. D. *J. Am. Chem. Soc.* **1983**, *105*, 1660–1661.
35. Walborsky, H. M.; Wüst, H. H. *J. Am. Chem. Soc.* **1982**, *104*, 5807–5808.
36. McMurry, J. E.; Fleming, M. P. *J. Org. Chem.* **1975**, *40*, 2555–2556.
37. McMurry, J. E.; Hoz, T. *J. Org. Chem.* **1975**, *40*, 3797–3798.

38. McMurry, J. E.; Melton, J. *J. Org. Chem.* **1973**, *38*, 4367–4373.
39. Timms, G. H.; Wildsmith, E. *Tetrahedron Lett.* **1971**, *2*, 195–198.
40. Ho, T-L.; Wong, C. M. *Synth. Commun.* **1973**, *3*, 37–38.
41. McMurry, J. E.; Silvestri, M. *J. Org. Chem.* **1975**, *40*, 2687–2688.
42. Mukaiyama, T.; Ogawa, Y.; Kuroda, K.; Matsuo, J.-I. *Chem. Lett.* **2004**, *33*, 1412–1413.
43. Eilitz, U.; Böttcher, C.; Sieler, J.; Gockel, S.; Haas, A.; Burger, K. *Tetrahedron*, **2001**, *57*, 3921–3925.
44. Detsi, A.; Koufaki, M.; Calogeropoulou, T. *J. Org. Chem.* **2002**, *67*, 4608–4611.
45. Eguchi, T.; Terachi, T.; Kakinuma, K. *J. Chem. Soc. Chem. Comm.* **1994**, *2*, 137–138.
46. Columbus, I.; Biali, S. *J. Org. Chem.* **1994**, *59*, 3402–3407.
47. McMurry, J. E.; Haley, G. J.; Matz, J. R.; Clardy, J. C.; Mitchell, J. *J. Am. Chem. Soc.* **1986**, *108*, 515–516.
48. Emond, S. J.; Debroy, P.; Rathore, R. *Org. Lett.* **2008**, *10*, 389–392.

1.3.2 The Nicholas Reaction

Kevin M. Shea

1.3.2.1 Description

The Nicholas reaction enables efficient substitution reactions of propargyl alcohols, ethers, and acetates. Prior to the substitution step, dicobalt octacarbonyl reacts with the alkyne to yield cobalt-alkyne complex **1**. The resulting organometallic complex reacts with inter- or intramolecular nucleophiles in the presence of a Lewis or protic acid to furnish desired substitution products **2**. The cobalt-complexed alkyne can be oxidatively removed after this step or used to further functionalize the Nicholas reaction products.[1] The stereoselective synthesis of chiral products using the title reaction is also possible.[2]

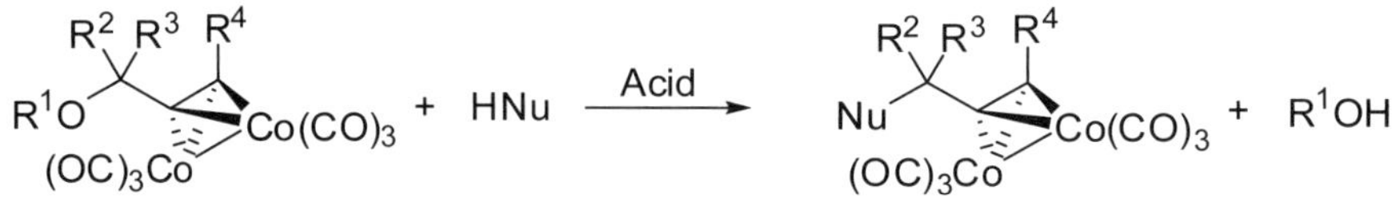

1 2

1.3.2.2 Historical Perspective

Kenneth M. Nicholas, currently a professor at the University of Oklahoma, discovered the reaction that bears his name while a professor at Boston College.[3] He began the study of cobalt-complexed alkynes during his graduate research at the University of Texas, Austin under the direction of Rowland Pettit. In 1971 Nicholas and Petit demonstrated the use of a cobalt–alkyne complex as an alkyne protecting group,[4] and one year later they reported the enhanced stability of carbocations adjacent to cobalt-alkyne complexes.[5] In 1977, along with his graduate student Rosa Lockwood, Nicholas described the first propargylic substitution reactions of propargyl alcohols complexed with dicobalt hexacarbonyl in the presence of strong protic acids and electron rich benzene derivatives. For example, propargyl alcohol **3** can be cobalt-complexed to yield cobalt–alkyne complex **4**. Addition of tetrafluoroboric acid yields stabilized carbocation **5** that undergoes a Friedel–Crafts reaction with anisole to furnish *para* disubstituted benzene **6**. Oxidative decomplexation with iron(III) provides the target alkyne **7** in 43% overall yield.[3]

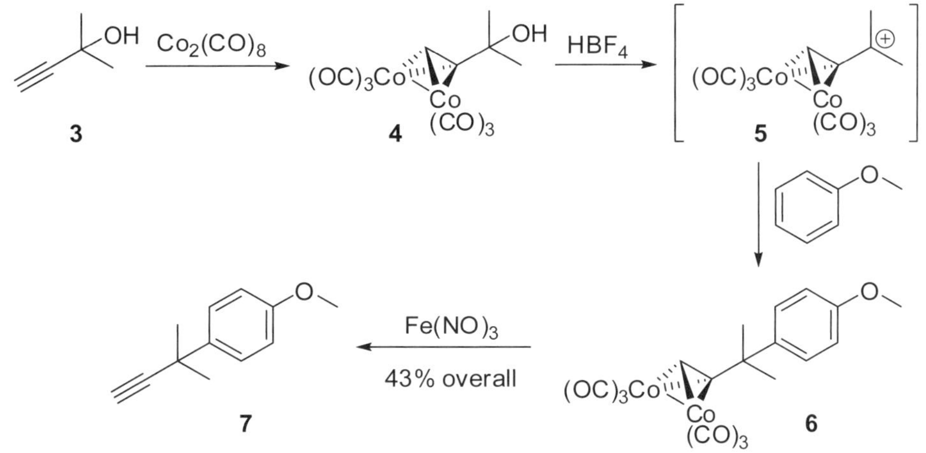

Nicholas continued his study of these propargylic substitution reactions for nearly the next twenty years[1d,e] and expanded his research to include the behavior of radicals stabilized by adjacent cobalt-complexed alkynes.[6]

1.3.2.3 Mechanism

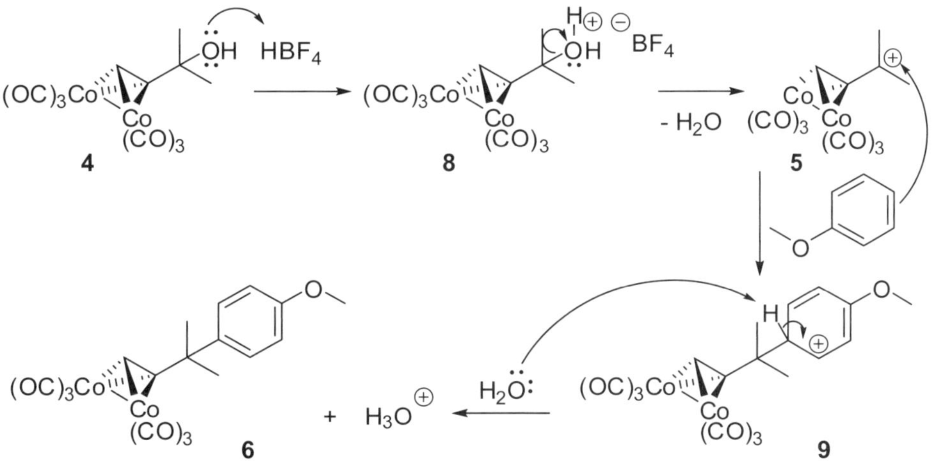

The mechanism of the Nicholas reaction is best described as an S_N1 process. Protonation of the alcohol in **4** followed by loss of water from cation **8** yields cobalt-stabilized carbocation **5**. Friedel–Crafts reaction of this electrophile with anisole provides resonance-stabilized carbocation **9** which, upon removal of a proton, furnishes the substitution product **6**. In addition to electron rich aromatics like anisole, a variety of neutral carbo- and heterocyclic nucleophiles react successfully with the carbocation intermediate. The most important advantage of this mechanism versus standard propargylic substitution reactions is that S_N1' and S_N2' reactions are impossible thus eliminating the pathway that yields allene byproducts.[1]

Stabilization of adjacent carbocations can be depicted by several resonance structures (**5**, **10**–**12**) which highlight opportunities for charge delocalization. Early NMR and IR studies by Nicholas on the isolable tetrafluoroborate salt of **5** and several analogs demonstrated measurable loss of electron density at cobalt (versus alcohol **4**), thus implicating structures like **10** and **12** in the stabilization of the positive charge.[7] Furthermore, Melikyan obtained an X-ray crystal structure of a carbocation stabilized by two adjacent cobalt-complexed alkynes that showed shortening of one cobalt α-carbon distance like that shown in compound **12**.[8]

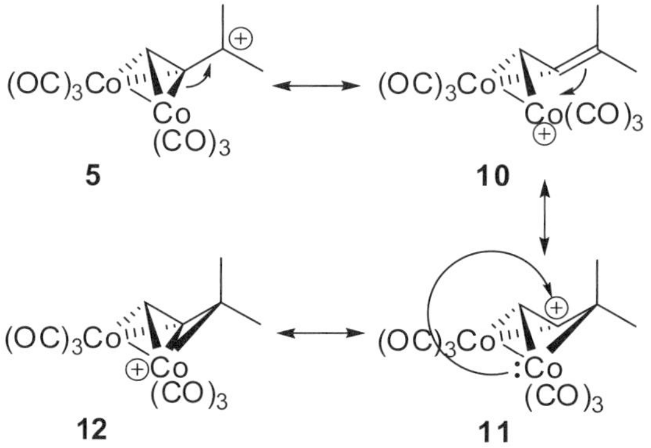

1.3.2.4 *Variations and Improvements*

Modifications to the standard Nicholas reaction generally fall into the following categories: asymmetric reactions, use of heteroatom nucleophiles, use of metals other than cobalt, reactions of neutral electrophiles, reactions of carbocations not in the α-position, cycloadditions, and rearrangements.

Numerous groups have investigated the asymmetric Nicholas reaction; studies focus on chiral nucelophiles, chiral cobalt-alkyne complexes, chiral cobalt ligands, and chirality transfer.[1b,2a] Two recent reports highlight examples of these strategies. Kann demonstrated that chiral phosphoramidite ligands effectively promote Nicholas reactions with moderate to good enantioselectivities.[9] Tyrrell reported a detailed study of asymmetric Nicholas reactions using a chiral auxiliary, chiral propargyl alcohols, and substrates derived from the chiral pool. In the best example, chiral non-racemic secondary alcohol **13** undergoes a boron trifluoride promoted intramolecular cyclization followed by cobalt decomplexation with ceric ammonium nitrate (CAN) to yield only one isomer of cyclic ether **14**.

Although the absolute stereochemistry is unclear, ^{1}H NMR analysis supports assignment of the relative stereochemistry as cis.[10]

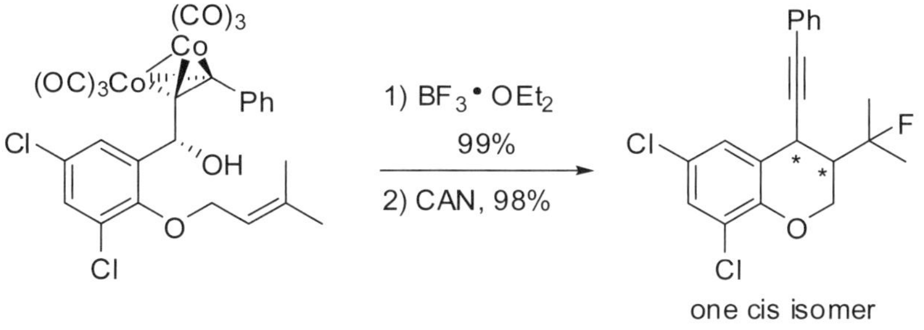

13 14

The use of heteroatom nucleophiles is a very popular variation of the Nicholas reaction.[1a] Alcohols[11] and amines[12] are most prevalent, while examples of azides,[13] thiols,[14] carboxylic acids,[15] epoxides,[16] hydrides,[17] and fluorides[18] are known. Isobe has made extensive use of alcohol nucleophiles in the preparation of various sized cyclic ethers. In studies directed toward the synthesis of ciguatoxin, Isobe converted alcohol **15** into eight-membered ring cyclic ether **16** using the Nicholas reaction.[19]

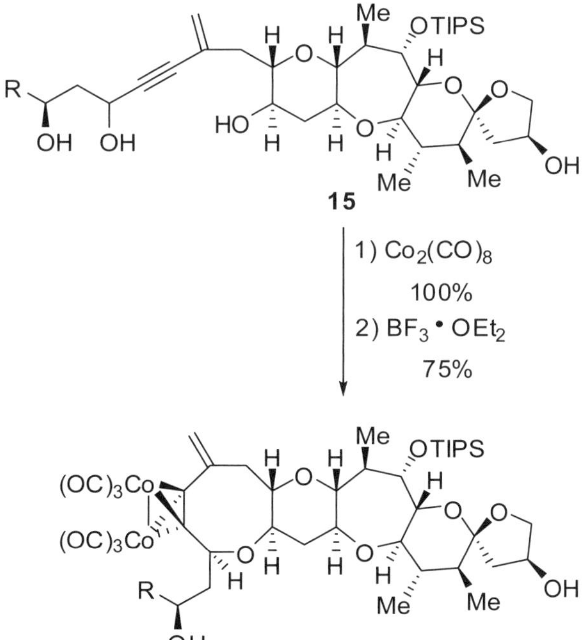

16

Many examples exist of alkynophylic metals other than cobalt that also promote propargylic substitution reactions; rhenium,[20] ruthenium,[21] and gold[22] can all be useful substitutes for cobalt. For example, Campagne synthesized enyne **18** upon treatment of propargyl alcohol **17** with allyltrimethylsilane and a Au(III) catalyst.[22]

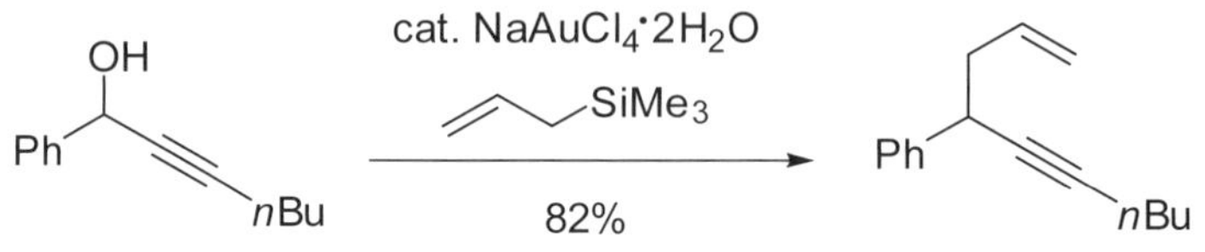

17 **18**

An interesting modification of the Nicholas reaction is an aldol reaction with an aldehyde adjacent to a cobalt-complexed alkyne. Chauvin reported a recent example of this strategy involving double addition of several carbon nucelophiles to cobalt-complexed acetylenedicarbaldehyde and comparison to the analogous Nicholas reactions with the corresponding diacetal. Depending on the reaction conditions, the products ranged from the expected addition product to unexpected oxygen heterocycles.[23]

Carbocations adjacent to cobalt-alkyne complexes that are also conjugated to π bonds enable a variety of conjugate addition options. Sierra and de la Torre reported examples for the synthesis of terpene-based hybrids. Treatment of alcohol **19** with boron trifluoride and furan furnished addition product **20** via an S_N1' mechanism.[24] Green employed a similar strategy for the synthesis of substituted cycloheptenes.[25]

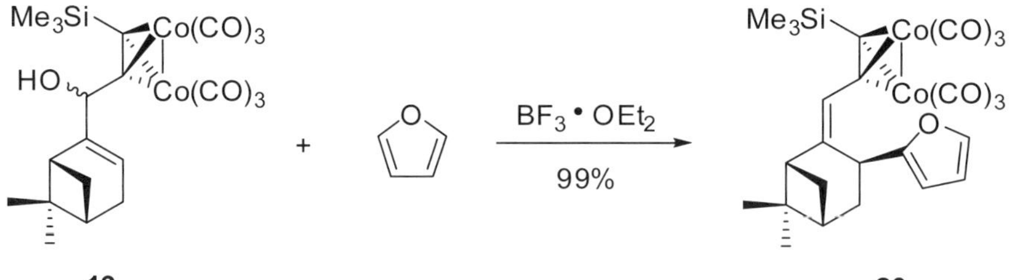

19 **20**

Christie and Jones first demonstrated in 2004 that appropriately substituted cyclopropanes adjacent to cobalt-alkyne complexes react with Lewis acids to yield 1,3-dipoles that are poised to participate in dipolar cycloadditions. Reaction of cyclopropane **21** with benzaldehyde and boron trifluoride provides tetrahydrofuran **22** in 83% yield.[26] Kerr subsequently applied this strategy to the synthesis of tetrahydro-1,2-oxazines upon combination of cyclopranes like **21** with a variety of nitrones.[27]

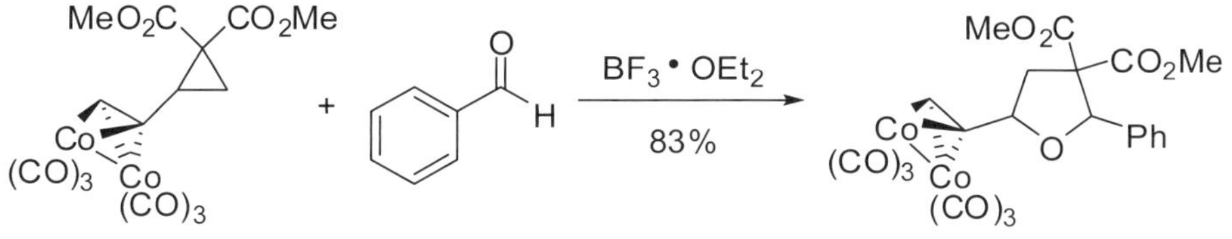

19 **20**

Harrity exploited the carbocation stabilizing ability of cobalt–alkyne complexes to promote a novel O→C rearrangement reaction. Exposure of cyclic enol ether **23** to diethylaluminum chloride promotes ionization of the C–O bond to yield the stabilized carbocation and an enolate. Bond rotation followed by C–C bond formation provides cyclohexanone product **24**.[28]

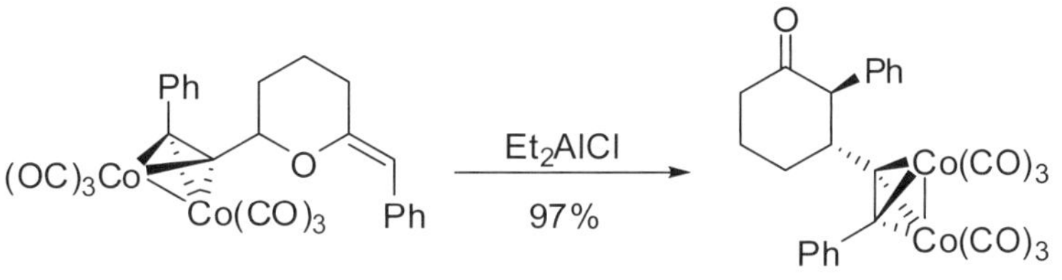

23 **24**

1.3.2.5 Synthetic Utility

Since its discovery just over thirty years ago, the Nicholas reaction has become a highly useful tool for the organic chemistry community. Applications of the Nicholas reaction fall into four categories: intermolecular reactions, endocyclic intramolecular reactions, exocyclic intramolecular reactions, and tandem reactions. For this discussion, endocyclic means that the cobalt-complexed alkyne is in the ring formed during the Nicholas reaction, while exocyclic indicates that the cobalt-alkyne complex is outside the newly generated ring.

Numerous carbon nucleophiles participate in the Nicholas reaction; the most popular are electron rich aromatics, enamines, enol ethers, ketene acetals, alkenes, allylsilanes, allylstannanes, and organoaluminum compounds.[1a,29] Upon completion of the substitution reaction, the cobalt complex can be oxidatively removed with I_2, Fe(III), or Ce(IV).[1a] Additional chemistry based on cobalt-complexed alkynes, like the Pauson-Khand reaction,[30] can also be performed to further increase molecular complexity.

Intermolecular Reactions

Konno reported a detailed study of the reaction of fluorine-containing propargyl acetates under Nicholas conditions with a variety of nucleophiles. Allylstannanes and allylsilanes provided moderate yields of the desired products, while enamines, silyl enol ethers, and silyl ketene acetals furnished the target compounds with excellent efficiency. In one example, cobalt-alkyne complex **25** reacts with silyl ketene acetal **26** in the presence of trimethylsilyl triflate to yield, after cobalt decomplexation, ester **27**.[31]

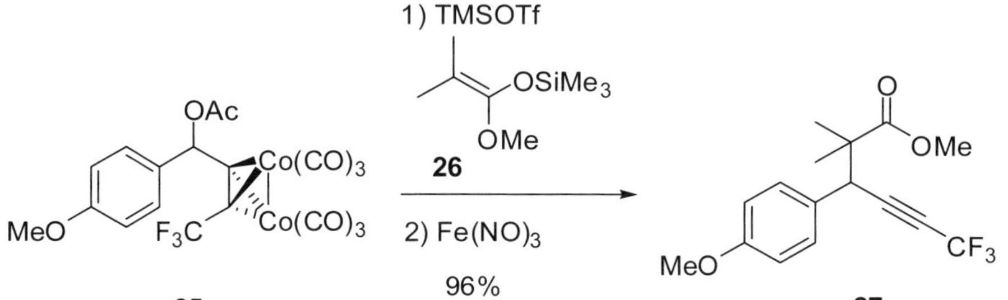

25 **27**

As part of his research into the synthesis of oxygen-bridged nine- and ten-membered cycloalkanes, Montaña introduced a key carbon fragment via the Nicholas reaction. Boron trifluoride promoted reaction between silyl enol ether **28** and cobalt–alkyne complexed propargyl alcohol **29** provides substitution product **30** in excellent yield.[32]

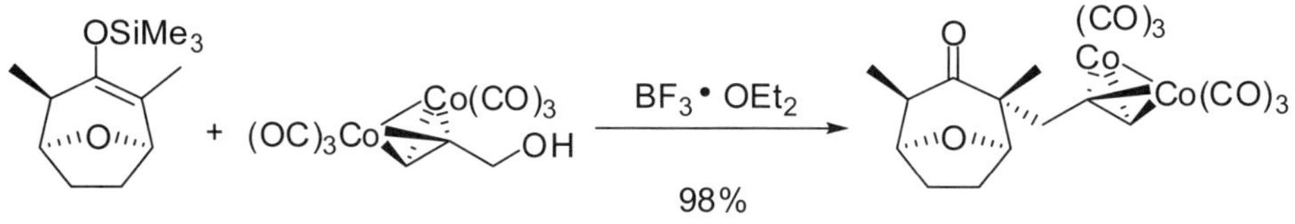

28 **29** **30**

Gómez and López reported the synthesis of *C*-ketosides using Nicholas reactions with several nucleophiles. In addition to the standard stabilization by the adjacent cobalt-alkyne complex, the carbocations generated in this study were further stabilized as oxonium ions (e.g., **32**). Reaction of galacto sugar **31** with 1-methylindole, followed by iodine promoted decomplexation, furnishes compound **33** in 64% overall yield. Interestingly, in the gluco series of cobalt-complexed alkynes, similar conditions led to double addition of several nucleophiles resulting in products substituted at both C-1 and C-4.[33]

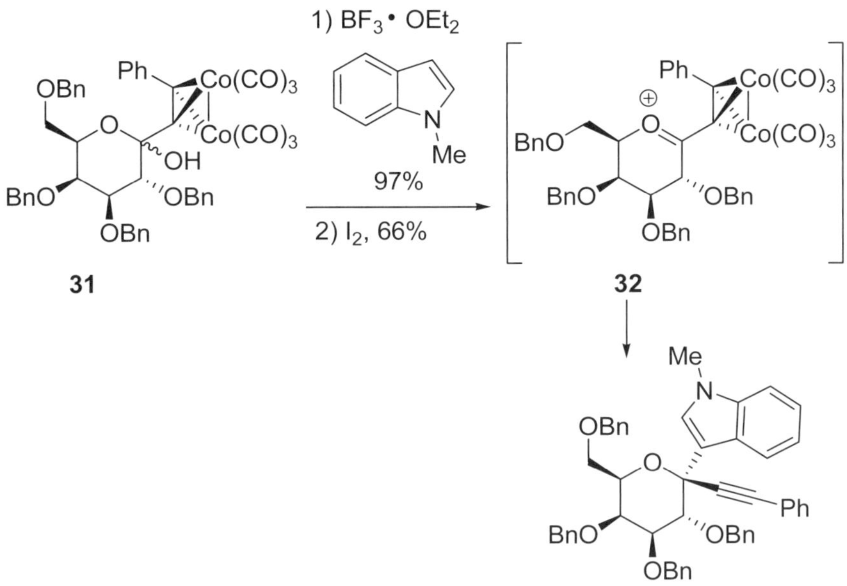

31 32 33

Kann recently investigated the intermolecular Nicholas reaction with substrates on the solid phase. In a detailed study with a variety of carbon and heteroatom nucleophiles, she demonstrated that the Nicholas reaction is a useful strategy for cleaving the substrate from the resin. Enyne **35** is available upon treatment of resin-bound cobalt–alkyne complex **34** with allyltrimethylsilane and boron trifluoride. Generation of the requisite carbocation results in cleavage from the polymer which is followed by a standard solution phase substitution reaction.[18] Subsequently, Kann applied this methodology for the synthesis of alkynylbenzyl galactosides which were evaluated as galectin inhibitors.[34]

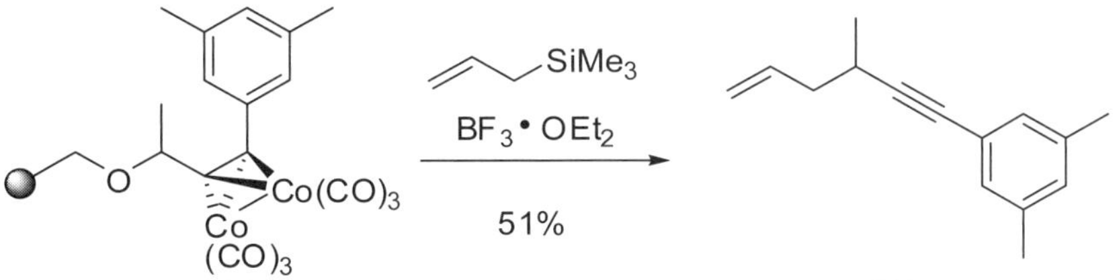

34 35

As highlighted in the *Variations and Improvements* section (**19→20**), de la Torre and Sierra use the Nicholas reaction to synthesize natural product hybrids. They reported standard transformations with electron rich aromatic nucleophiles in 2006.[35]

Endocyclic Intramolecular Reactions

One of Green's many applications of the chemistry of cobalt–alkyne complexes in synthesis involved formation of a seven-membered ring via an endocyclic intramolecular Nicholas reaction. The key step in his successful synthesis of allocolchicine NSC 51046 (**38**) was production of cyclic cobalt–alkyne complex **37** from reaction of acetate **36** and boron trifluoride. Hydrosilylation of the organometallic complex followed by desilylation yielded the corresponding alkene that was ultimately transformed into the target tricycle.[36]

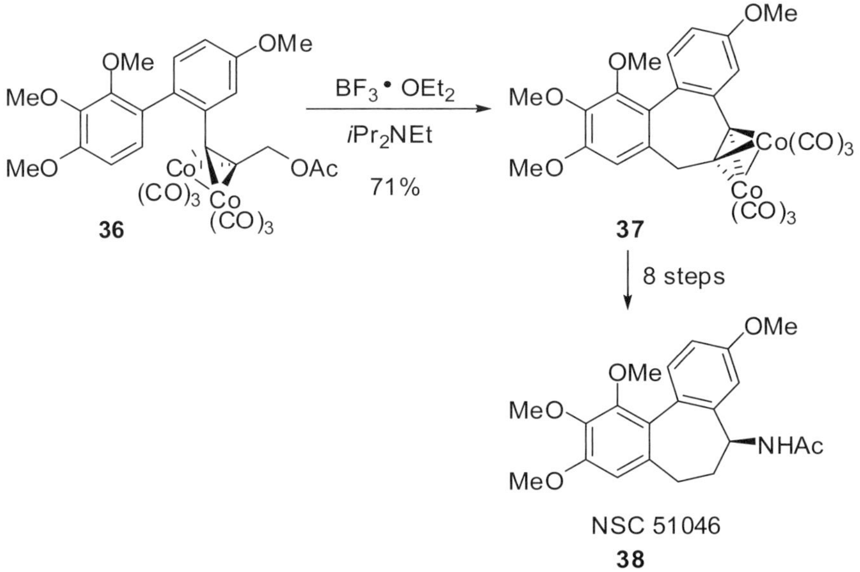

This type of Nicholas reaction has also been applied to the synthesis of ingenol and enediyne antitumor agents. Tanino and Kuwajima prepared the C ring of ingenol beginning with cobalt-alkyne complex **39**. Treatment with methylaluminum bis(2,6-dimethyl-4-nitrophenoxide) afforded target **40** which was ultimately transformed into ingenol (**41**).[37] Magnus targeted esperamicin, calicheamicin, dynemicin, and neocarzinostatin using the endocyclic intramolecular Nicholas reaction. For use in the syntheses of esperamicin and calicheamicin, he converted cobalt–alkyne complex **42** into enyne **43**. Subsequent exposure of **43** to iodine revealed the characteristic endiyne structure (**44**) common to all of these antitumor compounds.[38]

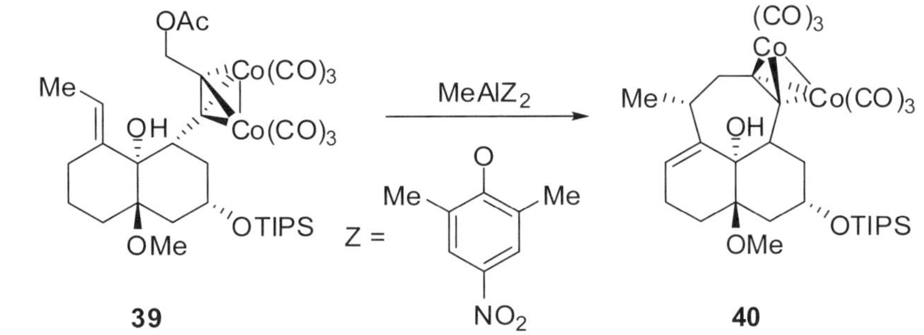

39 **40**

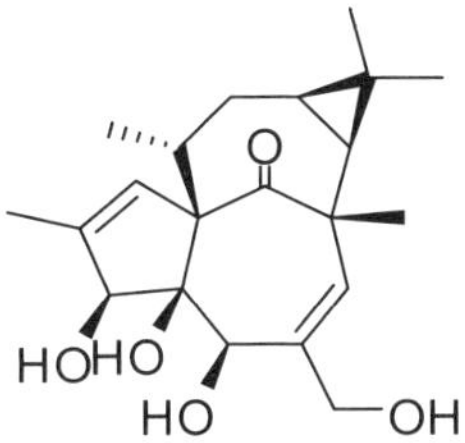

Ingenol
41

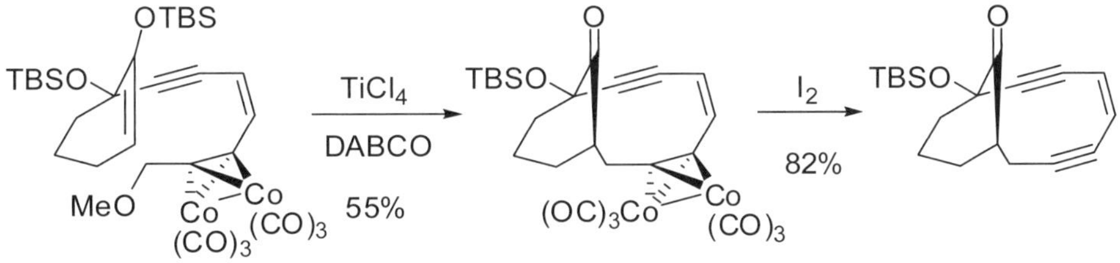

42 **43** **44**

Exocyclic Intramolecular Reactions

Bertrand recently generated six- and seven-membered rings using a terminal alkene nucleophile in exocyclic intramolecular Nicholas reactions. Depending on the nature of the Lewis acid, the carbocation intermediate formed upon cyclization (e.g., **46**) could be converted into a halide, amide, ester, or alkene. For example, alcohol **45** undergoes a 6-*endo*-cyclization to yield chlorocyclohexane **47** upon exposure to titanium tetrachloride.[39]

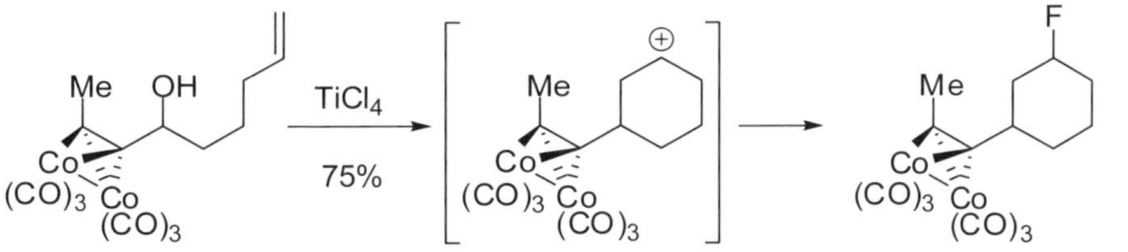

77:23 *cis:trans*

45 **46** **47**

In addition to her studies on the asymmetric Nicholas reaction mentioned previously (**13→14**), Tyrrell investigated standard intramolecular cyclizations to generate benzopyrans. An important advance in this report is that Tyrrell performed the cobalt complexation, Nicholas reaction, and cobalt decomplexation using a one-pot procedure. The conversion of propargyl alcohol **48** into benzopyran **49** highlights this strategy.[40]

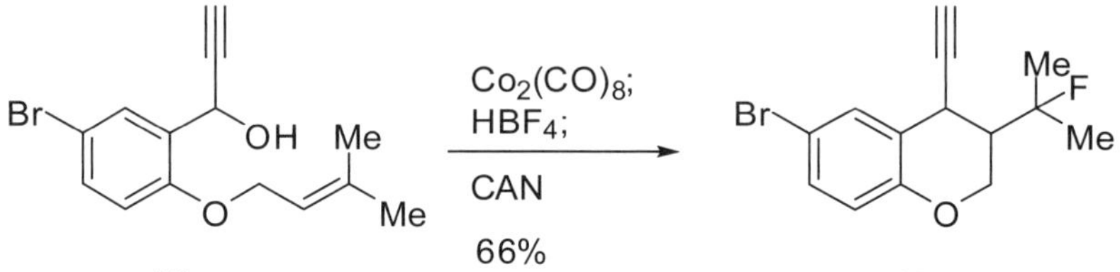

48 **49**

Tandem Reactions

Several reports describe a tandem intermolecular Nicholas/intramolecular Nicholas reaction sequence for the synthesis of cyclic compounds. Green employed this strategy for the synthesis of indolophanetetrayne cobalt complexes. Dimerization of substituted indole **50** with boron trifluoride furnishes target **51** in 55% yield.[41] Green also prepared a cobalt-complexed cycloheptyne via tandem intermolecular Nicholas/intramolecular Nicholas reactions. Boron trifluoride promotes combination of cobalt–alkyne complex **52** with allyltrimethylsilane to first yield intermolecular product **53** and then the desired intramolecular product **54**.[42]

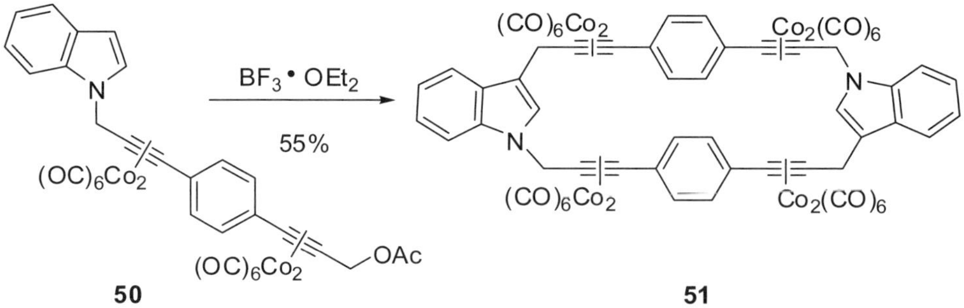

50 **51**

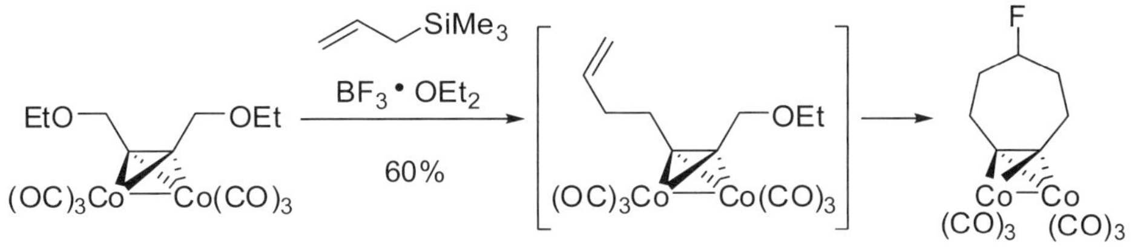

52 **53** **54**

Tyrrell demonstrated a three step tandem sequence involving an intermolecular Nicholas reaction, intramolecular Nicholas reaction, and a cationic cyclization. Treatment of silyl enol ether **55** with hexacarbonyl(propiolaldehyde diethyl acetal) dicobalt and boron trifluoride provides cobalt–alkyne complex **56**. Exposure of this material to tetrafluoroboric acid promotes an intramolecular Nicholas reaction to form the second six-membered ring. Alkyne decomplexation with ceric ammonium nitrate enables the final cyclization step to yield the target tricycle **57**.[43]

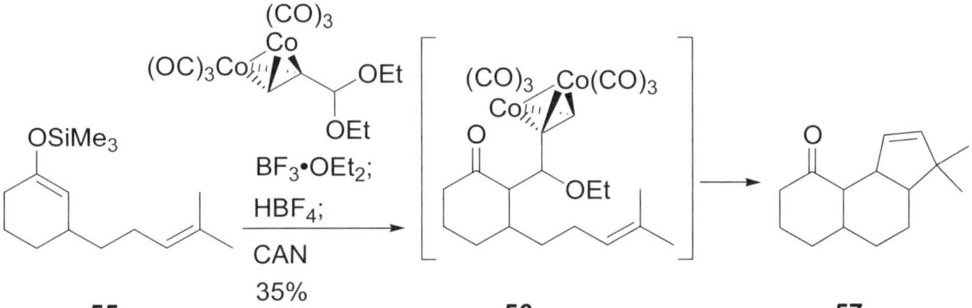

55 **56** **57**

The most impressive use of a tandem strategy involving the Nicholas reaction in a total synthesis project is Schreiber's preparation of (+)-epoxydictymene (**61**). Cobalt complexation of **58** followed by an endocyclic intramolecular Nicholas reaction with an allylsilane nucleophile yields Pauson–Khand precursor **59**. Treatment of **59** with *N*-methylmorpholine-*N*-oxide (NMO) promotes the Pauson–Khand reaction to furnish tetracycle **60** which was ultimately converted to the target natural product **61**.[44]

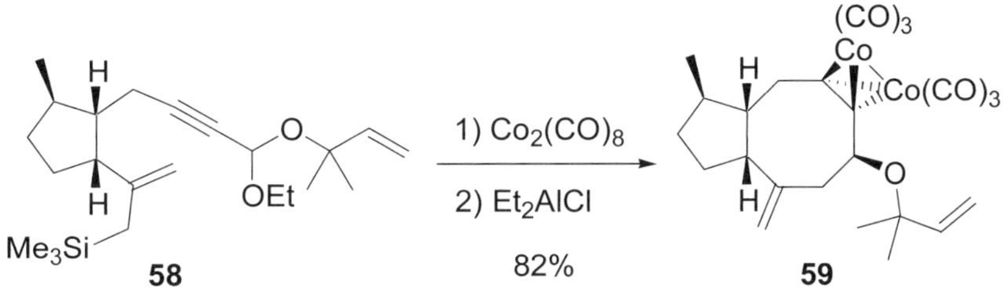

58 **59**

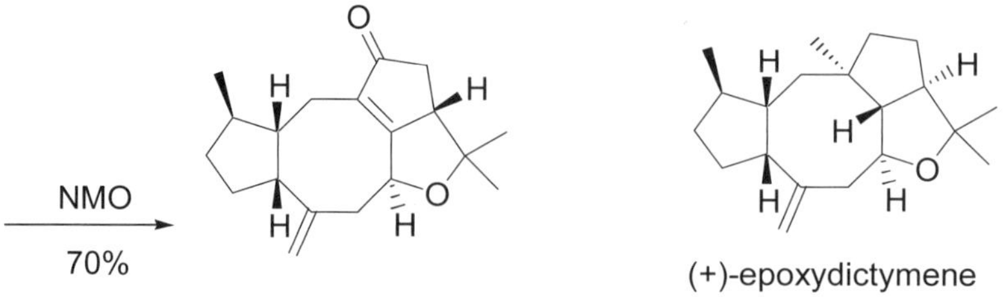

60

(+)-epoxydictymene
61

1.3.2.6 Experimental

Nicholas published an *Organic Syntheses* paper highlighting the preparation of 2-(1-methyl-2-propynyl)cyclohexanone (**65**) from the reaction of 1-trimethylsiloxycyclohexene (**62**) with hexacarbonyl (1-methyl-2-propynylium)dicobalt tetrafluoroborate (**63**) and subsequent decomplexation of cobalt complexed alkyne **64**. This report also includes procedures for the synthesis of the two starting materials (**62** and **63**).[45]

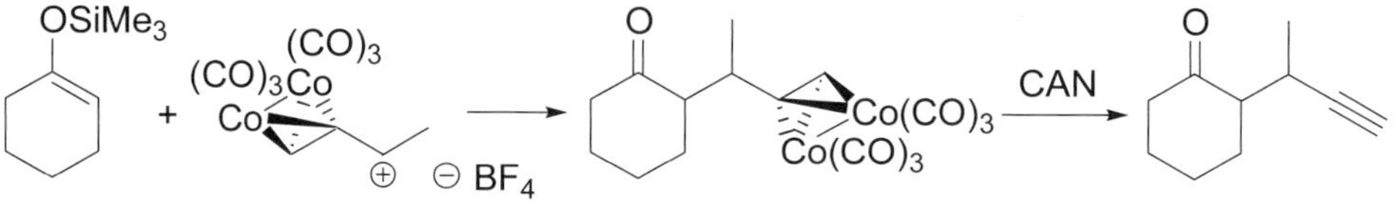

62 **63** **64** **65**

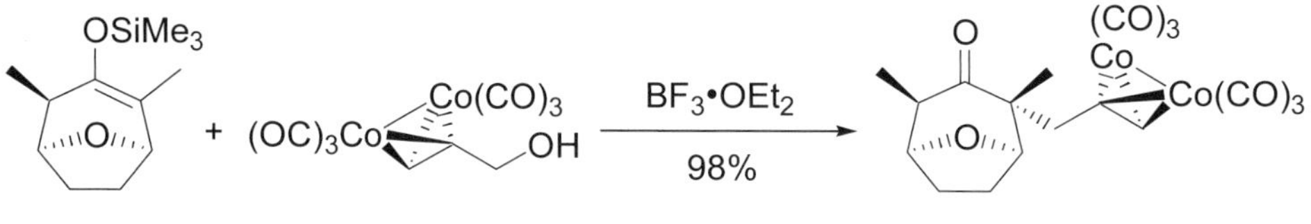

28 **29** **30**

Hexacarbonyl(μ-{η^4-[2,4-dimethyl-2-(prop-2-yn-1-yl)-8-oxabicyclo-[3.2.1]octan-3-one]})dicobalt(Co-Co):[32]
In a 50 mL flask fitted with a magnetic stirring bar, argon inlet, and septum, compound **29** (150 mg, 0.440 mmol) was dissolved in anhydrous dichloromethane (2 mL). Then, compound **28** (90 mg, 0.40 mmol), dissolved in anhydrous dichloromethane (2 mL), was added by cannula. The mixture was cooled to 0 °C, and BF_3•OEt_2 (300 μL, 2.0 mmol) was added all at once. The reaction mixture was maintained at 0 °C for 5 min and then allowed to warm to room temperature. After 2 h (monitored by TLC), the reaction mixture was quenched at 0 °C by the addition of triethylamine (400 μL, 2.5 mmol) and washed with ice water. The organic solution was dried with anhydrous $MgSO_4$, filtered and then percolated through a short pad of

activated neutral alumina, eluting with dichloromethane. The organic solution was concentrated to dryness in a rotary evaporator (without heating!), giving **30** as a dark red oil (190 mg, 98%).

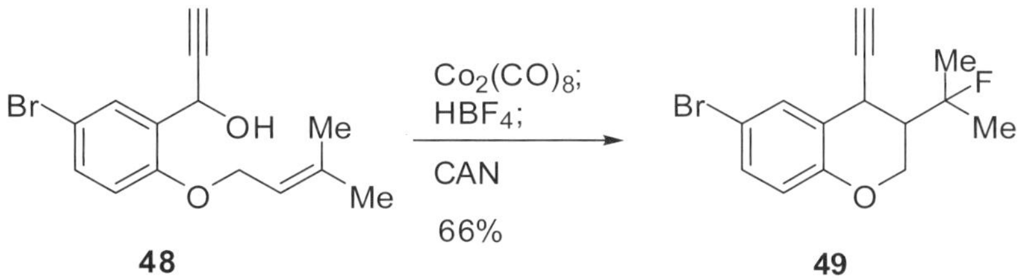

48 **49**

6-Bromo-4-ethynyl-3-(1-fluoro-1-methylethyl)chromane:[41]
To a solution of propynyl alcohol **48** (0.80 g, 2.7 mmol) in anhydrous dichloromethane (12 mL), under an atmosphere of nitrogen, was added octacarbonyldicobalt (1.02 g, 3.0 mmol) and the reaction was stirred at ambient temperature. The progress of the reaction was monitored by observing the evolution of carbon monoxide from the reaction mixture. TLC analysis, after fifteen minutes, showed the presence of a faster moving compound (R_f 0.45, 2:1 hexane:diethyl ether). The reaction mixture was then cooled to -10 °C whereupon tetrafluoroboric acid diethyl ether complex (0.52 mL, 3.0 mmol, 85% by volume) was added and the mixture left to stir. TLC analysis, after five minutes, showed the presence of a new compound (R_f 0.65, 2:1 hexane:diethyl ether). To the reaction mixture, maintained at -10 °C, was added dropwise methanolic ceric ammonium nitrate (CAN, 6.67 g, 12.20 mmol, 30 mL) until the evolution of carbon dioxide ceased and the yellow color of CAN persisted (about fifteen minutes). TLC analysis of the reaction mixture revealed the presence of a new compound (R_f 0.40, 3:1 hexane:diethyl ether). Residual methanol was removed *in vacuo* and the residue was partitioned between diethyl ether (25 mL) and water (25 mL). The aqueous phase was extracted with diethyl ether (3 × 20 mL) and the combined organic extracts were dried over anhydrous magnesium sulfate, filtered, and the solvent removed *in vacuo* to afford an oil. Purification was effected by column chromatography on silica (3 : 1 hexane:diethyl ether) to afford the desired compound **49** (0.53 g, 66%) as a yellow oil.

1.3.2.7 References

1. a) [R] Teobald, B. J. *Tetrahedron* **2002**, *58*, 4133–4170. b) [R] Green, J. R. *Curr. Org. Chem.* **2001**, *5*, 809–826. c) [R] Welker, M. E. *Curr. Org. Chem.* **2001**, *5*, 785–807. d) [R] Caffyn, A. J. M.; Nicholas, K. M. In *Comprehensive Organometallic Chemistry II*; Abel, E. W., Stone, F. G. A., Wilkinson, G., Eds.; Pergamon: Kidlington, 1995; Vol. 12, pp 685–702. e) [R] Nicholas, K. M. *Acc. Chem. Res.* **1987**, *20*, 207–214.
2. a) [R] Diaz, D. D.; Betancort, J. M.; Martin, V. S. *Synlett* **2007**, 343–359. b) [R] Muller, T. J. J. *Eur. J. Org. Chem.* **2001**, 2021–2033.

3. Lockwood, R. F.; Nicholas, K. M. *Tetrahedron Lett.* **1977**, 4163–4166.
4. Nicholas, K. M.; Pettit, R. *Tetrahedron Lett.* **1971**, *37,* 3475–3478.
5. Nicholas, K. M.; Pettit, R. *J. Organomet. Chem.* **1972**, *44,* C21–C24.
6. [R] Salazar, K. L.; Nicholas, K. M. *Tetrahedron* **2000**, *56,* 2211–2224.
7. Connor, R. E.; Nicholas, K. M. *J. Organomet. Chem.* **1977**, *125,* C45–C48.
8. Melikyan, G. G.; Bright, S.; Monroe, T.; Hardcastle, K. I.; Ciurash, J. *Angew. Chem., Int. Ed.* **1998**, *37,* 161–164.
9. Ljungdahl, N.; Pera, N. P.; Andersson, K. H. O.; Kann, N. *Synlett* **2008**, 394–398.
10. Tyrrell, E.; Millet, J.; Tesfa, K. H.; Williams, N.; Mann, A.; Tillett, C.; Muller, C. *Tetrahedron* **2007**, *63,* 12769–12778.
11. Quintal, M. M.; Closser, K. D.; Shea, K. M. *Org. Lett.* **2004**, *6,* 4949–4952.
12. Hernández, J. N.; Ramírez, M. A.; Rodríguez, M. L.; Martín, V. S. *Org. Lett.* **2008**, *10,* 2349–2352.
13. Shuto, S.; Ono, S.; Imoto, H.; Yoshii, K.; Matsuda, A. *J. Med. Chem.* **1998**, *41,* 3507–3514.
14. Mohamed, A. B.; Green, J. R.; Masuda, J. *Synlett* **2005**, 1543–1546.
15. Shea, K. M.; Closser, K. D.; Quintal, M. M. *J. Org. Chem.* **2005**, *70,* 9088–9091.
16. Pinacho Crisostomo, F. R.; Carrillo, R.; Leon, L. G.; Martin, T.; Padron, J. M.; Martin, V. S. *J. Org. Chem.* **2006**, *71,* 2339–2345.
17. Díaz, D. D.; Ramírez, M. A.; Martín, V. S. *Chem.–Eur. J.* **2006**, *12,* 2593–2606.
18. Gachkova, N.; Cassel, J.; Leue, S.; Kann, N. *J. Comb. Chem* **2005**, *7,* 449–457.
19. Hamajima, A.; Isobe, M. *Org. Lett.* **2006**, *8,* 1205–1208.
20. Ohri, R. V.; Radosevich, A. T.; Hrovat, K. J.; Musich, C.; Huang, D.; Holman, T. R.; Toste, F. D. *Org. Lett.* **2005**, *7,* 2501–2504.
21. [R] Nishibayashi, Y.; Uemura, S. *Curr. Org. Chem.* **2006**, *10*, 135–150.
22. Georgy, M.; Boucard, V.; Campagne, J.-M. *J. Am. Chem. Soc.* **2005**, *127,* 14180–14181.
23. Soleilhavoup, M.; Maurette, L.; Lamirand, C.; Donnadieu, B.; McGlinchey, M. J.; Chauvin, R. *Eur. J. Org. Chem.* **2003**, *2003,* 1652–1660.
24. Alvaro, E.; de la Torre, M. C.; Sierra, M. A. *Org. Lett.* **2003**, *5,* 2381–2384.
25. DiMartino, J.; Green, J. R. *Tetrahedron* **2006**, *62,* 1402–1409.
26. Christie, S. D. R.; Davoile, R. J.; Elsegood, M. R. J.; Fryatt, R.; Jones, R. C. F.; Pritchard, G. J. *Chem. Commun.* **2004**, 2474–2475.
27. Lebold, T. P.; Carson, C. A.; Kerr, M. A. *Synlett* **2006**, 364–368.
28. Meek, S. J.; Pradaux, F.; Carbery, D. R.; Demont, E. H.; Harrity, J. P. A. *J. Org. Chem.* **2005**, *70,* 10046–10056.
29. [R] Tyrrell, E. *Chem. Ind.* **2006**, 22–23.
30. [R] Brummond, K. M.; Kent, J. L. *Tetrahedron* **2000**, *56,* 3263–3283.
31. Konno, T.; Nagai, G.; Ishihara, T. *J. Fluorine Chem.* **2006**, *127,* 510–518.
32. Montana, A. M.; Ponzano, S.; Kociok-Koehn, G.; Font-Bardia, M.; Solans, X. *Eur. J. Org. Chem.* **2007**, 4383–4401.
33. Gomez, A. M.; Uriel, C.; Valverde, S.; Lopez, J. C. *Org. Lett.* **2006**, *8,* 3187–3190.
34. Bergh, A.; Leffler, H.; Sundin, A.; Nilsson, U. J.; Kann, N. *Tetrahedron* **2006**, *62,* 8309–8317.
35. Alvaro, E.; de la Torre, M. C.; Sierra, M. A. *Chem.–Eur. J.* **2006**, *12,* 6403–6411.
36. Djurdjevic, S.; Green, J. R. *Org. Lett.* **2007**, *9,* 5505–5508.
37. Tanino, K.; Onuki, K.; Asano, K.; Miyashita, M.; Nakamura, T.; Takahashi, Y.; Kuwajima, I. *J. Am. Chem. Soc.* **2003**, *125,* 1498–1500.
38. [R] Magnus, P. *Tetrahedron* **1994**, *50*, 1397–1418.
39. Olier, C.; Gastaldi, S.; Christie, S. D. R.; Bertrand, M. P. *Synlett* **2007**, 423–426.
40. Mann, A.; Muller, C.; Tyrrell, E. *J. Chem. Soc., Perkin Trans. 1* **1998**, 1427–1438.
41. Gibe, R.; Green, J. R.; Davidson, G. *Org. Lett.* **2003**, *5,* 1003–1005.
42. Lu, Y.; Green, J. R. *Synlett* **2001**, 243–247.
43. Tyrrell, E.; Tillett, C. *Tetrahedron Lett.* **1998**, *39,* 9535–9538.
44. Jamison, T. F.; Shambayati, S.; Crowe, W. E.; Schreiber, S. L. *J. Am. Chem. Soc.* **1997**, *119,* 4353–4363.
45. Varghese, V.; Saha, M.; Nicholas, K. M. *Org. Synth.* **1989**, *67,* 141–148.

1.3.3 Nozaki–Hiyama–Kishi Reaction

Larry Yet

1.3.3.1 Description

The one-pot Barbier-type addition of alkenyl, aryl, allyl, vinyl, propargyl, alkynyl, or allenylchromium compounds to aldehydes or ketones is known as the Nozaki–Hiyama–Kishi (NHK) reaction.[1–3] An excellent review by Fürstner published in 1999 detailed the exhaustive literature on the carbon–carbon bond formations involving organochromium(III) reagents. This chapter will present major developments and examples of recent carbon–carbon bond formation methodology and improvements as well as their use in natural products synthesis since 1999.

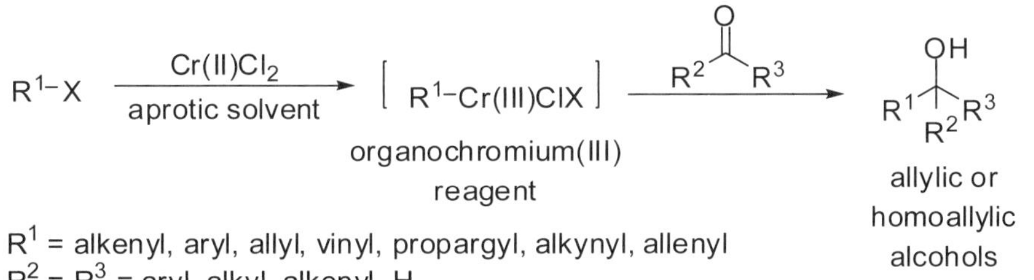

R^1 = alkenyl, aryl, allyl, vinyl, propargyl, alkynyl, allenyl
R^2 = R^3 = aryl, alkyl, alkenyl, H
X = Cl, Br, I, OTf
Solvent = DMF, DMSO, THF

The NHK reaction has become a powerful synthetic tool for the chemoselective formation of carbon-carbon bonds under very mild conditions. The most notable feature of these reactions include: 1) $Cr(II)Cl_2$ or $Cr(III)Cl_3$ are inexpensive commercially available reagents; 2) the broad range of substrates amenable to insertion of Cr(II) under mild conditions; 3) reactions can take place intermolecularly or intramolecularly where the thermodynamically driving force is the formation of a strong O–Cr(III) bond; 4) aldehydes or ketones can react; however it is chemoselective for the aldehyde if a ketone is present in the same molecule; 5) a low basicity of the organochromium reagents allow compatibility of a wide range of sensitive functional groups in the same molecule; 6) distinct stereochemical preference of *anti* products in reactions of crotylchromium reagents; and 7) simple set up and excellent reliability of the reaction. However, drawbacks of the NHK reactions include the toxic chromium salts and their need for greater than stoichiometric quantities.

1.3.3.2 Historical Perspective

From 1977–1983, Nozaki and Hiyama published the first seminal papers that started an explosive development of this particular branch of organometallic chemistry.[4] Both research groups demonstrated in great detail that stoichiometric quantities of $CrCl_2$ readily inserted into allyl-, alkynyl-, propargyl-, and aryl halides or sulfonates under aprotic conditions to give the corresponding organochromium(III) reagents, which reacted with aldehydes and ketones to give various alcohol products with high chemoselectivity and stereoselectivity. In 1986, Kishi and Nozaki independently and almost simultaneously discovered that traces of nickel salts exerted a catalytic effect on the formation of the C–Cr(III) bond.[5–6] This aided in the reactions with less reactive substrates such as vinyl or aryl halides or triflates.

1.3.3.3 Mechanism

The $CrCl_2/NiCl_2$ is the most widely employed synthetic tool in the Nozaki–Hiyama–Kishi reaction today and the generalized mechanism is outlined below.[1–3] In the nickel(II)-catalyzed NHK reaction, the first step in this cycle is the reduction of Ni(II) to Ni(0), which inserts into the carbon–halogen bond via an oxidative addition reaction to give organonickel species **1**. During this reduction, the Cr(II)Cl_2 is oxidized to Cr(III)Cl_3. This organonickel species **1** is then transmetallated with Cr(III)Cl_3 to form the organochromium(III) nucleophile **2**, which then reacts with the carbonyl compounds **3** to give **4**, where the final product is obtained after work-up. It should be noted that Cr(II)Cl_2 is a one-electron donor and therefore ≥ 2 equivalents is required per equivalent of organic halide is needed for the formation of any organochromium(III) nucleophile. Only low catalyst loading of $NiCl_2$ (0.1–2%) is needed.

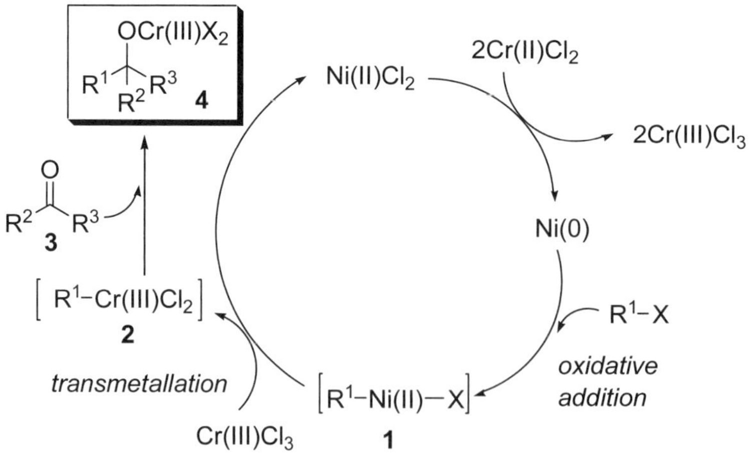

A chromium-catalyzed version has been developed by Fürstner which makes this process environmentally benign.[7] The key feature of this process uses chlorotrimethylsilane as an additive for the silylation of the chromium alkoxide species **5** in order to release the metal salt from the product **6**. The liberated Cr(III)Cl_2X can then be reduced to the active species Cr(II)Cl_2 by means of a stoichiometric and nontoxic reducing agent as a manganese(0) metal.

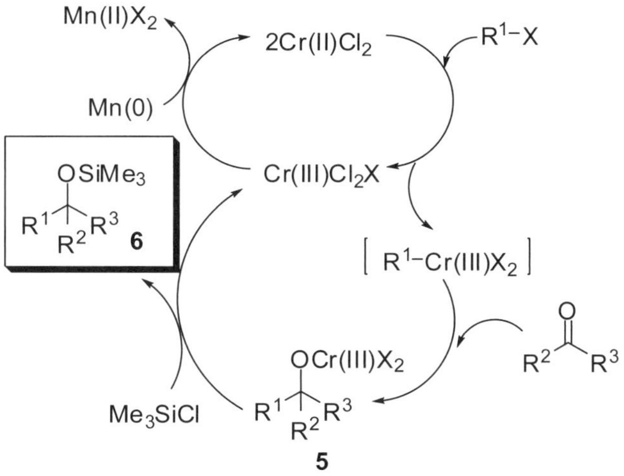

1.3.3.4 Variations and Improvements

The most useful improvement in the last decade has been the catalytic use of the NHK reaction developed by Fürstner.[7] Most of the improvements in the last few years involved the asymmetric NHK reaction with the special use of ligands with good enantioselectivities up to 98%.[8] This section will outlined the various ligands that have shown marked improvement in this area.

Bipyridyl Ligands

The first reports of the enantioselective NHK reaction was published by Kishi *et al.* in the chromium-mediated allylation and alkenylation reactions using specially designed bipyridyl ligands.[9] Bipyridyl itself inhibited the cross-coupling. It was, however, possible to tune the complexation capacity by introducing substituents at the 6-position, with the best ligand being **7**. However, the bipyridine derivative must be used in (over)stoichiometric amounts along with (over)stoichiometric amounts of $CrCl_2$, thus limiting the practicality of this method.

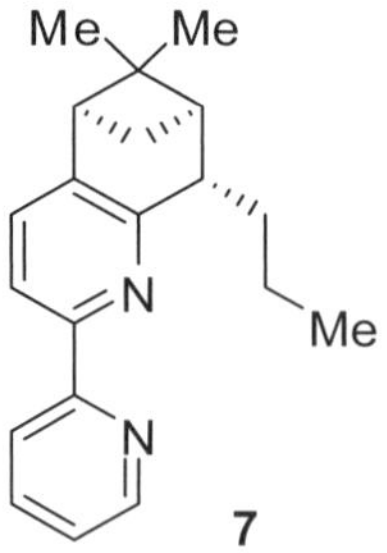

7

Chiral Salen Ligands

Umani–Ronchi adapted the Fürstner protocol to achieve the first catalytic, enantioselective variant of this reaction.[10] The chiral chromium salen complex was prepared from the *in situ* reduction of the anhydrous $CrCl_3$ to $CrCl_2$ with an excess of manganese metal, followed by complexation with the salen ligand **8** in the presence of catalytic triethylamine.[11] Then the addition of allylic chloride (**9**) to aldehydes **10** to give the allylic alcohols **11** in moderate yields and in up to 95% *ee*. The same groups employed the same conditions for the addition of 2-butenyl bromides to aldehydes to achieve up to 83:17 *syn/anti* of allylic alcohol products[12] and for the addition of 1,3-dichloropropene to aromatic aldehydes to obtain the syn chlorohydrin adduct in modest yield which were further converted to optically active vinyl epoxides.[13] The [Cr(salen)]-catalyzed addition of propargyl halides to aromatic aldehydes allowed the synthesis of enantiomerically enriched homopropargyl alcohols in moderate yields with up to 56% *ee*.[14]

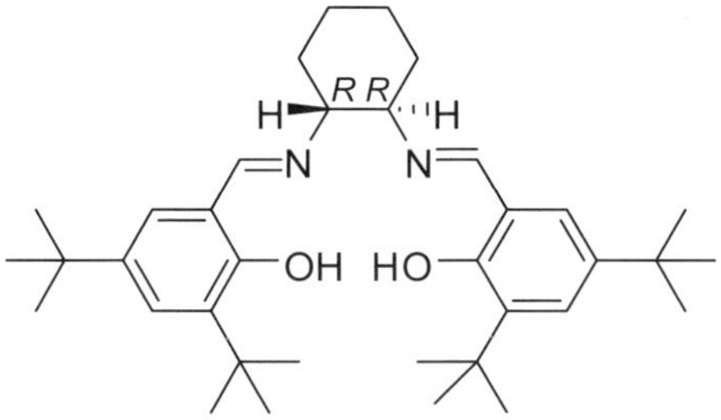

8

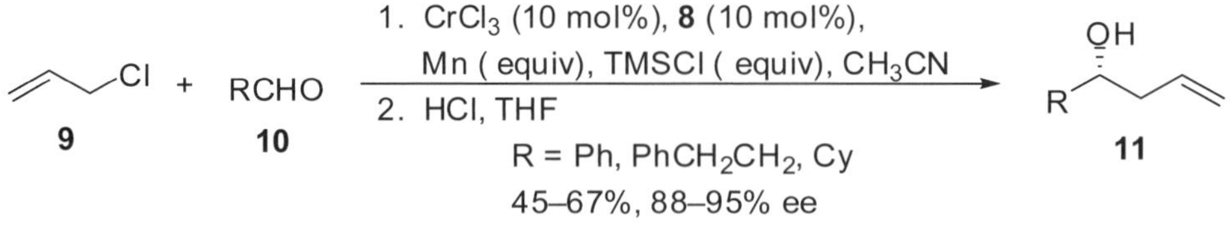

Berkessel *et al.* have demonstrated that salen ligand **12** based on *endo,endo*-2,5-diaminonorbornane (DIANANE) promoted efficient and highly enantioselective catalytic NHK reactions of various allylic and vinylic halides to aromatic and aliphatic aldehydes with up to 92% *ee*.[15]

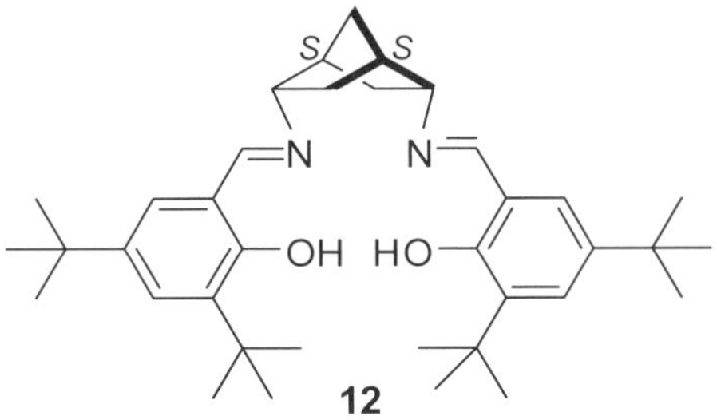

12

Other Ligands

Asymmetric Nozaki–Hiyama allylation of benzaldehyde with a multidentate proline-derived amino alcohol ligand **13** showed good enantioselectivity up to 90% *ee*.[16] Yamamoto *et al.* showed that tethered bis-(8-quinolinolato)-(TBOx) chromium complex **14** participated in catalytic enantioselective Nozaki–Hiyama allylation, crotylation and allenylation reactions of aliphatic and aromatic aldehydes up to 97% *ee*.[17]

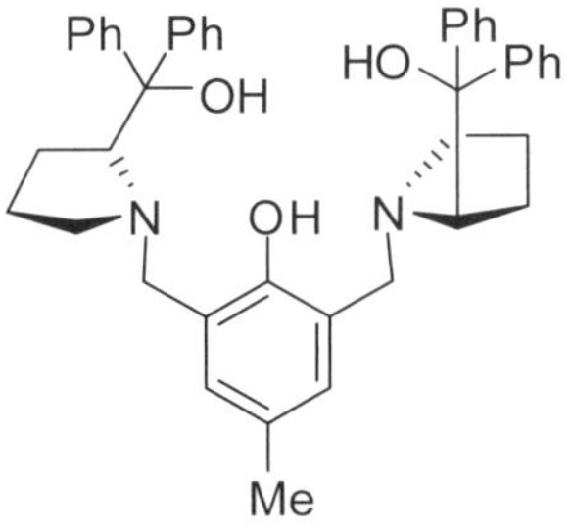

13

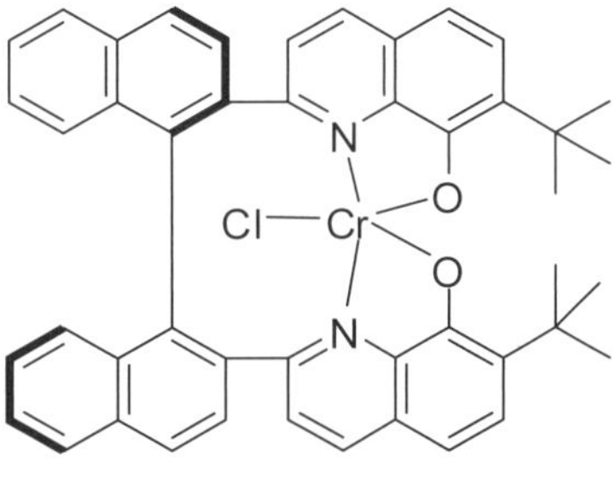

TBOxCr(III)Cl (**14**)

Chiral Oxazoline Ligands

The most developed ligands for the asymmetric Nozaki–Hiyama–Kishi reaction have the chiral oxazoline-type structures. Kishi was the first to show that sulfonamide ligands **15** can effect asymmetric induction under both stoichiometric and catalytic conditions.[18] Sulfonamide ligand **15a** was found to be a suitable ligand, after several screening experiments, for the Ni/Cr mediated stoichiometric enantioselective coupling of dihydrocinnamaldehyde (**16**) with various vinyl and allyl halides and triflates **17** to give the (*R*)-

alcohols **18**. The catalytic application of ligand **15a** and **15b** was exploited in the same set of reactions with manganese metal (2 equiv) and chlorotrimethylsilane (2 equiv) in propionitrile or THF as solvents.[19] Ligand **15c** was utilized in the Fe/Cr- and Co/Cr-mediated catalytic asymmetric 2-haloallylations of aldehydes.[20] Ligands **15c** and **15d** were further exploited in several new catalytic reactions: 1) Ni/Cr-mediated alkenylation, alkynylation, and arylation; 2) Co/Cr-mediated 2-haloallylation, alkylation, and propargylation; and 3) Cr-mediated allylation.[21]

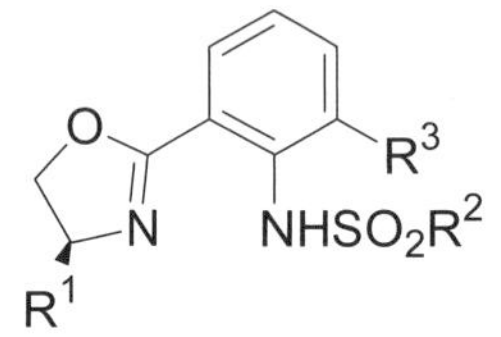

15a R^1 = *i*-Pr, R^2 = R^3 = Me
15b R^1 = *i*-Pr, R^2 = Me, R^3 = OMe
15c R^1 = *t*-Bu, R^2 = Bn, R^3 = H
15d R^1 = *t*-Bu, R^2 = 2-Naphth, R^3 = Me

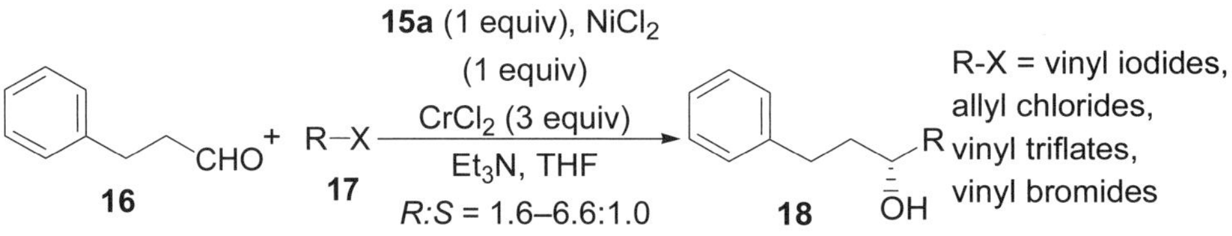

These discoveries were employed in the synthesis of the C14–C26 precursor **22** and C26–C38 segment of **24** of halicohondrins with stoichiometric and catalytic uses of ligand **15a**.[18–19]

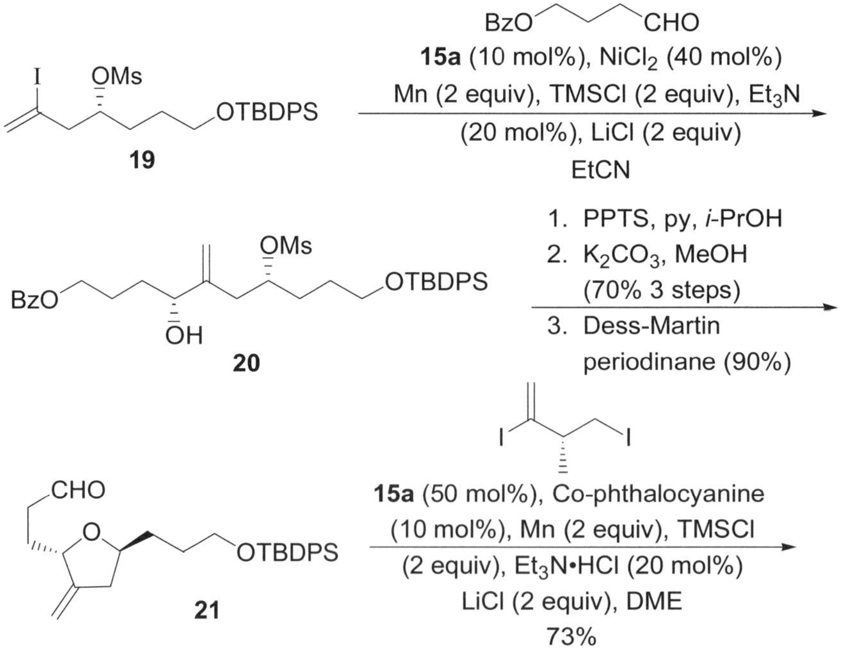

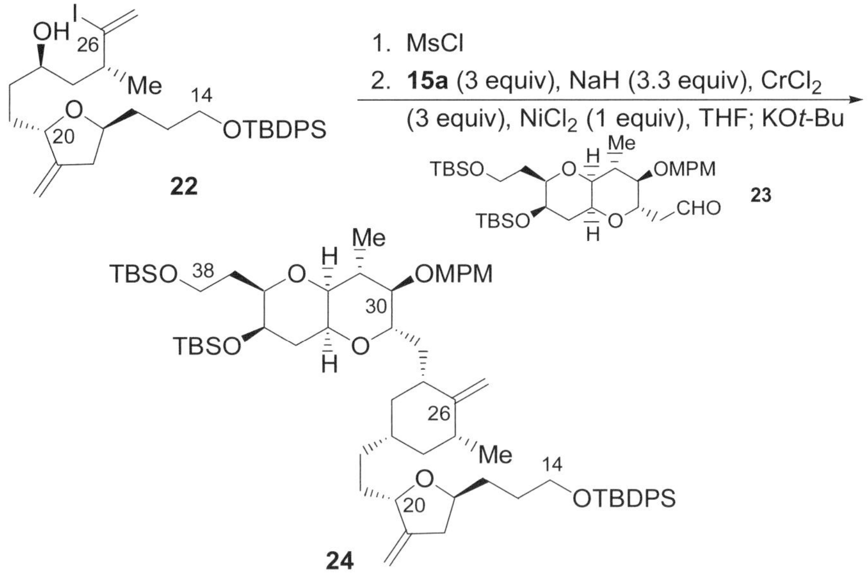

Nakada *et al.* designed and synthesized a C_2-symmetrical tridentate bis(oxazolinyl)carbazole ligand **25** for the asymmetric catalysis of Nozaki-Hiyama allylation and methallylation.[22] They reasoned that the allyl–Cr(III) ligand did not undergo significant dissociation due to the stabilization by three bonds: a σ-bond with the carbazole nitrogen and two coordination bonds with the oxazoline nitrogens, leaving a vacant coordination site at which an aldehyde can bind. Ligand **25**, $CrCl_2$, and Mn were all mixed in THF at room temperature; the Cr(II)-ligand **26** complex thus prepared *in situ* was then used for the enantioselective allylation.[23] Addition of allyl or methallyl halides afforded intermediate **26** which then reacted with aldehydes to give enantioenriched alcohols **27**. An allylation and methallylation reaction with this ligand was showcased in the enantioselective total synthesis of the potent HMG-CoA reductase inhibitor FR901512.[24] Furthermore, catalytic asymmetric Nozaki-Hiyama propargylation with ligand **25** proceeded with good to excellent enantioselectivity.[25] The first enantioselective NHK allenylation of terminally silylated propargyl halides using ligand **25** was also reported.[26]

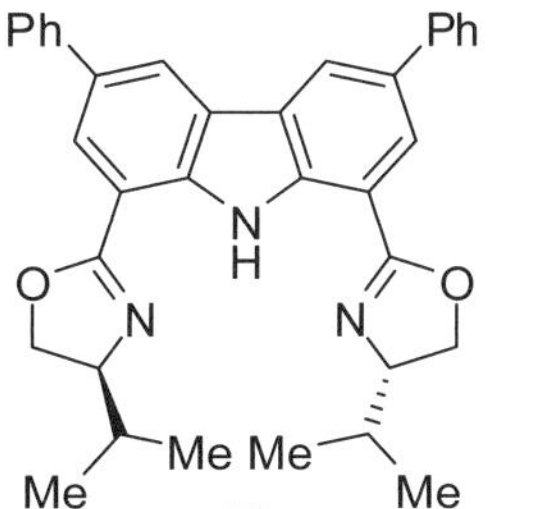

25

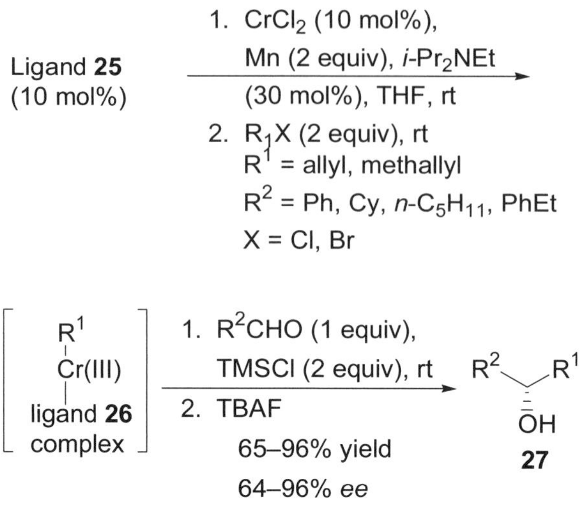

Sigman *et al.* identified a new set of stereochemically diverse oxazoline ligands derived from simple amino acids that promoted the Cr-catalyzed enantioselective addition of allylic halides to aldehydes in up to 95% *ee*.[27] Ligand diastereomer **28** was found to be the best general catalytic system for these reactions. Furthermore, diastereomer **29** was the ligand of choice for the enantioselective Cr-catalyzed addition of allyl bromide to ketones in up to 93% *ee*.[28]

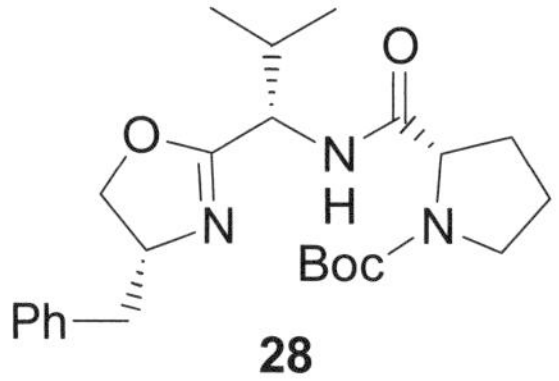

28

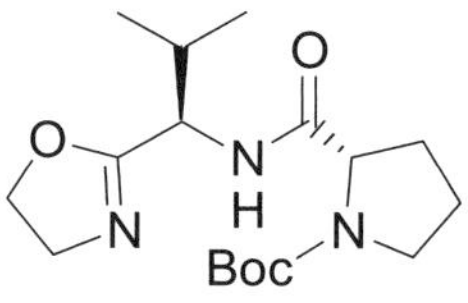

29

Guiry *et al.* found that the non-symmetric bis(oxazoline) ligand **30** was the optimal ligand in the allylation and crotylation of a range of aryl and aliphatic aldehydes.[29] The enantioselectivities obtained in the allylation reaction were up to 91% *ee* and in the crotylation reaction up to 92% *ee* with typical *syn*:*anti* ratios of up to 80:20. Sixteen members of a new ligand class incorporating an oxazoline ring linked by an amide bond to a chiral protected proline unit were reported; ligand **31** was found to be the best in the enantioselective allylation of benzaldehyde with 57% *ee*.[30]

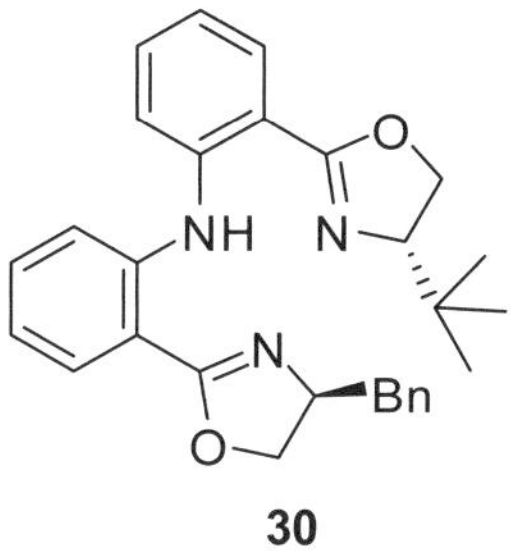

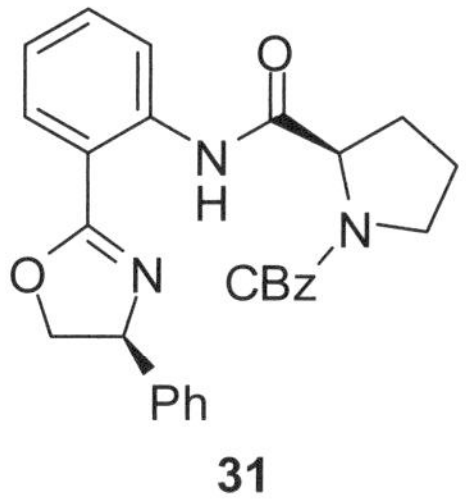

1.3.3.5 Synthetic Utility

Cyclization to Large-Sized Ring Systems

Popik reported the first example of triggering of the thermal Bergman cyclization by the photochemical ring contraction.[31] Thus, iododiyne **32** underwent NHK reaction in good yield to give cyclodeca-1,5-diyne **33**, which was further elaborated to 2-diazo-6,7-benzocycloundeca-4,8-diyn-1,3-dione (**34**). Upon irradiation **34** underwent Wolff rearrangement to produce reactive 10-membered enediynes followed by spontaneous Bergman cyclization. The synthesis and cytotoxicity of enediyne prodrugs with 3-hydroxy-4-arylmethylidene)cyclodeca-1,5-diyne scaffolds **36** were prepared from the NHK reaction of iododiyne **35** followed by esterification and silyl deprotection.[32] Pilli has reported the intramolecular NHK reaction for the construction of ten-membered lactones[33] and Bermejo has investigated the intramolecular NHK reaction for the preparation of the ten-membered ether of the eleutheside family of natural products.[34] Malacria has published an efficient preparation of a highly strained eleven-membered ring using the NHK reaction.[35]

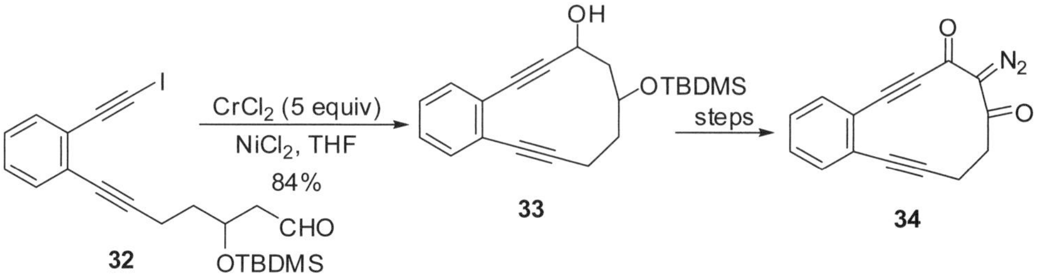

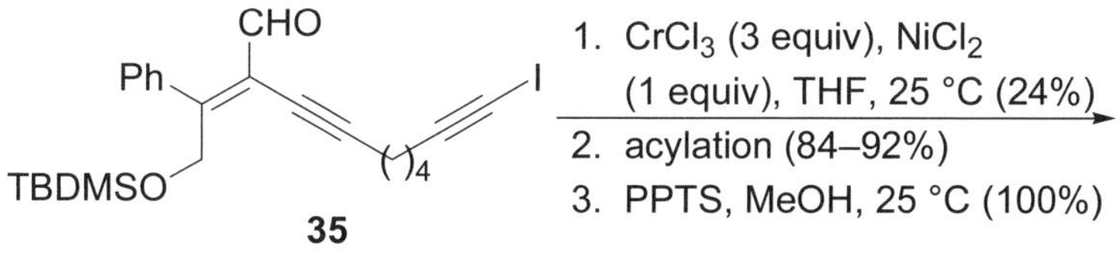

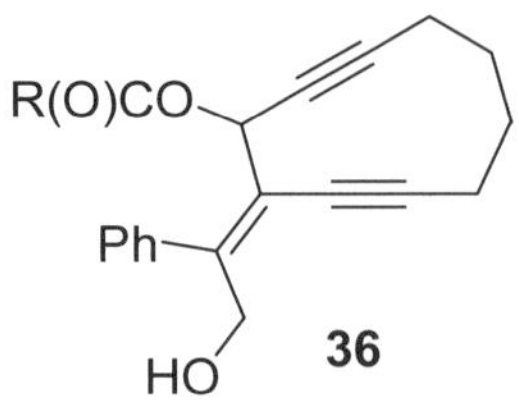

36

(Z)-Haloallylic Alcohols

Halogenated alkenols are versatile synthetic intermediates as well as critical structural units in a variety of biologically active natural products. Previous preparations of these structures are often hampered by multistep synthesis and by poor stereoselectivity. The Nozaki–Hiyama reaction has provided good synthetic entry to these structures. For example, a highly diastereoselective and straightforward synthesis for (*Z*)-2-fluoroallylic alcohols **39** via a Nozaki–Hiyama–Kishi type reaction with the corresponding bromofluoroolefins **37** with aliphatic and aromatic aldehydes **38** was developed.[36] Application of lithium, organoindium, organosamarium, organocopper, and organochromium carbenoid conditions failed to effect this transformation.

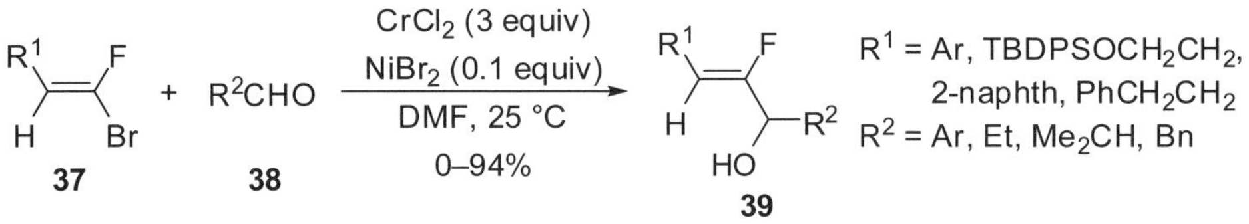

Mioskowski and Flack showed that (*Z*)-2-chloroalk-2-en-1-ols **42** were obtained in excellent yields from a wide variety of aldehydes by addition of (*E*)-chromium vinylidene carbenoids **41**, generated from trichloroalkanes **40** with $CrCl_2$ in THF at room temperature.[37] The same authors also reported $CrCl_2$-mediated condensations of χ-chloro-*gem*-trichloroalkanes with aldehydes to give homolallyic alcohols through a hydride rearrangement followed by a Nozaki–Hiyama allylation.[38]

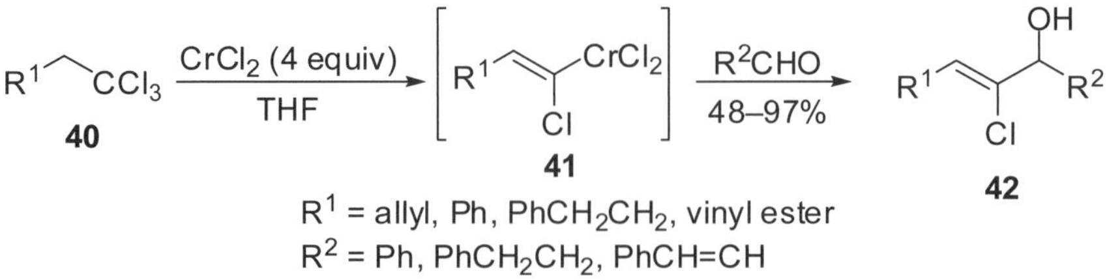

Miscelaneous Reactions

Cheng has published a convenient and synthetically useful alternative method to the NHK reaction for the arylation of aromatic aldehydes in a mild and selective way with nickel(II) bromide/zinc/dppe mediated protocol for the synthesis of diaryl carbinols.[39] Durandetti reported an electrochemical coupling of aryl halides with aldehydes for the synthesis of diaryl carbinols which was catalytic in chromium and nickel salts.[40] Comins utilized the NHK reaction to prepare 5-(1-hydroxyalkyl)-2,3-dihydro-4-pyridones, which were then explored in reductive, oxidative and substitutive reactions.[41] The first asymmetric catalytic synthesis of *syn*-alk-1-ene-3,4-diols was developed; the regio-, diastereo- and enantioselective addition of 3-chloropropenyl pivaloate to aldehydes was made possible by exploiting Salen–Cr(II) species in a catalytic version of the NHK reaction.[42]

Applications in Natural Products

The unique features of the Nozaki–Hiyama–Kishi reaction make this an attractive methodology in the application to natural products synthesis, particularly the pronounced chemoselectivity and remarkable compatability with an array of functional groups. However, there are very few reports published of the application of enantioselective NHK reactions in the total synthesis of natural products. This section is divided into three parts: 1) preparation of fragments with the NHK reaction in the synthesis of a portion of a natural product; 2) the NHK reaction in the final or second last step of a natural product; 3) the NHK reaction utilized anywhere in the total synthesis of a natural product. The next several pages show natural products where the NHK reaction have been employed in the syntheses of natural products since Fürstner's review published in 1999.[1]

The structures shown below are natural products where a portion of the target has been completed using the NHK reaction conditions. The swiggly lines denote the bond formed by the NHK reaction. Other information includes the natural products, the principal author, the halo or triflate precursor, the yields and ratio of isomers, enantiomers, or diastereomers from that step.

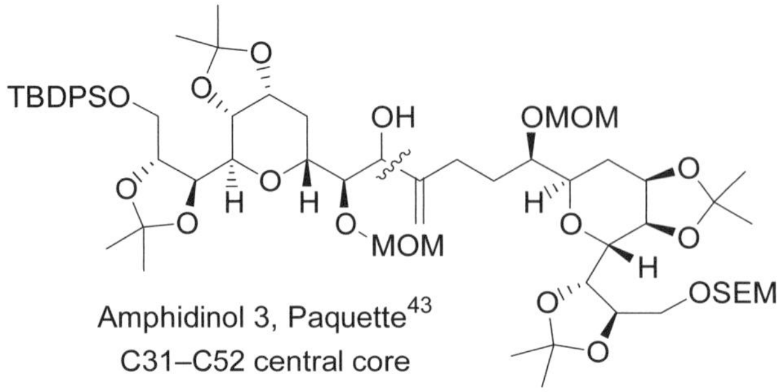

Amphidinol 3, Paquette[43]
C31–C52 central core
vinyl iodide, 62%, dr 1:1

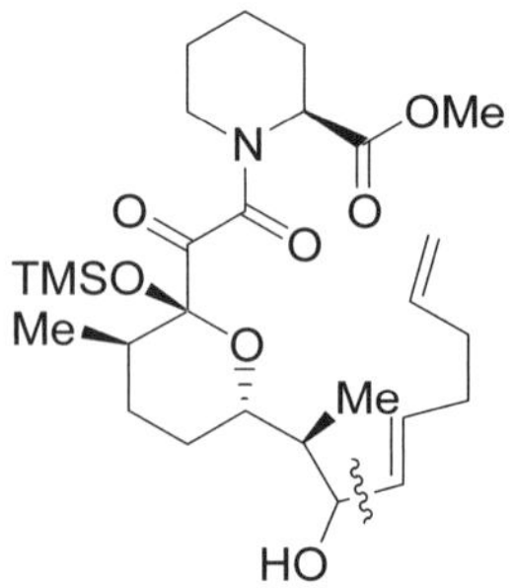

Antascomicin A, Chakraborty[44]
C1–C21 fragment
vinyl iodide, 62%

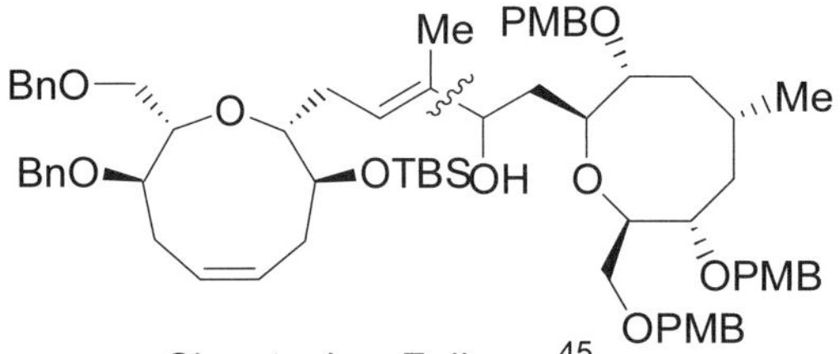

Ciguatoxins, Fujiwara[45]
FGHI ring
vinyl iodide, 85% of separable diastereomers

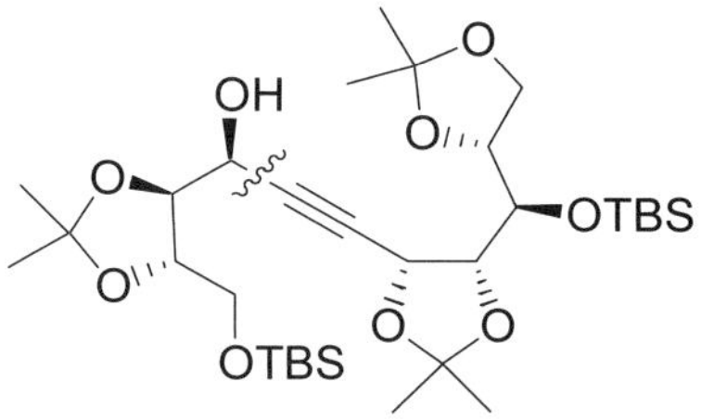

Hikizimycin, Furstner[46]
higher sugar core
iodoalkyne, 82%, > 95% *de*

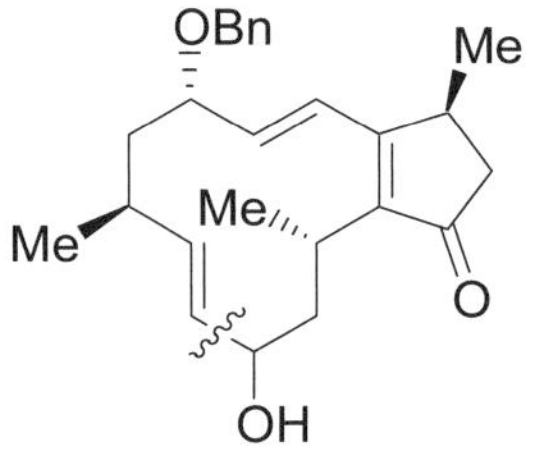

Mangicols, Uemura[47]
spirotetracyclic carbon core
vinyl iodide, 76%, 5:1 ratio

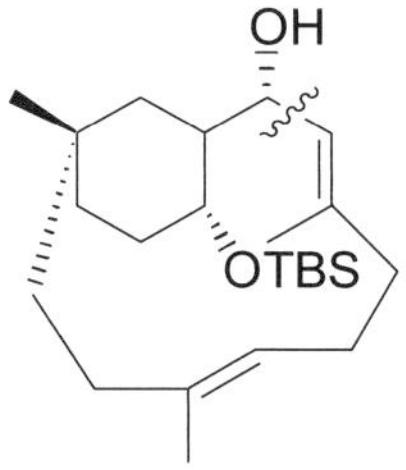

Phomactins, Maleczka[48]
octahydrochromene core
vinyl iodide, 60%, 1:1.3 *Z/E*

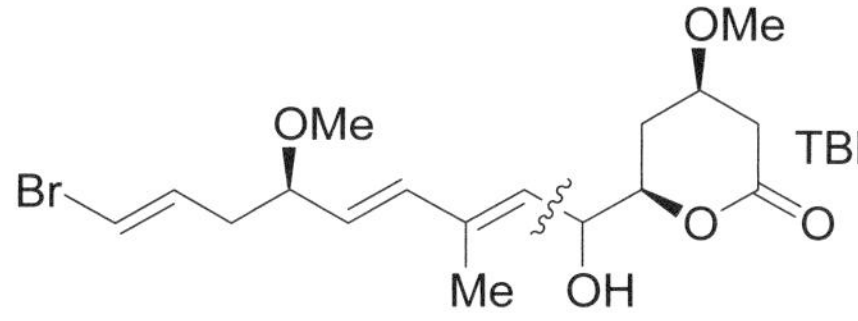

Phorboxazole B, Paterson[49]
C33–C46 fragment
vinyl iodide, 29%, *R/S* = 1,0:3:4

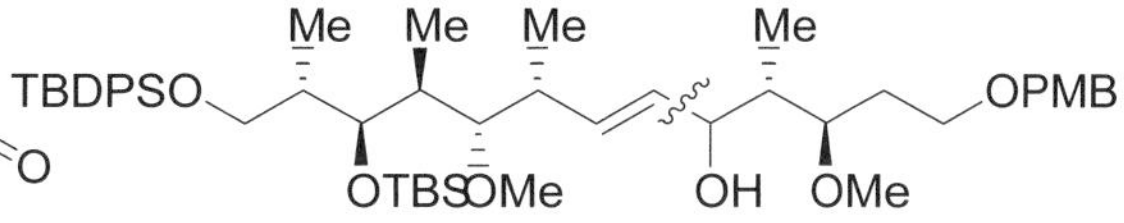

Reidispongiolide A, D'Auria[50]
C23–C35 fragment
vinyl iodide, 74%

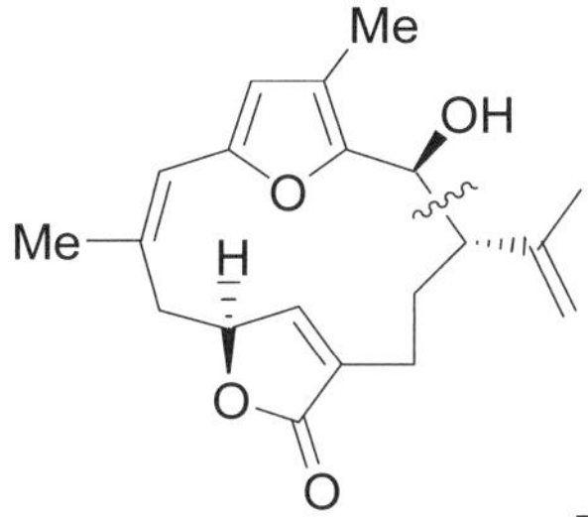

(+/-)-Bipinnatin J, Trauner[51]
vinyl bromide, 59%, *dr* > 9:1

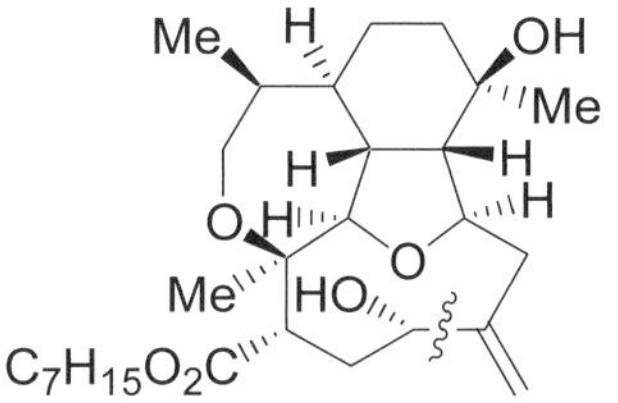

Briarellins E and F, Overman[52]
vinyl iodide, 79%

The structures below are natural products where the NHK reaction was used either in the second or last step of the synthesis where the natural product was completed.

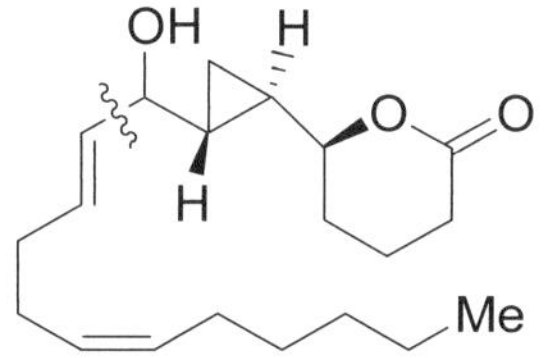

Eicosanoid 4, Mohapatra[53]
vinyl iodide, 82%

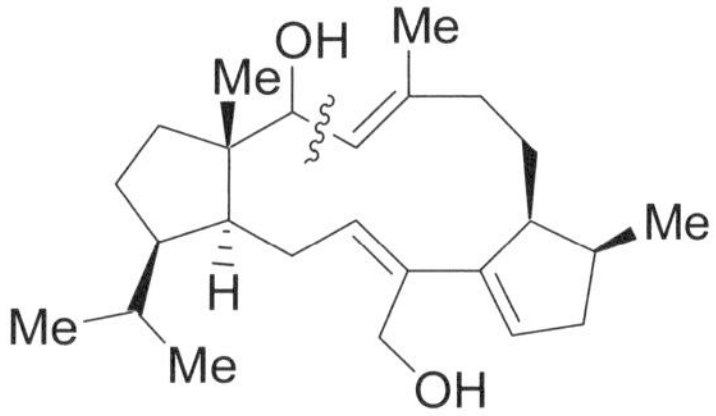

1,22-Dihydroxynitianes, Dake[54]
vinyl iodide, 70%
1:1 ratio of diastereomers

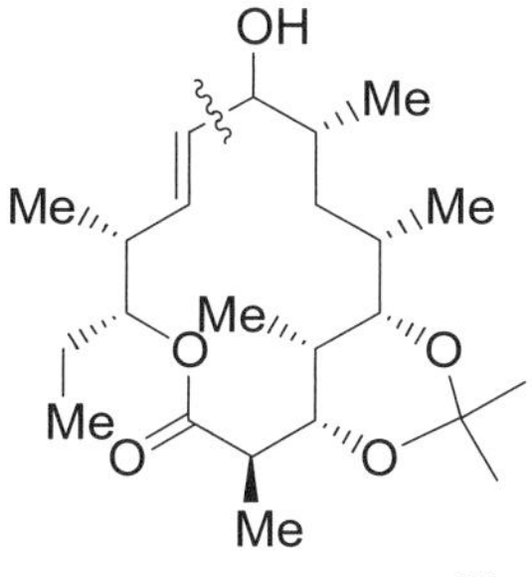

Narbonolide, Fecik[55]
vinyl iodide, 58%

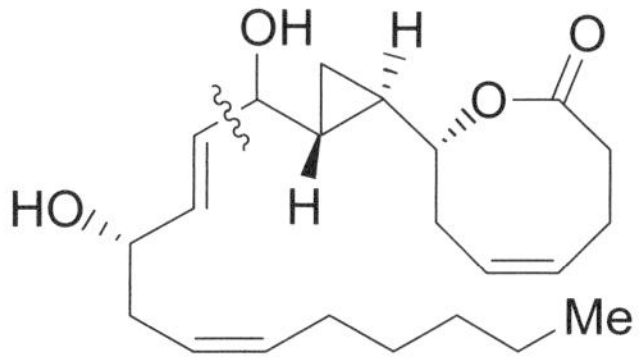

Solandelactones E and F, White[56]
vinyl iodide, 68%
3.5:1 ratio of diastereomers

The structures below are the completed total syntheses of natural products where the NHK reaction is utilized in any part of the synthetic scheme.

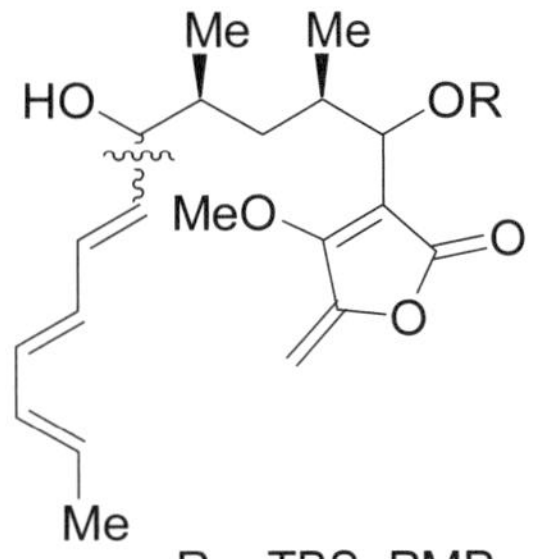

R = TBS, PMB

Abyssomicin C, Couladouros[57]

vinyl iodide, 40%

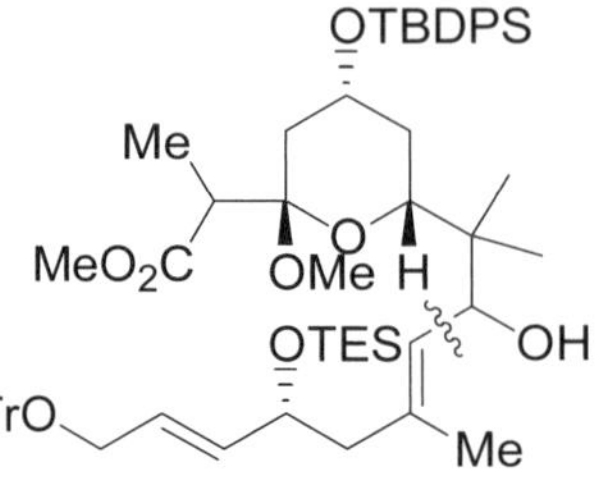

Aurisides A and B, Kigoshi[58]

vinyl iodide, 90%

2:1 ratio of diastereomers

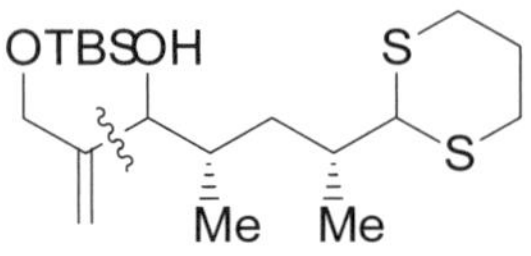

Azapriacid-1, Nicolaou[59]

vinyl iodide, 95%

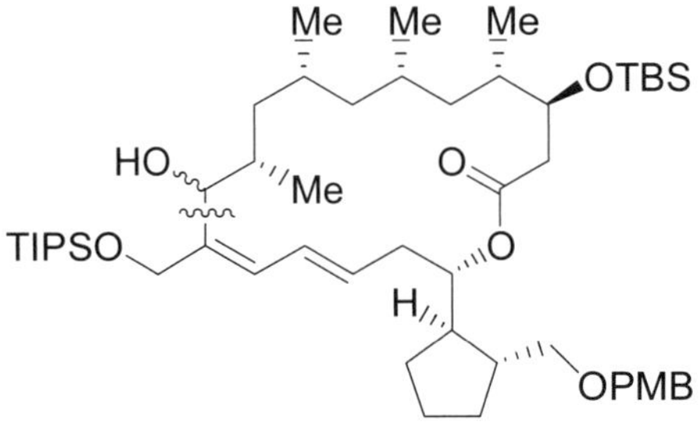

Borrelidin, Omura[60]

vinyl iodide, 55%; vinyl bromide, 13%

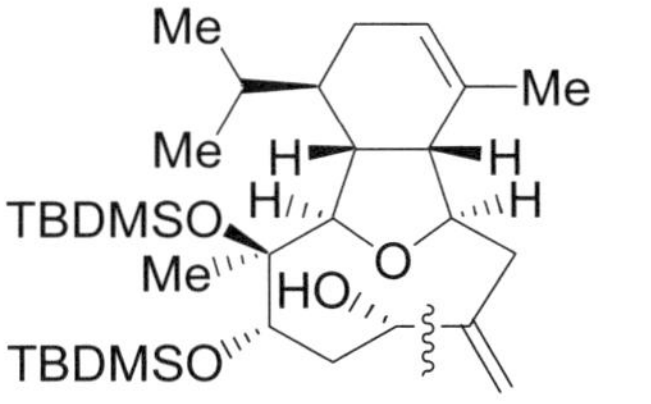

Deacetoxyalcyonin Acetate, Alcyonin, Overman[61]

vinyl iodide, 55%, >20:1 ratio

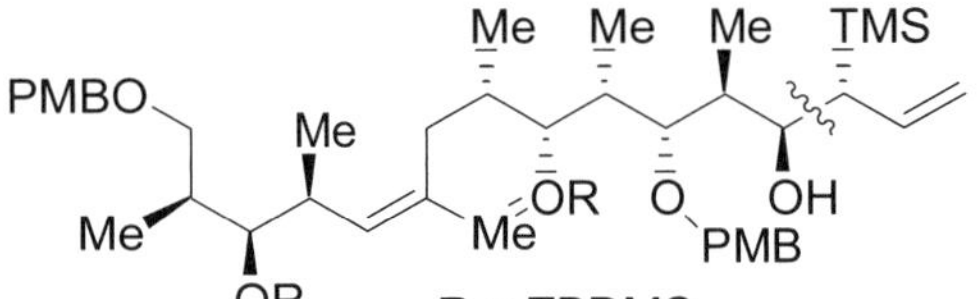

R = TBDMS

C7–C24 Fragment of (+)-Discodermolide, Mickel[62]

vinyl bromide, 81%

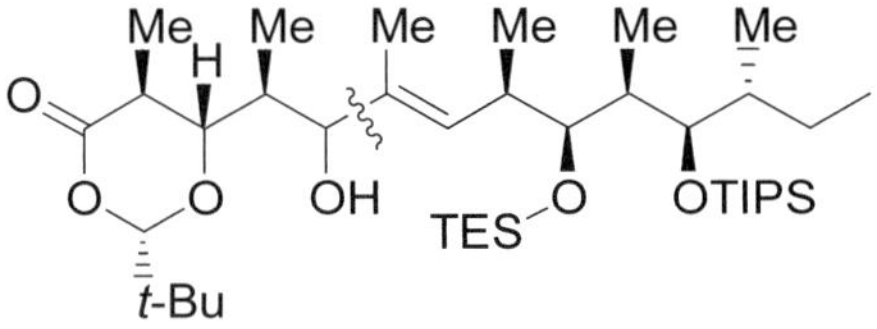

Ebelactone A, Fleming[63]

iodide, 41%, 77:23 ratio

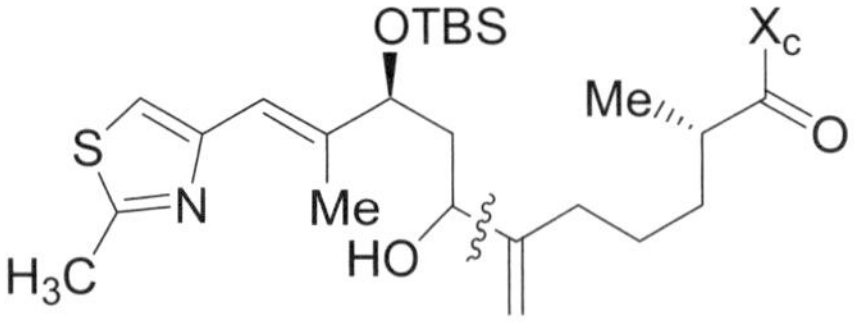

Epothilones B and D, Taylor[64]

vinyl iodide, 93%

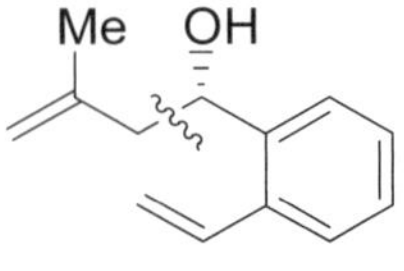

FR901512, Nakada[65]

methallyl chloride, 93%, 92% *ee* with Nakada catalyst **25**

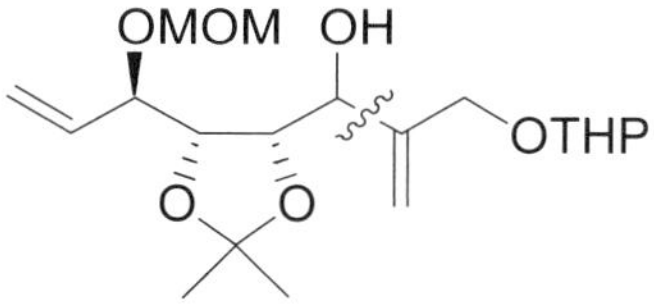

(-)-Gabosine precursor, Rao[66]
vinyl iodide, 84%

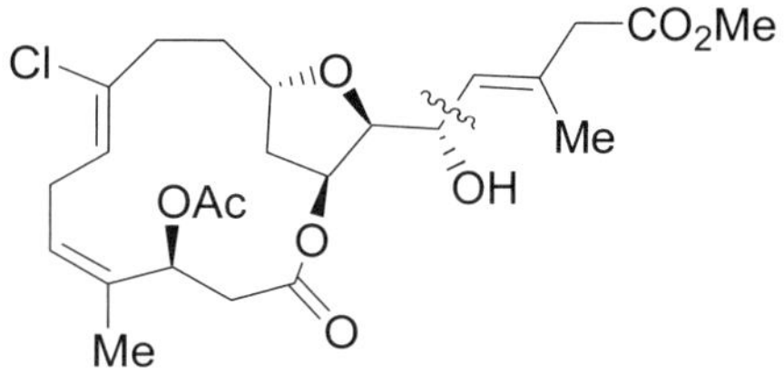

Hateruamalide, Kigoshi[67]
vinyl iodide, 57%, 11:1 ratio

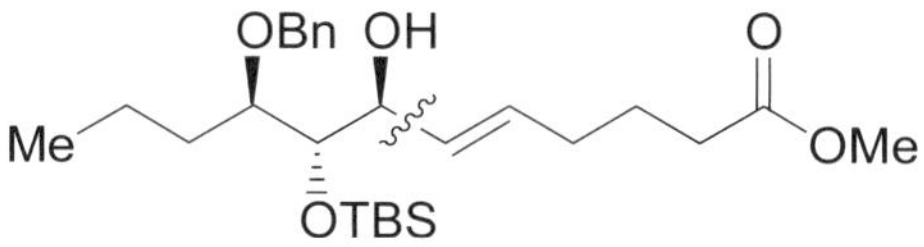

(+)-Herbarumin I, Pilli[68]
vinyl iodide, 76%, 3.4:1 ratio

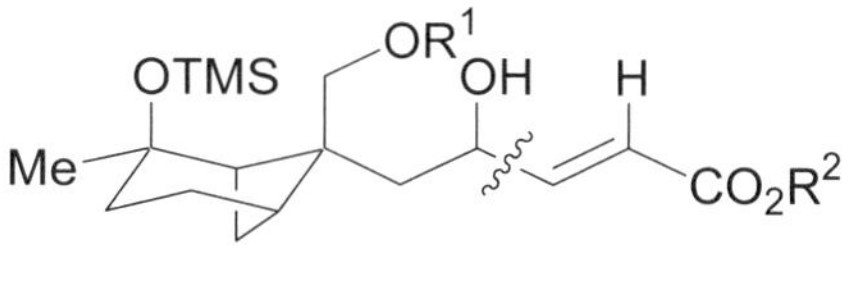

(+)-Massarinolidn B and
(+)-4-*epi*-Massarinolin B,
Bermejo[69]
vinyl iodide, 56–95%

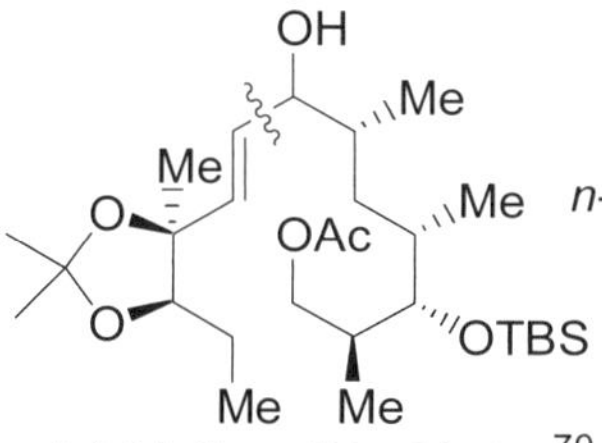

(+)-Methynolide, Yadav[70]
vinyl iodide, 65%

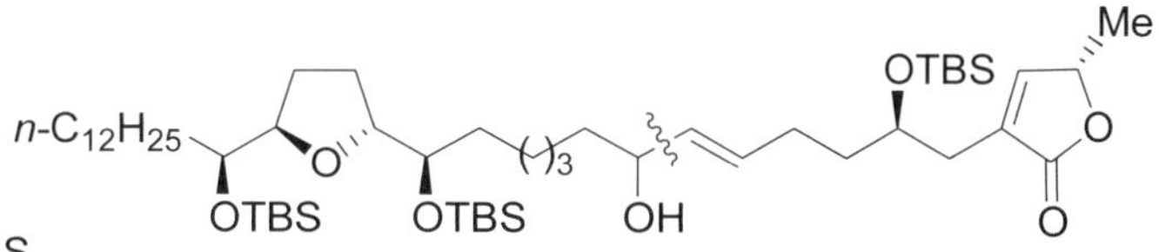

Mosin B, Tanaka[71]
vinyl iodide, 71%

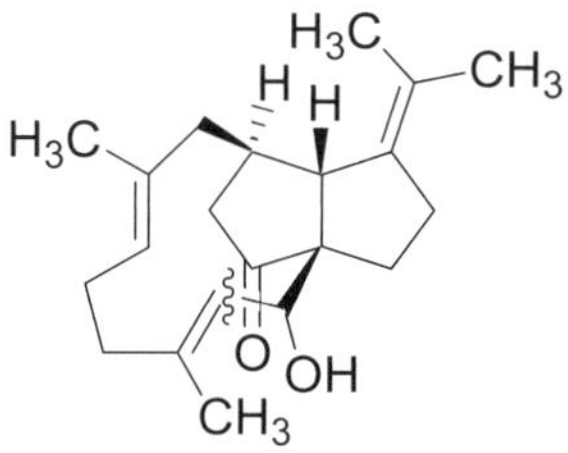

(–) and (+)-Negillamine A_2, Ready[72]
vinyl iodide, 73%, *dr* >10:1

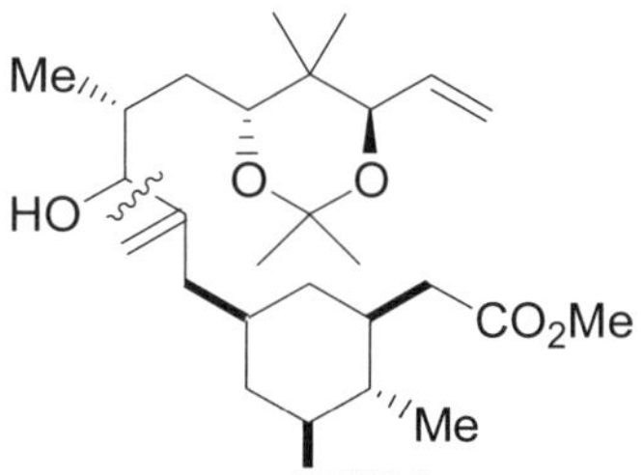

Polycavernoside A, White[73]
vinyl bromide, 79%
1:1 ratio of epimers

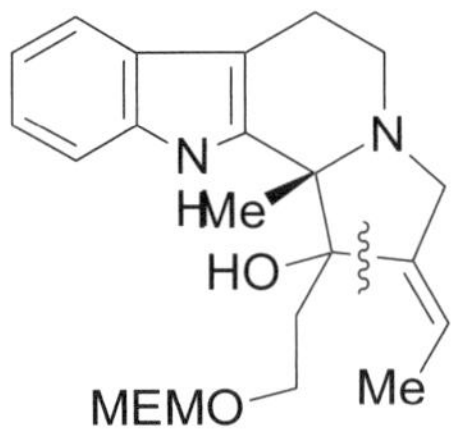

(–)-Subincanadines A and B, Takayama[74]
vinyl iodide, 88%
1.7:1 separable diastereomers

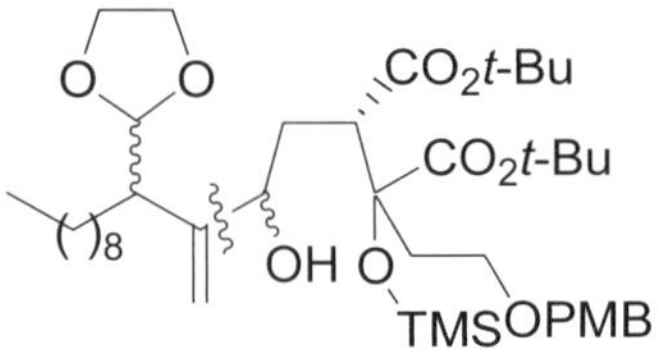

(±)-Trachyspic Acid, Hatakeyama[75]
vinyl triflate, 83%

NHK Reaction in Drug Discovery Analogues

The NHK reaction has also been reported in the syntheses of biologically active drug candidates such as analogues of laulimalide and epitholine B.

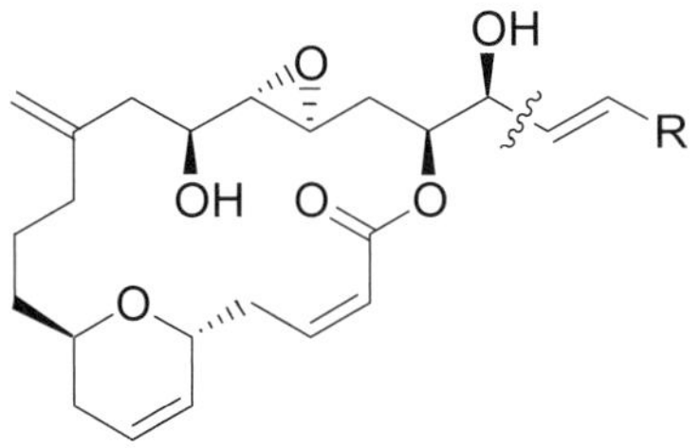

Laulimalide Analogues, Paterson[76]
vinyl iodide, DIANNE catalyst **12**

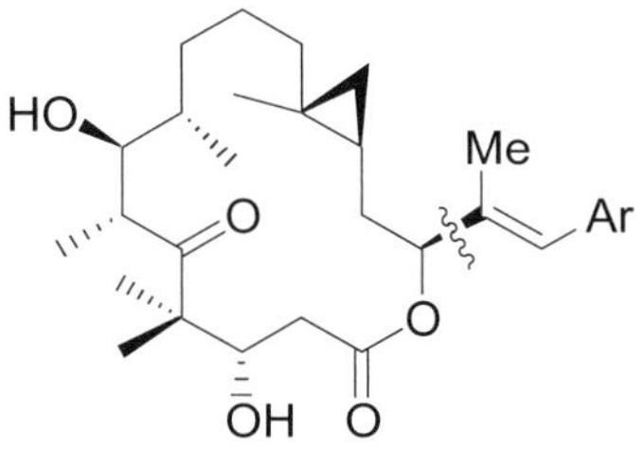

trans-12,13-cyclopropyl
Epothilone B analogues, Nicolaou[77]
vinyl iodide

1.3.3.6 Experimental

(*S*)-1-Phenyl-3-butene-1-ol (45)[22]

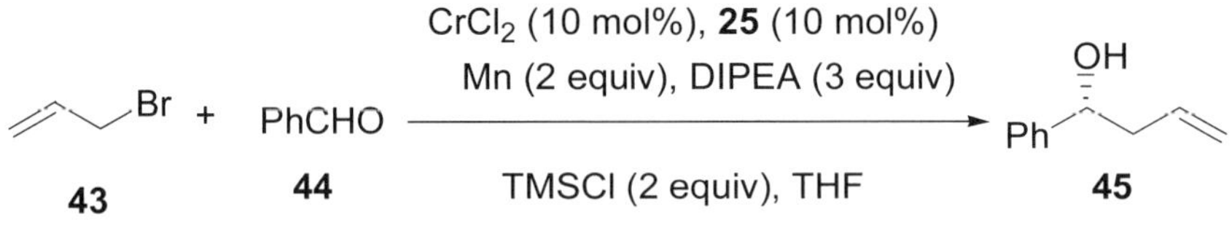

A mixture of ligand **25** (27.1 mg, 0.0500 mmol), $CrCl_2$ (6.3 mg, 0.0510 mmol), and Mn (53.8 mg, 0.979 mmol) was azeotroped three times with toluene and dried under high vacuum, and was suspended in THF (2 mL). The color of the suspension immediately turned to brown. To the stirred suspension was added DIPEA (0.026 mL, 0.15 mmol), and after 5 min to the resulting mixture was added allyl bromide (**43**, 0.086 mL, 0.99 mmol). After stirring for 30 min, the color of the mixture turned to greenish brown. To the stirred mixture was added benzaldehyde (**44**, 0.050 mL, 0.49 mmol), TMSCl (0.125 mL, 0.985 mmol) successively at 0 °C. After 12 h, the color of the reaction mixture turned to reddish brown. The reaction was quenched with saturated aqueous sodium bicarbonate (1 mL), filtered through a pad of Celite and the filtrate evaporated under vacuum. The crude product was dissolved in THF (2 mL) and the stirred mixture was treated with TBAF (1 mL, 1.0 mmol, 1.0 M in THF). The reaction was quenched with adding saturated aqueous ammonium chloride (2 mL), extracted with Et_2O (4 × 10 mL), dried over sodium sulfate, and evaporated. The residue was purified by flash chromatography (hexanes/ethyl acetate 10:1) to afford (*S*)-1-phenyl-3-butene-1-ol (**45**, 64.7 mg, 89%, 93% *ee*): $[\alpha]_D^{27}$ –46.2 (*c* 0.94, benzene), *ee* was determeined by HPLC (254 nm); Daicel Chiral Cell OD-H 0.46 cm × 25 cm; hexanes/isopropanol 19:1; flow rate = 0.3 mL/min; retention time: 26.4 min for (*R*)-1-phenyl-3-butene-1-ol, 28.7 min for (*S*)-1-phenyl-3-butene-1-ol.

***9-(Tert*-butyldimethylsilyloxy)-5,6,12,13-tetrahydro-8,9,10,11-tetrahydro-7*H*-benzo[11]annulen-7-ol** (33).[31]

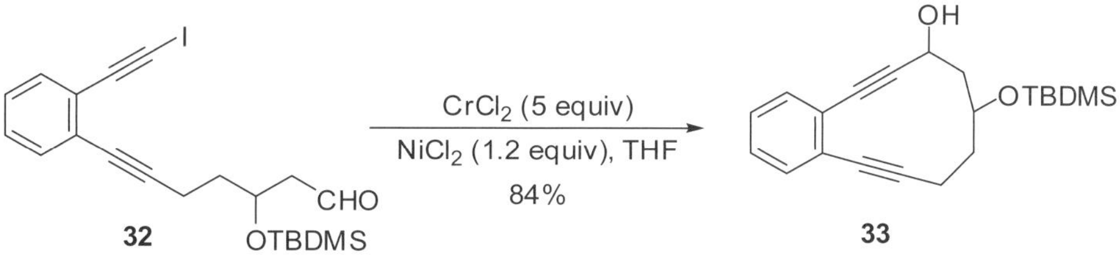

32 33

A solution of **32** (2.48 g, 5.33 mmol) in anhydrous THF (10 mL) was added to a vigorously stirred deoxygenated suspension of anhydrous $CrCl_2$ (3.28 g, 26.65 mmol) and anhydrous $NiCl_2$ (0.820 g, 6.33 mmol) in THF (700 mL) under argon and stirred for 4 h. The reaction mixture was concentrated to ~ 100 mL, diluted with ethyl acetate (300 mL) and washed with brine/saturated aqueous ammonium chloride (200 mL, 1:1) and brine (100 mL). The aqueous phase was back-extracted with ethyl acetate (100 mL). The combined organic phases were dried over magnesium sulfate and the solvents were removed under vacuum. The crude product was separated by column chromatography (ethyl acetate/dichloromethane 5:95) to yield two diastereomers, **33a** (0.364 g, 20%, R_f = 0.45) and **33b** (1.157 g, 64%, R_f = 0.40). **33a**: colorless solid.

1.3.3.7 References

1. [R] Fürstner, A. *Chem. Rev.* **1999**, *99*, 991.
2. [R] Denmark, S. E.; Fu, J. *Chem. Rev.* **2003**, *103*, 2763.
3. [R] Kishi, Y. *Pure Appl. Chem.* **1992**, *64*, 343.

4. a) Okude, Y.; Hirano, S.; Hiyama, T.; Nozaki, H. *J. Am. Chem. Soc.* **1977**, *99*, 3179. b) Okude, Y.; Hiyama, T.; Nozaki, H. *Tetrahedron Lett.* **1977**, 3829. c) Hiyama, T.; Okude, Y.; Kimura, K.; Nozaki, H. *Bull. Chem. Soc. Jpn.* **1982**, *55*, 561. d) Hiyama, T.; Kimura, K.; Nozaki, H. *Tetrahedron Lett.* **1981**, *22*, 1037. e) Takai, K.; Kimura, K; Kuroda, T.; Hiyama, T.; Nozaki, H. *Tetrahedron Lett.* **1983**, *24*, 5281.
5. Jin, H.; Uenishi, J; Christ, W. J.; Kishi, Y. *J. Am. Chem. Soc.* **1986**, *108*, 5644.
6. Takai, K.; Tagashira, M.; Kuroda, T.; Oshima, K.; Utimoto, K.; Nozaki, H. *J. Am. Chem. Soc.* **1986**, *108*, 6048.
7. a) Fürstner, A.; Shi, N. *J. Am. Chem. Soc.* **1996**, *118*, 12349. b) Fürstner, A.; Shi, N. *J. Am. Chem. Soc.* **1996**, *118*, 2533. c) Fürstner, A. *Chem. Eur. J.* **1998**, *4*, 567. d) Fürstner, A. *Pure Appl. Chem.* **1998**, *70*, 1071.
8. [R] Hargaden, G. C.; Guiry, P. J. *Adv. Synth. Catal.* **2007**, *349*, 2407.
9. Chen, C.; Tagami, K; Kishi, Y. *J. Org. Chem.* **1995**, *60*, 5386.
10. a) [R] Bandini, M.; Cozzi, P. G.; Umani-Ronchi, A. *Pure App. Chem.* **2001**, *73*, 325. b) [R] Bandini, M.; Cozzi, P. G.; Umani-Ronchi, A. *Chem. Commun.* **2002**, 919.
11. Bandini, M.; Cozzi, P. G.; Melchiorre, P.; Umani-Ronchi, A. *Agnew. Chem., Int. Ed.* **1999**, *38*, 3357.
12. a) Bandini, M.; Cozzi, P. G.; Umani-Ronchi, A. *Agnew. Chem., Int. Ed.* **2000**, *39*, 2327. b) Bandini, M.; Cozzi, P. G.; Umani-Ronchi, A. *Tetrahedron* **2001**, *57*, 835.
13. Bandini, M.; Cozzi, P. G.; Melchioore, P.; Morganti, S.; Umani-Ronchi, A. *Org. Lett.* **2001**, *3*, 1153.
14. Bandini, M.; Cozzi, P. G.; Melchiorre, P.; Tino, R.; Umani-Ronchi, A. *Tetrahedron: Asymmetry* **2001**, *12*, 1063.
15. Berkessel, A.; Menche, D.; Sklorz, C. A.; Schroder, M.; Paterson, I. *Agnew. Chem., Int. Ed. Engl.* **2003**, *42*, 1032.
16. Rozners, E.; Fontanez, J. *Lett. Org. Chem.* **2005**, *2*, 524.
17. a) Xia, G.; Yamamoto, H. *J. Am. Chem. Soc.* **2006**, *128*, 2554. b) Xia, G.; Yamamoto, H. *J. Am. Chem. Soc.* **2007**, *129*, 496. c) Yamamoto, H; Xia, G. *Chem. Lett.* **2007**, *36*, 1082.
18. Wan, Z.-K.; Choi, H.-w.; Kang, F.-A.; Nakajima, K.; Demeke, D.; Kishi, Y. *Org. Lett.* **2002**, *4*, 4431.
19. Choi, H.-w.; Nakajima, K.; Demeke, D. ; Kang, F.-A.; Jun, H.-S.; Wan, Z.-K.; Kishi, Y. *Org. Lett.* **2002**, *4*, 4435.
20. Kurosu, M.; Lin, M.-H.; Kishi, Y. *J. Am. Chem. Soc.* **2004**, *126*, 12248.
21. Namba, K.; Kishi, Y. *Org. Lett.* **2004**, *6*, 5031.
22. Inoue, M.; Suzuki, T.; Nakada, M. *J. Am. Chem. Soc.* **2003**, *125*, 1140.
23. Suzuki, T.; Kinoshita, A.; Kawada, H.; Nakada, M. *Synlett* **2003**, 570.
24. Inoue, M.; Nakada, M. *J. Am. Chem. Soc.* **2007**, *129*, 4164.
25. Inoue, M.; Nakada, M. *Org. Lett.* **2004**, *6*, 2977.
26. Inoue, M.; Nakada, M. *Angew. Chem., Int. Ed.* **2006**, *45*, 252.
27. Lee, J.-Y.; Miller, J. J.; Hamilton, S. S.; Sigman, M. S. *Org. Lett.* **2005**, *7*, 1837.
28. Miller, J. J.; Sigman, M. S. *J. Am. Chem. Soc.* **2007**, *129*, 2752.
29. a) McManus, H. A.; Guiry, P. J. *J. Org. Chem.* **2002**, *67*, 8566. b) McManus, H. A.; Cozzi, P. G.; Guiry, P. J. *Adv. Synth. Catal.* **2006**, *348*, 551.
30. Hargaden, G. C.; Muller-Bunz, H.; Guiry, P. J. *Eur. J. Org. Chem.* **2007**, 4235.
31. Karpov, G. V.; Popik, V. V. *J. Am. Chem. Soc.* **2007**, *129*, 3792.
32. a) Dai, W.-M.; Wu, A.; Hamaguchi, W. *Tetrahedron Lett.* **2001**, *42*, 4211. b) Dai, W.-M.; Tachi, Y.; Nishimoto, S.-i.; Zhong, X.; Guo, Z. *Lett. Drug Des. Discov.* **2004**, *1*, 69. c) Tachi, Y.; Dai, W.-M.; Tanabe, K.; Nishimoto, S.-i. *Bioorg. Med. Chem.* **2006**, *14*, 3199. d) Dai, W.-M.; Li, K. W.; Wu, A.; Hamaguchi, W.; Lee, M. Y. H.; Zhou, L.; Ishii, A.; Nishimoto, S.-i. *J. Med. Chem.* **2002**, *45*, 758.
33. Pilli, R. A.; Victor, M. M. *Tetrahedron Lett.* **2002**, *43*, 2815.
34. Sandoval, C.; Lopez-Perez, J. L.; Bermejo, F. *Tetrahedron* **2007**, *63*, 11738.
35. Elliott, M. R.; Dhimane, A.-L; Hamon, L.; Malacria, M. *Eur. J. Org. Chem.* **2000**, *1*, 155.
36. Dutheuil, G.; Lei, X.; Pannecoucke, X.; Quirion, J.-C. *J. Org. Chem.* **2005**, *70*, 1911.
37. a) Falck, J. R.; Barma, D. K.; Mioskowski, C.; Schlama, T. *Tetrahedron Lett.* **1999**, *40*, 2091. b) Barma, D. K.; Baati, R.; Valleix, A.; Mioskowski, C.; Falck, J. R. *Org. Lett.* **2001**, *3*, 4237.

38. Tisserand, S.; Bejot, R.; Billaud, C.; Li, D. R.; Falck, J. R.; Mioskowski, C. *Tetrahedron Lett.* **2006**, *47*, 5177.
39. Majumdar, K. K.; Cheng, C.-H. *Org. Lett.* **2000**, *2*, 2295.
40. Durandetti, M.; Nedelec, J.-V.; Perichon, J. *Org. Lett.* **2001**, *3*, 2073.
41. Comins, D. L.; Hiebel, A.-C.; Huang, S. *Org. Lett.* **2001**, *3*, 769.
42. Lombardo, M.; Licciulli, S.; Morganti, S.; Trombini, C. *Chem. Commun.* **2003**, 1762.
43. Bedore, M. W.; Chang, S.-K.; Paquette, L. A. *Org. Lett.* **2007**, *9*, 513.
44. Chakraborty, T. K.; Mohan, B. K. *Tetrahedron Lett.* **2006**, *47*, 4999.
45. a) Takizawa, A.; Fujiwara, K.; Doi, E.; Murai, A.; Kawai, H.; Suzuki, T. *Tetrahedron Lett.* **2006**, *47*, 747. b) Takizawa, A.; Fujiwara, K.; Doi, E.; Murai, A.; Kawai, H.; Suzuki, T. *Tetrahedron* **2006**, *62*, 7408.
46. Fürstner, A.; Wuchrer, M. *Chem. Eur. J.* **2006**, *12*, 76.
47. Araki, K.; Saito, K; Arimoto, H; Uemura, D. *Angew. Chem., Int. Ed.* **2004**, *43*, 81.
48. Mi, B.; Maleczka, R. E., Jr. *Org. Lett.* **2001**, *3*, 1491.
49. Paterson, I.; Steven, A.; Luckhurst, C. A. *Org. Biomol. Chem.* **2004**, *2*, 3026.
50. Zampella, A.; Sepe, V.; D'Orsi, R; D'Auria, M. D. *Lett. Org. Chem.* **2004**, *1*, 308.
51. Roethle, P. A.; Trauner, D. *Org. Lett.* **2006**, *8,* 345.
52. Corminboeuf, O.; Overman, L. E.; Pennington, L. D. *J. Am. Chem. Soc.* **2003**, *125*, 6650.
53. Mohapatra, D. K.; Yellol, G. S.; *Arkivoc* **2005**, *3*, 144.
54. Wilson, M. S.; Woo, J. C. S.; Dake, G. R. *J. Org. Chem.* **2006**, *71*, 4237.
55. Venkatraman, L; Aldrich, C. C.; Sherman, D. H.; Fecik, R. A. *J. Org. Chem.* **2005**, *70*, 7267.
56. White, J. D.; Martin, W. H. C.; Lincoln, C.; Yang, J. *Org. Lett.* **2007**, *9*, 3481.
57. Couladouros, E. A.; Bouzas, E. A.; Magos, A. D. *Tetrahedron* **2006**, *62*, 5272.
58. Suenaga, K; Hoshino, H.; Yoshii, T.; Mori, K.; Sone, H.; Bessho, Y.; Sakakura, A.; Hayakawa, I.; Yamada, K.; Kigoshi, H. *Tetrahedron* **2006**, *62*, 7687.
59. Nicolaou, K. C.; Pinko, P. M.; Bernal, F.; Frederick M. O.; Qian, W.; Uesaka, N.; Diedrichs, N.; Hinrichs, J; Koftis, T. V.; Loizidou, E.; Petrovic, G.; Rodriquez, M.; Sariah, D.; Zou, N. *J. Am. Chem. Soc.* **2006**, *128*, 2244.
60. Nagamitsu, T.; Takano, D.; Marumoto, K; Fukuda, T.; Furuya, K; Otoguro, K; Takeda, K.; Kuwajima, I.; Harigaya, Y.; Omura, S. *J. Org. Chem.* **2007**, *72*, 2744.
61. a) MacMillan, D. W. C.; Overman, L. E.; Pennington, L. D. *J. Am. Chem. Soc.* **2001**, *123*, 9033. b) Corminboeuf, O.; Overman, L. E.; Pennington, L. D. *Org. Lett.* **2003**, *5*, 1543.
62. Mickel, S. J.; Sedelmeier, G. H.; Niederer, D.; Schuerch, F.; Seger, M.; Schreiner, K.; Dacffler, R.; Osmania, A.; Bixel, D.; Loiseleur, Loiseleur, O.; Cercus, J.; Stettler, H.; Schaer, K.; Gamboni, R.; Bach, A.; Chen, G.-P.; Chen, W.; Geng, P. Lee, G. T.; Loeser, E.; Mckenna, J.; Kinder, F. R.; Konigsberger, K; Prasad, K.; Ramsey, T. M.; Reel, N.; Repic, O.; Rogers, L; Shich, W.-C.; Wang, R.-M.; Waykole, L; Xue, S. *Org. Proc. Res. Devel.* **2004**, *8*, 113.
63. Archibald, S. C.; Barden, D. J.; Bazin, J. F. Y.; Fleming, I.; Foster, C. F.; Mandal, A. K.; Mandal, A. K.; Parker, D.; Takaki, K.; Ware, A. C.; Williams, A. R. B.; Zwicky, A. B. *Org. Biomol. Chem.* **2004**, *2*, 1051.
64. Taylor, R. E.; Chen, Y. *Org. Lett.* **2001**, *3*, 2221.
65. Inoue, M; Nakada, M. *J. Am. Chem. Soc.* **2007**, *129*, 4164.
66. Ramana, G. V.; Rao, B. V. *Tetrahedron Lett.* **2005**, *46*, 3049.
67. Kigoshi, H.; Kita, M.; Ogawa, S.; Itoh, M.; Uemura, D. *Org. Lett.* **2003**, *5*, 957.
68. Sabino, A. A.; Pilli, R. A. *Tetrahedron Lett.* **2002**, *43*, 2819.
69. Lopez, M. R.; Bermejo, F. A. *Tetrahedron* **2006**, *62*, 8095.
70. Yadav, J. S.; Pratap, T. V.; Rajender, V. *J. Org. Chem.* **2007**, *72*, 5882.
71. a) Maezaki, N.; Kojima, N.; Sakamoto, A.; Iwata, C.; Tanaka, T. *Org. Lett.* **2003**, *3*, 429. b) Maezaki, N.; Kojima, N.; Sakamoto, A.; Tominaga, H.; Iwata, C.; Tanaka, T.; Monden, M.; Damdinsuren, B.; Nakamori, S. *Chem. Eur. J.* **2003**, *9*, 389.
72. Bian, J.; Van Wingerden, M; Ready, J. M. *J. Am. Chem. Soc.* **2006**, *128*, 7428.
73. a) White, J. D.; Blakemore, P. R.; Browler, C. C.; Hong, J.; Lincoln, C. M.; Nagomyy, P. A.; Robarge, L. A.; Wardrop, D. J. *J. Am. Chem. Soc.* **2001**, *123*, 8593. b) Blakemore, P. R.; Browder, C. C.; Hong, J.; Lincoln, C. M.; Nagornyy, P. A.; Robarge, L. A.; Wardrop, D. J.; White, J. D. *J. Org. Chem.* **2005**, *70*, 5449.
74. Suzuki, K.; Takayama, H. *Org. Lett.* **2006**, *8*, 4605.

75. Hirai, K.; Ooi, H.; Esumi, T.; Iwabuchi, Y.; Hatakeyama, S. *Org. Lett.* **2003**, *3*, 857.
76. a) Paterson, I.; Bergmann, H.; Menche, D.; Berkessel, A. *Org. Lett.* **2004**, *6*, 1293. b) Paterson, I.; Menche, D.; Hakansson, A.; Longstaff, A.; Wong, D.; Barasoain, I.; Buey, R. M.; Diaz, J. F. *Bioorg. Med. Chem. Lett.* **2005**, *15*, 2243.
77. a) Nicolaou, K. C.; Namoto, K.; Ritzen, A.; Ulven, T.; Shoji, M.; Li, J.; D'Amico, G.; Liotta, D.; French, C. T.; Wartmann, M.; Altmann, K.-H.; Giannakakou, P. *J. Am. Chem. Soc.* **2001**, *123*, 9313. b) Nicolaou, K. C.; Ritzen, A.; Namoto, K.; Buey, R. M.; Diaz, J. F.; Andreu, J. M.; Wartman,, M.; Altmann, K.-H.; O'Brate, A.; Giannakakou, P. *Tetrahedron* **2002**, *58*, 6413.

1.3.4 Tebbe Methylenation

Ji Zhang

1.3.4.1 Description

The Tebbe methylenation describes the conversion of carbonyl compound **1** into a terminal alkene **2** using titanium–aluminium complex **3** (the Tebbe reagent). Sterically encumbered carbonyl groups present in aldehydes, ketones, carboxylic acid derivatives (esters, lactones, and amides), and carbonates can be successfully methylenated utilizing this reagent. Several closely related titanium carbenoid reagents, including the Petasis, Takeda, and Takai reagents for carbonyl olefinations are also discussed here.[1–4]

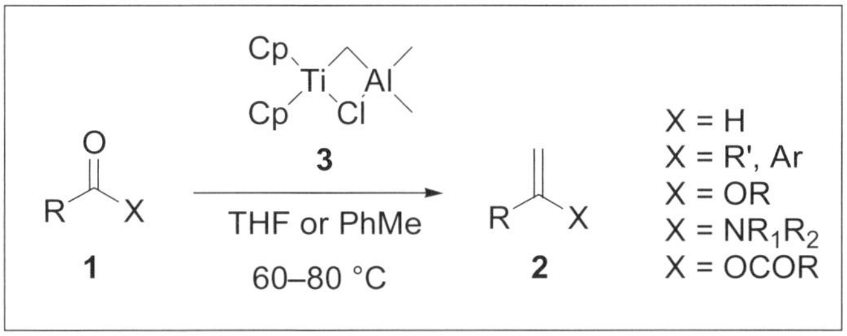

1.3.4.2 Historical Perspective

In 1978, Tebbe and co-workers at du Pont prepared the titanium–aluminum methylidene complex **3** from titanocene dichloride **4** and trimethylaluminium in toluene.[5] This complex **3** is an versatile methylene transfer reagent for conversion of ketones to terminal olefins when it was treated with a Lewis base such as pyridine or THF. Later on, Grubbs developed a facile method for the *in situ* generation of Tebbe reagent **3**.[6] In 1990, Pine and co-workers observed that the Tebbe reagent is unique in the carbonyl groups of carboxylic acid derivatives which are readily methylenated.[7] Examples include the formation of vinyl enol ethers from esters and enamines from amides. This reagent also provides an excellent alternative to the Wittig reagent for the methylenation hindered or base sensitive ketones, which have proven to be difficult reactants. In 1991, Pine compared the Tebbe reagent with the Wittig reagent for ketone methylenation and demonstrated for a variety of ketones, especially with hindered substrates, that the Tebbe reagent gives superior product yields.[8] It was also noted that the Tebbe reaction accomplishes methylenation in a non-basic medium, thus racemization does not take place on substrates with enolizable chiral centers. Today, this titanium-based alkylidenating reagent has been employed to prepare synthetic intermediates that cannot be access effectively using traditional alkenation methods.[9]

1.3.4.3 Mechanism

A highly reactive titanocene methylidene **5** is generated when the Tebbe reagent is treated with a Lewis base, even as mild as tetrahydrofuran (THF). Intermediate **5** efficiently methylenates a range of carbonyl groups, presumably via formation of oxatitanacyclobutane **7** by a [2 + 2] cycloaddition, which decomposes with elimination of $Cp_2Ti{=}O$ to afford alkene **8** in a matter of minutes at mild conditions. The driving force of this reaction is presumably the irreversible formation of the strong titanium oxygen double bond. Titanocene methylidene **5** is a typical Schrock carbene,[10] being an electron-deficient (16e) complex of an early transition metal in a high formal oxidation state. Such Schrock carbenes are nucleophilic at the carbene carbon atom and electrophilic at titanium, with their reactivity towards carbonyl groups being dominated by their high energy HOMOs. Thus, titanium alkylidenes would be expected to react preferentially with the most electrophilic carbonyl groups. It is known that the Tebbe reagent reacts more rapidly with amides than esters in the absence Lewis base because amide itself is a better Lewis base than ester, therefore, it generates the reactive titanium methylidene **5** more effectively.[7]

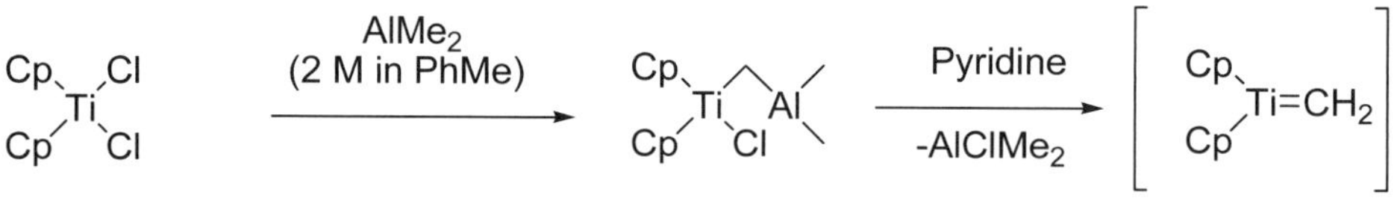

4 **3** **5**

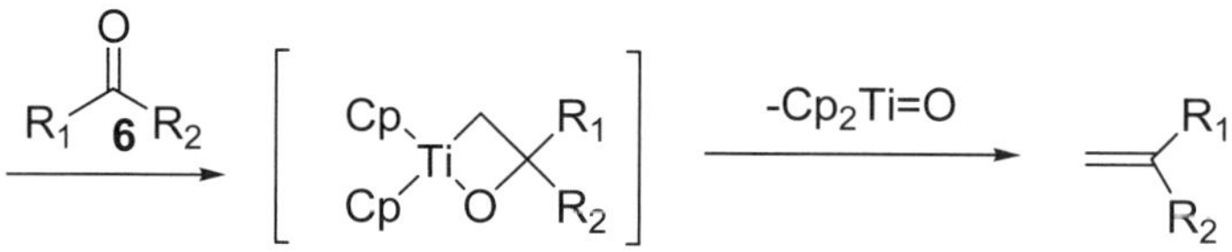

7 **8**

The main advantage of the Tebbe reagent **3** is for methylenating a range of carbonyl compounds and the reactive titanocene methylidene **5** is generated at mild conditions in the absence of strong base. The drawback of the Tebbe reagent is that it is capable for methylenation, to make $R_1R_2C{=}CH_2$, but unsuitable for the generation of other alkenes, such as $R_1R_2C{=}CHR$ ($R \neq H$). It should also be noted that the Tebbe reagent and the by-products formed by its decomposition are Lewis acidic and the handling of this reagent is hampered by its extreme sensitivity to air and moisture.

1.3.4.4 Variations and Improvements: Petasis olefination

Despite the great synthetic utility of the Tebbe reagent, the presence of aluminium in its structure results in several drawbacks, such as Lewis-acidic aluminium byproducts, poor stability of the Tebbe reagent in air and water, as well as the tedious treatment in the separation which limits its application in organic synthesis. A more superior reagent that exhibits similar reactivity is dimethyl titanocene **9** (Petasis reagent). In 1990, Petasis reported that dimethyltitanocene, readily prepared from titanocene dichloride **4** and methyllithium, or more preferably methylmagnesium chloride, methylenated carbonyl compound when the reaction mixture was heated to 60–75 °C in either THF or toluene.[11] The Petasis reagent tolerates brief exposure to air and water and is stable at room temperature when kept in solution in the dark. This reagent serves as a mild and practical alternative to the Tebbe reagent for carbonyl methylenation.

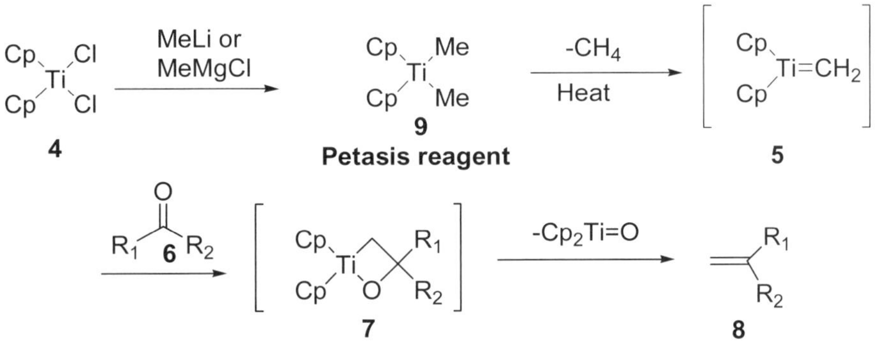

The common element between Tebbe and Petasis methylenation is that both share the same reactive intermediate, titanocene methylidene **5**, in the reaction pathway. Hughes has provided a strong evidence that the Petasis reaction proceeds by the rate-determining generation of titanocene methylidene **5** via an α-elimination to remove of methane, followed by a rapid reaction with the carbonyl compound.[12]

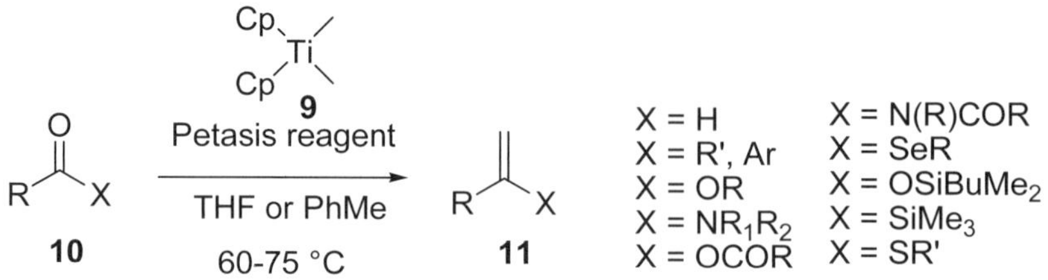

The α,β-unsaturated esters **12** and **14**, spirobislactone **16**, and vinylogous lactone **18** are smoothly methylenated by Petasis reagent. Silyl esters **22** and **24** are converted to silyl enol ethers **23** and **25**. Carbonate **20** can be methylenated to give ketene acetal **21**. Amide **26** and lactams can be methylenated, however the reaction is generally sluggish and the complete separation of Ti species is usually difficult. In a similar manner, thioester **28** and selenoester **30** are converted to alkenyl sulphide **29** and alkenyl selenide **31**, respectively. Additionally it has been demonstrated that acyl silanes can be converted to the corresponding alkenyl silanes.[13,14]

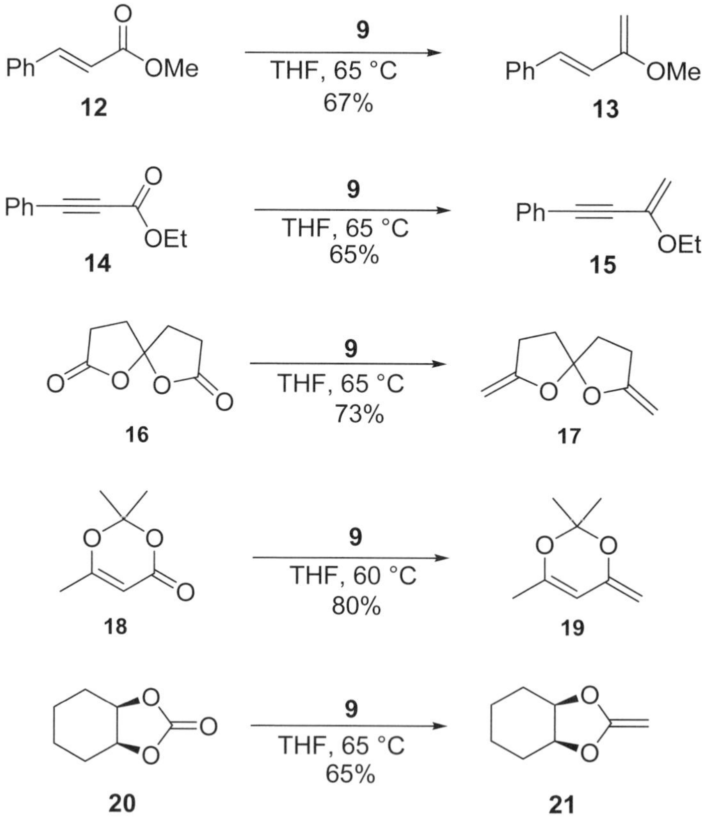

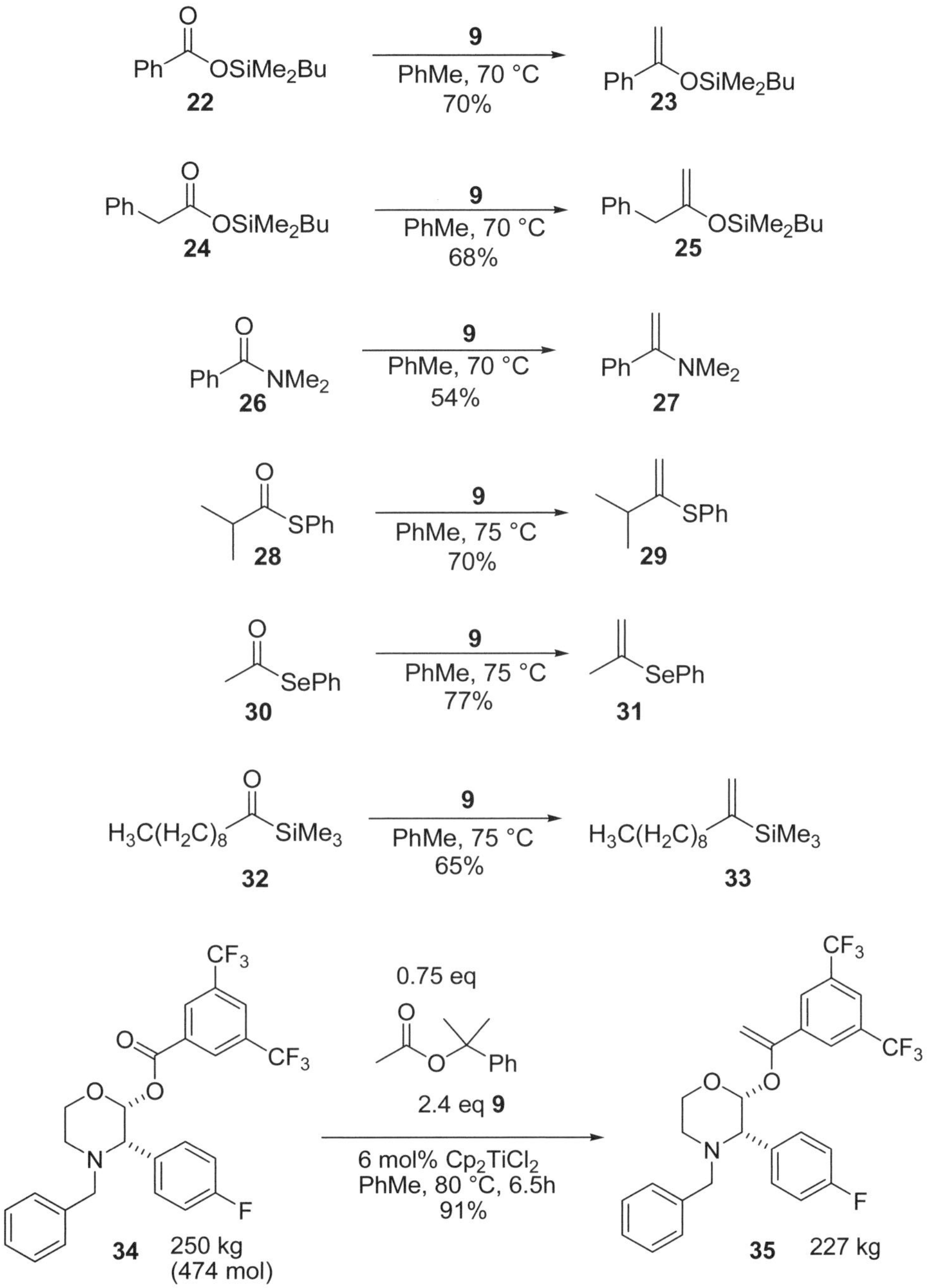

An elegant example of the preparation, storage, and use of the Petasis reagent for methylenation in multikilogram scale (250 kg, 474 mole) was provided by Payack and co-workers at Merck Process Chemistry Department in 2004.[15] The Petasis reagent should be stored refrigerated as a solution (in toluene or THF) because it is unstable in the solid state and decomposes, releasing heat and gas. Payack recommends quenching the methylenation reaction with a mixture of aqueous sodium bicarbonate and methanol while stirring at slightly elevated temperature (~ 40 °C) in order to convert titanium residues into insoluble waste that can be removed by filtration. On large scale it was demonstrated that ethanol can be used instead of methanol with the conversion being complete after stirring at 60 °C for 6 h. Payack and co-workers successfully developed a practical process for converting ester **34** to give enol ether **35** in excellent yield using the Petasis reagent. On the other hand, it was found that the utilization of a Tebbe methylenation generates only a 15% yield of enol ether **35**. This process was used to make hundreds of kilograms of an advanced intermediate of Aprepitant (Emend), which is employed as a therapy to prevent chemotherapy-induced nausea and vomiting.

Aldehydes and ketones can be selectively methylenated in the presence of less electrophilic carbonyl groups such as esters, amides and carbamates using the Petasis reagent. In Parsons' model study towards kainic acid, selective methylenation of ketone **36** was achieved using the Petasis reagent, avoiding epimerization of the neighbouring stereocenters and reaction with the methyl ester.[16] In Hegedus's route to α-alkyl-α-amino acid, key intermediate **39** was synthesized by the Petasis reagent.[17]

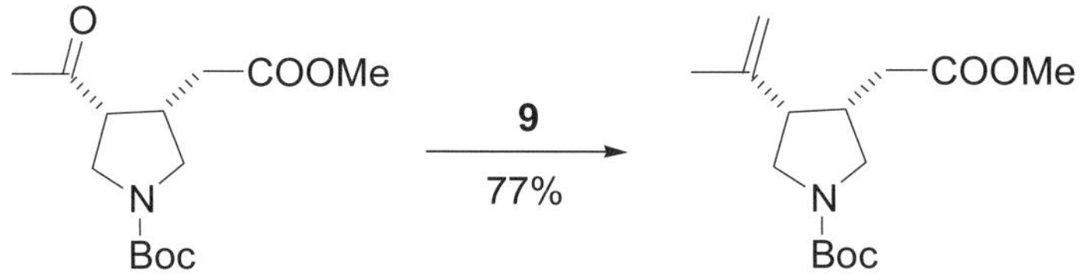

36 **37**

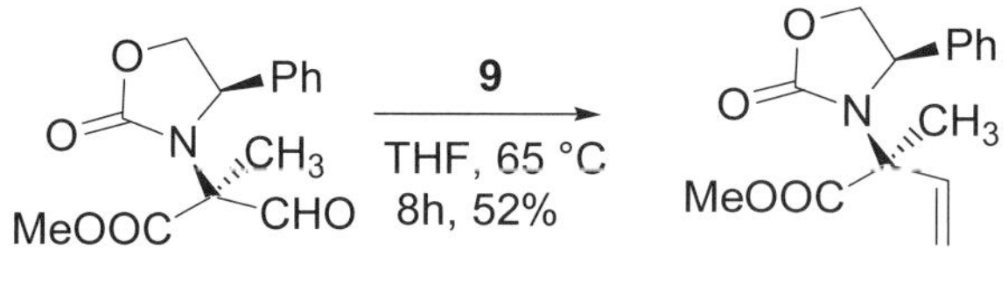

38 **39**

While homologated titanocene-aluminium complexes analogous to **3** are difficult to prepare; however, the corresponding dialkyl titanocenes of **5** can be readily prepared from titanocene dichloride and the appropriate organolithium or Grignard reagent. Therefore, Petasis-type reagents can be used not only for methylenation but also for olefination. Among these compounds that were shown to exhibit similar reactivity with **5** are the dibenzyl **40**,[18] bis(trimethyl-silymethyl) **41**, bis(cyclopropyl) **42** and other titanocenes **43**, **44** and **45**.[19-21] It should be noted that these dialkyl titanocenes are only useful for alkylidenation of carbonyl groups that are not capable of undergoing a prohibitively fast β-elimination process.

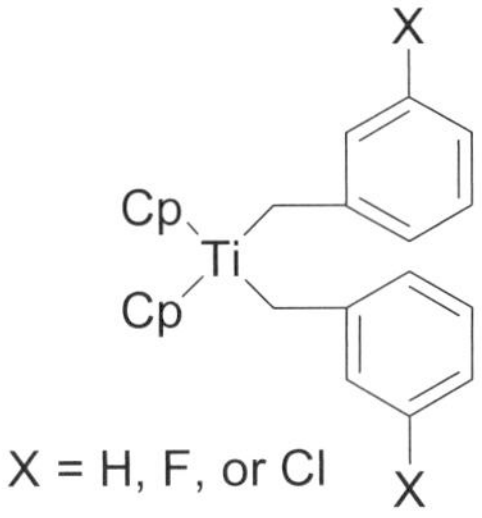

40

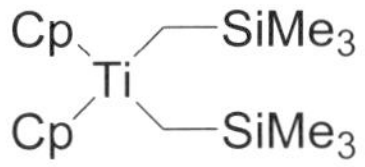

41

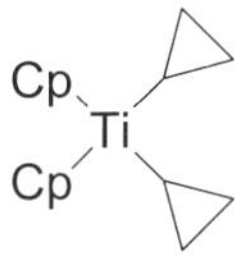

42

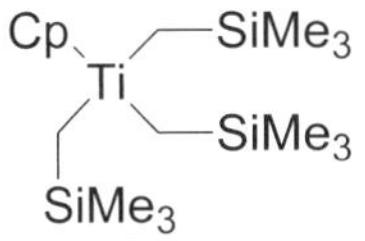

43

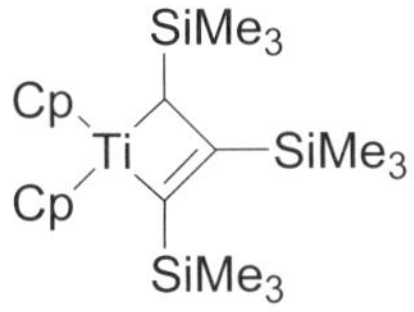

44

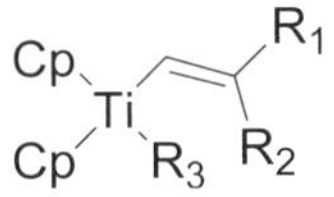

45

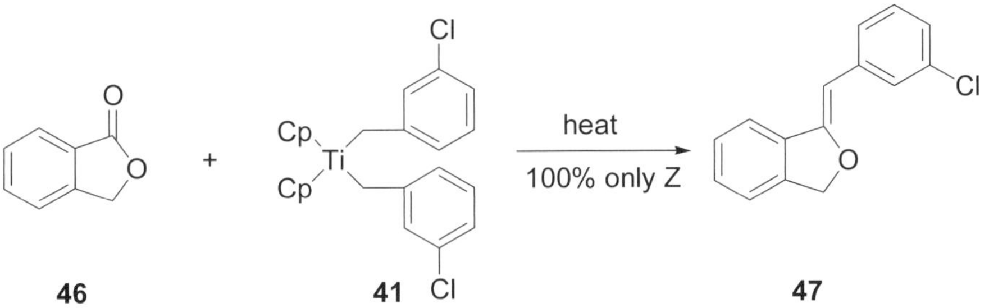

46 **41** **47**

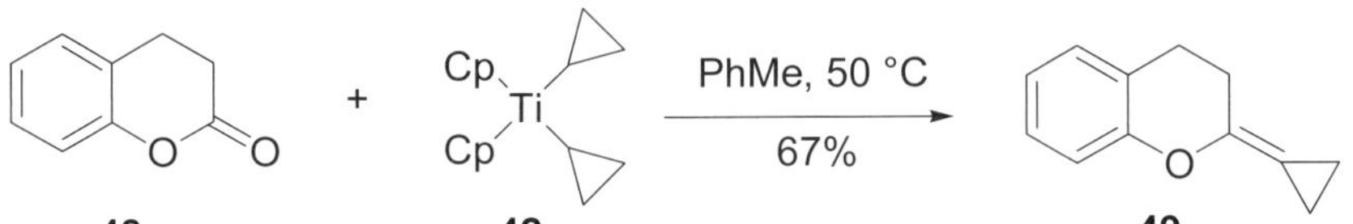

48 **42** **49**

Grubbs reagents, titanacycles **50** and **51** were prepared by the reaction of Tebbe reagent with a terminal alkene in the presence of a Lewis base.[22] When these complexes are heated, reactive titanocene methylidene **5** is regenerated and will methylenate carbonyl group (**49** to **53**).[23]

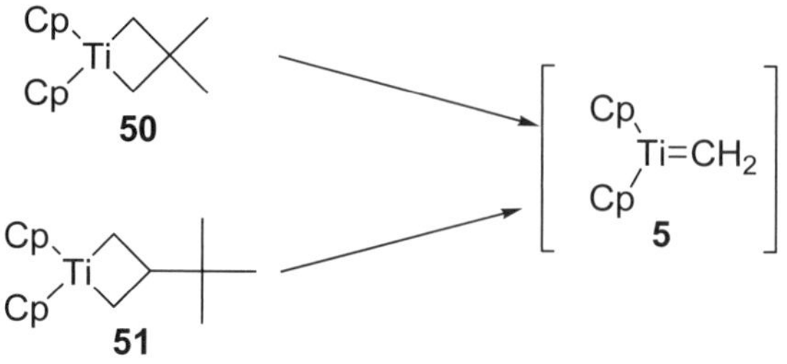

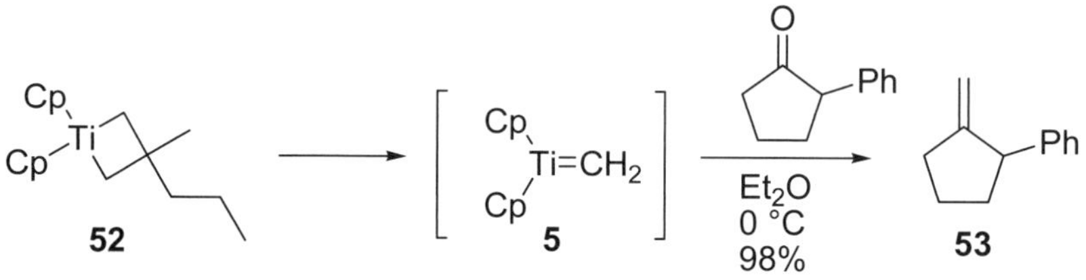

Takeda and co-workers discovered that the easily accessible thioacetals **54** and **56** can be reduced by a low-valent titanium reagent to give Schrock carbenes that are competent to alkylidenate aldehydes, ketones, esters, lactones, and thioesters.[24] This method has the advantage of tolerating hydrogen atoms on the β-carbon atom relative to the titanium atom of the Schrock carbene. The other distinct advantage of the Takeda olefination is that functionality can be incorporated into the titanium alkylidene complex from a range of functionalised thioacetals.[25]

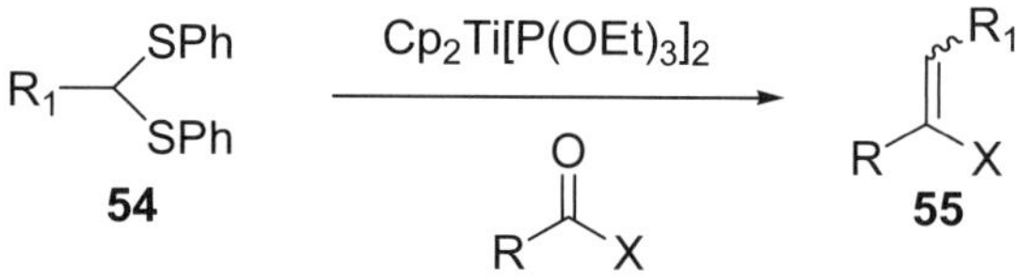

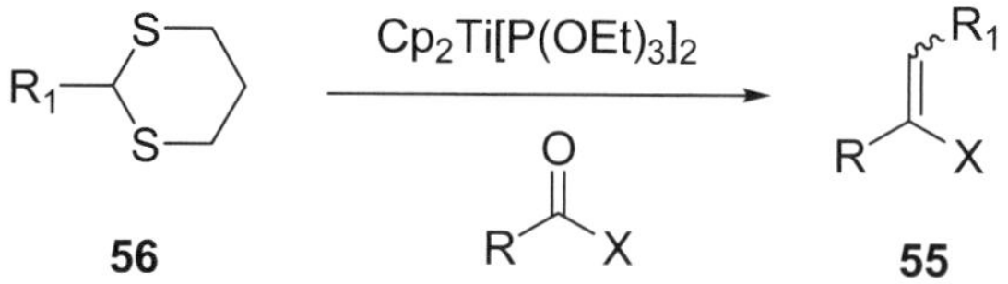

A plausible mechanism for the Takeda alkylidenation is given here.[26] The first oxidative addition to generate titanocene complex **57** is essentially instantaneous, while the second oxidative addition, giving bimetallic **58** or **59** is slower. By analogy with the Petasis reagent, the rate-determining step is likely to be the generation of the titanium(IV) alkylidene complex **60**, a Schrock carbene.

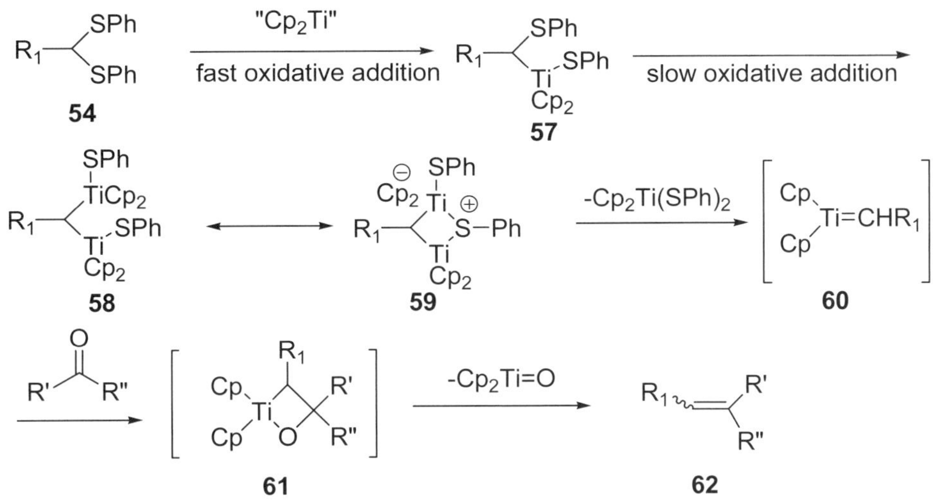

Takai reported a simple, general, and stereoselective method for the alkylidenation of ester to give Z-enol ethers.[27] The titanium carbene complex is easily prepared *in situ* by the reaction of $RCHBr_2$ with a low-valent titanium species generated by treatment of $TiCl_4$ with zinc and tetramethylenediamine (TMEDA) in THF. Without isolation, the complex is used for carbonyl alkenation.[28] It was reported that the presence of a small amount of lead in the zinc was crucial to the reaction.[29]

1.3.4.5 Synthetic Utility

The Tebbe reagent has been used extensively for the methylenation of a variety of carbonyl groups in the presence of other functional groups. For example, the aldehyde group in **63** was selectively methylenated in the presence of the ester without epimerisation of the adjacent chiral centers.[30] Without the protection of the hydroxy group, methylenation of readily enolizable ketone **65** gave alkene **66** in 61% isolated yield.[31] Ketone **67** possessing a β-halide atom was successfully methylenated, giving the substituted styrene **68** in 93% isolated yield.[32] Keto esters **69** and **71** are efficiently methylenated using the Tebbe reagent to afford olefins **70** and **72** respectively.[7,11] Additionally, the methylenation of carbonates with Tebbe reagent generates ketene acetals.

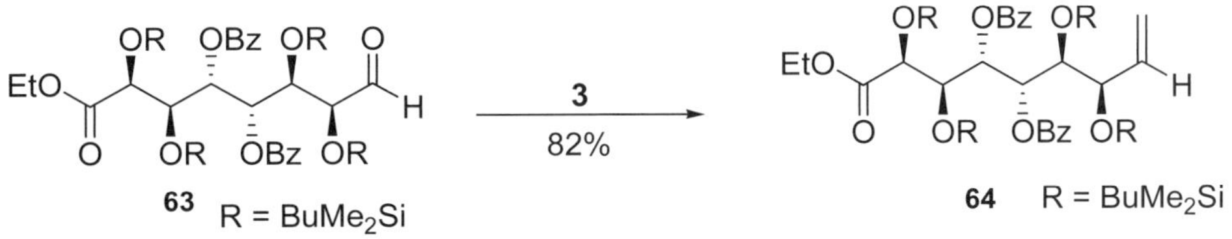

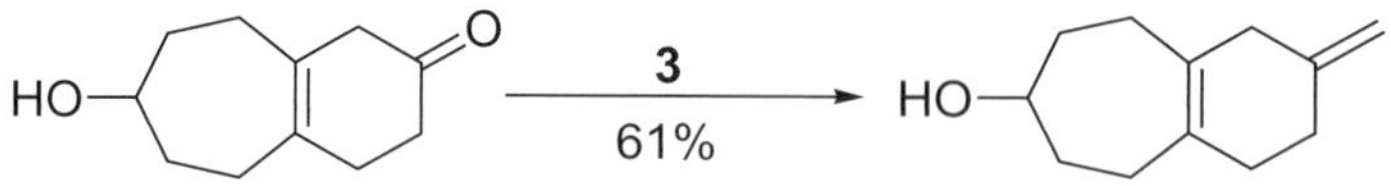

65 66

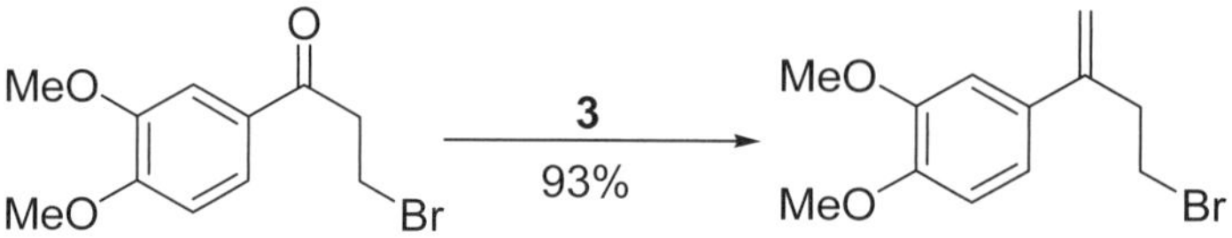

67 68

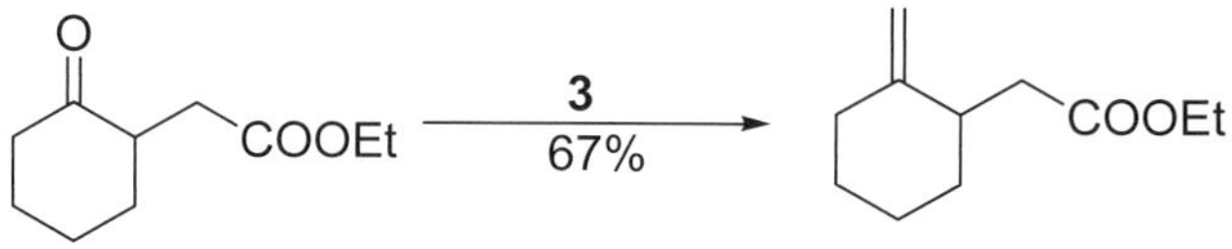

69 70

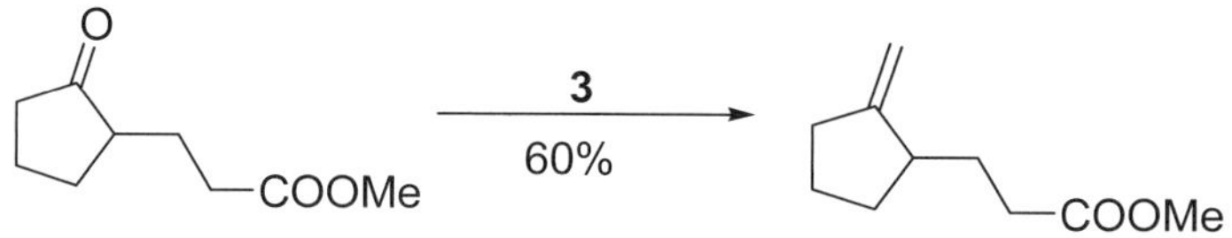

71 72

The Petasis reagent has been demonstrated to methylenate base-sensitive substrates without epimerization of sensitive stereocenters. For example the easily enolizable cyclopentanones **73** and **75** were converted to alkenes **74** and **76** in 77% and 81% isolated yield respectively.[33]

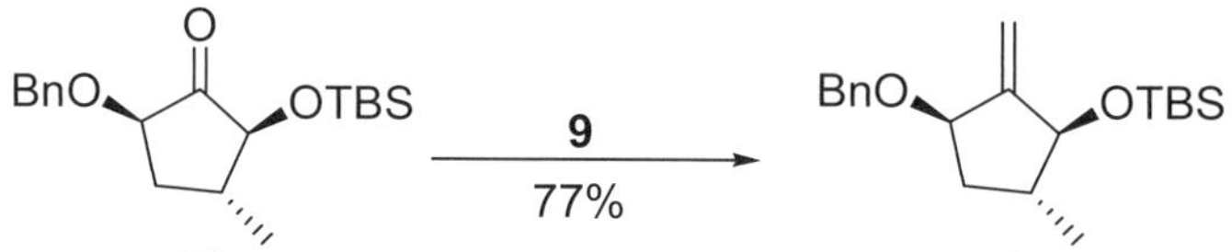

73 74

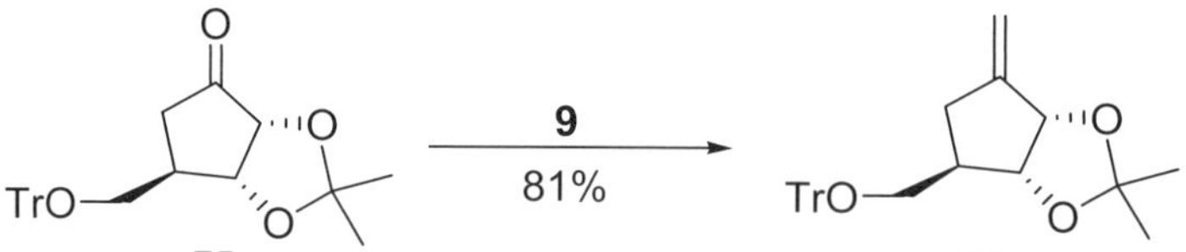

75 76

Using double Tebbe methylenation (**77** to **78**) followed by a Claisen rearrangement, Paquette and co-workers developed a concise, reliable, and efficient scheme for 4-cyclooctenones.[34] Using this methodology, Paquette reported an alternate enantioselective route to a key bicyclic intermediate, demonstrated the feasibility of a Tebbe–Claisen sequence for assembling the entire kalmanol backbone[35] and enantiomerically pure (+)-*cis*- and (+)-*trans*-lauthisan via a Claisen ring expansion.[36]

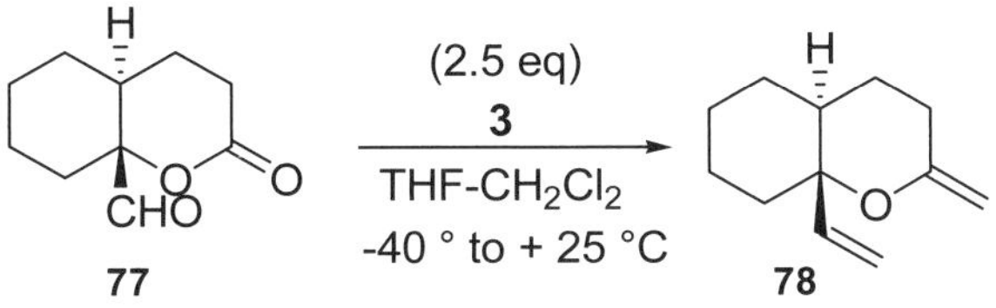

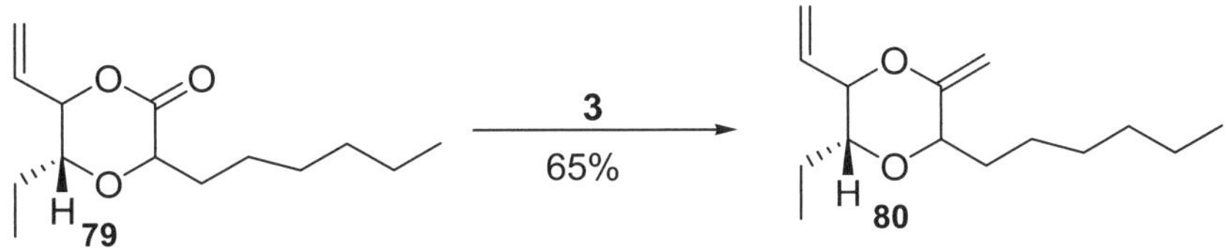

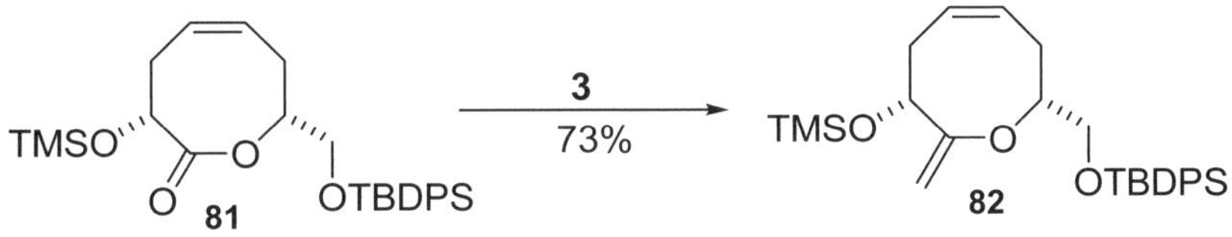

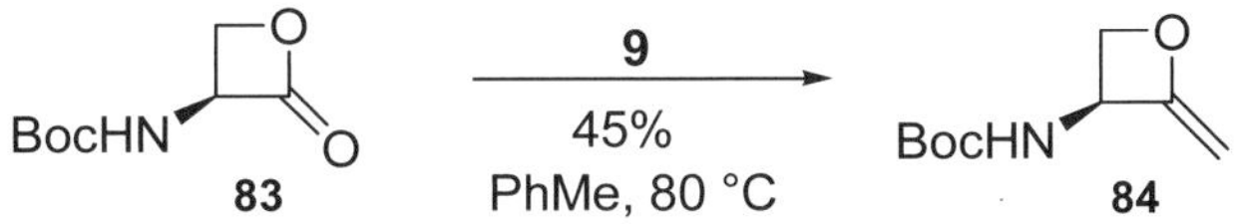

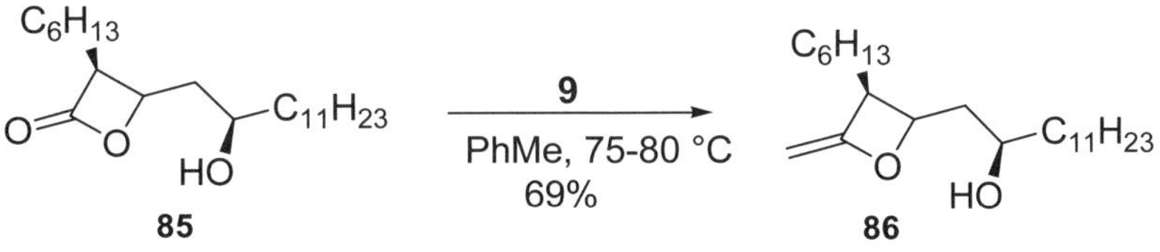

In the synthesis of (+)-laurencin, Holmes used Tebbe reagent converted ester to alkene in 71% yield.[37] Nicolaou developed a novel method for direct conversion of olefinic ester to cyclic enol ethers with Tebbe reagent.[38]

Methylenation of tertiary amides utilizing Petasis reagent, including *N*-acyl heterocycles, gives enamines.[13,39a] From the corresponding α-lactam **92**, and β-lactams **89**, methyleneaziridine **93** and methylenetidine **91** were synthesized via Petasis olefination.[39b]

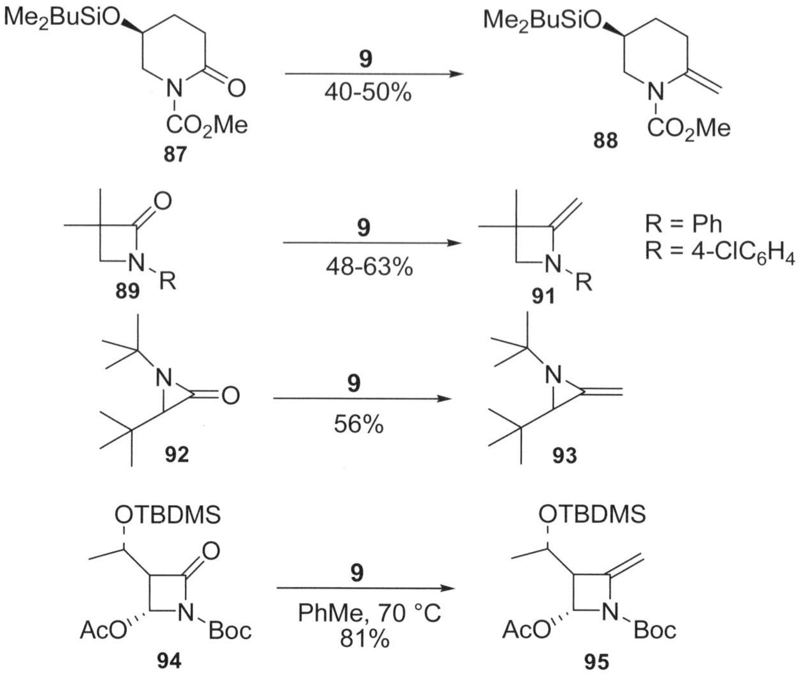

Petasis reagent was also effective for the methylenation of aldonolactones, **96**[40] and **98**.[41] Petasis olefination of unsymmetrical oxalates and oxalate monoesters or monoamides **100** under microwave-assisted, provides pyruvate-based enol ethers and enamines **101** in higher yields.[42]

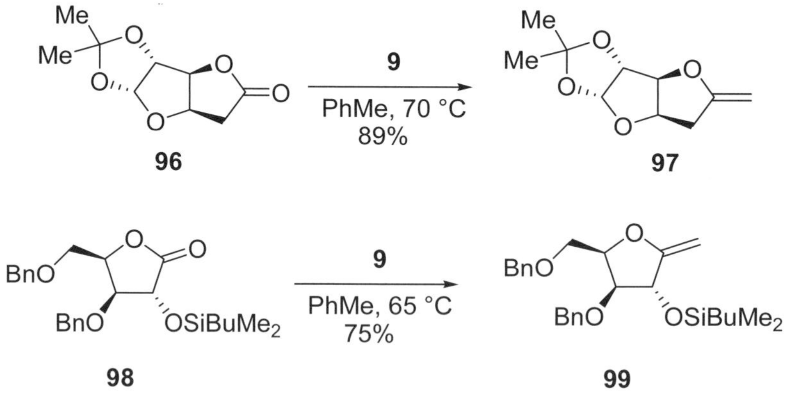

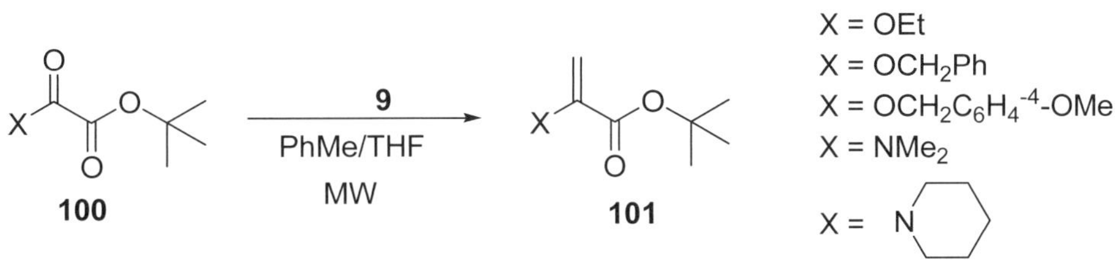

Depending on the number of equivalents of Petasis reagent used, anhydrides, as well as thioanhydrides and imides can be methylenated one or both carbonyl groups. In the case of cyclic substarates, the bis-methylenation provides a novel method for accessing functionalized furans or thiophenes, such as **104** *via* subsequent isomerization of the newly generated olefin.[43]

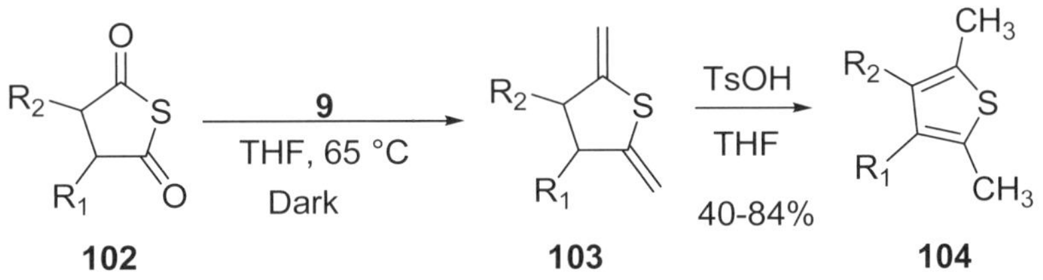

1.3.4.6 Experimental

The Tebbe Reaction; General Procedure:
To a solution of carbonyl (1.0 mmol) in THF (2–3 mL) at 0 °C is added a toluene solution of the Tebbe reagent (2 mL of 0.5 M solution, 1.0 mmol). The mixture is allowed to warm to rt, and after about 15 minutes, Et_2O (15–20 mL) is added. Then 5–10 drops of aq NaOH (0.1 M slowly added while stirring the mixture. After gas evolution ceases, the mixture is dried (Na_2SO_4) and filtered using a celite pad. Rotary evaporation of the solvent provides the crude product which is purified by column chromatography using alumina and an eluent of 2% Et_2O in pentane or petroleum ether.

The Petasis Reaction; General Procedure:
A 0.5 M toluene (or THF) solution of Cp_2TiMe_2 (2–3 equiv) was mixed with the carbonyl compound (1 mmol) and stirred under argon in the dark at 65 °C. After the reaction was completed (12–26 h), the mixture was diluted with petroleum ether. The resulting yellow-orange precipitate was removed by filtration, and the filtrate was concentrated. The residue was purified by flash column chromatography on silica or basic alumina (for vinyl ethers).

1.3.4.7 References

1. [R] Pine, S. H. *Org. React.* **1993**, *43*, 1.
2. [R] Hartley, R. C.; McKiernan, G. J. *J. Chem. Soc., Perkin Trans. 1* **2002**, 2763.
3. [R] Hartley, R. C.; Li, J.; Main, C. A.; McKiernan, G. J. *Tetrahedron* **2007**, *63*, 4825.
4. [R] Petasis, N. A. "*Transition Metals for Organic Synthesis*" Volume 1, Second Revised Ed. Beller M. and Bolm, C. Wiley-VCH, 2004, pp427–447.
5. Tebbe, F. N.; Parshall, G. W.; Reddy, G. S. *J. Am. Chem. Soc.* **1978**, *100*, 3611.
6. Cannizzo, L. F.; Grubbe, R. H. *J. Org. Chem.* **1985**, *50*, 2386.
7. Pine, S. H.; Pettit, R. J.; Geib, G. D.; Cruz, S. G.; Gallego, C. H.; Tijerina, T.; Pine, R. D. *J. Org. Chem.* **1985**, *50*, 1212.
8. Pine, S. H.; Shen, G. S.; Hoang, H. *Synthesis* **1991**, 165.
9. [R] (a) Maercker, A. *Org. React.* (N. Y.) **1965**, *14*, 270. [R] (b) House, H. O. "*Modern Synthetic Reactions*"; Benjamin; Menlo Park, CA, **1972**; pp682–709. [R] (c) Bestman, H. J.; Vostrowsky, O. *Top. Cur. Chem.* **1983**, *109*, 65.
10. [R] Schrock, R. R. *Chem. Rev.* **2002**, *102*, 145.
11. (a) Petasis, N. A.; Bzowej, E. I. *J. Am. Chem. Soc.* **1990**, *112*, 6392–6394. (b) Petasis, N. A.; Hu, Y. H.; Fu, D. K. *Tetrahedron. Lett.* **1993**, *36*, 6001.
12. Hughes, D. L.; Payack, J. F.; Cai, D. W.; Verhoeven, T. R.; Reider, P. J. *Organometallics* **1996**, *15*, 663.
13. Petasis, N. A.; Lu, S. P. *Tetrahedron. Lett.* **1995**, *36*, 2393.
14. Petasis, N. A.; Bzowej, E. I. *Tetrahedron. Lett.* **1993**, *34*, 943–946.
15. (a) Payack, J. F.; Huffman, M. A.; Cai, D. W.; Hughes, D. L.; Collins, P. C.; Johnson, B. K.; Cottrell, I. F.; Tuma, L. D. *Org. Process Res. Dev.* **2004**, *8*, 256. (b) Payack, J. F.; Hughes, D. L.; Cai, D.; Cottrell, I. F.; Verhoeven, T. R. *Org. Synth.* **2002**, *79*, 19.
16. Greenwood, E. S.; Hitchcock, P. B.; Parsons, P. J. *Tetrahedron* **2003**, *59*, 3307.
17. Colson, P. J.; Hegedus, L. S. *J. Org. Chem.* **1993**, *58*, 5918.
18. Petasis, N. A.; Bzowej, E. I. *J. Org. Chem.* **1992**, *57*, 1327.
19. Petasis, N. A.; Akritopoulou, I. *Synlett* **1992**, 665.
20. Petasis, N. A.; Staszewski, J. P.; Fu, D. K. *Tetrahedron Lett.* **1995**, *36*, 3619.
21. Petasis, N. A.; Bzowej, E. I. *Tetrahedron Lett.* **1993**, *34*, 943.
22. (a) Ott, K. C.; Grubbs, R. H. *J. Am. Chem. Soc.* **1981**, *103*, 5922. (b) Cannizzo, L. F.; Grubbs, R. H. *J. Org. Chem.* **1985**, 50, 2386. (c) Cannizzo, L. F.; Grubbs, R. H. *J. Org. Chem.* **1985**, *50*, 2316. (d) Stille, J. R.; Grubbs, R. H. *J. Am. Chem. Soc.* **1986**, *108*, 855.
23. (a) Clawson, L.; Buchwald, S. L.; Grubbs, R. H. *Tetrahedron Lett.* **1984**, *25*, 5733. (b) Anslyn, E. V.; Grubbs, R. H. *J. Am. Chem. Soc.* **1987**, *109*, 4880.
24. (a) Horikawa, Y.; Watanabe, M.; Fujiwara, T.; Takeda, T. *J. Am. Chem. Soc.* **1997**, *119*, 1127. (b) Takeda, T.; Saito, J.; Tsubouchi, A. *Tetrahedron Lett.* **2003**, *44*, 5571.
24. (a) Takeda, T.; Watanabe, M.; Rahim, M. A.; Fujiwara, T. *Tetrahedron Lett.* **1998**, *39*, 3753. (b) Rahim, A.; Taguchi, H.; Watanabe, M.; Fujiwara, T.; Takeda, T. *Tetrahedron Lett.* **1998**, *39*, 2153. (c) Takeda, T.; Sato, K.; Tsubouchi, A. *Synthesis* **2004**, 1457.
25. See reference 3, page 4830.
26. Okazoe, T.; Takai, K.; Oshima, K.; Utimoto, K. *J. Org. Chem.* **1987**, *52*, 4410.
27. [R] Tsuji, J. *Transition Metal Reagents and Catalysts, Innovations in Organic Synthesis*, John Wiley & Sons, **2000**, 326.
28. Takai, K.; Kakiuchi, T.; Kataoka, Y.; Utimoto, K. *J. Org. Chem.* **1994**, *59*, 2668.
29. Ikemoto, N.; Schreiber, L. S. *J. Am. Chem. Soc.* **1992**, 114, 2524.
30. Clawson, L.; Buchwald, S. L.; Grubbs, R. H. *Tetrahedron Lett.* **1984**, *25*, 5733.
31. Winkler, J. D.; Muller, C. L.; Scott, R. D. *J. Am. Chem. Soc.* **1988**, *110*, 4831.
32. (a) Marschner, C.; Penn, G.; Griengl, H. *Tetrahedron Lett.* **1993**, *49*, 5067. (b) Matsuura, T.; Nishiyama, S.; Yamamura, S. *Chem. Lett.* **1993**, 1503.
33. Philippo, C. M. G.; Vo, N. H.; Paquette, L. A. *J. Am. Chem. Soc.* **1991**, *113*, 2762.
34. Borrelly, S.; Paquette, L. A. *J. Am. Chem. Soc.* **1996**, *118*, 727.
35. Paquette, L. A.; Sweeney, T. J. *J. Org. Chem.* **1990**, *55*, 1703.
36. Robinson, R. A.; Clark, J. S.; Holmes, A. B. *J. Am. Chem. Soc.* **1993**, 115, 10400.
37. Nicolaou, K. C.; Postema, M. H. D.; Claiborne, C. F. *J. Am. Chem. Soc.* **1996**, *118*, 1565.

38. (a) Herdeis, C.; Heller, E. *Tetrahedron: Asymmetry*, **1993**, *4*, 2085. (b) Tehrani, K. A.; Kimpe, N. D. *Tetrahedron Lett.* **2000**, *41*, 1975.
39. Csuk, R.; Glanzer, B. I. *Tetrahedron* **1991**, *47*, 1655.
40. Faivre–Buet, V.; Eynard, I.; Nga, H. N.; Descotes, G.; Grouiller, A. *J. Carbohydr. Chem.* **1993**, *72*, 349.
41. Cook, M. J.; Fleming, D. W.; Gallagher, T. *Tetrahedron Lett.* **2005**, *46*, 297.
42. Kates, M. J.; Schauble, J. H. *J. Org. Chem.* **1994**, *59*, 494.

Chapter 2. Carbon-Chain Homologations

2.1	Arndt–Eistert homologation	336
2.2	Morita–Baylis–Hillman reaction	350
2.3	Benzoin condensation	381
2.4	Corey–Fuchs reaction	393
2.5	Henry reaction	404
2.6	Horner–Wadsworth–Emmons reaction	420
2.7	Julia–Lythgoe olefination	447
2.8	Knoevenagel condensation	474
2.9	Mukaiyama aldol reaction	502
2.10	Peterson olefination	521
2.11	Sakurai allylation reaction	539
2.12	Stetter reaction	576
2.13	Wittig reaction	588

2.1 Arndt–Eistert Homologation

Matthew J. Fuchter

2.1.1 Description

The conversion of a carboxylic acid to its homolog (one CH_2 group longer) in three stages *via* an α-diazo methylketone is known as the Arndt–Eistert homologation.[1–3] It is the best preparative method for the chain elongation of carboxylic acids.

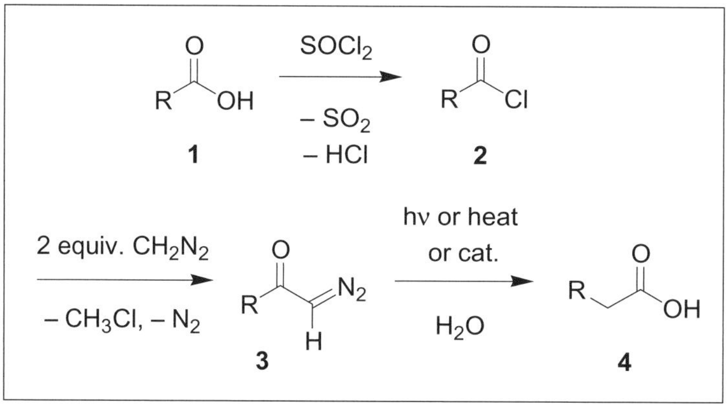

In the first step, the carboxylic acid **1** is converted to an acid chloride **2**. While classically thionyl chloride ($SOCl_2$) is used for this process, any procedure which converts a carboxylic acid to an acid chloride is suitable [$POCl_3$, $(COCl)_2$]. The second step involves formation of an α-diazo methylketone **3**, followed by a Wolff rearrangement[3] in the third step. Classically a silver catalyst (silver oxide/water, or silver benzoate/triethylamine) is used to mediate the Wolff rearrangement, although heat, photolysis or other metals (Pt, Cu) can be employed. In the presence of water the homologated acid **4** is isolated, however, on addition of alternative nucleophiles, other products can be produced. For example, the use of amine nucleophiles gives a homologated amide and alcohols give a homologated ester. The reaction tolerates a wide range of non-acidic functional groups (alkyl, aryl, olefins).

2.1.2 Historical Perspective

Ludwig Wolff made the discovery that diazo ketones can be converted into derivatives of an acid in 1915.[4] Wolff found, for example, that the treatment of ω-diazoacetophenone ($PhCOCHN_2$) with ethanolic ammonia and a silver catalyst gave phenylacetamide ($PhCH_2CONH_2$). Although the first examples of this rearrangement indicated a 1,2-shift had occurred, it was not appreciated how water was incorporated en route to carboxylic acids. Following Staudinger's seminal work on ketenes,[5] a revised mechanism was published by Wolff.[6] For over 20 years the Wolff rearrangement remained largely an oddity, however, once efficient syntheses of diazo ketones began to emerge, the situation changed. In 1935, Fritz Arndt and Bernd Eistert published their seminal paper on the homologation of carboxylic acids.[7] The first review on the Arndt–Eistert homologation appeared a mere 7 years later and included 76 recorded examples of its use.[2] The Arndt–Eistert homologation is now the best preparative method for the chain elongation of carboxylic acid derivatives, having proved itself as a flexible and scalable synthetic method.

2.1.3 *Mechanism*

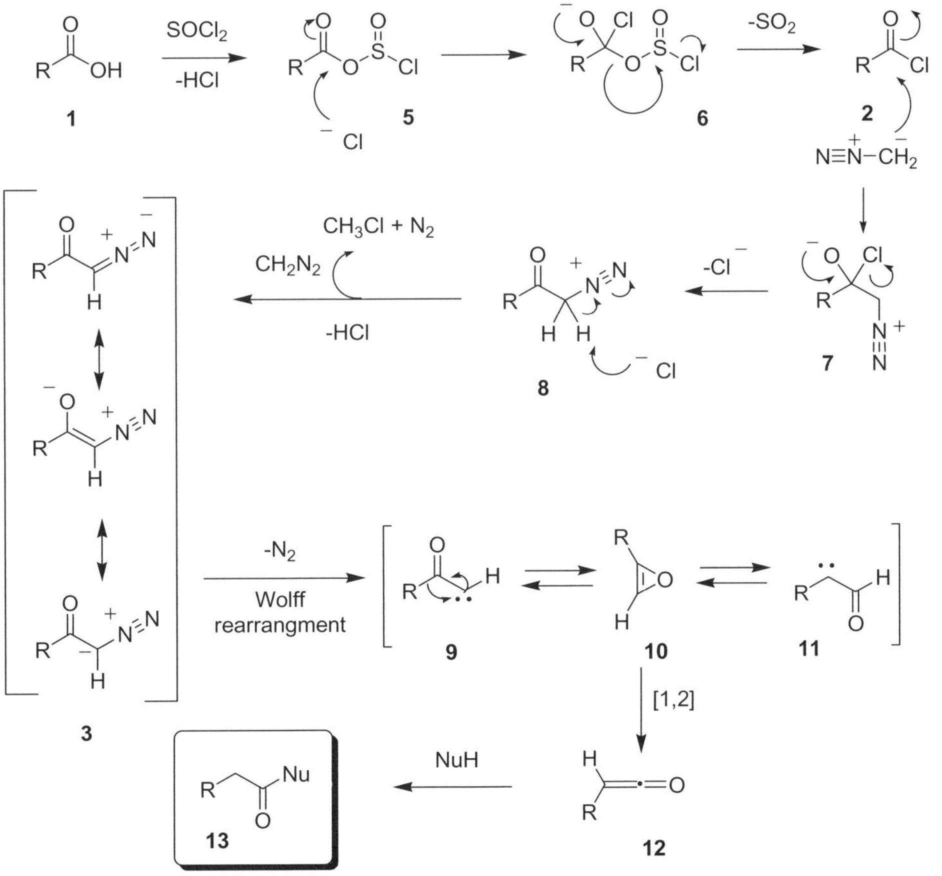

Exposure of carboxylic acid **1** to thionyl chloride (or analogous reagent) produces acid chloride **2**, and in the process sulfur dioxide and HCl are produced. Displacement of the chloride moiety by diazomethane furnishes diazo ketone **3**. Since HCl is a by-product of this reaction, two equivalents of diazomethane must be used to prevent side products (chloroketones). The HCl reacts with the second equivalent of diazomethane to form methyl chloride and nitrogen. Incidentally, Newmann and Beal reported a modification whereby triethylamine is added to capture the released HCl.[8] In these processes one equivalent of diazomethane is sufficient.

The diazo ketone **3** can exist in two conformations, namely the *s*-(*Z*) and *s*-(*E*) conformations (only *s*-(*Z*) shown), which arise from C–C bond rotation. The favoured conformation is *s*-(*Z*) due to an attractive interaction between the negatively-charged oxygen and positively-charged nitrogen and it is this conformation which preferentially undergoes the Wolff rearrangment.[3] Classically, a silver catalyst was employed to mediate the Wolff rearrangement, although the role of the catalyst is poorly understood. Loss of nitrogen from diazo ketone **3** gives a carbene **9** which can interconvert *via* a 1,2-oxygen shift, with oxirene **10** as an intermediate or transition state.[3] A rapid [1,2]-shift affords ketene **12**, which reacts readily with nucleophiles to give homologated acid derivatives **13**.

2.1.4 *Synthetic Utility*

β-Amino acid Synthesis

Perhaps the largest body of work regarding the application of the Arndt–Eistert homologation is in the chain-extension of α-amino acids. The produced β-amino acids are components of many biologically active natural products, including paclitaxel (Taxol®), HIV protease inhibitors and β-lactam antibiotics.[2,3]

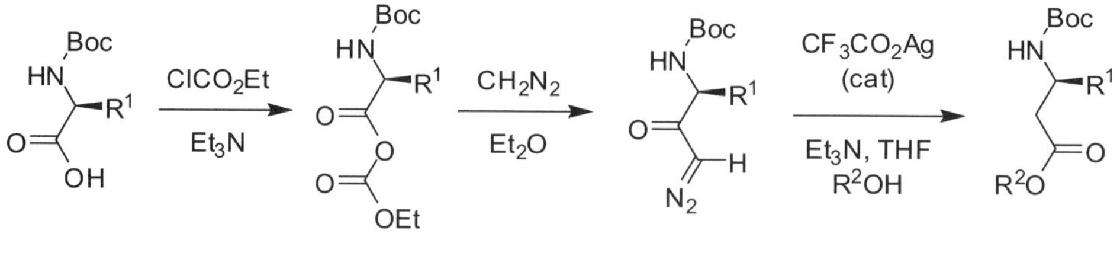

14 **15** **16** **17**

Seebach for example, has used the Arndt–Eistert homologation to prepare biologically relevant β-amino acids, from their α-amino acid counterparts.[9] Conversion of α-amino acid **14** to the mixed anhydride **15**, followed by formation of the diazo ketone **16** and Wolff rearrangement furnished the β-amino acid derivatives **17** in good yields. The oligomers of β-amino acids, as opposed to α-peptides, are of particular interest due to their ability to fold into well-defined secondary structures in solution as well as the solid state.

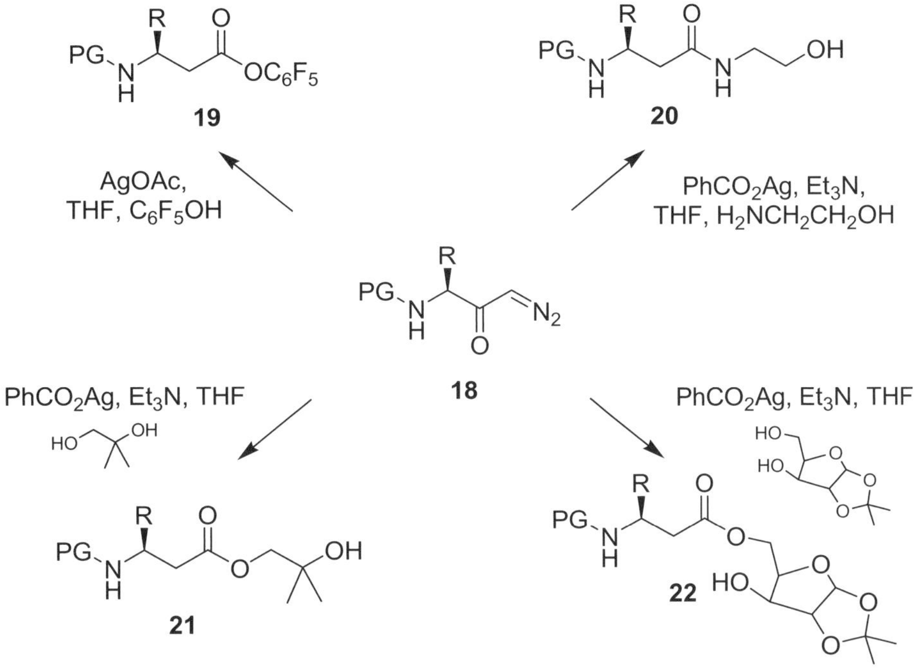

Although the Wolff rearrangement stage is most often performed in the presence of water or methanol, other nucleophiles can be employed to capture the transient ketene intermediate. For example, silver ion-catalyzed rearrangement of diazo ketone **18** in the presence of pentafluorophenol gave pentafluorophenyl esters of β-amino acids **19** in good yield and purity.[10] These activated esters are key intermediates in peptide synthesis. Good chemoselectivity was observed for reaction of diazo ketone **18** with 2-aminoethanol, furnishing amide **20** exclusively.[11] Likewise, the primary hydroxy group of 3-methylbutane-1,3-diol was acylated preferentially yielding ester **21**.[11] Discrimination between primary and secondary hydroxy groups was observed in the acylation of isopropylidene-*D*-xylofuranose and the product **22** was isolated in 55% and a selectivity of 19 : 1.[11]

Natural Product Synthesis

The practical applicability of the Arndt–Eistert reaction is evident from its employment in the preparation of numerous complex molecular architectures.

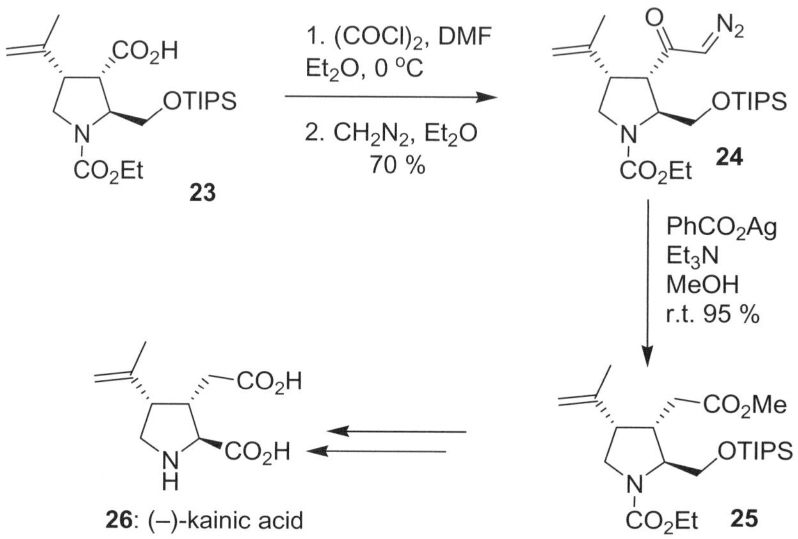

Gallagher and co-workers used the Arndt–Eistert homologation strategy in their synthesis of the potent anthelmintic agent (–)-kainic acid (**26**).[12] Functionalized pyrrolidine **23** was elaborated from *L*-aspartic acid. The presence of DMF was crucial to achieve high conversions of the acid **23** to the acid chloride **24**, and key intermediate **25** was obtained in good yield. This was subsequently transformed to (–)-kainic acid (**26**).

In his synthesis of the CP molecules, Nicolaou employed an Arndt–Eistert homologation of an advanced intermediate.[13] These fungal-derived natural products are inhibitors of farnesyl transferase, which could be used to control the level of serum cholesterol. Due to its sensitive nature, intermediate **27** was converted into diazo ketone **29** *via* the acyl mesylate **28** as opposed to an acid chloride. The diazo ketone **29** was immediately dissolved in DMF : H_2O and heated to 120 °C in the presence of silver oxide. This gave the homologated product **30** in 35% yield, which was subsequently transformed into CP molecule **31**.

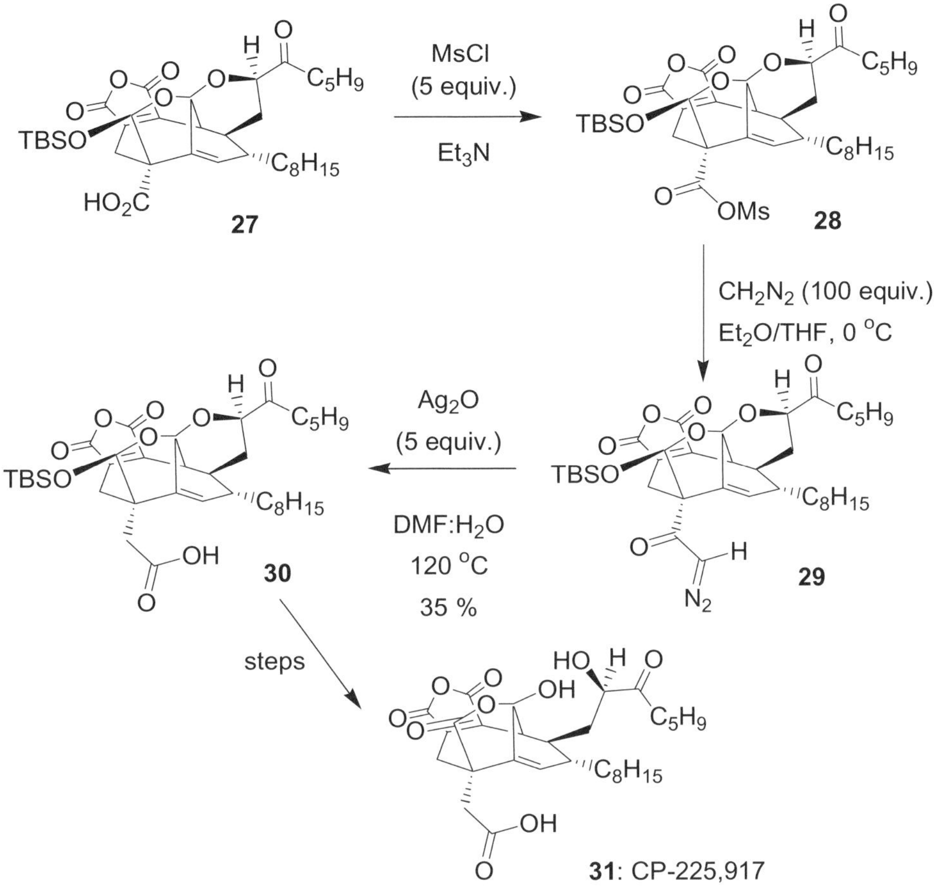

MsCl
(5 equiv.)
Et3N
27
28
CH2N2 (100 equiv.)
Et2O/THF, 0 °C
29
Ag2O
(5 equiv.)
DMF:H2O
120 °C
35 %
30
steps
31: CP-225,917

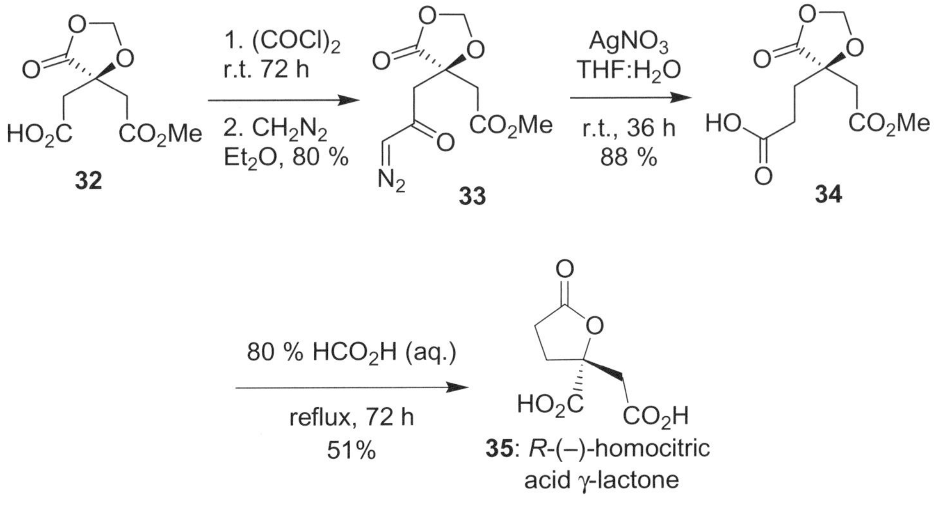

1. (COCl)2
r.t. 72 h
2. CH2N2
Et2O, 80 %
32
33
AgNO3
THF:H2O
r.t., 36 h
88 %
34
80 % HCO2H (aq.)
reflux, 72 h
51%
35: R-(–)-homocitric acid γ-lactone

Russell and co-workers developed a scalable synthesis of (*R*)-(–)-homocitric acid–γ-lactone (**35**) using an Arndt–Eistert homologation strategy.[14] Starting from citric acid derivative **32**, key intermediate **34** was produced in multi-gram quantities.

The Arndt–Eistert homologation is equally applicable to the synthesis of alkaloids. For example, Stoltz demonstrated its application to the first synthesis of the bis-indole alkaloid dragmacidin D (**38**).[15] In the final steps of the synthesis, carboxylic acid **36** was homologated to the diazo ketone, which was subsequently exposed to hydrobromic acid. This gave α-bromo ketone **37**, which was elaborated to racemic dragmacidin D (**38**).

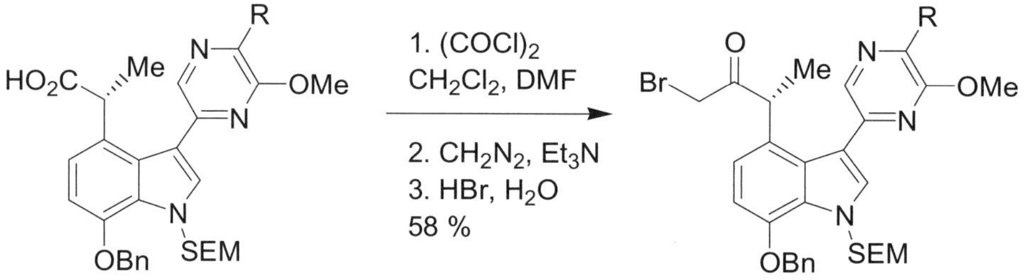

36 R = *N*-tosyl-6-bromo-3-indolyl **37**

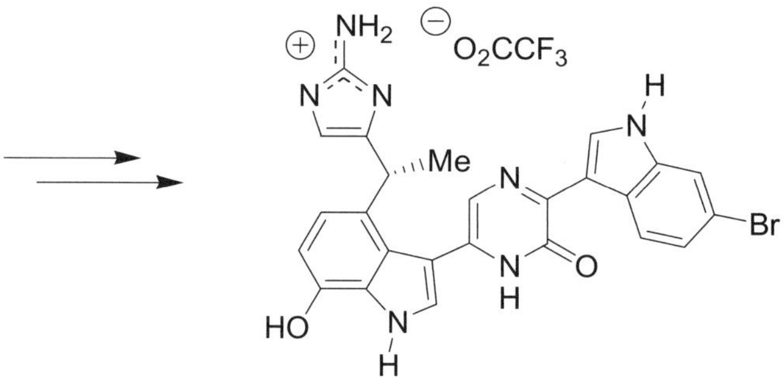

38: (±)-dragmacidin D

Rearrangement of α,β-Epoxy Diazomethyl Ketones

Oxiranecarboxylic acids **41** (glycidic acids) can be converted into α,β-epoxy diazomethyl ketones **42** via mixed anhydrides.[16] It was found that photolysis of these compounds in the presence of alcohols gave γ-hydroxy-α,β-unsaturated esters **44**.[17] It is thought that nucleophilic attack of the alcohol on the ketene **43** results in epoxide ring opening. The *E* olefin isomer is predominately formed, although small quantities of *Z* esters are also isolated (< 10%).[18] Conveniently non-racemic, chiral substrates are readily prepared via Sharpless asymmetric epoxidation of allylic alcohol **39**, followed by oxidation. The photo-rearrangement proceeds with retention of configuration at C-4.[18] This approach has been utilized in the synthesis of numerous naturally occurring compounds including rubrenolide,[19] cytochalasin B[20] and patulolide C.[21]

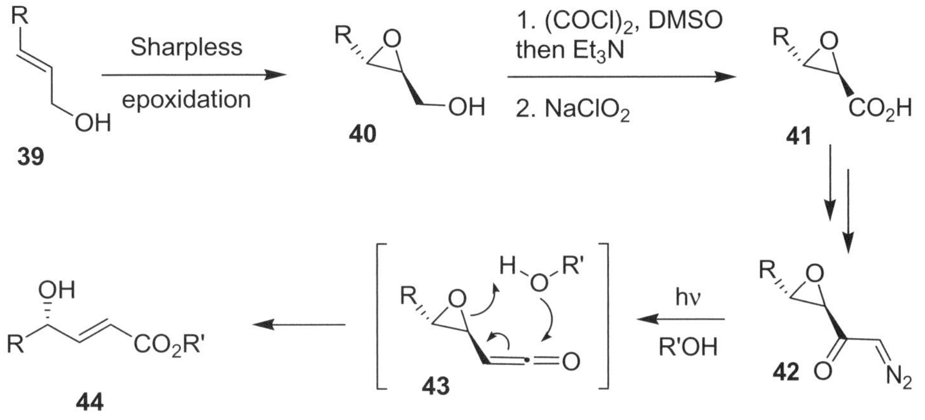

Miscellaneous Examples

Ring strain of the migrating group is compatible with the Wolff rearrangement and this was, for example, exploited in the homologation of cubanecarboxylic acid **45**.[22]

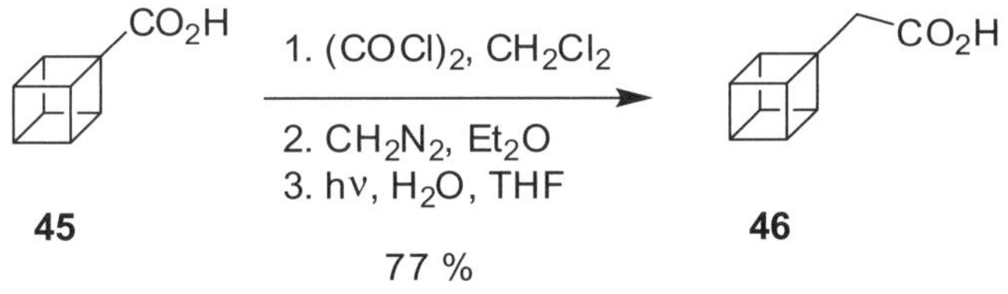

The presence of double bonds in the migrating group is also tolerated. In their synthesis of substituted cyclopentadiene derivatives, Zellweger and co-workers used several Arndt–Eistert homologations to elaborate substrate **47** to cyclopentadiene **50**.[23]

Whereas thermal and catalytic methods of mediating the Wolff rearrangement led instead to a cyclopropanation reaction, furnishing ketone **51**, the photochemical reaction provided desired homologated product **50** in good yield.

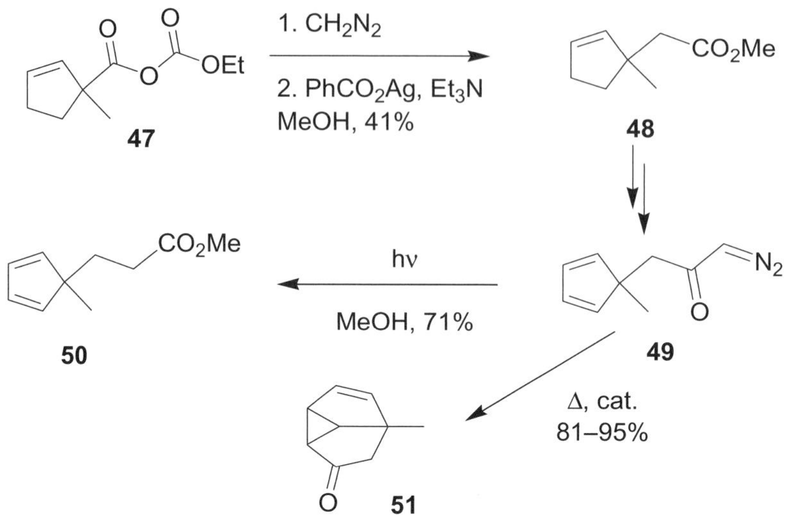

2.1.5 *Variations*

In general, only slight modifications to the Arndt–Eistert homologation have been reported. As mentioned previously, classically two equivalents of diazomethane are required due to the production of HCl upon reaction of diazomethane with the acid chloride. Newmann and Beal reported a modification whereby triethylamine is added to capture the released HCl and therefore only one equivalent of diazomethane is required.[8] Another major source of diversity in the reported Arndt–Eistert reactions is in the initial activation of the carboxylic acid. While thionyl chloride is classically used, other reagents that mediate the conversion of a carboxylic acid to an acid chloride are equally suitable.[1–3] As reported above, alternative activation methods such as the formation of mixed anhydrides and acyl mesylates are also be applicable.

Perhaps the most important variation strategy lies in the replacement of diazomethane. Diazomethane is highly toxic and explosive and therefore substitution of this reagent is desirable. Several reported efforts towards this goal are discussed below.

Diazo Group Transfer

One popular method in the synthesis of α-diazo ketones is in the base-catalyzed "diazo group transfer" reaction of sulfonyl azides with activated dicarbonyl compounds (*The Regitz Diazo Reaction*).[24] While direct diazo transfer to ketone enolates is usually not feasible, a two step "deformylative diazo transfer" strategy has been employed, whereby a ketone is first formylated under Claisen condensation conditions and then treated with a sulfonyl azide reagent.[25]

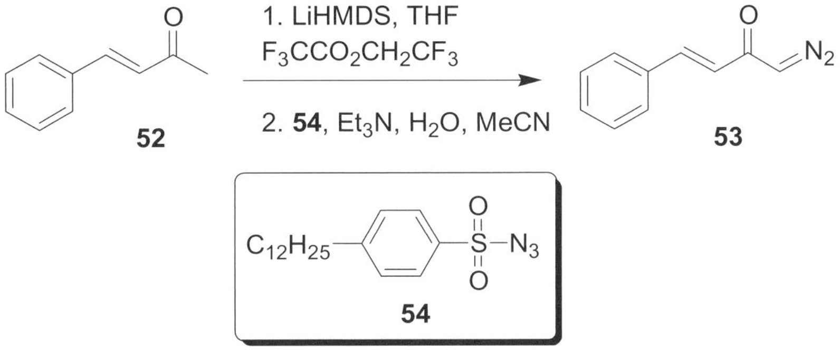

Several important α-diazo ketones cannot be prepared via this approach however due to their base-sensitive nature. Danheiser and co-workers reported a "detrifluoroacetylative diazo transfer" strategy to combat this shortcoming.[26] Deprotonation of substrate **52**, followed by a trifluoroethyl trifluoroacetate (TFEA) quench furnished the trifluoroacetylated product. This undergoes facile detrifluoroacetylative diazo transfer from sulfonyl azide **54**, to yield diazo ketone **53**. In general they found TFEA to be superior to other trifluoroacetylating agents.

Trimethylsilyldiazomethane

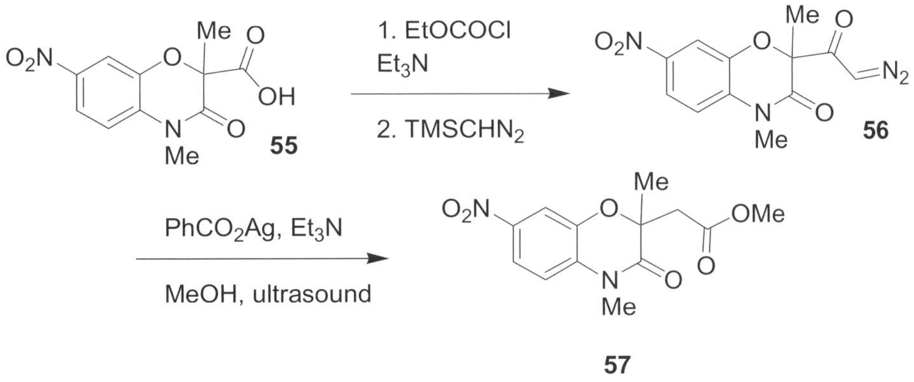

Another strategy for the safe preparation of α-diazo ketones is in the replacement of diazomethane with the thermally stable trimethylsilyl-diazomethane. Shioiri and co-workers were the first to report the use of trimethylsilyldiazomethane as an alternative, safer reagent for use in the Arndt–Eistert reagent for relatively simple substrates.[27] Dolenc more recently showed the ability to use mixed anhydrides rather than acid chlorides in combination with trimethylsilyldiazomethane.[28] For example, exposure of the mixed anhydride derivative of acid **55** to trimethylsilyldiazomethane gave α-diazo ketone **56** which subsequently underwent a Wolff rearrangement to give homologated product **57**.

Homologation via Ynolate Anions

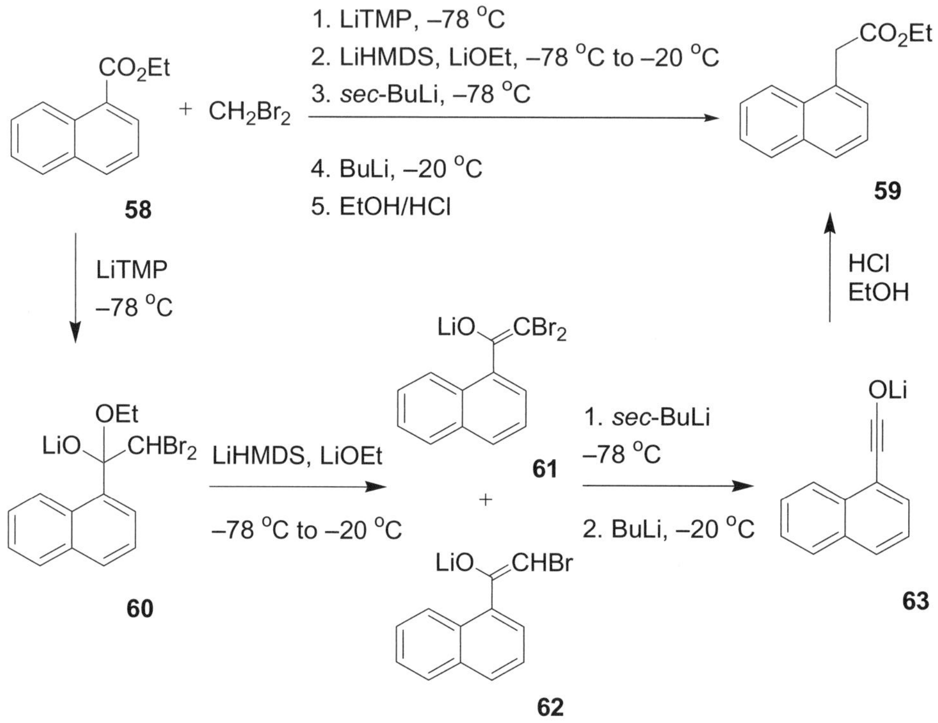

More elaborate replacement homologation reactions have also been developed. For example, Kowalshi and co-workers have reported a one-pot ester homologation via ynolate anions.[29] This alternative to the classical Arndt–Eistert reaction was used to mediate the conversion of naphthalene derivative **58** to its homolog **59** in good yield. The mechanistic steps involved in this rather complex reaction mixture have been reported as follows: Deprotonation of dibromomethane, followed by nucleophilic attack produces tetrahedral intermediate **60**. The addition of LiHMDS serves to deprotonate the mono and dibromoketones that are formed upon warming of the reaction to –20 °C. This gives enolate anions **61** and **62**. *sec*-Butyl lithium affects metal–halogen exchange and rearrangement of **61**, converting it to ynolate anion **63**. Subsequent addition of *n*-butyllithium at –20 °C deprotonates enolate **62**, also yielded ynolate anion **63**. An acidic ethanol quench yields the homologated product **59** in good yield.

2.1.6 *Experimental*

Homologation Using CF_3CO_2Ag via a Mixed Anhydride[30]

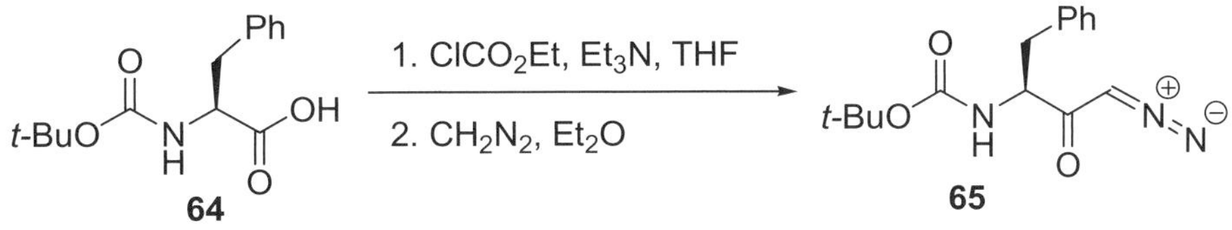

(*S*)-3-(*tert*-Butyloxycarbonylamino)-1-diazo-4-phenylbutan-2-one (65).[30] To an ice-cold, stirred solution of Boc-phenylalanine **64** (25.0 g, 94.2 mmol) in dry THF (250 mL) was added triethylamine (13.1 mL, 94.0 mmol) followed by ethyl chloroformate (9.45 mL, 94.0 mmol). After 15 min., an ethereal solution of diazomethane[31] (about 125 mL) was added. After a further 45 min. the remainder of the diazomethane solution is added (about 85 mL). The mixture was allowed to warm to ambient temperature and stirred for 3 h. 0.5 N Acetic acid (75 mL) was carefully added to destroy and excess diazomethane, followed by saturated aqueous sodium bicarbonate solution (75 mL). The aqueous layer was separated and the organic layer washed with brine (75 mL). The organic layer was dried over anhydrous magnesium sulfate, filtered and solvent evaporated under reduced pressure. The crude product **65** was dried *in vacuo* and used immediately.

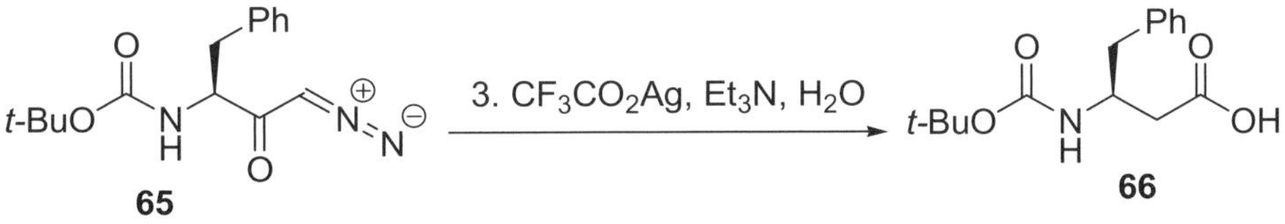

(*S*)-3-(*tert*-Butyloxycarbonylamino)-4-phenylbutanoic acid (66).[30]
A solution of silver trifluoroacetate (2.72 g, 12.3 mmol) in dry triethylamine (39 mL, 279 mmol) was added to a solution of the diazoketone **65** in THF (380 mL) and water (38 mL) at –25 °C in the dark. The resultant mixture was allowed to warm to ambient temperature overnight. The solvent was evaporated under reduced pressure and then the residue was stirred in saturated aqueous sodium bicarbonate (100 mL) for 1 h. The mixture was then partitioned between water (150 mL) and ethyl acetate (200 mL). The organic layer was washed with brine (30 mL), followed by saturated aqueous sodium bicarbonate (30 mL). The organic layer was further extracted with saturated aqueous sodium bicarbonate (3 × 30 mL) and the combined

aqueous layers washed with ethyl acetate (50 mL). This organic extract was washed with aqueous sodium bicarbonate (2 × 25 mL) and combined with the previous organic extracts. This process is repeated a second time. Congo Red indicator (10 drops) was added to the total combined aqueous layers, followed by ethyl acetate (100 mL). The mixture was cooled to 0 ºC and 5 N HCl was added drop-wise until the color of the indicator changed from red to blue. The organic layer was separated and the aqueous layer additionally extracted with ethyl acetate (3 × 100 mL). The total combined organic layers were dried over anhydrous magnesium sulfate, filtered and solvent evaporated under reduced pressure. The product was further dried *in vacuo*. The product crystallizes slowly to essentially pure material (17.1 g, 65 %), which was further recrystallized (Et_2O : petroleum ether, 1 : 1) to give the product **66** (12.1 g, 46 %).

Homologation Using $PhCO_2Ag$ via an Acid Chloride

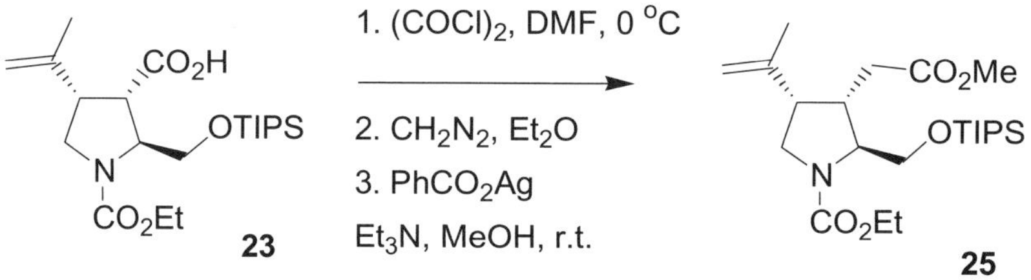

23 **25**

(2*S*,3*S*,4*S*)-3-Diazoacetyl-1-ethoxycarbonyl-4-isopropenyl-2-triisopropyl-silyloxymethylpyrrolidine.[12]

To an ice-cold, stirred solution of the pyrrolidinecarboxylic acid **23** (0.16 g, 0.39 mmol) in dry ether (3 mL) was added dry DMF (1 mL) followed by freshly distilled oxalyl chloride (61 mL, 0.695 mmol). After 0.5 h at 0 °C and 1.5 h without cooling, the solvent was evaporated and to a solution of the residual acid chloride in dry ether (3 mL) was added an excess of an ice-cold, dried (KOH) solution of diazomethane in ether. The resulting solution was left at ambient temperature overnight and then evaporated. The residue was chromatographed over silica gel eluted with 10% ether/light petroleum to give the diazoketone (0.12 g, 70%) as an oil.

Methyl (2*S*,3*S*,4*S*)-1-Ethoxycarbonyl-4-isopropenyl-2-triisopropyl-silyl-oxymethyl-pyrrolidin-3-ylacetate (25).[12]

A solution of silver benzoate (0.005 g) in dry triethylamine (0.05 mL) was added to a solution of the diazoketone (0.115 g, 0.26 mmol) in dry methanol (5 mL) and the resultant mixture stirred at ambient temperature for 16 h. It was then filtered through a Kieselguhr and evaporated. The residue was dissolved in ether (15 mL) and the resulting solution washed with saturated aqueous sodium hydrogen carbonate (3 mL) then dried and evaporated to leave the pyrrolidinylacetate **25** (0.110 g, 95%) as a colourless oil.

2.1.7 References

1. [R] Bachmann, W. E.; Struve, W. S. *Org. React.* **1942**, *1*, 38.
2. [R] Matthews, J. L.; Braun, C.; Guibourdenche, C.; Overhand, M.; Seeback, D. *Enantiosel. Synth. β-Amino Acids*, **1997**, 105.
3. [R] Kimse, W. *Eur. J. Org. Chem.* **2002**, 2193.
4. Wolff, L. *Ann.* **1912**, *394*, 25.
5. a) Staudinger, H. *Ber. Dtsch. Chem. Ges.* **1905**, *38*, 1735. b) Staudinger, H. *Ber. Dtsch. Chem. Ges.* **1907**, *40*, 1145. c) Staudinger, H. *Ber. Dtsch. Chem. Ges.* **1911**, *44*, 1619.
6. Wolff, L. *Justus Liebigs Ann. Chem.* **1912**, *394*, 23.
7. Arndt, F.; Eistert, B. *Ber. Dtsch. Chem. Ges. B.* **1935**, *68*, 200.
8. Newmann, M. S.; Beal, P. F. *J. Am. Chem. Soc.* **1950**, *72*, 5163.
9. Gademann, K.; Ernst, M.; Hoyer, D.; Seeback, D. *Angew. Chem., Int. Ed. Engl.* **1999**, *38*, 1223.
10. Gopi, H. N.; Babu, V. V. S. *Lett. Peptide Sci.* **2000**, *7*, 165.
11. Guibourdenche, C.; Podlech, J.; Seebach, D. *Liebigs Ann.* **1996**, *79*, 288.
12. Cooper, J.; Knight, D. W.; Gallagher, P. T. *J. Chem. Soc., Perkin Trans. 1*, **1992**, 553.
13. Nicolaou, K. C.; Baran, P. S.; Zhong, Y.-L.; Choi, H.-S.; Yoon, W. H.; He, Y.; Fong, K. C. *Angew. Chem., Int. Ed. Engl.* **1999**, *38*, 1669.
14. Ancliff, R. A.; Russell, A. T.; Sanderson, A. J. *Tetrahedron: Asymmetry*, **1997**, *8*, 3379.
15. Garg, N. K.; Sarpong, R.; Stoltz, B. M. *J. Am. Chem. Soc.* **2002**, *124*, 13179.
16. Zwanenburg, B.; Thijs, L. *Tetrahedron Lett.* **1974**, 2459.
17. a) Woolsey, N. F.; Khalil, M. H. *Tetrahedron Lett.* **1974**, 4309; b) Woolsey, N. F.; Khalil, M. H. *J. Org. Chem.* **1975**, *40*, 3521; c) van Haard, P. M. M.; Thijis, L.; Zwanenburg, B. *Tetrahedron Lett.* **1975**, 803.
18. Thijs, L.; Donnerholt, F. J.; Leemhuis, F. M. C.; Zwanenburg, B. *Tetrahedron Lett.* **1990**, *31*, 6589.
19. Saito, T.; Thijis, L.; Ettema, G.-J.; Zwanenburg, B. *Tetrahedron Lett.* **1993**, *34*, 3589.
20. Thijis, L.; Stokkingreef, E. H. M.; Lemmens, J. M.; Zwanenburg, B. *Tetrahedron,* **1985**, *41*, 2949.
21. a) Thijis, L.; Eggenberger, D. M.; Zwanenburg, B. *Tetrahedron Lett.* **1989**, *30*, 2153; b) Leemhuis, F. M. C.; Thijis, L.; Zwanenburg, B. *J. Org. Chem.* **1993**, *58*, 7170.
22. Eaton, P. E.; Yip, Y. C. *J. Am. Chem. Soc.* **1991**, *113*, 7692.
23. Burger, U.; Zellweger D. *Helv. Chim. Acta* **1986**, *69*, 676.
24. Regitz, M. *Synthesis* **1972**, 351.
25. Regitz, M.; Rüer, J.; Liedhegener, A. *Org. Synth. Coll. Vol. VI*, **1988**, 389.
26. Danheiser, R. L.; Miller, R. F.; Brisbois, R. G. *Org. Synth. Coll. Vol. IX*, **1998**, 197.
27. Aoyama, T.; Shiori, T. *Tetrahedron Lett.* **1980**, *21*, 4461.
28. Cesar, J.; Dolenc, M. S. *Tetrahedron Lett.* **2001**, *42*, 7099.
29. Rajarathnam, E. R. Kowalski, C. J. *Org. Synth. Coll. Vol. IX*, **1998**, 426.
30. Linder, M. R.; Steurer, S.; Podlech, J. *Org. Synth. Coll. Vol. X*, **2004**, 194.
31. De Boer, Th. J.; Backer, H. J. *Org. Synth. Coll. Vol. IV*, **1963**, 250.

2.2 Morita–Baylis–Hillman Reaction

Chris Limberakis

2.2.1 Description

The Morita–Baylis–Hillman or Baylis–Hillman reaction involves the reaction between an electrophile **1**, usually a carbonyl containing compound such as an aldehyde, ketone, or imine, and an activated alkene **2** in the presence of a catalyst such as an amine or phosphine to deliver an α-methylene-β-hydroxy carbonyl or α-methylene-β-amino carbonyl adduct **3**.

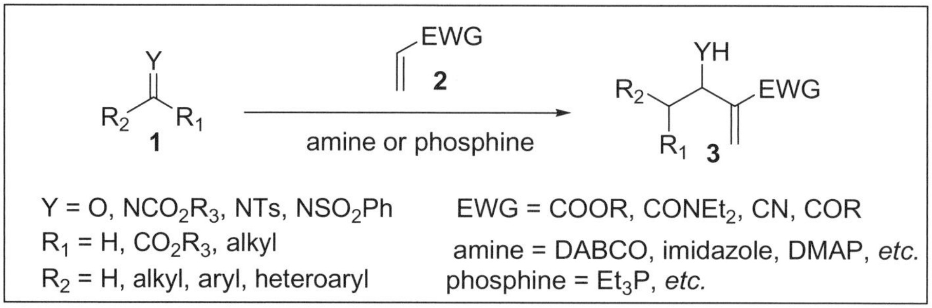

2.2.2 Historical Perspective

The genesis of the Morita–Baylis–Hillman (MBH) reaction began rather inauspiciously. In 1968, Morita described the reaction between an aldehyde and acrolein nitrile as well as methyl acrylate in the presence of tricyclohexylphosphine to afford 2-(1-hydroxyethyl) acrylonitrile and 2-(1-hydroxyethyl) methyl acrylate adducts in low yields (< 20%).[1] An example is shown below using methyl acrylate **4** and acetaldehyde **5**.

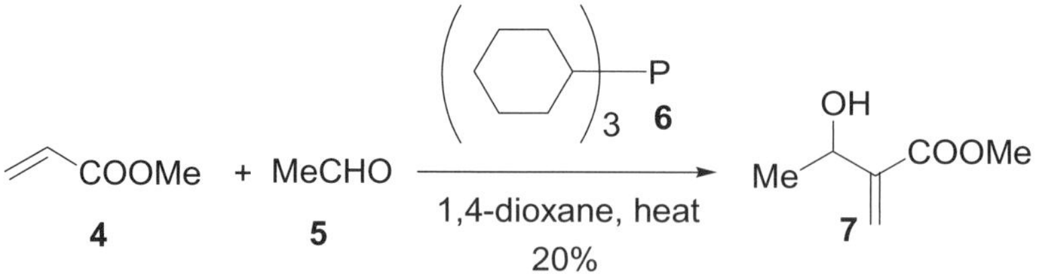

However, in 1972, Baylis and Hillman patented the synthesis of α-methylene-β-hydroxy carbonyl compounds between aldehydes and activated alkenes including α,β-unsaturated aldehydes, ketones, esters, amides, and nitriles in the presence of tertiary bases such as 1,4-diazabicyclo-[2.2.2]octane (DABCO) **9**, quinuclidine, and indolizine in high yield.[2]

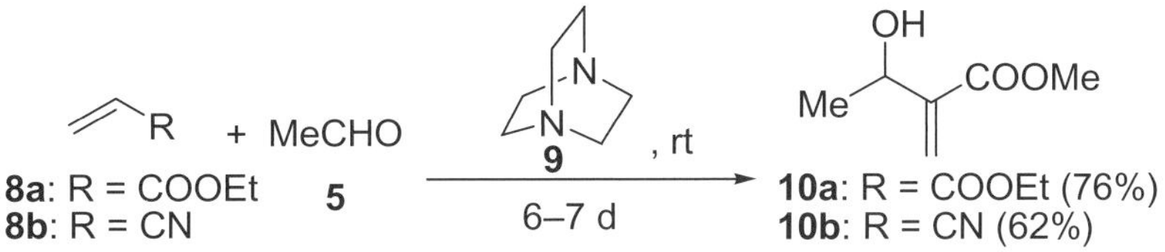

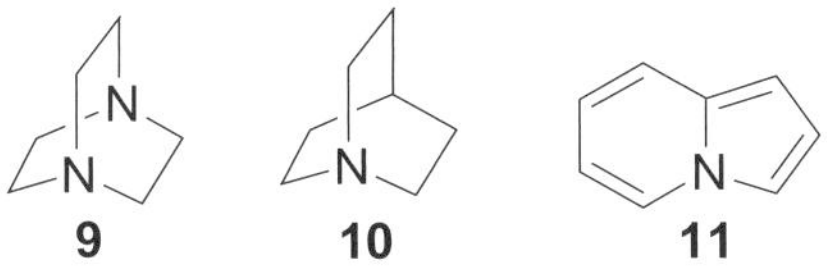

This improvement of the original Morita protocol provided a viable synthetic method, for this multi-component transformation formed a unique C–C bond and afforded functionality conducive to further derivatization in an atom economical manner albeit suffering long reaction times from days to weeks. In the early 1980s, the synthetic organic chemistry community realized the utility of this reaction[3] and has continued into the 21st century. Thus, the reaction has been the subject of numerous reviews.[3,4] Because of these recent reviews, this chapter will highlight only some literature examples of the last several years.

2.2.3 Mechanism

The most widely accepted MBH mechanism is depicted below involving methyl vinyl ketone **12** (MVK), benzaldehyde, and 1,4-diazabicyclo[2.2.2]octane **9** (DABCO).[4a] The first step is a Michael addition between DABCO and MVK to deliver the zwitterionic enolate **13**. With this intermediate in hand, two possible paths are possible — either an aldol condensation with benzaldehyde or another Michael addition with another molecule of MVK. Let us first proceed with the former route, since the aldol condensation leads to the major product. Hence, zwitterionic intermediate **13** reacts with benzaldehyde to deliver the zwitterionic aldol adduct **14a** which then under goes proton migration followed by an elimination reaction to give the α-methylene-β-hydroxy carbonyl adduct **16a** and DABCO **9**. It is believed that the rate determining step in path I is the aldol condensation and efforts to stabilize intermediate **14a** have been investigated. Alternatively following path II, zwitterionic enolate **13** could attack another molecule of MVK in a 1,4-fashion to afford the Michael adduct **14b**,and then proton migration followed by an elimination reaction of the thus afforded adduct gives Michael-type dimer **16b**. Although this mechanism is generally accepted, subsequent mechanistic studies have been conducted.[5–7]

Path I: Michael addition, aldol, elimination

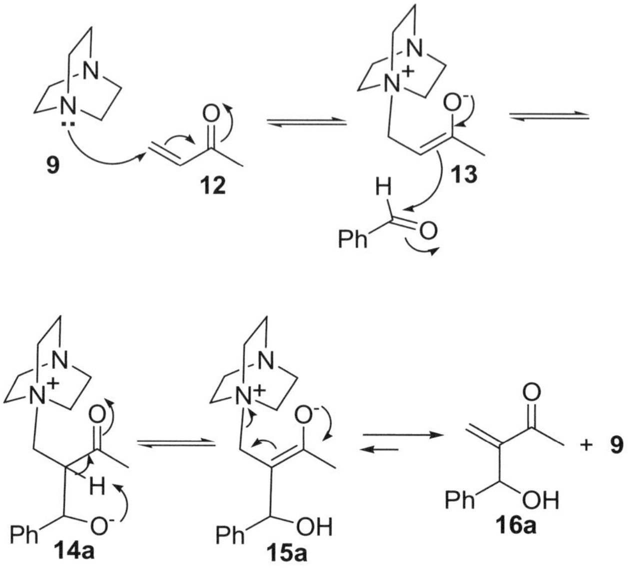

Path II: Michael, Michael, elimination

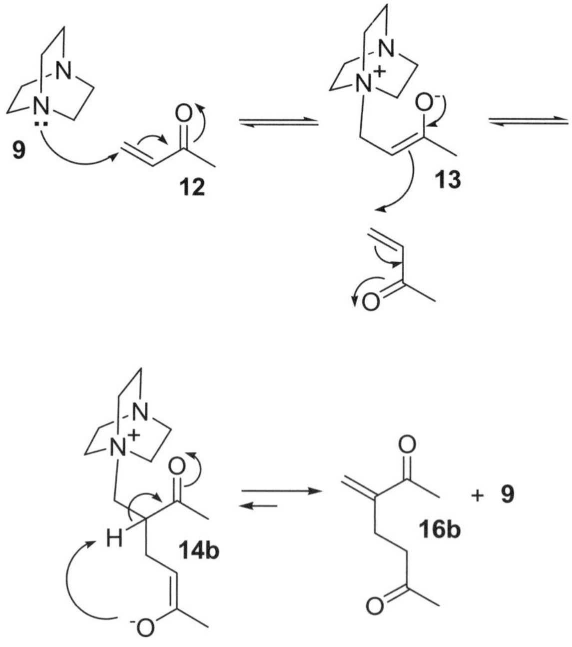

Aggarwal and coworkers showed a direct correlation between the pK_a of a series of quinuclidine catalysts and the reaction rate of the MBH.[5] In the study, the reaction between 2-pyridinecarboxylaldehyde and methyl acrylate in the presence of 5 mol% of catalyst was performed with no solvent. They determined the reaction rate increased as the pK_a of the catalyst increased. For instance, in the presence of quinuclidine **10** (pK_a = 11.3) a relative rate constant (k_{rel}) of 11.3 was reported; however, when the less basic quinuclidone **22** (pK_a = 6.3) was used the relative rate dropped significantly to 0.006 as shown in the table below.

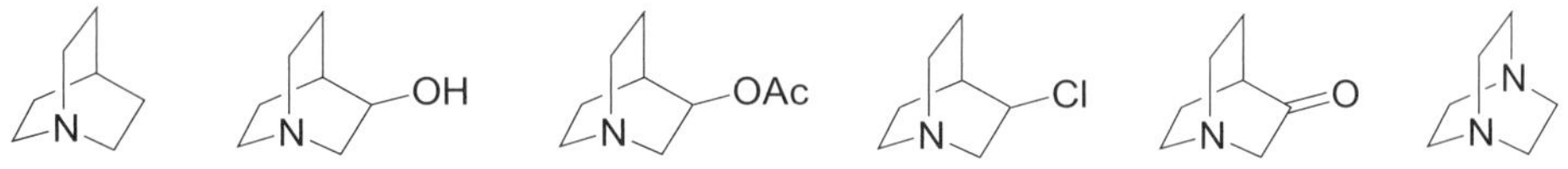

10 **19** **20** **21** **22** **9**

catalyst	pK_a	k_{rel}
10	11.3	9.0
19	9.9	4.3
20	9.3	0.15
21	8.9	0.04
22	6.9	0.006
9	8.5	1

Interestingly, DABCO (**9**, pK_a = 8.5), with relative rate of 1, accelerated the reaction more than 3-acetoxyquinuclidine **20** (pK_a = 9.3) or 3-chloroquinucilidine **21** (pK_a = 8.9). Presumably, DABCO was a more effective catalyst, because of the two basic nitrogen atoms.

There have also been proposed MBH mechanisms in the presence of protic solvents and aprotic additives.[6] For example, Aggarwal has suggested that a proton transfer from an alcohol to intermediate **23** delivers the six-membered transition state **24** which then leads to the protonated species **25**.

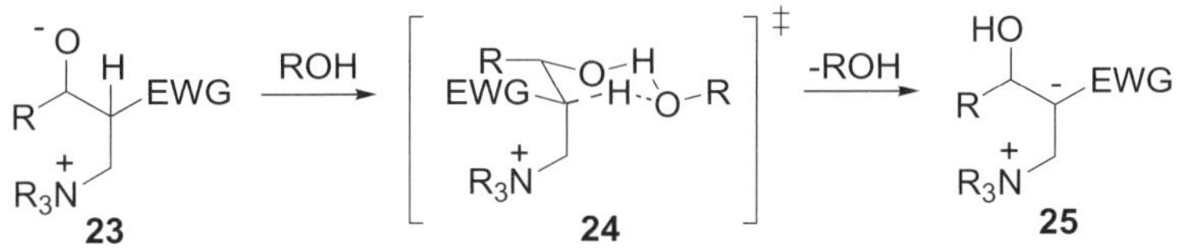

With respect to the aprotic conditions, McQuade and coworkers have studied this approach.[7] Hence, the zwitterion **26** attacks another molecule of aldehyde to afford **27**. Intermediate **27** then undergoes an intramolecular hydrogen transfer via the six-membered transition state **28** to deliver the acetal intermediate **29** upon elimination.

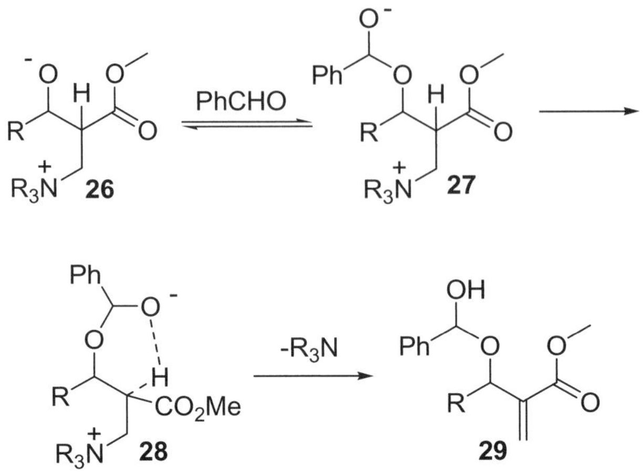

2.2.4 *Variations and Improvements*

In the last 20 years, the literature has been laden with variations and improvements of the MBH reaction as well as applications of MBH adducts to natural product synthesis and medicinal chemistry.[3,4] More specifically, all three components of the reaction including the activated alkene, electrophile, and catalyst have been addressed. Also, there are now aza-variants, asymmetric versions, and intra-molecular applications. Moreover, some very encouraging reports regarding the rate acceleration of this sluggish reaction have been reported using new catalysts, microwave irradiation, ultrasound, ionic liquids, polar solvents, *etc.* Although there remains ample space to improve and expand the scope of this reaction, it has been become a powerful carbon-carbon bond forming transformation.

Solvent effects

Although the MBH reaction can be accelerated in the presence of water[8], the product yield appears to be dependant on the water/co-solvent ratio.[9] For instance, the MBH reaction between methyl vinyl ketone **12** and *p*-nitrobenzaldehyde **30** in the presence of the co-catalyst system imidazole/proline in DMF/water was optimized when the DMF water ratio was 9:1. However, in a THF/water system, a 19:1 ratio was optimal.

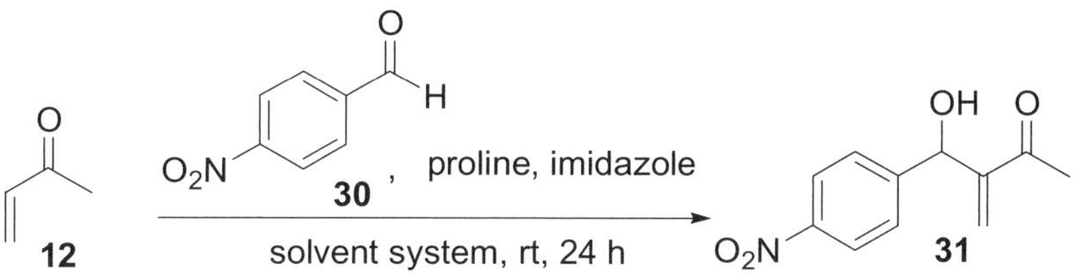

DMF:water	% yield of product
1:1	30
7.5:2.5	40
9:1	80
19:1	63
just DMF	61

THF:water	% yield of product
1:1	32
9:1	30
19:1	62
just THF	27

Although polar protic solvents, such as water and methanol, accelerate the rate of the MBH reaction, Chong and coworkers showed that longer chain alcohols were superior in affording higher yields.[10] For example, the MBH reaction of acetaldehyde **5** and methyl vinyl ketone **12** in the presence of DABCO in octanol gave a 65% yield of MBH adduct **32** and 8% yield of the Michael dimer **33**. Alternatively in MeOH, the yield of **32** was only 28% and **33** was delivered in 12% yield. Interestingly, the chain length of octanol was optimal, since shorter and longer chained alcohols gave lower yields. The scope of the methodology was extended to a variety of aliphatic and aromatic aldehydes while using methyl vinyl ketone. Also, octanol's role was changed from solvent to additive. Typically, two equivalents of octanol were required to achieve the desired rate acceleration, and reaction yields were typically above 70%.

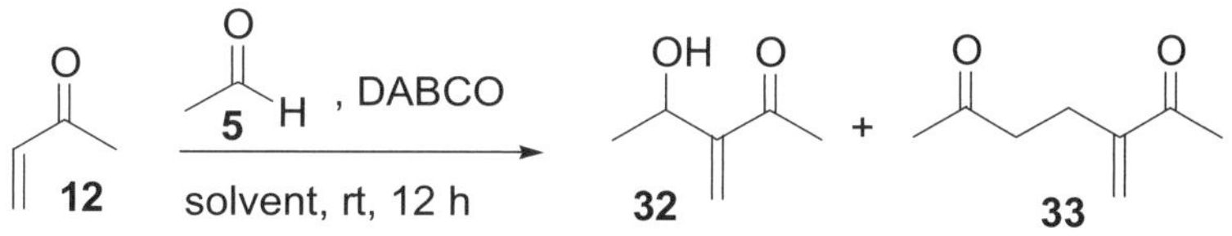

solvent	MBH adduct	dimer
H_2O	0	0
CH_3OH	28	12
$CH_3(CH_2)_3OH$	50	10
$CH_3(CH_2)_4OH$	52	22
$CH_3(CH_2)_7OH$	65	8
$CH_3(CH_2)_{11}OH$	50	9

Krishna and coworkers reported a rate acceleration of the MBH reaction in an aprotic polar solvent, sulfolane, when compared to 1,4-dioxane/water, methylene chloride, or dimethylsulfoxide. The MBH adducts **35** and **36** were afforded in 74–80% yield from aldehydes **34**.

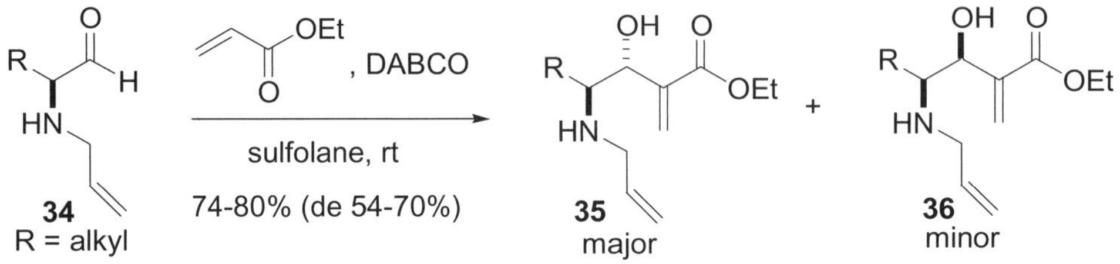

Ionic liquids have also played a central role in the rate acceleration of the MBH reaction.[12,13] For example, when *n*-butylpyridinium tetrafluoroborate was used as the solvent in the MBH reaction between a series of arylcarboxaldehydes **37a–d** and activated alkenes in the presence of urotropine **38** (HMTA), the reaction times were dramatically reduced from 16–24 h to 4–8 h with equal or better yields. Presumably, the ionic liquid medium stabilized the polar intermediates found in the reaction pathway.

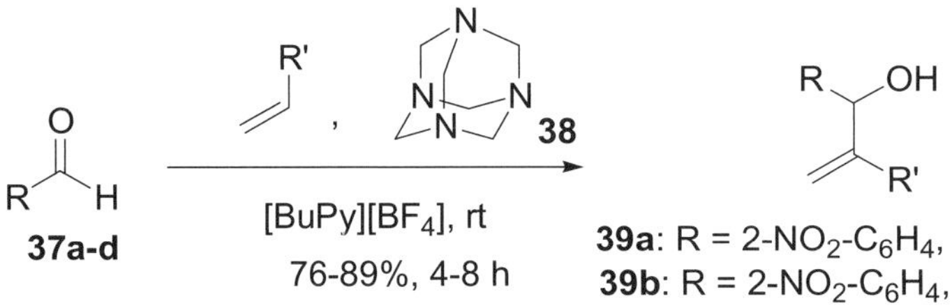

39a: R = 2-NO_2-C_6H_4, R' = CO_2CH_3
39b: R = 2-NO_2-C_6H_4, R' = CO_2NH_2
39c: R = 3-NO_2-C_6H_4, R' = CO_2CH_3
39d: R = 4-NO_2-C_6H_4, R' = CO_2CH_3

Compounds	Yield and reaction time: dioxane:H_2O (1:1)	Yield and reaction time: [BuPy][BF_4]
39a	82%, 24 h	81%, 8 h
39b	72%, 16 h	76%, 8 h
39c	80%, 24 h	85%, 6 h
39d	78%, 24 h	89%, 4 h

Catalyst

Aggarwal showed that 1,8-diazabicycloundec-7-ene (DBU) was also a very effective catalyst for the MBH reaction and significantly reduced the reaction time when compared to DABCO.[14] In addition, reaction rates were also shortened when compared to other catalysts such as dimethylaminopyridine (DMAP) and 3-hydroxyquinuclidine.

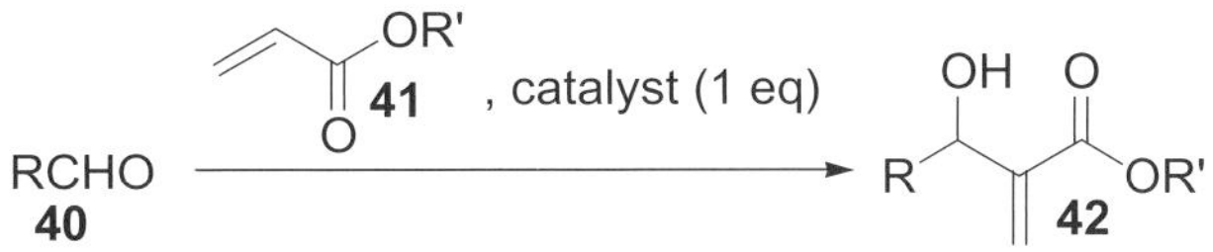

catalyst	R	R'	time (h)	yield
DBU	Ph	Me	6	89%
	Ph	*t*-Bu	72	74%
	4-OMePh	Me	48	62%
DABCO	Ph	Me	120	91%
	Ph	*t*-Bu	672	65%
	4-OMePh	Me	480	90%

As mentioned in the previous section, urotropine (HMTA) is an inexpensive alternative to DABCO. More specifically, MBH reactions between aromatic aldehydes and methyl acrylate and acrylonitrile using HMTA, 0.1 eq or 1 eq, were run in a variety of solvents including tetrahydrofuran, methanol, dimethylsulfoxide, acetonitrile, and dioxane:water (1:1).[15] In addition, the reactions could be run neat.

Moreover, *N*-methylmorpholine (NMM, pK_a =7.5) has been used successfully.[16] Because of its low basicity, NMM could be an alternative to the traditional catalysts if substrate racemization is an issue.

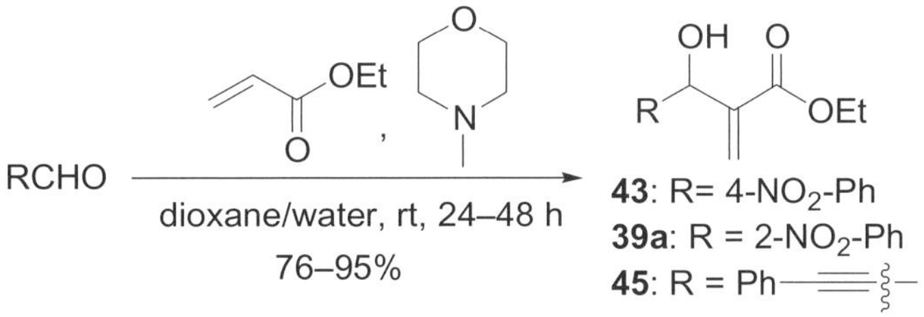

Alternatively, the reaction time of the MBH reaction between cylcopenten-2-one **46** and *p*-chlorobenzaldehyde **47** in the presence of an imidazole catalyst or L-histidine in a 1 M sodium bicarbonate solution is another mild medium.[17] A reaction rate increase was observed in 1 M sodium bicarbonate versus a THF/water mixture. With imidazole as the catalyst, the reaction time was reduced from 40 to 11 h with the yield improving from 74% to 90%.

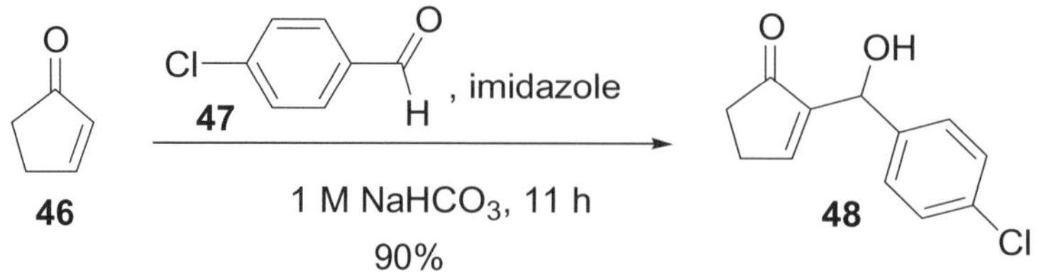

Although most of the MBH reactions rely on tertiary amine bases, Cheng and coworkers demonstrated that alkoxides can deliver excellent results.[18] For example, the MBH reactions of cyclopenten-2-one **46** and cyclohexen-2-one **49** with a variety of aromatic aldehydes in the presence of 50 mol% sodium methoxide in MeOH afforded products **48–51** in 60–94% yield with reaction times under one hour.

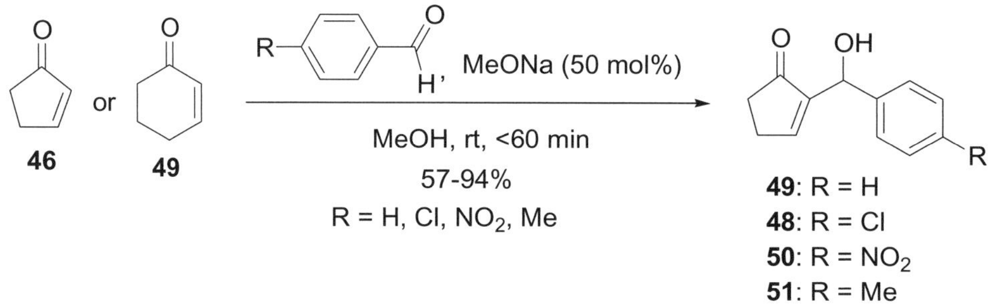

There has also been interest in the titanium-mediated MBH.[4] In a recent example, Liu and coworkers used a titanium-mediated MBH between the *S*,*S*-acetals **52a–c** and 2-pyridinecarboxylaldehyde **37d** and 2-quinolonecarboxaldehyde **55** as a route to di- and tri-substituted indolizines.[19] The MBH reactions were run in acetonitrile at 0 °C in the presence of titanium tetrachloride to deliver the adducts **53a–c** and **56a–c** in 67–85% yield. The products were then converted to the indolizines **54a–c** and **57a–c**, respectively, in hot acetic anhydride in good yields.

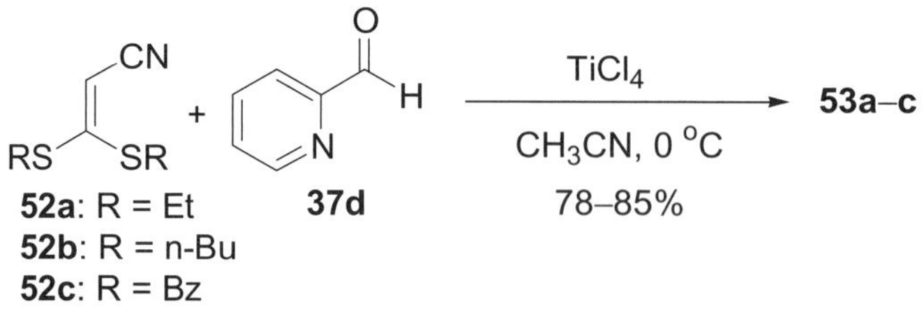

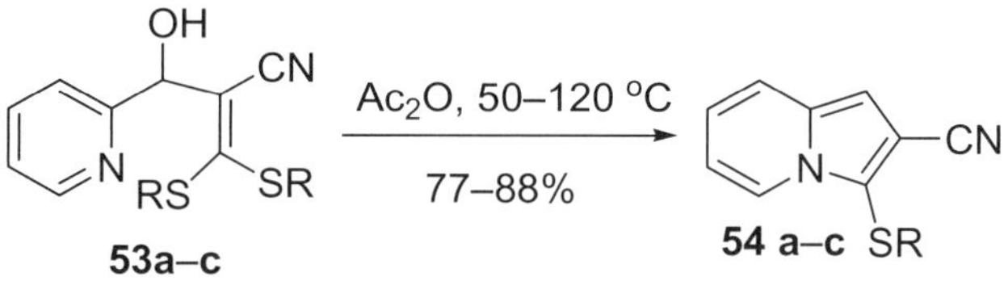

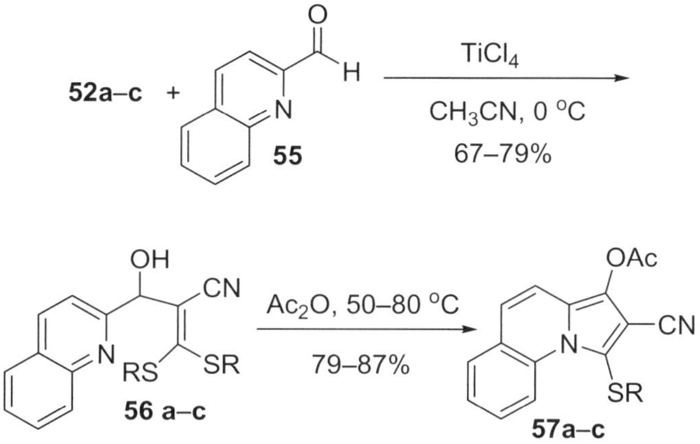

Miscellaneous

In a recent report using the environmentally friendly high speed ball milling (HSBM) technique, Mack and Shumba described a dramatic increase in reaction rate of the MBH reaction between *p*-nitrobenzaldehyde **30** and methylacrylate **4** in the presence of DABCO.[20] In the event, all reagents were placed in a stainless steel vial and milled with a stainless steel ball-bearing for 30 minutes to afford the MBH adduct **58** in high yield. Previous reports using solvent-free conditions required three to four days and delivered the product in 70–87% yield.

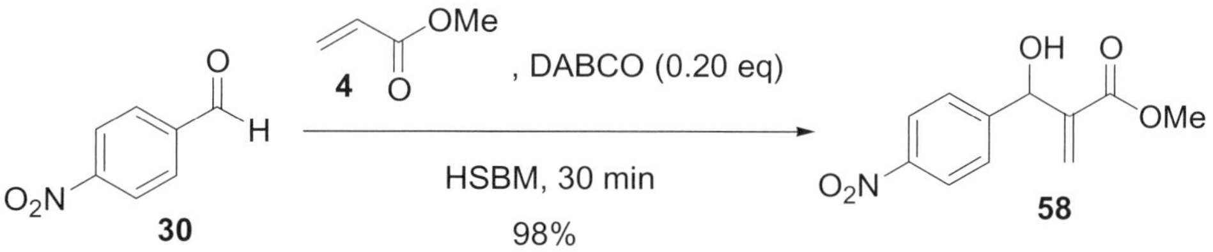

Although the MBH reaction typically relies on tertiary amines and phosphines as the catalysts, Basavaiah and Brière expanded the chlacogenide MBH reaction by using thiolane in the presence of TBDMSOTf.[21] This variant of the MBH was run using 2-cyclohex-1-one **49** and acetals **59** with stoichiometric amounts of thiolane and TBDMSOTf to afford the products **60a–e** in moderate to good yield.

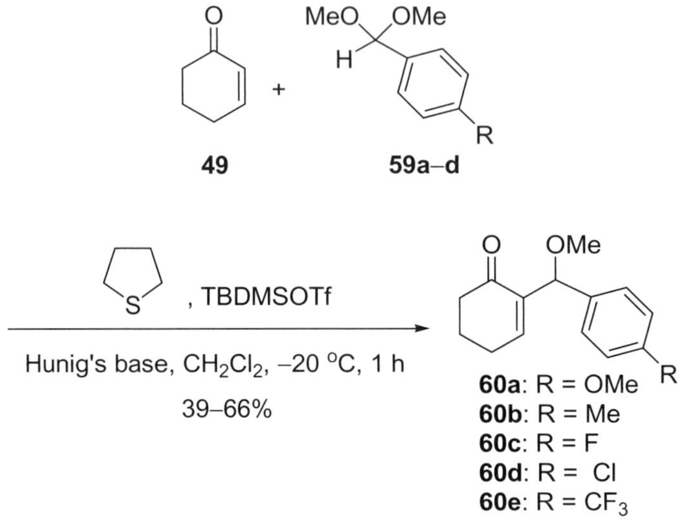

Asymmetric Versions

Traditional

For example, Schaus and Rodgen reported an asymmetric version of the MBH reaction involving cyclohexenone **49** and a variety of aliphatic aldehydes **61a–d** in the presence of the bis-phenol catalyst **63**.[22] The MBH adducts **62a–d** were delivered in good to excellent yields with enantioselectivities ranging from 86–98%.

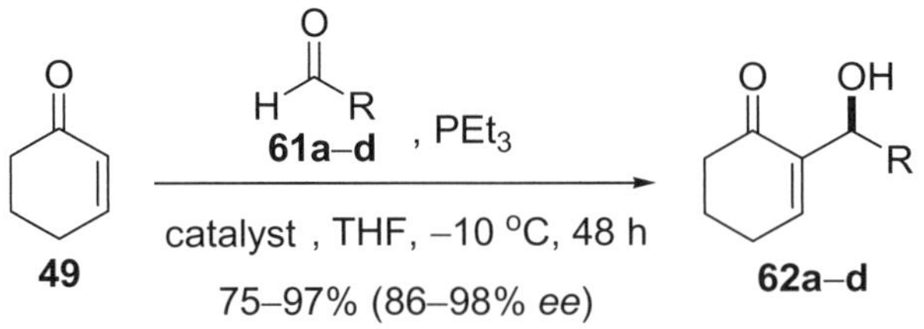

Moreover, Shi and Liu achieved good to excellent enantioselectivity when they utilized the bis(thio) urea catalyst **64** derived from *R*-(–)-5,5′,6,6′,7,7′,8,8′-octahydro-1,1′-binaphtyl-2,2′-diamine (H8-BINAM) to deliver the MBH adducts.[23] This asymmetric variant was carried out using cyclohexenone **49** and aromatic aldehydes in the presence of a catalytic amount of DABCO and catalyst **64** to deliver adducts **65a–e** in excellent yields with good enantioselectivities.

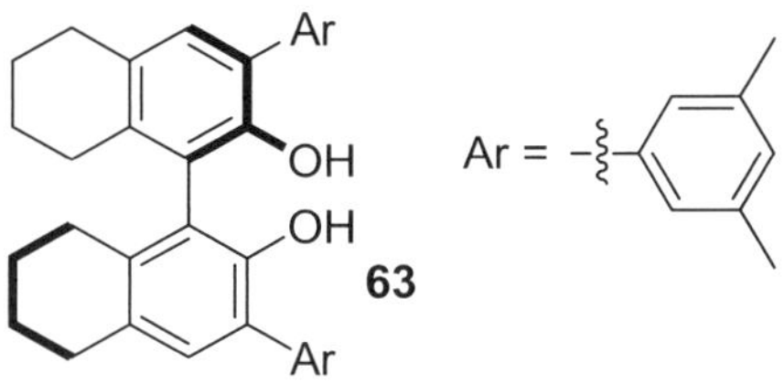

Compound	R	% Yield	% ee
62a	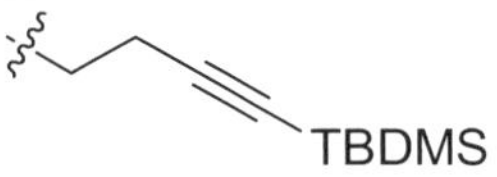 TBDMS	75%	86%
62b	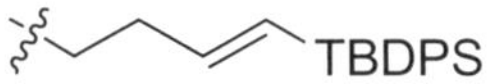 TBDPS	94%	98%
62c	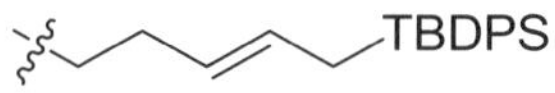 TBDPS	80%	90%
62d	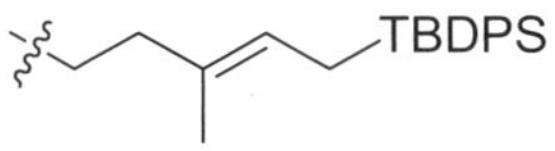 TBDPS	97%	98%

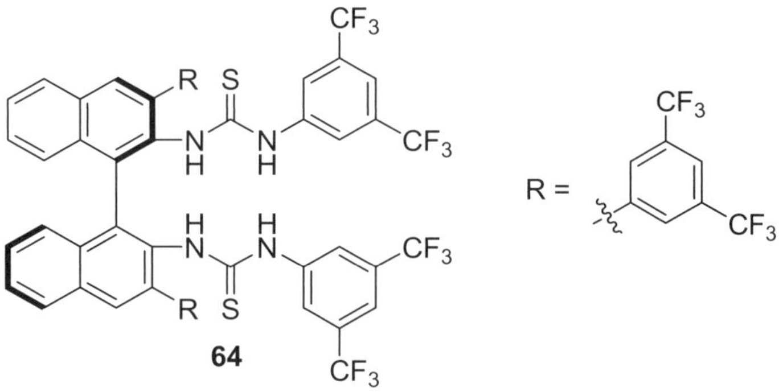

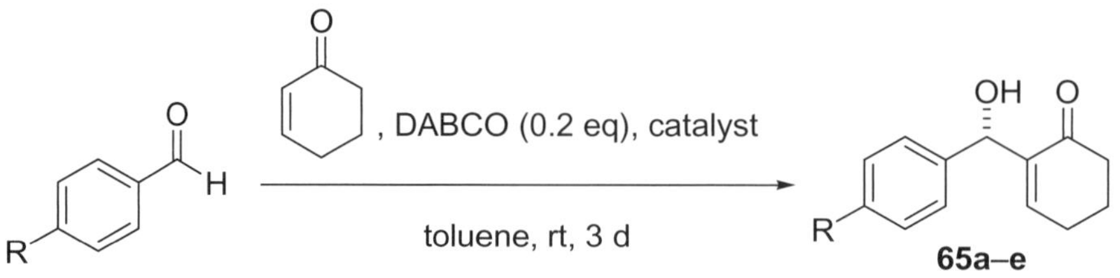

Compound	R	% yield	% *ee*
65a	H	99	81
65b	F	88	84
65c	Cl	90	85
65d	CH_3	85	76
65e	NO_2	94	72

Aza variant

The aza variant has also seen some advancements.[4, 24–27]

For example, Sasai and coworkers reported the metal-free bifunctional Binol derivative **67** as an effective catalyst for an enantio-selective aza MBH reaction.[24] In the presence of 10 mol% of **67**, methyl vinyl ketone and a series of tosylamines **66a–e** were converted to adducts **68a–e** in high yields with 88–95% *ee*. It has been rationalized that both the 2-hydroxy and pyridine moieties of **67** contributed to this highly enantioselective transformation.

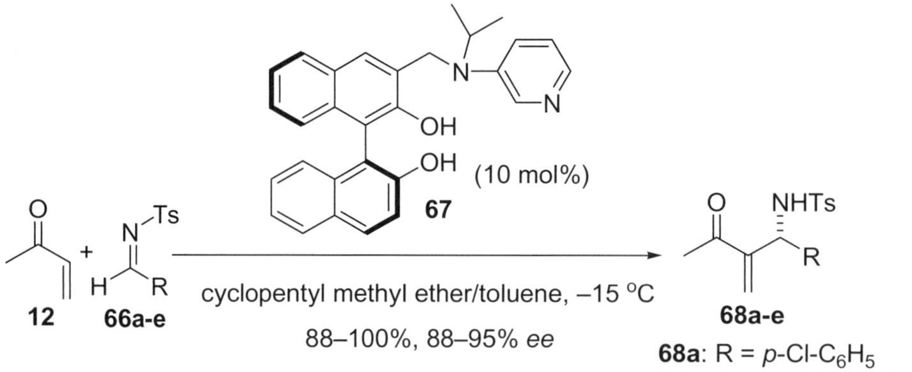

68a: R = *p*-Cl-C_6H_5
68b: R = p-OMe-C_6H_5
68c: R = p-NO_2-C_6H_5
68d: R = 2-furyl
68e: R = 2-naphthyl

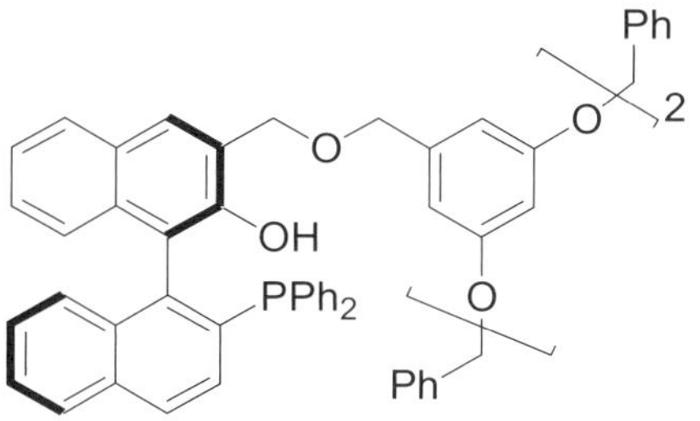

69

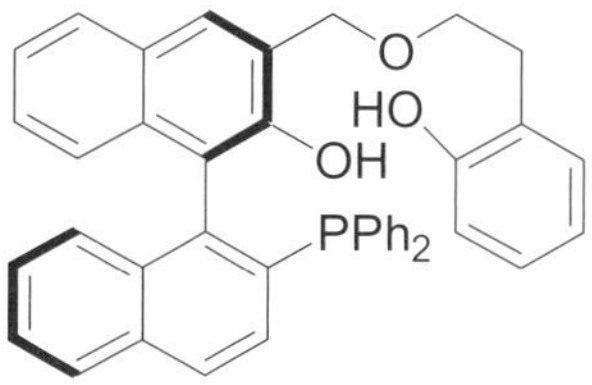

70

68b,c 71-73

product	catalyst **69**	catalyst **70**
71: R = C_6H_5	95% (91% ee)	84% (96% ee)
72: R = *p*-F-C_6H_5	84% (93% ee)	100% (95% ee)
68b: R = *p*-OMe-C_6H_5	90% (95% ee)	97% (95% ee)
68c: R = *p*-NO_2-C_6H_5	99% (97% ee)	98% (94% ee)
73: R = cinnamyl	94% (90% ee)	71% (87% ee)

Furthermore, two recent reports were published concerning bifunctional chiral phosphine catalysts **69**[26] and **70**.[27] In both cases, *N*-sulfonated imines and methyl vinyl ketone in the presence of either catalyst afforded the (*S*)-adducts **68b**,**c** and **71–73** in high yields with excellent enantioselectivities.

2.2.5 Synthetic Utility

The MBH reaction has been applied to both natural product synthesis[28] and medicinal chemistry.[11,29]

Clive and coworkers successfully utilized a MBH reaction to construct the hexahydroquinolizine fragment **75** of the marine natural product halichlorine (**74**).[28a] The aldehyde **76** was treated methyl acrylate in the presence of DABCO and scandium triflate to deliver a diastereomeric mixture of alcohols **77**. The alcohols were then acetylated with acetyl chloride to afford the acetates **78** in 71% yield over two steps. The acetates were then converted in two steps to the hexahydroquinolizine fragment **75**.

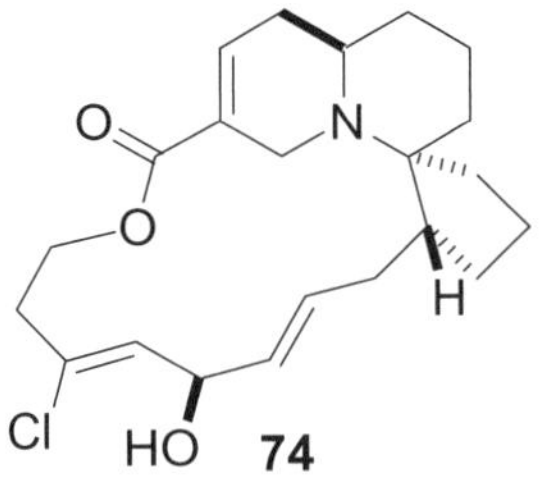

74

Halichlorine

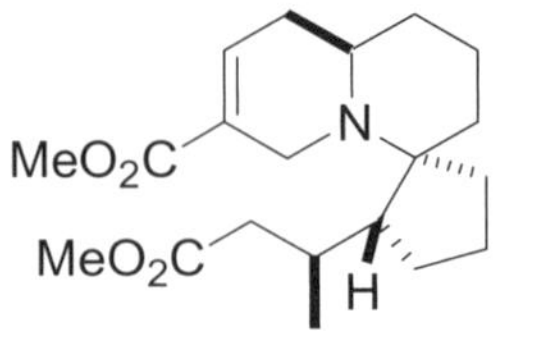

75

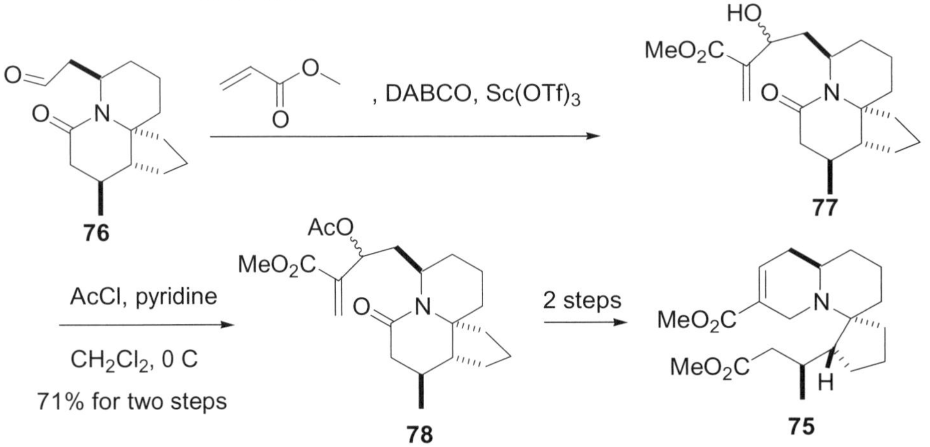

Ubukata and coworkers synthesized the potent antimicrobial natural product 6-tuliposide B (**80**) via a MBH reaction between *tert*-butyldimethylsilyloxy acetaldehyde **82** and the glucoside acrylate **81** in the presence of 3-hydroxyquinucolidine **19** to deliver a diastereomeric mixture of alcohols **83** in 40% yield.[28b] The alcohols were then separated using chiral chromatography, and the desired diastereomer **83a** was then treated with TFA to cleave both silyl protecting groups to afford the 6-tuliposide B (**80**).

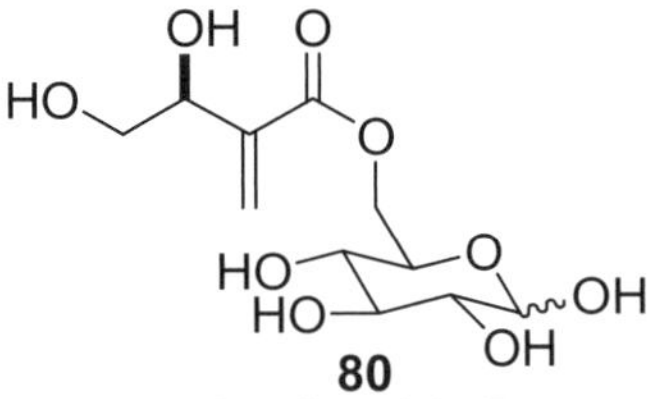

80

6-tuliposide B

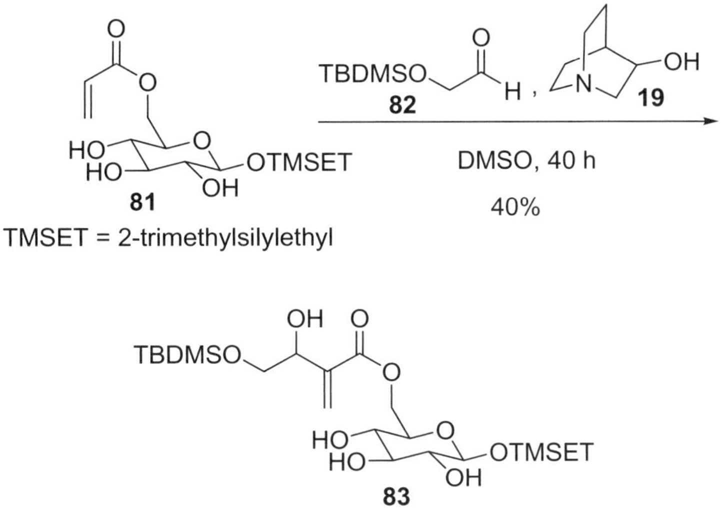

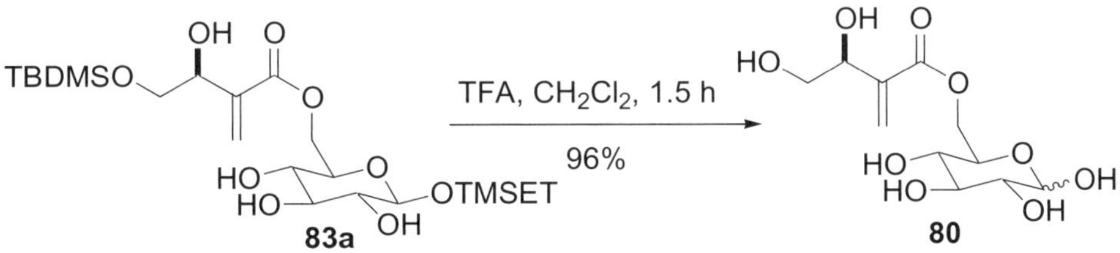

An MBH reaction was used to introduce key functionality to form a challenging oxacycle ring system.[28c] Aldehyde **85** was treated with ethyl acrylate in the presence of DABCO in a mixture of dioxane and water to deliver a diastereomeric mixture of alcohols **86a** and **86b**. After chromatographic separation of the diastereomers, each one was then converted to the corresponding oxacycle via a radical cyclization reaction. The thus afforded alcohols were acetylated to deliver acetate **87a** and acetate **87b** in 64% and 71%, respectively, over two steps.

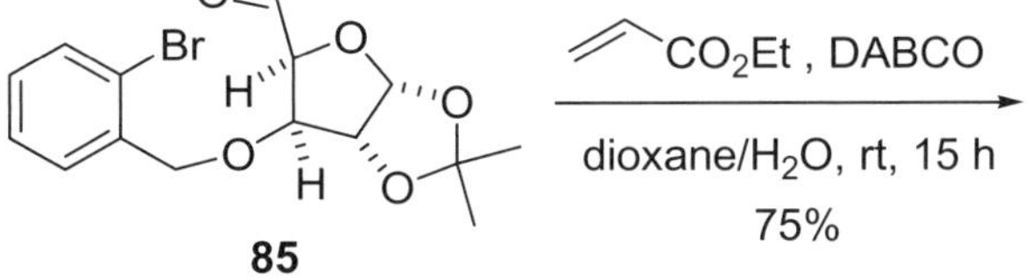

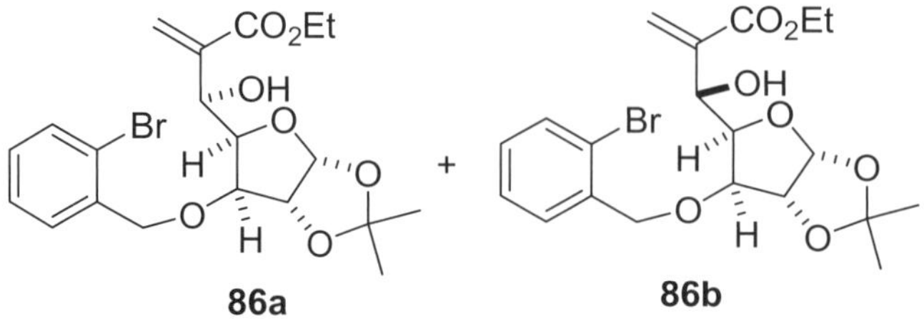

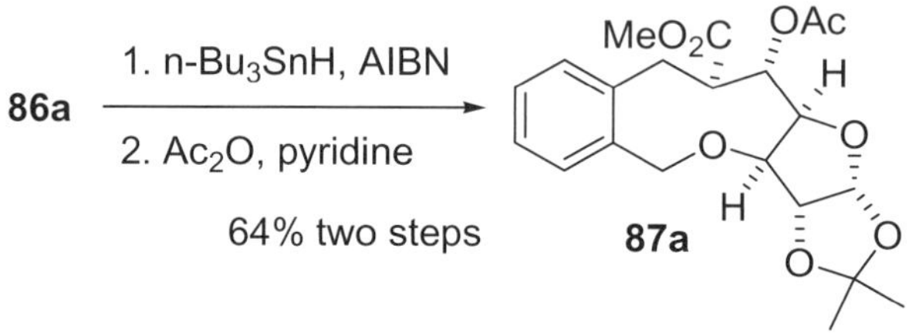

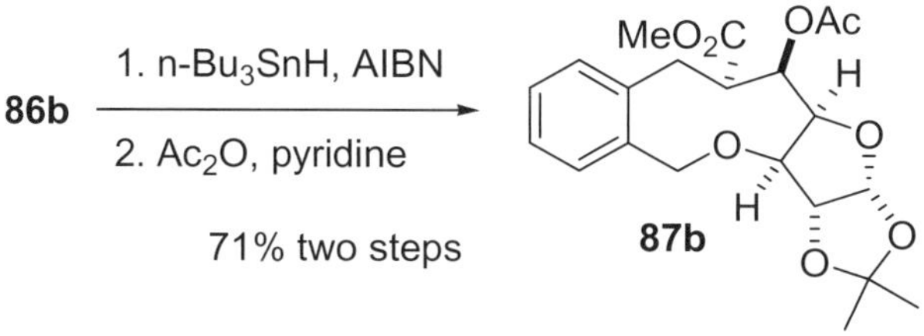

Krishna and coworkers synthesized syributin 1 (**92a**) and 2 (**92b**), which have agricultural applications, using an MBH reaction.[28d] The transformation was done using the aldehyde **88** and ethyl acrylate in a dioxane/water mixture in the presence of DABCO. The MBH adducts were delivered in 72% yield with a de of 80%. Interestingly, when this reaction was run under high pressure, no diastereoselectivity was observed. The MBH adducts **89** were then easily converted to the diene **90** which then underwent a ring closing metathesis to give the butenolide **91**. Further derivatization afforded **92a** and **92b**.

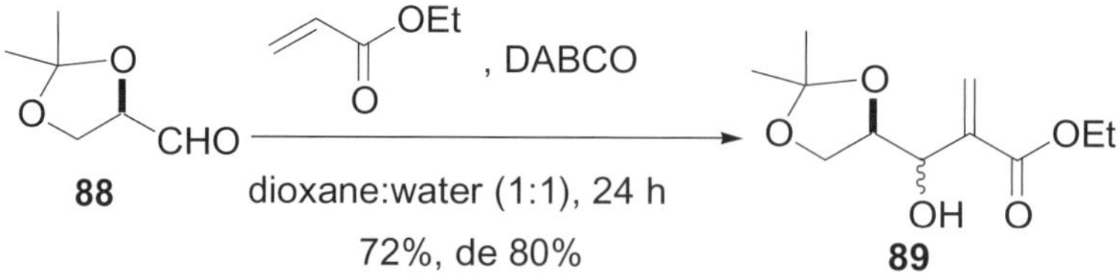

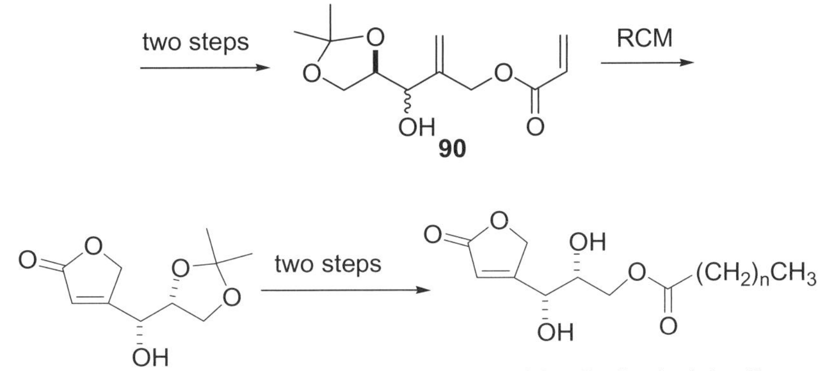

90

91

92a: Syributin 1 (n=4)
92b: Syributin 2 (n=6)

A MBH reaction played a pivotal role in the construction of the C1–C9 fragment of caribenolide I.[28e] Despite occurring in the first step of the sequence the MBH, the reaction between the aldehyde **94** and methyl acrylate in the presence of 3-hydroxyquinuclidine afforded the MBH adduct **95** in 62% yield. This transformation not only produced the C5–C9 portion of the molecule but established key sites of derivatization which included the α,β-unsaturated aldehyde and the C9–OPMB moieties. Both of these moieties were directly elaborated to aldehyde **98** which compromised the C1–C11 fragment of carbenolide 1 (**93**).

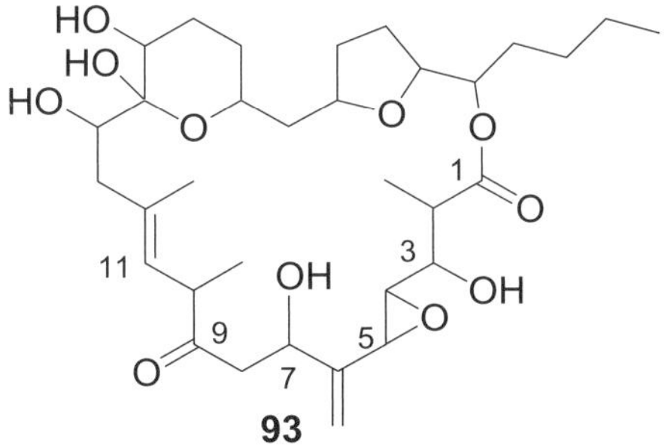

93
carbenolide I

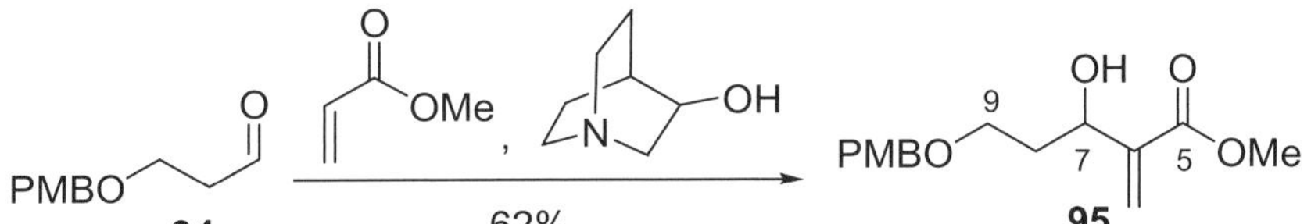

94

95

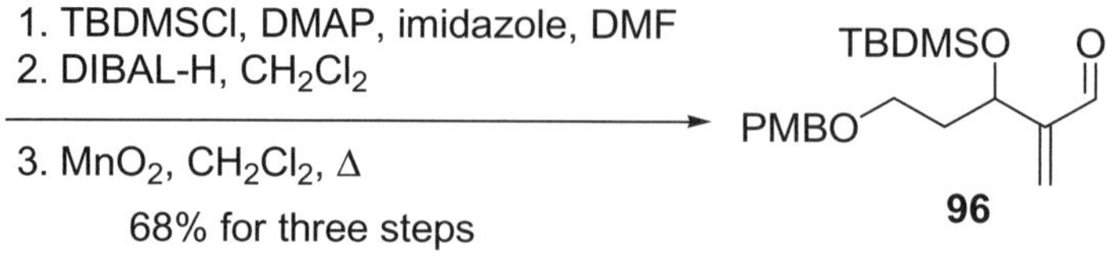

96

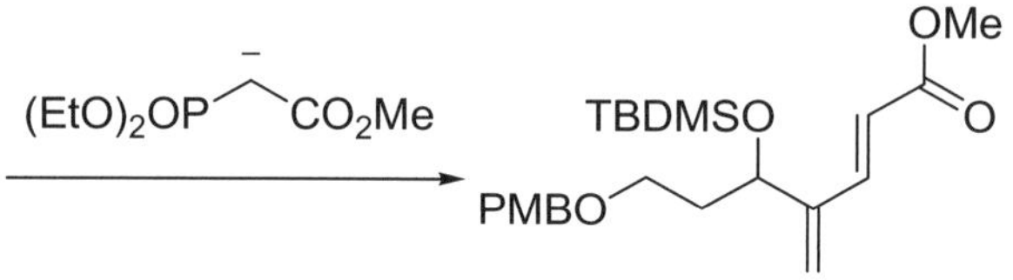

97

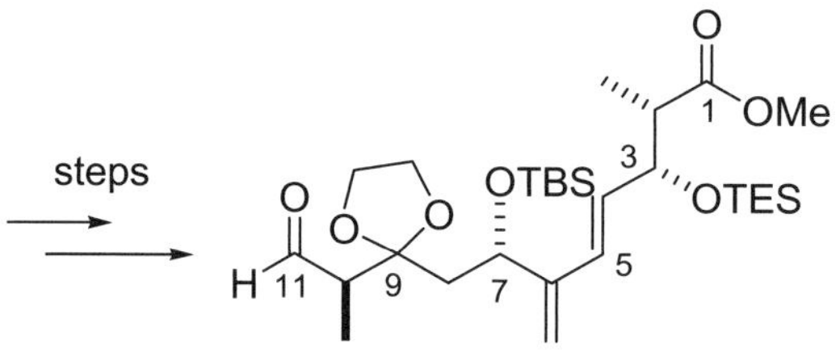

98

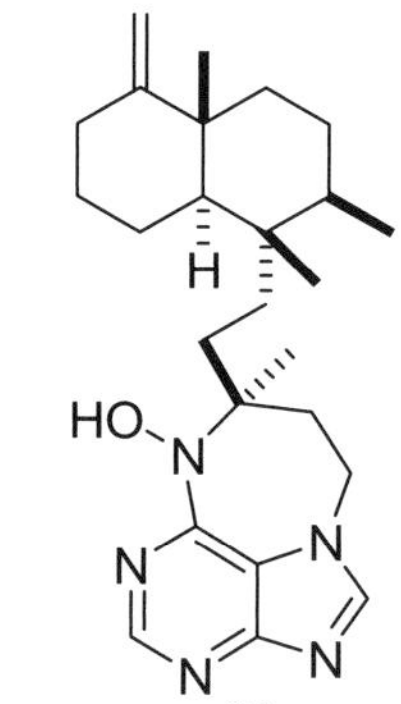

99

asmarine A

In an asymmetric variant of the MBH, Schaus and Rodgen constructed the penultimate intermediate of the decalin subunit of asmarine A (**99**).[28f] The transformation was accomplished by allowing cyclohexenone **49** to react with the chiral aldehyde **100** in the presence of triethylphosphine and a catalytic amount of the chiral bis-napthol **63** to deliver the chiral alcohol **101** in 99% yield with an enantiomeric excess of 99%. The alcohol was then exposed to boron trifluoride to afford the decalin **102** in 91% with high diastereoselectivity.

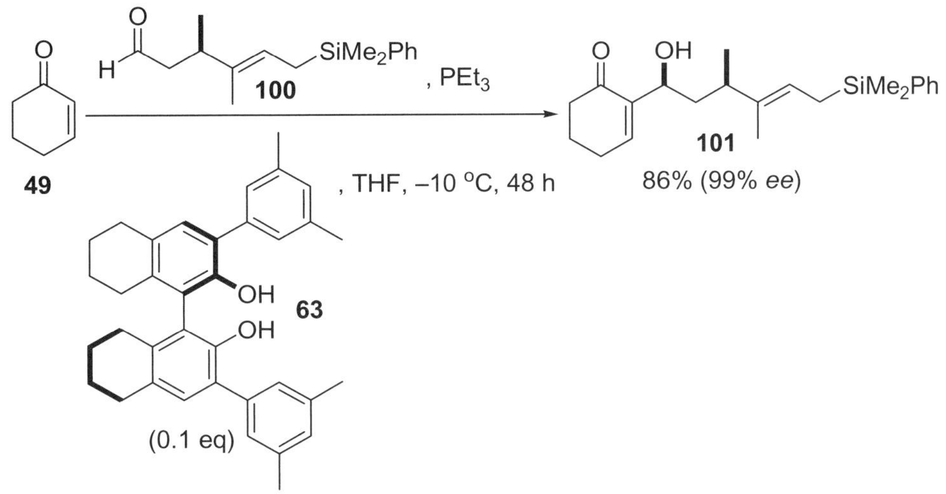

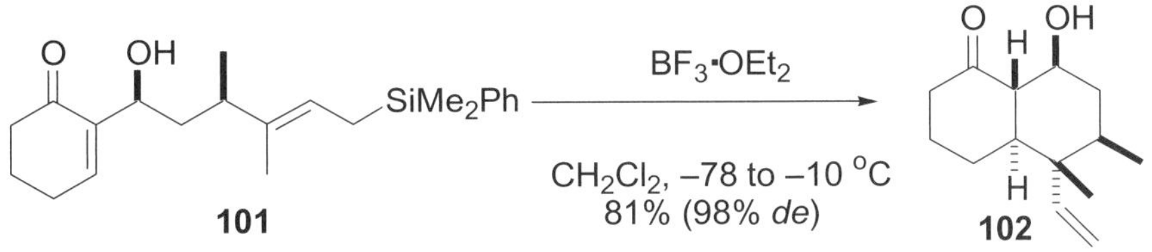

Coehlo and Lopes used a non-traditional approach in their construction of the dihydroisoquinolin5(6H)-core of the amaryllidaceae alkaloid plicamine (**103**).[28g] The MBH reaction was performed between 6-bromopiperonal **106** and methyl acrylate using ultrasound in the presence of the ionic liquid 1-methyl methylimidazolium to deliver the adduct **107** in 80% yield. Again, the MBH reaction provided the requisite functionality to allow further derivatization to afford the dihydroisoquinoline subunit **105** after several chemical steps.

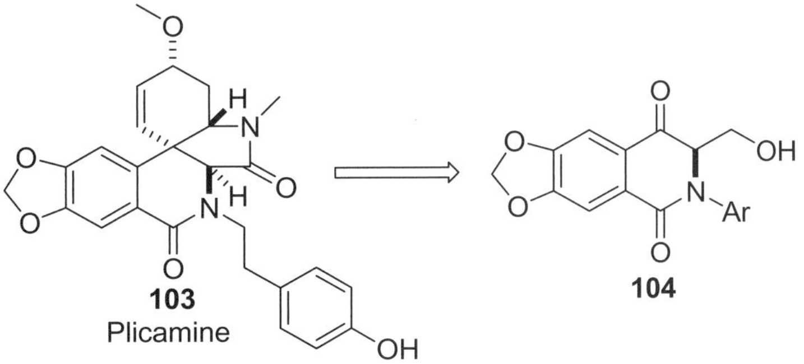

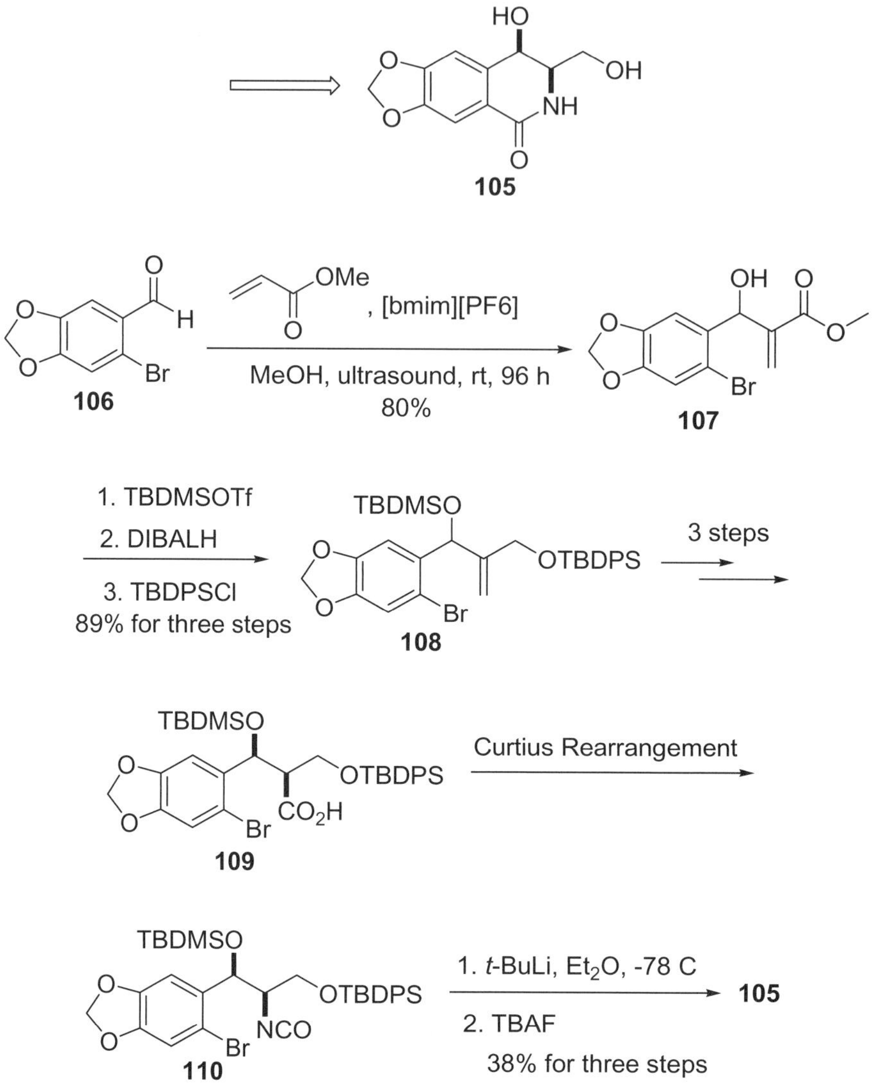

Roush and coworkers reported an intramolecular MBH on a highly functionalized intermediate which resulted in the total synthesis of (–)-spinosyn A (**111**).[28h,i] Under dilute conditions in *tert*-amyl alcohol, compound **112** underwent a vinylogous MBH reaction in the presence of trimethyl phosphine to afford the tetracyclic structure **113** as the major product. Subsequent reductive dehalogentaion and cleavage of the PMP group delivered the spinosyn A pseudoglycon **114** in 73% yield from compound **112**. Glycosidation of **114** and subsequent transformations yielded (-)–spinosyn A (**111**).

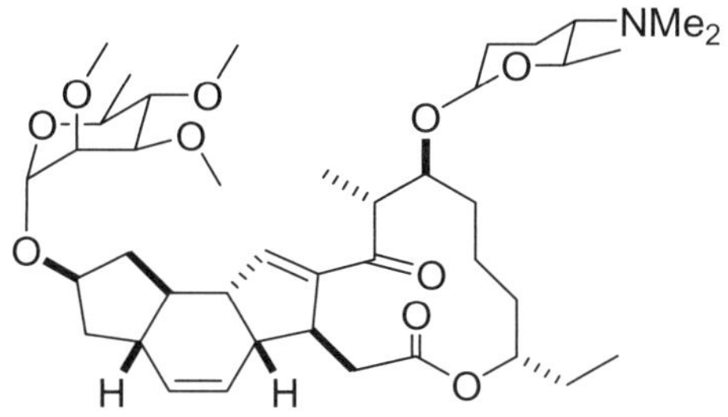

111
(–)-spinosyn A

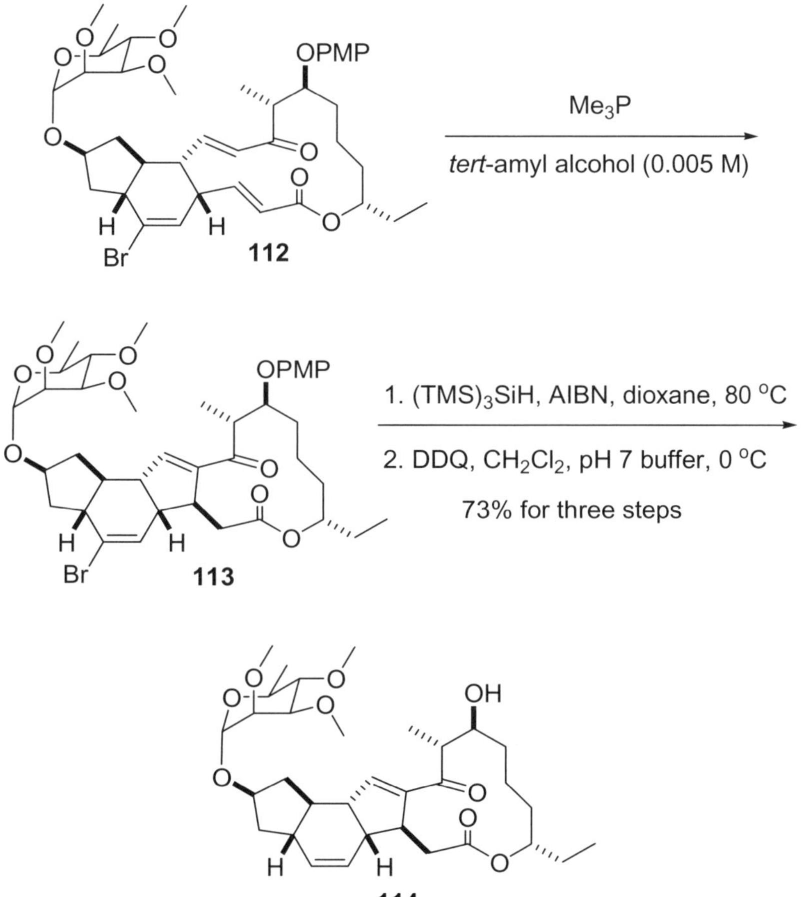

In another intramolecular MBH report, Corey and coworkers synthesized a key intermediate in their synthesis of salinosporamide A (**117**).[28j] The transformation was accomplished by treating the α,β-unsaturated amide **115** with quinuclidine in DME to deliver a diastereomeric mixture (9:1) of γ-lactams **116a** and **116b** in a combined yield of 90%. The major diastereomer **116a** was then converted in several steps to the natural product.

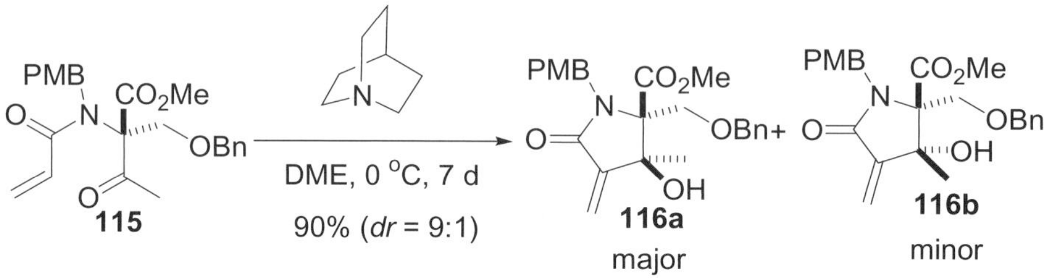

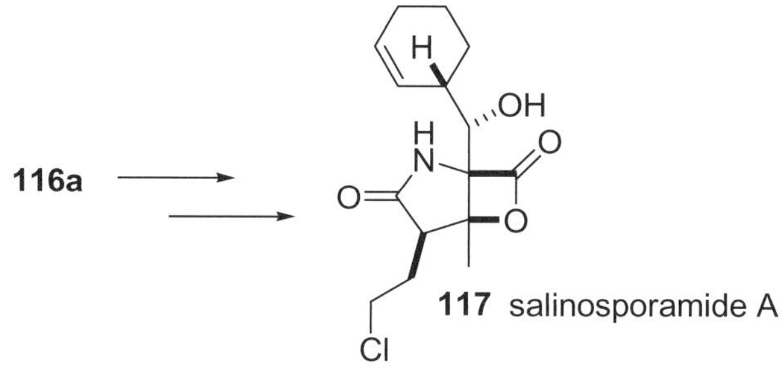

Medicinal Chemistry

Coehlo and coworkers synthesized the carboxylic acid precursor of (+)-efaroxan (**118**), an α_2 adrenoreceptor antagonist, which is a treatment for neurodegenerative disease, migraine, and type II diabetes.[29a] The synthesis opened with an MBH reaction which afforded key moieties. Thus, 2-fluorobenzaldehyde **120** was converted to the MBH adduct **121** under ultrasound conditions in 90% yield. After acetylation of the hydroxyl group, the acetylated MBH adduct was treated with dimethyl cuprate to furnish the trisubstituted olefin **122** via an S_N2' reaction. The methyl ester of **122** was reduced to the allylic alcohol which then underwent a Sharpless asymmetric oxidation to furnish the epoxide **123**. In several steps, the epoxy alcohol **123** was then converted to the carboxylic acid **124**. Upon treatment with sodium hydride the α-hydroxy carboxylic acid, **124** cyclized to deliver the hydrobenzofuran **119** in 65% yield and hence the precursor of (+)-efaroxan (**118**).

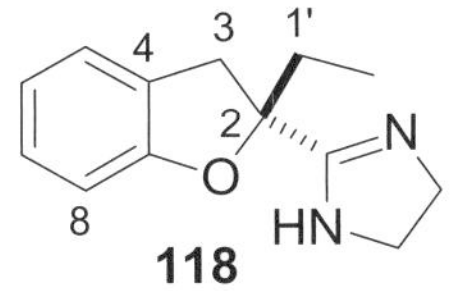

118

(R)-efaroxan

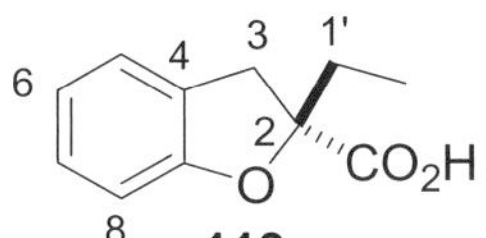

119

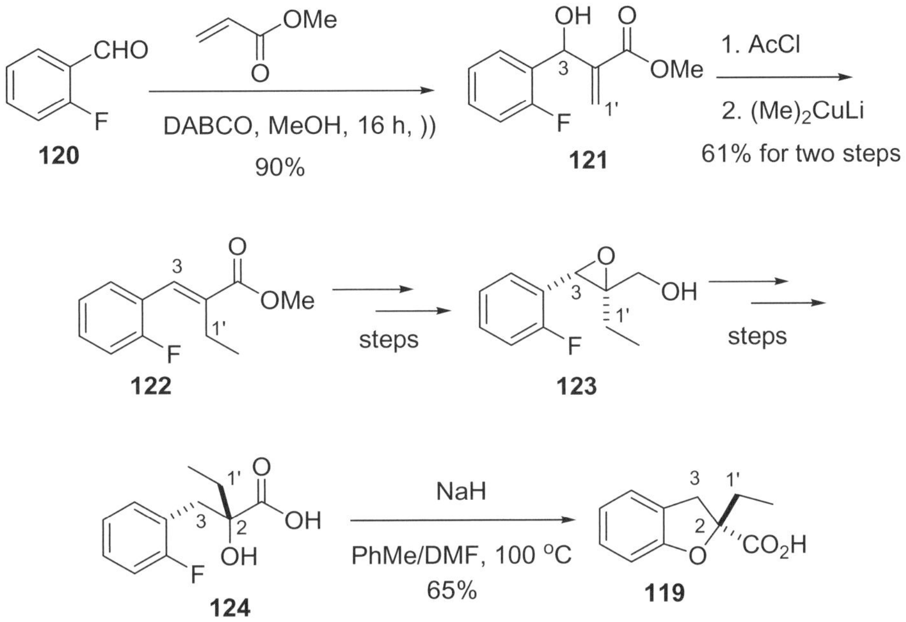

Although the syntheses of the germane broad-spectrum fluoroquinolone antibiotics are well precedented,[30] Lee and Hong developed a MBH route.[29b] The approach strategy involved using the appropriate arylaldehyde and methyl propiolate in the presence of tetrabutylammnonium iodide and zirconium chloride. This methodology was used to synthesize the penultimate intermediate of ciprofloxacin (**130**). The sequence opened with a MBH/zirconation reaction to form the β-iodo-α-hydroxyl-alkylacrylate **127** in 84% yield. Oxidation of the allylic alcohol **127**, using the Dess–Martin reagent, afforded an intermediate keto-ester. Treatment of the keto–ester with cyclopropyl amine delivered the β-amino intermediate **128** which then cyclized using sodium hydride to afford the fluoroquinolone core **129**. The conversion of **129** to ciprofloxacin is well precedented and ultimately involves displacement of the C7–F with piperidine.[30,31]

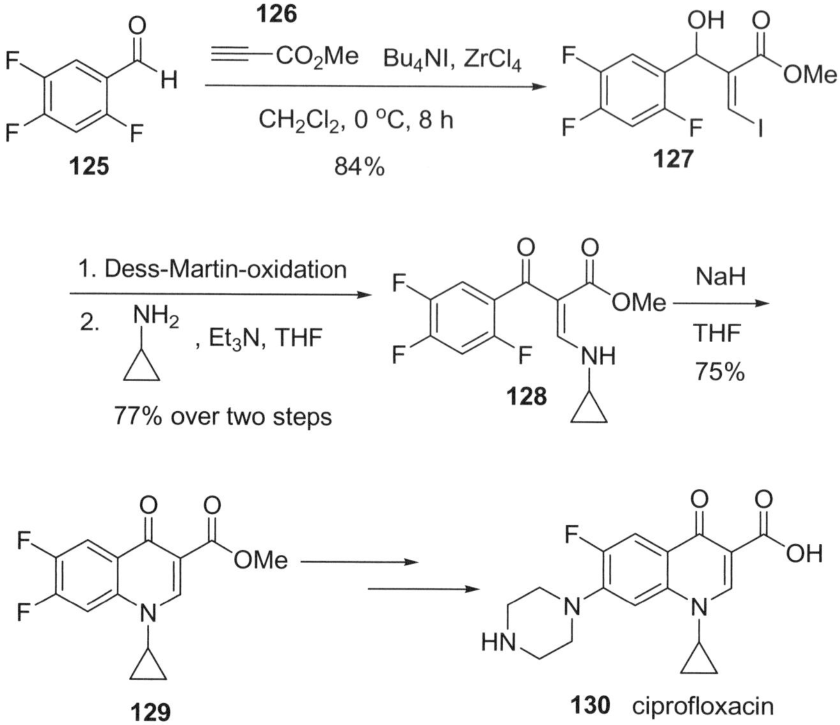

Prabhakar and coworkers used an MBH reaction to synthesize a variety of C3-alkyl/aryl-2,3-dideoxy hex-2-enopyranosides that exhibited in vitro activity against *Mycobacterium* tuberculosis.[29c,d] Using a titanium chloride/tetrabutyl ammonium iodide mediated MBH, pyranosides **131a** and **131b** were converted to the MBH adducts **132a** and **132b**, respectively.

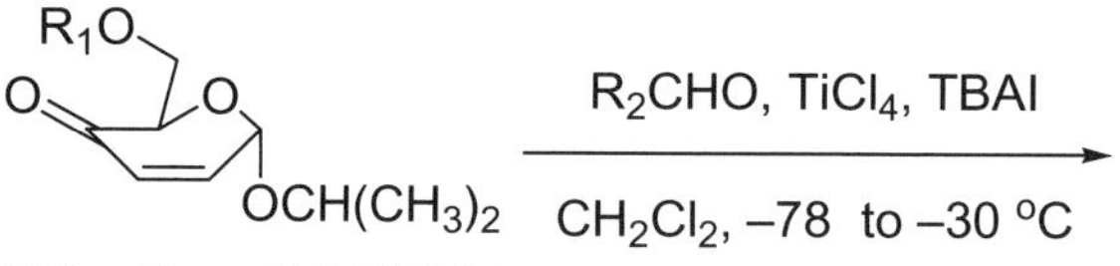

131a: R_1 = -COC(CH_3)$_3$
131b: R1 = -$COCH_3$

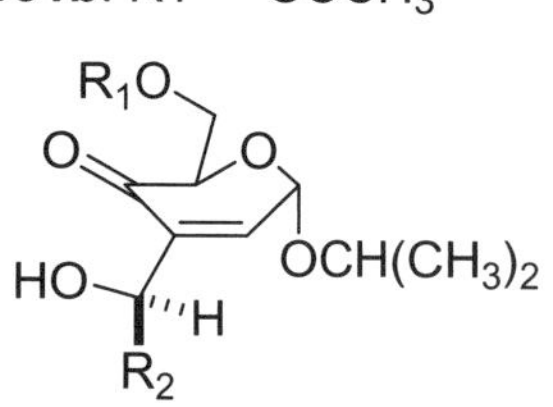

132a R_1 = -COC(CH_3)$_3$,R_2 = *p*-CNPh, -(CH_2)$_8$$CH_3$
132b R1 = -$COCH_3$, *p*-NO_2Ph

A series of antithrombotic compounds were synthesized in two steps. First, the isoxazolescarboxaldehydes **133a**,**b** were converted to the MBH adducts **134a–d** which then underwent a Michael addition with *N*-methylpiperazine to afford compounds **135a–d** in 51–65% yield over two steps.[28e] Compounds were also further derivatized to afford other analogs.

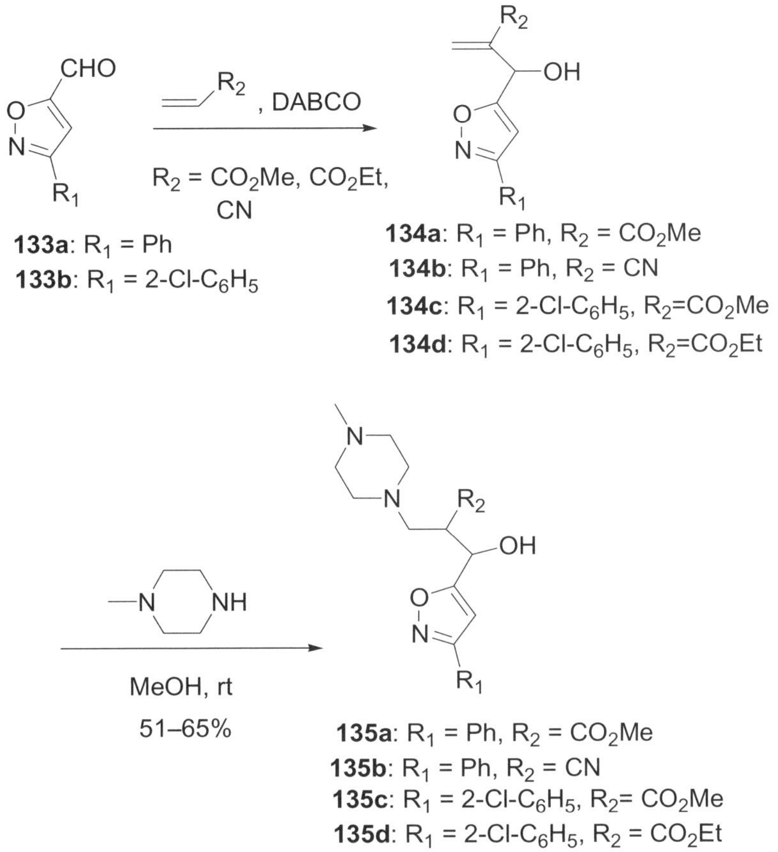

Rao and co-workers synthesized a number of compounds that exhibited activity against a number of Gram-positive and Gram-negative bacterial strains as well as antifungal properties.[29f] A central intermediate to these derivatives were a set of acetoxy MBH adducts. Thus, the chloropyridine carboxyaldehydes **136a–d** were converted to the adducts **137a–d** in 85–96% yield using cyclohexenone in the presence of imidazole. Subsequent acetylation of **137a–d** gave the pivotal intermediates **138a–d**. The acetates were then converted to quinolone derivatives **139a–d** in the presence of nitroethane and potassium carbonate in 85–94% yield.

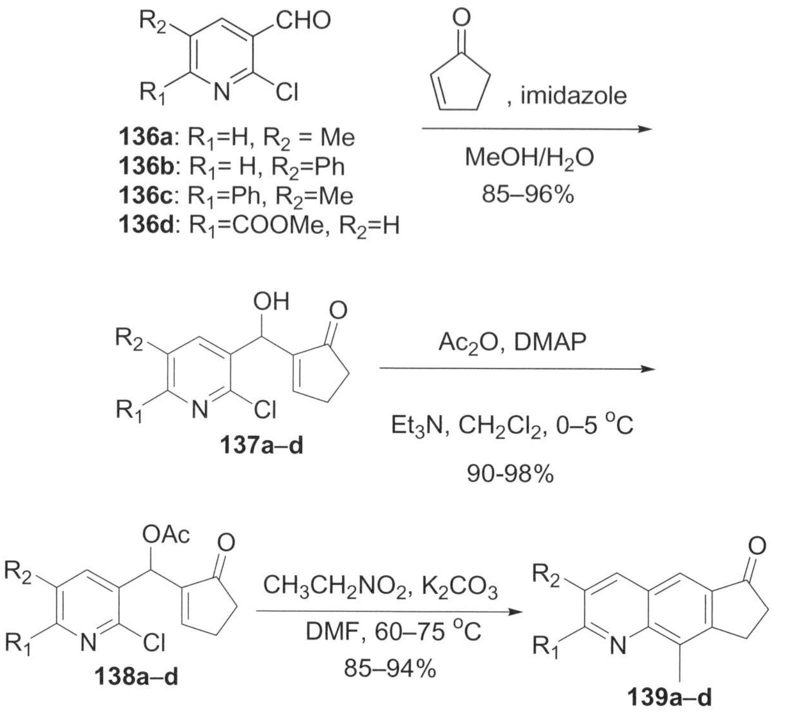

2.2.6 *Experimental*

DABCO mediated intermolecular MBH[11a]

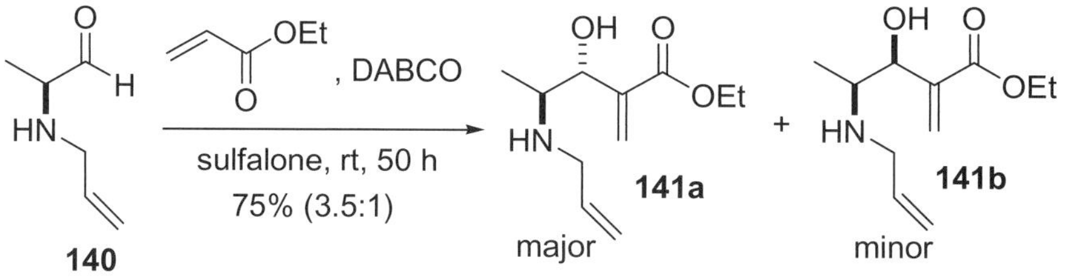

A solution of *tert*-butyl-*N*-allyl-[(1*S*)-1-methyl-2-oxoethyl]-carbamate **140** (1.0 g, 4.7 mmol) was treated with ethyl acrylate (1.02 mL, 9.39 mmol) and DABCO (0.63 g, 5.63 mmol) in sulfolane at room temperature for 50 h. Then, the reaction mixture was diluted with water (2 × 20 mL) and extracted with ether (2 × 30 mL). The combined organic layers were washed with brine (25 mL), dried (Na_2SO_4), and concentrated under reduced pressure to get a residue. The residue was purified by column chromatography (silica gel 60–120 mesh, *n*-hexane/EtOAc, 94:6) to afford 0.86 g (58.5%) of acrylate

141a and 0.24 g (16.5%) of acrylate **141b**. The combined yield was 1.10 g (75%).

Titanium-mediated intermolecular MBH[29d]

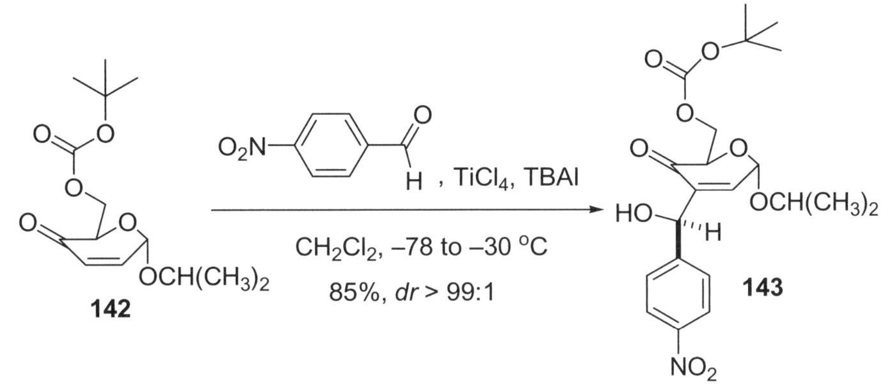

To a stirred solution of tetrabutylammonium iodide (TBAI, 0.2 mmol) in dry dichloromethane (5 mL) at –78 ºC was added titanium tetrachloride (1.5 mmol) dropwise. After stirring for two minutes, a mixture of the enone **142** (1 mmol) and *p*-nitrobenzaldehyde (2 mmol) in dry dichloromethane (5 mL) was added. The reaction mixture was slowly warmed to –30 ºC and kept at this temperature for 8 h. A saturated aqueous solution of sodium bicarbonate was added, followed by filtration through a celite pad. The organic layer from the filtrate was separated, and the aqueous layer was extracted with ethyl acetate. The combined organic layers were washed with brine, dried over sodium sulfate, filtered, and concentrated. The crude product was then purified by column chromatography eluting with hexane:ethyl acetate (80:20) to deliver the product **143** as an oil in 85% yield.

Intramolecular MBH[28j]

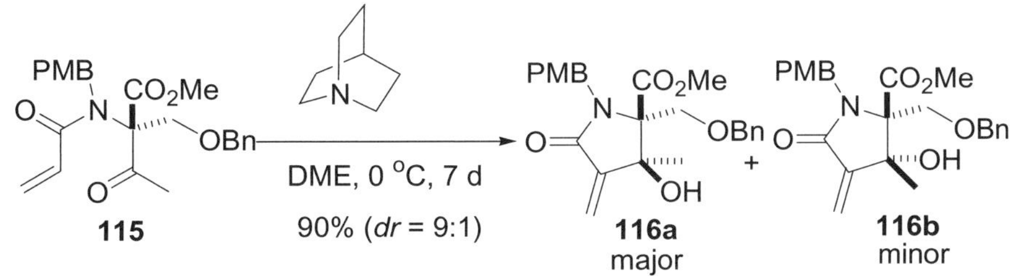

A mixture of the keto-amide **115** (8.5 g, 20.0 mmol) and quinuclidine (2.22 g, 20.0 mmol) in DME (10 mL) was stirred rapidly for 7 d at 0 ºC. After completion, the reaction mixture was diluted with ethyl acetate (50 mL)

washed with 2N HCl, followed by water and dried over sodium sulfate. The solvent was removed *in vacuo* to give the crude adducts **116a,b** (7.65 g, 90%, 9:1 mixture of diastereomers) as a viscous oil.

IntermolecularAza-MBH[32]

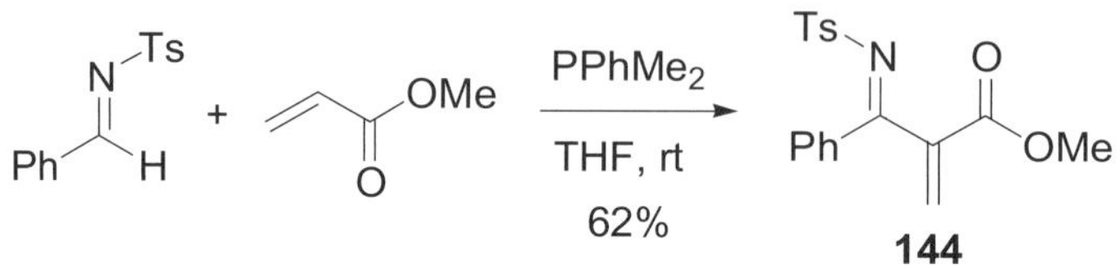

144

To a solution of *N*-benzylidene-4-methylbenzenesulfonamide (129 mg, 0.5 mmol) and methyldiphenylphosphine (9 uL, 0.05 mmol) in THF (1.0 mL) at room temperature was added methyl acrylate (89 mg, 0.60 mmol), and the reaction mixture was further stirred at room temperature for 6 h. The reaction was monitored by TLC. When the *N*-tosylated imine disappeared, the solvent was removed under reduced pressure, and the residue was purified by flash chromatography [SiO_2, EtOAc:petroleum ether (1:5)] to afford 108 mg (63%) of the adduct **144** as a colorless solid.

2.2.7 References

1. Morita, K. Suzuki, Z.; Hirose, H. *Bull. Chem. Soc. Jpn.* **1968**, *41*, 2815.
2. (a) Baylis, A. B.; Hillman, M. E. D. *Ger. Offen.* **1972**, *2*, 155,133; Chem Abstr. **1972**, *77*, 34174q; Hillman, M. E. D.; Baylis, A. B. U. S. Patent **1973**, 3, 743,669.
3. [R] Basavaiah, D.; Rao, K. V.; Reddy. R. J. *Chem. Soc. Rev.* **2007**, *36*, 1581–1588.
4. (a) [R] Singh, V.; Batra, S. *Tetrahedron* **2008**, *64*, 4511–4574. (b) [R] Masson, G.; Housseman, C.; Zhu, J. *Angew. Chem. Int. Ed.* **2007**, *46*, 4614–4628. (aza) (c) [R] Shi, Y-L.; Shi, M. *Eur. J. Org. Chem.* **2007**, 2905–2916. (aza) (d) [R] Basavaiah, D.; Rao, A. J.; Satyanarayana,T. *Chem. Rev.* **2003**, *103*, 811–891. (e) [R] Langer, P. *Angew. Chem., Int. Ed.* **2000**, *39*, 3049–3052. (f) [R] Ciganek, E. *Org. React.* **1997**, *51*, 201–478. (g) [R] Basavaiah, D.; Rao, A. J.; Hyma, R. S. *Tetrahedron* **1996**, *52*, 8001–8062. (f) [R] Fort,Y.; Berthe, M. C. Caubere, P. *Tetrahedron* **1992**, *48*, 6371–6384.
5. Aggarwal, V. K.; Emme, I.; Fulford, S. Y. *J. Org. Chem.* **2003**, *68*, 692–700.
6. Robiette, R.; Aggarwal, V. K.; Harvey, J. N. *J. Am. Chem. Soc.* **2007**, *129*, 15513–15525.
7. (a) Price, K. E.; Broadwater, S. J.; Jung, H. M.; McQuade, D. T. *Org. Lett.* **2005**, *7*, 147–150. (b) Price, K. E.; Broadwater, S. J.; Walker, B. J.; McQuade, D. T. *J. Org. Chem.* **2005**, *70*, 3980–3987. (c) Roy, D.; Suno, R. B. *Org. Lett.* **2007**, *9*, 4873–4876
8. (a) Cai, J.; Zhou, Z.; Zhao, G.; Tang, C. *Org. Lett.* **2002**, *4*, 4723–4725. (b) Aggarwal, V. K.; Dean, D. K.; Mereu, A.; Williams, R. *J. Org. Chem.* **2002**, *67*, 510–514.
9. Davies, H. J.; Ruda, A. M.; Tomkinson, N. C. O. *Tetrahedron Lett.* **2007**, **48**, 1461–1464.
10. (a) Park, K-S.; Kim, J.; Choo, H.; Chong, Y. *Synlett* **2007**, 395–398. (b) Kang, S.-Y.; Park, K-S.; Kim, J.; Choo, H.; Chong, Y. *Bull. Kor. Chem. Soc.* **2007**, *28*, 179–180.
11. (a) Krishna, P. R.; Reddy, P. S. *J. Comb. Chem.* **2008**, *10*, 426–435. (b) Krishna, P. R.; Manjuvani, P.; Kannan, V.; Sharma, G. V. M. *Tetrahedron Lett.* **2004**, *45*, 1183–1185.
12. Zhao, S.-H.; Bie, H.-Y. Chen, Z.-B. *Org. Prep. Proc. Int.* **2005**, *37*, 231–237.
13. (a) Gong, H.; Cai, C.-q.; Yang, N.-f.; Yang, L.-w.; Zhang, J.; Fan, Q-h. *J. Mol Cat A.* **2006**, *249*, 236–239. (b) Mi, X.; Luo, S.; Cheng, J.-P. *J. Org. Chem.* **2005**, 2338–2341. (c) Mi, X.;

Luo, S.; Xu, H.; Zhang, L.; Cheng, J.-P. *Tetrahedron* **2006**, *62*, 2537–2544. (d) Lin, Y.-S.; Lin, C.-Y.; Liu, C. W.; Tsai, T. Y. R. *Tetrahedron* **2006**, *62*, 872–877. (e) Pègot, B.; Vo-Thanh, G.; Gori, D.; Loupy, A. *Tetrahedron Lett.* **2004**, *45*, 6425–6428. (f) Hsu, J.-C.; Yen, Y.-H.; Chu, Y.-H. *Tetrahedron Lett.* **2004**, *45*, 4673–4676. (g) Kumar, A.; Pawar, S. S. *J. Mol. Cat A.* **2004**, *211*, 43–47.

14. Aggarwal, V. K.; Mereu, A. *Chem. Commun.* **1999**, 2311–2312.
15. de Souza, R. O. M. A.; Meireles, B. A.; Aguiar, L. C. S.; Vasconcellos, M. L. A. A. *Synthesis* **2004**, 1595–1600.
16. Krishna, P. R.; Sekhar, E. R.; Kannan, V. *Synthesis* **2004**, 857–860.
17. Luo, S.; Mi, X.; Xu, H.; Wang, P. G.; Cheng, J.-P. *J. Org. Chem.* **2004**, *69*, 555–558.
18. Luo, S.; Mi, X.; Xu, H.; Wang, P. G.; Cheng, J.-P. *J. Org. Chem.* **2004**, *69*, 8413–8422.
19. Sun, S.; Wang, M.; Deng, H.; Liu, Q. *Synthesis* **2008**, 573–583.
20. Mack, J.; Shumba, M. *Green Chemistry* **2007**, *9*, 328–330.
21. Rao, J. S.; Brière, J-F.; Metzner, P.; Basavaiah, D. *Tetrahedron Lett.* **2006**, *47*, 3553–3556.
22. Rodgen, S. A.; Schaus, S. E. *Angew. Chem. Int. Ed.* **2006**, *45*, 4929–4932.
23. Shi, M.; Liu, X-G. *Org. Lett.* **2008**, *10*, 1043–1046.
24. Matsui, K.; Takizawa, S.; Sasai, H. *J. Am. Chem. Soc.* **2005**, *127*, 3680–3681.
25. (a) Qi, M.-J.; Ai, T.; Shi, M.; Li, Guigen *Tetrahedron* **2008**, *64*, 1181–1186. (b) Shi, M.; Xu, Y.-M.; Shi, Y.-L. *Chem. Eur. J.* **2005**, *11*, 1794–1802. (c) Shi, M.; Ma, G.-N.; Gao, J. *J. Org. Chem.* **2007**, *72*, 9779–9781. (d) Zhao, L.-J.; Kwong, C. K.-W.; Shi, M.; Toy, P. H. *Tetrahedron* **2005**, *61*, 12026–12032.
26. Liu, Y-h.; Shi, M. *Adv. Synth. Catal.* **2008**, *350*, 122–128.
27. Ito, K.; Nishida, K.; Gotanda, T. *Tetrahedron Lett.* **2007**, *48*, 6147–6149.
28. (a) Clive, D. L. J.; Yu, M.; Li, Z. *Chem. Commun.* **2005**, 906–908. (b) Shigetomi, K. Kishimoto, T. Shoji, K.; Ubukata, M. *Tetrahedron: Asymmetry* **2008**, *19*, 1444–1449. (c) Majhi, T. P.; Neogi, A.; Ghosh, S.; Mukherjee, A. K.; Chattopadhyay, P. *Tetrahedron* **2006**, *62*, 12003–12010. (d) Krishna, P. R.; Narsingam, M.; Kannan, V. *Tetrahedron Lett.* **2004**, *45*, 4773–4775. (e) Seck, M.; Franck, X.; Seon-Meniel, B.; Hocquemiller, R.; Figadère, B. *Tetrahedron Lett.* **2006**, *47*, 4175–4180. (f) Rodgen, S. A.; Schaus, S. E. *Angew. Chem. Int. Ed.* **2006**, *45*, 4929–4932. (g) Lopes, E. C. S.; Coelho, F. *J. Braz. Chem. Soc.* **2007**, *18*, 1415–1438. (h) Winbush, S. M.; Mergott, D. J.; Roush, W. R. *J. Org. Chem.* **2008**, *73*, 1818–1829. (i) Mergott, D. J.; Frank, S. A.; Roush, W. R. *Proc. Nat. Acad. Sci. USA* **2004**, *101*, 11955–11959. (j) Reddy, L. R.; Saravanan, P.; Corey, E. J. *J. Am. Chem. Soc.* **2004**, *126*, 6230–6231. (k) Amarante, G. W.; Rezende, P.; Cavallaro, M.; Coelho, F. *Tetrahedron Lett.* **2008**, *49*, 3744–3748. (l) König, C. M.; Harms, K.; Koert, U. *Org. Lett.* **2007**, *9*, 4777–4779. (m) Lee, S. I.; Hwang, G.-S.; Shin, S. C.; Lee, T. G.; Jo, R. H.; Ryu, D. H. *Org. Lett.* **2007**, *9*, 5087–5089. (n) Das, B.; Banerjee, J.; Chowdhury, N.; Majhi, A.; Mahender, G. *Helv. Chim. Acta* **2006**, *89*, 876–883. (o) Ho, A.; Cyrus, K.; Kim, K.-B. *Eur. J. Org. Chem.* **2005**, 4829–4834. (p) Kabalka, G. W.; Venkataiah, B. *Tetrahedron Lett.* **2005**, *46*, 7325–7328. (q) Chapuis, C.; Büchi, G. H.; Wüest, H. *Helv. Chim. Acta* **2005**, *88*, 3069–3088. (r) Heim, R.; Wiedemann, S.; Williams, C. M.; Bernhardt, P. V. *Org. Lett.* **2005**, *7*, 1327–1329.
29. (a) de Carvalho e Silveira, G. P.; Coelho, F. *Tetrahedron Lett.* **2005**, *46*, 6477–6481. (b) Hong, W. P.; Lee, K-J. *Synthesis* **2006**, 963–968. (c) Saquib, M.; Gupta, M. K.; Sagar, R.; Prabhakar, Y. S.; Shaw, A. K.; Kumar, R.; Maulik, P. R.; Gaikwad, A. N.; Sinha, S.; Srivastava, A. K.; Chaturvedi, V.; Srivastava, R.; Srivastava, B. S. *J. Med. Chem.* **2007**, *50*, 2942–2950. (d) Sagar, R,; Pant, C. S.; Pathak, R.; Shaw, A. K. *Tetrahedron* **2004**, *60*, 11399–11406. (e) Batra, S.; Roy, A. K.; Patra, A.; Bhaduri, A. P.; Surin, W. R.; Raghavan, S. A. V.; Sharma, P.; Kapoor, K.; Dikshit, M. *Bioorg. Med. Chem.* **2004**, *12*, 2059–2077. (f) Narender, P.; Srinivas, U.; Ravinder, M.; Rao, B. A.; Ramesh, Ch.; Harakishore,K.; Gangadasu, B.; Murthy, U. S. N.; Rao, V. J. *Bioorg. Med. Chem.* **2006**, *14*, 4600–4609. (g) Ma, D. Yu, S.; Li, B.; Chen, L.; Chen, R.; Yu, K.; Zhang, L.; Chen, Z.; Zhong, D.; Gong, Z.; Wang, R.; Jiang, H.; Pei, G. *ChemMedChem* **2007**, *2*, 187–193. (h) Mohan, R.; Rastogi, N.; Namboothiri, I. N. N.; Mobin, S. M.; Panda, D. *Bioorg. Med. Chem.* **2006**, *14*, 8073–8085.

(i) Nag, S.; Pathak, R.; Kumar, M.; Shukla, P. K.; Batra, S. *Bioorg. Med. Chem. Lett.* **2006**, *16*, 3824–3828. (j) Kohn, L. K.; Pavam, C. H.; Veronese, D.; Coelho, F.; De Carvalho, J. E.; Almeida, W. P. *Eur. J. Med. Chem.* **2006**, *41*, 738–744. (k) Frezza, M.; Balestrino, D.; Soulère, L.; Reverchon, S.; Queneau, Y.; Forestier, C.;Doutheau, A. *Eur. J. Org. Chem.* **2006**, 4731–4736. (l) Dadwal, M.; Mohan, R.; Panda, D.; Mobin, S. M.; Namboothiri, I. N. N. *Chem. Comm.* **2006**, 338–340. (m) Narender, P.; Srinivas, U.; Gangadasu, B.; Biswas, S.; Rao, V. J. *Bioorg. Med. Chem. Lett.* **2005**, *15*, 5378–5381. (n) Mateus, C. R.; Coelho, F. *J. Braz. Chem. Soc.* **2005**, *16*, 386–396.

30. (a) Mitscher, L. A. *Chem. Rev.* **2005**, *105*, 559–592. (b) Da Silva, A. D.; De Almeida, M. V.; DeSouza, M. V. N.; Couri, M. R. C. *Curr. Med. Chem.* **2003**, *10*, 21–39. (c) Limberakis, C., Chapter 4. Quinolones: Levofloxacin (Levaquin®), Moxifloxacin (Avelox®), Gemifloxacin (Factive®), and Garenoxacin (T-3811), In *The Art of Drug Synthesis*; Li, J. J.; Johnson, D. S., eds. John Wiley & Sons: Hoboken, NJ, 2007, pp. 39–69.
31. Li, J. J.; Johnson, D. S.; Sliskovic, D. R.; Roth, B. D. *Contemporary Drug Synthesis.* John Wiley & Sons: Hoboken, NJ, 2004, pp. 75–83.
32. Xu, Y.-M.; Shi, M. *J. Org. Chem.* **2004**, *69*, 417–425.

2.3 Benzoin condensation

Victor J. Cee

2.3.1 *Description*

The benzoin condensation is the coupling of two aldehyde molecules to give an α-hydroxyketone (acyloin) product. In its most classical form, the homocoupling of benzaldehyde gives the parent benzoin (R = R′ = Ph). Contemporary modifications of the benzoin condensation include the use of acylsilanes as well as imine derivatives in place of one aldehyde partner.[1]

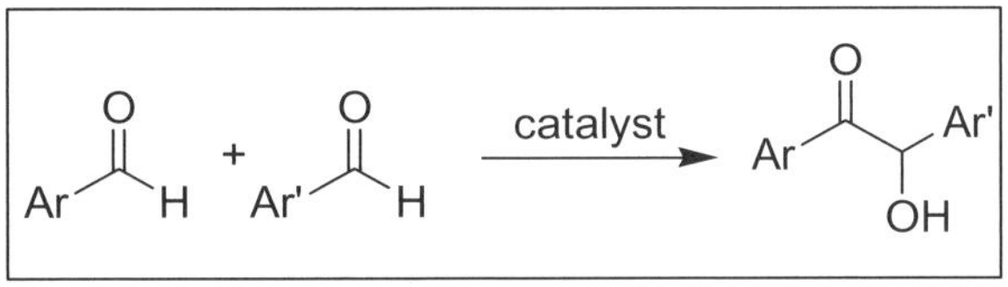

Metal cyanides[2] and heterocyclic carbenes[3] are commonly employed catalysts for the benzoin condensation. Symmetrical acyloins can be prepared by dimerization of a wide range of aromatic and aliphatic aldehydes. Chiral heterocyclic carbene catalysts have also been developed that provide symmetrical acyloin products with high levels of enantiomeric purity.

The cross-benzoin reaction between two different aldehydes typically produces a statistical mixture of products, although in some cases a single thermodynamic product predominates.[2] A number of approaches have been developed to circumvent the limitations of the cross-benzoin reaction. In one approach, a thiamine diphosphate-dependent enzyme is used to promote a selective cross-benzoin reaction, often with high levels of asymmetric induction. In other approaches, one aldehyde coupling partner is replaced with a selective acyl donor. Cyanohydrin derivatives have proven to be ideal preformed acyl donors, and their use constitutes a stepwise benzoin condensation that is stoichiometric in cyanide. The discovery that acylsilanes can serve as cyanohydrin precursors has led to the development of a highly selective cyanide-catalyzed cross-benzoin condensation. By employing a chiral metallophosphite catalyst instead of potassium cyanide, good to excellent levels of asymmetric induction are possible.

Modified benzoin condensations in which the acyl acceptor is not an aldehyde constitute a variation of the classical cross-benzoin condensation. Aldehydes and ketones can be coupled in an intramolecular annulation reaction to give tertiary α-hydroxyketones. The selective cross-coupling of aldehydes and imines to give α-aminoketones has also been realized in an aza-benzoin condensation.

2.3.2 Historical Perspective

Wöhler and Liebig are generally credited with the first published report of a cyanide-catalyzed benzoin condensation.[4] In 1903, Lapworth published the accepted mechanism for the reaction, in which an intermediate aldehyde cyanohydrin is deprotonated to generate an acyl anion equivalent with inverted reactivity at the carbonyl carbon.[5] As early as 1943, it was reported that the combination of a thiazolium salt and a base could also effectively catalyze the benzoin condensation.[6] This work, along with studies of reactions catalyzed by thiamine-dependent enzymes,[7] indicated that acyl anion equivalents are also likely intermediates in these thiazolium/base-catalyzed processes. Breslow provided the generally accepted mechanism for these transformations in 1958 suggesting that the thiazolium salt and base generate an active thiazole-2-ylidine carbene catalyst.[8] A resurgence of interest in acyl anion chemistry has led to significant improvements over the classical cyanide-catalyzed benzoin condensation. Asymmetric benzoin dimerization, regio- and enantioselective cross-benzoin condensations, as well as benzoin annulations and aza-benzoin condensations have been developed.

2.3.3 Mechanism

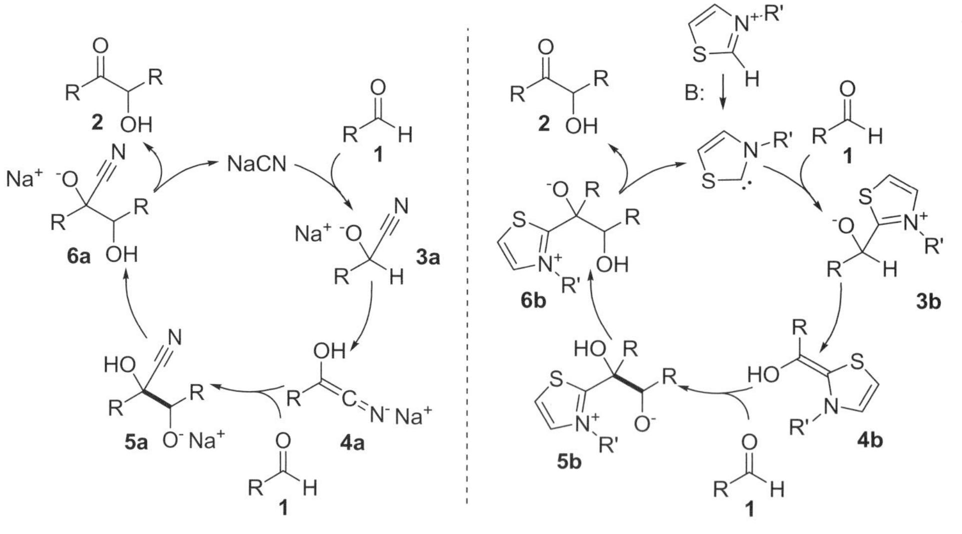

The widely accepted Lapworth mechanism[5] for the cyanide-catalyzed condensation and Breslow mechanism[8] for the heterocyclic carbene-catalyzed condensation contain many parallels and are illustrated together. The reaction is initiated by nucleophilic addition of the catalyst to the aldehyde, giving intermediates **3a/b**. For the thiazolium-catalyzed condensation, the thiazolium salt serves as a pre-catalyst, with deprotonation generating the active thiazole-2-ylidine catalyst. The nitrile and thiazolium groups are sufficiently electron-withdrawing to render what was the formyl hydrogen acidic, and proton transfer leads to the enamine-enol species **4a/b**. These intermediates exhibit inverted reactivity at the acyl carbon, and react as nucleophiles with another molecule of aldehyde to give **5a/b**. Following proton transfer, the benzoin product **2** is produced upon collapse of tetrahedral intermediate **6a/b**, with concomitant release of the catalyst.

2.3.4 Synthetic Utility

Benzoin Dimerization

The cyanide-catalyzed benzoin dimerization exhibits a somewhat limited substrate scope for the preparation of symmetrical aryl-aryl′ benzoin products, and is not suitable for the preparation of symmetrical alkyl-alkyl′ acyloin products.[2] The development of heterocyclic carbene catalysts resulted in a greatly expanded substrate scope, including many substituted aromatic aldehydes, heteroaromatic aldehydes, and aliphatic aldehydes. Despite the improvements offered by heterocyclic carbenes, the electron rich 4-*N*,*N*-dimethylaminobenzaldehyde and the electron deficient 4-nitrobenzaldehyde still remain very challenging substrates.

Enantioselective Benzoin Dimerization

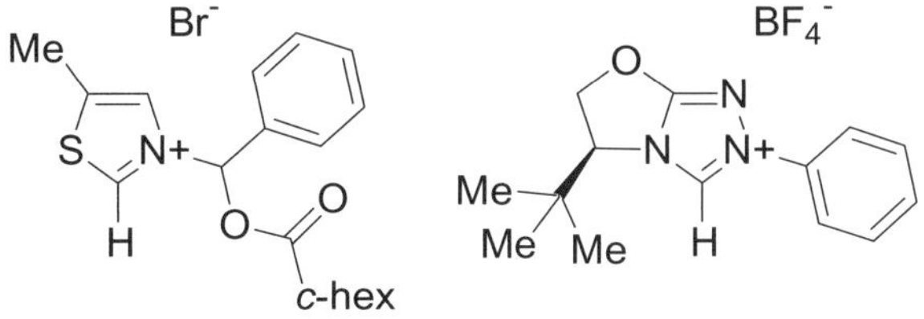

(+)-7 benzoin: 0.8% *ee*

8 benzoin: 90% *ee*

The pioneering work of Sheehan[9] showed as early as 1966 that the use of the chiral thiazolium precatalyst **(+)-7** imparted low but measurable enantioselectivity in the conversion of benzaldehyde to benzoin. Since this report, many groups have developed chiral carbene catalysts that effect asymmetric benzoin condensation reactions. The best catalyst identified to date with respect to enantioselectivity is the triazolium pre-catalyst **8** reported by Enders and co-workers, which is capable of producing benzoin in 90% *ee*.[10]

Müller and co-workers recently developed an enantioselective benzoin dimerization using purified enzymes from *Pseudomonas*. The thiamine diphosphate (ThDP) dependent enzymes benzaldehyde lyase (BAL) and benzoylformate decarboxylase (BFD) were found to catalyze the reversible benzoin condensation of aromatic aldehydes. The reaction is driven in the forward direction by the poor solubility of the benzoin products in aqueous media.[11] A wide variety of aromatic aldehydes are accepted by BAL, and products of the (*R*)-configuration are produced in excellent yield and enantiomeric purity.[12] The (*S*)-enantiomer of benzoin is also available in high enantiomeric purity from a BAL-catalyzed kinetic resolution of rac-benzoin. In the presence of excess acetaldehyde, BAL selectively converts (*R*)-benzoin into (*R*)-2-hydroxy-1-phenylpropanone, while the (*S*)-benzoin enantiomer is not a substrate for the enzyme. At 49% conversion, (*S*)-benzoin is resolved to > 99% *ee*. BFD can produce (*R*)-benzoin from benzaldehyde in comparable yield and enantiomeric purity with respect to BAL, but the substrate scope appears more limited.[13]

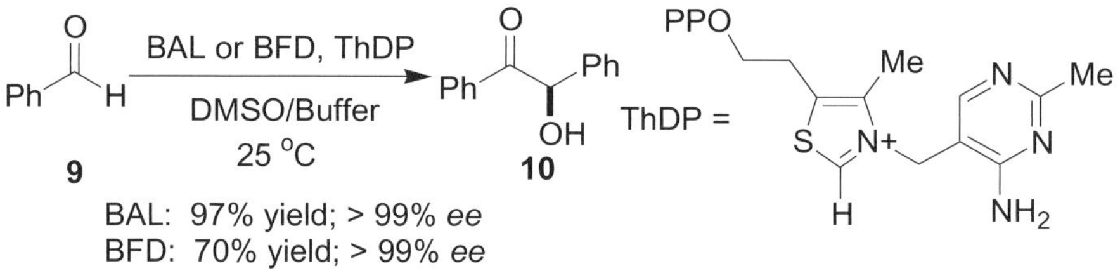

Classical Cross-Benzoin Condensation

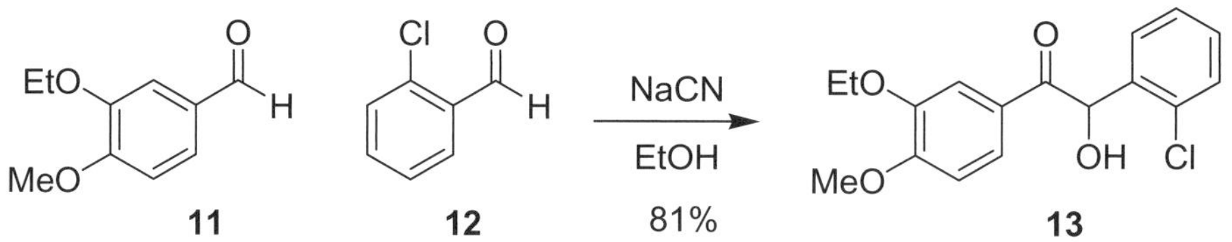

The cross-benzoin condensation often results in the production of mixtures of products, although a small number of specific aldehyde pairs have been identified to participate in selective cross-benzoin condensations. For example, the reaction of 3-ethoxy-4-methoxybenzaldehyde (**11**) and 2-chlorobenzaldehyde (**12**) in 1 : 1 stoichiometry provides 2′-chloro-3-ethoxy-4-methoxybenzoin **13** in high yield.[14] These selective reactions are generally believed to be under thermodynamic control, with the predominant product containing the electron-rich aryl ring adjacent to the ketone.[15] Investigations of cross-benzoin condensations between aryl/aliphatic[16] as well as aliphatic/aliphatic[17] aldehydes have also identified selective coupling partners or preparatively useful conditions, but these approaches suffer from limited generality.

Modified Cross-Benzoin Condensation

Two related modifications have been developed to provide cross-benzoin products not accessible under classical benzoin condensation conditions. In the first approach, one aldehyde partner is converted to a cyanohydrin derivative which serves as the stoichiometric equivalent of intermediate **3a** on the catalytic cycle of the cyanide-catalyzed benzoin condensation. A number of cyanohydrin derivatives can be used, with the most popular being the *O*-silyl cyanohydrin **14**, first reported by Hünig and Wehner.[18] In a subsequent step, the cyanohydrin derivative is treated with stoichiometric base followed by the second aldehyde to give the silylated cross-benzoin product **16** in high yield.

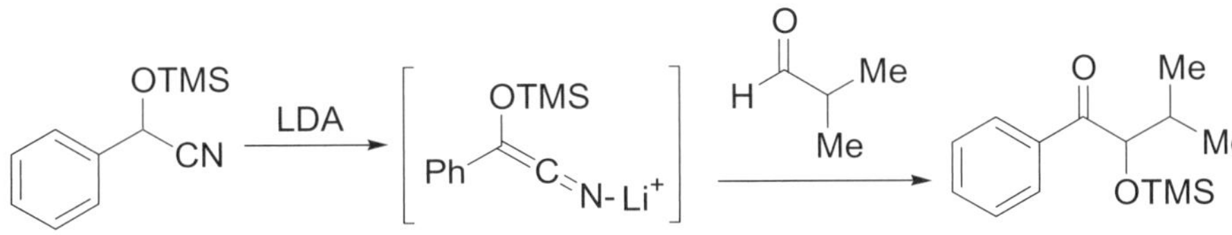

14 **15** **16** 100%

The observation that acylsilanes such as **17** could directly provide intermediate **18** via cyanide addition and [1,2]-Brook rearrangement has led to an improved process that is catalytic in cyanide. The reaction produces *O*-silylated aryl–aryl′, aryl–alkyl′, and alkyl–aryl′ acyloin products in moderate to high yield.[19] Further work by Johnson and co-workers established the superiority of lanthanum tricyanide for this process, which allows selective access to even alkyl–alkyl′ acyloin products.[20]

Enantioselective Cross-Benzoin

Chiral metallophosphites were found to be effective catalysts of the acylsilane/aldehyde cross-benzoin condensation in work reported by Johnson and co-workers.[21] The TADDOL-derived catalyst **21** is very effective for the preparation of aryl-aryl' cross benzoin products, with yields from 65–87% and enantiomeric purities from 81–91% *ee*. The more difficult aryl–alkyl′, and alkyl–aryl′ cross-benzoin products can also be generated, with yields from 72–88% and enantiomeric purities from 41–73% *ee*.

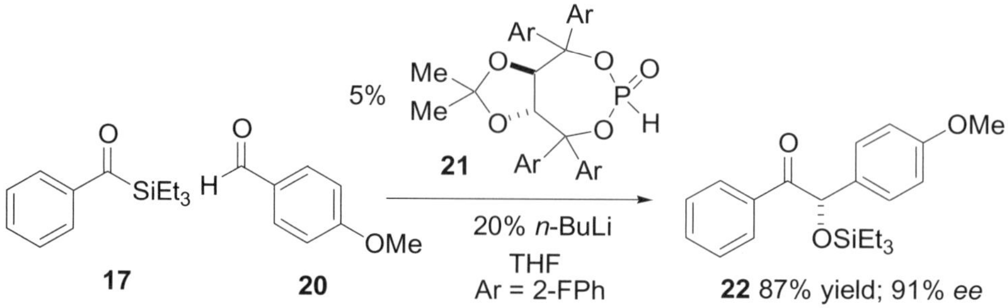

The enzymes benzaldehyde lyase (BAL) and benzoylformate decarboxylase (BFD) have also been shown to catalyze enantioselective cross-benzoin reactions. Aryl–aryl′ as well as aryl–alkyl′ products are produced in high yield and enantiomeric purity. In the case of aryl–aryl′ products, the success of the reaction depends on the empirical identification of suitable donor/acceptor pairs. Aldehydes containing *ortho*-substituents were found to be ideal acceptors in BAL and BFD-catalyzed cross-benzoin reactions.[22] The preparation of (*R*)-1-(4-bromophenyl)-2-(2-chlorophenyl)-2-hydroxyethanone (**25**) from 4-bromobenzaldehyde (**23**) and 2-chlorobenzaldehyde (**24**) highlights the high conversion and selectivity possible in this transformation. As with *rac*-benzoin, *rac*-1-(4-bromophenyl)-2-(2-chlorophenyl)-2-hydroxyethanone is easily resolved by BAL to give the (*S*)-enantiomer with high enantiomeric purity.

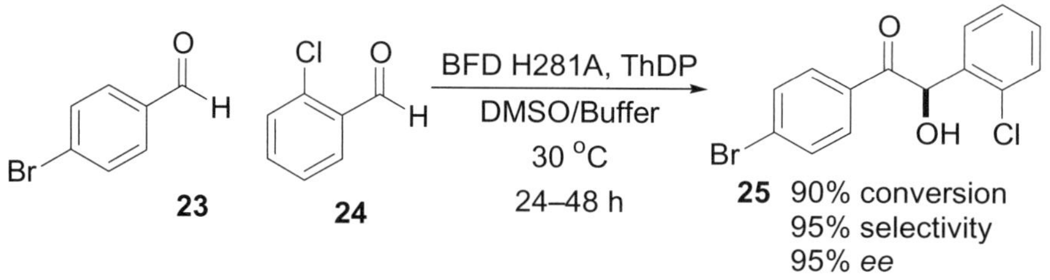

Müller and co-workers have also shown that BAL is effective for the selective preparation of aryl–alkyl′ acyloin products. The aliphatic aldehyde must be present in excess, and longer reaction times are necessary. Under these optimized conditions, (*R*)-2-hydroxy-1-phenylpropan-1-one (**26**) is available in excellent yield and enantiomeric purity from the BAL-catalyzed coupling of benzaldehyde and acetaldehyde.[12] This reaction has recently been optimized for large-scale production.[23]

1 equiv 3 equiv **26** 94% yield > 99% *ee*

Benzoin Annulations

A conceptually unique annulation process is available through the intramolecular benzoin condensation of dialdehyde or ketoaldehyde substrates. In one of the few examples of a dialdehyde annulation, Wong and co-workers reported the cyanide-catalyzed intramolecular benzoin reaction of dialdehyde **27** to give benzoxepinone **28** in modest yield.[24] Very recently, Mennen and Miller described macrocyclization reactions of dialdehydes catalyzed by triazolium carbene catalysts to give 11–14-membered rings in 16–43% yield.[25]

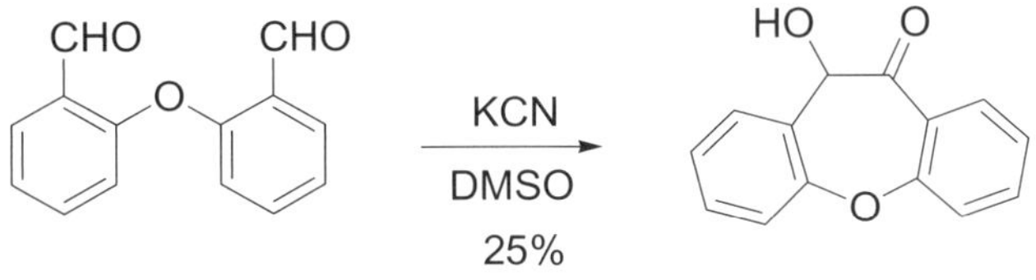

27 **28**

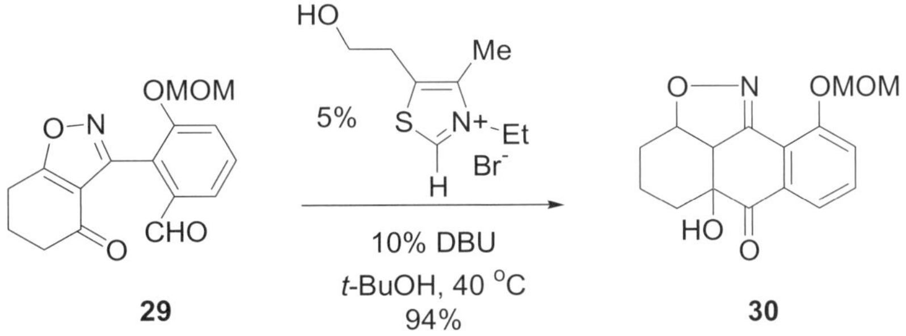

29 **30**

The intramolecular benzoin condensation between an aldehyde and ketone was virtually unknown until 2003, when Suzuki and co-workers established that thiazolylidine catalysts could effect this transformation in good to excellent yield for the synthesis of functionalized preanthraquinones such as **30**.[26] Subsequent to this work, reports of annulation reactions of a variety of ketoaldehyde substrates, as well as asymmetric variants, have appeared in the literature.[1b]

Aza-Benzoin Condensation

The heterocyclic carbene-catalyzed coupling of an aldehyde with an imine constitutes a variation of the cross-benzoin condensation called the aza-benzoin condensation. α-Amino ketones are obtained in generally good yields, and asymmetric induction is also possible. The reaction was first reported by López-Calahorra and co-workers for the coupling of aromatic and aliphatic aldehydes with iminium salts.[27] Katritzky and co-workers later showed that aliphatic aldehydes do not give azabenzoin products as initially reported, but simple Mannich addition products.[28] The substrate scope of the reaction was improved by Murry, Frantz, and co-workers, who found that arylsulfonylamides are ideal iminium precursors.[29] Reaction of benzaldehyde with phenylsulfonylcarbamate **31** under the optimized conditions provides the Boc-protected α-aminoketone **32** in good yield. Notably, classical benzoin dimerization of the aldehyde is not observed under these conditions. Additional studies by other researchers have established optimal conditions for the aza-benzoin condensation of aldehydes and *N*-phenyl imines[30] as well as acylsilanes and *N*-phosphinoyl imines.[31]

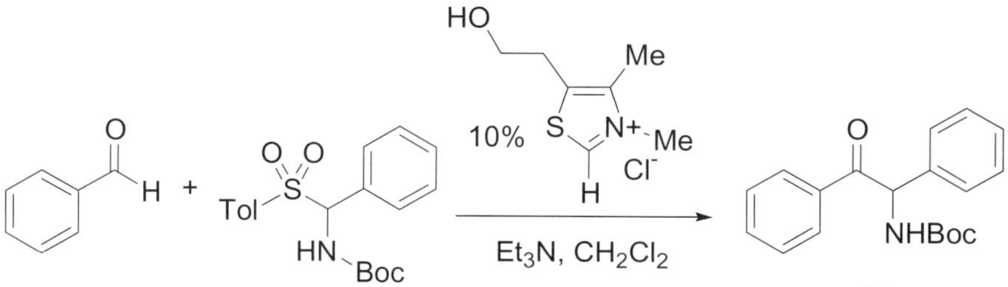

9 **31** **32**

A number of challenges were overcome by Miller and co-workers in the development of an asymmetric aza-benzoin condensation.[32] Using the general reaction partners developed by Murry, Frantz, and co-workers, it was found that the thiazolium dipeptide derivative **33** is an effective catalyst for the asymmetric reaction, providing enantiomeric purities from 75–87% *ee* for the coupling of a limited group of substituted benzaldehydes and arylsulfonylamides. Racemization of the product α-aminoketone under the reaction conditions is a significant concern; short reaction times, reactive electron-deficient aldehydes, and the hindered base pentamethylpiperidine are all necessary to achieve both good conversion and high enantiomeric excess.

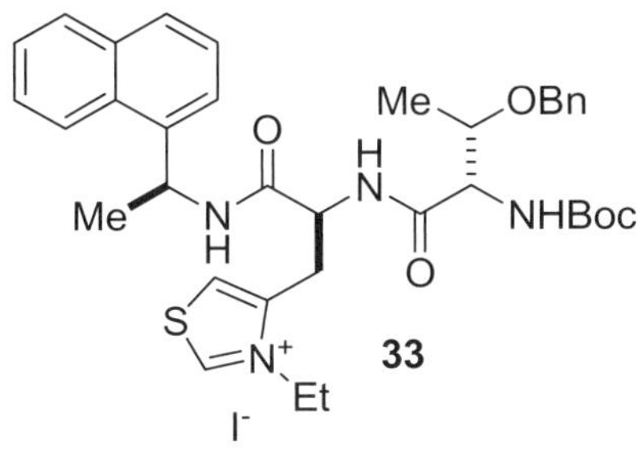

33

2.3.5 *Experimental*

NaCN-Catalyzed Dimerization of Benzaldehyde[33]

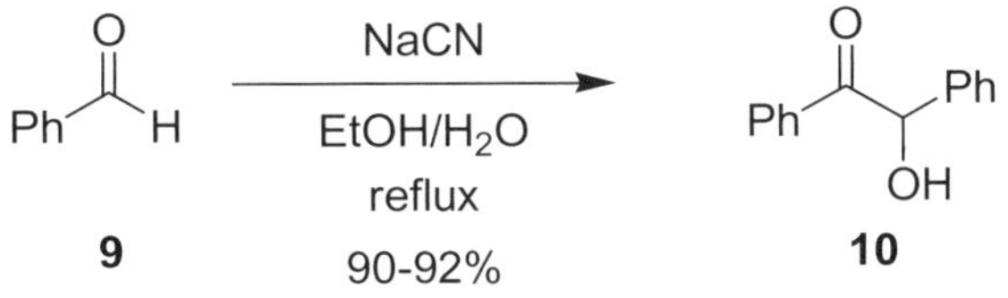

2-Hydroxy-1,2-diphenylethan-1-one (10).
In a 3 L round-bottomed flask fitted with a reflux condenser are placed 625 cc. of 95% alcohol, 500 mL of water, 500 g (476 mL, 4.7 moles) of pure benzaldehyde (**9**), and 50 g. of sodium cyanide (96–98%). The mixture is then heated and kept boiling for one-half hour. In the course of about twenty minutes, crystals begin to separate from the hot solution. At the end of the thirty minutes, the solution is cooled, filtered with suction, and washed with a little water. The yield of dry crude benzoin (**10**), which is white or light yellow, is 450–460 g. (90–92% of the theoretical amount). In order to obtain it completely pure, the crude substance is recrystallized from 95% alcohol, 90 g of crude material being dissolved in about 700 mL of boiling alcohol; upon cooling, a yield of 83 g of white, pure benzoin which melts at 129 °C is obtained.

BAL-Catalyzed Asymmetric Dimerization of Benzaldehyde[12]

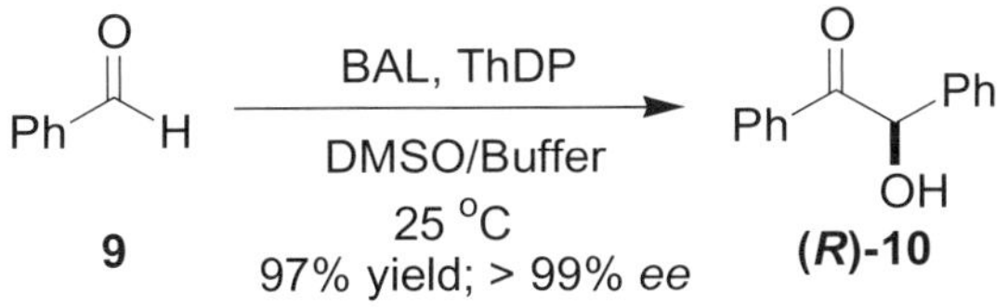

(*R*)-2-Hydroxy-1,2-diphenylethan-1-one ((*R*)-10).
Benzaldehyde (**9**, 318 mg, 3 mmol) was dissolved in a mixture of dimethyl sulfoxide (20 mL) and potassium phosphate buffer [80 mL, 50 mM, pH 7.0, containing $MgSO_4$ (2.5 mM) and ThDP (0.15 mM)]. After addition of BAL (20 U) the reaction mixture was allowed to stand at 25 °C for 48 h before a further 20 U of BAL were added. After 62 h no more benzaldehyde was detected (GC-MS). The reaction mixture was extracted with dichloromethane (250 mL) and the organic layer washed with water (25 mL) and brine (25 mL) and dried over Na_2SO_4. Evaporation of the solvent and purification of the crude product by crystallization afforded (*R*)-2-hydroxy-1,2-diphenyl-ethan-1-one (**(*R*)-10**) as a colorless solid; yield: 305 mg (96%, 99% *ee*).

Cross-Benzoin Reaction: Coupling of Acylsilane and Aldehyde[20]

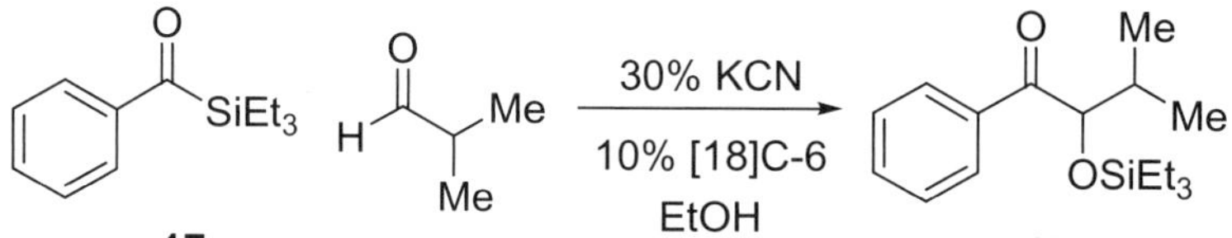

17 **19** 66%

3-Methyl-1-phenyl-2-(triethylsilyloxy)butan-1-one (19).
A dry round-bottom flask with a magnetic stir bar was charged with 106 mg (0.48 mmol) of acylsilane **17**, 9.5 mg (0.15 mmol) of KCN, 14 mg (0.048 mmol) of 18-crown-6 and 4 mL of Et_2O. To the resulting mixture, 48 μL (0.53 mmol) of isobutyraldehyde was added as an Et_2O solution (6 mL) *via* cannula over 2 h, and the reaction was kept for another 1 h at 25 °C. To the reaction mixture was added 15 mL of H_2O. The mixture was stirred for 5 min before the organic layer was separated and the aqueous layer was extracted with three 15 mL portions of Et_2O. The organic extracts were combined, dried ($MgSO_4$), and the solvent was removed with a rotary evaporator. The crude product was purified by flash chromatography with 45:1 hexanes/EtOAc to afford 93 mg (66%) of 3-methyl-1-phenyl-2-(triethylsilyloxy)butan-1-one (**19**) as a clear oil.

Aza-Benzoin Reaction[29]

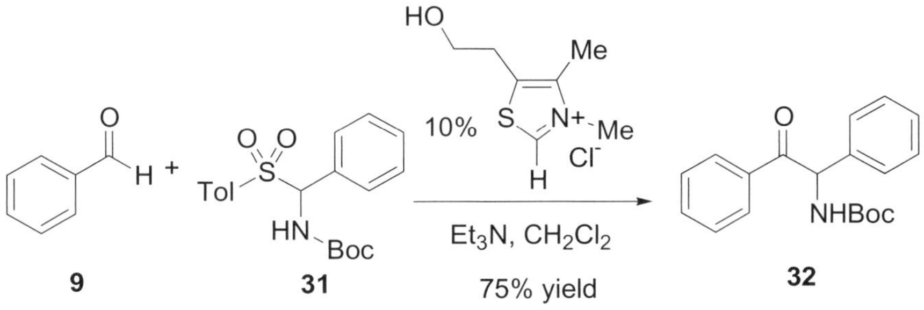

9 **31** **32**

***N*-(3-Benzyloxy-2-oxo-1-phenyl-propyl)-carbamic acid *tert*-butyl ester (32).**

A 50mL flask was charged with the tosyl-amide **31** (3.0 mmol, 1.0 equiv) and the thiazolium catalyst (0.3 mmol, 0.1 equiv) and purged with nitrogen for 15 min. To the flask was added CH_2Cl_2 (20 mL) followed by the aldehyde **9** (3.3 mmol, 1.1 equiv) and the resulting mixture stirred and heated to 35 °C. Triethylamine (45 mmol, 15 equiv) was added in one portion via syringe and the corresponding reaction was monitored by HPLC analysis for consumption of the tosyl-amide. After the reaction was complete, it was cooled to 25 °C and water was added (20 mL). The resulting layers were separated and the aqueous layer was extracted with CH_2Cl_2 (10 mL). The combined organic layers were washed with brine (10 mL), dried over Na_2SO_4, and concentrated in vacuo. The product was isolated from the crude reaction mixture by crystallization from (1 : 1) methylene chloride/hexanes to give *N*-(3-benzyloxy-2-oxo-1-phenyl-propyl)-carbamic acid *tert*-butyl ester (**32**) in 75% yield.

2.3.6 References

1 [R] (a) Johnson, J. S. *Angew. Chem. Int Ed.* **2004**, *43*, 1326–1328. (b) Johnson, J. S. *Curr. Opin. Drug. Disc. Devel.* **2007**, *10*, 691–703.

2 [R] Ide, W. S.; Buck, J. S. *Org. React.* **1948**, *4*, 269–304.

3 [R] (a) Enders, D.; Balensiefer, T *Acc. Chem. Res.* **2004**, *37*, 534–541. (b) Marion, N.; Diez-González, S.; Nolan, S. P. *Angew. Chem., Int. Ed.* **2007**, *46*, 2988–3000. (c) Enders, D.; Niemeier, O.; Henseler, A. *Chem. Rev.* **2007**, *107*, 5605–5655.

4 Wöhler, F.; Liebig, J. *Ann. Pharm.* **1832**, *3*, 249–282.

5 Lapworth, A. *J. Chem. Soc.* **1903**, *83*, 995–1005.

6 Ukai, T.; Tanaka, R.; Dokawa, T. *J. Pharm. Soc. Jpn.* **1943**, *63*, 296–300.

7 [R] Jordan, F. *Nat. Prod. Rep.* **2003**, *20*, 184–201.

8 (a) Breslow, R. *J. Am. Chem. Soc.* **1958**, *80*, 3719–3726. (b) Breslow, R.; Kim, R. *Tetrahedron Lett.* **1994**, *35*, 699–702.

9 Sheehan, J. C.; Hunneman, D. H. *J. Am. Chem. Soc.* **1966**, *88*, 3666–3667.

10 Enders, D.; Kallfass, U. *Angew. Chem., Int. Ed.* **2002**, *41*, 1743–1745.

11 Demir, A. S.; Pohl, M.; Janzen, E.; Müller, M. *J. Chem. Soc., Perkin Trans. 1* **2001**, 633–635.

12 Demir, A. S.; Sesenoglu, Ö; Eren, E.; Hosrik, B.; Pohl, M.; Janzen, E.; Kolter, D.; Feldmann, R.; Dünkelmann, P.; Müller, M. *Adv. Synth. Catal.* **2002**, *344*, 96–103.

13 Demir, A. S.; Dünnwald, T.; Iding, H.; Pohl, M.; Müller, M. *Tetrahedron: Asym.* **1999**, *10*, 4769–4774.

14 Buck, J. S.; Ide, W. S. *J. Am. Chem. Soc.* **1932**, *54*, 3302–3309

15 Jenkins, S. S. *J. Am. Chem. Soc.* **1931**, *53*, 3115–3122.

16 Stetter, H.; Dämbkes, G. *Synthesis* **1977**, 403–404.

17 Heck, R.; Henderson, A. P.; Köhler, B.; Rétey, J.; Golding, B. T. *Eur. J. Org. Chem.* **2001**, 2623–2627.

18 Hünig, S.; Wehner, G. *Synthesis* **1975**, 391–392.

19 Linghu, X.; Johnson, J. S. *Angew. Chem., Int. Ed.* **2003**, *42*, 2534–2536.

20 Linghu, X.; Bausch, C. C.; Johnson, J. S. *J. Am. Chem. Soc.* **2005**, *127*, 1834–1840.

21 Linghu, X.; Potnick, J. R.; Johnson, J. S. *J. Am. Chem. Soc.* **2004**, *126*, 3070–3071.

22 Dünkelmann, P.; Kolter-Jung, D.; Nitsche, A.; Demir, A. S.; Siegert, P.; Lingen, B.; Baumann, M.; Pohl, M.; Müller, M. *J. Am. Chem. Soc.* **2002**, *124*, 12084–12085.

23 Stillger, T.; Pohl, M.; Wandrey, C.; Liese, A. *Org. Process Res. Dev.* **2006**, *10*, 1172–1177.

24 Yang, Z.; Wong, H. N. C.; Hon, P. M.; Chang, H. M.; Lee, C. M. *J. Org. Chem.* **1992**, *57*, 4034–4037.

25 Mennen, S. M.; Miller, S. J. *J. Org. Chem.* **2007**, *72*, 5260–5269.

26 Hachisu, Y; Bode, J. W.; Suzuki, K. *J. Am. Chem. Soc.* **2003**, *125*, 8432–8433.

27 Castells, J.; López-Calahorra, F.; Bassedas, M.; Urrios, P. *Synthesis* **1988**, 314–315.

28 Katritzky, A. R.; Cheng, D.; Musgrave, R. P. *Heterocycles* **1996**, *42*, 273–280.

29 Murry, J. A.; Frantz, D. E.; Soheili, A.; Tillyer, R.; Grabowski, E. J. J.; Reider, P. J. *J. Am. Chem. Soc.* **2001**, *123*, 9696–9697.

30 Li, G.-Q.; Dai, L.-X.; You, S. L. *Chem. Comm.* **2007**, 852–854.

31 Mattson, A. E.; Scheidt, K. A. *Org. Lett.* **2004**, *6*, 4363–4366.

32 Mennen, S. M.; Gipson, J. D.; Kim, Y. R.; Miller, S. J. *J. Am. Chem. Soc.* **2005**, *127*, 1654–1655.

33 Adams, R.; Marvel, C. S. *Org. Synth. Coll. Vol. I*, **1941**, 94.

2.4 Corey–Fuchs Reaction

Xiaojun Han

2.4.1 Description[1]

Corey–Fuchs reaction is a two step reaction converting an aldehyde to an alkyne by one-carbon homologation of the aldehyde. The Wittig-like reaction of aldehyde **1** and dibromocarbene forms dibromoalkene **2**. The treatment of dibromoalkene **2** with two equivalent of *n*-BuLi form a lithium alkynylide, quenched by eletrophiles, such as proton, CO_2, aldehydes, ketones, and alkyl halides to form alkyne **3**.

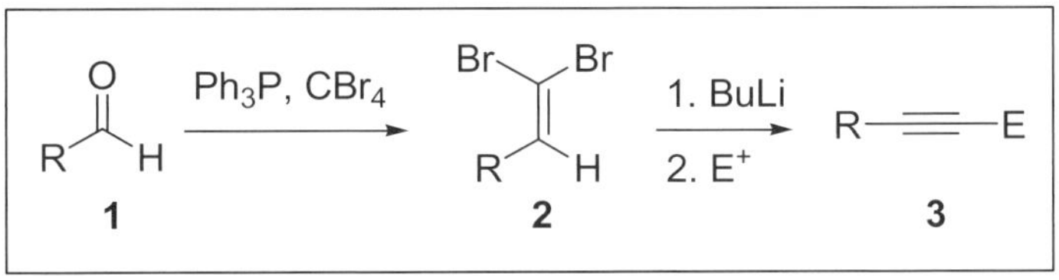

2.4.2 Historical Perspective[1]

This transformation was first disclosed by Corey and Fuchs in 1972. Three aldehydes, an aromatic, a linear aliphatic and a cyclic aliphatic aldehyde were chosen to show the generality of this reaction. For all three aldehydes, the reaction proceeded well.

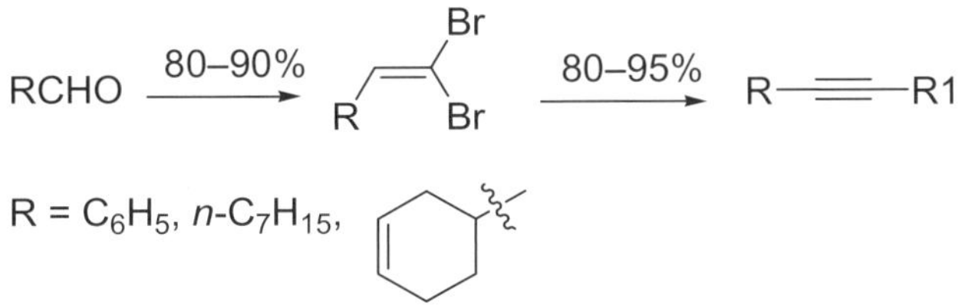

Two procedures were developed for the first step. The aldehyde (1 equiv) was added to a mixture of PPh_3 (4 equiv) and CBr_4 (2 equiv) in CH_2Cl_2 at 0 °C with a reaction time of 5 min. Alternatively, the aldehyde (1 equiv) was added to the mixture, prepared by the reaction of Zn dust (2 equiv), PPh_3 (2 equiv) and CBr_4 (2 equiv) at 23 °C for 24–30 h, for a reaction time of 1–2 h. The advantage of the first procedure is very short total reaction time; for the second procedure, the isolation procedure is simpler, especially for those non-polar alkynes, since less PPh_3 is used and unreacted

PPh_3 could eluent together with non-polar alkynes during flash column chromatography.

Corey–Fuchs alkyne synthesis can be seen as a cousin of the Wittig olefin synthesis.

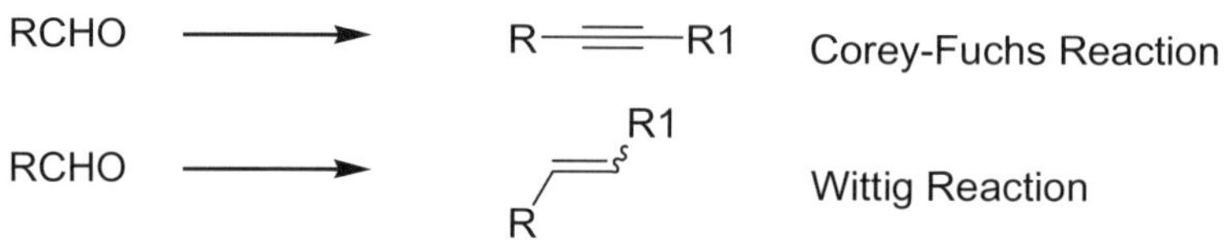

2.4.3 *Mechanism*

Dibromo ylide **4** is generated by the reaction of PPh_3 and CBr_4.[2] The reaction of aldehyde **1** and ylide **4** follows the pathway of Wittig reaction[3] to afford dibromoalkene **2**, which reacting with 2 equiv of *n*-BuLi affords lithium alkynalide **5**. The reaction of alkynalide **5** and eletrophiles produces alkyne **3**.

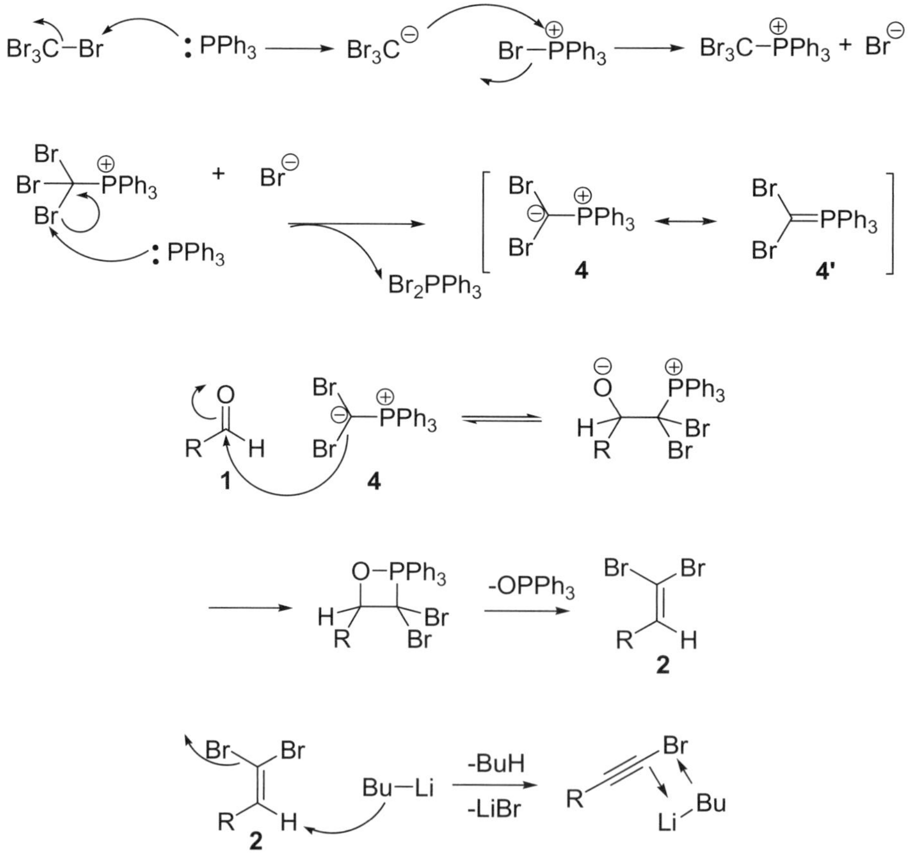

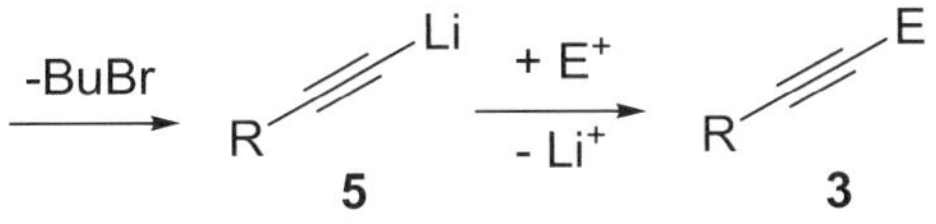

There are no mechanistic studies carried out for this reaction. However, from our knowledge of the reaction between PPh_3 and CBr_4,[2] and Wittig reaction,[3] the above proposed mechanism[4] seems reasonable.

2.4.4 *Variation and Improvement*

For acid labile substrates, a base such as PyH[5] or Et_3N[6] could be added to neutralize the formed HBr in the first step of dibromoalkene formation. $P(NMe_2)_3$[7] was also used for this transformation as both a phosphine source and a base. Solvents such as MeCN,[8] hepatane,[9] and THF[7] worked equally well as the most widely used solvent CH_2Cl_2.

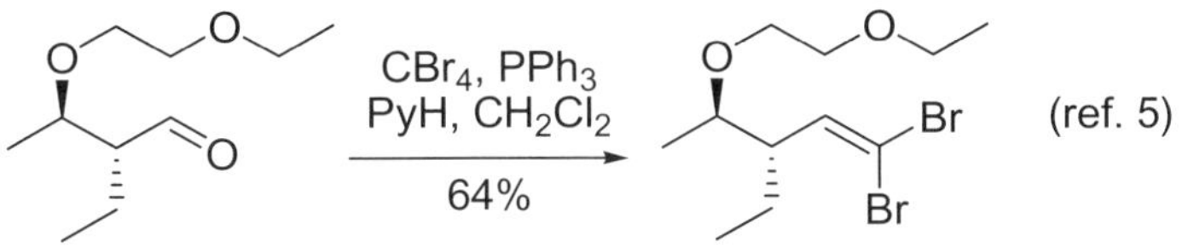

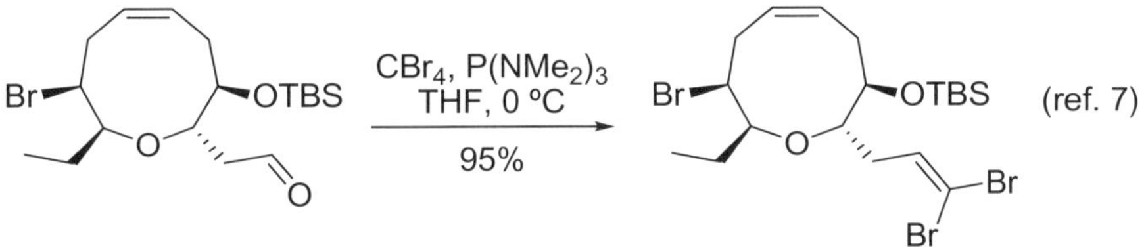

Dibromoalkenes could also be produced by the reaction of hydrazones and CBr_4 in the presence of catalytic amount of CuCl.[10] This approach does not involve PPh_3, but its synthetic utility needs to be seen.

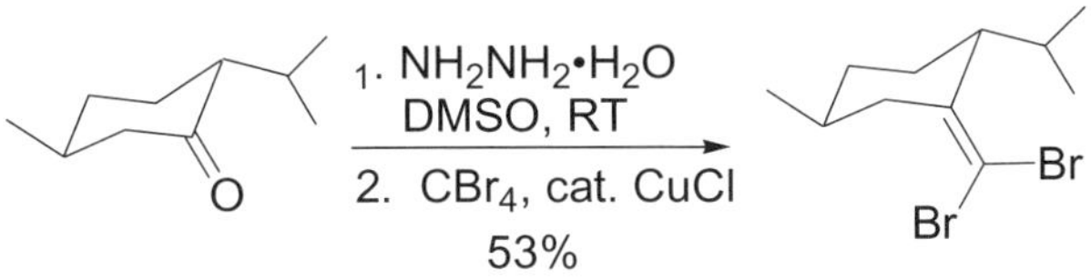

To use less *n*-BuLi, *t*-BuOK[11] or NaHMDS[12] was reported to effect the transformation of dibromoalkenes to alkynyl bromide, which was then converted to terminal alkynes by *n*-BuLi.

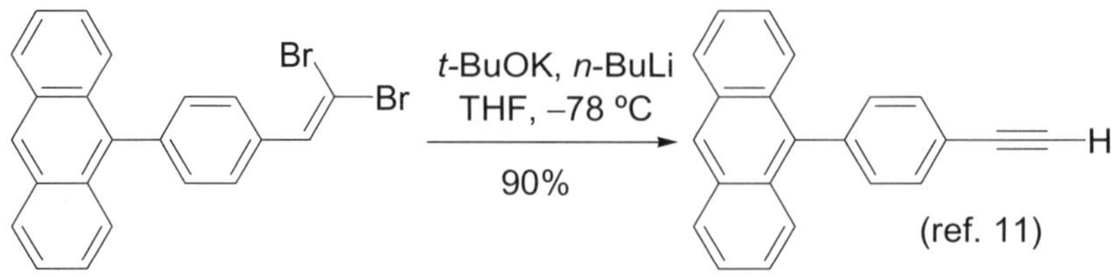

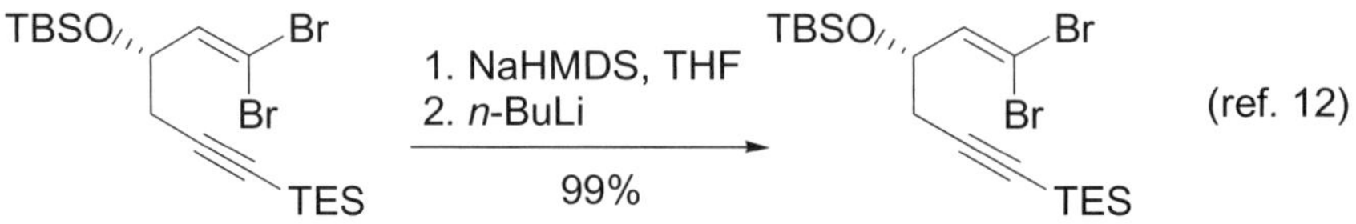

Other basic conditions, such as magnesium turning in refluxing THF,[13] KOH in *i*-PrOH/H_2O under reflux,[14] small alkyl magnesium bromides[6,9], and LDA[15] can all effect this reaction. These conditions avoid the use of *n*-BuLi and this affords better functional group tolerance.

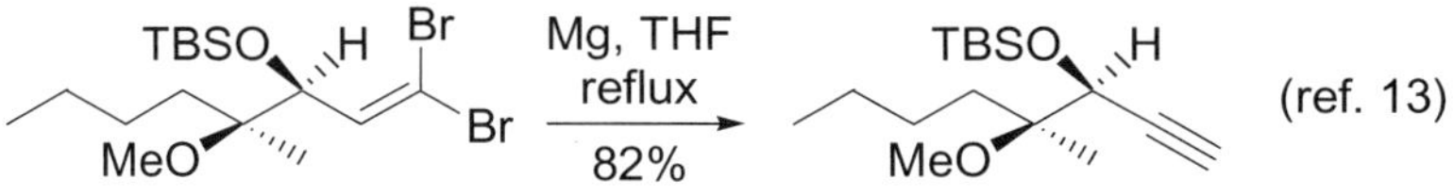

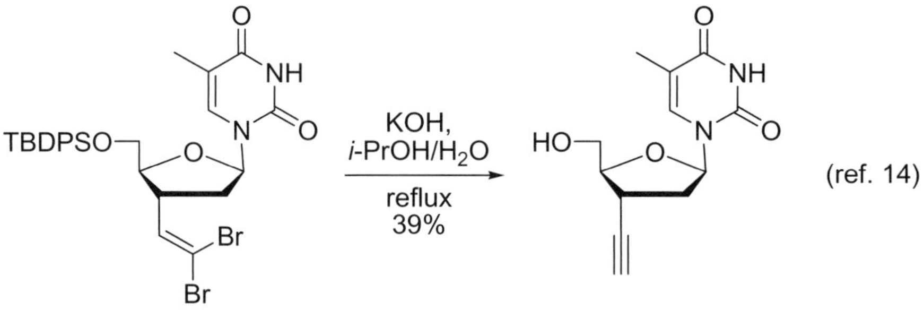

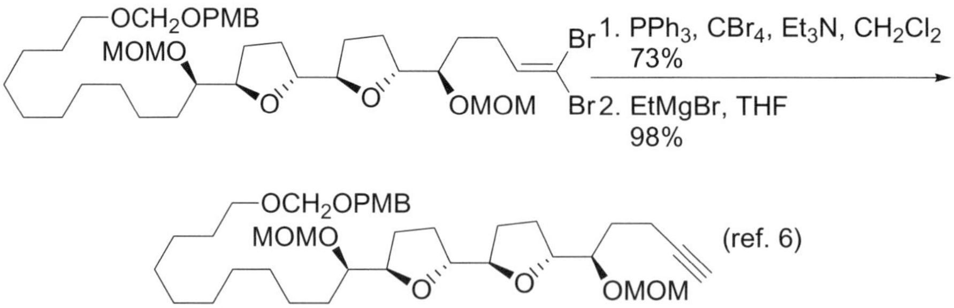

Lithium alkynylides can also be trapped by reactive eletrophiles other than protons, such as alkyl halides, acyl halides[9] and Weinreb amide.[16]

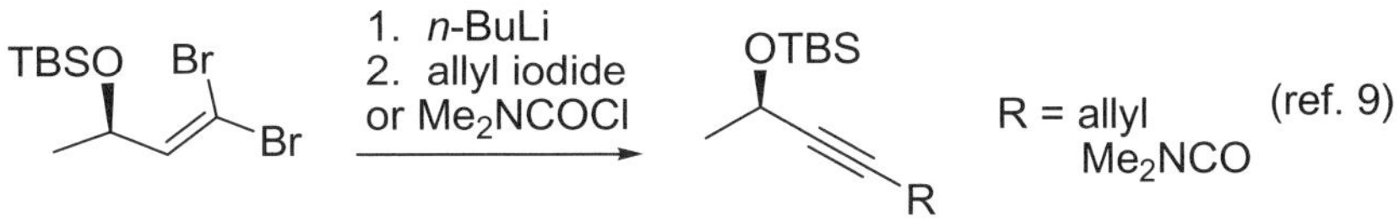

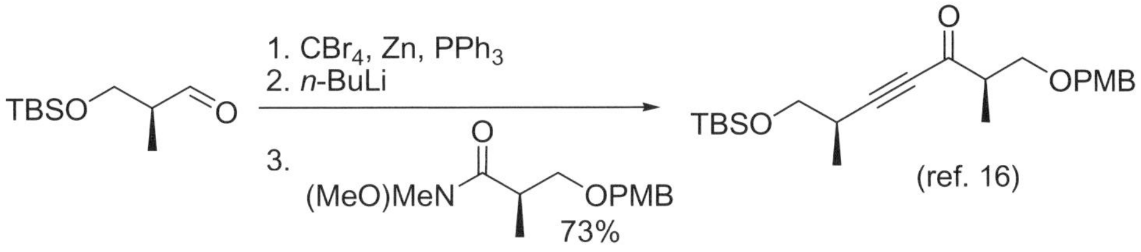

2.4.5 *Synthetic Utility*

Fürstner[17] prepared acetylene **6** by the titled reaction, which isomerized to phenanthrene **7** in the presence of 5% $PtCl_2$ in PhMe at 80 ºC by the $PtCl_2$ catalyzed 6-*endo*-dig cyclization. Heteroaryl containing acetylenes could also perform this transformation.

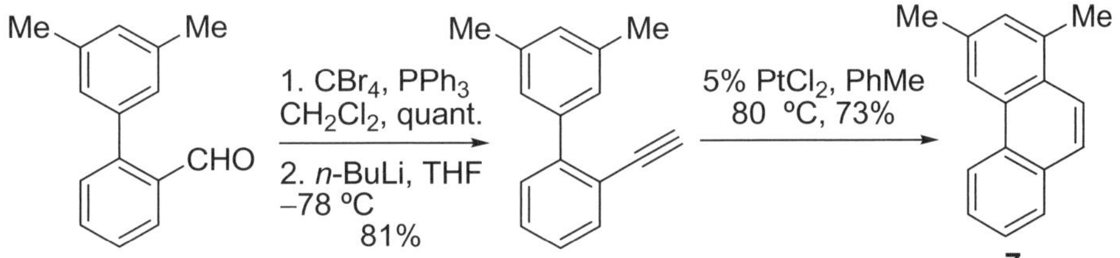

6 **7**

In their total synthesis of 6,7-dehydrostipiamide, Negishi[18] prepared acetylene **8** (C_8–C_{18} subunit) by the Corey–Fuchs reaction.

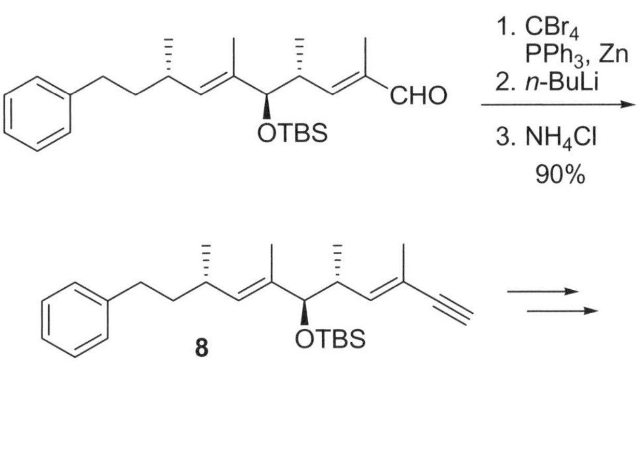

8

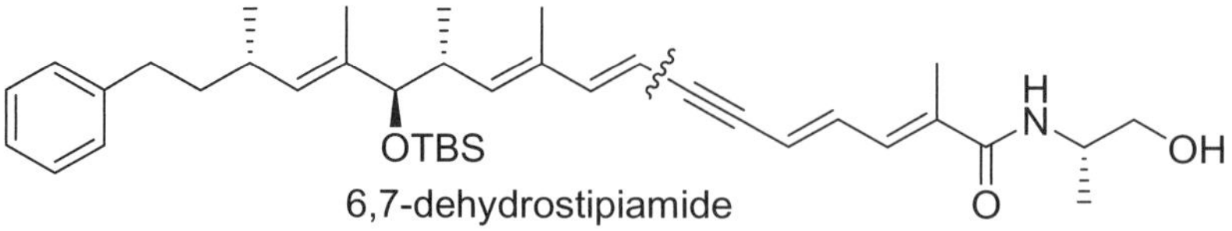

Alkyne **13** was prepared from aldehyde **9** by Corey–Fuchs reaction (**10**), alkylation (**11**), 1,2-Wittig rearrangement (**12**) and silylation reaction. The reaction of alkyne **13** and Bu_3SnH afforded compound **14** via a 5-*exo-dig* cyclization and the cleavage of Si–O bond by MeLi.[19]

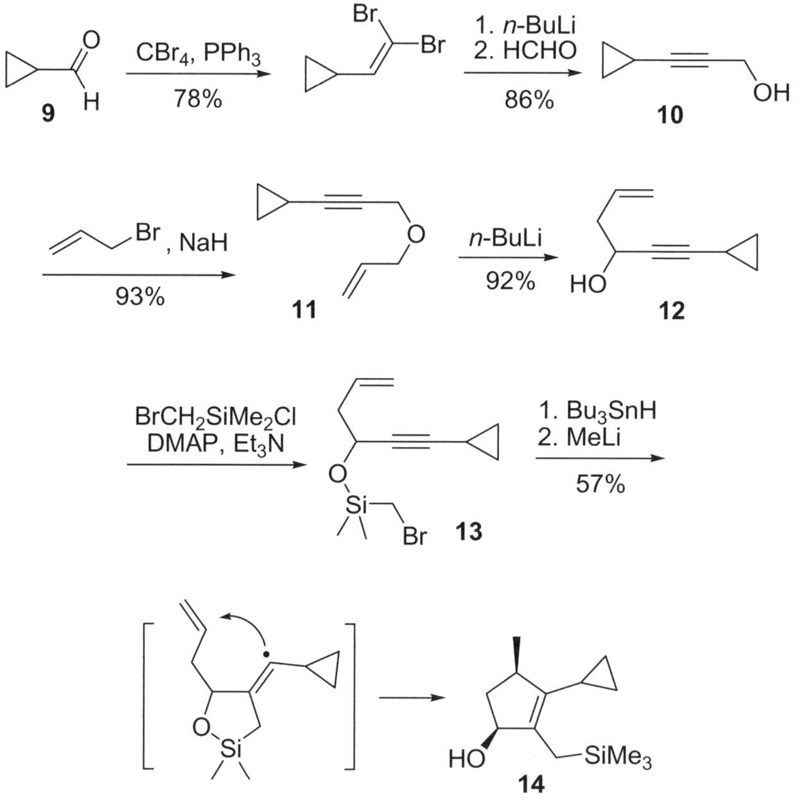

Propargylsilane (**16**) was prepared by Corey–Fuchs reaction via dibromoalkene **15**. (*Z*)-Allylsilanes (**17**) was synthesized by cis hydrogenation of C≡C of propargylsilane **16** by treating it with $Ni(OAc)_2 \cdot 4H_2O$, $NaBH_4$, H_2 and ethylenediamine.[20]

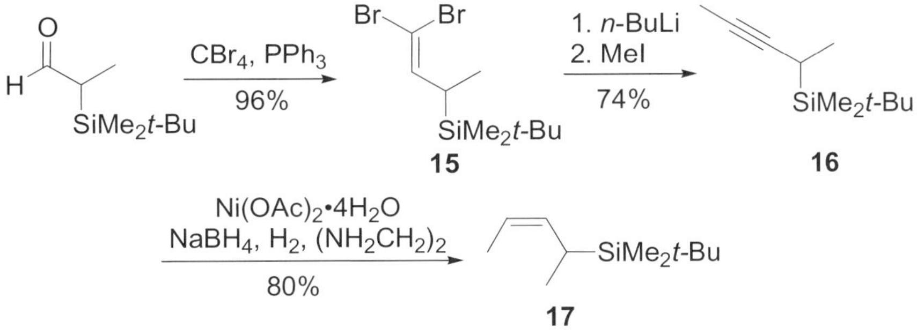

In the effort towards the synthesis of olivine, acetylene **19** was made from aldehyde **18** by the reaction of CBr_4 and PPh_3 to form a dibromoalkene, hydroboration/oxidation to form an alcohol, and treatment of the resulting dibromoalkene with *n*-BuLi.

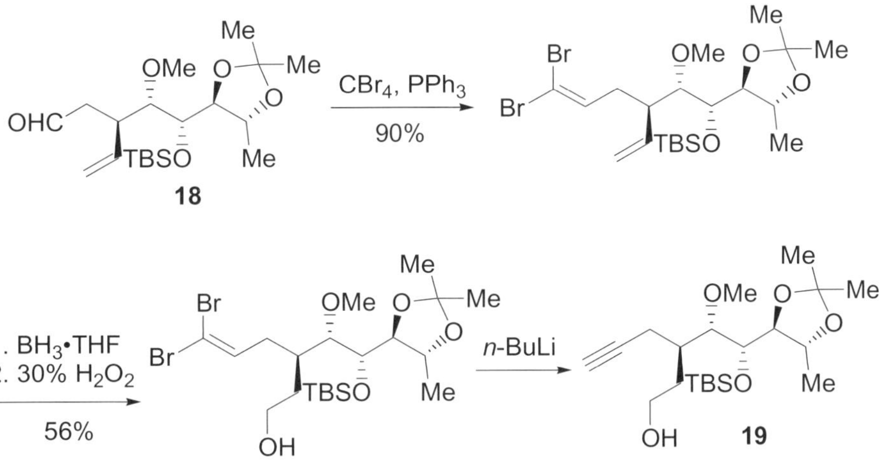

Corey–Fuchs reaction has also been employed to make pharmaceutically interesting compounds. The treatment of bromo-acetylene **20** with *t*-BuLi and the quench of resulting lithium alkylanide with phenyl isocyanate produced amide **21**. After SEM removal, NMDA receptor antagonist **22** for stoke therapy was obtained.[22]

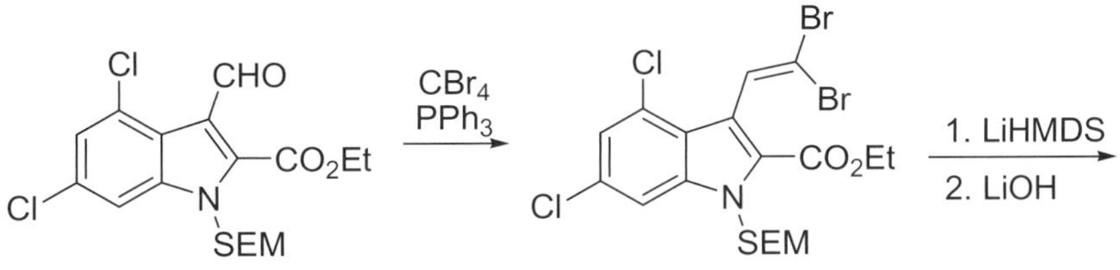

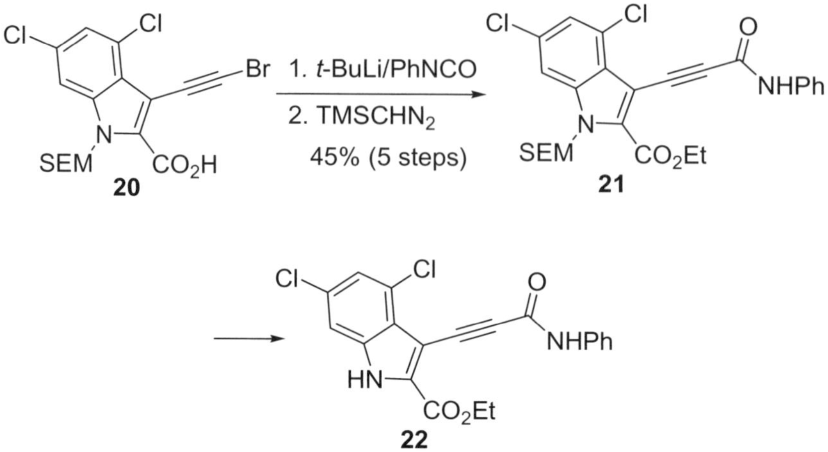

The enantiomerically pure alkynylogous amino acids **24** were prepared from aldehyde **23** by Corey–Fuchs reaction.[23]

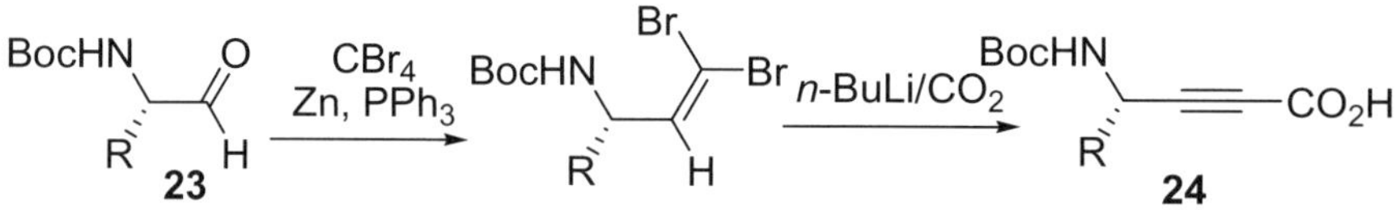

Acetylene **25** was converted to compound **26**, which was a subtype selective muscarinic agonist, studied for the treatment of Alzheimer's disease.[24]

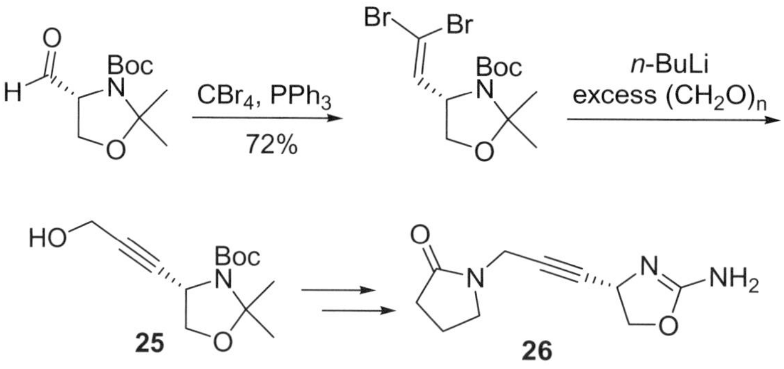

There are also wide spread application of the Corey–Fuchs reaction in the material science, since dibromoalkenes and terminal acetylenes are well suited for Pd-catalyzed reactions to make highly conjugated linear and branch compounds.[25] For example, acetylene **27** was made by the Corey–

Fuchs reaction, which was employed for Sonogashira reaction to make compounds **28**–**31**. These compounds were studied as potential redox-active molecular wires.[26]

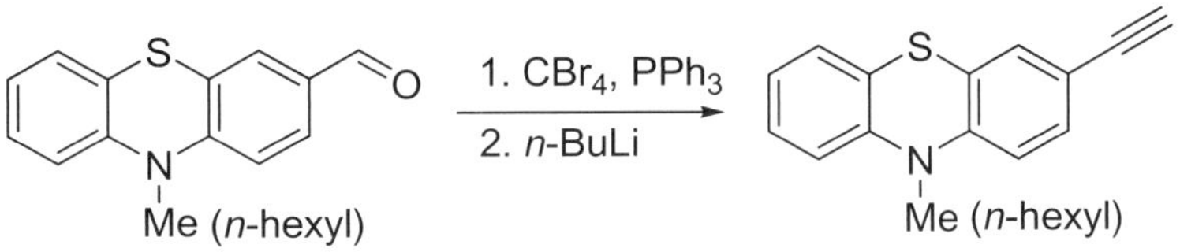

27

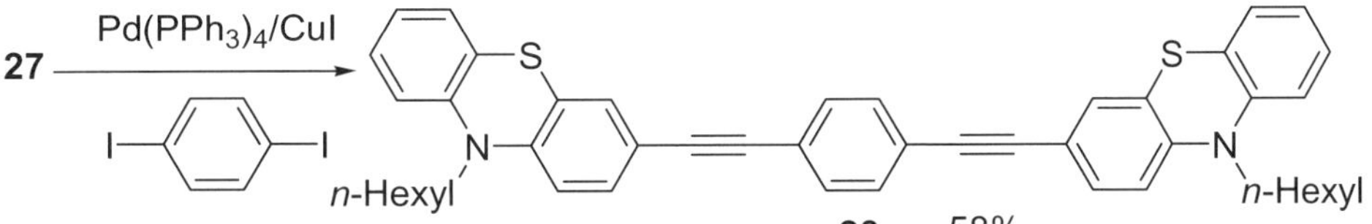

28 58%

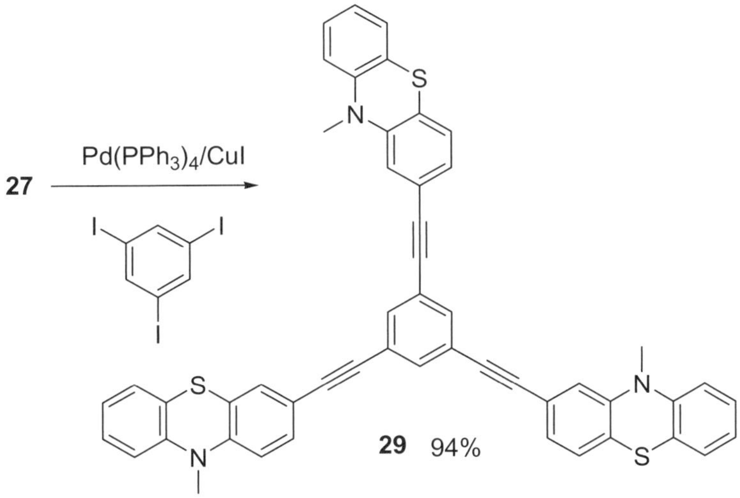

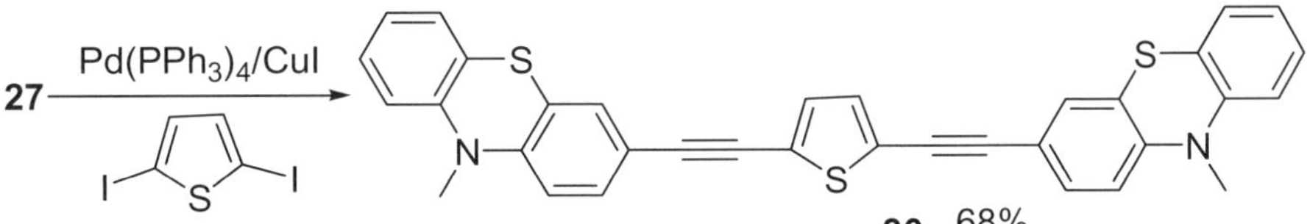

30 68%

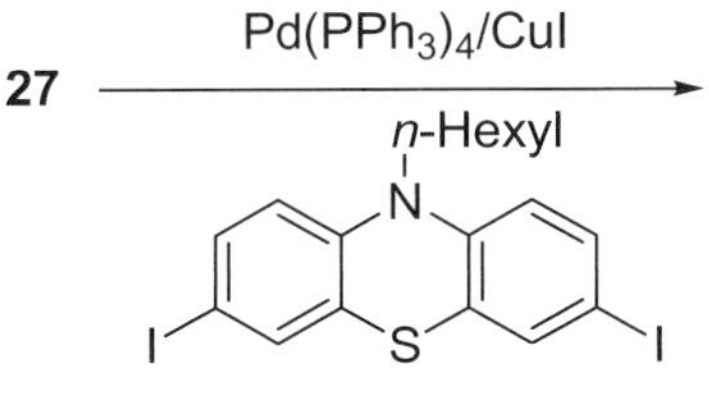

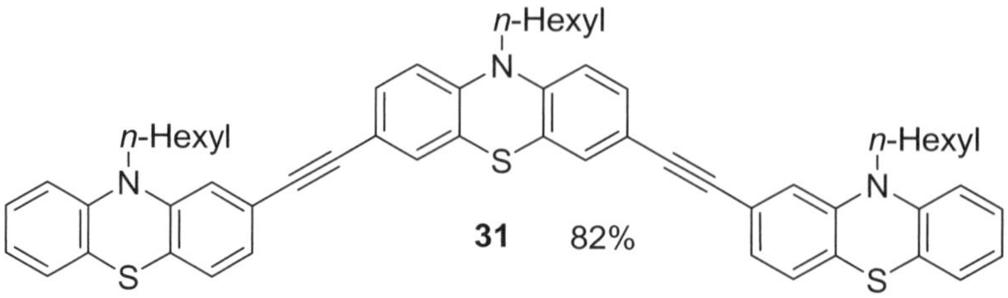

31 82%

2.4.6 *Experimental*[20]

***tert*-Butyl(4,4-dibromobut-3-en-2-yl)dimethylsilane** (15):

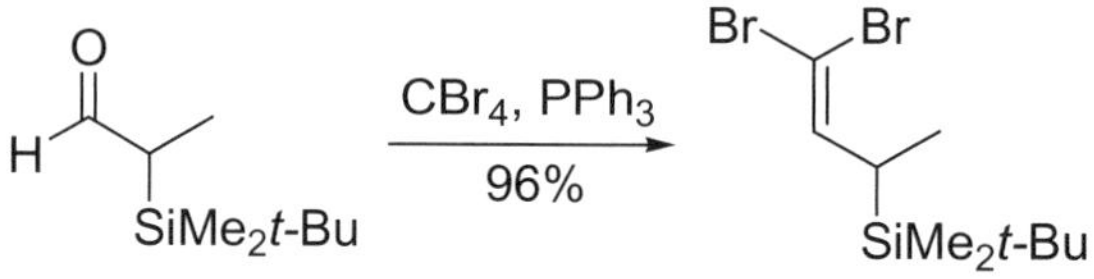

15

To a stirred solution of CBr_4 (2.8 equiv) in freshly distilled CH_2Cl_2 (1 M), was added a solution of PPh_3 (5.7 equiv) in CH_2Cl_2 (2 M) at 0 °C. After 30 min at 0 °C, a solution of 2-(*tert*-butyldimethylsilyl)propanal (1.0 equiv) in CH_2Cl_2 (0.5 M) was added. It was stirred at 0 °C for 2 h. Ether (4 mL/mmol of aldehyde) was added; the precipitate was filtered off and washed with ether (3 × 2 mL/mmol of aldehyde). The combined filtrates were washed with H_2O (3 mL/mmol of aldehyde), saturated aq. $NaHCO_3$ (3 mL/mmol of aldehyde), saturated aq. NH_4Cl (3 mL/mmol of aldehyde), brine (3 mL/mmol of aldehyde), dried over $MgSO_4$ and evaporated *in vacuo*. The residue was purified by column chromatography on silica gel eluenting with pentane to afford the title compound in 96% yield.

***tert*-Butyldimethyl(pent-3-yn-2-yl)silane** (16):

15 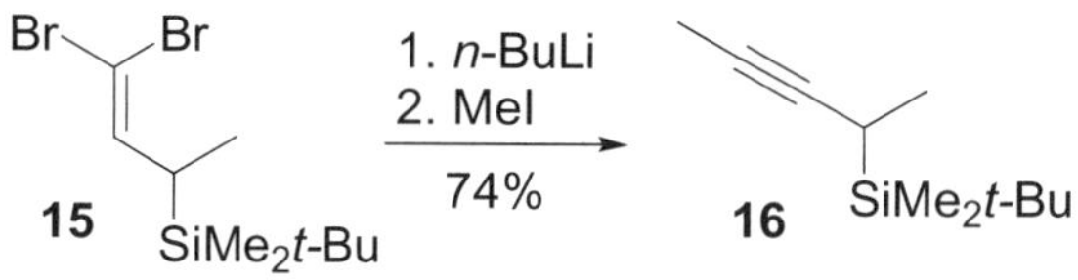 **16**

To a solution of compound **15** (1.00 equiv) in THF (0.25 M) at −78 °C, *n*-BuLi (2.14 equiv, 2.36 M) was added. After stirring for 30 min at −78 °C, MeI (1.20 equiv) was added. The mixture was heated at 40 °C for 2 h and then it was stirred at room temperature for 18 h. It was cooled to −78 °C and quenched with ether/H_2O (1:1 v/v; 10 mL/mmol of **15**). The combined organic layers were washed with brine (20 mL/mmol of **15**), dried over $MgSO_4$, evaporated *in vacuo*. The residue was purified by column chromatography on silica gel eluenting with pentane to afford the title compound in 74% yield.

2.4.7 References

1. Corey, E. J.; Fuchs, P. L. *Tetrahedron Lett.* **1972**, 3769–3772.
2. Hodge, P.; Khoshdel, E. *J. Chem. Soc., Perkin Trans. I* **1984**, 195–198
3. (a) Yamataka, H.; Nagareda, K.; Takatsuka, T.; Ando, K.; Hanafusa, T.; Nagase, S. *J. Am. Chem. Soc.* **1993**, *115*, 8570–8576. (b) Vedejs, E.; Marth, C. F. *J. Am. Chem. Soc.* **1990**, *112*, 3905–3909.
4. http://www.organic-chemistry.com/namedreactions/corey-fuchs-reaction.shtm
5 Nakahata, N.; Fujimura, S.; Kuwahara, S. *Chem. Eur. J.* **2006**, *12*, 4584–4593.
6 Han, H.; Sinha, M. K.; D'Souza, L. J.; Keinan, E.; Sinha, S. *Chem. Eur. J.* **2004**, *10*, 2149–2158.
7 Uemura, T.; Suzuki, T.; Onodera, N.; Hagiwara, H.; Hoshi, T. *Tetrahedron Lett.* **2007**, *48*, 715–719.
8 Mayor, M.; Buschel, M.; Fromm, K. M.; Lehn, J.-M.; Daub, J. *Chem. Eur. J.* **2001**, *7*, 1266–1272.
9 Ku, Y.-Y.; Patel, R. R.; Elisseou, E. M.; Sawick, D. P. *Tetrahedron Lett.* **1995**, *36*, 2733–2736.
10 Korotchenko, V. N.; Shastin, A. V.; Nenajdenko, V. G.; Balenkova, E. S. *Org. Biomol. Chem.* **2003**, *1*, 1906–1908.
11 Bolletta, F.; Fabbri, D.; Lombardo, M.; Prodi, L.; Trombini, C.; Zaccheroni, N. *Organometallics*, **1996**, *15*, 2415–2417.
12 Ji, N.; O'Dowd, H.; Rosen, B. M.; Myers, A. G. *J. Am. Chem. Soc.* **2006**, *128*, 14825–14827.
13 Hijfte, L. V.; Kolb, M.; Witz, P. *Tetrahedron Lett.* **1989**, *30*, 3655–3656.
14 Yan, J.; Naeslund, C.; Al-Madhoun, A. S.; Wang, J.; Ji, W.; Cosquer, G. Y.; Johnsamuel, J.; Sjoberg, S.; Eriksson, S.; Tjarks, W. *Bioorg. Med. Chem. Lett.* **2002**, *12*, 2209–2212.
15 Sankaraman, S.; Venkataramana, G.; Varghese, B. *J. Org. Chem.* **2008**, *73*, 2404–2407.
16 Mulzer, J.; Berger, M. *J. Org. Chem.* **2004**, *69*, 891–898.
17. Mamane, V.; Hannen, P.; Fürstner, A. *Chem. Eur. J.* **2004**, *10*, 4556–4575.
18. Zeng, X.; Zeng, F.; Negishi, E. *Org. Lett.* **2004**, *6*, 3245–3248.
19. Mainetti, E.; Fensterbank, L.; Malacria, M. *Synlett* **2002**, 923–926.
20. Tietze, L. F.; Neumann, T.; Kajino, M.; Pretor, M. *Synthesis* **1995**, 1003–1006.
21. Gilbert, A. M.; Miller, R.; Wulff, W. D. *Tetrahedron* **1999**, *55*, 1607–1630.
22. Giacobbe, S. A.; Fabio, R. D.; Baraldi, D.; Cugola, A.; Donati, D. *Syn. Commun.* **1999**, *29*, 3125–3135.
23. Reetz, M. T.; Strack, T. J.; Kanand, J.; Goddard, R. *Chem. Commun.* **1996**, 733–734.
24. Chung, J. Y. L.; Wasicak, J. T. *Tetrahedron Lett.* **1990**, *31*, 3957–3960.
25. (a) Nierengarten, J.-F.; Zhang, S.; Gégout, A.; Urbani, M.; Armaroli, N.; Marconi, G.; Rio, Y. *J. Org. Chem.* **2005**, *70*, 7550–7557. (b) Nierengarten, J.-F.; Gu, T.; Hadziioannou, G.; Tsamouras, D.; Kransnikov, V. *Helv. Chim. Acta* **2004**, *87*, 2948–2966. (c) Kaafarani, B. R.; Pinkerton, A. A.; Neckers, D. C. *Tetrahedron Lett.* **2001**, *42*, 8137–8139
26. Krämer, C. S.; Müller, T. J. J. *Eur. J. Org. Chem.* **2003**, 3534–3548.

2.5 Henry Reaction

Alan X. Wang

2.5.1 Description

The Henry reaction, or nitroaldo reaction, is one of the classic carbon-chain formation methods utilized in organic synthesis.[1] It involves the condensation of nitroalkanes with aldehydes or ketones in the presence of bases (often catalytic amount) to afford the mixtures of diastereomeric 2-nitroalcohols, which in turn can be converted into other useful synthetic intermediates, such as 2-aminoalcohols, α-hydroxyketones, homologous ketones, and perhaps most importantly, nitroalkenes[2] through various functional transformations.

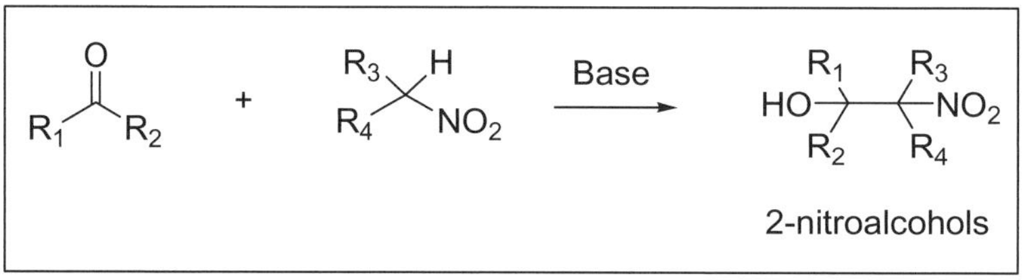

The vast majority of the Henry reactions involve the condensation of aliphatic and aromatic aldehydes with nitroalkanes. Cautions need to be exercised since the condensation product 2-nitroalcohols can undergo dehydration to form α-nitroalkenes, which may polymerize under strong acidic or basic conditions. Aldo condensation and Cannizzaro reaction may occur also under strong basic conditions and need to be minimized. Nitroaldo condensation with aromatic aldehydes is especially prone to dehydration. Except for those cases using nitromethane and occasionally primary nitroalkanes, nitroaldo reaction of ketones[3] is sluggish in general and the yield is often less satisfactory[4] primarily due to the unfavorable steric effect and prevalent retro-Henry reaction. Dehydration of the products of the notroaldo reactions with ketones is also more difficult to avoid. Nitromethane and primary nitroalkanes are more frequently used in the Henry reaction. Reactions with secondary nitroalkanes are generally less robust than with primary and nitromethane, while tertiary nitroalkanes do not undergo the nitroaldo reaction at all due to the lack of α-proton. Since the nitro group has a strong inductive effect stabilizing its adjacent carbanion, a wide range of bases can be chosen from to deprotonate a nitroalkane. Alkali[5] and alkaline earth metal[6] alkoxides, hydroxides in alcoholic or aqueous solvent seem to be popular choices for substrates with limited functionality. Aluminum and titanium alkoxides[7] also serve the nitroaldo reaction well. Organic bases, such as primary, secondary, and tertiary amines have also been extensively used in the literatures.[8] Anion-exchange resins and fluoride are also useful catalysts for the Henry reaction. In some cases, higher yield and greater selectivity can be achieved by employing mild and solvent free media such as alumina and alumina-supported fluoride for substrates bearing sensitive functional groups.[9]

2.5.2 Historical Perspective

In the late 19th century, V. Meyer[10] reported the first synthesis of mononitroalkane, nitropentane, by reacting amyl iodide with silver nitrite. Soon after, the synthesis of nitromethane was reported by H. Kolbe.[11] Their seminal works had aroused great deal of interests in nitroalkanes.[12] Merely two decades later, Louis Henry made the discovery that nitoalkanes can be added to aldehydes to form vicinal nitroalcohols.[13] This transformation, later known in the literature as the Henry, or nitroaldo reaction, provides a general method of construction of carbon-carbon bond and spontaneous introduction of vicinal functional groups. Therefore, it has been widely applied in organic synthesis[14] and natural product synthesis. However, the niroaldo reaction suffers from an inherent drawback, that is, lack of stereo-selectivity primarily due to the facts that the nitroaldo process is reversible (retro-Henry) and the stereogenic center of reactive intermediate nitronates is epimerized during the protonation. In the late 1970's, using of silyl nitronates,[15] lithionitrates, and doubly lithiated nitronates[16] was reported to achieve much improved synthetic yields and stereo-selectivities. Since then, the Henry reaction had enjoyed resurgence in the synthetic field. Ever increasing demand of optically pure materials in the late 20th century has challenged chemists in this field to develop novel reagents to promote asymmetric nitroaldo condensation. In 1992, M. Shibasaki[17] reported the first example of catalytic asymmetric Henry reaction using the combination of (*S*)-(–)-binaphthol and lanthanum *tert*-butoxide as the catalyst. An ever growing list of robust chiral catalytic systems have since been developed and found their applications in asymmetric synthesis of pharmaceutical intermediates and natural products.[18–22]

2.5.3 Mechanism

It is generally accepted that the condensation of nitroalkanes **1** with aldehydes proceeds with the nitronates **2** as the intermediates. The role of the base is to shift the tautomeric equilibrium towards the formation of nitronic acids, or *aci*-nitroalkanes **3**. Since **3** are much stronger acids (with a p*K*a range of 2~6) than **1** (with a p*K*a range of 9~10)[1], they are more readily deprotonated with the base. After the deprotonation, the formed nucleophilic nitronates **2** add to the carbonyl group of aldehydes or ketones (though less commonly seen) to form nitroalkoxy anions **4**. Subsequent protonation of **4** furnishes the final condensation products 2-nitroalcohols **5** and regenerates the base in the process to complete the catalytic cycle.

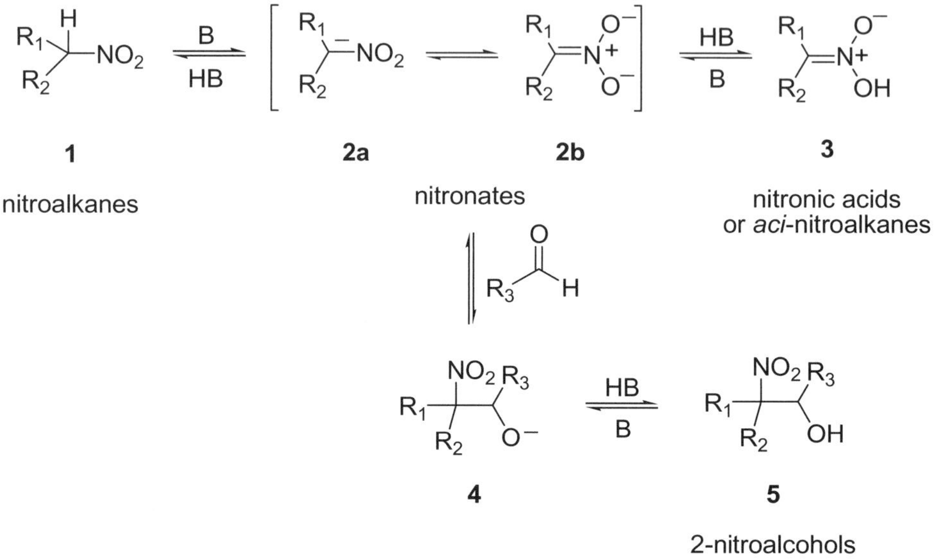

2.5.4 *Synthetic Utility*

There are very few examples of nitroalkanes found in nature products and biologically active pharmaceutical products. However, thanks to readily available transformations of the nitro group into other functional groups, nitroalkane has found great utility in organic synthesis. The synthetic application of the Henry reaction can be illustrated in the following scheme. Dehydration of 2-nitroalcohols **6** leads to the formation of very versatile synthetic intermediates, conjugated nitroalkenes **7**. Reduction of the nitro group of **6** affords vicinal aminoalcohols **10**, which play important roles in amino sugar, antibiotics, and active pharmaceutical intermediates synthesis. Oxidation of **6** yields another useful intermediates α-nitroketones **11**, while conversion of the nitro group in **6** to carbonyl provides α-hydroxyketones **12**. Direct denitration of **6** furnishes alcohols **13**. When R_2 = H, **6** can be directly converted into α-hydroxycarboxylic acids **15**.

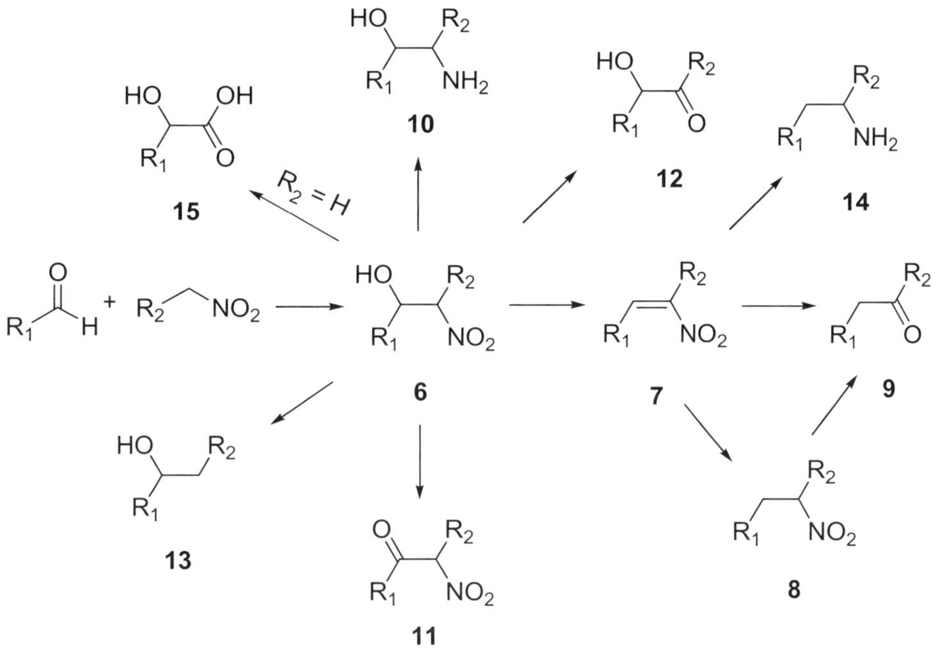

Nitroalkenes

Conjugated nitroalkenes **7** are probably the most useful intermediates derived from the nitroaldo reaction products **6**. Wide range of mild reagents including dicyclohexylcarbodiimide (DCC), pivaloyl chloride, methanesulfonyl chloride, and phthalic anhydride can be used for the dehydration of **6**. For instance, it is reported that DCC in the presence of catalytic copper(I) chloride in diethyl ether converted **6** into **7** in good to excellent yields.[23] A list of nucleophiles including silyl enol ethers, carbonyl enolates, sulfur-, oxygen-, and nitrogen-centered nucleophiles readily add to Michael receptors **7**.[2,24,25] Nitroalkenes **7** are also widely used as dienophiles in the Diels–Alder reactions.[26] Reduction of **7** under mild conditions such as sodium borohydride in THF[27] affords saturated nitroalkanes **8**, which can be further converted into ketones **9** (Nef reaction[28]). Direct transformation from **7** to **9** can be made with various reducing reagents such as Zn,[29] Raney nickel and sodium hypophosphite.[30] Besides lithium aluminum hydride, catalytic hydrogenation,[31] excess borane[32] can also be used to reduce **7** to amines **14**.

Synthetic utility of the nucleophilic addition of **7** can be exemplified by transformation using enol silanes as nucleophiles to generate 1,4-dicarbonyl compounds. Model substrate **16** reacts with 2-nitropropene in dichloromethane at –78 °C in the presence of $TiCl_4$ or $SnCl_4$ to give the addition product **17a** and/or **17b**, which is hydrolyzed during work up to furnish 1,4-dicarbonyl **18** in good yield.[24]

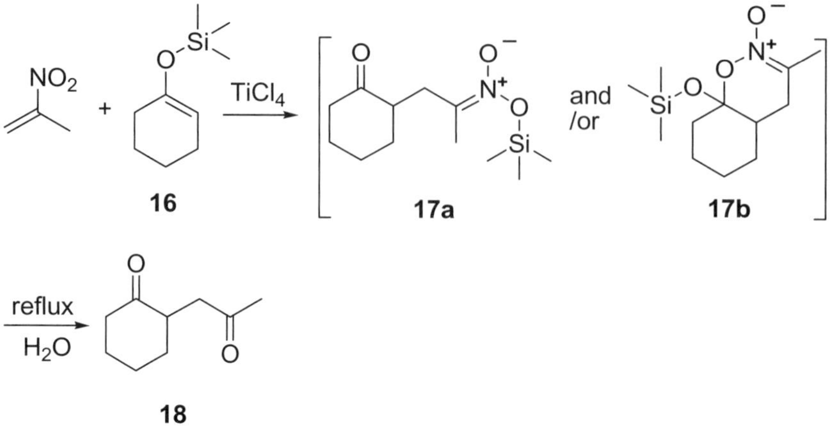

Functionalized conjugated nitroalkenes, such as 2-nitro-2-propen-1-ol pivalate (**NPP**), deserve special considerations. They can serve as extremely versatile multiple coupling reagents by sequential coupling with various nucleophiles and electrophiles due to the unique feature of three reactive sites in one molecule.[33]

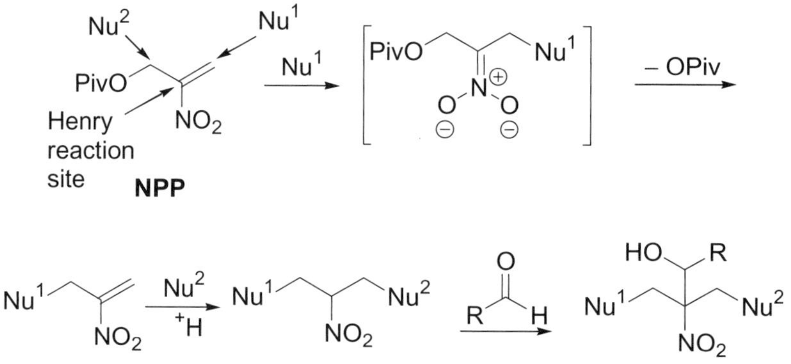

For instance, it is reported that a chiral copper catalyst (*R,R*)-**L** promotes asymmetric conjugate addition of dialkyl zinc to α,β-unsaturated ketone **19** to form homochiral zinc enolate **20**. This intermediate is then trapped *in situ* with **NPP** as electrophile, without the need of additional palladium catalysis. Good yield, high *trans/cis* (95/5) ratio, and excellent enantioselectivity (99%) are obtained.[34] Moreover, the multi-functionalized nature of **21** makes it a versatile intermediate for further elaboration.

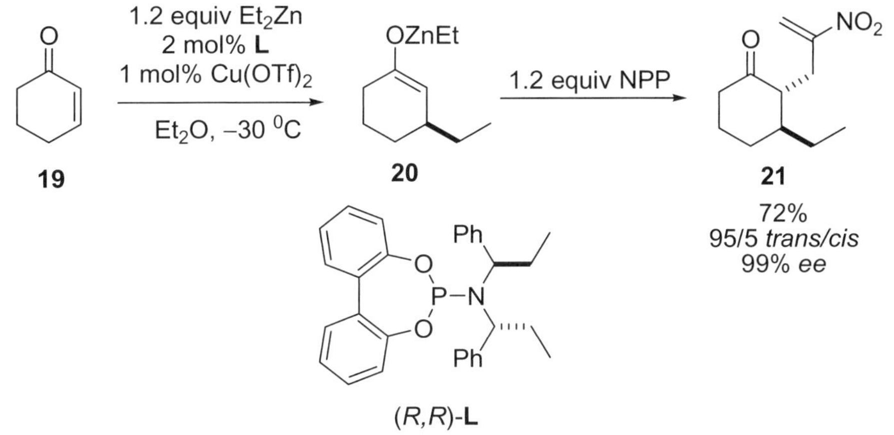

β-Amino alcohols

Hydrogenation[35] catalyzed by palladium, platinum, or Raney nickel of **6** retains the stereogenic center of vicinal carbon, thus provides a complimentary access to chiral vicinal amino alcohols to the conventional routes such as asymmetric epoxide ring opening with amines and their related precursors. Employing lanthanum-(*R*)-binaphthol complex to promote asymmetric nitroaldo reaction, followed by the reduction of nitro group, to form amino alcohols serves as the perfect example of this synthetic strategy.[36]

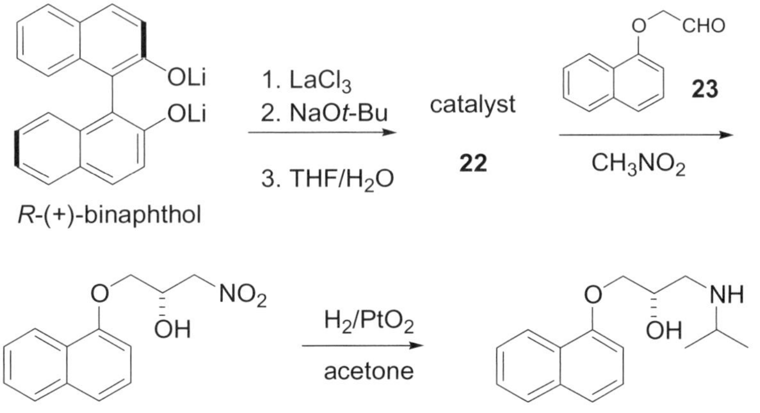

24
80% (92% *ee*)

25
S-(–)-propanolol

Reaction of *R*-(+)-binaphthol with lanthanum chloride heptahydrate and sodium *tert*-butoxide in THF and water forms the chiral catalyst system

22, which is used *in situ* to promote the condensation between aldehyde **23** and nitromethane in THF. 2-Nitroalcohol **24** is obtained in good yield with excellent *ee*. Platinum oxide catalyzed hydrogenation of **24** in the presence of acetone completes the reduction and alkylation sequence in one-pot to furnish the final product **25**.

A dinuclear zinc complex **26** of a novel chiral ligand is a very efficient chiral catalyst system[37] to promote asymmetric Henry reaction. It is employed as the key step in the synthesis of (–)-denopamine, a selective β_1-adrenoceptor agonist, clinically effective in treating congestive cardiomyopathy. Aldehyde **27** condenses with nitromethane in the presence of 10% catalyst **26** in THF to afford vincinal nitroalcohol **28** in 88% yield with 90% *ee*. Hydrogenation of **28** catalyzed by palladium on carbon in ethanol furnishes aminoalcohol **29**, which is then converted into the final product **30**.

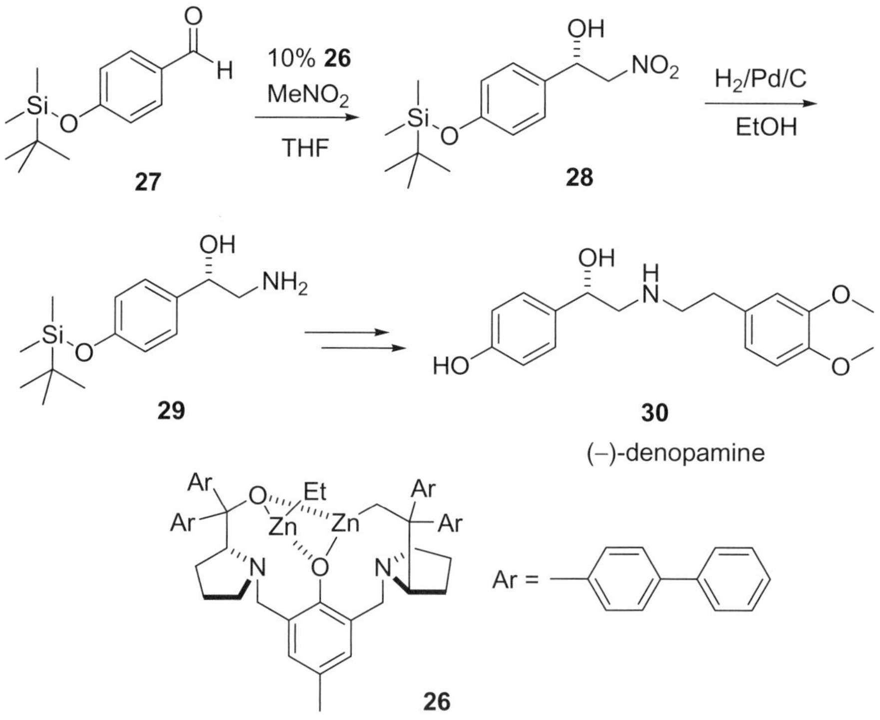

The key step of making Amprenavir **35**, an HIV protease inhibitor successfully marketed by Vertex, involves asymmetric Henry reaction catalyzed by a chiral quaternary salt **31**.[38] 2-Nitroalcohol **33** is made with great diastereoselectivity (17:1), while the same reaction promoted by achiral potassium fluoride merely affords 4:1 ratio. Reduction of **33** is accomplished by $NiCl_2/NaBH_4$ to give 2-aminoalcohol **34** as the key precursor to the synthesis of **35**.

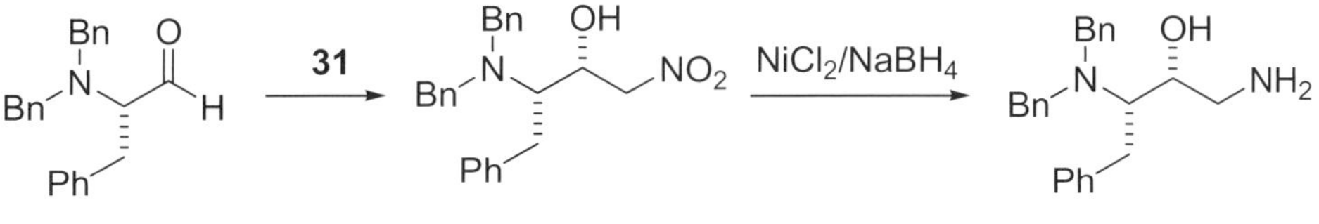

17:1 *de*

32 **33** **34**

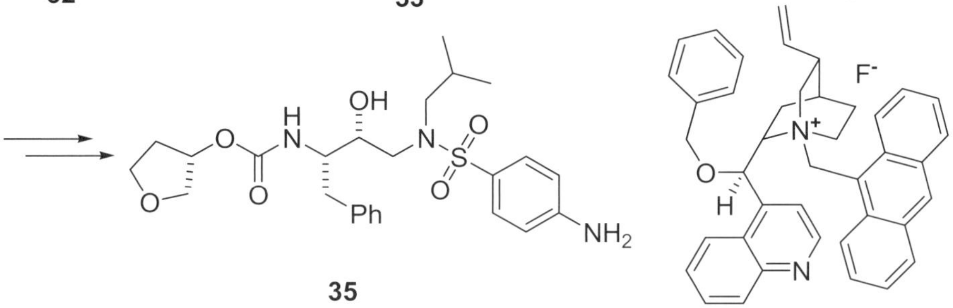

35 **31**

α-Nitroketones

Mild oxidizing reagents such as pyridium chlorochromate[39] readily convert **6** to **11**. One-pot method involving solvent-free nitroaldo reaction on alumina followed by *in situ* oxidation[40] also provides **11** in satisfactory yield. Denitration under free radical conditions with tin hydride/AIBN[41] readily reduces the secondary and tertiary nitro group in α-nitroketones **11** to yield ketones. Michael addition of **11** to α,β-unsaturated carbonyl derivatives can be performed under both basic conditions[42] and on slightly acidic solid catalyst such as silica gel.[43] Synthetic application of these versatile α-nitroketones **11** can be found in a well-written review article published recently.[44] Nevertheless, synthesis of 1,5-dicarbonyl compounds **37** exemplifies the strategy utilizing both reduction and the Michael addition reaction of α-nitroketones.**11**.

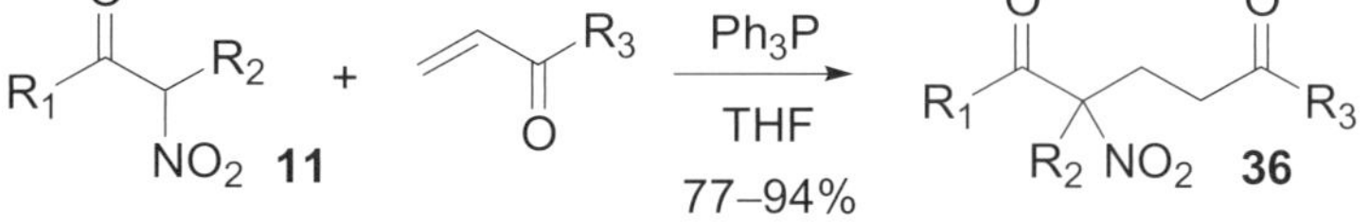

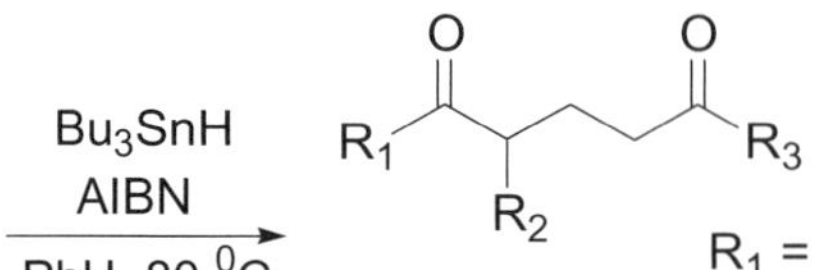

37 71–90%

R_1 = Ph, n-C_7H_{15}, n-Pr, n-C_4H_9
R_2 = H, Me, Et
R_3 = H, Me

The addition of **11** to methyl vinyl ketone or acrylaldehyde proceeds in the presence of triphenylphosphine with good to excellent yields in THF at room temperature. The subsequent denitration of **36** is accomplished with tin hydride and free radical initiator AIBN in refluxing bezene with good yields to form 1,5-dicarbonyl compounds **37**.[45]

α-Hydroxyketones

Another important synthetic application of **6** involves elaborating the nitro group into the carbonyl group[28] to furnish useful building blocks, α-hydroxycarbonyl compounds **12**. Sequential treatment of 2-nitroalcohol **38** with base and acid provides aldehyde **39** in good yield.[46] A slightly modified method converts 2-nitroalcohol **40** into α-hydroxyketone **41** in modest yield.[47]

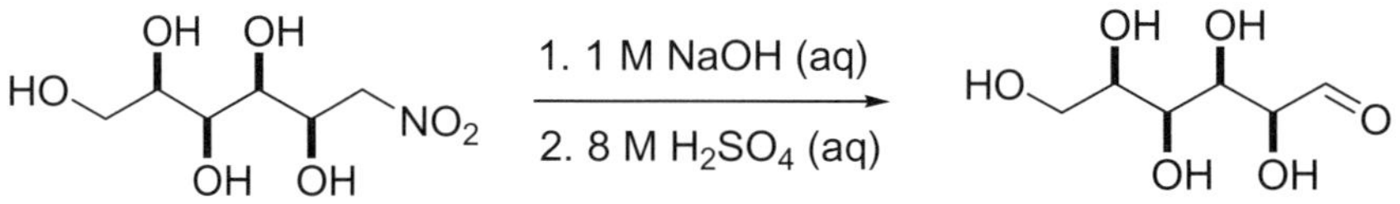

38 **39** 68%

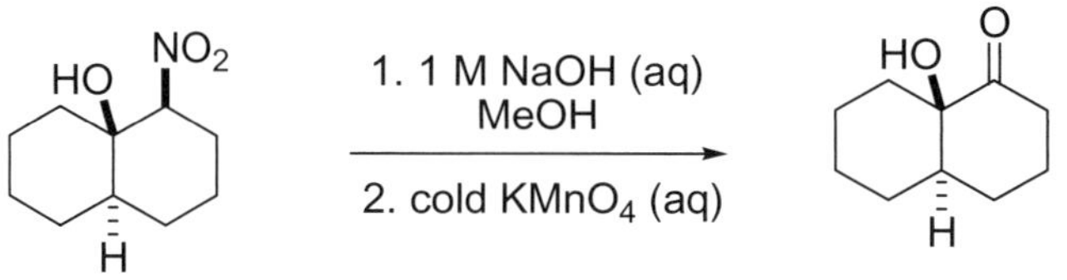

40 **41** 45%

Alcohols

The free radical condition[41] with n-Bu_3SnH/AIBN in benzene has been employed to replace the nitro group with the hydrogen atom in 2-nitroalcohol **42** with excellent diastereoselectivity.[48] On the other hand, secondary benzylic nitro group of 2-nitroalcohol **44** can be reduced to afford homobenzylic alcohol **45** by hydrogenation catalyzed by Pearlman's catalyst

$(Pd(OH)_2/C)$ in ethanol at elevated temperature with good yield and the retention of the adjacent stereogenic center.[49]

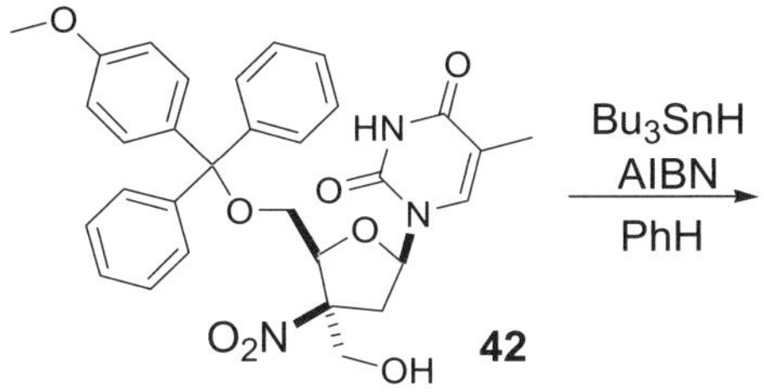

42

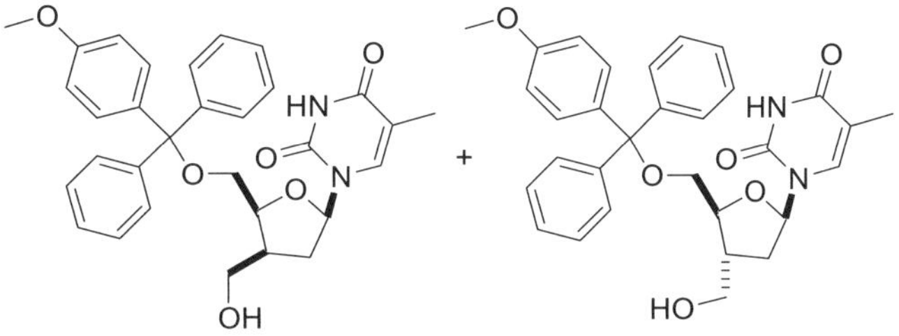

43a
82%

43b
9%

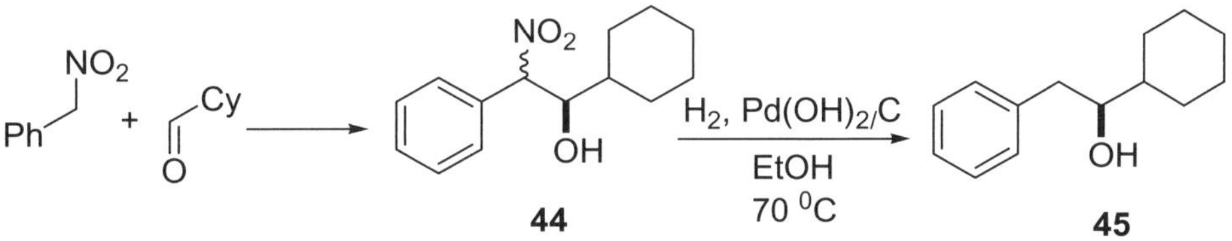

44

45
83%, 94% *ee*

α-Hydroxycaboxylic acids

Direct oxidation of primary nitroalkanes to carboxylic acids is rarely useful since strong reaction conditions are needed and often result in low yields. Lately, a mild and efficient condition using sodium nitrite and acetic acid in DMSO is developed to directly transform the nitro group to carboxylic acid **46**.[50] Under similar conditions, 2-nitroalcohol **47** is transformed into carboxylic acid methyl ester **48** in good yield[21] without epimerization of the adjacent stereogenic center.

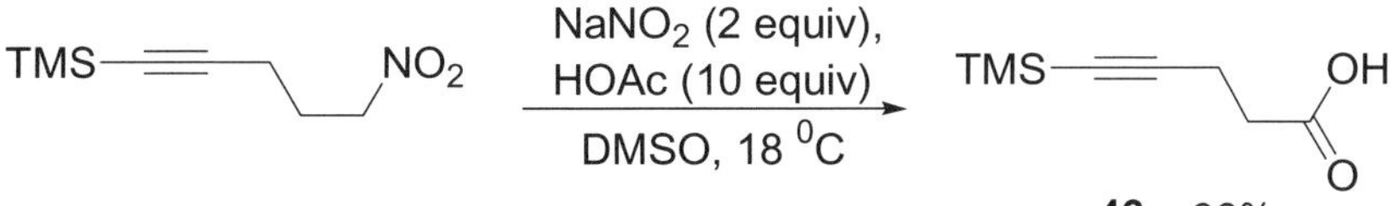

46 98%

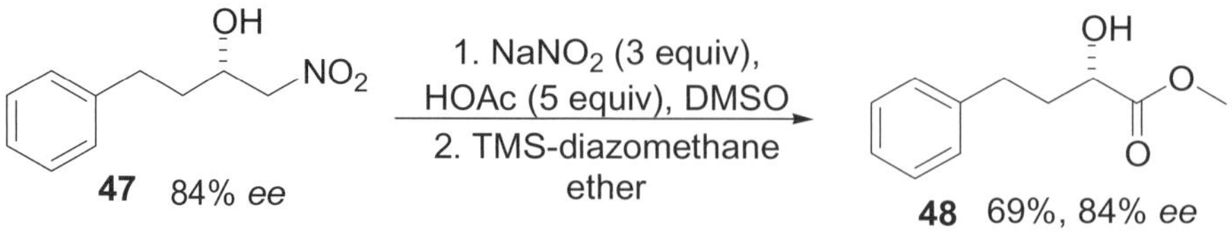

2.5.5 *Variation and Improvements*

Various modifications have been made to the original Henry reaction to either expand its synthetic application (aza-Henry reaction[51]), or to improve its yield and selectivity (silylnitronates and lithionitronates. condensations[54,55]).

Aza-Henry Reaction

It is reported that catalytic amount of Lewis acid promotes nitroaldo condensation of imine **49** with nitronates **50** to form 2-nitroamines **51**, which can be reduced to diamines **52** with 15:1 *syn/anti* ratio after removal of the protection group.[52] This seminal work just sets up the stage for the development of catalytic asymmetric aza-Henry reaction. For instance, copper based chiral Lewis acid with *C2*-symmetry **53** is used to promote addition of nitropropane to imine **54** to form intermediate 2-nitroamine **55** with high diastereo- and enantio-selectivities, which can be further elaborated to afford chiral diamine **56**.[53]

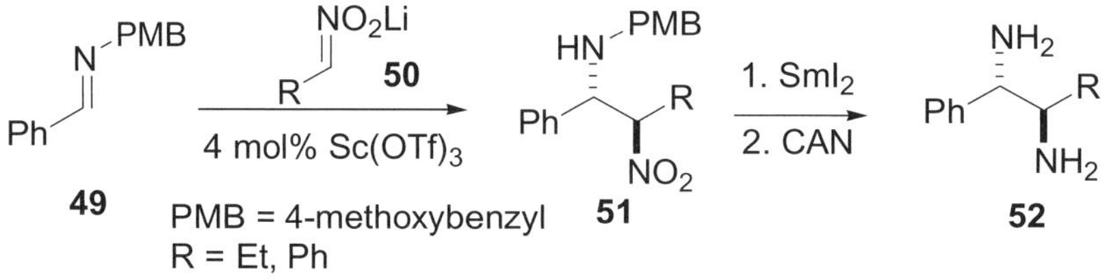

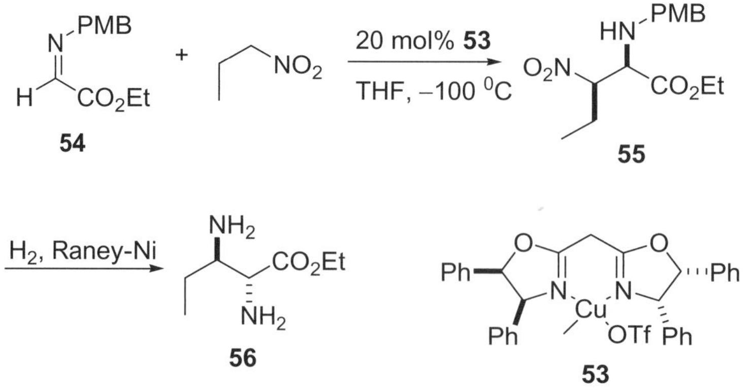

Nitronates condensation

Sterochemical control in the Henry reaction had been essentially neglected until lithionitronates, doubly-lithiated nitronates and silylnitronates methodologies were explored. Treatment of nitropropane with *n*-BuLi furnishes nitronate **57**, which then reacts with 4-nitrobenzaldehyde in the presence of Ti(*i*-OPr)Cl_3 in THF to provide nitroalcohol **55** in high *erythro* (*anti*) selectivity[7]. Silyl nitronates in the presence of fluorides also prefers *erythro (anti)* products[54] while dilithiated nitronates reacts with aldehydes to afford 2-nitroalcohols in high *threo* (*syn*) selectivity.[55]

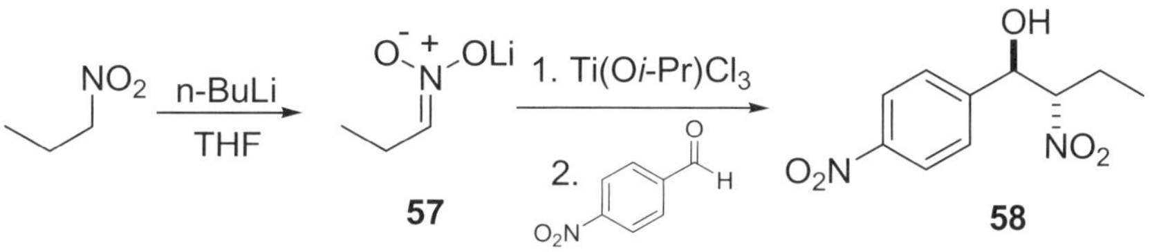

Intramolecular Henry Reaction

Intramolecular Henry reaction is employed as the key step to construct the formal synthesis of indolizidine 167B, which was first isolated from neo-tropical frog genera *Dendrobates*. The Michael addition of pyrrolidinone **59** to nitroalkene **60** is promoted by KO*t*-Bu with good yield and high diastereoselectivity. When the formed product nitroaldehyde **61** is treated with DBU in acetonitrile, intramolecular nitroaldo condensation takes place to afford β-nitroalcohol **62** with good yield and high diastereoselectivity. Reduction with Bu_3SnH/AIBN furnishes denitro compound **63** as the precursor to indolizidine 167B.[56]

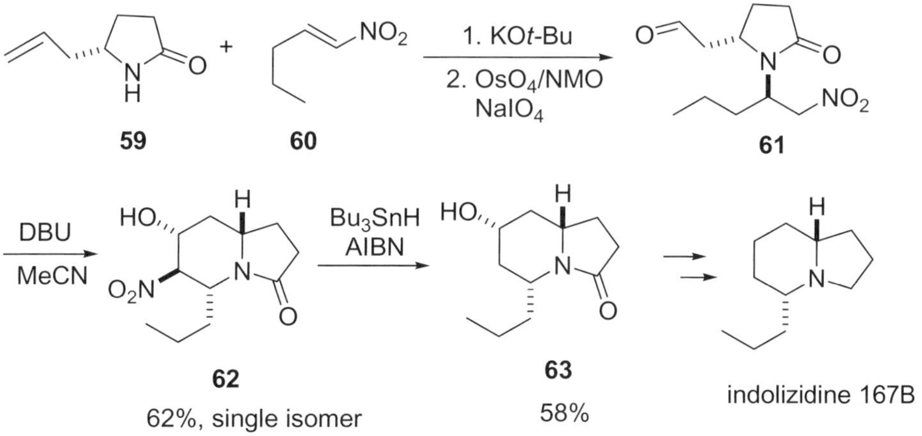

Retro-Henry Reaction

Anhydrous copper (II) sulfate adsorbed on silica gel in refluxing benzene is reported to effect the C–C bond cleavage between the hydroxy group and the nitro group of 2-nitroalcohol **64** to from nitroketone **65** in good yield as the key step in the synthesis of pharacantholide I.[57]

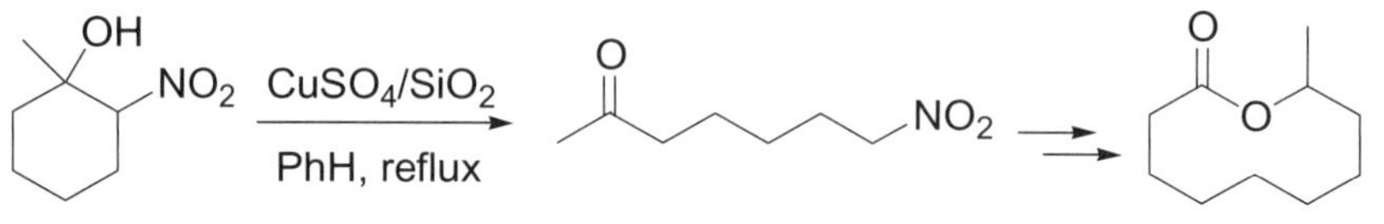

64 **65** 67% (±)-phoracantholide I

2.5.6 *Experimental*

A typical Henry reaction procedure using K_3PO_4 as the base in CH_3CN[58]

To a stirred solution of nitroethane (0.83 g, 11 mmol) in acetonitrile (15 mL) was added anhydrous potassium phosphate (0.10 g, 0.46 mmol) followed by mnitrobenzaldehyde (1.51 g, 10 mmol) and stirring continued. On completion of reaction (TLC monitoring), water (30 mL) was added to it and the reaction mixture was extracted with diethyl ether (50 mL). The ether extract was washed with water (3 × 20 mL) and dried over anhydrous Na_2SO_4. Removal of the solvent gave a residue which was filtered through a short column of silica gel to afford 1-(3-nitrophenyl)-2-nitropropan-1-ol (**2**) (2.2 g, 93%).

A typical asymmetric Henry reaction procedure using a chiral copper catalyst[20]

A mixture of $Cu(OAc)_2{\bullet}H_2O$ (100 mg, 0.50 mmol) and the ligand **2** (197 mg, 0.55 mmol) in EtOH (23 mL) was stirred at room temperature for 1 h, giving a deep blue colored solution. Nitromethane (27.1 mL, 500 mmol) was added, followed by *o*-anisaldehyde (6.95 g, 50 mmol) in one portion. The resulting green solution was stirred at room temperature for 56 h, after which it was diluted with 20% EtOAc/hexanes (50 mL). The mixture was filtered through a short SiO_2 plug (3.5 × 4.0 cm), eluting with 20% EtOAc/hexanes (200 mL). The filtrate was concentrated *in vacuo* and the residue was purified by column chromatography (20% EtOAc/hexanes) to give the title compound **3c** (*R*)-1-(2-methoxyphenyl)-2-nitroethanol (9.05 g, 92%) as a yellow oil in 94% *ee*.

A typical Henry reaction procedure involving lithionitrate and aldehydes and ketones[55]

To a cooled (–90 °C) stirred solution of 50 mL of THF, 10 mL of HMPA or 45 mL of THF and 15 ml of DMPU and 10 mmol THP-protected nitroalcohol was added 15.2 mL (22 mmol) of *n*-butyllithium (1.45 M in hexane). The resulting yellow mixture was allowed to warm to –40 °C during 3 h, and at –90 °C the electrophile (10 mmol) was slowly added. After the reaction mixture had warmed to –60 °C within 90 min (aldehydes, ketones) or to –40 °C within 2 h (esters, alkylhalides), the mixture was cooled again to –90 °C and quenched with 3 mL (50 mmol) of acetic acid. The clear cold reaction solution was combined with 100 mL of ether and washed successively with two 25 mL portions of cold saturated aqueous $NaHCO_3$. Each aqueous phase was extracted with one 100 mL portion of ether. The combined organic phases were washed 4 times with water and once with saturated aqueous NaCl, dried with anhydrous $MgSO_4$, filtered, and concentrated. Chromatography of the resulting residue on silica gel with ether/hexane afforded the pure nitro compound. The deprotection of the THP group was carried out with 20 mL of MeOH and 0.3 g of Amberlyst at 45 °C for 2 h, except in the case of nitro ketones and nitro esters which were deprotected at room temperature. Distillation or recrystallization provided analytically pure samples.

A typical Henry reaction procedure involving silylnitronate and aldehyde[59]

Preparation of (*t*-butyl)dimethylsilyl ester of *aci*-nitro(*p*-methoxyphenyl)methane.

To a solution of 5.4 mmol lithium diisopropylamide in 15 mL THF stirred at –78 °C under argon were added 0.84 g (5 mmol) *p*-methoxyphenylnitromethane, and after 30 min, 0.90 g (6.0 mmol) (*t*-butyl)dimethylsilyl chloride dissolved in 2 mL THF. The temperature was allowed to rise to 20 °C overnight. The solvent removed *in vacuo*, and the residue triturated with 25 ml pentane, filtered through celite and concentrated. The 0.61 g (43.6%) of crude silyl nitronate thus obtained was recrystallized from pentane to furnish crystals suitable for X-ray structure analysis.

Silyl nitroaldo addition.

General procedure.

To a solution of 160 mg (0.5 mmol, pre-dried by heating for 4 h at 90 °C/0.1 Torr) of tetra(*n*-butyl)ammonium fluoride in 15 ml of THF at –78 °C was added the aldehyde (10 mmol) followed by the silyl nitronate (11 mmol). The resulting solution was stirred at –78 °C for 3 h, then allowed to warm up to RT overnight. It was then poured into 200 ml of hexane, and washed with 3 × 20 ml of water, and dried with $MgSO_4$. Evaporation of solvent and bulb-to-bulb distillation gave the desired protected or free vicinal nitroalcohol.

*6-Nitro-7-trimethylsilyloxytridecane (**1a**).*

Reaction of heptanal with the trimethylsilyl ester of 1-aci-nitrohexane gave 1a as a colorless oil in 80% yield.

2.5.7 References

1. [R] Rosini, G. In *Comprehensive organic Synthesis*, B. Trost, I. Flemming, Eds. Pergamon, **1991**, *2*, 321–340.
2. [R] Barret, A. G.; Graboski, G. G. *Chem. Rev.*, **1986**, *86*, 751–762.
3. Matsumoto, K. *Angew. Chem., Int. Ed.* **1984**, *23* (8), 617–618.
4. Nightingale, D. V.F.; Erickson, B.; Knight, N. C. *J. Org. Chem.* **1950**, *15*, 782–784.
5. [R] Fieser, L. F.; Fieser, M.; In *Reagents for Organic Synthesis*, Wiley, **1967**, *1*, 739–741.
6. Vanderbilt, B.; Hass, H. *Ind. Eng. Chem.* **1940**, *32*, 34–38.
7. Barrett, A.; Robyr, C.; Spilling, C. *J. Org. Chem.* **1989**, *54*, 1233–1234.
8. Forsyth, A.; Paton, R.; Watt, I. *Tetrahedron Lett.* **1989**, *30*, 993–996.
9. Hanessian, S.; Kloss, J. *Tetrahedron Lett.* **1985**, *26*, 1261–1264.
10. Meyer, V.; Stüber, O. *Chem. Ber.* **1872**, *5*, 399, 514.
11. Kolbe, H. *J. Prakt. Chem.* **1872**, *5*, 427–432.
12. Reynolds, H.; Adkins, H. *J. Am. Chem. Soc.* **1939**, *51*, 279–287.
13. Henry, L. *Compt. Rend.* **1895**, *120*, 1265–1268.
14. [R] Hass, H.; Riley, E. *Chem. Rev.* **1943**, *32*, 373–430.
15. Colvin, E. W.; Seebach, D. *J. Chem. Soc., Chem. Commun.* **1978**, 689–691.
16. Seebach, D.; Lehr, F. *Angew. Chem.* **1976**, 88, 540–541.
17. Sasai, H.; Suzuki, T.; Arai, S.; M. Shibasaki, M. *J. Am. Chem. Soc.* **1992**, *114*, 4418–4420.
18. [R] Luzzio, F. A. *Tetrahedron* **2001**, *57*, 915–945
19. [R] Palomo, C.; Oiarbide, M.; Laso, A. *Eur. J. Org. Chem.* **2007**, 2561–2574.
20. Shotome, Y.; Hashimoto, Y.; Nagasawa, K. *Eur. J. Org. Chem.* **2006**, 2894–2897.
21. Trost, B. M.; Yeh, V. S. C. *Angew. Chem., Int. Ed.* **2002**, *41*, 861–863.
22. Evans, D.; Seidel, D.; Rueping, M.; Lam, W.; Shaw, J.; Downey, W. *J. Am. Chem. Soc.* **2003**, *125*, 12692–1263.
23. Knochel, P.; Seebach, D. *Synthesis* **1982**, 1017–1018.
24. [R] Yoshikoshi, A.; M. Miyashita, M. *Acc. Chem. Res.* **1985**, *18*, 284–290.
25. Berner, O. M.; Tedeschi, L.; D. Enders, D. *Eur. J. Org Chem.* **2002**, *12*, 1877–1894.
26. [R] Denmark, S. E.; and A. Thorarensen, A. *Chem. Rev.* **1996**, *96*, 137–165.
27. Hallinan, E. A.; Dorn, C. R.; Moore, W. M.; Jerome, G. M.; Manning, P. T.; Pitzele, B. S. *Tetrahedron Lett.* **2003**, *44*, 7345–7347.
28. [R] Ballini, R.; Petrini, M. *Tetrahedron* **2004**, *60*, 1017–1047.
29. Ghosh, A. K.; Lei, H. *J. Org. Chem.* **2002**, *67*, 8783–8788.
30. Monti, D.; Gramatica, P.; Speranza, G.; Manitto, P. *Tetrahedron Lett.* **1983**, *24*, 417–418.
31. Cainelli, G.; Galletti, P.; Giacomini, D.; Licciulli, S.; Quintavalla, A. *Eur. J. Org. Chem.* **2007**, *15*, 2526–2533.
32. Kabalka, G. W.; Guindi, L. H. M.; Varma, R. S. *Tetrahedron* **1990**, *46*, 7443–7457.
33. Seebach, D.; Konchel, P. *Helv. Chim. Acta* **1984**, 67, 261–283.
34. Rathgeb, X.; March, S.; Alexakis, A. *J. Org. Chem.* **2006**, *71*, 5737–5742.
35. Sasai, H.; Tokunaga, T.; Watanabe, S.; Suzuki, T.; Itoh, N.; Shibasaki, M *J. Org. Chem.* **1995**, *60*, 7388–7389.
36. Sasai, H.; Itoh, N.; Suzuki, T.; Shibasaki, M. *Tetrahedron Lett.* **1994**, *35*, 6123–6126.

37. Trost, B. M.; Yeh, V. S. C.; Ito, H.; Bremeyer, N. *Org. Lett.* **2002**, *4*, 2621–2623.
38. Corey, E. J.; Zhang, F. *Angew. Chem., Int. Ed.* **1999**, *38*, 1931–1934.
39. Dauzonne, D.; Grandjean, C. *Synthesis* **1992**, *7*, 677–680.
40. Rosini, G.; Ballini, R.; Sorrenti, P.; Petrini, M. *Synthesis* **1984**, 607–608.
41. Tormo, J.; Hays, D. S.; Fu, G. C. *J. Org. Chem.* **1998**, *63* 5296–5297.
42. [R] Rosini, G.; Ballini, R.; Petrini, M.; Marotta, E.; Rughi, P. *Org. Prep. Proc. Int.* **1990**, *22*, 707–746.
43. Ballini, R.; Fiorini, D.; Gil, M. V.; Palmieri, A. *Green Chem.* **2003**, *5*, 475–476.
44. [R] Ballini, R.; Bosica, G.; Fiorini, D.; Palmieri, A. *Tetrahedron* **2005**, *61*, 8971–8993.
45. Ono, N.; Miyake, H.; Kaji, A. *J. Chem. Soc., Chem. Comm.* **1983**, 875–876.
46. Dromowicz, M,; Koell, P. *Carbohydr. Res.***1998**, *308*, 169–171.
47. Chung, W. K.; Chiu, P. *Synlett* **2005**, 55–58.
48. Gard, N.; Plavec, J.; Chattopadhyaya, J. *Tetrahedron* **1993**, *49*, 5189–5202.
49. Fessard, T. C.; Motoyoshi, H.; Carreira, E. M. *Angew. Chem., Int. Ed.* **2007**, *46*, 2078–2081.
50. Matt, C.; Wagner, A.; Mioskowski, C. *J. Org. Chem.* **1997**, *62*, 234–235.
51. [R] Westermann, B. *Angew. Chem., Int. Ed.* **2003**, *42*, 151–153.
52. Adams, H.; Anderson, J. C.; Peace, S.; Pennell, A. M. K. *J. Org. Chem.* **1998**, *63*, 9932–9943.
53. Knudsen, K. R.; Risgaard, T.; Nishiwaki, N.; Gothelf, K. V.; Jørgenson, K. A. *J. Am. Chem. Soc.* **2001**, *123*, 5843–5844.
54. Marti, R. E.; Heinzer, J.; Seebach, D. *Liebigs Ann.* **1995**, 1193–1215.
55. Eyer, M.; Seebach, D. *J. Am. Chem. Soc.* **1985**, *107*, 3601–3606.
56. Kamimura, A.; Nagata, Y.; Kadowaki, A.; Uchidaa, K.; Uno, H. *Tetrahedron* **2007**, *63*, 11856–11861.
57. Saikia, A. K.; Hazarika, M. J.; Barua, N. C.; Bezbarua, M. S.; Sharma, R. P.; Ghosh, A. C. *Synthesis*, **1996**, 981–985.
58. Desai, U. V.; Pore, D. M.; Mane, R. B.; Solabannavar, S. B.; Wadgaonkar, P. P. *Synth. Commun.* **2004**, *34*, 19–24.
59. Colvin, E. W.; Beck, A. K.; Seebach, D. *Helv. Chim Acta* **1981**, *64*, 2264–2271.

2.6 Horner–Wadsworth–Emmons Reaction

Frank Rong

2.6.1 *Description*

The formation of alkene **2** from phosphine oxide carbanion and aldehydes is referred to as the Horner reaction. When using phosphonate carbanion, the reaction is known as the Horner–Wadsworth–Emmons (HWE) reaction.[1–5]

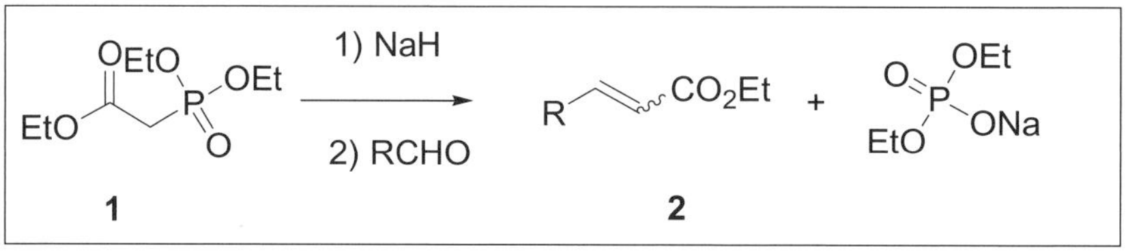

First, the diethyl phosphate **1** or phosphine oxide was treated with strong base to produce the phosphoryl-stabilized carbanions, which was further reacted with aldehyde to form the olefin **2**. The reaction often forms a mixture of (*Z*)- and (*E*)-olefins depending upon the substrates and the reaction condition. This reaction mainly used for the homologation of aldehydes to give (*E*)-α,β-unsaturated esters and related system. The phosphate by-product can be washed away with water. This gives more advantageous than the corresponding Wittig reaction.

Strong bicyclic guanidine base, such as 1,5,7-triazabicyclo[4.4.0]dec-5-ene (TBD) and its methyl derivatives MTBD, acts as promoters for HWE reactions.[6] The reaction promoted by TBD and MTBD worked very well with both aromatic and aliphatic aldehydes **3** and **5**, respectively. The reaction gave mainly or only the (*E*)-products, **4** and **6**, selectively. However, very poor (or none) yield was obtained when reacting with ketones in the presence of these base. The presence of LiCl did not affect positively the kind of reactions.

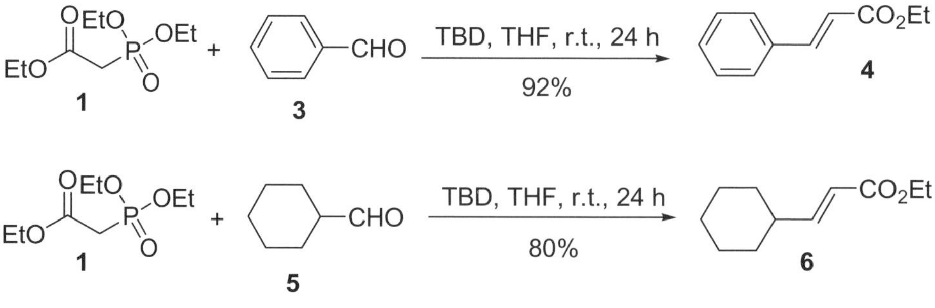

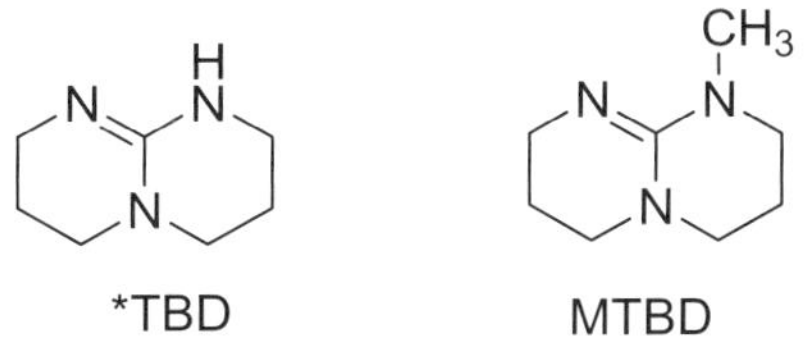

*TBD: 1,5,7-triazabicyclo[4.4.0]dec-5-ene

In recent years, a great attention has been devoted to variants of the HWE reaction which produce (*Z*)-α,β-unsaturated esters and related systems with high degree of stereocontrol.[7,8] Still–Gennari phosphonate reaction is one of the important variants of the Horner–Wadsworth–Emmons (HWE) reaction using bis(trifluoroethyl)phosphonate **7** to give (*Z*)-olefins selectively, **9**:**10** ≥ 50:1.[7] The trifluoroethylphosphonoester **7** was readily prepared from commercially available trimethylphosphonoacetate and trifluoroethanol. The reaction was conducted at –78 °C except for the case of using K_2CO_3 as base at 0 °C. Among the base examined, KHMDS/18-crown-6 was found to be particularly effective and allowed highly steroselective formation of (*Z*)-unsaturated esters **11** with aldehydes. The K_2CO_3 procedure is also effective in many instances and has the virtue of operating under mild, weakly basic conditions.

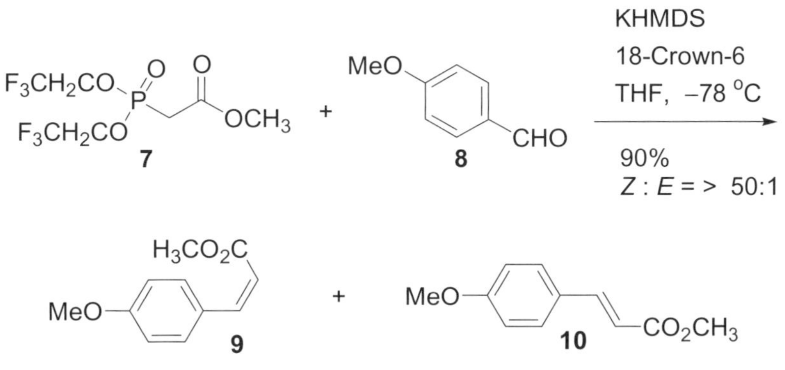

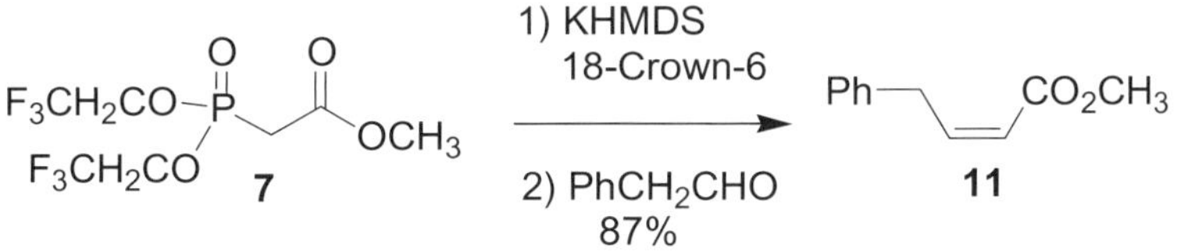

Ando reported a new HWE reaction procedure using ethyl (diarylphosphono)acetates **12**.[9,10] The reagents were prepared from triethyl phosphonoacetate, PCl_3, and the corresponding phenols. The reaction of **12** with several kinds of aldehydes in the presence of Triton B, NaH or KHMDS in THF solvent revealed that these reagents are useful for the synthesis of (*Z*)-unsaturated esters, **13**:**4** = 99:1. Additionally this procedure has lower overall cost.

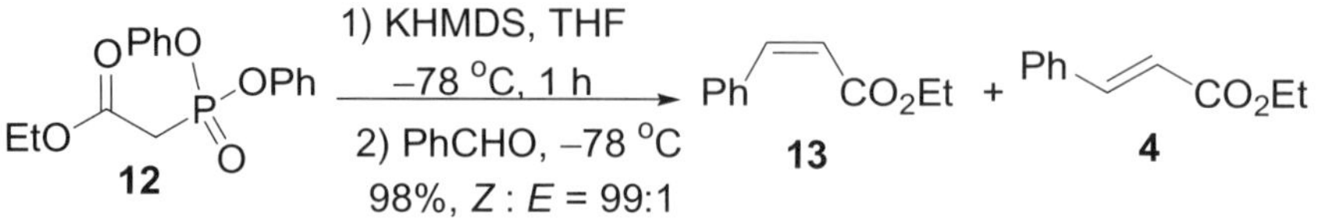

2.6.2 *Historical Perspective*[4]

Horner and his colleagues in 1958 were the first to react phosphoryl-stabilized carbanions with aldehydes and ketones to produce olefins.[11,12] The carbanions used were generated from either diethyl benzylphosphonate or diphenylphosphine oxides. The benzylic carbanions were found to react with benzophenone to give the triphenyl olefin in good yield. However, the special advantages of phosphonates or phosphine oxides were not utilized broadly in alkene synthesis for years. Since 1961 Wadsworth and Emmons published their results in *the Journal of American Chemical Society*, the utility of the phosphonates in the alkene preparation started becoming popular.[13,14] In the ensuing years, there has been confusion about whom to credit for this class of reaction, as the names "Horner", "Wadsworth", "Emmons", "Wadsworth–Emmons", and "Horner–Wittig" have appeared as descriptors with regularity. Horner was the first to use phosphine oxides;[11,12] however, since his group[12] only examined a single phosphonate reagent, Wadsworth and Emmons can also claim to develop the phosphonate modification of the Wittig reaction.[13] It will be fair to refer the phosphonate-mediated olefinations as the "Horner–Wadsworth–Emmons" (more concisely "HWE") reaction. And the phosphine oxide variant will be called the "Horner" reaction. Since Still–Gennari modified HWE olefination has been applied broadly and efficiently for last twenty years, it will be called as Still–Gennari modified HWE olefination reaction if the bis(trifluoroethyl)-phosphonate be used as variant.

2.6.3 *Mechanism*

Horner–Wadsworth–Emmons reaction

The mechanism for the HWE reaction is shown in above for an aldehyde condensation. The phosphoryl-stabilized carbanion **15** attacks the carbonyl in a stepwise manner, to give oxyanion intermediate **16**, which then decomposes via a transient four-centered intermediate, **17A** or **17B**, to yield olefin, **18** or **19**. The stereochemistry is determined by a combination of the steroselectivity in the initial carbon-carbon bond-formation step, and perhaps, reversibility of intermediates (**16** and **17**). Warren and colleagues have investigated in detail of the reaction of phosphine oxides.[15–17] Their results demonstrated that *erythro-* and *thero-***16** can be captured by protonation and isolated as stable β-hydroxy phosphine oxides, example of which have been independently and sterospecifically decomposed to the respective (*Z*)- and (*E*)-alkenes.

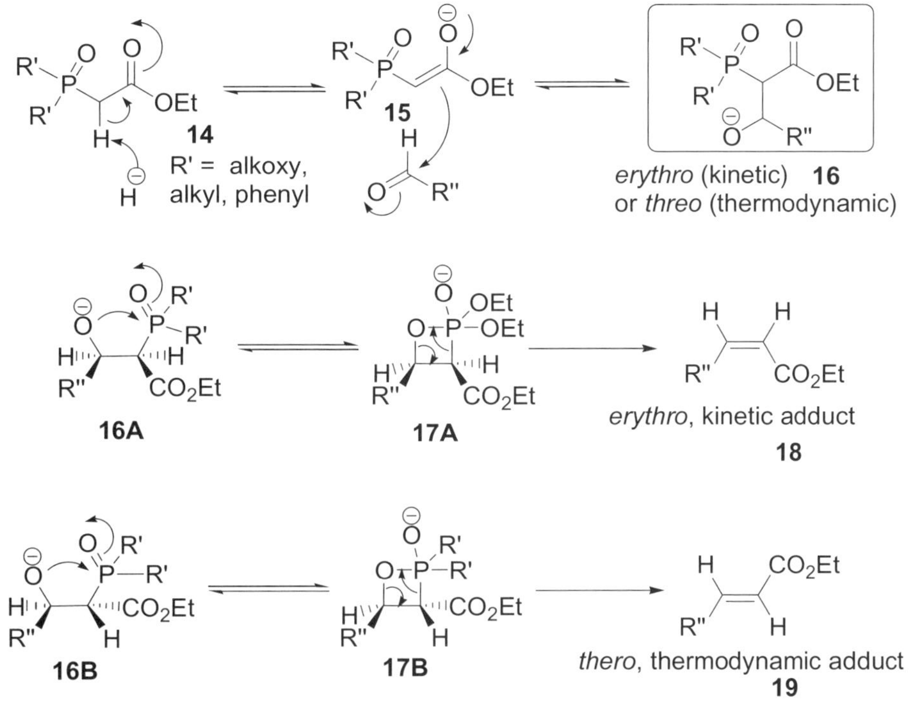

The HWE reaction is generally restricted to phosphonates bearing a *α*-substituent that can stabilize a carbanion (e.g., CO_2^-, CO_2Me, CN, aryl, vinyl, SO_2R, $P(O)(OR)_2$, SR, OR, and NR_2). The absence of such groups usually results in poor yields of alkene products. Larsen and Aksnes have done the reaction rate studies on the HWE reaction.[18,19] The reaction of several phosphonate reagents with sodium ethoxide and *para-* and *meta-*substituted banzaldehydes were studied by monitoring levels of aldehyde and alkene by UV spectroscopy. The reaction was found to be first order in aldehyde, ethoxide, and phosphonate, and third order overall, with the rate-limiting step being the initial condensation of phosphonate with aldehyde. Cyclic phosphonate **20** reacted about 20 times faster than acyclic counterpart **21**.

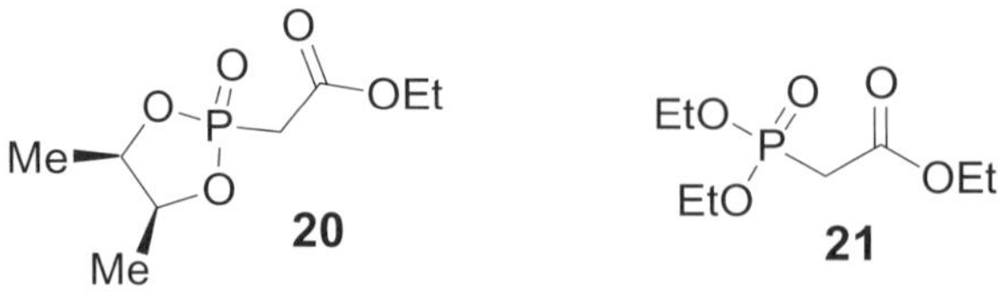

20 21

The enhancement was attributed to a more pronounced release of ring strain on convention from the tetrahedral to the pent coordinate state at phosphorus with canbanions derived from **20**.[18] An order of magnitude decrease in rate was seen with phosphinate Et(O)(Ph)P(O)CH_2CO_2Et, relative to $(Et)_2P(O)CH_2CO_2Et$, and phosphine oxide $Ph_2P(O)CH_2CO_2Et$ reacted generally 35 times slower than the phosphinate. This may be explained by the relative ease with which the reaction from the phosphinate can achieve a penta coordinated state relative to the phosphine oxide.

Still–Gennari HWE reaction[7,8]

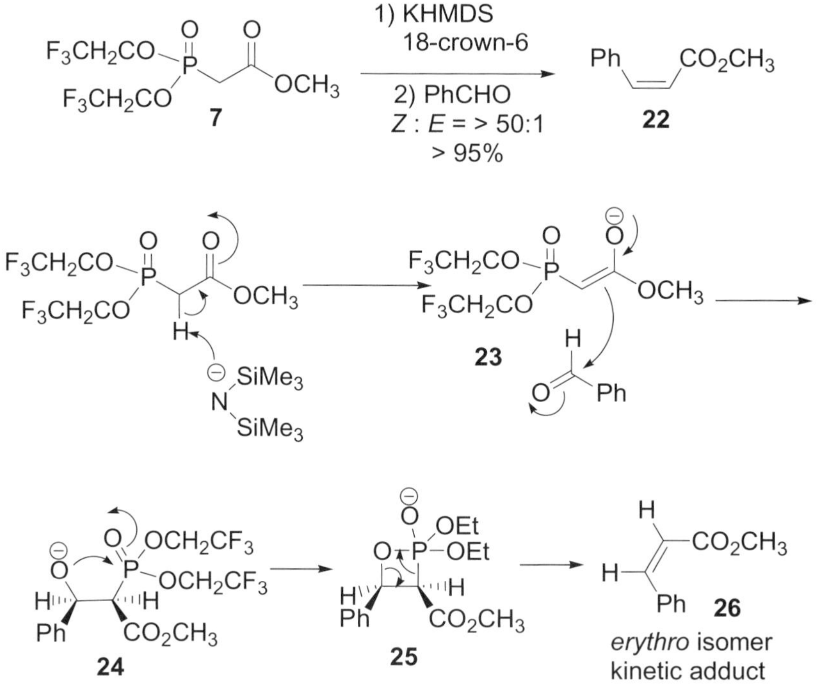

The mechanism for the Still–Gennari HWE reaction is shown in above for an aldehyde condensation. The phosphoryl-stabilized carbanion **23** attacks the carbonyl to give oxyanion intermediate **24**, which then decomposes via a transient four-centered intermediate **25** to yield olefin **26**. The stereochemistry is in favour of erythro isomer, a kinetic adduct.

2.6.4 Synthetic Utility

The HWE reactions were used broadly in alkene preparation for last thirty years. It also achieved in the *Solid-Phase Organic Synthesis* (SPOS). Polymer-bound HWE regents may offer advantages over the soluble regents in cases where the products are water soluble and difficult to separate from the anionic phosphors side product. Reiser and Jauch have applied HWE reaction to the Solid Phase Organic Synthesis of a key intermediate for combinatorial synthesis of mniopetals, kuehneromycins and marasmanes.[20] First, (*E*)-4-bromo-butenol was loaded onto Wang resin via in situ prepared Wang trichloro acetimidate in the presence of BF_3, diethyl ether complex. Then, the polymer bound bromide **27** was transformed into the polymer bound phosphonate **28** in a Michaelis–Arbuzov reaction. After deporotonation of the phosphonate, it was treacted immediately with the free aldehyde to give the triene **29**. This makes easier in the synthesis of mniopetals, inhibitors of HIV reverse transcriptase.

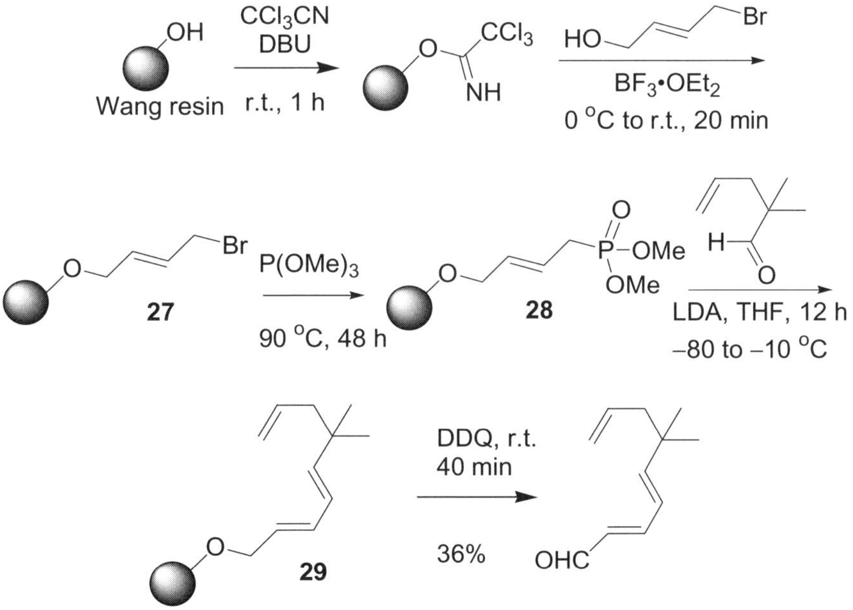

HWE olefination has applied in the synthesis of simple and functionalized α,β-unsaturated nitriles. Scettri and co-workers have developed a practical procedure for HWE olefination of aldehydes **3** and ketones **33** with α-cyano phosphonates **30** using lithium hydroxide as a mild base.[21] The (*E*)-steroselectivity is predominant and the olefins are obtained in good to high yields. Activated 4 Å molecular sieve have been shown to shorten reaction times with ketones, but they are not necessary to secure good yields of the final products, rendering the methodology even simpler. A complete control of the stereoselectivity has been achieved in the case of α-hydroxy ketone **33**, where the directing effect of the hydroxyl group is reasonably considered to be responsible of the observed exclusive formation of the (*E*)-isomer **34**.

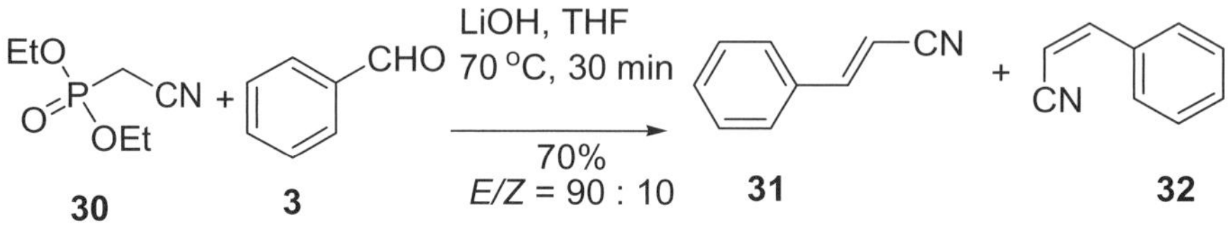

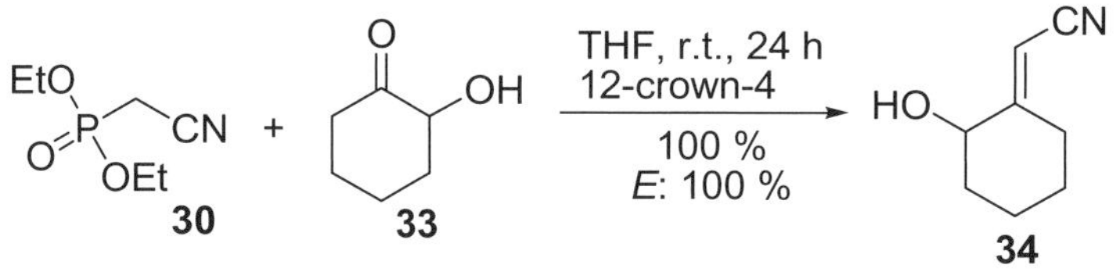

Kurihara reported imidazomethyl phosphonates **35** as another HWE-type reagents.[22] The **35** reacted readily with various aldehydes and ketons in the presence of *t*-BuOK to produce (*E*)-vinylimiazoles **36** in good to excellent yields. The reactions of **35** with all substrates, aliphatic aldehydes, aromatic and heteroaromatic aldehydes, provided substituted (*E*)-vinylimidazoles. The synthetic utility of the reagent **35** was demonstrated by the efficient preparation of few histamine H_3 ligands (**37**) by simple hydrogenation of intermediate **36**.

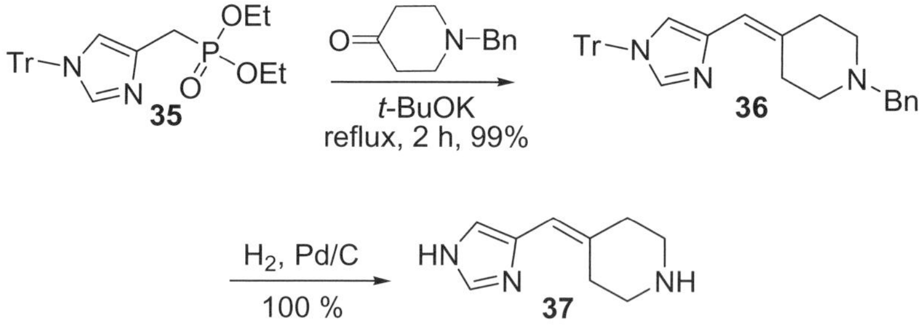

Inter- and intramolecular HWE reactions of 5-(diethoxyphosphoryl)-1-acyl-2-alkyl(aryl)-2,3-dihydro-4-pyridones showed very interesting results by Comins and Ollinger.[23] Treating phosphornate **38** with *t*-BuOK forms anion, followed by addition of the aldehyde, and raising the reaction temperature to 35 °C gave good yields of the desired enones **39**. An intramolecular HWE olefination was also carried out under similar conditions in below scheme. Dihydropyridone **40** was added to Grignard reagent **41** and CuBr•SMe_2 in THF to provide an 84% yield of the *trans*-piperidone **42**. Hydrolysis of the acetal **42** using aqueous oxalic acid in THF and water produced the crude aldehyde **43** in near quantitative yield. Intramolecular HWE reaction of **43** in the presence of Et_3N and LiCl in THF allowed a facile preparation of *trans*-bicyclic enone **44** in a highly stereocontrolled fashion. The stereochemistry of **44** was confirmed by single crystal X-ray analysis.

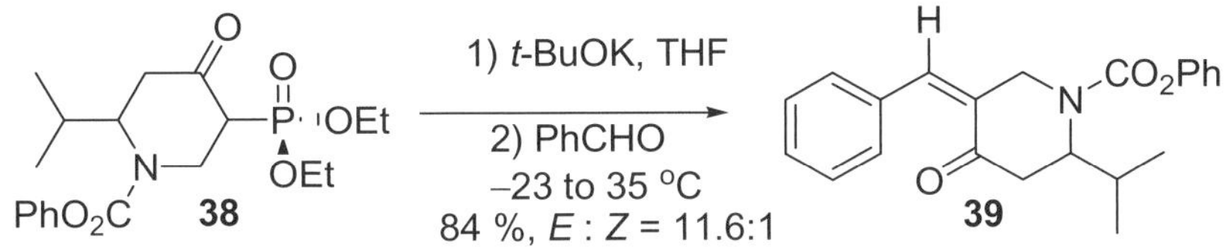

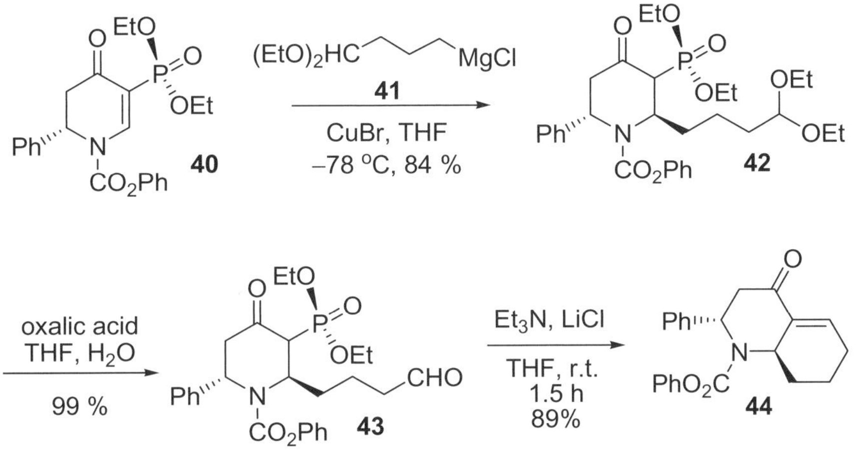

The Horner–Emmons olefination shows a preference for the formation of the more stable (*E*)-olefins. On the other hands, Still and Gennari modified HWE olefination shows high (*Z*)-stereoselectivity and high yield in the preparation of unsaturated esters, **45**(*Z*) : **46**(*E*) = > 50:1, from a variety of aromatic, saturated and unsaturated aliphatic aldehydes.[7,24]

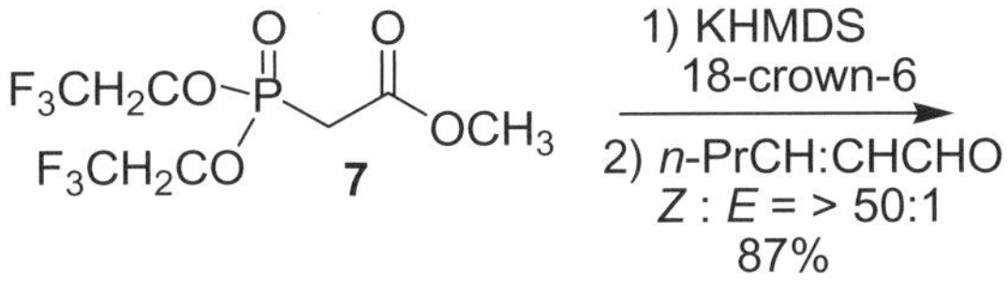

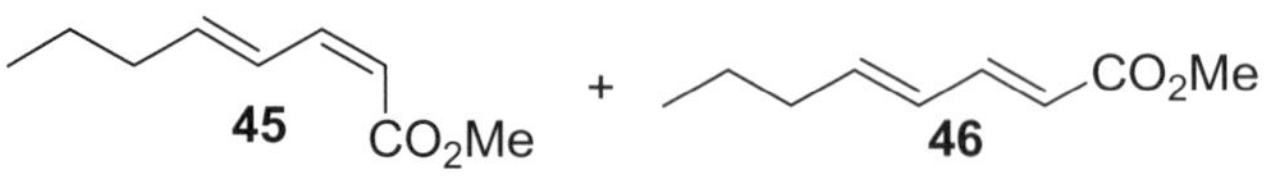

The (*Z*)-monolignols (hydroxycinnamyl alcohols) **49** were found along with their predominant (*E*)-isomers in a number of plants. Still–Gennarri reaction was successful applied to the preparation of (*Z*)-unsaturated esters **48** from aldehyde **47**. The ester **48** was further converted to (*Z*)-coniferyl alcohol **49** in the synthesis of (*Z*)-monolignols.[25]

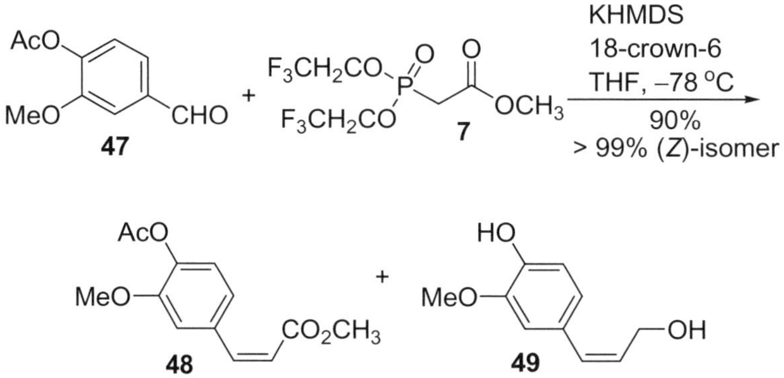

Excellent (*Z*)- or (*E*)-selectivity was observed by Nagao and co-workers in the Still–Gennarri HWE reactions of methyl bis-(2,2,2-trifluoroethyl)phosphonoacetate **7** or ethyl 2-fluoro-2-(diethylphosphono)-acetate **55** with aryl alkyl ketones bearing substituents on an aromatic moiety employing $Sn(OSO_2CF_3)_2$ as mediator in the presence of *N*-ethylpiperidine.[26] It was noticed that $Sn(OSO_2CF_3)_2$ mediated HWE reaction was superior to the NaH condition. Example, *o*-tolyl methyl ketone has no reaction with phosphornate **7** under NaH condition, whereas the reaction in the presence of $Sn(OSO_2CF_3)_2$ gave 84% yield with *E*/*Z* ratio of 4:96. In the case of aromatic compounds, ArCOMe **50**, having electron–donating methyl or methoxy group(s), the *E* : *Z* stereoselectivity was excellent. No reaction occurred in the case of ketone with a bulky mesityl group. In contrast, the similar reaction of ketones having an electron-withdrawing nitro or trifluoromethyl group, resulted in a slightly lower stereoselectivity.

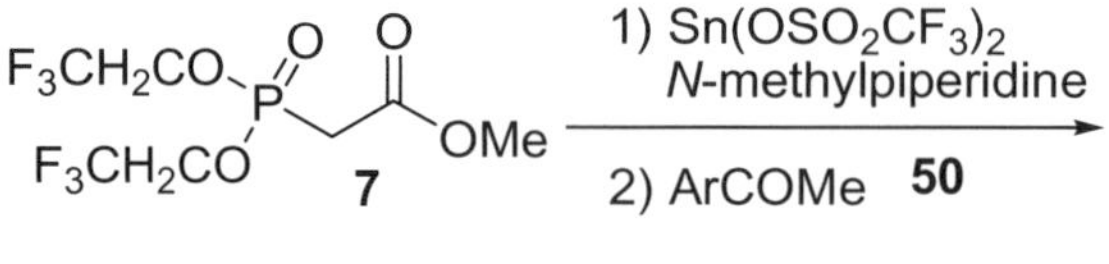

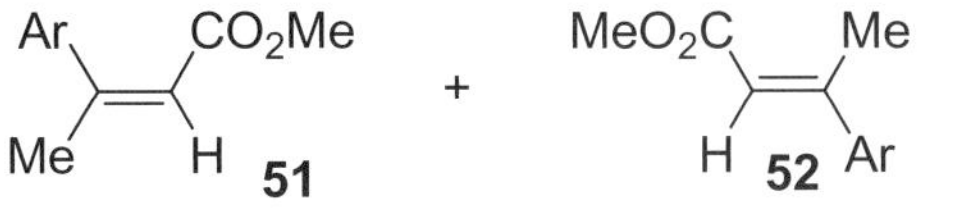

ArCOMe:
50

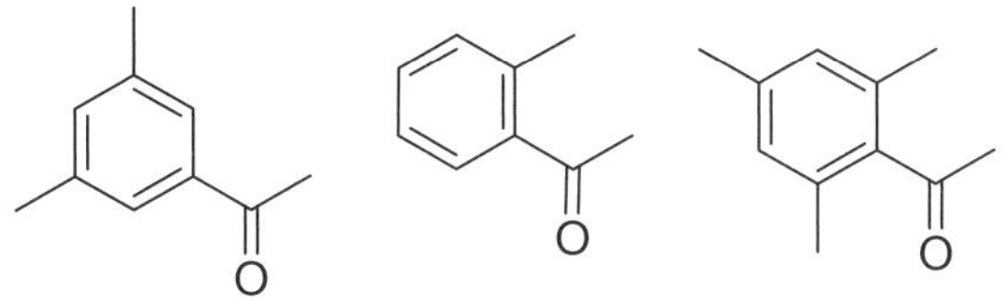

Yield:	55%	84%	-
Alkene (*E*/*Z*):	2 : 98	4 : 96	-

ArCOMe:
50

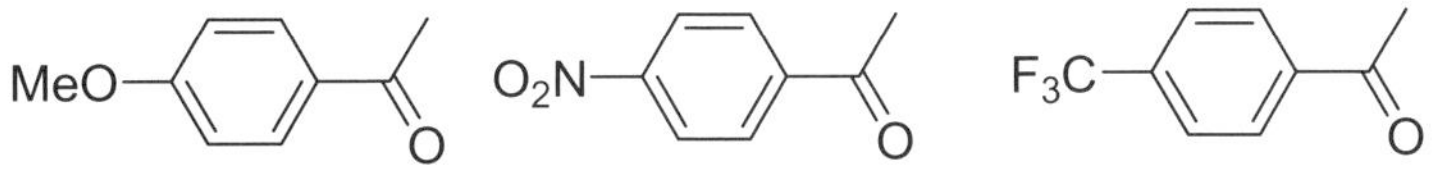

Yield:	95%	100%	97%
Alkene (*E*/*Z*):	2:98	17:83	9:91

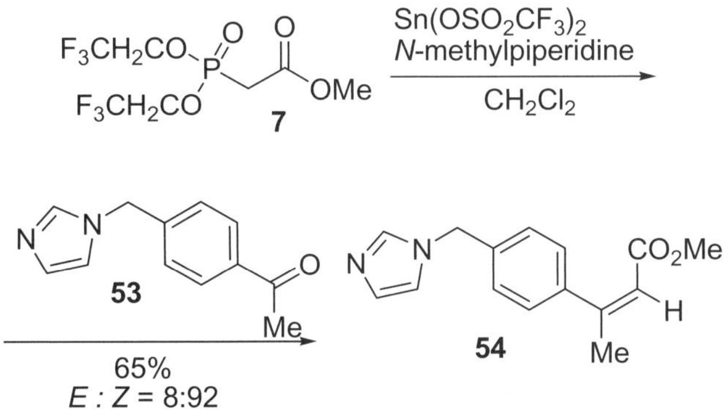

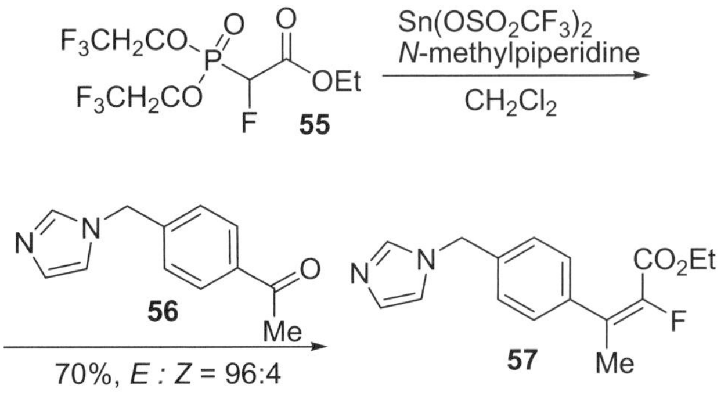

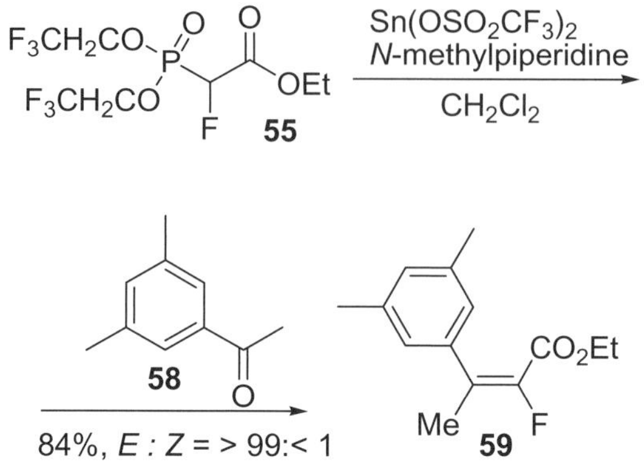

Taylor and co-workers discovered that Still–Gennari HWE reaction gives unexpected (*E*)-isomeric product, **62**(*E*) : **61**(*Z*) = 90:10, in the reaction of (*Z*)-3-stannylpropenal **63**. By contrast, the Ando procedure provides reasonable (*Z*)-stereoselectivity, **62**(*E*) : **61**(*Z*) = 13 : 87, in most cases including those with a *cis*-orientated stannyl substituent.[27]

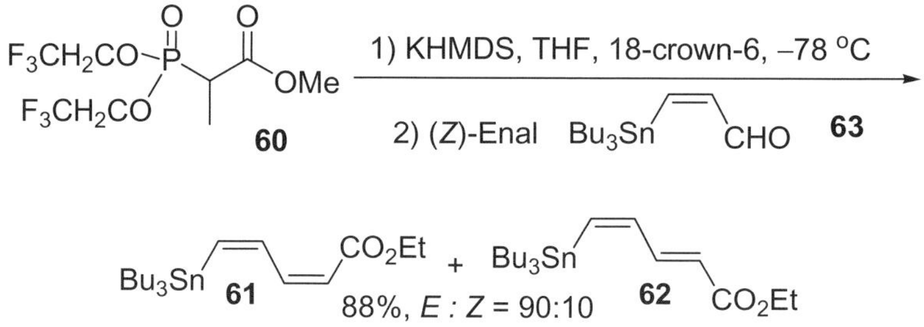

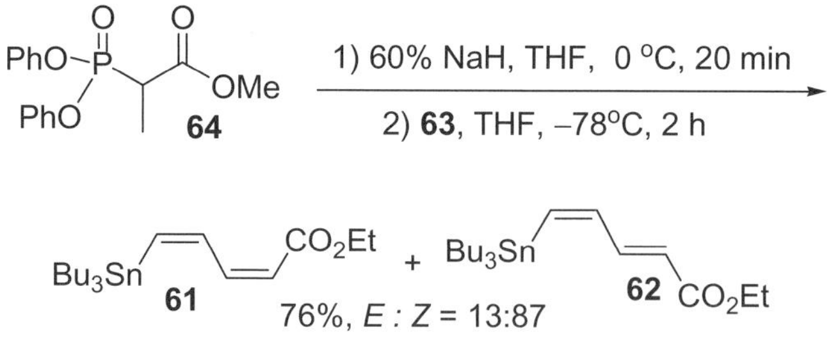

2.6.5 *Natural Product Synthesis*

The practical applicability of the HWE reaction and Still–Gennari varianted HWE olefination reaction is evident from its employment in the total-synthesis of numerous complex natural products. Macrolactin A, **68**, is a 24-membered polyene macrolide. It displays strong cytotoxic activity *in vitro* on

B16–F10 murine melanoma cells (IC_{50} = 3.5 μg/mL as well as powerful antiviral activity against *Herpes simplex* types I and II and against human HIV-1 virus replication. Bonini and co-workers reported a concise and efficient stereoselective synthesis of the C_1–C_{11} fragment of macrolactin A, **67,** utilizing Still–Gennari HWE reaction condition.[28] Aldehyde **65** prepared from corresponding alcohol, was reacted with phosphonate **7** in the presence of KHMDS and 18-crown-6 affording (2*Z*,4*E*)-diene **66** with excellent stereoselectivity (*Z,E*/*E,E* = 95:5) in 70% yield.

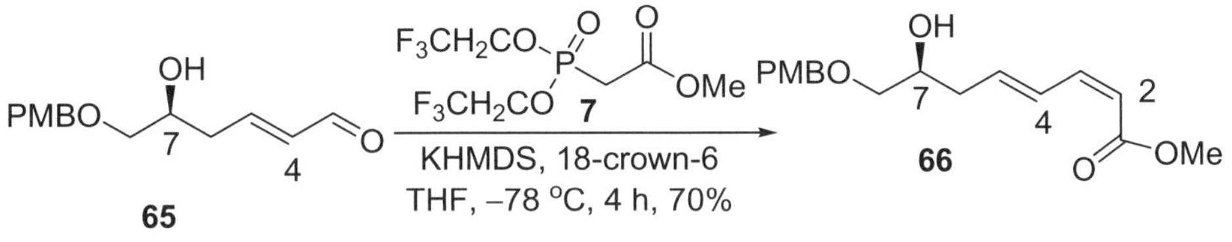

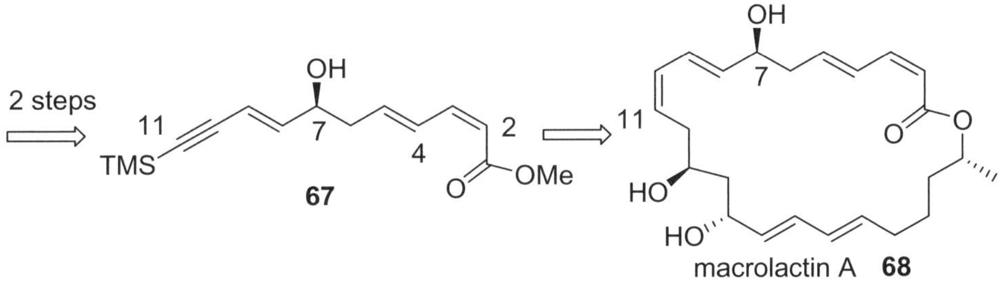

The epothilones A–E, **72–73**, have shown their eminent cytotoxic activity against tumor cells, taxol-like mitose inhibition and toxity against multiple drug-resistant tumor cell lines. Öhler and co-workers have developed an easy access to epothilones A–D, in which the intermediate **71** was obtained in 89% yield from aldehyde **69** and only the (*Z*)-isomer, C_{12}–C_{13}, was formed by Still–Gennari olefination condition.[29]

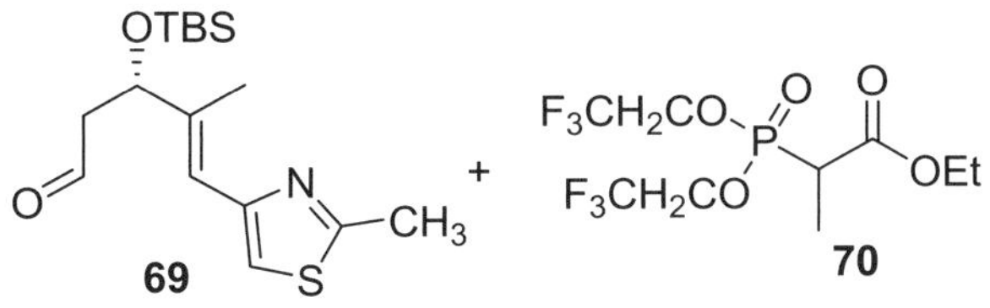

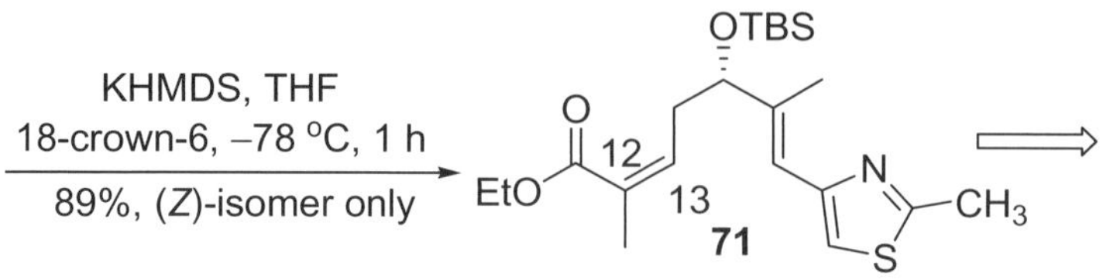

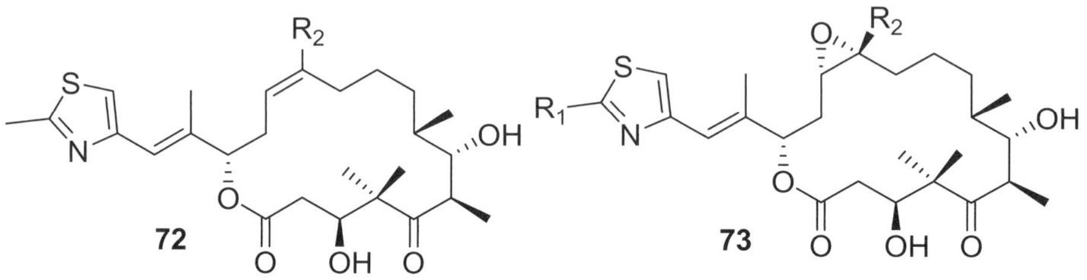

R₂ = H epothilone C
R₂ = Me epothilone D

R₁ = Me, R₂ = H epothilone A
R₁ = Me, R₂ = Me epothilone B
R₁ = CH₂OH, R₂ = H epothilone E

Tedanolide **77** has showen ED_{50}'s of 250 *pg*/mL (vs. the KB human carcinoma cell line) and 16 *pg*/mL (*vs*. PS lymphocytic leukemia). 13-Deoxytedanolide **78** has shown a T/C of 189% at a dose of 125 μg/kg vs p388 cell lines. Many synthetic approaches have been demonstrated for the total synthesis of these two compounds. Jung and co-workers have developed a nonaldol aldol process in the synthesis of a C_1–C_{11} fragment.[30] Treatment of the key intermediate **74** with trimethylsilyl triflate, followed by Still–Gennari olefination conditions afforded the desired C_1–C_{11} (*Z*)-conjugated ester **75** in 80% yield as a 1:1 mixture with the pyran derivative **76**.

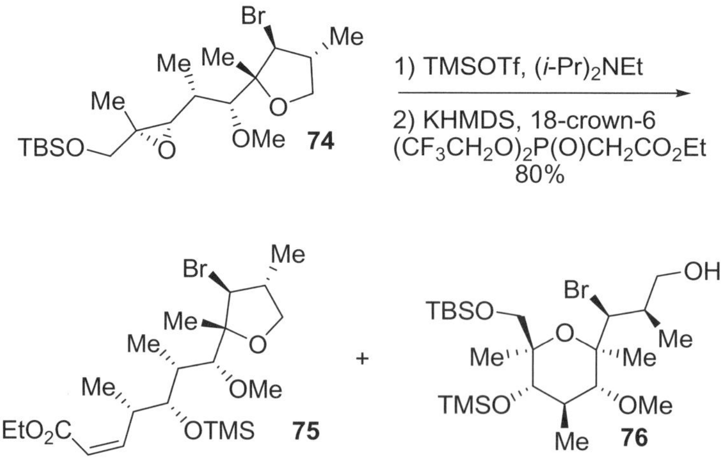

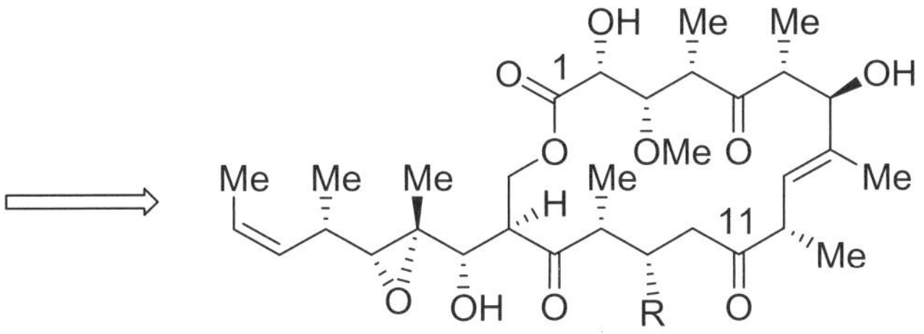

77 R = OH
78 R = H

Dictyostatin **83** is a potent cytotoxic macrolide. It displays powerful growth inhibitory activity against a number of murine and human cancer cells at low nanomolar concentrations, and retains activity against taxol-resistant cancer cells that express active P-glycoprotein (PGP). Paterson and co-workers disclosed the first total synthesis of dictyostatin employing (*Z*)-selective intermolecular Still–Gennari olefination.[31] The acid **79** was converted to acid chloride by using the Ghosez reagent ($Me_2C{=}C(Cl)NMe_2$). Then, the acid chloride was added to a solution of $(F_3CCH_2O)_2P(O)CH_2Li$ in THF at –100 °C to afford phosphonate **80** containing the required functionality for the pivotal fragment assembly. When a mixture of aldehyde **81** and phosphonate **80** was subjected to an excess of K_2CO_3 in the presence of 18-crown-6 in toluene, the Still–Gennari HWE coupling proceeded smoothly to produce (*Z*)-enone **82** efficiently, with good selectivity.

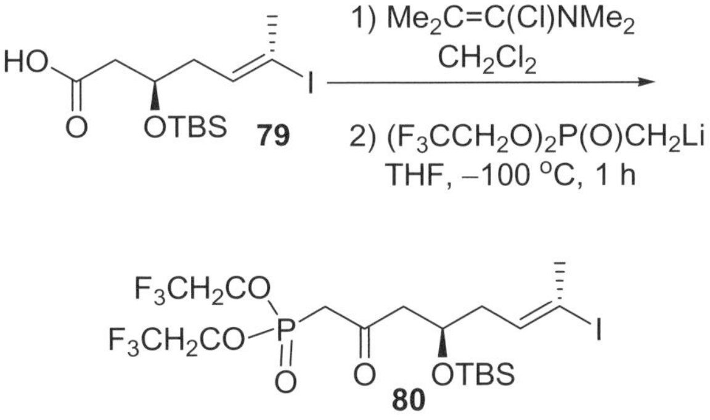

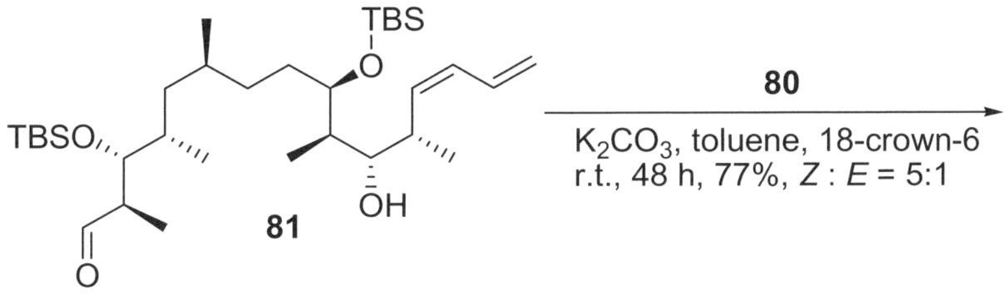

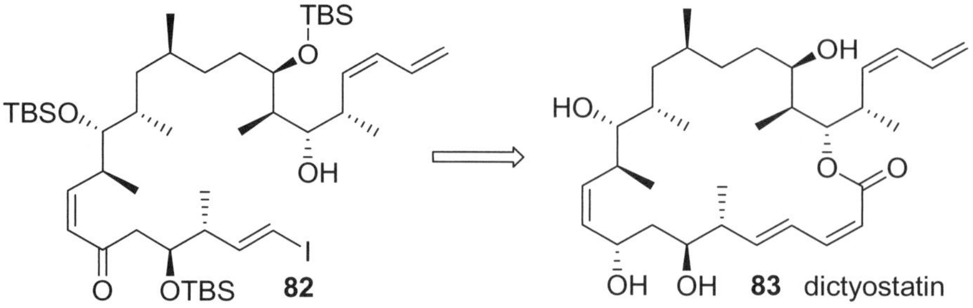

SNF 4435 C and SNF4435 D are two natural products. Both compounds have been shown to selectively block induced B-cell proliferation versus induced T-cell proliferation and show potent immunosuppressive activity in vitro. The total synthetic approaches toward the immunosuppressant SNF C and SNF4435 D utilizing Still–Gennari reaction and HWE olefination failed. However, both Still–Gennari reaction targeting intermediate **86** and HWE olefination targeting compound **89** were demonstrated nicely, with good yield and high stereoselectivity.[32]

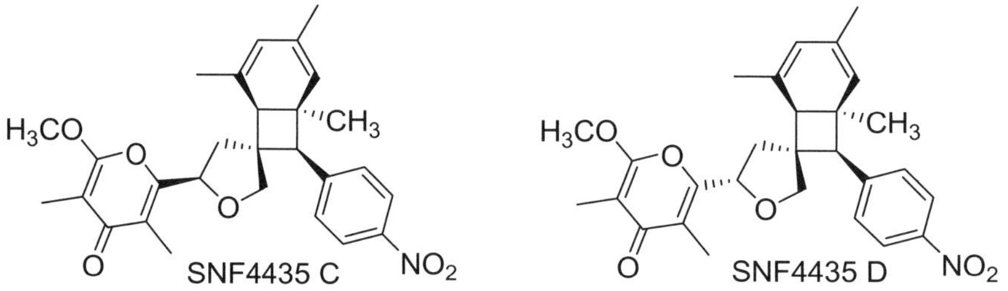

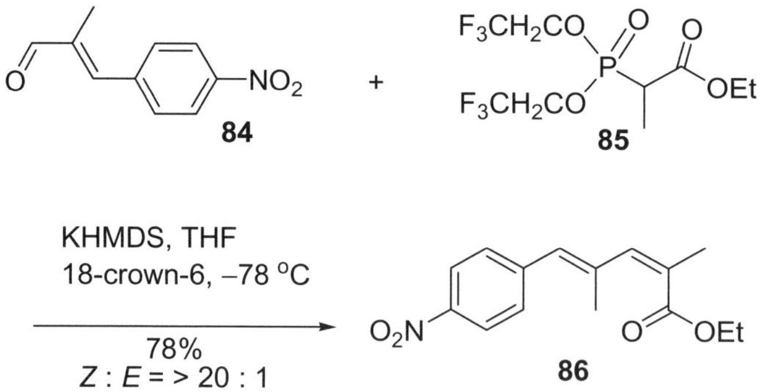

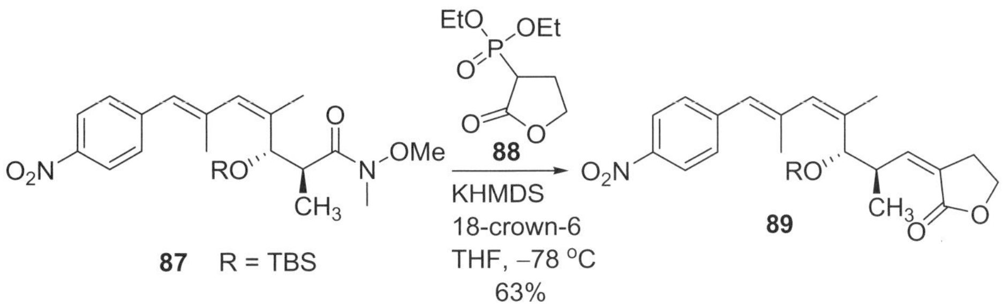

Discodermolide **92** is a unique polyketide isolated from the Caribbean deep sea sponge *Discodermia dissolute*. It inhibited T-cell proliferation with an IC_{50} of 9 nM and graft versus host disease in transplanted mice. The growth of Taxol-resistant ovarian and colon cancer cells is inhibited by discodemolide with an IC_{50} of < 2.5 nM. Paterson and co-workers has completed the stereocontrolled total synthesis of (+)-discodermolide **92** in 10.3% overall yield (23 steps longest linear sequence).[33] Still–Gennari phosphonate reaction provides high yield (87%) and high selectivity in the formation of (*Z*)-C_8–C_9 olefin, **91**, from alcohol **90** in the total synthesis of discodemolide **92**.

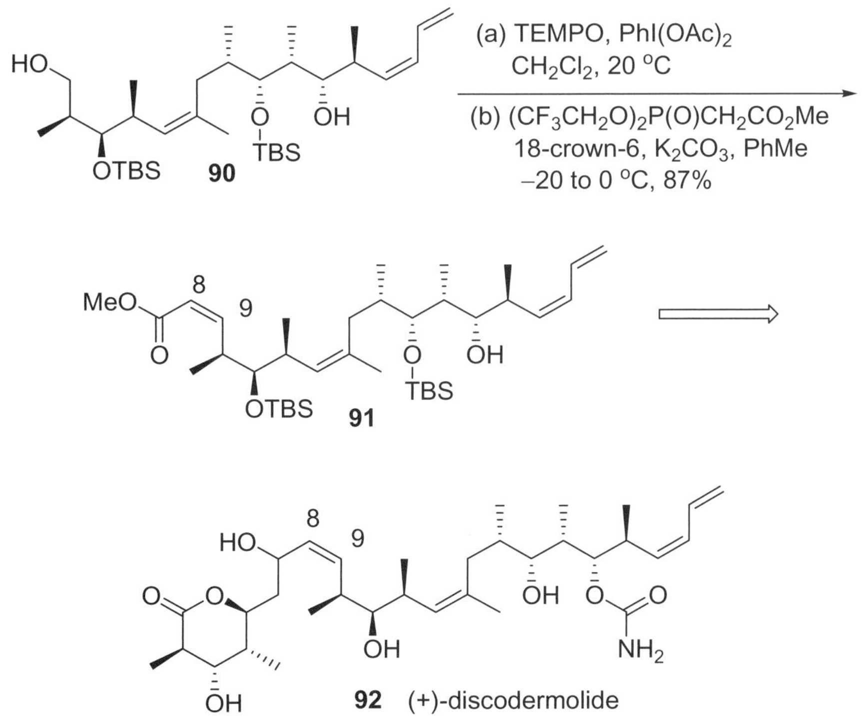

Mickel and co-workers at Novartis Pharma was very successful in large-scale synthesis of (+)-discodermolide **92**, Still–Gennari HWE reaction in the formation of C_8–C_9 (*Z*)-olefin especially.[34] The compound **91** was obtained in 76% yield (622 g) from **90** by two step reactions, and KHMDS was used as base in the olefination reaction.

In the effort of total synthesis of discodemolide, the Still–Gennari type HWE reaction was also used by Paterson and Lyothier to form the C_8–C_9 olefin.[35] Phosphonate **94** was treated with NaH in THF at 0 °C for 30 min prior to the addition of aldehyde **93**. The crude product showed a mixture of (*Z*)- and (*E*)-olefins, **95** and **96,** in a 5:1 ratio. The desired (*Z*)-isomer **96** could be isolated by flash chromatography in 73% yield. The (*Z*)-enone **96** was further treated with K-Selectride at –25 °C for 24 h. A single diastereomeric alcohol **97** was formed selectively and isolated in 59% yield. This alcohol processing (*7R*)-configuration was leading to 7-*epi*-discodermolide **98** as final product by four additional reactions.

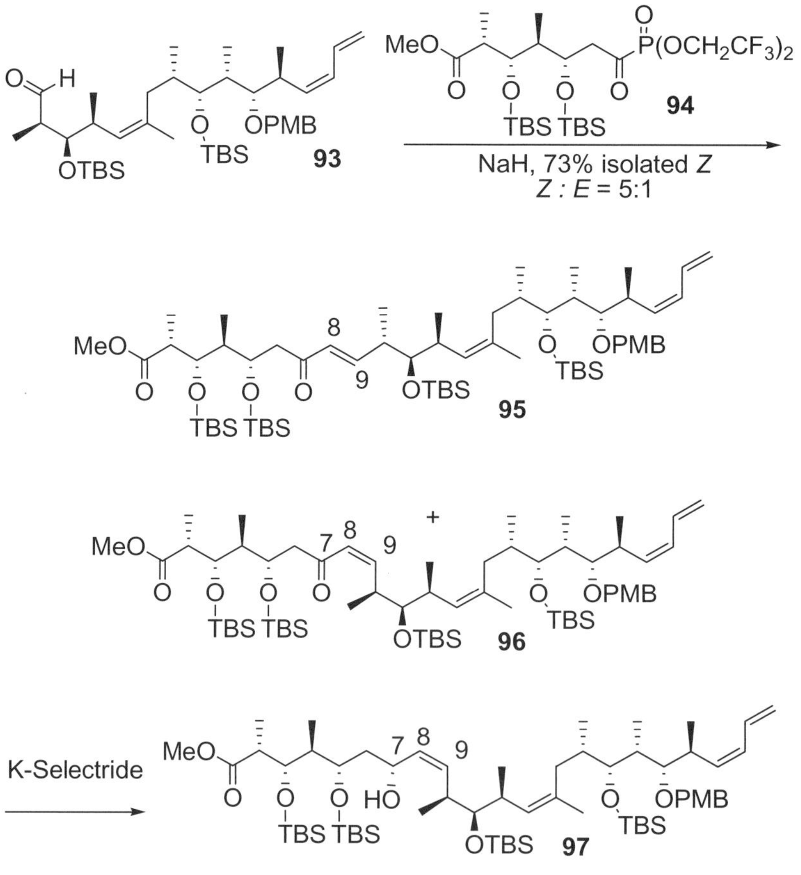

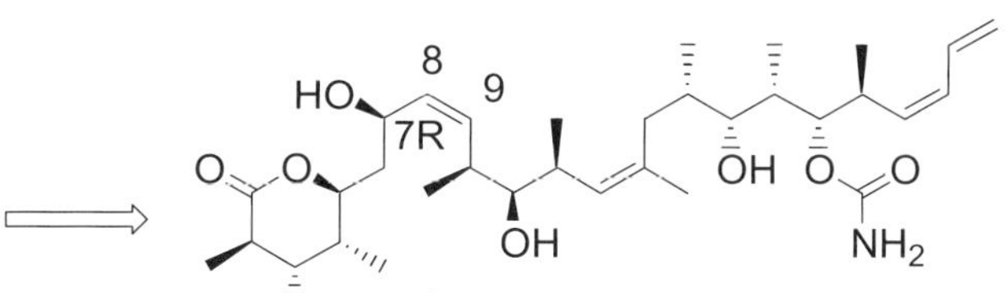

98 7-*epi*-discodermolide

Paterson and Lyothier have made more effort on the C_7 reduction in the total synthesis of (+)-discodermolide by adjustment of protecting groups.[35] The 3,5-diol protected by cyclopentylidene acetal was a good choice. Utilizing the optimized Still–Gennari HWE olefination conditions developed previously, phosphonate **99** was treated with NaH in THF at –10 °C for 30 min and then reacted with aldehyde **93**. A clean olefination reaction occurred that was now more selective than obtained with bis-TBDMS-protected phosphonate **94** (*Z/E* = 10 : 1), with the (*Z*)-enone **100** isolated by simple flash chromatography on silica gel in 74% yield. The (*Z*)-enone **100** was further treated with DDQ, and with $Cl_3CC(O)NCO/K_2CO_3$ in MeOH to provide enone **102**. The enone **102** was therefore treated with K-Selectride in toluene at –78 °C. The reduction proved to be clean and efficient, completed within 3 h at –78 °C, and provided a single diasterenmeric alcohol **103** with (*7S*)-configuration, in 90% yield. This proceeded cleanly using HF•py, with concomitant *δ*-lactonization, providing (+)-discodermilide in 84% yield. This is called as third-generation total synthesis of (+)-discodermilide.

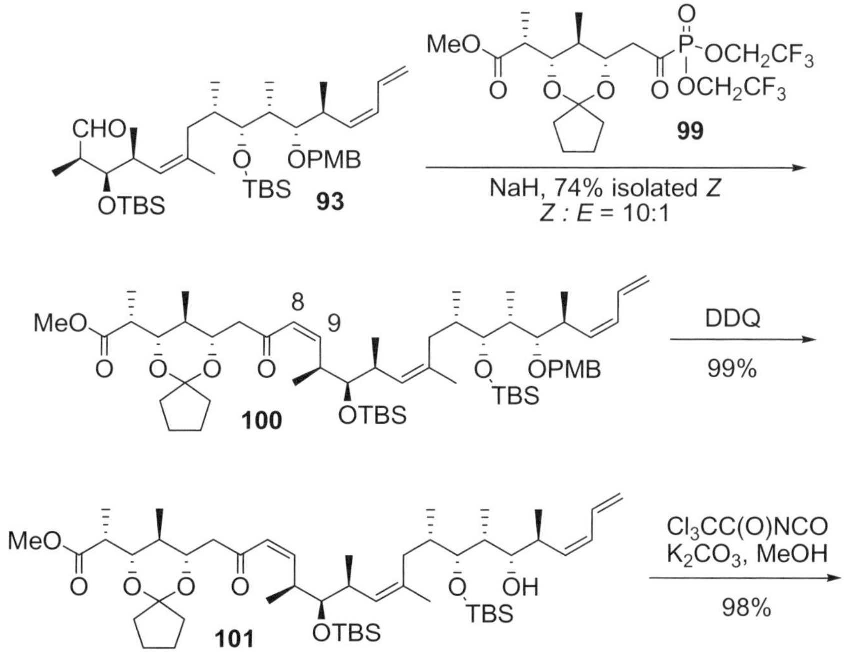

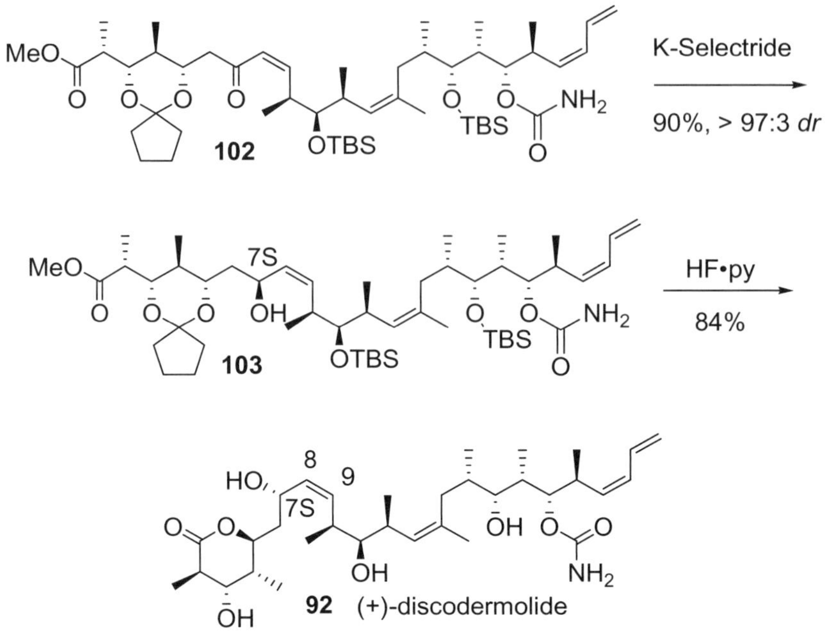

White and co-workers reported a convergent and efficient route to the synthesis of marine toxin polycavernoside A, **106**.[36] In which, condensation of aldehyde **104** with the Still–Gennari phosphonate gave the expected *cis*-α,β-unsaturated ester, (*Z*)-C_2–C_3 olefin **105** in 94% yield, which is important toward the synthesis of polycavernoside A, **106**.

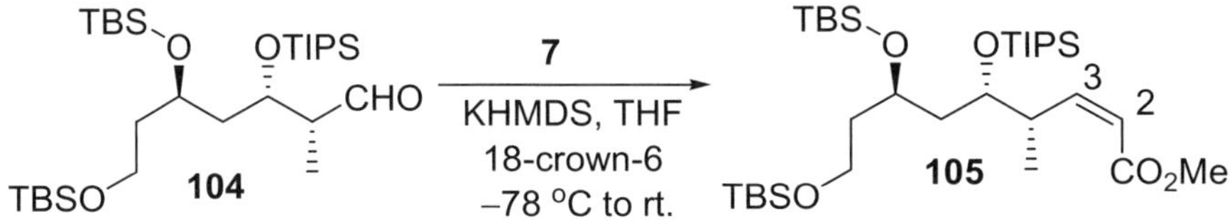

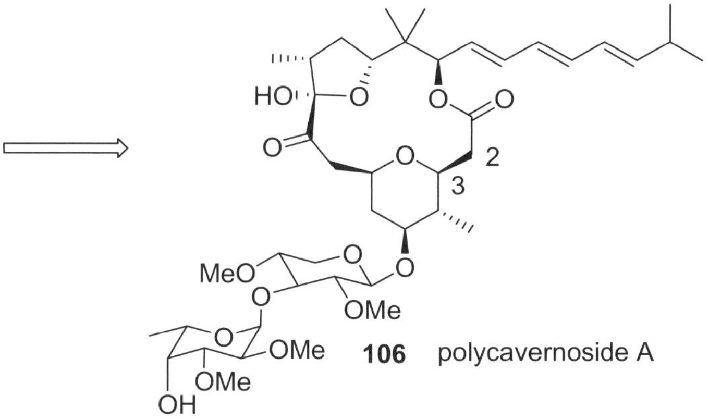

106 polycavernoside A

Laulimalide **109,** also known as fijianolide B, is a 20-membered macrolide that was isolated from various marine sponges. Laulimalide **109** is a potent inhibitor of cellular proliferation, with IC_{50} values against numerous drug sensitive cell lines in the low nanomolar range (KB cell line, 15 ng/mL; MCF-7, 7 *n*g/mL; P388, A549, HT29, MEL28, 10–50 *n*g/mL; MDA-MB-435, 6 *n*g/mL; SK-OV-3, 12 *n*g/mL). Couple approaches have been developed for the total synthesis of laulimalide. One is utilizing Still–Gennari olifination to make the macrocyclization in the total synthesis. It was expected to give C_2–C_3-olefin with high (*Z*)-selectivity. However, it finished as an 1.8:1 *E/Z* mixture of the macrolactones **108**.[37] Isolation of (*Z*)-**108** from the mixture by chromatography, followed by deprotection with dimethylboron bromide and exposure with (+)-diisopropyl tartrate (DIPT) provided laulimalide **109**.

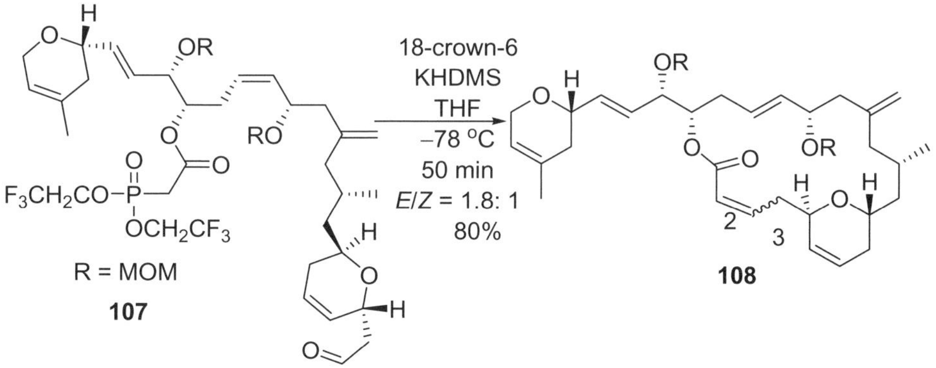

107 **108**

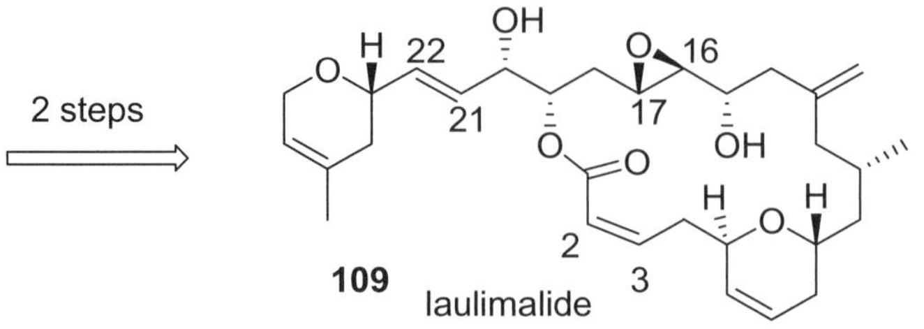

Mulzer and co-workers have developed three different routes for the total synthesis of deoxylaulimalide, which is the immediate precursor of the marine sponge metabolite laulimalide. One of the efficient routes was empolying HWE olefination in the formation of C_{16}–C_{17} **116**, by reacting phosphate **115** with an aldehyde **114**, and C_{21}–C_{22} **113**, by reacting phosphate **111** with aldehyde **112**, in the total synthesis of laulimalide **109**.[38]

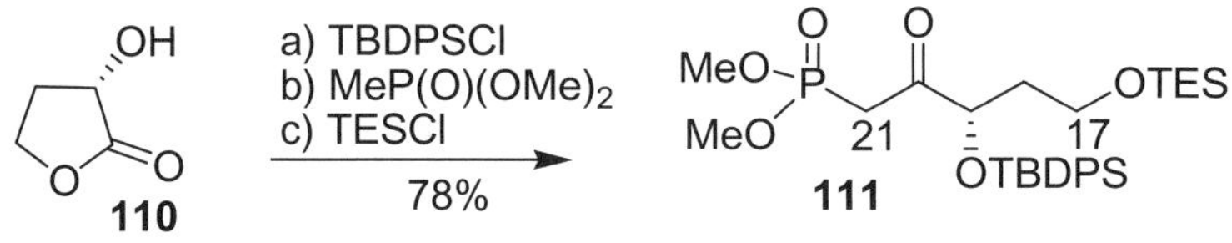

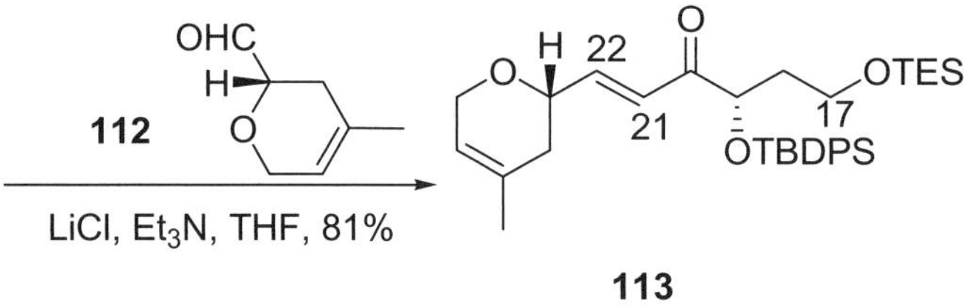

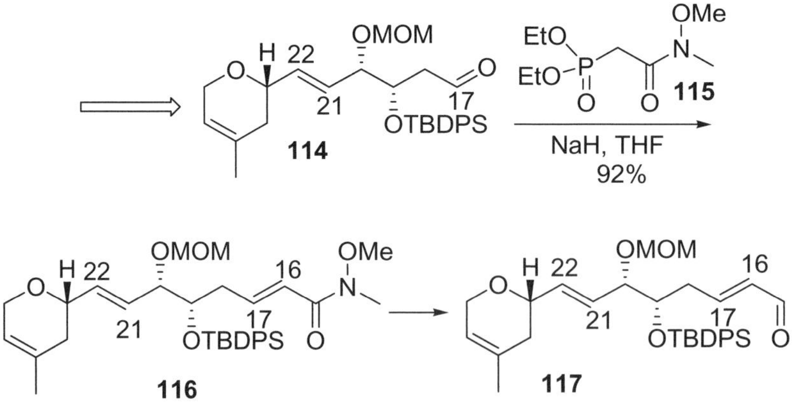

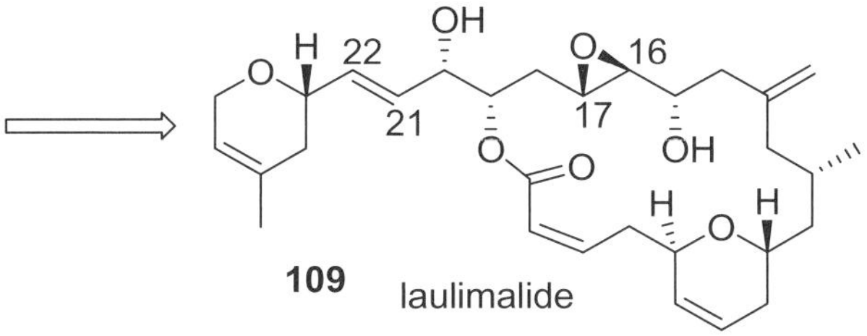

109 laulimalide

Leucascandrolide A, **118**, is a doubly *O*-bridged 18-membered macrolide. It exhibits high cytotoxicity in vitro against human KB and p388 tumor cell lines displaying low IC_{50}'s of 0.05 and 0.26 μg/mL, respectively. It also shows potent antifungal activity against Candida albicans, a pathogenic yeast that attacks AIDS patents and other immunocompromised individuals. Panek and co-workers reported a retrosynthesis of leucascandrolide A.[39] It reveals two principal fragments: the 18-membered macrolide **119** and the $C_{1'}$–$C_{11'}$ oxazole side chain **120**. An efficient and convergent synthesis of the $C_{1'}$–$C_{11'}$ side chain **120** was achieved. The bond connection is made through the use of a palladium(0) catalyzed Sonogashira cross-coupling between trifloy oxazole **122** and **121**. Installation of the (*Z*)-olefin using a Lindlar reduction, followed by removal of the silyl protecting group using TBAF buffered with HOAc gave oxazile **122** in 80% yield for the two steps. Completion of the fragment was accomplished by using a Dess–Martin oxidation followed by a Still–Gennari olefination to give the completed $C_{1'}$–$C_{11'}$ fragment **125**.

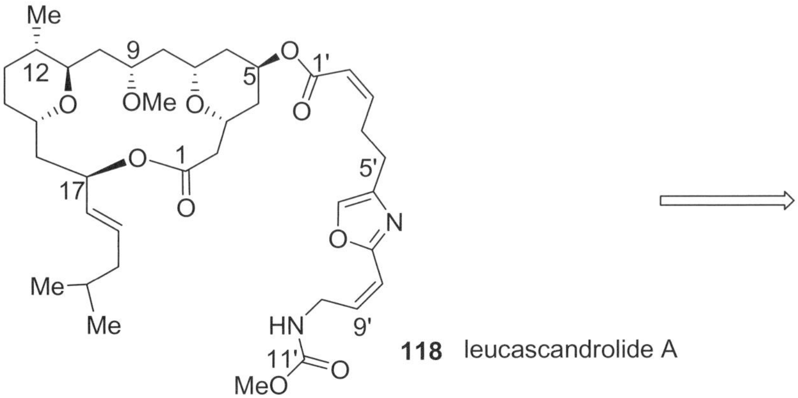

118 leucascandrolide A

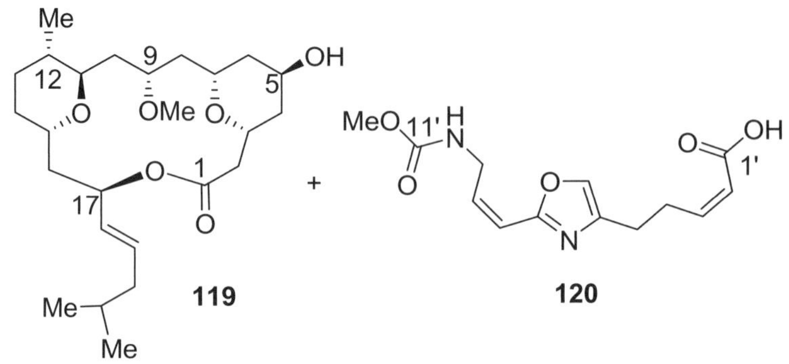

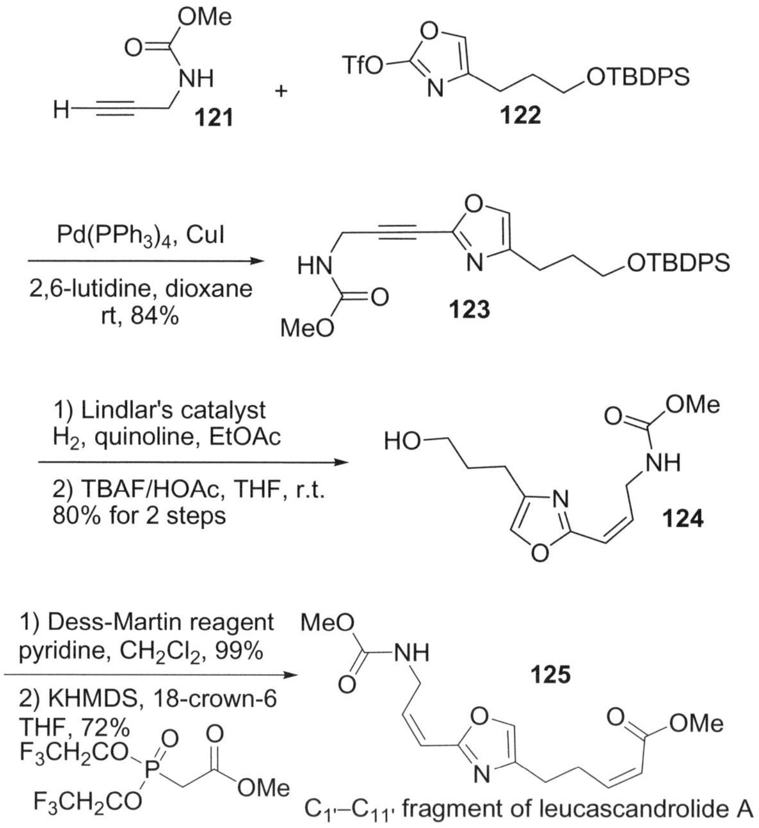

Natural product phorboxazoles have demonstrated potent antifungal activity against *Candida albicans* and *Saccharomyces carlsbergensis*, and exceptional inhibition of cell growth. These metabolites showed extraordinary potency (mean $GI_{50} \leq 1.6 \times 10^{-9}$ M; GI_{50}: 50% inhibition of cell growth) when bioassayed for 60 human tumor cell strains at the national cancer Institute (NCI). Lin and co-workers reported a total synthesis of phorboxazole B, **129**.[40] The Still–Gennari olefination is one of the key strategies in the total synthesis. Intramolecular olefination of **127**, prepared from intermediate **126,** in the presence of K_2CO_3 and 18-crown-6 in toluene effected a (*Z*)-selective macrolization to afford a mixture of macrocycles (*Z*/*E* = 5:1), which were easily separated on silica gel to give (*Z*)-macrolide **128** in 56% yield. Then, cleavage of the silyl ethers and the mixed methyl acetal by sequential treatment of **128** with TBAF and 6% aqueous HCl/THF produced phorboxazole B (**129**).[40]

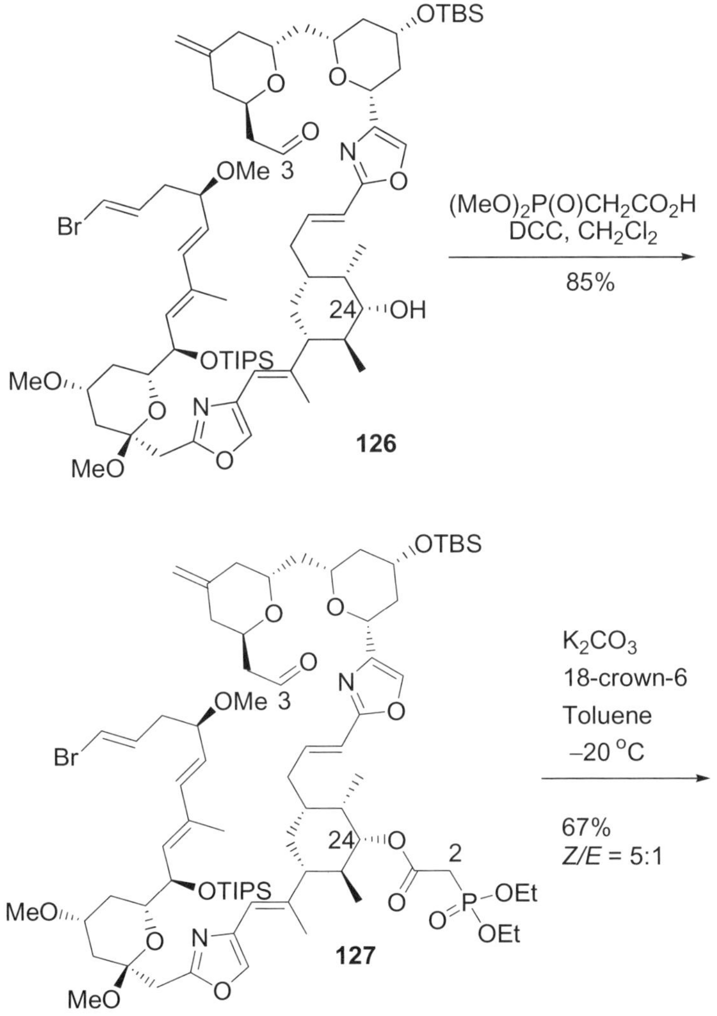

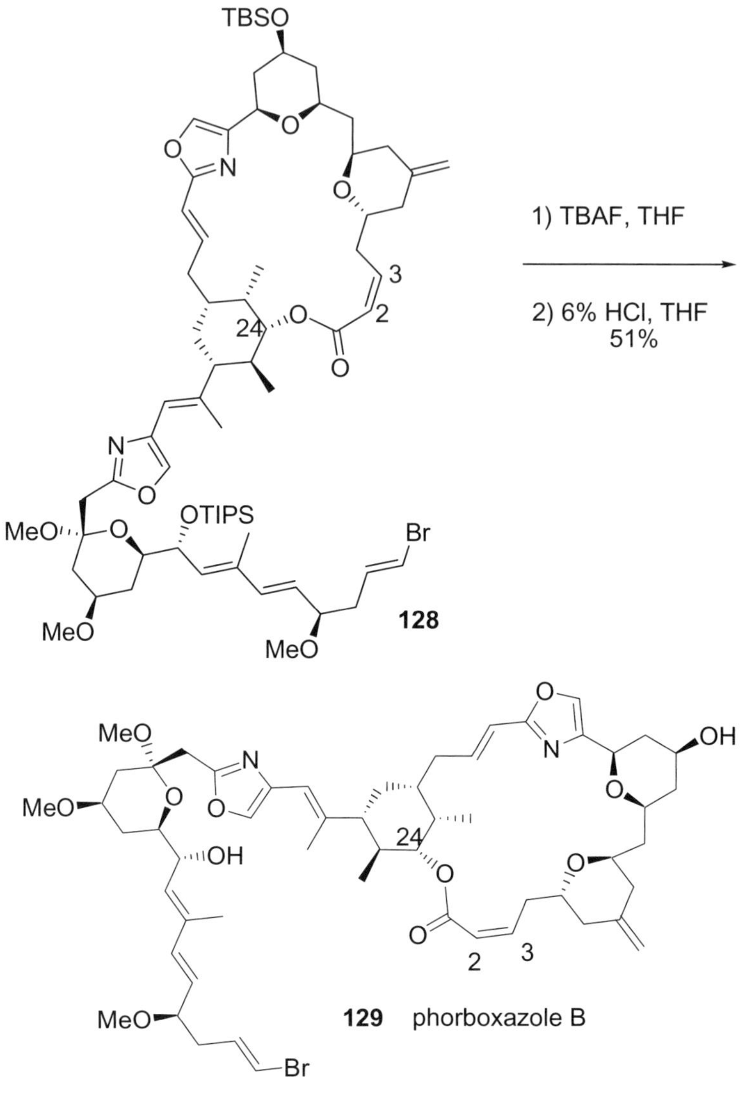

2.6.6 *Experimental*

Preparation of (methyl 4-*O*-acetyl-3-methoxy-(*Z*)-cinnamate (48).[25]
A solution of bis(2,2,2-trifluoroethyl)-(methoxycarbonylmethyl) phosphonate **7** (3.3 g, 10.3 mmol), 18-crown-6 (10.9 g, 41.2 mmol) in anhydrous THF was cooled to –78 °C under nitrogen and treated with KHMDS (20.6 mL, 0.5 M in toluene, 10.3 mmol). 4-*O*-acetyl-3-methoxybenzaldehyde **47** (2 g, 10.3 mmol) was then added, and the resulting mixture was stirred for 30 min. at –78 °C. [Note: if the reaction is kept strictly at low temperature, the selectivity is > 99% for the (*Z*)-isomer; if the temperature is allowed to rise during the

addition, 5% or more of the (*E*)-isomer can result even at –60 °C]. Satd. NH_4Cl solution was added and the product was extracted into ethyl ether (3 × 50 mL). The organic phase was dried over $MgSO_4$ and the solvents removed to give an oily product (2.34 g, 90%).

Preparation of methyl (2*R*,3*S*,4*S*,5*S*,8*Z*,10*S*,11*S*,12*S*,13*Z*,16*S*,17*R*,18*S*,19*S*,20*S*,21*Z*)-11,17-bis(*tert*-butyldimethylsilyloxy)-3,5-cyclopentylidene acetal-19-(4-methoxybenzyloxy)-2,4,10,12,14,16,18,20-octamethyl-7-oxo-tetracosa-8,13,21,23-tetraenoate (100).[35]
Phosphonate **99** (10 mg, 18.4 μmol) was dissolved in THF (1 mL) and cooled to –10 °C. Sodium hydride (60% in mineral oil, 0.8 mg, 19 μmol) was added, and the reaction mixture was stirred at –10 °C for 30 min. Aldehyde **93** (40 mg, 55.3 μmol) was added in solution in THF (2 mL), and the reaction mixture was allowed to warm to room temperature slowly and stirred for 2.5 days. Aqueous NH_4Cl (10 mL) was added, the mixture was extracted with dichloromethane (4 × 5 mL), and the combined organic extracts were dried over $MgSO_4$ and concentrated under reduced pressure. The *Z*/*E* ratio was determined by ^1H NMR spectroscopy as being 10 : 1. Purification by flash chromatography (10% AcOEt in hexane) afforded the (*Z*)-enone **100** as a colorless oil (13.5 mg, 74%).

2.6.7 References

1. Horner, L.; Hoffmann, H.; Wippel, H. G.; Klahre, G. *Chem. Ber.* **1959**, *92*, 2499.
2. Wadsworth, W. S., Jr.; Emmons, W. D. *J. Am. Chem. Soc.* **1961**, *83*, 1733–1738.
3. Wadsworth, D. H.; Schupp, O. E.; Seus, E. J.; Ford, J. A., Jr. *J. Org. Chem.* **1965**, *30*, 680–685.
4. [R] (a) Maryanoff, B. E.; Reitz, A. B. *Chem. Rev.* **1989**, *89*, 863–927; (b) Maercker, A. *Org. React.* **1965**, *14,* 270–490. (c) Murphy, P. J.; Brennan, J. *Chem. Soc. Rev.* **1988**, *17,* 1–30; (d) Vedejs, E.; Peterson, M. J. *Top. Stereochem.* **1994**, *21,* 1; (e) Deagostino, A.; Prandi, C.; Tonachini, G.; Venturello, P. *Trends Org. Chem.* **1995**, *5*, 103.
5. Shair, M. D.; Yoon, T. Y.; Mosny, K. K.; Chou, T. C.; Danishefsky, S. J. *J. Am. Chem. Soc.* **1996**, *118*, 9509–9525.
6. Simoni, D.; Rossi, M.; Rondanin, R.; Mazzali, A. Baruchello, R.; Malagutti, C.; Roberti, M.; Invidiata, F. P. *Org. Lett.* **2000**, *2,* 3765–3768.
7. Still, W. C.; Gennari, C. *Tetrahedron Lett.* **1983**, *24*, 4405–4408.
8. Nicolaou, K. C.; Nadin, A.; Leresche, J. E.; LaGreca, S.; Tsuri, T.; Yue, E. W.; Yang, Z. *Angew. Chem., Int. Ed. Engl.* **1994**, *33*, 2187.
9. Ando, K. *J. Org. Chem.* **1997**, *62*, 1934–1939.
10. Ando, K. *J. Org. Chem.* **1999**, *64*, 6815–6821.
11. Horner, L.; Hoffmann, H.; Wippel, H. G. *Chem. Ber.* **1958**, *91*, 61.
12. Horner, L.; Hoffmann, H.; Wippel, H. G.; Klahre, G. *Chem. Ber.* **1959**, *92*, 2499.
13. Wadsworth, W. S., Jr.; Emmons, W. D. *J. Am. Chem. Soc.* **1961**, *83*, 1733–1738.
14. Wadsworth, W. S., Jr. *Curr. Cont./Phys. Chem. Earth. Sci.* **1983**, *32*, 18.
15. Buss, A. D.; Warren, S. *J. Chem. Soc., Perkin Trans. 1* **1985**, 2307.
16. Buss, A. D.; Cruse, W. B.; Kennard, O.; Warren, S. *J. Chem. Soc., Perkin Trans. 1* **1984**, 243.
17. Buss, A. D.; Warren. S. *Tetrahedron Lett.* **1983**, *24*, 3931.
18. Larsen, R. O.; Aksnes, G. *Phosphorus Sulfur* **1983**, *15*, 219.
19. Larsen, R. O.; Aksnes, G. *Phosphorus Sulfur* **198**3, *15*, 229.

20. Reiser, U.; Jauch, J. *Synlett* **2001**, *1*, 90–92.
21. Lattanzi, A.; Orelli, L. R.; Barone, P.; Massa, A.; Iannece, P.; Scettri, A. *Tetrahedron Lett.* **2003**, *44*, 1333–1337.
22. Harusawa, S.; Koyabu, S.; Inoue, Y.; Sakamoto, Y.; Araki, L.; Kurihara, T. *Synthesis* **2002**, *8*, 1072–1078.
23. Comins, D.; Ollinger, C. G. *Tetrahedron Lett.* **2001**, *42*, 4115–4118.
24. Bodnarchuk, N. D.; Malovik, V. V.; Derkach, G. I. *Zh. Obshch. Khim.* **1970**, *40*, 1210.
25. Ralph J.; Zhang Y. *Tetrahedron* **1998**, *54*, 1349–1354.
26. Sano, S.; Yokoyama, K.; Shiro, M.; Nagao, Y. *Chem. Pharm. Bull.* **2002**, *50*, 706–709.
27. Franci, X.; Martina, S. L. X.; McGrady, J. E.; Webb, M. R.; Donald, C.; Taylor, R. J. K. *Tetrahedron Lett.* **2003**, *44*, 7735–7740.
28. Bonini, C.; Chiummiento, L.; Videtta, V.; Colobert, F.; Solladie, G. *Synlett* **2006**, *15*, 2427–2430.
29. Mulzer, J.; Mantoulidis, A.; Öhler, E. *Tetrahedron Lett.* **1998**, *39*, 8633–8636.
30. Jung, M. E.; Marquez, R. *Org. Lett.* **2000**, *2*, 1669–1672.
31. Paterson, I.; Britton, R.; Delgado, O.; Meyer, A.; Poullennec, K. G. *Angew. Chem., Int. Ed.* **2004**, *43*, 4629–4633.
32. Beaudry, C. M.; Trauner, D. *Org. Lett.* **2002**, *4*, 2221–2224.
33. Paterson, I.; Florence, G. G.; Gerlach, K.; Scott, J. P.; Sereinig. N. *J. Am. Chem. Soc.* **2001**, *123*, 9535–9544.
34. Mickel, S. J.; Sedelmeier, G. H.; Niederer, D.; Schuerch, F.; Seger, M.; Schreiner, K.; Daeffler, R.; Osmani, A.; Bixel, D.; Loiseleur, O.; Cercus, J.; Stettler, H.; Schaer, K.; Gamboni, R. *Org. Proc. Res. Dev.* **2004**, *8*, 113–121.
35. Paterson, I. Lyotheir, I. *J. Org. Chem.* **2005**, *70*, 5494–5507.
36. White, J. D.; Blakemore, P. R.; Browder, C. C.; Hong, J.; Lincoln, C. M.; Nagornyy, P. A.; Robarge, L. A.; Wardrop, D. J. *J. Am. Chem. Soc.* **2001**, *123*, 8593–8595.
37. Mulzer, J.; Öhler, E. *Angew. Chem., Int. Ed.* **2001**, *40*, 3842–3846.
38. Ahmed, A.; Hoegenauer. E. K.; Enev, V. S.; Hanbauer, M.; Kaehlig, H.; Öhler, E.; Mulzer, J. *J. Org. Chem.* **2003**, *68*, 3026–3042.
39. Dakin, L. A.; Langille, N. F.; Panek, J. S. *J. Org. Chem.* **2002**, *67*, 6812–6815.
40. Li, D.-R.; Zhang, D.-H.; Sun, C.-Y.; Zhang, J.-W.; Yang, L.; Chen, J.; Liu, B.; Su, C.; Zhou, W.-S.; Lin, G.-Q. *Chem. Eur. J.* **2006**, *12*, 1185–1204.

2.7 Julia–Lythgoe Olefination

Frank Rong

2.7.1 Description

Julia–Lythgoe olefination refers to the formation of predominantly *trans*-olefins **3** via the addition of phenyl sulfones to aldehydes or ketones, followed by alcohol functionalization and subsequent reductive elimination.[1–6] The yield for the transformation from carbonyl to alkene is extremely high; usually greater than 80% overall. The selectivity can be also high for the preparation of disubstituted (*E*)-alkenes. The Na–Hg amalgam was used initially as a reducing agent in the reductive vicinal elimination. However, because of the high toxicity of Hg, this step was improved by replacing Na–Hg with other reducing agents such as lithium naphthalenide,[7] SmI_2,[8] Mg,[9] and electro reductive reaction.[10]

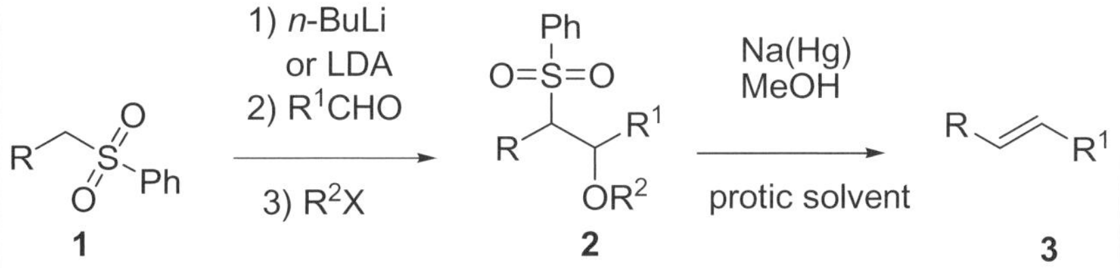

$R^2 = COCH_3$, X = $OCOCH_3$; R^2 = COPh, SO_2CH_3, X = Cl

Kocienski and Lythgoe found the reductive elimination could best be carried out with the acetoxy or benzoyloxy sulfones **2**.[5,6] If the lithio sulfone derivative is used for addition to the carbonyl, the reaction can be worked up with acetic anhydride or benzoyl chloride to obtain the alkene precursor. In the case where enolization of the carbonyl is a complication, the magnesium derivative can frequently be used successfully. Methanol, ethyl acetate/methanol or THF/methanol were the solvents of choice and a temperature of –20 °C was effective at suppressing the undesired elimination of the acetoxy group to produce the vinyl sulfone.

The olefination of ketones to prepare trisubstituted alkenes employing Na–Hg affords moderate yields, unpredictable stereoselectivities and large amounts of retro-aldol products from the intermediate *β*-alkoxy sulfones. High yields and moderate stereoselectivities of trisubstituted alkenes are obtained by a modification of the Julia–Lythgoe olefination reaction, involving the *in situ* capture of the intermediate *β*-alkoxy sulfones with a suitable oxophilic electrophile and the employment of SmI_2/HMPA to promote the reductive elimination at low temperatures.[2,11,12]

Modified one-port Julia–Lythgoe olefination to give predominantly (*E*)-olefins from heteroarylsulfones **4** and aldehydes is called as Julia–Kocienski olefination.[13–16]

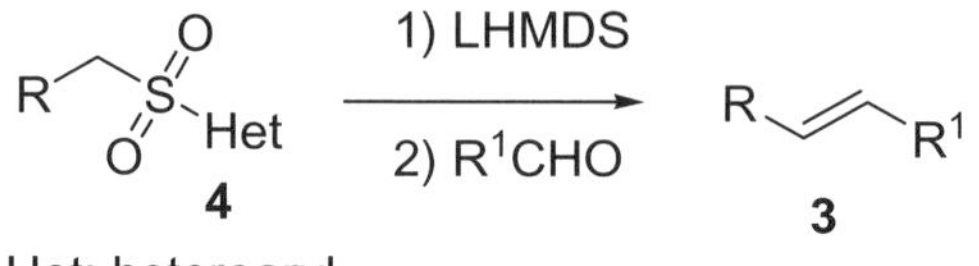

4 **3**

Het: heteroaryl

The one-port olefination of Sylvestre Julia is operationally simpler and more amenable to scale up than the classical ¾-step variant originally reported by Marc Julia. This reaction consists of the replacement of the phenyl sulfone moiety traditionally in the classical reaction, with different heteroaryl sulfones, such as benzothiazol-2-yl (BT, **5**) sulfone. This allows the direct olefination process and eliminates the sulfone reduction step. The stereochemistry of the reaction in the synthesis of 1,2-disubstituted alkenes is dependent on the base and solvent.

Kocienski and co-workers discovered that metallated 1-phenyl-1*H*-tetrazol-5-yl sulfones (PT, **6**) gave much better yields compared to its BT-substituted counterpart suggesting that the PT sulfone anions are less prone to self-condensation.[15] Other heterocyclic derivatives, such as pyridine-2-yl (PYR, **7**), 1-*tert*-butyl-1*H*-tetrazol-5-yl (TBT, **8**), and 3,5-bis(trifluoromethyl)phenyl (BTFP, **9**) sulfones[17–19] have also provided useful levels of stereoselectivity in the one-pot Julia–Kocienski olefination reaction.

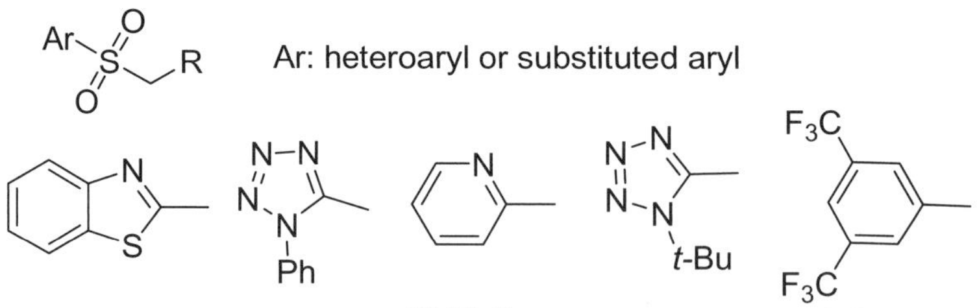

BT, 5 **PT, 6** **PYR, 7** **TBT, 8** **BTFP, 9**

2.7.2 *Historical Perspective*[4]

Marc Julia and Paris invented this methodology for the preparation of (*E*)-olefin in the synthesis of Liaisons in 1973.[1] The Julia coupling was applied to the synthesis of mono-, di- and tetra-substituted alkenes in the original communication. Kocienski and Lythgoe first demonstrated the *trans* selectivity of the Julia coupling process in 1978.[5,6] Their contribution of reductive elimination with the acetoxy or benzoyloxy sulfones gave a tremendous improvement and broad application of Julia olefination.

Sylvestre Julia and co-workers discoveried in 1991 a direct synthesis of olefins by reaction of carbonyl compounds with lithio derivatives of 2-[alkyl- or (2′-alkenyl)- or benzyl-sulfonyl]-benzothiazoles (BT, **5**).[13] Since the initial study of the reaction of metallated BT sulfone **5** with carbonyl compounds, the versatility of these derivatives has been fully demonstrated through their application in the total synthesis of a large number of nature products. Kocienski and co-workers found in 1998 that 1-phenyl-1*H*-tetrazol-5-yl sulfone (PT, **6**) is a better olefination partner comparing to BT sulfones.[15] This allowed the one-port Julia–Lythgoe olefination to be employed more efficiently and broadly, especially in the synthesis of nature products.

2.7.3 Mechanism

Julia–Lythgoe olefination reaction[2]

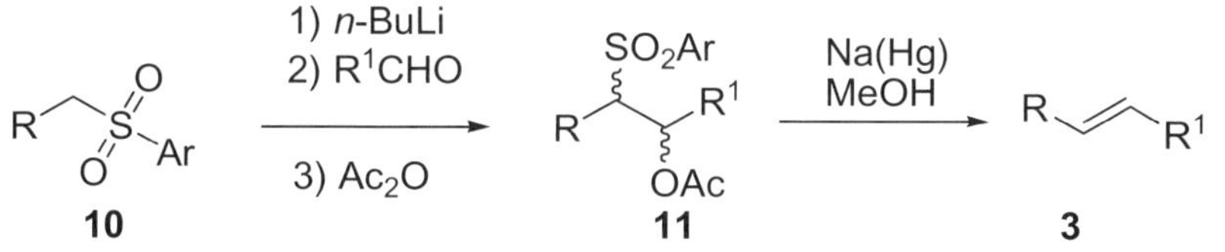

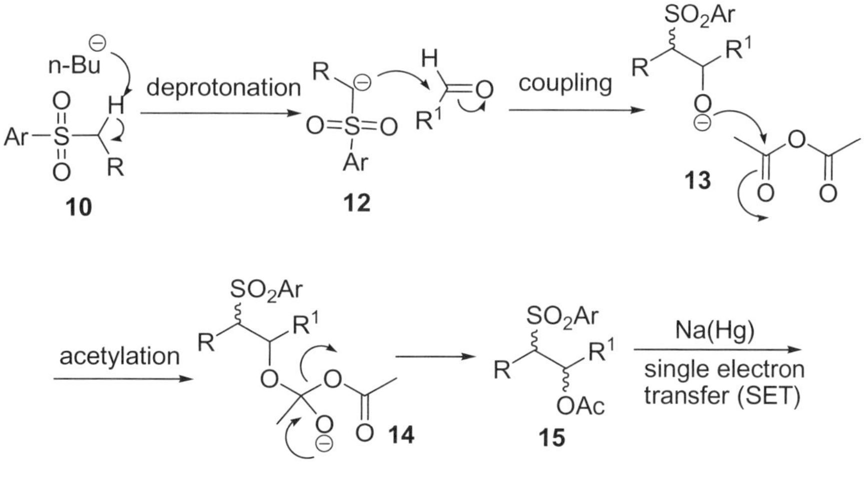

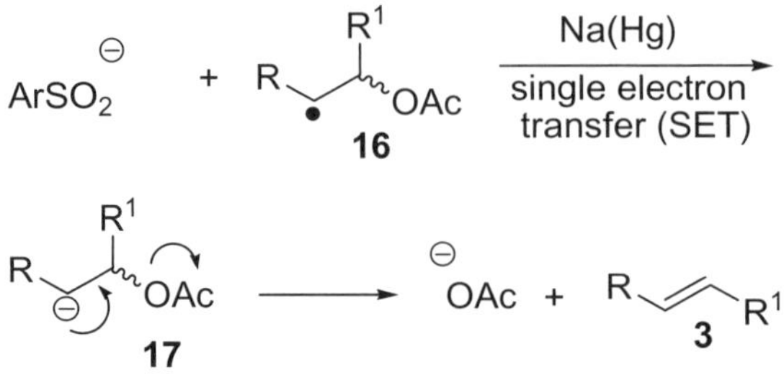

The sulfone derivative is metallated **12** and attached to the carbonyl **13**, followed by functionalization **15**, and reductive elimination, to produce the alkene **3**. The diastereoisomeric *erythro*- and *thre*o-acetoxy sulfones **15** could be separated and both isomers were converted to the *trans*-alkene. It was hypothesized that the (*E*)-selectivity is derived from the reductive removal of the arylsulfonyl group, generating an anion **17** that assumes the low energy *trans*-configuration before loss of the acylate anion. As demonstrated by numerous examples, the mechanism for reductive elimination is consistent with the finding that the alkenes obtained are the thermodynamic mixture and that increased branching at the site of elimination should, for steric reasons, increase the *trans* selectivity.

Julia–Kocienski olefination[13–16,20]

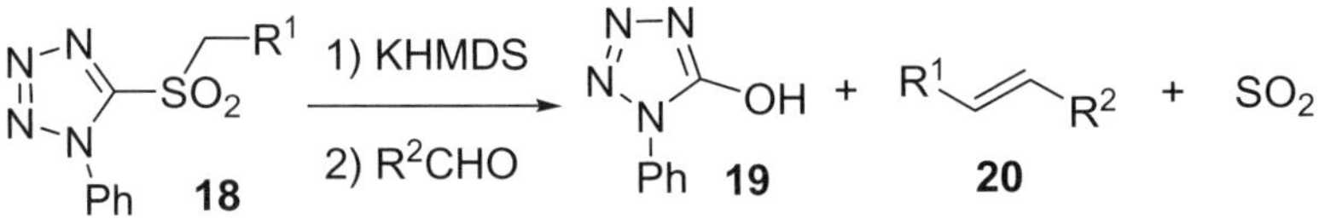

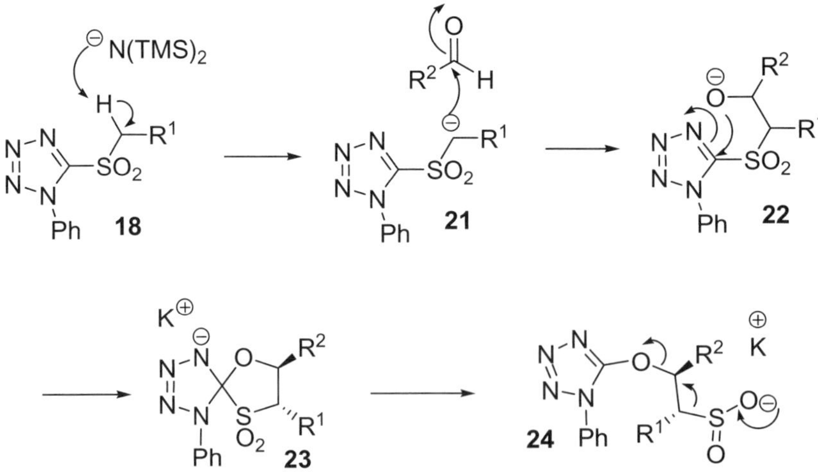

The heteroaryl sulfone **18** is deprotonated to give the carbanion **21**. Addition of canbanion to the carbonyl compound gives adduct **22**, which then undergoes a series of transformations resulting in the expulsion of sulphur dioxide and the potassium derivative of PT alcohol **25** with concomitant formation of the alkene **20**. The use of larger counterion (such as K^+) and polar solvents (such as DME) favors an open transitin state **26** (PT = phenyltetrazolyl):

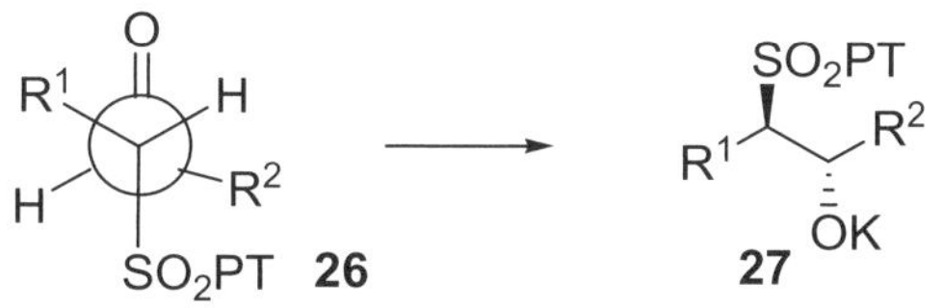

2.7.4 Synthetic Utility

Since Julia and Pairs invented the methodology for the preparation of (*E*)-olefin, it has been employed to solve many challenging synthetic problem. More importantly it had also been modified by many ways, such as changing precursors, reducing agents and conditions, to improve the yield, stereoselectivity and convenient operation. For example, Keck and co-workers reported the use of samarium diiodide as an alternative to sodium/mercury amalgam in the Julia–Lythgoe olefination.[21] The optimum protocol developed utilizes SmI_2 reduction of vinyl sulfones in the presence of DMPU and MeOH and gives generally high yields with good to excellent *E* stereoselectivity. For example, treating **28** with *n*-BuLi, then added to PhCHO, followed by acetylation gave **29** in 82% yield. **29** was further treated with DBU gave **30** in 97% yield with *E*/*Z* ratio of 4:1. The intermediate **30** was reduced by SmI_2 in THF gave **31** in 97% yield with *E*/*Z* ratio of 25:1.

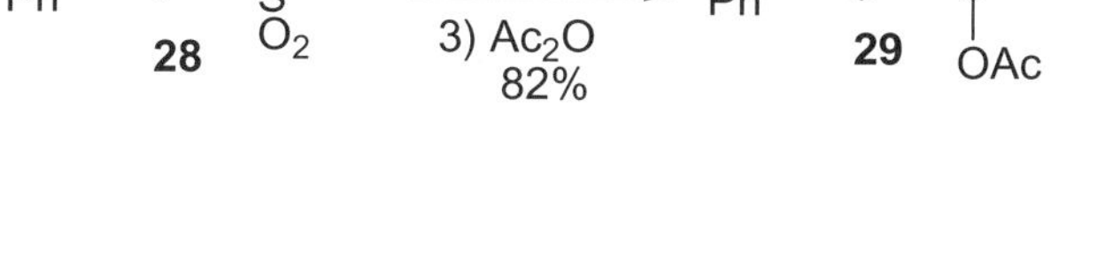

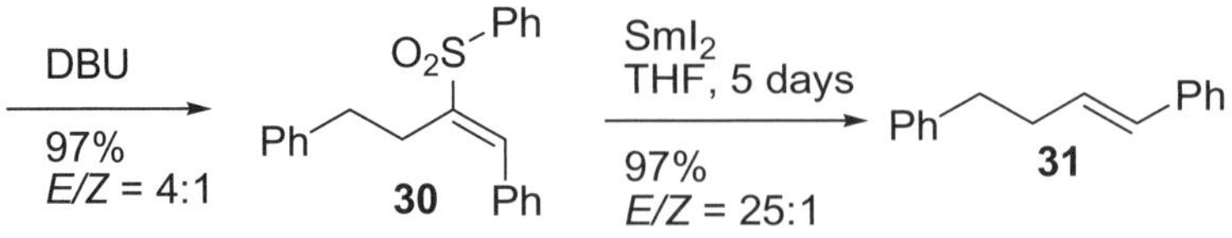

The Julia–Lythgoe olefination suffers from the production of mixtures of *E*/*Z* olefinic isomers and the poor yields of trisubstituted alkenes obtained during attempted coupling between ketones and primary sulfone anions. Marko and co-workers developed a novel variant of the Julia–Lythgoe olefination reaction, involving the *in situ* capture of intermediate *β*-alkoxy-sulfones by benzoyl or trimethylsilyl chloride, followed by SmI_2-mediated reductive elimination.[22] The sulfone **32** was treated with *n*-BuLi, and then attached to aldehyde **33**, followed by benzolation gave intermediate **34** in 89% yield, which was further reduced by SmI_2 gave **35** as *E* product in 64% yield. This method also allows the efficient preparation of trisubstituted alkenes directly from ketones and primary sulfones. This protocol also provides a connective preparation of dienyl ethers, which are important partners in Diels–Alder cycloadditions. The intermediate is a mixture of *syn*- and *anti*-diastereoisomers **37**. However, the geometric integrity of the C–C double bond of the substrate aldehydes was retained in the product. Reductive elimination of benzoate **37** using SmI_2 in THF in the presence of DMPU affording the desired dienyl ethers in 82% yield as a 1.7 : 1 mixture of **38**(*E*):**39**(*Z*)-isomers at the newly formed enol-ether double bond. No isomerisation of the styrenyl double bond occurred under these conditions.

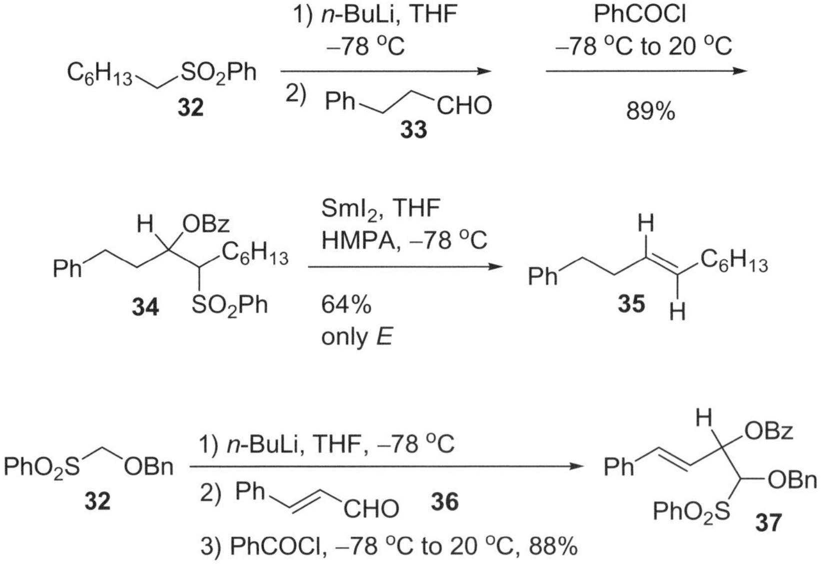

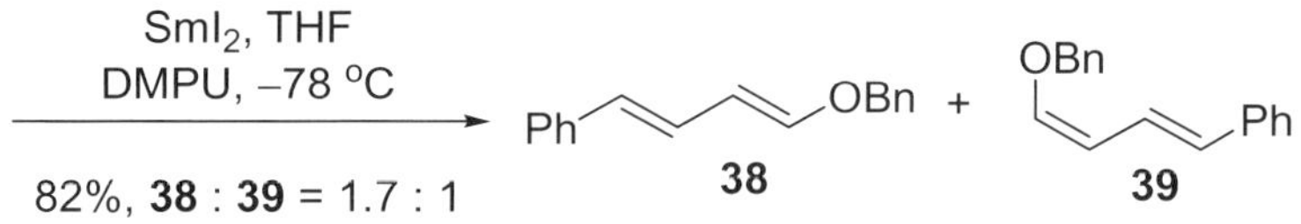

Modification of the Julia–Lythgoe olefination reaction between ketones and primary sulfones leads to trisubstituted alkenes in good overall yields reported by Marko and co-workers.[23] Samarium diiodide/HMPA shown to play a crucial role in the reductive elimination step. Starting from sulfone **40**, key intermediate **42** was produced in 93% yield, which was further converted to **43** in 73% yield with *E/Z* ratio of 2:1. More complicated intermediate **44** was similarly reduced by SmI_2 to **45** in 68% yield.

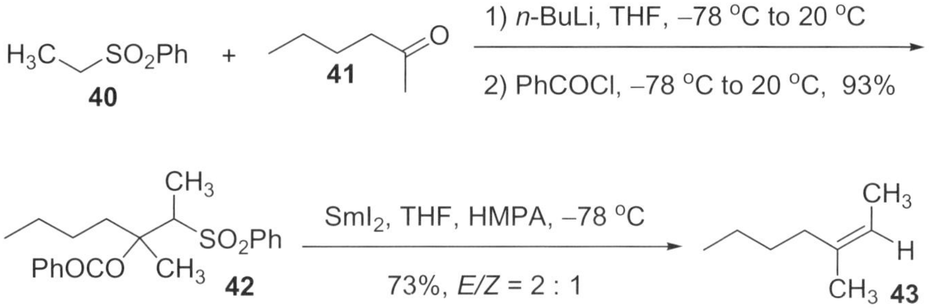

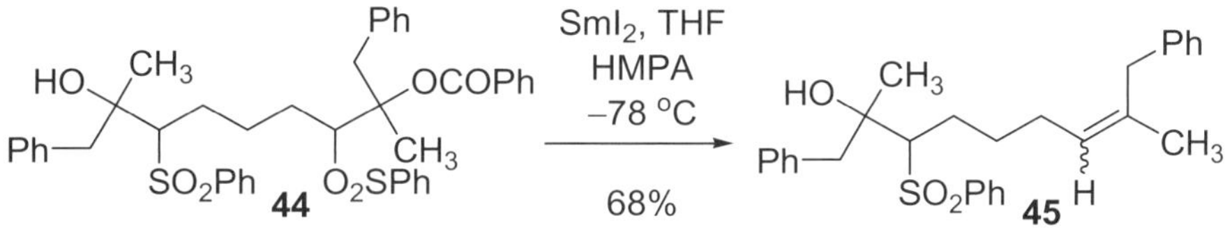

Aryl sulfoxides can be also performed as precursors in the classical Julia–Lythgoe olefination reaction. Satoh and co-workers offered a sulfoxide version of the Julia–Lythgo olefination.[24] Reaction of *β*-mesyloxy (or acetoxy) sulfoxides, drived from alkyl (or arylmethyl) phenyl sulfoxides and carbonyl compounds in two steps, with alkylmetals (*n*-BuLi, *t*-BuLi, or EtMgBr) at low temperature gave olefins in good to excellent yields. For example, starting from sulfoxide **46,** the intermediate **48** was prepared in 88% yield, which was further treated with *n*-BuLi furnished the alkene **49** in 90% yield with *E/Z* ratio of 3:1. When the *β*-hydroxy sulfoxides derived from arylaldehydes were treated with mesyl chloride in the presence of triethylamine, the sulfoxides directly gave (*E*)-olefins in good yields. Example, **52** was converted to alkene **53** in 90% yield within 30 min at room temperature. The reductive vicinal elimination was found to take place through the direct sulfoxide–metal exchange. The stereoselectivity depends on the substrates.

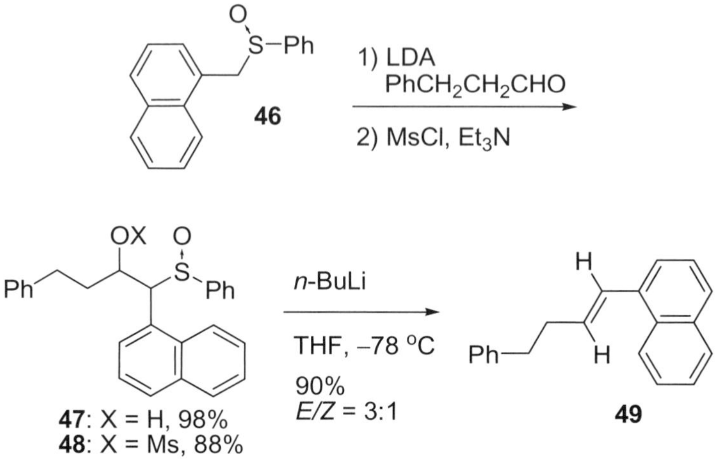

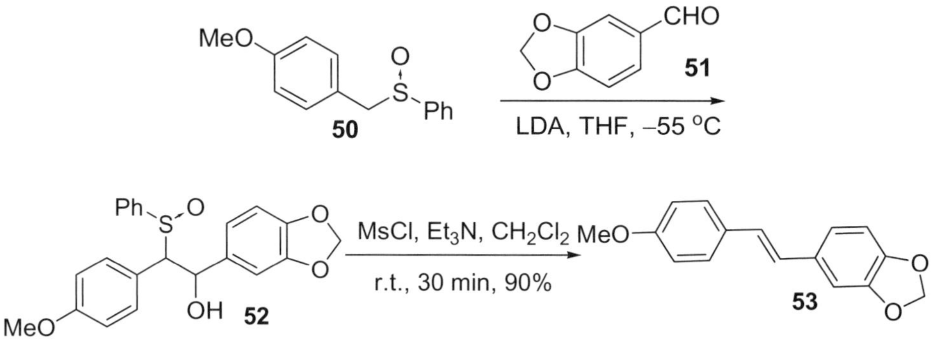

The tri-substituted olefin was also prepared by sulfoxide version of the Julia–Lythgoe olefination.[24] The α-branched sulfoxide, cyclohexyl *p*-tolyl sulfoxide **54** was treated with LDA at –55 °C, then added to the aldehyde in THF to give the adduct **55** in 78% yield, which was a mixture of two diastereomers (L : P = 66 : 34; L is less polar isomer and P is the more polar isomer on silica gel TLC). The mixture of **55** was separated by silica gel column chromatography. The isomers were acetylated separately to give **56** in almost quantitative yield. The **56** was treated with *n*-BuLi to provide final product **57**. The P-isomer gave better yields (90%) than the L-isomer (43%).

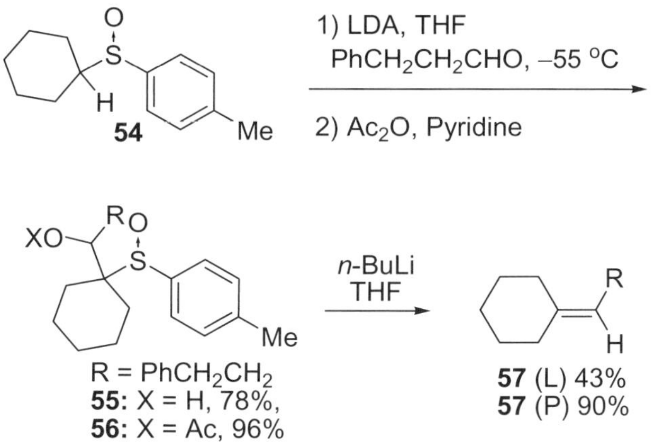

The use of the Julia–Lythgoe olefination with cyclopropylsulfones and cyclopropylsulfoxides for the synthesis of alkylidenecyclopropanes was reported by Bernard and co-workers.[25] The adduct sulfide **60**, prepared from sulfide **58**, was oxidized to sulfone **61** and sulfoxide **64** by controlling oxidation temperature. Both intermediate **61** and **64** can be further converted to trisubstituted alkene **63** under different reductive condition.

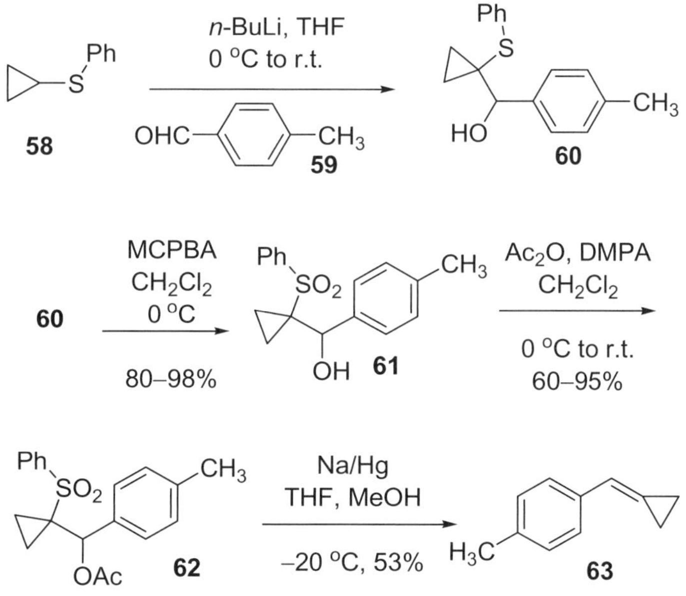

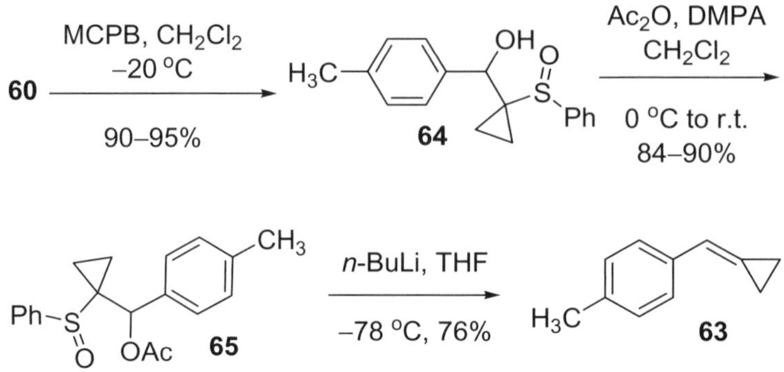

Marko and co-workers also made an efforts on the modification of the classical Julia–Lythgoe olefination using sulfoxides instead of sulfones.[26] The modified reaction, after *in situ* benzoylation and SmI_2/HMPA- or DMPU-mediated reductive elimination, provides 1,2-di-, tri-, and tetra-substituted olefins in moderate to excellent yields and *E/Z* selectivity. For example benzoated sulfoxide **67** was obtained by addition of sulfoxide **66** to aldehyde **33**, which was subsequently reduced into alkene **68** in 67% yield with *E/Z* ratio of > 95:1.

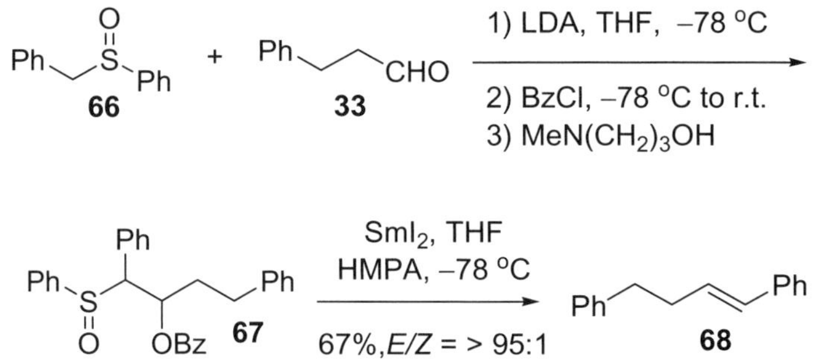

The Julia–Kocienski olefination is more convenient in the synthesis of alkenes than the classical Julia olefination. A direct synthesis of olefins by reaction of carbonyl compounds with lithio derivatives of BT sulfones was first reported by Julia and co-workers.[13] The lithium alkoxides formed as intermediates undergo an intramolecular addition to the neighboring C=N group followed an S to O benzothiazole transfer and simultaneous extrusion of sulphur dioxide and ejection of 2(3*H*)-benzothiazolone anion. The sulfone **69** was treated with LDA, and then reacted with acetone gave alkene **70** in 51% yield.

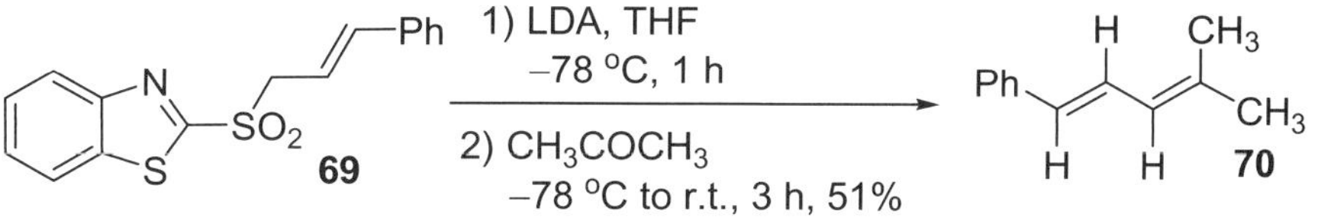

A stereoselective construction of the *D-erythro*-azidosphingosine characteristic *trans* double bond was accomplished by Panza and co-workers by condensation reaction between tetradecanol and a heterocyclic sulfone derived from diethyl *D*-tartrate, following the Kocienski modification of the Julia–Lythgoe olefination.[27] Alcohol **71** was first converted into the 1-phenyl-1*H*-tetrazole-5-yl thioether under Mitsunobu conditions and then oxidized to **73** in 80% yields. A solution of sulfone **73** in DME at –55 °C was treated with KHMDS to give a stable anion of compound **73**, which was then reacted with tetradecanal to give compound **74** in 53% yield. The compound **74** can be efficiently transformed into the target, 3-*O*-(4-methoxybenzyl)-azidosphingosine, with reported procedure.

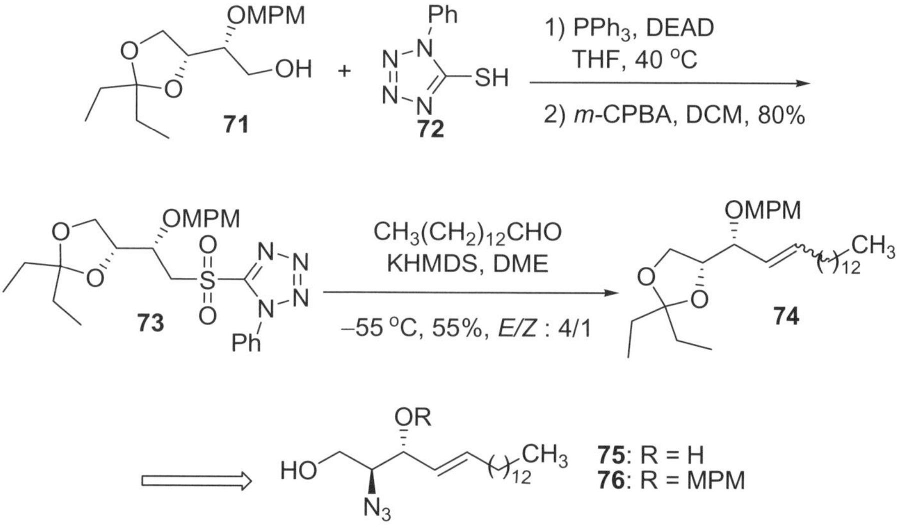

The synthesis of *E,Z*-dienes, **79** and **80**, was developed by Charette and co-workers from *α,β*-unsaturated aldehydes, **77**, and heteroarylsulfones, **78,** using the Julia–Lythgoe olefination reaction.[28] The selectivity of the olefination reaction under optimal conditions is better than 88:12 when a pyridylsuofone was used as the precursor.

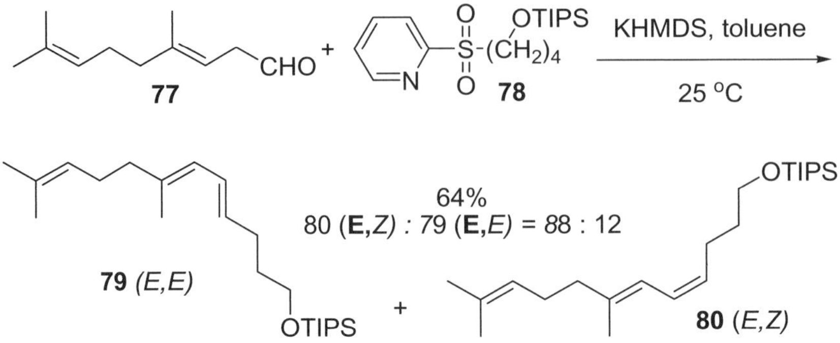

The Julia–Lythgoe reaction precursors, heteroarylsulfones, were prepared efficiently by using H_2O_2/Na_2WO_4 combination oxidation of heteroarylthioethers in the presence of alkenes.[28] Coupling **82** with **81** gave heteroarylthioether in 84% yield, which was subsequently oxidized to heteroarylsulfone **83** in > 95% yield at room temperature.

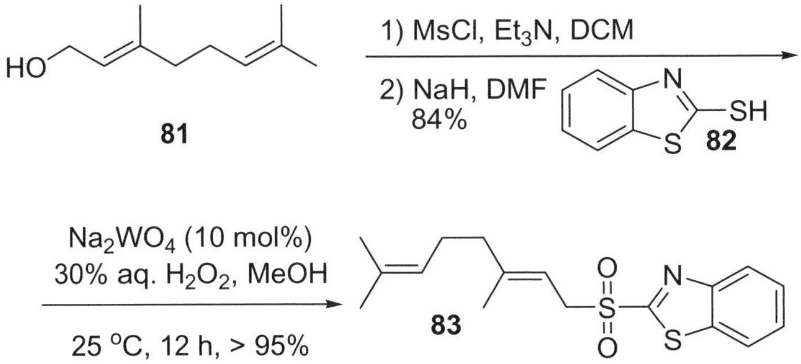

Najera and co-workers reported the application of 3,5-bis(trifluoromethyl)phenyl (BTFP, **9**) sulfones in the modified Julia olefination to the synthesis of various stilbenes such as resveratrol.[19] This one-pot protocol can be performed using KOH at room temperature or the phosphazene base P4–*t*-Bu at –78 °C, and given a high yielding and good stereoselectivity. For example, the sulfone **85**, prepared from **84** by two-steps reaction, was reacted with aldehyde **86** in the presence of P4–*t*-Bu in THF finished the alkene **87** in 74% yield with *E*/*Z* ratio of 98:2.

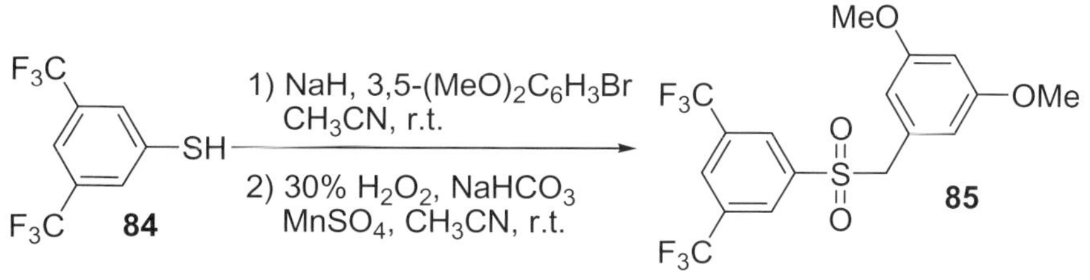

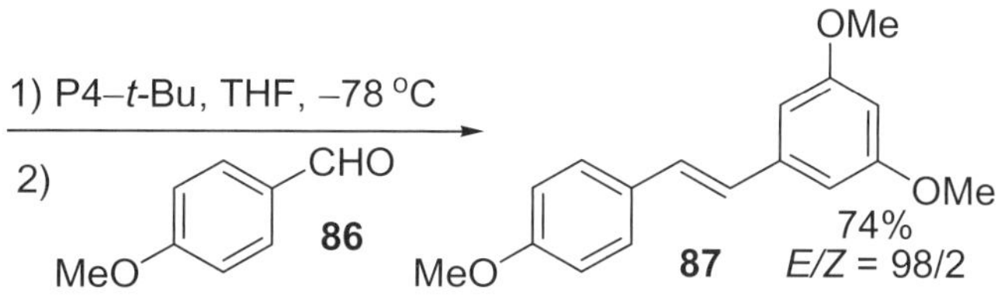

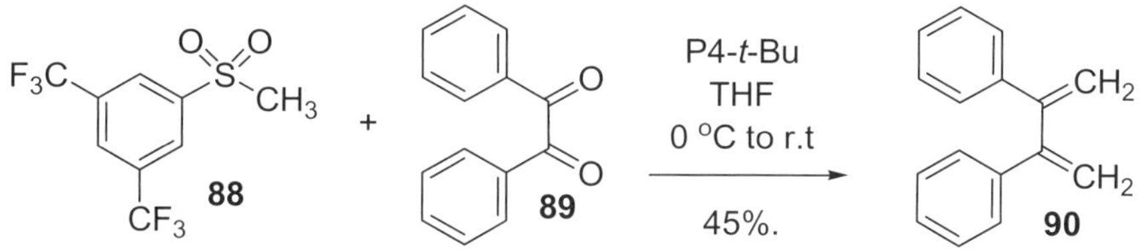

The synthesis of terminal alkenes from carbonyl compounds is a very important reaction in organic synthesis. Najera and co-workers found out that BTFP methyl sulfone is a good reagent to perform methylation reactions of carbonyl compounds through the Julia–Kocienski protocol.[18] The reaction proved to be much more efficient when Barbier-type conditions were used. Example, alkene **90** was obtained in 45% yield by reacting sulfone **88** with diketone **89** in the presence of P4–*t*-Bu in THF. For example, the sulfone **92**, prepared from **84** by two-steps reaction in 81% yield, was converted to alkene **94** in 95% yield and **96** in 71% yield respectively.

Najera and co-workers have also successfully employed BTFP sulfones in the synthesis of tri- and tetrasubstituted olefins by reacting with carbonyl compounds in the presence of phosphazene base P4–*t*-Bu at room temperature in THF.[17]

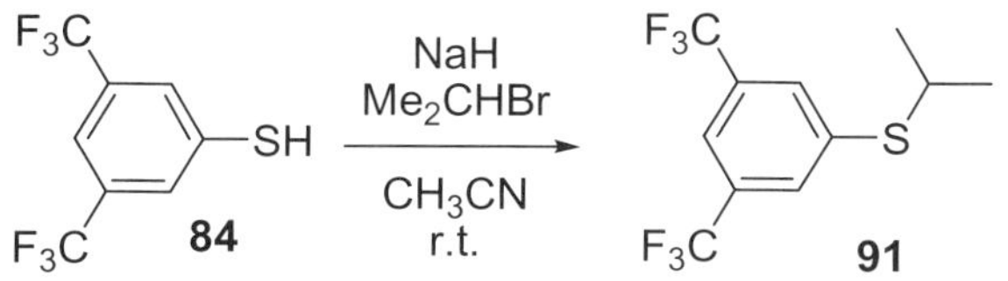

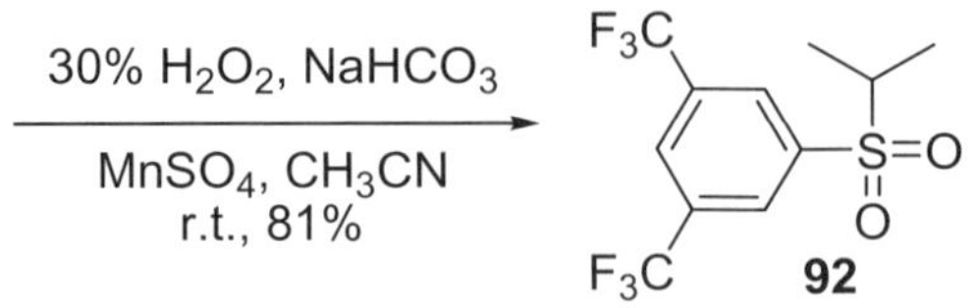

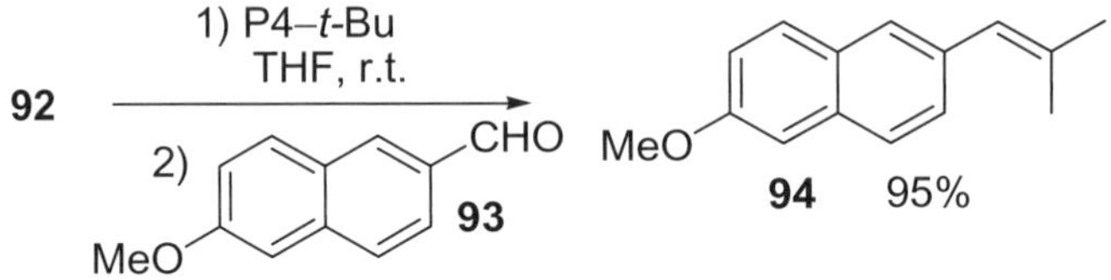

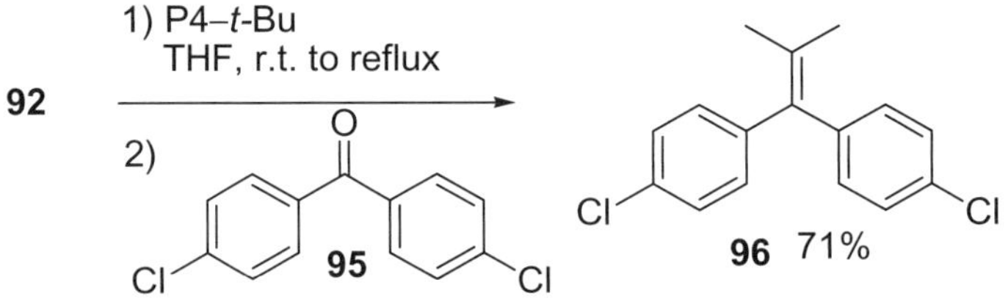

2.7.5 *Natural Product Synthesis*

The Julia–Lythgoe olefination and Kocienski modification have applied broadly in the synthesis of nature products. Isoprostane of A_2 and J_2 are isomeric of the cyclopentenone prostaglandins A_2 and J_2, respectively, which are reported to exert unique biological effects. Prostaglandins of A_2 and J_2 series have been reported to be active against a wide variety of DNA and RNA viruses, including HIV-1 and influenza virus. They also possess a potent anti-inflammatory activity due to the inhibition and modification of the subunit IKKβ of the enzyme I*k*B kinase. Zanoni and co-workers reported the first total synthesis of A_2 Isoprostane **101** employed a stereoselective Julia–Lythgoe olefination in the formation of C_{13}–C_{14} double bond.[29] The intermediate **99**, obtained in 87% yield by addition of sulfone **97** to aldehyde **98**, was reduced by Na(Hg) to alkene **100** in 81% yield.

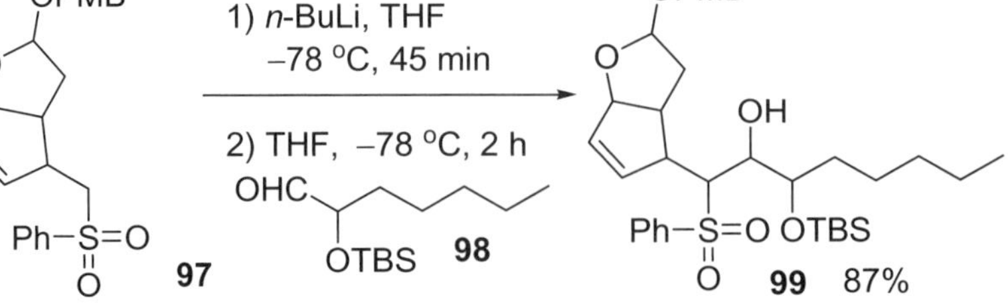

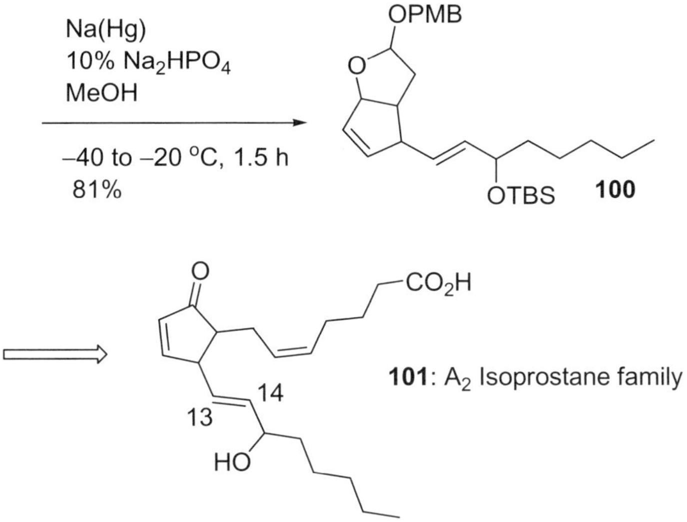

The macrolactins are a structurally diverse class of secondary metabolites isolated from a deep-sea bacterium. Macrolactin A exhibits a broad spectrum of activity with significant antiviral and cancer cell cytotoxic properties including inhibition of B16–F10 murine melanoma cell replication with *in vitro* IC_{50} values of 3.5 μg/mL. It also has implications for controlling human HIV replication and is a potent inhibitor of *Herpes simplex* types I and II. Marino and co-workers reported a stereocontrolled total synthesis of (–)-macrolactin A **107**. Julia–Lythgoe olefination was employed in the formation of C_{18}–C_{19} double bond.[30] The intermediate **105** was obtained in 78% yield by addition of sulfone **103** to aldehyde **102**, benzolation of the adduct intermediate, and treatment with KO*t*-Bu. The alkene **106** was obtained in 56% yield with *E/Z* ratio of 8:2 by reduction of **105** using SmI_2 in the presence of DMPU.

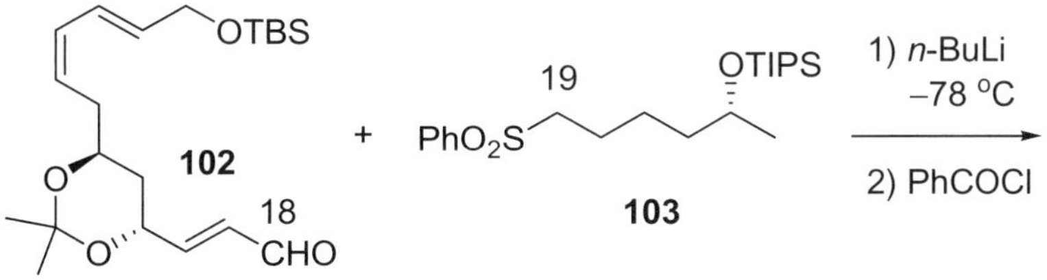

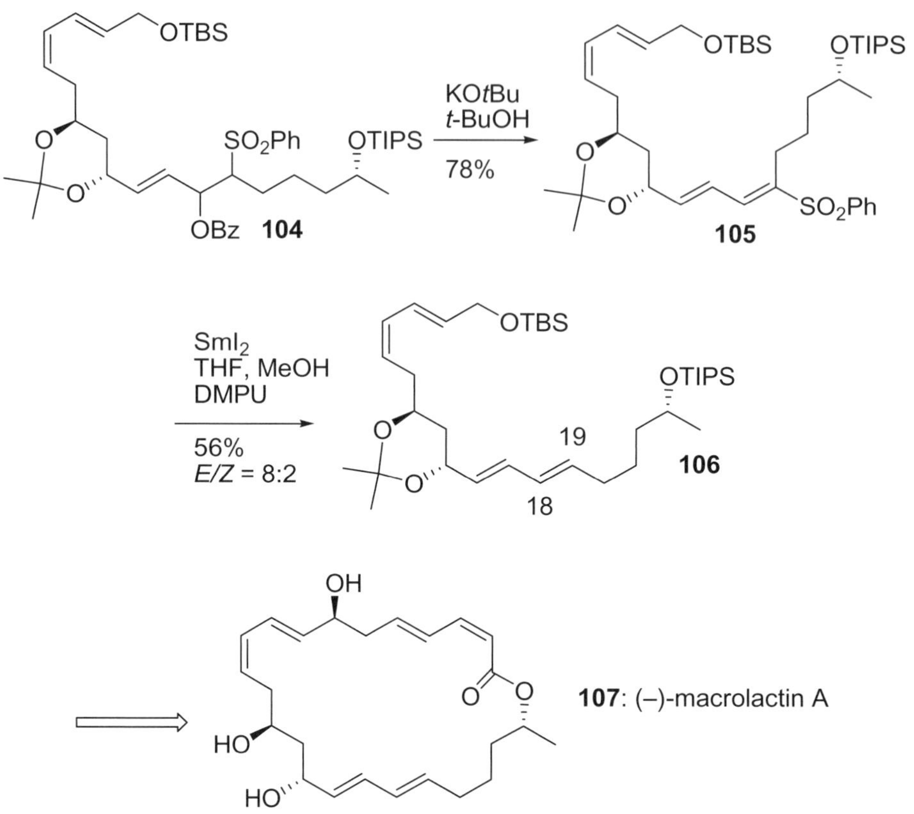

Moenocinol is a part of moenomycin antibiotics, and was synthesized by Huang and Yang by 10 linear steps in 12% overall yield. Julia–Kocienski olefination occurred in a stereospecific manner to give the desired 6*E*-configratin of moenocinol.[31] The intermediate **111**, obtained in 58% yield by reacting sulfone **109** with aldehyde **110**. The final product **112** was obtained in 86% yield by de-protection of **111** using TBAF in THF.

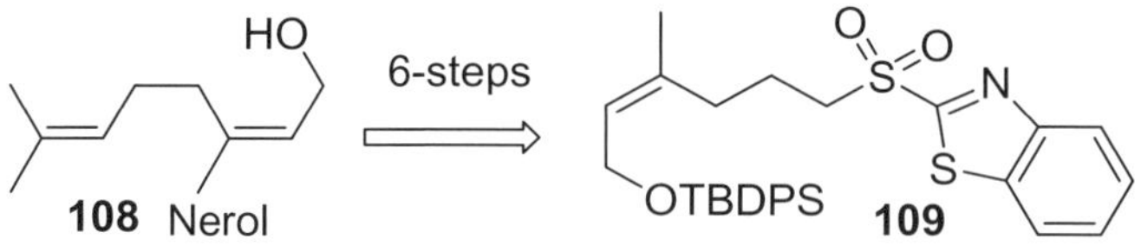

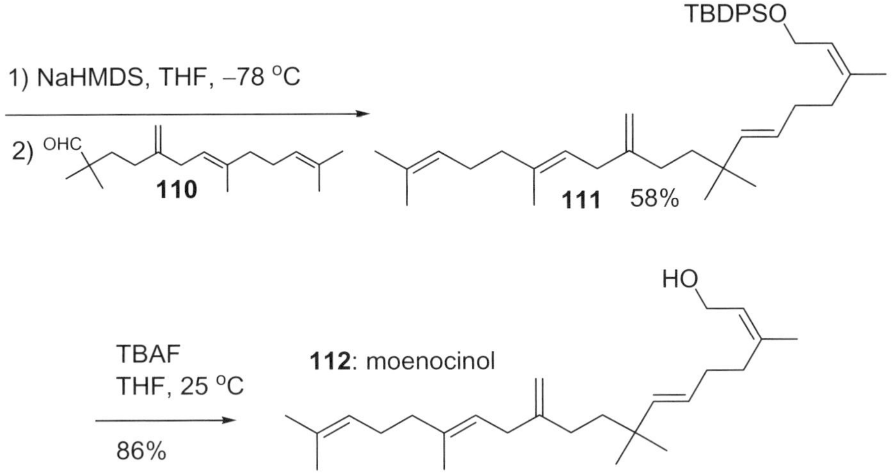

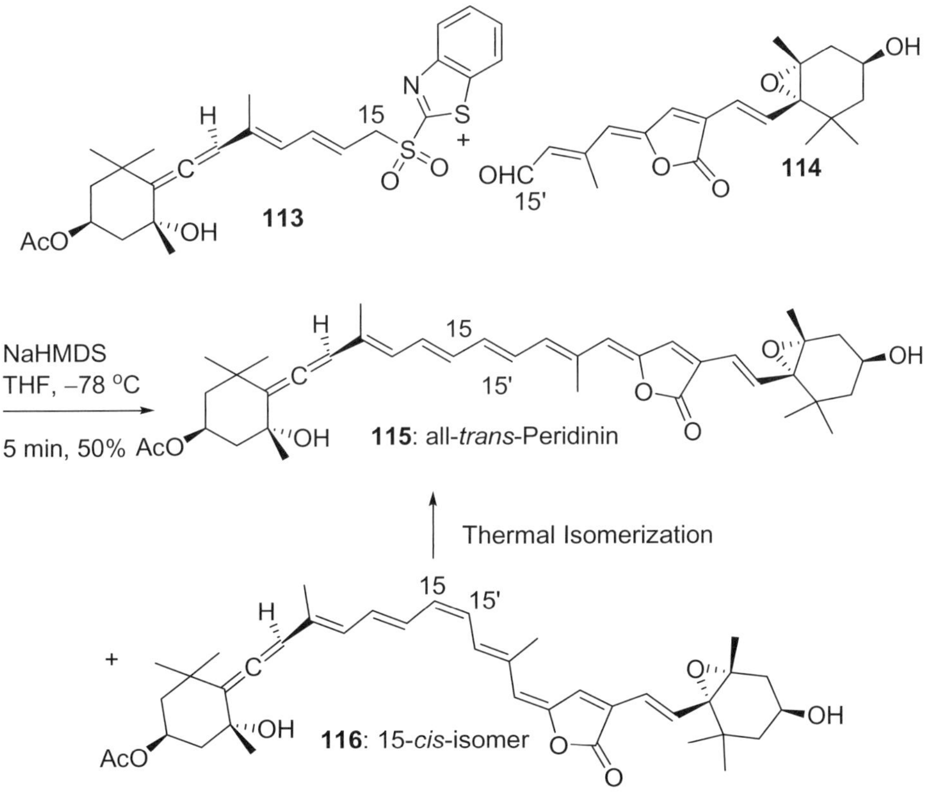

Peridinin, a polyfunctional carotenoid, was isolated from the planktonic algae dinoflagellates causing red tides and shown antitumor and anticarcinogenic activities. Katsumura and co-workers reported a stereocontrolled total synthesis of Peridinin **115**. The modified Julia–Kocienski olefination was used to form the C_{15}–$C_{15'}$ double bond.[32] Addition of sulfone **113** to aldehyde **114** gave a mixture of all *trans*-peridinin **115** and 15-*cis*-isomer **116** which was converted back to **115** by thermal isomerisation.

Danishefsky and co-workers reported a total synthesis of antibiotic *dl*-indolizomycin **121**.[33] In which Julia olefination was successful employed to prepare the triene moiety. The intermediate **119**, prepared by addition of **118** to aldehyde **117**, was reduced to **120** by using 5% Na(Hg).

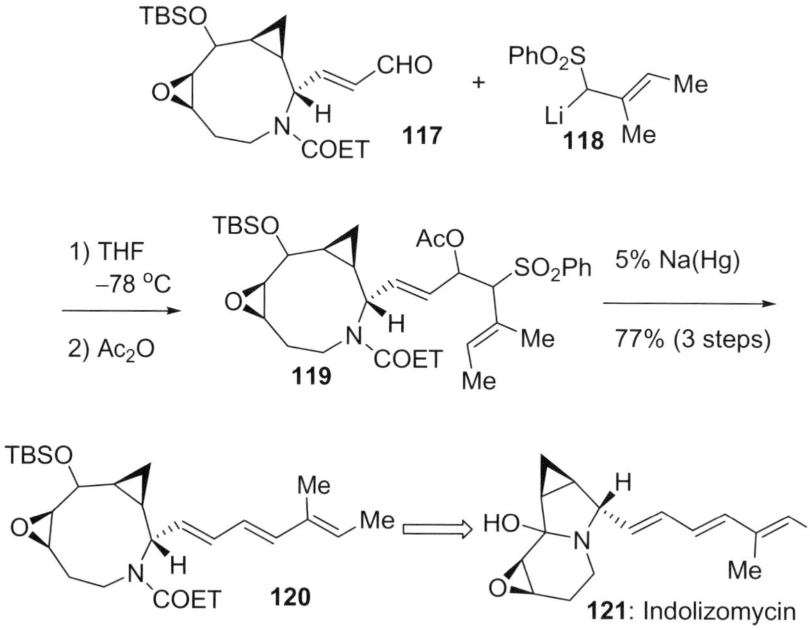

(+)-Thiazinotrienomycin, novel ansamycin antibiotics produced by *Streptomyces* sp. MJ672-m3, shown significant *in vitro* cytotoxicity against human cancer cell lines derived from the cervix, stomach, colon, and breast. Smith and Wan reported a first total synthesis of (+)-thiazinotrienomycin E **128**.[34] Both the Julia protocol and the Kocienski modified Julia–Lythgoe olefination were used to elaborate the *E,E,E*-triene subunit in a stereo-controlled fashion. The alkene **126** was obtained in 35% yield with *E/Z* ratio of 1.5:1 by addition of sulfone **124**, prepared from **123**, to aldehyde **125**. However, the **126** was obtained in 85% yield with *E/Z* ratio of 10:1 by addition of **127**, prepared from 1-phenyl-1*H*-tetrazole **72**, to aldehyde **125**. It is very obvious that the Kocienski modified Julia–Lythgoe olefination gave better results than Julia protocol in term of yield and *E*-selectivity.

Importantly no isomerisation at either the C(4,5) or C(6,7) olefins was observed.

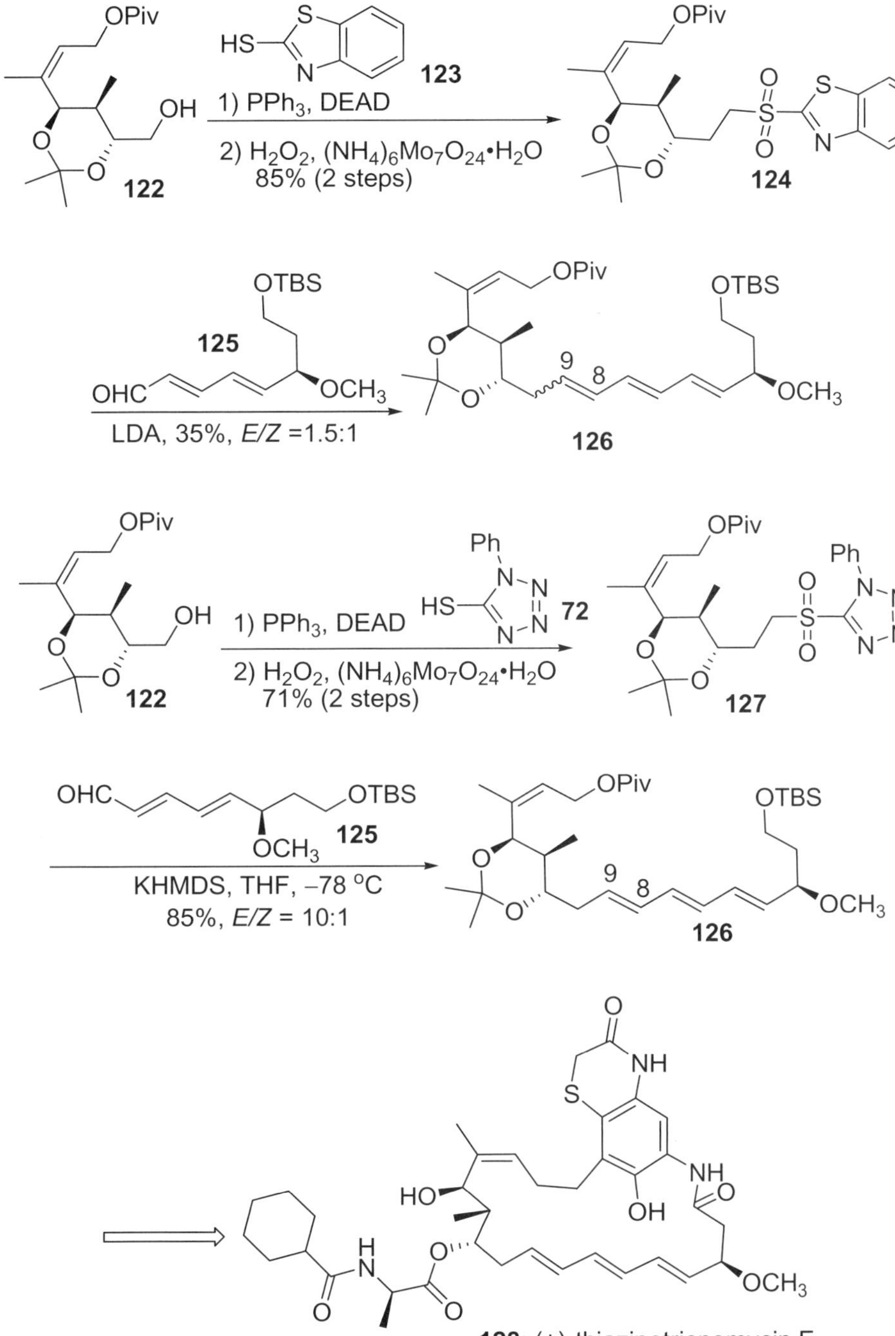

128: (+)-thiazinotrienomycin E

Natural product (+)-U-106305 was a potent inhibitor of the *in vitro* cholesteryl ester transfer protein reaction. Charette and Lebel reported an enantioselective total synthesis of (+)-U-106305 **134**.[35] Julia olefination was used for the double bond formation. The yield of **131** was > 90% with *E/Z* ratio of 3.5:1 by reacting sulfone **129** with aldehyde **130** under optimized condition using 1.0 M solution of NaHMDS in THF. Similarly alkene **133** was obtained in 92% yield with *E/Z* ratio of 4:1 by reacting aldehyde **132** with sulfone **129** and followed by TBAF de-protection of silyl moiety.

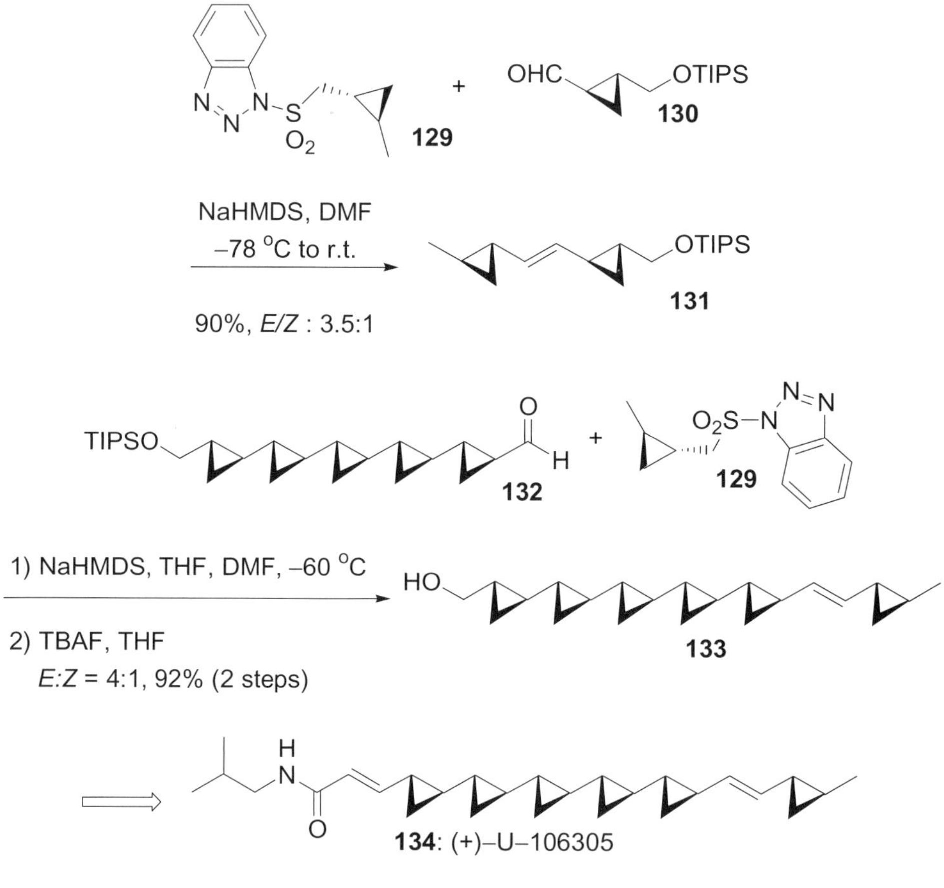

Ambruticin is a novel antifungal agent. It exhibits pronounced activity against systemic medical pathogens, and displays potent inhibitory activity against the yeast strain *Hansenula anomala* with an MIC of 0.03 ug/mL. Liu and Jacobsen reported a concise and highly stereocontralled total synthesis of Ambruticin **138**.[36] The total synthesis was accomplished in 16 steps and 12% yield in the longest linear sequence (21 steps, overall). Kocienski–Julia olefination was used in the formation of C_8–C_9 double bond. It was found that the high selectivity for either double bond isomer could be obtained. *Z*-alkene of **137** was produced as major product (*E/Z* = 1:8) by reacting sulfone **136** with aldehyde **135** if using NaHMDS in THF at –78 °C. However, *E* isomer of **137** almost exclusively (*E/Z* = > 30:1) was obtained if utilizing LiHMDS at –35 °C in polar solvents, such as DMF/DMPU (1:1 v/v).

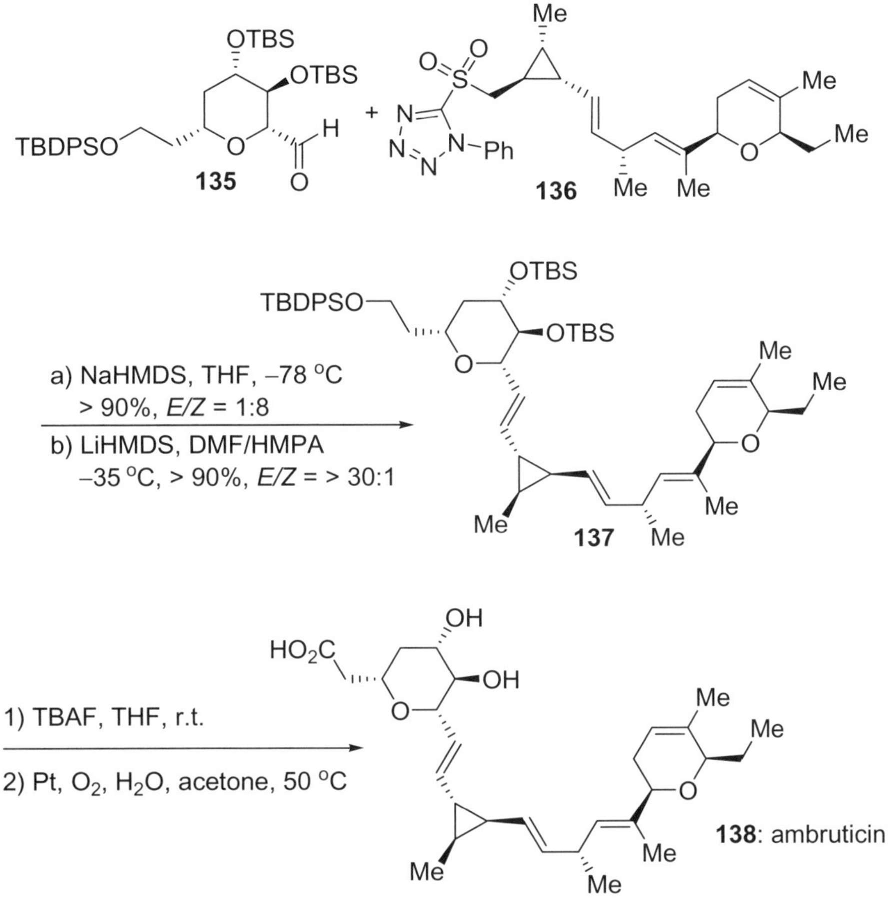

138: ambruticin

Hennoxazole A displays potency against *Herpes Simplex* virus type 1 and peripheral analgesic activity comparable to that of indomethacin. Williams and co-workers reported a total synthesis of (–)-hennoxazole A **141**.[37] The Kocienski modification of the Julia–Lythgoe olefination was very successfully employed in the formation of C_{17}–C_{18} alkene in 85% yield with excellent *E*-selectivity (*E/Z* = 91:9) by reacting sulfone **140** with aldehyde **139**. Hydrolysis of the C_4 pivaloate ester (LiOH in aqueous THF/MeOH) provided synthetic hennoxazole A (**141**) in 72% yield.

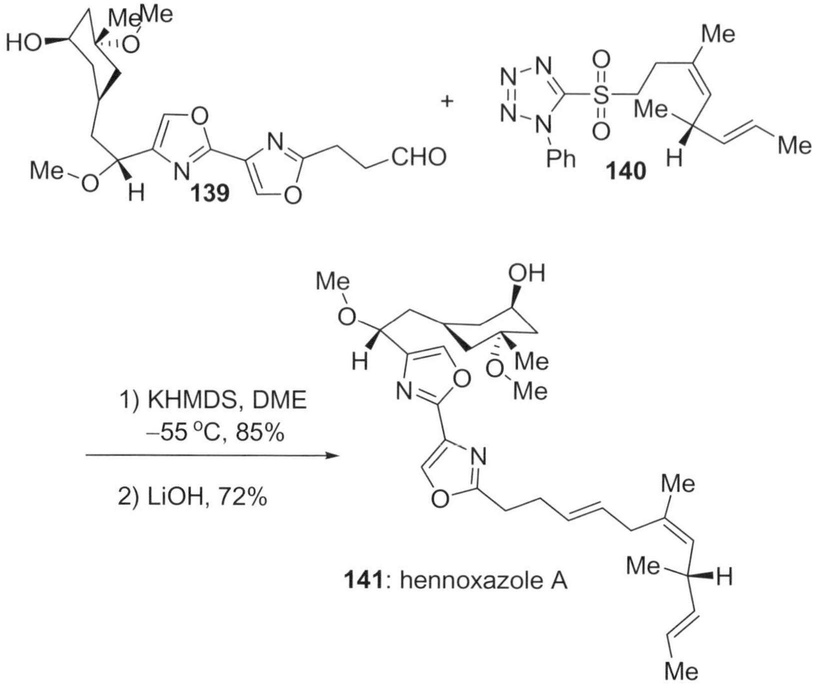

Spirotryprostatin A **142** and B **143** are two powerfully bioactive indole alkaloids. Both compounds inhibit the cell cycle in the G2/M phase, and **143** shows cytotoxic activity on the growth of human leukemia cell lines. Meyers and Carreira reported a total synthesis of **143**.[38] The Kocienski modified Julia olefination was used in the formation of trisubstituted olefin without scrambling at C_{18}. The alkene **146** was prepared in 78% yield by reacting sulfone **145** with aldehyde **144**. The final product **143** was obtained by four-step reaction from the intermediate **146**.

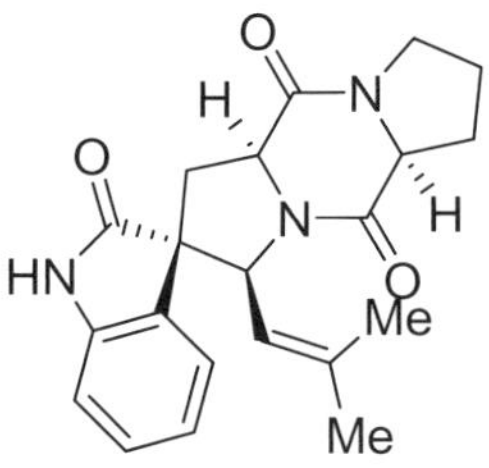

142: spirotryprostatin A

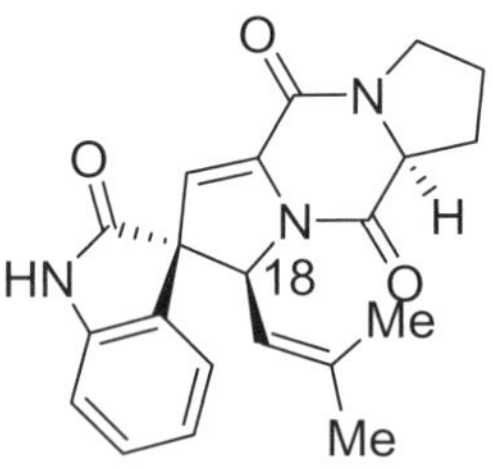

143: spirotryprostatin B

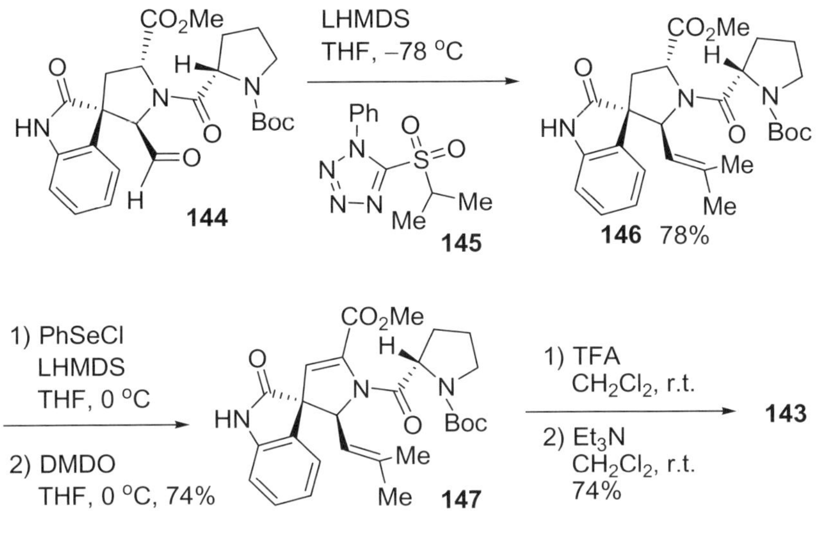

Herboxidiene **154** shows potent herbicidal activity and up-regulates gene expression of low density lipoprotein receptors. Kocienski and co-workers synthesized herboxidiene **154** successfully from two key fragments using a modified Julia olefination based on the benzothiazolyl sulfone activator.[20] The yield was 81% with excellent selectivity (10*E*:*Z* = 91:9) by one-port reaction between sulfone **152** and the aldehyde **151**. 1-Phenyl-1*H*-tetrazolyl sulfone **149** was used as activator in the synthesis of the C_1–C_{10} oxane fragment **151**. Addition of KHMDS to the mixture of sulfone **149** and aldehyde **148** in 1,2-dimethoxyethane at –60 °C gave a 93% yield of the alkene **151** with good stereoselectivity (*E*:*Z* = 93:7).

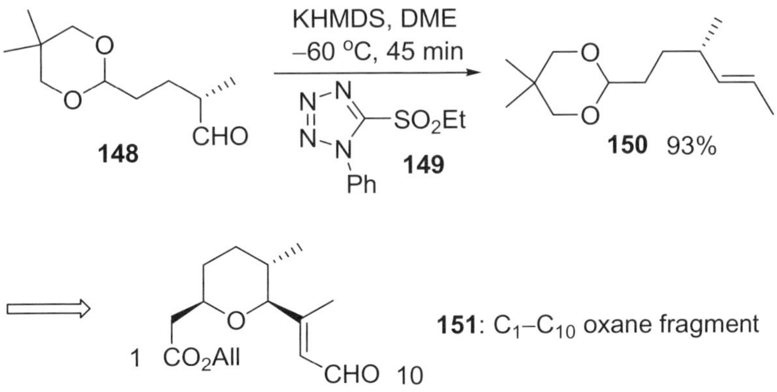

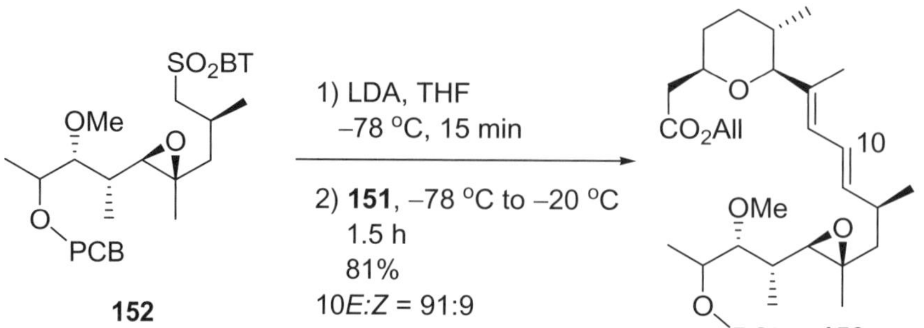

152 **153**

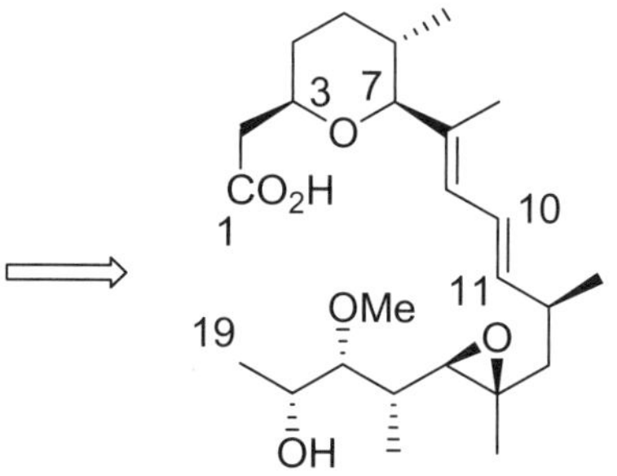

154: herboxidiene

2.7.5 *Experimental*

Procedure for the coupling of the alkyl sulfone and the aldehyde to generate the α-alkoxy sulfone.

Preparation of 1-acetoxy-1,4-diphenyl-2-(phenylculfonyl)butane (29).[21]
A solution of alkyl sulfone **28** (1.00 g, 3.85 mmol) in THF (35 mL) was cooled to −78 °C, *n*-BuLi (1.88 mL of 2.25 M in hexanes, 4.24 mmol) was added dropwise, and the bright yellow clear solution was mixed for 30 min. Benzaldehyde (429 mg, 0,411 mL, 4.039 mmol) in 4 mL of THF was added dropwise via cannula, and the solution was washed in with 2 mL of THF. The mixture was kept at −78 °C for 3 h before Ac_2O (786 mg, 0.730 mL, 7.70 mmol) was added dropwise via syringe. The bath was maintained at −78 °C for 1 h before the bath was allowed to expire, bringing the solution to room temperature overnight. The reaction was quenched with saturated NH_4Cl, the solution was extracted with CH_2Cl_2 (1 × 50 mL, 2 × 15 mL) and with ether (1 × 20 mL), and the combined organics were dried over $MgSO_4$, filtered through Celite (0.5 cm) and silica gel (2 cm) and concentrated *in vacuo*. The crude product was purified by RPLC (4 mm plate) loaded with 2 mL of 5% EtOAc/hexanes and eluted with 100 mL of hexanes, 100 mL of 5% EtOAc/hexanes, 100 mL of 10% EtOAc/hexanes, 100 mL of 15% EtOAc/hexanes, 150 mL of 20% EtOAc/hexanes, and 150 mL of 35%

EtOAc/hexanes to yield 1.325 g (85%) of **29** as a clear slightly yellow liquid: R_f 0.14 (20% EtOAc/hexane); major isomer.

Procedure for the conversion of acetoxy sulfones to vinyl sulfones with 1,8-diazobicyclo[5.4.0]-undec-7-ene.

Preparation of 1,4-diphenyl-2-(phenylsulfonyl)-1-butene (30**).**[21]
To a solution of acetoxy sulfone **29** (1.47 g, 3.607 mmol) in THF (50 mL) was added DBU (3.30 g, 3.24 mL, 21.65 mmol) dropwise via syringe. After 18 h, the reaction was judged complete by TLC and quenched by dilution with either (20 mL) and brine (10 mL). The aqueous phase was re-extracted with methylene chloride (3 × 60 mL). The combined organics were extracted with brine (10 mL), dried over $MgSO_4$, filtered through Celite (0.5 cm) and silica gel (2.0 cm), and concentrated *in vacuo*. The crude product was purified by RPLC on a 4 mm plate in two passes loaded with 5% EtOAc/hexanes and eluted with 50 mL of hexanes, 75 mL of 5% EtOAc/hexanes, 100 mL of 10% EtOAc/hexanes, 150 mL of 15% EtOAc/hexanes, and 150 mL of 20% EtOAc/hexanes to yield 1.26 g (97%) of the product **30** as a clear colorless solid (mp 87–90 °C): R_f 0.38 (20% EtOAc/hexane); major isomer.

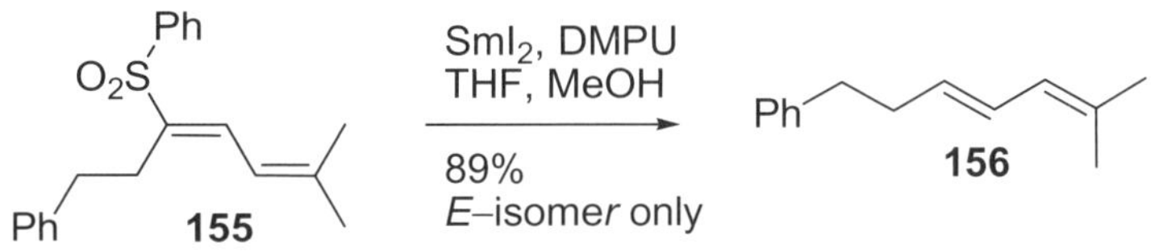

Procedure for the reductive cleavage of a vinyl sulfone to afford the olefin.

Preparation of 2-methyl-7-phenyl-2,4-heptadiene (156**).**[21]
To a suspension of samarium (249 mg, 1.660 mmol) in THF (15 mL) was added iodine (373 mg, 1.470 mmol) in one portion. The mixture was then heated at 65 °C (bath temperature) for 90 min over which time the reaction went from a cloudy brick red color to yellow and then blue green in color characteristic of SmI_2. The mixture was cooled to rt, and the vinyl sulfone **155** (60.0 mg, 0.184 mmol), DMPU (236 mg, 223 μL, 1.84 mmol), and MeOH (61.0 mg, 77 μL, 1.84 mmol) in 2 mL of THF were added dropwise via cannula and washed in with 1 mL of THF. The mixture went purple over the course of the addition. After 30 min, the reaction was judged complete by TLC and quenched with saturated aqueous Na_2SO_3, mixed for 30 min, and then extracted with ether (1 × 20 mL). The organic phase was washed once with 10 mL of brine, and the aqueous phase was then re-extracted once with 10 mL of pentane. The combined organic phase was dried over $MgSO_4$,

filtered through Celite (0.5 cm) and silica gel (1.0 cm), concentrated *in vacuo*, and purified via RPLC on a 2 mm plate loaded with 2 mL of 5% EtOAc/hexanes and run with 150 mL of hexanes, 150 mL of 5% EtOAc/hexanes, and 150 mL of 10% EtOAc/hexanes to yield the product **156** as a clear colorless liquid (31 mg, 89%).

Preparation of Triene (+)-126.[34]
A solution of (+)-**127** (1.9 g, 3.65 mmol) in THF (20 mL) at –78 °C was treated with potassium bis(trimethylsilyl)amide (0.5 M in toluene, 9.5 mL, 4.75 mmol). The resultant yellow solution was stirred for 20 min before (–)-**125** (1.1 g, 3.84 mmol) in THF (10 mL) was introduced via cannula. The mixture was stirred for 1 h at –78 °C, warmed at room temperature and stirred for an additional 1 h. The mixture was poured into brine (100 mL) and extracted with ether (3 × 100 mL). The combined organic phases were dried over $MgSO_4$, filtered, and concentrated. Flash chromatography (hexanes/ethyl acetate, 20:1) provided (+)-**126** (1.8 g, 85% yield) as a colorless oil.

2.7.6 References

1. Julia, M.; Paris, J. M. *Tetrahedron Lett.* **1973**, *14*, 4833–4836.
2. Kocienski, P. *Phosphorus and Sulfur* **1985**, *24*, 97–127 (review).
3. Kelly, S. E. Comp. *Org. Syn.* **1991**, *1*, 792–806 (review).
4. Kocienski, P.; Lythgoe, B; Waterhouse, I. *J. Chem. Soc., Perkin Trans I* **1980**, 1045–1050.
5. Kocienski, P.; Lythgoe, B. Ruston, S. *J. Chem. Soc., Perkin Trans I* **1978**, 829.
6. Kocienski, P.; Lythgoe, B.; Roberts, D. A. *J. Chem. Soc., Perkin Trans I* **1978**, 834–837.
7. Jones, A. B.; Villalobos, A.; Linde II, G. R.; Danishefsky, S. J. *J. Org. Chem.* **1990**, *55*, 2786.
8. Ihara, M.; Suzuki, S.; Taniguchi, T.; Tokunaga, Y.; Fukumoto, K. *Synlett.* **1994**, 859.
9. Lee, G. H.; Lee. H. K.; Choi, E. B.; Kim, B. T.; Pak, C. S. *Tetrahedron Lett.* **1995**, *36*, 5607.
10. Shono, T.; Matsumura, Y.; Kashimura, S. *Chem. Lett.* **1978**, 69.
11. Julia, M.; Launay, M.; Stacino, J.-P.; Verpeaux, J.-N. *Tetrahedron Lett.*, **1982**, 23, 2465.
12. Fabre, J.-L.; Julia, M.; Verpeaux, J.-N. *Tetrahedron Lett.* **1982**, *23*, 2469.
13. Baudin, J. B.; Hareau, G.; Julia, S. A.; Ruel, O. *Tetrahedron* **1991**, *32*, 1175–1178.
14. Baudin, J. B.; Hareau, G.; Julia, S. A.; Ruel, O. *Bull. Soc. Chim. Fr.* **1993**, *130*, 336–357.
15. Blakemore, P. R.; Cole, W. J.; Kocienski, P. J.; Morley, A. *Synlett.* **1998**, 26–28.
16. Kocienski, P. J.; Bell, A.; Blakemore, P. R. *Synlett.* **2000**, *3*, 365–366.
17. Alonso, D. A.; Fuensanta, M.; Najera, C. Eur. *J. Org. Chem.* **2006**, 4747–4754.
18. Alonso, D.; Fuensanta, .; Najera, C.; Varea, M. *J. Org. Chem.* **2005**, 70, 6404–6416.
19. Alonso, D. A.; Najera, C.; Varea, M. *Tetrahedron Lett.* **2004**, *45*, 573–577.
20. Blakemore, P. R.; Kocienski, P. J.; Morley, A.; Muir, K. *J. Chem. Soc., Perkin Trans. I* **1999**, 955–968.
21. Keck, G.; Savin, K. A.; Weglarz, M. A. *J. Org. Chem.* **1995**, *60*, 3194–3204.
22. Marko, I. E.; Murphy, F.; Kumps, L.; Ates, A.; Touillaux, R.; Craig, D.; Carballares, S.; Dolan, S. *Tetrahedron*, **2001**, *57*, 2609–2619.
23. Marko, I. E.; Murphy, F.; Dolan, S. *Tetrahedron Lett.* **1996**, *12*, 2089–2092.
24. Satoh, T.; Hanaki, N.; Yamada. N.; Asano, T. *Tetrahedron* **2000**, *56*, 6223–6234.
25. Bernard, A. M.; Frongia, A.; Piras, P. P.; Secci, F. *Synlett* **2004**, *6*, 1064–1068.
26. Pospíšil, J.; Pospíšil, T, Markó, I. E. *Org. Lett.* **2005**, *7*, 2373–2376.
27. Compostella, F.; Franchini, L.; Panza, L.; Prosperi, D. *Tetrahedron* **2002**, *58*, 4425–4428.
28. Charette, A. B.; Berthelette, C.; St-Martin, D. *Tetrahedron Lett.* **2001**, *42*, 5149–5153.

29. Zanoni, G.; Porta, A.; Vidari, G. *J. Org. Chem.* **2002**, *67*, 4346–4351.
30. Marino, J. P.; McClure, M. S.; Holub, D. P. *J. Am. Chem. Soc.* **2002**, *124*, 1664–1668.
31. Huang, H.-J.; Yang, W.-B. *Tetrahedron Lett.* **2007**, *48*, 1429–1433.
32. Furuichi, N.; Hara, H.; Osaki, T.; Nakano, M.; Mori, H.; Katsumura, S. *J. Org. Chem.* **2004**, *69*, 7949–7959.
33. Kim, G.; Chu–Moyer, M. Y.; Danishefsky, S. J. *J. Am. Chem. Soc.* **1990**, *112*, 2003–2005.
34. Smith, III, A. B.; Wan, Z. *J. Org. Chem.* **2000**, *65*, 3738–3753.
35. Charette, A. B.; Label, H. *J. Am. Chem. Soc.* **1996**, *118*, 10327–10328.
36. Liu, P.; Jacobsen, E. N. *J. Am. Chem. Soc.* **2001**, *123*, 10772–10773.
37. Williams, D. R.; Brooks, D. A.; Berliner, M. A. *J. Am. Chem. Soc.* **1999**, *121*, 4924–4925.
38. Meyers, C.; Carreira, E. M. *Angew. Chem., Int. Ed.* **2003**, *42*, 694–696.

2.8 Knoevenagel Reaction

Ivar M. McDonald

2.8.1 Description

The Knoevenagel reaction is the condensation of an active methylene compound with an aldehyde or ketone, to give an α,β-unsaturated dicarbonyl compound.(**1** + **2** → **3**).[1,2]

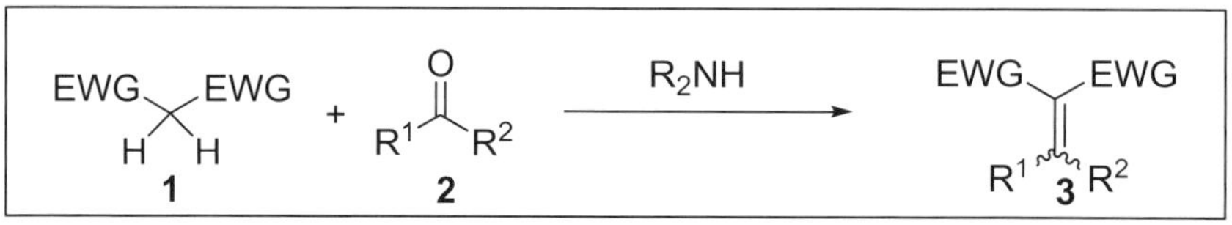

The Knoevenagel condensation is distinguished from the related aldol condensation in that the active methylene component (the nucleophile) must be doubly-activated by two electron-withdrawing groups (EWGs) such as NO_2, CN, COR, CO_2H, CO_2NR_2, SO_2OR, SO_2NR_2, SO_2R, SOR, SR, $PO(OR)_2$, aryl, and heteroaryl. The double activation of the active methylene unit allows the Knoevenagel condensation to take place under much milder conditions than the related aldol condensation. Thus, the Knoevenagel condensation is usually promoted by weak bases, such as an amine or salt thereof. As with prior reviews of the Knoevenagel condensation,[1,2] reactions promoted by strong bases such as alkoxides or metal amides are not included in the scope of this chapter.

2.8.2 Historical Perspective

In 1885, Arthur Hantzsch showed that the condensation of ethylacetoacetate, benzaldehyde and ammonia provided the symmetrical dihydropyridine **6**.[3,4] 1,5-Diketone **7** was also isolated from the reaction mixture and it was shown that it could be converted to **6** by treatment by resubjecting it to ethanolic ammonia. Subsequent experiments showed that if a primary amine such as methyl, ethyl or allyl amine was used, **7** was isolated as the sole product.[4] Hantzsch showed that the amine was an essiential part of the condensation, as no reaction occurred between ethylacetoacetate and benzaldehde in the absence of amine, but even a catalytic amount of amine was an effective promoter of this condensation. Hantzsch's synthesis of **7** was probably the first reported instance of what was to become known as the Knoevenagel condensation.

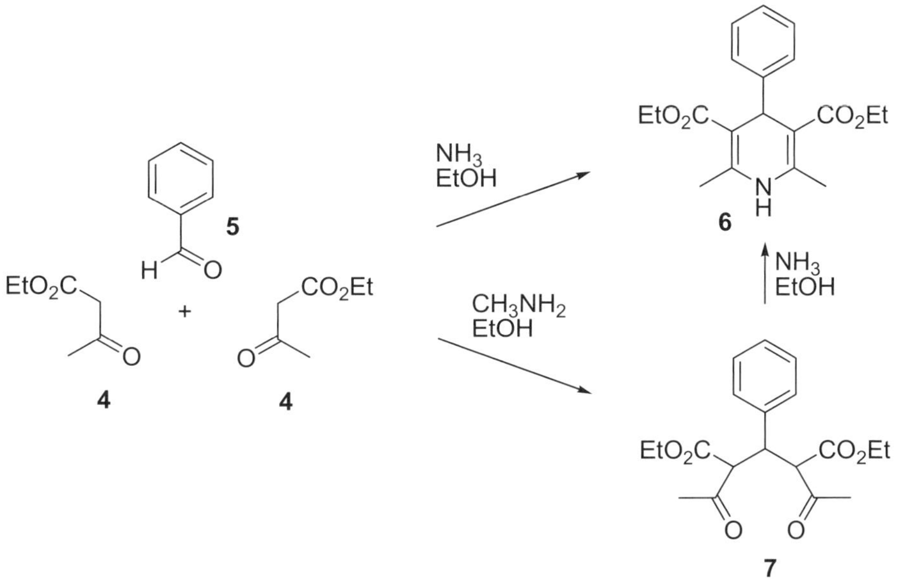

In 1894, Emil Knoevenagel expanded on Hantzsch's work in two significant ways.[5] First, he showed that this reaction could be promoted by any primary or secondary amine, not just the few primary amines demonstrated by Hantzsch. (In Knoevenagel's hands, tertiary amines, such as pyridine, quinoline, diethylaniline or dimethylaniline were ineffective, although this was to be later revisited by Hann and Lapworth, *vide infra*).

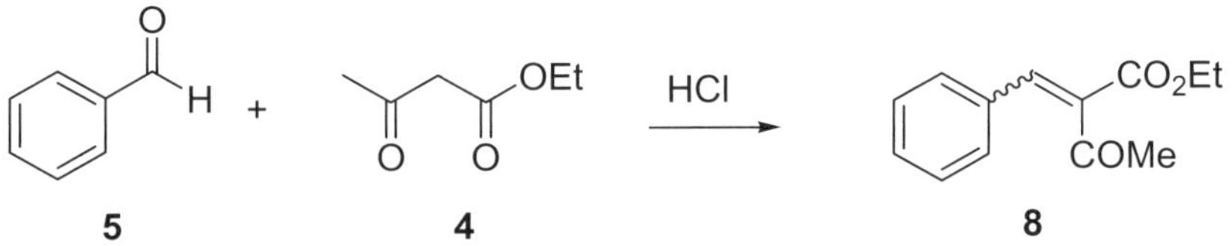

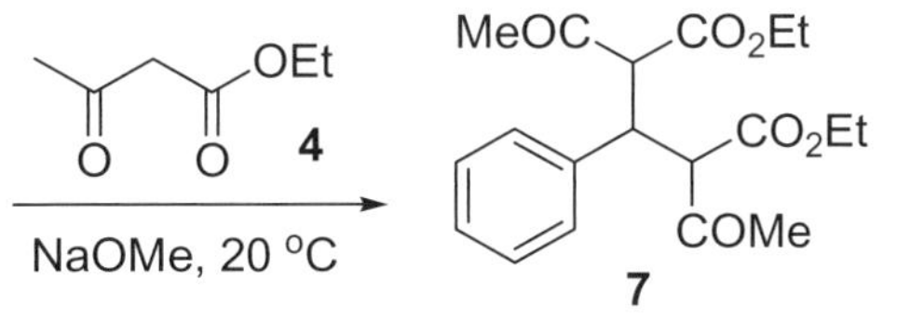

Knoevenagel also showed that the synthesis of **7** could be carried out in a stepwise manner, first by condensation of ethylacetoacetate with benzaldehyde, followed by Michael addition of an additional equivalent of acetoacetate. This method allowed for the possibility of synthesizing unsymmetrical pyridines, which were not otherwise available using the

Hantzsch procedure. In his initial work, Knoevenagel used Claisen's method (anhydrous HCl) for the synthesis of the alkylidine acetoacetates, and used sodium ethoxide to catalyze the Michael addition of the second equivalent of acetoacetate.

Later that same year, Knoevenagel expanded the scope of this transformation in showing that diethylmalonate (**10**) could take the place of ethylacetoacetate as the active methylene component in Hantzsch's procedure, condensing with formaldehyde in the presence of diethylamine to form a bis-adduct, **11**.[6]

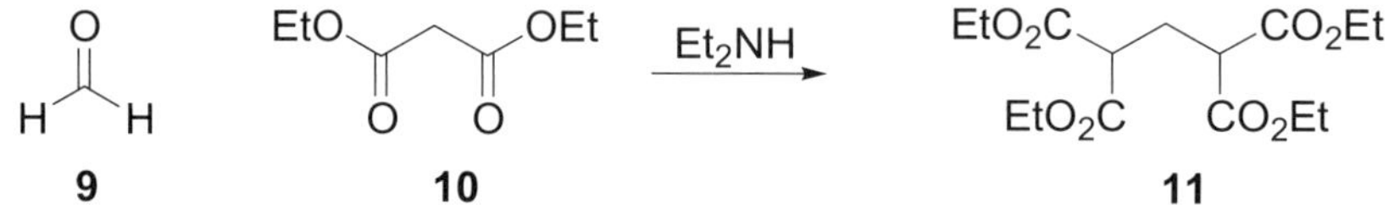

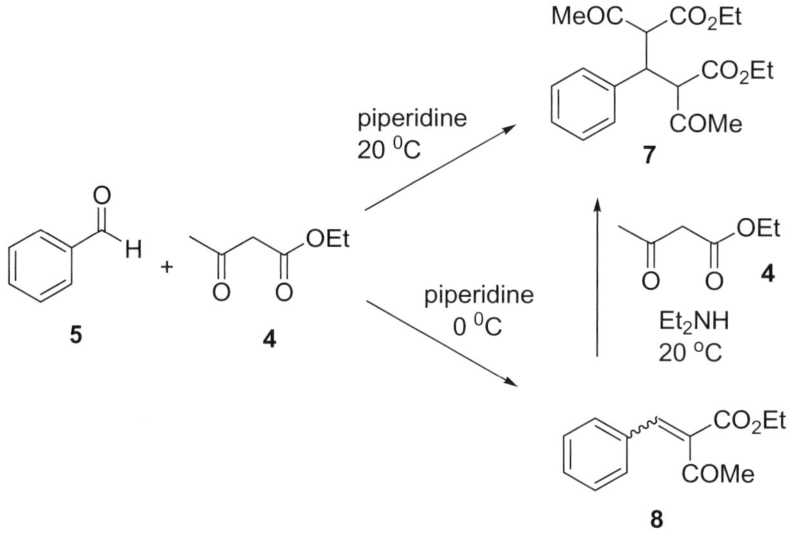

Two years after these initial reports, Knoevenagel was again investigating the condensation between benzaldehyde and ethylacetoacetate, which, when carried out with excess ethylacetoacetate in the presence of catalytic piperidine at room temperature or 0 °C yielded bis-adduct **11**.[7] When an assistant serendipitously performed the reaction with more efficient cooling, however, the reaction did not proceed to completion as before, but gave a mixture of bis-adduct **11** and benzylidine acetoacetate **8**, the condensation product of one equivalent each of ethylacetoacetate and benzaldehyde. Later experiments showed that if the temperature was controlled at 0 °C and the ethylacetoacetate and benzaldehyde were used in equimolar amounts, the benzylidine acetoacetate **8** could be isolated in quantitative yield. This is the transformation that later generally became known as the Knoevenagel condensation. Subsequent experiments showed that benzylidine product **8** could be converted to the bis-adduct **7** by resubjecting it to similar reaction conditions, thus establishing the intermediacy of the benzylidine compound **8** as an intermediate in the formation of the bis-adduct **7**.

While the benzylidine (or alkylidine) acetoacetates and related compounds are sometimes referred to as Knoevenagel products, it is important to note that Knoevenagel was by no means the first to produce these products by the condensation of ethylacetoacetate and aldehydes, as this had been done at least ten years earlier by Claisen and others.[8] The important contribution of Knoevenagel was the discovery of a method that allowed for these condensations to take place under much milder conditions (i.e., catalytic piperidine at 0 °C) than was allowed by the previous methods (anhydrous HCl, sodium alkoxides *etc.*).

2.8.3 Mechanism

The Knoevenagel reaction is a base-catalyzed aldol-type reaction that can occur through two possible mechanisms, depending on the type of base used. When Emil Knoevenagel made his initial discovery of this reaction, it was already known that benzaldehyde could condense with two equivalents of piperidine to provide the benzylidine bispiperidine aminal **12**.[9] Therefore, he proposed the intermediacy of an aminal (or imine) in the condensation.

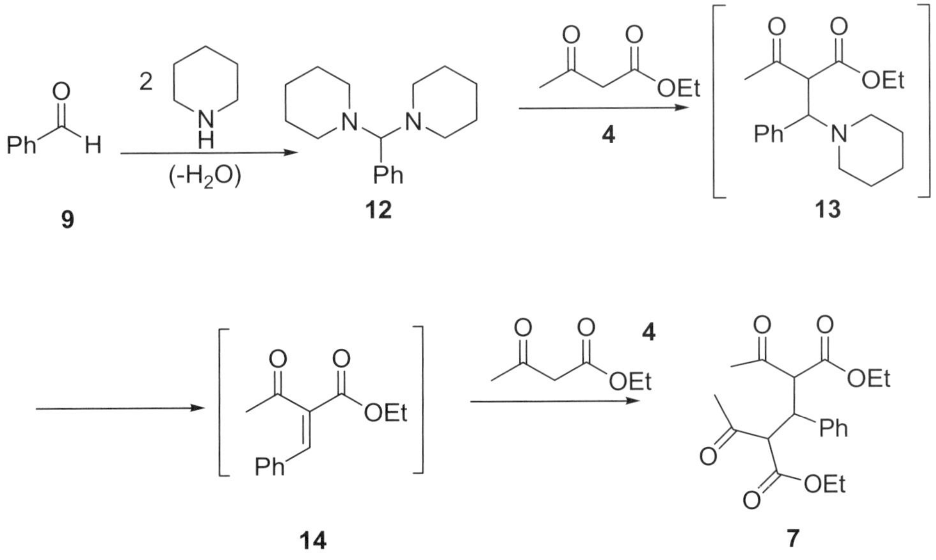

Indeed, when the pre-formed aminal **12** was treated with ethyl acetoacetate in ethanol, the bis-adduct **7** was isolated in good yield.[10] Knoevenagel's proposal for the mechanism of this reaction therefore was first, condensation of the amine and aldehyde to give the aminal **12**, followed by attack of this by the acetoacetate to produce a presumed β-amino dicarbonyl intermediate **13**, which would then eliminate piperidine to give the α,β-unsaturated dicarbonyl compound **14**, which in this case underwent a final Michael addition with an additional equivalent of ethyl acetoacetate to provide the bis-adduct **7**.

A few years after Knoevenagel proposed his mechanism for the reaction that came to bear his name, Hann and Lapworth showed that primary or secondary amines were not required for the transformation to take place, as Knoevenagel had claimed, but that tertiary amines could indeed promote the reaction between menthylacetoacetate **15** and aldehydes (although they were less efficient).[11] The use of a tertiary amine obviously precludes the possibility of an aminal or imine intermediate, therefore Hann and Lapworth proposed direct reaction between the enolized dicarbonyl compound and the aldehyde (Hann–Lapworth mechanism). This mechanism would proceed similarly to that proposed by Knoevenagel, but without aminal formation, and proceeding through a β-hydroxy dicarbonyl compound such as **18**, rather than a β–amino dicarbonyl compound (e.g., **13**).

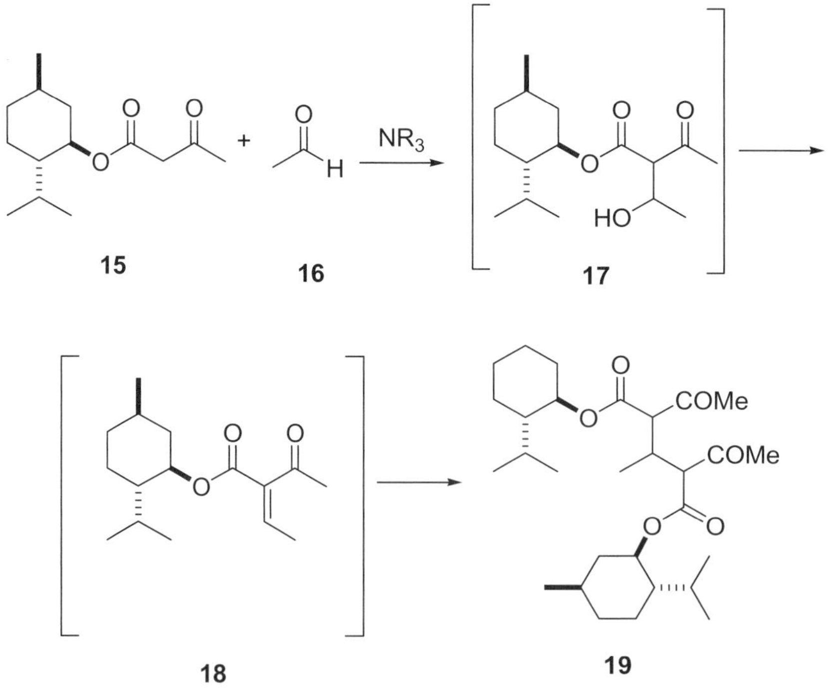

Clear evidence for both mechanisms exists,[1,2] as both the β-amino dicarbonyl[12] and β-hydroxy dicarbonyl[13] intermediates been isolated in some cases. It can be safely assumed that if the reaction is promoted by a tertiary base, the Hann–Lapworth mechanism is operative, but in the presence of primary or secondary amines, either mechanism may occur.

2.8.4 Variations and Improvements

Verley–Doebner Modification

In 1898 Knoevenagel showed that benzylidine aminal **20** condensed with malonic acid proceeded with decarboxylation to give cinnamic acid (**22**).[14] This was found to be a fairly general method for the condensation of malonic acid with aromatic aldehydes, but was not efficient for aliphatic aldehydes.

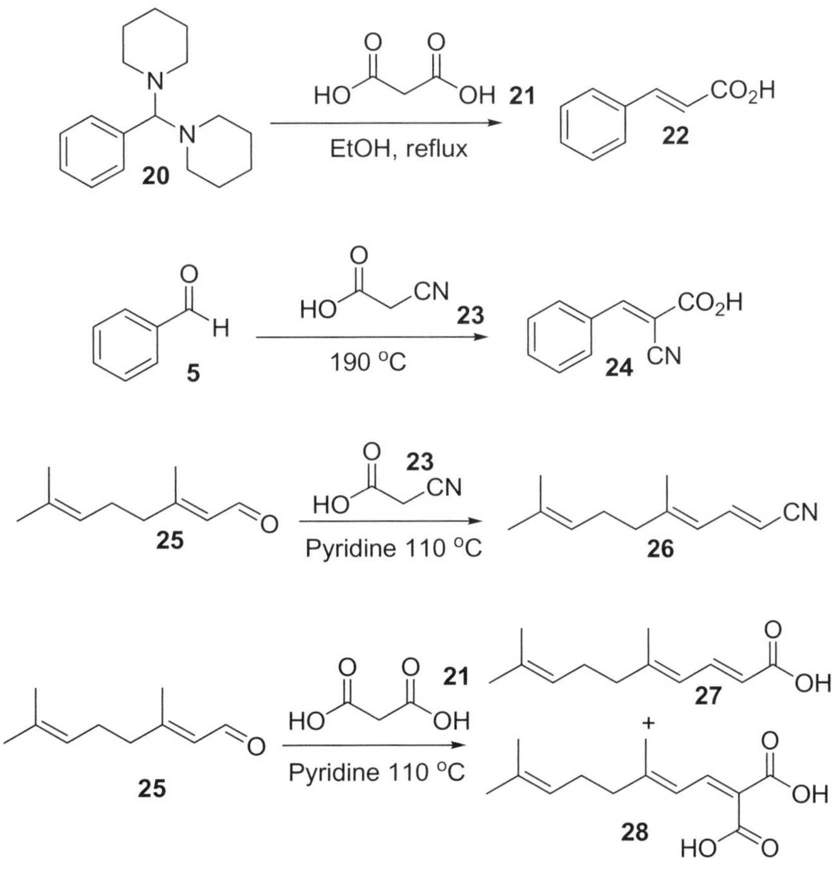

The uncatalzyed condensation of cyanoacetic acid with benzaldehyde to furnish α-cyanocinnamic acid **24** was demonstrated by Fiquet in 1893.[15] In 1899, Albert Verley was investigating the properties of citral (**25**), and attempted to condense it with cyanoacetic acid under these same conditions, but was unable to successfully promote this condensation, as polymerization of the product predominated. When Verley added a tertiary amine base, such as pyridine, condensation took place under much milder conditions and proceeded with concomitant decarboxylation to give the citrylidine acetonitrile **26** in reasonable yield.[16] Subsequent analogous reaction between malonic acid and citral provided the citrylidine acetic acid **27**, along with a slightly greater amount of the citrylidine malonic acid **28**.[18] These same conditions were employed one year later by Oscar Doebner in the condensation of malonic acid and crotonaldehyde to produce sorbic acid. The use of excess pyridine in this reaction is interesting, because pyridine usually not an effective promoter of the Knoevenagel condensation. The condensation of malonic acid with aldehydes is often referred to as the Doebner modification, but, as pointed out by Boxer and Linstead,[19] priority should be assigned to Verley, and more accurately should be referred to as the Verley–Doebner modification of the Knoevenagel condensation.

One potential complication in the Verley–Doebner modification is the propensity of some substrates to form β,γ-unsaturated isomers in addition to the α,β-unsaturated compounds. Many amines have been found to catalyze the condensation of malonic acid with aldehydes, but pyridine has been found to be unusual in producing high yields of the α,β-isomers; many other amines give significant amounts of the β,γ-isomers in addition to the α,β-isomers.[19]

Boxer and Linstead developed a method for selectively obtaining the β,γ-isomers by using triethanolamine as the base, which, although effective, suffered from very poor yields. Valentine Ragoussis showed that for unbranched aldehydes, the β,γ-unsaturated products could be obtained in excellent yield and selectivity by using a catalytic amount of piperidinium acetate in DMSO at 100 °C (**29** + **21** → **30**).[20,21] This transformation was shown to be general for unbranched aldehydes, but neither ketones nor aldehydes that were α-branched condensed with malonic acid under these conditions.

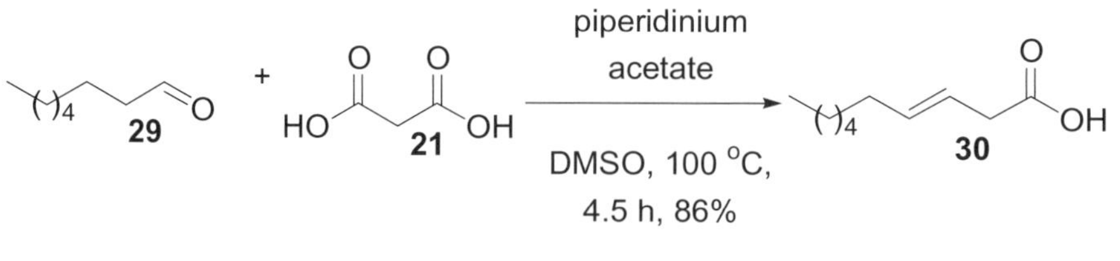

Since pyridine is the base in the Verley–Doebner modification, a mechanism similar to the Hann–Lapworth mechanism takes place, with the addition of a decarboxylation step. In 1952, E. J. Corey studied the mechanism of decarboxylation in the Verley–Doebner modification of the Knoevenagel condensation by employing a number of model systems, including ethyl hydrogen isopropylidine malonate **31**, which was found to give a mixture of β,β-dimethylacrylate **32** and a greater amount of the β,γ-unsaturated isomer **33** upon refluxing in pyridine.[22] This result was deduced to be due to a mechanism wherein the first step is tautomerization of **34** to **35**. Decarboxylation of the β,γ-unsaturated isomer **34** is facile as the resultant anion is resonance stabilized, and this stabilized anion can be trapped at either the α- or γ-position to yield **33** and **32**, respectively. Separate experiments showed that ethyl hydrogen mesitylene malonate **36** did not undergo decarboxylation in refluxing pyridine even though the corresponding aldehyde **38** smoothly provided the cinnamic acid **37** when treated with malonic acid under the same conditions.

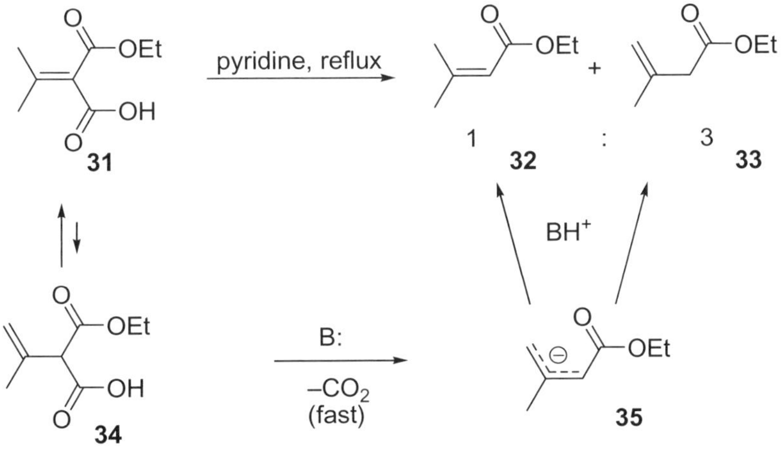

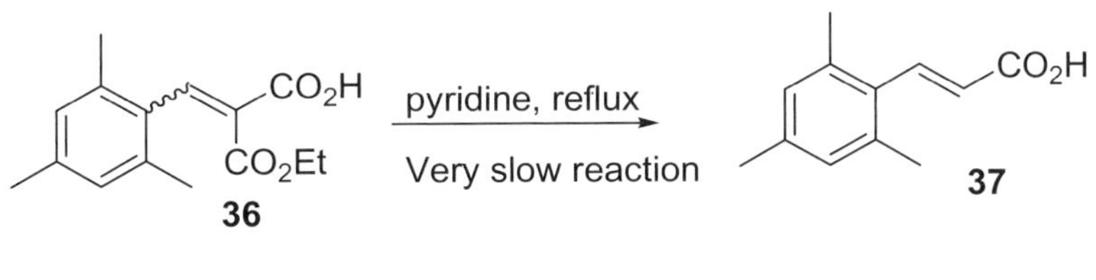

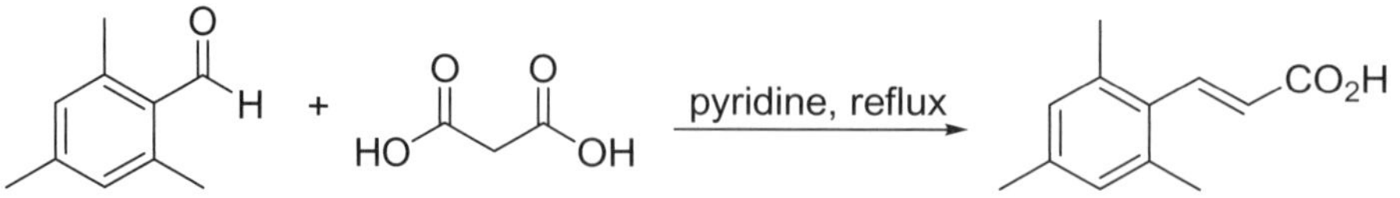

38 21 37

Corey therefore proposed that the decarboxylation step would only occur in situations where the α-carbon was not linked by a double bond (i.e., when the α-carbon was sp^3 hybridized), and concluded that the decarboxylation likely occurred in a concerted manner with the elimination of water (**38** + **21** → **40** *via* **39**).

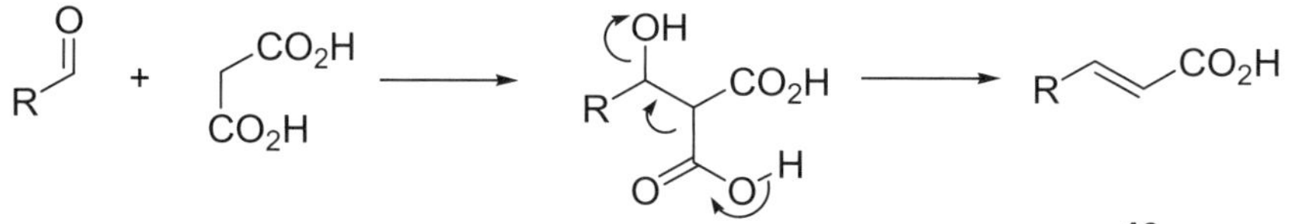

38 21 39 40

Nokami hydroxylative Knoevenagel reaction

Another important variant of the Knoevenagel reaction is the Nokami hydroxylative Knoevenagel reaction, which employs a phenylsulfinyl group as one of the electron-withdrawing groups in the active methylene component (e.g., **42**). The initial product of condensation is a vinylsulfinate, such as **43**, which then isomerizes to the allylsulfinate **44** which then undergoes a Mislow–Evans [2,3]-sigmatropic rearrangement to give the γ-hydroxylated, α,β-unsaturated compounds **46**.[23–26]

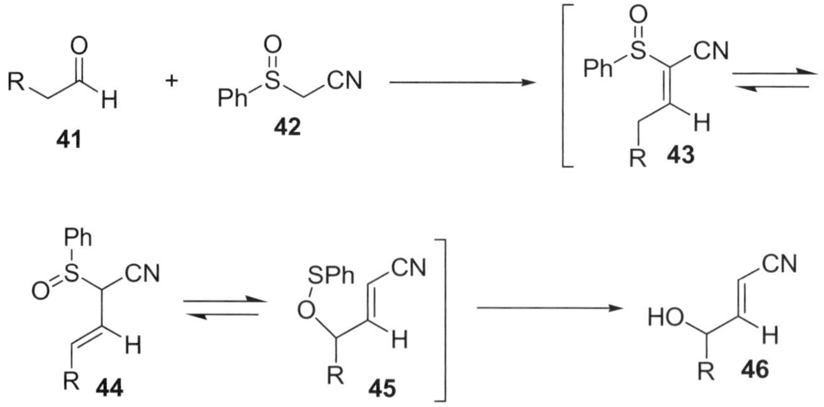

41 42 43 44 45 46

The Nokami hydroxylative Knoevenagel reaction has been expanded from the initial report using phenylsulfinylacetonitrile **42** to the use of phenylsulfinylacetates and phenylsulfinylketones. The scope of the carbonyl compound has also been expanded to include ketones as well as aldehydes, but the use of ketones with available α-hydrogens on both sides adds the obvious potential complications in terms of mixtures of products.

Standard reaction conditions

Initial reports from Knoevenagel used secondary amines such as piperidine or diethylamine as the catalyst, usually in an alcohol as solvent, and at temperatures between 0 °C and room temperature. The range of possible promoters of the Knoevenagel condensation was soon expanded by Hann and Lapworth to include tertiary amines, and soon thereafter, many, many modified reaction conditions were discovered, some of which had clear advantages over the standard conditions, while the benefits of others are less clear. This section will focus on those modifications that have shown some advantage over the standard conditions.

Cope Modification

Several research groups, including Knoevenagel's,[27] employed amine salts as catalysts for the condensation of aldehydes with active methylene compounds, and it was proposed that the salts may be more efficient catalysts than the corresponding free bases because the acidic component assisted in promoting the elimination of water from the aldol-like intermediates.[28] Arthur Cope expanded on these findings and found that by using amine salts such as ammonium acetate or piperidinium acetate in acetic acid solution, excellent yields could be obtained if the water was continuously removed from the reaction by distillation.[29] These conditions were later modified to use benzene as a solvent, which still allowed for the removal of water by azeotropic distillation, but provided for milder overall conditions; these are the conditions referred to as the Cope modification of the Knoevenagel reaction.[30]

Catalysis by amino acids

In 1909, British–American biochemist H. D. Dakin showed that amino acids could efficiently promote both the Knoevenagel condensation and the Verley–Doebner modification of the Knoevenagel condensation.[31] This method was somewhat forgotten for many years, until it was rediscovered by Haley and Maitland in 1951 and Prout in 1953.[32,33] Prout carried out reactions with a number of amino acids and found that four were particularly useful: *p*-aminophenol, α-aminophenylacetic acid, β-alanine, and ε-aminocaproic acid. The use of amino acids in the catalysis of the Knoevenagel condensation has again increased in recent years, as organocatalysis with amino acids such as proline has gained increased attention.

Lehnert Modification

In 1970, Willy Lehnert showed that in the presence of titanium tetrachloride and a tertiary amine such as pyridine or triethylamine, the Knoevenagel condensation of both aldehydes and ketones with diethylmalonate proceeded under mild conditions (0 °C to room temperature) to provide the condensation products in good yield (e.g., **47→48**).[34] The scope of the reaction was therefore increased as hindered ketones could be coupled with malonate esters in reasonable yield.[35]

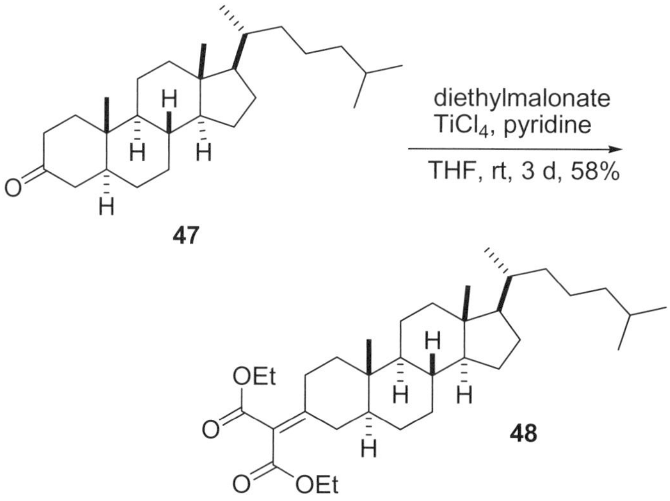

Other promoters of the Knoevenagel condensation

Along with the conditions previously mentioned, a plethora of other reagents have been shown to promote the Knoevenagel condensation. Few of these have been utilized to any degree, and the reader is directed to previous reviews of the Knoevenagel condensation for a fuller discussion of some of these conditions. Conditions reported include alkali metal fluorides, with KF being the most generally employed due to its adequate solubility and low cost.[36] Many inorganic solids have been reported to promote the Knoevenagel reaction including basic alumina,[37,38] magnesium and zinc oxides,[38] aluminium phosphate/aluminium oxide,[39] xonotlite,[40] hydrotalcite,[41] diammonium hydrogenphosphate[42] and calcium vanadate apatite.[43] Other solid or solid-supported reagents include functionalized silica gel[44] and ion-exchange resins Amberlite IR-4B[45] and Dowex-3.[46] Lewis acid promoters include $TiCl_4$(without added base),[47] $Ti(Oi\text{-}Pr)_4$,[48] $ZnCl_2$,[30] $Zn(OAc)_2$,[49] $CeCl_3{\bullet}7H_2O$,[50] $Mg(ClO_4)$[51] and $CuCl_2$.[52] Yet more conditions include I_2/K_2CO_3,[53] tetrabutylammonium hydroxide,[54] tungsten phosphoric acid,[55] ionic liquids,[56–59] reactions in water,[60] solvent-free conditions,[61] microwave promoted reactions,[59,62] reactions promoted by phase-transfer catalysts,[61] and finally, for some very reactive substrates, uncatalyzed reactions.[60] Unfortunately for the practicing chemist, few of these conditions are well worked out enough or understood well enough to have become common, and almost all examples of the Knoevenagel condensation in the literature employ an amine base, or a salt thereof (Cope conditions, or amino acid catalysis) or $TiCl_4$/Base(Lehnert conditions).[1,2]

2.8.5 Synthetic Utility

Scope of the active methylene component (nucleophile)

As described earlier, the Knoevenagel condensation requires two electron withdrawing groups to be present on the active methylene component. Almost any combination of EWGs may be employed, with the caveat that reactivity generally decreases in the order $NO_2 > CN > COCH_3 > COPh > CO_2R >$ Phenyl,[1] thus if both electron–withdrawing groups are of lesser reactivity, harsher reaction conditions may be required, or reactions with less reactive carbonyl compounds may be difficult.

Malonic Acid

Malonic acid condenses with most aldehyde and some very reactive ketones to give either the alkylidine (or arylidine) malonic acids or the corresponding monocarboxylic acrylic or cinnamic acids.[1,2] The alkylidine (or arylidine) malonic acids are best produced using ammonia in alcohol at 70 °C or below.[63] Pyridine is the solvent and base of choice (Verley–Doebner modification) for the synthesis of α,β-unsaturated acrylic or cinnamic acids,[19] while piperidinium acetate in refluxing xylenes is preferred for the selective synthesis of the β,γ-isomers.[21] With malonic acid, formation of bis-adducts by Michael addition of an additional equivalent of the active methylene is rare.[1,2]

Malonic Esters

Acyclic malonic esters react with most aldehydes under fairly mild conditions (secondary amines with warming), but ketones and less reactive aldehydes require conditions such as the Lehnert conditions ($TiCl_4$/amine). Malonic esters have a propensity to provide the bis-adducts via Michael addition. Cyclic malonic esters (such as Meldrum's acid) are more reactive than the corresponding acyclic versions, with reactions with simple aldehydes proceeding without catalyst in DMF or DMSO.[64] Formation of the bis-adduct is also more common with cyclic esters, especially in the coupling with unbranched aliphatic aldehydes.[1,2]

Malononitrile

The nitrile group is one of the most reactive electron withdrawing groups in the active methylene component, and with two nitriles, malononitrile is very reactive and condenses with almost all aldehydes and many ketones under mild conditions.[1,2,65,66] Even sterically hindered ketones such as pinacolone condense with malononitrile in 48% yield using β-alanine as a catalyst.[33] Formation of the bis-adduct is less of a problem with malononitrile than with malonic esters.

Cyanoacetic acid, Cyanoacetic esters

The nitrile in cyanoacetic acid and cyanoacetic esters makes them more reactive than the corresponding malonic acid or esters. As with malonic acid, cyanoacetic acid can condense to give the alkylidine cyanoacetic acid or the decarboxylated α,β-unsaturated nitriles. Cyanoacetic acid reacts in a similar fashion as malonic acid, and often gives mixtures of the α,β- and β,γ-unsaturated compounds.[67] Cyanoacetic esters usually react to give the α,β-unsaturated cyanoacetates, but the bis-adducts may be isolated if excess cyanoacetic ester is used.

β-Keto esters

Condensations of β-ketoesters (such as ethylacetoacetate) are often complicated by formation of the bis-adducts.[1,2] The mono-addition products can often be isolated by performing the reaction at 0 °C with a secondary amine catalyst; at room temperature, the bis-adducts usually predominate.

1,3-Diketones

Condensations with 1,3-diketones can be problematic, with aromatic aldehydes reacting cleanly to give the α,β-unsaturated products, but with aliphatic aldehydes and ketones giving mixtures of α,β- and β,γ- unsaturated compounds.[1,2,68,69] Michael addition to form the bis-adducts can also be an issue, especially with cyclic 1,3-diketones, for which the mono-addition products can often not be isolated.[70]

Aryl and heteroaryl activating groups

Aryl activating groups, especially those with electron-withdrawing substituents, are often good activating groups for the Knoevenagel condensation. Examples of arylacetic acid, arylacetic esters, arylacetonitrile are well-known, and often occur under standard Knoevenagel conditions (piperidine in refluxing alcohol).[1,2] In many cases, diarylmethane derivatives condense with aldehydes, but these transformations often require strong bases unless the aryl groups are sufficiently electron-deficient.[71,72] Electron-poor heterocycles are often very effective electron-withdrawing groups in the Knoevenagel condensation.

Heterocycles

A number of heterocycles that might not be readily recognized as active methylene compounds react in Knoevenagel condensations under standard conditions. Oxindole **49**,[73] butenolide **50**,[74] thiazolidinedione **51**,[75] hydantoin **52,**[76] pyrazolone **53**,[77] isoxazolidinone **54**,[77] hydroxypyrone **55**[78] have all been shown to participate in Knoevenagel condensations.

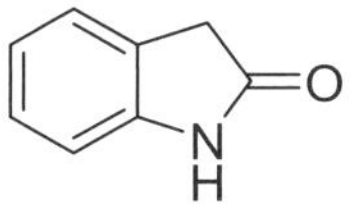

49

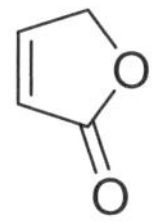

50

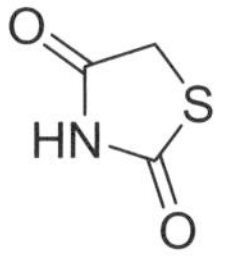

51

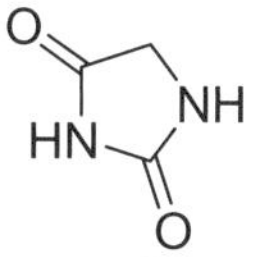

52

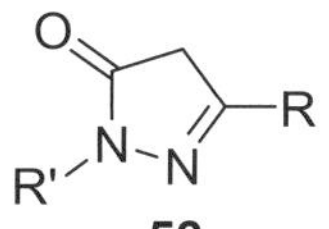

53

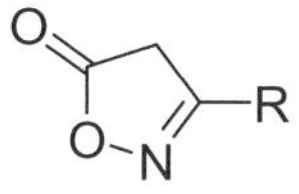

54

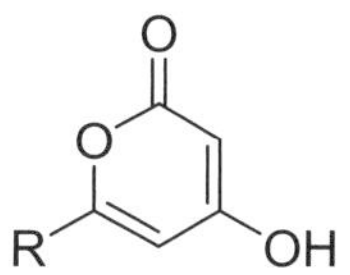

55

Scope of the electrophile

Aldehydes and ketones are by far the most common examples of electrophiles used in the Knoevenagel condensation, with reactivity tracking with steric bulk. Aldehydes therefore react much more quickly than ketones, and highly substituted ketones react only very sluggishly, if at all. Reactivity generally decreases in the order aldimine (**58**) > aldehyde (**56**) > ketimine (**57**) > enamine (**63**) > ketone (**57**).[2] A few other surrogates for aldehydes and ketones have been employed in the Knoevenagel condensation, the first of which was the aminal **12** used by Knoevenagel in his mechanistic explorations.[10] Imines (**58**, **59**),[79] aminals (**60**), α-aminoethers (**61**),[80] enamines (**63**),[81] acetals (**62**)[82] and dibromomethane derivatives (**64**)[83] have all been used in the Knoevenagel condensation to a limited extent.

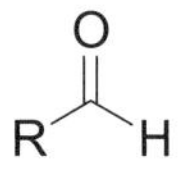

56
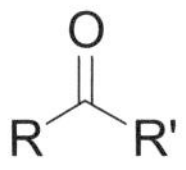

57
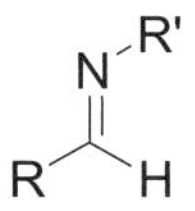

58
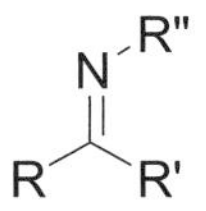

59

60

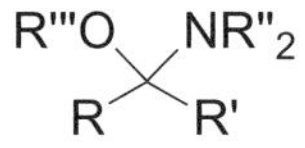

61

62
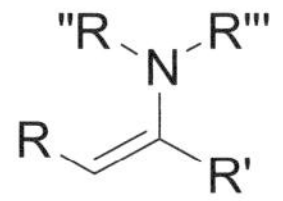

63
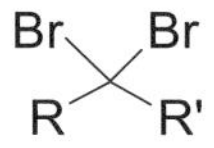

64

Stereochemical considerations

Stereochemistry in the Knoevenagel condensation is driven by sterics, with the thermodynamic products being formed. If the two EWGs in the active methylene compound are different in size, excellent selectivity can be obtained in the condensation with aldehydes, or with ketones that are also sufficiently different on each side.[1,2]

Tandem reactions

The two electron-withdrawing groups attached to the olefin of the Knoevenagel products makes the LUMO quite low,[2] and makes them excellent substrates for further transformations. This reactivity was first observed in Knoevenagel's initial report in which the bis-adducts were isolated, a transformation that was later understood to involve sequential Knoevenagel condensation followed by Michael addition.

A number of tandem transformations taking advantage of the reactivity of the Knoevenagel products have been described, including tandem Knoevenagel–Michael, Knoevenagel–ene, Knoevenagel–Diels–Alder, and

Knoevenagel reactions followed by sigmatropic rearrangements or electrocyclizations.[2,84]

Knoevenagel–ene reaction

In his synthesis of the cadinane–sesquiterpene veticadinol (**69**), Lutz Tietze employed a sequence of Knoevenagel condensation of (*R*)-citronellal (**66**) and dimethylmalonate followed by an ene reaction provide **68** with the requisite trans stereochemistry in what was to become the decalin core of veticadinol **69**.[2,85]

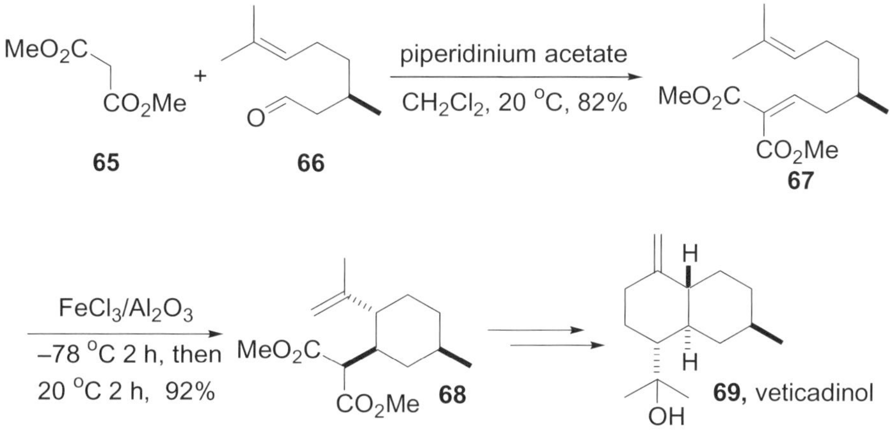

Tandem Knoevenagel–Diels–Alder reactions

The tandem Knoevenagel–Diels–Alder sequence is well known, and has been extensively studied by Tietze and others.[2] The Knoevenagel products have been shown to be active as dienophiles in normal-electron-demand Diels–Alder reactions, as well as heterodienes in inverse-electron-demand hetero-Diels–Alder reactions. One of the most elegant examples of a tandem Knoevenagel–hetero-Diels–Alder sequence is Tietze's synthesis of hirsutine **76**.[86] In this sequence, a mixture of aldehyde **70**, Meldrum's acid **71**, enol ether **73** (a mixture of isomers) and a catalytic amount of ethylenediamine diacetate were sonicated in Benzene for 12 h at 60 °C. This effected a cascade sequence that began with Knoevenagel condensation to form the heterodiene **72**, followed by intermolecular hetero-Diels–Alder reaction with enol ether **73** to lead to **74**, which underwent *in situ* hydrolysis and decarboxylation to yield intermediate **75** in 84% yield.

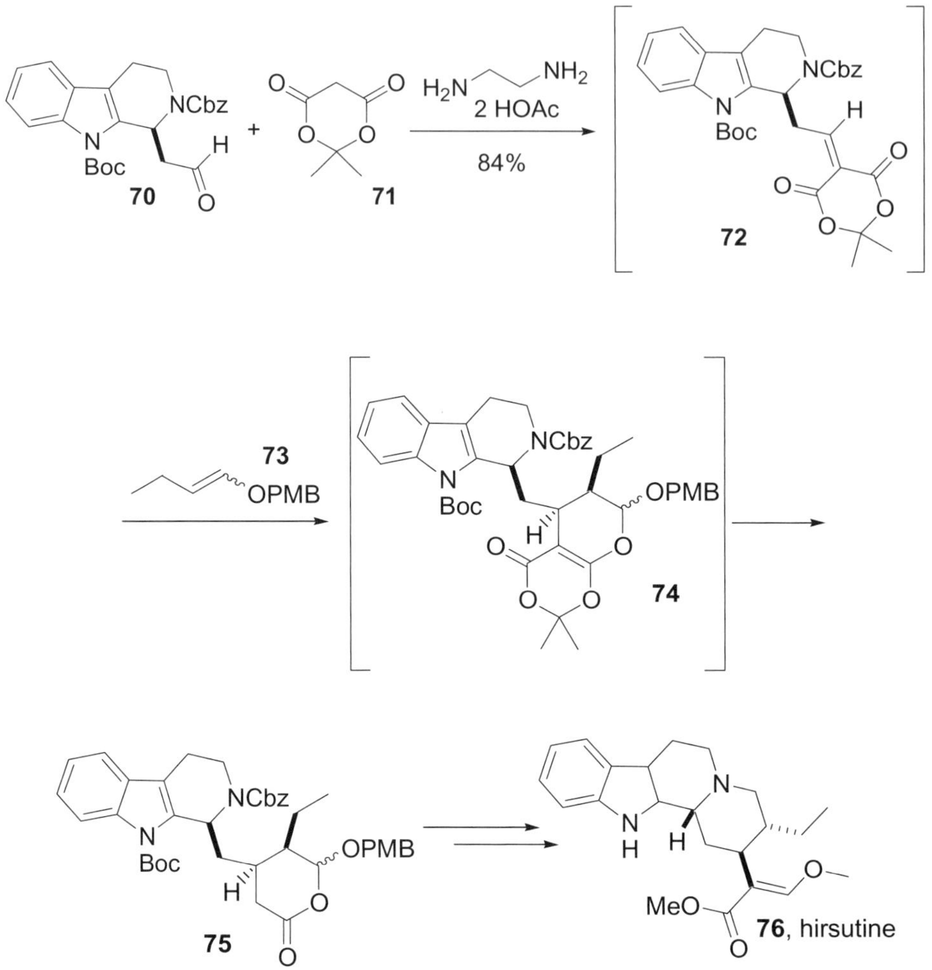

Tandem Knoevenagel–Electrocyclization

Tandem sequences involving Knoevenagel condensation followed by electrocyclization have not been employed as extensively as other tandem reactions (such as the tandem Knoevenagel–Diels–Alder reaction), but have been used in a few cases with great success, rapidly constructing complex polycyclic cores.

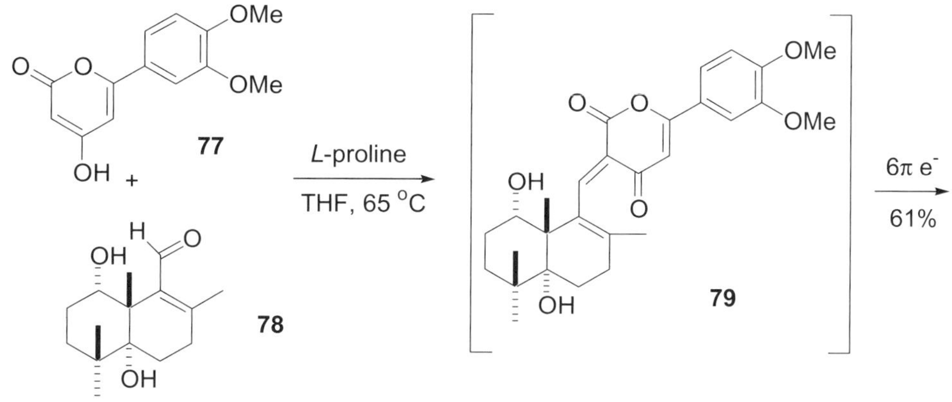

In 2002, Satoshi Ōmura's group used a proline-catalyzed Knoevenagel condensation between α,β-unsaturated aldehyde **78** and 4-hydroxy-2-pyrone **77** to form dienone **79**, which underwent a spontaneous 6π-electron electrocyclization to form **80**, which contains the pentacyclic core of arisugacin A (**81**).[78]

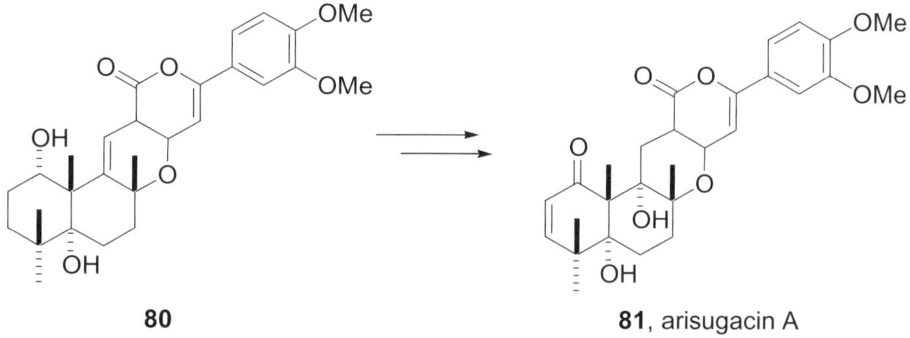

Tandem Knoevenagel–Mislow–Evans

The tandem Knoevenagel–Mislow–Evans sequence is also known as a hydroxylative Knoevenagel reaction, and has been described previously. An elegant example employing this sequence was in Trost's synthesis of allocyanthin B2.[87] Reacting aldehyde **82** with phenylsulfinyl acetonitrile and piperidine effects a Knoevenagel condensation followed by a [2,3]-sigmatropic rearrangement to form an intermediate γ-hydroxy-α,β-unsaturated nitrile that undergoes spontaneous lactonization to yield tricycle **83**. Reduction of the olefin, nitrile and lactone moieties provides **84** and sets the stage for a final intramolecular aldol condensation to afford allocyanthin B2 (**85**).

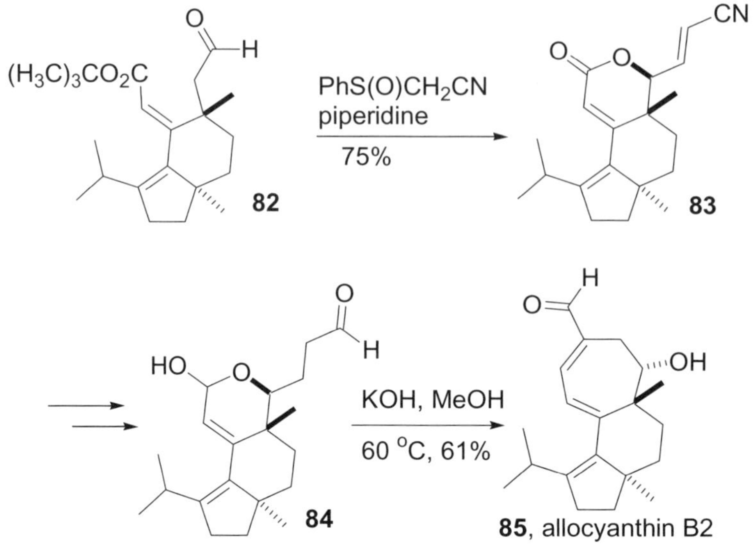

Heterocycle syntheses

The Knoevenagel condensation also plays a part in a number of prominent heterocycle syntheses including the Knorr pyrrole synthesis,[88] Hantzsch dihydropyridine synthesis,[89] Feist–Bénary furan synthesis[90] and the Gewald reaction,[91] which is used to make 2-aminothiophenes.

Synthesis of natural products and biologically active molecules

In 1928, Ben Corson and Roger Stoughton were studying the condensation of malononitrile with aromatic aldehydes when they made a surprising discovery regarding the physiological properties of some of these benzalmalononitriles.[92] They reported that "certain of these dinitriles have the effect of sneeze and tear gases. They are harmless when wet, but to handle the dry powder is disastrous...In sneezing caused by *o*-chlorobenzalmalononitrile **88**, the face smarts, especially if damp. The smarting is intensified by washing. Most of the discomfort can be avoided if a gas mask is worn when handling the dry powder."

Dinitrile **88** was eventually named for the men who surely suffered in its discovery, and was called CS gas after the initials of their last names. CS gas (which is not a gas but rather an aerosol) has gained widespread use as a tear gas and riot control agent.

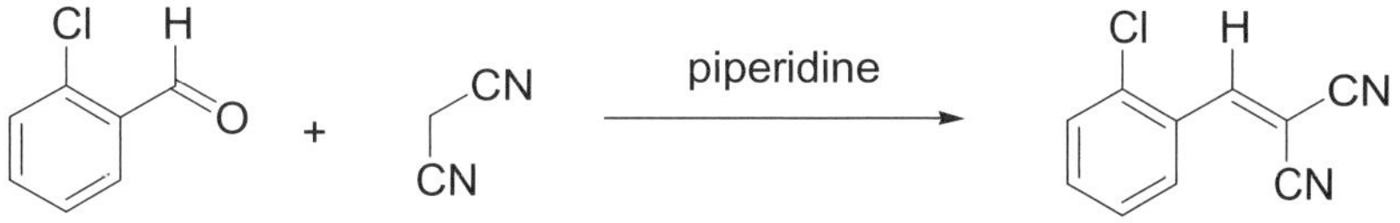

86 **87** **88**, CS gas

In their total synthesis of gelsemine (**94**), Tohru Fukuyama and Gang Liu used a Knoevenagel condensation of a cyclopropyl carboxaldehyde **89** to later set the stage for a key divinylcyclopropane–tropinone rearrangement.[93] Initial efforts employing oxindole as the active methylene component provided a mixture of olefin isomers favoring the undesired *E*-isomer. By switching to 4-iodooxindole **90** as the active methylene component, they found that the selectivity in the condensation was reversed and the desired *Z*-isomer **91** was formed isolated in 88% yield.

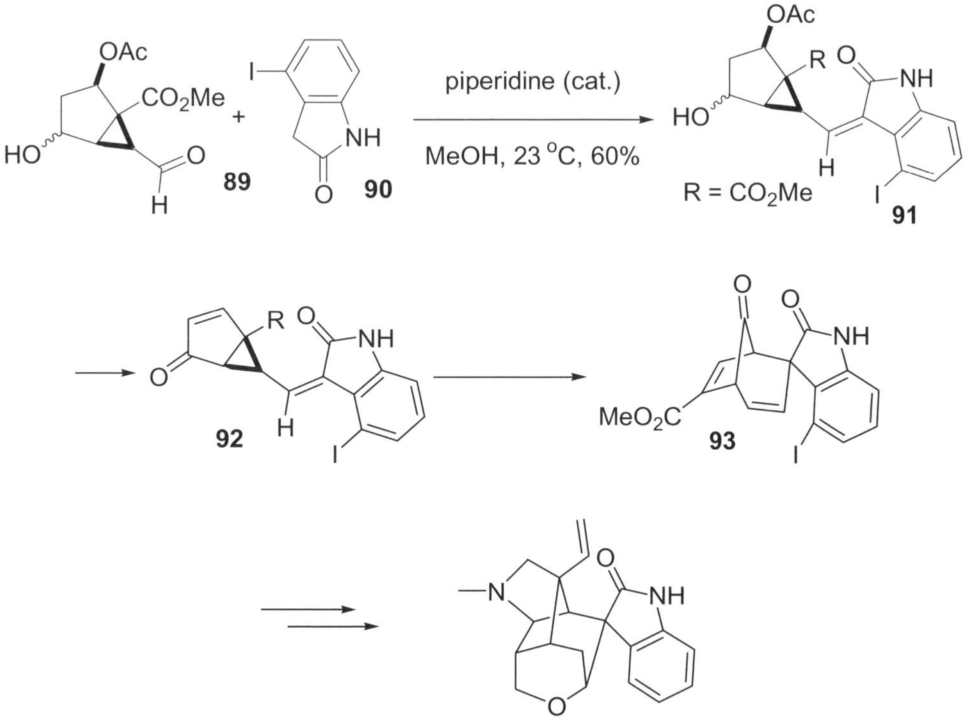

89 **90** **91** **92** **93** **94**, gelsemine

In Goverdhan Mehta's synthesis of guanacastepene C (**97**), an intramolecular Knoevenagel condensation (**95** → **96**) was employed to construct the six-membered ring of the guanacastepene core.[94]

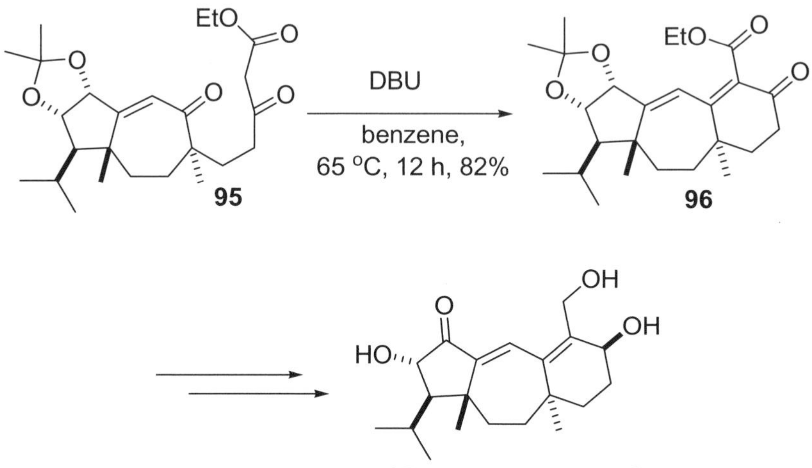

97, guanacastepene C

The antidiabetic thiazolidinedione activators of peroxisome proliferator-activated receptors (PPARs) include rosiglitazone (**100**, Avandia®), pioglitazone (Actos®) and troglitazone (Rezulin®), and all include a benzyl-linked thiazolidinedione ring which is constructed via a Knoevenagel condensation between thiazolidinedione **51** and a substituted benzaldehyde such as **98** under Cope conditions.[75]

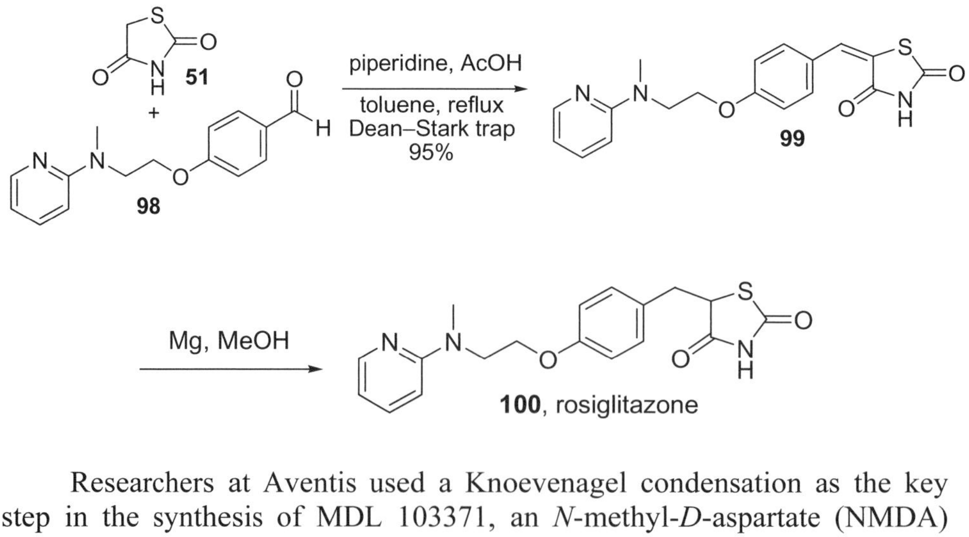

100, rosiglitazone

Researchers at Aventis used a Knoevenagel condensation as the key step in the synthesis of MDL 103371, an *N*-methyl-*D*-aspartate (NMDA) antagonist, a compound for the potential treatment of stroke.[95] Condensation of 3-formylindole **101** with phenylacetonitrile **102** as the active methylene component provided **103** solely as the undesired (but expected, see Section 2.8.5.2, Stereochemical considerations) *Z*-isomer. Isomerization to the *E*-isomer was accomplished following reduction of the aromatic nitro group and hydrolysis of the nitrile to the acid to give **104**.

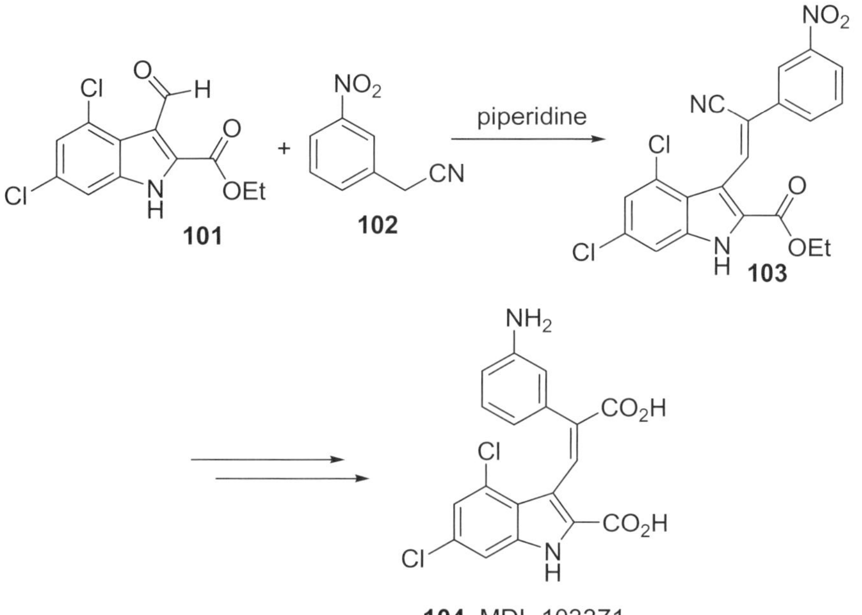

104, MDL 103371

Merck described the synthesis of PDE4 inhibitor **107** as a potential anti-inflammatory drug, the synthesis of which involved a Knoevenagel condensation between aldehyde **105** and benzyl oxadiazole **106** as the key step.[72] The researchers were concerned about the possibility of the formation of olefin isomers in the condensation, due to the similar sterics of the phenyl and ox diazole rings. They found however, that by performing the reaction in isopropanol (in which both the starting aldehyde **105** and product **107** have limited solubility) the initially-formed mixture of olefin isomers equilibrates, and the desired product crystallizes out of the mixture to provide exclusively the desired isomer.

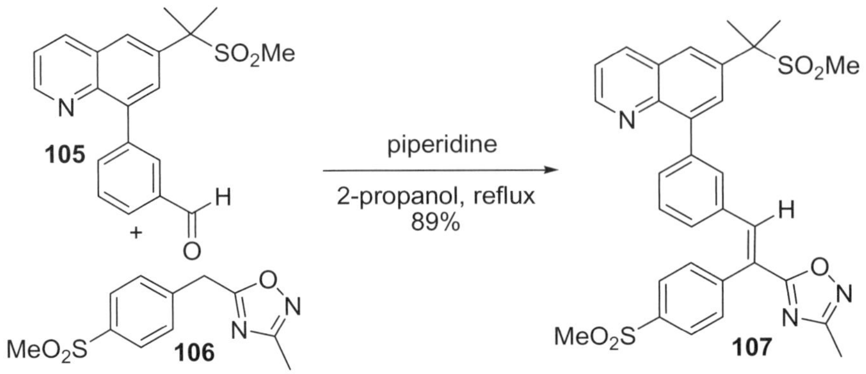

2.8.6 *Experimental*

Ethyl *n*-butylidineaceteoacetate (Knoevenagel Method):[30]

A mixture of ethylacetoacetate (65 g, 0.5 mol) and freshly distilled n-butyraldehyde (36 g, 0.55 mol) was cooled to -5 °C. Piperidine (0.5 g) in 1 g of alcohol was added slowly over 5–10 minutes so that the reaction temperature did not rise above +5 to 10 °C. The mixture was cooled to 0 °C and placed in a refrigerator for 12–24 h, after which the ester was washed with water containing a few drops of acetic acid (3 × 100 mL). The washings were extracted with ether, and the combined ether extract and ester layer were distilled in vacuum through a Widmer column. The yield of ethyl butylidineaceteoacetate was 71.4 g, 81%.

Ethyl (1-ethylpropylidine)cyanoacetate (Cope Modification):[96]

In a 500 mL round-bottomed flask equipped with a Dean–Stark constant water separator which is connected to a reflux condenser are placed ethyl cyanoacetate (67.8 g, 0.60 mol), diethylketone (56.8 g, 0.66 mol), ammonium acetate (9.2 g, 0.12 mol), glacial acetic acid (40 g, 0.48 mol) and 100 mL benzene. The flask is heated on an oil bath at 160–165 °C, and the water that distils out of the mixture with the refluxing benzene is removed from the separator at intervals. Refluxing is continued for 24 h.

The reaction is cooled and washed with three 25 mL portions of 10% NaCl, after which the benzene is removed by distillation under reduced pressure. The residue is transferred to a 1 L bottle, a solution of sodium bisulfite (78 g, 0.75 mol) in water (310 mL) and the mixture is shaken on a mechanical shaker for 2 h. The turbid solution is diluted with water (500 mL) and extracted with benzene (3 × 50 mL). The extracts are discarded. The bisulfite solution is then cooled in an ice bath, and an ice-cold solution of NaOH (32 g, 0.8 mol) in water (130 mL) is added dropwise with mechanical stirring. The ester layer which separates is extracted at once with benzene (4 × 25 mL) The benzene solution is washed with 1% HCl (50 mL), dried for a short time over sodium sulfate (20 g), filtered into a 250 mL modified Claisen flask and distilled under reduced pressure to obtain ethyl (1-ethylpropylidine)cyanoacetate (65.4–75 g, 60.5–68% yield).

Ethyl (1-methylpropylidine)cyanoacetate (Dakin-Prout modification; amino-acid catalysis):[97]

In a 1 L round-bottom flask fitted with a 24/40 joint are placed β-alanine (0.45g, 5.1 mmol), ethyl cyanoacetate (106 mL, 0.1 mol), butanone (108 mL, 1.2 mol), glacial acetic acid (20 mL) and benzene (100 mL). A Barrett-style Dean–Stark trap and condenser are attached to the flask, and the mixture is heated briskly under reflux until water ceases to be collected in the trap (7–12 h).

The reaction mixture is decanted into a 500 mL round-bottom flask, which is attached to a fractionating column (60 cm Vigreaux). The solvent is removed at atmospheric pressure while the oil bath is heated finally at 160 °C. The residue is distilled at reduced pressure to furnish four fractions: (a) acetic acid and other materials boiling below 95 °C/16 mmHg; (b) ethyl cyanoacetate, bp 95–110 °C/16 mmHg, (c) mixed fractions bp 110–124 °C/16 mmHg, and (d) ethyl (1-methylpropylidine)cyanoacetate bp 124–126 °C/16 mmHg. Fraction (d) amounts to 117–122 g, and refractionation of (c) gives an additional 18–24 g, total yield 135–146 g (81–87.5% yield).

Diethyl 5α-Cholestan-3-ylidinemalonate (48**; Lehnert Modification):**[34]

A solution of $TiCl_4$ (5 mL, 45.6 mmol) in CCl_4 (22 mL) was added dropwise to 40 mL THF with stirring under ice cooling and under nitrogen. To this yellow emulsion, diethyl malonate (3.1 mL, 20.4 mmol) and 5α-cholestan-3-one **47** (3.81 g, 9.9 mmol) were added, and then pyridine (3.25 mL, 40.2 mmol) in THF (7 mL) was added dropwise. The resulting brick-red emulsion was stirred at room temperature under a nitrogen atmosphere for three days. Water was added to the reaction mixture, and the whole was stirred until all the solids had dissolved, and then extracted with ether. The ethereal extract was washed with 2 N HCl, 5% $NaHCO_3$, sat. NaCl, dried over Na_2SO_4 and evaporated to dryness. The residue was chromatographed on silica gel. Elution with benzene gave 3.87 g of crude product. Recrystallization from ethanol gave diethyl 5α-cholestan-3-ylidinemalonate (3.07 g, 58.9% yield) as colorless plates.

m-Nitrocinnamic acid (Verley–Doebner Modification):[98]

In a 1 L round bottom flask fitted with a reflux condenser are placed *m*-nitrobenzaldehyde (151 g, 1 mol), malonic acid (115 g, 1.1 mol), 95% ethanol (250 mL) and pyridine (25 mL). The mixture is heated on a steam bath under gentle reflux for 6–8 h and then cooled. The large masses of crystals are broken up with a spatula, and the reaction mixture is cooled on an ice bath. The solid is collected on a Büchner funnel, and the residue is washed with cold ethanol (100 mL) and then ether (2 × 100 mL). The crude product is suspended in ethanol (300 mL) and digested on a steam plate for 2–3 h. The mixture is cooled and filtered, and the solid is air-dried. *m*-nitrocinnamic acid is isolated as a light yellow solid, 144–155 g (75–80% yield).

(*E*)-Dec-3-enoic acid (Ragoussis Modification)[21]

In a 1 L round-bottom flask, equipped with a condenser and a bubbler filled with DMSO, a solution of malonic acid (104.1 g, 1.0 mol), piperidinium acetate (from 0.85 g piperidine and 0.6 g acetic acid, 0.01 mol) and *n*-octanal (64.0 g, 0.5 mol) in DMSO (500 mL) was stirred under nitrogen at room temperature for 20 min. Then the nitrogen was removed and the solution was heated on a steam bath, with stirring. A rapid evolution of CO_2 was observed. Heating was maintained until the evolution of CO_2 ceased (4 h). The solution was cooled to room temperature, poured into 1 L of cold water and extracted with diethyl ether. The combined extracts were washed with water, dried over anhydrous sodium sulfate and evaporated under reduced pressure. The crude product was purified by distillation under vacuum, using a simple, 30 cm Vigreaux column to give (*E*)-dec-3-enoic acid (73.1 g, 86% yield).

2.8.7 Reference

1. [R] Jones, G. *Org. React.* **1967,** *15*, 204–599.
2. [R] Tietze, L. F.; Beifuss, U. In *Comprehensive Organic Synthesis*; Trost, B. M., Ed.; Pergamon: Elmsford, NY, **1991**, Vol. 2, 341–394.
3. a) Hantzsch, A. *Justus Liebigs Ann. Chem.* **1882,** *215*, 1–82. b) Hantzsch, A. *Ber. Dtsch. Chem. Ges.* **1883,** *16*, 1946–1948.
4. Hantzsch, A. *Ber. Dtsch. Chem. Ges.* **1885,** *18*, 2579–2586.
5. Knoevenagel, E. *Justus Liebigs Ann. Chem.* **1894,** *281*, 25–126.
6. Knoevenagel, E. *Ber. Dtsch. Chem. Ges.* **1894,** *27*, 2345–2346.
7. Knoevenagel, E. *Ber. Dtsch. Chem. Ges.* **1896,** *29*, 172–174.
8. a) Claisen, L. *Ber. Dtsch. Chem. Ges.* **1881,** *14*, 345–349. b)) Claisen, L.; Crismer, L. *Justus Liebigs Ann. Chem.* **1883,** *218*, 129–144. c) Carrick, J. T. *J. Prakt. Chem.* **1890,** *42*, 159–160. d) Carrick, J. T. *J. Prakt. Chem.* **1892,** *45*, 500–515. e) [R] Johnson, J. R. *Org. React.* **1942,** *1*, 210–265.

9. Ehrenberg, A. *J. Prakt. Chem.* **1887,** *36*, 117–131.
10. Knoevenagel, E. *Ber. Dtsch. Chem. Ges.* **1898,** *31*, 2585–2595.
11. Hann, A. C. O.; Lapworth, A. *J. Chem. Soc.* **1904,** *85*, 46–56.
12. Tanikaga, R.; Tamura, T.; Nozaki, Y.; Kaji, A. *J. Chem. Soc., Chem. Comm.* **1984,** 87–88.
13. Iwasaki, G.; Saeki, S.; Hamana, M. *Chem. Lett.* **1986,** *15*, 173–176.
14. Knoevenagel, E. *Ber. Dtsch. Chem. Ges.* **1898,** *31*, 2596–2619.
15. a) Fiquet, E. *Ann. Chim. Phys.* **1893,** *29*(6) 438–504. b) Fiquet, E. *Bull. Soc. Chim. Fr.***1892,** *7*(iii), 11–13. c) Fiquet, E. *Bull. Soc. Chim. Fr.* **1892,** *7*(iii), 767–771.
16. Verley, A. *Bull. Soc. Chim. Fr.* **1899,** *21*(iii), 413–414.
17. Verley, A. *Bull. Soc. Chim. Fr.* **1899,** *21*(iii), 414–418.
18. Doebner, O. *Ber. Dtsch. Chem. Ges.* **1900,** *33*, 2140–2142.
19. Boxer, S. E.; Linstead, R. P. *J. Chem. Soc.* **1931,** 740–751.
20. Ragoussis, N. *Tetrahedron Lett.* **1987,** *28*, 93–96.
21. Ragoussis, N.; Ragoussis, V. *Perkin Trans. 1.* **1998,** 3529–3533.
22. Corey, E. J. *J. Am. Chem. Soc.* **1952,** *74*, 5987–5905.
23. Nokami, J.; Mandai, T.; Imakura, Y.; Nishiuchi, K.; Kawada, M.; Wakabayashi, S. *Tetrahedron Lett.* **1981,** *22*, 4489–4490.
24. Ono, T.; Tamaoka, T.; Yuasa, Y.; Matsuda, T.; Nokami, J.; Wakabayashi, S. *J. Am. Chem. Soc.* **1984,** *106*, 7890–7893.
25. Trost, B.; Mallart, S. *Tetrahedron Lett.* **1993,** *34*, 8025–8028.
26. Nokami, J.; Kataoka, K.; Shiraishi, K.; Osafune, M.; Hussain, I.; Sumida, S. *J. Org. Chem.* **2001,** *66*, 1228–1232.
27. Knoevenagel, E. DE 162,281, **1904.**
28. Blanchard, K. C.; Klein, D. L.; MacDonald, J. *J. Am.Chem. Soc.* **1931,** *53*, 2809–2810.
29. Cope, A. C. *J. Am. Chem. Soc.* **1937,** *59*, 2327–2330.
30. Cope, A. C. Hofmann, C. M.; Wyckoff, C.; Hardenbergh, E. *J. Am. Chem. Soc.* **1941,** *63*, 3452–3456.
31. Dakin, H. D. *J. Biol. Chem.* **1909,** *7*, 49–55.
32. Haley, C. A. C.; Maitland, P. *J. Chem. Soc.* **1951,** 3155–3174.
33. Prout, F. S. *J. Org. Chem.* **1953,** *18*, 928–933.
34. a) Lehnert, W. *Tetrahedron Lett.* **1970,** *54*, 4723–4724. b) Lehnert, W. *Tetrahedron* **1973,** *29*, 635–638. c) Lehnert, W. *Synthesis* **1974,** 667–669.
35. Kano, K.; Hayashi, K.; Mitsuhashi, H. *Chem. Pharm. Bull.* **1982,** *30*, 1198–1203.
36. a) [R] Clark, J. H. *Chem. Rev.* **1980,** *80*, 429-452. b) Yakabson, G. G.; Akhmetova, N. E. *Synthesis* **1983,** 169–184.
37. Texier-Boullet, F.; Villemin, D.; Ricard, M.; Moison, H.; Foucaud, A. *Tetrahedron* **1985,** *41*, 1258–1266.
38. Moison, H.; Texier-Boullet, F.; Foucaud, A. *Tetrahedron* **1987,** *43*, 537–542.
39. Cabello, J. A.; Campelo, J. M.; Garcia, A.; Luna, D.; Marinas, J. M. *J. Org. Chem.* **1984,** *49*, 5195–5197.
40. Chalais, S.; Laszlo, P.; Mathy, A. *Tetrahedron Lett.* **1985,** *26*, 4453–4454.
41. Ebitani, K.; Motokura, K.; Mori, K.; Mizugaki, T. Kaneda, K. *J. Org. Chem.* **2006,** *71*, 5440–5447.
42. a) Balalaie, S.; Bararjanian, M.; Hekmat, S.; Salehi, P. *Synth. Commun.* **2006,** *36,* 2549–2557. b) Balalaie, S.; Bararjanian, M.; Hekmat, S.; Salehi, P. *Synth. Commun.* **2006,** *36,* 3703–3711. c) Balalaie, S.; Bararjanian, M.; Sheikh-Ahmadi, M.; Hekmat, S.; Salehi, P. *Synth. Commun.* **2007,** *37,* 1097–1108.
43. Hara, T.; Kanai, S.; Mori, K.; Mizugaki, T.; Ebitani, K.; Jitsukawa, K.; Kaneda, K. *J. Org. Chem.* **2006,** *71*, 7455–7562.
44. Lai, G.; Peng, J.; Li, J.; Qiu, H.; Jiang, J.; Jiang, K.; Shen, Y. *Tetrahedron Lett.* **2006,** *47*, 6951–6953.
45. Mastagli, P.; Lambert, P.; Andric, N. *Bull. Soc. Chim. Fr.* **1956,** 796–798.
46. Hein, R. W.; Astel, M. J.; Shelton, J. R. *J. Org. Chem.* **1961,** *26*, 4874–4878.
47. Mastagli, P, Gnanadickam, C.; Lambert, P. *Compt. Rend.***1962,** *254*, 1295–1297.

48. Yamashita, K.; Tanaka, T.; Hayashi, M. *Tetrahedron* **2005**, *61*, 7981–7985.
49. Homann, W. K. DE 3,423,243, **1986**.
50. Bartoli, G.; Beleggia, R.; Giuli, S.; Giuliani, A.; Marcantoni, E.; Massaccesi, M.; Paoletti, M. *Tetrahedron Lett.* **2006,** *47*, 6501–6504.
51. Bartoli, G.; Bosco, M.; Carlone, A.; Dalpozzo, R.; Galzerano, P.; Melchiorre, P.; Sambri, L. *Tetrahedron, Lett.* **2008,** *49*, 2555–2557.
52. Attanasi, O.; Filippone, P.; Mei, A. *Synth. Commun.* **1983,** *13*, 1203–1208.
53. Ren, Y.-M.; Cai, C. *Synth. Commun.* **2007,** *37*, 2209–2213.
54. Balalie, S.; Bararjanaian, M. *Synth. Commun.* **2006,** *36*, 533–539.
55. Oskooie, H. A.; Heravi, M. M.; Derikvand, F.; Khorasani, M.; Bamoharram, F. F. *Synth. Commun.* **2006,** *36*, 2819–2823.
56. Ranu, B. C.; Jana, R. *Eur. J. Org. Chem.* **2006,** 3767–3770.
57. a) Darvatkar, N. B.; Deorukhkar, A. R. Bhilare, S. V.; Salunkhe, M. M. *Synth. Commun.* **2006,** *36*, 3043–3051. b) Forbes, D. C.; Law, A. M.; Morrison, D. W. *Tetrahedron Lett.* **2006,** *47*, 1699–1703.
58. Chowdhury, S.; Mohan, R. S.; Scott, J. L. *Tetrahedron* **2007,** *63*, 2363–2389.
59. Valizadeh, H.; Shockravi, A.; Gholipur, H. *J. Heterocyclic Chem.* **2007,** *44*, 867–870.
60. a) Bigi, F.; Conforti, M. L.; Maggi, R.; Piccinno, A.; Sartori, G. *Green Chem.* **2000,** *2*, 101–103. b) McCluskey, A.; Robinson, P. J.; Hill, T.; Scott, J. L.; Edwards, J. K. *Tetrahedron Lett.* **2002,** *43*, 3117–3120.
61. Rong, L.; Li, X.; Wang, H.; Shi, D.; Tu, S.; Zhuang, Q. *Synth. Commun.* **2006,** *36*, 2407–2412.
62. Abd El-Rahman, N. M.; El-Kateb, A. A.; Mady, M. F. *Synth. Commun.* **2007,** *37*, 3961–3970.
63. Boehm, T. *Arch. Pharm.* **1929,** *267*, 702–714.
64. Hedge, J. A.; Kruse, C.W.; Snyder, H. R. *J. Org. Chem.* **1961,** *26*, 3166–3170.
65. [R] Freedman, F. *Chem. Rev.* **1980,** *80*, 329–350.
66. Jones, G. R. N. *Nature* **1972,** *235*, 257–261.
67. Baerts, G. *Bull. Soc. Chim. Belges.* **1932,** *41*, 314–317.
68. Lehnert, W. *Synthesis* **1974,** 667–669.
69. Martin, D. F.; Shamma, M.; Fernelius, W. C. *J. Am. Chem. Soc.* **1958,** *80*, 5851–5856.
70. Swoboda, G.; Schuster, P. *Monatsh. Chem.* **1964,** *95*, 398–401.
71. Buetler, U.; Fuenfschilling, P. C.; Steinkemper, A. *Org. Process Res. Dev.* **2007,** *11*, 341–345.
72. Conlon, D. A.; Drahus-Paone, A.; Ho, G.-J.; Pipik, B.; Helmy, R.; McNamara, J. M.; Shi, Y.-J.; Williams, J. M.; MacDonald, D.; Deschênes, D.; Gallant, M.; Mastracchio, A.; Roy, B.; Scheigetz, J. *Org. Process Res. Dev.* **2006,** *10*, 36–45.
73. Walker, G. N. *J. Am. Chem. Soc.* **1955,** *77*, 3844–3850.
74. Chavan, S. P.; Pathak, A. B.; Pandey, A.; Kalkote, U. R. *Synth. Commun.* **2008,** *37*, 4253–4263.
75. Cantello, B. C. C.; Cawthornre, M. A.; Cottam, G. P.; Duff, P. T.; Haigh, D.; Hindley, R. M.; Lister, C. A.; Smith, S. A. Thurlby, P. L. *J. Med. Chem.* **1994,** *37*, 3977–3985.
76. Balavoine, F.; Malabre, P.; Alleaume, T.; Rey, A.; Cherfils, V.; Jeanneton, O.; Seigneurin-Venin, S.; Revah, F. *Bioorg. Med. Chem. Lett.* **2007,** *17*, 3754–3759.
77. Tietze, L. F.; Brumby, T.; Pretor, M.; Remberg, G. *J. Org. Chem.* **1988,** *53*, 810–820.
78. Sunazuka, T.; Handa, M.; Nagai, K.; Shirahata, T.; Harigaya, Y.; Otoguro, K.; Kuwajima, I.; Omura, S. *Org. Lett.* **2002,** *4*, 367–369.
79. Robertson, D. N. *J. Org. Chem.* **1960,** *25*, 47–49.
80. Benzing, M.; Vilsmaier, E. *Chem. Ber.* **1987,** *120*, 1873–1880.
81. a) Alt, G. H.; Gallegos, G. A. *J. Org. Chem.* **1971,** *36*, 1000–1002. b) Prout, F. S. *J. Org. Chem.* **1973,** *38*, 399–400.
82. Letif, N.; Mishriky, N. *Can. J. Chem.* **1966,** *44*, 1271–1274.
83. Augustine, J. K.; Naik, Y. A.; Mandal, A. B.; Chowdappa, N.; Praveen, V. B. *J. Org. Chem.* **2007,** *72*, 9854–9856.

84. Enders, D.; Grondal, C.; Hüttl, M. R. M. *Angew. Chem., Int. Ed.Engl.* **2007,** *46*, 1570–1581.
85. Tietze, L. F.; Beifuss, U.; Antel, J.; Sheldrick, G. M. *Angew. Chem., Int. Ed. Engl.* **1988,** *27*, 703–705.
86. Tietze, L. F.; Zhou, Y. *Angew. Chem., Int. Ed.* **1999,** *38*, 2045–2047.
87. Trost, B. M.; Dong, L.; Schroeder, G. M. *J. Am. Chem. Soc.* **2005,** *127*, 2844–2845.
88. Gribble, G. W. Knorr and Paal–Knorr Pyrrole Synthesis. In *Name Reactions in Heterocyclic Chemistry*; Li, J. J., Ed.; John Wiley & Sons: Hoboken, NJ, 2005; pp 79–88.
89. Shea, K. M. Feist-Bénary Furan Synthesis. In *Name Reactions in Heterocyclic Chemistry*; Li, J. J., Ed.; John Wiley & Sons: Hoboken, NJ, 2005; pp 160–167.
90. Galatsis, P. Hantzsch (Dihydro)-Pyridine Synthesis. In *Name Reactions in Heterocyclic Chemistry*; Li, J. J., Ed.; John Wiley & Sons: Hoboken, NJ, 2005; pp 304–322.
91. Tinsley, J. M. Gewald Aminothiophene Synthesis. In *Name Reactions in Heterocyclic Chemistry*; Li, J. J., Ed.; John Wiley & Sons: Hoboken, NJ, 2005; pp 193–1198.
92. Corson, B. B.; Stoughton, R. W. *J. Am. Chem. Soc.* **1928,** *50*, 2825–2837.
93. Fukuyama, T.; Liu, G. *J. Am. Chem. Soc.* **1996,** *118*, 7426–7427.
94. Mehta, G.; Pallavi, K.; Umarye, J. D. *Chem. Commun.* **2005,** 4456–4458.
95. Watson, T. J.; Horgan, S. W.; Shah, R. S.; Farr, R. A.; Schnettler, R. A.; Nevill, C. R. Jr.; Weiberth, F. J.; Huber, E.W.; Baron, B. M.; Webster, M. E.; Mishra, R.K.; Harrison, B. L.; Nyce, P. L.; Rand, C. L.; Goralski, C. T. *Org. Proc. Res.Dev.* **2000,** *4*, 477–480.
96. Cope, A.C.; Hancock, T.M. *Org. Syntheses Coll. Vol. 3*, **1955,** 399.
97. Prout, F. S.; Hartman, R. J.; Huang, P.-Y.; Korpics, C. J. . *Org. Syntheses Coll. Vol. 4*, **1963,** 93.
98. Wiley, R. H.; Smith, N. R. *Org. Synth. Coll. Vol. 4*, **1963,** 731.

2.9 Mukaiyama Aldol Reaction

Richard J. Mullins and Michael T. Corbett

2.9.1 Description

The Mukaiyama aldol reaction is the nucleophilic addition of a trimethylsilyl enol ether **1** to either an aldehyde **2** or a ketone in the presence of a Lewis acid to form a β-hydroxyketone **3**.

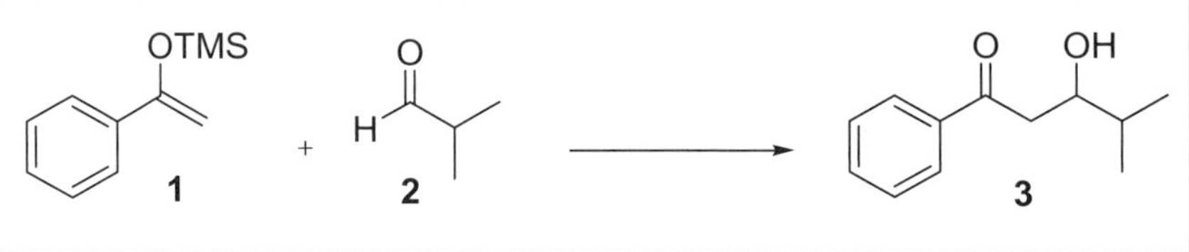

2.9.2 Historical Perspective

Since its identification in 1872 by Charles-Adolphe Wurtz and Alexander Borodin, the aldol reaction has found immense synthetic utility in the formation of carbon–carbon bonds. The utility of the aldol reaction, however, was typically limited due to the formation of various condensation adducts. The desire to reduce the formation of unwanted byproducts led researchers to investigate modifications to the classical aldol model. One theme which emerged from studies to overcome the limitations to the aldol reaction was the incorporation of more powerful lithium amide bases for the production of kinetic and/or thermodynamic lithium enolates.[1,2]

Studies done by Mukaiyama and coworkers in 1973 investigated the control of the aldol reaction when Lewis acids, such as $TiCl_4$, were employed to mediate the cross-aldol addition.[4] Investigations concluded that when silyl enol ethers were reacted with ketones or aldehydes, the cross-aldol adduct was obtained in good yield with no self-addition or condensation products. The synthetic utility of the new reaction was further investigated in the subsequent year when studies done by Mukaiyama examined the role of different Lewis acids in mediating aldol product formation.[5] A wide range of metal halide Lewis acids were utilized as mediators in the reaction of benzaldehyde with isopropenyl acetate demonstrating the reaction's tolerance to a wide range of reaction conditions.

In the following year, the Mukaiyama aldol reaction was extensively studied, covering a wide range of factors associated with the reaction.[6] It was found that $TiCl_4$ was the most effective Lewis acid due to its ability to activate the carbonyl carbon thus making it susceptible to nucleophilic reactions. As might be expected, it was also noted that aldehydes reacted more readily at –78 °C, whereas ketones reacted slowly at low temperatures and therefore necessitated a reaction temperature of 0 °C. Mukaiyama also confirmed that the silyl enol ethers of both ketones and aldehydes shared similar reactivity since all previous studies were conducted solely with the enolates of ketones. This reactivity, however, allows for significant functional group tolerance. A reactant comprised of an aldehyde and a ketone will selectively react through the aldehyde when at –78 °C. Similarly, ketones react preferentially when esters are also present in the compound at room temperature.

2.9.3 Mechanism

Numerous in-depth mechanistic studies have been performed on the Mukaiyama aldol reaction.[6–9] Although various mechanisms exist in the literature that take into account the various roles of the numerous catalysts used for the enantio- and diastereoselective Mukaiyama aldol reaction, the commonly accepted mechanism accounting for bond formation is shown below.[10] The reaction begins with the coordination of a Lewis acid with aldehyde **4** to form complex **5**. Due to its enhanced electrophilicity, complex **5** is attacked by the π-bond of the enol silane **6**, giving rise to resonance stabilized cation **7**. At this point, either intermolecular silyl cleavage upon hydrolysis or intramolecular silyl transfer to the product hydroxyl group occurs to give products such as **8** or **9**.

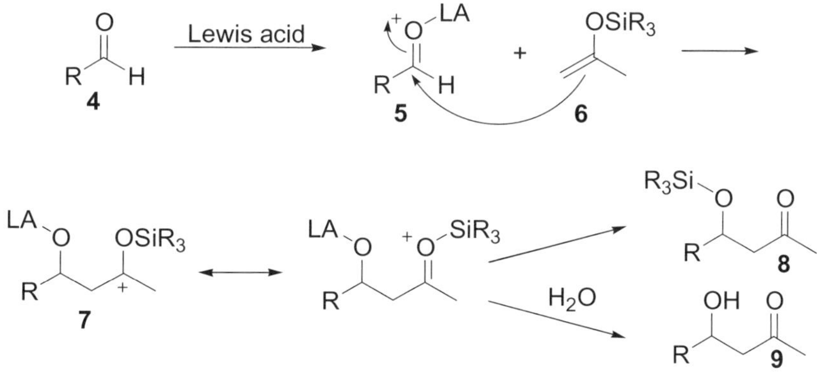

While the order of silyl transfer or cleavage is inconsequential to bond formation, it is one of the more important and hotly debated aspects of the mechanism owing to its importance in the development of catalytic enantioselective variants of the Mukaiyama aldol reaction. Intramolecular silyl transfer, as shown in the formation of **10**, would regenerate the chiral, non-racemic metal catalyst. On the other hand, intermolecular reaction with, for example, a triflate ion, would generate TMSOTf, a powerful, and most importantly, achiral Lewis acid, which could itself catalyze the Mukaiyama reaction, resulting in diminished enantioselectivity in the overall process.

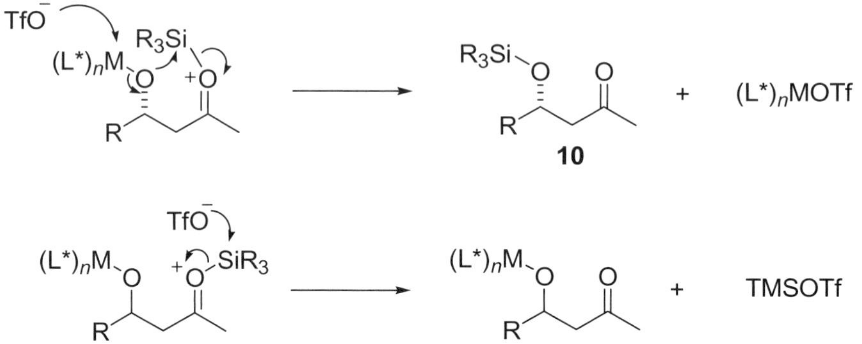

Alternatively, a Friedel–Crafts mechanism has been proposed to account for bond formation via the Mukaiyama aldol reaction.[11] As stated, attack of the enol silane **11** on the activated aldehyde **12** provides carbocation **13**. Prior to silyl group transfer or outright silyl cleavage seen in the mechanism above, removal of the α-hydrogen regenerates the enol silane **14**. While highly dependent on specific reaction conditions, the isolation of **15** leads to the suggestion of **14** as a potential intermediate in the Mukaiyama aldol reaction.

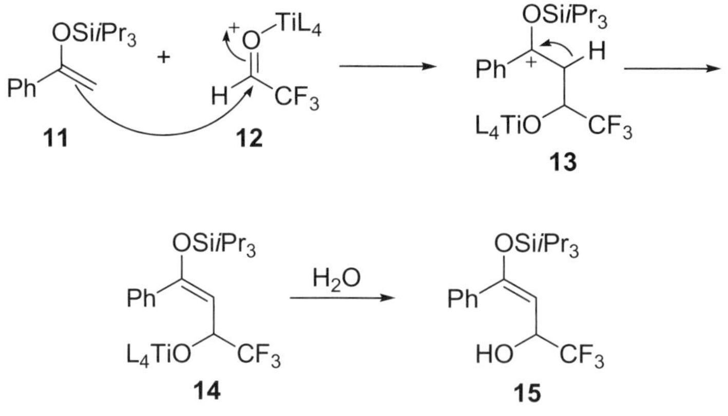

Another mechanistic interpretation of the Mukaiyama aldol invokes an ene reaction pathway accounting for bond formation.[12] Following complexation, the ene reaction occurs via the organized chair-like transition state **16**, providing the *syn*-stereochemistry observed in the adduct **17**. Steric interactions between the methyl group and the glyoxylate ester result in the positioning of the ester in the pseudo-axial position. When an *E*-enol silane is utilized in the reaction, the stereochemical outcome is the same, via an alternate chair-like transition state. The isolation of products, such as **17**, under standard Mukaiyama conditions lends some credibility to this ene pathway as a potential mechanism.

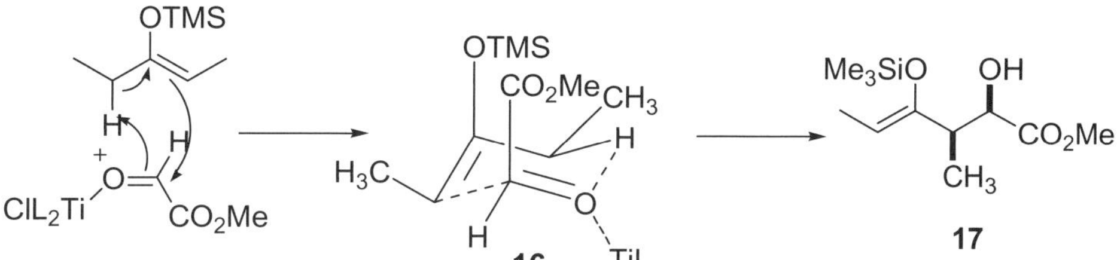

16 **17**

Finally, stereochemical aspects of the mechanism have been extensively studied by Evans and coworkers during their construction of a model to map the diastereoselective induction observed in the course of the aldol reaction.[13] Similarly, mechanistic studies were performed by the Evans group with regards to their bis(oxazolinyl)-pyridine (pybox)-copper(II) complex, a catalyst for the enantioselective Mukaiyama reaction.[14]

2.9.4 Synthetic Utility

Diastereoselective Synthesis

The applicability of the Mukaiyama aldol reaction has been far reaching. As demonstrated in the area of complex natural product synthesis, the reaction has been found to proceed very efficiently under mild conditions with high functional group tolerance. Importantly, these reactions also proceed with high stereoselectivity, even when carried out in the coupling of very complex fragments. An impressive example of this was demonstrated in the Evans synthesis of 6-deoxyerythronolide B,[15] a biosynthetic precursor to the erythromycin antibiotics. The highly convergent Mukaiyama aldol coupling of two fragments was performed via the reaction of silyl enol ether **19** and aldehyde **18** in the presence of the catalyst BF_3•OEt_2 to yield the single aldol adduct **20** in high yield (83%). The excellent diastereoselectivity was attributed to the double stereodifferentiating effects of the α-methyl and β-alkoxy substituents. The 1,2-Felkin–Anh stereochemical result is matched with 1,3-*anti*-diastereofacial selectivity resulting in the highly selective production of **20**. This general phenomenon has been termed by Evans and coworkers "merged 1,2- and 1,3-asymmetric induction" and is exquisitely described in a full report.[16]

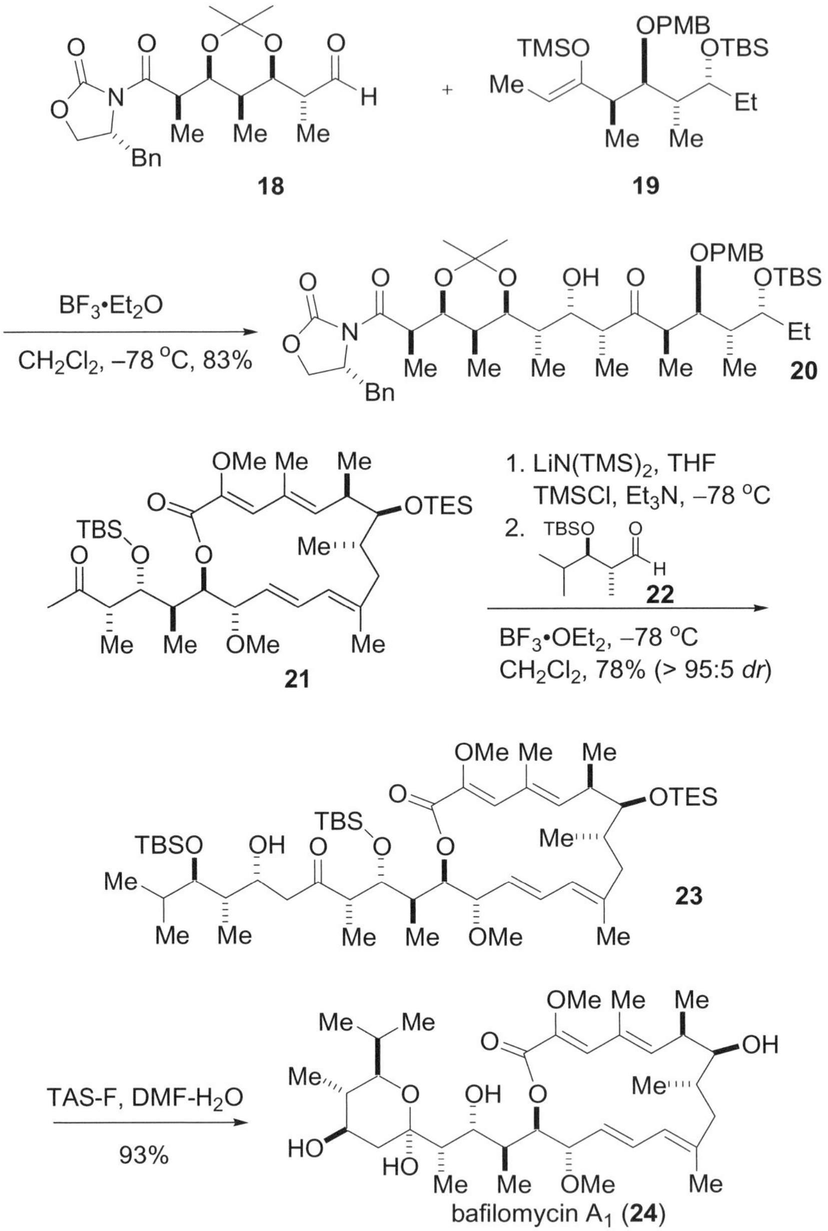

Another impressive demonstration of this phenomenon was seen in the late stage coupling reaction utilized for the synthesis of (–)-bafilomycin A_1 (**24**) by Roush and coworkers.[17] Following conversion of the methyl ketone **21** to the enol silane, coupling with aldehyde **22** proceeded with excellent stereoselectivity to give **23**, the Felkin addition product. Global deprotection resulting in hemiacetal formation completed the synthesis of the macrolide antibiotic. The successful application of this reaction was anticipated based on previous successes in using 2,3-*anti*-β-hydroxy aldehydes as electrophiles in the Mukaiyama aldol reaction.[18]

In a similar manner, the synthesis of the immunosuppressant (–)-pironetin (PA48153C) featured the Mukaiyama aldol as a key step in the later stages.[19] Consistent with the experimental results and the model proposed by Evans,[16] this reaction resulted in the preparation of the single diastereomer **25**, a key intermediate in the Keck synthesis.

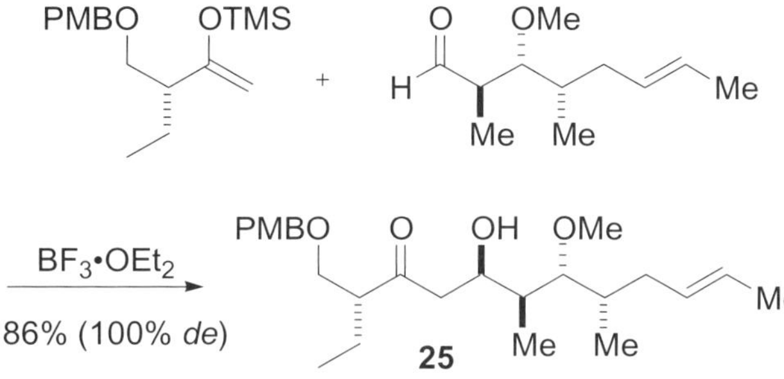

The outstanding diastereoselectivity in these reactions is further demonstrated in Paterson's total synthesis of the potential antiviral agent (+)-concanamycin F.[20] Taking advantage of the double asymmetric induction afforded by the 2,3-*anti*-β-hydroxy aldehyde **27**, the coupling of **26** and **27**, promoted by BF_3•OEt_2, produced the single isomer **28** in high yield.

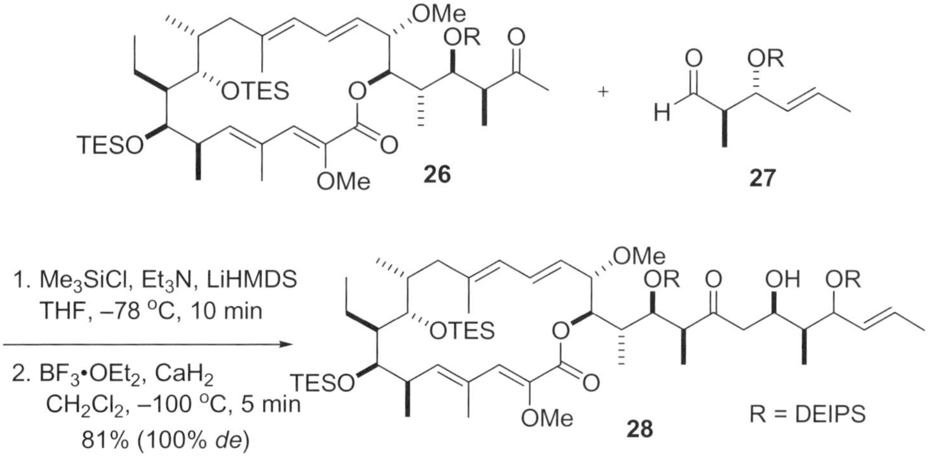

In contrast, the Mukaiyama aldol reaction used in the Heathcock synthesis of the C_{29}–C_{44} fragment of spongistatin[21] proceeded with comparatively reduced diastereoselectivity. The stereochemically complex enol silane **30** was coupled to **29**, a 2,3-*syn*-β-alkoxy aldehyde, resulting in

31 in modest yield and diastereoselectivity. As compared to previous examples, the α- and β- stereochemistry of the aldehyde component provides for the stereochemically mismatched case, resulting in diminished selectivity.

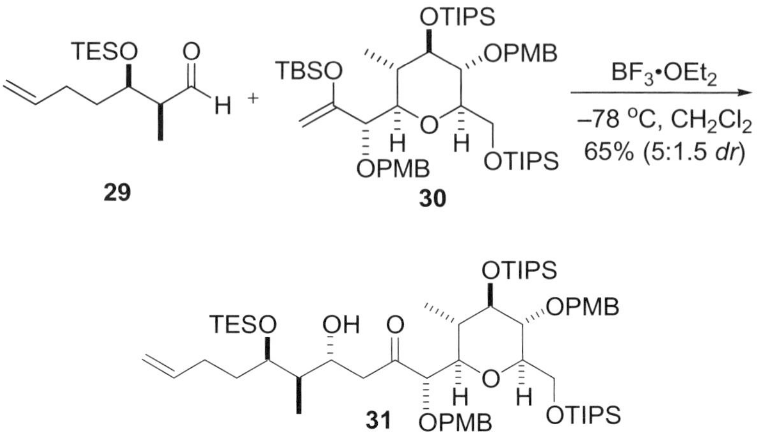

Moderate levels of diastereoselectivity in the Mukaiyama aldol reaction can also be achieved using the β-alkoxy substituent to serve as the lone stereodirecting agent. In contrast to the high levels of double asymmetric induction observed in the cases above, during the course of their asymmetric synthesis of the marine macrolide bryostatin, the Evans group[22] observed the reaction of β-alkoxyaldehyde **32** with the bis(trimethylsilyl)-dienol ether **33** in the presence of the alkoxytitanium Lewis acid $TiCl_2(Oi\text{-}Pr)_2$, which yielded the aldol adduct **34** in good yield (83%) and with good diastereoselectivity (94:6 *dr*).

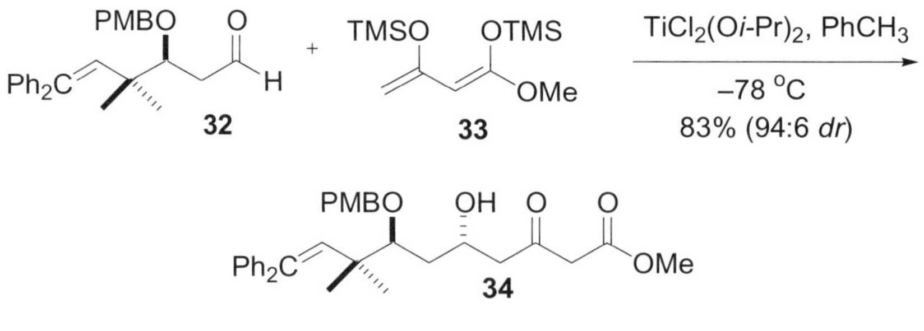

Similar results have been obtained in the coupling of aldehyde **35** and enol silane **36**, which resulted in a 9:1 ratio of diastereomers favoring **37**, a key intermediate in the synthesis of the antiviral marine natural product (–)-hennoxazole.[23,24] In a similar manner, coupling of **38** and **39** resulted in **40**, providing an alternate route to the same molecule.

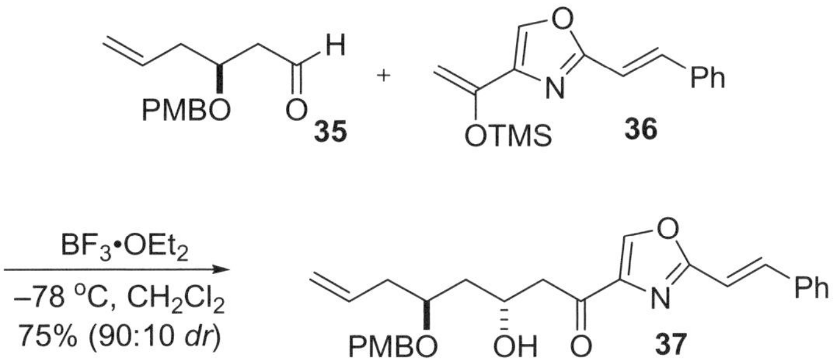

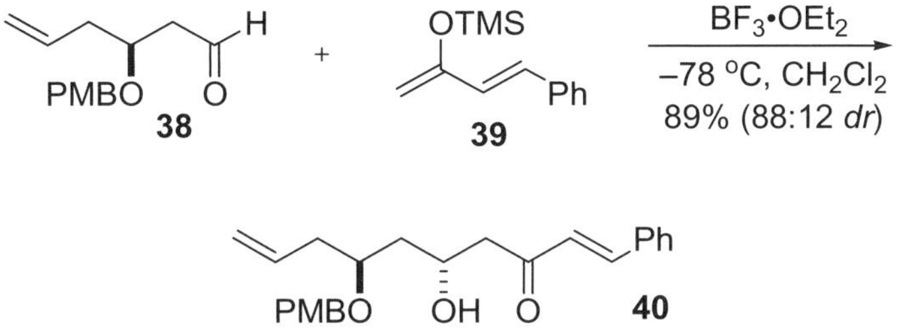

The Mukaiyama aldol reaction has also been shown to proceed in a stereoselective manner as a result of chelation control as demonstrated by Oikawa and coworkers during their total synthesis of the antifungal antibiotic tautomycin.[25] The key-step of their synthesis was the aldol coupling of the enol silane derived from ketone **41** to aldehyde **42** to provide the *anti*-Felkin aldol adduct **43** as a single stereoisomer in good yield (54%). The ability to couple two highly functionalized subunits utilizing the Mukaiyama aldol is noteworthy, with the reaction stereoselectivity presumed to be the result of the chelation between the Lewis acid and the methoxy group α to the carbonyl.

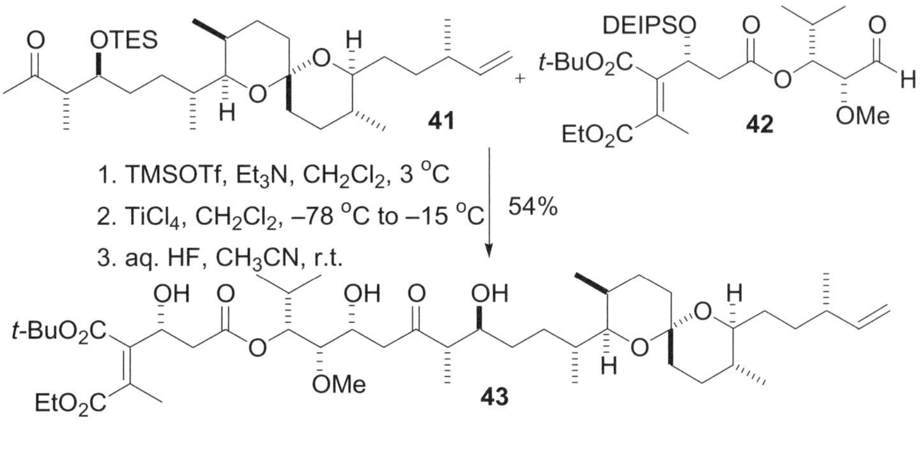

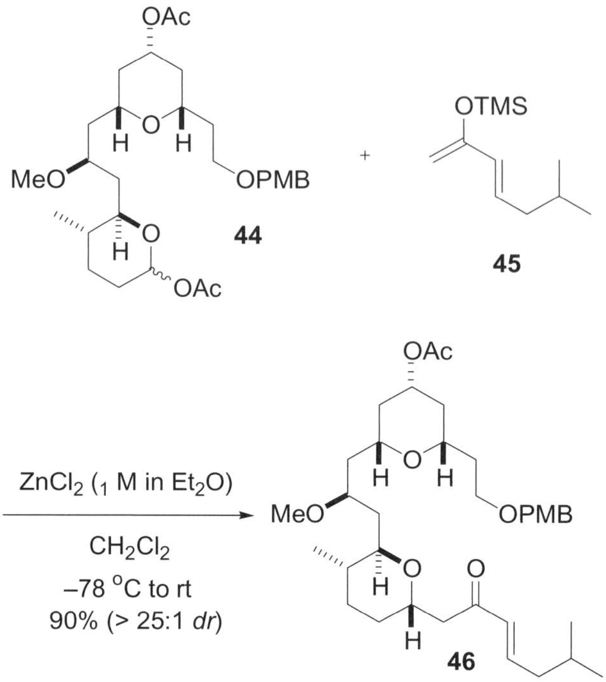

Acetals can be effectively utilized as electrophiles for enol silane substrates under Lewis acidic conditions. As demonstrated by Williams and coworkers, the diastereoselective $ZnCl_2$-catalyzed Mukaiyama aldol process between enol silane **45** and acetal **44** occurred to produce **46**. Thus, the *E*-unsaturated ketone of the marine macrolide leucascandrolide A was directly installed with a high level of diastereoselectivity resulting from axial addition to the resulting 6-membered oxocarbenium ion.[26] Similar results were demonstrated in the Paterson synthesis of lecuascandrolide.[27]

Vinylogous Mukaiyama Aldol Reaction

The integration of silyl dienol ethers into Mukaiyama aldol reactions in place of silyl enol ethers was pioneered by Paterson and coworkers.[28] The vinylogous Mukaiyama aldol reaction is a Lewis acid promoted addition of a silyl dienol ether to an aldehyde to yield solely γ-attack, achieving diastereoselective and/or enantiomeric aldol adducts in high yield. High selectivity is achieved using both boron and titanium Lewis acids; although, BF_3•OEt_2 provided the highest levels of diastereoselectivity. In the syntheses of the marine macrolides swinholide A and scytophycin C, the silyl dienol ether **48** was added to aldehyde **47** to produce aldol adduct **49** in high yield (78%) and with the expected high 1,3-*anti*-diastereoselectivity (90:10 *de*). Notably, only the *E*-isomer of the conjugated aldehyde was produced.

Similar results have been obtained for enantioselective vinylogous Mukaiyama aldol reactions using both metal[29,30] and non-metal[31] containing catalysts.

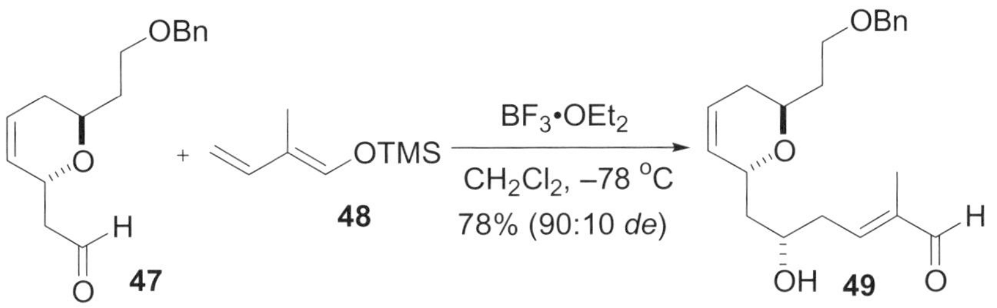

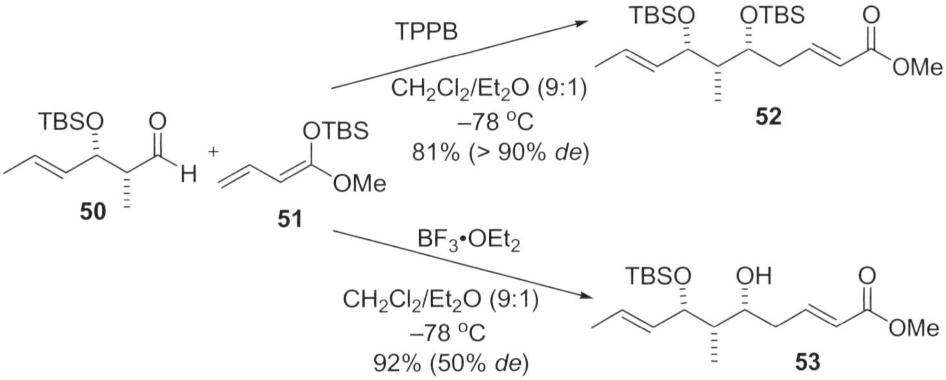

Efforts to increase the diastereoselectivity of the vinylogous Mukaiyama aldol reaction were continued by Kalesse and coworkers utilizing tris(pentafluorophenyl)borane (TPPB) catalysts in lieu of BF_3•OEt_2. By using TPPB, a bulky equivalent of BF_3•OEt_2, the Felkin–Anh stereoselectivity became the predominant factor in the Evans' merged 1,2- and 1,3-asymmetric induction model. As applied in the synthesis of ratjadone, the addition of the diene **51** to the aldehyde **50** yielded, with relatively poor stereocontrol, the aldol adduct **53** when utilizing BF_3•OEt_2. In contrast, the desired product **52** was obtained using TPPB, occurring with complete transfer of the silyl ether.[32] Although the yields are slightly lower when using TPPB, the increase in *de* from 50% to > 90% is substantial. Similar diastereoselectivities were obtained using $B(C_6H_5)_3$ as the Lewis acid catalyst, but without transfer of the silyl group. Thus, complementary methods exist depending on the desire for silyl transfer in the Mukaiyama aldol reaction. In a series of subsequent publications, Kalesse and coworkers expand on the breadth of application of the TPPB catalysts in numerous natural product syntheses,[33,34] including the preparation of **54** and **55**

featuring the use of silylketene acetals toward the synthesis of oleandolide[35] and amphidinolide H2,[36] respectively.

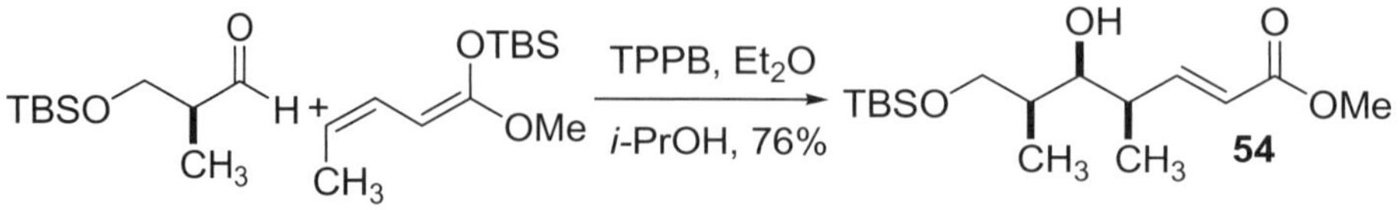

54

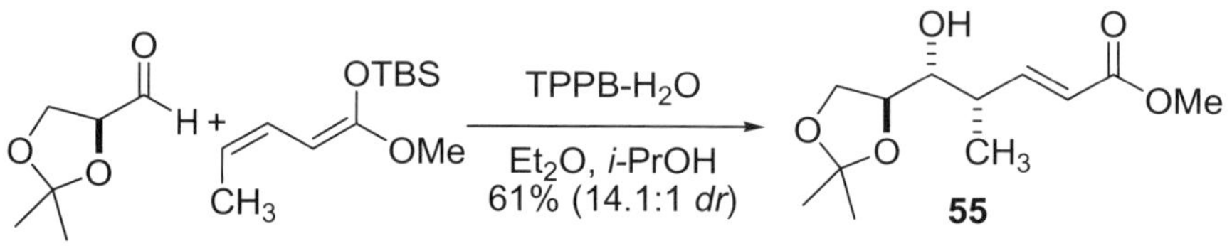

55

Remote asymmetric induction can be obtained through the use of chiral auxiliaries, such as valine derived oxazolidinones, within the framework of the vinylogous Mukaiyama aldol reaction. During the synthesis of khafrefungin, an antifungal agent, Kobayashi and coworkers reacted the vinylketene silyl *N,O*-acetal **56** with the aldehyde **57** to yield the *anti*-aldol adduct **58** in excellent yield (98%) and high diastereoselectivity (> 20:1).[37]

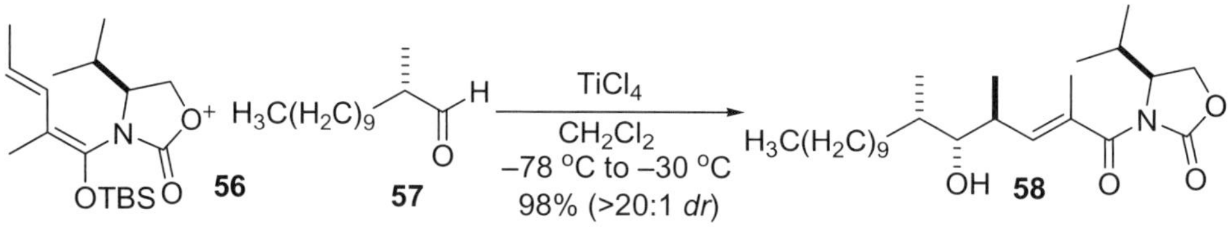

Tandem Reactions

The synthetic utility of the Mukaiyama aldol reaction is broadened when coupled with other reactions in a tandem fashion. Romo and coworkers developed a highly diastereoselective tandem Mukaiyama aldol–lactonization (TMAL) reaction sequence for the synthesis of β-lactones.[38] Despite the relatively high diastereoselectivity, the $ZnCl_2$ mediated aldol reaction was only optimized when smaller silyl and thiol groups of the ketene acetyl were employed for the improved efficiency of the subsequent lactonization. The methodology was utilized in the synthesis of (–)-panicilin D, a pancreatic lipase inhibitor, where the ketene thioacetal **60** was added to aldehyde **59**, upon which lactonization followed by deprotection yielded the β-lactone **61** as a diastereomeric mixture (9.3:1) in modest yield.

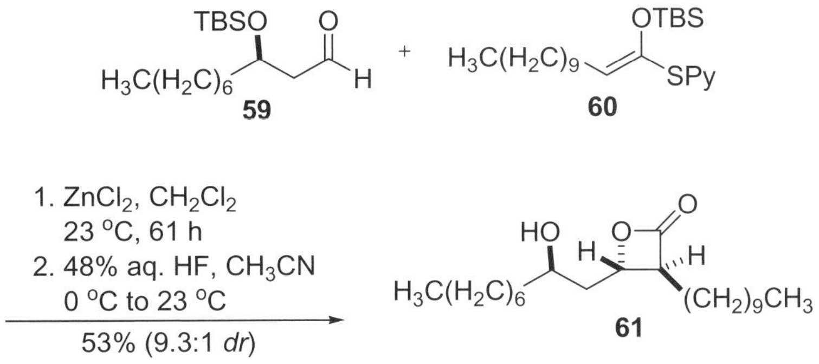

Incorporation of an epoxide-opening prior to a vinylogous Mukaiyama aldol reaction allows for the formation of a stereotriad along an aliphatic chain, which can be used to rapidly produce polyketide frameworks. Utilizing the non-aldol, aldol sequence pioneered by Jung,[39] Kalesse and coworkers reacted the epoxy alcohol **62** with TBSOTf to initiate both the epoxide ring-opening/hydride transfer and the subsequent vinylogous Mukaiyama aldol reaction with the resulting aldehyde to produce ester **64** in high yield (72%) and good *de* (> 10 : 1) following protection of the resulting diol.[40]

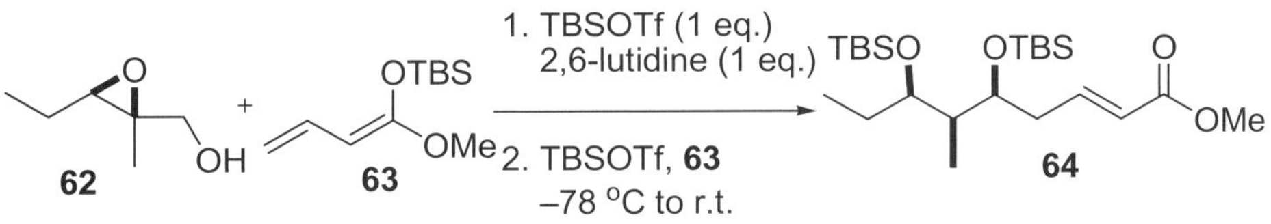

Enantioselective Variants

While substantial utility has been demonstrated for the Mukaiyama aldol reaction in diastereoselective natural product syntheses, more recent research efforts have been focused on the development of catalytic enantioselective variants of the reaction.[41] These enantioselective variants of the reaction have provided creative solutions to problems associated with stereocontrolled syntheses of molecules of polyacetate origin. A wide range of chiral Lewis acid and Lewis base catalysts have been developed that exhibit high levels of enantioselectivity in the Mukaiyama aldol reaction. While the list is certainly not exhaustive, some such catalysts are shown below (**65**–**72**).

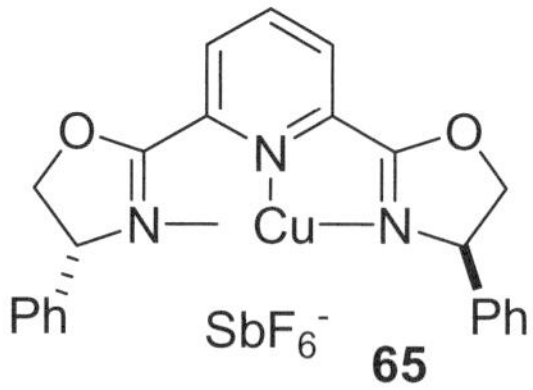

65

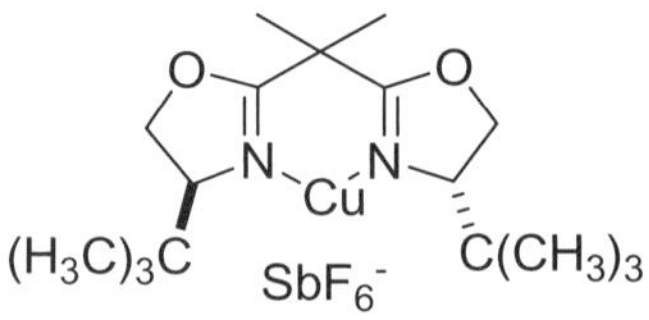

66

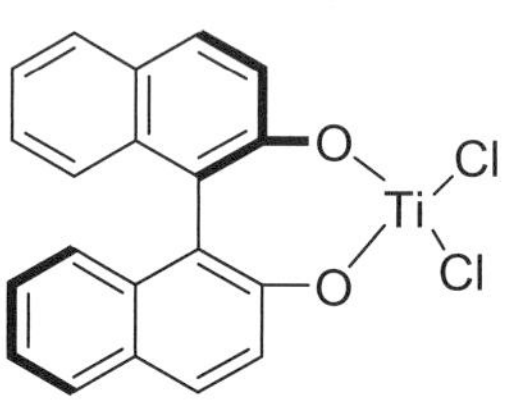

67

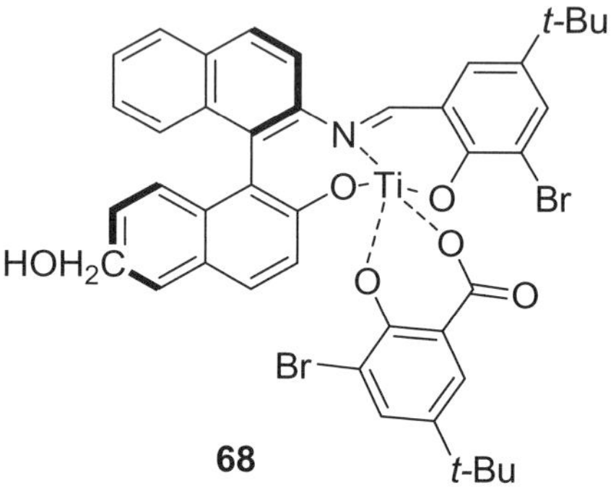

68

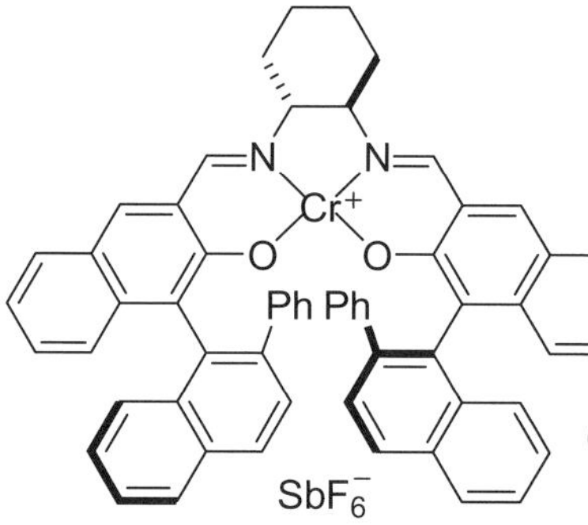

69

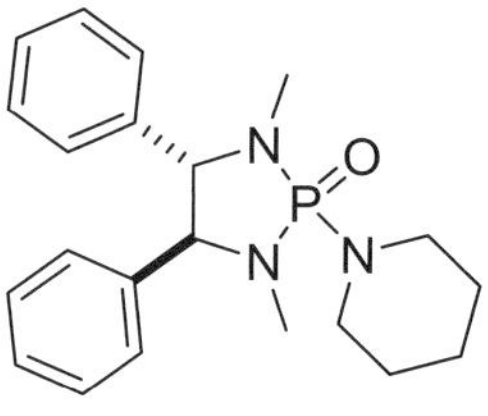

70

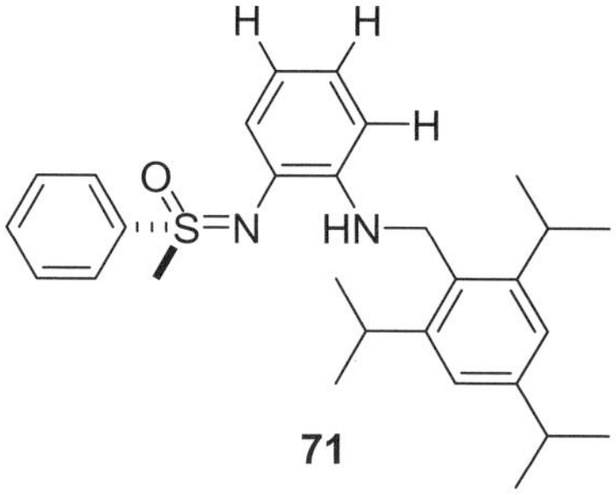

71

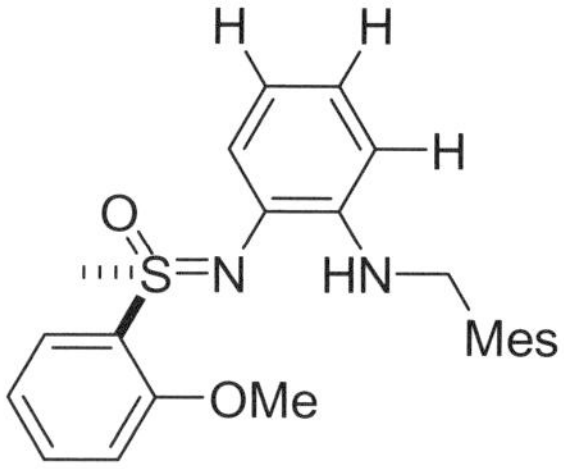

72

The development of C_2-symmetric bis(oxazolinyl)–Cu(II) complexes **65** and **66** by the Evans group greatly contributed to the advancement of the enantioselective Mukaiyama aldol reaction.[42] In addition, these chiral, non-racemic catalysts have found utility in a variety of enantioselective C–C bond forming processes.[14,43] One application of catalyst **65** was demonstrated in the total synthesis of the marine macrolide phorboxazole B where the bis(trimethylsilyl)dienol ether **74** was added to (benzyloxy)acetaldehyde (**73**) to yield the enantiopure (> 99% *ee*) δ-hydride-β-ketoester **75** in high yield (85%).[44,45] While the majority of Mukaiyama aldol reactions have aldehydes as the electrophilic partner, Evans and coworkers have also demonstrated the highly enantioselective reaction between ketone **76** and silylketene acetal **77** using the catalyst **66**.[42] Finally, the vinylogous aldol reaction between 2-(trimethylsiloxy)furan (**80**) and pyruvate ester **79** resulted in **81** with impressive enantio- and diastereoselectivity. These complexes have also found substantial utility in the development of an aqueous, catalytic version of the Mukaiyama aldol.[46–51] Other metals, such as zinc(II) and indium(III), have been utilized with the pybox ligand; however, the copper(II) complexes have been found to produce higher yield and enantioselectivity.[52,53] Finally, polymer bound pybox catalysts have been prepared, allowing for easier recovery of the catalyst following the reaction.[54,55]

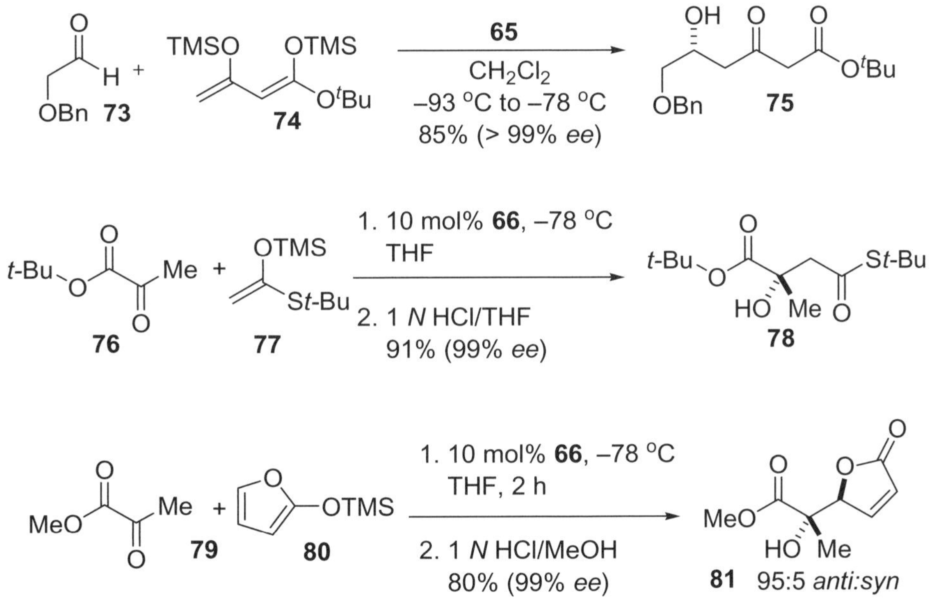

Highly enantioselective asymmetric Mukaiyama aldol reactions can also be carried out utilizing a chiral binaphthol (BINOL)-derived titanium(IV) Lewis acid catalyst **67**.[56] Mikami, Keck, and Carreira have extensively investigated the influence of the BINOL-based titanium(IV) catalyst and its ability to achieve high enantioselectivity.[57–59] For example, **82** and **83** were efficiently coupled to give **84** with excellent enantioselectivity.[58]

A similar reaction was used for preparation of the remote stereocenter in **85**, an intermediate in Paterson's synthesis of aurisides A and B. The asymmetric catalysis of the aldol reaction with fluoral has also been investigated by Mikami for highly enantioselective syntheses of fluorine-containing aldols.[60,61] The reaction of the ketene trimethylsilyl acetal **86** with fluoral (**87**) in the presence of the BINOL-catalyst **67** yielded the aldol adduct **88** in moderate yield (56%) and high enantioselectivity (90% *ee*).

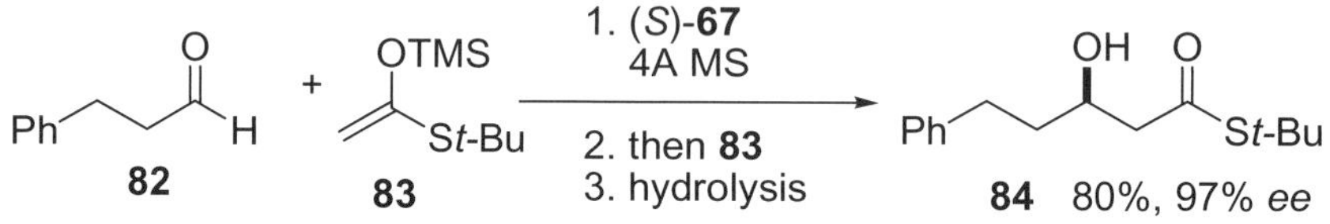

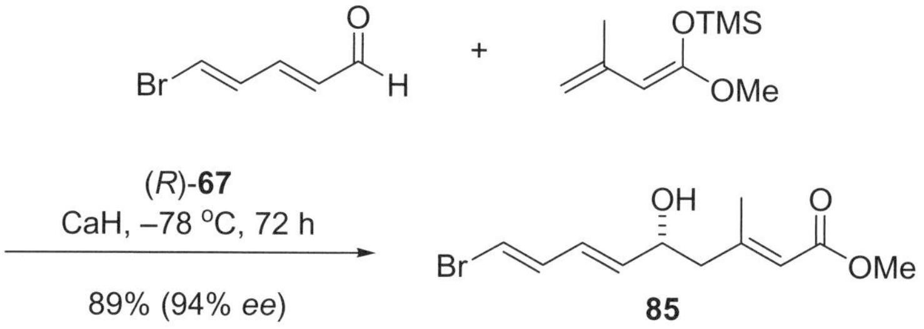

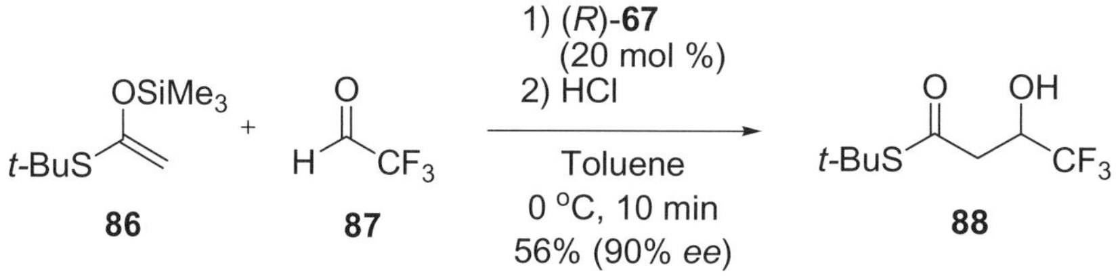

Although a majority of the catalytic complexes employed in the aldol reaction are bidentate, Carreira and coworkers published the synthesis of a new chiral tridentate chelating ligand for the efficient asymmetric induction of stereochemistry in aldol adducts.[59] The Ti(IV) complex **68**, an analog of the BINOL catalyst previously mentioned, was further stabilized by 3,5-di-*tert*-butylsalicyclic acid as a counterion to increase the yields, selectivity, and efficiency of the asymmetric reaction. This new catalyst is particularly effective in the addition of either *O*-trimethylsilyl, or *O*-ethyl, or *O*-methyl ketene to both aliphatic and aromatic aldehydes enantioselectively to obtain the respective aldol adduct. For example, the reaction of the silylketene acetal **90** with the aromatic aldehyde **89** in the presence of **68** obtains the aldol adduct **91** in high yield (91%) and excellent enantioselectivity (97% *ee*).

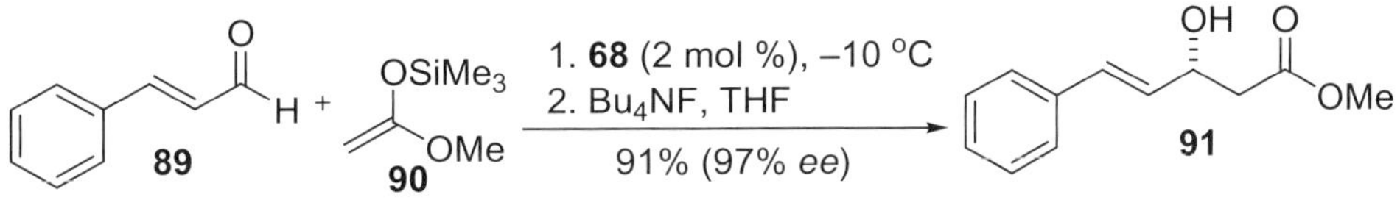

The Mukaiyama aldol reaction has also been used in the enantioselective asymmetric synthesis of δ-hydroxy-β-keto ester derivatives. Katsuki and coworkers utilized the chiral Cr(salen) complex **69** for the catalysis of the aldol reaction in the presence of water or alcohol.[62] The presence of alcohol in the reaction greatly increases the enantioselectivity of the aldol addition. In the reaction of the silyl enol ether **92** with the aldehyde **93** in the presence of the catalyst **69**, triethylamine, and isopropanol yielded the aldol adduct **94** in high yield (87%) and excellent enantioselectivity (97% *ee*).

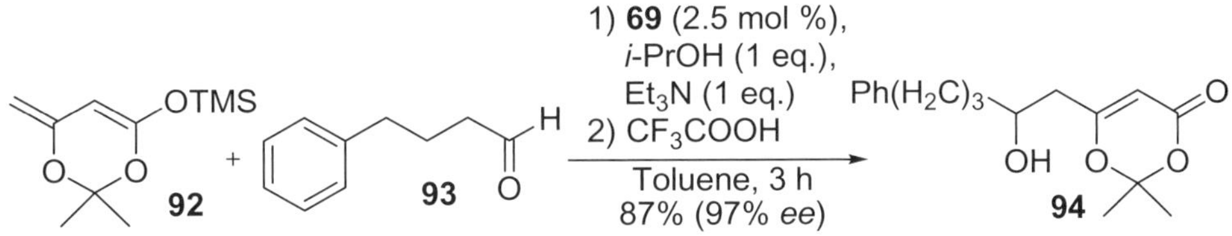

As an alternative to the BINOL-ligands utilized by Carreira, Yamamoto and coworkers developed a group of BINAP ligands for use in a catalyzed Mukaiyama aldol reaction with AgOTf, KF, and 18-crown-6.[63] The reaction of the trimethylsilyl ether **96** with the aldehyde **95** yielded the aldol adduct **97** in high yield (84%) and excellent enantioselectivity (97% *ee*). Unlike other Mukaiyama aldol reactions, this particular example appears to proceed through a closed transition state as the diastereoselectivity of the reaction is dependant on the geometry of the starting enol silane.

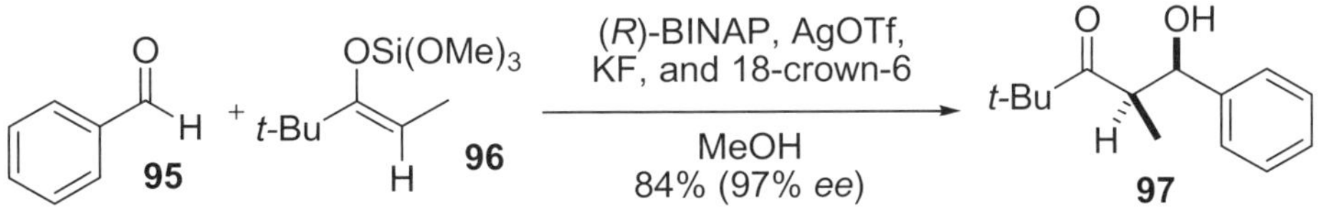

Bolm and coworkers have developed a synthetic procedure for the efficient synthesis of enantiomerically enriched alcohols from pyruvates.[64] As an alternative to the popular PYBOX and BUBOX complexes pioneered by Evans, new C_1-symmetric aryl-bridged aminosulfoximene ligands **71** and **72** were synthesized for use in the copper(II)-catalyzed Mukaiyama aldol reaction. By varying the stereochemistry of the sulfoximine, it is possible to obtain almost enantiopure aldol adducts. For example, the reaction of the enol ether **98** with the aldehyde **99** in the presence of (*S*)-**71** yielded the aldol adduct (*R*)-**100** in high yield (90%) and excellent stereoselectivity (99% *ee*). Similarly, the reaction of the enol ether **101** with the aldehyde **102** in the presence of (*R*)-**72** produced the aldol adduct (*S*)-**103** in high yield (82%) and excellent stereoselectivity (98% *ee*).

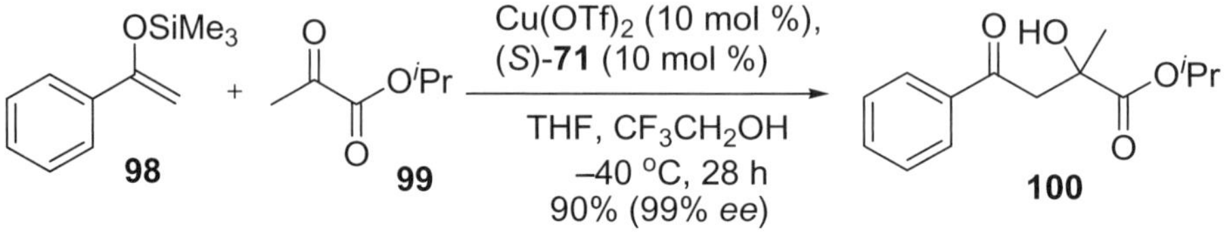

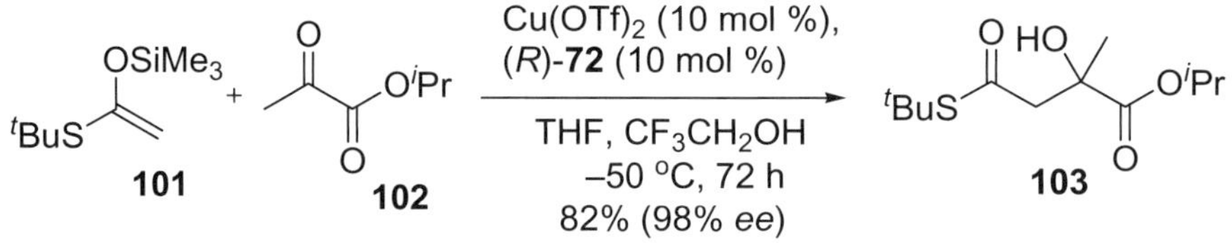

2.9.6 *Experimental*

(3*S*,4*R*,5*R*,6*R*,7*S*,10*R*,11*S*,12*R*)-6,12-di[(*tert*-butyldimethylsilyl)oxy]-3-methoxy-4-(benzoyl)-5,7,11,13-tetramethyl-tetradeca-1-ene (106).[17]

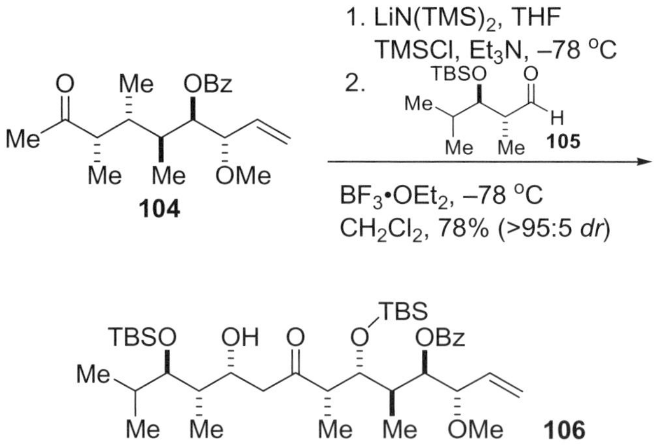

To a –78 °C solution of methyl ketone **104** (42 mg, 0.09 mmol) in THF (2.0 mL) was added a freshly prepared solution of TMSCl and Et_3N (1:1 (v/v), 0.100 mL) and LiHMDS (0.180 mL of a 1.0 M solution in toluene, 0.18 mmol). The mixture was stirred for 30 min, at which point TLC analysis showed that all of the starting methyl ketone **104** had been consumed. The reaction was quenched with 3 mL of pH 7 buffer and diluted with pentane (5 mL). The aqueous layer was extracted with pentane (3 × 5 mL) and then the combined organic layers were washed with pH 7 buffer and brine, then dried over $MgSO_4$, filtered, and concentrated *in vacuo*. The crude TMS enol ether was dried under high vacuum for 2 h. ^{1}H NMR analysis of this crude material showed > 95% purity; this material was used immediately in the next step.

To a –78 °C solution of aldehyde **105** (38 mg, 0.16 mmol) and the TMS silyl enol ether prepared in the preceding experiment (theoretically 0.09 mmol) in 1.5 mL of CH_2Cl_2 was added BF_3•OEt_2 (0.030 mL, 0.24 mmol). After 1 h, an additional 0.040 mL of BF_3•OEt_2 was added. After another 30 min, the reaction was quenched by adding saturated aq. $NaHCO_3$ (1.0 mL). The reaction was diluted with EtOAc (3.0 mL) and the reaction was warmed to room temperature. The layers were separated and the aqueous phase was extracted with EtOAc (3 × 5 mL) and then the combined organic layers were washed with brine, dried over $MgSO_4$, and concentrated *in vacuo*. The 500 MHz ^{1}H NMR spectrum of the crude aldol reaction mixture indicated that aldol **106** was the only product (> 95:5 selectivity). This crude yellow oil was purified by column chromatography (SiO_2, 15–20% Et_2O/hexanes) to yield 12 mg of starting methyl ketone **104** and 43 mg of aldol **106** as a colorless oil (78%, 93% based on recovered **104**).

2.9.7 References

1. Wittig, G.; Hesse, A. *Org. Syn.* **1970**, *50*, 66–71.
2. House, H. O.; Crumrine, O. S. *J. Am. Chem. Soc.* **1973**, *95*, 3310–3324.
3. Mukaiyama, T.; Narasaka, K.; Banno, K. *Chem. Lett.***1973**, 1011–1014.
4. Mukaiyama, T.; Izawa, T.; Saigo, K. *Chem. Lett.***1974**, 323–326.
5. Mukaiyama, T.; Banno, K.; Narasaka, K. *J. Am. Chem. Soc.* **1974**, *24*, 7503–7509.
6. Ohkouchi, M.; Masui, D.; Yamaguchi, M.; Yamagishi, T. *J. Mol. Cat. A: Chem.* **2001**, *170*, 1–15.
7. Cozzi, P. G.; Floriani, C. *Organometallics* **1994**, *13*, 2131–2134.
8. Hiraiwa, Y.; Ishihara, K.; Yamamoto, H. *Eur. J. Org. Chem.* **2006**, 1837–1844.
9. Wong, C. T.; Wong, M. W. *J. Org. Chem.* **2007**, *72*, 1425–1430.
10. Hollis, T. K.; Bosnich, B. *J. Am. Chem. Soc.* **1995**, *117*, 4570–4581.
11. Ishii, A.; Kojima, J.; Mikami, K. *Org. Lett.* **1999**, *12*, 2013–2016.
12. Mikami, K.; Matsukawa, S. *J. Am. Chem. Soc.* **1993**, *115*, 7039–7040.
13. Evans, D. A.; Dart, M. J.; Duffy, J. L.; Yang, M. G. *J. Am. Chem. Soc.* **1996**, *118*, 4322–4343.
14. Evans, D. A.; Kozlowski, M. C.; Murry, J. A.; Burgey, C. S.; Campos, K. R.; Connell, B. T.; Staples, R. J. *J. Am. Chem. Soc.* **1999**, *121*, 669–685.
15. Evans, D. A.; Kim, A. S. *Tetrahedron Lett.***1997**, *1*, 53–56.
16. Evans, D. A.; Dart, M. J.; Duffy, J. L.; Yang, M. G. *J. Am. Chem. Soc.* **1996**, *118*, 4322–4343.
17. Scheidt, K. A.; Bannister, T. D.; Tasaka, A.; Wendt, M. D.; Savall, B. M.; Fegley, G. J.; Roush, W. R. *J. Am. Chem. Soc.* **2002**, *124*, 6981–6990.
18. Paterson, I.; Cumming, J. G.; Smith, J. D.; Ward, R. A. *Tetrahedron Lett.* **1994**, *35*, 441–444.
19. Keck, G. E.; Knutson, C. E.; Wiles, S. A. *Org. Lett.* **2001**, *5*, 707–710.

20. Paterson, I.; Doughty, V. A.; McLeod, M. D.; Trieselmann, T. *Angew. Chem., Int. Ed.* **2000**, *39*, 1308–1312.
21. Wallace, G. A.; Scott, R. W.; Heathcock, C. H. *J. Org. Chem.* **2000**, *65*, 4145–4152.
22. Evans, D. A.; Carter, P. H.; Carreira, E. M.; Prunet, J. A.; Charette, A. B.; Lautens, M. *Angew. Chem., Int. Ed.* **1998**, *37*, 2354–2359.
23. Yokokawa, F.; Asano, T.; Shioiri, T. *Tetrahedron* **2001**, *57*, 6311–6327.
24. Yokokawa, F.; Asano, T.; Shioiri, T. *Org. Lett.* **2000**, *26*, 4169–4172.
25. Oikawa M.; Ueno, T.; Oikawa, H. *J. Org. Chem.* **1995**, *60*, 5048–5068.
26. Williams, D. R.; Patnaik, S.; Plummer, S. V. *Org. Lett.* **2003**, *26*, 5035–5038.
27. Paterson, I.; Tudge, M. *Angew. Chem., Int. Ed.* **2003**, *42*, 343–347.
28. Paterson, I.; Smith, J. D. *J. Org. Chem.* **1992**, *57*, 3261–3264.
29. Paterson, I.; Florence, G. J.; Heimann, A. C.; Mackay, A. C. *Angew. Chem., Int. Ed.* **2005**, *44*, 1130–1133.
30. De Rosa, M.; Soriente, A.; Scettri, A. *Tetrahedron: Asymmetry* **2000**, *11*, 3187–3195.
31. Gondi, V. B.; Gravel, M.; Rawal, V. H. *Org. Lett.* **2005**, *25*, 5657–5660.
32. Christmann, M.; Kalesse, M. *Tetrahedron Lett.***2001**, *42*, 1269–1271.
33. Hassfeld, J.; Kalesse, M. *Tetrahedron Lett.***2002**, *43*, 5093–5095.
34. Hassfeld, J.; Eggert, U.; Kalesse, M. *Synthesis* **2005**, *7*, 1183–1199.
35. Hassfeld, J.; Kalesse, M. *Synlett* **2002**, *12*, 2007–2010.
36. Liesener, F. P.; Jannsen, U.; Kalesse, M. *Synthesis* **2006**, *15*, 2590–2602.
37. Shirokawa, S.; Shinoyama, M.; Ooi, I.; Hosokawa, S.; Nakazaki, A.; Kobaiyashi, S. *Org. Lett.* **2007**, *5*, 849–852.
38. Yang, H. W.; Zhao, C.; Romo, D. *Tetrahedron* **1997**, *48*, 16471–16488.
39. Jung, M. E.; D'Amico, D. C. *J. Am. Chem. Soc.* **1993**, *115*, 12208–12209.
40. Rahn, N.; Kalesse, M. *Synlett* **2005**, *5*, 863–865.
41. Gröger, H.; Vogl, E. M.; Shibasaki, M. *Chem. Eur. J.* **1998**, *4*, 1137–1141.
42. Evans, D. A.; Burgey, C. S.; Kozlowski, M. C.; Tregay, S. W. *J. Am. Chem. Soc.* **1999**, *121*, 686–699.
43. [R] Johnson, J. S.; Evans, D. A. *Acc. Chem. Res.* **2000**, *33*, 325–335.
44. Evans, D. A.; Fitch, D. M.; Smith, T. E.; Cee, V. J. *J. Am. Chem. Soc.* **2000**, *122*, 10033–10046.
45. Evans, D. A.; Wu, C.; Masse, C. E.; MacMillan, D. W. C. *Org. Lett.* **2002**, *20*, 3379–3382.
46. Jankowska, J.; Mlynarski, J. *J. Org. Chem.* **2006**, *71*, 1317–1321.
47. Jankowska, J.; Paradowska, J.; Mlynarski, J. *Tetrahedron Lett.* **2006**, *47*, 5281–5284.
48. Jankowska, J.; Paradowska, J.; Rakiel, B.; Mlynarski, J. *J. Org. Chem.* **2007**, *72*, 2228–2231.
49. Yang, B.-Y.; Chen, X.-M.; Deng, G.-J.; Zhang, Y.-L.; Fan, Q.-H. *Tetrahedron Lett.* **2003**, *44*, 3535–3538.
50. Kobayashi, S.; Nagayama, S.; Busujima, T. *Tetrahedron* **1999**, *55*, 8739–8746.
51. Benaglia, M.; Cinquini, M.; Cozzi, F.; Celentano, G. *Org. Biomol. Chem.* **2004**, *2*, 3401–3407.
52. Fu, F.; Teo, Y.-C.; Loh, T.-P. *Tetrahedron Lett.* **2006**, *47*, 4267–4269.
53. Evans, D. A.; Kozlowski, M. C.; Tedrow, J. S. *Tetrahedron Lett.* **1996**, *37*, 7481–7484.
54. Fraile, J. M.; Pérez, I.; Mayoral, J. A.; Reiser, O. *Adv. Synth. Catal.* **2006**, *348*, 1680–1688.
55. Orlandi, S.; Mandoli, A.; Pini, D.; Salvadori, P. *Angew. Chem.* **2001**, *113*, 2587–2589.
56. Matsukawa, S.; Mikami, K. *Tetrahedron: Asymmetry* **1995**, *6*, 2571–2574.
57. Mikami, K.; Matsukawa, M. *J. Am. Chem. Soc.* **1994**, *116*, 4077–4078.
58. Keck, G. E.; Krishnamurthy, D. *J. Am. Chem. Soc.* **1995**, *117*, 2363–2364.
59. Carreira, E. M.; Singer, R. A.; Lee, W. *J. Am. Chem. Soc.* **1994**, *116*, 8837–8838.
60. Mikami, K.; Takasaki, T.; Matsukawa, S.; Maruta, M. *Synlett* **1995**, 1057–1058.
61. Mikami, K.; Yajima, T.; Takasaki, T.; Matsukawa, S.; Terada, M.; Uchimaru, T.; Maruta, M. *Tetrahedron* **1996**, *52*, 85–98.
62. Shimada, Y.; Matsuoka, Y.; Irie, R.; Katsuki, T. *Synlett* **2001**, *1*, 57–60.
63. Wadamoto, M.; Ozasa, N.; Yanagisawa, A.; Yamamoto, H. *J. Org. Chem.* **2003**, *68*, 5593–5601.
64. Langner, M.; Rémy, P.; Bolm, C. *Chem. Eur. J.* **2005**, *11*, 6254–6265.

2.10 Peterson Olefination

Nadia M. Ahmad

2.10.1 Description

The Peterson olefination concerns the construction of double bonds from trialkylsilyl-substituted organometallics and carbonyls. The reaction involves the formation of an α-hydroxysilane, which then undergoes elimination to give the alkene. Elimination can take place under either acidic or basic conditions.

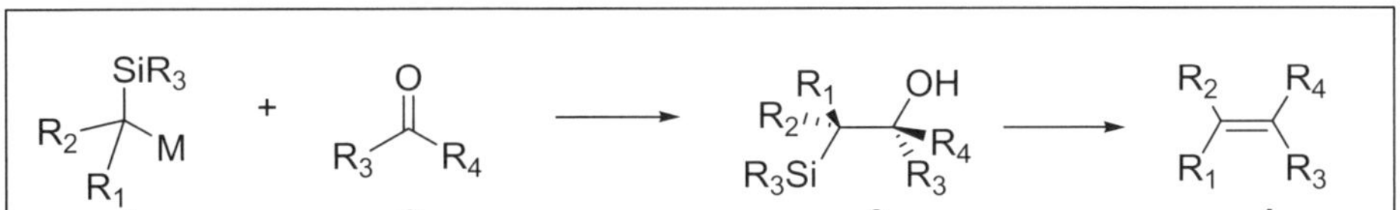

1 **2** **3** **4**

2.10.2 Historical Perspective

Prior to the 1967 report by Donald Peterson,[1] olefination reactions, and in particular, methylenation reactions, could be carried out with a variety of reagents, most notably with phosphorus-substituted anions such as the Wittig,[2a,b] Horner,[3] and Wadsworth–Emmons[4] protocols. In addition to these phosphorus-stabilised carbanions, Corey and Kwiatkowski introduced α-lithiophosphonic acid bisamides and α-lithiophosphonothioate esters.[5,6] At the same time, boron– and sulfur-substituted carbanions were also reported in the literature as methylenating agents.[7,8] Peterson and coworkers considered the mechanism of action for the reactions of carbonyl compounds with phosphorus-stabilised anions and determined that silicon-substituted anions should also be able to effect such homologations. Silicon (i) is readily attacked by alkoxides, (ii) forms strong bonds with oxygen, and the resulting silanoate is a good leaving group, (iii) has d orbitals therefore pentavalent bond formation can take place. The hypothesis was supported by a reaction reported several years earlier by Gilman and coworkers in which alkene **8** was synthesised with concommitant loss of the silanoate anion **9**.[9] Gilman did not extend the chemistry to substrates not containing phosphorus.

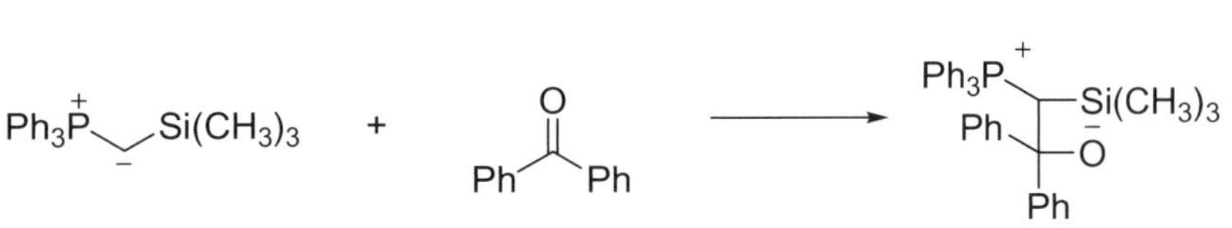

5 **6** **7**

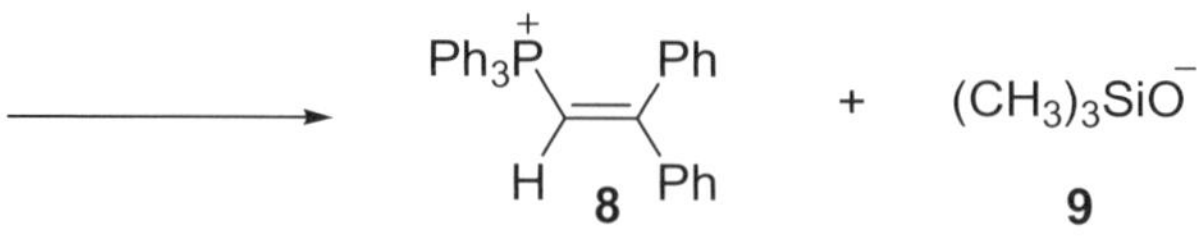

Peterson went on to describe reactions of several lithiated silanes with carbonyls compounds, all giving the desired alkenes in good yields, albeit with very little stereoselectivity. In 1975 however, Peterson and Hudrlik published their studies on the stereoselective elimination of the β-hydroxyalkylsilyls.[10] The reduction of 5-trimethylsilyl-4-octanone **10** was carried out with DIBAL-H to give one diastereoisomer, **11**. The authors found that elimination with sodium or potassium hydride gave *trans*-4-octene as the major isomer **12**, while elimination under acidic conditions resulted in predominantly *cis*-4-octene **13**. Mild conditions were employed, affording stereochemical purity of up to 95% with excellent yields.

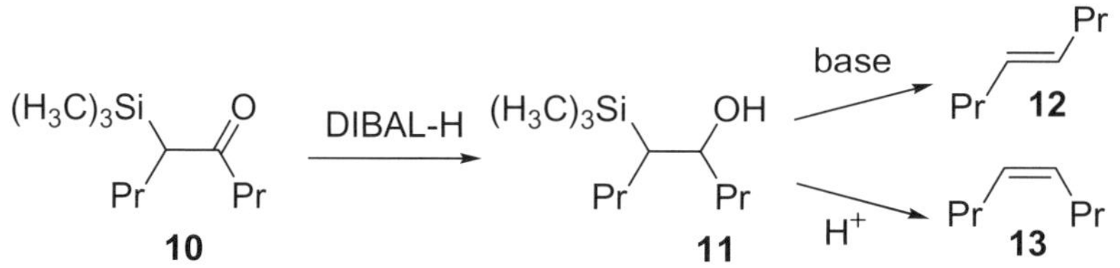

2.10.3 Mechanism

Due to the mechanism of action, the Peterson olefination is a very versatile method of forming carbon–carbon double bonds. In particular, the reaction can be carried out under basic or acidic conditions giving rise to either the *trans* or *cis* isomer respectively. In cases where the stereoselectivity of the new double bond is irrelevant, the ability to employ either acid or basic conditions means that potentially sensitive functionalities in the remainder of the molecule can be suitably accommodated.

The mechanism begins with the addition of a silyl-substituted carbanion **14** to a carbonyl compound **15**; an aqueous work-up then leads to a diastereomeric mixture of β-hydroxyalkylsilanes, often isolable and sometimes separable. The stereo-selectivity of the reaction can be controlled by the steric demands of the silyl group; the use of more sterically demanding silyl groups results in the *erythro* isomer as the major product.

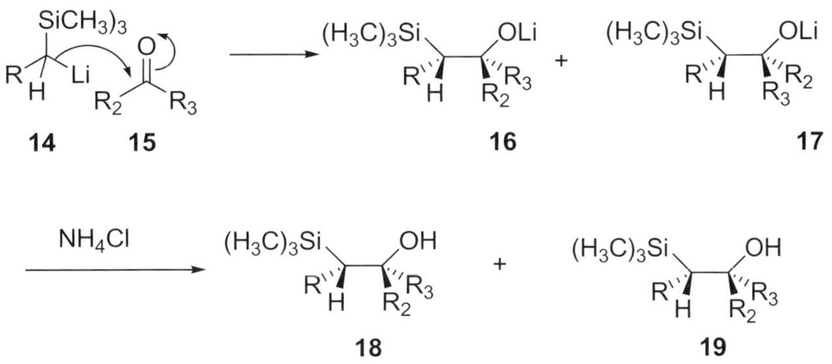

A subsequent basic elimination can proceed *via* two pathways: (i) deprotonation of the hydroxyl followed by a silyl 1,3-shift **20**, (ii) deprotonation followed by formation of a penta-coordinate 1,2-oxasiletanide **21** which then collapses to give the alkene product, **23**.

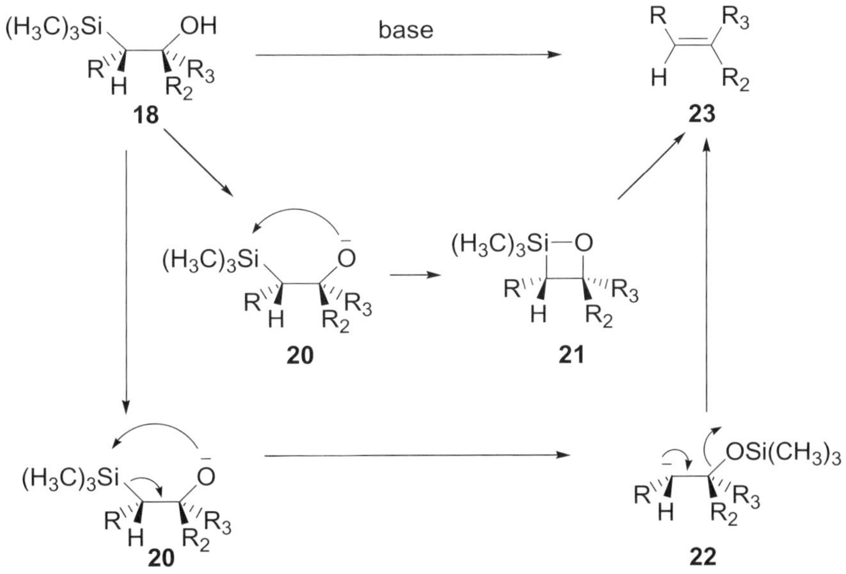

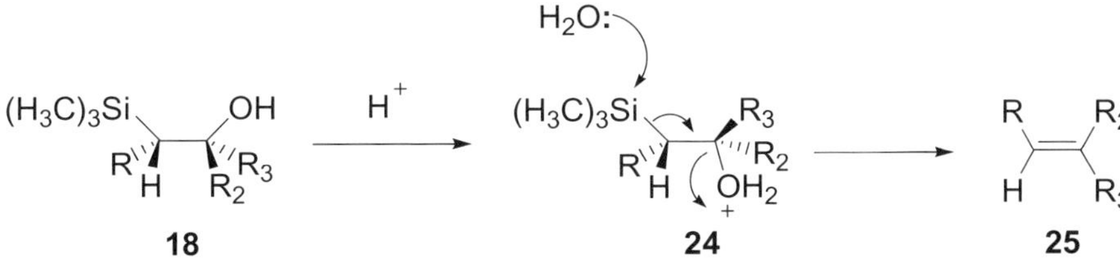

On the other hand, acid hydrolysis proceeds *via anti*-elimination to give the *cis*-isomer.

2.10.4 Variations and Improvements

Several modified Peterson protocols exist for the methylenation reactions of carbonyl compounds.[11] Under the usual conditions, methylenation reactions using silylithium reagents suffer from poor chemoselectivity as the silyllithiums also act as strong bases. However, Johnson and Tait discovered that the equimolar addition of cerium chloride to the silyllithium followed by addition of the carbonyl compound resulted in the corresponding β-hydroxysilanes **27** in good yields.[12] The same reactions carried out in the absence of $CeCl_3$ gave significantly lower yields.

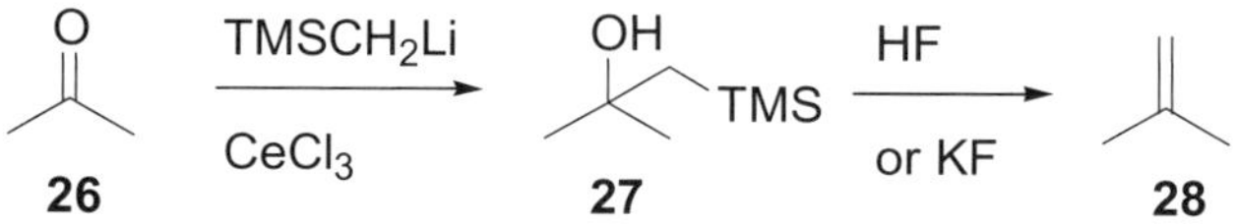

26 **27** **28**

The use of the fluoride ion or zinc bromide as a catalyst in the Peterson reaction with bis(trimethylsilyl) methyl derivatives had previously been reported.[13a–c] Bellassoued and Ozanne went on to report the one-pot synthesis of (2*E*,4*E*) dienals from the corresponding aldehydes using a catalytic amount of cesium fluoride in DMSO.[14a,b] The method obviates the need for an acidic or basic work up.

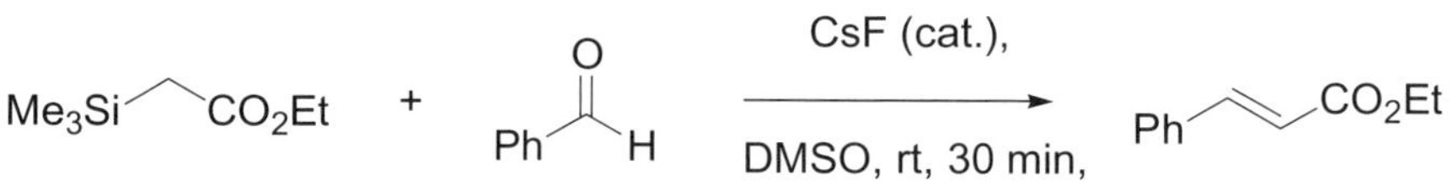

29 **30** **31**

Chiral tridendate ligands have been utilized to effect asymmetric Peterson reactions, resulting in olefins with axial chirality in high yields and enantioselectivities.[15] This is reported as the first example of its kind, namely the enantioselective Peterson olefination of α-silanyl ester enolates **33** with substituted cyclohexanones **32**, facilitated by an external chiral ligand **34**.

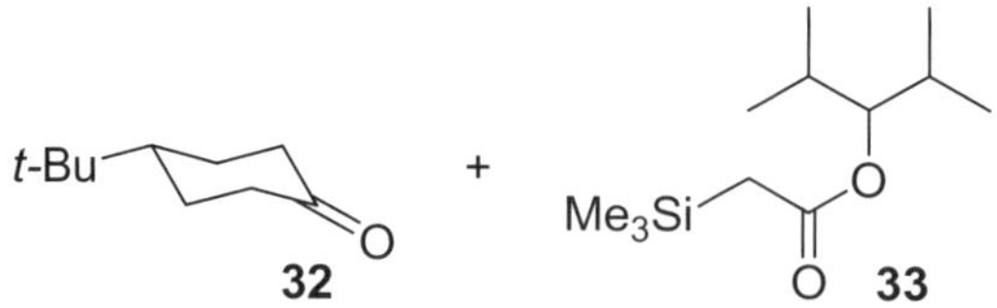

32 **33**

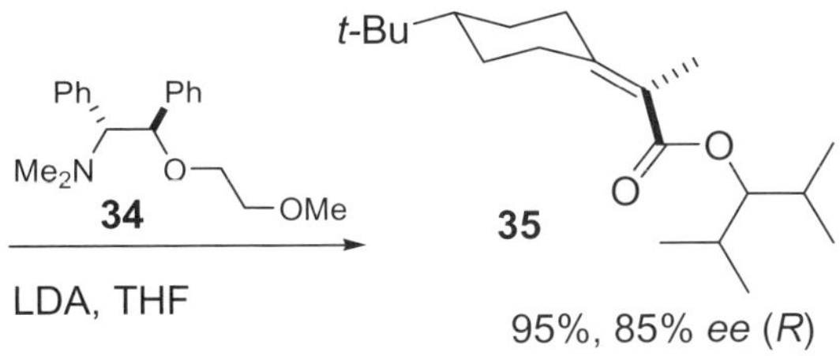

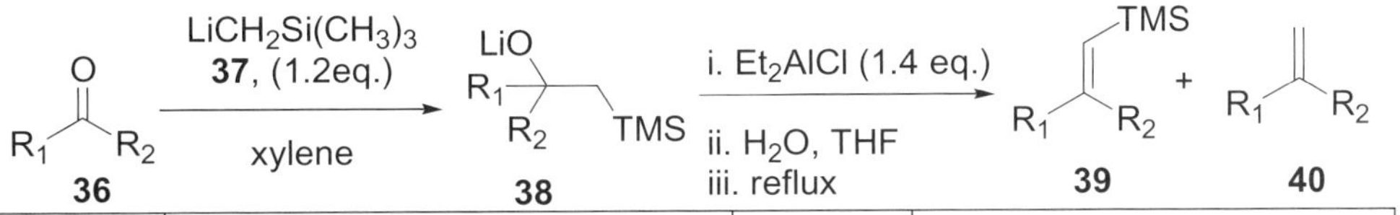

Entry	Substrate	Yield	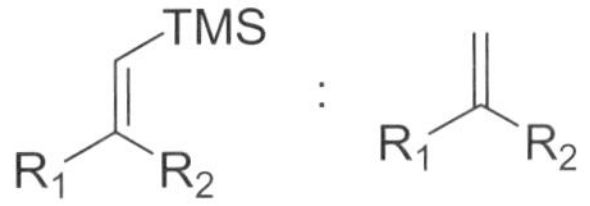
a	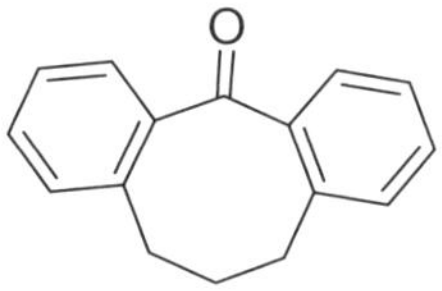	76	10:1
b	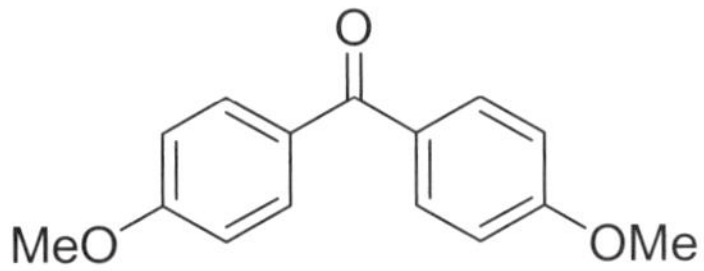	83	1:0 (not detected)
c	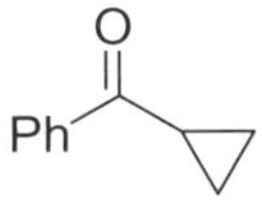	79	78:1 (*E*:*Z* 1:19)
d	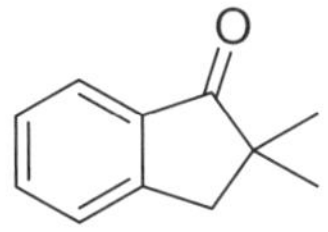	61	1:0 (not detected) 60:1
e	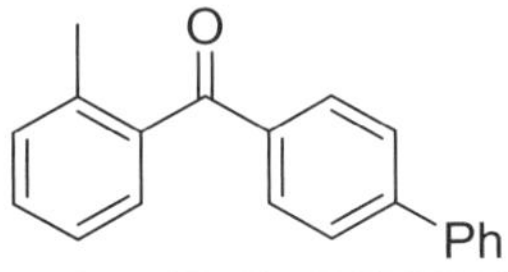	85	11:1 (*E*:*Z* 2:1)

Kwan and Battiste have employed the Peterson protocol to synthesize vinylsilanes **39/40** in a one-pot process.[16] The procedure involved the addition of trimethylsilylmethyl lithium to non-enolizable aromatic ketones followed by addition of diethylaluminium chloride, then water. The tabulated examples illustrate the rare use of organoaluminum reagents to promote elimination. Good chemo- and stereoselectivity was observed.

This methodology has been further expanded to include an aldehyde selective Peterson methylenation reagent. Abedi and Battiste have illustrated the use of tris(trimethylsilylmethyl)alane (TTMA) which selectively methylates aldehydes in the presence of ketones.[17] Notable side products from the reaction include the Meerwein–Pondorff–Verely reduction of the starting aldehyde along with Oppenauer oxidation of the trimethylsilylmethyl alcohol. The reagent can be used neat or as a complex with lithium bromide; yields were found to be higher when the lithium salt was used, with fewer or no side products.

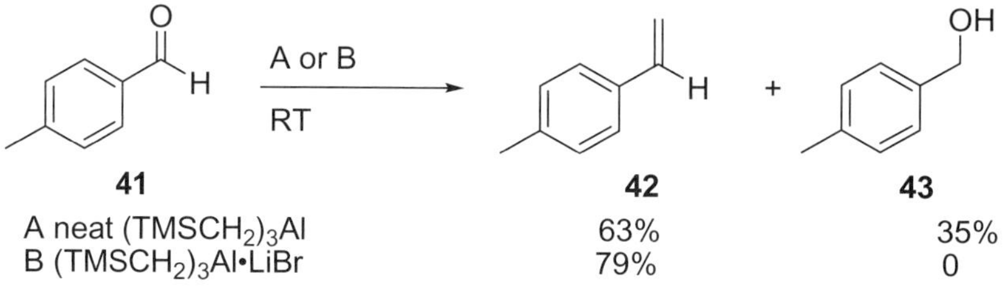

Iqbal and Evans have described their efforts in the synthesis of cross–conjugated dienones (**46**) by utilizing a one-pot conjugate–addition Peterson olefination reaction.[18a–c] This was followed by a retro-Diels–Alder to unmask the endocyclic carbon–carbon double bond in 73% yield.

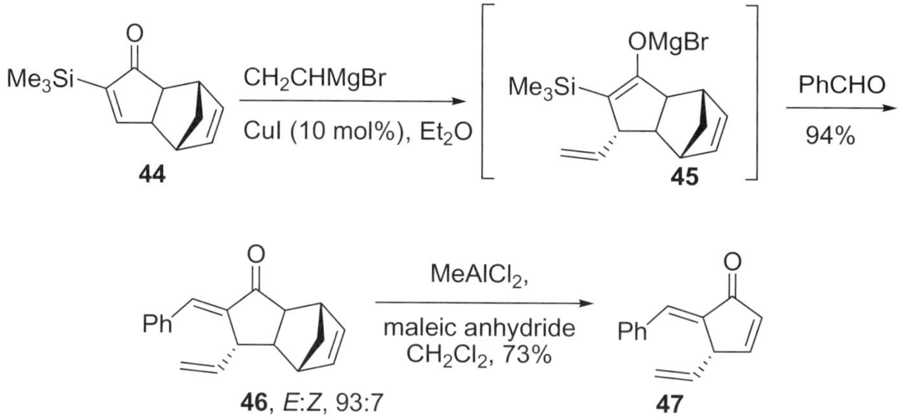

2.10.5 *Synthetic Utility*

Pulido and coworkers have reported the highly diastereoselective synthesis of *erythro*-β-hydroxysilanes *via* the reaction of α-*tert*-butyldiphenylsilyl carbonyl compounds with organometallics.[19] It had been shown previously that the formation of β-hydroxysilanes does not generally proceed with high diastereoselectivity and can vary according to the silyl group employed.[20] However, the steric bulk of the *tert*-butyldiphenylsilyl in this case was used effectively to synthesis stereodefined *β*-hydroxysilanes which were then used in the stereoselective synthesis of di- and tri-substituted olefins *via* the Peterson olefination reaction. In addition, the use of the *tert*-butyldiphenylsilyl group enables the isolation of the *α*-aldehydes in contrast with the corresponding *α*-trimethylsilyl aldehydes which are also more difficult to synthesise.[21a,b]

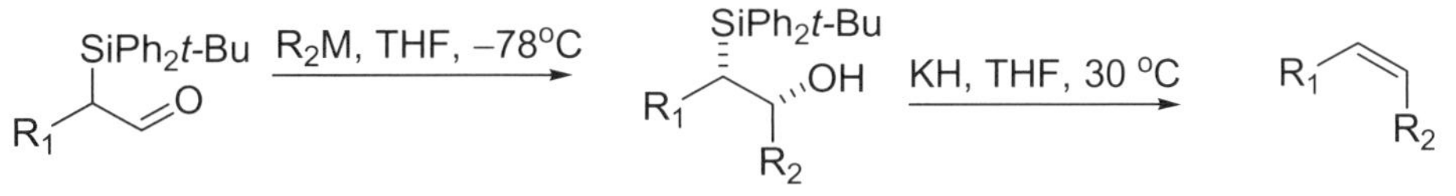

48 **49** **50**

Entry	R_1	R_2M	Yield %	Yield % (*Z*:*E*)
a	Ph	MeLi	87	86 (92:8)
b	Ph	MeLi	90	91 (85:15)
c	Ph	vinylMgBr	83	93 (95:5)
d	Bu	PhLi	70	80 (96:4)

Trisubstituted alkenes were also obtained by the reaction of *β-tert*-butyldiphenylsilylketone **51** with lithium reagents, with appreciable stereoselectivity following the Felkin–Ahn model. The *β-tert*-butyldiphenylsilylketones were found to be unstable, converting to the corresponding silyl enol ethers on standing at room temperature for several hours and therefore had to be used soon after synthesis. However, the Peterson olefination to the trisubstituted alkenes could take place directly without isolation of the *β*-hydroxysilane.

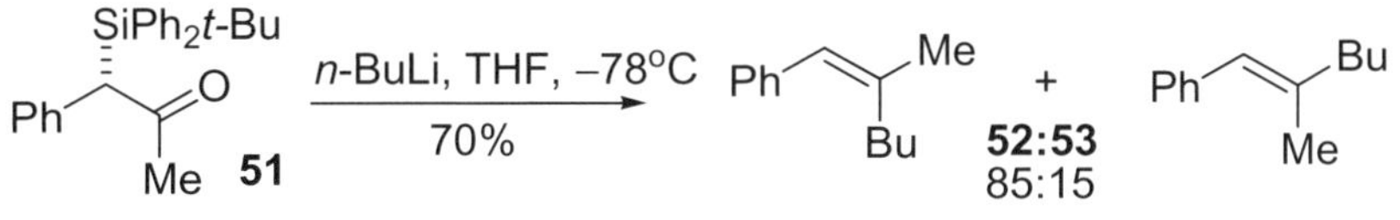

The Peterson olefination has been utilized for the formation of double bonds in the presence of numerous functionalities, thus illustrating the wide scope of this homologation reaction. These include benzyl carbamates,[22a–c] *α*-fluorovinylphosphonates,[23] *γ*-substituted allylsilanes,[24] and fluoro-alkenoates.[25]

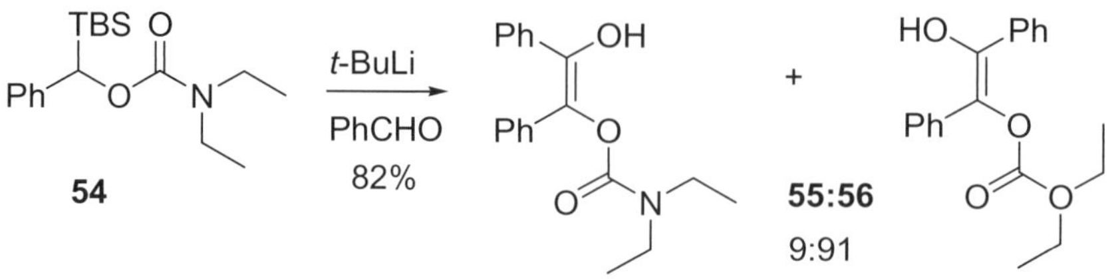

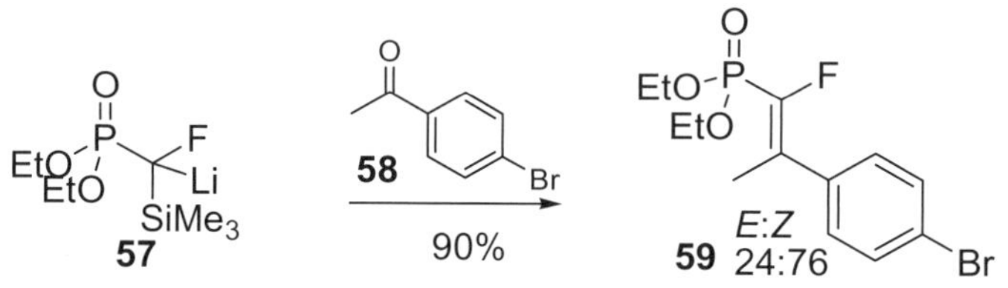

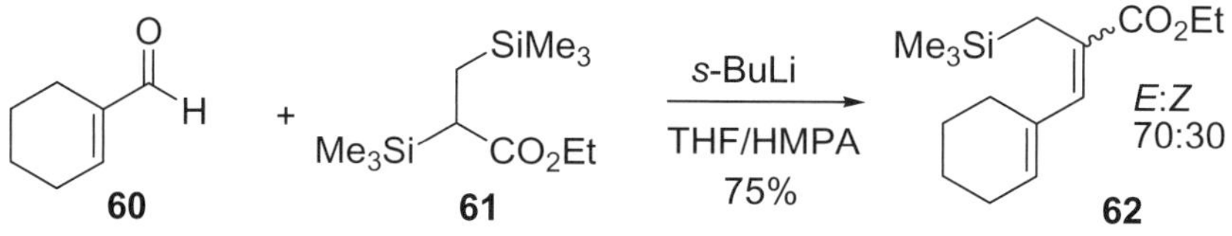

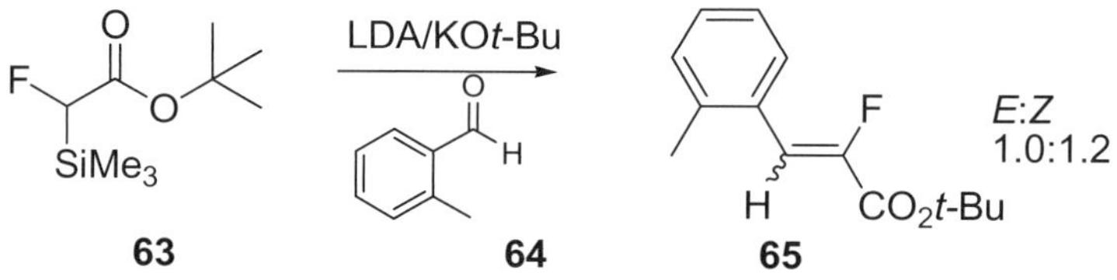

The Peterson reaction has been employed to synthesize various functional groups containing olefins. For example, Fürstner and coworkers have developed a one-pot *N*-acylation/Peterson elimination procedure to furnish enamides.[26] An attractive feature of the Peterson reaction utilised in this procedure is that basic conditions are employed for the elimination instead of acidic conditions which in this case would lead to the hydrolysis of the enamide. The reaction proceeds in good yields over the two steps (*N*-acylation/Peterson elimination) and all products were obtained as single diastereoisomers, thus, with complete stereoselective conversion of the β-hydroxysilane **66**/**68** to the enamide product **67**/**69**.

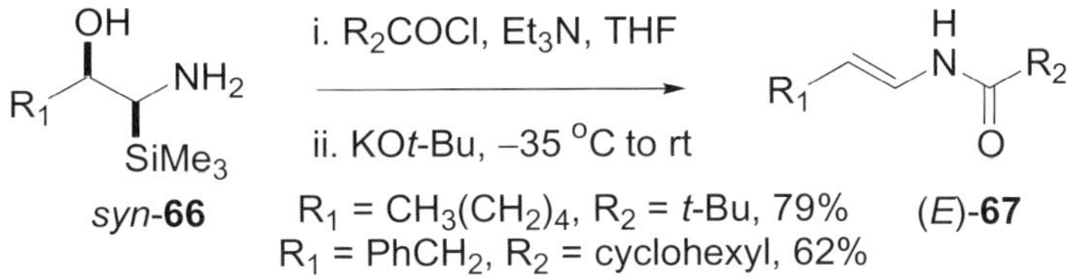

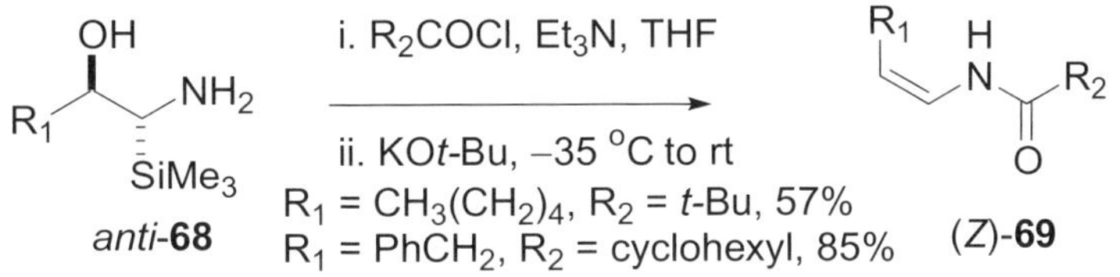

In a similar manner, but without the use of either acidic or basic conditions, Chakraborty and Laxman have carried out a stereoselective *in situ* Peterson olefination as the last step in their total synthesis of (+)-crocacin A **71**.[27] In a particularly elegant step, the deprotection of a silyl ether **70** unmasked an oxy-anion which then enabled the Peterson elimination to take place. The skipped diene present in the natural product was formed in 86% yield.

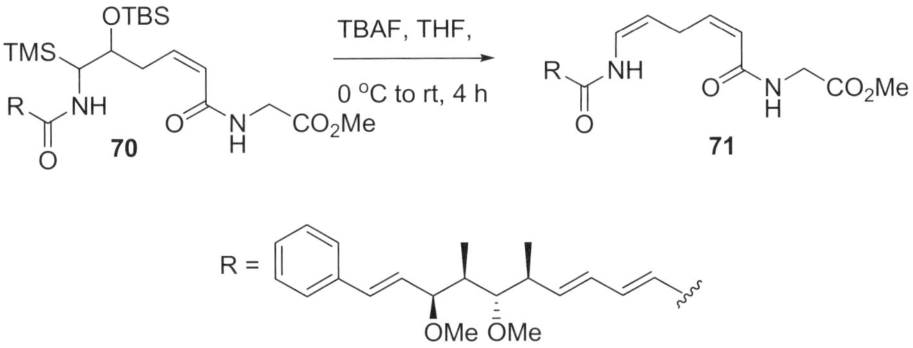

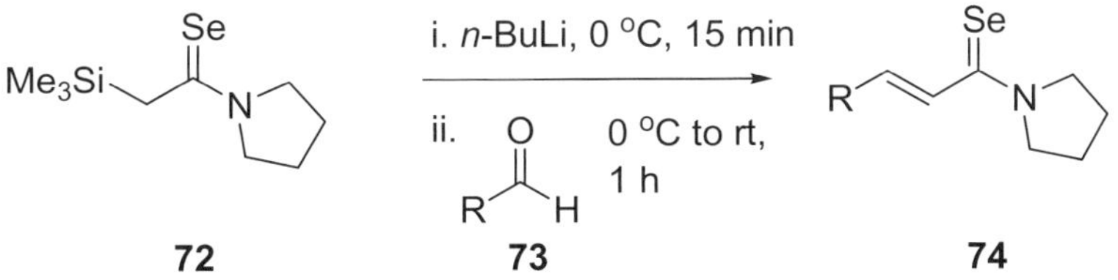

Entry	RCHO	Yield %
a	2-$MeOC_6H_4$	91
b	4-NCC_6H_4	76
c	4-FC_6H_4	83
d	C_6H_5	98

In an analogous fashion, unsaturated selenoamides **74** have also been generated *via* the Peterson olefination reaction.[28] The reaction proceeded with excellent yields.

A modified Peterson reaction has been used to generate silenes which have then been converted to diols and lactones.[29] The reaction involves the formation of the sila-Grignard reagents under standard conditions, followed by treatment with isobutyraldehyde to give the silene precursor **76**. Silacyclohexene **76** was then produced by reaction with 1,3-pentadiene. High diastereoselectivity was observed and confirmed by 2D NMR experiments on the subsequent reduction and oxidation products. Thus, standard conditions converted silane **78** into diol **79** which was then oxidised to give lactone **80**. The protocol can be expanded to achieve further functionalisation, for example in the synthesis of homoallylic alcohol **82**.

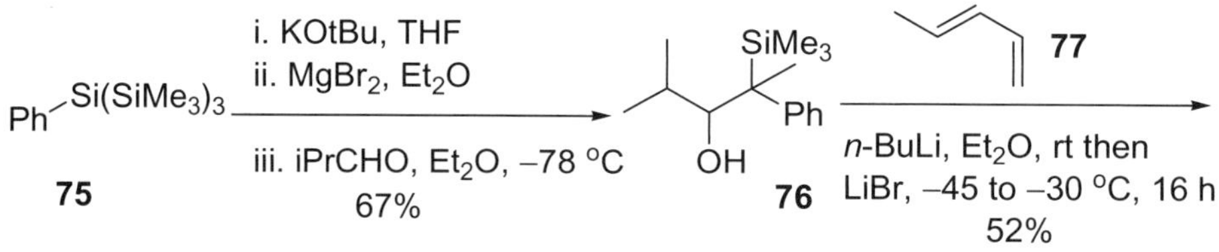

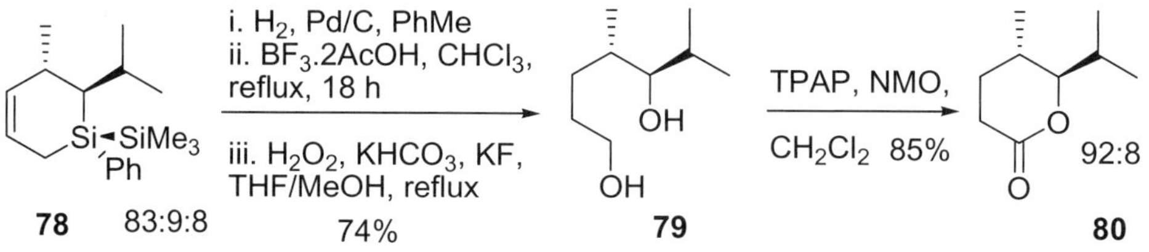

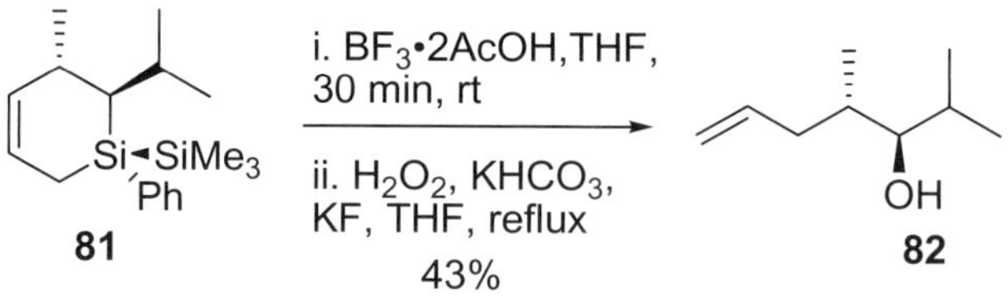

The preparation of 1-(1-alkenyl)benotriazoles **84** by the Peterson olefination is a particularly facile process due to the ability of the benzotriazolyl moiety to stabilise an anion through resonance stabilisation.[30]

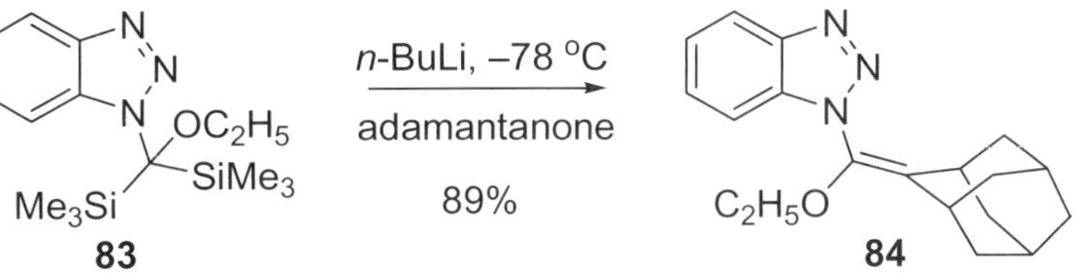

The synthesis of α-fluorovinylsulfones by the use of the Peterson olefination has been described by Usuki *et al.* who reacted silyl lithiums **86** generated *in situ* from the corresponding α-fluoro-α-silyl sulfones **85** and TBSCl, with carbonyl compounds **87**.[31] They found that the reaction proceeded in better yield when an aldehyde was used rather than a ketone. The phenylsulfonyl moiety could be reductively removed using a sodium amalgam in NaH_2PO_4–$NaHPO_4$.

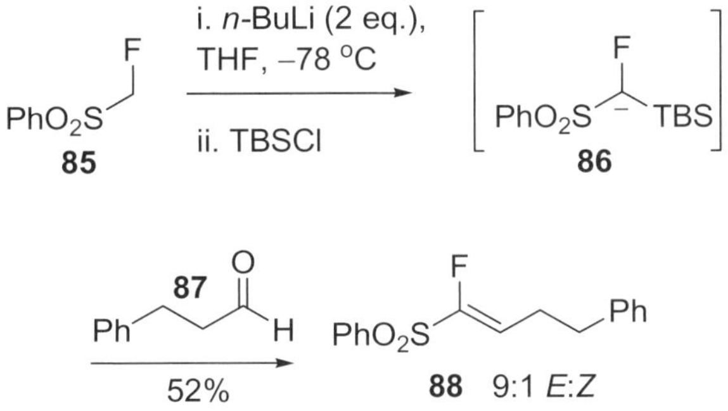

The Peterson olefination has been used to synthesise monofluoromethylenephosphonates **91** which were then used as potential phosphate mimics in biological systems.[32] Although the reaction was not stereoselective, the geometric isomers were separable by chromatography.

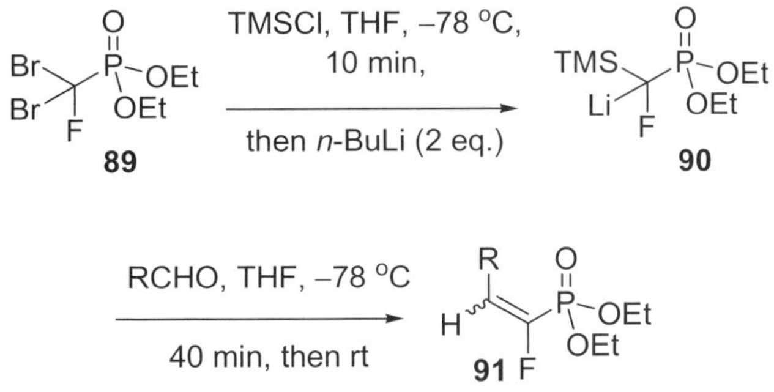

Van Vranken and coworkers have illustrated an elegant use of convergent stereocontrol in Peterson olefinations.[33] In their synthesis of (±)-3-hydroxybakuchiol, the reaction between a neopentyl α-silyl alkyllithium intermediate and an aryl aldehyde generated a mixture of *syn*- and *anti*-β-silyl alkoxides. The mixture was treated under basic, kinetic conditions to give stereoselective elimination of the *syn*-β-silyl alkoxide thus affording an *E* alkene **93**. Subsequent heating of the reaction mixture and addition of acid caused stereospecific elimination of the *anti*-β-silyl alkoxide resulting in the same *E* alkene *via* the complementary cationic pathway (see section 2.10.3). Excellent selectivity was obtained.

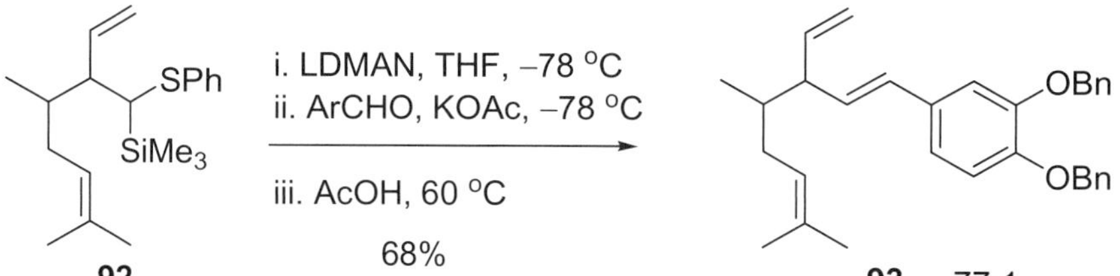

92 **93** 77:1

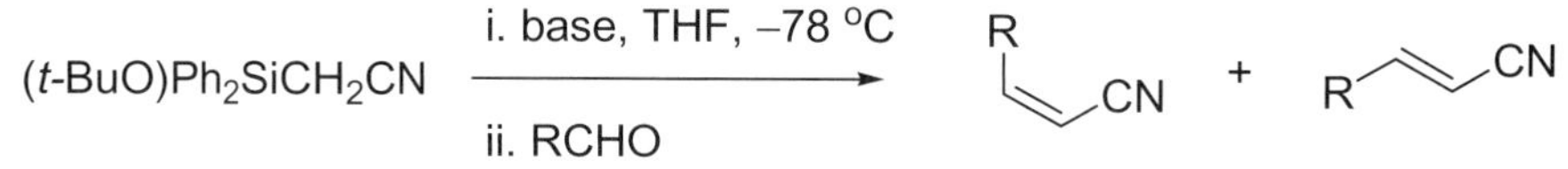

94 **95** **96**

Entry	Aldehyde	Base	Product	*Z*:*E*[a]	Yield %[b]
a	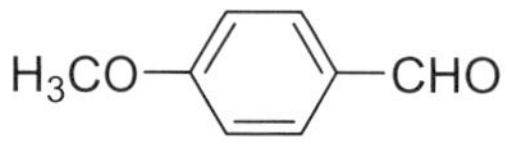	KHMDS	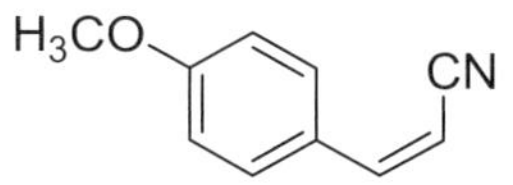	96:4	94
b	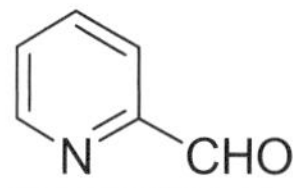	KHMDS	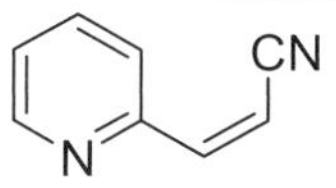	92:8	88
c	TMS—≡—CHO	NaHMDS		93:7	98
d	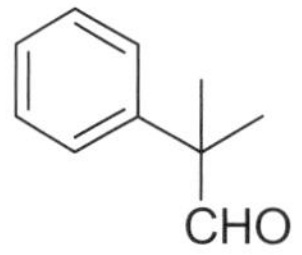	*n*-BuLi	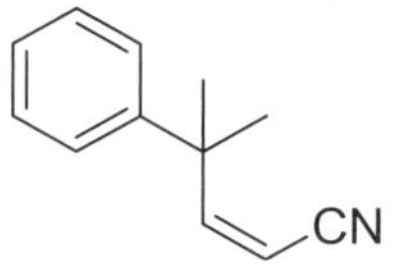	95:5	87

[a]Determined by ^{1}H NMR of crude reaction mixture. [b]Combined isolated yield of (*Z*)- and (*E*)-olefins.

The development of a Peterson reagent, (*t*-BuO)Ph$_2$SiCH$_2$CN, for the preparation of (*Z*) -β-monosubstituted-α,β-unsaturated cyanides **95/96** has been described by Kojima *et al.*[34] This method allows for the selective formation of cyanides which are not otherwise easily attainable.[35a–d] Examples of the transformations, which can be carried out with a variety of bases, reported arc shown.

Roush and co-workers have effectively applied the Peterson olefination conditions to synthesise cyclitol derivatives as intermediates in their approach towards the conduritol natural products.[36] Starting from common precursors **97**, both acidic and basic conditions were employed to give the desired alkenes **98**/**100** in excellent yields; removal of the protecting groups then furnished the natural products **99**/**101**.

Floreancig and coworkers employed a sequential Peterson olefination and Prins cyclisation reaction in their total synthesis of (+)-dactylolide hence neatly illustrating the versatility of the Peterson reaction.[37] The *β*-hydroxysilane **103** was synthesised *in situ* by the double addition of the necessary Grignard reagent onto ester **102**. Treatment of the resulting tertiary alcohol with pyridinium triflate and magnesium sulfate then prompted the Peterson olefination and subsequent Prins cyclisation to occur, affording tetrahydropyran **106** in 75% yield.

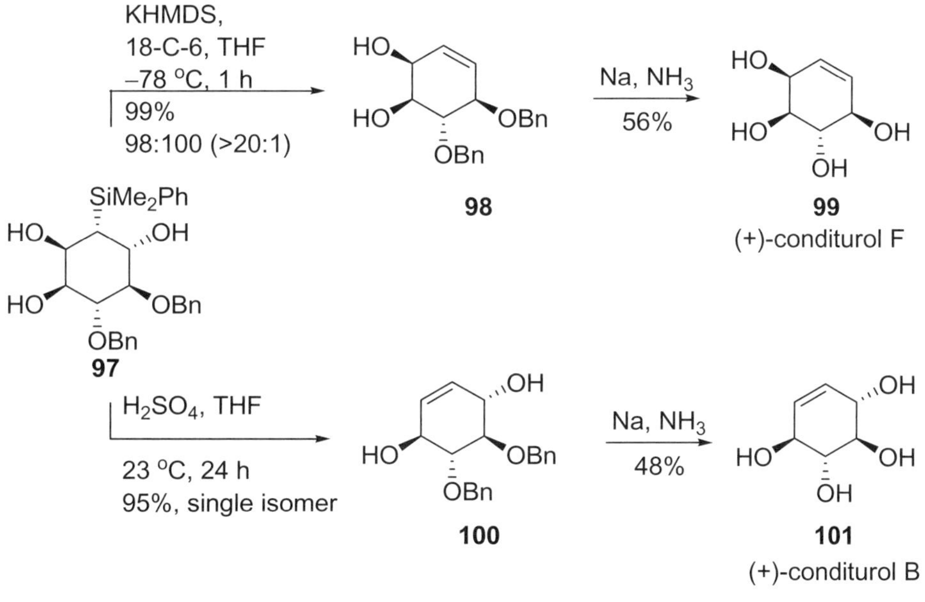

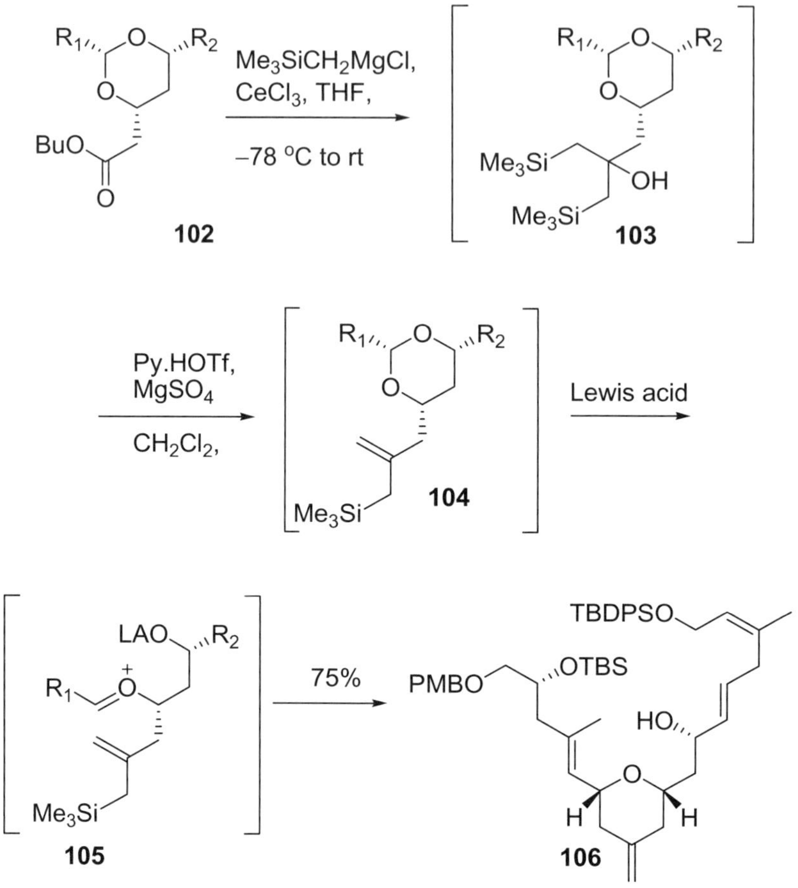

The condensation reaction of γ-(trimethylsilyl)allenylboranes **107** with conjugated allenic aldehydes **108** followed by Peterson olefination has been carried out by Wang *et al.*[38] The procedure resulted in the formation of enyne–allenes **111/112** which were then used in intramolecular transformations *via* radical reactions. The reaction proceeded with excellent stereoselectivity depending on whether acidic or basic conditions were used for the elimination of the β-hydroxysilane. Wang then applied this protocol to form σ-isotoluenes and diene–allenes.[39]

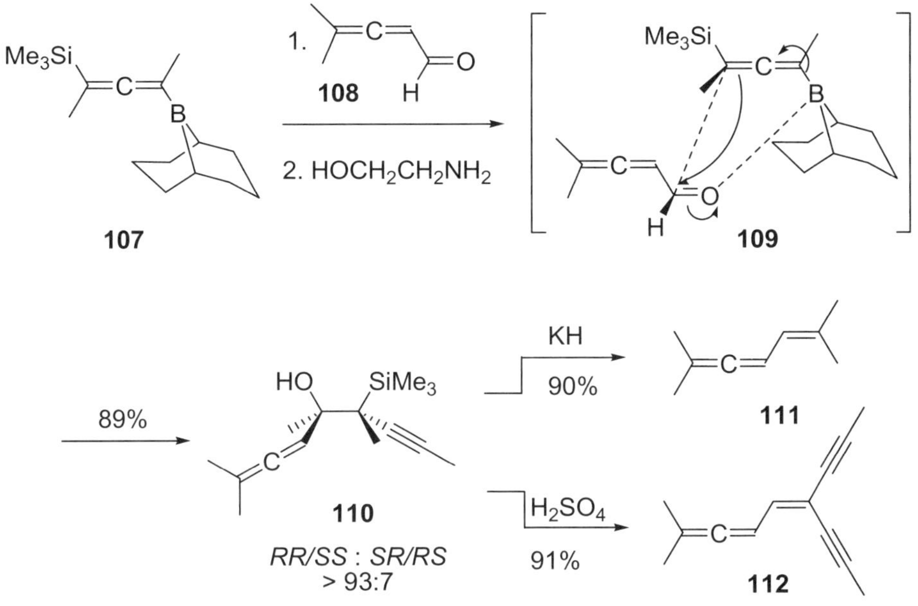

The synthesis of 1,3-dicyclohexadienes **114** can be carried out in a number of ways,[40a–c] however, a particular efficient method reported by Lautens and co-workers employs a tandem ring-opening–Peterson elimination.[41]

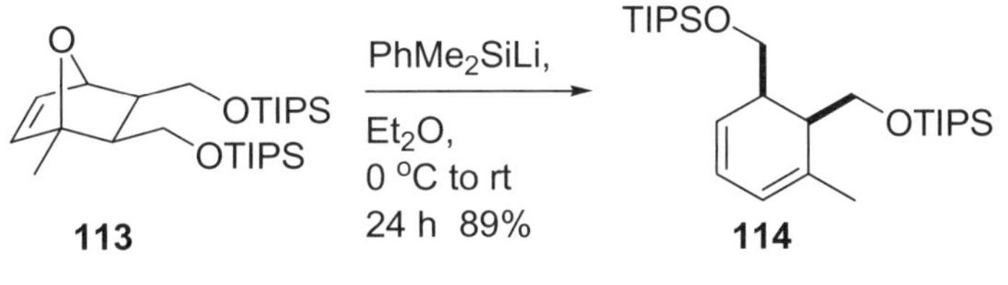

The reaction proceeds in good yield and a variety of oxabicyclic substrates are suitable for the transformation.

2.10.6. Experimental

(1*S*,2*S*,5*R*,6*R*)-5,6-Dibenzyloxy-cyclohex-3-ene-1,2-diol (98)[36]

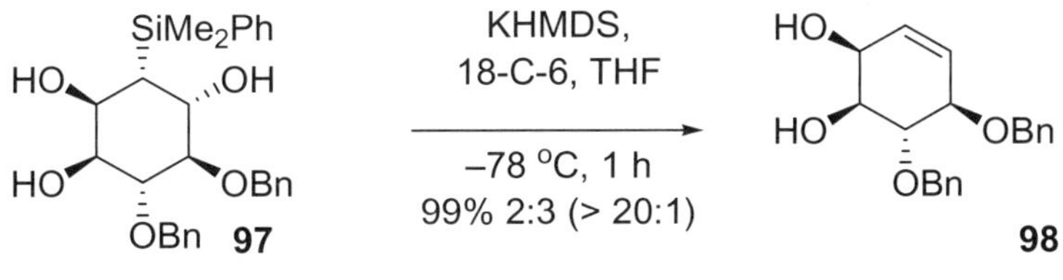

To a –78 °C solution of triol **97** (17.0 mg, 0.03 mmol) and 18-crown-6 (9.0 mg, 0.03 mmol) in THF (0.2 mL) was added a –78 °C solution of KHMDS (37 mg, 0.19 mmol in 0.3 mL of THF) via cannula. After being stirred at –78 °C for 1 h, the mixture was quenched with saturated aqueous NH_4Cl solution, diluted with H_2O, saturated aqueous NaCl, and saturated aqueous Rochelle salt solution and allowed to warm to 23 °C. The mixture was stirred for 1 h, then extracted with EtOAc (2 × 50 mL). The combined organic extracts were dried (Na_2SO_4), filtered and concentrated *in vacuo*. The crude oil was purified by gradient flash chromatography (10 g SiO_2, 40–50% EtOAc/hexanes) to yield 11.5 mg (99%) of cyclohexene **98** with > 20:1 *dr*.

(1*S*,4*S*,5*R*,6*R*)-5,6-Dibenzyloxy-cyclohex-2-ene-1,4-diol (100)[36]

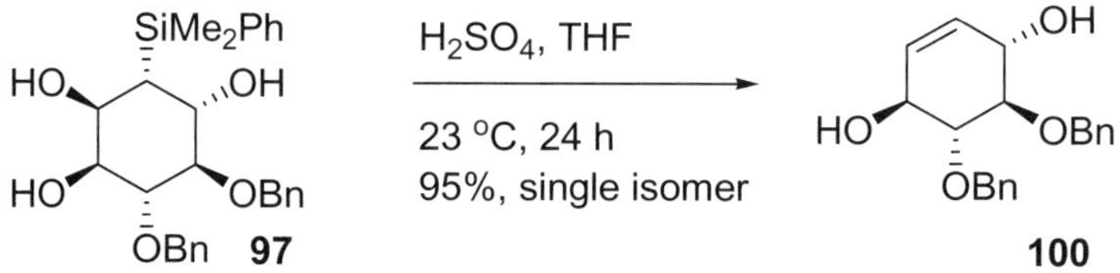

To a solution of triol **97** (30.0 mg, 0.06 mmol) in THF (2 mL) at 23 °C was added H_2SO_4 (20 μL of a 0.3 M THF solution). The mixture was stirred at 23 °C for 12 h, then a second portion of H_2SO_4 (20 μL of a 0.3 M THF solution) was added and the reaction was stirred for an additional 12 h. The reaction mixture was then poured into an aq. saturated $NaHCO_3$ solution, diluted with H_2O, saturated aqueous NaCl, and then extracted with Et_2O (2 × 50 mL). The combined organic extracts were dried Na_2SO_4), filtered and concentrated *in vacuo* to yield 19 mg (95%) of conduritol B derivative **100** as a light yellow solid.

(2*S*,3*E*,7*E*,11*S*)-13-methoxy-2-(methoxymethoxy)-4,8,11-trimethyl-15-methylenebicyclo[9.3.1.]pentadeca-1(14),3,7,13-tetraene (116)[42]

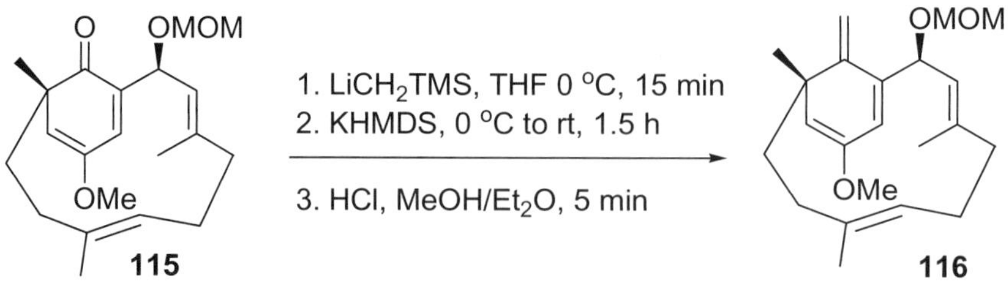

Trimethylsilylmethyllithium (1.0 M in THF, 0.12 mL) was added drop-wise to a solution of compound **2b** (20.7 mg, 0.0598 mmol) in 0.3 mL of THF at 0 °C. The solution was stirred for 15 minutes at 0 °C, and then KHMDS (0.24 mL, 0.12 mmol, 0.5 M in toluene) was added. The reaction mixture was stirred for 1.5 hours at room temperature before quenching with H_2O (5 mL). The aqueous phase was separated and extracted with Et_2O (3 × 10 mL). The combined organic extracts were washed with brine (20 mL), dried over Na_2SO_4 and concentrated *in vacuo* to give an unstable enol ether intermediate. The crude enol ether was dissolved in MeOH (1 mL), and then treated with aqueous HCl (1 mL, 1%) and stirred at room temperature for 5 min. The mixture was diluted with Et_2O (5 mL) then neutralized with saturated aqueous $NaHCO_3$ (10 mL). The aqueous phase was separated and then extracted with Et_2O (2 × 10 mL). The combined organic extracts were washed with brine (10 mL), dried (Na_2SO_4) and concentrated *in vacuo*. The residue was purified by flash column chromatography on silica gel (9:1 hexanes/Et_2O) to give ketone **116** (14.6 mg, 0.0442 mmol, 74%).

2.10.7 References

1. Peterson, D. J. *J. Org. Chem.* **1967**, *32*, 780–784.
2. (a) Wittig, G.; Schölkopf, U. *Chem. Ber.* **1954**, *87*, 1318–1330. (b) Wittig, G.; Haag, W. *Chem. Ber.* **1955**, *88*, 1654–1666.
3. Horner, L.; Hoffman, H.; Wippel, H. G.; Klahre, G. *Chem. Ber.* **1959**, *92*, 2499–2505.
4. Wadsworth Jr. W. S.; Emmons, W. D. *J. Am. Chem. Soc.* **1961**, *83*, 1733–1738.
5. Corey, E. J.; Kwiakowski, G. T. *J. Am. Chem. Soc.* **1966**, *88*, 5652–5653.
6. Corey, E. J.; Kwiakowski, G. T. *J. Am. Chem. Soc.* **1966**, *88*, 5654–5656.
7. Cainelli, G.; Bello, G. D.; Zubiani, G. *Tetrahedron Lett.* **1966**, 4315–4318.
8. Corey, E. J.; Durst, T. *J. Am. Chem. Soc.* **1966**, *88*, 5656–5657.
9. Gilman, H.; Tomasi, R. A.; *J. Org. Chem.* **1962**, *27*, 3647–3649.
10. Hudrlik, P. F.; Peterson, D. *J. Am. Chem. Soc.* **1975**, *97*, 1464–1468.
11. [R] van Staden, L. F.; Gravestock, D.; Ager, D. J. *Chem. Soc. Rev.* **2002**, *31*, 195–200.
12. Johnson, C. R.; Tait, B. D. *J. Org. Chem.* **1987**, *52*, 281–283.
13. (a) Gaudemar, M.; Bellassoued, M. *Tetrahedron Lett.* **1990**, *31*, 329–352. (b) Bellassoued, M.; Majidi, A. *J. Org. Chem.* **1993**, *58*, 2517–2522. (c) Palomo, C.; Aizpurua, J. M.; Garcia, J. M.; Ganboa, I.; Cossio, F. P.; Lecea, B. Lopez, C. *J. Org. Chem.* **1990**, *55*, 2498–2503.
14. (a) Bellassoued, M.; Salemkour, M. *Tetrahedron Lett.* **1993**, *34*, 5281–5284. (b) Bellassoued, M.; Ozanne, N. *J. Org. Chem.* **1995**, *60*, 6582–6584.
15. Iguchi, M.; Tomioka, K. *Org. Lett.* **2002**, *4*, 4329–4331.
16. Battiste, M. A. *Tetrahedron Lett.* **2002**, *43*, 8765–8768.
17. Abedi, A.; Battiste, M. A. *J. Organometallic Chem.* **2004**, *689*, 1856–1859.
18. (a) Iqbal, M.; Evans, P.; *Tetrahedron Lett.* **2003**, *44*, 5741–5745. (b) Tsuge, O.; Kanemasa, S.; Ninomiya, Y. *Chem. Lett.* **1984**, *11*, 1993–1996. (c) Cooke, M. P. Jr.; Pollock, C. M. *J. Org. Chem.* **1993**, *58*, 7474–7481.
19. Barbero, A.; Blanco, Y.; García, C.; Pulido, F. J. *Synthesis*, **2000**, *9*, 1223–1228.
20. Obayashi, M.; Utimoto, K.; Nonaki, H. *Bull. Chem. Soc. Jpn.* **1979**, *52*, 1760–1764.
21. (a) Hudrlik, P. F.; Hudrlik, A. M.; Misra, R. N.; Peterson, D.; Withers, G. P.; Kulkarni, A. K.; *J. Org. Chem.* **1980**, *45*, 4444–4448. (b) Sato, T.; Abe, T.; Kuwajima, I. *Tetrahedron Lett.* **1978**, *19*, 259–262.

22. (a) Mason, P. H.; Yoell, D. K.; Van Staden, L. F.; Emslie, N. D. *Synth. Commun.* **1995**, 3347–3350. (b) Van Staden, L. F.; Bartels–Rahm, B.; Emslie, N. D. *Tetrahedron Lett.* **1997**, *38*, 1851–1852. (c) Van Staden, L. F.; Bartels–Rahm, B.; Field, J. S.; Emslie, N. D. *Tetrahedron Lett.* **1998**, *54*, 3255–3278.
23. Waschbüsch, R.; Carran, J.; Savignac, P. *Tetrahedron*, **1996**, *52*, 14199–14216.
24. Suzuki, H.; Ohta, S.; Kuroda, C. *Synth. Commun.* **2004**, *34*, 1383–1392.
25. Lin, J.; Welch, J. T. *Tetrahedron Lett.* **1998**, *39*, 9613–9616.
26. Fürstner, A.; Brehm, C.; Cancho–Grande, Y. *Org. Lett.* **2001**, *3*, 3955–3957.
27. Chakraborty, T. K.; Laxman, P. *Tetrahedron Lett.* **2003**, *44*, 4989–4992.
28. Murai, T.; Fujishima, A.; Iwamoto, C.; Kato, S. *J. Org. Chem.* **2003**, *68*, 7979–7982.
29. Berry, M. B.; Griffiths, R. J.; Sanganee, M. J.; Steel, P. G.; Whelligan, D. K. *Tetrahedron Lett.* **2003**, *44*, 9135–9138.
30. Pleynet, D. P. M.; Dutton, J. K.; Johnson, A. P. *Tetrahedron*, **1999**, *55*, 11903–11926.
31. Asakura, N.; Usuki, Y.; Iio, H. *J. Fluorine Chem.* **2003**, *124*, 81–88.
32. Keeney, A.; Nieschalk, J.; O'Hagan, D. *J. Fluorine Chem.* **1996**, *80*, 59–62.
33. Perales, J. B.; Makino, N. F.; Van Vranken, D. L. *J. Org. Chem.* **2002**, *67*, 6711–6717.
34. Kojima, S.; Fukuzaki, T.; Yamakawa, A.; Murai, Y. *Org. Lett.* **2004**, *6*, 3917–3920.
35. (a) Zhang, T. Y.; O'Toole, J. C.; Dunigan, J. M.; *Tetrahedron Lett.* **1998**, *39*, 1461–1464. (b) Yamakado, Y.; Ishiguro, M.; Ikeda, N.; Yamamoto, H. *J. Am. Chem. Soc.* **1981**, *103*, 5568–5570. (c) Crowe, W. E.; Goldberg, D. R.; *J. Am. Chem. Soc.* **1995**, *117*, 5162–5163. (d) Randl, S.; Gessler, S.; Wakamatsu, H.; Blechert, S. *Synlett* **2001**, 430–432.
36. Heo, J.-N.; Holson, E. B.; Roush, W. R. *Org. Lett.* **2003**, *5*, 1697–1700.
37. Aubele, D.; Wan, S.; Floreancig, P. E. *Angew. Chem., Int. Ed. Engl.* **2005**, *44*, 3485–3488.
38. Andemichael, Y. W.; Gu, Y. G.; Wang, K. K. *J. Org. Chem.* **1992**, *57*, 794–796.
39. Andemichael, Y. W.; Wang, K. K. *J. Org. Chem.* **1992**, *57*, 796–798.
40. (a) Pearson, A. J. *Acc. Chem. Res.* **1980**, *13*, 463–469. (b) Gibson, D. T.; Hensley, M. Yoshioka, H. Mabry, T. J. *Biochemistry*, **1970**, *9*, 1626–1630. (c) Kundig, E. P.; Simmons, D. P. *J. Chem. Soc., Chem. Commun.* **1983**, 1320–1322.
41. Lautens, M.; Ma, S.; Belter, R. K.; Chiu, Leschziner, A. *J. Org. Chem.* **1992**, *57*, 4065–4066.
42. Huang, J.; Wu, C.; Wulff, W. D. *J. Am. Chem. Soc.* **2007**, *129*, 13366.

2.11 Sakurai allylation reaction

Marco A. Biamonte

2.11.1 Description

The Sakurai reaction can be defined as the allylation of a carbonyl compound, or equivalent thereof, performed with an allylsilane and promoted by a Lewis acid.

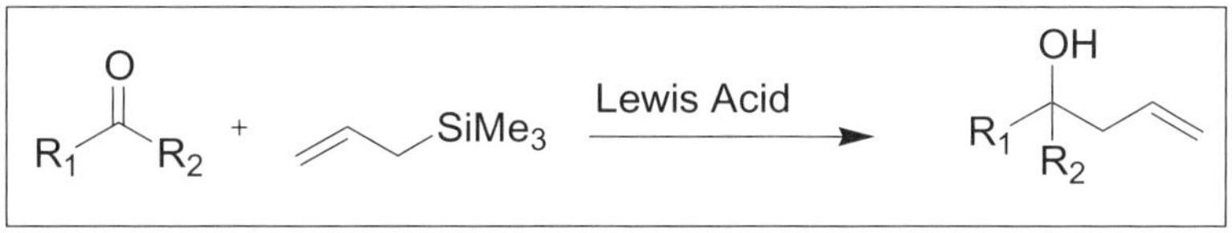

This book chapter is limited to Lewis acid-mediated reactions, and does not discuss the important field of Lewis base-mediated allylations,[1,2] nor does it describe the reactions of allylsilanes with other electrophiles such as epoxides, imines, and allyl–X (X = –Cl, –OR, –OAc). The Sakurai reaction has been covered under different forms in reviews focusing on: "The Stereochemistry of the Sakurai reaction",[3] "Intramolecular Addition Reactions of Allylic and Propargylic Silanes",[4] "Selective Reactions Using Allylic Metals",[5] "Catalytic Enantioselective Addition of Allylic Organometallic Reagents to Aldehydes and Ketones",[6] and "Modern Carbonyl Chemistry".[7]

The Sakurai reaction is most commonly performed in dichloromethane at –78 °C with one equivalent of Lewis acid. Diethyl ether and tetrahydrofuran are rarely used as solvents, because they sequester the Lewis acid as ethereal complexes.

The most popular promoters are $BF_3{\bullet}OEt_2$, $TiCl_4$, and $SnCl_4$. They are followed, arguably, by $AlCl_3$, $AlCl_2Et$, and TMSOTf. Protic acids are used more rarely since they tend to cause an unwanted protodesilylation. BF_3, $TiCl_4$, and $SnCl_4$ have somewhat different properties. BF_3 does not react with allylsilanes directly, and has only one coordination site. $TiCl_4$ is also inert towards allylsilanes, even at room temperature, and has two coordination sites. $TiCl_4$ forms 1 : 1 or 1 : 2 adducts with aldehydes, depending on the stoichiometry of the reagents. It can also form chelates with α- and β-alkoxyaldehydes, which influences the stereochemistry of the reaction (chelation control). $TiCl_4$ is best used at low temperatures (–78 °C) to prevent the formation of unwanted chlorination products that appear at room temperature. The major characteristic of $SnCl_4$ is that, in the presence of an allylsilane, it undergoes a transmetallation within minutes at –80 °C. Hence, with $SnCl_4$, the real allyl donor is more often than not an allylstannane. In addition, $SnCl_4$ has two coordination sites, and forms exclusively 1 : 2 adducts with aldehydes, regardless of the stoichiometry. $SnCl_4$ does not give chlorinated side-products at room temperature.[8] Aldehydes, ketones, and enones react at –78 °C upon catalysis with $TiCl_4$ or $SnCl_4$, sometimes instantaneously as in the case of cyclohexenone.

Allylsilanes can be prepared by a wide array of methods, including (1) the reaction of allyl metals with $ClSiR_3$, (2) the reaction of silylanions ($MSiR_3$) with allylic substrates, (3) the Kumada coupling of Me_3SiCH_2MgBr with vinyl halides, catalyzed by Pd or Ni species, (4) the Wittig reaction of β-silylated Wittig reagents, (5) the cross-metathesis of olefins with allylsilanes, and (6) the reductive silylation of unsaturated compounds.[9]

2.11.2 Historical Perspective

In 1976, Akira Hosomi and Hideki Sakurai of Tohoku University in Sendai, Japan, published a letter entitled "*Syntheses of γ,δ-Unsaturated Alcohols From Allylsilanes And Carbonyl Compounds In The Presence of Titanium Tetrachloride*".[10] The letter describes the reaction depicted in equation 1, in which allyl silanes react with aldehydes or ketones to provide homoallylic alcohols. The following year, Hosomi and Sakurai extended their finding to ketals, which provide homoallylic ethers (eq. 2),[11] and to α,β-unsaturated ketones (eq. 3),[12] in which case the addition occurs in a 1,4-fashion and becomes a valuable method to generate quaternary centers.

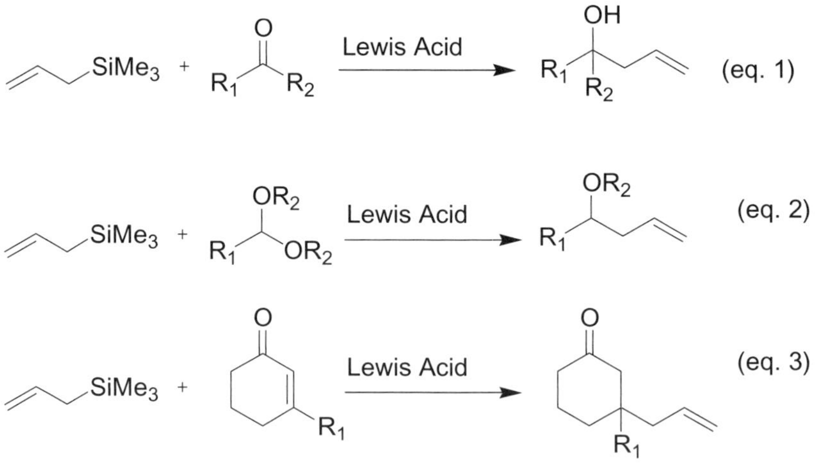

At the time of the reports of Hosomi and Sakurai, allylsilanes possessed the remarkable feature of being the only known allylmetal species that were stable to air and moisture. Furthermore, allylsilanes were configurationally stable. They did not undergo the π-allyl rearrangement characteristic of allylmagnesium halides and allyllithium reagents. Even crotyltrimethyltins undergo a 1,3-metallotropic shift which renders the preparation of α-methylallylstannanes difficult.[13] However, organosilanes could only be used in specific cases, and Hosomi and Sakurai, by demonstrating the utility of $TiCl_4$ catalysis, were the first to provide conditions applicable to a wide range of substrates, thus paving the way for the use of allylsilanes in organic synthesis. As an added benefit, the reaction proved to be regiospecific, occurring exclusively at the γ-carbon of the allyl group, regardless of how sterically hindered it is.[14]

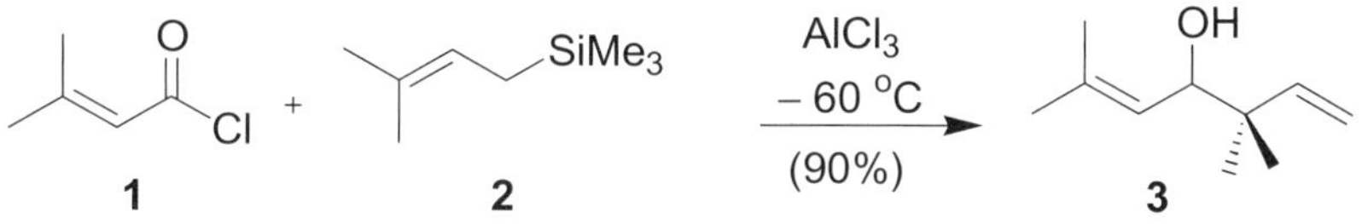

Since then, researchers have continuously refined the scope of the reaction. In the 1970's and 1980's the focus was to understand the stereochemical issues associated with the reaction. Researchers then gradually turned their attention to asymmetric Lewis acids, catalytic reactions, catalytic and enantioselective reactions, reactions catalyzed by Lewis bases, and reactions that can be performed in aqueous solvents.

2.11.3 Mechanism

General Considerations

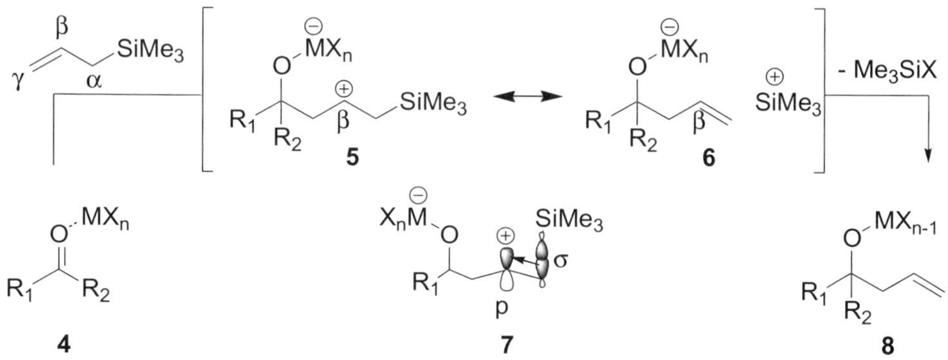

The recurrent theme in the above reactions is the activation of a carbonyl compound **4** (or equivalent) by a Lewis acid, followed by the addition of the allylsilane. The reaction proceeds through a β-silyl cation **5** that is stabilized by hyperconjugation, as represented by the resonance form **6**. The hyperconjugation can also be depicted by structure **7**, which is another way to indicate that the electrons forming the C–Si bond are partially delocalized into the vacant p orbital of the carbocation. This effect accelerates the reaction by a factor of 30,000–200,000 ($\Delta\Delta G^*$ =4.2 kcal/mol).[15,31] The subsequent elimination of the silyl group generates a new double bond.

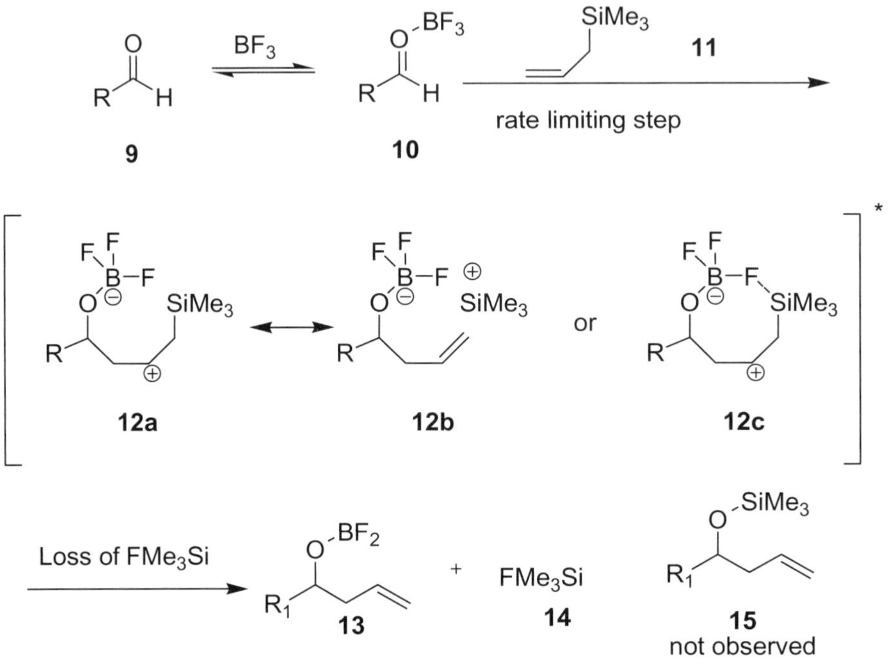

The mechanism can be further illustrated by using BF_3 as the representative Lewis acid. The reaction starts with the equilibrium leading to the formation of the aldehyde•BF_3 complex **10**. Complex **10** then reacts with the allylsilane **11** in what is the rate limiting step of the process. It is generally assumed that the transition state is acyclic (**12a**, **12b**). The 8-membered-ring transition state **12c** was proposed by Bottoni based on computational studies[16] and stereochemical analyses. Although the 8-membered ring may be operative in some cases, it has been clearly ruled out in intramolecular cases by Denmark.[6] A fluoride is then delivered from the boron to the silicon atom (unclear if intra or intermolecularly) to produce difluoroboron ether **13** and fluorotrimethylsilane **14**. The latter (**14**) is the only silicon species detected by ^{29}Si NMR spectroscopy between –80 to 20 °C. Silyl ether **15** is not observed. Similarly, the use of $TiCl_4$ as Lewis acid results in the formation of the corresponding titanium alkoxide and chlorotrimethylsilane. Hence, the Lewis acid not only activates the carbonyl group but also provides a nucleophilic halide.

The Lewis Acid–Carbonyl Complex

Understanding the physical chemistry of Lewis acids is important in selecting the most appropriate Lewis acid. The Lewis acid is involved in two distinct processes. It binds to the carbonyl group, and as result activates it. Normally, the stronger the acid binds to the carbonyl, the more reactive the acid–carbonyl complex becomes. Yet, it is interesting to analyze binding and reactivity separately. This section deals with bond strengths, and the next section will address the reactivity of the complex.

Table 1: Equilibrium constants at room temperature (0.23M).[17]

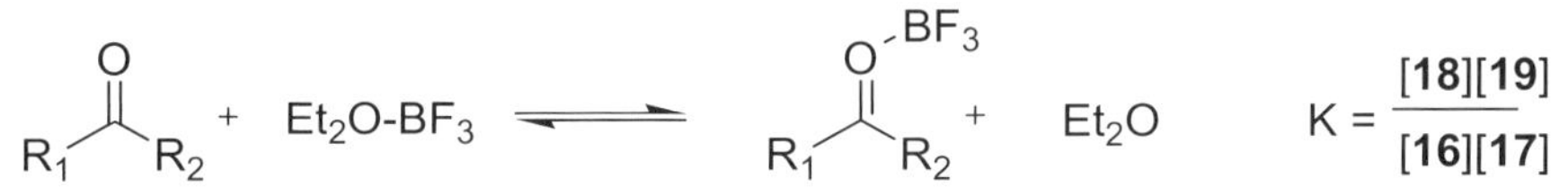

16 17 18 19

Entry	*Compound*	*Solvent*	*K*	*[16]:[18]*	*ΔG [kcal/mol]*
1	Benzaldehyde	$CDCl_3$	0.208±0.011	69:31	0.93
2	*p*-Anisaldehyde	$CDCl_3$	7.24	27:73	–1.1
3	*p*-Tolualdehyde	$CDCl_3$	0.838	52:48	0.10
4	*p*-Chlorobenzaldehyde	$CDCl_3$	0.112	75:25	1.29
5	*p*-Nitrobenzaldehyde	$CDCl_3$	0.009	92:08	2.79
6	Cyclohexanone	$CDCl_3$	0.276	66:34	0.76
7	Isobutyraldehyde	$CDCl_3$	0.204	69:31	0.94
8	Benzaldehyde	CD_2Cl_2	0.16±0.015	71:29	1.08
9	Benzaldehyde	C_6D_6	0.15	72:28	1.12

Equilibrium constants with BF_3 adducts have been determined by NMR at room temperature (Table 1), and the corresponding ΔG values can be derived.[17,18] The diethyl ether present in the commercial Et_2O•BF_3 complex competes with the carbonyl group effectively. At room temperature in $CDCl_3$, BF_3 rapidly interchanges its position on ether and benzaldehyde, and only 31% of benzaldehyde is found in the complexed form **18** (entry 1). Most aromatic and aliphatic aldehydes are predominantly in the uncomplexed form at room temperature (8–48% of complex), which translates into a positive ΔG. Therefore, counterintuitively, cooling the mixture only shifts the equilibrium towards the left, and as a result there is less complex present at low temperatures. For benzaldehyde (ΔG = 0.93 kcal/mol) the proportion of complex **18** is calculated to decrease from 31% at room temperature to 23% at –78 °C. *p*-Anisaldehyde (entry 2) is the only exception among the examples reported in Table 1, where 73% of the *p*-anisaldehyde is actually coordinated to BF_3.

The equilibrium is further affected by steric factors. The bond strength of BF_3 adducts has been measured by calorimetry (in the absence of diethyl ether, Table 2).[19,20] ΔH appears to be very sensitive to steric hindrance, with a difference of nearly 10 kcal/mol between di-*tert*-butyl ketone and acetone (entries 1 and 7). Sterically unhindered aldehydes, ketones, and esters bind BF_3 with ΔH = –13.6 to –18.2 kcal/mol (entries 3–8). Compared to these carbonyl compounds, diethyl ether forms a slightly stronger bond (ΔH = –18.8 kcal/mol, entry 9), and tetrahydrofuran forms an even stronger bond (–21.6 kcal/mol). This explains why ether and tetrahydrofuran are rarely used as solvents: BF_3 binds to ethereal solvents more effectively than to the intended aldehyde.

Table 2: Selected Enthalpies of Complex Formation with Boron Trifluoride in Dichloromethane at 298 K. For a more complete list, see reference 19.

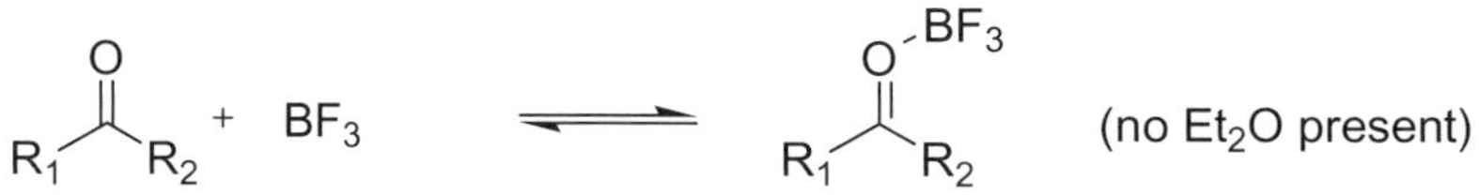

20 21 22

Entry	*Compound*	*ΔH [kcal/mol]*
1	di tert-butyl ketone	–7.49±0.09
2	acetonitrile	–14.4±0.11
3	acetaldehyde	–16.6±0.29
4	acetophenone	–17.8±0.03
5	benzaldehyde	–17.9±0.23
6	ethyl acetate	–18.0±0.07
7	acetone	–18.1±0.05
8	cylcohexanone	–18.2±0.19
9	diethyl ether	–18.8±0.09
10	tetrahydrofuran	–21.6±0.06
11	dimethylsulfoxide	–25.2±0.08
12	*N,N*-dimethylformamide	–27.0±0.08

Table 3: The enthalpy of complexation (ΔH) of crotonaldehyde with various Lewis Acids in dichloromethane at 295 K.[21]

Entry	*Acid*	*ΔH [kcal / mol]*
1	BCl_3	–24.4±0.3
2	$SbCl_5$	–21.2±0.4
3	$TiCl_4$	–14.0±0.6
4	$SnCl_4$	–12.2±0.4

In absolute terms, $TiCl_4$ and $SnCl_4$ do not form bonds quite as strong as BF_3 (ΔH = –14.0 and –12.2 kcal/mol respectively; Table 3, entries 3 and 4). Yet the absence of diethyl ether competing for complexation allows the aldehyde to be fully complexed. No free aldehyde was detected by NMR studies the presence of 0.5 eq of $TiCl_4$ or $SnCl_4$, using 1:1 $CD_2Cl_2/CDCl_3$ as solvent.[8]

The tridimensional structure of carbonyl–Lewis acid complexes has been determined by X-ray studies. The crystal structure of the benzaldehyde•BF_3 complex, shows that the BF_3 group coordinates *anti* to the Ph group as in **23**.[22] Likewise, the crystal structure of (4-(-*t*-butyl)benzaldehyde)$_2$•$SnCl_4$ shows an *anti* coordination of $SnCl_4$.[23] Importantly, a formyl hydrogen bond has been postulated by Corey as an organizing element to rationalize the outcome of a number of reactions involving aldehyde-metal complexes (**25**).[24–27]

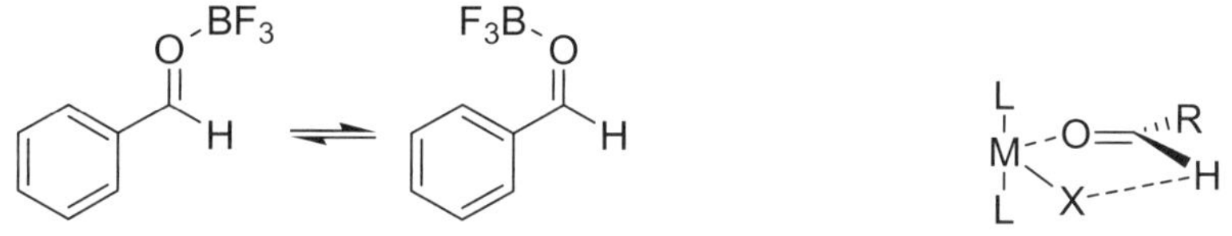

23 *Anti* coordination

24 *Syn* coordination

25 Formyl Hydrogen Bond

Effect of the Lewis Acid on the reactivity of the acid-carbonyl complex

Once the acid is complexed to the aldehyde, the aldehyde becomes an excellent electrophile. The electrophilicity depends directly on how low is the LUMO energy of the complex. To establish a scale of Lewis acidity, the LUMO energy of crotonaldehyde•Lewis acid complexes was calculated by MNDO (gas phase, Table 4).[28,29] The calculations allow one to rank order the acids, and BCl_3 proved to be the strongest acid, and $SnCl_4$ the mildest one:

$$BCl_3 > BBr_3 > AlCl_3 > EtAlCl_2 > BF_3 > Et_2AlCl > SnCl_4.$$

Several other acidity scales have been proposed.[21] One of them consists in analyzing the complexation of a Lewis acid to crotonaldehyde by ^1H or ^{13}C NMR spectroscopy.[30] This approach is attractive because it is experimentally simple and takes into account solvent effects. Addition of a Lewis acid to a 0.3M solution of crotonaldehyde in CD_2Cl_2 at –20 °C causes a downfield shift ($\Delta\delta$) of the 2-, 3-, and 4-proton resonances (and an erratic shift of the 1-proton). The shift is commensurate with the electron withdrawing effect of the acid, and therefore to the energy of the LUMO (π^*). This gives rise to another scale of Lewis acidity, based on NMR shifts. The NMR-based scale parallels very closely the LUMO-based scale, and the Lewis acids are ranked in a nearly identical order:[28]

$BBr_3 > BCl_3 > AlCl_3 > EtAlCl_2 > BF_3 > EtAlCl_2 > TiCl_4 > Et_2AlCl > SnCl_4$

Table 4: Scale of Lewis Acidity of crotonaldehyde•Lewis acid complexes based on (i) the calculated energy of the LUMO or (ii) ^{1}H NMR shift of H-3.

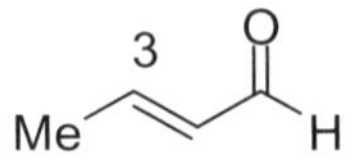

crotonaldehyde

	Energy of LUMO calculated by MNDO [(a)] *π* [eV]*	*NMR* [(b)] *Δδ [ppm]*
BCl_3	–2.52	1.35
BBr_3	–2.50	1.49
$SiCl_4$	–2.43	
$AlBr_3$	–2.37	
$AlCl_3$	–2.31	1.23
$EtAlCl_2$	–2.03	1.15
BF_3	–1.93	1.17
Et_2AlCl	–1.82	0.91
AlF_3	–1.77	
$Me(OPh)_2Al$	–1.76	
$HgCl_2$	–1.74	
$HgBr_2$	–1.72	
HgF_2	–1.71	
HgI_2	–1.66	
Et_3Al	–1.62	0.63
$SnCl_4$	–1.58	0.87

(a) Calculated energy of the LUMO. 1 eV = 23.06 kcal/mol. (b) Shift of chemical shift of crotonaldehyde H-3 complexes in CD_2Cl_2.

Effect of aldehyde substituents on the reaction rate

For the Sakurai reaction, we are not aware of studies in which the electronic effect of aldehyde substituents has been systematically investigated.

Effect of silane substituents on the reaction rate

The rate of the Sakurai reaction depends not only on the Lewis acid, but also on the ability of the silyl group to stabilize the intermediate carbocation. The cation is further stabilized when the silyl group carries electron-donating substituents. This becomes obvious when considering the mesomeric form **3**. The relative nucleophilicity of allylsilanes has been quantified (see Figure 1).[31] The method uses a diaryl cation (Ar–CH$^+$–Ar) as probe. Diarylcations are easily generated from the corresponding halide and a Lewis acid such as BCl_3. They are colored, and their disappearance upon reaction with allylmetal species is conveniently monitored spectroscopically. The Me_3Si

group accelerates the addition to the cation by a factor of 1.99×10^5 compared to no substituents (H). The less electron-rich Ph_3Si group is two orders of magnitude less nucleophilic than the Me_3Si group. For analogous reasons, a Me group in the β-position of the allylsilane (isobutenyl) accelerates the reaction by a factor of about 10^4.

Figure 1: Relative nucleophilicity of allylsilanes.[31]

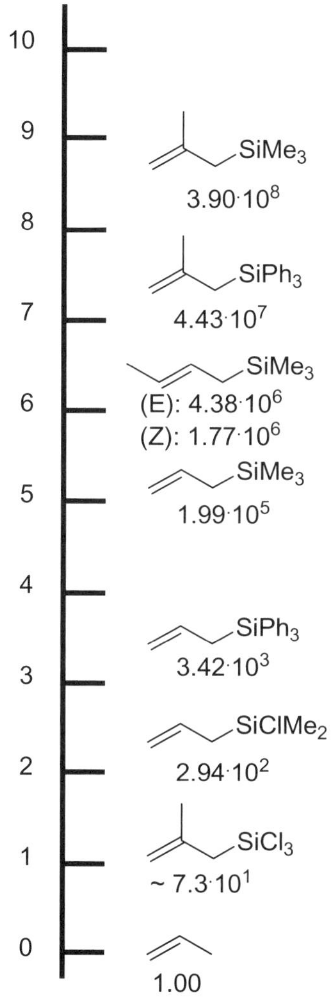

2.11.4 Variations and Improvements

A feature of the Sakurai reaction that has contributed to its popularity is that the stereochemical outcome can be controlled. In the course of the reaction, one sp^2 and two sp^3 centers are formed (**28**). If appropriately substituted, the

sp^2 center can give rise to regioisomers having an *E* or *Z* configuration. Similarly, each sp^3 center can give rise to a stereogenic center (*R* or *S*), and therefore to enantiomeric pairs of *syn*/*anti* diastereroisomers. We will first discuss the *E*/*Z* isomerism arising from the formation of a double-bond, and then move the *syn*/*anti* isomerism arising from the two vicinal sp^3 centers.

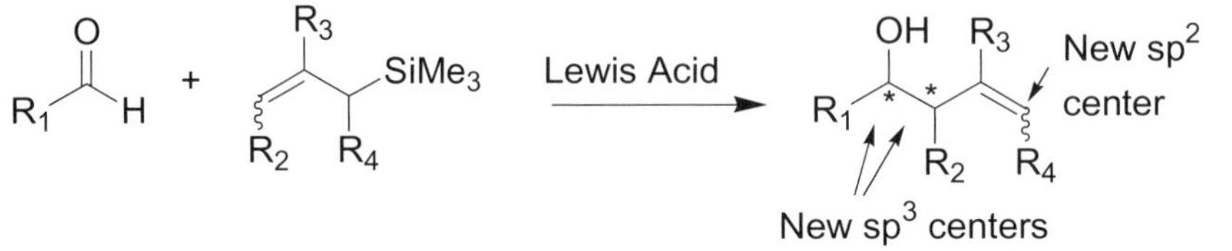

26 **27** **28**

E/Z Isomerism

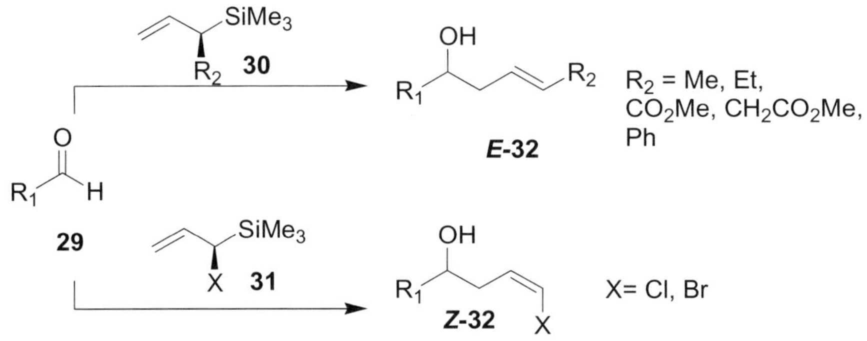

E/*Z* isomers are produced when the allyl silane is further substituted at the allylic positions as in **30** and **31**. In the majority of cases, the *E* olefin predominates. This occurs when the allylic substituent R_2 is Me,[32] Et,[33] CO_2Me,[34] CH_2CO_2Me,[35] or Ph.[36] In contrast, α-halogen substituted allylsilanes give *Z* olefins with high selectivity.[37]

Syn/Anti *Isomerism*

Syn/*anti* diastereoisomers can only be produced if the allylic γ-carbon is substituted, as in the case of crotylsilane **34**. Crotyltrimethylsilanes are not commercial and must be prepared.[38,39] As a rule of thumb, both *E* and *Z*-crotylsilanes **34** give predominantly the *syn* isomer, with the *E*-crotylsilane giving better selectivities, as shown in Table 5.[40,41] There are, however, exceptions.

Table 5: Diastereoselectivity in the reaction of carbonyl compounds with *E*- or *Z*-crotyltrimethylsilane in CH_2Cl_2 with 1.2 eq $TiCl_4$ in CH_2Cl_2 at –78 °C.[40]

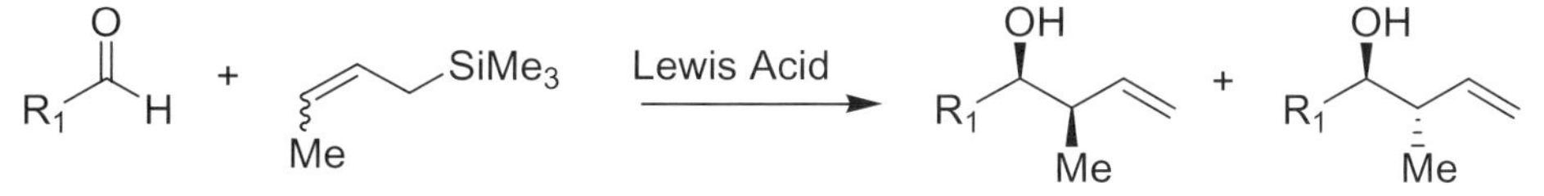

33 **34** **35 (*Syn*)** **36 (*Anti*)**

Entry	*R1*	*Acid*	*Syn:anti* *From E-silane*	*Syn:anti* *From Z-silane*
1	Et	$TiCl_4$	95:5	65:35
2	*i*-Pr	$TiCl_4$	97:3	64:36
3	*t*-Bu	$TiCl_4$	>99:1	69:31

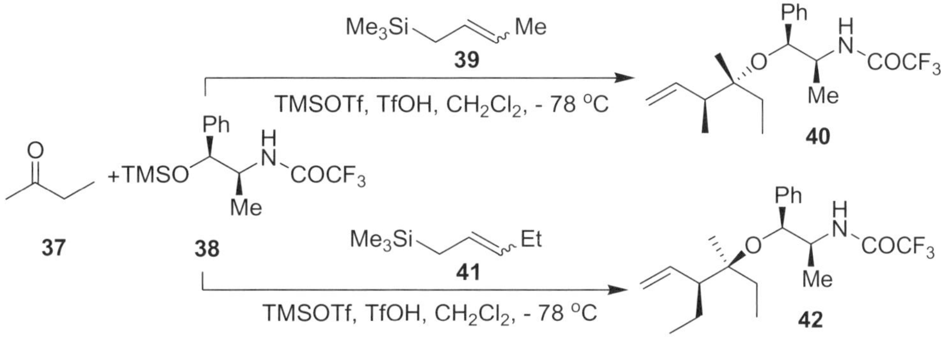

One exception is found when using norpseudoephedrine derivative **38** as chiral auxilary. Both *E*- and *Z*-crotylsilanes **39** gave the *anti* adduct **40** as the major product. However, the nominally more hindered *E*- and *Z*-pentenylsilane **41** gave the *syn* adduct **42** as the major product, testifying to the delicate balance of non-bonded interactions occurring in the transition state.[38]

Another exception is found in a multicomponent reaction where the carbonyl compound **43** is premixed with TMSOMe to give ketal **44** *in situ* (Table 6).[42] The reaction proceeds via methyloxenium ion **45**. *E*-crotyltrimethylsilane gives in all cases (Table 6, entries 1–3) the *syn*-adduct as expected. However, *Z*-crotyltrimethylsilane, gives either the *syn*- or the *anti*-product, depending on the aldehyde used.

Table 6: Diastereoselectivities of the Lewis-Acid promoted multicomponent reaction of ketones, Me_3SiOMe, and crotyltrimethylsilanes at –78 °C.

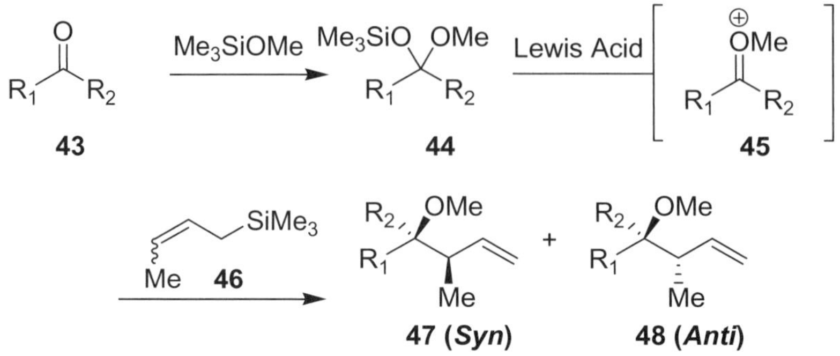

Entry	*R1*	*R2*	*Acid*	*Syn:anti From E-silane*	*Syn:anti From Z-silane*
1	Me	H	Me_3SiOTf	82:18	31:69
2	Et	H	Me_3SiOTf	89:11	45:55
3	tBu	H	Me_3SiOTf	98:2	95:5
4	Et	Me	TfOH	26:74	27:73

Rationalizing the syn/anti selectivity is not easy, and requires sophisticated computational methods. The simple structure of the intermediate methyloxenium **45** yields itself well to calculations. Computations were performed at a high level of theory (DFT with solvent effects) for the TMSOTf-catalyzed reaction of dimethylketals.[42] The methyloxenium ion can have either a *E* or *Z* configuration. If one assumes that the ion has the *E* configuration, and that all transitions states must have staggered conformations to minimize non-bonded interactions, there are six possible transition states (**TS1**–**TS6**). Three of them lead to the *syn* diastereomer, and three others to the *anti* diastereomer. The double bond of the crotylsilane can approach the double bond of the oxenium ion either in a synclinal or antiperiplanar manner. For acetaldehyde, the synclinal transition state TS1 is preferred. The next best transition is actually not truly staggered. It is halfway between TS2 and TS3, with the C=C bond nearly eclipsing the C–H bond (dihedral angle O=C–C=C = 104.8°). This second transition is energetically close to the first one ($\Delta\Delta G$ = 0.55 kcal/mol). Both transition states yield the *syn* adduct. The situation for pivaldehyde is somewhat different. The preferred transition state is the one hybrid between TS2 and TS3, and is followed by TS1 (ΔG=1.01 kcal/mol). Thus, even in highly diastereoselective reactions such as the one of acetaldehyde and pivaldehyde with *E*-crotyltrimethylsilane (Table 6, entry 3), there can be two operating transition states. It is not possible to single out a “default” transition state that accounts for all cases.

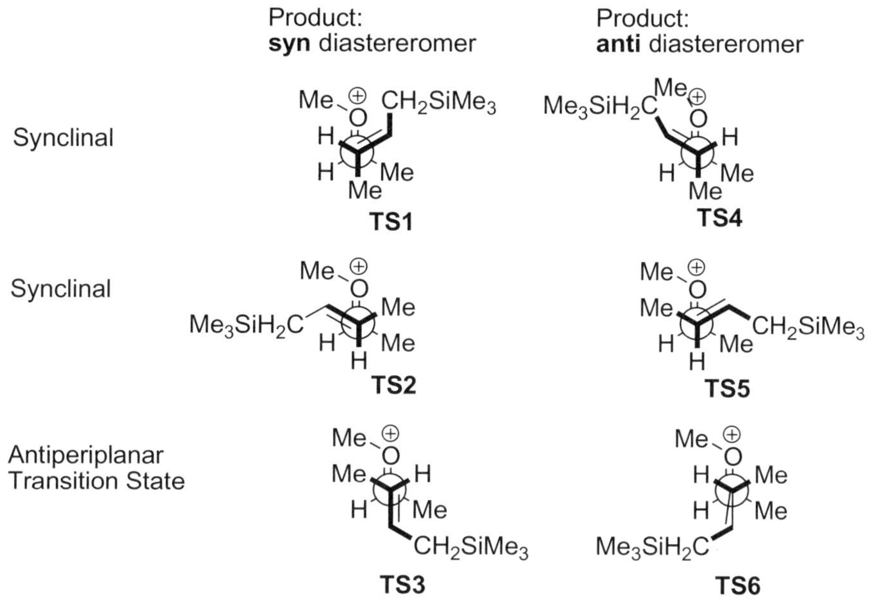

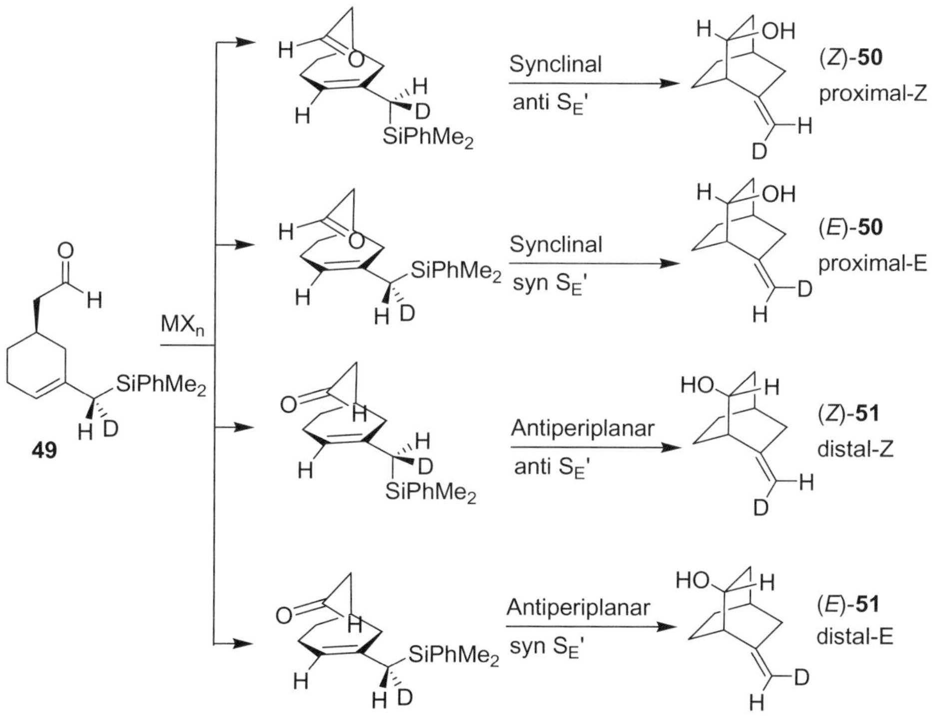

Furthermore, in intermolecular reactions, it is impossible to determine experimentally if the transition state is actually synclinal or antiperiplanar.

However, it is possible to do so for intramolecular reactions. Probe **49** was designed by Denmark to assess if the transition state is synclinal or antiperiplanar.[6] The probe also allows to analyze whether, once the β-silyl cation is formed, the nucleofuge (Me_2PhSi group) is oriented towards or away from the oxygen atom in the elimination step yielding the olefin. The results reported in Table 7 show that the synclincal transition state is always favored (≥ 60:40). The antiperiplanar geometry is still significant in the reactions using BF_3•OEt_2 or $SnCl_4$, but less so with CF_3COOH or $SiCl_4$. In the elimination step, the Me_2PhSi group is always orientated away from (*anti* to) the oxygen atom.

Table 7: Effect of Lewis acid on the geometry of the transition state of the intramolecular Sakurai reaction with probe 49

Entry	*Lewis acid*	*Synclinal/ Antiperiplanar (**50**/**51**)*	*anti* S_E'/ *syn* S_E' *(E-**50**/Z-**50**)*	*anti* S_E'/ *syn* S_E' *(E-**51**/Z-**51**)*
1	BF_3•OEt_2	75/25	94/4	94/6
2	$SnCl_4$	60/40	91/9	94/6
3	CF_3SO_3H	95/5	93/7	94/6
4	$SiCl_4$	98/2	95/5	–

In summary, the transition state of a Sakurai reaction is acyclic. There is no bond that maintains the carbonyl oxygen atom and the silicon atom in close proximity. The loose nature of the transition state complicates the control of the stereochemistry (and the analysis of the transition state). Nonetheless, good levels of diastereoselection can be achieved in the reaction of *E*-crotyltrimethylsilane with aldehydes, which affords products carrying a vicinal OH and a Me group, with a strong preference for the 1,2-*syn* diastereoisomer. It is not possible to obtain the 1,2-*anti* isomer with high selectivities.

1,2-Asymmetric induction

Asymmetric centers located at the α- and β-positions of the aldehyde influence the facial selectivity of the Sakurai reaction (1,2- and 1,3-asymmetric induction, respectively). 2-Phenylpropanal **52** is an example of α-methyl substituted aldehyde. It reacts with trimethylsilane to give mostly the *syn* adduct **53**, as predicted by the Cram and Felkin–Anh models (Table 8).[43] The selectivities are modest (≤ 2.7:1)

Table 8: Effect of an α-methyl substituent on the diastereoselectivity of the Sakurai reaction.[44,45]

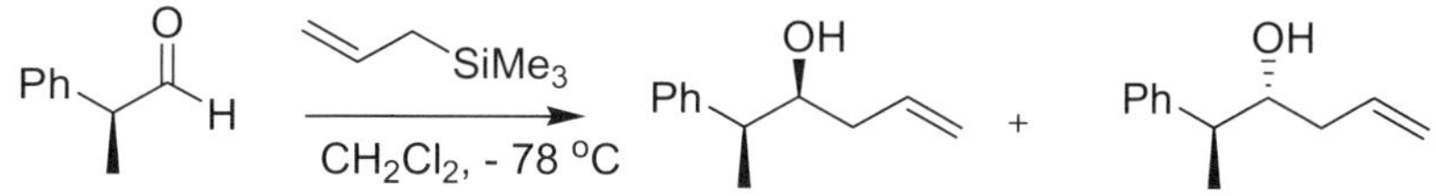

52 **53 (*Syn*)** Cram **54 (*Anti*)** anti-Cram

Acid	*syn / anti*
BF_3	2 : 1
$TiCl_4$	1.6 : 1
$SnCl_4$	2.2 : 1
$AlCl_3$	2.7 : 1

As a reminder, the Felkin–Ahn model can be illustrated with aldehyde **55**. The largest substitutent (*R*) is perpendicular to the C=O bond, and the nucleophile attacks the carbonyl group at an angle (Bürgi–Dunitz trajectory), eclipsing the sterically least demanding bond (H–C) as in transition state **56**.

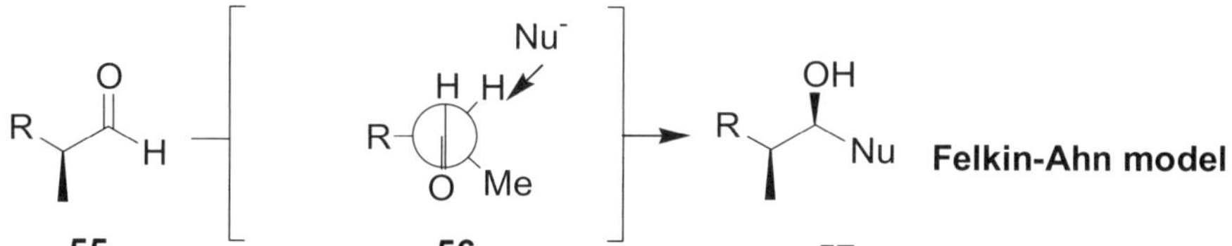

55 **56** **57**

If the aldehyde carries a chelating substituent (e.g., BnO–) in the α-position, as in **58**, other models, such as the chelation or Cornforth models may apply.[46] The latter is used with Lewis acids incapable of forming chelates, e.g., BF_3. The most stable conformer is the one in which dipoles are antiparallel to each other (**61**), and the nucleophile attacks from the least hindered face.

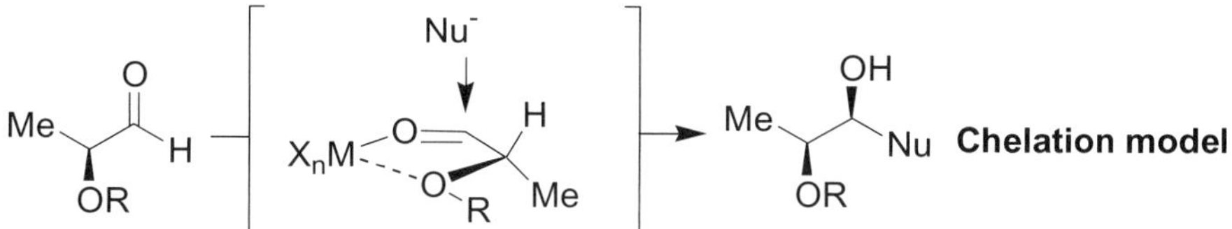

58 **59** **60**

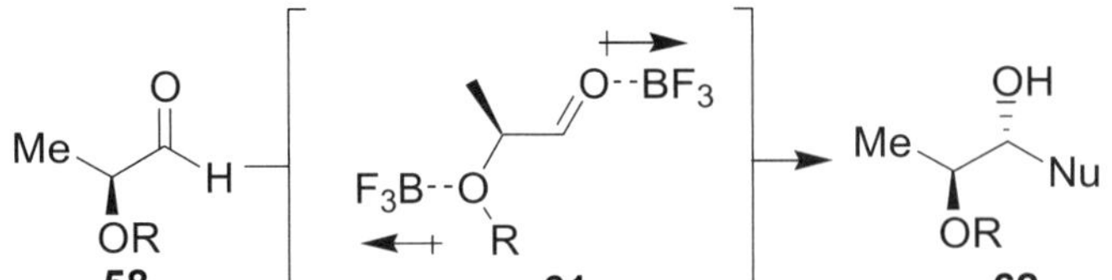

58 61 62 **Cornforth's model**

Table 9: Effect of α-benzyloxy substituents on the diastereoselectivity of the Sakurai reaction.[44,47]

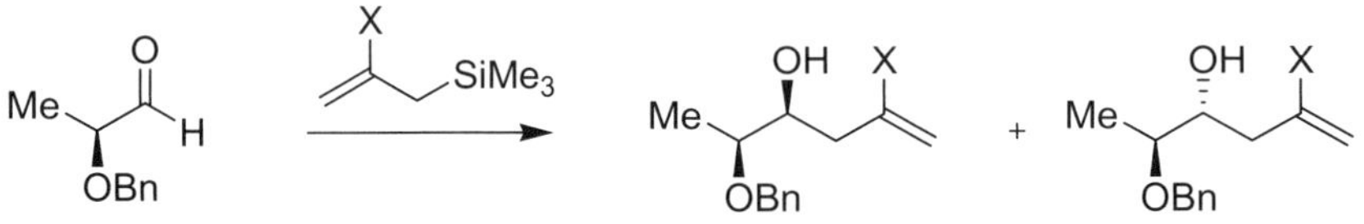

63 **64 (*Syn*)** Chelation **65 (*Anti*)** non-Chelation

Entry	*X*	*Acid*	*syn : anti*	*Yield*
1	H	$SnCl_4$	35:1	94
2	H	$BF_3 \cdot OEt_2$	1:1.5	50
4	Me	$SnCl_4$	45:1	81
5	Me	$BF_3 \cdot OEt_2$	1: 2.6	40

Thus, 2-benzyloxyaldehyde **63** provides the *syn* adduct **64** with good selectivities when using $SnCl_4$, in line with the chelation model (Table 9). In contrast, BF_3 promotes the *anti* stereoselection predicted by the Cornforth model, but in modest excess. $TiCl_4$ gives undesired, chlorinated products.[44]

Chelation control is also feasible in the case of α-methoxycyclohexanone **66** where the nucleophile attacks from the equatorial face. The approach from the axial face is hindered by the two axial hydrogens.[48]

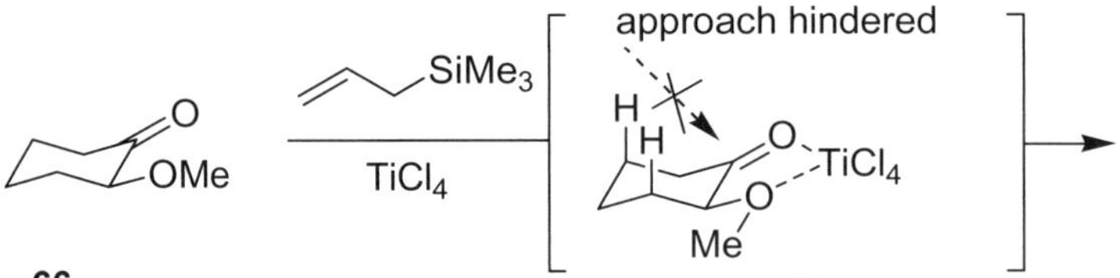

66 **67**

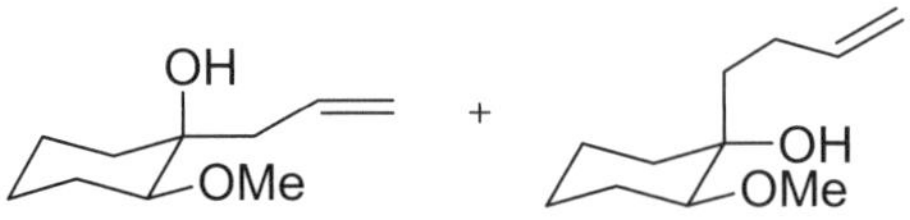

68 (*syn*) Chelation > 99 : 1 **69 (*anti*)**

1,3-Asymmetric induction

β-Alkoxy groups are also effective at promoting 1,3-asymmetric induction (Table 10). Interestingly, the levels of induction obtained in the Sakurai reaction are higher than when using RMgX, RLi, or R_2CuLi.[50]

Table 10: Effect of a β-alkoxy substituent on the diastereoselectivity of the Sakurai reaction.

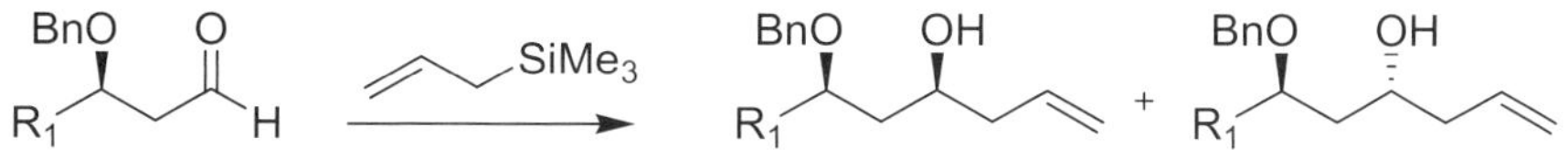

70 **71 (*syn*)** **72 (*anti*)** Chelation

Entry	*Acid*	*syn:anti*	*Ref*
1	$SnCl_4$	10:90	49
2	$TiCl_4$	5:95	50

The $MgBr_2$-mediated allylation of aldehyde **73** provided exclusively the 1,2-*syn* adduct **74** showing that 1,2-chelation overrides 1,3-chelation.[51]

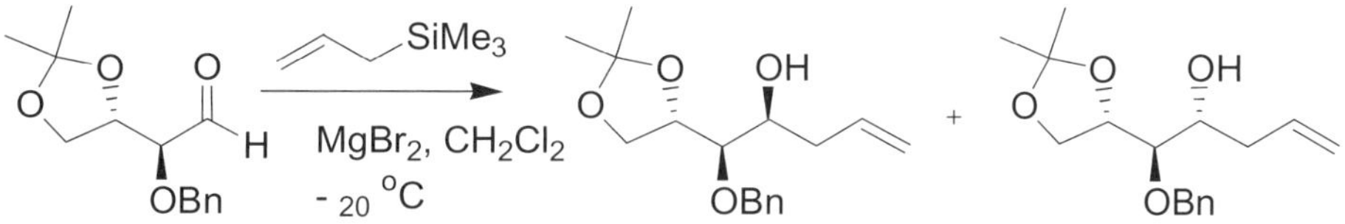

73 **74 (1,2-*Syn*)** 1,2-chelation **75 (1,3-*Syn*)** 1,3-chelation

To summarize this section, the Sakurai reaction obeys the usual rules of asymmetric induction. An α-methyl group favors the Cram-Felkin-Anh product (1,2-*syn*). An α-alkoxy group can lead to the chelation product (1,2-*syn*) or the non-chelation product (1,2-*anti*), depending on the Lewis acid. In the presence of chelating Lewis acids ($TiCl_4$, $SnCl_4$), β-alkoxyaldehydes provide the 1,3-*anti*-product preferentially. Only the reactions performed under chelation control with alkoxyaldehydes afford good level of diastereoselection.

Chirality transfer from chiral silanes

The use of chiral silanes has been reviewed.[52] Chiral silanes **76–78** have the asymmetric center located on the α-carbon. Reaction of **76** with aldehydes in the presence of $TiCl_4$ gave *ee* values of 91–95%.[36] The $TiCl_4$-mediated reaction of crotylsilanes **77** and **78** gave *syn*-homoallylic alcohols. The *E*-

olefin **77** gave excellent diastereoselectivities (92:8 to > 99:1) and *ee* values (86–92%), while *Z*-olefin **78** gave low diastereo- and enantioselectivities.

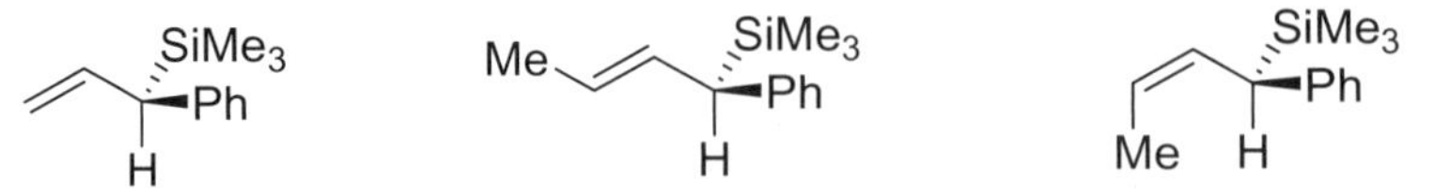

76 **77** **78**

Silane **80** has an asymmetric silicon atom. The chiral center is further away from the reaction center than for **76–78**, and **80** gave nearly racemic material in Sakurai reactions.[53] Allylsilane **83**, in which the asymmetric center is even more remote from the double bond, gave modest, but somewhat better *ee* values (21–46%).[54]

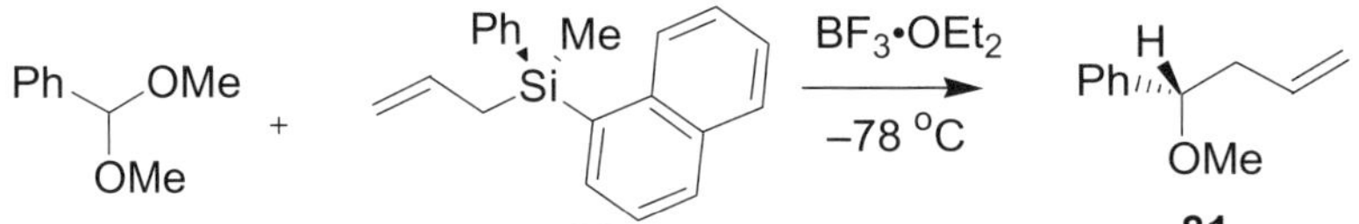

79 **80** **81**

ee = 3.9–5.5%
yield = 20%

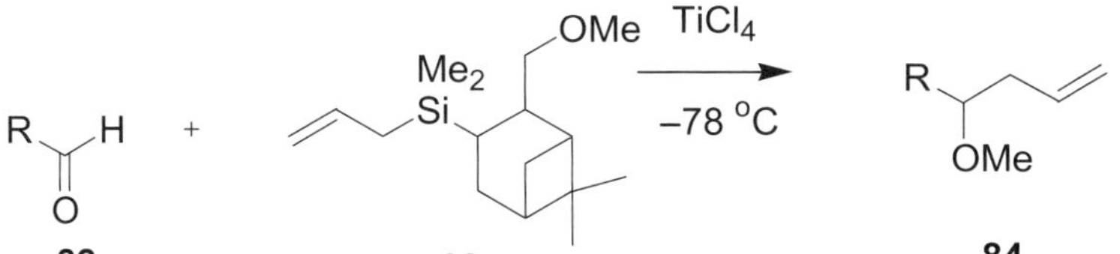

82 **83** **84**

ee = 21–46%

Asymmetric Catalytic Reactions

The development of asymmetric catalytic Sakurai reactions offers major challenges. As mentioned earlier, the BF_3-catalyzed allylation stops at the boron alkoxide **86** and $FSiMe_3$ (**87**). The silyl ether **88** is not formed, nor is BF_3 regenerated. Likewise, the use of $TiCl_4$ yields the titanium alkoxide **89** and $ClSiMe_3$ (**90**). The silyl ether **88** is not produced and $TiCl_4$ is not released. For the reaction to be catalytic, however, the alkoxide must react with the halotrimethylsilane (as in **91** with **92**) to regenerate the catalyst MX_n. Furthermore, the acyclic nature of the transition state makes any form of asymmetric control difficult to achieve.

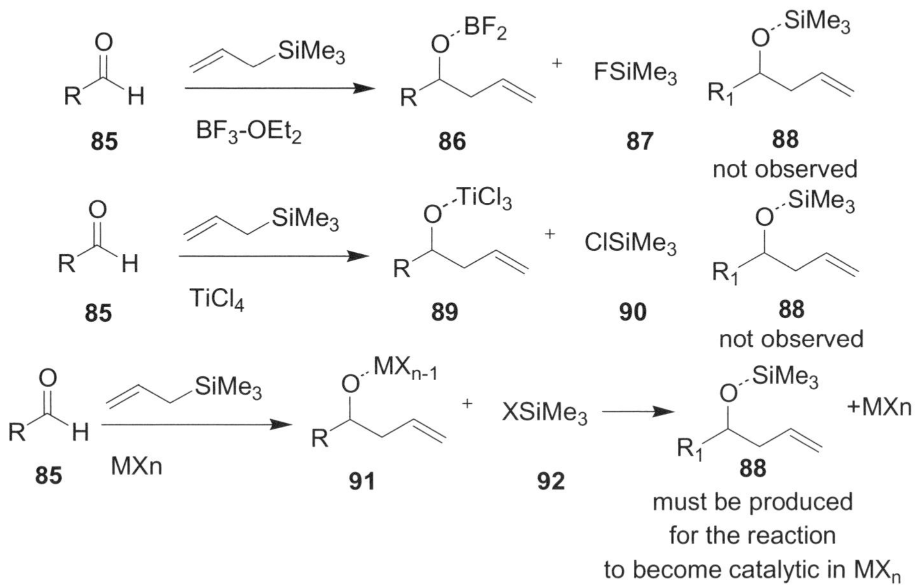

The first example of catalytic, enantioselective Sakurai reaction was effected by means of Yamamoto's chiral acyloxyborane (CAB) catalyst **96** (Table 11).[55] The CAB catalysts are derived from tartartic acid, and **97** gave a better yield than **96**.[56]

Table 11: Effect of chiral promoters 96 and 97 on the enantioselectivity of the Sakurai reaction

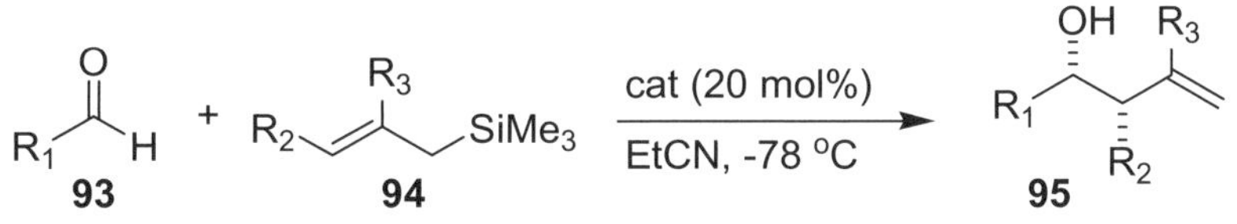

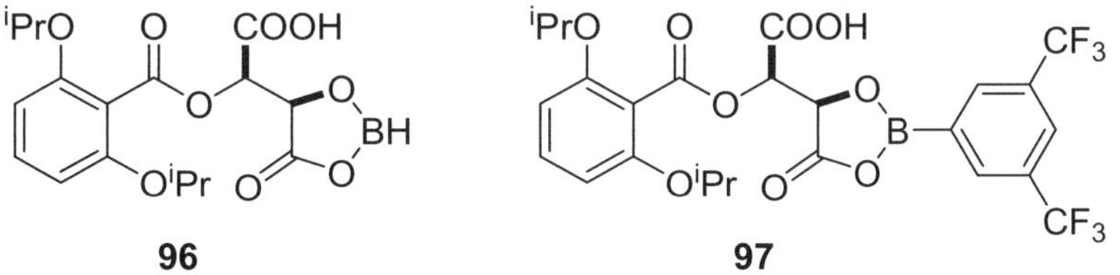

Entry	R_1	R_2	R_3	*cat*	*Yield [%]*	*Syn:Anti*	*er*
1	Ph	Me	Et	**96**	74	97:3	98:2
2	n-Pr	Me	Et	**96**	36	95:5	93:7
3	Ph	H	Me	**96**	68	–	91:9
4	Ph	H	H	**96**	46	–	78:22
5	Ph	H	Me	**97**	99	–	94:6

BINOL/titanium catalysts have also been reported (BINOL=1,1′-binaphthalene-2,2′-diol) in the allylation of glyoxaldehyde **98**. The addition of molecular sieves is extremely important for high reactivity and selectivity.[57]

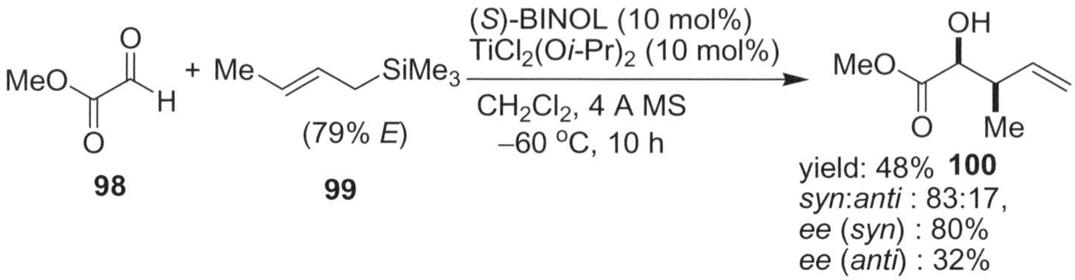

The $TiCl_2(Oi\text{-}Pr)_2$/BINOL system is only weakly acidic, and therefore limited to reactive aldehydes such as glyoxaldehyde. To circumvent the problem, Carreira introduced the TiF_4/BINOL system, which provides a more Lewis acidic, and more reactive catalyst (Table 12). In addition, the greater strength of the Ti–F bond compared to the Si–F bond assists the catalyst turnover.[58,59]

Table 12: Sakurai reaction promoted by BINOL/TiF_4

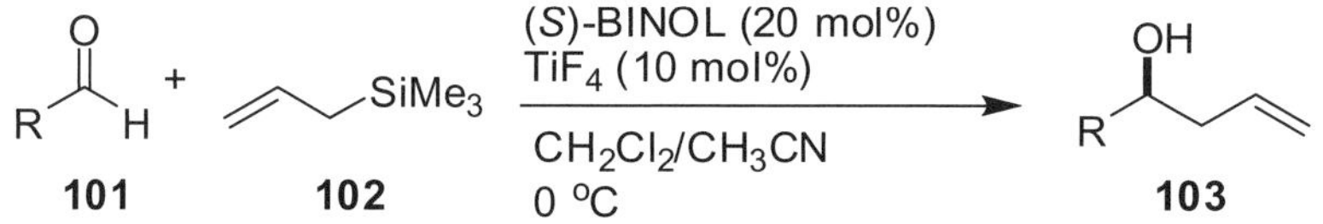

Entry	*R*	*Yield [%]*	*er*	*ee [%]*
1	Me_3C–	91	97:3	94
2	Ph–	85	90:10	80
3	*c*-C_6H_{11}–	72	80:20	60
4	$PhCH_2CH_2$–	69	80:20	60

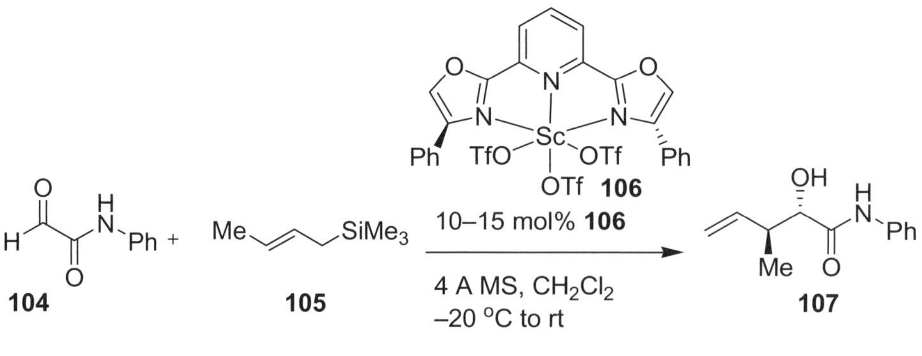

Recently, Evans found the Pybox catalyst **106** to be effective in promoting the reaction of allylsilanes with glyoxamides (91–99% *ee*).[60]

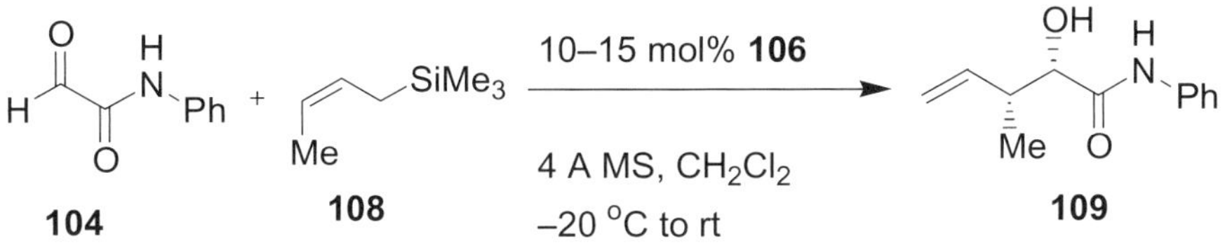

The BINAP/AgF system catalyzes the reaction of allyltrimethoxysilanes (Table 13). ^{1}H NMR studies suggest a fast transmetallation and equilibration to the *E*-crotylsilver species, followed presumably by a six-membered, chair-like transition state. The reaction is limited to aromatic aldehydes.[61]

Table 13: Sakurai reaction promoted by BINAP/AgF

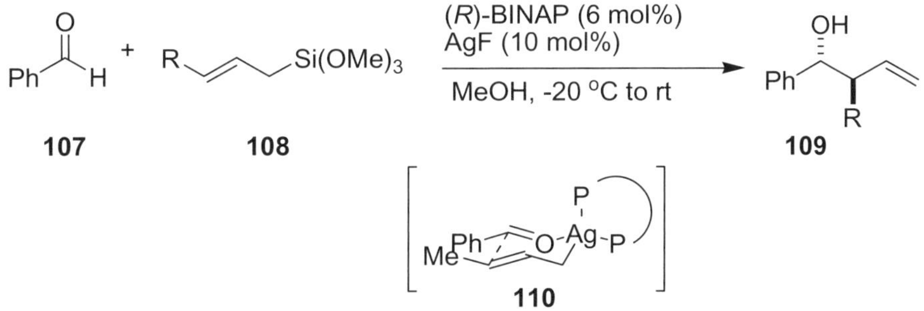

Entry	*R*	*E:Z*	*Yield [%]*	*Anti(er) : syn (er)*
1	H	–	84	*er* = 97:3
2	Me	83:17	77	92 (98:2) : 8 (81:19)
3	Me	<1:99	82	94 (97:3) : 6 (80:20)

Shibasaki employed $F_2Ph_3Si^-$ to activate crotyltrimethoxysilane **111**, together with CuCl to activate acetophenone **112**. The reaction could be made modestly enantioselective (61% *ee*) by adding BINAP.[62] Yamamoto employed instead AgF with the more sophisticated ligand **114**.[63] In both cases the course of the reaction did not significantly depend on the *E/Z* configuration of the starting crotylsilane, and in both cases the reaction may well proceed through an allylmetal (M = Cu, Ag) species, thus hardly qualifying these as Sakurai reactions.

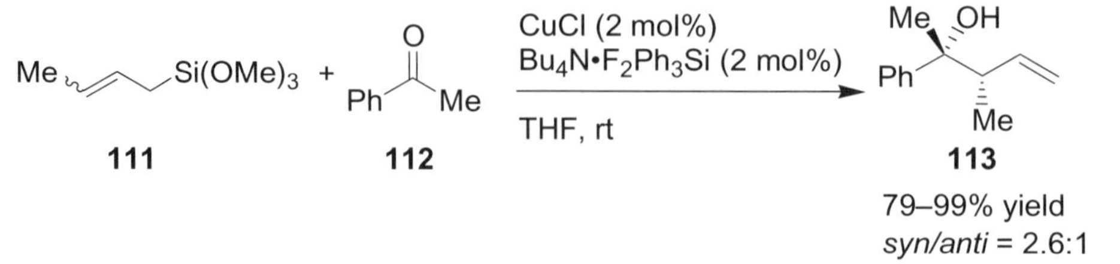

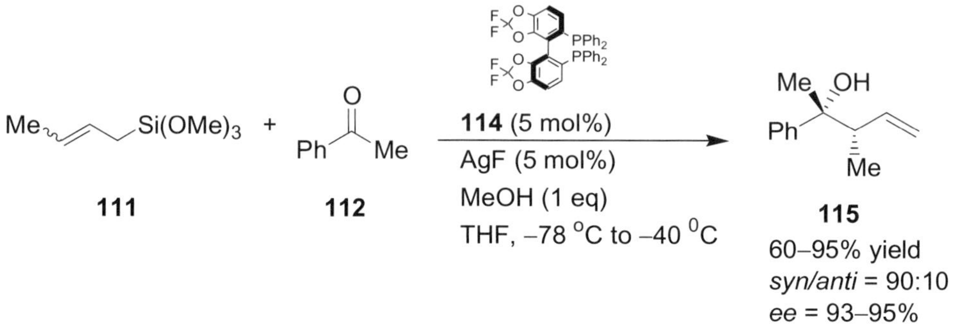

Some Synthons of Interest

3,3-Bis(trimethylsilyl)-1-propene **116** is an interesting allyl synthon because it can be functionalized in both directions.[64] Propene **116** is obtained by hydrogenation of the corresponding propyne over $Pd/BaSO_4$. Reaction of **116** with an aldehyde yields vinyl silane **117** that can be reacted with electrophiles.

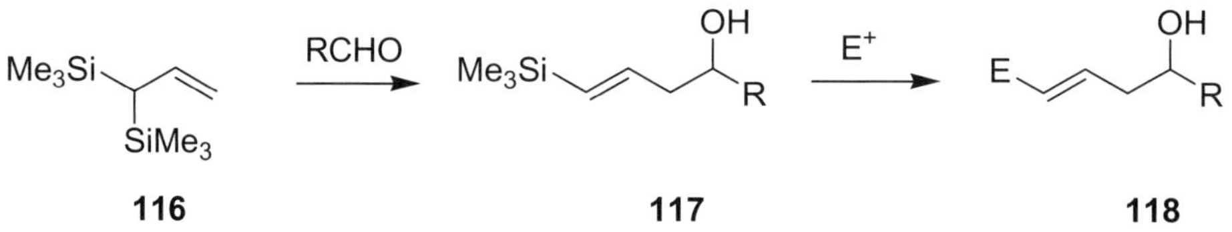

Analogous bis-silanes are readily accessible by a Kumada coupling.[65] For instance, the reaction of bis-silylated Grignard reagent **119** with 2-bromopropene **120** gave 3,3-bis(trimethylsilyl)-2-methyl-1-propene **121**. Like other methallylsilanes, **121** is prone to undergo a Sakurai reaction as well a competitive ene reaction. Using "mildly acidic" conditions (BF_3•OEt_2 in Et_2O) the ene reaction predominates. With "more acidic" conditions ($SnCl_4$ in CH_2Cl_2), the Sakurai reaction becomes the predominant pathway.[65] The authors ascribed the difference in reactivity to the varying acidity of the medium. They did not discuss the possibility that $SnCl_4$, unlike BF_3, may react with **121** to generate *in situ* a reactive allylstannane.

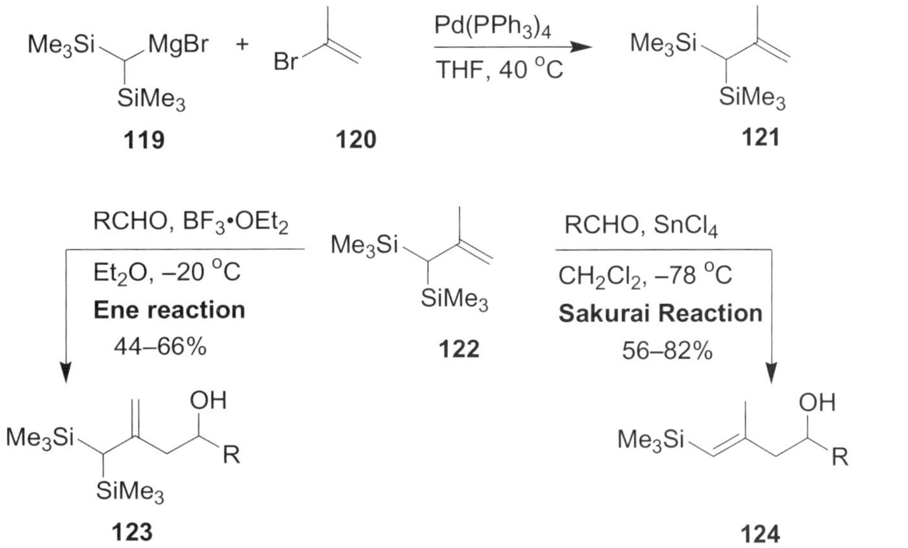

The fact that methallylsilanes are prone to ene reactions is also evident with the readily available silane[66] **125**, which reacts with crotonaldehyde **126** in the presence of 1.5 equivalents of Et_2AlCl to produce the allylic alcohol **127** in 55% yield.[67]

2.11.5 Synthetic Utility

This section attempts to give a sense for how the Sakurai reaction is used in modern total synthesis.

A straightforward application of the Sakurai reaction is found in a study by Crews pertaining to the synthesis of amphidinolide B1.[68,69] Methallylsilane **128** was reacted with α-alkoxyaldehyde **129** in the presence of BF_3•OEt_2 to give the non-chelation-controlled product **130** in 90% yield and > 95 : 5 selectivity. The MOM and PMB protecting groups were compatible with BF_3. The Sakurai reaction was thus used to generate an early synthetic intermediate.

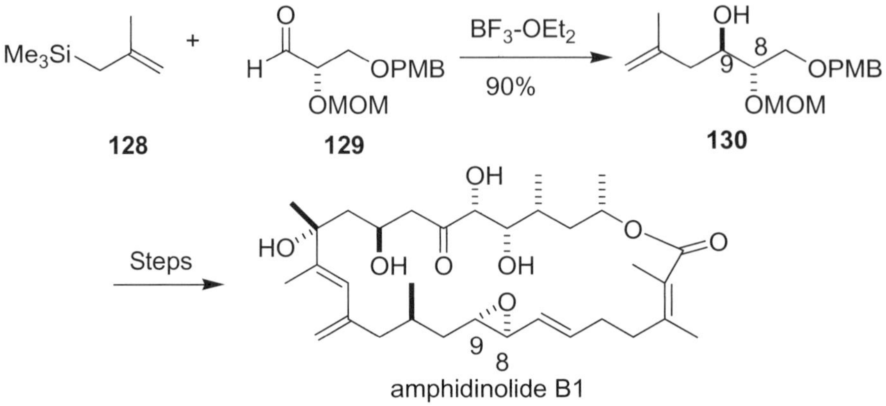

The Sakurai reaction is a good tool to generate C-glycosides. This was illustrated by Crimmins, in the construction of the GHIJ fragment of brevetoxin A. Treatment of ketal **131** with allyltrimethylsilane and TMSOTf provided pyrane **132**. The allylic substituent was introduced in the equatorial position with > 10:1 diastereoselectivity.[70] The olefin was then subjected to a hydroboration, a typical way of elaborating on a Sakurai reaction, to give alcohol **133**, which was converted into GHIJ rings of brevetoxin A.

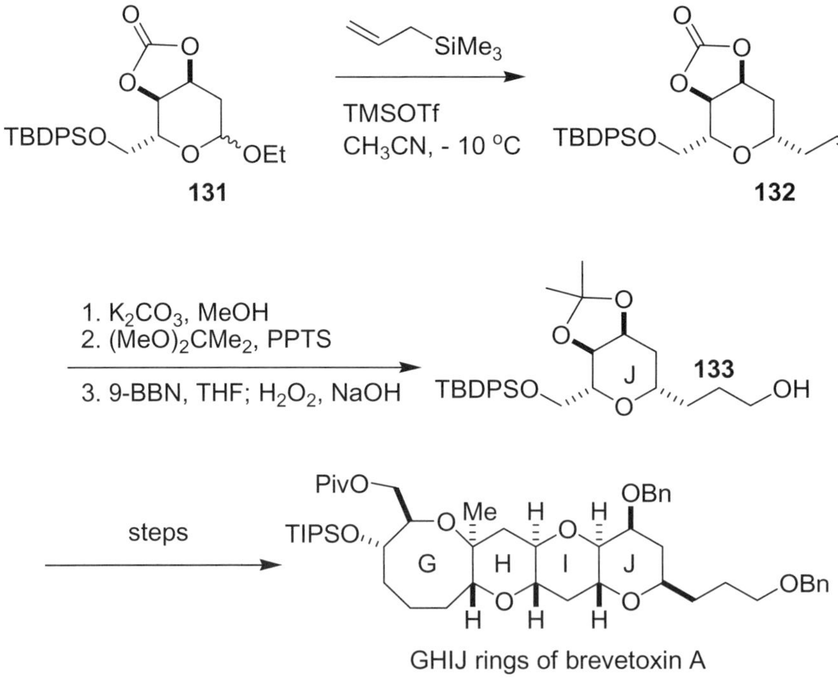

The chemical stability of organosilanes sets them apart from most other organometallic species. This allows the introduction of a silyl group early in the synthetic scheme. The complexity of the silane can be gradually increased, and the Sakurai reaction can be triggered at the appropriate time. This strategy was used by Schaus to construct the decalin core of Clerodane.[71] A 1,2-addition of $PhMe_2SiMgCl$ to tiglic aldehyde (**134**) introduced the silyl group, without the subsequent Si elimination characteristic of a regular Peterson olefination, to yield the racemic allylic alcohol **135**. The racemic **135** was oxidized, and then reduced with Corey's CBS catalyst to give the enantiomerically pure alcohol **136**. A Claisen rearrangement was next used to force the olefin of **136** to migrate, so as to unveil the allyl silane functional group. Thus, treatment of **136** with catalytic $Hg(OAc)_2$ in refluxing ethyl vinyl ether gave a vinyl ether that spontaneously rearranged upon chromatography to give allyl silane **137**. Next, an asymmetric Morita–Baylis–Hilman reaction catalyzed by biphenyl **138** gave cyclohexenone **139**, containing both an enone and an allylsilane moiety. The molecule was thus poised to undergo an intramolecular Sakurai reaction, which was promoted by BF_3•OEt_2 to yield the clerodane core **140**.

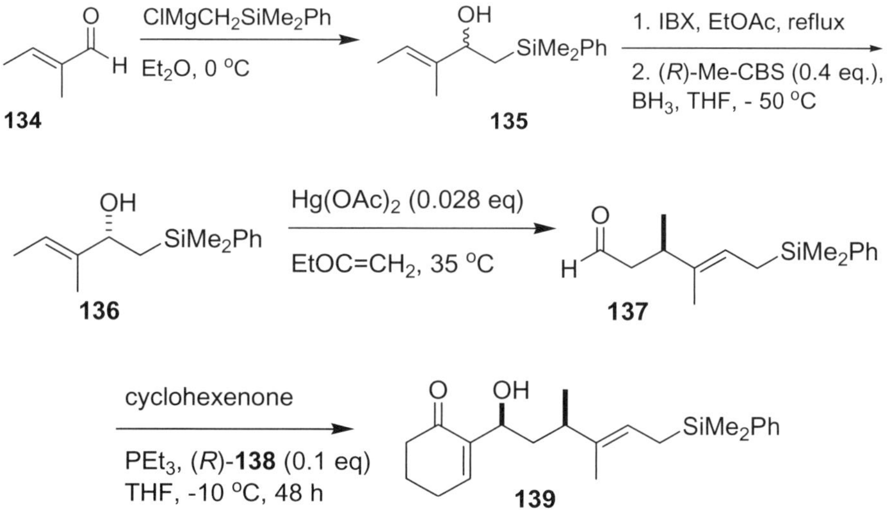

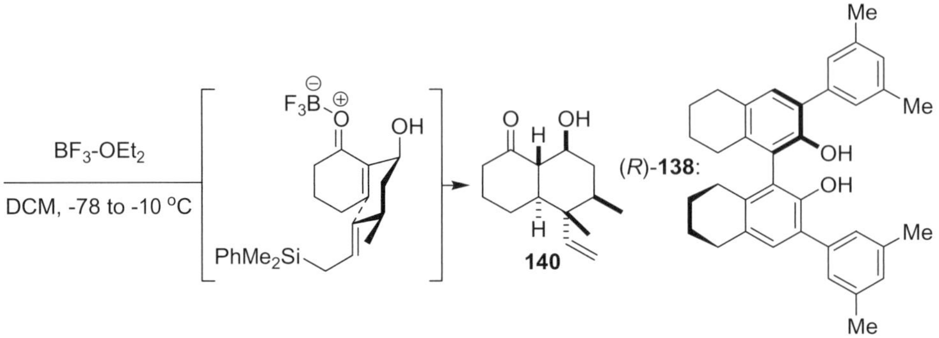

In a sophisticated development of the Sakurai reaction, Markó has developed a three component reaction involving an aldehyde, an alcohol, and an allylsilane.[72] The reaction was used in the synthesis of jerangolid D.[73] Reaction of aldehyde **141** and allyl alcohol **142** provided an hemiacetal which, upon TMSOTf catalysis underwent a Sakurai reaction with allylsilane **143** to give allyl ether **144**. The sequence built a new stereogenic center (under non-chelating conditions). A ring-closing metathesis of **144** then provided the desired pyrane ring **145**.

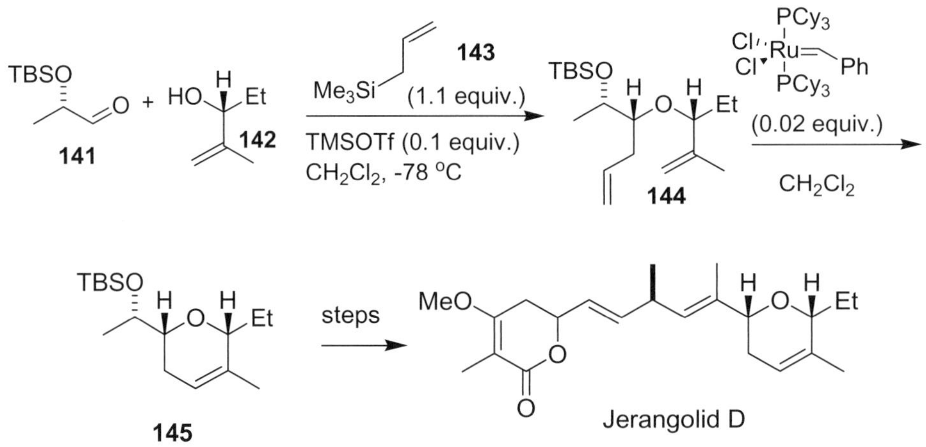

By connecting the alcohol and allylsilane moieties into a single molecule, the above Sakurai reaction becomes a two-component system (instead of three components as in the previous scheme). This enables the formation of the pyrane ring in a single step, as illustrated in studies directed at polycavernoside A.[74] Vinyl iodide **146** was subjected to a Pd-catalyzed Kumada coupling with Me_3SiCH_2MgCl to provide allylsilane **147**. Then **147**, which also carries an OH group, was reacted with aldehyde **148** in the presence of 1 equivalent of $BF_3{\bullet}OEt_2$ to provide the intermediate oxonium

149, which underwent an intramolecular Sakurai reaction at low temperature. Warming the reaction mixture to room temperature cleaved the TBS silyl ether and gave tetrahydropyran **150** in 91% yield. The Sakurai reaction effectively controlled the relative configuration of two new stereogenic centers. Variations on the same theme are found in Panek's syntheses of apicularen A[75] and (+)-leucascandrolide A.[76]

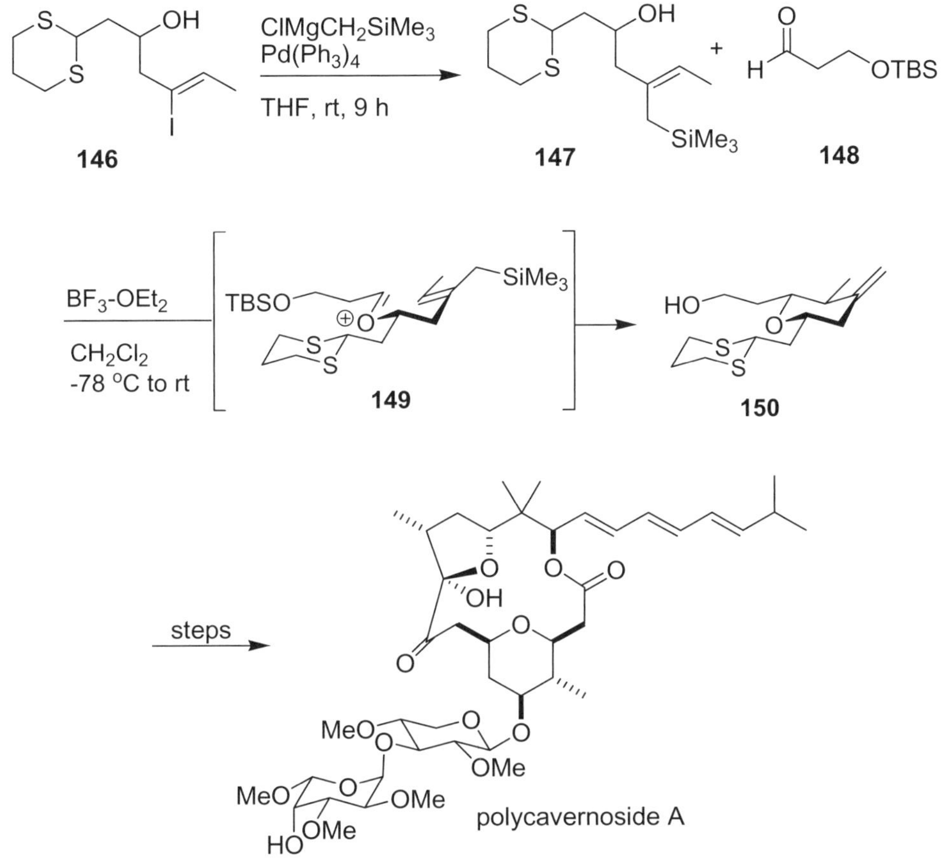

The synthesis of methyl monate C constitutes a similar example, in which the intramolecular Sakurai reaction between ketal **151** and hydroxylated allyl silane **152** gave pyrane **154**, setting up two stereogenic centers, and positioning all the pyrane substituents in equatorial positions.[67]

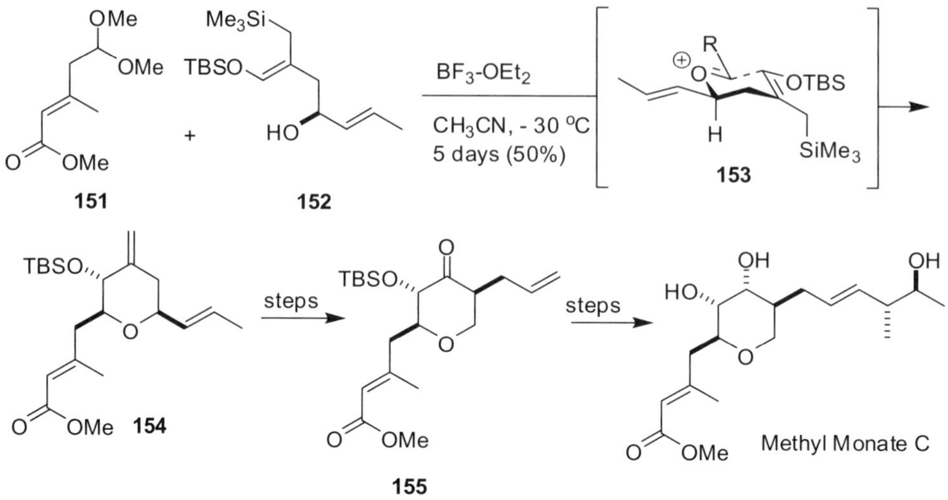

Tetrahyropyranes could also be obtained by performing sequentially Sakurai, Prins, and Ritter reactions in a single pot.[77] The starting 4-acetoxy-1,3-dioxolane **156** contains two acetals that are sufficiently differentiated in their rate of ionization. TMSOTf promoted the Sakurai reaction between ketal **156** and allylsilane **157** at –45 °C in CH_2Cl_2:CH_3CN to give **158**. Addition of TfOH and warming the mixture to –15 °C gave oxonium **159** and triggered the Prins cyclization to give carbocation **160** which was intercepted by CH_3CN in a Ritter reaction to yield 4-acetamidotetrahydropyran **161**. This recent methodology should soon find applications in new total syntheses as it allows one to build four new asymmetric centers in a controlled manner, as in the preparation of **164**.

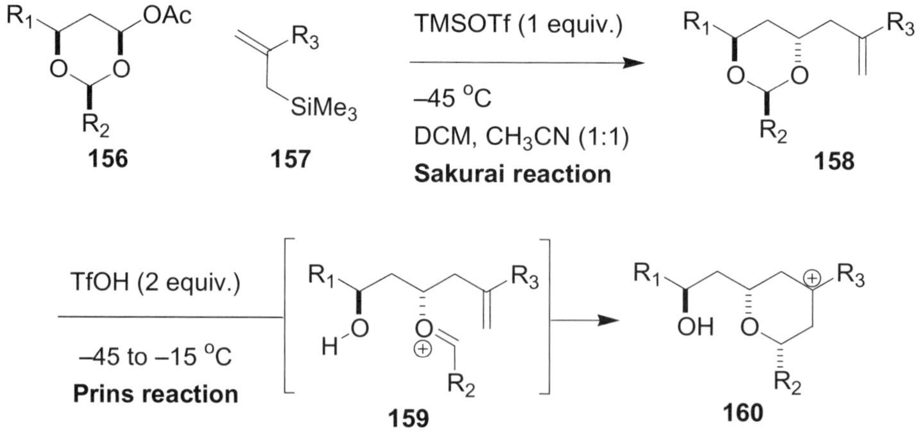

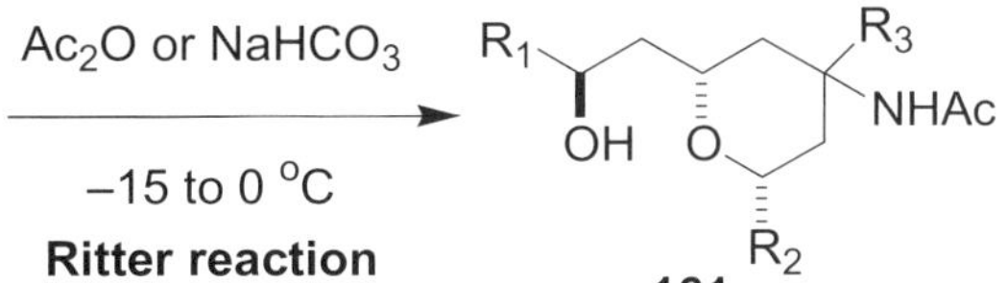

161

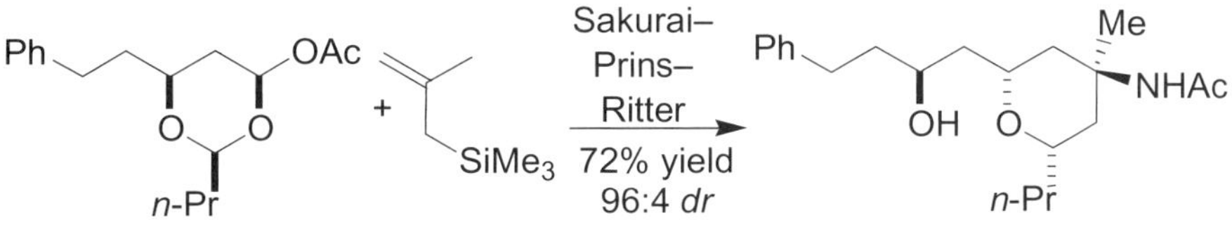

162 **163** **164**

Finally, one of the most noteworthy examples of Sakurai reactions is found in Wender's total synthesis of (–)-laulilamide.[78] The synthesis of the allylsilane group is instructive *per se*. Treatment of methyl ester **165** with $TMSCH_2MgBr$ provided methallylsilane **166** in a simple, single-step reaction. The reaction most likely occurred by a double addition of the Grignard reagent, followed by elimination of Me_3SiOH (Peterson olefination). The most remarkable feature, however, is that the synthetic plan relied on an asymmetric Sakurai reaction as the key reaction to bring together two advanced intermediates. The C_3–C_{14} fragment **166** and the C_{15}–C_{27} fragment **167** were assembled by means of Yamamoto's chiral acyloxyborane **97** to provide alcohol **168**. It should be noted in passing that fragment **167** is an α,β-unsaturated aldehyde, and that the addition of the allyl group proceeded in a 1,2-fashion, rather than in a 1,4-manner as would have been the case with an α,β-unsaturated ketone. In summary, Wender provided the ultimate proof of the robustness of the asymmetric Sakurai reaction by electing it as a key step in the late stage of a total synthesis.

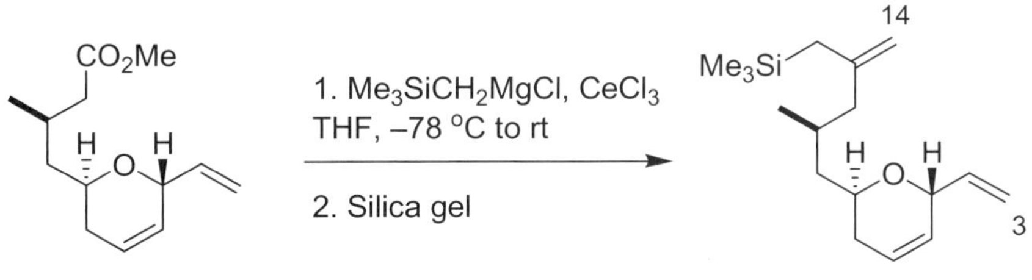

165 **166**

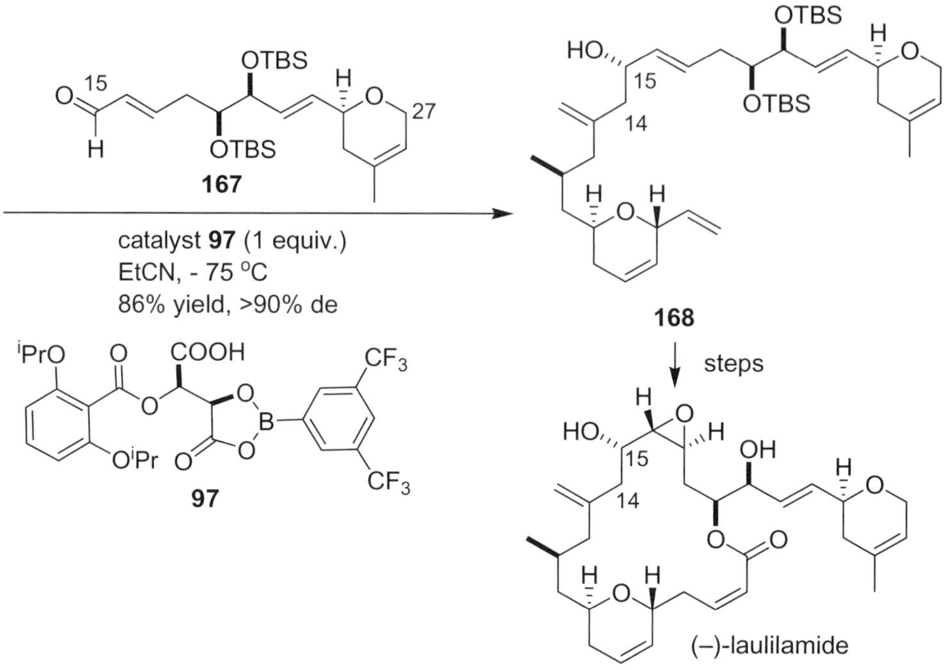

In conclusion, the Sakurai reaction is a powerful method to introduce a nucleophilic allyl synthon. Similar allylations include the Keck reaction of allylstannanes, and the Roush reaction of allylboranes. What sets the Sakurai reaction apart is the exceptional stability of allylsilanes, which are stable to air and water, and can easily be chromatographed, unlike most other allylmetal species. In addition, allylsilanes are easily accessible by a variety of reactions. These factors have allowed the Sakurai reaction to find widespread use, from basic methodology to total synthesis.

2.11.6 Experimental

We report here the preparation of *E*-crotylsilane. The same procedure can be applied to the preparation of *Z*-crotylsilane. Next, three typical Sakurai reaction conditions are described, promoted by $TiCl_4$, BF_3, and $SnCl_4$ respectively. An example illustrating the limitations of $TiCl_4$ towards acid-labile protecting groups is provided. Finally, two recent protocols using milder Lewis Acids ($ZnCl_2$, $FeCl_3$) are given.

E-Crotyltrimethylsilane[79]

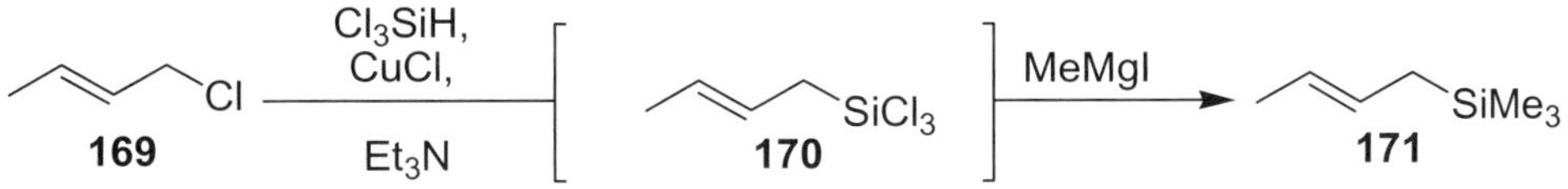

Commercial (*E*)-crotyl chloride (2.98 g, 33 mmol) was dissolved in dry ether (10 mL) under nitrogen and magnetic stirring. After cooling to 0 °C, trichlorosilane (4.47 g) was added followed by triethylamine (3.33 g, 33 mmol) and copper(I) chloride (0.98 g, 10 mmol). The mixture was stirred at room temperature for 12 h. After addition of dry ether (50 mL) the white precipitate formed was filtered under nitrogen and the ethereal solution, containing trichlorocrotylsilane, was slowly added at 0 °C to a solution of MeMgI (120 mL of a solution containing approximately 140 mmol of the Grignard reagent) under mechanical stirring. The mixture was refluxed for 8 h and, after cooling to 0 °C, quenched with a pre-cooled solution of NH_4Cl. The ethereal layer was rapidly separated and dried over anhydrous Na_2SO_4. The ether was evaporated using a Vigreux column at atmospheric pressure (avoid the use of any vacuum apparatus) and the residual product distilled to give *E*-crotyltrimethylsilane **171** (2.75 g, 65% yield).

A somewhat simplified version of this procedure was extended to the preparation of *Z*-crotyltrimethylsilane, as well as to a variety of γ-substituted allyl silanes.[60] For additional preparations of allyl silanes, see references 80–82.

An example of $TiCl_4$-promoted Sakurai reaction[83]

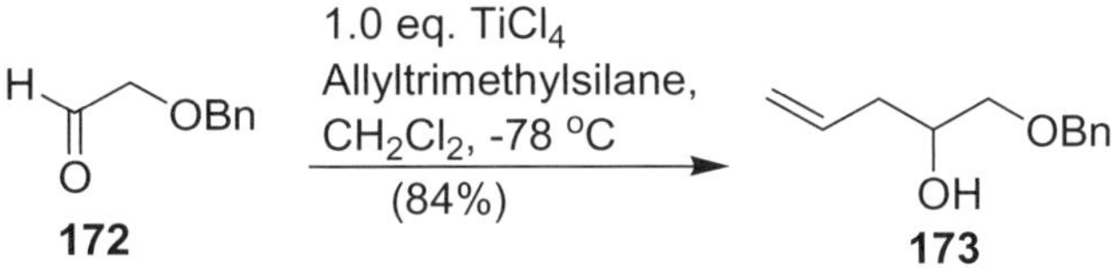

To a stirred solution of benzyloxyacetaldehyde (**172**, 2.5 g, 16.6 mmol) in dichloromethane (200 mL) was added titanium tetrachloride (1.8 mL, 16.6 mmol) at –78 °C. The resulting mixture was stirred for 10 min, allyltrimethylsilane was added, and the mixture was stirred at –78 °C for 30 min. The reaction was then quenched with H_2O (50 mL). The layers were separated, and the aqueous layer was extracted with diethyl ether (2 × 100 mL). The combined organic layers were successively washed with saturated aqueous $NaHCO_3$ solution and brine and dried over anhydrous $MgSO_4$. Evaporation of the solvent under reduced pressure provided a residue that was purified by column chromatography on silica gel (10% EtOAc in hexanes as the eluent) to afford alcohol **173** (2.7 g, 84%).

An example of BF_3-promoted Sakurai reaction[84]

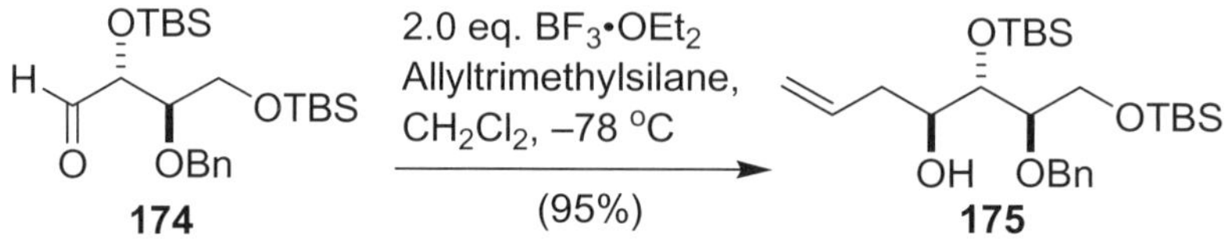

To a solution of **174** (3.5 g, 7.19 mmol) in CH_2Cl_2 (30 mL) were added allyltrimethylsilane (2.29 mL, 14.4 mmol) at –78 °C followed by BF_3•Et_2O (1.81 mL, 14.4 mmol) dropwise at –78 °C. The solution was stirred at –78 °C for 5 h and quenched with sat. aq. NH_4Cl (20 mL) at –78 °C. The mixture was warmed up to room temperature and the organic phase was separated. The aqueous fraction was extracted with CH_2Cl_2 (3 × 20 mL) and the organic phases were combined, dried over Na_2SO_4, and concentrated under reduced pressure. The crude residue was purified by flash column chromatography on silica gel [Gradient eluent: 2% to 10% EtOAc in hexanes] to provide the alcohol **175** (3.28 g, 95%) as a colorless oil.

An example of $SnCl_4$-promoted Sakurai reaction[44]

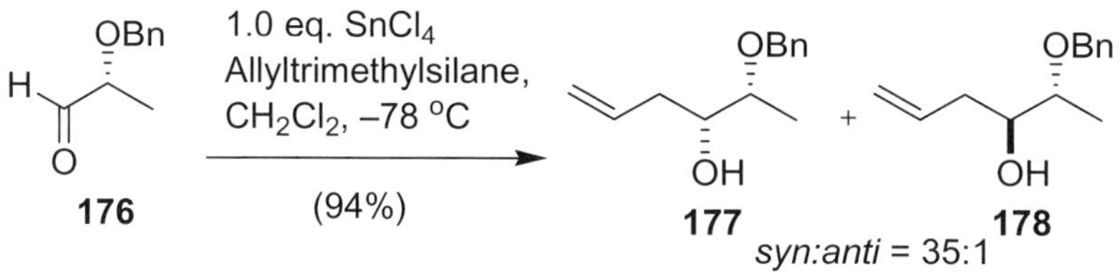

A solution of 0.117 mL (260 mg, 1 mmol) of stannic chloride in 4 mL of dry CH_2Cl_2 was cooled to –78 °C. To the solution was added dropwise a solution of 164 mg (1 mmol) of 2-(phenylmethoxy)propanal in 0.5 mL of dry CH_2Cl_2 with a syringe over a 2-min period. The solution was stirred for 3 min and 0.175 mL (125 mg, 1.1 mmol) of allytrimethylsilane was added in one portion. After stirring at –78 °C for 15 min, the clear reaction mixture was quenched with water. The reaction mixture was allowed to warm to room temperature and was extracted with ether. The ethereal layer was dried over Na_2SO_4. After evaporation of the solvent, the residue was purified by flash chromatography (9 g silica gel, 1:15 ethyl acetate/hexanes) to afford 193 mg

(94%) of a clear, colorless liquid, which was a 35:1 mixture of the diastereoisomers **177** and **178**, respectively.

Limitations of $TiCl_4$ as a promoter towards acid-sensitive protecting groups[85]

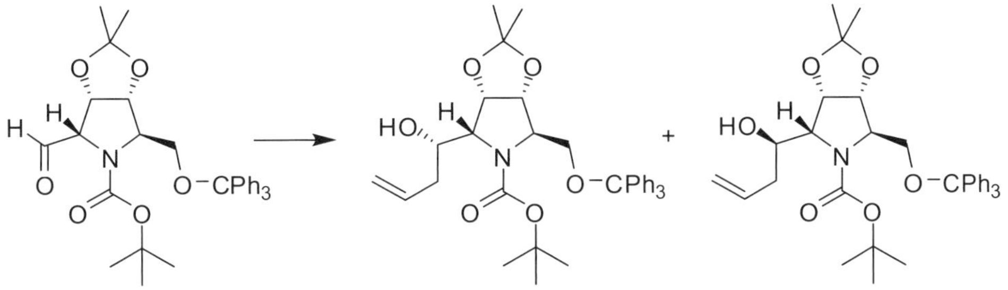

179 **180** **181**

The reaction of **179** with allyltrimethylsilane and $TiCl_4$ illustrates the limit of the Sakurai reaction in the presence of acid-sensitive protecting groups (Boc, trityl, acetonide). Ikota obtained a remarkable 21% yield (it could have been worse) by quenching the reaction at –78 °C with NaOH. The Sakurai reaction was not an optimal choice in this case. The reaction proceeded better using allylmagnesium chloride (**180**: 60%, **181**: 24%) or allyllithium (**180**: 68%, **181**: 13%) in THF at –78 °C. The interest of the experimental procedure is the work-up that minimizes the unwanted deprotections.

A solution of $TiCl_4$ (160 mg, 0.843 mmol) in CH_2Cl_2 (1 mL) was added to a solution of crude allyltrimethylsilane (115 mg, 1.0 mmol) and crude aldehyde **179** (obtained by Swern oxidation, 230 mg, 0.42 mmol) in CH_2Cl_2 (2.5 mL) at –78 °C over a period of 5 min. After being stirred for 10 min at –78 °C, the mixture was basified with 10% aqueous NaOH and extracted with AcOEt. The AcOEt extracts were washed with half-saturated aqueous NaCl. Drying followed by evaporation and purification of the residue by column chromatography (AcOEt:hexane 1:4) gave **180** (52 mg, 21%)

$ZnCl_2$ as promoter

The natural product artemisinin is a widely used antimalarial. Here, a mild Lewis acid ($ZnCl_2$) is used to promote a Sakurai reaction with a ketone.[86]

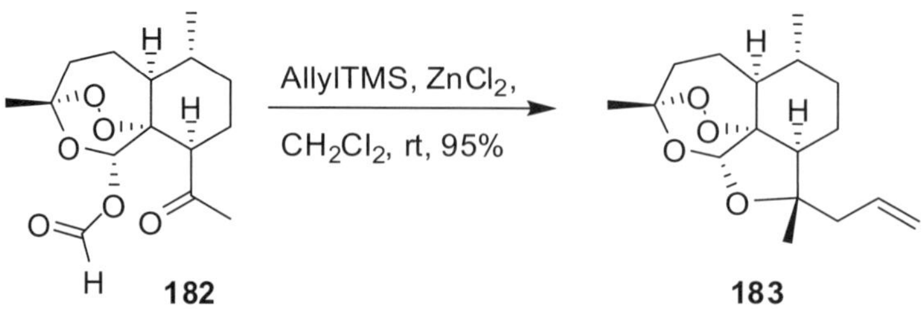

182 183

To a solution of allyltrimethylsilane (0.20 mmol) in dried CH_2Cl_2 (10 mL) was added anhydrous $ZnCl_2$ (16 mg, 0.12 mmol) and 4 Å molecular sieves under argon. To the allyltrimethylsilane solution was added a solution of **182** (30 mg, 0.10 mmol) in CH_2Cl_2 (2 mL) at 0 °C. After stirring overnight at room temperature, the reaction mixture was diluted with diethyl ether (20 mL) and quenched with a saturated NH_4Cl solution. The organic layer was collected, and the aqueous layer was extracted with diethyl ether (3 × 20 mL). The combined organic extracts were washed with brine, dried over anhydrous $MgSO_4$, and concentrated under reduced pressure. The residue was purified by flash column chromatography to give **183** (28 mg, 95%).

$FeCl_3$ in $MeNO_2$ as promoter[87]

The following procedure uses inexpensive $FeCl_3$ and $MeNO_2$. Nitromethane is an unusually polar solvent for Sakurai reactions. No reaction occurred in MeCN or THF. The protocol was also effective for non-aromatic aldehydes such as **187** (rt, 1 h, 92%) or the easily enolizable **188** (rt, 1 h, 99%). Electron-withdrawing groups are well tolerated, such as in p-nitrobenzaldehyde **189** (–20 °C, 0.5 h, 99%) but electron-donating substituents such as in *p*-anisaldehyde **190** promote over-alkylation to give solely **191**, and none of the desired adduct (–20 °C, 0.5 h, 0%).

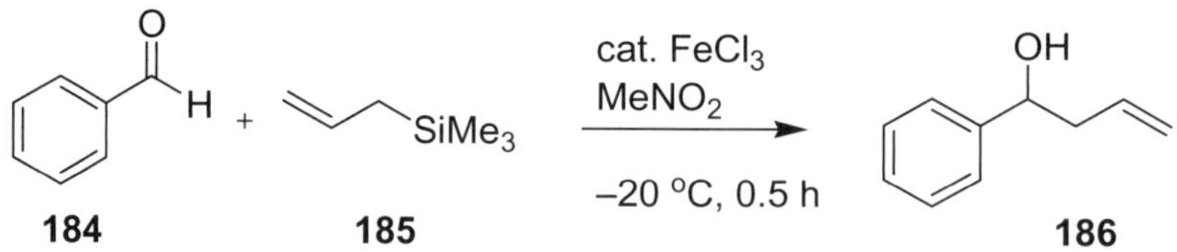

184 185 186

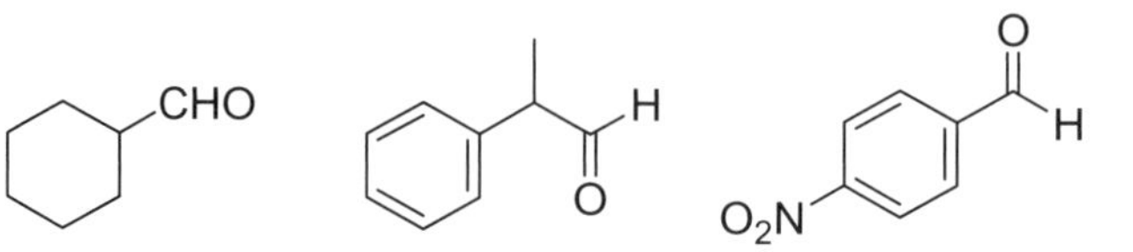

187 188 189

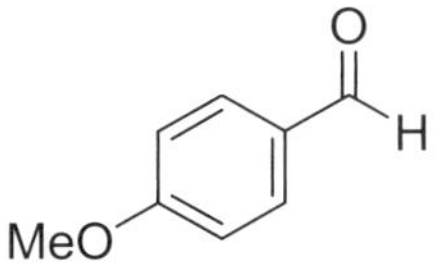

190

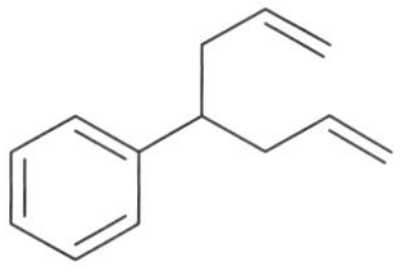

191

To a solution of $FeCl_3$ (5.4 mg, 0.033 mmol) in $MeNO_2$ (1.0 mL) were added allyltrimethylsilane (159 μL, 1.00 mmol) and a solution of benzaldehyde (76.3 mg, 0.66 mmol) in $MeNO_2$ (1.0 mL) at –20 °C under argon atmosphere. After stirring the reaction mixture for 0.5 h, 1 N HCl (2.0 mL) was added and the reaction was allowed to warm to room temperature. The resultant mixture was stirred for an additional 0.5 h and poured into saturated sodium hydrogenocarbonate. The organic materials were extracted with Et_2O and dried over anhydrous magnesium sulfate. The solvent was evaporated and 1-phenyl-3-buten-1-ol (90.0 mg) was isolated by thin-layer chromatography on silica gel (ether : hexane = 1 : 1).

2.11.7 References

1. [R] For hypervalent silicon as a reactive site in selective bond-forming processes, see: Rendler, S.; Oestreich, M. *Synthesis* **2005**, *11*, 1727.
2. [R] For Chiral Lewis Bases as Catalysts, see: Kočovský, P.; Malkov, A. I. in *Enantioselective Catalysis – Reactions and Experimental Procedures*, Dalko, P. I. Ed., Wiley, **2007**.
3. [R] Yamamoto, Y., Sasaki, N., in *Chemical Bonds – Better Ways to Make Them and Break Them*, Bernal, I. Ed., Elsevier, **1989**.
4. [R] Schinzer, D. *Synthesis* **1988**, 263.
5. [R] Yamamoto, Y.; Asao, N. *Chem Rev.* **1993**, *93*, 2207.
6. [R] Denmark, S. E.; Fu, J. *Chem. Rev.* **2003**, *103*, 2763.
7. [R] Modern Carbonyl Chemistry, Otera, J. Ed. Wiley-VCH, Weinheim, **2000**.
8. [R] Denmark, S. E.; Almstead, N. G. *Tetrahedron* **1992**, *48*, 5565.
9. [R] (a) Fleming, I. *Chem Soc. Rev.* **1981**, *10*, 83. [R] (b) Chan, T. H.; Fleming, I. *Synthesis* **1979**, 761. (c) Hosomi, A.; Iguchi, H.; Sakurai, H. *Chem. Lett.* **1982**, 223.
10. Hosomi, A.; Sakurai, H. *Tetrahedron Lett.* **1976**, *17*, 1295.
11. Hosomi, A.; Sakurai, H. *Chem. Lett.* **1976**, 941.
12. Hosomi, A.; Sakurai, H. *J. Am. Chem. Soc.* **1977**, *99*, 1673.
13. Hoffmann, R. W. *Angew. Chem., Int. Ed. Engl.* **1982**, *21*, 555. Yamamoto, Y.; Maruyama, K. *Heterocycles*, **1982**, *18*, 357. Also, crotyldiallylboron rearrange rapidly at room temperature, while crotylboronates (crotyl-B$(OR)_2$) derivates are sufficiently thermally stable.
14. Pillot, J.-P.; Dunogues, J.; Calas, R. *Tetrahedron Lett.* **1976**, *17*, 1871.
15. Mayr, H.; Pock, R.; *Tetrahedron* **1986**, *42*, 4211.
16. The 8-membered ring was suggested to arise from an electrostatic interaction between the partially charged fluorine and silicon atoms (–0.40 and +0.36, respectively). Bottoni, A.; Costa, A. L.; Di Tommaso, D.; Rossi, I.; Tagliavini, E. *J. Am. Chem. Soc.* **1997**, *119*, 12131.
17. Gajewski, J. J.; Ngernmeesri, P. *Org. Lett.* **2000**, *2*, 2813.
18. The equilibrium constant K = [**18**][**19**]/[**16**][**17**] = exp(– ΔG/RT), where R is the gas constant (0.00199 kcal K^{-1} mol^{-1}) and T the absolute temperature. To calibrate the ideas, for the equilibrium to shift 10x to the right, i.e., 10 × more of each **18** and **19**, K needs to shift by 100x, and ΔG by 1.78 kcal/mol (at –78 °C = 195 °K).
19. Maria, P.-C.; Gal, J.-F. *J. Phys. Chem.* **1985**, *89*, 1296.

20. Bond enthalpies alone do not allow calculation of equilibrium constants. The entropic term TΔS also needs to be known to calculate ΔG = ΔH – TΔS and the corresponding equilibrium constant. However, for the particular case of equilibria with borontrifluoride etherate, the term TΔS appears to be negligible, and ΔG ≈ ΔH. For benzaldehyde competing with diethyl ether, the difference in binding energies is calculated to be ΔH = 0.9 ± 0.3 kcal/mol (Table 2, entry 5 minus entry 9). This is within experimental error of the ΔG = 0.93 kcal/mol reported in the previous table (Table 1, entry 1).
21. Childs, R. F.; Mulholland, D. L.; Nixon, A. *Can. J. Chem.* **1982**, *60*, 809.
22. Reetz, M. T.; Hüllmann, M.; Massa, W.; Berger, S.; Rademacher, P.; Heymanns, P. *J. Am. Chem. Soc.* **1986**, *108*, 2405.
23. Denmark, S. E; Henke, B. R. Weber, E. J. *J. Am. Chem. Soc.* **1987**, *109*, 2512.
24. Corey, E. J.; Lee, T.W. *Chem. Commun.* **2001**, 1321.
25. Corey, E. J.; Rohde, J. J.; Fisher, A.; Azimoioara, M. D. *Tetrahedron Lett.* **1997**, *38*, 33.
26. Corey, E. J.; Rohde, J. J. *Tetrahedron Lett.* **1997**, *38*, 37.
27. Corey, E. J.; Barnes-Seeman, D.; Lee, T. W. *Tetrahedron Lett.* **1997**, *38*, 1699.
28. Laszlo, P.; Teston, M. *J. Am. Chem. Soc.* **1990**, *112*, 8750.
29. It has not been experimentally demonstrated that the LUMO energy of the complex correlates with the rate of a Sakurai reaction. However, the scale correlated very well with the ability of the Lewis acid to catalyze the ene reaction between β-pinene and methyl acrylate. Laszlo, P.; Teston-Henry, M. *Tetrahedron Lett.* **1991**, *32*, 3837.
30. Childs, R. F.; Mulholland, D. L.; Nixon, A. *Can. J. Chem.* **1982**, *60*, 801.
31. Hagen, G.; Mayr, H. *J. Am. Chem. Soc.* **1991**, *113*, 4954. In the presence of Lewis Acids such as BCl_3.
32. Hayashi, T.; Iwamura, H.; Uozomi, Y. *Tetrahedron Lett.* **1994**, *35*, 4813.
33. Greese, T. A.; Hutchinson. K. D.; Overman, L. E. *J. Org. Chem.* **1993**, *58*, 2468.
34. Naruta, Y.; Uno, H.; Maruyama, K. *Chem. Lett.* **1982**, 961.
35. (a) Lambert, W. T.; Burke, S. D. *Org. Lett.* **2003**, *5*, 515. (b) Lambert, W. T.; Hanson, G. H.; Benayoud, F.; Burke, S. D. *J. Org. Chem.* **2005**, *70*, 9382.
36. (a) Hayashi, T.; Konishi, M.; Kumada, M. *J. Org. Chem.* **1983**, *48*, 281. (b) Hayashi, T.; Konishi, M.; Okamoto, Y.; Kabeta, K., Kumada, M. *J. Org. Chem.* **1986**, *51*, 3772.
37. Hosomi, A.; Ando, M.; Sakurai, H. *Chem. Lett.* **1984**, 1385.
38. Tietze, L. F.; Volkel, L.; Wulff, C.; Weigand, B.; Bittner, C.; McGrath, P.; Johnson, K.; Schäfer, M. *Chem. Eur. J.* **2001**, *7*, 733.
39. Kamachi, T.; Kuno, A.; Matsuno, C.; Okamoto, S. *Tetrahedron Lett.* **2004**, *45*, 4677.
40. Hayashi, T.; Kabeta, K.; Hamachi, I.; Kumada, M. *Tetrahedron Lett.* **1983**, *24*, 2865.
41. (a) Sakurai, H. *Pure Appl. Chem.* **1982**, *54*, 1. (b) Sakurai, H. *Pure Appl. Chem.* **1985**, *57*, 1759.
42. Tietze, L. F.; Kinzel, T.; Schmatz, S. *J. Am. Chem. Soc.* **2006**, *128*, 11483.
43. For a summary of Cram's, Felkin's and Anh's work, see Lodge E. P.; Heathcock, C. H. *J. Am. Chem. Soc.* **987**, *109*, 3353. See also Wong, S. S.; Paddon-Row, M. N. *J. Chem. Soc., Chem. Commun.* **1990**, 456.
44. Heathcock, C. H.; Kiyooka, S.; Blumenkopf, T. A. *J. Org. Chem.* **1984**, *49*, 4214.
45. Yamamoto, Y.; Komatsu, T.; Maruyama, K. *J. Organomet. Chem.* **1985**, *285*, 31.
46. Reetz, M. T. *Angew. Chem., Int. Ed. Engl.* **1984**, *23*, 556.
47. Kiyooka, S.; Heathcock, C. H. *Tetrahedron Lett.* **1983**, *24*, 4765.
48. Reetz, M. T.; Kesseler, K.; Schmidtberger, S., Wenderoth, B.; Steinbach, R. *Angew. Chem., Int. Ed. Engl.* **1983**, *22*, 989.
49. Heathcock, C. H.; Flippin, L. A. *J. Am. Chem. Soc.* **1983**, *105*, 1667.
50. Reetz, M. T.; Jung, A. *J. Am. Chem. Soc.* **1983**, *105*, 4833.
51. Williams, D. R.; Klingler, F. D. *Tetrahedron Lett.* **1987**, *28*, 869.
52. [R] Masse, C. E.; Panek, J. S. *Chem. Rev.* **1995**, *95*, 1293.
53. Hathaway, S. J.; Paquette, L. A. *J. Org. Chem.* **1983**, *48*, 3351.
54. Coppi, L.; Mordini, A.; Taddei, M. *Tetrahedron Lett.* **1987**, 28*,* 969.
55. Furuta, K.; Mouri, M.; Yamamoto, H. *Synlett* **1991**, 561.
56. Ishihara, K.; Mouri, M.; Gao, Q.; Maruyama, T.; Furuta, K.; Yamamoto, H. *J. Am. Chem. Soc.* **1993**, *115*, 11490.
57. Aoki, S.; Mikami, K.; Terada, M.; Nakai, T. *Tetrahedron* **1993**, *49*, 1783.

58. Gauthier, D. R.; Carreira, E. M. *Angew. Chem., Int. Ed. Engl.* **1996**, *35*, 2363.
59. Bohde, J. W.; Gauthier, D. R.; Carreira, E. M. *Chem. Commun.* **2001**, 2560.
60. Evans, D. A.; Aye, Y.; Wu, J. *Org. Lett.* **2006**, *8*, 2071.
61. Yanagisawa, A.; Kageyama, H.; Nakatsuta, Y.; Asakawa, K.; Matsumoto, Y.; Yamamoto, H. *Angew. Chem., Int. Ed. Engl.* **1999**, *38*, 3701.
62. Yamasaki, S.; Fujii, K.; Wada, R.; Kanai, M.; Shibasaki, M. *J. Am. Chem. Soc.* **2002**, *124*, 6536
63. Wadamoto, M.; Yamamoto, H. *J. Am. Chem. Soc.* **2005**, *127*, 14556.
64. Princet, B.; Anselme, G.; Pornet, *J. Synth. Commun.* **1999**, *29*, 3326.
65. Williams, D. R.; Morales-Ramos, A. I.; Williams, C. M. *Org. Lett.* **2006**, *8*, 4396.
66. Trost, B. M.; Chan, D. M. T.; Naninga, T. N. *Org. Synth.* **1984**, *62*, 58.
67. Van Innis, L.; Plancher, J. M.; Markó, I. E. *Org. Lett.* **2006**, *8*, 6111.
68. Mandal, A. K.; Schneekloth, J. S.; Kuramochi, K.; Crews, C. M. *Org. Lett.* **2006**, *8*, 427.
69. Reetz, M. T.; Kesseler, K. *J. Org. Chem.* **1985**, *50*, 5434.
70. Crimmins, M. T.; Zuccarello, J. L.; Cleary, P. A.; Parrish, J. D. *Org. Lett.* **2006**, *8*, 159.
71. Rodgen, S. A; Schoss, S. E. *Angew. Chem., Int. Ed. Engl.* **2006**, *45*, 4929.
72. Pospíšil, J.; Kumamoto, T.; Markó, I. E. *Angew. Chem., Int. Ed. Engl.* **2006**, *45*, 3357.
73. Pospíšil, J.; Markó, I. E. *J. Am.Chem. Soc.* **2007**, *129*, 3516.
74. Pérez-Balado, C.; Markó, I. E. *Tetrahedron* **2006**, *62*, 2331.
75. Su, Q.; Panek, J. S. *J. Am. Chem. Soc.* **2004**, *126*, 2425.
76. (a) Su, Q.; Panek, J. S. *Angew. Chem., Int. Ed. Engl.* **2005**, *44*, 1223 (b) Su, Q.; Dakin, L. A.; Panek, J. S. *J. Org. Chem.* **2007**, *72*, 2.
77. Epstein, O. L.; Rovis, T. *J. Am. Chem. Soc.* **2006**, *128*, 16480.
78. Wender, P. A.; Hegde, S. G.; Hubbard, R. D.; Zhang, L. *J. Am. Chem. Soc.* **2002**, *124*, 4956.
79. D'Aniello, F.; Falorni, M.; Mann, A.; Taddei, M. *Tetrahedron: Asym.* **1996**, *7*, 1217.
80. Kamachi, T.; Kuo, A.; Matsuno, C.; Okamoto, S. *Tetrahedron* **2004**, *45*, 4677.
81. Thibaudeau, S.; Gouverneur, V. *Org. Lett.* **2003**, *5*, 4891 and references therein.
82. Soderquist, J. A., Santiago, B. *Tetrahedron Lett.* **1989**, *30*, 5696.
83. Ghosh, A. K.; Lei, H. *J. Org. Chem.* **2000**, *65*, 4779.
84. Liu, J.; Hsung, R. P. *Org. Lett.* **2005**, *7*, 2273.
85. Ikota, N.; Nakagawa, H.; Ohno, S.; Noguchi, K.; Okuyama, K. *Tetrahedron* **1998**, *54*, 8985.
86. Liu, Y.; Wong, V. K.-W.; Ko, B. C.-B.; Wong, M.-K.; Che, C.-M. *Org. Lett.* **2005**, *7*, 1561.
87. Watahaki, T.; Oriyama, T. *Tetrahedron Lett.* **2002**, *43*, 8959.

2.12 Stetter Reaction

Victor J. Cee

2.12.1 *Description*

A highly efficient olefin hydroacylation known as the Stetter reaction resulted from the discovery that an activated olefin could intercept the putative acylanion intermediate of the classical benzoin reaction.[1] Metal cyanides and heterocyclic carbenes are commonly employed catalysts for the Stetter reaction. Chiral heterocyclic carbenes as well as chiral metallophosphites have been developed as catalysts to provide 1,4-dicarbonyl compounds with high levels of enantiomeric purity.[2]

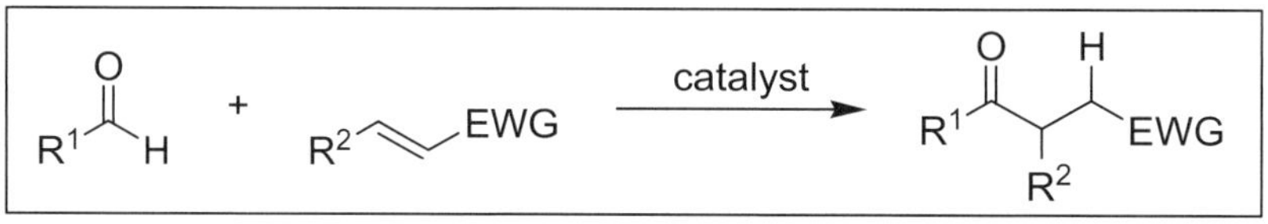

2.12.2 *Historical Perspective*

In a series of publications beginning in 1973, Hermann Stetter and co-workers reported that activated olefins could intercept the putative acylanion intermediate of the benzoin reaction.[3] Typical catalysts for the benzoin reaction, sodium cyanide and thiazolylidine carbenes, were found to perform well in this new reaction. Stetter also established that the success of the reaction is due to the reversible nature of the benzoin condensation relative to the irreversible formation of 1,4-dicarbonyl products.[1] As a consequence, benzoins or aldehydes can be used interchangeably as reactants. The reaction has proven to be a highly efficient method for the synthesis of 1,4-dicarbonyl compounds and 4-oxonitriles. A resurgence of interest in acyl anion chemistry has resulted in many new discoveries, including alternative acyl donors, as well as catalysts capable of highly enantioselective intra- and intermolecular Stetter reactions.[2]

2.12.3 *Mechanism*

The widely accepted Breslow mechanism[4] for carbene-catalyzed acylanion chemistry is illustrated for the Stetter reaction. The thiazolium precatlyst is deprotonated by base to give the active thiazolylidine carbene, and the reaction is initiated by nucleophilic addition of the catalyst to aldehyde **1**, giving intermediate **4**. The thiazolium substituent is sufficiently electron-withdrawing to render what was the formyl hydrogen acidic, and proton transfer leads to the enamine-enol species **5**. This intermediate exhibits inverted reactivity at the acyl carbon, and reacts with Michael acceptor **2** to give **6**. Proton transfer and expulsion of catalyst provides the 1,4-dicarbonyl product **3**. Intermediate **5** may in principle react with a molecule of aldehyde **1** to give the benzoin product **9**, but the reversibility of this process, compared with the irreversible formation of **3**, ensures that the 1,4-dicarbonyl product can generally be produced in high yield and purity. Catalysis by metal cyanide is believed to follow essentially the same course of the illustrated carbene-catalyzed reaction.[5]

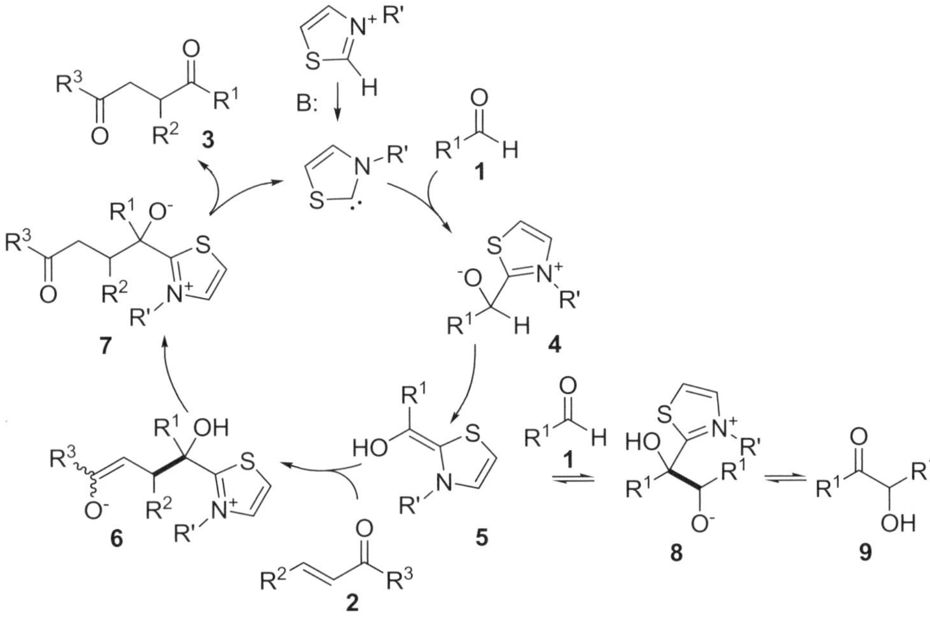

2.12.4 *Synthetic Utility*

Intermolecular Stetter Reaction

Initial work in the early 1970s by Stetter and co-workers established the scope of the intermolecular reaction.[1] Aromatic and heteroaromatic aldehydes are smoothly coupled to α,β-unsaturated ketones, esters and nitriles under sodium cyanide or thiazolylidine catalysis (entries 1–5, **3a–d**). In contrast, the coupling of aliphatic aldehydes and activated olefins (entry 6, **3e**) is successful only under thiazolylidine catalysis.[3b]

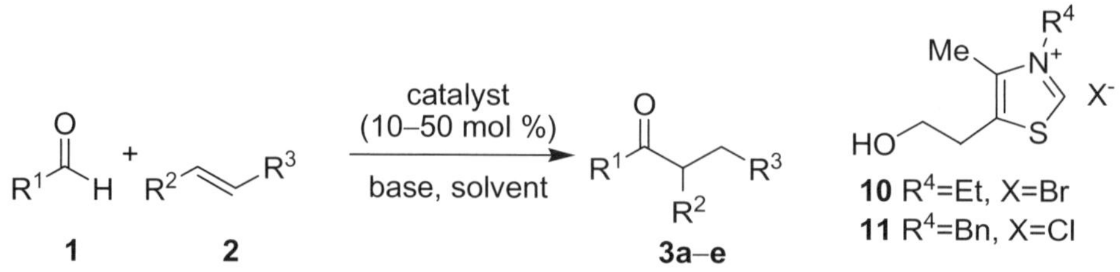

Entry	R^1	R^2	R^3	Catalyst	Base	Solvent	Product	Yield (%)
1	Ph	H	COMe	NaCN	--	DMF	**3a**	82
2	Ph	H	COMe	**10**	TEA	--	**3a**	65
3	Ph	Ph	COMe	NaCN	--	DMF	**3b**	80
4	Ph	H	CN	NaCN	--	DMF	**3c**	80
5	Ph	H	CO_2Et	NaCN	--	DMF	**3d**	55
6	*n*-Hex	H	COMe	**11**	TEA	EtOH	**3e**	75

Simple and efficient, the Stetter hydroacylation reaction quickly became a popular route for the production of 1,4-dicarbonyl compounds as precursors for the synthesis of pyrrole, furan, and pyridazine heterocycles. The Parke–Davis route to atorvastatin (Lipitor®) illustrates the use of a Stetter/Paal–Knorr sequence to access a high value pharmaceutical intermediate (**16**) from 4-fluorobenzaldehyde **12** and benzylidine amide **13**.[6]

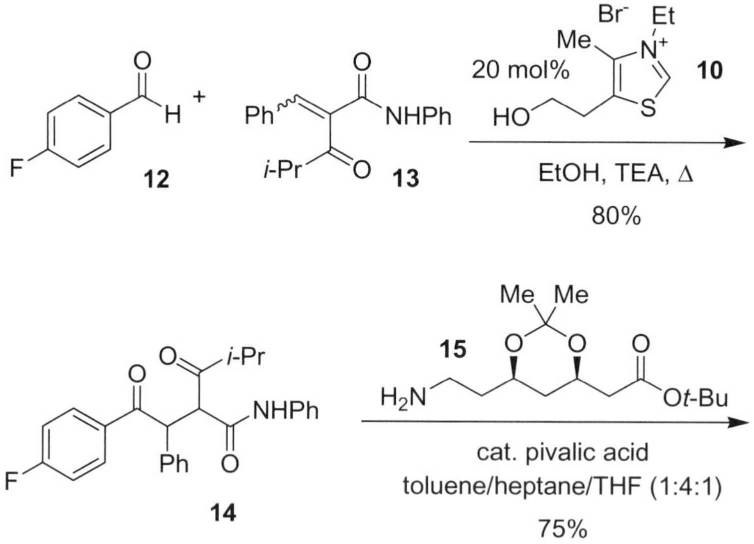

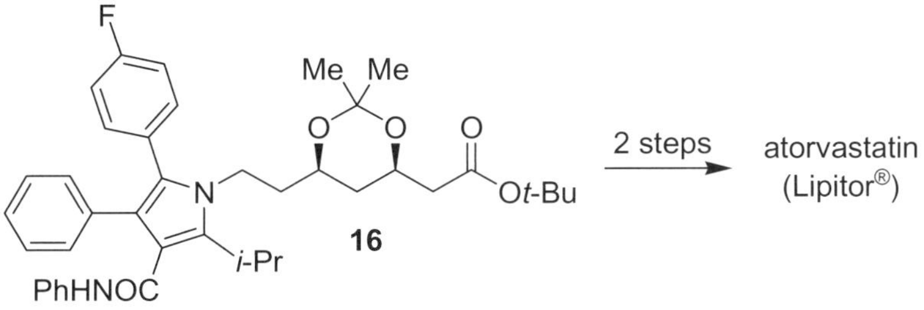

Intermolecular Biomimetic Stetter Reaction

Stetter and Lorenz reported in 1985 that a thiazolylidine catalyst could promote the addition of α-ketoacid derivatives (**17**) to activated olefins (**18**) with loss of CO_2 to give 1,4-diketones (**19**).[7] The reaction has been called biomimetic because of its resemblance to the family of biochemical transformations effected by thiamine diphosphate-dependent enzymes acting on pyruvate.[8] While the original report was limited to alkyl vinyl ketones, more recent work by Scheidt and co-workers has greatly expanded the scope of this reaction by employing α,β-unsaturated 2-acylimidazoles as the activated olefin component.[9]

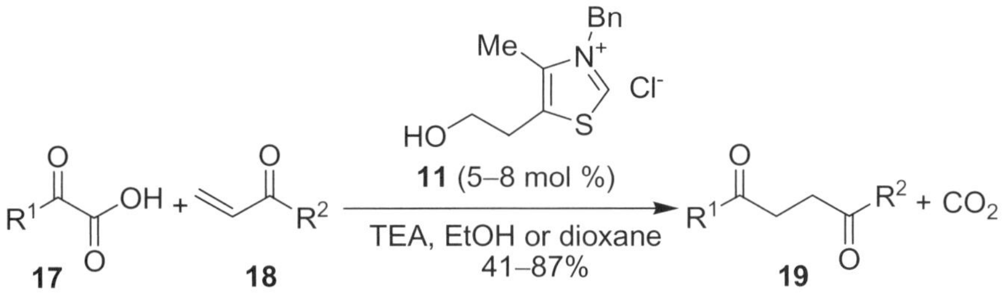

Intermolecular Sila-Stetter Reaction

Degl'Innocenti and co-workers established in 1987 that acylsilanes could react with activated olefins in a cyanide-catalyzed sila-Stetter reaction,[10] and recent work by Scheidt and co-workers has established the utility of thiazolylidine catalysts in this Stetter variation.[11] The reaction is presumed to occur via a catalyst-initiated [1,2]-Brook rearrangement, which serves to provide the silylated acylanion donor intermediate **23**. The use of excess 2-propanol in the reaction facilitates silyl transfer and catalyst turnover. Compared to the classical Stetter approach, this variation provides comparable yields of 1,4-diketone products from aromatic (**20**) and aliphatic (**21**) acylanion precursors.

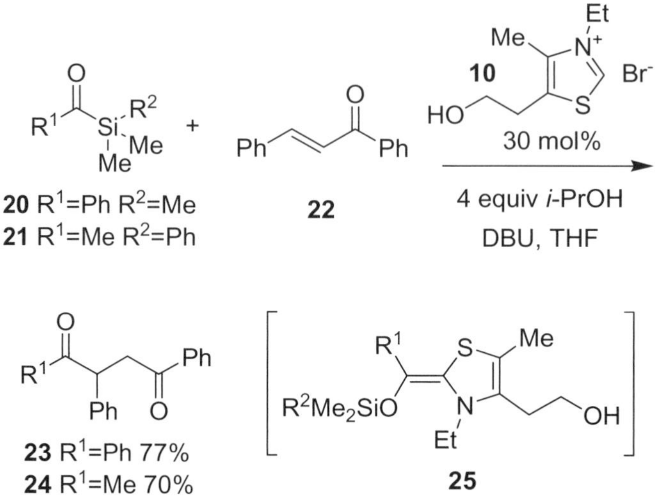

Enantioselective Intermolecular Stetter Reaction

Enders and co-workers were the first to report enantioselectivity in an intermolecular Stetter reaction promoted by a chiral *N*-heterocyclic carbene.[2d] The chiral thiazolium precatalyst **27** effects a Stetter reaction between butanal (**26**) and chalcone (**22**) to give diketone **28** with low efficiency but measurable enantiomeric excess.

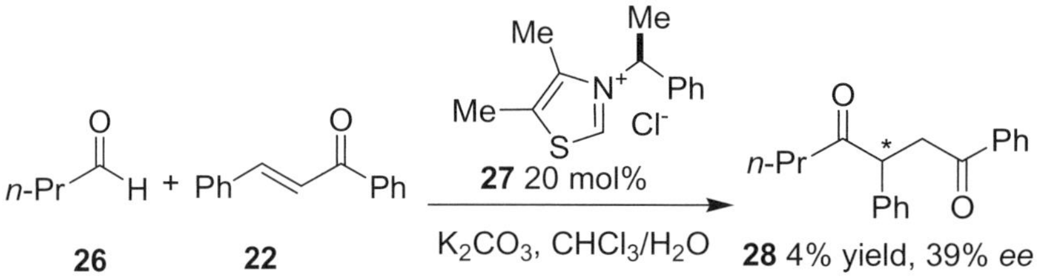

Despite significant research into new chiral carbene catalysts, few improvements in the direct asymmetric intermolecular addition of aldehydes to activated olefins have been possible. An innovative alternative was reported by Johnson and co-workers, who showed that chiral metallophosphite **31** could catalyze an asymmetric sila-Stetter reaction with high enantioselectivity.[12] Due to the absence of a silyl scavenger, catalyst turnover occurs by a [1,4]-retro-Brook reaction, and the initially isolated α-silylamide product is recrystallized to improve the enantiomeric purity from 90 to 99% *ee*. Desilylation completes the three-step procedure to give γ-ketoamide **32** in good yield and excellent enantiomeric purity. The reaction is currently limited to aromatic acylsilanes and α,β-unsaturated amides, but the *p*-methoxyphenyl ketone may be oxidatively converted to the corresponding *p*-methoxyphenyl ester to allow access to a variety of enantiomerically enriched succinic acid derivatives.

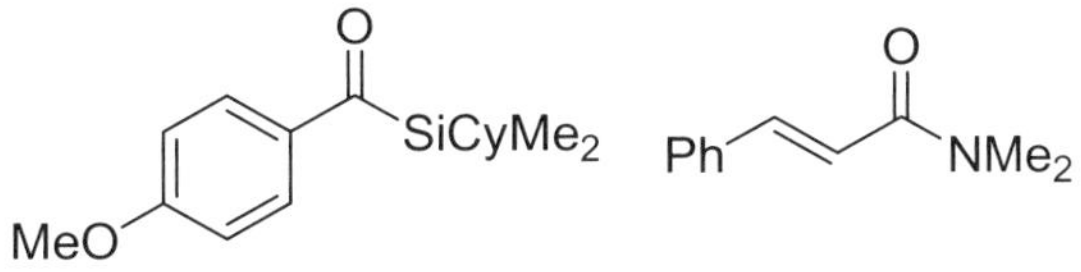

29 **30**

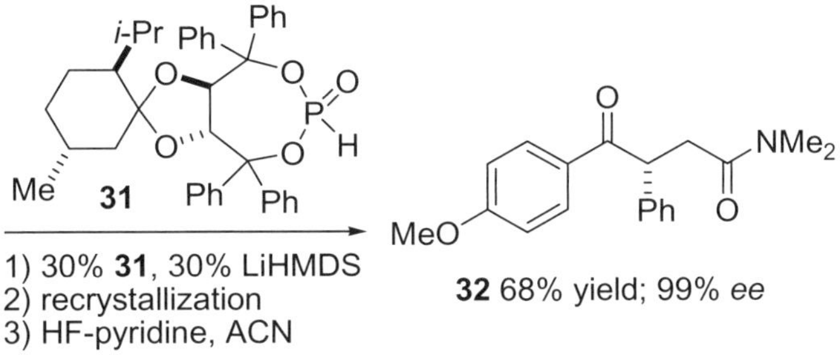

Intramolecular Stetter Reaction

Trost and co-workers are credited with the first intramolecular Stetter reaction in their total synthesis of hirsutic acid C.[13] An excess of thiazolium precatalyst **34** is employed to promote the intramolecular cyclization of bicycle **33**, which provides the key C_2–C_3 bond of the natural product in the tricyclic intermediate **35**.

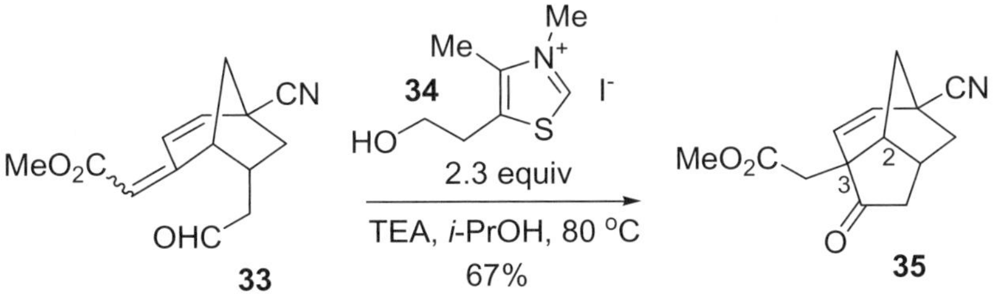

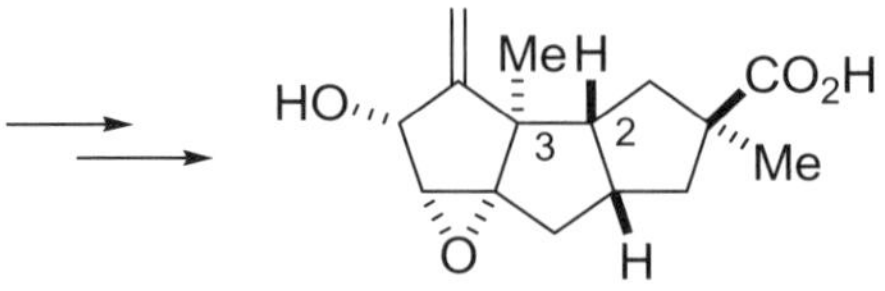

hirsutic acid C

Further work has established the utility of the intramolecular Stetter reaction for the synthesis of dihydrobenzofuranone and chroman-4-one products (entry 1, **37**),[14] as well as fused enediones of varying ring sizes (entry 2, **39**).[15] In the latter reaction, elimination of the neighbouring acetate group under the reaction conditions installs the tetrasubstituted olefin. An intramolecular Stetter reaction promoted by triazolium salt **41** has been reported to afford the *cis*-decalin ring system of **42** (entry 3), a key intermediate in a formal total synthesis of the natural product platensimycin.[16]

Entry	Substrate	Precatalyst	Product
1	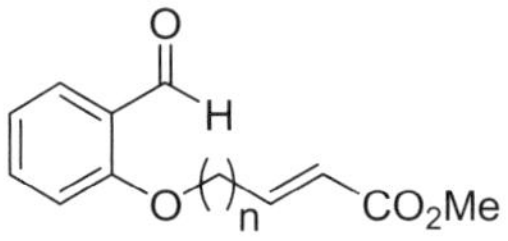 **36** n=0–1	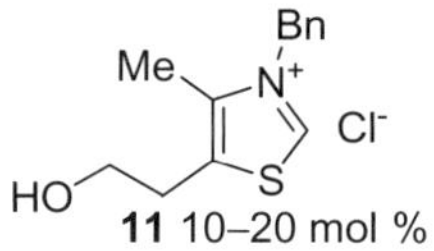 **11** 10–20 mol %	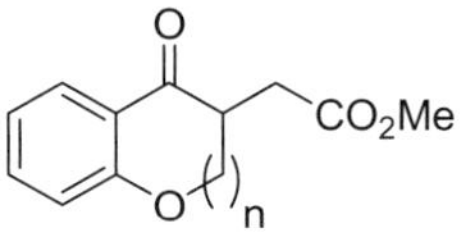 **37** n=0–1; 39–86%
2	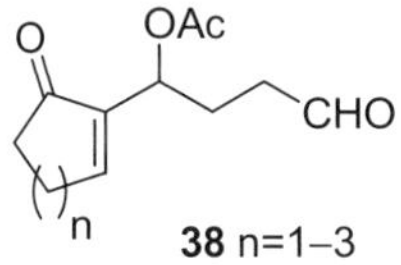 **38** n=1–3	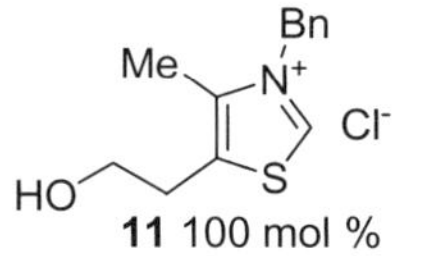 **11** 100 mol %	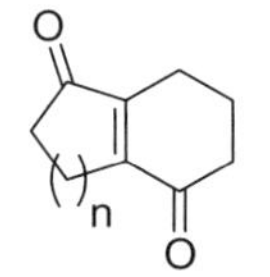 **39** n=1–3; 50–80%
3	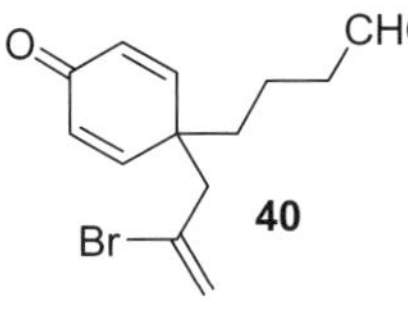 **40**	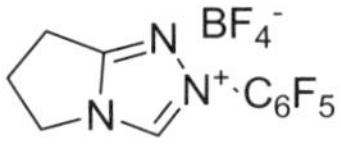 **41** 100 mol %	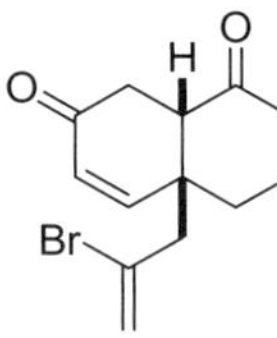 **42** 64%

Enantioselective Intramolecular Stetter Reaction

In contrast to the limited success in achieving direct asymmetric intermolecular Stetter reactions, there have been an abundance of reports detailing highly enantioselective intramolecular Stetter annulations. The first reported example was published by Enders and co-workers in 1996, in which the Ciganek reaction was rendered modestly asymmetric through the use of chiral triazolium precatalyst **45**.[17] More recently, Rovis and co-workers have developed a family of highly active and selective triazolium precatalysts, including **46–48**, that currently represent the state of the art for the catalytic asymmetric intramolecular Stetter reaction.[18]

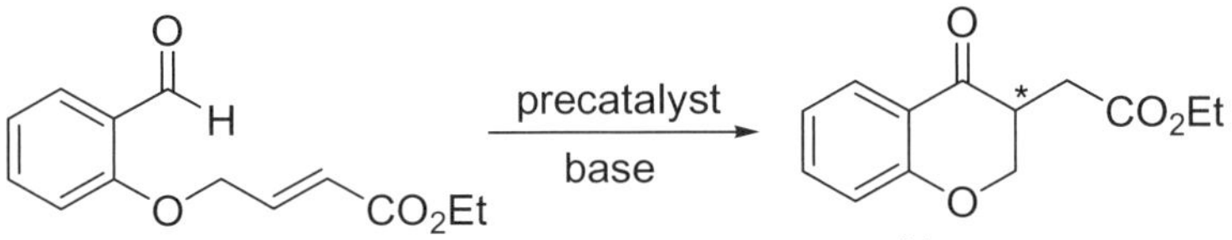

43 **44**

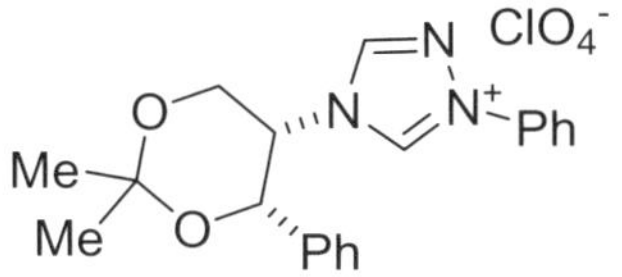

45 (20 mol %)
69% yield, 56% *ee* (*R*)

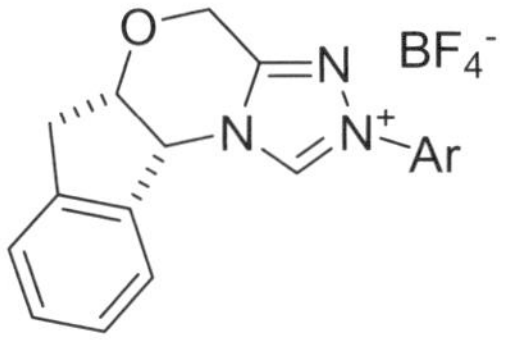

46 Ar = *p*-anisyl (20 mol %)
94% yield, 94% *ee* (*R*)

47 Ar =Ph-F$_5$ (20 mol %)
92% yield, 94% *ee* (*R*)

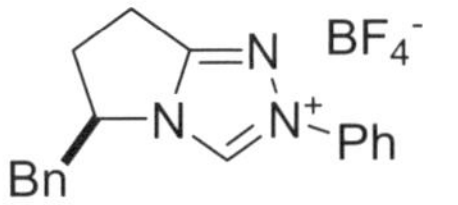

48 (20 mol %)
94% yield, 90% *ee* (*S*)

Rovis and co-workers have utilized their improved chiral triazolium catalysts to greatly expand the scope of the asymmetric intramolecular Stetter reaction. Key advances in Ciganek-type reactions include the use of tri-substituted olefin acceptors to introduce additional structural complexity. Contiguous stereocenters are set in high diastereomeric ratios provided care is taken to remove the hexamethyldisilylamine resulting from deprotonation of the chiral triazolium pre-catalyst (entry 1).[19] Products containing quaternary stereogenic centers (entry 2) can also be produced in excellent yield and *ee*.[20] Cyclization of aliphatic aldehydes onto diverse electron-deficient olefins has also been explored, providing monocyclic ketones (entry 3, **56**),[18b,19,20] hydrobenzofuranones (entry 4, **58**),[21] and spirofuranone-lactams (entry 5, **60**)[22] in high yield and with excellent levels of enantiomeric excess.

Entry	Substrate	Precatalyst	Product	%Yield (%*ee*; *dr*)
1	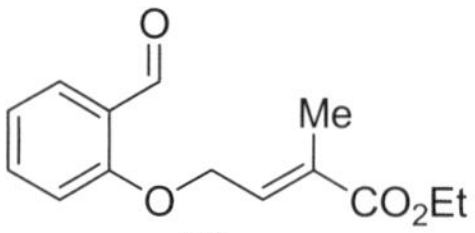 **51**	**46** (20 mol %)	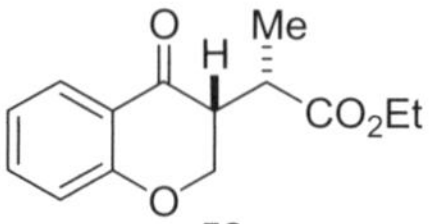 **52**	80 (97; 150:1)
2	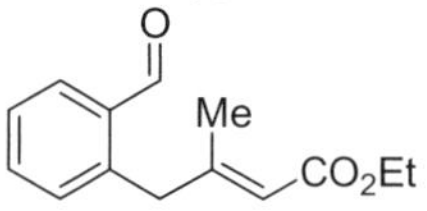 **53**	**47** (20 mol %)	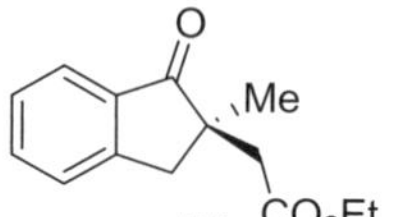 **54**	95 (99)
3	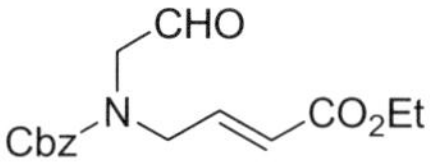 **55**	**48** (20 mol %)	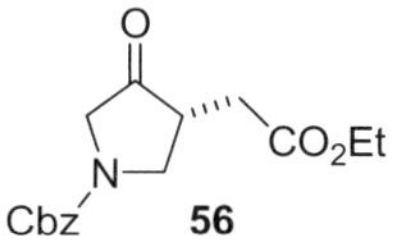 **56**	80 (99)
4	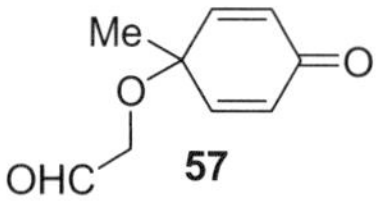 **57**	**46** (20 mol %)	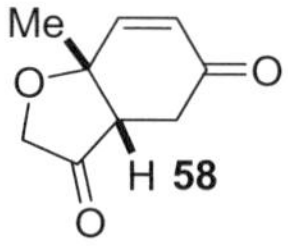 **58**	90 (92; >95:05)
5	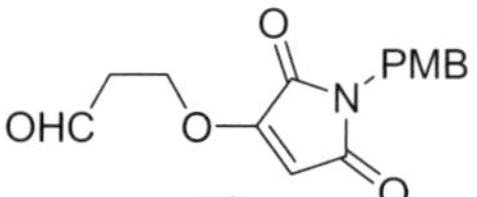 **59**	***ent*-47** (20 mol %)	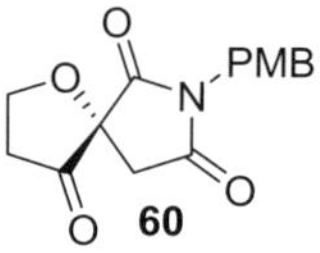 **60**	88 (99)

2.12.5 Experimental

NaCN-Catalyzed Intermolecular Stetter Reaction[1a]

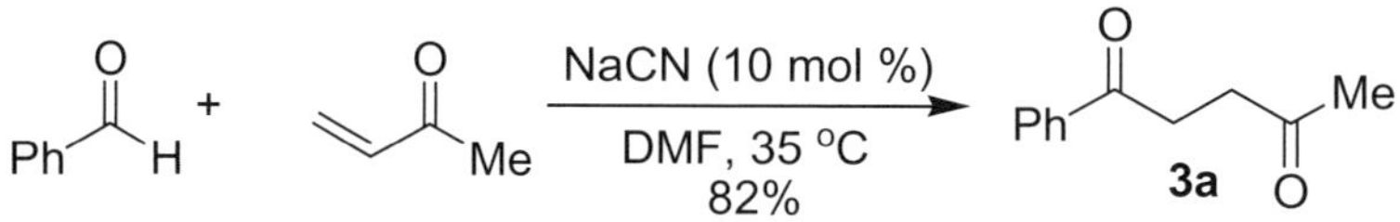

1-Phenyl-1,4-pentanedione (3a).

A solution of freshly distilled benzaldehyde (10.6 g, 100 mmol) and anhydrous DMF (50 mL) is added dropwise within 10 min to a stirred mixture of sodium cyanide (0.49 g, 10 mmol) and DMF (50 mL) at 35 °C. After 5 minutes' stirring, a solution of freshly distilled methyl vinyl ketone (5.3 g, 75 mmol) in DMF (100 mL) is added at 35 °C over 20 min. Stirring is continued for 1 h at the same temperature and the reaction mixture then treated with twice the amount of water. After repeated extraction with chloroform the combined extracts are washed with dilute hydrochloric acid (pH 2), then with sodium hydrogen carbonate solution, and finally with water. After removal of solvent, the residue is vacuum distilled. Yield 10.8 g (82%).

Thiazolylidine-Catalyzed Intermolecular Stetter Reaction[3b]

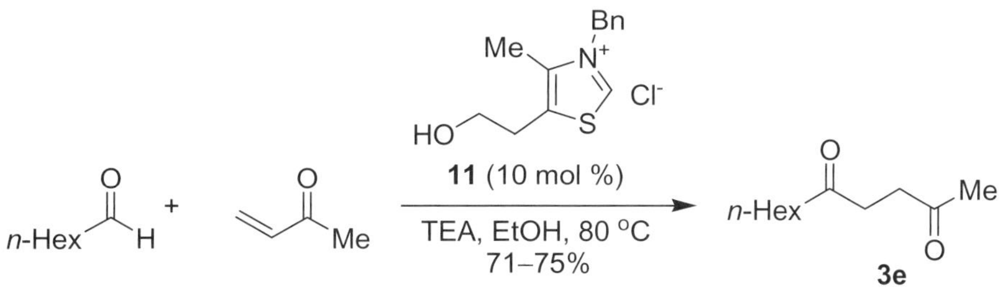

2,5-Undecanedione (3e)**.**
A 1000-mL, three-necked, round-bottomed flask equipped with a mechanical stirrer, a short gas inlet tube, and an efficient reflux condenser fitted with a potassium hydroxide drying tube is charged with 26.8 g (0.1 mol) of 3-benzyl-5-(2-hydroxyethyl)-4-methyl-1,3-thiazolium chloride (**11**), 500 mL of absolute ethanol, 77.2 g (1.1 mol) of 3-buten-2-one, 60.6 g (0.6 mol) of triethylamine, and 114.2 g (1.0 mol) of heptanal. A slow stream of nitrogen is started and the mixture is stirred and heated in an oil bath at 80 °C. After 16 hr the reaction mixture is cooled to room temperature and concentrated by rotary evaporation. Then 500 mL of chloroform is added to the residue and the mixture is washed with 200 mL of dilute hydrochloric acid (5%), 200 mL of saturated sodium hydrogen carbonate solution, and, finally, with two 200-mL portions of water. After the solution is dried with anhydrous magnesium sulfate, the chloroform is distilled off and the residue is fractionated under reduced pressure through a 30-cm Vigreux column. The main fraction is collected at 80–82 °C/0.3 mm. The yield for **3e** is 130–138 g (71–75% based on heptanal) of a colorless distillate, which solidifies on standing at room temperature, mp 33–34 °C.

Intermolecular Sila-Stetter Reaction[11]

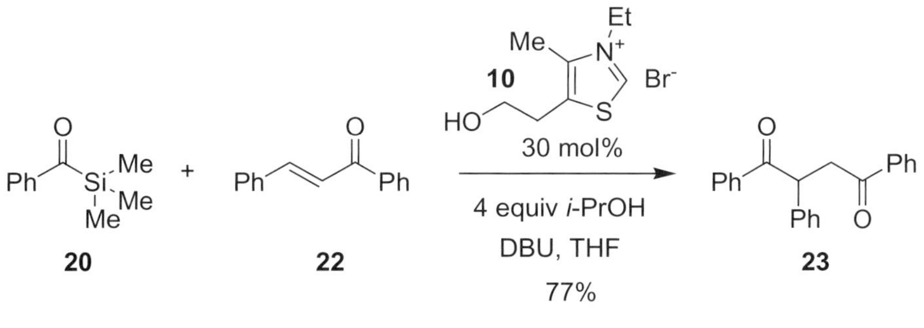

1,2,4-triphenylbutane-1,4-dione (23).

A screw-capped test tube was charged with the thiazolium salt **10** (30 mg, 0.119 mmol) in a nitrogen-filled drybox. The test tube was removed from the box and placed under a positive pressure of nitrogen. Benzoyltrimethylsilane **20** (140 mg, 0.768 mmol) in THF (0.25 mL) was added by syringe to the test tube followed by the addition of DBU (17 μL, 0.119 mmol). The reaction mixture was heated to 70 °C after which **22** (0.384 mmol) in THF (0.25 mL) was added by syringe followed by the addition of isopropanol (120 μL, 1.56 mmol). The reaction was allowed to stir at 70 °C for 24 hours. Upon completion by TLC (40% ether/hexanes), the reaction was cooled to room temperature, diluted with ethyl acetate (20 mL) and washed with water (20 mL). The aqueous layer was washed with ethyl acetate (3 × 30 mL) and the combined organic extracts were dried over sodium sulfate, filtered, and concentrated *in vacuo*. The resulting residue was purified by flash column chromatography on silica gel (10% ether/hexanes) to give 1,2,4-triphenylbutane-1,4-dione **23** (0.093 g, 77%) as a colorless oil.

Asymmetric Intramolecular Stetter Annulation[18b]

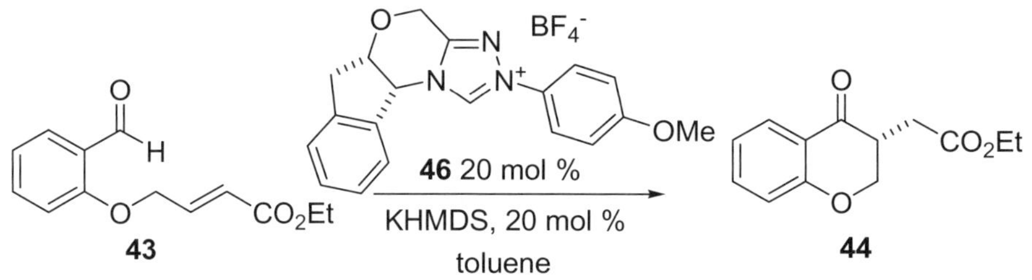

(*R*)-Ethyl 2-(4-oxo-3,4-dihydro-2*H*-chromen-3-yl)acetate (44).

A flame dried round-bottom flask was charged with triazolium salt **46** (0.2 equiv) and toluene (5 mL). To this solution was added KHMDS (0.5 M in toluene prepared prior to use from 0.05 g of KHMDS in 0.5 mL of toluene) (0.2 equiv) via syringe, and the solution was stirred at ambient temperature for 5 min. A solution of **43** (1 equiv, 0.12 mmol) in toluene (2 mL) was added. The resulting solution was allowed to stir at ambient temperature and monitored by TLC. The reaction mixture was placed directly onto a silica gel column. The desired product was purified by flash column chromatography, eluted with a suitable solution of hexane and ethyl acetate (typically 4:1). Evaporation of solvent afforded analytically pure product **44**.

2.12.6 References

1. [R] (a) Stetter, H. *Angew. Chem., Int. Ed. Engl.* **1976**, *15*, 639–647. (b) Stetter, H.; Kuhlmann, H. *Org. React.* **1991**, *40*, 407–496.
2. [R] (a) Enders, D.; Balensiefer, T. *Acc. Chem. Res.* **2004**, *37*, 534–541. (b) Johnson, J. S. *Angew. Chem. Int Ed.* **2004**, *43*, 1326–1328. (c) Christmann, M. *Angew. Chem., Int. Ed.* **2005**, *44*, 2632–2634. (d) Enders, D.; Niemeier, O.; Henseler, A. *Chem. Rev.* **2007**, *107*, 5605–5655. (e) Johnson, J. S. *Curr. Opin. Drug. Disc. Devel.* **2007**, *10*, 691–703. (f) Marion, N.; Diez-González, S.; Nolan, S. P. *Angew. Chem., Int. Ed.* **2007**, *46*, 2988–3000.
3. (a) Stetter, H.; Schreckenberg, M. *Angew. Chem., Int. Ed. Engl.* **1973**, *12*, 81. (b) Stetter, H.; Kuhlmann, H.; Haese, W. *Org. Synth.* **1987**, *65*, 26–31.
4. Breslow, R. *J. Am. Chem. Soc.* **1958**, *80*, 3719–3726. (b) Breslow, R.; Kim, R. *Tetrahedron Lett.* **1994**, *35*, 699–702.
5. Lapworth, A. *J. Chem. Soc.* **1903**, *83*, 995–1005.
6. Baumann, K. L.; Butler, D. E.; Deering, C. F.; Mennen, K. E.; Millar, A.; Nanninga, T. N.; Palmer, C. W.; Roth, B. D. *Tetrahedron Lett.* **1992**, *33*, 2283–2284.
7. Stetter, H.; Lorenz, G. *Chem. Ber.* **1985**, *118*, 1115–1125.
8. Jordan, F. *Nat. Prod. Rep.* **2003**, *20*, 184–201.
9. Myers, M. C.; Bharadwaj, A. R.; Milgram, B. C.; Scheidt, K. A. *J. Am. Chem. Soc.* **2005**, *127*, 14675–14680.
10. Degl'Innocenti, A.; Ricci, A.; Mordini, A.; Reginato, G.; Colotta, V. *Gazz. Chim. Ital.* **1987**, *117*, 645–648.
11. (a) Mattson, A. E.; Bharadwaj, A. R.; Scheidt, K. A. *J. Am. Chem. Soc.* **2004**, *126*, 2314–2315. (b) Mattson, A. E.; Bharadwaj, A. R.; Zuhl, A. M.; Scheidt, K. A. *J. Org. Chem.* **2006**, *71*, 5715–5724..
12. Nahm, M. R.; Linghu, X.; Potnick, J. R.; Yates, C. M.; White, P. S.; Johnson, J. S. *Angew. Chem., Int. Ed.* **2005**, *44*, 2377–2379. (b) Nahm, M. R.; Potnick, J. R.; White, P. S.; Johnson, J. S. *J. Am. Chem. Soc.* **2005**, *128*, 2751–2756.
13. Trost, B. M.; Shuey, C. D.; DiNinno Jr., F.; McElvain, S. S. *J. Am. Chem. Soc.* **1979**, *101*, 1284–1285.
14. Ciganek, E.; *Synthesis* **1995**, 1311–1313.
15. Markó, I. E.; de Merode, T.; Wasnaire, P. *Chem. Comm.* **2007**, 4755–4757.
16. Nicolaou, K. C.; Tang, Y.; Wang, J. *Chem. Comm.* **2007**, 1922–1923.
17. Enders, D; Breuer, K.; Runsink, J. *Helv. Chim. Acta* **1996**, *79*, 1899–1902.
18. (a) Kerr, M. S.; Read de Alaniz, J.; Rovis, T. *J. Am. Chem. Soc.* **2002**, *124*, 10298–10299. (b) Read de Alaniz, J.; Kerr, M. S.; Moore, J. L.; Rovis, T. *J. Org. Chem.* **2008**, *73*, 2033–2040.
19. Read de Alaniz, J.; Rovis, T. *J. Am. Chem. Soc.* **2005**, *127*, 6284–6289.
20. Kerr, M. S.; Rovis, T. *J. Am. Chem. Soc.* **2004**, *126*, 8876–8877.
21. Liu, Q.; Rovis, T. *J. Am. Chem. Soc.* **2006**, *128*, 2552–2553.
22. Orellana, A.; Rovis, T. *Chem. Comm.* **2008**, 730–732.

2.13 Wittig Reaction

Paul Galatsis

2.13.1 Description

The Wittig reaction,[1] not to be confused with the Wittig rearrangement,[2] is defined by the coming together of a phosphorous ylide and an aldehyde or ketone moiety within a molecule giving rise to an alkene functional group that links the two reacting species with concomitant loss of the corresponding phosphine oxide.

The synthetic power of the Wittig reaction is in its control of alkene geometry, which remains an ongoing challenge in organic synthesis. The high selectivity for (*Z*)- or (*E*)-alkenes is dependent upon all the various reaction parameters, but the type of ylide has a great influence on this outcome. The phosphorous ylides can be categorized based on their general reactivity. Substituents that can strongly stabilize the carbanionic nature of the ylide by conjugation (e.g., $-CO_2R$, –CN, or $-SO_2R$) are considered "stabilized" ylides and typically favor the formation of (*E*)-alkenes. Substituents, like phenyl or allyl, which can only moderately stabilize the negative charge of the ylide are called "semi-stabilized" or "moderately" stabilized ylides and often do not give any great preference of alkene geometry. If there are no substituents that can interact with the negative charge then these are considered "non-stabilized" ylides and usually favor formation of (*Z*)-alkenes.

2.13.2 Historical Perspective

Georg Wittig and coworkers began publishing on what would become the Wittig reaction in 1953.[3] The discovery of this reaction was not planned as Wittig did not favor this method of experimentation.[4] It was, however, "recognized" by Wittig as an interesting observation in a series of experiments he was conducting on the chemistry of bonding arrangements of the Group V elements by examining their anionic counterparts. It was during his investigation on the chemistry of pentacoordinate phosphorous that he noted the uniqueness and generality of the method. Other groups saw the value in this particular reaction and it was quickly adopted by academic and industrial[5] chemists. Ultimately, the Nobel Committee also recognized its utility and in 1979 Wittig shared the Nobel Prize in Chemistry with H. C. Brown (hydroboration reaction) for their respective contributions.[6]

2.13.3 Mechanism

The mechanism[1] for the Wittig reaction has become more controversial over the past couple of decades. Up to the early 1970s, the classical mechanism began facing greater challenges as more experimental data, focused on understanding the stereochemical outcome of this reaction began to be reported, with improvements in technology and instrumentation. As a result, there is probably no unified, all encompassing mechanism to explain all examples of this reaction. The diversity of structural and stereoelectronic variations within the ylide and the carbonyl group has resulted in a potential continuum of possible mechanistic pathways.

As indicated (*vide supra*), the classical mechanism for the Wittig reaction involves the initial nucleophilic addition of the ylide carbanion (resonance structures **1** and **2**) to the electrophilic carbon of the carbonyl in **3** to afford betaine intermediate **4**. Rotation about the central C–C bond provides for ring closure to 1,2-oxaphosphetane **5**. These species are thermally unstable and readily decompose, via a concerted electrocyclic process, to generate the corresponding phosphine oxide **8** and the (*E*)-alkene **6** or (*Z*)-alkene **7**.

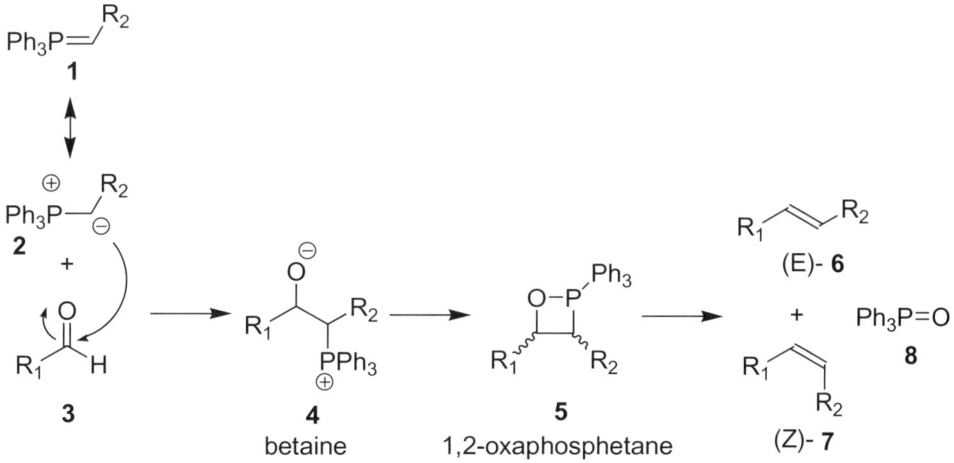

By this process, the alkene geometry is set during the initial 1,2-carbonyl addition and a rationalization for the observed stereochemistry could be made. Kinetic addition of ylide **1** to carbonyl compound **3** would give rise to a distribution of betaines favoring the *erythro*, kinetic adduct **9** over the thermodynamic, *threo* adduct **10**. Once formed the rapid loss of **8** via the oxaphosphetanes **13** and **14** would give rise to the alkene **7** over the more stable **6**, respectively.

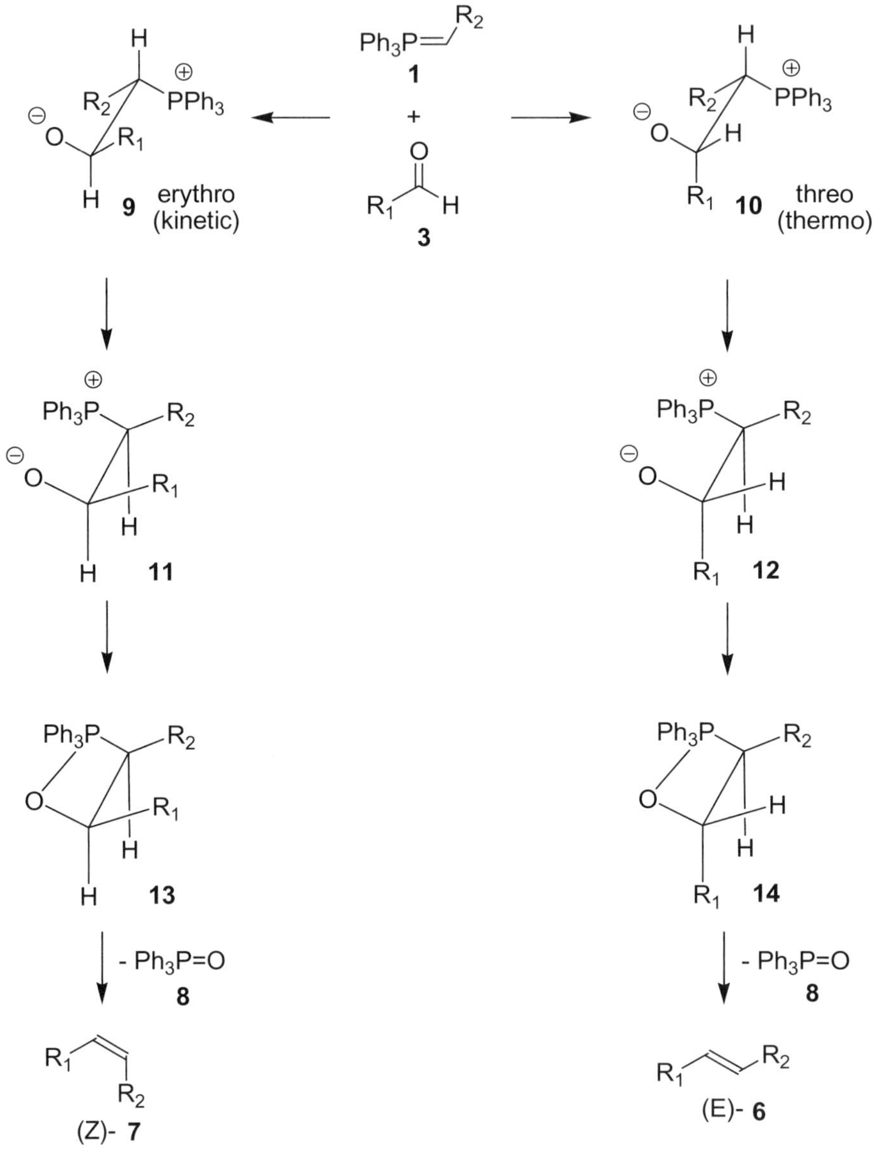

It has been observed that the Wittig reaction is greatly influenced by the presence of soluble salts.[1,7] The degree to which the *erythro*/*threo* adducts (**9** and **10**) are able to equilibrate via reversible formation of the starting material, is significantly affected by the inclusion of, or protocols to exclude, soluble salts, in particular lithium. Coordination of lithium to the betaine (**11** and **12**)/oxaphosphetane (**13** and **14**) intermediates retards the rate of loss of phosphine oxide **8**, thus allowing more time for equilibration. For example, consider the reaction of benzaldehyde **15** with the ylide **16** to generate alkene **17**.[7a] The addition of lithium salts had a dramatic effect on the overall yield of the reaction. Not only was the yield negatively impacted, the product distribution was affected. Under salt free conditions, the product distribution favored the *cis* over *trans*. However, the addition of various forms of lithium salts resulted in an erosion of the selectivity to 50 to 50.

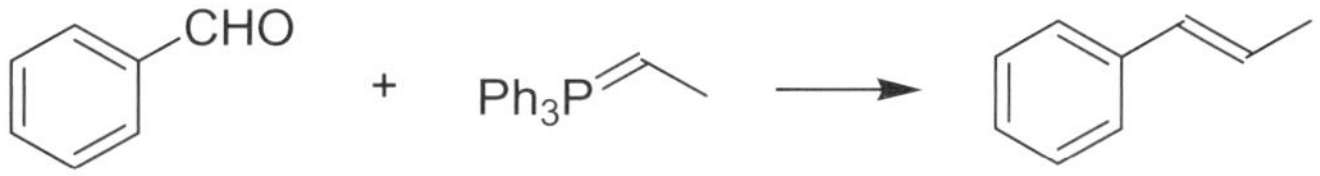

15 **16** **17**

Additive	Yield (%)	*Z*:*E*
salt free	98	87:13
LiCl	70	81;19
LiBr	68	61:39
LiI	76	58:42
$LiBPh_4$	63	50:50

Experimental evidence that supported the existence of oxaphosphetanes as intermediates in the Wittig reaction was obtained in 1973, when Vedejs and coworkers were able to determine their presence by ^{31}P NMR. Subsequently, the Maryanoff[1,9] and Vedejs[10] labs both spent the next 15 years probing deeper into the mechanistic details surrounding the origin of stereochemistry in this reaction. At the time, betaines had never been observed directly and this raised the question of what part in the reaction sequence they played, if any. As a consequence, a mechanism invoking an initial [2 + 2] cycloaddition was put forward. Therefore, in a reaction between carbonyl **19** and ylide **18** the expected adducts would be **20** and **21**, respectively. These adduct could then lose **8** to afford the corresponding alkenes **22** and **23**, respectively. In the presence of lithium salts, coordination to the oxygen center of the oxaphosphetane could facilitate a ring-opening to the metal-coordinated betaines.

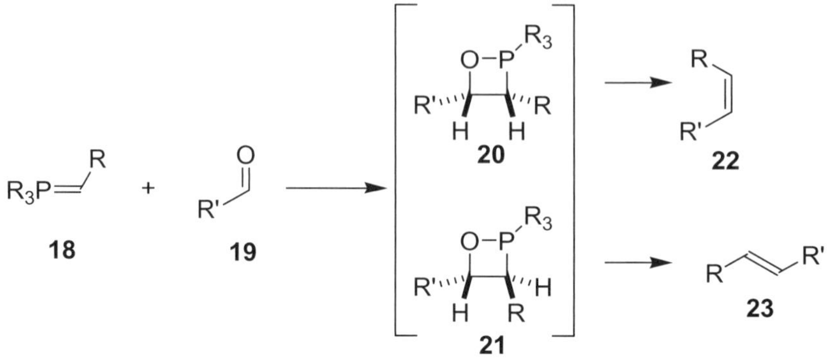

18 19 20 21 22 23

Stereochemistry could be rationalized by which mode this cycloaddition progressed. The two modes proposed included a *cis*-selective, puckered 4-membered TS **24** and a *trans*-selective, planar TS **25**.

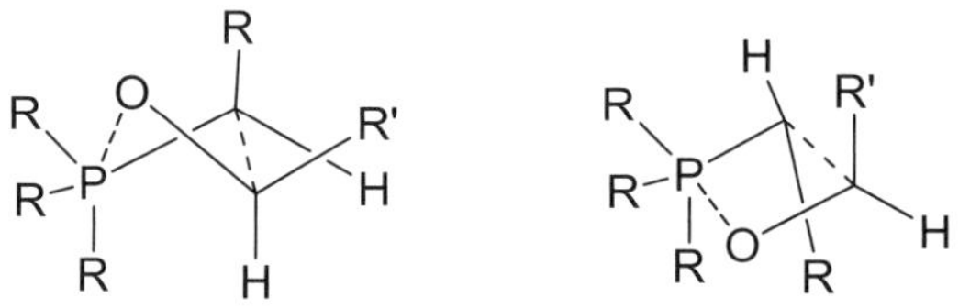

puckered TS
24

planar TS
25

More recently, the observation of a betaine was reported.[11] The system was biased to stabilize the betaine by coordination to pyridine nitrogens. The addition of ketone **27** and ylide **26** allowed for formation of oxaphosphetane **28** which was observable in the ^{31}P NMR. Upon addition of lithium bromide, betaine **29** could be resolved. It is interesting to note that once the lithium ion was sequestered in a crown ether, the oxaphosphetane **28** was regenerated.

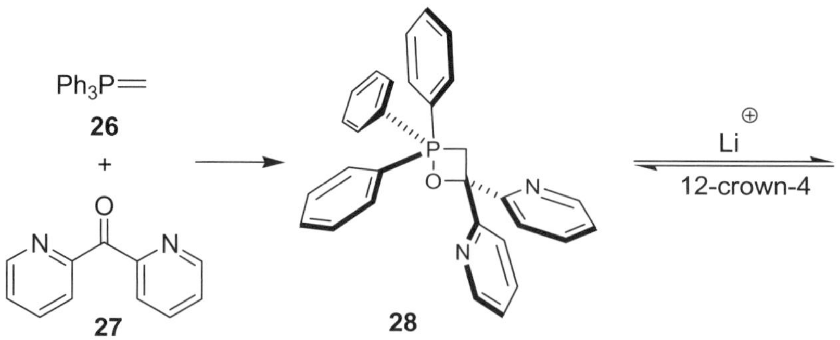

26 27 28

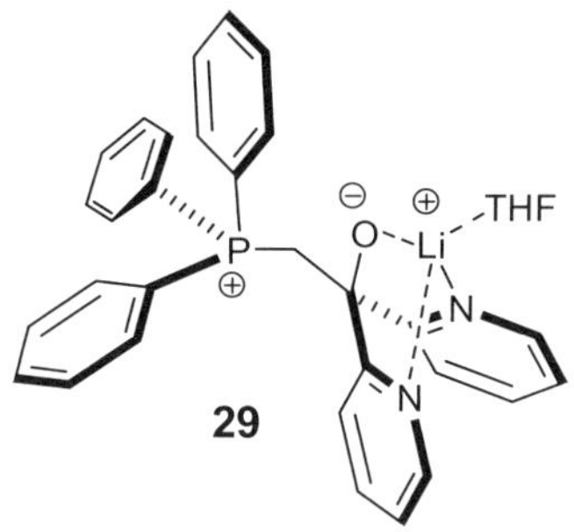

29

2.13.4 *Variations, Improvements or Modifications*

One of the first variations of the Wittig reaction was initially reported by Horner and coworkers[12] and rapidly followed by an initial report by Wadsworth and Emmons.[13] These examples made use of phosphine oxide/phosphonate derivatives of the ylides first reported by Wittig and are now collectively known as the Horner–Wadsworth–Emmons reaction (HWE).[1] Ylide formation occurs upon deprotonation of dialkoxy phosphonate **31** and alkene **32** is formed from carbonyl compound **30** with loss of the corresponding phosphate derivative **33**. The use of this variation has advantages over the conventional version: a) phosphonate carbanions are known to be more nucleophilic due to decreased stabilization by valence shell expansion of the phosphorous atom, thus are able to react with a wider diversity of carbonyl compounds. b) the phosphorous-based product of the reaction, a water-soluble phosphate, allows for a greater ease of reaction work-up. c) the enhanced reactivity of the phosphonate permits direct derivitization of the reagent. d) the Arbuzov reaction allows for ready preparation of the desired phosphonate.

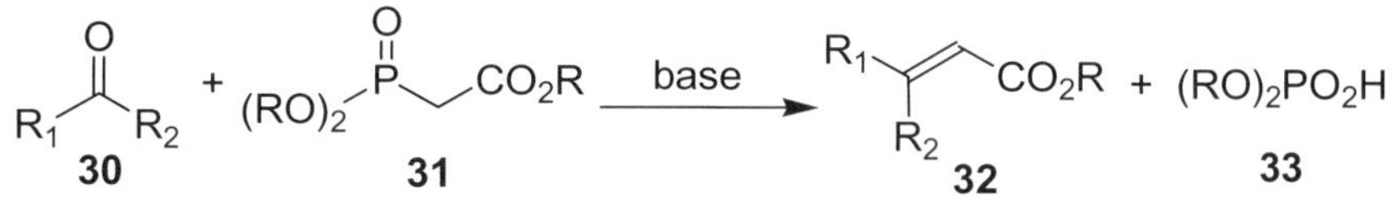

Mechanistically, this variation is similar to the classical Wittig reaction.[1] The only departure stems from the physico-chemical nature of the intermediates. The addition of ylide **34** to carbonyl compound **3** occurs as before to set up a mixture of *erythro* **35** and *threo* **37** adducts. Now the betaines are more acidic, which facilitates the interconversion of **35** and **37**. Deprotonation of **35**, *in situ*, produces **36**. Kinetic protonation occurs from the least sterically hindered face, as shown in **38**. With the equilibration to

37 complete, oxaphosphetane **39** can be formed which can then undergo loss of **8** to form exclusively the thermodynamic product **40**.

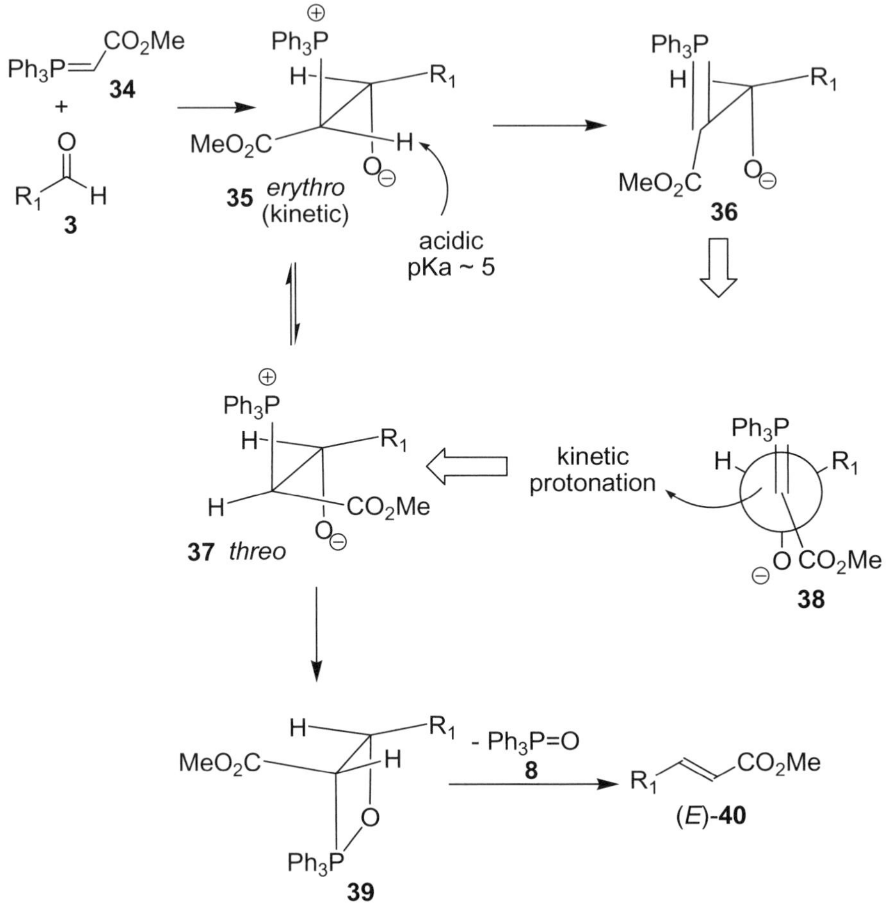

As greater use and understanding of the Wittig reaction grew, more reports of modifications of the ylide were published. Initial investigations focused on phosphonamides.[14] By this method, alkenes **43** could be formed via thermal decomposition of **42** with loss of **44**. This intermediate could be obtained upon deprotonation of **41** and exposing the resultant ylide to carbonyl compound **30**. This reagent was found to be complementary to the Witting and HWE reactions. The advantages were found to included: a) ready direct elaboration of the carbon framework, b) improved stereochemical control, c) ease of purification, d) alternative preparative routes, and e) low cost of goods.

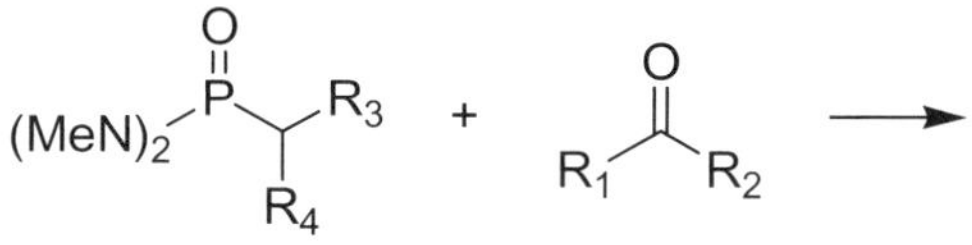

41 **30**

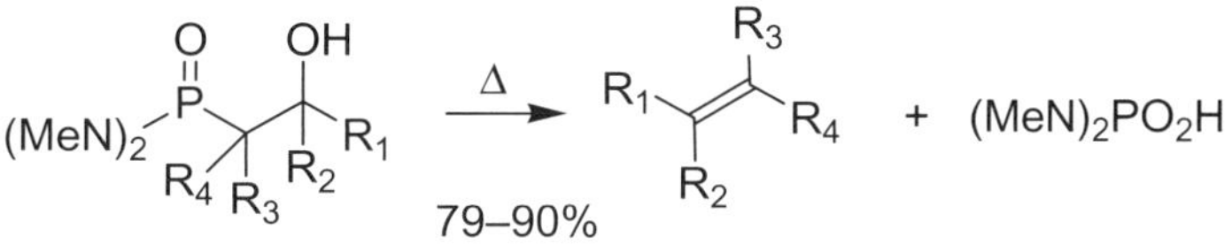

42 **43** **44**

The thiophosphonate was also found to be able to undergo alkene formation chemistry.[15] The ylide of **45** reacts with **30** to generate the adduct **46**. This then undergoes a similar decomposition as **42** to afford alkene **43**.

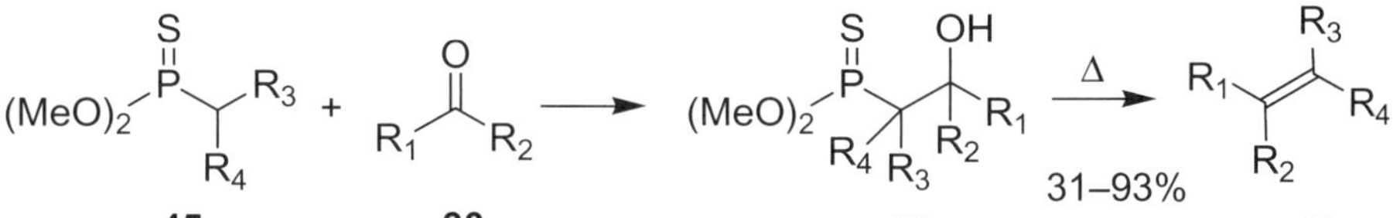

45 **30** **46** **43**

Combination of these two modifications was determined to be effective, particularly when the corresponding phosphonium ylide was inert or sluggish.[16] Thus reacting the anion of **47** with carbonyl compound **30** produced alcohol **48**. Methylation of this intermediate was required for the preparation of alkene **49** with concomitant loss of **50**.

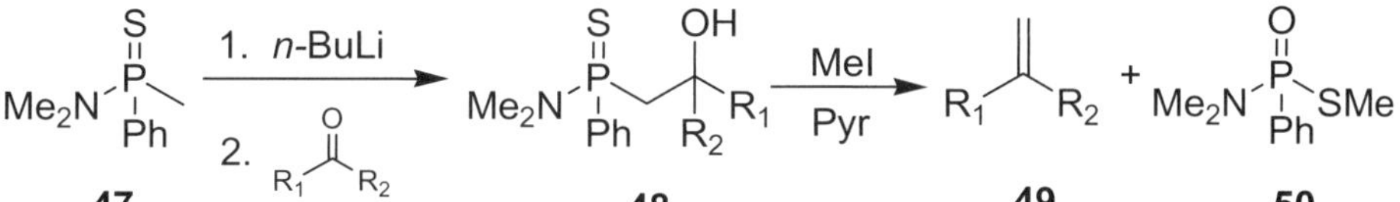

47 **30** **48** **49** **50**

Modifications directly to the alkoxy phosphonate were examined. The Still–Gennari modification[17] of the HWE reaction provided access to (*Z*)-selective alkenes. In this example, the substituents on the phosphonate were modified as in **51**. Executing the standard reaction conditions with aldehyde **52** now resulted in **53**. It was postulated that this modification accelerates the elimination of the initial adduct, thus preventing equilibration to the more stable *threo* intermediate (*vide supra*).

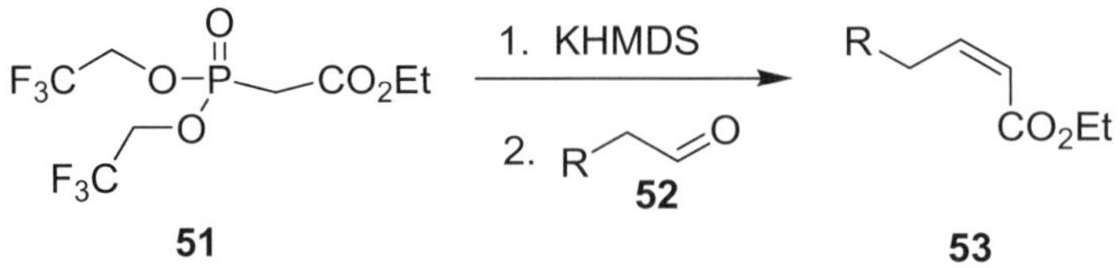

Substitution of the phosphorus atom, itself, has also been reported.[18] Substituted arsonium ylides **54** were found to react in an analogous manner with aldehydes **3** to afford alkene **55**. This reaction proceeded best with aldehydes, as cyclohexanone only gave a 15% yield of the corresponding alkene.

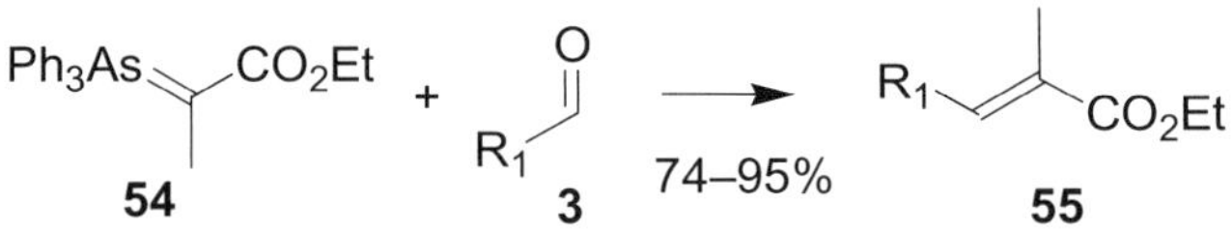

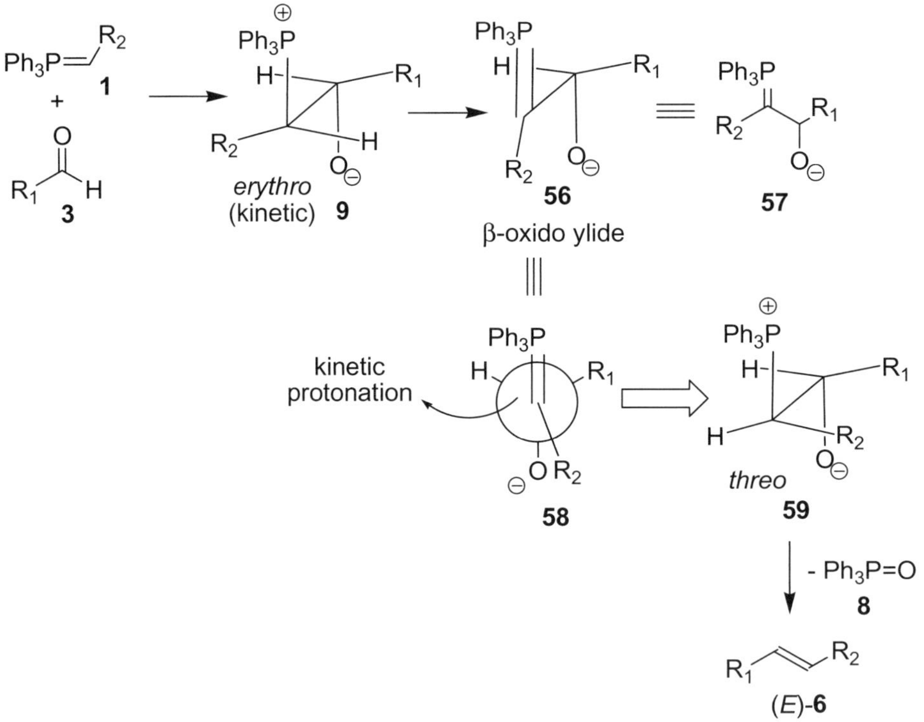

In 1966, Schlosser and Christmann,[19] as a way of controlling the stereochemistry of the reaction, published a variation that provided support for the classical mechanism of the Wittig reaction. The Schlosser modification allowed for a *trans*-selective outcome for the alkene geometry.

This was accomplished by treating the putative betaine intermediate **9**, generated from ylide **1** and aldehyde **3**, with a strong base (PhLi or *n*-BuLi) in ether/THF (1:1). This deprotonation afforded β-oxido ylide **56**, also illustrated by **57**. In a manner similar to the equilibration occurring in HWE (*vide supra*), one can add an external electrophile which kinetically adds to the sterically less hindered face of **58**. The resultant **59** can then undergo loss of **8** to produce **6**, in a controlled equilibration of the *erythro/threo* betaines.

Electrophilic species, other than a proton, can be used, thus adding to the synthetic utility of this variation. One such example is the end-game solution to the total synthesis of α-santalol **61**.[20] The advanced intermediate **60** was treated with ylide **16**. Formation of the β-oxido ylide with *n*-BuLi and addition of formaldehyde provided a one-step procedure for alkene formation and elaboration to **61**.

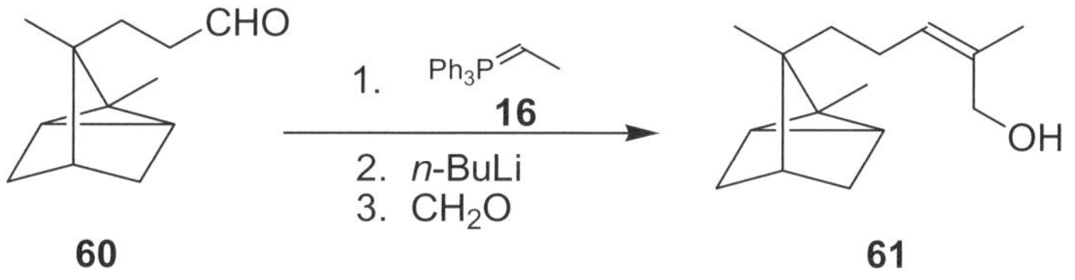

As the complexity of substrates being submitted to the Wittig reaction dramatically increased, gentler reactions conditions were required to accommodate greater inter-functional group compatibility. Thus the nature of the base necessary to form the ylide was studied. Roush and coworkers[21] reported the successful use of lithium chloride with amine bases (DBU or DIPEA) for base sensitive aldehydes or phosphonates. One example cited involved the coupling of phosphonate **62** with aldehyde **63** to afford triene **64** in good yield. This stands in sharp contrast to when NaH was used and no desired product was observed.

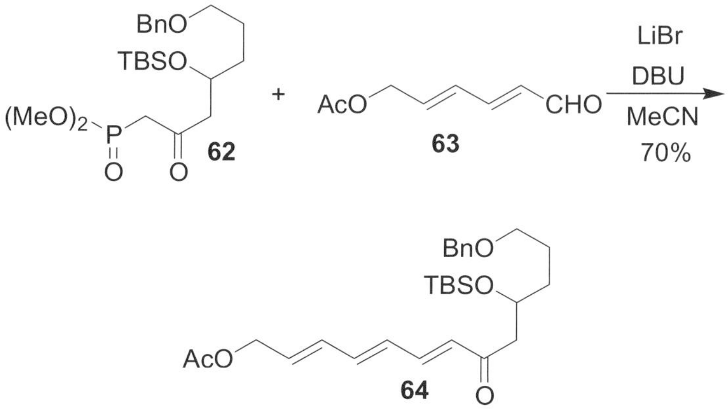

Rathke and Nowak[22] extended this chemistry by showing magnesium salts could also be used with triethylamine as the base. Additionally, strong bicyclic guanidine bases[23] have been employed in these reactions without the need for the addition of salts. One example makes use of **66** to facilitate the reaction of **65** with **15** to generate **67**.

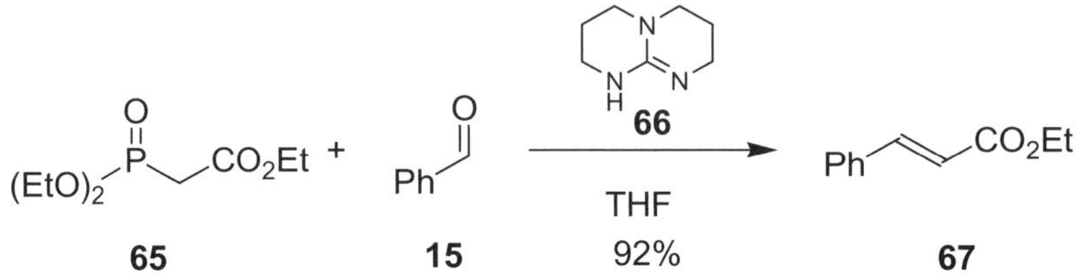

65 **15** **67**

In cases where the aldehyde is epimerizable, the recent report citing the use of lithium hexafluoroisopropoxide has been found to be very useful.[24] Comparing these reaction conditions with the protocol of Roush shows the utility in the transformation of **68** into **69**. Shorter reaction times, improved alkene stereochemistry and optical purity were observed.

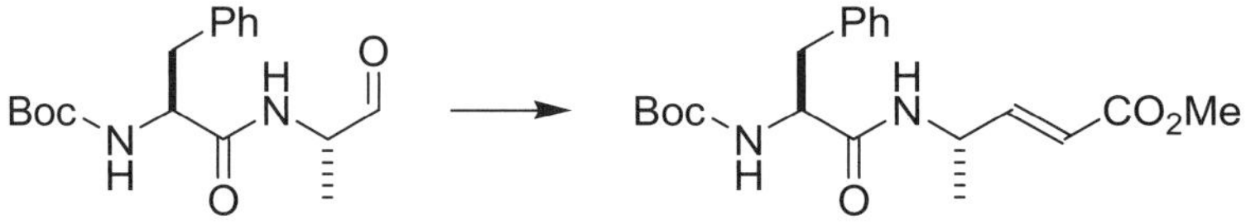

68 **69**

	time (h)	Product (de)	E:Z	Yield (%)
$LiOCH(CF_3)$	0.8	95	> 19:1	88
LiCl, DBU	18	80	7:1	84

The first report of the use of solid-supported reagents in the Wittig reaction dates back to 1971 and disclosed the use of polymeric diphenylphosphine **70**.[25] This reagent could be transformed into ylide **71** and reacted with benzaldehyde **15** to afford alkene **72**. Work-up was made easier since the phosphine oxide by-product was now immobilized on the support as **73**.

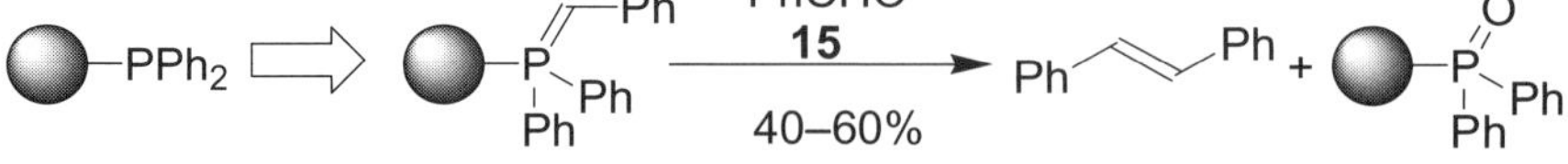

70 **71** **72** **73**

Alternative methods of linking to the solid support were also examined. Johnson[26] showed that amino polymers could be used to generated phosphonates **74**. HWE reactions with aldehydes could produce polymer-bound **75** and treatment with TFA would give access to the isolated amide **76**.

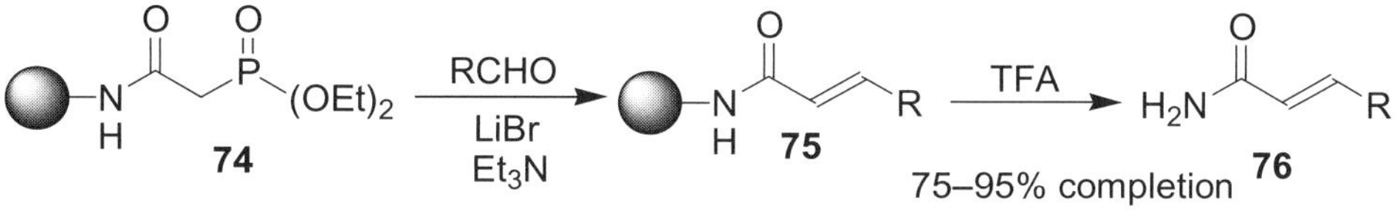

74 **75** **76**

Linking through the aldehyde can also be accomplished.[27] A library of tamoxifen analogs was prepared by treating immobilized aldehyde **77** with ylide **78**. Cleavage from the resin with TFA then afforded the desired product **79**.

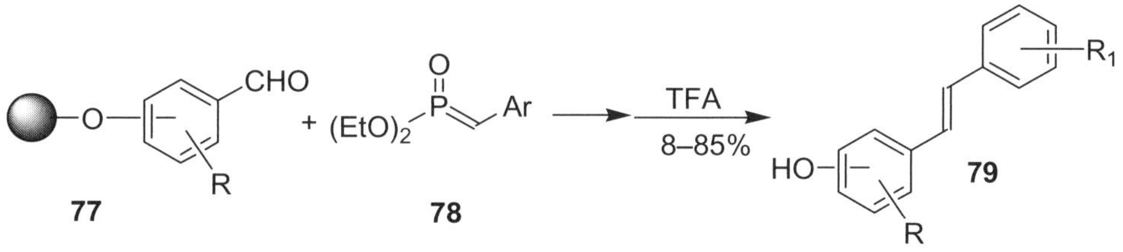

77 **78** **79**

New technologies have been applied to this chemistry. Rate acceleration induced by microwave irradiation occurs in examples of "difficult" reactions. Using solvent-free conditions,[28] ylide **81** was treated with acetophenone **80** to afford **82**. This process occurred more efficiently than traditional heating.

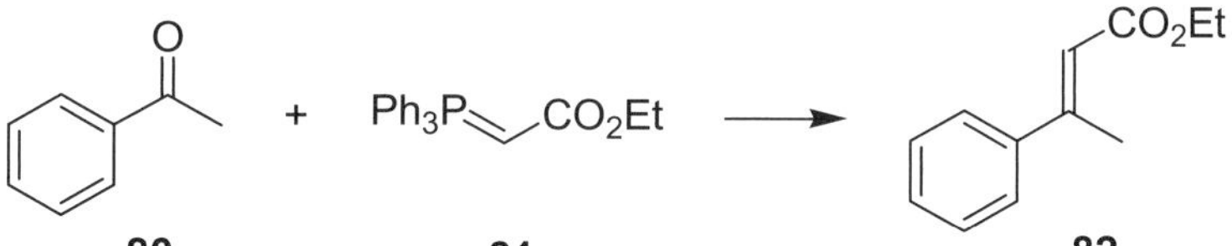

80 **81** **82**

Δ	10 h	170°	58%
μW	20 min	210°	67%

The combination of solid-support and microwave irradiation directed towards the Witting reaction was investigated in the formation of **85**.[29] Multifuntionalized aldehyde **84** could be induced to react with solid-supported ylide **83** in a very efficient manner.

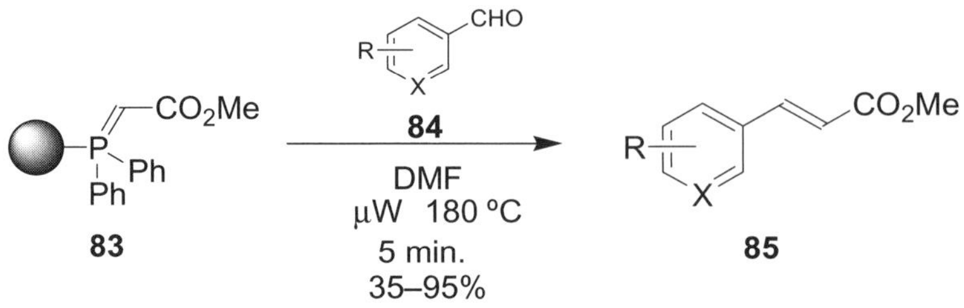

Upon initial examination, how one could control chirality in a reaction where the product is an alkene (no new sp^3 centers are formed) which is inherently achiral could be asked. Despite this impression, asymmetric variations of the Wittig reaction have been reported. One approach is to use a chiral auxiliary in the ester moiety of a phosphonate. The first example of a chiral Witting made use of menthol as a chiral auxiliary.[30] Reaction of the ketone **86** with the chiral HWE reagent **87** gave rise to **88**. However, the levels of chiral induction were not reported.

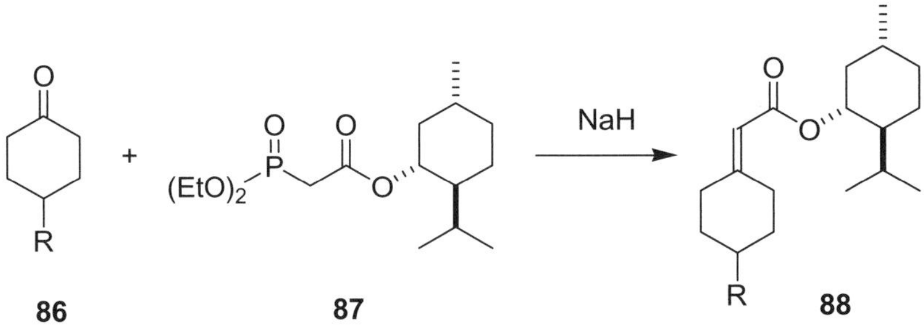

Carbohydrate-derived alcohols (isomannide and isosorbide) have been examined.[31] In one example, the isomannide derivative **89** was able to react with ketone **90** to afford **91** in excellent yield.

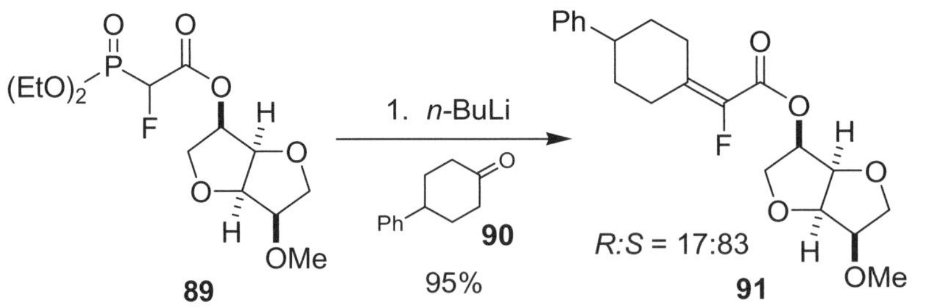

The first example of a chiral phosphorous-atom ylide was communicated by Bestmann.[32] The chiral reagent **93** underwent the Witting reaction with **92** to afford **94** but with poor chiral induction.

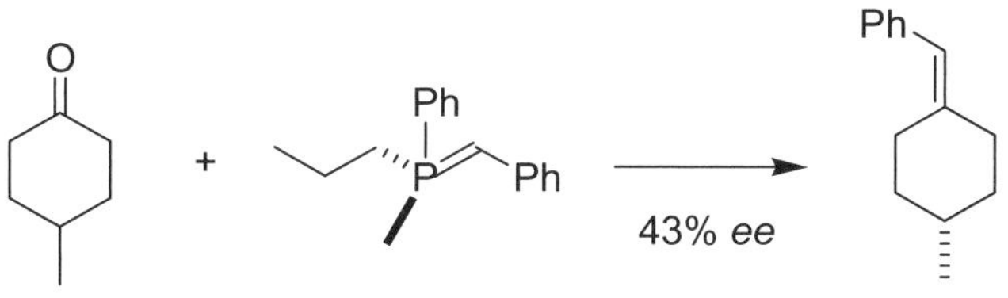

92 **93** **94**

Hanessian and coworkers reported an improved variation of this approach.[33] They were able to prepare the chiral phosphorous reagent **95**. Upon deprotonation to form the ylide and exposure to ketone **96**, the desired product **97** could be formed in good optical purity.

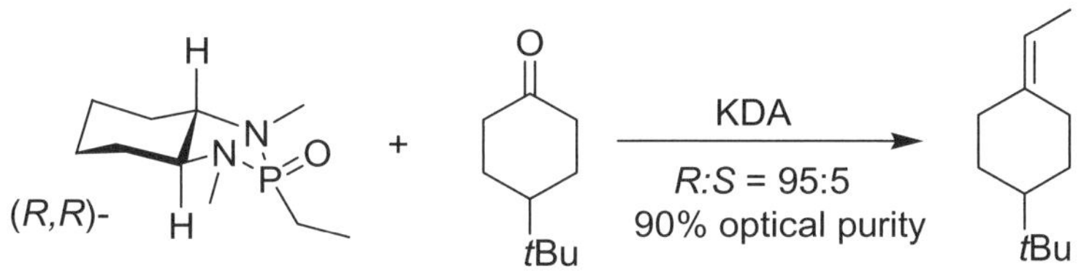

95 **96** **97**

Additional examples of this approach made use of other "ligands" on the phosphorous atom. For example, Denmark made use of bicyclic [2.2.1] system **98**.[30] Reaction of the corresponding ylide with ketone **96** generated ester **99** in good yield. The level of chiral induction could be modulated by the substituent on the nitrogen atom.

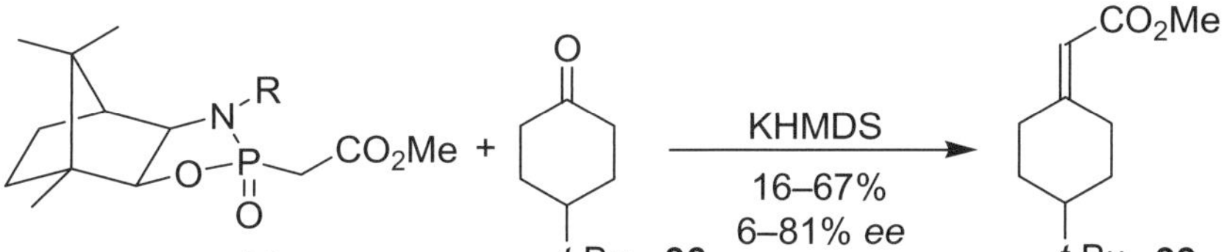

98 **96** **99**

Whereas, Fuji made use of a binol auxiliary in the preparation of allenes.[35] Ketene intermediates could be generated by base induced elimination from **100**. Reaction of this intermediate with the ylide formed from **101** was then able to generate allene **102** in good yield but with a varying amount of chiral induction.

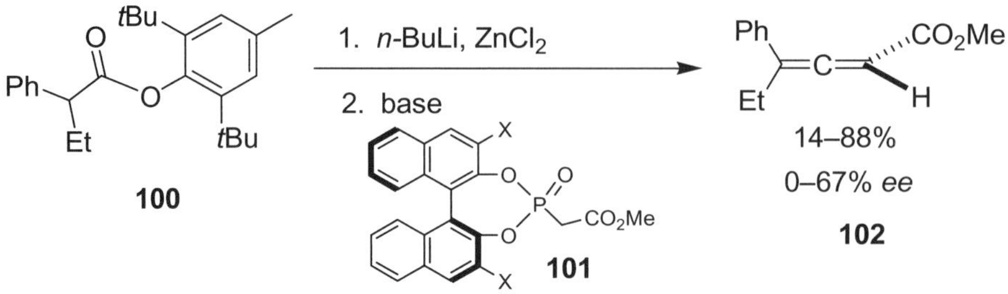

Phosphazenes are the nitrogen version of the Wittig ylide and give rise to the aza-Wittig reaction.[36] Thus **103** is able to react with carbonyl compound **30** to generate the corresponding imine **104** with concomitant loss of phosphine oxide.

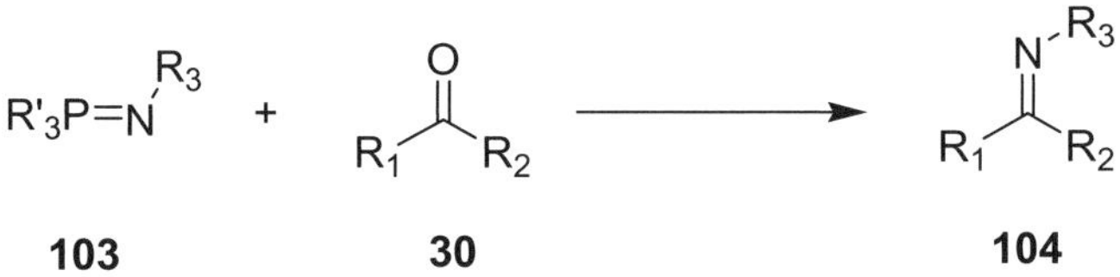

One example of the utility of this variation of the reaction can be illustrated in a synthesis of the alkaloid (–)-dendrobine **108**.[37] The tricyclic azide **105** was exposed to triphenylphosphine to generate the phosphazene **106** *in situ*. The intramolecular aza-Wittig then produced **107**, an advanced intermediate towards **108**.

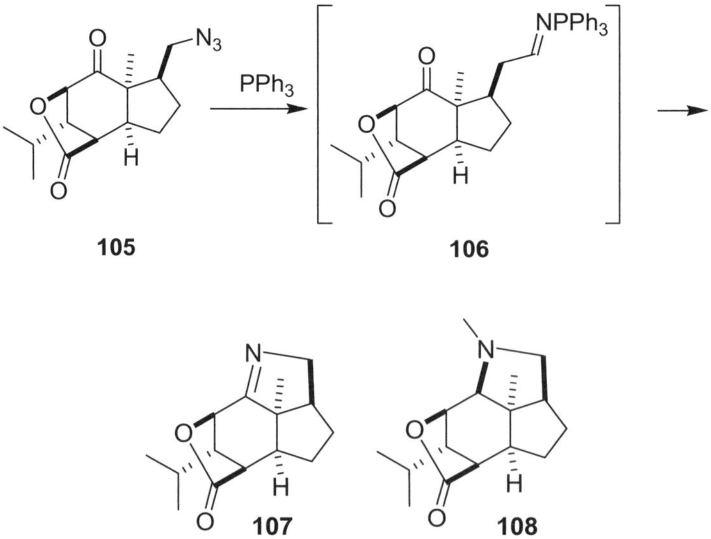

While the Wittig reaction is traditionally thought to occur predominately with aldehydes and ketone, non-classical Wittig reactions are known to occur with esters, lactones, thiol esters, amides, anhydrides, and other functional groups.[38] One such example can be illustrated using the synthesis of (–)-quinocarcin **111**.[39] Exposing phosphonium bromide **109** to potassium *tert*-butoxide at elevated temperatures generated the corresponding ylide which was able to regioselectively react with the imide carbonyl to generate **110**. This product was readily transformed into the antitumor antibiotic **111**.

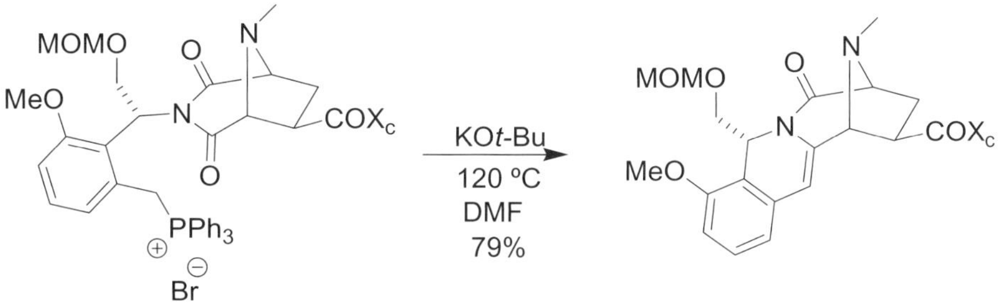

109 **110**

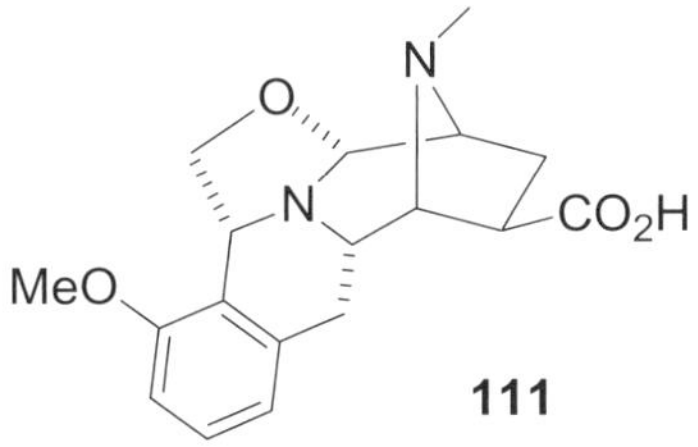

111

Other modifications or variations of the Wittig reaction exist. Some are more analogous than others. For example, sulfur ylides have also been investigated and the Corey–Chaykovsky reaction employs their use.[40] However, olefination is not the outcome of this reaction sequence. The nucleophilic addition of ylide **112** to the carbonyl compound **30** gives rise to the corresponding epoxides **113**. Furthermore, reaction with imines produces aziridines and cyclopropanes are formed with α,β-unsaturated carbonyl systems.

112 **30** **113**

While alkenes are produced in the Peterson reaction,[41] the key reactive species is α-silylcarbanion **114**. Reacting **114** with **30** using either acidic or basic work-up conditions gives rise to the corresponding alkene **43**.

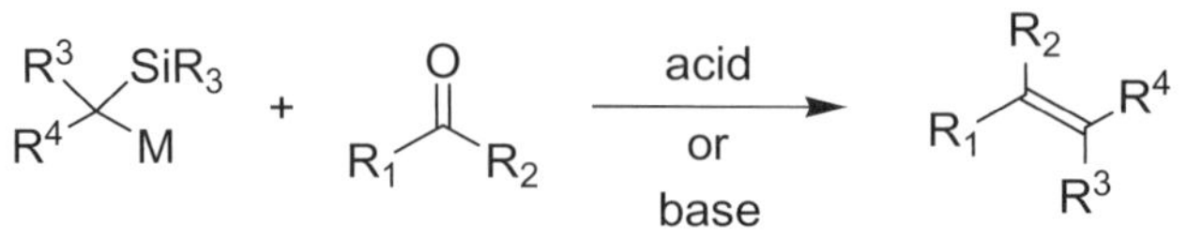

114 **30** **43**

The Tebbe reaction makes use of an organometallic reagent to carry out the olefination process.[42] The reactive species **116** is formed from the Tebbe reagent ($Cp_2TiCH_2AlCl_2$). Treating carbonyl compound **30** with **116** affords alkene **115**. When exposing ester or amide **117** to **116** the corresponding enol or enamine is formed, respectively. Mechanistically, the reaction proceeds via a titana-oxetane which would be analogous to the oxaphosphetane intermediate in the Wittig.

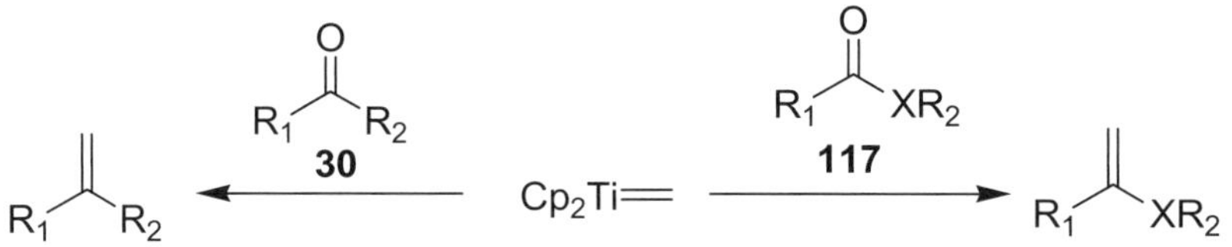

115 **116** **118**

2.13.5 *Synthetic Utility*

While the use of this reaction for the preparation of alkenes is well documented, and discussed (*vide infra*), there are examples that extend the use of this chemistry. Dauben[43] reported the synthesis of cyclohexadienes using allyl phosphonates. Thus exposure of ketone **119** to ylide **120** afforded intermediate **121** by an initial Michael addition. Regeneration of the ylide to form **122**, catalyzed by the resultant enolate, could then facilitate the ring closure to form the desired product **123**.

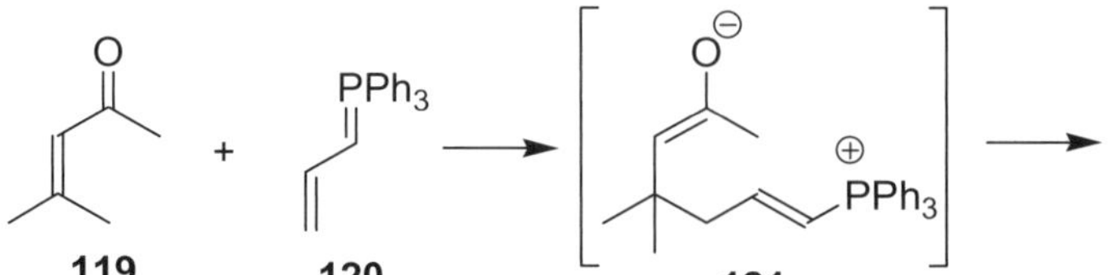

119 **120** **121**

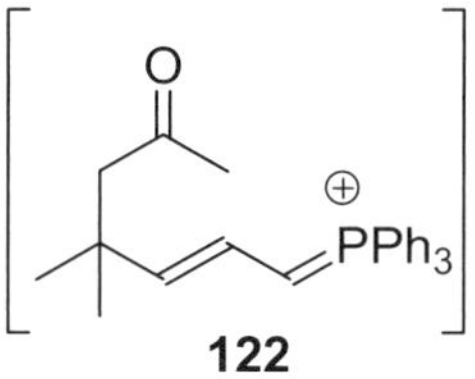

122 **123**

In a similar manner, Fuchs[44] showed that enolate **124** could undergo a 1,5-addition to butadienyl phosphonium ion **125** to generate ylide **126** *in situ* that could then give cyclohexadiene **127** by Wittig ring closure.

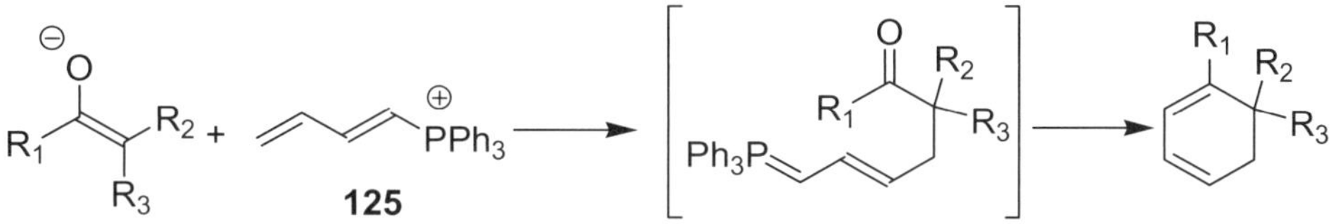

124 **126** **127**

Direct intramolecular Wittig or HWE reactions are possible but one needs to be aware of sensitive functionality and the formation of the ylide. The Roush method (*vide supra*) provided one way to detour past this potential road block. In small systems, for example, Woodward[45] was able to bypass this problem. His preparation of penem antibiotics involved ylide **128** reacting, intramolecularly, to form the bicyclic ring framework **129**.

128 **129**

More elaborate systems require greater finesse. In their approach to amphoteronolide B **132**, Nicolaou[46] was able to use a HWE reaction to carry out the alkene formation, ring-closure reaction very late in the total synthesis. Thus treatment of **130** with potassium carbonate produced **131** in 70% yield.

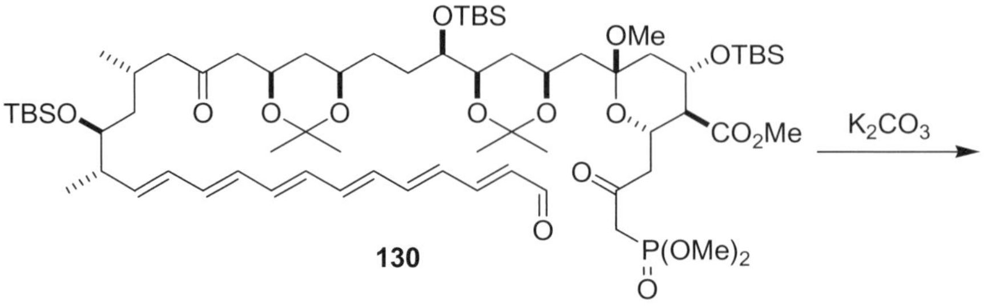

130

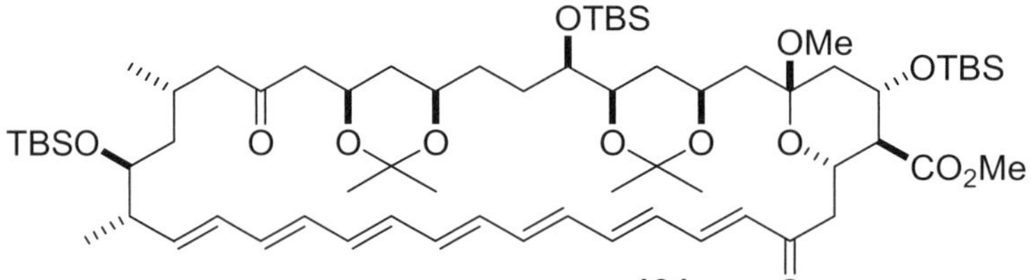

131

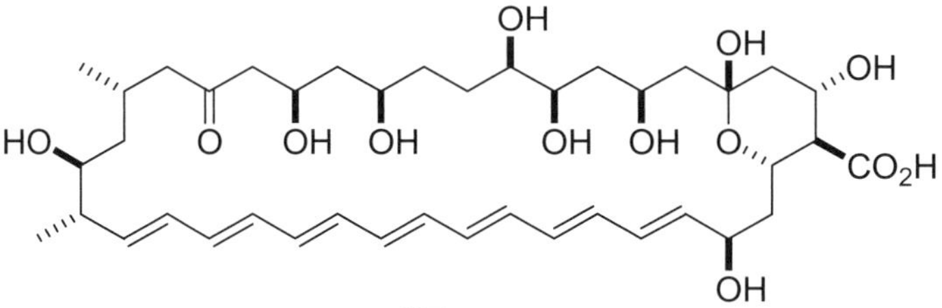

132

Nicolaou also made use of the Wittig reaction in his synthesis of brevetoxin B **136**.[47] The western **133** and eastern **134** portions of the molecule were linked in 75% yield using a classical Wittig reaction to generate **135**.

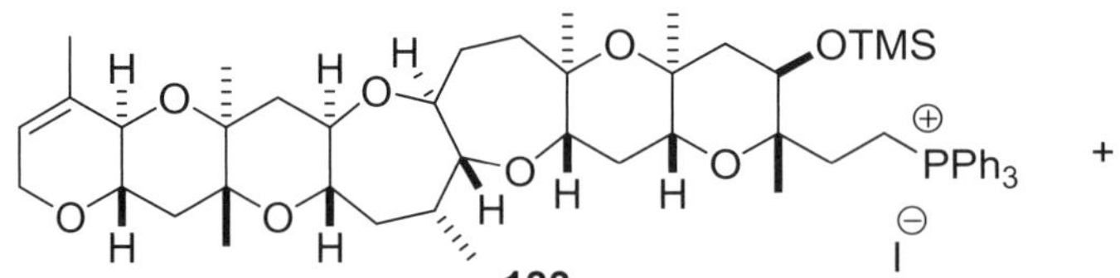

133

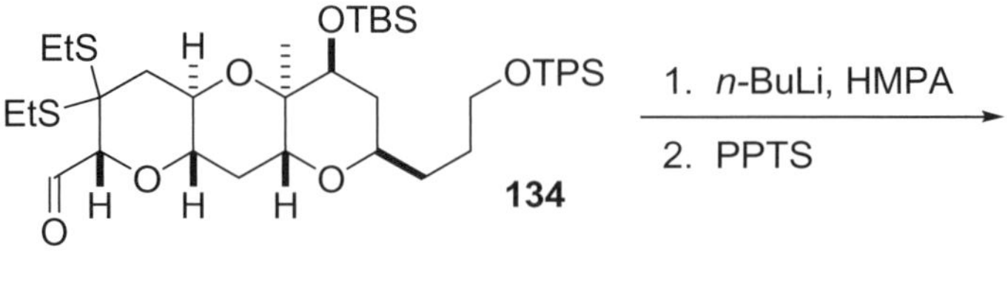

134

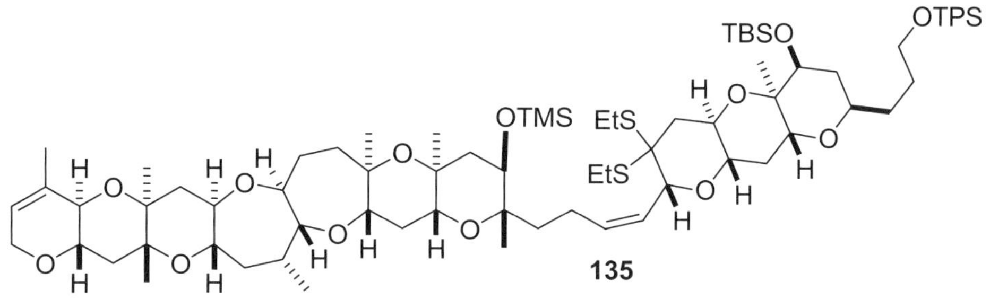

135

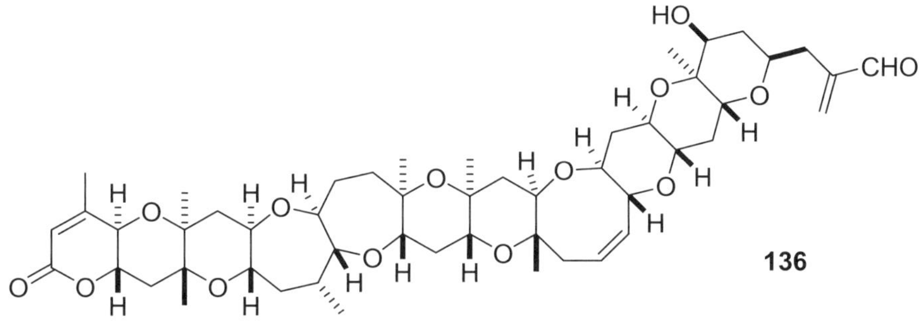

136

Molander has recently reported the compatibility of organotrifluoroborates with the Wittig and HWE reactions.[48] The value in this transformation can be realized in light of the increased use of this functional group in palladium cross-coupling reactions. This approach allows one to structurally elaborate a relatively sensitive moiety before conducting, for example, a Suzuki reaction. Thus, aldehyde **137** afforded alkene **139** when exposed to ylide **138**.

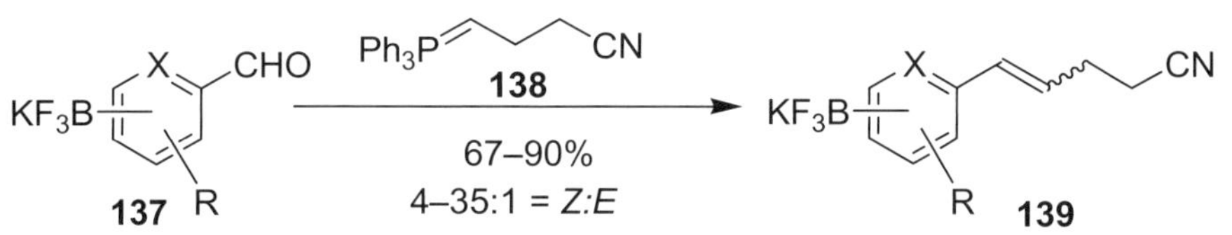

A synthetic approach to the anti-influenza agent, A-215675 (**143**) made use of an embedded Weinreb amide.[49] The chiral aldehyde **141** was treated with functionalized ylide generated from **140** to afford **142**. The Weinreb amide moiety was then used to elaborate **142** to the desired target **143**.

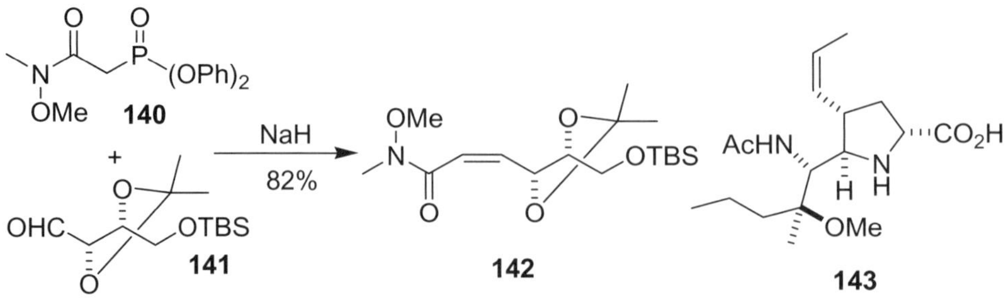

A model system for the marine sponge metabolite, spiculoic acid **147** was generated by an intramolecular Diels–Alder reaction.[50] The requisite triene **146** was prepared using a Wittig reaction. To this end, aldehyde **144** was coupled with ylide **145**. The resultant **146** was not isolated and underwent an intramolecular Diels–Alder reaction to generate a mixture of adducts, **148** and **149**.

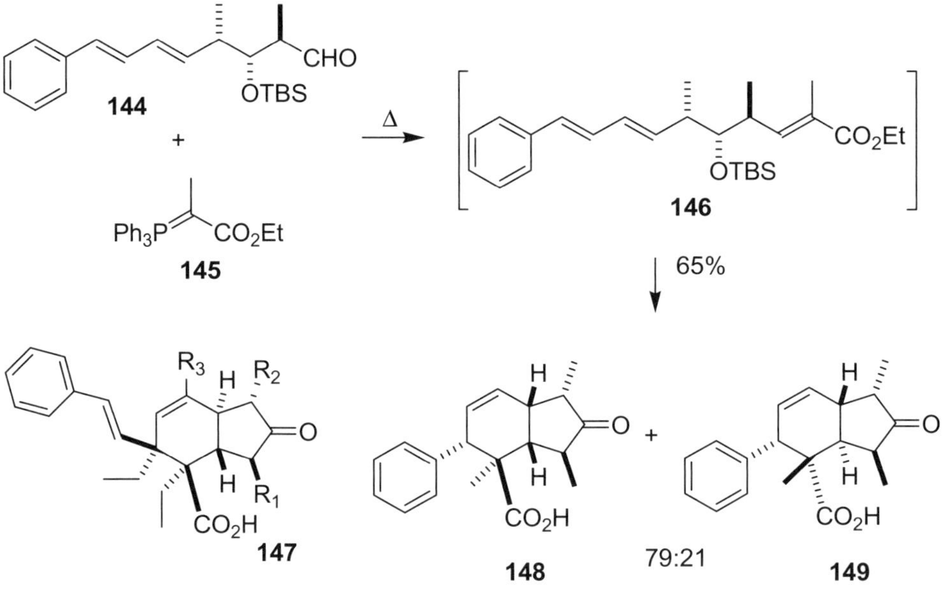

Applications of these reactions to the construction of heterocyclic ring systems are wide and varied.[51] Ring frameworks of all sizes and heteroatoms including oxygen, nitrogen, and sulfur are fully exemplified. One rapid access to isoquinoline ring systems is shown in the reaction of dialdehyde **150** with the bis-functionalized reagent **151** to produce **152** in 55% yield.[52]

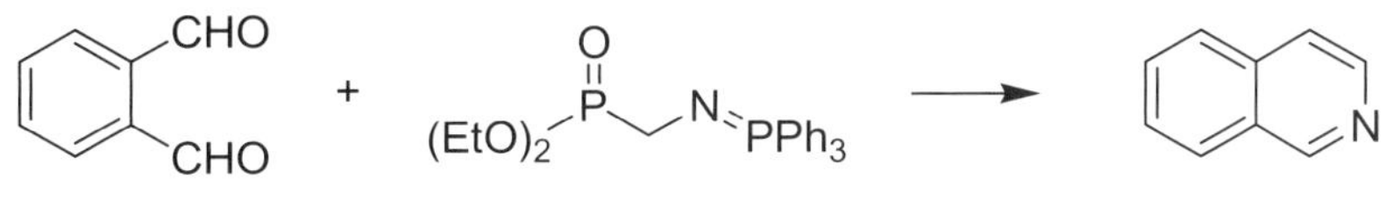

150 151 152

An example of ring sizes that are readily accessed was reported by Vogtle.[53] The pyridinophane **155** was assembled by treating bi-pyridine **153** with bis-phosphonium salt **154** in the presence of base followed by hydrogenation of the resultant alkenes.

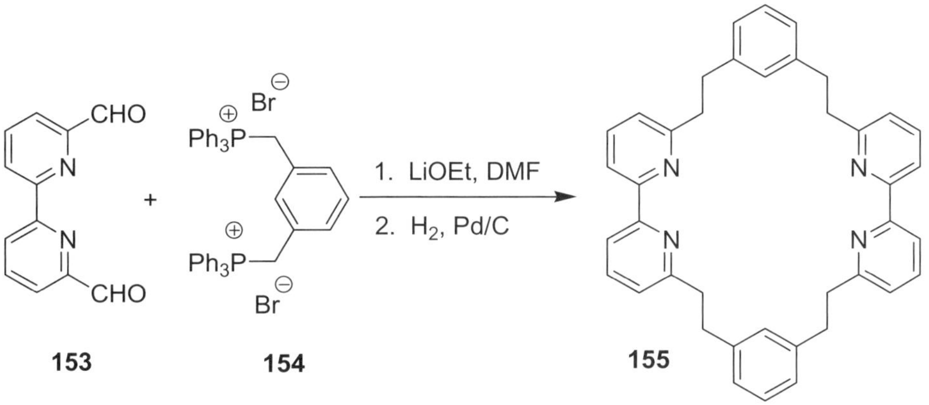

153 154 155

2.13.6 Experimental

Classical Wittig Reaction[1b,54]

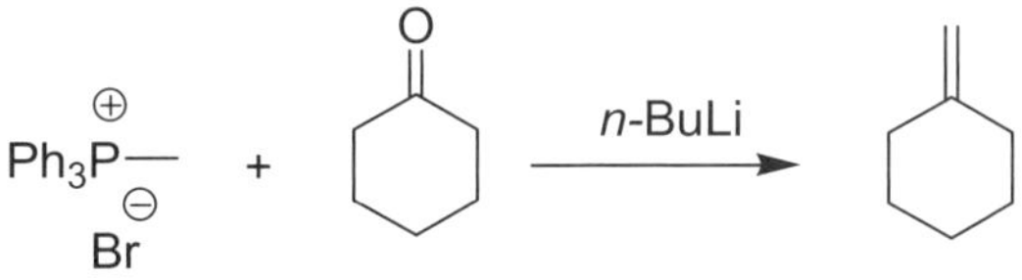

156 157 158

Methylenecyclohexane (158)

To a solution of *n*-BuLi (100 mmol) and 200 mL of anhydrous ether is cautiously added triphenylmethylphosphonium bromide **156** (35.7 g, 100 mmol) over 5 min. The reaction was stirred at room temperature for 4 h before adding cyclohexanone **157** (10.8 g, 110 mmol) dropwise. A white precipitate forms and the reaction is heated under reflux overnight. Upon cooling to room temperature, the precipitate is removed by suction and washed with ether. The combined organic phases are washed with water and dried with $CaCl_2$. Fractionation of the crude product, after the ether is carefully removed, afforded **158** (3.8 g, 40%) as a liquid.

Horner–Wadsworth–Emmons Reaction[13,55]

157 **65** **159**

Ethyl Cyclohexylideneacetate (159)

To a slurry of 50% sodium hydride (2.4 g, 0.05 mol) in 100 mL of dry 1,2-dimethoxyethane at room temperature was added triethyl phosphonoacetate **65** (11.2 g, 0.05 mol) dropwise. Once addition was complete, the reaction was stirred an additional 60 min. until all gas evolution had ceased. Cyclohexanone **157** (4.9 g, 0.05 mol) was added at such a rate as to keep the reaction temperature below 30 °C. The reaction was stirred for 15 min after addition was completed before taking-up the viscous, semi-solid in a large excess of water. The aqueous solution was extracted with ether and the organic phase was dried with $MgSO_4$, filtered, and concentrated to give 5.8 g (70%) of **159** as a liquid.

2.13.7 References

1. [R] a) Trippet, S. *Q. Rev. Chem. Soc.* **1963**, *17*, 406–440. b) Wittig, G. *Pure Appl. Chem.* **1964**, *9*, 245–254. c) Maercker, A. *Org. Reac.* **1965**, *14*, 270–490. d) Schlosser, M. *Top. Stereochem.* **1970**, *5*, 1–30. e) Treucroft, J.; Sammes, P. G. *Q. Rev. Chem. Soc.* **1971**, *25*, 135–169. f) Boutagy, J.; Thomas, R. *Chem. Rev.* **1974**, *74*, 87–99. g) Wadsworth, W. S. Jr. *Org. Reac.* **1977**, *25*, 73–253. h) Bestmann, H. J. *Pure Appl. Chem.* **1979**, *51*, 515–533. i) Bestmann, H. J. *Pure Appl. Chem.* **1980**, *52*, 771–788. j) Becker, K. B. *Tetrahedron* **1980**, *36*, 1717–1745. k) Bestmann, H. J.; Vostrosky, O. *Top. Curr. Chem.* **1983**, *109*, 85–163. l) Stec, W. *Acc. Chem. Res.* **1983**, *16*, 411–417. m) Murphy, P. J. *Chem Soc. Rev.* **1988**, *17*, 1–30. n) Maryanoff, B. E.; Reitz, A. B. *Chem. Rev.* **1989**, *89*, 863–927. o) Minami, T.; Motohoshiya, J. *Synthesis*, **1992**, 333–349. p) Nicolaou, K. C.; Harter, M. W.; Gunzner, J. L.; Nadin, A. *Liebigs Ann.* **1997**, 1283–1301. q) Rein, T.; Pederson, T. M. *Synthesis* **2002**, 579–593.
2 For a recent review see: [R] Hiersemann, M.; Abraham, L.; Pollex, A. *Synlett* **2003**, 1088–1095.
3. a) Wittig, G.; Geissler, G. *Liebigs Ann.* **1953**, *580*, 44–57. b) Wittig, G.; Schollkopf, U. *Liebigs Ann.* **1954**, 1318– 1330. c) Wittig, G.; Haag, W. *Liebigs Ann.* **1955**, 1654–1666.
4. [R] Hoffmann, R. W. *Angew. Chem., Int. Ed.* **2001**, *40*, 1411–1416.
5. Pommer, H. *Angew. Chem., Int. Ed.* **1977**, *16*, 423–492.
6. [R] a) Vedejs, E. *Science* **1980**, *207*, 42–44. [R] b) Wittig, G. *Science* **1980**, *210*, 600–604.
7. a) Schlosser, M.; Christmann, K. F. *Liebigs Ann.* **1967**, 1–35. b) Schlosser, M.; Schaub, B.; de Oliveira-Neto, J.; Jeganathan, S. *Chimia* **1986**, *40*, 244–245. c) Reitz, A. B.; Nortey, S. O.; Jordan, A. D. Jr.; Mutter, M. S.; Maryanoff, B. E. *J. Org. Chem.* **1986**, *51*, 3302–3308. d) Wang, Q.; Deredas, D.; Huynh, C.; Schlosser, M. *Chem. Eur. J.* **2003**, *9*, 570–574.
8. Vedejs, E. ; Snobble, K. A. J. *J. Am. Chem. Soc.* **1973**, *95*, 5778–5780.

9. Maryanoff, B. E.; Reitz, A. B.; Mutter, M. S.; Inners, R. R.; Almond, H. R. Jr.; Whittle, R. R.; Olofson, R. A. *J. Am. Chem. Soc.* **1986**, *108*, 7664–7678.
10. a) Vedejs, E.; Meier, G. P.; Snobble, K. A. J. *J. Am. Chem. Soc.* **1981**, *103*, 2823–2831. b) Vedejs. E.; Marth, C. F.; Ruggeri, R. *J. Am. Chem. Soc.* **1988**, *110*, 3940–3948. c) Vedejs, E.; Marth, C. F. *J. Am. Chem. Soc.* **1988**, *110*, 3948–3958. d) Vedejs, E.; Marth, C. F. *J. Am. Chem. Soc.* **1990**, *112*, 3905–3909. e) Vedejs, E.; Peterson, M. J. *Adv. Carbanion Chem.* **1996**, *2*, 1–85.
11. Neumann, R. A.; Berger, S. *Eur. J. Org. Chem.* **1998**, 1085–1087.
12. a) Horner, L.; Hoffmann, H.; Wippel, H. G. *Chem. Ber.* **1958**, *91*, 61–63. b) Horner, L.; Hoffmann, H.; Wippel, H. G.; Klahre, G. *Chem. Ber.* **1959**, *92*, 2499–2505.
13. Wadsworth, W. S, Jr.; Emmons, W. D. *J. Am. Chem. Soc.* **1961**, *83*, 1733–1738.
14. a) Corey, E. J.; Kwiatkowski, G. T. *J. Am. Chem. Soc.* **1966**, *88*, 5652–5653. b) Corey, E. J.; Kwiatkowski, G. T. *J. Am. Chem. Soc.* **1966**, *88*, 5653–5654. c) Corey, E. J.; Kwiatkowski, G. T. *J. Am. Chem. Soc.* **1968**, *90*, 6816–6821.
15. Corey, E. J.; Kwiatkowski, G. T. *J. Am. Chem. Soc.* **1966**, *88*, 5654–5656.
16. Johnson, C. R.; Elliott, R. C. *J. Am. Chem. Soc.* **1982**, *104*, 7041–7044.
17. Still, W. C.; Gennari, C. *Tetrahedron Lett.* **1983**, *24*, 4405–4408.
18. Castells, J.; Lopez-Calahorra, F.; Yu, Z. *Tetrahedron* **1994**, *50*, 13765–13774.
19. a) Schlosser, M.; Christmann, K. F. *Angew. Chem., Intl. Ed* **1966**, *5*, 126. b) Schlosser, M.; Christmann, K. F.; Piskala, A. *Chem. Ber.* **1970**, *103*, 2814–2820.
20. Corey, E. J.; Yamamoto, H. *J. Am. Chem. Soc.* **1970**, *92*, 226– 228.
21. Blanchette, M. A.; Choy, W.; Davis, J. T.; Essenfeld, A. P.; Masamune, S.; Roush, W. R.; Sakai, T. *Tetrahedron Lett.* **1984**, *25*, 2183–2186.
22. Rathke, M. W.; Nowak, M. *J. Org. Chem.* **1985**, *50*, 2624–2636.
23. Simoni, D.; Rossi, M.; Rondannin, R.; Mazzali, A.; Baruchello, R.; Malagutti, C.; Roberti, M.; Invidiata, F. P. *Org. Lett.* **2000**, *2*, 3765–3768.
24. Blasdel, L. K.; Myers, A. G. *Org. Lett.* **2005**, *7*, 4281–4283.
25. Camps, F.; Castells, J.; Font, J.; Vela, F. *Tetrahedron Lett.* **1971**, 1715–1716. See also: a) Heitz, W.; Michels, R. *Angew. Chem., Intl. Ed.* **1972**, *11*, 298–299. b) Bolli, M. H.; Ley, S. V. *J. Chem. Soc., Perkin Trans. 1* **1998**, 2243–2246.
26. Johnson, C. R.; Zhang, B. *Tetrahedron Lett.* **1995**, *36*, 9253–9256.
27. Williard, R.; Jammalamadaka, V.; Zava, D.; Benz, C. C.; Hunt, C. A.; Kushner, P. J.; Scanlan, T. S. *Chem. Biol.* **1995**, *2*, 45–51.
28. Spinella, A.; Fortunati, T.; Soriente, A. *Synlett* **1997**, 93–94.
29. Westman, J. *Org. Lett.* **2001**, *3*, 3745–3747.
30. Tomoskozi, I.; Janzso, G. *Chem. Ind.* **1962**, 2085–2086.
31. Sano, S.; Teranishi, R.; Nakano, F.; In, K.; Takeshige, H.; Ishii, T.; Shiro, M.; Nagao, Y. *Heterocycles* **2003**, *59*, 793–804.
32. Bestmann, H. J.; Heid, E.; Ryschka, W.; Lienert, J. *Liebigs Ann.* **1974**, 1684–1687.
33. Hanessian, S.; Delorme, D.; Beaudoin, S.; Leblanc, Y. *J. Am. Chem. Soc.* **1984**, *106*, 5754–5756.
34. Denmark, S. E.; Rivera, I. *J. Org. Chem.* **1994**, *59*, 6887–6889.
35. Tanaka, K.; Otsubo, K.; Fuji, K. *Tetrahedron Lett.* **1996**, *37*, 3735–3738.
36. [R] For a recent review see: Palacios, F.; Alonso, C.; Aparicio, D.; Rubiales, G.; de los Santos, J. M. *Tetrahedron* **2007**, *63*, 523–575. For examples of phospha-variations see; Shah, S.; Protasiewicz, J. D. *Coord. Chem. Rev.* **2000**, *210*, 181–201.
37. Sha, C.-K.; Chiu, R.-T.; Yang, C.-F.; Yao, N.-T.; Tseng, W.-H.; Liao, F.-L.; Wang, S.-L. *J. Am. Chem. Soc.* **1997**, *119*, 4130–4135.
38. For a recent review see: Murphy, P. J.; Lee, S. E. *J. Chem. Soc., Perkin Trans.* 1 1999, 3049–3066.
39. Garner, P.; Ho, W. B.; Shin, H. *J. Am. Chem. Soc.* **1993**, *115*, 10742–10753.
40. [R] See entry from this series: Li, J. J. *Name Reactions in Heterocyclic Chemistry* **2005**, pp 2–14, John Wiley & Sons, Inc., Hoboken, NJ.
41. Recent reviews see: [R] a) van Staden, L. F.; Gravestock, D.; Ager, D. J. *Chem. Soc. Rev.* **2002**, *31*, 195–200. [R] b) Kawashima, T.; Okazaki, R. *Synlett* **1996**, 600–608. [R] c) Ager, D. J. *Org. Reac.* **1990**, *38* 1–223.

42. For recent reviews see: [R] a) Pine, S. H. Org. React. 1993, 43, 1–91. [R] b) Beadham, I.; Micklefield, J. *Curr. Org. Syn.* **2005**, *2*, 231–259.
43. Dauben, W. G.; Hart, D. J.; Ipaktschi, J.; Kozilowski, A. P. *Tetrahedron Lett.* **1973**, 4425–4428.
44. Fuchs, P. L. *Tetrahedron Lett.* **1974**, 4055–4058.
45. Ernest, I.; Gosteli, J.; Greengrass, C. W.; Holick, W.; Jackman, D. E.; Pfaendler, H. R.; Woodward, R. B. *J. Am. Chem. Soc.* **1978**, *100*, 8214–8222.
46. Nicolaou, K. C.; Daines, R. A.; Chakroborty, T. K.; Ogawa, Y. *J. Am. Chem. Soc.* **1988**, *110*, 4685–4696.
47. Nicolaou, K. C.; *Angew. Chem., Intl. Ed.* **1996**, *35*, 589–607.
48. Molander, G. A.; Figueroa, R. *J. Org. Chem.* **2006**, *71*, 6135–6140.
49. Momose, T.; Hama, N.; Higashino, C.; Sato, H.; Chida, N. *Tetrahedron Let*t. **2008**, *49*, 1376–1379.
50. Crossman, J. S.; Perkings, M. V. *Tetrahedron* **2008**, *64*, 4852–4967.
51. a) Heron, B. M. *Heterocycles* **1995**, *41*, 2357–2386. b) Schobert, R.; Gordon, G. J. *Curr. Org. Chem.* **2002**, *6*, 1181–1196.
52. Katritzky, A. R.; Jiang, J.; Greenhill, J. V. *J. Org. Chem.* **1993**, *58*, 1987–1988.
53. Vogtle, F.; Hochberg, R.; Kochendorfer, F.; Windscheif, P.-M.; Volkmann, M.; Jansen, M. *Chem. Ber.* **1990**, *123*, 2181–2185.
54. Wittig, G.; Schoellkopf, U. *Org. Syn.* **1960**, *40*, 66–68.
55. Wadsworth, W. S., Jr.; Emmons, W. D. *Org. Syn.* **1965**, *45*, 44–47.

Chapter 3. Radical Chemistry

3.1	Barton–McCombie deoxygenation	614
3.2	Barton nitrite photolysis	633
3.3	Sandmeyer reaction	648
3.4	Wohl–Ziegler reaction	661

Note from the editor:

More appropriately, these four radical-related name reactions belong to the previous volume (Volume 2) on *Name Reactions for Functional Group Transformations*. They were, regrettably, not included in that volume due to logistical reasons. Because of their importance in organic synthesis, instead they are added to this volume (Volume 3) on *Name Reactions for Homologations-1*. After all, it is better to be late than never.

3.1 Barton-McCombie Deoxygenation

John Mancuso

3.1.1 *Description*

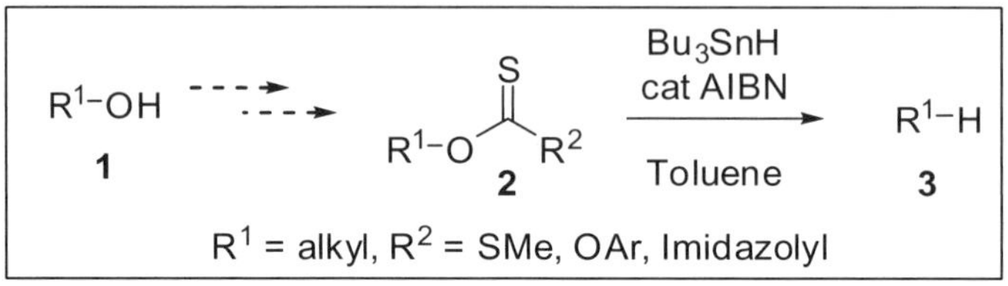

The Barton–McCombie reaction describes the overall reduction of a 1°, 2°, or 3° alcohol (**1**) to the parent alkane (**3**), effectively replacing –OH with –H. Reduction is achieved via pre-activation of the hydroxyl group as its thiocarbonyl derivative (**2**) followed by reaction with a reducing agent such as tributyltin hydride. The reductive process involves a radical-mediated transformation.

3.1.2 *Historical Perspective*

In their seminal paper, Barton and McCombie describe a method to deoxygenate secondary alcohols through a radical chain mechanism.[1] This process was presented as an alternative to the standard conditions of derivatization of an alcohol to a tosylate/mesylate followed by reduction. Such polar processes are problematic with hindered carbon centers and can lead to rearrangements and/or eliminations if carbocationic intermediates are involved.

Inspired by the mechanism of thiobenzoate pyrolysis, whereby the driving force of the fragmentation is the conversion of a thiocarbonyl radical to a more stable carbonyl radical, a series of *O*-alkylthioesters were prepared from steroid derivatives bearing a secondary alcohol function. A variety of thiocarbonyl groups were originally screened for the reduction of *O*-ergosteryl derivatives and it was determined that *S*-methylxanthates (**4**), thionobenzoates (**5**) and thiocarbonyl imidazolides (**6**) could be readily reduced with tributyltin hydride. The neutral conditions of the Barton-McCombie reaction allow functional transformations to occur while preserving labile groups, thus the reaction has proven successful for use on the preparation of a variety of natural products, saccharides, and terpene derivatives.[2]

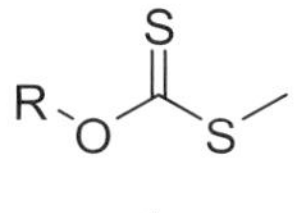

4

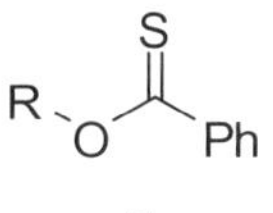

5

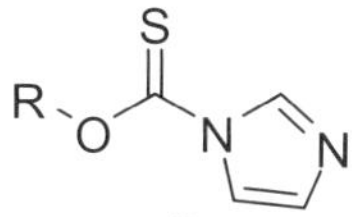

6

3.1.3 Mechanism

A radical initiator, such as azobisisobutyronitrile (AIBN, **7**) decomposes homolytically under reaction conditions to generate **8** which abstracts a hydrogen from tributyltin hydride (**9**) creating a tin-centered radical **10**. The tin radical attacks at the sulphur atom of the thiocarbonyl derivative **2**, generating intermediate **11** which collapses leaving tributyltin xanthate **12** (which can further decompose to O=C=S and **13**) and alkyl radical **14**. Intermediate **14** can abstract another atom of hydrogen from tributyltin hydride to generate the desired alkane product **3**, and replenish the pool of tin radical **10**.

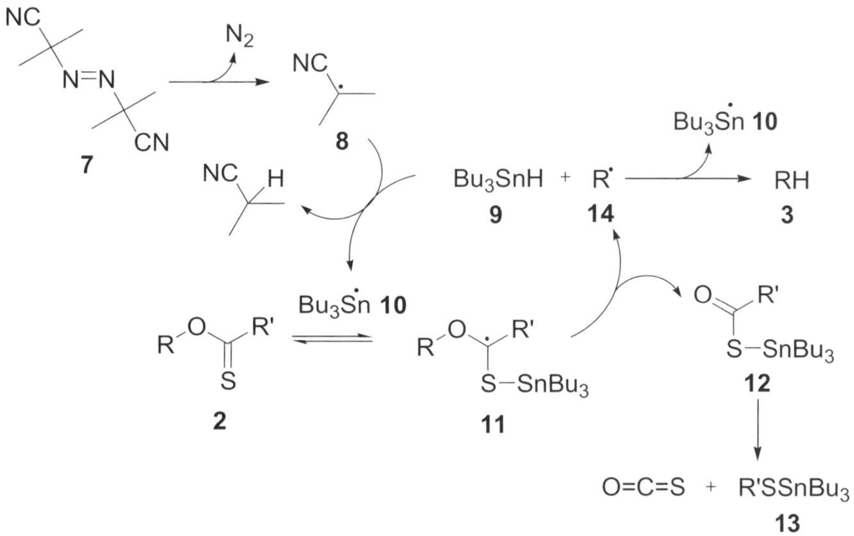

Fragmentation of **11** is known to be the rate determining step and, depending on its lifetime, this radical can be captured intramolecularly if a proximal acceptor group is present.[3]

The driving force of the reaction is the formation of a very stable sulfur–tin bond, ultimately yielding **13** and oxycarbon sulfide. The reaction mechanism has been probed by ^{119}Sn NMR spectroscopy.[4] Since the reaction proceeds through free-radical intermediates, the absence of charged entities ensures a low solvation and thus the reaction exhibits a higher degree of insensitivity to steric factors.

Product distribution is highly dependant on the substitution pattern on the radical center of **11**. For cyclic thiocarbonate **15** containing primary and secondary centers, fragmentation gives the more stable 2° radical intermediate, thus generating **16** as the sole product. For cyclic carbonate **17** with two secondary centers, a mixture of two regiomeric deoxygenated products **18** and **19** is obtained.[5]

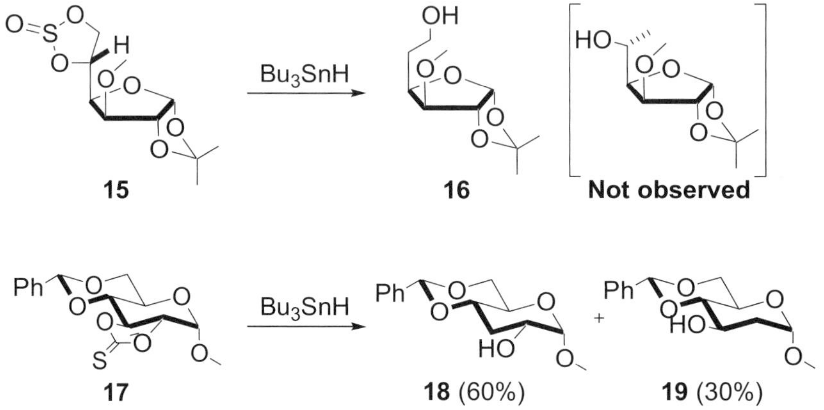

Ziegler and Zheng have investigated a substrate containing a cyclic thiocarbonate of primary and secondary hydroxyl functions (**20**) in which deoxygenation of the primary center is preferred.[6] In this case, relief of bond angle bending strain is responsible for the regiochemistry observed, thus fragmentation gives **21** as the sole product.

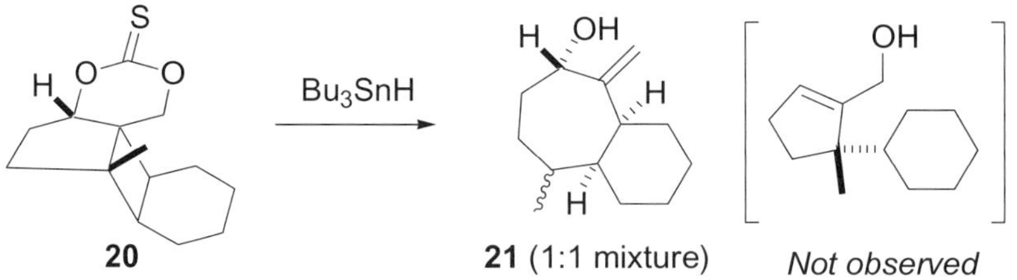

Dideoxygenation via 1,2-dixanthate elimination is also possible and, in acyclic systems, will usually generate the more stable *trans* olefin.[7] For example, treatment of the bis(xanthate) **22** with tin hydride in refluxing toluene gave selective formation of the *E*-alkene **23**.

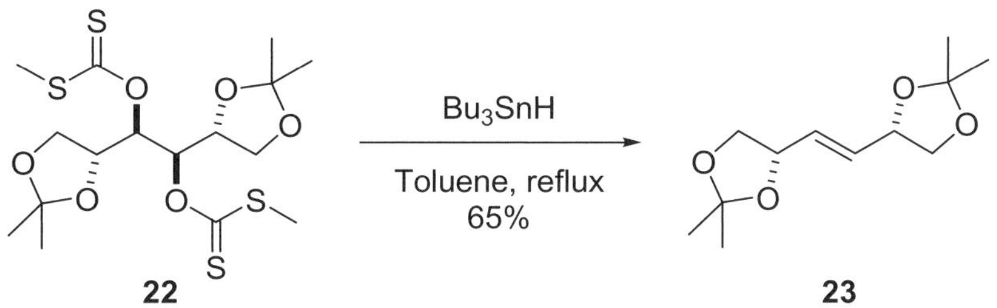

22 23

3.1.3 *Variation and Improvements*

Robins later introduced the (phenoxy)thiocarbonate **24** as a more reactive derivative[8] for deoxygenation of secondary centers. Subsequently, Barton and Jaszberenyi showed that introduction of electron-withdrawing groups on the phenyl ring increased the speed of the desired fragmentation.[9] In particular, use of the 2,4,6-trichlorophenoxy (**25**) or perfluorophenoxy (**26**) derivatives allowed reduction to proceed within 15 minutes in an AIBN/Bu_3SnH system at 110 °C. In addition to the simpler procedure for preparation of these thionocarbonate derivatives and their deoxygenation, the formation of side products for the latter step was minimized.

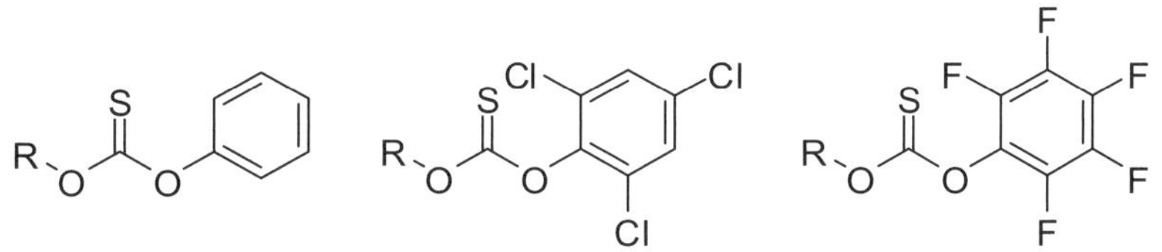

24 25 26

Optimization of the deoxygenation of primary alcohols, and in particular, neopentyl alcohols, for which the deoxygenation under neutral conditions should be much easier as opposed to ionic displacement and reduction, was examined.[10] It was determined that slow addition of tributyltin hydride to the thiocarbonyl derivative could allow deoxygenation to occur without competing Chugaev elimination. Formation of the primary radical was achieved at lower temperatures through the use of thionobenzoate (**5**) and thiocarbonylimidazolide (**6**) derivatives instead of xanthates (**4**). Compound **27** could be converted to **28** at 130 °C as opposed to 150 °C. Further temperature reduction (i.e., refluxing benzene or toluene) could be achieved through the use of halogenated derivatives **25** and **26**.[11]

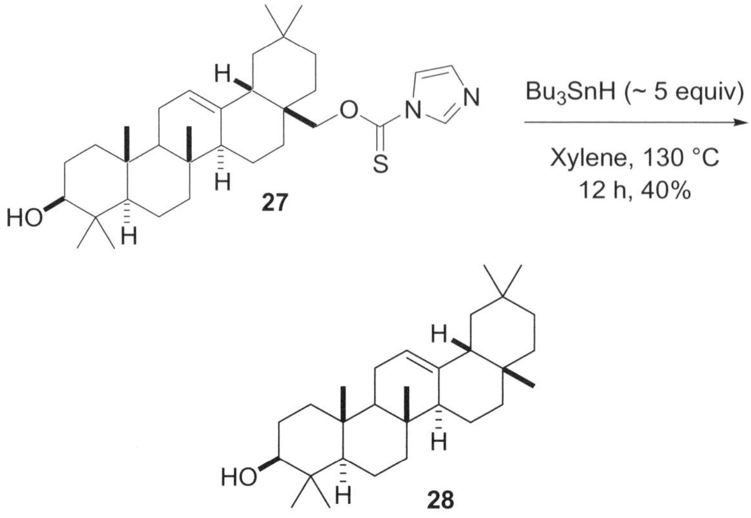

Deoxygenation of tertiary alcohols was examined, as an alternative and potentially superior procedure to protocols involving dehydration followed by hydrogenation.[12] The difficulty with the use of thiocarbonyl derivatives for this class of compounds lies in their propensity to eliminate readily to the olefin during preparation. Thioformates were investigated, even though these functional groups posed problems for secondary alcohol deoxygenations. Fission proceeds readily as scission of the C–O bond leads to the formation of a very stable tertiary radical.

Thioformate synthesis required that the tertiary alcohol **29** was reacted first with *p*-dimethylaminophenyl isocyanide in the presence of copper oxide to generate an isoamide intermediate (**30**). This was then converted to thioformate **31** through action of hydrogen sulfide and was subsequently reduced to **32** using standard Barton–McCombie conditions.

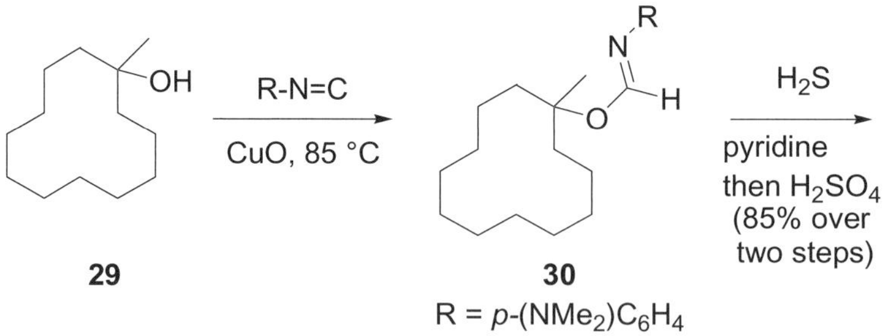

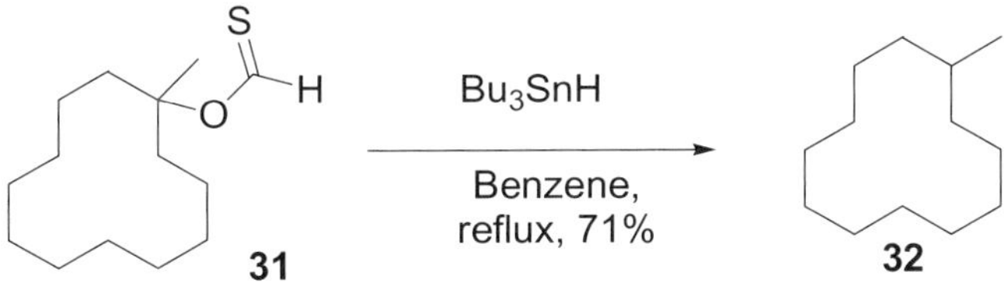

31 **32**

The oxalate derivatives of *tert*-butylsulfide and *N*-hydroxythiopyridone have also been used as deoxygenation precursors for tertiary alcohols. In these cases, the resultant radical intermediate was intercepted by activated olefins instead of a reducing agent.[13]

3.1.6 *Alternative Reducing Agents*

Tin-based Reductants

As a solution to the problem of completely removing tin residues during product purification, Neumann and Peterseim introduced an insoluble polystyrene-bound di-*n*-butyltin hydride reagent.[14] The reagent can be filtered from the product mixture and recycled for further use. A variety of secondary alcohols and diols could be deoxygenated in good yields (65–98%).

Fu and coworkers developed a variant of the Barton–McCombie deoxygenation which employs a catalytic amount of the tin component with an alternate stochiometric reductant.[15] The reduction of a thiocarbonate **33** by tin hydride (**9**) yields product **34** and O=C=S. The Bu_3Sn-OPh by-product **35**, can be reduced by a stochiometric silyl hydride **37**, such as poly(methyl)hydrosiloxane (PMHS), to regenerate **9** and generate PMHS–OPh **36**.

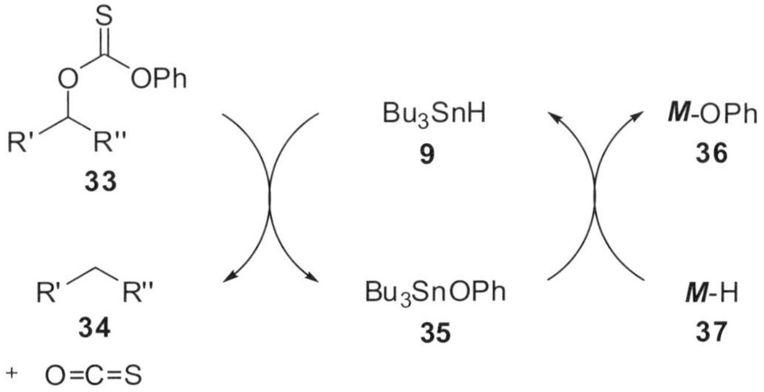

Optimal conditions involved the use of a catalytic amount (7.5 mol%) of $(Bu_3Sn)_2O$ with an excess (5 equiv) of PMHS. Yields were comparable to stoichiometric tin reactions. However, the catalytic tin protocol required addition of an excess of *n*-BuOH for efficient catalyst turnover. This additive was found to aid in the initial generation of Bu_3SnH from the oxide as well as facilitate regeneration of the hydride from Bu_3SnOPh, and displayed a 10-fold reaction acceleration over reactions without *n*-BuOH.

Dumartin and coworkers extended Fu's catalytic protocol by replacing $(Bu_3Sn)_2O$ with Neumann's supported tin hydride polymer.[16] However, PMHS was found to be inefficient, potentially due to problems with PMHS diffusion into the polystyrene matrix of the supported tin reagent. The less bulky trimethoxysilane was more successful at promoting tin hydride turnover. Although yields were comparable and residual tin contamination was negligible, it should be noted that reaction times (24–36 h) were much longer than the use of supported tin hydride alone or Fu's original $(Bu_3Sn)_2O$ protocol (9–16 h).

Clive and Wang have introduced a water-soluble variant of triphenyltin hydride that can be used for the Barton–McCombie reaction.[17] Work-up involves basic or acidic work-up by washing the reaction products with an H_2O/THF solution of LiOH or TsOH, respectively, to remove all tin residues.

Silicon-based Reductants

Since organotin compounds exhibit toxicity, introduce purification difficulties, present disposal problems and are generally expensive reagents, efforts have been made to develop tin-free variants of the Barton–McCombie deoxygenation.[18] Although simple, low-molecular weight reagents such as Et_3SiH would be ideal, the greater strength of the Si–H bond results in a slower reaction compared to Bu_3SnH, and chain propagation is short-lived at moderate temperatures. However, the advantage of employing the silane is that the addition of the R_3Si• radical to the thiocarbonyl group is much less likely to be reversible in comparison to R_3Sn·because of the greater strength of the resultant Si–S bond.

Barton and co-workers examined a variety of silanes as potential reducing agents: triethylsilane, triethoxysilane, phenylsilane, diphenylsilane and triphenylsilane.[19] These reagents were compared using a system involving triethylborane-air reduction of cyclododecyl xanthates at room temperature. Of all the silanes tested, diphenylsilane was found to be the most reactive. Yields of a variety of secondary xanthate substrates were excellent at room temperature and did not increase substantially with elevated temperatures. For primary xanthates, reduction was problematic at room temperature and could be improved by elevating reaction temperature to 80 °C. Tris(trimethylsilyl)silane[20] was also compared to diphenylsilane and was found to give a similar reactivity profile, however, the authors note the reduced cost of the latter reagent.

Schummer and Höfle also investigated tris(trimethylsilyl)silane, a reagent previously shown by Chatgilialoglu to reduce organic halides, selenides and xanthates, for the reduction of phenylthionocarbonates (**24**).[21] Reduction was possible at 80 °C in toluene with the advantage that by-products were very lipophilic and could be easily removed by column chromatography. Barton and colleagues also demonstrated that primary and secondary 4-fluoro or (2,4,6-trichlorophenyl)thionocarbonates (**25**) could be readily reduced with phenylsilane or diphenylsilane, the former reagent being advantageous over the latter for its lower cost and volatility, thereby simplifying product purification.[22]

An alternative procedure for the reduction of xanthates, as proposed by Roberts and co-workers, was to employ a catalytic amount of a thiol, in combination with triethylsilane.[23] Using 1,1-di-*t*-butyl peroxide (DTBP) as radical initiator, *t*-dodecanethiol as an "acceptor" polarity-reversal catalyst, followed by triethylsilane as stoichiometric reductant. The thiol serves to transfer H-atom from the silane reductant to the intermediate radical species (R·).[24]

Since the deoxygenation of compound **38** was well characterized using classical Barton–McCombie conditions, Roberts chose this same model substrate to illustrate the viability of silane/thiol reduction protocol. Reduction with this system gave a good yield of **39**, however, it was observed that minor diasteromer **40**, resulting from radical-induced epimerization at C-5, could also be isolated. The rationale for the formation of **38** was potentially due to the thiyl catalyst abstracting H• from the substrate faster than from the silane. The authors propose that the use of a better donor, such as diphenylsilane, could prevent such side-products from occurring.

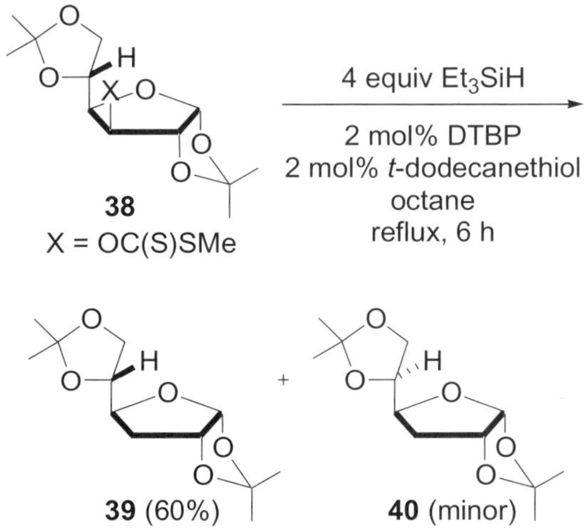

Chatgilialoglu and co-workers examined 5,10-dihydrosilanthrene (**41**) as a reducing agent for the deoxygenation of steroid **42** to give **43**.[25] It was found that phenylthionocarbonate derivatives (**24**) were the only substrate class that was compatible; xanthates (**4**), thiocarbonyl imidazolides (**6**) and *N*-phenyl thionocarbamates gave incomplete conversions and lower yields. It was believed that decomposition of these substrate classes generate intermediates which inhibit the reductant. Using a free-radical clock experiment, the reactivity of **41** was determined to be less than tris(trimethylsilyl)silane but greater than that of diphenylsilane.

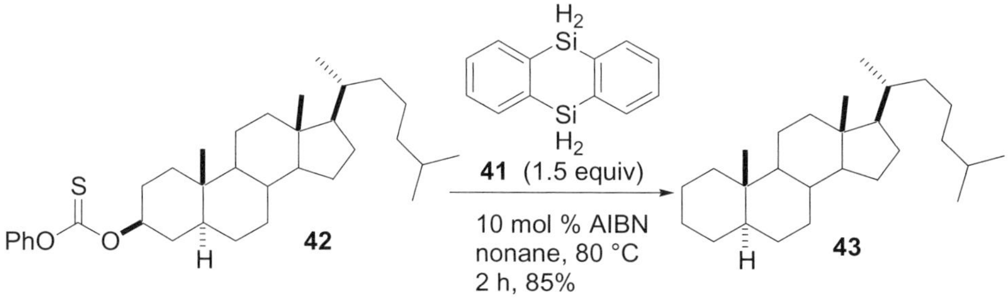

Studer and co-workers have used silylated cyclohexadienes as radical chain reducing agents for the reduction of **44** to **45**.[26] Release of $R_3Si\bullet$ followed by re-aromatization is the driving force of the reaction. These reagents can be easily prepared and initiation can be performed with AIBN, triethylborane/O_2 or by simply using an air atmosphere.

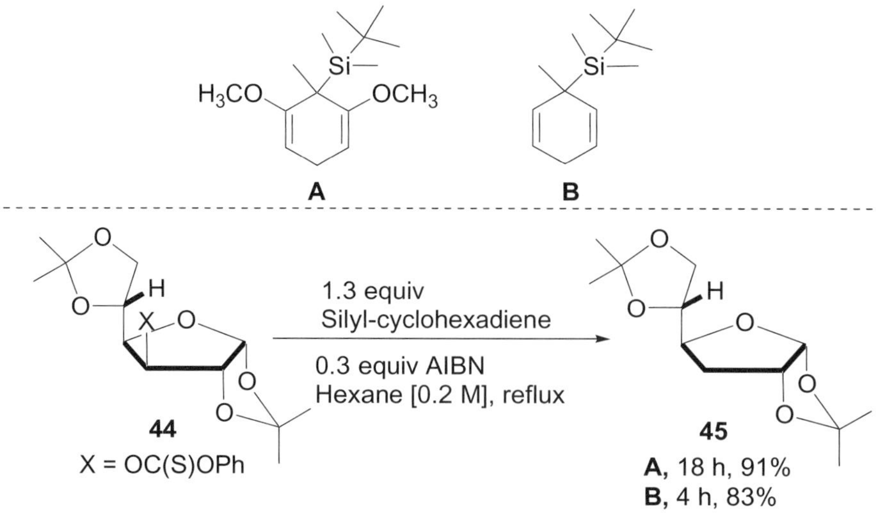

Phosphorous-based Reductants

Phosphorus-based hydrogen donors were investigated with the goals of having a low-cost, metal-free reagent amenable to large scale deoxygenations of xanthates and thionocarbonates.[27] Dimethyl and diethylphosphite were successfully used for the reduction of xanthate **38** to **39,** however, AIBN could not be used as initiator and was replaced with benzoyl peroxide.

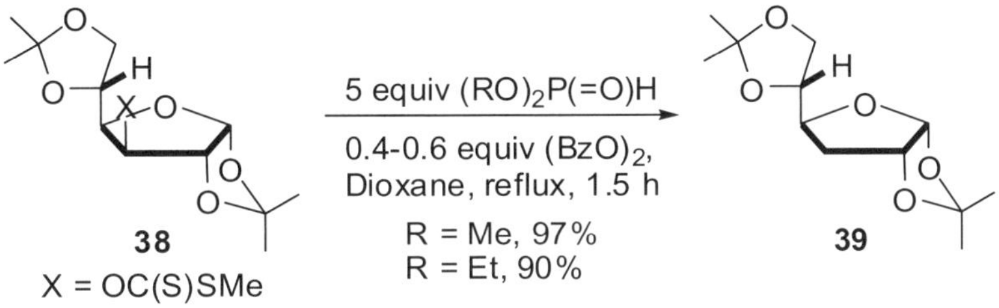

Although dialkylphosphites exhibited a lower reactivity profile compared to silanes, it was found that deoxygenations could be performed using the dialkylphosphite as solvent. For example, conversion of **38** to **39** was achieved in 90% yield within an hour in neat, refluxing diethylphosphite. Excess reagents and phosphorous-containing residues could be washed out during reaction work-up.

Hypophosphorous acid was also investigated as reductant. This reagent was compatible with AIBN but required the addition of an amine such as triethylamine to keep reaction conditions neutral. With the exception of (pentafluorophenyl)thionocarbonates (**24**), a variety of (4-fluorophenyl)thionocarbonates and xanthates (**4**) could be deoxygenated easily. Hypophosphoric acid is only available commercially as a 50% aqueous solution, which precludes its use for water-sensitive substrates, therefore a convenient alternative was to use a crystalline, anhydrous salt of hypophosphoric acid with *N*-ethylpiperidine.[28] This system was used to readily reduce the fucose derivative **46** to **47** in excellent yield.

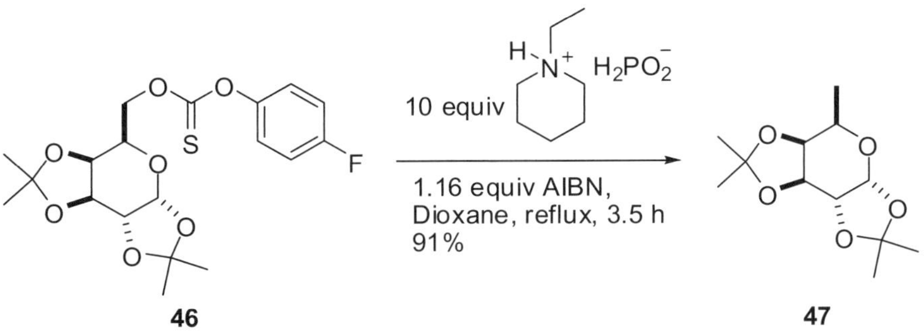

Vicinal diols could also be deoxygenated via their xanthate derivates, however a sacrificial olefin, such as 1-dodecene, was a necessary additive to prevent the phosphorous-centered radical from adding to the desired product.

Phosphine-boranes can also be used as reducing agents.[29] These reagents are usually easily-handled, non-toxic, crystalline, air-stable solids. Efficiency as a chain propagator was found to be higher in ethereal solvents as opposed to benzene or toluene. Optimal conditions for the deoxygenation of **38** employed 2 equivalents of tri-*n*-butylphosphine-borane and 0.2 equivalents AIBN in refluxing dioxane for an hour allowing **39** to be isolated in 89% yield. Under these conditions, it was possible to also isolate the side product of the radical deoxygenation reaction, n-$Bu_3P{\bullet}BH_2SC(O)SMe$ (85% yield), thus proving that the radical chain mechanism follows a similar pathway as that of tributyltin hydride.

It was also determined that these phosphine-borane reagents show specific reactivity towards reduction of xanthates over bromide and chloride groups in a substrate. Bis(xanthates) could also be reduced to olefins without the necessity of adding a sacrificial olefin acceptor to protect the deoxygenation product.

Miscellaneous Reductants

Kim and colleagues studied the use of the one-electron oxidant tetrabutylammonium peroxydisulfate as initiator in conjunction with formate as reductant.[30] This protocol was designed to be more appropriate for bulk chemical deoxygenations, as the use of AIBN or peroxides can be dangerous on large scale. Tetrabutylammonium peroxydisulfate is soluble in organic solvents such as acetonitrile, DMF and chlorinated solvents.

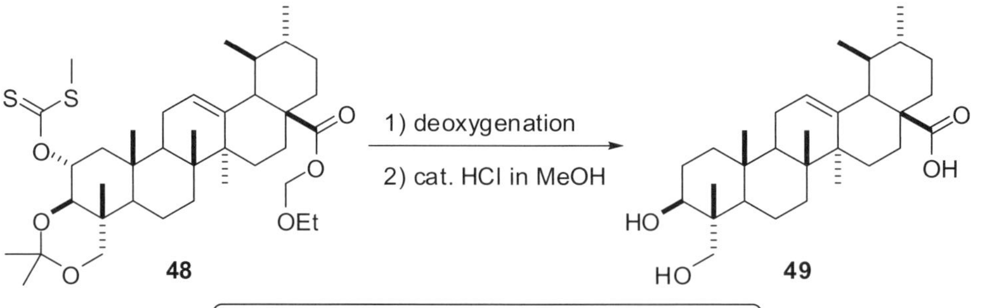

Deoxygenation conditions (yield of **49** after deprotection):
- Bu_3SnH, Toluene, 110°C (76%)
- Ph_2SiH_2, AIBN, Toluene, 110°C (70%)
- $(Bu_4N)_2S_2O_8$, HCO_2Na, DMF, 50°C, 4h (95%)

The protocol can be run at lower temperatures and simplifies product purification by avoiding metal residues. For example, conversion of **48** to **49** was achieved in greater yield and under milder conditions as compared to the use of tin or silicon-based reducing agents. Dideoxygenation of **22** to **23** also proceeded in good yield under mild conditions.

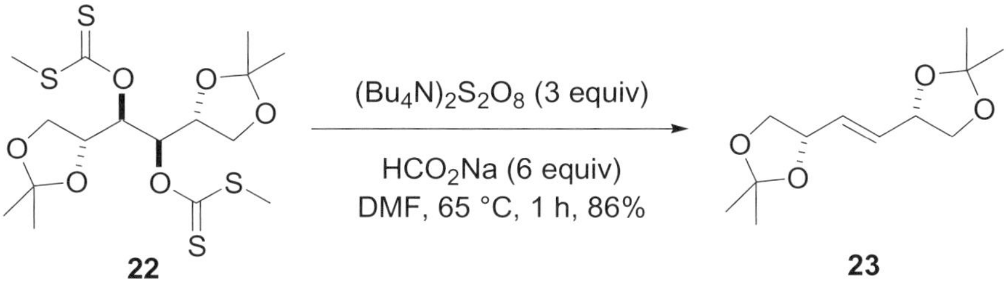

The reaction is believed to propagate via transfer of a single electron to the thiocarbonyl group from CO_2 radical anion. When the reaction is conducted in DMSO as solvent using $(Bu_4N)_2S_2O_8/DCO_2Na$, 91% incorporation of deuterium was observed in a sample substrate, proving that formate is the hydrogen source.

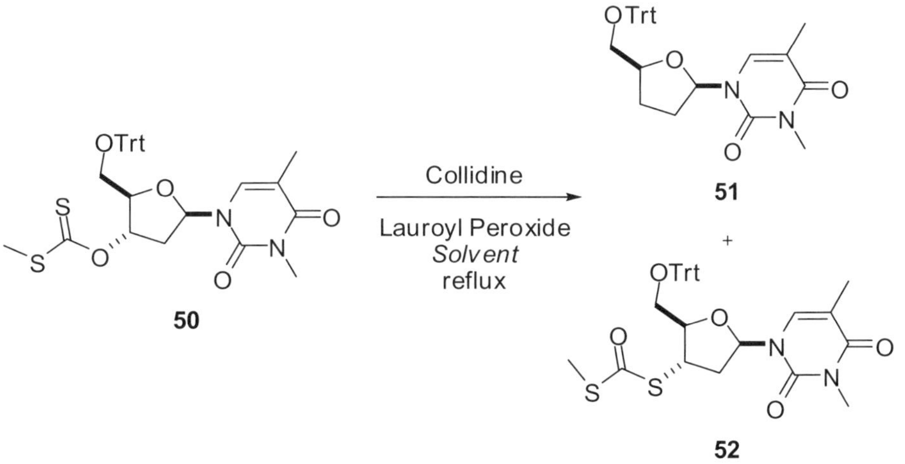

Solvent (yield of **51** and **52**)
2-Propanol (45%, 41%)
Benzene (11%, 60%)

Zard and Quiclet–Sire described the use of 2-propanol as solvent and hydride donor in conjunction with lauroyl peroxide as radical initiator for xanthate deoxygenation.[31] Collidine is a necessary additive to prevent hydrolysis of acid sensitive functionalities by the lauric acid by-product.

Depending on the stability of the intermediate radical formed, competing radical rearrangement of the substrate (**50**) can occur, giving mixture of deoxygenated product **51** and *O*- to *S*-rearranged xanthate **52**. Removing the hydrogen donor by replacing the solvent with benzene gave higher yields of rearranged material.

The use of complexes of borane ligated to *N*-heterocyclic carbenes (NHC) as radical hydrogen atom donors has been reported recently.[32] Complexes such as **53** and **54** have been postulated to be superior hydrogen donors than amine- and phosphine-boranes due to π-conjugation, resulting in a weakened B–H bond.

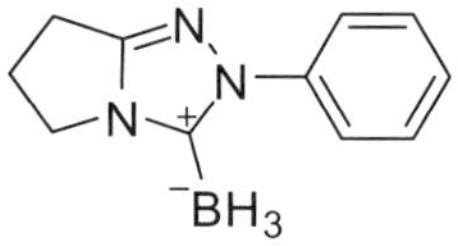

53

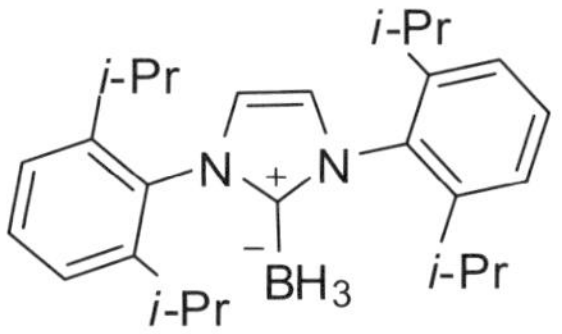

54

These NHC-borane complexes are easily-handled, air-stable, white crystalline solids. Using 1–2 equivalents of **53**, substrate **38** could be deoxygenated to **39** in refluxing benzene in 2 hours in 60% yield (using 0.5 equivalents AIBN or 1 equivalent of triethylborane/air atmosphere).

3.1.5 Synthetic Utility

Kehler and Hansen used the Barton–McCombie deoxygenation to prepare a series of *sec*-alkylmethylphosphinates.[33] Very few compounds in this class have been reported in the literature due to difficulties in preparation. The tertiary alcohol in **55** was unreactive to chlorothionoformates and thus was derivatized as the methyl oxalate ester (**56**) which was then deoxygenated under classical conditions to generate **57**.

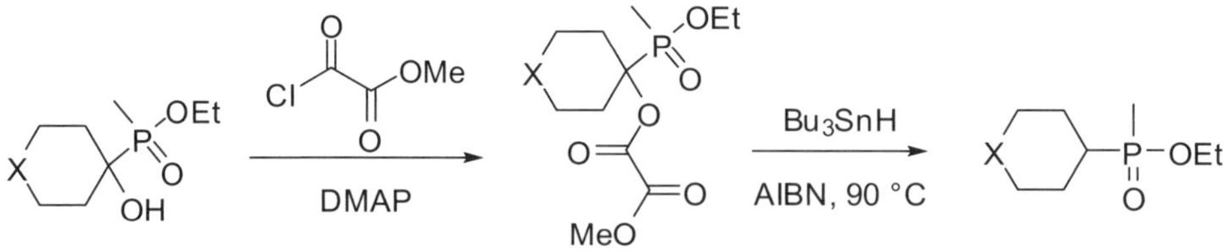

55 **56** **57**

X= CH_2, 80%
X= N(CO)OEt, 75%

X= CH_2, 68%
X= N(CO)OEt, 90%

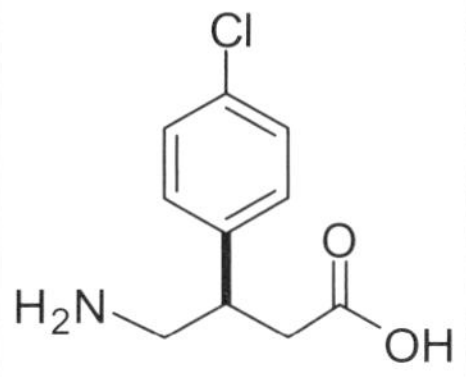

58 (*R*)-Baclofen

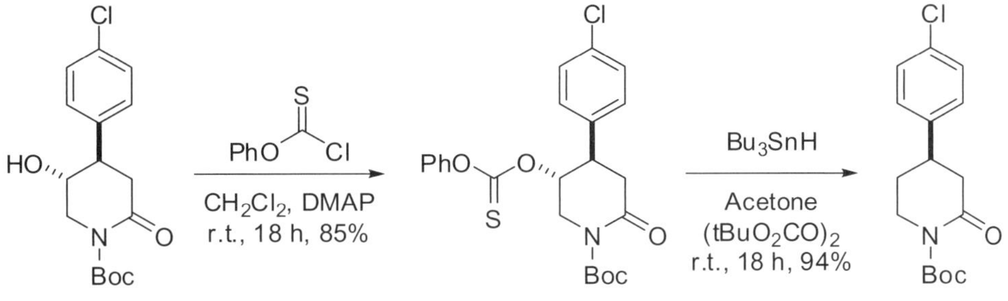

59 **60** **61**

A series of (*R*)-baclofen (**58**) homologues was prepared by Krogsgaard–Larsen and co-workers to probe their activity against the 4-aminobutanoic $acid_B$ ($GABA_B$) receptor.[34] Removal of the hydroxyl group in **59** was a key step of the synthesis and could not be attained using zinc-based reducing agents or platinum-based hydrogenation conditions. Deoxygenation of the thiocarbonyl derivative **60** was accomplished using tributyltin hydride with di-*tert*-butyl peroxyoxalate to give **61**. The use of AIBN as initiator led to decomposition of the starting material.

Danishefsky and co-workers disclosed the total synthesis of a neurotropic illicinone, tricycloillicinone (**64**), a natural product which enhances the action of choline acetyltransferase, thereby increasing acetylcholine levels and is therefore of interest for the treatment of Alzheimer's disease.[35] Due to the steric environment around the secondary alcohol, activation of **62** to **63** under neutral conditions failed and required conversion first to its potassium salt using KHMDS at low temperature. Deoxygenation to generate the desired compound **60** then proceeded in moderate yield.

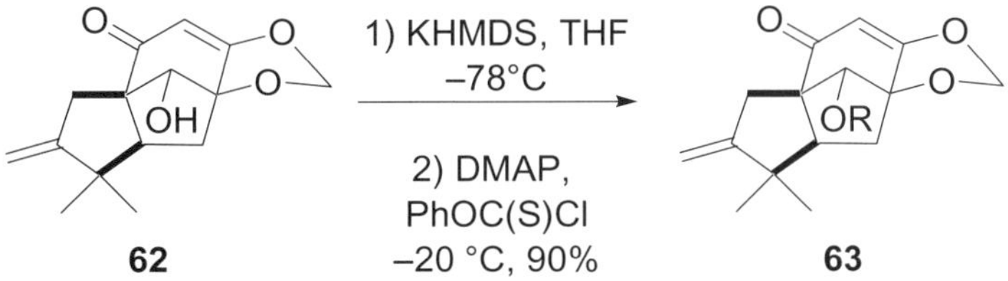

62 **63**
R = C(S)OPh

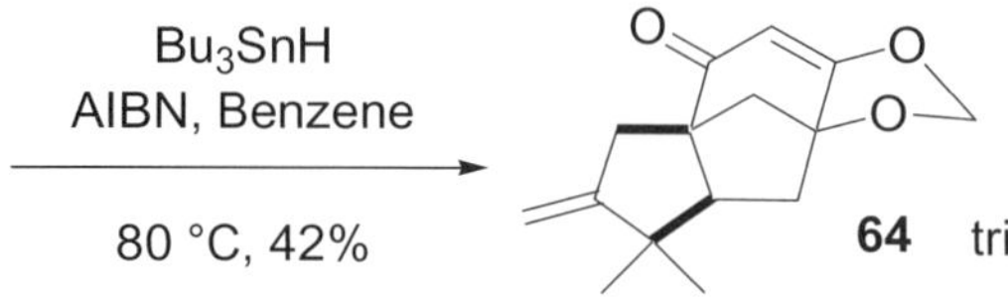

64 tricycloillicinone

Audran and colleagues described two syntheses of furanosesquiterpene natural products employing similar synthetic pathways, culminating in the use of the Barton-McCombie deoxygenation as a key step.[36] The ester unit in bicyclic lactones **65** and **66** is used to set stereochemistry of the C-2 center. After further elaboration, the ester stereocontrol element is then reductively cleaved and cyanated to yield the diols **67** and **68**, from **65** and **66**, respectively. Then, in a one-pot, two-step process, activation and double deoxygenation was performed to yield reduced compounds **69** and **70**, which are the intermediates in the synthesis of (+)-ricciocarpin A (**71**) and (+)-ancistrofuran (**72**), respectively.

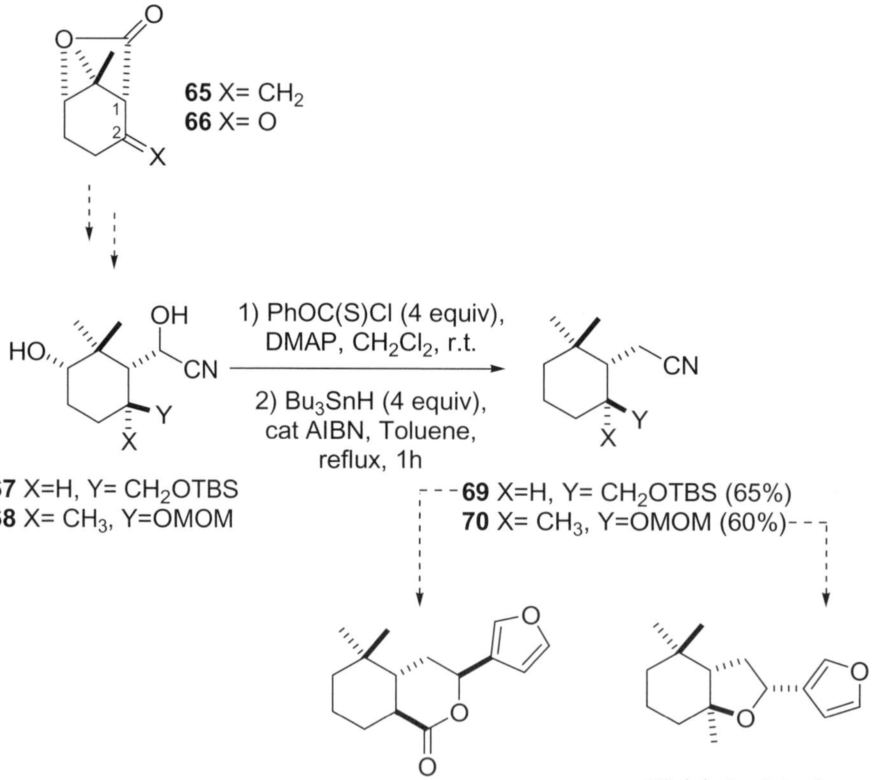

71 (+)-Ricciocarpin A

72 (+)-Ancistrofuran

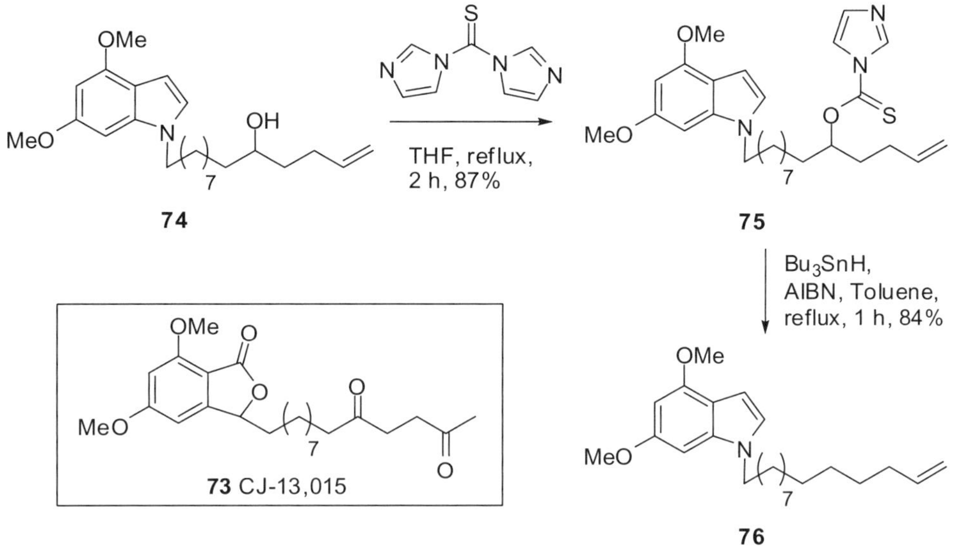

Brimble and co-workers published the synthesis of a series of analogues of the known inhibitor CJ-13,015 (**73**), which possesses antibiotic activity against *Helicobacter pylori*.[37] The 4,6-dimethoxyindole ring was chosen to replace the 5,7-dimethoxyphthalide ring system as an attempt to increase activity against *H. pylori*. However, this indole ring system was known to be very sensitive and required neutral conditions to activate the alcohol **74** to the thiocarbonyl imidazolide compound **75**. Deoxygenation was then attained using the classical protocol to give **76** in good yield, followed by further elaboration to generate a series of analogues.

The expedient total synthesis of tetracyclic core of the enantiomer of (–)-nakadomarin A (**77**) was reported by Zhai and co-workers recently.[38] This marine natural product, related to manzamine, was found to be toxic to murine lymphoma L1210 cells and displayed antimicrobial activity. The functionalized tetracycle **78** contained a trisubstituted alkene moeity which could not be hydrogenated selectively due to the presence of the more reactive furan ring. This alkene was unresponsive to treatment with TFA/Et_3SiH, and hydroboration followed by protonation (using HBr or HOAc) was also unsuccessful. This quandary was resolved by conversion to the secondary alcohol **79** (using a hydroboration/oxidation procedure) which was then converted to xanthate **80**, which could be readily deoxygenated in good yield to **81**.

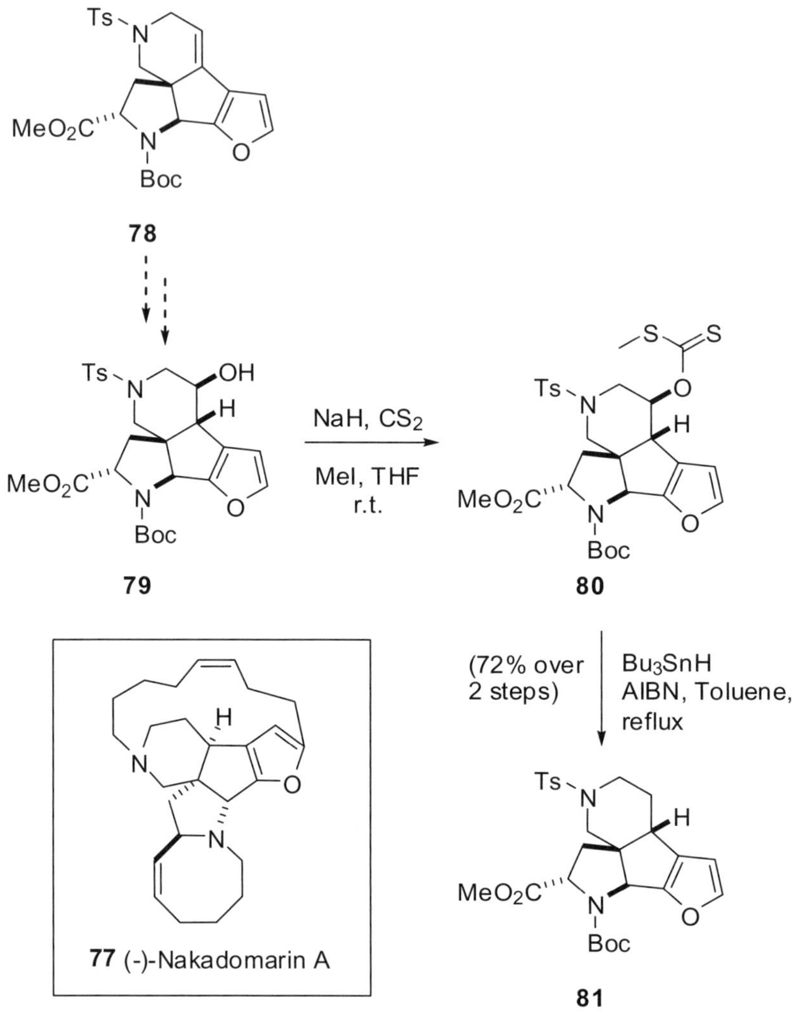

3.1.6 *Experimental*

*4,6-Dimethoxy-1-(10'-(*N*-imidazolylthiocarbonyloxy)-tetradec-13'-enyl)-1*H*-indole 75.*[37]

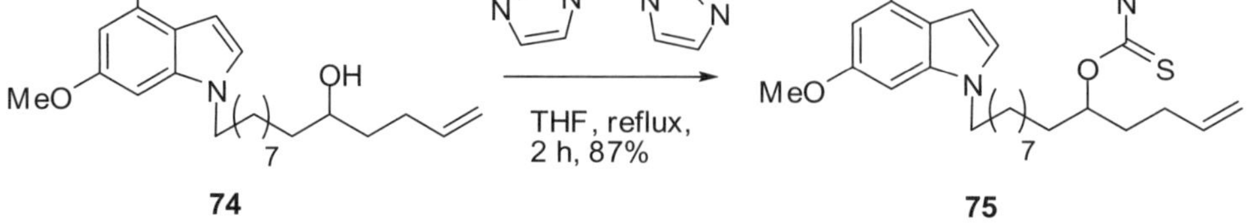

1,1′-Thiocarbonyldiimidazole (0.10 g, 0.6 mmol) was added to a solution of alcohol **74** (0.088 g, 0.2 mmol) in tetrahydrofuran (1.5 mL) and the mixture heated to reflux for 2 h. The reaction mixture was poured into a separating funnel containing water (5 mL) and shaken. The aqueous layer was extracted with dichloromethane (6 × 5 mL) and the organic extracts were combined and washed with saturated sodium bicarbonate solution (10 mL) and brine (10 mL), then dried over magnesium sulfate. The solvent was removed *in vacuo* and the resulting bright yellow oil purified via flash chromatography using hexane–ethyl acetate (4:1, R_f 0.31) as eluent to afford the title compound **75** (98 mg, 87%) as a colorless oil.

*4,6-Dimethoxy-1-(tetradec-13′-enyl)-1*H*-indole 76.*[37]

75 **76**

Tributyltin hydride (0.053 mL, 0.2 mmol) and azobisisobutyronitrile (catalytic) were added to a solution of Barton–McCombie precursor **75** in toluene (2 mL) and the mixture heated to reflux for 1 h. The reaction was then filtered through a column of silica gel containing powdered potassium fluoride (10% w/w) and washed with dichloromethane (100 mL). The solvent was removed *in vacuo* and the resultant oil dissolved in diethyl ether (10 mL) and washed with saturated sodium bicarbonate (5 mL), water (5 mL) and brine (5 mL). The combined aqueous layers were then extracted with diethyl ether (4 × 5 mL) before the organic layers were combined and dried over magnesium sulfate and the solvent removed *in vacuo*. The resulting bright yellow oil was purified via flash chromatography using hexane–ethyl acetate (4:1, R_f 0.90) eluent to afford the title compound **76** (62 mg, 84%) as a colorless oil.

3.1.7 References

1. Barton, D. H. R.; McCombie, S. W. *J. Chem. Soc. Perkin I* **1975**, 1574–1585.
2. [R] Selected reviews on the Barton-McCombie reaction: a) Hartwig, W. *Tetrahedron* **1983**, *39*, 2609–2645. b) Crich, D.; Quintero, L. *Chem. Rev.* **1989**, *89*, 1413–1432.
3. a) Rhee, J. U.; Bliss, B. I., RajanBabu, T. V. *J. Am. Chem. Soc.* **2003**, *125*, 1492–1493. b) Rhee J. U.; Bliss, B. I. RajanBabu, T. V. *Tetrahedron Asymmetry* **2003**, *14*, 2939–2959.
4. Barton, D. H. R.; Jang, D. O.; Jaszberenyi, J. Cs. *Tet. Lett.* **1990**, *31*, 3991–3994.
5. Barton, D. H. R.; Subramanian, R. *J. Chem. Soc. Perkin I* **1977**, 1718–1723.
6. Ziegler, F. E.; Zheng, Z.-l. *J. Org. Chem.* **1990**, *55*, 1416–1418.

7. Barrett, A. G. M.; Barton, D. H. R.; Bielski, R. *J. Chem. Soc., Perkin Trans. 1*, **1979**, 2378-2381.
8. Robins, M. J.; Wilson, J. S.; Hansske, F. *J. Am. Chem. Soc.* **1983**, *105*, 4059–4065.
9. Barton, D. H. R.; Jaszberenyi, J. Cs. *Tetrahedron Lett.* **1989**, *30*, 2619–2622.
10. Barton, D. H. R.; Motherwell, W. B.; Stange, A. *Synthesis* **1981**, 743–745.
11. Barton, D. H. R.; Blundell, P.; Dorchak, J.; Jang, D.O.; Jaszberenyi, J. Cs. *Tetrahedron* **1991**, *47*, 8969–8984.
12. Barton, D. H. R.; Hartwig, W.; Hay-Motherwell, R. S.; Motherwell, W. B.; Stange, A. *Tetrahedron Lett.* **1982**, *23*, 2019–2022.
13. Barton, D. H. R.; Crich, D. *Tetrahedron Lett.* **1985**, *26*, 757–760.
14. Neumann, W. P.; Peterseim, M. *Synlett* **1992**, 801–802.
15. a) Lopez, R. M.; Hays, D. S.; Fu, G. C. *J. Am. Chem. Soc.* **1997**, *119*, 6949–6950. b) Tormo, J.; Fu, G. C. *Org. Syn.* **2002**, *78*, 239–243.
16. Boussaguet, P.; Delmond, B.; Dumartin, G.; Peyere, M. *Tetrahedron Lett.* **2000**, *41*, 3377–3380.
17. Clive, D. L. J.; Wang, J. *J. Org. Chem.* **2002**, *67*, 1192–1198.
18. [R] For a review on tin hydride substitutes in radical chain reactions, see: Studer, A; Amrein, S. *Synthesis* **2002**, *7*, 835–849.
19. [R] Barton, D. H. R.; Jang, D. O.; Jaszberenyi, J. Cs. *Tetrahedron Lett.* **1990**, *31*, 4681–4684.
20. [R] Chatgilialoglu, C. *Acc. Chem. Res.* **1992**, *25*, 188–194.
21. Schummer, D.; Höfle, G. *Synlett* **1990**, 705–706.
22. Barton, D. H. R.; Jang, D. O.; Jaszberenyi, J. Cs. *Synlett* **1991**, 435–438.
23. Cole, S. J.; Kirwan, J. N.; Roberts, B. P.; Willis, C. R. *J. Chem. Soc. Perkin Trans I* **1991,** 103–112.
24. Cai, Y.; Roberts, B. P. *Tetrahedron Lett.* **2001**, *42*, 763–766.
25. Gimisis, T.; Ballestri, M.; Ferreri, C.; Chatgilialoglu, C.; Boukherroub, R.; Manuel, G. *Tetrahedron Lett.* **1995**, *36*, 3897–3900.
26. a) Studer, A.; Amrein, S. *Angew. Chem., Int. Ed.* **2000**, *39*, 3080–3082. b) Studer, A.; Amrein, S.; Schleth, F.; Schulte, T.; Walton, J. C. *J. Am. Chem. Soc.* **2003,** *125*, 5726–5733.
27. Barton, D. H. R.; Jang, D. O.; Jaszberenyi, J. Cs. *J. Org. Chem.* **1993**, *58*, 6838–6842.
28. Commercially available from Aldrich, Inc.: 1-Ethylpiperidine hypophosphite, CAS#: 145060-63-5.
29. Barton, D. H. R.; Jacob, M. *Tetrahedron Lett.* **1998**, *39*, 1331–1334.
30. Park, H. S.; Lee, H. Y.; Kim, Y. H. *Org. Lett.* **2005**, *7*, 3187–3190.
31. Quiclet-Sire, B.; Zard, S. Z. *Tetrahedron Lett.* **1998**, *39,* 9435–9438.
32. Ueng, S.-H.; Brahmi, M.-M.; Derat, É.; Fensterbank, L.; Lacôte, E.; Malacria, M.; Curran, D. P. *J. Am. Chem. Soc.* **2008**, *130*, 10082–10083.
33. Hansen, H. I.; Kehler, J. *Synthesis* **1999**, *11*, 1925–1930.
34. Karla, R.; Ebert, B.; Thorkildsen, C.; Herdeis, C.; Johansen, T. N.; Nielsen, B.; Krogsgaard-Larsen, P. *J. Med. Chem.* **1999**, *42*, 2053–2059.
35. Pettus, T. R. R.; Inoue, M.; Chen, X.-T.; Danishefsky, S. J. *J. Am. Chem. Soc.* **2000**, *122*, 6160–6168.
36. a) Palombo, E.; Audran, G.; Monti, H. *Synlett* **2005**, *13*, 2104–2106. b) Palombo, E.; Audran, G.; Monti, H. *Tetrahedron* **2005**, *61*, 9545–9549.
37. Wilson, Z. E.; Heapy, A. M.; Brimble, M. A. *Tetrahedron* **2007**, *63,* 5379–5385.
38. Deng, H.; Yang, X.; Tong, Z.; Li, Z.; Zhai, H. *Org. Lett.* **2008**, *10,* 1791–1793.

3.2 Barton Nitrite Photolysis Reaction

Timothy J. Hagen

3.2.1 Description

The Barton nitrite photolysis reaction involves the conversion of a nitrite ester **1** to a γ-oximino-alcohol **3** by photolysis involving the homolytic cleavage of a nitrogen oxygen bond followed by hydrogen abstraction.[1]

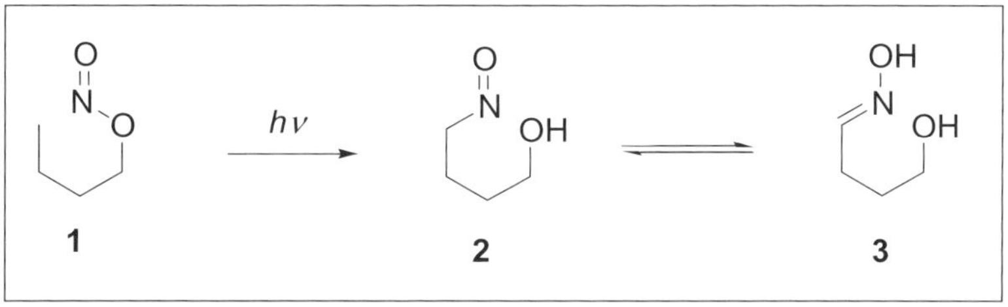

3.2.2 Historical Perspective

There were two important factors that were critical in the invention of the Barton nitrite photolysis reaction. The first was Sir Derek H. R. Barton's creativity and willingness to invent a new chemical reaction and the second was his solid understanding of conformational analysis. Barton's attitude toward reaction invention can be summarized with a quote from a review that he wrote on the invention of chemical reactions. "The invention of chemical reactions is not a popular academic activity. There are probably no graduate courses on this subject. Perhaps the reason for the lack of interest is the widespread belief that all new chemical reactions are discovered by accident." "For several decades now I have argued that it is possible to invent new chemical reactions of synthetic significance. The first step, of course, is to recognize which reaction you would like to invent."[2] The second factor critical to the invention of the Barton nitrite photolysis was Barton's knowledge and insight of conformational analysis. In 1950 Barton published a seminal paper on "The Conformation of the Steroid Nucleus" in which he described his theory of conformational analysis.[3] Every organic chemist that has followed him has employed these principles to their work, in one way or another. In 1969 Barton and Hassel would be awarded the Nobel Prize for "their contributions to the development of the concept of conformation and its application in chemistry".[4]

In 1960, Barton took a "vacation" in Cambridge, Massachusetts where he worked in a small outfit called the Research Institute for Medicine and Chemistry. Barton was working on the synthesis of the adrenocortical hormone aldosterone (**7**). Ever since the structure of aldosterone was determined by Reichstein and his colleagues[5] by a remarkable feat of classical degradation, tremendous effort was expended on its total and partial synthesis. The masked aldehyde function at C-18 made partial synthesis from another steroid improbable, so there was a need to invent a new reaction for substitution of unactivated neopentylic methyl groups. In order to make aldosterone, Barton invented a novel nitrite photolysis reaction by simply writing down on a piece of paper what he thought would be an ideal process. Barton's choice of nitrites was based on the hypothesis that they would afford alkoxy radicals on photolysis, that the axial alkoxy radical would abstract a δ-hydrogen and the resulting alkyl radical would capture NO, the other photolysis product.[2] The nitroso-methyl group was postulated to isomerise to the oxime, a perfect aldehyde precursor. His skilled collaborator, Dr. John Beaton, was able to quickly reduce his idea to practice. The readily available corticosterone acetate (**4**) was smoothly transformed to its nitrite **5** and on photolysis, this afforded the crystalline oxime **6**. Treatment of the latter with nitrous acid readily gave pure aldosterone acetate (**8**).[6] Although the overall yield was not high (~ 20%), this new reaction did permit the synthesis of 60 grams of aldosterone acetate (**8**) when the rest of the world supply was in milligram quantities.[2] This dramatic increase in supply allowed the biology of aldosterone to be fully evaluated. It was said that Barton considered it his most satisfying piece of work.[7]

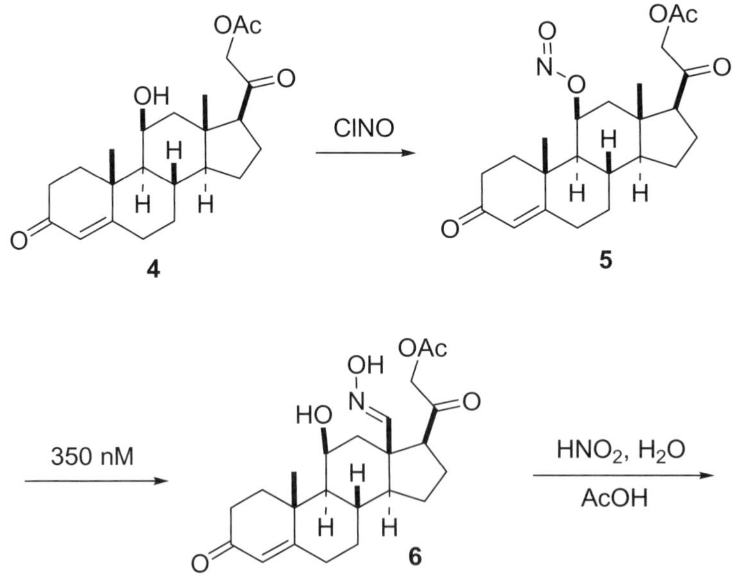

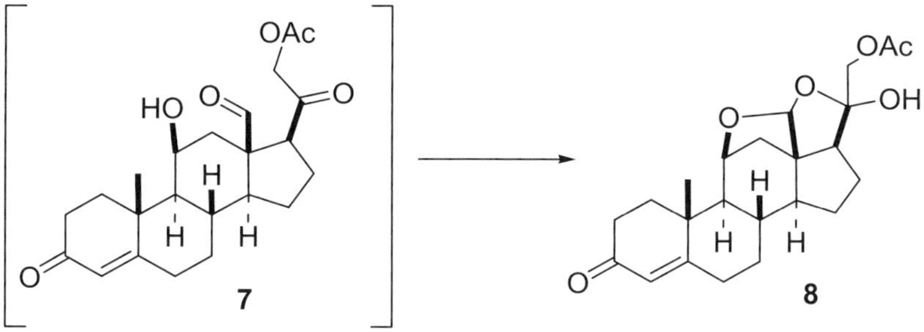

3.2.3 *Mechanism*

Photolysis of some steroidal nitrites in either benzene or toluene produces an alkoxy radical, which then undergoes an intramolecular isomerization by hydrogen-atom transfer involving a six-membered ring transition state. This is then followed by the trapping of this second alkyl radical by the nitric oxide produced in the initial photolysis, resulting in a net 1,5-migration of NO from oxygen to carbon.[8]

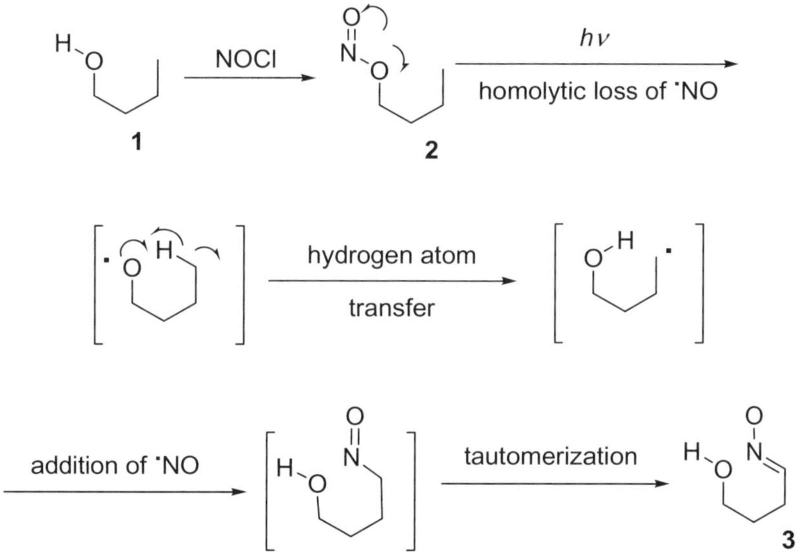

3.2.4 *Synthetic Utility*

The Barton nitrite photolysis reaction has played an important role in the synthesis of natural products and has been previously reviewed.[9,10] The invention of this reaction enabled chemists to convert remote, un-activated, sp^3 hybridized carbon atoms on conformationally constrained hydrocarbons, efficiently and predictably, into atoms of higher oxidation states and thus allowing for further functionalization of these carbon atoms. Evidence of the power of this reaction comes from the Barton labs conversion of corticosterone acetate (**4**) to aldosterone acetate (**5**). Skeletons with several fused rings, such as steroids, are well suited to these remote functionalizations. There are numerous examples for the synthesis of steroids. The numbering system for the steroid core is shown below.

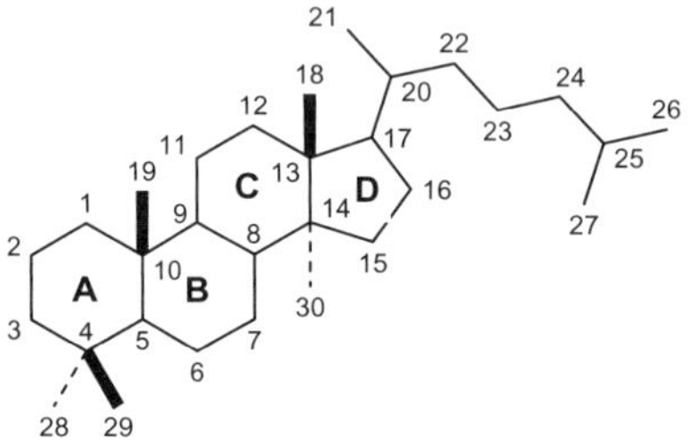

Functionalizaton of C-18 in the Steroid Core

The Barton nitrite photolysis was used in the conversion of nitrite of corticosterone acetate (**5**) to **6** during the synthesis of aldosterone actetate.[1]

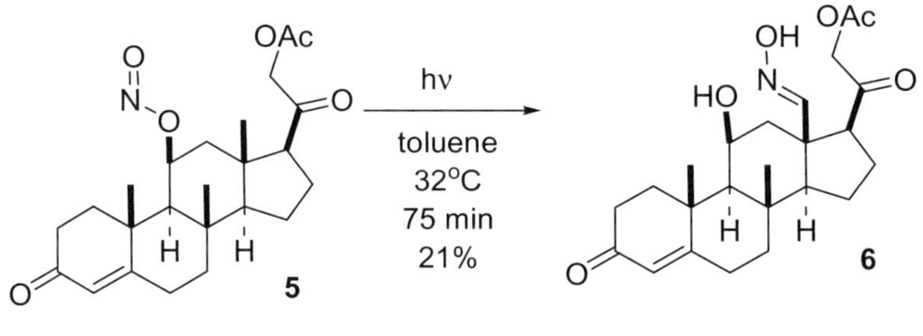

Suginome and coworkers prepared **10** by photolysis of steroidal 20α-ol nitrite **9** which was then converted into 12-deoxofukujusonorone in ten-steps.[11]

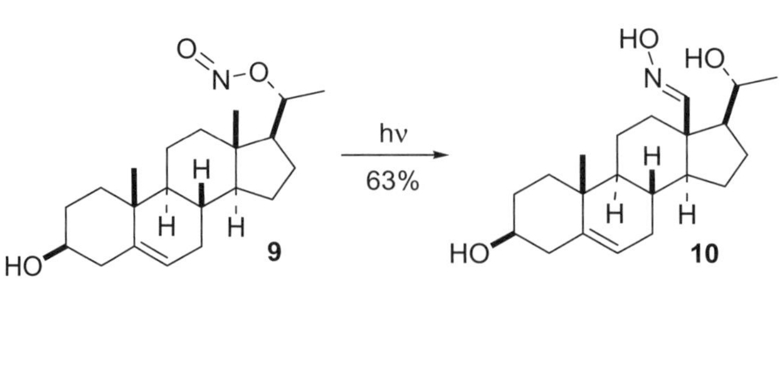

The synthesis of seychellogenin by Habermehl utilized a Barton nitrite photolysis reaction for the conversion of **11** to **12**.[12]

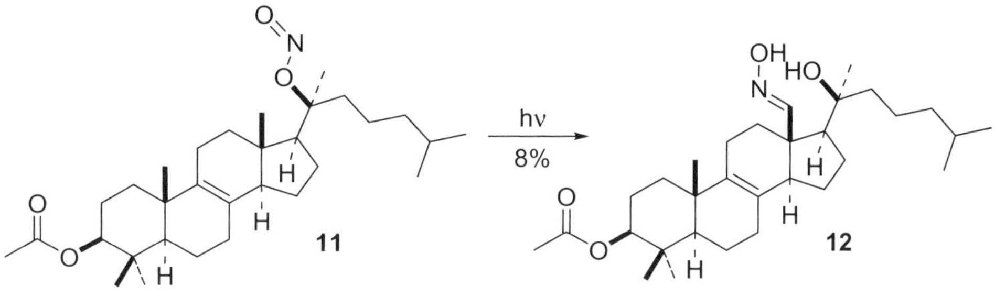

Functionalizaton of C-19 in the Steroid Core

Functionalization of the methyl groups in 4,4-dimethyl-5α-androstanes was investigated.[13] Photolysis of **13** yielded the product **14** preferentially. The authors concluded that the conformation of **13** is compatible with a "flattened chair"[13] for ring A in the androstane series.

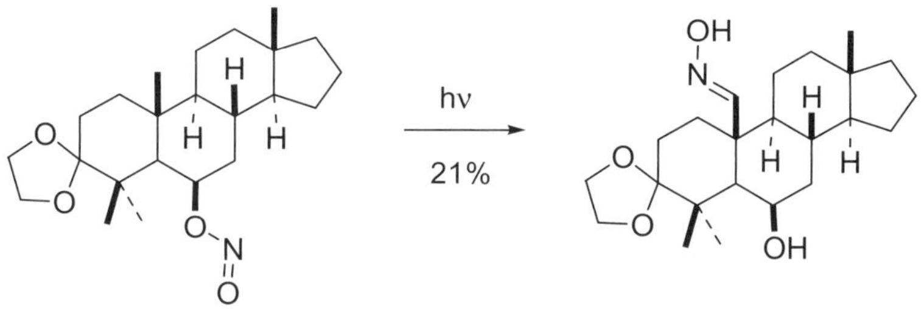

13 **14**

The synthesis of di- and tetrahydroaldosterone derivatives and the C-19 position isomer of 3α,5β-tetrahydroaldosterone was accomplished by Harnik and coworkers.[14] Photolysis of the nitrite **15** provided the oxime **16** that was then converted to the 19-oxygenated compound.

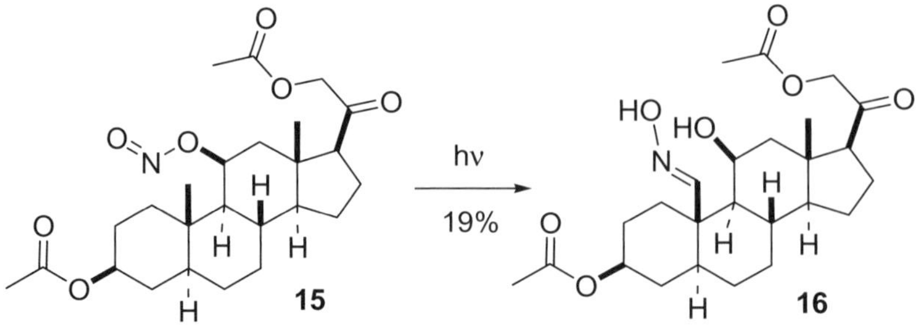

The nitrite photolysis reaction was utilized by Barton to prepare 19-substituted lanostane derivatives.[15] The photolysis of 3β-acetoxylanostan-11β-yl nitrite **17** afforded the 3β-acetoxy-19-hydroxy-iminolanostan-11β-ol **18**.

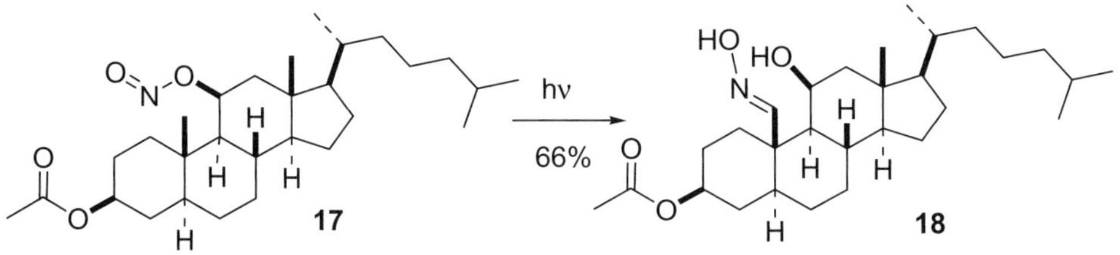

Photolysis of the nitrite ester derived from 5α-cholestan-2β-ol (**19**) by Wolff and coworkers gave 19-oximino-5α-cholestan-2β-ol **20** which was then converted to 17β-hydroxy-1β,19-cyclo-5α-androstan-2-one.[16]

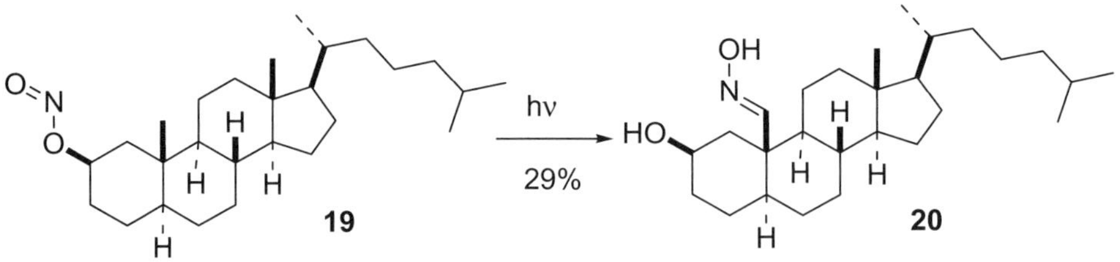

Suginone reported the photolysis of 6β-nitrocholest-4-ene **21** to yield hydroxyimino-cholestenol **22**.[17]

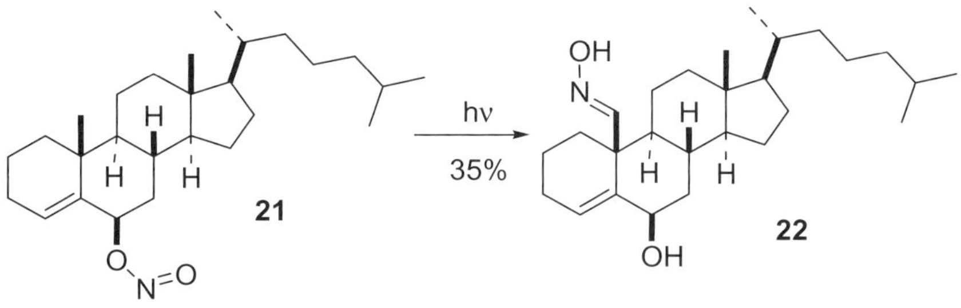

Functionalizaton of C-28 in the Steroid Core

Photolysis of the nitrite **23** yielded the ring A C-28 oxidized product (**24**) preferentially for ring A in the androstane series.[13]

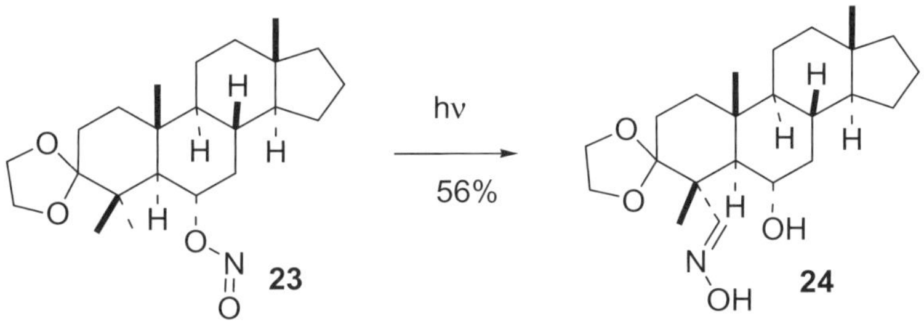

Functionalizaton of C-30 in the Steroid Core

The synthesis of 32-functionalized lanostane derivatives was accomplished by the photolysis of **25** to yield **26**.[18]

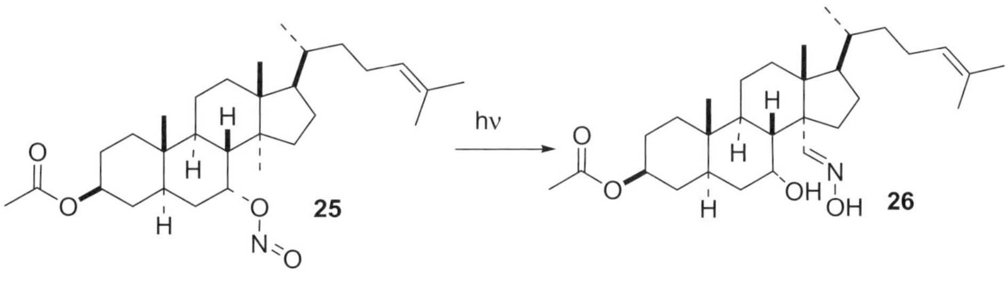

Functionalization of non-steroidal rigid systems

The Barton nitrite photolysis reaction also had an important role in terpene synthesis and the elucidation of terpene structure. For example, in Corey's syntheses of azadiradione[19] and perhydrohistrionicotoxin[20] selective functionalization by means of the nitrite photolysis reaction was the crucial step. The yields for the nitrite photolysis step were low (28% and 20%, respectively), but it would be challenging to imagine a better way to functionalize these un-activated positions in **27** and **29**.

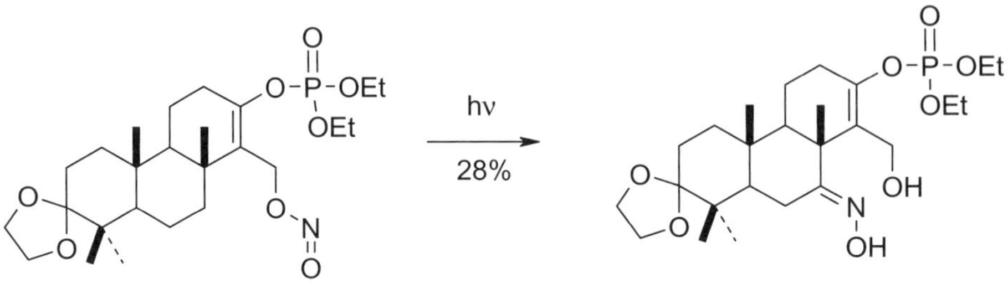

27 **28**

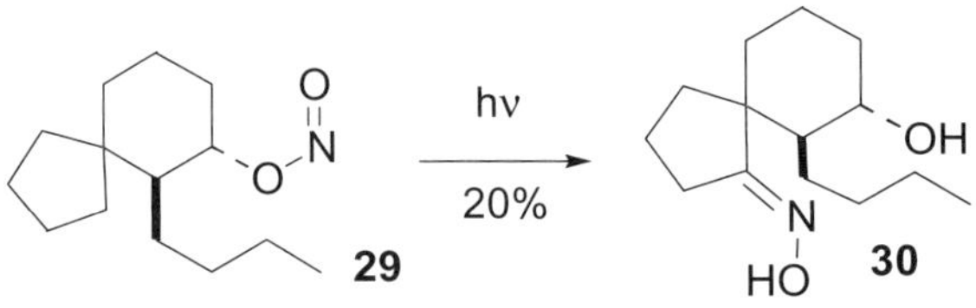

29 **30**

Masamune and co-workers oxidized the angular methyl group of **31** using the nitrite photolysis reaction; this facilitated the synthesis of rishitin.[21]

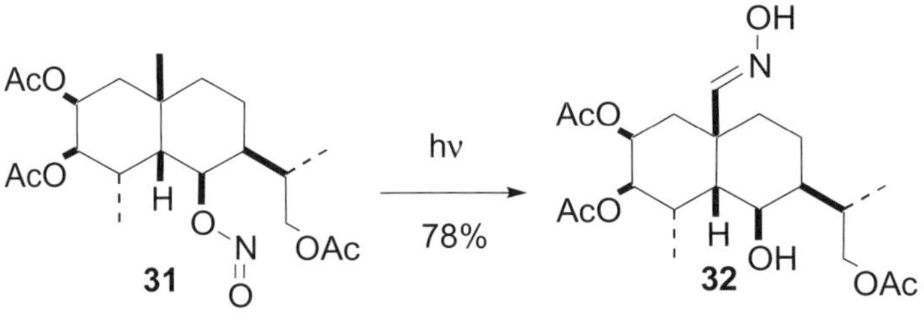

31 **32**

The Barton reaction does not always provide a single major product. For example, during the functionalization of compound **33**, reaction at either (or both) the C-4 and C-10 methyls was expected, but oxidation of the C-8 methyl was not expected.[22] This remote functionalization occurred via two consecutive 1,5-H-atom transfers.

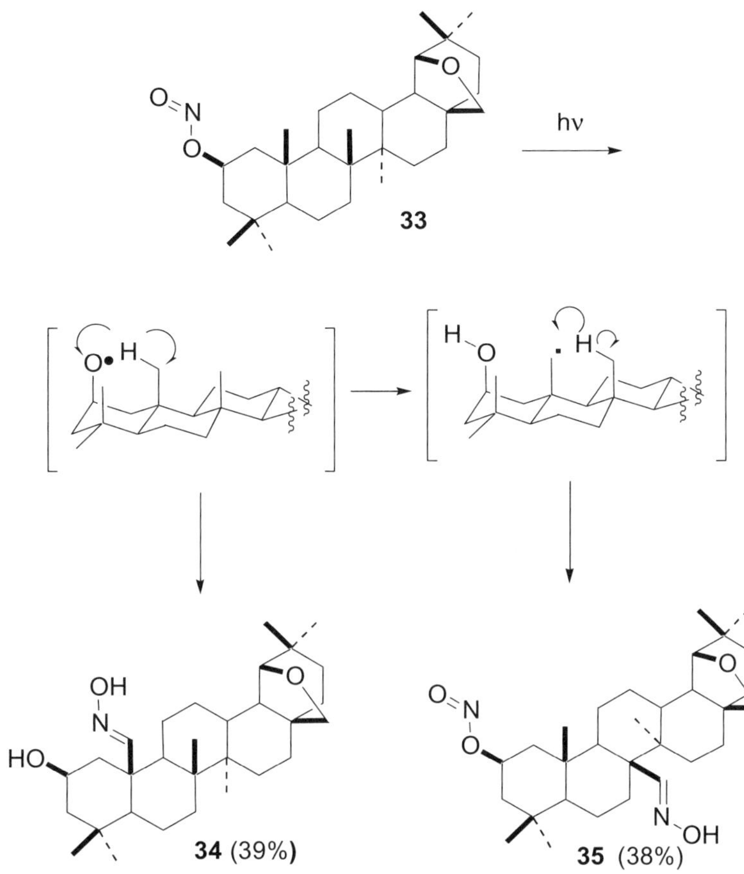

The remote functionalization in alkaloids is usually associated with the Hofmann–Loffler–Freytag reaction however the Barton protocol may also be useful. Two examples are shown below, where photolysis of the nitrite yielded the corresponding oximes in good yields.[23,24]

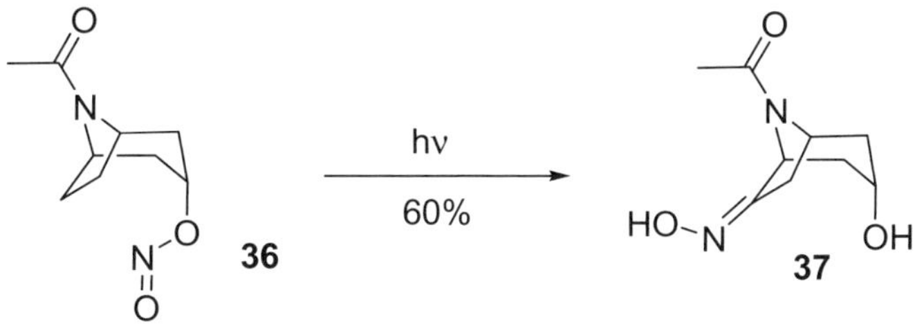

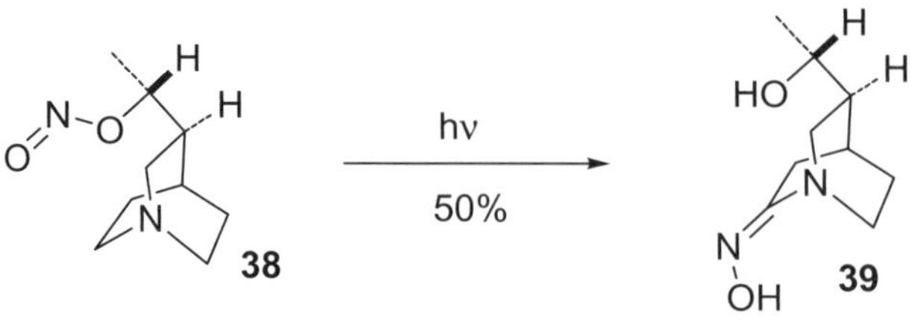

Recently, Nakano and co-authors have utilized the Barton nitrite photolysis reaction in the synthesis of chiral ligands for catalysts.[25] A pseudoenantiomer of β-isocupreidine (β-ICD), was synthesized from quinine by employing a Barton nitrite photolysis reaction of the nitrite **40** to form the oxime **41**.

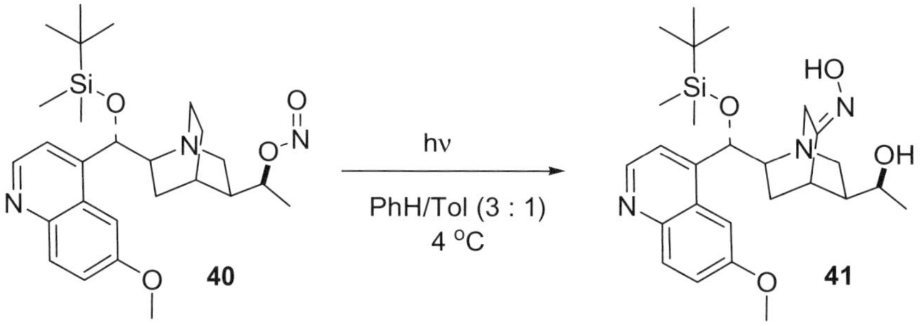

1α-Amino-1,6,9-trideoxy forskolin was recently synthesized starting from drimenal and an isoprenoid C_5 unit.[26] A tricyclic labdane with the entire forskolin skeleton was available in only four steps. Barton's nitrite photolysis reaction was applied to functionalize C-1.

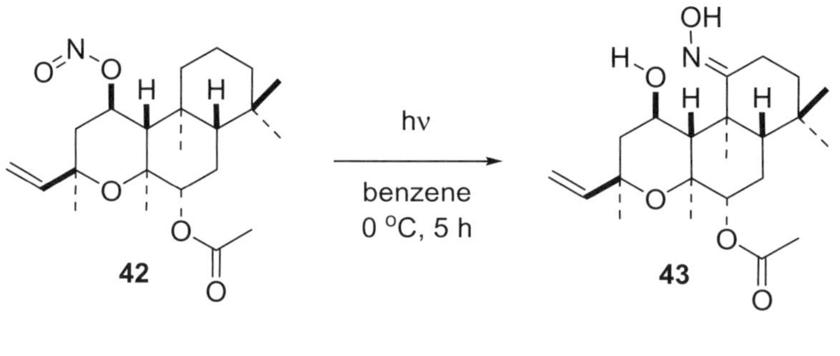

In the synthesis of 1α,25-dihydroxy-18-norvitamin D_3, Sicinski and co-workers utilized a Barton nitrite photolysis reaction.[27]

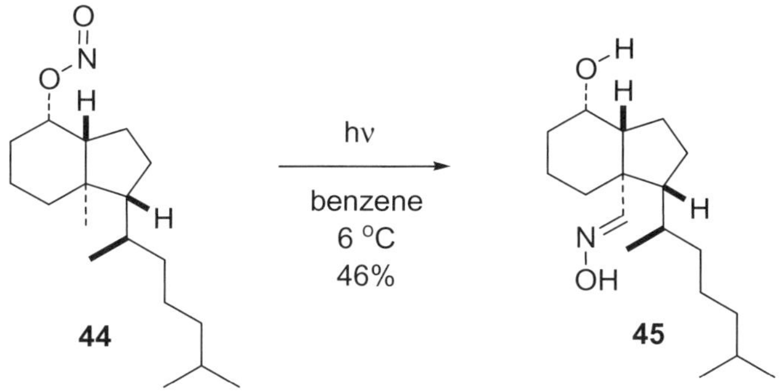

Functionalization of Flexible Molecules

Although Barton reactions occur more easily with rigid structures flexible molecules can also be functionalized by this method. The reaction of the acyclic system **46** shown below must involve a rapid hydrogen atom abstraction from C-5, since cyclic products derived from the addition of the oxygen radical to the double bond are not observed.[28]

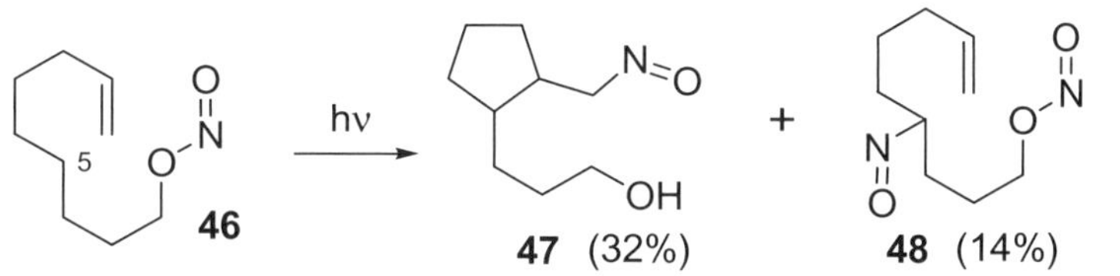

3.2.5 Variations and Improvements

Boar and coworkers synthesized cycloartenol via 19-oxygenated lanostanes by photolysis of the lanostenyl nitrite **49** in CCl_4 in the presence of oxygen to yield the nitrate **50** instead of the usual oxime. This compound (**50**) was then converted into cycloartenol in a 62% yield.[29]

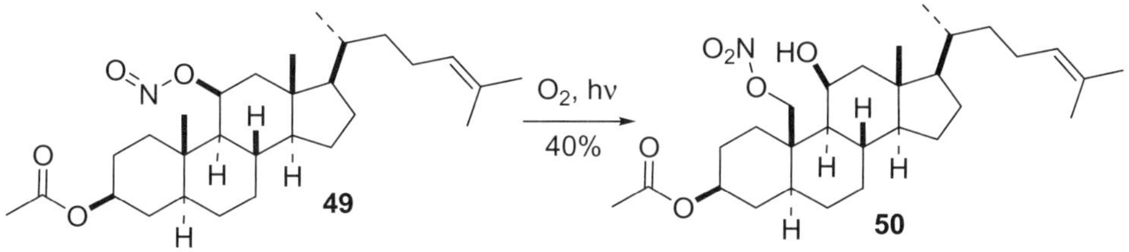

Researchers from Searle converted the 11β-nitrite ester **51** which was then irradiated in benzene in the presence of iodine, to give the intermediate 18-iodo derivative, which immediately cyclized on refluxing with NaOAc in MeOH to give Δ1-18-deoxyaldosterone acetate **52**.[30]

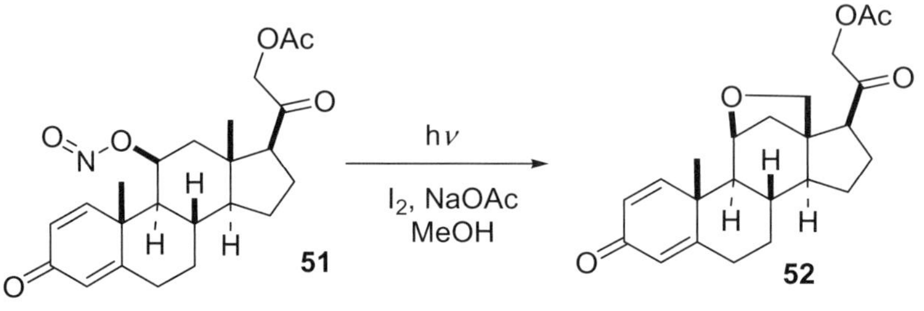

In the improved synthesis of aldosterone by Barton, a key step involved photolysis of the nitrite **53** to give the desired attack at C-18, which then reacted with the hydroxyl at C-11 to produce a hydroxylamine group that then cyclized to the nitrone.[31] Further work showed that the yield of the nitrone varied with the choice of solvent used in the photolysis. A systematic study of the photolysis of 11β-nitrites showed that for this reaction the solvent order of MeCN > THF > PhCl > PhMe > benzene > acetone and addition of base (1% solvent volume) gave improved yields.

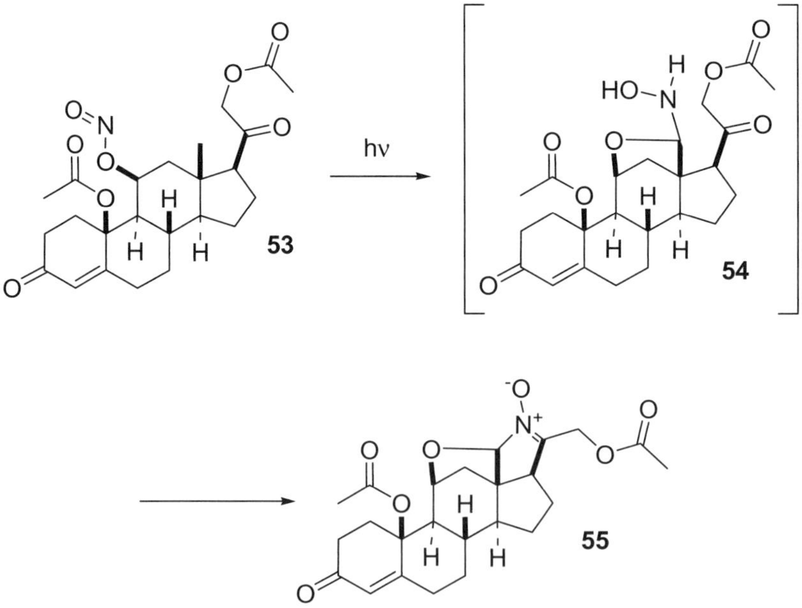

The Barton nitrite photolysis reaction of **56**, to give **57**, a key intermediate for the synthesis of an endothelin receptor antagonist, was successfully carried out in a continuous microflow system using a pyrex glass-covered stainless-steel microreactor having a microchannel (1000 μm width, 107 μm depth, 2.2 m length). The authors found that a 15 W black light (peak wavelength: 352 nm) as the light source, suffices for the Barton reaction, creating a compact photo-micro reaction system. Multi-gram scale production was attained using two serially connected, multi-lane microreactors.[32]

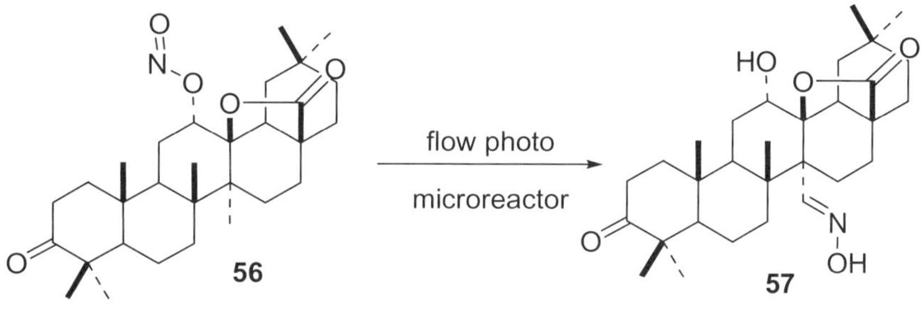

A recent patent application disclosed a method for making polyurethanes made from nitrite esters and di/poly-isocyanates with the help of UV-irradiation for use as coatings.[33] The application disclosed polymerizable and curable compositions and processes of polymerizing and curing polyurethanes involving an isocyante compound and a nitrite ester compound. Irradiation of the nitrite ester compound with UV generates an active hydrogen compound by the Barton nitrite photolysis reaction that reacts with an isocyanate compound to produce and/or crosslink a polyurethane.

3.2.6 Experimental

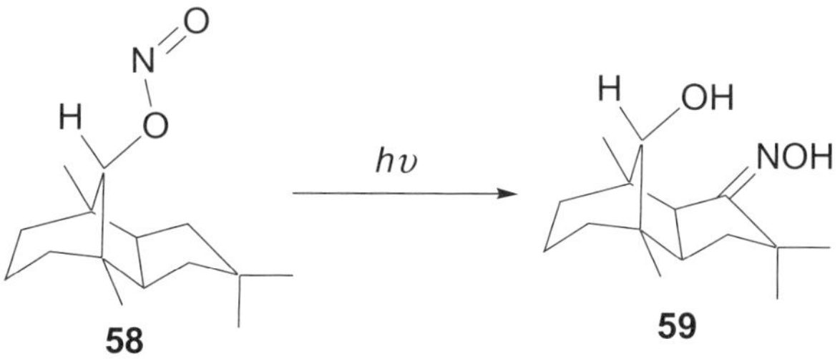

9-Hydroxy-2,2,4,8-tetramethyl-octahydro-4,8-methano-azulen-1-one oxime (59)[34]

A solution of α-nitrite **58** (7.3 g) in thiophene free benzene (200 mL) under nitrogen in a Pyrex flask equipped with a condenser was irradiated externally with the filament unit of a 250-W Hg lamp (General Electric H250-A5) placed as close as possible so that its heat brought and maintained the solution at reflux. A solid began to appear after 40 min, and the irradiation was continued for a total of 75 min. The photoproduct (mp 224–226 °C dec, 4.2 g) was collected and the filtrate was concentrated and irradiated 30 min to give a second crop (mp 223–225°C dec, 0.7 g); total 4.9 g (67%). Recrystallization from dioxane gave the photoproduct **59** as shining needles.

3.2.7 References

1. Barton, D. H. R.; Beaton, J. M.; Geller, L. E.; Pechet, M. M. *J. Am. Chem. Soc.* **1960**, *82*, 2640–2641.
2. [R] Barton, D. H. R. *Aldrichimica Acta* **1990**, *23*, 3–10.
3. [R] Barton, D. H. R. *Experientia* **1950**, *6*, 316–329.
4. See brochure of Nobel Committees for Physics and Chemistry, The Royal Swedish Academy of Sciences, List of Nobel Prize Laureates 1901–1994, Almquist & Wiksell Tryckeri: Uppsala, Sweden.
5. Simpson, S. A.; Tait, T. F.; Wettstein, A.: Neher, R.; Euw, J.; Schindler, O.; Reichstein, T. *Helv. Chim. Acta* **1954**, *37*, 1163–1200.
6. Barton, D. H. R.; Beaton, J. M.; Geller, L. E.; Pechet, M. M. *J. Am. Chem. Soc.* **1961**, *83,* 4076–4083.
7. Barton, D. H. R. *Some Recollections of Gap Jumping*, American Chemical Society: Washington D.C., 1991, pp75–81.
8. Akhtar, M.; Barton, D. H. R.; Sammes, P. G. *J. Am. Chem. Soc.* **1964**, *86*, 3394–3395
9. [R] Majetich, G.; Wheless, K. *Tetrahedron* **1995**, *51*, 7095–7129.
10. [R] Suginome, H. *CRC Handbook of Organic Photochemistry and Photobiology* 2nd edn. **2004**, 102/1–102/16.
11. Suginome, H.; Takakuwa, K.; Orito, K. *Chem. Lett.* **1982**, 1357–1360.
12. Habermehl, G.; Seib, K. *Naturwissenschaften* **1978**, *65*, 155–156.
13. Midgley, J. M.; Parkin, J. E.; Whalley, W. B. *J. Chem. Soc., Chem. Commun.* **1970**, *13*, 789–790.
14. Harnik, M.; Lederman, Y.; Szpigielman, R.; Herling, J. *Tetrahedron* **1976**, *32*, 1001–1004.
15. Barton, D. H. R.; Budhiraja, R. P.; McGhie, J. F. *J. Chem. Soc., Part C* **1969**, 336–338.
16. Wolff, M. E.; Morioka, T. *J. Org. Chem.* **1965**, *30*, 2553–2557.
17. Suginome, H.; Nakayama, Y.; Senboku, M. *J. Chem. Soc., Perkin Trans. 1* **1992**, *14*, 1837–1842.
18. Batten, P. L.; Bentley, T. J.; Boar, R. B.; Draper, R. W.; McGhie, J. F.; Barton, D. H. R. *J. Chem. Soc., Perkin Trans. 1* **1972**, *5*, 739–748
19. Corey, E. J.; Hahl, R. W. *Tetrahedron Lett.* **1989**, *30*, 3023–3026.
20. Corey, E. J.; Arnett, J. F.; Widiger, G. N. *J. Am. Chem. Soc.* **1975**, *97,* 430–431.
21. Murai, A.; Nishizakura, K.; Katsui, N.; Masamune, T. *Tetrahedron Lett.* **1975**, 4399–4402.
22. Sejbal, J.; Klinot, J.; Vystrcil, A. *Collect. Czech. Chem. Commun.* **1988**, *53,* 118–131.
23. Khuong-Huu, F.; Bennett, C. R.; Fouche, P.; Goutarel, *R. C. R. Acad. Sci. Paris* **1972**, *275,* 499–502.
24. Stotter, P. L.; Hill, K. A.; Friedman, M. D. *Heterocycles* **1987**, *25,* 259–262.
25. Nakano, A.; Ushiyama, M.; Iwabuchi, Y.; Hatakeyama, S. *Adv. Synth. Catal.* **2005**, *347*, 1790–1796.

26. Anikin, A.; Maslov, M.; Sieler, J.; Blaurock, S.; Baldamus, J.; Hennig, L.; Findeisen, M.; Reinhardt, G.; Oehme, R.; Welzel, P. *Tetrahedron* **2003**, *59*, 5295–5305.
27. Sicinski, R. R.; Perlman, K.L.; Prahl, J.; Smith, C.; DeLuca, H. F. *J. Med. Chem.* **1996***, 22,* 4497–4506.
28. Cekovic, Z.; Ilijev, D. *Tetrahedron Lett.* **1988**, *29,* 1441–1444.
29. Boar, R. B.; Copsey, D. B. *J. Chem. Soc., Perkin Trans. 1* **1979**, *2*, 563–564.
30. Chinn, L. J..; Desai, B. N.; Lenz, G. R. *J. Chem. Res., (S)* **1978**, *8*, 284–285.
31. Barton, D. H. R.; Basu, N. K.; Day, M. J.; Hesse, R. H.; Pechet, M. M.; Starratt, A. N. *J. Chem. Soc., Perkin Trans. 1* **1975**, *22*, 2243–2251
32. Sugimoto, A.; Sumino, Y.; Takagi, M.; Fukuyama, T.; Ryu, I. *Tetrahedron Lett.* **2006**, *47*, 6197–6200.
33. Taylor, A.; Todd, J. L. PCT Int. Appl. WO 2008078104A1, 3 July **2008**.
34. Nickon, A; Iwadare, T.; McGuire, T. J.; Mahajan, J. R; Narang, S. A.; Umezawa, B. *J. Am. Chem. Soc.* **1970** *92*, 1688–1696.

3.3 Sandmeyer Reaction

Jeremy M. Richter

3.3.1 Description

The conversion of an aromatic diazo compound, generated from the corresponding aniline, to the aryl halide (chloride or bromide) or cyanide is known as the Sandmeyer reaction.[1,2]

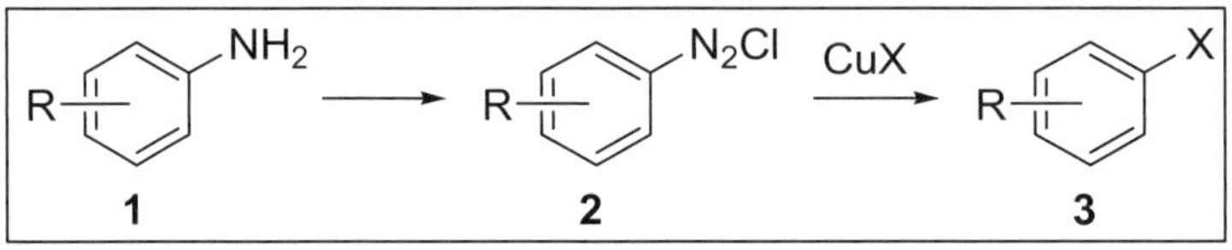

The reaction begins with conversion of the aniline (**1**) to the diazo compound (**2**). Usually in the same pot, **2** is then treated with the cuprous halide or cyanide to form the aryl halide or cyanide (**3**).

3.3.2 Historical Perspective

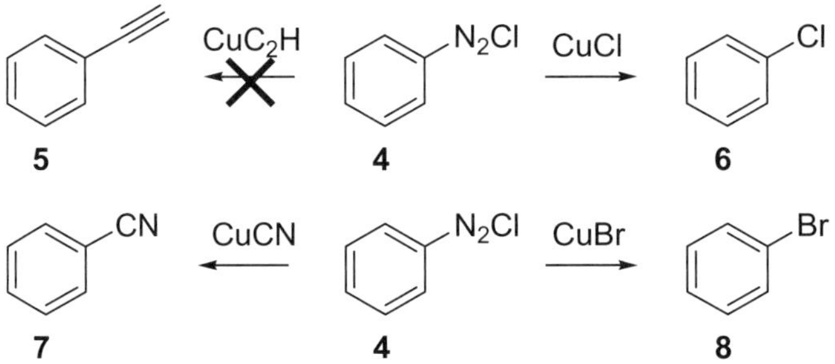

During an unsuccessful attempt to prepare phenylacetylene (**5**) from diazobenzene (**4**), Sandmeyer discovered that chlorobenzene (**6**) could be efficiently prepared from **4** under these reaction conditions.[1] While investigating the formation of this unexpected product, he discovered that the active reagent was cuprous chloride, formed in hydrochloric acid solution. He further postulated that this particular reaction was unique to cuprous chloride, since cupric and ferrous chloride did not exhibit a similar behavior. It would be many years, however, before a more complete understanding of this process could be secured. Sandmeyer's subsequent studies demonstrated that, when reacted with diazobenzene (**4**), cuprous bromide produced bromobenzene (**8**) and cuprous cyanide produced cyanobenzene (**7**).[2] Strictly speaking, these are the only three transformations that bear the "Sandmeyer" name; however there are several other reaction classes of similar mechanism which are routinely categorized as Sandmeyer reactions (*vide infra*). These seminal investigations provided the impetus for further study of this reaction in the ensuing decades.[3–7]

3.3.3 Mechanism

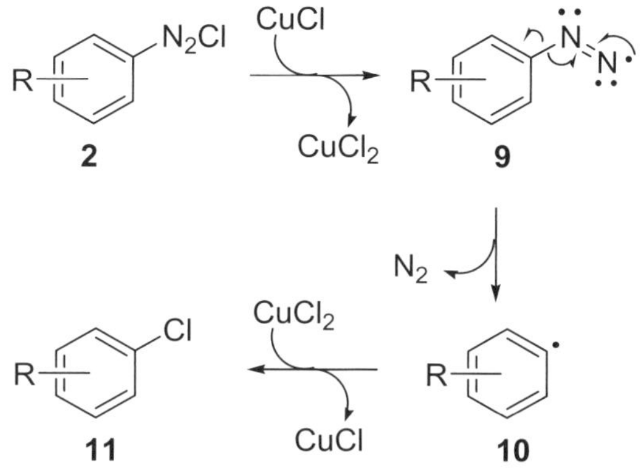

The Sandmeyer reaction has been the focus of numerous mechanistic studies since its discovery in 1884,[8–27] and the exact course of the transformation is still debated in the literature. For almost 60 years, it was commonly thought that the Sandmeyer reaction was an ionic process, until the work of Waters definitively demonstrated that the reaction actually proceeds through a radical mechanism.[9] Furthermore, Kochi provided evidence that the reaction proceeds through inner-sphere electron transfer,[16] however this conclusion has come under scrutiny in recent years. It is currently believed that the mechanism may vary between inner- and outer-sphere, depending on the substituents on, and therefore redox potential of, the diazobenzene.[27] Regardless of the specific nature of the electron transfer event (the rate determining step[23]), the reaction begins with transfer of an electron from cuprous chloride to diazobenzene (**4**), generating fleeting radical intermediate **9**. This intermediate decomposes with the expulsion of dinitrogen, thus providing aryl radical **10**. Interception of this radical with chlorine, from cupric chloride, gives the final product **11**. The exact nature of this atom transfer event is also poorly understood: it could either be *via* the bridging chlorine atom or an alternate S_H2 process, or perhaps a continuum of both, depending on the substituents on the aromatic ring.[24,25] Further studies have demonstrated that this reaction is indeed catalytic in copper;[24] however, addition of extra cupric halide increases the efficiency of the halogenation event.[6,21]

3.3.4 Variations

Since its discovery in 1884, the Sandmeyer reaction has undergone numerous subtle variations and improvements and it is beyond the scope of this chapter to discuss each of these. Rather, representative procedures will be given in section 3.3.6, which provides state-of-the-art protocols for accomplishing these transformations. However, a few of these modifications are worth noting. There have been several reports of initiation of the Sandmeyer reaction by reductants other than cuprous halides, however cupric halide is still required to effect halide transfer to radical **10**.[21,28,29] Therefore, it is unclear whether these reductants are indeed reducing **4**, or if they are reducing cupric halide to cuprous halide. Several other metals ($ZnCl_2$, $FeCl_2$, CoCl) are capable of catalyzing both the reduction and ligand transfer steps.[11,30] When copper metal is utilized as the initiating reductant, the reaction is commonly known as the Gattermann reaction.[31] It has also been reported that the reaction can proceed efficiently if the arenediazonium tetrachlorocuprate salt is prepared prior to halogenation.[32,33] Alternate halide sources have been utilized, some of which allow for the direct conversion to the aryl halides from the anilines, without isolation of the intermediate diazoarenes.[34–36] Finally, performing the reaction in polyethylene glycol/dichloromethane was found to improve the efficiency of these transformations.[37]

There have been many reports of reactions that are broadly classified as Sandmeyer-like, yet they do not strictly belong to this specific category. Several groups have reported that triazines of type **12** can be used to mask the diazo, then allow for direct Sandmeyer-type reaction to form the substituted products (**3**).[38–42] The iodination of diazoarenes is commonly called a Sandmeyer reaction, however it is unclear whether this reaction is even proceeding *via* the same mechanism. Indeed, the uncatalyzed version of this reaction predates Sandmeyer's seminal investigations by nearly 20 years.[43] Subsequently, much work has been done to render this transformation (**2** to **13**) more efficient.[3,7,8,34,44–49] As with iodination, the hydroxylation of diazoarenes to form phenols (**14**) was reported 20 years prior to Sandmeyer's seminal investigations.[50] These particular reactions usually proceed *via* uncatalyzed, cationic mechanisms,[51-54] however strict control of the reaction conditions can allow for catalyzed, radical-based hydroxylations to occur.[24,55] This hydroxylation method is commonly known as the Sandmeyer Hydroxylation reaction. The conversion of diazoarenes to nitroarenes (**15**) has also been reported under a variety of conditions,[3,4,6,56–63] as well as the conversion to a myriad of sulfur-substituted arenes (**16**).[4,35,64–69] The decomposition of diazoarene tetrafluoroborates (**17**) to arylfluorides (**18**) has been reported and is commonly known as the Balz–Schiemann reaction.[70–74] Finally, intercepting the aryl radical formed from the

Sandmeyer reaction (**10**) with a variety of olefinic coupling partners, followed by halogenation of the resulting alkyl radical, leading to **19**, is commonly known as the Meerwein reaction, many modifications of which are known.[5,6,75]

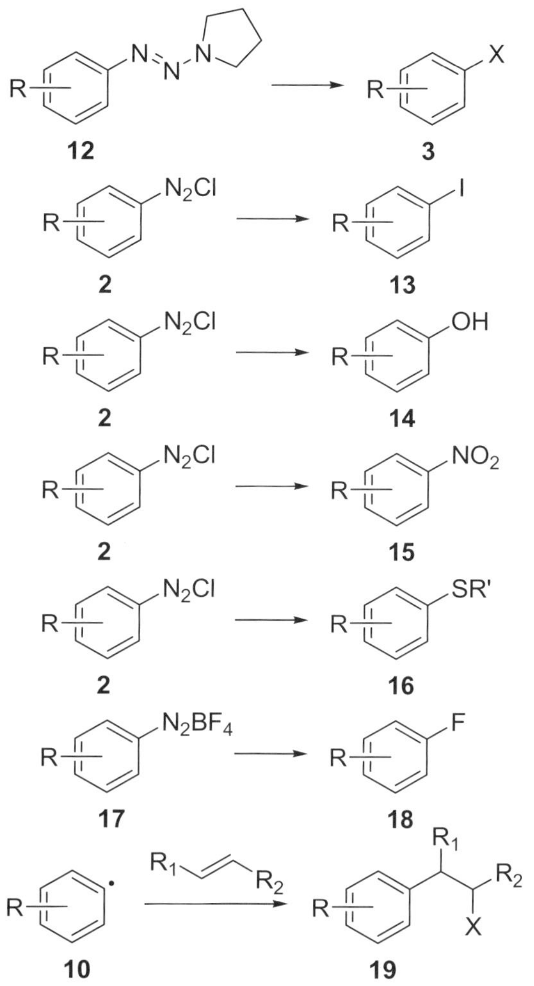

3.3.5 *Synthetic Utility*

Natural Product Total Synthesis

The power of the Sandmeyer reaction to accomplish the functionalization of aromatic rings has led to this methodology being utilized in countless natural product total syntheses. Banerjee and coworkers used a Sandmeyer bromination reaction in the synthesis of dichroanal B (**22**),[76] efficiently converting aniline **20** into aryl bromide **21** in 88% yield. This particular synthesis highlights the utility of the Sandmeyer reaction on electron-rich aromatic rings with a variety of substituents.

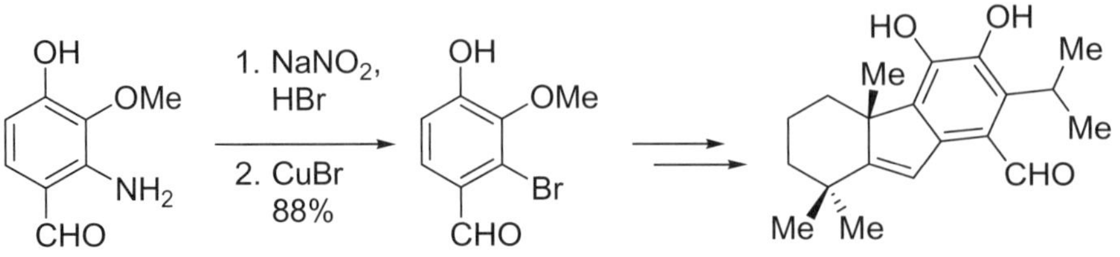

20 **21** dichroanal B (**22**)

Weinreb also utilized a Sandmeyer bromination in his approach towards the total synthesis of chartelline A.[77] He was able to effect the conversion of aniline **23** into the tri-bromide **24** in 88% yield, even in the presence of a potentially sensitive spiro-β-lactam.

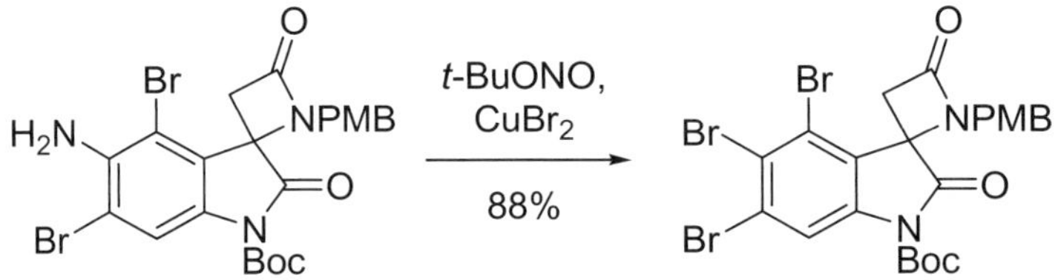

23 **24**

The Sandmeyer reaction is not only useful for the preparation of arylbromides. The Boger group used a Sandmeyer reaction to install both a hydroxyl and chloro substituent in their synthesis of the teicoplanin aglycon (**29**).[78] Diazotization of **25** followed immediately by hydroxylation with cuprous oxide (halide-free/radical conditions) furnished phenol **26** in excellent yield, even with several sensitive functionalities present in the molecule. Several steps were then required to convert **25** into aniline **27**. At this point, a Sandmeyer chlorination proceeded in moderate yield to furnish the aryl chloride **28**. This particular reaction is noteworthy, as the reaction proceeded well, in the presence of a variety of functional groups, many of which were unprotected.

The Sandmeyer-like iodination has also found application in the synthesis of numerous natural products. Fukuyama utilized this reaction to install an aryl iodide at an early stage in his synthesis of eudistomin C (**32**).[79] Diazotization of **30** followed by the addition of potassium iodide in the same pot produced aryl iodide **31** in 62% yield.

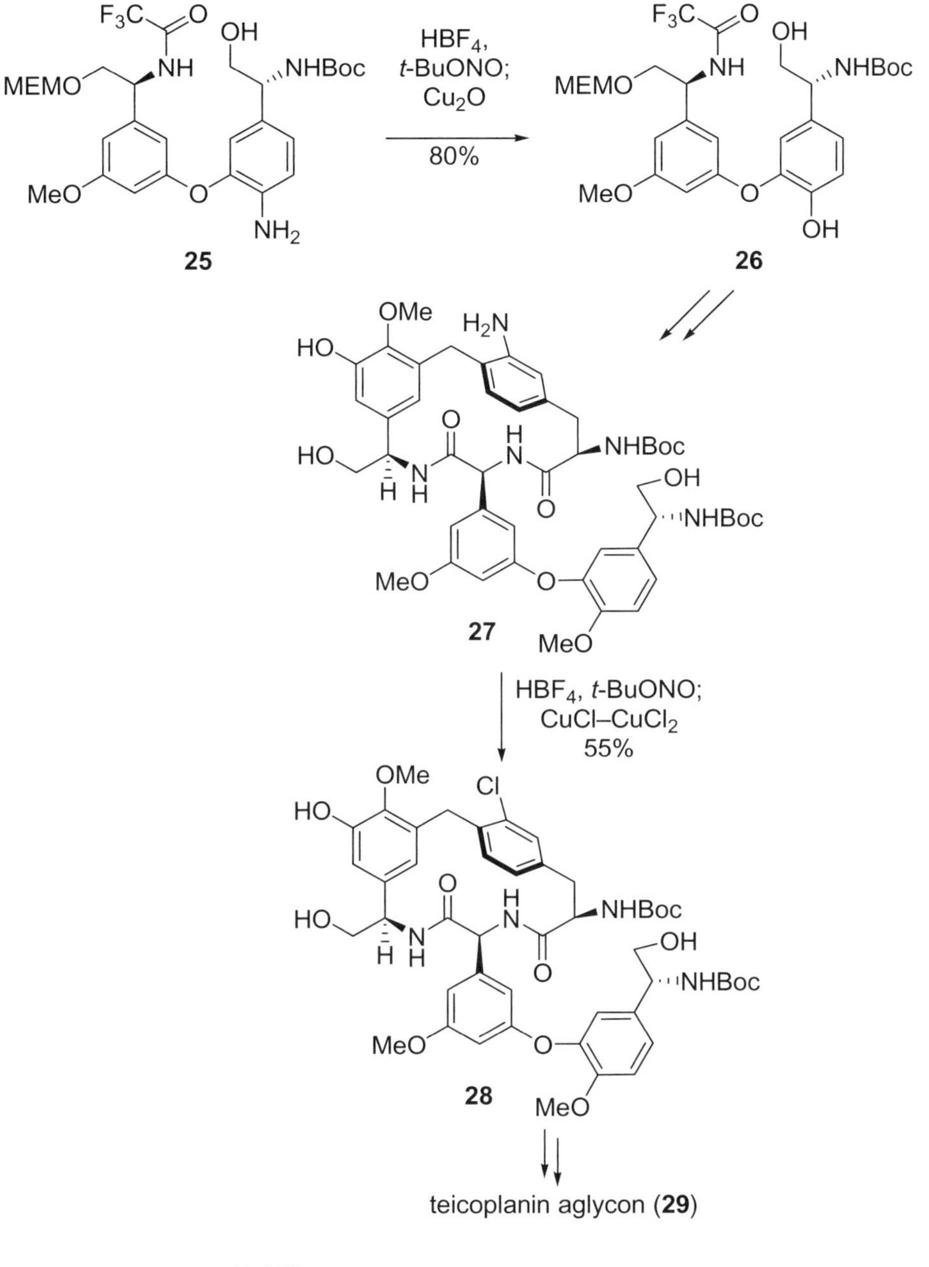
HBF$_4$, t-BuONO; Cu_2O
80%
25
26
27
HBF$_4$, t-BuONO; CuCl–$CuCl_2$
55%
28
teicoplanin aglycon (29)

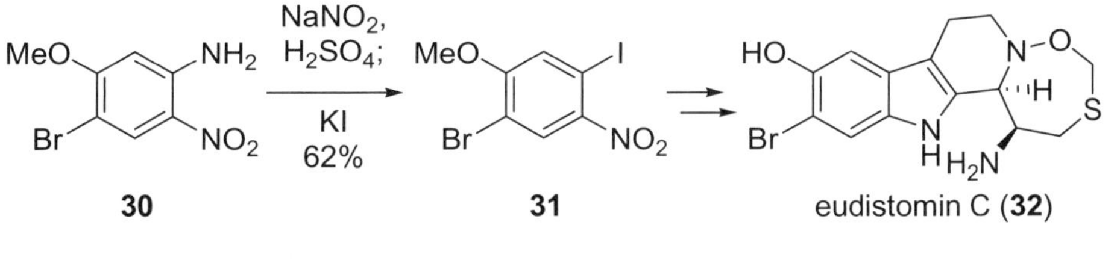
$NaNO_2$, H_2SO_4; KI
62%
30
31
eudistomin C (32)

As an additional example, Harmata used a modified Sandmeyer-like iodination in his synthesis of the natural product pseudopteroxazole (**35**).[80] Triazine formation on **33** proceeded quantitatively, after which exposure to diiodomethane produced the desired aryl iodide (**34**) in good yield.

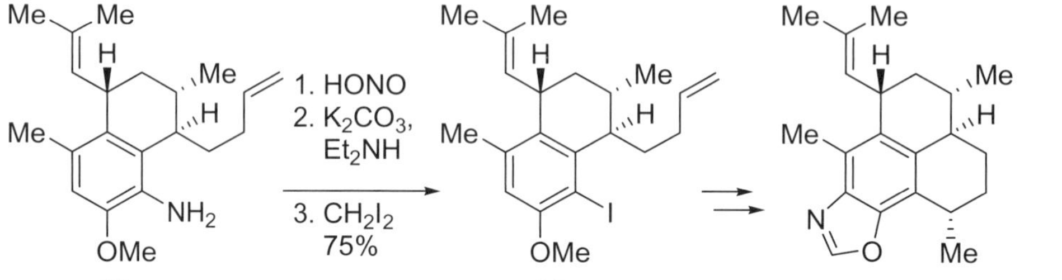

33 **34** pseudopteroxazole (**35**)

Medicinal and Heterocyclic Chemistry

The Sandmeyer reaction has not only been useful in the total synthesis of natural products, but the preparation of medicinally relevant compounds and heterocyclic products has been greatly simplified by this procedure. The H. Lundbeck & Co. has utilized the Sandmeyer bromination in the preparation of dopamine-uptake inhibitors (**38**), *via* Sandmeyer reaction on anthranilic acid **36** to produce aryl bromide **37** in excellent yield.[81]

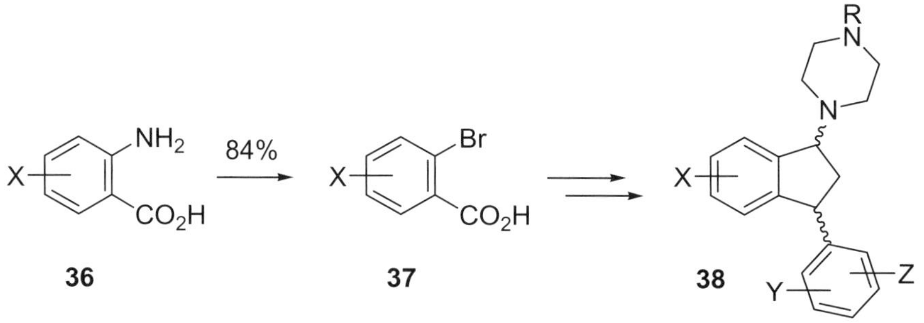

36 **37** **38**

The Dainippon Pharmaceutical Company has utilized a Sandmeyer reaction in the preparation of antibacterial pyridonecarboxylic acids.[82] The diazonaphthyridone (**39**) was either chlorinated or cyanated under Sandmeyer conditions to provide the desired products (**40** and **41** respectively) in good yield. These products were eventually processed to the active pharmaceutical agents (**42**).

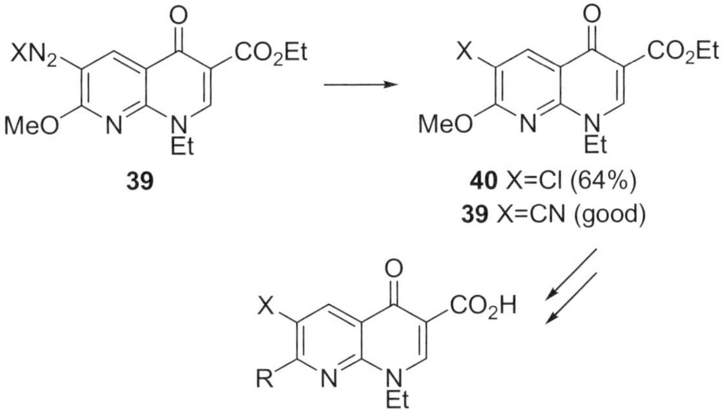

39 → 40 X=Cl (64%), 39 X=CN (good) → 42

Investigators at SmithKline Beecham subsequently utilized the Sandmeyer cyanation for modification of the potassium channel activator cromakalim (**45**).[83] Diazo formation on the aniline **43** followed immediately by treatment with cuprous cyanide and potassium cyanide in the same flask led to the desired aryl cyanide (**44**) in 73% yield.

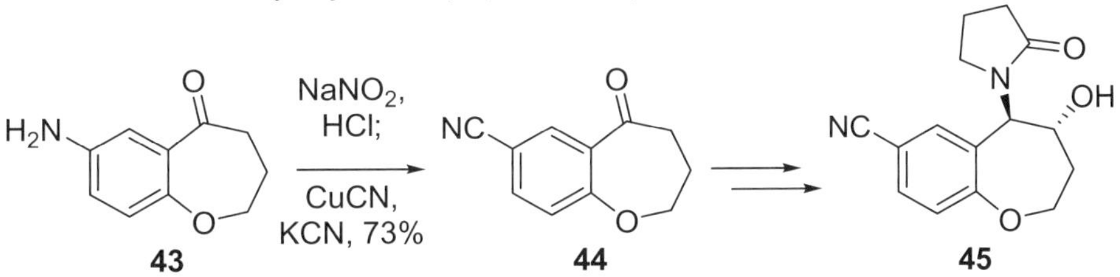

43 44 45

Finally, Somei reported the use of a Sandmeyer bromination to access variably substituted indole heterocycles.[84] Anilinic compounds **46** and **47** were efficiently diazotized and brominated to provide the tribromides **48** and **49** in 81% and 78% yields, respectively.

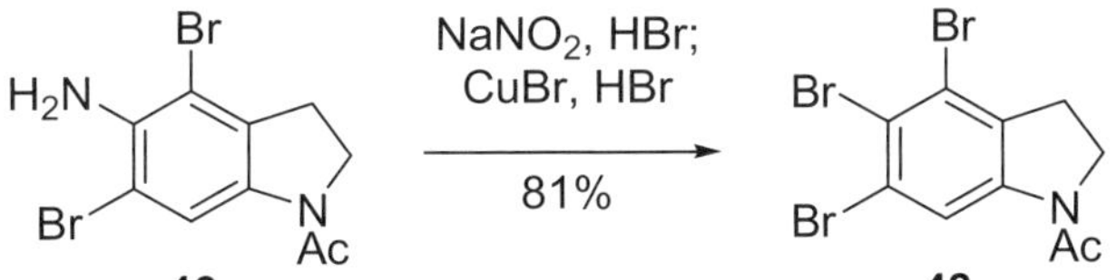

46 48

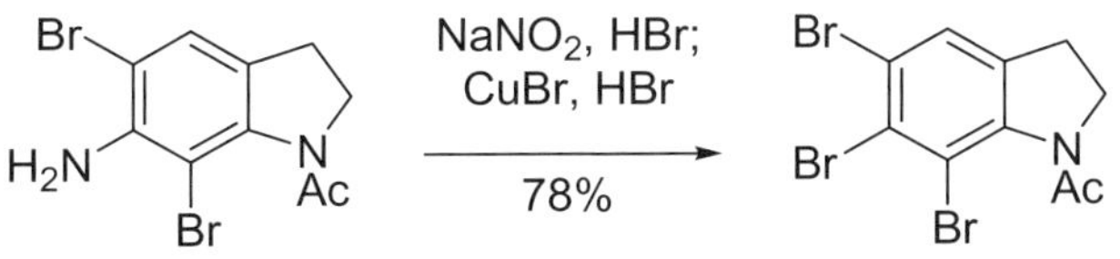

47 49

3.3.6 Experimental

Bromination[77]

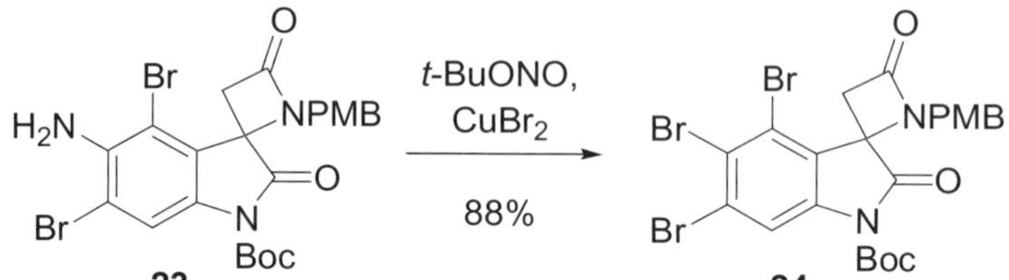

23 24

To a solution of dibromo aniline [**23**] (1.25 g, 2.20 mmol) in 55 mL of MeCN was added $CuBr_2$ (2.46 g, 11.0 mmol), followed by *t*-BuONO (437 μL, 3.31 mmol), and the solution was then heated at 50 °C for 1 h. The reaction mixture was diluted with saturated $NaHCO_3$ solution, extracted with EtOAc, dried over MgSO4, and concentrated *in vacuo*. The residue was purified by flash column chromatography (20–50% EtOAc/hexanes) to yield tribromide [**24**] as a white solid (1.22 g, 88%).

Chlorination[78]

A solution of crude aniline in CH_3CN (50 μL) was treated with HBF_4 (0.08 M solution in CH_3CN, 13 μL, 1.04 μmol) at 0 °C, and the reaction mixture was stirred at 0 °C for 5 min before the addition of *t*-BuONO (0.08 M solution in CH_3CN, 13 μL, 1.04 μmol). The resulting mixture was stirred at 0 °C for 5 min. The reaction mixture was cooled to –20 °C and immediately added to an aqueous solution (0.1 mL) containing CuCl (2.7 mg, 27.3 μmol) and $CuCl_2$ (15.0 mg, 111.5 μmol) at 0 °C. The mixture was stirred vigorously at 25 °C for 1 h and added to a solution of EtOAc (3 mL) and H_2O (3 mL)]. The water layer was extracted with EtOAc (3 × 5 mL), and the combined organic solution was washed with saturated aqueous NaCl (10 mL), dried (Na_2SO_4), and concentrated *in vacuo*. PTLC (SiO_2, 3% CH_3OH-$CHCl_3$) afforded [**28**] (0.45 mg, 0.8 mg [theoretical], 55%).

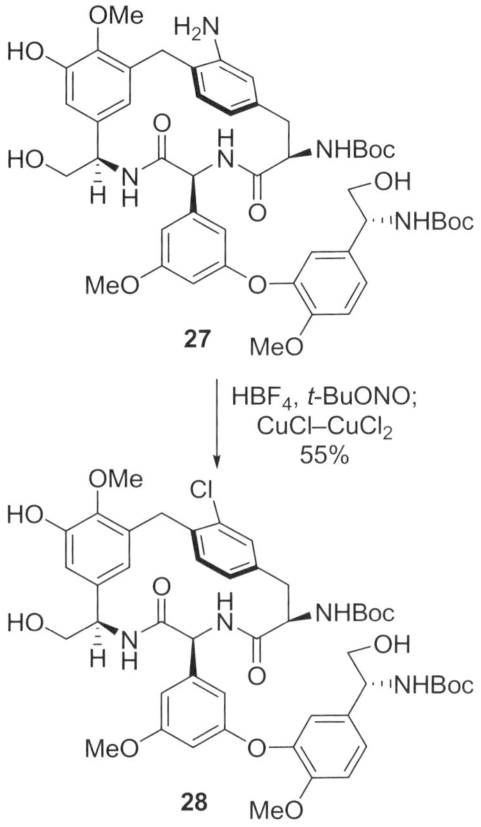

Cyanation[83]

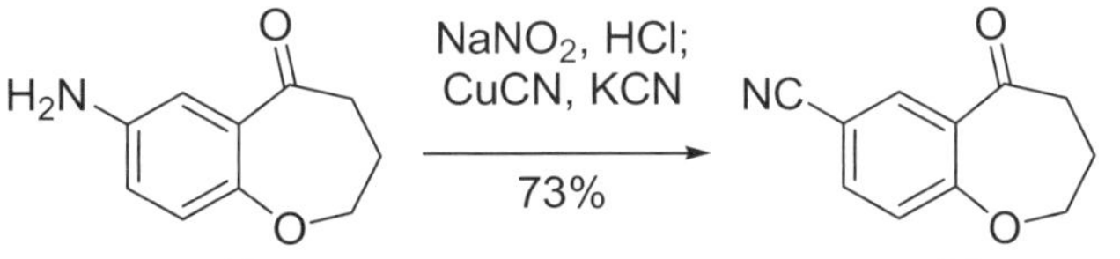

43 44

Aqueous sodium nitrite (1.26 g, 18.3 mmol in 20 cm^3) was added dropwise to a solution of the 7-aminobenzoxepine [**43**] (3.1 g, 17.5 mmol) in (1:2) aq. ethanol (60 cm^3) containing conc. HCl (4.4 cm^3) at 0 ºC until an immediate reaction was observed with starch iodide paper. The solution was stirred for 15 min and then added dropwise to a solution of copper(I) cyanide (14.8 g, 0.166 mmol) and potassium cyanide (10.36 g, 0.159 mmol) in water (50 cm^3) at 100 ºC at such a rate that the temperature did not drop below 70 ºC. The mixture was heated at 100 ºC for a further 0.5 h, cooled, and extracted with ethyl acetate. The combined extracts were, dried, filtered, and evaporated to give a brown solid, which was chromatographed (dichloromethane) to give the *title compound* (2.39 g, 73%).

Hydroxylation[78]

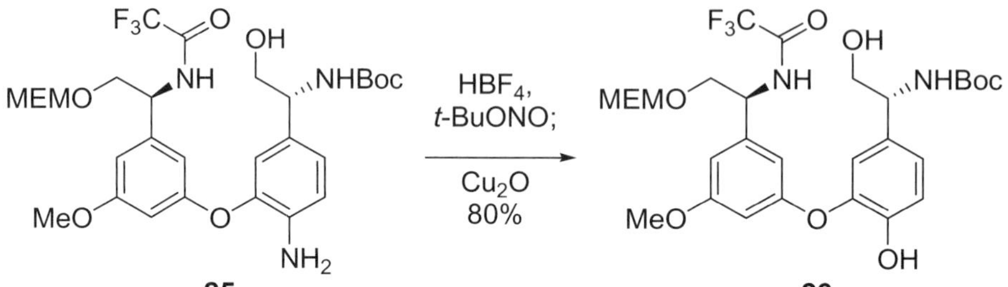

25 **26**

[**25**] was dissolved in CH_3CN (10 mL) and cooled with an ice-bath. HBF_4 (48% solution in water, 1.1 equiv, 220 μL) was added and the solution was stirred at 0 ºC for 10 min. *t*-BuONO (1.1 equiv. 200 μL) dissolved in CH_3CN (1 mL) was added dropwise and the solution was stirred at 0 ºC for 10 min. The solution was cooled to –15 ºC and H_2O (20 mL) was added. After 10 min, the solution was added to a cold solution (0 ºC) of $Cu(NO_3)_2$ (200 equiv, 80 g) and Cu_2O (45 equiv, 10 g) in H_2O (200 mL). The mixture was vigorously stirred at 0 ºC for 1 h and extracted with EtOAc (2 × 100 mL). The organic phases were combined, washed with saturated aqueous NaCl, dried (Na_2SO_4), filtered and concentrated. Flash chromatography (SiO_2, 5 × 30 cm, 60% EtOAc-hexanes) provided [**26**] (840 mg, 1.05 g theoretical, 80%).

Iodination[79]

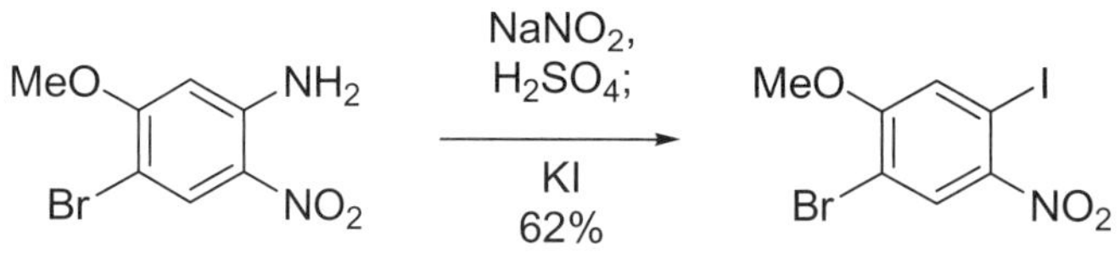

30 **31**

To a solution of 4-bromo-5-methoxy-2-nitrobenzenamine ([**30**]) (7.00 g, 28.3 mmol) and H_2SO_4 (7.00 g) in CH_3CN (250 mL) was added $NaNO_2$ (3.91 g, 56.7 mmol) in H_2O (25 mL) at 0 ºC. After stirring for 15 minutes, KI (18.8 g, 115 mmol) in H_2O (25 mL) was added to the reaction mixture at 0 ºC. The reaction mixture was then allowed to warm up to room temperature and was stirred for 15 minutes. The mixture was poured into sat. $Na_2S_2O_3$ and extracted with EtOAc. The extracts were washed with brine, dried over $MgSO_4$, filtrated, and evaporated.

3.3.7 References

1. Sandmeyer, T. *Ber.* **1884**, *17*, 1633–1635.
2. Sandmeyer, T. *Ber.* **1884**, *17*, 2650–2653.
3. [R] Hodgson, H. H. *Chem. Rev.* **1947**, *40*, 251–277.
4. Cowdrey, W. A.; Davies, D. S. *Quart. Revs.* **1952**, *6*, 358–379.
5. Nonhebel, D. C. *Special Publication – Chemical Society*. **1970**, *24*, 409–437.
6. [R] Galli, C. *Chem. Rev.* **1988**, *88*, 765–792.
7. M[R] erkushev, E. B. *Synthesis* **1988**, 923–937.
8. Hodgson, H. H.; Birtwell, S.; Walker, J. *J. Chem. Soc.* **1941**, 770–776.
9. Waters, W. A. *J. Chem. Soc.* **1942**, 266–270.
10. Hodgson, H. H.; Birtwell, S.; Walker, J. *J. Chem. Soc.* **1942**, 720–723.
11. Hodgson, H. H.; Birtwell, S.; Walker, J. *J. Chem. Soc.* **1944**, 18–19.
12. Hodgson, H. H.; Sibbald, D. D. R. *J. Chem. Soc.* **1944**, 393–395.
13. Hodgson, H. H.; Sibbald, D. D. R. *J. Chem. Soc.* **1945**, 545–546.
14. Hodgson, H. H.; Sibbald, D. D. R. *J. Chem. Soc.* **1945**, 819–821.
15. Hodgson, H. H.; Mahadevan, A. P. *J. Chem. Soc.* **1947**, 173–174.
16. Kochi, J. K. *J. Am. Chem. Soc.* **1957**, *79*, 2942–2948.
17. Nonhebel, D. C.; Waters, W. A. *Proc. Roy. Soc. London. Series A.* **1957**, *242*, 16–27.
18. Dickerman, S. C.; Weiss, K.; Ingberman, A. K. *J. Am. Chem. Soc.* **1958**, *80*, 1904–1911.
19. Rømming, C.; Wœrstad, K. *Chem. Commun.* **1965**, 299–300.
20. Dickerman, S. C.; DeSouza, D. J.; Jacobson, N. *J. Org. Chem.* **1969**, *34*, 710–713.
21. Galli, C. *J. Chem. Soc. Perkin Trans. I.* **1981**, 1459–1461.
22. Galli, C. *J. Chem. Soc. Perkin Trans. II.* **1982**, 1139–1141.
23. Galli, C. *J. Chem. Soc. Perkin Trans. II.* **1984**, 897–902.
24. Hanson, P.; Jones, J. R.; Gilbert, B. C.; Timms, A. W. *J. Chem. Soc. Perkin Trans. II.* **1991**, 1009–1017.
25. Hanson, P.; Hammond, R. C.; Gilbert, B. C.; Timms, A. W. *J. Chem. Soc. Perkin Trans. II.* **1995**, 2195–2202.
26. Hanson, P. Rowell, S. C.; Taylor, A. B.; Walton, P. H.; Timms, A. W. *J. Chem. Soc. Perkin Trans. II.* **2002**, 1126–1134.
27. Hanson, P.; Jones, J. R.; Taylor, A. B.; Walton, P. H.; Timms, A. W. *J. Chem. Soc. Perkin Trans. II.* **2002**, 1135–1150.
28. Doyle, M. P.; Guy, J. K.; Brown, K. C.; Mahapatro, S. N.; VanZyl, C. M.; Pladziewicz, J. C. *J. Am. Chem. Soc.* **1987**, *109*, 1536–1540.
29. Gilbert, B. C.; Hanson, P.; Jones, J. R.; Whitwood, A. C.; Timms, A. W. *J. Chem. Soc. Perkin Trans. II.* **1992**, 629–636.
30. Hodgson, H. H.; Foster, C. K. *J. Chem. Soc.* **1942**, 581–583.
31. Gattermann, L. *Ber.* **1890**, *23*, 1218.
32. Obushak, M. D.; Lyakhovych, M. B.; Ganushchak, M. I. *Tetrahedron Lett.* **1998**, *39*, 9567–9570.
33. Obushak, N. D.; Lyakhovich, M. B.; Bilaya, E. E. *Russ. J. Org. Chem.* **2002**, *38*, 38–46.
34. Oae, S.; Shinhama, K.; Kim, Y. H. *Chem. Lett.* **1979**, 939–942.
35. Lee, J. G.; Cha, H. T. *Tetrahedron Lett.* **1992**, *33*, 3167–3168.
36. Doyle, M. P.; Siegfried, B.; Dellaria, J. F. Jr. *J. Org. Chem.* **1977**, *42*, 2426–2431.
37. Suzuki, N.; Azuma, T.; Kaneko, Y.; Izawa, Y.; Tomioka, H. *J. Chem. Soc. Perkin Trans I.* **1987**, 645–647.
38. Foster, N. I.; Heindel, N. D.; Burns, H. D.; Muhr, W. *Synthesis* **1980**, 572–573.
39. Ku, H.; Barrio, J. R. *J. Org. Chem.* **1981**, *46*, 5239–5241.
40. Satyamurthy, N.; Barrio, J. R. *J. Org. Chem.* **1983**, *48*, 4394–4396.
41. Heindel, N. D.; Van Dort, M. *J. Org. Chem.* **1985**, *50*, 1988–1990.
42. Gross, M. L.; Blank, D. H.; Welch, W. M. *J. Org. Chem.* **1993**, *58*, 2104–2109.
43. Griess, J. P. *Ann. Chem., Justus Liebigs.* **1866**, *137*, 39–91.
44. Friedman, L.; Chlebowski, J. F. *J. Org. Chem.* **1968**, *33*, 1636–1638.
45. Korzeniowski, S. H.; Gokel, G. W. *Tetrahedron Lett.* **1977**, *18*, 3519–3522.
46. Citterio, A.; Arnoldi, A. *Synth. Commun.* **1981**, *11*, 639–642.

47. Coudret, C. *Synth. Commun.* **1996**, *26*, 3543–3547.
48. Suzuki, H.; Nonoyama, N. *Tetrahedron Lett.* **1998**, *39*, 4533–4536.
49. Ozeki, N.; Shimomura, N.; Harada, H. *Heterocycles.* **2001**, *55*, 461–464.
50. Griess, P.; Hofmann, A. W. *Phil. Trans. Roy. Soc. London* **1864**, *164*, 667–731.
51. Lewis, E. S. *J. Am. Chem. Soc.* **1958**, *80*, 1371–1373.
52. Horning, D. E.; Ross, D. A.; Muchowski, J. M. *Can. J. Chem.* **1973**, *51*, 2347–2348.
53. Swain, C. G.; Sheats, J. E.; Harbison, K. G. *J. Am. Chem. Soc.* **1975**, *97*, 783–790.
54. Swain, C. G.; Sheats, J. E.; Gorenstein, D. G.; Harbison, K. G. *J. Am. Chem. Soc.* **1975**, *97*, 791–795.
55. Cohen, T.; Dietz, A. G. Jr.; Miser, J. R. *J. Org. Chem.* **1977**, *42*, 2053–2058.
56. Sandmeyer, T. *Ber.* **1887**, *20*, 1494–1497.
57. Hantzsch, A.; Blagden, G. W. *Ber.* **1900**, *33*, 2544–2556.
58. Orton, K. J. P. *J. Chem. Soc.* **1903**, *83*, 796–814.
59. Vesely, V.; Dvorak, K. *Bull. Soc. Chim.* **1922**, *31*, 421–424.
60. Bucherer, H. T.; van der Recke, G. *J. Prakt. Chem.* **1931**, *132*, 113–144.
61. Hodgson, H. H.; Marsden, E. *J. Chem. Soc.* **1944**, 22–24.
62. Clark, K. J.; Fray, G. I. *J. Chem. Soc.* **1960**, 894.
63. Singh, P. R.; Kumar, R.; Khanna, R. K. *Tetrahedron Lett.* **1982**, *23*, 5191–5194.
64. Leukardt. *J. Prakt. Chem.* **1890**, *41*, 179.
65. Gattermann, L.; Haussknecht, W. *Ber.* **1890**, *23*, 738–739.
66. Gattermann, L. *Ber.* **1899**, *32*, 1136–1159.
67. Korczynski, A. *Bull. Soc. Chim.* **1921**, *29*, 283–290.
68. Kaufmann, H. P.; Küchler, K. *Ber.* **1934**, *67*, 944–948.
69. Jenkins, C. L.; Kochi, J. K. *J. Org. Chem.* **1971**, *36*, 3095–3102.
70. Balz, G.; Schiemann, G. *Ber.* **1927**, *60*, 1186–1190.
71. [R] Roe, A. *Org. React.* **1949**, *5*, 193–228.
72. Sellers, C.; Suschitzky, H. *J. Chem. Soc.* **1965**, 6186–6188.
73. Suschitzky, H. *Adv. Fluorine Chem.* **1965**, *4*, 1–27.
74. Sharts, C. M. *J. Chem. Educ.* **1968**, *45*, 185–192.
75. Meerwein, H.; Buchner, E.; van Emster, K. *J. Prakt. Chem.* **1939**, *152*, 237–266.
76. Banerjee, M.; Mukhopadhyay, R.; Achari, B.; Banerjee, A. K. *Org. Lett.* **2003**, *5*, 3931–3933.
77. Sun, C.; Lin, X.; Weinreb, S. M. *J. Org. Chem.* **2006**, *71*, 3159–3166.
78. Boger, D. L.; Kim, S. H.; Mori, Y.; Weng, J.-H.; Rogel, O.; Castle, S. L.; McAtee, J. J. *J. Am. Chem. Soc.* **2001**, *123*, 1862–1871.
79. Yamashita, T.; Kawai, N.; Tokuyama, H.; Fukuyama, T. *J. Am. Chem. Soc.* **2005**, *127*, 15038–15039.
80. Harmata, M.; Hong, X. *Org. Lett.* **2005**, *7*, 3581–3583.
81. Bøgesø, K. P. *J. Med. Chem.* **1983**, *26*, 935–947.
82. Matsumoto, J.-i.; Miyamoto, T.; Minamida, A.; Nishimura, Y.; Egawa, H.; Nishimura, H. *J. Med. Chem.* **1984**, *27*, 292–301.
83. Buckle, D. R.; Eggleston, D. S.; Houge-Frydrych, C. S. V.; Pinto, I. L.; Readshaw, S. A.; Smith, D. G.; Webster, R. A. B. *J. Chem. Soc. Perkin Trans. I.* **1991**, 2763–2771.
84. Ohta, T.; Somei, M. *Heterocycles*. **1989**, *29*, 1663–1667.

3.4 Wohl–Ziegler Reaction

Timothy T. Curran

3.4.1 Description

The Wohl–Ziegler reaction is the reaction of an allylic or benzylic substrate **1** with *N*-bromosuccinimide (**2**, NBS) under radical initiating conditions to provide the corresponding allylic or benzylic bromide **3**. Conditions used to promote the radical reaction are typically radical initiators, light and/or heat and carbon tetrachloride (CCl_4) is typically utilized as the solvent.

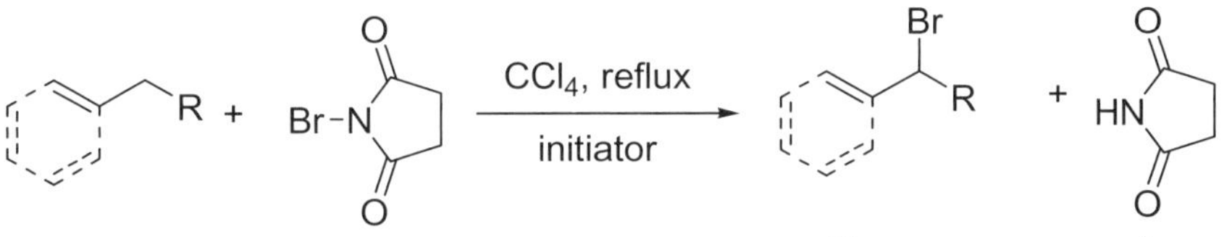

1 2 3 4

Although the transformation is typically non-stereoselective, the reaction results in the selective functionalization of the allylic or benzylic position due to the nature of the reaction conditions which results in primarily attack of weak C–H bonds with insertion of a halide.

3.4.2 Historical Perspective

In 1919 Wohl reported the reaction of *N*-bromoacetamide with tetramethylethylene which provided a brominated substance. This work was followed-up in a 1921 report in which Wohl and Jaschinowski further studied the halogenating properties of *N*-bromoacetamide.[1] Several years later Wolfe and Awang[2] critically evaluated and re-investigated portions of Wohl's two reports, determining that *N*-bromoacetamide reaction with some alkenes did not result in free radical bromination but instead gave products resulting from electrophilic addition of bromine to the alkene. Wohl's work remained in the literature seemingly unnoticed by many, until Ziegler reported the reaction of NBS with **5** during his synthetic studies toward the cantharidins.[3]

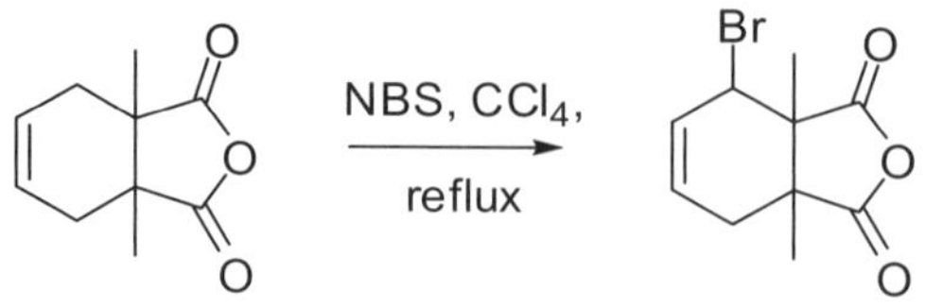

5 6

In a subsequent report, Ziegler and co-workers[4] systematically studied the halogenation of several allylic substrates using a variety of reagents including *N*-bromo and *N*-chloro pthalimide, chloramine-T, *N*-bromo and *N*-chlorosaccharin, *N*-chloro-*N*-benzoyl-*p*-toluenesulfonamide, *N*-chloro-di-*p*-toluylsulfonamide, and most notably, *N*-bromo and *N*-chlorosuccinimide. Successful allylic bromination was reported for several substrates, among which were the conversion of cyclohexene (**7**) into bromocyclohexene (**8**) and the conversion of phenyl propene **9** into cinnamyl bromide (**10**).

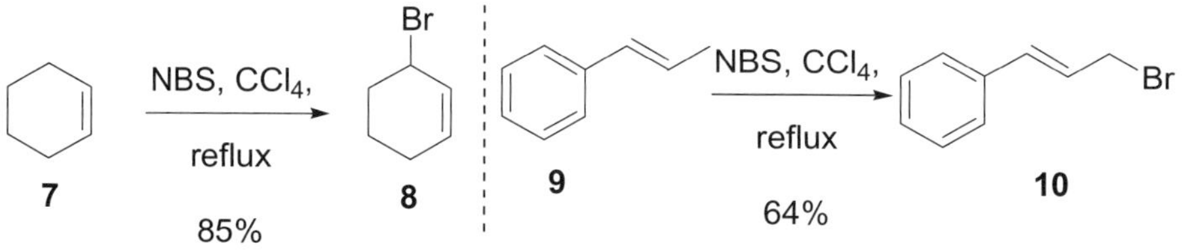

7 8 9 10

3.4.3 Mechanism

The mechanism of the Wohl–Ziegler reaction was initially proposed by Bloomfield[5] who suggested that the succinimidyl radical was the dominant chain-carrier, and later revised by Goldfinger[6] who proposed the bromine radical to be the key, chain-carrying species. One key to the Goldfinger mechanism is the formation and maintenance of a low concentration of Br_2 in the reaction medium. This particular situation occurs most reliably when CCl_4 is used as solvent, as the solubility of NBS is low. Support for the mechanism has come in several studies showing that the selectivity of bromine at low concentration and NBS provide very similar ratios of bromination products. For example, work by Incremona and Martin[7] on allylic substrates, which compared bromination rates and selectivities using NBS and Br_2, provided strong evidence that bromine was the chain carrier. In addition, Walling and Reiger[8] have completed studies on relative rates which compared imide brominating reagents (three different bromohydantoins with NBS) again proposing that a common chain carrier was at play and was proposed to be the bromine radical. Similarly, Desmond and co-workers[9] studied the relative reactivity of NBS, Br_2 and $BrCCl_3$ with a series of benzyl brominations and showed that while NBS and Br_2 were similar, $BrCCl_3$ gave a different reaction profile when substrates **11** were subjected to photo-initiated Wohl–Ziegler conditions.

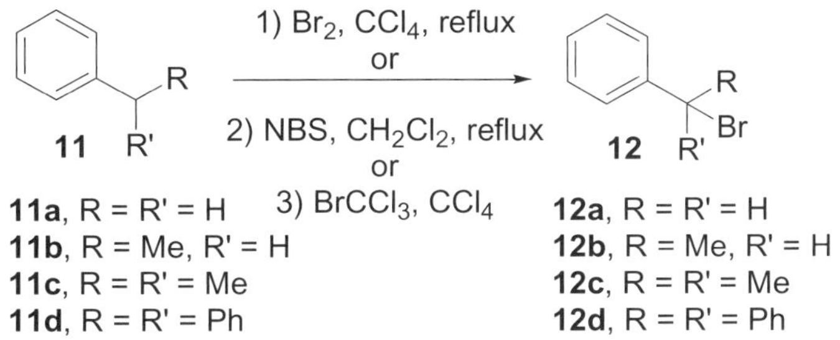

The initiation step provides a radical source by thermal or photochemical dissociation of initiators, which then provides bromine radicals by reaction with Br_2. Initiators are sometimes present in the alkene as allyl hydroperoxides which may be present due to inadvertent, prior auto-oxidation. Bromine or HBr may be present in trace amounts in NBS. Reaction of the bromine radical **20** with the substrate **1** proves selective for allylic or benzylic hydrogens due to the near thermoneutral nature of the reaction which breaks the C–H bond and forms the H–Br bond. Reaction of the formed carbon-centered radical **21** with Br_2 provides the desired bromide **3** and $Br^{\bullet}$ **20**. Hydrogen bromide **17** reacts with NBS to form succinimide **4** and resupplies the required low concentration of Br_2. Alternatively, reaction of substrate radical **21** with NBS **2** provides product **3** and succinimidyl radical **22** ($S^{\bullet}$). Due to energy and kinetics considerations, abstraction of the allylic hydrogen by the $S^{\bullet}$ should be slower than abstraction of bromine from NBS by an allyl radical.[10] In using solvents in which NBS, succinimide **4** or it's radical **22** are not very soluble, $S^{\bullet}$ is not the key chain-carrier. By-products and side-reactions can occur with $S^{\bullet}$.

Alternate solvents, however, in which the $S^{\bullet}$ has increased solubility, like CH_2Cl_2 or MeCN, are capable of providing a different reaction manifold; and, therefore different selectivities and reactivities are observed. The different selectivites observed were attributed to different chain carriers, and for a time, it was thought that there were two electronically different $S^{\bullet}$ (excited succinimidyl σ- and ground state, succinimidyl π-radical). This hypothesis was set forth in part due to the formation of β-bromopropionyl isocyanate (**24**, β-bpic), a fragmentation product of $S^{\bullet}$. Based on orbital symmetry arguments, ring opening of the excited state σ-succinimidyl radical for the formation of the observed impurity β-bpic **24** would be preferred and

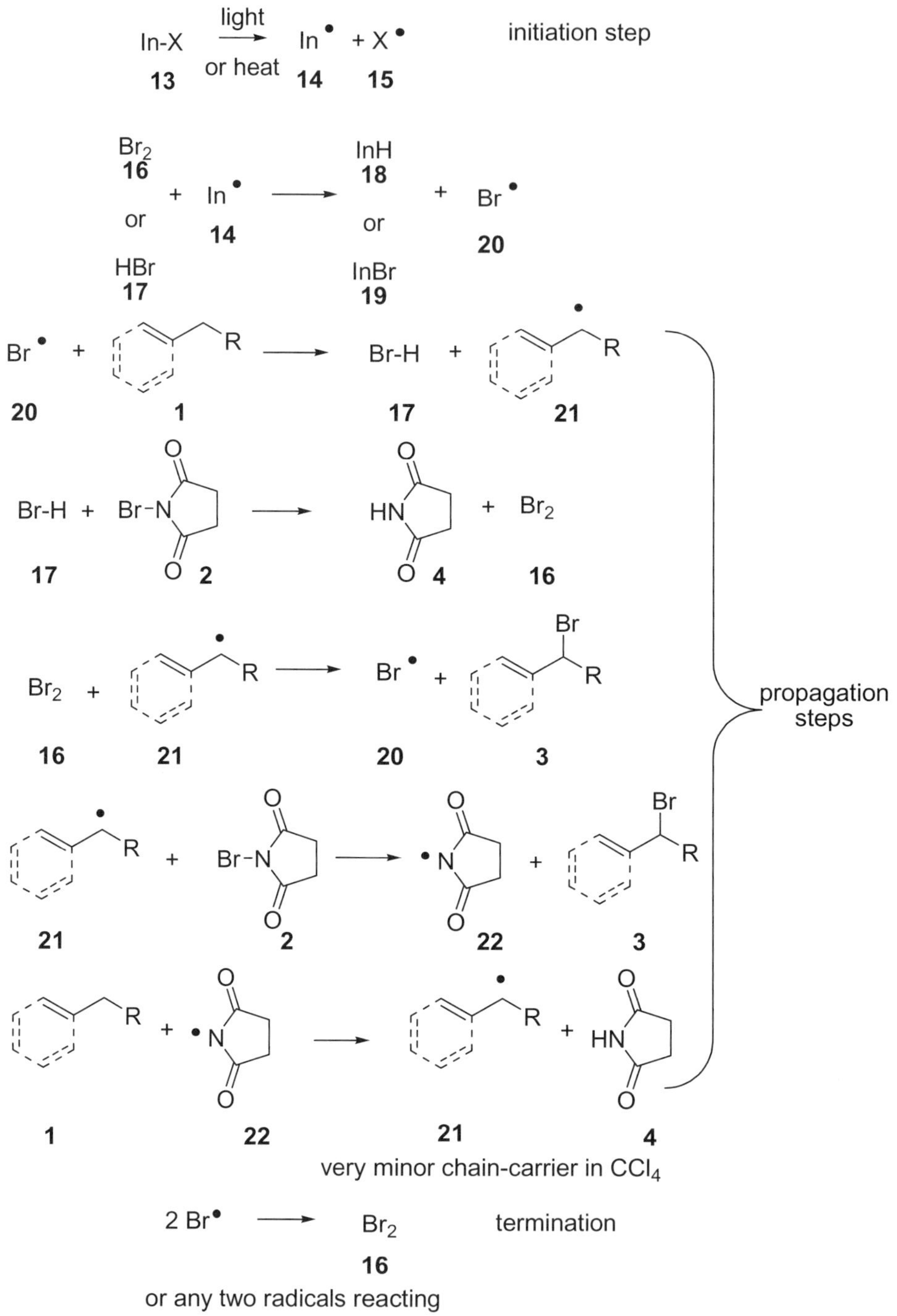

the experimentally supported radical data suggests that S$^{\bullet}$ is a π-succimidyl radical.[11] Data supports the notion of more than two chain carriers for this reaction and a complex of some type has been proposed as a composite radical[12] ($SBr_2^\bullet$), as the third chain carrier and the suggestion of the σ-excited succinimidyl radical was abandoned yet questions remain on the formation of **24**.[13]

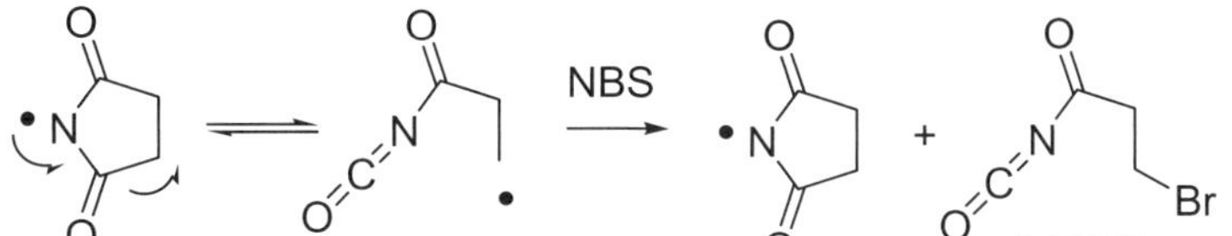

β-BPIC

22 **23** **22** **24**

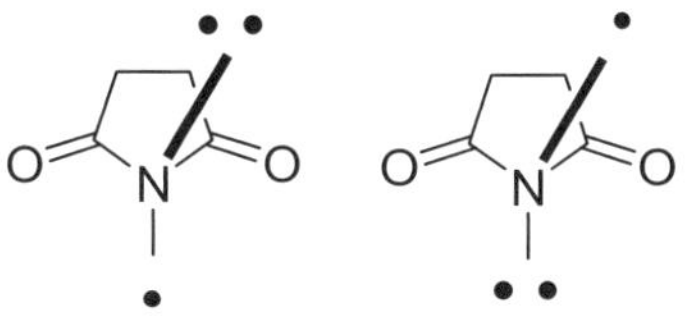

σ π

Additives may also impact the mechanism. Accelerators to the reaction were found to be 1) radical initiators, 2) bromine generators and 3) *t*-amines.[10] The importance of initiation and the fact that bromine plays a crucial role in the reaction mechanism makes it easy to understand the impact of such additives. However, one must realize that the Br_2 concentration must be kept relatively low or alkene addition products may be observed. The impact of *t*-amines as accelerators is somewhat less obvious, but most likely arises from amine-induced, homolytic decomposition of NBS or hydroperoxide. Retarders of the reaction are common inhibitors of radical formation or substances that will intercept the chain-carriers like tetrabromo-*p*-benzoquinone, *s*-trinitrobenzene, *m*-dinitrobenzene, or hydroquinone.

Regioselectivity and Stereoselectivity

Selectivity of bromination is dependent upon the ease of radical formation. In general, tertiary benzylic or allylic > secondary benzylic or allylic > primary benylic or allylic taking into account the stability of the radical or the energy required to abstract the hydrogen. For this reason, the conditions for the Wohl–Ziegler reaction prove to be very selective for the removal of $H^\bullet$ for benzylic and allylic hydrogens. Initially, the more substituted radical was thought not to form,[1] but this has been proven false. For example, Dauben[14] has shown that clean and reproducible formation of the tertiary bromide **26** was not only possible but was achieved in good yield.

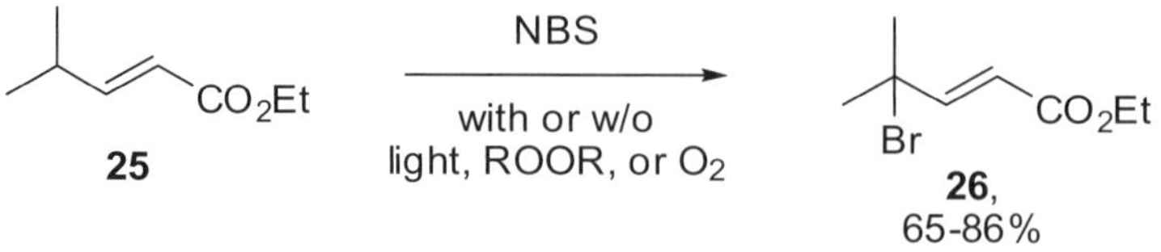

25 → 26, 65-86%

When unsymmetric alkenes are reacted under the Wohl–Ziegler conditions, product mixtures can be obtained. For example, reaction of 1-octene (**27**) gave about a 3:1 mixture of primary and secondary bromination products **28** and **29**.[15]

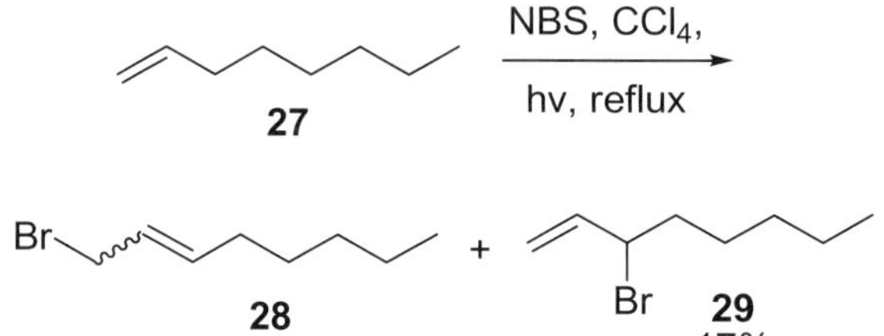

27 → **28** 44% + **29** 17%

While a mixture of products was observed in bicyclic systems, the product ratios were different, favoring the secondary bromide over the primary. Substrate **30** was brominated to afford a 3 : 2 ratio of the secondary and primary bromo products **31** and **32**, respectively, in high yield. These terminal alkenes show that the delocalization of the radical can occur to provide reactive radicals **33** and **34**. No bridgehead bromination occurred.[16]

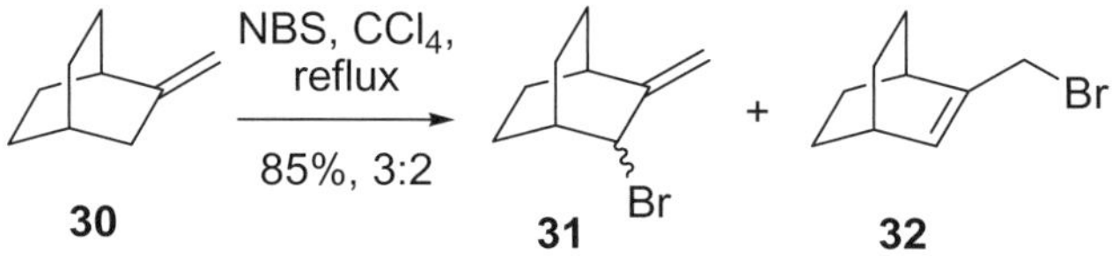

30 → **31** + **32**

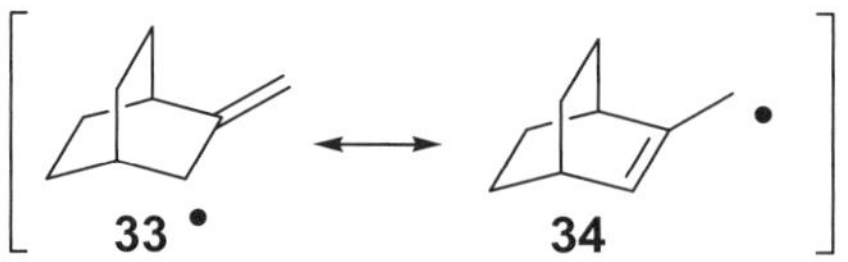

33 ↔ **34**

As noted by Ziegler,[4] secondary radicals reacted more rapidly than primary radicals; for example, 2-methyl-2-butene required 16 h to finish while 2-methyl-2-hexene required only 10 min. An additional example is shown for the reaction of 2-heptene wherein the secondary bromide is formed in 76%.[15]

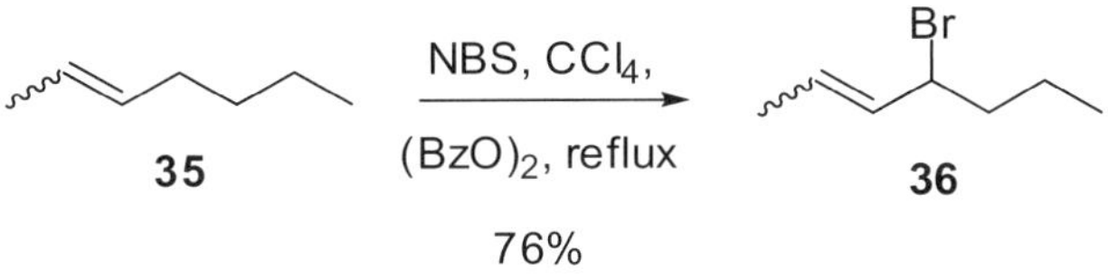

35 → 36

Bromination products are not typically formed when the allyl radical can incorporate a carbon atom with an attached heteroatom. For example, bromination of **37** provided only **38** (PhthN = phthalimido) in 77% yield.[17]

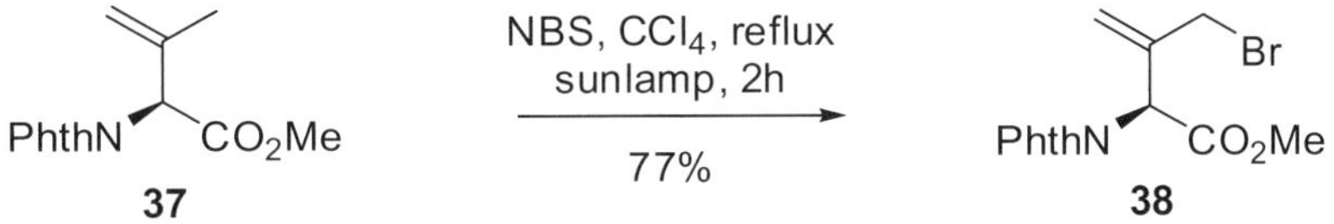

37 → 38

Another example of primary bromide formation in which the halide addition occurs away from the heteroatom was reported in the work by Stoodley and co-workers.[18] This work reported the functionalization of a precursor to 4π components for enantioselective Diels–Alder reactions.

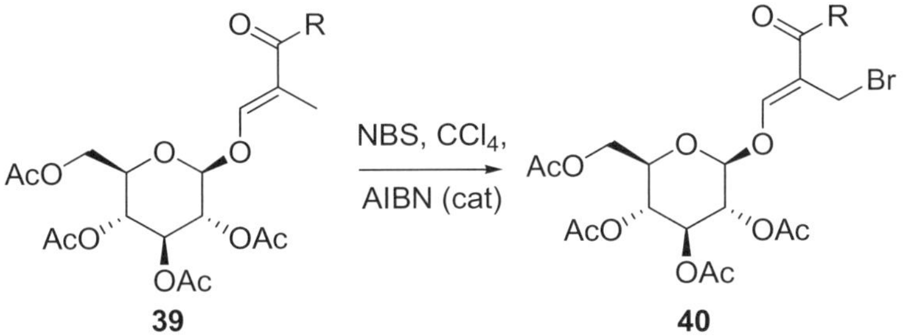

39 → 40

R	Yield %
Me	72
Et	79
OMe	61
OEt	63

From a stereochemical point of view, Dauben[19] reported the bromination of enriched α-deutero-ethyl benzene (**41**) and showed it to be totally non-selective. This suggests that without other mitigating circumstances, the radical is flat.

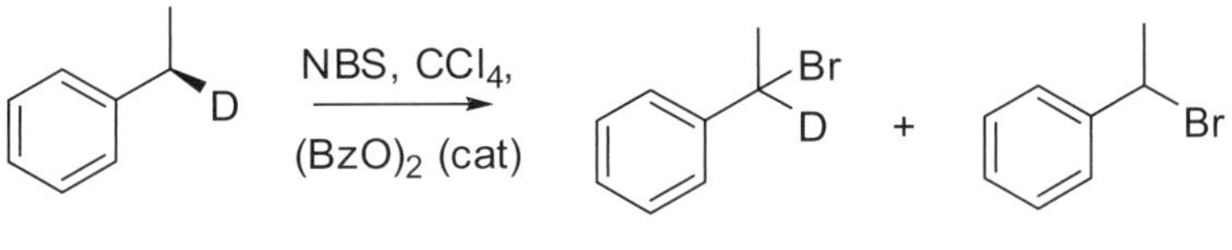

enriched **41** → **42** + **43**

racemic!

However there are several examples in which, due to conformational constraints, the stereochemistry of the radical generated can be selective. For example, Macomber[20] has shown that bromination of chiral oxaphospholenes produced tertiary allylic bromides in high selectivity. The authors suggest that the intermediate radical is locked into a conformation keeping the phenyl and the C5 *t*-butyl groups *trans* to one another. Additionally, interaction between the intermediate radical and phenyl group may be another stabilizing factor. Thus, a mixture of isomers **44** provided selectively **45**. Furthermore, enantio-enriched **46** provided **47** cleanly and in very high yield.

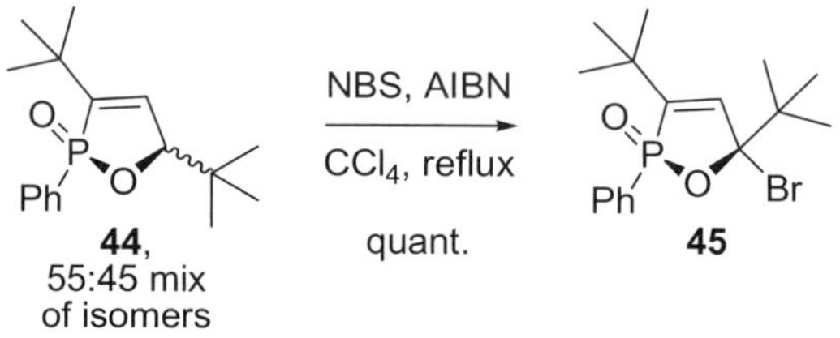

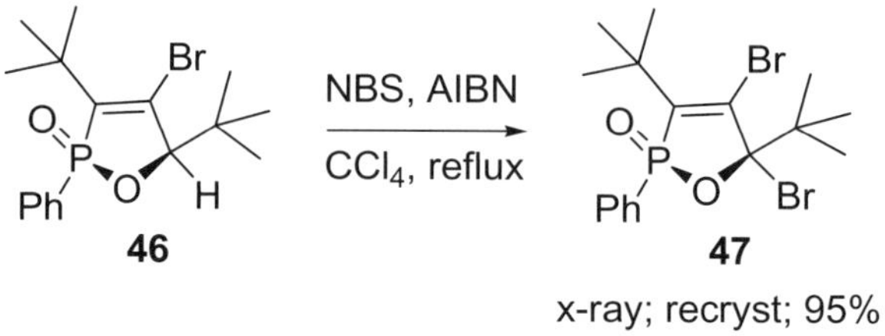

Similarly, bromination of a dioxabicyclo-octene **48** or **49** gave good yield of **50** in 98% purity. Neither the origin of the high stereoselectivity nor the stabilization of the requisite *exo*-radical **51** over the other radicals was explained.[21]

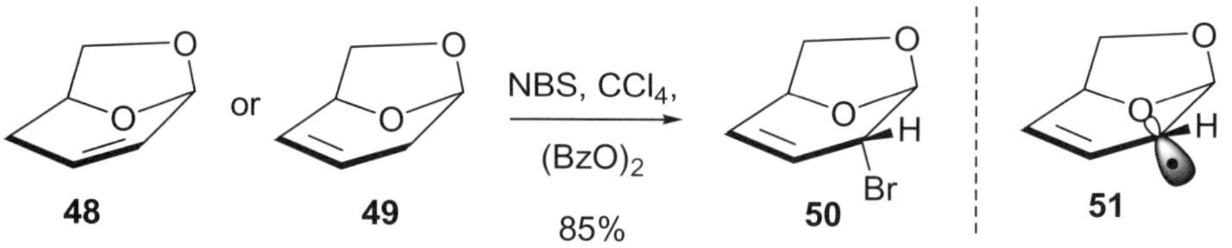

Radical bromination of the V-shaped diaryl **52** gave the single *exo*-bis-bromo derivative **53**. The explanation for the high selectivity was attributed to the requirement of the formed radical orbital to be delocalized in the aromatic plane, thereby acting as a radical conformational anchor.[22]

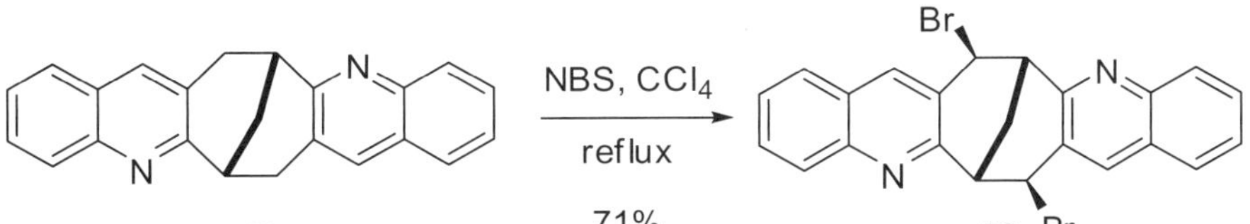

52 53

3.4.4 *Synthetic Utility*

Recently, the Wohl–Ziegler reaction conditions have been applied to the benzylic bromination of *N*-benzyl amides to promote deprotection. Interestingly, oxygen and light seem to be key in promoting this room-temperature reaction. *N*-Bromination of *N*-methyl acetamide is proposed to be the first step, followed by Br$^{\bullet}$ generation by the action of oxygen and light. Radical formation at the benzylic position of **54** followed by bromination provides **55** which upon work-up provides the desired deprotected amide **56** with benzaldehyde **57** as a by-product. Several examples were provided showing the utility in producing secondary amides from the corresponding tertiary benzyl amides.[23]

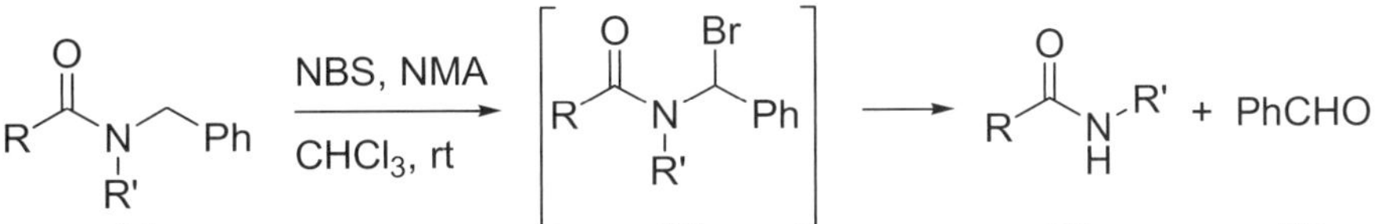

54 55 56 57

R	R'	Yield %
Ph	CH_2CMe_3	76
$(CH_2)_{12}Me$	Me	87
CH_2CH_2Ph	Me	43
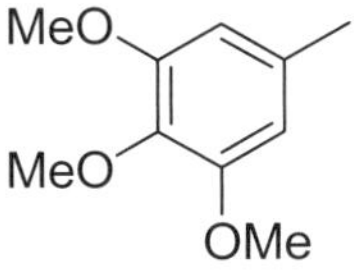	Me	80
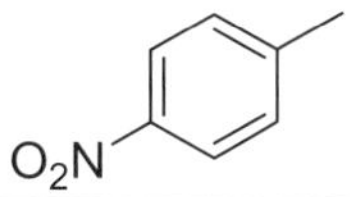	Me	74

The Wohl–Ziegler reaction has been used as a step in total syntheses on the academic and industrial scale. It has been applied to a broad range of substrates and delivered on a variety of scales. Argade reported[24] using these conditions as a key step in the synthesis of gymnoascolide A **60** in which bromination of anhydride **58** using NBS with catalytic benzoyl peroxide provided primary bromide **59** in 80% yield. Bromide **59** was used as a leaving group for a carbon nucleophile.

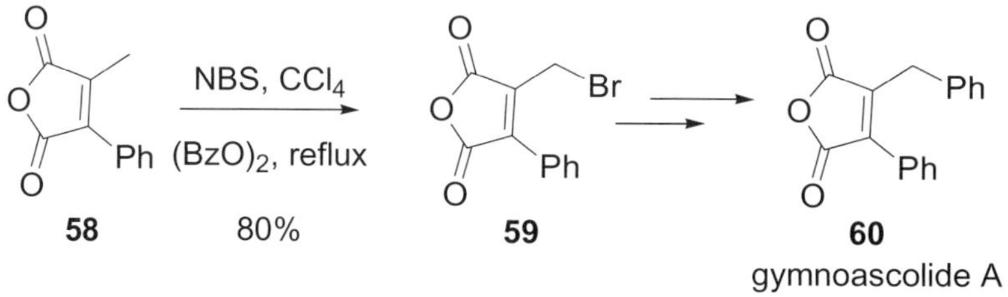

Corey and co-workers[25] reported the stereoselective bromination of **61** in their total synthesis of the anti-influenza neuramidase inhibitor Tamiflu® to occur in 95% yield and excellent stereoselectivity providing (±)-**62**. Here, Corey used the allyl bromide system as a precursor to a diene.

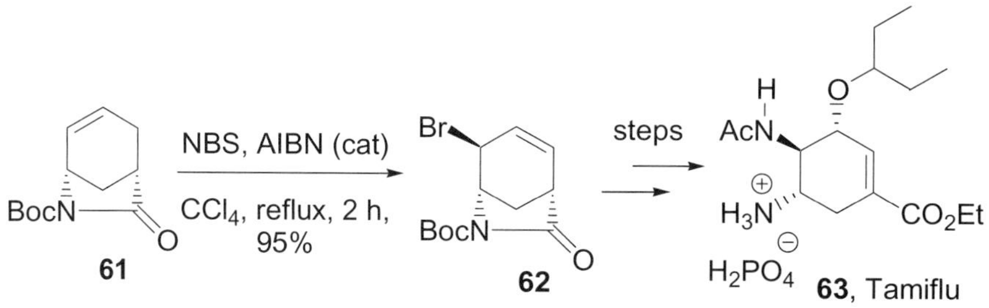

The bromination of pregnene **64** provided allyl bromide **65** that served as a key intermediate for the synthesis of the antipsoriatic vitamin D3 analogue, maxacalcitol which is used in the treatment of antihyperparathyroidism. The authors use the allyl bromide generated by reaction of NBS with catalytic AIBN in heptane as a precursor to generate the diene.[26] This reaction sequence is reported on the 65 kg scale. The diene system is key to the last reaction sequence which is the photo-promoted, conrotatory, ring-opening of the cyclohexadiene system and triene isomerization delivering maxacalcitol **66**.

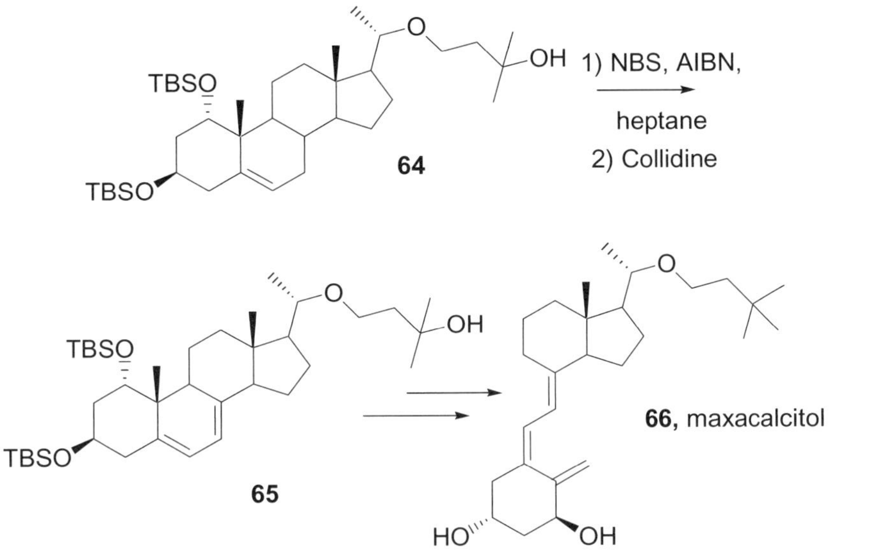

Benzylic bromination of atropisomers has proven important in the preparation of a variety of dinapththazepine catalysts, like Maruoka's chiral, quaternary ammonium salt used for PTC. Amine **69** is considered a key building block for the preparation of this type of chiral catalyst. Bis-bromination of **67** provided **68** in 54% yield using NBS and catalytic AIBN in cyclohexane.[27] Notably, the pure, bis-bromide **68** precipitates out of solution. Again, the benzyl bromide system was utilized as a leaving group.

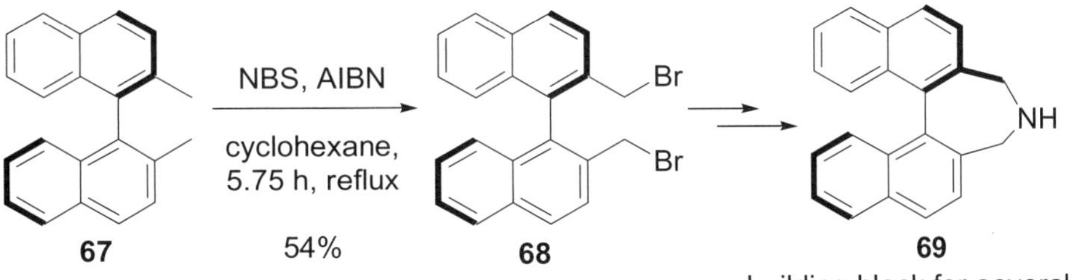

building block for several chiral catalysts

3.4.5 Variations and Improvements

Initiators

A variety of initiators have been used to promote this reaction. Peroxides and azo-compounds are thermally unstable and provide radicals. From these two classes, AIBN and benzoyl peroxide have been the most commonly used chemical initiators. The use of other diazo compounds (derivatives of AIBN, e.g., 1, 1′-azobiscyanocyclohexane) or other peroxides (*t*-butyl peroxide) have also been described. Irradiation is also a popular alternative. Photolytic cleavage of bromine provides two $Br^{\bullet}$. Light will also cleave peroxides as well as acetone.[28] As discussed in the mechanistic section, peroxides generated in small quantities by ene reaction of O_2 with the alkene can serve to initiate the reaction.

Microwave promoted reactions (no chemical initiators) have been reported. An interesting result from this work was the reported 45% yield of the bis-bromo compound **71** an electron rich toluene **70** which typically provides aromatic substitution.[29]

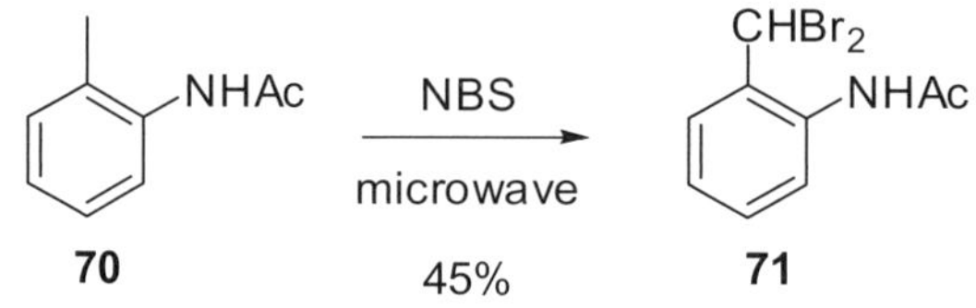

70 **71**

Other Halogenating Agents

As cited earlier, Ziegler[4] has utilized a variety of halogenating reagents and several more have been investigated. While some of these are of academic interest, most have not seen the utility over the years as NBS has. For example, tribromo-quinoline (**72**) was shown to undergo free radical bromination of cyclohexene **7** to provide 61% of 3-bromocyclohexene (**8**) along with 78% yield of the dibromo-quinoline **73**.[30]

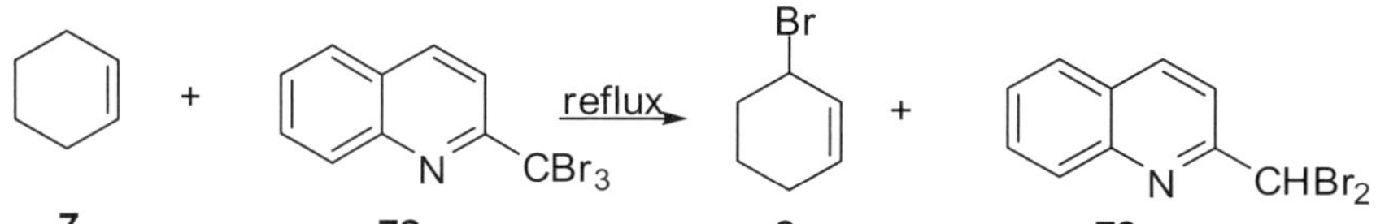

7 **72** **8**, 61% **73**, 78%

In general, bromohydantoins, most other *N*-halo imides (*N*-bromo glutarimide, *N*-bromo-phthalimide, tetramethyl-*N*-bromosuccinimide, etc…except for *N*-iodosuccinimide), Br_2, $BrCCl_3$, and *N*-bromocaprolactam[31] are suitable reagents for radical brominations of allylic and benzylic substrates. Of all of these reagents, the most used reagents are NBS and Br_2.

Alternative Solvents

A solvent study has been published on the benzylic bromination of three toluenes (4-nitrotoluene, 4-methylbiphenyl and 4-*t*-butyltoluene) in several solvents. A tension was observed in ring vs. radical bromination and it was determined to avoid ring bromination, a dielectric constant <10 for the solvent was selected. Improved reactivity was observed in purified CCl_4.[32] Benzylic bromination in water has been reported.[33] With electron rich aromatics (polyalkyl, NHAc or OR substituted), ring bromination was observed. Bromination of **74** and **75** with NBS in water using a 40w incandescent light provided **76** and **77** in 83% and 52% yield, respectively.

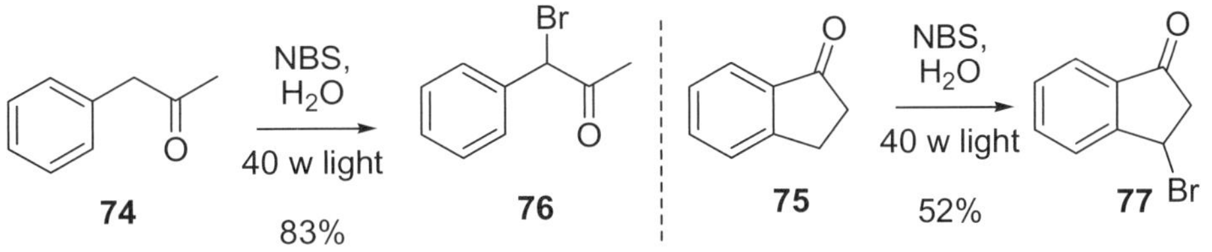

The Wohl–Ziegler reaction has been reported in the ionic liquid, 1-butyl-3-methylimidazolium hexafluorophasphate ([bmim]PF_6), and compared with results from solvent free reactions.[34] Again, electron rich benzylic substrates gave rise to ring brominated products, but electron deficient aromatics gave good yield of the desired benzyl bromide.

3.4.6 Experimental

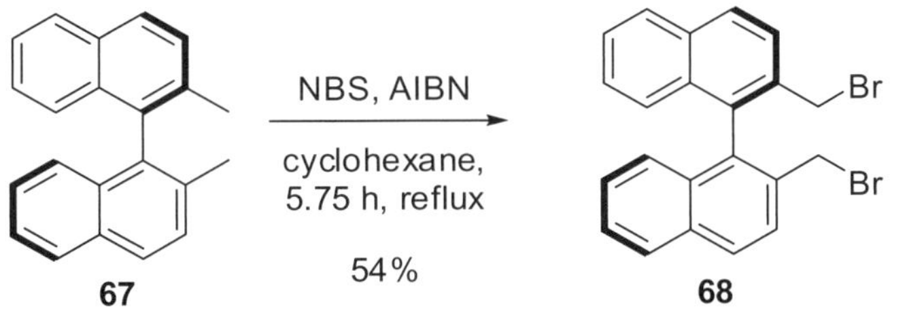

Preparation of (*R*)-2,2′-Bis(bromomethyl)-1,1′-Binaphthyl (68)[27]

To a suspension of (*R*)-2,2′-dimethyl-1,1′-binaphthyl (24.4 g, 86.4 mmol) in cyclohexane (170 mL) was added *N*-bromosuccinimide (NBS, 33.8 g, 190 mmol) followed by 2,2′-azobisisobutyronitrile (AIBN, 0.70 g, 4.26 mmol) at room temperature (20–25 °C). The mixture was stirred and heated at reflux for 2 h, during which the progress of the reaction was monitored by TLC [AcOEt/*n*-hexane (1:10); *Rf* = 0.60 for **67**, 0.49 for **68**]. On consumption of **67**, the mixture was allowed to cool to room temperature (20–25 °C), and AcOEt (56 mL) was added with stirring. The mixture was poured into H_2O (350 mL). The biphasic mixture was stirred for not longer than an hour by which solids had ceased to precipitate. Crystalline solids were collected by filtration and air-dried overnight (8–12 h) to give **68** (20.6 g, 54.3%).

3.4.7 References

1. [R] Djerassi, C. *Chem. Rev.* **1948,** *43,* 271.
2. Wolfe, S.; Awang, D. V. C. *Can. J. Chem.* 1971, *49*, 1384.
3. Ziegler, K.; Schenck, G.; Krockow, E. W.; Siebert, A.; Wenz, A.; Weber, H. *Justus Liebigs Ann. Der Chem.* **1942**, *552,* 1.
4. Ziegler, K.; Spaeth, A.; Schaaf, E.; Schumann, W.; Winkelmann, E. *Justus Liebigs Ann. Der Chem.* **1942**, *552,* 80.
5. Bloomfield, G. F. *J. Chem. Soc.* **1944**, 114.
6. Adam, J.; Gosselain, O. A.; Goldfinger, P. *Nature* **1953**, *171*, 704.
7. Incremona, J. H.; Martin, J. C. *J. Am. Chem. Soc.* **1970**, *92*, 627.
8. Walling, C.; Rieger, A. L. *J. Am. Chem. Soc.* **1963**, *85*, 3134.
9. Russell, G. A.; DeBoer, C.; Desmond, K. M. *J. Am. Chem. Soc.* **1963**, *85*, 365.
10. Dauben, H. J.; McCoy, L. L. *J. Am. Chem. Soc.* **1959**, *81*, 4863.
11. [R] Skell, P. S.; Day, J. C. *Accts. Chem. Res.* **1978**, *11*, 381.
12. Eberson, L. *Pure Appl. Chem.* **1991**, *63*, 205.
13. Skell, P. S.; Luning, U.; McBain, D. S.; Tanko, J. M. *J. Am. Chem. Soc.* **1986**, *108*, 121.
14. Dauben, H. J.; McCoy, L. L. *J. Org. Chem.* **1959**, *24*, 1577.
15. [R] Volhardt, K. P. C. and Shore, N. E. *Organic Chemistry: Structure and Function, 3rd Edition*, pps. 572–573. W. H. Freeman; New York, 1999 (3rd edition).
16. Davies, D. I.; Parfitt, L. T. *Tetrahedron Lett.* **1969**, *5*, 293.
17. Easton, C. J.; Edwards, A. J.; McNabb, S. B.; Merrett, J. C.; O'Connell, J. L.; Simpson, G. W.; Simpson, S. S.; Willis, A. C. *Org. Biomol. Chem.* **2003**, *1*, 2492.
18. Esteves, A. P.; Freitas, A. M.; Raynor, C. M.; Stoodley, R. J. *J. Chem Soc., Perkin Trans.* **2000**, *1*, 1753.
19. Dauben, H. J.; McCoy, L. L. *J. Am. Chem. Soc.* **1959**, *81*, 5404.
20. Macomber, R. S.; Rardon, D. E.; Ho, D. M. *J. Org. Chem.* **1992**, *57*, 3874.
21. Ranganayakulu, K.; Brown, R. K. *J. Org. Chem.* **1974**, *39*, 3941.
22. Marjo, C. E.; Bishop, R.; Craig, D. C.; Scudder, M. L. *Medeleev Commun.* **2004**, 278.
23. Kuang, L.; Zhou, J.; Chen, S.; King, K. *Synthesis* **2007**, *20*, 3129.
24. Baag, M. M., Argade, N. P. *Synthesis* **2008**, *1*, 26.
25. Yeung, Y.-Y.; Hong, S.; Corey, E. J. *J. Am. Chem. Soc.* **2006**, *128*, 6310.
26. Shimizu, H.; Shimizu, K.; Kubodera, N.; Mikami, T.; Tsuzaki, K.; Suwa, H.; Harada, K.; Hiraide, A.; Shimizu, M.; Koyama, K.; Ichikawa, Y.; Hirasawa, D.; Kito, Y.; Kobayashi, M.; Kigawa, M.; Kato, M.; Kozono, T.; Tanaka, H.; Tanabe, M.; Iguchi, M.; Yoshida, M. *Org. Process Res. Dev.* **2005**, *9*, 278.
27. Ikunaka, M.; Maruoka, K.; Okuda, Y.; Ooi, T. *Org. Process Res. and Dev.* **2003**, *7*, 644.
28. [R] Sykes, P. *A Guidebook to Mechanism in Organic Chemistry*, sixth ed., pp299–328. Longman Scientific & Technical, Essex, England and John Wiley and Sons, Inc., New York, 1986.
29. Goswami, S.; Dey, S.; Jana, S.; Adak, A. K. *Chem. Lett.* **2004**, *33*, 916.
30. Brown, B. R.; Grice, J. O.; Hammick, D. L. L.; Thewlis, B. H. *J. Chem. Soc.* **1951**, 1149.
31. Taub, B.; Hino, J. B. *J. Org. Chem.* **1960**, *25*, 263.
32. Offermann, W.; Vogtle, F. *Angew. Chem., Int. Ed. Engl.* **1980**, *19*, 464.
33. Podgorsek, A.; Stavber, S.; Zupan, M.; Iskra, J. *Tetrahedron Lett.* **2006**, *47*, 1097.
34. Togo, H.; Hirai, T. *Synlett* **2003**, 702.

Appendixes

Appendix 1
Table of Contents for Volume 1: ***Name Reactions in Heterocyclic Chemistry***
Published in 2005

PART 1 THREE- AND FOUR-MEMBERED HETEROCYCLES 1

Chapter 1 Epoxides and Aziridines 1

1.1 Corey–Chaykovsky reaction 2
1.2 Darzens glycidic ester condensation 15
1.3 Hoch–Campbell aziridine synthesis 22
1.4 Jacobsen–Katsuki epoxidation 29
1.5 Paterno–Büchi reaction 44
1.6 Sharpless–Katsuki epoxidation 50
1.7 Wenker aziridine synthesis 63

PART 2 FIVE-MEMBERED HETEROCYCLES 69

Chapter 2 Pyrroles and Pyrrolidines 69

2.1 Barton–Zard reaction 70
2.2 Knorr and Paal–Knorr pyrrole syntheses 79
2.3 Hofmann–Löffler–Freytag reaction 90

Chapter 3 Indoles 99

3.1 Bartoli indole synthesis 100
3.2 Batcho–Leimgruber indole synthesis 104
3.3 Bucherer carbazole synthesis 110
3.4 Fischer indole synthesis 116
3.5 Gassman indole synthesis 128
3.6 Graebe–Ullman carbazole synthesis 132
3.7 Hegedus indole synthesis 135
3.8 Madelung indole synthesis 140
3.9 Nenitzescu indole synthesis 145
3.10 Reissert indole synthesis 154

Chapter 4	**Furans**	
4.1	Feist–Bénary furan synthesis	160
4.2	Paal–Knorr furan synthesis	168
Chapter 5	**Thiophenes**	**183**
5.1	Fiesselmann thiophene synthesis	184
5.2	Gewald aminothiophene synthesis	193
5.3	Hinsberg synthesis of thiophene derivatives	199
5.4	Paal thiophene synthesis	207
Chapter 6	**Oxazoles and Isoxazoles**	**219**
6.1	Claisen isoxazole synthesis	220
6.2	Cornforth rearrangement	225
6.3	Erlenmeyer–Plöchl azlactone synthesis	229
6.4	Fischer oxazole synthesis	234
6.5	Meyers oxazoline method	237
6.6	Robinson–Gabriel synthesis	249
6.7	van Leusen oxazole synthesis	254
Chapter 7	**Other Five-Membered Heterocycles**	**261**
7.1	Auwers flavone synthesis	262
7.2	Bucherer–Bergs reaction	266
7.3	Cook–Heilbron 5-amino-thiazole synthesis	275
7.4	Hurd–Mori 1,2,3-thiadiazole synthesis	284
7.5	Knorr pyrazole synthesis	392
PART 3	**SIX-MEMBERED HETEROCYCLES**	**301**
Chapter 8	**Pyridines**	**302**
8.1	**Preparation via condensation reactions**	**303**
8.1.1	Hantzsch (dihydro)-pyridine synthesis	304
8.1.1.1	Description	304
8.1.1.2	Historical perspective	304
8.1.1.3	Mechanism	305
8.1.1.4	Variations	307
8.1.1.4.1	Guareschi-Thorpe pyridine synthesis	307
8.1.1.4.2	Chichibabin (Tschitschibabin) pyridine synthesis	308
8.1.1.4.3	Bohlmann–Rahtz pyridine synthesis	309
8.1.1.4.4	Kröhnke pyridine synthesis	311

8.1.1.4.5	Petrenko–Kritschenko piperidone synthesis	313
8.1.1.5	Improvements or modifications	314
8.1.1.6	Experimental	320
8.1.1.6.1	Three-component coupling	320
8.1.1.6.2	Two-component coupling	320
8.1.1.7	References	321
8.2	**Preparation via cycloaddition reactions**	**323**
8.2.1	Boger reaction	323
8.3	**Preparation via rearrangement reactions**	**340**
8.3.1	Boekelheide reaction	340
8.3.2	Ciamician–Dennstedt rearrangement	350
8.4	**Zincke reaction**	**355**

Chapter 9 Quinolines and Isoquinolines **375**

9.1 Bischler–Napieralski reaction 376
9.2 Camps quinoline synthesis 386
9.3 Combes quinoline synthesis 390
9.4 Conrad–Limpach reaction 398
9.5 Doebner quinoline synthesis 407
9.6 Friedländer synthesis 411
9.7 Gabriel–Colman rearrangement 416
9.8 Gould–Jacobs reaction 423
9.9 Knorr quinoline synthesis 437
9.10 Meth–Cohn quinoline synthesis 443
9.11 Pfitzinger quinoline synthesis 451
9.12 Pictet–Gams isoquinoline synthesis 457
9.13 Pictet–Hubert reaction 465
9.14 Pictet–Spengler isoquinoline synthesis 469
9.15 Pomeranz–Fritsch reaction 480
9.16 Riehm quinoline synthesis 487
9.17 Skraup/Doebner–von Miller reaction 488

Chapter 10 Other Six-Membered Heterocycles **495**

10.1 Algar–Flynn–Oyamada reaction 496
10.2 Beirut reaction 504
10.3 Biginelli reaction 509
10.4 Kostanecki–Robinson reaction 521
10.5 Pinner pyrimidine synthesis 536
10.6 von Richter cinnoline reaction 540

Appendix 2, Table of Contents for Volume 2: *Name Reactions for Functional Group Transformations*

Published in 2007

Chapter 1 Asymmetric Synthesis 1
1.1 CBS reduction 2
1.2 Davis chiral oxaziridine reagents 22
1.3 Midland reduction 40
1.4 Noyori catalytic asymmetric hydrogenation 46
1.5 Sharpless asymmetric hydroxylation reactions 67

Chapter 2 Reduction 85
2.1 Eschweiler–Clark reductive alkylation of amines 86
2.2 Gribble reduction of diaryl ketones 93
2.3 Luche reduction 112
2.4 Meerwein–Ponndorf–Verley reduction 123
2.5 Staudinger reaction 129
2.6 Wharton reaction 152

Chapter 3 Oxidation 159
3.1 Baeyer–Villiger oxidation 160
3.2 Brown hydroboration reaction 183
3.3 Burgess dehydrating reagent 189
3.4 Corey–Kim oxidation 207
3.5 Dess–Martin periodinane oxidation 218
3.6 Tamao–Kumada–Fleming oxidation 237
3.7 Martin's sulfurane dehydrating reagent 248
3.8 Oppenauer oxidation 265
3.9 Prilezhaev reaction 274
3.10 Rubottom oxidation 282
3.11 Swern oxidation 291
3.12 Wacker–Tsuji oxidation 309
3.13 Woodward *cis*-dihydroxylation 327

Chapter 4 Olefination 333
4.1 Chugaev elimination 334
4.2 Cope elimination reaction 343
4.3 Corey–Winter olefin synthesis 354
4.4 Perkin reaction (cinnamic acid synthesis) 363
4.5 Perkow vinyl phosphate synthesis 369
4.6 Ramberg–Bäcklund olefin synthesis 386
4.7 Shapiro reaction 405

4.8 Zaitsev elimination 414

Chapter 5 Amine Synthesis 423
5.1 Fukuyama amine synthesis 424
5.2 Gabriel synthesis 438
5.3 Leuckart–Wallach reaction 451

Chapter 6 Carboxylic Acid Derivatives Synthesis 457
6.1 Fischer–Speier esterification 458
6.2 Mukaiyama esterification 462
6.3 Ritter reaction 471
6.4 Strecker amino acid synthesis 477
6.5 Yamada coupling reagent 500
6.6 Yamaguchi esterification 545

Chapter 7 Miscellaneous Functional Group Manipulations 551
7.1 Balz–Schiemann reaction 552
7.2 Buchwald–Hartwig reactions 564
7.3 Haloform reaction 610
7.4 Hunsdiecker reaction 623
7.5 Japp–Klingemann hydrazone synthesis 630
7.6 Krapcho decarboxylation 635
7.7. Nef reaction 645
7.8 Prins reaction 653
7.9 Regitz diazo synthesis 658
7.10 Sommelet reaction 689

Appendix 3
Table of Contents for Volume 4: *Name Reactions for Homologations-2*

Chapter 1.	**Rearrangements**	**1**
Section 1.1	***Concerted rearrangement***	***2***
1.1.1	Alder ene reaction	2
1.1.2	Claisen and related rearrangements	33
1.1.3	Cope and related rearrangements	87
1.1.4	Curtius rearrangement	135
1.1.5	Hofmann rearrangement	163
1.1.6	Lossen rearrangement	199
1.1.7	Overman rearrangement	210
1.1.8	[1,2]-Wittig rearrangement	226
1.1.9	[2,3]-Wittig rearrangement	241
1.1.10	Wolff rearrangement	257
Section 1.2	***Cationic rearrangement***	***274***
1.2.1	Beckmann rearrangement	274
1.2.2	Demjanov rearrangement	293
1.2.3	Meyer–Schuster rearrangement	305
1.2.4	Pinacol rearrangement	319
1.2.5	Pummerer rearrangement	334
1.2.6	Schmidt rearrangement	353
1.2.7	Wagner–Meerwein rearrangement	373
Section 1.3	***Anionic rearrangement***	***395***
1.3.1	Benzilic acid rearrangement	395
1.3.2	Brook rearrangement	406
1.3.3	Favorskii rearrangement	427
1.3.4	Grob fragmentation	453
1.3.5	Neber rearrangement	464
1.3.6	Payne rearrangement	474
1.3.7	Smiles rearrangement	489
1.3.8	Stevens rearrangement	516
Chapter 2.	**Asymmetric C–C bond formation**	**531**
2.1	Evans aldol reaction	532
2.2	Hajos–Wiechert reaction	554
2.3	Keck stereoselective allylation	583
2.4	Roush allylboronation	613

Chapter 3. Miscellaneous homologation reactions 640

3.1 Bamford–Stevens reaction 642
3.2 Mannich reaction 653
3.3 Mitsunobu reaction 671
3.4 Parham cyclization 749
3.5 Passerini reaction 765
3.6 Ugi reaction 786

Appendix 4
Table of Contents for Volume 5: *Name Reactions for Ring Formations*
Due in 2010

Chapter 1 Three-Membered Carbocycles
1.1 Freund reaction
1.2 Kishner cyclopropane synthesis
1.3 Kulinovich cyclopropanol synthesis
1.4 Pfau–Plattner azulene synthesis
1.5 Simmons–Smith reaction

Chapter 2 Five-Membered Carbocycles
2.1 Danheiser annulation
2.2 Dieckmann condensation
2.3 Nazarov cyclization
2.4 Pauson–Khand reaction
2.5 Weiss reaction

Chapter 3 Six-Membered Carbocycles
3.1 Bergman cyclization
3.2 Bradsher reaction
3.3 Bogert–Cook reaction
3.4 Bradsher cyclization (Bradsher cycloaddition)
3.5 Darzens synthesis of tetralin derivatives
3.6 Diels–Alder reaction
3.7 Dötz reaction
3.8 Elbs reaction
3.9 Favorskii rearrangement rearrangement
3.10 Fujimoto–Belleau reaction
3.11 Haworth reaction
3.12 Haworth phenanthrene synthesis
3.13 Myers–Moore cyclization
3.14 Parham cyclization
3.15 Pchorr reaction
3.16 Robinson annulation
3.17 Scholl reaction
3.18 Stork enamine reaction

Chapter 4 Large-Ring Carbocycles
4.1 Buchner method of ring enlargement
4.2 de Mayo reaction
4.3 Ring-closing metathesis (RCM)

4.4 Ruzicka large ring synthesis
4.5 Tiffeneau–Demjanov rearrangement
4.6 Thorpe–Ziegler reaction

Chapter 5 Transformation of Carbocycles

5.1 Blanc chloromethylation reaction
5.2 Friedel–Crafts reaction
5.3 Gattermann–Koch reaction
5.4 Houben–Hoesch reaction
5.5 Kolbe–Schmitt reaction
5.6 Vilsmeier–Haack reaction
5.7 von Richter reaction

Appendix 5
Table of Contents for Volume 6: *Name Reactions in Heterocyclic Chemistry-2*
Due in 2011

PART 1 THREE- AND FOUR-MEMBERED HETEROCYCLES

Chapter 1 Epoxides and Aziridines
1.1 Blum aziridine synthesis
1.2 Gabriel–Heine aziridine isomerization
1.3 Graham diziririne synthesis
1.4 Hassner aziridine-azirine synthesis
1.5 Neber reaction
1.6 Scheiner aziridine synthesis
1.7 Shi epoxidation

PART 2 FIVE-MEMBERED HETEROCYCLES

Chapter 2 Pyrroles and Pyrrolidines
2.1 Clauson–Kass pyrrole synthesis
2.2 Ehrlich reaction of pyrroles
2.3 Houben–Hoech acylation of pyrroles
2.4 Overman pyrrolidine synthesis
2.5 Padawa pyrroline synthesis
2.6 Trifimov synthesis of pyrroles

Chapter 3 Indoles
3.1 Bischler indole synthesis
3.2 Borsche–Drechsel cyclization
3.3 Gassman oxindole synthesis
3.4 Hinsberg(–Stollé) oxindole and oxiquinoline synthesis
3.5 Larock indole synthesis
3.6 Matinet dioxindole synthesis
3.7 Mori–Ban indole synthesis
3.8 Neber–Bosset oxindole synthesis
3.9 Sandmeyer diphenylurea isatin synthesis
3.10 Sandmeyer isonitrosoacetanilide isatin synthesis
3.11 Sommelet–Hauser rearrangement (indole)
3.12 Stollé synthesis

Chapter 4 Furans
4.1 Jeger tetrahydrofuran synthesis
4.2 Nierenstein reaction

4.3 Perkin rearrangement
4.4 Ueno–Stork radical cyclization

Chapter 5 Thophenes
5.1 Paal thiophene synthesis
5.2 Volhard–Erdmann cyclization

Chapter 6 Oxazoles and Isoxazoles
6.1 Davidson oxazole synthesis
6.2 Fischer oxazole synthesis
6.3 Huisgen 4-isoxazoline synthesis
6.4 Japp oxazole synthesis
6.5 Robinson–Gabriel synthesis
6.6 Schöllkopf oxazole synthesis

Chapter 7 Other five–membered heterocycles
7.1 Asinger thiazoline synthesis
7.2 Bamberger imidazole synthesis
7.3 Bredereck imidazole synthesis
7.4 Dimroth triazole synthesis
7.5 Finegan tetrazole synthesis
7.6 Hantsch thiazole synthesis
7.7 Huisgen tetrazole rearrangement
7.8 Knorr pyrazole synthesis
7.9 Pechmann pyrazole synthesis
7.10 Marckwald imidazole synthesis
7.11 Urech hydantoin synthesis
7.12 Wallach imidazole synthesis
7.13 Weidenhagan imidazole synthesis

PART 3 SIX-MEMBERED HETEROCYCLES

Chapter 8 Pyridines
8.1 Baeyer pyridine synthesis
8.2 Katrizky reaction

Chapter 9 Quinolines and Isoquinolines
9.1 Betti reaction
9.2 Bernthsen acridine synthesis
9.3 Dötz hydroquinone synthesis
9.4 Lehmstedt–Tanasescu reaction
9.5 Niementowski quinoline synthesis

Chapter 10 Pyrimidines
10.1 Niementowski quinazolone synthesis
10.2 Remfry–Hull pyrimidine synthesis
10.3. Shaw pyrimidine synthesis

Chapter 11 Other Six-Membered Heterocycles
11.1 Allan–Robinson reaction
11.2 Baker–Venkataraman rearrangement
11.3 Bargellini reaction
11.4 Borsche cinnoline synthesis
11.5 Ferrario–Akermann thiocyclization
11.6 Gastaldi pyrazine synthesis
11.7 Gutknecht pyrazine synthesis
11.8 König benzoxazine synthesis
11.9 Lehmsted–Tanasrescu acridone synthesis
11.10 Niementowski quinazoline synthesis
11.11 Pechmann pyrone synthesis
11.12 Pechmann pyrazole synthesis
11.13 Robinson–Schöpf reaction
11.14 Simonis chromone cyclization
11.15 Wesseley–Moser rearrangement
11.16 Widman–Stoermer cinnoline synthesis
11.17 Wichterle reaction

Chapter 12 Potpourri
12.1 ANRORC mechanism
12.2 Boulton–Katritzky rearrangement
12.3 Chichibabin amination reaction
12.4 Dimroth rearrangement
12.5 Einhorn–Brunner triazole synthesis
12.6 Hantzsch synthesis of pyrroles, pyridines, and thiazoles
12.7 King–Ortoleva reaction
12.8 Traube purine synthesis

Subject Index

Abyssomicin C, Nozaki-Hiyama-Kishi reaction, 312
Acetals, Mukaiyama aldol reaction, 509
Acetonitrile, Henry reaction, potassium phosphate-based procedure, 415
Acetophenone, McMurry coupling, 275–276
Acetylene
  Corey-Fuchs reaction, 396–401
  Glaser coupling, organocopper reagents, 241–252
  Sonogashira reaction, 109–110
Acetylide formation, Castro-Stephens reaction, catalytic mechanism, 218–220
Acetyl sulfones, Julia–Lythgoe olefination, vinyl sulfone conversion, 470
Acid chloride, Arndt-Eistert homologation, 347–348
Aci-nitro(*p*-methoxy-phenyl)methane preparation, (*t*-butyl)Dimethylsilyl ester, 416
Acylsilane, cross-benzoin condensation, 389
Additives, Stille coupling, 140–142
Alcohols
  Barton-McCombie deoxygenation, 616–617
  Henry reaction, 411–412
  Tsuji-Trost reaction and oxidation, 204–205
  Wittig reaction variants, 599
Aldehydes
  benzoin condensation
    annulations, 386
    basic principles, 380–381
    cross-coupling reaction, 389
  Corey-Fuchs reaction, 392–402
  Henry reaction, 403–415
    lithionitrate and ketone, 416
    silylnitronate, 416
  Horner-Wadsworth-Emmons reaction, 421–423
  Knoevenagel reaction, electrophilic scope, 487
  McMurry coupling, 269–272
  Mukaiyama aldol reaction, 504–505, 507–508
  Sakurai allylation reaction, Lewis acid-carbonyl complex, 544–546
  Tebbe methylenation, 324–327
Aldo condensation, Henry reaction and, 403
Aldol reaction
  Knoevenagel mechanism, 476–478
  Mukaiyama aldol reaction
    basic principles, 501
    diastereoselective synthesis, 504–509
    enantioselective variants, 512–517
    historical perspective, 501–502
    mechanism, 502–504
    synthetic utility, 504–517
    tandem reactions, 511–512
    TMS compounds, 517–518
    vinologous reaction, 509–511
  Nicholas reaction modification, 288–289
Aldonolactones, Tebbe methylenation, 330–331
Aldosterone, Barton nitrite photolysis reaction, 633
  experimental compounds, 643–645
  steroid core functionalization, 635–638
Alkaline earth metals, McMurry coupling, 272
Alkaloids, Arndt-Eistert homologation, 341
Alkenes
  Julia–Kocienski olefination, 455–459
  Julia–Lythgoe olefination, 451–459
  McMurry coupling, 270–271

Alkenes *(continued)*
Negishi cross-coupling reaction, 83–86
Peterson olefination, 526
Wittig reaction, 587–589
variations, 592–603
Wohl-Ziegler reaction, 664–668
1-(1-Alkenyl)benzotriazoles, Peterson olefination, 529–530
Alkenyl chloride reagents, Kumada cross-coupling reactions, 52–59
Alkoxides, Morita-Baylis-Hillman homologation, 357
Alkoxy phosphonate, Wittig reaction, 594–595
β-Alkoxy substituent, Sakurai allylation reaction, 554
Alkyl bromides, Negishi cross-coupling reactions, 74–75
Alkyl halides
Kumada cross-coupling reactions, 63–64
Stille coupling, 147–149
Fu's procedure, 157
Suzuki coupling reactions, unactivated compounds, 173–174
Alkynes
Castro-Stephens reaction, silylated alkynes, 230
Corey-Fuchs reaction, 397–401
Glaser coupling, organocopper reagents, 238–241
Negishi cross-coupling reaction, 83–86
Sonogashira reaction, 105–106
homocoupling, 109–110
nontraditional coupling partners, 128
Alkynylboronates, Glaser coupling, organocopper reagents, 241
Alkynyltrialkylsilanes, Glaser coupling, organocopper reagents, 240–241
Allenes, Wittig reaction modification, 600–601
Allocolchicine, Nicholas reaction, 292–293
Allylic compounds
Tsuji-Trost reaction, hydrogenolysis, 205–206
Wohl-Ziegler reaction, 665–668
Allylic halides, Castro-Stephens coupling, 229–230
Allylsilanes, Sakurai allylation reaction, 538–539
nucleophilicity, 546
Aluminum compounds, Tebbe methylenation
basic properties, 319–321
Petasis olefination, 321–327
Alzheimer's therapy
Barton-McCombie deoxygenation, 626–629
Corey-Fuchs reaction, 399–400
Ambruticin, Julia–Kocienski/Julia–Lythgoe olefination, 465–466
Amides
Corey-Fuchs reaction, 398–399
Negishi cross-coupling reaction, 89
Tebbe methylenation, 329–331
Wohl-Ziegler reaction, 668–672
Amine nucleophiles
Knoevenagel reaction
historical perspective, 473–475
mechanism, 476–478
tertiary amines, 482
Verley-Doebner modification, 478–481
Tsuji-Trost reaction, 199–201
β-Amino acid synthesis, Arndt-Eistert homologation, 337–338
Amino acids
Knoevenagel reaction, catalysis by, 482–483
Dakin-Prout modification, 496
Negishi cross-coupling reaction, 87–89
β-Amino alcohols, Henry reaction, 408–410
Aminosulfoximene ligands, Mukaiyama aldol reaction, enantioselectivity, 516–517
Amphidinol 3, Nozaki-Hiyama-Kishi reaction, 310
Amphidinolide B1, Sakurai allylation reaction, 560–561
Amprenavir, Henry reaction, 409–410
(+)-Ancistrofuran, Barton-McCombie deoxygenation, 627
Ando procedure, Horner-Wadsworth-Emmons reaction, synthetic procedures, 429
Androstane series, Barton nitrite photolysis reaction, 636–638
Anhydrides
Arndt-Eistert homologation, 346–347
Tebbe methylenation, 331
Anilides, asymmetric intramolecular Heck reaction, 13–17
Anion-exchange resins, Henry reaction, 404
Anionic pathways
Heck reaction, 7–9
Hiyama cross-coupling reaction, 34–35
Annulation
benzoin condensation, 386

Stetter reaction, asymmetric annulation, 585
Annulene, Glaser coupling, organocopper reagents, 246–252
Antascomicin A, Nozaki-Hiyama-Kishi reaction, 310
Antibacterial agents
Morita-Bayless-Hillman reaction, 374–375
Sandmeyer reaction, 653–654
Antibiotic agents
Barton-McCombie deoxygenation, 628
Julia–Kocienski/Julia–Lythgoe olefination, 463–469
Mukaiyama aldol reaction, 508–509
Anti coordination, Sakurai allylation reaction, 544
syn/anti isomerism, 547–551
Antidiabetic compounds, Knoevenagel reaction, 493
Anti-inflammatory drugs, Knoevenagel reaction, 494
Anti-influenza agents
Wittig reaction, 606–607
Wohl-Ziegler reaction, 669
Antithrombotic compounds, Morita-Bayless-Hillman reaction, 374
Anti-tubercular drugs, Morita-Bayless-Hillman reaction, 373
Anti-tumor agents, Horner-Wadsworth-Emmons reaction, 430–443
Antiviral agents, Mukaiyama aldol reaction, 506–509
Aqueous solvents
Heck reactions, 25–26
Sonogashira reaction, 127–128
Arndt-Eistert homologation, carbon-chain homologations
acid chloride homologation using $PhCO_2Ag$, 347–348
β-amino acid synthesis, 337–338
basic principles, 335
diazo group transfer, 343–344
α,β-epoxy diazomethyl ketone rearrangement, 341–342
historical perspective, 335–336
mechanisms, 336–337
mixed anhydride homologation using CF_3CO_2Ag, 346–347
natural product synthesis, 339–341
synthetic utility, 337–343
trimethylsilyldiazomethane, 344–345
Wolff rearrangement, 342–343
ynolate anions, 345–346
Aromatic aldehydes, Nozaki-Hiyama-Kishi reaction, 309
Artemisinin, Sakurai allylation reaction, 570–571
Aryl bromides
Sandmeyer reaction, 651
Suzuki reaction, 167–168
Aryl chlorides
Sonogashira reaction, 124
Stille coupling, 145–147
Suzuki coupling reactions, 169–170
Aryl compounds
aryl tosylates, Kumada cross-coupling reactions, 60
Hiyama cross-coupling reaction, 36–37
Knoevenagel reaction, 486
Kumada cross-coupling reactions, 52–59
Negishi cross-coupling reactions, 73
Aryl halides
Nozaki-Hiyama-Kishi reaction, 309
Sonogashira reaction, 124
Arylmagnesium reagents, Negishi cross-coupling reactions, 80
4-Arylpiperidines, Negishi cross-coupling reactions, 81
Aryl sulfoxides, Julia–Lythgoe olefination, 452–453
Aryl triflates, Suzuki reaction, 167–168
Arylzinc chlorides and bromide-vinylic tellurides, Negishi cross-coupling reaction, 83
Asmarine A, Morita-Bayless-Hillman reaction, 367–368
Asymmetric allylic alkylation (AAA), Tsuji-Trost reaction, 188–189
amine nucleophiles, 199–201
carbon nucleophiles, 196–197
metal catalysts, 206–208
oxygen nucleophiles, 198–199
sulfur nucleophiles, 201
Asymmetric reactions
BAL-catalyzed benzaldehyde dimerization, 388–389
Heck reactions
intermolecular reaction, 17
intramolecular reaction, 13–17

Asymmetric reactions *(continued)*
Henry reaction, chiral copper catalyst, 415
Kumada cross-coupling, 55–57
Morita-Baylis-Hillman homologation, 359–361
Mukaiyama aldol reaction
enantioselectivity, 514–517
merged 1,2- and 1,3-asymmetric induction, 504–509
Nicholas reaction, 286–289
Nozaki-Hiyama-Kishi reaction, chiral oxazoline ligands, 305–307
Sakurai allylation reaction
1,2-asymmetric induction, 551–553
1,3-asymmetric induction, 554
catalytic reactions, 555–559
Tsuji-Trost reaction, allylic alkylation, 188–190
Ullman coupling, 265–266
Atorvastatin, Stetter reaction, 577–578
1-(*Z*)-Atractylodinol, Negishi cross-coupling reaction, 83–84
Atropisomers, Wohl-Ziegler reaction, 670
Aurisides A and B
Mukaiyama aldol reaction, enantioselectivity, 515–517
Nozaki-Hiyama-Kishi reaction, 312
Aza-benzoin condensation, 387–388, 390
Azadiradione, Barton nitrite photolysis reaction, 638–639
Aza-Henry reaction, variation and improvements, 413
Aza Morita-Baylis-Hillman reaction
basic principles, 361–362
intermolecular reactions, 377
Azapriacid-1, Nozaki-Hiyama-Kishi reaction, 312
Azidosphingosine, Julia–Kocienski olefination, 456–459
Azobisisobutyronitrile (AIBN)
Barton-McCombie deoxygenation, 614–616
(*R*)-baclofen homologues, 626
4,6-dimethoxy-1-(tetradec-13′-enyl)-1*H*-indole, 630
miscellaneous reductants, 623–625
phosphorous-based reductants, 622–623
Wohl-Ziegler reaction, 668–671

(*R*)-Baclofen homologues, Barton-McCombie deoxygenation, 626
(–)-Bafilomycin A_1, Mukaiyama aldol reaction, 505–506
Bafilomycin A1, Stille coupling, 154
Balz-Schiemann reaction, 649–650
Barton-McCombie deoxygenation
basic principles, 613
4,6-dimethoxy-1-(tetradec-13′-enyl)-1*H*-indole, 630
historical perspective, 613–614
mechanisms, 614–616
miscellaneous alternative reductants, 623–625
phosphorous-based reductants, 622–623
silicon-based reductants, 619–621
synthetic utility, 625–629
1,1′-thiocarbonyldiimidazole, 629–630
tin-based reductants, 618–619
variation and improvements, 616–618
Barton nitrite photolysis reaction
basic principles, 632
flexible molecule functionalization, 642
historical perspective, 632–634
9-hydroxy-2,2,4,8-tetramethyl-octahydro-4,8-methano-azulen-1-one oxime, 645
mechanisms, 634
non-steroidal rigid system functionalization, 638–642
steroid core functionalizations, 635–638
synthetic utility, 634–642
variations and improvements, 642–644
Baylis-Hillman reaction. *See* Morita-Baylis-Hillman reaction
Benzaldehyde
Knoevenagel reaction, 474–478
Verley-Doebner modification, 478–481
NaCN-catalyzed dimerization, 388
Nozaki-Hiyama-Kishi reaction, 303
Benzaldehyde lyase (BAL)
asymmetric benzaldehyde dimerization, 388–389
benzoin condensation, enantioselective dimerization, 383–385
Benzalmalononitriles, Knoevenagel reaction, 491–495
Benzofurans, Negishi cross-coupling reactions, 90
Benzoin condensation
carbon-chain homologations

annulations, 386
aza-benzoin condensation, 387, 390
BAL-catalyzed benzaldehyde dimerization, 388–389
basic principles, 380–381
cross-benzoin acylsilane-aldehyde coupling, 389
cross-benzoin condensation, 383–385
dimerization, 382–383
enantioselective cross-benzoin condensation, 384–385
enantioselective dimerization, 382–383
historical perspective, 381
mechanisms, 381–382
NaCN-catalyzed benzaldehyde dimerization, 388
synthetic utility, 382–387
Stetter reaction, 575–576
Benzolyl peroxide, Wohl-Ziegler reaction, 670–671
Benzopyrans, Nicholas reaction, 294
Benzoylformate decarboxylase (BFD), benzoin condensation, enantioselective dimerization, 383–385
α-Benzyloxy substituents, Sakurai allylation reaction, 553
1-(Benzyloxy)-4-iodopyrazole, Negishi cross-coupling reactions, 77
N-(3-Benzyloxy-2-oxo-1-phenyl-propyl)-carbamic acid *tert*-butyl ester, benzoin condensation, 390
1-(4-Benzylphenyl)ethanone, Suzuki coupling reaction, 181
Bergman cyclization, Nozaki-Hiyama-Kishi reaction, 307
Betaines, Wittig reaction, 591
Biaryl compounds
enantioselective cross-benzoin condensation, 385
Ullman coupling reaction, 263–265
Bicyclic guanidine base, Horner-Wadsworth-Emmons reaction, 419–421
BINAP compounds
Heck reaction
asymmetric intermolecular reactions, 17
neutral pathways, 6–7
Mukaiyama aldol reaction, enantioselectivity, 516–517
Binaphthyl (BINOL) ligands
Mukaiyama aldol reaction, enantioselectivity, 515–517
Sakurai allylation reaction, chiral promoters, 557
(*R*)-BINOL, Kumada cross-coupling, asymmetric reactions, 57
Bioconjugates, Sonogashira reaction, 121–122
Biologically active molecules, Knoevenagel reaction, 491–495
Biomimetic Stetter reaction, 578
Biopolymers, Sonogashira reaction, 115–117
Bipinnatin J, Nozaki-Hiyama-Kishi reaction, 311
Bipyridines, Negishi cross-coupling reactions, 97
Bipyridyl ligands, Nozaki-Hiyama-Kishi reaction, 301–302
(*R*)-2,2′-Bis(bromomethyl)-1,1′-binaphthyl, Wohl-Ziegler reaction, 672
6,6-Bis-phenylsulfonyl-4-(*tert*-butyldimethylsilanyloxymethyl)-1-(4-methoxy-benzyloxy)-hex-3-en-2-ol, Tsuji-Trost reaction, 209
Bis(trifluoromethyl)phenyl (BTFP), Julia–Lythgoe olefination, 457–459
Borane compounds, Barton-McCombie deoxygenation, 625
Boron trifluoride promoter, Sakurai allylation reaction
basic properties, 538–539
example, 569
Lewis acid-carbonyl complex, 542–543
Borrelidin, Nozaki-Hiyama-Kishi reaction, 312
Breslow mechanism, Stetter reaction, 575–576
Brevenal, Suzuki coupling reactions, 179–180
Brevetoxins
Sakurai allylation reaction, 561–562
Wittig reaction, 605–606
Briarellins E and F, Nozaki-Hiyama-Kishi reaction, 311
Bromination
Sandmeyer reaction, 655
Wohl-Ziegler reaction, mechanisms, 662–668
6-Bromo-4-ethynyl-3-(1-fluoro-1-methylethyl)chromate, Nicholas reaction, 297
N-Bromosuccinimide (NB*S*), Wohl-Ziegler reaction, 660–672
Bronze, Ullmann coupling, 261
[1,2]-Brook rearrangement, benzoin condensation, 384–385

Bryostatin, Mukaiyama aldol reaction, 507
Bulky ligands
  Sonagashira reaction, 129–130
  Suzuki coupling reactions, 170
tert-Butyl(4,4-dibromobut-3-en-2-yl)dimethylsilane, Corey-Fuchs reaction, 401
tert-Butyldimethyl(pent-3-yn-2-yl)silane, Corey-Fuchs reaction, 401–402

Cadinane-sesquiterpene veticadinol, tandem Knoevenagel-ene reaction, 488
Cadiot-Chodkiewicz reaction, Castro-Stephens reaction and
  organocopper reagents, 221–223
  related processes, 231
Camptothecin, Negishi cross-coupling reactions, 94
Cannizzaro reaction, Henry reaction and, 403
Carbazole synthesis, Sonogashira reaction, 120–121
Carbenes, Heck reactions, 23
Carbocations, Nicholas reaction, 286–289
Carbohydrates, Negishi cross-coupling reaction, 87
Carbonates, Barton-McCombie deoxygenation, 615
Carbon–carbon bond formation
  Heck reaction, 2–3
  Mukaiyama aldol reaction, enantioselectivity, 514–517
  Negishi cross-coupling reactions, 75–76
    organozinc reagents, 89
  Nicholas reaction, 289
  Peterson olefination, 521–522
  Sonogashira reaction, 100–102
  Stille coupling, sigma carbon bonds, 144–145, 148–149, 152–156
  Suzuki reaction, 163
    mechanisms, 165–166
Carbon-chain homologations
  Arndt-Eistert homologation
    acid chloride homologation using $PhCO_2Ag$, 347–348
    β-amino acid synthesis, 337–338
    basic principles, 335
    diazo group transfer, 343–344
    α,β-epoxy diazomethyl ketone rearrangement, 341–342
    historical perspective, 335–336
    mechanisms, 336–337
    mixed anhydride homologation using CF_3CO_2Ag, 346–347
    natural product synthesis, 339–341
    synthetic utility, 337–343
    trimethylsilyldiazomethane, 344–345
    Wolff rearrangement, 342–343
    ynolate anions, 345–346
  benzoin condensation
    annulations, 386
    aza-benzoin condensation, 387, 390
    BAL-catalyzed benzaldehyde dimerization, 388–389
    basic principles, 380–381
    cross-benzoin acylsilane-aldehyde coupling, 389
    cross-benzoin condensation, 383–385
    dimerization, 382–383
    enantioselective cross-benzoin condensation, 384–385
    enantioselective dimerization, 382–383
    historical perspective, 381
    mechanisms, 381–382
    NaCN-catalyzed benzaldehyde dimerization, 388
    synthetic utility, 382–387
  Corey-Fuchs reaction
    description, 392
    historical perspective, 392–393
    mechanisms, 393–394
    synthetic utility, 396–401
    tert-butyl(4,4-dibromobut-3-en-2-yl)dimethylsilane, 401
    tert-butyldimethyl(pent-3-yn-2-yl)silane, 401–402
    variation and improvement, 394–396
  Henry reaction
    alcohols, 411–412
    β-amino alcohols, 408–410
    asymmetry procedure, chiral copper catalyst, 415
    aza-Henry reaction, 413
    basic principles, 403–404
    historical perspective, 404
    α-hydroxycarboxylic acids, 412–413
    α-hydroxyketones, 411
    intramolecular reaction, 414
    lithionitrate, aldehydes and ketones, 416

mechanism, 404–405
nitroalkenes, 406–408
α-nitroketones, 410–411
nitronates condensation, 414
6-nitro-7-trimethylsilyloxytridecane, 417
potassium phosphate-based acetonitrile procedure, 415
retro-Henry reaction, 415
silylnitronate and aldehyde, 416–417
synthetic utility, 405–413
Horner-Wadsworth-Emmons reaction
basic principles, 419–421
historical perspective, 421
mechanism, 421–423
methyl 4-*O*-acetyl-3-methoxy-(*Z*)-cinnamate, 443–444
natural product synthesis, 429–443
Still-Gennari variation, 423
synthetic utility, 424–429
Julia–Kocienski olefination
mechanism, 449–450
natural product synthesis, 459–469
synthetic utility, 450–459
Julia–Lythgoe olefination
acetoxy sulfone-vinyl sulfone conversion, 470
alkyl sulfone-aldehyde coupling, 469–470
basic principles, 446–447
historical perspective, 447–448
mechanism, 448–449
natural product synthesis, 459–469
synthetic utility, 450–459
vinyl sulfone reductive cleavage, 470–471
Knoevenagel reaction
active methylene nucleophile, 484
amino acid catalysis, 482–483, 496
aryl/heteroaryl activating groups, 486
basic principles, 473
condenstion promoters, 483–484
Cope modification, 482, 495–496
cyanoacetic acid and esters, 485
Dakin-Prout modification, 496
diethyl 5α-cholestan-3-ylidinemalonate, 496
1,3-diketones, 486
electrophile scope, 487
ene reaction, 488
ethyl (1-ethylpropylidine)cyanoacetate, 495–496
ethyl *n*-butylidineacetoacetate, 495
heterocycles, 486, 491
historical perspective, 473–476
β-keto esters, 485
Lehnert modification, 483, 496
malonic acid, 484
malonic esters, 485
malononitrile, 485
mechanism, 476–478
natural products and bioactive molecules, 491–495
Nokami hydroxylative variant, 481–482
Ragoussis modification, 497
standard conditions, 482
stereochemistry, 487
synthetic utility, 484–495
tandem reactions, 487–488
Diels-Alder reaction, 488–489
electrocyclization, 489–490
Knoevenagel-Mislow-Evans, 490–491
Verley-Doebner modification, 478–481, 497
Morita-Baylis-Hillman reaction
asymmetric reactions, 359–362
aza variant, 361–362
basic properties, 349
catalyst, 355–358
DABCO mediated intermolecular reaction, 375–376
high speed ball milling technique, 358–359
historical perspective, 349–350
intermolecular aza-MBH, 377
intramolecular reaction, 376–377
mechanism, 350–353
medicinal chemistry, 371–375
solvent effects, 353–355
synthetic utility, 362–375
titanium-mediated intermolecular reaction, 376
Mukaiyama aldol reaction
basic principles, 501
diastereoselective synthesis, 504–509
enantioselective variants, 512–517
historical perspective, 501–502
mechanism, 502–504
synthetic utility, 504–517
tandem reactions, 511–512
TMS compounds, 517–518
vinologous reaction, 509–511

Carbon-chain homologations *(continued)*
  Peterson olefination
    basic principles, 520
    experimental compounds, 534–536
    historical perspective, 520–521
    mechanism, 521–522
    synthetic utility, 525–534
    variations and improvements, 523–525
  Sakurai allylation reaction
    1,2-asymmetric induction, 551–553
    1,3-asymmetric induction, 554
    asymmetric catalytic reactions, 555–559
    basic principles, 538–539
    BF_3 promoter, 569
    chiral silane chirality transfer, 554–555
    crotylsilanes, 567–568
    E/Z isomerism, 547
    $FeCl_3$ promoter in $MeNO_2$, 571–572
    historical perspective, 539–540
    mechanism, 540–546
      aldehyde substituents, 545
      Lewis acid-carbonyl complex, 542–545
      silane substituents, 545–546
    $SNCl_4$ promoter, 569–570
    syn/anti isomerism, 547–551
    synthetic utility, 560–567
    synthons, 559–560
    $TiCl_4$ promoter, 568–570
    $ZnCl_3$ promoter, 570–571
  Stetter reaction
    asymmetric intramolecular annulation, 585
    basic principles, 575
    biomimetic reactions, 578
    enantioselective reactions, 579–583
    historical perspective, 575
    intermolecular reactions, 576–580, 583–585
    intramolecular reactions, 580–582, 584
    mechanism, 575–576
    NaCN-catalyzed intermolecular reaction, 583–584
    Sila-Stetter reaction, 578–579, 584–585
    synthetic utility, 576–582
    thiazolylidine-catalyzed intermolecular reaction, 584
  Wittig reaction
    basic principles, 587
    ethyl cyclohexylideneacetate, 609
    historical perspective, 587–588
    mechanism, 588–592
    methylenecyclohexane, 608–609
    synthetic utility, 603–608
    variations, improvements, and modifications, 592–603
Carbon nucleophiles, Tsuji-Trost reaction
  asymmetric allylic alkylation, 196–197
  "hard" nucleophiles, 195–196
  "soft" nucleophiles, 190–193
Carbonylation
  Stille coupling, 137–138, 158
  Tsuji-Trost reaction, 202–203
Carbonyl compounds
  Mukaiyama aldol reaction, 501–502
  Sakurai allylation reaction, 538–539
    mechanism, 540–546
    syn/anti isomerism, 547–551
  Tebbe methylenation, 327–331
  Wittig reaction, 594
Carboxylates, Tsuji-Trost reaction, asymmetric allylic alkylation, 198–199
Caribenolide I, Morita-Bayless-Hillman reaction, 366
Cascade cyclizations, Heck zipper reactions, 18
Castro-Stephens reaction
  organocopper reagents
    allylic halides, 229–230
    aryl halide alkynylation, 224–225
    basic principles, 212–214
    catalytic mechanism and *in situ* acetylide formation, 218–220
    Catiot-Chodkiewicz reaction, 221–223, 231
    classical mechanism, 217–218
    copper-catalyzed Sonogashira, 229
    copper(I) acetylide preparation, 231–232
    copper nanoclusters - heterogeneous catalysis, 228–229
    heterocycle synthesis, 226–227, 233
    historical perspective, 214–216
    ligand effects, 220–221
    macrocycle synthesis, 225–226
    methyl 2-(2-(5-(methoxycarbonyl)benzofuran-2-yl)ethyl)benzofuran-5-carboxylate, 234
    2-phenylfuro[3,2-b]pyridine, 233–234
    regioselectivity and stereoselectivity, 220
    Rosenmund-von Braun reaction, 223–224, 232
    silylated alkynes, 230
    in situ modification, 234

synthetic utility, 224–227
palladium-free Sonogashira reaction, 128
Sonogashira reaction, 102–103
Catalytic reactions
Castro-Stephens reaction, acetylide formation, 218–220
Heck reaction, 3–12
Knoevenagel reaction, amino acids, 482–483
Morita-Baylis-Hillman homologation, 355–358
Sakurai allylation reaction, asymmetric reactions, 555–559
Stille coupling
alkyl halides, 147–149
aryl chloride case, 145–147
Suzuki coupling reactions, 168–169
Catenanes, Glaser coupling, organocopper reagents, 249–252
Cationic pathways, Heck reaction, 5–7
Ceric ammonium nitrate (CAN), Nicholas reaction, 286–289
Cerium chloride, Peterson olefination, 523–525
Cesium fluoride, Peterson olefination, 523–525
CF_3CO_2Ag, Arndt-Eistert homologation of mixed anhydrides, 346–347
Chair-like transition state, Mukaiyama aldol reaction, 503–504
Chalcogenide, Morita-Baylis-Hillman homologation, 358–359
Chartelline A, Sandmeyer reaction, 651
Chelation model, Sakurai allylation reaction, 552
Chiral acyloxyborane (CAB) catalyst, Sakurai allylation reaction, 556
Chiral binaphthalenes, Negishi cross-coupling reactions, 81–82
Chiral copper catalysts, Henry reaction, 407–408
asymmetric reactions, 415
Chiral ligands
Barton nitrite photolysis reaction and, 641
Wittig reaction variants, 599–603
Chiral oxazoline ligands, Nozaki-Hiyama-Kishi reaction, 303–307
Chiral salen ligands, Nozaki-Hiyama-Kishi reaction, 302–303
Chiral silanes, Sakurai allylation reaction, chirality transfer, 554–555
Chiral thiazolium precatalyst, benzoin condensation, enantioselective dimerization, 382–383
Chlorination, Sandmeyer reaction, 655–656
p-Chlorobenzaldehyde, Morita-Baylis-Hillman homologation, 356–357
Cholestanol derivatives, Barton nitrite photolysis reaction, 637–638
Chroman-4-one, intramolecular Stetter reaction, 581
Chromium catalysts, Nozaki-Hiyama-Kishi reaction, 300–301
Ciganek-type reactions, Stetter reaction, 581–583
Ciguatoxins, Nozaki-Hiyama-Kishi reaction, 310
Cis-trans isomerization, Negishi cross-coupling reactions, 72
Claisen rearrangement
Arndt-Eistert homologation, diazo group transfer, 343–344
Knoevenagel reaction, 475
Sakurai allylation reaction, 562
Tebbe methylenation, 328–331
Clavulone analogs, Sonogashira reaction, 113–114
Cobalt
Kumada cross-coupling reactions, 64–66
Sandmeyer reaction, 649–650
Collidine, Barton-McCombie deoxygenation, 624–625
Combinatorial library synthesis, Sonogashira reaction, 113–114
Complex fragment coupling, Sonogashira reaction, 111–114
Concanamycin F, Mukaiyama aldol reaction, 506
Condensation promoters, Knoevenagel reaction, 483–484
Conformational analysis, Barton nitrite photolysis reaction, 632–633
Conjugated systems
Heck reactions, 19
Hiyama cross-coupling reaction, additions, 40–41
Peterson olefination, 525
Tsuji-Trost reaction, diene formation, 203–204
Cope modification, Knoevenagel reaction, 482
Copper acetylide
Cadiot-Chodkiewicz reaction, 221–223
silylated alkynes, 230
Castro-Stephens reaction

Copper acetylide *(continued)*
classical mechanism, 217–218
experimental preparations, 232–233
historical perspective, 215–216
in situ catalytic mechanism, 218–220
Copper catalysts. *See also* Organocopper reagents
Henry reaction, 407–408
asymmetric reactions, 415
Mukaiyama aldol reaction, enantioselectivity, 513–517
Sandmeyer reaction, 649–650
Sonogashira reaction
basic properties, 100–102
Castro-Stephens reaction and, 229
high-activity ligands, 123–125
historical perspective, 102–103
mechanisms of, 104–106
Stille coupling, 140–142
vinylstannanes, 155–156
Copper-free Sonogashira reaction
bulky phosphine ligand, 129
historical perspective, 103
mechanism of, 106–107
Copper(I) phenylacetylide
Castro-Stephens reaction, 233
Glaser coupling, basic principles, 236–237
Corey-Chaykovsky reaction, Wittig reaction modification, 602
Corey-Fuchs reaction, carbon-chain homologations
description, 392
historical perspective, 392–393
mechanisms, 393–394
synthetic utility, 396–401
tert-butyl(4,4-dibromobut-3-en-2-yl)dimethylsilane, 401
tert-butyldimethyl(pent-3-yn-2-yl)silane, 401–402
variation and improvement, 394–396
Corey's synthesis, Barton nitrite photolysis reaction, 638–639
Cornforth model, Sakurai allylation reaction, 553
Corticosterone acetate, Barton nitrite photolysis reaction, 635–636
Cristatic acid, Tsuji-Trost reaction, 193
(+)-Crocacin, Peterson olefination, 528
Cromakalim, Sandmeyer reaction, 654
Cross-benzoin condensation, 384
Cross-coupling reactions
benzoin condensation
acylsilane and aldehyde, 389
basic principles, 380–381
Castro-Stephens reaction
allylic halides, 229–230
classical mechanism, 217–218
Hiyama cross-coupling, 33–45
basic principles, 33
conjugate additions, 40–41
experimental compounds, 45
heterocycle examples, 39–40
historical perspective, 33–34
improvements, 36–38
mechanisms, 34–35
natural product applications, 43–44
organosiloxane preparations, 35–36
organosilyl precursors, 38–39
synthetic utility, 38–44
tandem reactions, 41–42
Kumada cross-coupling
alkyl halides, 63–64
aryl tosylates, 60
asymmetric cross-coupling, 55–57
basic principles, 47–48
highly-active nickel-catalyzed procedure, 67
historical perspective, 48–49
Knochel-type Grignard reagents and triarylmagnesiates, 60–63
mechanisms of, 49–52
metallic catalysts, 64–66
natural product synthesis, 58–59
standard nickel-catalyzed procedure, 66–67
synthetic utility, 52–59
McMurry coupling, historical perspective, 268–272
Negishi cross-coupling
alkenes, alkynes, polyenes, enynes, 83–86
amides and esters, 89
amino acids, 87–89
basic principles, 70
bipyridine, 97
carbohydrates, 87
carbon-carbon bond constructions, 76–77
catalytic improvements, 74–75
enantioselectivity, 75–76
five-membered heterocycles, 77–79
furanoenyne, 96
historical background, 71

mechanisms, 71–72
microwave irradiation, 73–74
natural products, 90–95
phosphine ligands, 72–73
six-membered rings, 79–82
synthetic utility, 77–96
Sonogashira reaction
alkyne homocopuling and acetylene reactivity, 109–110
aqueous solvent reactions, 127–128
basic principles, 100–102
bioconjugates, 121–122
bulky phosphine ligand, copper-free conditions, 129–130
complex fragment coupling, 111–114
copper co-catalyst, 104–106
copper-free mechanism, 106–107
cylindramide intermediate, 129
electronic and optical properties of compounds, 117–118
fatty acids and polyunsaturated systems, 114–115
heterocycle synthesis, 118–121
heterogeneous catalysts, 109
historical perspective, 102–103
ligand activity, 123–125
ligand-free catalysts, 108, 125–127
mechanisms, 107–108
metal catalysts, 128
non-traditional coupling partners, 128
polymers, 122–123
regioselectivity and stereoselectivity, 108
self-assembly subunits, dendrimers, and biopolymers, 115–117
synthetic utility, 110–123
Suzuki reaction
aryl chlorides, 169–170
basic principles and trends, 163
1-(4-benzylphenyl)ethanone, 181
comparisons with other cross-coupling reactions, 163–165
enantioselective reactions, 174–176
heteroaromatic compounds, 170–171
highly active catalysts, 168–169
historical perspective, 165–166
mechanism, 166–168
5-(2′-methylphenyl)-1*H*-indole, 180–181
organotrifluoroborate reagents, 171–173
synthetic utility, 176–180
unactivated alkyl halides and sulfonates, 173–174
Ullmann coupling
asymmetric copper coupling, 265–266
basic principles and historical perspective, 258
intramolecular palladium coupling, 266
mechanisms, 258–259
synthetic utility, 262–265
variations, improvements, or modifications, 259–261
Crotylsilanes, Sakurai allylation reaction, 567–568
Crotyltrimethoxysilane, Sakurai allylation reaction, 558–559
CS gas, Knoevenagel reaction, 491–492
Cyanation, Sandmeyer reaction, 656–657
Cyanide complexes
benzoin condensation, 381
modified cross-condensation, 384–385
Peterson olefination, 531–532
Cyanoacetic acid and esters, Knoevenagel reaction, 485
Cyclic compounds, Nicholas reaction, tandem reactions, 294–296
Cyclin-dependent kinases (CDKs), Negishi cross-coupling reactions, 95
Cyclitol derivatives, Peterson olefination, 532
Cyclization, Nozaki-Hiyama-Kishi reaction, large-sized ring systems, 307
Cycloalkanes, Nicholas reaction, 290–291
Cyclohexadienes, Barton-McCombie deoxygenation, silicon-based reductants, 621
Cyclopentadiene, asymmetric intramolecular Heck reaction, 14–17
Cyclopenten-2-one, Morita-Baylis-Hillman homologation, 356–357
Cyclopropanes, Nicholas reaction, 288–289
Cyclopropyl carboxaldehyde, Knoevenagel reaction, 492
Cyclopropylsulfoxides, Julia–Lythgoe olefination, 454–459
Cylindramide intermediate, Sonogashira reaction, 129

DABCO (1,4-diazobicyclo-[2.2.2]octane)
Castro-Stephens reaction, ligand effects, 221
Morita-Baylis-Hillman reaction

DABCO (1,4-diazobicyclo-[2.2.2]octane) *(continued)*
asymmetric reactions, 359–361
development of, 350
intermolecular reactions, 375–376
mechanisms, 350–353
solvent effects, 354–355
Sonogashira reaction, ligand-free catalysts, 125–127
(+)-Dactylolide, Peterson olefination and Prins cyclization reaction, 532
Dakin-Prout modification, Knoevenagel reaction, 496
Deacetoxyalcyonin acetate, alcyonin, Nozaki-Hiyama-Kishi reaction, 312
Decarboxylation, Knoevenagel reaction, Verley-Doebner modification, 481
(*E*)-Dec-3-enoic acid, Knoevenagel reaction, 497
[18]-Dehydrodesoxyepothilone B, Kumada cross-coupling reactions, 59
Dendrimers, Sonogashira reaction, 115–117
(–)-Dendrobine, Wittig reaction modification, 601
12-Deoxofukujusonorone, Barton nitrite photolysis reaction, 635–636
Deoxygenation reactions
Barton-McCombie deoxygenation
basic principles, 613
4,6-dimethoxy-1-(tetradec-13′-enyl)-1*H*-indole, 630
historical perspective, 613–614
mechanisms, 614–616
miscellaneous alternative reductants, 623–625
phosphorous-based reductants, 622–623
silicon-based reductants, 619–621
synthetic utility, 625–629
1,1′-thiocarbonyldiimidazole, 629–630
tin-based reductants, 618–619
variation and improvements, 616–618
McMurry coupling, 273–276
Deprotonation, Peterson olefination, 522
Deuterium, Barton-McCombie deoxygenation, miscellaneous reductants, 624–625
Diacetylenes, Glaser coupling, organocopper reagents, 242–252
Dialkylphosphites, Barton-McCombie deoxygenation, 622–623
endo, *endo*-2,5-Diaminonorbornane (DIANANE), Nozaki-Hiyama-Kishi reaction, 303
9,9′-Dianthryldiacetylene (Eglinton modification), Glaser coupling, 253
Diarylcations
Sakurai allylation reaction, Lewis acid-carbonyl complex, 545–546
Wohl-Ziegler reaction, 667–668
Diastereoselective synthesis
Mukaiyama aldol reaction, 504–509
vinylogous variant, 510–511
Sakurai allylation reaction
β-alkoxy substituent, 554
syn/anti isomerism, 547–551
1,8-Diazabicycloundec-7-ene (DBU), Morita-Baylis-Hillman reaction, 355–356
(2*S*,3*S*,4*S*)-3-Diazoacetyl-1-ethoxycarbonyl-4-isopropenyl-2-trilospropyl-silyloxymethylpyrrolidine, Arndt-Eistert homologation, 347
Diazo group transfer
Arndt-Eistert homologation, 343–344
Sandmeyer reaction, 647
variations, 649–650
Diazomethane, Arndt-Eistert homologation, 343–347
Diazonium salts, Heck reactions, 27
(1*S*,2*S*,5R,6*R*)-5,6-Dibenzyloxy-cyclohex-3-ene-1,2-diol, Peterson olefination, 534–535
(1*S*,4*S*,5R,6*R*)-5,6-Dibenzyloxy-cyclohex-2-ene-1,4-diol, Peterson olefination, 535
Dibromoalkene formation, Corey-Fuchs reaction, 394–396
Dibromo ylide, Corey-Fuchs reaction, 393–394
1,4-Dicarbonyl compounds, Stetter reaction, 577–578
2,4-Dichloropurine, Negishi cross-coupling reaction, 82
Dichloroquinoline, Negishi cross-coupling reaction, 82
Dictyostatin, Horner-Wadsworth-Emmons reaction, 432
1,3-Dicyclohexadienes, Peterson olefination, 534
Dicyclohexylcarbodiimide (DCC), Henry reaction, 406–408
Diels-Alder reaction
Glaser coupling, organocopper reagents, 244–252
Peterson olefination, 525

tandem Knoevenagel reaction, 488–489
Tsuji-Trost reaction, carbon nucleophiles, 193
Wittig reaction, 607
Dienes, Julia–Lythgoe olefination, 456–459
Dienones, Peterson olefination, 525
Diethyl 5α-cholestan-3-ylidinemalonate, Knoevenagel reaction, 496
Diethylmalonate, Knoevenagel reaction, 475
Dihydrobenzofuranone, intramolecular Stetter reaction, 581
5,7-Dihydro-dibenzo[a,c]cycloheptane-6,6-dicarboxylic acid diethyl ester, Ullman coupling, 266
1,22-Dihydroxynitianes, Nozaki-Hiyama-Kishi reaction, 311
1,3-Diketones, Knoevenagel reaction, 486
Dimerization
BAL-catalyzed asymmetric benzaldehyde dimerization, 388–389
benzoin condensation, 382–383
NaCN-catalyzed benzaldehyde, 388
4,6-Dimethoxy-1-(10′-(*N*-imidazolylthiocarbonyloxy)-tetradec-13′-enyl)-1*H*-indole, Barton-McCombie deoxygenation, 629–630
4,6-Dimethoxy-1-(tetradec-13′-enyl)-1*H*-indole, Barton-McCombie deoxygenation, 630
Dimethylaminopyridine (DMAP), Morita-Baylis-Hillman homologation, 356
(*R*)-2,2′-Dimethyl-1,1′-binaphthyl, Kumada cross-coupling reaction, 67
2,6-Dimethylbiphenyl, Hiyama cross-coupling reaction, 45
Dimethylfuran (DMF), Castro-Stephens reaction, 213–214
N,N-Dimethylglycine (DMG), Heck reactions, 24–25
(*E*)-3-(2,6-Dimethylphenyl)-3-methyl acrylic acid methyl ester, Heck reaction, 30
(*t*-butyl)Dimethylsilyl ester, *aci*-nitro(*p*-methoxy-phenyl)methane preparation, 416
Dinitrile, Knoevenagel reaction, 491–492
Dioxabicyclo-octene, Wohl-Ziegler reaction, 667
Diphenylacetylene, Sonogashira reaction, 129
1,4-Diphenyl-2-(phenylsulfonyl)-1-butene, Julia–Lythgoe olefination, 470
Diphenylphosphine, Wittig reaction variants, 597–598
1,3-bis(Diphenylphosphino)propane (dppp)
Heck reaction, neutral pathways, 6–7
Kumada cross-coupling reactions, 53–59
1,4-Di-(2′-quinolyl)-1,3-butadiyne (Cameron modification), Glaser coupling, 252–253
(+)-Discodermolide, Nozaki-Hiyama-Kishi reaction, 312
Discodermolide, Horner-Wadsworth-Emmons reaction, 434–437
2′,4-Disubstituted-2,4′-bithiazoles, Negishi cross-coupling reactions, 91–92
1,1-Di-t-butyl peroxide (DTBP), Barton-McCombie deoxygenation, silicon-based reductants, 620–621
1,2-Dixanthate, Barton-McCombie deoxygenation, 615–616
Diynes, Negishi cross-coupling reaction, 86
Dopamine-uptake inhibitors, Sandmeyer reaction, 653–654
Dragmacidin D, Arndt-Eistert homologation, 341
Drug discovery analogues, Nozaki-Hiyama-Kishi reaction, 314
Dynamic kinetic asymmetric transformation (DYKAT), Tsuji-Trost reaction
asymmetric allylic alkylation, 190
molybdenum-catalyzed reaction, 207

Ebelactone A, Nozaki-Hiyama-Kishi reaction, 312
(+)-Efaroxan, Morita-Bayless-Hillman reaction, 371–372
Eglinton conditions, Glaser coupling, organocopper reagents, 240–241
Eicosnoid 4, Nozaki-Hiyama-Kishi reaction, 311
Electrocyclization, tandem Knoevenagel reaction, 489–490
Electronic compounds, Sonogashira reaction, 117–118
Electrophiles
Knoevenagel reaction, 487
Wittig reaction, 596–603
Enamides, Peterson olefination, 526–527
Enantioselectivity
benzoin condensation
cross-benzoin reaction, 384–385
dimerization, 382–383
Mukaiyama aldol reaction, 512–517
Sakurai allylation reaction, chiral promoters, 556–559

Enantioselectivity *(continued)*
  Stetter reaction
    intermolecular reactions, 579–580
    intramolecular reactions, 581–583
  Suzuki coupling reactions, 174–176
Endocyclic intramolecular reactions, Nicholas reaction, 292–293
Endothelin receptor antagonist, Barton nitrite photolysis reaction, 644
Enediyne antitumor agent, Nicholas reaction, 292–293
Ene reactions
  Mukaiyama aldol reaction, 503–504
  tandem Knoevenagel reaction, 488
Enolates, Wittig reaction, 604
Enol silane, Mukaiyama aldol reaction, 507–509
Enthalpies, Sakurai allylation reaction, Lewis acid-carbonyl complex, 543
Enynes, Negishi cross-coupling reaction, 83–86
Epothilones
  Horner-Wadsworth-Emmons reaction, 430–443
  Nozaki-Hiyama-Kishi reaction, 312
Epoxides, Mukaiyama aldol reaction, 512
α,β-Epoxy diazomethyl ketone rearrangement, Arndt-Eistert homologation, 341–342
Equilibrium constants, Sakurai allylation reaction, Lewis-Acid carbonyl complex, 542–543
Erythro/threo adducts, Wittig reaction, 589–590
  variations, improvements or modifiactions, 592–603
Esters
  Negishi cross-coupling reaction, 89
  Tebbe methylenation, Petasis olefination, 322–327
Ethylacetoacetate, Knoevenagel reaction, 475–476
Ethyl *n*-butylidineacetoacetate, Knoevenagel reaction, 495
Ethyl cyclohexylideneacetate, Wittig reaction, 609
Ethyl(diarylphosphono)acetates, Horner-Wadsworth-Emmons reaction, 420–421
Ethyl (1-ethylpropylidine)cyanoacetate, 495–496
Ethyl (1-ethylpropylidine)cyanoacetate, Knoevenagel reaction, 495–496
Ethyl (1-methylpropylidine)cyanoacetate, Knoevenagel reaction, 496
(*R*)-Ethyl 2-(4-oxo-3,4-dihyro-2*H*-chromen-3-yl)acetate, Stetter reaction, 585
N-Ethynylzinc tosylamides, Negishi cross-coupling reaction, 85
Evans synthesis, Mukaiyama aldol reaction, 504–509
Exocyclic intramolecular reactions, Nicholas reaction, 293–294
E/Z isomerism, Sakurai allylation reaction, 547

Farnesyl transferase, Arndt-Eistert homologation, 339–341
Fatty acids, Sonogashira reaction, 114–115
Felkin-Ahn model
  Mukaiyama aldol reaction, 504–506
  Peterson olefination, 526
  Sakurai allylation reaction, 552
Ferrocenylphosphine, Kumada cross-coupling, asymmetric reactions, 57
Fischer carbene complex, Hiyama cross-coupling reaction, 38
Five-membered heterocycles, Negishi cross-coupling reactions, 77–79
"Flattened chair" conformation, steroid core, Barton nitrite photolysis reaction, 636–637
Flexible molecules, Barton nitrite photolysis reaction, 642
1-Fluoro-1-haloalkenes, Negishi cross-coupling reaction, 84
Fluoroquinolone antibiotics, Morita-Bayless-Hillman reaction, 372–373
Fluorous systems, Heck reactions, 27
α-Fluorovinylsulfones, Peterson olefination, 530
Formates, Barton-McCombie deoxygenation, miscellaneous reductants, 624–625
Formyl hydrogen bond, Sakurai allylation reaction, 544
Forskolin compounds, Barton nitrite photolysis reaction, 641
FR901512, Nozaki-Hiyama-Kishi reaction, 312
Friedel-Crafts reaction
  Mukaiyama aldol reaction, 503
  Nicholas reaction, organometallic reagents, 285–286
Furanoenyne, Negishi cross-coupling reactions, 96
Furanosesquiterpene natural products, Barton-McCombie deoxygenation, 627

(*Z*)-[4-(Furan-2-yl)buten-3-en-1-ynyl]trimethylsilane, Negishi cross-coupling reactions, 96
Fu's protocol
  Barton-McCombie deoxygenation, tin-based reductants, 618–619
  Heck reactions, 30
  Hiyama cross-coupling reaction, 37–38
  Stille coupling
    aklyl halides, 157
    sigma carbon bonds, 144–145, 148–149

Gabosine precursor, Nozaki-Hiyama-Kishi reaction, 313
Gelsemine, Knoevenagel reaction, 492
GJIJ ring fragments, Sakurai allylation reaction, 561–562
Glaser coupling, organocopper reagents
  basic principles, 236
  9,9′-dianthryldiacetylene (Eglinton modification), 253
  1,4-di-(2′-quinolyl)-1,3-butadiyne (Cameron modification), 252–253
  2,4-hexadiyne-1,6-diol, 252
  historical perspective, 236
  mechanism, 236–237
  N,N′-phenylbuta-1,3-diyne-1,4-tosylamide (Hay modification), 253
  synthetic utility, 241–252
  variations and improvements, 238–241
C-Glycosides, Sakurai allylation reaction, 561–562
Grignard reagents
  Kumada cross-coupling
    asymmetric reactions, 56–57
    basic principles, 47–48
    Knochel-type reagents, 60–63
    mechanisms of, 52–59
    metal catalysis, 64–66
  Suzuki reaction, 164–165
Grubbs reagents, Tebbe methylenation, 325–327
Guanacastepene C, Knoevenagel reaction, 493
Gymnoascolide A, Wohl-Ziegler reaction, 669

Halichlorine, Morita-Bayless-Hillman reaction, 362
Halides
  Sandmeyer reaction, 649–650
  Sonogashira reaction, 128
  Wohl-Ziegler reaction, 666–668
(*Z*)-haloallylic alcohols, Nozaki-Hiyama-Kishi reaction, 308
Halogenated alkenols, Nozaki-Hiyama-Kishi reaction, 308
Halogenating agents, Wohl-Ziegler reaction, 671
Hammett studies, Castro-Stephens reaction, classical mechanism, 218
Hann-Lapworth mechanism, Knoevenagel reaction, 477–478
Hantzsch's synthesis, Knoevenagel reaction, 473–474
"Hard" nucleophiles, Tsuji-Trost reaction
  basic mechanism, 187–188
  carbon nucleophiles, 195–196
Hateruamalide, Nozaki-Hiyama-Kishi reaction, 313
Hay conditions, Glaser coupling, 238
Heck reaction, palladium chemistry, 2–31
  alternative solvents, 25–27
  anionic pathways, 7–9
  aqueous media, 25–26
  asymmetric intermolecular reaction, 17
  asymmetric intramolecular reaction, 13–17
  basic principles, 2
  carbenes, 23
  cationic *vs.* neutral pathways, 5–7
  conjugated systems, 19
  fluorous systems, 27
  Fu's experimental conditions, 30
  high-activity ligands, 20–23
  historical perspective, 2–3
  ionic liquids, 26–27
  Jeffery's experimental conditions, 30–31
  leaving groups, 27
  ligand-free catalysts, 11–12, 23–25
  mechanisms, 3–12
  metal catalysts, 28
  palladacycles, 22–23
  phosphines, 21–22
  polymers, 20
  reductive reactions, 28–29
  regioselectivity and stereoselectivity, 9–10
  supercritical fluids, 27
  supported catalysts, 23–25
  synthetic utility, 12–20
  zipper reactions, 18

Helicobacter pylori antibiotics, Barton-McCombie deoxygenation, 628
(–)-Hennoxazole, Mukaiyama aldol reaction, 507–508
Hennoxazole A, Julia–Kocienski/Julia–Lythgoe olefination, 466–467
Henry reaction
  carbon-chain homologations
    alcohols, 411–412
    β-amino alcohols, 408–410
    asymmetry procedure, chiral copper catalyst, 415
    aza-Henry reaction, 413
    basic principles, 403–404
    historical perspective, 404
    α-hydroxycarboxylic acids, 412–413
    α-hydroxyketones, 411
    intramolecular reaction, 414
    lithionitrate, aldehydes and ketones, 416
    mechanism, 404–405
    nitroalkenes, 406–408
    α-nitroketones, 410–411
    nitronates condensation, 414
    potassium phosphate-based acetonitrile procedure, 415
    retro-Henry reaction, 415
    silylnitronate and aldehyde, 416–417
    synthetic utility, 405–413
  6-nitro-7-trimethylsilyloxytridecane, 417
(*E*)-(Hept-1-enyl)naphthalene, Hiyama cross-coupling reaction, 45
(+)-Herbarumin I, Nozaki-Hiyama-Kishi reaction, 313
Herboxidiene, Julia–Kocienski/Julia–Lythgoe olefination, 468–469
Herrmann's catalyst, Heck reactions, palladacycles, 22–23
Heteroaromatic compounds, Suzuki coupling reactions, 170–171
Heteroaryl activating groups, Knoevenagel reaction, 486
Heteroatomic nucleophiles, Nicholas reaction, 287–289
  intermolecular reactions, 291
Heterocouplings
  Glaser coupling, organocopper reagents, 244–252
  McMurry coupling, 269–272
Heterocyclic chemistry
  Castro-Stephens reaction, 233–234
  Hiyama cross-coupling reaction, 39–40
  Knoevenagel reaction, 486, 491
  Negishi cross-coupling reactions, five-membered heterocycles, 77–79
  Sandmeyer reaction, 653–654
  Sonogashira reaction, 118–121
  Wittig reaction, 607–608
Heterogeneous catalysis
  Sonogashira reaction, 109
  Suzuki coupling reactions, 168–169
Hexacarbonyl(μ-(η^4-[2,4-dimethyl-2-{prop-2-yn-1-yl}-8-oxabicyclo-[3.2.1.]octan-3-one]))dicobalt, Nicholas reaction, 296–297
2,4-Hexadiyne-1,6-diol, Glaser coupling, 252
Hexylbenzene, Kumada cross-coupling reaction, 67
Highest occupied molecular orbitals (HOMOs), Tebbe methylenation, 320–321
High speed ball milling (HSBM) technique, Morita-Baylis-Hillman homologation, 358–359
Hikizimycin, Nozaki-Hiyama-Kishi reaction, 310
Hirsutic acid C, intramolecular Stetter reaction, 580–581
HIV inhibitors, Horner-Wadsworth-Emmons reaction, 424–429
Hiyama cross-coupling reaction, palladium chemistry, 33–45
  basic principles, 33
  conjugate additions, 40–41
  experimental compounds, 45
  heterocycle examples, 39–40
  historical perspective, 33–34
  improvements, 36–38
  mechanisms, 34–35
  natural product applications, 43–44
  organosiloxane preparations, 35–36
  organosilyl precursors, 38–39
  synthetic utility, 38–44
  tandem reactions, 41–42
Hofmann-Loffler-Freytag reaction, Barton nitrite photolysis reaction and, 640–641
(*R*)-(–)-Homoctric acid-γ-lactone, Arndt-Eistert homologation, 341
Homocoupling reactions
  alkynes, Sonogashira reaction, 109–110

Glaser coupling, organocopper reagents, 239–241
Ullman coupling, 264–265
Homogeneous catalysts, Suzuki coupling reactions, 168
Hopf's synthesis, McMurry coupling, 270–271
Horner-Wadsworth-Emmons reaction
carbon-chain homologations
basic principles, 419–421
historical perspective, 421
mechanism, 421–423
methyl 4-*O*-acetyl-3-methoxy-(*Z*)-cinnamate, 443–444
natural product synthesis, 429–443
Still-Gennari variation, 423
synthetic utility, 424–429
Peterson olefination, 520–521
Wittig reaction variations, 592–603, 604–605
β-Hydride elimination, Heck reaction, regioselectivity and stereoselectivity, 9–10
Hydrogenolysis, Tsuji-Trost reaction, allylic compounds, 205–206
(±)-3-Hydroxybakuchiol, Peterson olefination, 531
α-Hydroxycarboxylic acids, Henry reaction, 412–413
2-Hydroxy-1,2-diphenylethan-1-one, benzoin condensation, 388
(*R*)-2-Hydroxy-1,2-diphenylethan-1-one, benzoin condensation, 389
α-Hydroxyketones, Henry reaction, 411
Hydroxylation, Sandmeyer reaction, 657
β-Hydroxysilanes, Peterson olefination, 520–528
19-Hydroxytaxol, McMurry coupling, 278–279
9-Hydroxy-2,2,4,8-tetramethyl-octahydro-4,8-methano-azulen-1-one oxime, Barton nitrite photolysis reaction, 645
Hypophosphorous acid, Barton-McCombie deoxygenation, 622–623

Imidazomethyl phosphonates, Horner-Wadsworth-Emmons reaction, 425–429
Immunosuppressive compounds, Mukaiyama aldol reaction, 506
Indium complexes, Mukaiyama aldol reaction, enantioselectivity, 514–517
Indole compounds
Barton-McCombie deoxygenation, 628
4,6-dimethoxy-1-(10′-(*N*-imidazolylthiocarbonyloxy)-tetradec-13′-enyl)-1*H*-indole, 629–630
4,6-dimethoxy-1-(tetradec-13′-enyl)-1*H*-indole, 630
McMurry coupling, 281
Sandmeyer reaction, 654
Indolizomycin, Julia–Kocienski/Julia–Lythgoe olefination, 463
Ingenol antitumor agent, Nicholas reaction, 292–293
Initiators, Wohl-Ziegler reaction, 670–671
Intermolecular reactions
Horner-Wadsworth-Emmons reaction, 425–429
Morita-Bayless-Hillman reaction, 375–377
Nicholas reaction, 290–291
Sakurai allylation reaction, *syn/anti* isomerism, 550–551
Stetter reaction, 576–580, 583–585
biomimetic reactions, 578
enantioselectivity, 579–580
NaCN-catalyzed reaction, 583–584
Sila-Stetter variant, 578–579, 584–585
thiazolylidine-catalyzed reaction, 584
Intramolecular reactions
Heck reaction, 13–17
Henry reaction, 414
Hiyama cross-coupling reaction, 43–44
Horner-Wadsworth-Emmons reaction, 425–429
Morita-Baylis-Hillman homologation, 376–377
Nicholas reaction
endocyclic compounds, 292–293
exocyclic reactions, 293–294
Sakurai allylation reaction, *syn/anti* isomerism, 551
Stetter reaction, 580–583
asymmetric annulation, 585
Suzuki coupling reactions, 178–179
Tsuji-Trost reaction, allylation, carbon nucleophiles, 192–193
Ullman coupling, 266
Wittig reactions, 604
Iodides, Negishi cross-coupling reaction, 85
Iodination, Sandmeyer reaction, 657
5-Iodomidazole, Negishi cross-coupling reactions, 78

Ionic liquids
Heck reactions, 26–27
Morita-Baylis-Hillman reaction, 355
Iridium, Tsuji-Trost reaction, asymmetric allylic alkylation, 206–208
Iron complexes
Kumada cross-coupling reactions, 64–66
Sakurai allylation reaction, $FeCl_3$ promoter, 571–572
Sandmeyer reaction, 649–650
Isocoumarin, Sonogashira reaction, 120
Isomerization
Knoevenagel reaction, Verley-Doebner modification, 479–481
Sakurai allylation reaction
E/Z isomerism, 547
syn/anti isomerism, 547–551

Jatrophone compound, Stille-Hegedus synthesus, 153
Jefferey's conditions, Heck reactions, 19, 30–31
ligand-free and supported catalysts, 23–25
Jerangolid D, Sakurai allylation reaction, 563–564
Julia–Kocienski olefination, carbon-chain homologations
mechanism, 449–450
natural product synthesis, 459–469
synthetic utility, 450–459
Julia–Lythgoe olefination, carbon-chain homologations
acetoxy sulfone-vinyl sulfone conversion, 470
alkyl sulfone-aldehyde coupling, 469–470
basic principles, 446–447
historical perspective, 447–448
mechanism, 448–449
natural product synthesis, 459–469
synthetic utility, 450–459
vinyl sulfone reductive cleavage, 470–471

(–)-Kainic acid, Arndt-Eistert homologation, 339
Keck synthesis
Mukaiyama aldol reaction, 506
Sakurai allylation reaction, 567
β-Keto esters, Knoevenagel reaction, 485
Keto esters, McMurry coupling, 277–278
Ketones
Arndt-Eistert homologation, 339–341
α,β-epoxy diazomethyl ketone rearrangement, 341–342
benzoin condensation, 380–381
Henry reaction, 416
Horner-Wadsworth-Emmons reaction, 425–429
Julia–Lythgoe olefination, 451–459
Knoevenagel reaction
electrophilic scope, 487
Lehnert modification, 483
McMurry coupling, 269–272
Negishi cross-coupling reaction, 85
Sakurai allylation reaction, *syn/anti* isomerism, 548–551
Tebbe methylenation, 324–327
Wittig reaction modification, 597–603
Knochel-type Grignard reagents, Kumada cross-coupling, 60–63
Knoevenagel reaction, carbon-chain homologations
active methylene nucleophile, 484
amino acid catalysis, 482–483, 496
aryl/heteroaryl activating groups, 486
basic principles, 473
condenstion promoters, 483–484
Cope modification, 482, 495–496
cyanoacetic acid and esters, 485
Dakin-Prout modification, 496
diethyl 5α-cholestan-3-ylidinemalonate, 496
1,3-diketones, 486
electrophile scope, 487
ene reaction, 488
ethyl (1-ethylpropylidine)cyanoacetate, 495–496
ethyl *n*-butylidineacetoacetate, 495
heterocycles, 486, 491
historical perspective, 473–476
β-keto esters, 485
Lehnert modification, 483, 496
malonic acid, 484
malonic esters, 485
malononitrile, 485
mechanism, 476–478
natural products and bioactive molecules, 491–495
Nokami hydroxylative variant, 481–482
Ragoussis modification, 497
standard conditions, 482
stereochemistry, 487
synthetic utility, 484–495

tandem reactions, 487–488
Diels-Alder reaction, 488–489
electrocyclization, 489–490
Knoevenagel-Mislow-Evans, 490–491
Verley-Doebner modification, 478–481, 497
Knots, Glaser coupling, organocopper reagents, 249–252
Kumada cross-coupling reactions, palladium chemistry
alkyl halides, 63–64
aryl tosylates, 60
asymmetric cross-coupling, 55–57
basic principles, 47–48
highly-active nickel-catalyzed procedure, 67
historical perspective, 48–49
Knochel-type Grignard reagents and triarylmagnesiates, 60–63
mechanisms of, 49–52
metallic catalysts, 64–66
natural product synthesis, 58–59
Negishi cross-coupling reactions vs., 79
standard nickel-catalyzed procedure, 66–67
synthetic utility, 52–59

Lactonamycin, Negishi cross-coupling reactions, 93
β-Lactones, Mukaiyama aldol reaction, 511–512
Lanostane derivatives, Barton nitrite photolysis reaction, 637–638, 642–643
Lanthanum-(*R*)-binaphthol complex, Henry reaction, 408–410
Lapworth mechanism, benzoin condensation, 381–382
Large scale Stille coupling, procedures for, 159
Laulimalide analogues
Horner-Wadsworth-Emmons reaction, 438–440
Nozaki-Hiyama-Kishi reaction, 314
Sakurai allylation reaction, 566–567
(+)-Laurencin, Tebbe methylenation, 329–331
Leaving groups, Heck reactions, 27
Lehnert modification, Knoevenagel reaction, 483
diethyl 5α-cholestan-3-ylidinemalonate, 496
Leucascandrolide
Horner-Wadsworth-Emmons reaction, 440
Mukaiyama aldol reaction, 509
Lewis acid-mediated reactions, Sakurai allylation reaction, 538–539
mechanism, 540–546
Ligand effects
Castro-Stephens reaction, 220–221
Heck reactions, 20–21
Hiyama cross-coupling reaction, 37–38
Kumada cross-coupling, asymmetric reactions, 55–57
Sonogashira reaction, 101–102
high-activity ligands, 123–125
mechanisms of, 105–106
Stille coupling, 139–140
Suzuki coupling reactions, 169–170
Ligand-free catalysts
Heck reaction
basic mechanisms, 11–12
supported catalysts, 23–25
Sonogashira reaction, 108, 125–127
Lithionitrate, Henry reaction, 416
Lithium salts, Wittig reaction, 590, 596–603
Lowest unoccupied molecular orbital (LUMO), Sakurai allylation reaction, Lewis acid-carbonyl complex, 544–546

Macrolactins, Julia–Kocienski/Julia–Lythgoe olefination, 460–461
Macrolides
Horner-Wadsworth-Emmons reaction, 431–443
Mukaiyama aldol reaction, 507
enantioselectivity, 514–517
vinylogous variant, 509–511
Magnesium compounds
Corey-Fuchs reaction, 395–396
Wittig reaction, 597
Magnesium-halogen exchange, Kumada cross-coupling reactions, 61–63
Malonic acid, Knoevenagel reaction, 484
Malonic esters, Knoevenagel reaction, 485
Malononitrile, Knoevenagel reaction, 485
Manganese catalysts
Nozaki-Hiyama-Kishi reaction, 304–307
Stille coupling, 142
Mangicols, Nozaki-Hiyama-Kishi reaction, 310
Mannich addition products, aza-benzoin condensation, 387–388
(+)-Massarinolin, Nozaki-Hiyama-Kishi reaction, 313
McMurry coupling, organometallic reagents
basic principles, 268
historical perspective, 268–272

McMurry coupling, organometallic reagents *(continued)*
indole, 281
mechanism, 273–276
receptor, 281–282
synthetic utility, 278–281
variations, improvements, modifications, 276–278
MDL compound, Knoevenagel reaction, 493–494
Medicinal chemistry
Morita-Bayless-Hillman reaction, 371–375
Sandmeyer reaction, 653–654
Meerwein-Pondorff-Verley reduction, Peterson olefination, 525
Meerwein reaction, 650
Merged 1,2- and 1,3-asymmetric induction, Mukaiyama aldol reaction, 504–509
Metal catalysts
Heck reactions, 28
Henry reaction, β-amino alcohols, 408–410
Kumada cross-coupling reactions, 64–66
Sonogashira reaction, 126–129
Stille coupling, 141–142
Tsuji-Trost reaction, 206–208
(2*S*,3E,7E,11*S*)-13-Methoxy-2-(methoxymethoxy)-4,8,11-trimethyl-15-methylenebicyclo[9.3.1]pentadeca-1(14),3,7,13-tetraene, Peterson olefination, 535–536
Methyl 4-*O*-acetyl-3-methoxy-(*Z*)-cinnamate, Horner-Wadsworth-Emmons reaction, 443–444
Methyl 6,6′-dimethyl-2,2′-bipyridine-4-carboxylate, Negishi cross-coupling reactions, 97
Methylenation reactions, Peterson olefination, 523–525
Methylenecyclohexane, Wittig reaction, 608–609
Methylene nucleophiles, Knoevenagel reaction, 484
Methyl (2*S*,3*S*,4*S*)-1-Ethoxycarbonyl-4-isopropenyl-2-triisopropyl-silyl-oxymethyl-pyrrolidin-3-ylacetate, Arndt-Eistert homologation, 347–348
Methyl ketones, Mukaiyama aldol reaction, 517–518
Methyl 2-(2-(5-(methoxycarbonyl)benzofuran-2-yl)ethyl)benzofuran-5-carboxylate, Castro-Stephens reaction, 234
2-Methyl-7-phenyl-2,4-heptadiene, Julia–Lythgoe olefination, 470–471
5-(2′-Methylphenyl)-1*H*-indole, Suzuki coupling reaction, 180
3-Methyl-1-phenyl-2-(triethylsilyloxy)butan-1-one, benzoin condensation, 389
α-Methyl substituent, Sakurai allylation reaction, 551–553
Methyl *tert*-butyl ether (MTBE), Stille coupling, 148–149
Methyl vinyl ketone (MVK), Morita-Baylis-Hillman reaction, 350–353
(+)-Methynolide, Nozaki-Hiyama-Kishi reaction, 313
Michael addition and elimination
Henry reaction, α-nitroketones, 410–411
Knoevenagel reaction, 474–475
Morita-Baylis-Hillman reaction, 351–353
Wittig reaction, 603–604
Michaelis-Arbuzov reaction, Horner-Wadsworth-Emmons reaction, 424–429
Microwave irradiation
McMurry coupling, 272
Negishi cross-coupling reactions, 73–74
Stille coupling, toxicity reduction, 150–151
Ullman coupling, 264–265
Wittig reaction variants, 598
Mislow-Evans [2,3]sigmatropic rearrangement, Nokami hydroxylative Knoevenagel reaction, 481–482
Moenocinol, Julia–Kocienski/Julia–Lythgoe olefination, 461–462
Molybdenum, Tsuji-Trost reaction, 206–208
Monofluoromethylenephosphonates, Peterson olefination, 530
Morita-Baylis-Hillman reaction, carbon-chain homologations
asymmetric reactions, 359–362
aza variant, 361–362
basic properties, 349
catalyst, 355–358
DABCO mediated intermolecular reaction, 375–376
high speed ball milling technique, 358–359
historical perspective, 349–350
intermolecular aza-MBH, 377
intramolecular reaction, 376–377
mechanism, 350–353

medicinal chemistry, 371–375
solvent effects, 353–355
synthetic utility, 362–375
titanium-mediated intermolecular reaction, 376
Mosin B, Nozaki-Hiyama-Kishi reaction, 313
Mukaiyama aldol reaction
basic principles, 501
diastereoselective synthesis, 504–509
enantioselective variants, 512–517
historical perspective, 501–502
mechanism, 502–504
synthetic utility, 504–517
tandem reactions, 511–512
TMS compounds, 517–518
vinologous reaction, 509–511

NaCN-catalyzed dimerization
benzaldehyde, 388
intermolecular Stetter reaction, 583–584
(–)-Nakadomarin A, Barton-McCombie deoxygenation, 628–629
Nanostructures, Sonogashira reaction, 116–117
Naphthyl-naphthyl Negishi cross-coupling reaction, 81
Narbononlide, Nozaki-Hiyama-Kishi reaction, 311
Natural products
Arndt-Eistert homologation, 339–341
Barton-McCombie deoxygenation, 626–629
Barton nitrite photolysis reaction, 634–642
Hiyama cross-coupling reaction, 43–44
Homer-Wadsworth-Emmons reaction, 429–443
Horner-Wadsworth-Emmons reaction, 429–443
Julia–Lythgoe olefination, 459–469
Knoevenagel reaction, 491–495
Kumada cross-coupling reactions, 58–59
Morita-Bayless-Hillman reaction, 362–375
Mukaiyama aldol reaction, diastereoselective synthesis, 504–509
Negishi cross-coupling reactions, 90–96
Nozaki-Hiyama-Kishi reaction, 309–314
Sandmeyer reaction, 650–653
Sonogashira reaction, 112–114
Suzuki coupling reactions, 176–177
Tsuji-Trost reaction, carbon nucleophiles, 193
(–) and (+)-Negillamine A_2, Nozaki-Hiyama-Kishi reaction, 313
Negishi cross-coupling reactions, palladium chemistry
alkenes, alkynes, polyenes, enynes, 83–86
amides and esters, 89
amino acids, 87–89
basic principles, 70
bipyridine, 97
carbohydrates, 87
carbon-carbon bond constructions, 76–77
catalytic improvements, 74–75
enantioselectivity, 75–76
five-membered heterocycles, 77–79
furanoenyne, 96
historical background, 71
mechanisms, 71–72
microwave irradiation, 73–74
natural products, 90–95
phosphine ligands, 72–73
six-membered rings, 79–82
synthetic utility, 77–96
Neuramidase inhibitors, Wohl-Ziegler reaction, 669
Neutral pathways, Heck reaction, 5–7
N-Heterocyclic carbenes (NHC), Barton-McCombie deoxygenation, 625
Nicholas reaction, organometallic reagents
basic principles, 284
6-bromo-4-ethynyl-3-(1-fluoro-1-methylethyl)chromane, 297
endocyclic intramolecular reactions, 292–293
exocyclic intramolecular reactions, 293–294
historical perspective, 284–285
intermolecular reactions, 290–291
mechanisms, 285–286
synthetic utility, 289–296
tandem reactions, 294–296
variations and improvements, 286–289
Nicholas reaction, organometallic reagents, hexacarbonyl(μ-(η^4-[2,4-dimethyl-2-{prop-2-yn-1-yl}-8-oxabicyclo-[3.2.1.]octan-3-one]))dicobalt, 296–297
Nickel catalysts
Kumada cross-coupling
alkyl halides, 63–64
asymmetric reactions, 55–57
basic principles, 47–48
mechanisms, 49–52
standard procedures, 66–67

Nickel catalysts *(continued)*
synthetic applications, 52–59
Negishi cross-coupling reactions
basic principles, 70
microwave irradiation, 73–74
Nozaki-Hiyama-Kishi reaction, 300–301
chiral oxazoline ligands, 303–307
Sonogashira reaction, 126–127
Stille coupling, 142
alkyl halide reactions, 148–149
Tsuji-Trost reaction, asymmetric allylic alkylation, 208
Ullmann coupling, 261
NIckel(II)-phosphine complex, Kumada cross-coupling reactions, 49, 53–59
Nitroaldo condensation, Henry reaction, 403–404
Nitroalkanes, Henry reaction, 403–415
Nitroalkenes, Henry reaction, 406–408
m-Nitrocinnamic acid, Knoevenagel reaction, Verley-Doebner modification, 497
Nitrogen nucleophiles, Tsuji-Trost reaction, 194
α-Nitroketones, Henry reaction, 410–411
Nitronates, Henry reaction, 414
Nitrone derivatives, Barton nitrite photolysis reaction, 643–644
2-Nitro-2-propen-1-ol pivalate (NPP), Henry reaction, 407–408
6-Nitro-7-trimethylsilyloxytridecane, Henry reaction, 417
N-methylmorpholine (NMM), Morita-Baylis-Hillman homologation, 356
Nokami hydroxylative Knoevenagel reaction, mechanisms, 481–482
Non-steroidal rigid systems, Barton nitrite photolysis reaction, 638–641
Norpseudoephedrine derivative, Sakurai allylation reaction, *syn/anti* isomerism, 548–551
Nozaki-Hiyama-Kishi reaction, organometallic reagents
basic principles, 299–300
bipyridyl ligands, 301–302
chiral oxazoline ligands, 303–304
chiral salen ligands, 302–303
drug discovery analogs, 314
(*Z*)-haloallylic alcohols, 308
historical perpsective, 300
large-sized ring system cyclization, 307–308
mechanism, 300–301
miscellaneous reactions, 309
multidentate ligands, 303–307
natural products, 309–314
(*S*)-1-phenyl-3-butene-1-ol, 314–315
synthetic utility, 307–314
9-(*tert*-butyldimethylsilyloxy)-5,6,12,13-tetrahydro-8,9,10,11-tetrahydro-7*H*-benzo[11]annulen-7-ol, 315

Olefins
Horner-Wadsworth-Emmons reaction, 424–429
Julia–Lythgoe olefination
acetoxy sulfone-vinyl sulfone conversion, 470
alkyl sulfone-aldehyde coupling, 469–470
basic principles, 446–447
historical perspective, 447–448
mechanism, 448–449
natural product synthesis, 459–469
synthetic utility, 450–459
vinyl sulfone reductive cleavage, 470–471
Negishi cross-coupling reaction, 84
Peterson olefination
basic principles, 520
experimental compounds, 534–536
historical perspective, 520–521
mechanism, 521–522
synthetic utility, 525–534
variations and improvements, 523–525
Stetter reaction, basic principles, 575
Tebbe methylenation, 321–327
Olivine, Corey-Fuchs reaction, 398
Organoboron compounds, Suzuki reaction
basic principles, 164–165
mechanisms, 165–168
Organochromium(III) nucleophile, Nozaki-Hiyama-Kishi reaction, 300–301
Organocopper reagents
Castro-Stephens reaction
allylic halides, 229–230
aryl halide alkynylation, 224–225
basic principles, 212–214
catalytic mechanism and *in situ* acetylide formation, 218–220
Catiot-Chodkiewicz reaction, 221–223, 231
classical mechanism, 217–218
copper-catalyzed Sonogashira, 229

copper(I) acetylide preparation, 231–232
copper nanoclusters - heterogeneous catalysis, 228–229
heterocycle synthesis, 226–227, 233
historical perspective, 214–216
ligand effects, 220–221
macrocycle synthesis, 225–226
methyl 2-(2-(5-(methoxycarbonyl)benzofuran-2-yl)ethyl)benzofuran-5-carboxylate, 234
2-phenylfuro[3,2-b]pyridine, 233–234
regioselectivity and stereoselectivity, 220
Rosenmund-von Braun reaction, 223–224, 232
silylated alkynes, 230
in situ modification, 234
synthetic utility, 224–227
Glaser coupling
basic principles, 236
9,9′-dianthryldiacetylene (Eglinton modification), 253
1,4-di-(2′-quinolyl)-1,3-butadiyne (Cameron modification), 252–253
2,4-hexadiyne-1,6-diol, 252
historical perspective, 236
mechanism, 236–237
N,N′-phenylbuta-1,3-diyne-1,4-tosylamide (Hay modification), 253
synthetic utility, 241–252
variations and improvements, 238–241
Ullmann coupling
asymmetric copper coupling, 265–266
basic principles and historical perspective, 258
intramolecular palladium coupling, 266
mechanisms, 258–259
synthetic utility, 262–265
variations, improvements, or modifications, 259–261
Organohalides, Negishi cross-coupling reactions, 75
Organometallic reagents
Knoevenagel reaction, 483–484
McMurry coupling
basic principles, 268
historical perspective, 268–272
indole, 281
mechanism, 273–276
receptor, 281–282
synthetic utility, 278–281
variations, improvements, modifications, 276–278
Nicholas reaction
basic principles, 284
6-bromo-4-ethynyl-3-(1-fluoro-1-methylethyl)chromane, 297
endocyclic intramolecular reactions, 292–293
exocyclic intramolecular reactions, 293–294
historical perspective, 284–285
intermolecular reactions, 290–291
mechanisms, 285–286
synthetic utility, 289–296
tandem reactions, 294–296
variations and improvements, 286–289
Nozaki-Hiyama-Kishi reaction
basic principles, 299–300
bipyridyl ligands, 301–302
chiral oxazoline ligands, 303–304
chiral salen ligands, 302–303
drug discovery analogs, 314
(*Z*)-haloallylic alcohols, 308
historical perpsective, 300
large-sized ring system cyclization, 307–308
mechanism, 300–301
miscellaneous reactions, 309
multidentate ligands, 303–307
natural products, 309–314
(*S*)-1-phenyl-3-butene-1-ol, 314–315
synthetic utility, 307–314
9-(*tert*-butyldimethylsilyloxy)-5,6,12,13-tetrahydro-8,9,10,11-tetrahydro-7*H*-benzo[11]annulen-7-ol, 315
Tebbe methylenation
basic principles, 319
historical perspective, 319–320
mechanisms, 320–321
Petasis olefination, 321–327
Petasis-Tebbe procedural comparisons, 331
synthetic utility, 327–331
Organometallic reagents, Nicholas reaction, hexacarbonyl(μ-(η^4-[2,4-dimethyl-2-{prop-2-yn-1-yl}-8-oxabicyclo-[3.2.1.]octan-3-one]))dicobalt, 296–297
Organosilanes, Sakurai allylation reaction, 562
Organosiloxanes, Hiyama cross-coupling reaction, 35–36

Organosilyl precursors, Hiyama cross-coupling reaction, 38–39
Organostannane, Stille coupling and activation of, 142–145
  toxicity reduction, 149–151
Organotrifluoroborate reagents
  Suzuki coupling reactions, 171–178
  Wittig reaction, 606
Organozinc reagents, Negishi cross-coupling reactions, 70
  carbon–carbon bond formation, 75–76, 89
  five-membered heterocycles, 77–79
Oxacycle ring system, Morita-Bayless-Hillman reaction, 364
Oxalate derivatives, Barton-McCombie deoxygenation, 618
Oxaphosphetanes, Wittig reaction, 590
Oxaphospholenes, Wohl-Ziegler reaction, 667
Oxazoles, Negishi cross-coupling reactions, 78
Oxazoline ligands, Nozaki-Hiyama-Kishi reaction, 303–307
Oxidative addition, Stille coupling, mechanism, 136–139
Oxygen nucleophiles, Tsuji-Trost reaction, 194–195
  asymmetric allylic alkylation, 198–199

Palladacycles, Heck reactions, 22–23
Palladium-carbon heterogeneity, Heck reactions, 25
Palladium chemistry
  Heck reaction, 2–31
    alternative solvents, 25–27
    anionic pathways, 7–9
    aqueous media, 25–26
    asymmetric intermolecular reaction, 17
    asymmetric intramolecular reaction, 13–17
    basic principles, 2
    carbenes, 23
    cationic *vs.* neutral pathways, 5–7
    conjugated systems, 19
    fluorous systems, 27
    Fu's experimental conditions, 30
    high-activity ligands, 20–23
    historical perspective, 2–3
    ionic liquids, 26–27
    Jeffery's experimental conditions, 30–31
    leaving groups, 27
    ligand-free catalysts, 11–12, 23–25
    mechanisms, 3–12
    metal catalysts, 28
    palladacycles, 22–23
    phosphines, 21–22
    polymers, 20
    reductive reactions, 28–29
    regioselectivity and stereoselectivity, 9–10
    supercritical fluids, 27
    supported catalysts, 23–25
    synthetic utility, 12–20
    zipper reactions, 18
  Hiyama cross-coupling reaction, 33–45
    basic principles, 33
    conjugate additions, 40–41
    experimental compounds, 45
    heterocycle examples, 39–40
    historical perspective, 33–34
    improvements, 36–38
    mechanisms, 34–35
    natural product applications, 43–44
    organosiloxane preparations, 35–36
    organosilyl precursors, 38–39
    synthetic utility, 38–44
    tandem reactions, 41–42
  Kumada cross-coupling reactions
    alkyl halides, 63–64
    aryl tosylates, 60
    asymmetric cross-coupling, 55–57
    basic principles, 47–48
    highly-active nickel-catalyzed procedure, 67
    historical perspective, 48–49
    Knochel-type Grignard reagents and triarylmagnesiates, 60–63
    mechanisms of, 49–52
    metallic catalysts, 64–66
    natural product synthesis, 58–59
    standard nickel-catalyzed procedure, 66–67
    synthetic utility, 52–59
  Negishi cross-coupling reactions
    alkenes, alkynes, polyenes, enynes, 83–86
    amides and esters, 89
    amino acids, 87–89
    basic principles, 70
    bipyridine, 97
    carbohydrates, 87
    carbon-carbon bond constructions, 76–77
    catalytic improvements, 74–75
    enantioselectivity, 75–76
    five-membered heterocycles, 77–79

furanoenyne, 96
historical background, 71
mechanisms, 71–72
microwave irradiation, 73–74
natural products, 90–95
phosphine ligands, 72–73
six-membered rings, 79–82
synthetic utility, 77–96
Sonogashira reaction
alkyne homocopuling and acetylene reactivity, 109–110
aqueous solvent reactions, 127–128
basic principles, 100–102
bioconjugates, 121–122
bulky phosphine ligand, copper-free conditions, 129–130
complex fragment coupling, 111–114
copper co-catalyst, 104–106
copper-free mechanism, 106–107
cylindramide intermediate, 129
electronic and optical properties of compounds, 117–118
fatty acids and polyunsaturated systems, 114–115
heterocycle synthesis, 118–121
heterogeneous catalysts, 109
historical perspective, 102–103
ligand activity, 123–125
ligand-free catalysts, 108, 125–127
mechanisms, 107–108
metal catalysts, 128
non-traditional coupling partners, 128
polymers, 122–123
regioselectivity and stereoselectivity, 108
self-assembly subunits, dendrimers, and biopolymers, 115–117
synthetic utility, 110–123
Stille coupling
additives, 140–142
alkyl halide catalytic development, 147–149
aryl chloride catalyst development, 145–147
basic principles, 133–134
carbonylative coupling, 158
historical perspective, 134–136
large-scale coupling, 159
ligand influences, 139–140
mechanisms of, 136–139
organostannane activation, 142–145
π-allyl Stille coupling, 158–159
Stille-Kelly reaction, 151–152
synthetic utility, 152–156
tin compound toxicity, 149–151
toxicity management, 156–157
unactivated secondary alkyl halides, 157
Suzuki reaction
aryl chlorides, 169–170
basic principles and trends, 163
1-(4-benzylphenyl)ethanone, 181
comparisons with other cross-coupling reactions, 163–165
enantioselective reactions, 174–176
heteroaromatic compounds, 170–171
highly active catalysts, 168–169
historical perspective, 165–166
mechanism, 166–168
5-(2′-methylphenyl)-1*H*-indole, 180–181
organotrifluoroborate reagents, 171–173
synthetic utility, 176–180
unactivated alkyl halides and sulfonates, 173–174
Tsuji-Trost reaction
alcohol oxidation, 204–205
allylic compound hydrogenolysis, 205–206
amine nucleophiles, 199–201
asymmetric allylic alkylation, 188–190
basic principles, 185
6,6-bis-phenylsulfonyl-4-(*tert*-butyldimethylsilanyloxymethyl)-1-(4-methoxy-benzyloxy)-hex-3-en-2-ol, 209
carbon nucleophiles, 192–193, 196–197
carbonylation, 202–203
conjugated diene formation, 203–204
"hard" carbon nucleophiles, 195–196
historical perspective, 186
mechanisms, 186–190
metal catalysts, 206–208
nitrogen nucleophiles, 194
oxygen nucleophiles, 194–195, 197–198
(2*S*,5*S*)-2-[3′-phenylsulfonyl-*N*′-(4-methoxybenzyl)succinimid-3′-yl]-5-benzoyloxy-2,5-dihydrofuran, 209
"soft" carbon nucleophiles, 190–193
sulfur nucleophiles, 201
synthetic utility, 190–201
2-(trimethylsilylmethyl)allyl acetate, 201–202
Palladium-free Sonogashira reaction, 128, 229
Palominol, Stille coupling, 154

Pearlman's catalyst, Henry reaction, alcohols, 411–412
(*S*)-[6,6′-(2*R*,4*R*-Pentadioxyl)]=(2,2′)-bis(diphenylphosphoryl)-(1,1′)-biphenyl, Ullman coupling, 265–266
PEPPSI-Ipr precatalyst, Negishi cross-coupling reactions, 74–75
Perhydrohistrionicotoxin, Barton nitrite photolysis reaction, 638–639
Peridinin, Julia–Kocienski/Julia–Lythgoe olefination, 461–462
Peroxisome proliferator-activated receptors (PPARs), Knoevenagel reaction, 493
Petasis reagent, Tebbe methylenation, 321–327
  carbonyl compounds, 328–331
  general procedure, 331
Peterson olefination
  basic principles, 520
  experimental compounds, 534–536
  historical perspective, 520–521
  mechanism, 521–522
  synthetic utility, 525–534
  variations and improvements, 523–525
  Wittig reaction modification, 603
$PhCO_2Ag$, Arndt-Eistert homologation via acid chloride, 347–348
(Phenoxy)thiocarbonate derivatives, Barton-McCombie deoxygenation, 616
N,N′-Phenylbuta-1,3-diyne-1,4-tosylamide (Hay modification), Glaser coupling, 253
(*S*)-1-Phenyl-3-butene-1-ol, Nozaki-Hiyama-Kishi reaction, 314–315
2-Phenylfuro[3,2-b]pyridine, Castro-Stephens reaction, 233–234
1-Phenyl-1,4-pentanedione, intermolecular Stetter reaction, 583–584
Phenylsulfinyl group, Nokami hydroxylative Knoevenagel reaction, 481–482
(2*S*,5*S*)-2-[3′-Phenylsulfonyl-*N*′-(4-methoxybenzyl)succinimid-3′-yl]-5-benzoyloxy-2,5-dihydrofuran, Tsuji-Trost reaction, 209
Phomactins, Nozaki-Hiyama-Kishi reaction, 310
Phorboxazoles
  Homer-Wadsworth-Emmons reaction, 441–443
  Nozaki-Hiyama-Kishi reaction, 311
Phosphazenes, Wittig reaction modification, 601
Phosphine-boranes, Barton-McCombie deoxygenation, phosphorous-based reductants, 623
Phosphines
  Castro-Stephens reaction, catalytic mechanism, 220
  Heck reactions, 21–22
  Negishi cross-coupling reactions, 72–73
  Sonogashira reaction
    bulky ligand condition, 129
    high-activity ligands, 124–125
  Stille coupling, 139–140
    alkyl halide reactions, 148–149
  Wittig reaction, 592–603
Phosphodiesterase inhibitor, Knoevenagel reaction, 494
Phosphonamides, Wittig reaction, 593–603
Phosphonate 99, Horner-Wadsworth-Emmons reaction, 444
Phosphonates
  Horner-Wadsworth-Emmons reaction, 422–423
  Wittig reaction variants, 598
Phosphorous compounds
  Barton-McCombie deoxygenation, phosphorous-based reductants, 622–623
  Peterson olefination, 520–521
  Wittig reaction modification, 595–603
Pinacol step, McMurry coupling, 273–276
  variations and improvements, 276–278
π-allyl Stille coupling, 158–159
(–)-Pironectin, Mukaiyama aldol reaction, 506
Platinum catalysts, Tsuji-Trost reaction, asymmetric allylic alkylation, 208
Plicamine, Morita-Bayless-Hillman reaction, 368
Polyacetylenes, Glaser coupling, organocopper reagents, 246–252
Polycavernoside A
  Horner-Wadsworth-Emmons reaction, 437–438
  Nozaki-Hiyama-Kishi reaction, 313
Polyenes
  asymmetric intramolecular Heck reaction, cyclization, 16–17
  Negishi cross-coupling reaction, 83–86
  Suzuki coupling reactions, 176–177
Polymers
  Barton-McCombie deoxygenation, tin-based reductants, 619
  Heck reactions, 20

Sonogashira reaction, 122–123
Polymethylhydrosiloxane (PMHS), Stille coupling, 151
Poly(methyl)hydrosiloxane (PMHS), Barton-McCombie deoxygenation, tin-based reductants, 618–619
Polyunsaturated systems, Sonogashira reaction, 114–115
Polyurethanes, Barton nitrite photolysis reaction, 644
Potassium organotrifluoroborate salts, Suzuki coupling reactions, 177–178
Potassium phosphate, Henry reaction, acetonitrile, 415
P(*o*-Tol), Heck reactions, 21–22
Precatalysts, Heck reaction, anionic pathways, 7–9
Pregnene, Wohl-Ziegler reaction, 669
Prins cyclization reaction, Peterson olefination and, 532
Promoters
Knoevenagel reaction, 483–484
Sakurai allylation reaction
chiral promoters, 556–557
overview, 538–539
2-Propanol, Barton-McCombie deoxygenation, 624–625
Propargyl alcohol, Castro-Stephens reaction, Cadiot-Chodkiewicz reaction and, 231
Propargylsilane, Corey-Fuchs reaction, 397–398
Prostaglandins, Julia–Kocienski/Julia–Lythgoe olefination, 459–460
Protein kinase C (PKC), Negishi cross-coupling reactions, 95
Protic solvents, Morita-Baylis-Hillman reaction, 352–353
Pseudohalides, Negishi cross-coupling reactions, 75
Pseudopteroxazole, Sandmeyer reaction, 653
Pybox catalyst, Sakurai allylation reaction, 558
Pyridazine, Negishi cross-coupling reactions, 80–81
Pyridines
Castro-Stephens reaction, 213–214
Knoevenagel reaction, Verley-Doebner modification, 480–481
Negishi cross-coupling reactions, 80
Pyridonecarboxylic acids, Sandmeyer reaction, 653–654
Quadrigemine, asymmetric intramolecular Heck reaction, 13–17
(–)-Quinocarcin, Wittig reaction modification, 602
Quinuclidine catalysts, Morita-Baylis-Hillman reaction, 352–353

Radical chemistry
Barton-McCombie deoxygenation
basic principles, 613
4,6-dimethoxy-1-(tetradec-13′-enyl)-1*H*-indole, 630
historical perspective, 613–614
mechanisms, 614–616
miscellaneous alternative reductants, 623–625
phosphorous-based reductants, 622–623
silicon-based reductants, 619–621
synthetic utility, 625–629
1,1′-thiocarbonyldiimidazole, 629–630
tin-based reductants, 618–619
variation and improvements, 616–618
Barton nitrite photolysis reaction
basic principles, 632
flexible molecule functionalization, 642
historical perspective, 632–634
9-hydroxy-2,2,4,8-tetramethyl-octahydro-4,8-methano-azulen-1-one oxime, 645
mechanisms, 634
non-steroidal rigid system functionalization, 638–642
steroid core functionalizations, 635–638
synthetic utility, 634–642
variations and improvements, 642–644
Sandmeyer reaction
basic principles, 647
bromination, 655
chlorination, 655–656
cyanation, 656–657
historical perspective, 647–648
hydroxylation, 657
iodination, 657
mechanism, 648
medicinal and heterocyclic chemistry, 653–654
natural product synthesis, 650–653
variations, 649–650
Wohl-Ziegler reaction
alternative solvents, 671–672

Radical chemistry *(continued)*
basic principles, 660
(*R*)-2,2′-bis(bromomethyl)-1,1′-binaphthyl, 672
halogenating agents, 671
historical perspective, 660–661
initiators, 670–671
mechanisms, 661–668
regioselectivity and stereoselectivity, 664–668
synthetic utility, 668–670
Raf kinase, Negishi cross-coupling reactions, 95
Ragoussis modification, Knoevenagel reaction
(*E*)-dec-3-enoic acid, 497
Verley-Doebner modification, 479–481
Rapamycin, Stille coupling in, 152–153
Reactivity pathway, copper-free Sonogashira reaction, 106–107
Receptor structures, McMurry coupling, 281–282
Redox-active molecular wires, Corey-Fuchs reaction, 400–401
Reductive cleavage, Julia–Lythgoe olefination, vinyl sulfones, 470–471
Reductive Heck reactions, palladium compounds, 28–29
Regioselectivity
Castro-Stephens reaction, 220
Heck reaction, 9–10
cationic pathways, 6–7
Sonogashira reaction, 108
Tsuji-Trost reaction, 187–188
carbon nucleophiles, 191–193
Wohl-Ziegler reaction, 664–668
Reidispongiolide A, Nozaki-Hiyama-Kishi reaction, 311
Retro-Henry reaction, 415
(+)-Ricciocarpin A, Barton-McCombie deoxygenation, 627
Ring systems
Horner-Wadsworth-Emmons reaction, 423
Nozaki-Hiyama-Kishi reaction, cyclization, 307
Sandmeyer reaction, natural product synthesis, 650–653
Rishitin, Barton nitrite photolysis reaction, 639
Rosenmund-von Braun reaction, Castro-Stephens reaction and
organocopper reagents, 223–229
related processes, 232
Roseophilin, Negishi cross-coupling reactions, 92–93
Rotaxanes, Glaser coupling, organocopper reagents, 249–252
Roush reaction
Sakurai allylation reaction, 567
Wittig reaction variants, 596–603, 604

Sakurai allylation reaction
1,2-asymmetric induction, 551–553
1,3-asymmetric induction, 554
asymmetric catalytic reactions, 555–559
basic principles, 538–539
BF_3 promoter, 569
chiral silane chirality transfer, 554–555
crotylsilanes, 567–568
E/Z isomerism, 547
$FeCl_3$ promoter in $MeNO_2$, 571–572
historical perspective, 539–540
mechanism, 540–546
aldehyde substituents, 545
Lewis acid-carbonyl complex, 542–545
silane substituents, 545–546
$SNCl_4$ promoter, 569–570
syn/anti isomerism, 547–551
synthetic utility, 560–567
synthons, 559–560
$TiCl_4$ promoter, 568–570
$ZnCl_3$ promoter, 570–571
Salen ligands
Mukaiyama aldol reaction, enantioselectivity, 516–517
Nozaki-Hiyama-Kishi reaction, 302–303
Salinosporamide A, Morita-Bayless-Hillman reaction, 370–371
Sandmeyer hydroxylation reaction, 649–650
Sandmeyer reaction
basic principles, 647
bromination, 655
chlorination, 655–656
cyanation, 656–657
historical perspective, 647–648
hydroxylation, 657
iodination, 657
mechanism, 648
medicinal and heterocyclic chemistry, 653–654
natural product synthesis, 650–653

variations, 649–650
Santiagonamine, Ullman coupling reaction, 263
Scyphostatin, Negishi cross-coupling reactions, 91
Sec-alkylmethylphosphinates, Barton-McCombie deoxygenation, 625
Selenoamides, Peterson olefination, 529
Self-assembly subunits, Sonogashira reaction, 115–117
Seychellogenin, Barton nitrite photolysis reaction, 636
Silane substituents
Barton-McCombie deoxygenation, silicon-based reductants, 619–621
Sakurai allylation reaction
chirality transfer, 554–555
Lewis acid-carbonyl complex, 545–546
synthons, 559–560
Sila-Stetter reaction, 578–579
intermolecular variants, 584–585
Silenes, Peterson olefination, 529
Silicon compounds
Barton-McCombie deoxygenation, silicon-based reductants, 619–621
Peterson olefination, 520–521
Silver ions, McMurry coupling, 280–281
Silylated alkynes, Castro-Stephens reaction, 230
α-Silylcarbanion, Wittig reaction modification, 603
Silyl enol ether, Mukaiyama aldol reaction, 504–505
Silyl nitro addition, Henry reaction, 416–417
Silylnitronate, Henry reaction, 416
Silyl transfer
Mukaiyama aldol reaction, 502–503
Sakurai allylation reaction, Lewis acid-carbonyl complex, 545–546
Six-membered rings, Negishi cross-coupling reactions, 79–82
SNF 4435 compounds, Horner-Wadsworth-Emmons reaction, 433
"Soft" nucleophiles, Tsuji-Trost reaction
basic mechanism, 187–188
carbon nucleophiles, 190–193
Solandelactones E and F, Nozaki-Hiyama-Kishi reaction, 311
Solid-phase chemistry, Stille coupling, toxicity reduction, 149–150
Solid-phase organic synthesis (SPO*S*), Horner-Wadsworth-Emmons reaction, 424–429
Solvents
Heck reactions, 25–27
Morita-Baylis-Hillman reaction, 353–358
Stille coupling, 148–149
Wohl-Ziegler reaction
alternative solvents, 671–672
mechanisms, 662–668
Sonication, McMurry coupling, 272
Sonogashira coupling reaction
Castro-Stephens reaction and, 213–214
Corey-Fuchs reaction, 400–401
palladium chemistry
alkyne homocopuling and acetylene reactivity, 109–110
aqueous solvent reactions, 127–128
basic principles, 100–102
bioconjugates, 121–122
bulky phosphine ligand, copper-free conditions, 129–130
complex fragment coupling, 111–114
copper co-catalyst, 104–106
copper-free mechanism, 106–107
cylindramide intermediate, 129
electronic and optical properties of compounds, 117–118
fatty acids and polyunsaturated systems, 114–115
heterocycle synthesis, 118–121
heterogeneous catalysts, 109
historical perspective, 102–103
ligand activity, 123–125
ligand-free catalysts, 108, 125–127
mechanisms, 107–108
metal catalysts, 128
non-traditional coupling partners, 128
polymers, 122–123
regioselectivity and stereoselectivity, 108
self-assembly subunits, dendrimers, and biopolymers, 115–117
synthetic utility, 110–123
Spiculoic acid, Wittig reaction, 607
(–)-Spinosyn A, Morita-Bayless-Hillman reaction, 368
Spirobislactone, Tebbe methylenation, 322–327
Spirotyrptostatin A, Julia–Kocienski/Julia–Lythgoe olefination, 467

Spongistatin, Mukaiyama aldol reaction, 506–507
S• radicals, Wohl-Ziegler reaction, 662–668
Stannic chloride ($SnCl_4$) promoter, Sakurai allylation reaction
  basic properties, 538–539
  example, 569–570
  Lewis acid-carbonyl complex, 543–544
Stereochemistry
  Knoevenagel reaction, 487
  Mukaiyama aldol reaction, 504
  Wittig reaction, 591
    Schlosser modification, 595–596
Stereoselectivity
  Castro-Stephens reaction, 220
  Heck reaction, 9–10
  Kumada cross-coupling reactions, 49–52
  Mukaiyama aldol reaction, 508–509
  Sonogashira reaction, 108
  Wohl-Ziegler reaction, 664–668
Steric factors, Sakurai allylation reaction, Lewis acid-carbonyl complex, 543
Steroidal nitrites, Barton nitrite photolysis reaction, mechanism, 634–638
Steroid synthesis, Barton nitrite photolysis reaction, 635–638
Stetter reaction
  asymmetric intramolecular annulation, 585
  basic principles, 575
  biomimetic reactions, 578
  enantioselective reactions, 579–583
  historical perspective, 575
  intermolecular reactions, 576–580, 583–585
  intramolecular reactions, 580–582, 584
  mechanism, 575–576
  NaCN-catalyzed intermolecular reaction, 583–584
  Sila-Stetter reaction, 578–579, 584–585
  synthetic utility, 576–582
  thiazolylidine-catalyzed intermolecular reaction, 584
Stille coupling, palladium chemistry
  additives, 140–142
  alkyl halide catalytic development, 147–149
  aryl chloride catalyst development, 145–147
  basic principles, 133–134
  carbonylative coupling, 158
  historical perspective, 134–136
  large-scale coupling, 159
  ligand influences, 139–140
  mechanisms of, 136–139
  organostannane activation, 142–145
  π-allyl Stille coupling, 158–159
  Stille-Kelly reaction, 151–152
  synthetic utility, 152–156
  tin compound toxicity, 149–151
  toxicity management, 156–157
  unactivated secondary alkyl halides, 157
Stille-Hegedus synthesus, jatrophone compound, 153
Stille-Kelly reaction, 151–152
Still-Gennari reaction
  Horner-Wadsworth-Emmons reaction
    mechanisms, 423–424
    natural products, 429–443
    synthetic procedures, 429
  Wittig reaction modification, 594–595
Still-Gennari reactions, Horner-Wadsworth-Emmons reaction, basic principles, 420–421
"Stitching" techniques, Stille coupling, 152–156
Stroke therapy
  Corey-Fuchs reaction, amide compounds, 398–399
  Knoevenagel reaction, 493–494
(–)-Subincanadines A an B, Nozaki-Hiyama-Kishi reaction, 314
Sulfonates, Suzuki coupling reactions, 173–174
Sulfones, Julia–Lythgoe olefination, 451–459
Sulfoxides, Julia–Lythgoe olefination, 451–459
Sulfur nucleophiles, Tsuji-Trost reaction, 201
Sulfur-tin bonds, Barton-McCombie deoxygenation, 614–616
Supercritical fluids, Heck reactions, 27
Suzuki reaction
  palladium chemistry
    aryl chlorides, 169–170
    basic principles and trends, 163
    1-(4-benzylphenyl)ethanone, 181
    comparisons with other cross-coupling reactions, 163–165
    enantioselective reactions, 174–176
    heteroaromatic compounds, 170–171
    highly active catalysts, 168–169
    historical perspective, 165–166
    mechanism, 166–168
    5-(2′-methylphenyl)-1*H*-indole, 180–181
    organotrifluoroborate reagents, 171–173

synthetic utility, 176–180
unactivated alkyl halides and sulfonates, 173–174
Ullmann coupling and, 262
syn-β-Hydride elimination, Heck reaction, regioselectivity and stereoselectivity, 10
Syn coordination, Sakurai allylation reaction, 544
syn/anti isomerism, 547–551
Synthons, Sakurai allylation reaction, 559–560
Syributin 1, Morita-Bayless-Hillman reaction, 365

TADDOL-derived catalyst, enantioselective cross-benzoin condensation, 384–385
Takeda alkylidenation, Tebbe methylenation, 326–327
Tamoxifen compounds
McMurry coupling, 279–280
Wittig reaction variants, 598
Tandem reactions
Hiyama cross-coupling reaction, 41–42
Knoevenagel reaction, 487–488
Diels-Alder reaction, 488–489
electrocyclization, 489–490
Knoevenagel-Mislow-Evans, 490–491
Mukaiyama aldol reaction, 511–512
Nicholas reaction, 294–296
Tautomycin, Mukaiyama aldol reaction, 508–509
Tebbe methylenation
organometallic reagents
basic principles, 319
historical perspective, 319–320
mechanisms, 320–321
Petasis olefination, 321–327
Petasis-Tebbe procedural comparisons, 331
synthetic utility, 327–331
Wittig reaction modification, 603
Tedanolide, Horner-Wadsworth-Emmons reaction, 431
Teicoplanin aglycon, Sandmeyer reaction, 651
Terpene synthesis, Barton nitrite photolysis reaction, 638–641
β-*tert*-Butyldiphenylsilylketone, Peterson olefination, 526
9-(*tert*-Butyldimethylsilyloxy)-5,6,12,13-tetrahydro-8,9,10,11-tetrahydro-7*H*-benzo[11]annulen-7-ol, Nozaki-Hiyama-Kishi reaction, 315
(*S*)-3-(*tert*-Butyloxycarbonylamino)-1-diazo-4-phenylbutan-2-one, Arndt-Eistert homologation, 346
(*S*)-3-(*tert*-Butyloxycarbonylamino)-4-phenylbutanoic acid, Arndt-Eistert homologation, 346–347
Tethered bis-(8-quinolinolato)-(TBOx) chromium complex, Nozaki-Hiyama-Kishi reaction, 303
Tetrabutylammonium peroxydisulfate, Barton-McCombie deoxygenation, 623–624
Tetraethynylethene, Negishi cross-coupling reaction, 86
Tetrahydrocannabinol, Kumada cross-coupling reactions, 59
Tetrahydropyranes, Sakurai allylation reaction, 565–566
THF-*N*-ethylpyrrolidinone (NEP) solvent, Negishi cross-coupling reactions, 74
Thiamine diphosphate (ThDP), benzoin condensation, enantioselective dimerization, 383
(+)-Thiazinotrienomycin, Julia–Kocienski/Julia–Lythgoe olefination, 463–464
Thiazolium precatalyst, Stetter reaction, 575–576
Thiazolylidine, intermolecular Stetter reaction, 584
2-Thiazolylzinc species, Negishi cross-coupling reactions, 78–79
Thioacetals, Tebbe methylenation, 326–327
Thiocarbonates, Barton-McCombie deoxygenation, 615–618
4,6-dimethoxy-1-(10′-(*N*-imidazolylthiocarbonyloxy)-tetradec-13′-enyl)-1*H*-indole, 629–630
miscellaneous reductants, 624–625
tin-based reductants, 618–619
Thioformate synthesis, Barton-McCombie deoxygenation, 617–618
Thionyl chloride, Arndt-Eistert homologation, 335–337
Thiophosphonate, Wittig reaction, 594–603
Tin compounds, Barton-McCombie deoxygenation
sulfur-tin bonds, 614–616
tin-based reductants, 618–619
Titanium-aluminum complex, Tebbe methylenation
basic properties, 319

Titanium-aluminum complex, Tebbe methylenation *(continued)*
historical perspective, 319–320
Titanium carbene complex, Tebbe methylenation, 327
Titanium compounds
McMurry coupling, 271–272
mechanistic pathways, 274–276
variations and improvements, 277–278
Morita-Baylis-Hillman homologation, 357–358
intermolecular reactions, 376
Mukaiyama aldol reaction, enantioselectivity, 515–517
Titanium tetrachloride ($TiCl_4$) promoter, Sakurai allylation reaction
basic properties, 538–539
example, 568–569
Lewis acid-carbonyl complex, 543–544
limitations, 570
Titanocene complexes, Tebbe methylenation, 324–327
Titanocene methylidene, Tebbe methylenation, 320–321
Toxicity reduction, Stille coupling
techniques for, 156–157
tin compounds, 149–151
(±)-Trachyspic acid, Nozaki-Hiyama-Kishi reaction, 314
Transmetalation pathways, Stille coupling, 139
Triarylmagnesiates, Kumada cross-coupling, 60–63
Triazines, Sandmeyer reaction, 649–650
Tricycloillicinone, Barton-McCombie deoxygenation, 626
Tridentate ligands, Mukaiyama aldol reaction, enantioselectivity, 515–517
Triene (+)-126, Julia–Lythgoe olefination, 471
Trimethylsilyl-dizaomethane, Arndt-Eistert homologation, 344–345
γ-(Trimethylsilyl)allenylboranes, Peterson olefination, 533–534
2-(Trimethylsilylmethyl)allyl acetate, Tsuji-Trost reaction, 201–202
1,2,4-Triphenylbutane-1,4-dione, Stetter reaction, 585
Triphenylphosphine-based catalysts, Suzuki coupling reactions, 168–169
Tris(pentafluorophenyl)borane (TPPB), Mukaiyama aldol reaction, vinylogous variant, 510–511
Tris(trimethylsilylmethyl)alane (TTMA), Peterson olefination, 525
Triynes, Negishi cross-coupling reaction, 86
Tsuji-Trost reaction, palladium chemistry
alcohol oxidation, 204–205
allylic compound hydrogenolysis, 205–206
amine nucleophiles, 199–201
asymmetric allylic alkylation, 188–190
basic principles, 185
6,6-bis-phenylsulfonyl-4-(*tert*-butyldimethylsilanyloxymethyl)-1-(4-methoxy-benzyloxy)-hex-3-en-2-ol, 209
carbon nucleophiles, 196–197
carbonylation, 202–203
conjugated diene formation, 203–204
"hard" carbon nucleophiles, 195–196
historical perspective, 186
mechanisms, 186–190
metal catalysts, 206–208
nitrogen nucleophiles, 194
oxygen nucleophiles, 194–195, 197–198
(2*S*,5*S*)-2-[3′-phenylsulfonyl-*N*′-(4-methoxybenzyl)succinimid-3′-yl]-5-benzoyloxy-2,5-dihydrofuran, 209
"soft" carbon nucleophiles, 190–193
sulfur nucleophiles, 201
synthetic utility, 190–201
2-(trimethylsilylmethyl)allyl acetate, 201–202
6-Tuliposide B, Morita-Bayless-Hillman reaction, 363–364
Turn-over number (TON)
Heck reactions, 21–22
Sonogashira reaction, 107–108
polymers, 122–123

Ullmann coupling, organocopper reagents
asymmetric copper coupling, 265–266
basic principles and historical perspective, 258
intramolecular palladium coupling, 266
mechanisms, 258–259
synthetic utility, 262–265
variations, improvements, or modifications, 259–261
2,5-Undecanedione, intermolecular Stetter reaction, 584
Urea catalysts, Morita-Baylis-Hillman homologation, 359–360

Urotropine, Morita-Baylis-Hillman homologation, 356
(+)-U-106305, Julia–Kocienski/Julia–Lythgoe olefination, 465

VEGFR kinase inhibitor, Stille coupling, 156
Verley-Doebner modification, Knoevenagel reaction, 478–481, 497
  amino acid catalysis, 482–483
  m-nitrocinnamic acid, 497
Vinblastine
  Sonogashira reaction, 112–113
  Ullmann coupling, 261
Vincristine, Ullmann coupling, 261
Vinylogous lactone, Tebbe methylenation, 322–327
Vinylogous Mukaiyama aldol reaction, 509–511
Vinylsilanes, Peterson olefination, 524–525
Vinylstannanes, Stille coupling, 155
Vinyl sulfone, Julia–Lythgoe olefination, acetoxy sulfone conversion, 470
Vitamin D derivatives
  Barton nitrite photolysis reaction, 641–642
  Negishi cross-coupling reactions, 94–95

Weinreb amides, Wittig reaction, 606–607
Wittig reaction
  carbon-chain homologations
    basic principles, 587
    ethyl cyclohexylideneacetate, 609
    historical perspective, 587–588
    mechanism, 588–592
    methylenecyclohexane, 608–609
    synthetic utility, 603–608
    variations, improvements, and modifications, 592–603
  Corey-Fuchs reaction, 393
  Peterson olefination, 520–521
  Sonogashira heterocycle coupling, 120–121
Wohl-Ziegler reaction
  alternative solvents, 671–672
  basic principles, 660
  (*R*)-2,2′-bis(bromomethyl)-1,1′-binaphthyl, 672
  halogenating agents, 671
  historical perspective, 660–661
  initiators, 670–671
  mechanisms, 661–668
  regioselectivity and stereoselectivity, 664–668
  synthetic utility, 668–670
Wolff rearrangement, Arndt-Eistert homologation
  β-amino acid synthesis, 337–338
  basic principles, 335–337
  miscellaneous compounds, 342–343

Xanthates, Barton-McCombie deoxygenation
  phosphorous-based reductants, 623
  silicon-based reductants, 620–621
Xerulinic acid, Negishi cross-coupling reactions, 93–94

Ylide carbanion structures, Wittig reaction, 588
  variations, improvements or modifiactions, 592–603
Ynolate anions, Arndt-Eistert homologation, 345–346

(*S*)-Zearalenone, Stille coupling and, 1550
Ziegler-Ullman reaction, organocopper reagents, 262–263
Zinc complexes
  Henry reaction, 409
  Mukaiyama aldol reaction, enantioselectivity, 514–517
  Sakurai allylation reaction, $ZnCl_4$ promoter, 570–571
  Sandmeyer reaction, 649–650
Zipper reactions, Heck cascade, 18
Zwitterion compounds, Morita-Baylis-Hillman reaction, 351–353